ENVIRONMENTAL RISK ASSESSMENT OF SOIL CONTAMINATION

Edited by **Maria C. Hernández-Soriano**

Environmental Risk Assessment of Soil Contamination
http://dx.doi.org/10.5772/57086
Edited by Maria C. Hernández-Soriano

CBS, Edition 2016

Contributors

Maria C. Hernandez Soriano, Jose C. Jimenez-Lopez, Sunday Paul Bako, Alysson Roberto Baizi E Silva, Fábio Camilotti, Juan Carlos Durán Álvarez, Blanca Jiménez, Emmanuel Onweremadu, Erika Bustos, Muhammad Aqeel Ashraf, Mohd. Jamil Maah, Ismail Yusoff, Ng Tham Fatt, Anna Lenart-Boroń, Piotr Boron, Thierry Woignier, Florence Clostre, Philippe Cattan, Yves Marie Cabiddoche, Megalie Lesueur Jannoyer, Joseph Levillain, Carmen Cristina Elekes, Anna Białk-Bielińska, Joanna Maszkowska, Alan Puckowski, Piotr Stepnowski, Wisley Moreira Farias, Eder Souza Martins, José Camapum De Carvalho, Fabrício Bueno Da Fonseca Cardoso, Edi Mendes Guimarães, Geraldo Rezende Boaventura, Anders Jonsson, Henrik Haller, Adebola Abosede Adeyi, Oludele Michael Michael Omidiran, Oladele Osibanjo, Bosiacki Maciej, Tomasz Kleiber, Bartosz Markiewicz, Jesús Pastor, Maria Jesus Gutierrez Gines, Ana Jesus Hernandez, Carmen Bartolomé, Dinora Vazquez-Luna, Anca Maria Moldoveanu, Sebastian Meyer, Bruno Glaser, Daniel Fischer, Yves Noël, Peter Quicker, Georg Kuffer, Affonso Celso Gonçalves Jr., Herbert Nacke, Daniel Schwantes, Gustavo Ferreira Coelho, Normaniza Osman, Mohamad Saifuddin, Aimee Halim, Paulo Favas, Joao Pratas, Mayank Varun, Rohan D'Souza, Manoj Stephen Paul, Santosh Kumar Sarkar, Dibyendu Rakshit, Francisco José Martin Peinado, Mariano Simon, Ines Garcia Fernandez, Veronica Gonzalez, Ana Romero, Marija Romic, Youssef F. F Lawgali, Andrew Meharg, Yousef Elgimat

Published by InTech

Janeza Trdine 9, 51000 Rijeka, Croatia

Notice

Publishing Process Manager Iva Lipovic

Technical Editor InTech DTP team

Cover InTech Design team

Additional hard copies can be obtained from orders@intechopen.com

Environmental Risk Assessment of Soil Contamination, Edited by Maria C. Hernández-Soriano

p. cm.

ISBN 978-953-51-1235-8

INTECH

open science | open minds

Contents

Preface

"Continuing change is the dominant factor in society today. No sensible decision can be made without taking into account not only the world as it is, but the world as it will be." Isaac Asimov.

The natural resources of the planet face enormous challenges in meeting food and energy demands of an increasing population, expected at 9 billion by 2050. Production challenges are exacerbated by unpredictable impacts on climate and natural ecosystems. Soil contamination and soil pollution constitute an essential issue to be addressed if we are to secure the life quality of present and future generations. Soil contamination is the presence of elevated concentrations of substances in the environment above the natural background level for the area and for the organism. Therefore, soil contamination is defined as the build-up of persistent toxic compounds in the natural soil environment, which have adverse effects on plant growth and animal health. More specifically, pollution is the introduction by man, directly or indirectly, of substances or energy into the environment resulting in deleterious effects to living resources and hazards to human health. The Environmental Protection Agency (EPA, USA) considers risk "to be the chance of harmful effects to human health or to ecological systems resulting from exposure to an environmental stressor". From historically contaminated areas to recent environmental disasters, the quality of world soils is constantly jeopardized, with an immediate associated risk for all living organisms linked to the affected soil area. According to a recent report from the European Commission (EEA, 2013), approximately three million sites in the EU have supported potential polluting activities, with 250,000 sites needing urgent remediation. The most frequent contaminants are heavy metals and mineral oil and the main causes of pollution are past and present industrial and commercial activities, agricultural chemicals, and the improper disposal and treatment of waste, but these categories vary widely worldwide. The Cooperative Research Centre for Contamination Assessment and Remediation reported in 2013 that Australia has 160,000 potentially contaminated sites, less than 1 per cent remediated, and most of the sites being in urban areas. According to data discussed at the 13th International Conference on Environmental Science and Technology (Athens, 2013), the number of contaminated sites exceeds 500,000 in the USA, 30,000 in Canada and 500,000 in Japan, while there are no official available data for several other areas such as Eastern Europe, Africa and South America and for countries such as China and India.

The protection of soil as the irreplaceable resource that sustains life in the planet requires integrated efforts from researchers and policy makers to develop sound risk assessment procedures and policies as well as feasible soil management and remediation strategies. To date, the relevance of soil contamination in environmental research and policy remains considerably belittled. Despite the substantial body of published research currently available, circa 3.5 million records returned from the web of science for "soil" search (650,000 records for "soil pollution" search), official guidelines and risk assessment procedures are frequently obsolete. Moreover, the advance of legislation concerning soil protection from the potential impact of human activities fails to meet the urgent need of developing sustainable land management practices. Overall, an increased effort to assess soil contamination and develop remediation and quality preservation strategies has become essential for the present and future of world soils.

Environmental risk assessment of soil contamination is a book designed by InTech - Open Access Publisher in collaboration with recognized international authors, unifying biological, chemical and agronomic approaches to research on soil contamination and pollution.

This book is intended to provide a wide depiction of current research in soil contamination and risk assessment, encompassing broad reviews and case studies on soil pollution by heavy metals and organic pollutants. Moreover, the book introduces several innovative approaches for soil remediation and risk assessment, including recent advances in phytoremediation. Therefore, this volume will be a valuable reading to a global audience of scientists, researchers, environmental educators, policy makers, technicians, managers, students and the general public.

The first section comprises nine chapters corresponding to broad studies on soil contamination and pollution, including a wide diversity of environmental scenarios and general revisions of soil pollution assessment and remediation strategies. Section 2 introduces seven chapters dedicated to advanced methodological approaches for risk assessment and soil remediation, including several up to date chemical and biological techniques. Section 3 specifically addresses progresses in phytoremediation, which is traditionally considered a top strategy for soil reclamation. A collection of 11 case studies on soil contamination with organic pollutants (section 4) and heavy metals (section 5) provides a worldwide depiction of soil contamination and pollution, highlighting the variability of contaminated sites and assessment approaches across the world.

The 31 chapters of this book can be read independently, but like different soil quality indicators they mutually benefit and enrich each other, providing a valuable composition that depicts current status of research in soil contamination and advances on risk assessment from a multidisciplinary approach.

For their excellent work, special thanks to the technical editors and sincere appreciation to Ms Iva Lipović, Deputy Head of Book Publishing Department, who coordinated the publication process of this volume and assisted me and the authors in completing our tasks smoothly and in a timely manner.

Dr. Maria C. Hernández-Soriano
The University of Queensland
Australia

Section 1

Risk Assessment: Soil Contamination Scenarios and Strategies

Chapter 1

Soil Contamination, Risk Assessment and Remediation

Muhammad Aqeel Ashraf, Mohd. Jamil Maah and Ismail Yusoff

Additional information is available at the end of the chapter

http://dx.doi.org/10.5772/57287

1. Introduction

Environment pollution is a burning topic of the day. Air, water and soil are being polluted alike. Soil being a "universal sink" bears the greatest burden of environmental pollution. It is getting polluted in a number of ways. There is urgency in controlling the soil pollution in order to preserve the soil fertility and increase the productivity. Pollution may be defined as an undesirable change in the physical, chemical and biological characteristics of air, water and soil which affect human life, lives of other useful living plants and animals, industrial progress, living conditions and cultural assets. A pollutant is something which adversely interfere with health, comfort, property or environment of the people. Generally most pollutants are introduced in the environment by sewage, waste, accidental discharge or else they are by-products or residues from the production of something useful. Due to this our precious natural resources like air, water and soil are getting polluted.

The basis of agriculture is Soil. All crops for human food and animal feed depend upon it. We are losing this important natural resource by the accelerated erosion 10 some extent. In addition to this the enormous quantities of man-made waste products, sludge and other product" from new waste treatment plants even polluted water are also causing or leading to soil pollution. In order to preserve the fertility and the productivity of the soil, control measures are to be taken in a herculean manner, thereby improving the health of all living beings.

Assessing the ecological risk of contaminated soil, pesticide application, sewage sludge amendment, and other human activities leading to exposure of the terrestrial environment to hazardous substances is a complicated task with numerous associated problems. Not only is terrestrial ecological risk assessment a relatively new field of science that has developed rapidly only since the mid-1980s, but it is also complicated by the fact that soil, in contrast to most aquatic environments, is very often on private lands and traded as real estate. Professional and economic divergence between the interests of scientists, stakeholders, authorities, engineers, managers, lawyers, nongovernment organizations (NGOs) and regulators is therefore not unusual. Even neglecting those aspects, a number of unresolved problems exist in the way we currently assess risk and manage the impact of anthropogenic substances in the terrestrial environment.

This chapter does not intend to present a comprehensive review of all published data from ecological studies at contaminated sites. Instead, the observations from all case studies are used in the discussion and form the basis for the final conclusion. In each case, we try to answer the following questions:

1. What is soil pollution and how it occurs?
2. How to determine the ecological risk assessment of the soil?
3. To what extent do soil screening levels (over)estimate risk?
4. Do bioassays represent a more realistic risk estimate?
5. Is it possible to make sound field surveys, or do we lack suitable reference situations?
6. What are the possible soil management methods for the polluted soils?

2. Soil pollution

Soil pollution is defined as the build-up in soils of persistent toxic compounds, chemicals, salts, radioactive materials, or disease causing agents, which have adverse effects on plant growth and animal health [1].

Soil is the thin layer of organic and inorganic materials that covers the Earth's rocky surface. The organic portion, which is derived from the decayed remains of plants and animal, is concentrated in the dark uppermost topsoil. The inorganic portion made up of rock fragments, was formed over thousands of years by physical and chemical weathering of bedrock. Productive soils are necessary for agriculture to supply the world with sufficient food [2].

There are many different ways that soil can become polluted, such as:

- Seepage from a landfill
- Discharge of industrial waste into the soil
- Percolation of contaminated water into the soil
- Rupture of underground storage tanks
- Excess application of pesticides, herbicides or fertilizer
- Solid waste seepage

The most common chemicals involved in causing soil pollution are:

- Petroleum hydrocarbons
- Heavy metals
- Pesticides
- Solvents

2.1. Inorganic toxic compounds

Inorganic residues in industrial waste cause serious problems as regards their disposal. They contain metals which have high potential for toxicity. Industrial activity also emits large amounts of arsenic fluorides and sulphur dioxide (SO_2) [3]. Fluorides are found in the atmosphere from superphosphate, phosphoric acid, aluminium, steel and ceramic industries. Sulphur dioxide emitted by factories and thermal plants may make soils very acidic. These metals cause leaf injury and destroy vegetation.

Copper, mercury, cadmium, lead, nickel, arsenic are the elements which can accumulate in the soil, if they get entry either through sewage, industrial waste or mine washings. Some of the fungicides containing copper and mercury also add to soil pollution. Smokes from automobiles contain lead which gets adsorbed by soil particles and is toxic to plants. The toxicity can be minimized by building up soil organic matter, adding lime to soils and keeping the soil alkaline [4].

2.2. Organic wastes

Organic wastes of various types cause pollution hazards. Domestic garbage, municipal sewage and industrial wastes when left in heaps or improperly disposed seriously affect health of human beings, plants and animals [5-7]. Organic wastes contain borates, phosphates, detergents in large amounts. If untreated they will affect the vegetative growth of plants. The main organic contaminants are phenols and coal.

Asbestos, combustible materials, gases like methane, carbon dioxide, hydrogen sulphide, carbon monoxide, sulphur dioxide, petrol are also contaminants. The radioactive materials like uranium, thorium, strontium etc. also cause dangerous soil pollution. Fallout of strontium mostly remains on the soil and is concentrated in the sediments [8]. Decontamination procedures may include continuous cropping and use of chelate amendments. Other liquids wastes like sewage, sewage sludge, etc. are also important sources of soil problems.

a. Sewage and sewage sludge

Soil pollution is often caused by the uncontrolled disposal of sewage and other liquid wastes resulting from domestic uses of water, industrial wastes containing a variety of pollutants, agricultural effluents from animal husbandry and drainage of irrigation water and urban runoff [9-10]. Irrigation with sewage water causes profound changes in the irrigated soils. Amongst various changes that are brought about in the soil as an outlet of sewage irrigation include physical changes like leaching, changes in humus content, and porosity etc., chemical changes like soil reaction, base exchange status, salinity, quantity and availability of nutrients like nitrogen, potash, phosphorus, etc. Sewage sludges pollute the soil by accumulating the metals like lead, nickel, zinc, cadmium, etc. This may lead to the phytoxicity of plants.

b. Heavy metal pollutants

Heavy metals are elements having a density greater than five in their elemental form. They mostly find specific absorption sites in the soil where they are retained very strongly either on the inorganic or organic colloids. They are widely distributed in the environment, soils, plants, animals and in their tissues. These are essential for plants and animals in trace amounts. Mainly urban and industrial aerosols, combustion of fuels, liquid and solid from animals and human beings, mining wastes, industrial and agricultural chemicals etc. are contributing heavy metal pollution. Heavy metals are present in all uncontaminated soils as the result of weathering from their parent materials. Concentration of heavy metals in soils and plants is given in Table 1.

Sl.No	Heavy metal	Hithosphere	Soil range	Plants
1	Cadmium (Cd)	0.2	0.01-0.7	0.2-0.8
2	Cobalt (Co)	40	1-40	0.05-0.5
3	Chromium (Cr)	200	5-3000	0.2-1.0
4	Copper (Cu)	70	2-100	4-15
5	Iron (Fe)	50,000	7000-5,50,000	140
6	Mercury (Hg)	0.5	0.01-0.3	0.015
7	Manganese (Mn)	1000	100-4000	15-100
8	Molybdenum (Mo)	2.3	0.2-5	1-10
9	Nickel (Ni)	100	10-1000	1
10	Lead (Pb)	16	2-200	0.1-10
11	Tin (Sn)	40	2-100	0.3
12	Zinc (Zn)	80	10-300	8-100

Table 1. Heavy metal concentration in the hithosphere, soils and plants (Ug/gm dry matter)

In agricultural soils, however, the concentration of one or more of these elements may be significantly increased in several ways, like through applications of chemicals, sewage sludge, farm slurries, etc. Increased doses of fertilizers, pesticides or agricultural chemicals, over a period, add heavy metals to soils which may contaminate them. Certain phosphatic fertilizers frequently contain trace amounts of cadmium which may accumulate in these soils. Likewise, some fertilizers when applied to soils, they add certain heavy metals which are given in Table 2.

The range of heavy metal contents in sludges is given in Table 3.

Sl.No	Fertilizer	Co	Cr	Cu	Mn	Mo	Ni	Pb	Zn
1	Nitrochalk	-	-	22	24	-	2	-	15
2	Calcium	0.1	Traces	Traces	Traces	-	-	-	1
3	Nitrate	-	-	To 10	To 5	-	-	-	-
4	Ammonium sulphate	<5	<5	0.800	0.80	<0.05 to 0.22	<5	Traces to 200	0.800
5	Super phosphate	0.02-13	0-1000	Traces to 1000	Traces to 2842	Traces to 35	Traces to 32	Traces to 92	70-3000
6	Potassium chloride	001	-	0-10	Traces-8	<0.05	<1	<1	0-3
7	Potassium sulphate	<5	<5	0-300 to 80	Traces to . 33	0.09	<5	<50	<50

Table 2. Heavy metal content of fertilizers (ug/gm)

Sl.No	Heavy metal	Range (ppm)
1	Cadmium	< 60-1500
2	Cobalt	2-260
3	Chromium	40-8800
4	Copper	200-8000
5	Iron	6000-62,000
6	Manganese	150-2500
7	Molybdenum	2-30
8	Nickel	20-5300
9	Lead	120-3000
10	Zinc	700-49,000

Table 3. Heavy metal contents in sludges (ppm)

The fate of heavy metals in soil will be controlled by physical and biological processes acting within the soil. Metal ions enter the soil solution from these various forms of combination in different rates they may either remain in solution or pass into the drainage water or be taken up by plants growing on the soil or be retained by the soil in sparingly soluble or insoluble forms. The organic matter of these soil have great affinity to heavy metals cations which form stable complexes thereby leading to reduced nutrient content [11-12].

2.3. Organic pesticides

Pesticides are quite frequently used to -control several types of pests now-a-days. Pesticides may exert harmful effects to micro-organisms, as a result of which plant growth may be affected. Pesticides which are not rapidly decomposed may create such problems. Accumulation is residues of pesticides in higher concentrations are toxic. Pesticides persistence in soil and movement into water streams may also lead to their entry into foods and create health hazards. Pesticides particularly aromatic organic compounds are not degraded rapidly and therefore, have a long persistence time which can be seen in Table 4.

Sl.No	Pesticide	Persistence time
1	BHC	11 yrs
2	DDT	10 yrs
3	2,4-D	2-8 weeks
4	Aldrin	9 yrs
5	Diuron	16 months
6	Atrazine	18 months
7	Siwazine	17 months
8	Chlordane	12 yrs
9	2,3 6-Trichlorobenzene (TBA)	2-5 yrs

Table 4. Persistence time for some selected pesticides

Mercury, cadmium and arsenic are common constituents of pesticides and all these heavy metals are toxic. At present DDT and a number of organochlorine compounds used as pesticides have been declared harmful and banned in U.S.A. and England [13-14]. It is due to the persistence of their residues in soils for considerable time without losing their toxicity. This has led to higher concentration of these pesticides in vegetation, in animal flesh and milk. Eventually man has been affected. In view of their demerits, organochlorines have been replaced by organophosphate pesticides which are more toxic, but do not leave any residue. They do not pollute the soil. The rodenticides too add to soil pollution. A major method of checking this pesticidal pollution is to increase the organic matter content of the sol and choose such pesticides which are non-persistent and leave no harmful residue.

2.4. Types of soil pollution

- Agricultural Soil Pollution

i. pollution of surface soil

ii. pollution of underground soil

- Soil pollution by industrial effluents and solid wastes

i. pollution of surface soil

ii. disturbances in soil profile

- Pollution due to urban activities

i. pollution of surface soil

ii. pollution of underground soil

2.5. Sources of soil pollution

The sources which pollute the soil are twofold: Agricultural sources and non-agricultural sources. Figure 1 shows the different sources for the soil pollution.

a. Agricultural sources

Soil pollution comes from different sources including agriculture and animal husbandry. Some of the agricultural practices lead to soil pollution. They are animal wastes, use of long lived pesticides, herbicides, fungicides, nematocides, etc. fertilizers and some agricultural practices.

b. Non-agricultural sources

Soil pollution by non-agricultural sources is usually the direct result of urban sprawl caused by rapidly increasing population and a rapidly per capita output of waste related to our modem way of life. Its materials that find their entry into the soil system have long persistence and accumulate in toxic concentration and thus become sources of pollution. Some of those most important soil pollutants are inorganic toxic compounds.

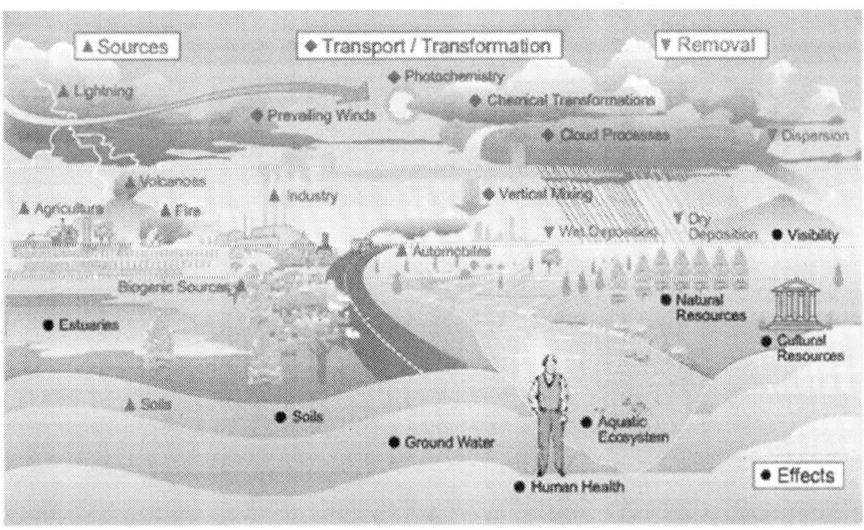

Figure 1. Sources of Soil Pollution [5]

2.6. Causes of soil pollution

Soil pollution is caused by the presence of man-made chemicals or other alteration in the natural soil environment. This type of contamination typically arises from the rupture of underground storage links, application of pesticides, and percolation of contaminated surface water to subsurface strata, oil and fuel dumping, leaching of wastes from landfills or direct discharge of industrial wastes to the soil. The most common chemicals involved are petroleum hydrocarbons, solvents, pesticides, lead and other heavy metals. This occurrence of this phenomenon is correlated with the degree of industrialization and intensities of chemical usage. A soil pollutant is any factor which deteriorates the quality, texture and mineral content of the soil or which disturbs the biological balance of the organisms in the soil. Pollution in soil has adverse effect on plant growth.

Pollution in soil is associated with

- Indiscriminate use of fertilizers
- Indiscriminate use of pesticides, insecticides and herbicides
- Dumping of large quantities of solid waste
- Deforestation and soil erosion

a. Indiscriminate use of fertilizers

Oxygen from air and water but other necessary nutrients like nitrogen, phosphorus, potassium, calcium, magnesium, sulfur and more must be obtained from the soil. Farmers generally use fertilizers to correct soil deficiencies. Fertilizers contaminate the soil with impurities, which come from the raw materials used for their manufacture. Mixed fertilizers often contain ammonium nitrate (NH_4NO_3), phosphorus as P_2O_5, and potassium as K_2O. For instance, As, Pb and Cd present in traces in rock phosphate mineral get transferred to super phosphate fertilizer. Since the metals are not degradable, their accumulation in the soil above their toxic levels due to excessive use of phosphate fertilizers becomes an indestructible poison for crops. The over use of NPK fertilizers reduce quantity of vegetables and crops grown on soil over the years. It also reduces the protein content of wheat, maize, grams, etc., grown on that soil. The carbohydrate quality of such crops also gets degraded [15]. Excess potassium content in soil decreases Vitamin C and carotene content in vegetables and fruits. The vegetables and fruits grown on over fertilized soil are more prone to attacks by insects and disease.

b. Indiscriminate use of pesticides, insecticides and herbicides

Plants on which we depend for food are under attack from insects, fungi, bacteria, viruses, rodents and other animals, and must compete with weeds for nutrients. To kill unwanted populations living in or on their crops, farmers use pesticides. The first widespread insecticide use began at the end of World War II and included DDT (dichlorodiphenyltrichloroethane) and gammaxene. Insects soon became resistant to DDT and as the chemical did not decompose readily, it persisted in the environment. Since it was soluble in fat rather than water, it biomagnified up the food chain and disrupted calcium metabolism in birds, causing eggshells to be thin and fragile. As a result, large birds of prey such as the brown pelican, ospreys, falcons and eagles became endangered. DDT has been now been banned in most western countries. Ironically many of them including USA, still produce DDT for export to other developing nations whose needs outweigh the problems caused by it [16].

c. Dumping of solid wastes

In general, solid waste includes garbage, domestic refuse and discarded solid materials such as those from commercial, industrial and agricultural operations. They contain increasing amounts of paper, cardboards, plastics, glass, old construction material, packaging material and toxic or otherwise hazardous substances. Since a significant amount of urban solid waste tends to be paper and food waste, the majority is recyclable or biodegradable in landfills. Similarly, most agricultural waste is recycled and mining waste is left on site. The portion of solid waste that is hazardous such as oils, battery metals, heavy metals from smelting industries and organic solvents are the ones we have to pay particular attention to. These can in the long run, get deposited to the soils of the surrounding area and pollute them by altering their chemical and biological properties [17].

d. Deforestation

Soil Erosion occurs when the weathered soil particles are dislodged and carried away by wind or water. Deforestation, agricultural development, temperature extremes, precipitation including acid rain, and human activities contribute to this erosion. Humans speed up this process by construction, mining, cutting of timber, over cropping and overgrazing. It results in floods and cause soil erosion. Forests and grasslands are an excellent binding material that keeps the soil intact and healthy. They support many habitats and ecosystems, which provide innumerable feeding pathways or food chains to all species. Their loss would threaten food chains and the survival of many species. During the past few years quite a lot of vast green land has been converted into deserts. The precious rain forest habitats of South America, tropical Asia and Africa are coming under pressure of population growth and development (especially timber, construction and agriculture). Many scientists believe that a wealth of medicinal substances including a cure for cancer and aids, lie in these forests. Deforestation is slowly destroying the most productive flora and fauna areas in the world, which also form vast tracts of a very valuable sink for CO_2 [18].

e. Pollution due to urbanization

Pollution of surface soils materials (like vegetables, animal wastes, papers, wooden pieces, carcasses, plant twigs, leaves, cloth wastes as well as sweepings) and many non-biodegradable materials (such as plastic bags, plastic bottles, plastic wastes, glass bottles, glass pieces, stone / cement pieces) [19-20]. On a rough estimate Indian cities are producing solid city wastes to the tune of 50,000 - 80,000 metric tons every day. If left uncollected and decomposed, they are a cause of several problems such as;

- Clogging of drains: Causing serious drainage problems including the burst / leakage of drainage lines leading to health problems.

- Barrier to movement of water: Solid wastes have seriously damaged the normal movement of water thus creating problem of inundation, damage to foundation of buildings as well as public health hazards.
- Foul smell: Generated by dumping the wastes at a place.
- Increased microbial activities: Microbial decomposition of organic wastes generate large quantities of methane besides many chemicals to pollute the soil and water flowing on its surface
- When such solid wastes are hospital wastes they create many health problems: As they may have dangerous pathogen within them besides dangerous medicines, injections.

f. Pollution of underground soil

Underground soil in cities is likely to be polluted by

- Chemicals released by industrial wastes and industrial wastes
- Decomposed and partially decomposed materials of sanitary wastes

Many dangerous chemicals like cadmium, chromium, lead, arsenic, selenium products are likely to be deposited in underground soil. Similarly underground soils polluted by sanitary wastes generate many harmful chemicals. These can damage the normal activities and ecological balance in the underground soil

2.7. Effects of soil pollution

a. Agricultural

- Reduced soil fertility
- Reduced nitrogen fixation
- Increased erodibility
- Larger loss of soil and nutrients
- Deposition of silt in tanks and reservoirs
- Reduced crop yield
- Imbalance in soil fauna and flora

b. Industrial

- Dangerous chemicals entering underground water
- Ecological imbalance
- Release of pollutant gases
- Release of radioactive rays causing health problems
- Increased salinity

http://dx.doi.org/10.5772/57287

- Reduced vegetation

c. Urban

- Clogging of drains
- Inundation of areas
- Public health problems
- Pollution of drinking water sources
- Foul smell and release of gases
- Waste management problems

d. Environmental

- Soil becomes unavailable to grow food
- If contaminated soil is used to grow food, the land will usually produce lower yields
- Can cause even more harm because a lack of plants on the soil will cause more erosion
- The pollutants will change the makeup of the soil and the types of microorganisms that will live in it.
- Thus it's possible for soil pollution to change whole ecosystems

2.8. Control of soil pollution

The following steps have been suggested to control soil pollution. To help prevent soil erosion, we can limit construction in sensitive area. In general we would need less fertilizer and fewer pesticides if we could all adopt the three R's: Reduce, Reuse, and Recycle. This would give us less solid waste.

a. Extraction and separation techniques

In solvent extraction, the contaminated & oil is mixed with an extracting agent in general (an aqueous solution but preferably an organic solvent). Potential applications include the removal of metals such as cadmium, copper, zinc, nickel, chromium, arsenic, antimony and lead using a mineral solution, zinc lead, organo-metallic compounds and some cyanides using sodium hydroxide solution. Hydrocarbons and halogenated hydrocarbons can also be removed [21]. Contamination is often preferentially present in the finer or coarser fraction of the soil or the organic components (ex. humus), contamination can therefore, be removed in some cases using a process which separates the soil into fractions on the basis of specific gravity or particle size or settling velocity.

b. Thermal methods

In thermal methods, there are two ways of heat treatment; removal of contaminants by evaporation either by direct heat transfer from heated air or an open flame or by indirect heat transfer, and destruction of the contaminants directly or indirectly at an appropriate temperature. The gas leaving the heating appliance must be treated to destroy or remove any contaminants or unwanted products of combustion. A related process is stream stripping in which steam is injected into soil to aid evaporation of relatively volatile contaminants which may be water soluble or insoluble.

c. Chemical methods

Treatment of the soil in suspension in a suitable liquid and without sludging is the two possible methods. In these, intimate, contact between soil and chemical is essential and should be frequently done so that the process of detoxification is complete.

d. Microbial treatment methods

The microbial treatment methods appear to be more promising which can deal with whole range of organic contaminants including phenol, polychlorinated hydrocarbons, oil and oil products, dioxins, etc. There are two different ways of approaching the problems.

1. A community of microbes already existing on the site is collected and cultured in the laboratory.

2. Strains of microbes are developed in the laboratory that is capable of metabolizing particular chemicals.

Excavation of the soil prior to treatment offers the greatest scope for creating optimum conditions. The excavated soil can be placed on thin layers to various depth using standard earth moving techniques and microbes and nutrients applied using standard agricultural techniques such as fertilizing, ploughing, harrowing, etc.

e. Reducing chemical fertilizer and pesticide use

Applying bio-fertilizers and manures can reduce chemical fertilizer and pesticide use. Biological methods of pest control can also reduce the use of pesticides and thereby minimize soil pollution.

f. Reusing of materials

Materials such as glass containers, plastic bags, paper, cloth etc. can be reused at domestic levels rather than being disposed, reducing solid waste pollution.

g. Recycling and recovery of materials

This is a reasonable solution for reducing soil pollution. Materials such as paper, some kinds of plastics and glass can and are being recycled. This decreases the volume of refuse and helps in the conservation of natural resources. For example, recovery of one tonne of paper can save 17 trees.

h. Reforesting

Control of land loss and soil erosion can be attempted through restoring forest and grass cover to check wastelands, soil erosion and floods. Crop rotation or mixed cropping can improve the fertility of the land.

i. Solid waste treatment

Proper methods should be adopted for management of solid waste disposal. Industrial wastes can be treated physically, chemically and biologically until they are less hazardous. Acidic and alkaline wastes should be first neutralized; the insoluble material if biodegradable should be allowed to degrade under controlled conditions before being disposed. As a last resort, new areas for storage of hazardous waste should be investigated such as deep well injection and more secure landfills. Burying the waste in locations situated away from residential areas is the simplest and most widely used technique of solid waste management.

2.9. Soil monitoring

The main objective of soil monitoring is to prevent and mitigate contamination by substances with the potential to exert an adverse effect on the soil itself, and on air, water and organisms that may contact the soil. Soil monitoring, within the approvals program, is directed primarily to the assessment of contaminants that have been released to the soil surface. Thus, subsurface facilities are generally not the reason for soil monitoring, but may be the reason for groundwater monitoring [22]. However, where soil contamination is known or suspected to originate from subsurface sources such as underground tanks or pipes, an assessment will be required. Where the above considerations indicate soil monitoring is required as a condition of an Approval, the proponent is required to carry out the following, as specified in the Soil Monitoring Directive: > prepare a soil monitoring proposal; > execute the approved soil monitoring plan; > interpret and report the results of the soil monitoring; and > prepare and execute a soil management plan where indicated by the results of soil monitoring. This guideline provides a background for the soil monitoring program and a description of soil management program requirements.

a. Legislative background

The soil monitoring program mostly developed under the Environmental Protection in support of the following principles: > development must be sustainable, meaning that the use of resources and the environment today must not impair prospects for their use by future generations; > the environmental impact of development must be prevented or mitigated; > polluters should bear the responsibility of paying for the costs of their actions; > remediation costs should be incorporated into financial planning so that adequate funds are available for site remediation and planners can know the true costs and benefits of source reduction programs. Recognizing that under the environmental protection is a shared responsibility, it follows that both the approval holder and the Department must have a means to assess environmental performance with respect to the above principles and requirements.

b. Soil quality standards

Environmental Protection expects that approval holders will manage their operations to prevent substance releases to soil. Substance releases to soil do occur, however, and contaminants are often present above background concentrations at industrial facilities. In view of this, Environmental Protection should have soil quality standards to guide assessment and remediation of soil contamination. Facilities that are currently uncontaminated have the opportunity to maintain conditions that allow unrestricted land-use. For these facilities, the minimum standards will be determined by the Tier I criteria or equivalent objectives. Older facilities, however, were often operated under different standards and environmental management practices than are currently acceptable.

2.10. Soil pollution risks

There is an increasing use of risk-oriented policies to deal with the local effects of soil pollution. The risks that such policies deal with are: human health risks and can also include eco-toxicological risks. These risks are expressed in terms of negative effects and chances between 0 and 1 that such negative effects will occur. Examples of areas where risk-oriented policies are applied to soil pollution include the United States of America [23], Canada [24] and countries in the European Union [25]. Historically, these risk oriented policies have followed the abandonment of policies aimed at restoring soils to their original 'clean' state.

Risk-based criteria or standards, developed in the framework of risk oriented policies, are applied to risks estimated with deterministic methodologies, following the steps of hazard characterization, appraisal of exposure and risk characterization, while using exposure-risk relations established beforehand. Risk-based criteria have been applied to decisions about soil remediation in the form of soil clean-up standards [26], to the use of soils for specific purposes and in the United States also to sediment management [27]. The risk-oriented policies considered here [28], assume that background exposure to pollutants carries no risk and that a specified level of soil pollution carries a maximum tolerable or maximum acceptable risk for organisms living locally. The latter is the main basis for standard setting.

In part, risk-oriented soil pollution legislation includes policy goals that are qualitative [29]. For instance, the primary UK legislation on contaminated soil defines land as contaminated in need of risk management 'if significant harm is being caused or there is a significant possibility of such harm being caused' [29]. Mostly, however policies have resulted in specific quantitative values for maximum tolerable or acceptable soil pollution. The analysis of such values used in different industrialized countries has shown that there are very large differences, roughly up to a factor [30]. According to Provoost et al. [31], these differences to a large extent originate in different political choices (e.g. including or excluding ecotoxicity) and in different assumptions as to the modeling of exposure to soil pollutants, including site related factors, such as soil type and building constructions [31].

2.10.1. Risks related to one soil pollutant

In practice, there are several matters which are at variance with the proper establishment of actual risk related to one soil pollutant. These are: the absence of standards for pollutants, neglect of background exposure, and neglect of routes of exposure to soil pollution, neglect of available dose- effect studies and neglect of biological availability. These will now be discussed in more detail.

a. Absence of quality standards

When data regarding soil pollutants are available, they should be compared with quality standards reflecting maximum tolerable risk of exposure. However, such standards are not always in place. For instance, of the volatile organic carbon compounds detected in groundwater samples by the US Geological Service, were unregulated- with no standards in place [32]. Similarly Patterson et al. [33] found a variety of brominated ethenes in Australian groundwater, all lacking standards.

b. Neglect of background exposure

For a proper estimate of soil pollution related risks, exposure to specific soil pollutants should be evaluated in combination with exposure to the same substance that is not related to local soil contamination. Several countries, such as Canada, Germany, Spain and Belgium, do indeed establish soil clean-up standards while considering background dietary and inhalatory exposure but others, e.g. Sweden, Norway and the Netherlands, do not [34]. Neglecting background exposure or specific types of background exposure may have implications for risk estimates.

c. Neglect of routes of exposure to soil pollution

In evaluating exposure to soil pollutants, assumptions regarding exposure routes are important. In this respect difference between countries may be noted. Soil clean-up standards for lead of Norway and Sweden differ in part because in Sweden the dominant exposure route is assumed to be by drinking water and in Norway it is thought to be by drinking water and ingestion of soil [34].

Inhalation of household dust and soil particles is not always taken into account in governmental decision making about risks of soil pollution. For instance, in the Netherlands inhalation of soil particles has been neglected as an exposure route, but in e.g. Spain it is not [34]. Neglect of inhalation would seem at variance with existing studies. Nawrot et al. [35] have studied the effects of cadmium pollution in soil (around former thermal zinc plants) and found a significant increase in lung cancer risk correlated with cadmium exposure. They plausibly explain this in terms of exposure of lung tissue to cadmium present in inhaled soil and household dust particles.

Household dust particles have also been found to be important in the exposure of children to pesticides in agricultural settings.

d. Neglect of available dose-effect studies

Akesson et al. [37] have analyzed the effects of low environmental cadmium exposure in an epidemiological study of Swedish women in the Lund area, being 53-64 years of age, excluding women from areas with soils heavily polluted by cadmium. Akesson et al. [37] found associations between the internal dose of cadmium and tubular and glomerular kidney effects, which may represent early signs of adverse effects. Women with diabetes seemed to be at increased risk of experiencing such early signs. In view of these data it seems plausible that at a background exposure that is common in Sweden, old women in the general population may be at risk for adverse cadmium effects [34] and that even a modest increase in cadmium exposure due to polluted soil may lead to added risk. However, when establishing soil clean-up standards in Sweden this background exposure has been neglected [34]. Nawrot et al. [35] have studied the relation between mortality and cadmium body burden in Belgium. They obtained evidence that total mortality and non-cardiovascular mortality may be elevated at cadmium body burdens which can be found among the population not living on soils that are currently considered to be a health risk. Similarly there are now strong indications that the negative effects of lead on the neurophysiologic and sexual development may well be found at the level of background exposure common in Western European and US cities [37], though soil pollution policy, at least in European counties, assumes that such background exposure is safe [14].

2.10.2. Ecotoxicological risks

Maximum acceptable or maximum tolerable ecotoxicological risks are usually derived from a limited number of studies concerning single species under laboratory conditions. Laboratory conditions may be very different from actual conditions in the field, and thus findings in the field are often at variance with laboratory studies [35]. In field studies it has been found that several factors which tend to be neglected in laboratory studies may strongly impact toxic effects of soil pollutants. These include among others: density and adaptability of populations of affected organisms, the presence of other environmental stress factors and the presence or absence of specific landscape elements such as buffer strips [35].

2.10.3. Biological availability

Biologically available pollutants determine risk [3]. Biological availability may vary strongly for different types of organisms [36]. Biological availability of a compound in a specific soil is also dependent on physical, chemical and biological and spatial factors [35]. Examples of such factors are pH, the amount and nature of organic and mineral compounds also present and the presence of organisms that can mobilize soil pollutants [37-39]. In practice, biological availability may be much at variance with total concentrations [40].

2.10.4. Combination effects

a. Limited accounting of combination effects

As to the overall risk of soil pollutants, cumulative effects of the combination of substances present in soils should be considered. However actual standard setting practice has largely focused on criteria relating to one element or compound. In some cases there are criteria for groups of compounds [34]. Such criteria limit the amount (in g/kg soil) of groups of compounds but often do not address the possibility that the risk per unit of weight may be different for different compounds. An exception to this is criteria for the presence of halogenated dioxins and benzofurans and planar biphenyls. The establishment of risk in case of exposure to these compounds uses addition on the basis of equivalent toxicity [40]. This is a major improvement, though it has been pointed out that this approach may still underestimate the risk of neuro-development effects [41].

b. Importance of combination effects

Combination effects may be important in two respects. Firstly, coexisting soil contaminants may impact each others' biological availability [43]. Secondly, exposure to a combination of pollutants may be associated with antagonistic, synergistic and additive interactions of these pollutants, impacting their effect on organisms [44-47]. Some risks of pollutant mixtures can be predicted on the basis of existing knowledge. For instance there is a fair chance that there will be dose additivity when effects are receptor mediated [48]. Also in case of narcotic effects, joint-mixture ecotoxicological effects may be predicted [48]. If responses are dissimilar, response addition may be used [49]. A methodology to deal with the ecotoxicity of mixtures giving rise to both dose-additive and response-additive effects has been proposed [48]. This two step model evaluates mixture toxicity for the same mode of action with concentration additivity and the toxicity for different modes of action with response additivity. For determining the severity of ecotoxicological effects in case of heavily polluted soils (in which legal maximum tolerable levels for one or more substances are exceeded), a systematic approach to combination effects based on a mixture of concentration addition and response addition has been proposed [41].

2.10.5. Remedies for shortcomings

Remedies would seem possible which would allow for a significant improvement in risk estimates. Unregulated substances can get standards. Standards may be regularly updated on the basis of new dose-effect studies. Risk estimates can include both background exposure and all exposure routes for local soil pollution. Estimates of biological availability can be integrated in risk assessments and improved by better testing of bioavailability or by in-vivo monitoring [51]. The deficiencies in taking account of combination effects in ecotoxicity, discussed in section may be addressed by directly testing of ecotoxicity, when the focus is on ecosystem functioning [52,53]. However it should be noted that small effects on the functioning of ecosystems may have large effects over time [35]. This necessitates large numbers of replicate tests that may well be beyond routine practice [35].

In determining combination effects on human health, direct testing on humans is an 'unethical option'. However biomarker-based monitoring of some aspects of soil pollution relevant to humans may be an option. For instance Roos et al. [54] have applied a biomarker based test to original and remediated soils that were contaminated by a variety of polycyclic aromatic hydrocarbons (PAH). They tested the expression profile of cytochromes P 450 [54]. Xiao et al. [55] have measured genotoxic risk of soil contamination using an in-vitro assay with Salmonella. Though the relation between such biomarker-based data gathered and the in-vivo risks awaits further elucidation, the application of tests based on biomarkers for soil pollution is an interesting option in dealing with combination effects on humans.

Also, estimates of risk may be derived from biomarkers which may be monitored in people exposed to soil pollution. Such biomarkers have emerged from epidemiological studies considering the combined effect of substances. An illustration thereof is the study by Lee et al. [56] which found a graded association of the concentration of blood lead and urinary cadmium concentrations with oxidative stress related markers in the US population. This suggests that oxidative stress may be useful as a biomarker for combination effects. It has furthermore been proposed to evaluate effects of exposure to nitroarenes by measuring haemoglobin adducts [57], and of mixtures of volatile organochlorines by measuring glutathione conjugative metabolites [58]. Bioassays based on aryl hydrocarbon (Ah) receptor mediated mechanisms have been proposed which will allow a better alternative to the measurement of polyhalogenated aromatic hydrocarbons [41]. Another option is to estimate risks to human health by taking into account cumulative combination effects in line with established cause-effect relations and research into the effects of actual combinations. It has been shown that risks of compounds with the same targets and the same modes of action may be estimated on the basis of concentration addition, while including toxicity equivalence factors for the compounds involved [59].

This has been shown to apply to receptor-mediated-and reactive mechanisms of toxicity, provided that no chemical reactions occur between the components of the mixture considered [60]. Currently this approach is applied to halogenated dioxins, benzofurans and planar polybiphenyls, though non-linear interactions are not completely absent in this category of compounds [61], and neurodevelopment effects may be underestimated, as pointed out before [41]. Extension of this approach is possible to e.g. polycyclic aromatics, including heterocyclic polycyclic aromatics [18,62] organophosphates that inhibit the enzyme cholinesterase [44,63], compounds that bind to estrogen receptors [64-66], carcinogens [67], a variety of petroleum products [68] and compounds that inhibit the MXR efflux pump [69].

3. Ecological risk assessment

Ecological risk assessment (ERA) is a process of collecting, organizing, and analyzing environmental data to estimate the risk or probability of undesired effects on organisms, populations, or ecosystems caused by various stressors associated with human activities. The basic principles of ecological risk assessment are described in numerous papers [70-72]. All varieties of ERA are associated with uncertainties. The value or usefulness of the different ERA methodologies depends on the uncertainty, predictability, utility, and costs. There are typically two major types of ERA. The first is predictive and is often associated with the authorization and handling of hazardous substances such as pesticides or new and existing chemicals in the European Union. This kind of ERA is ideally done before environmental release. The second type of ERA could be described as an impact assessment rather than a risk assessment, as it is the assessment of changes in populations or ecosystems in sites or areas already polluted. The predictive method is based on more or less generic extrapolations from laboratory or controlled and manipulated semi field studies to real-world situations. The descriptive method is more site specific as it tries to monitor ecosystem changes in historically contaminated soils such as old dumpsites or gas facilities or in field plots after amendment with pesticides or sewage sludge, for example.

Often ERA is performed in phases or tiers, which may include predictive as well as descriptive methods. The successive tiers require, as a rule of thumb, more time, effort, and money. The paradigm or schemes for ERA may vary considerable from country to country, but often consist of an initial problem formulation based on a preliminary site characterization, and a screening assessment, a characterization of exposure, a characterization of effects, and a risk characterization followed by risk management. Although exposure assessment is often just as or even more important, this chapter primarily considers effect assessment. In most European countries, ERA of contaminated soils consists of rather simplified approaches including soil screening levels (SSL) (a.k.a. quality objectives, quality criteria, benchmarks, guideline values) and simple bioassays for a first screening of risk [73-75]. National research or remediation programs have led to the development of a large variety of guideline values.

Although hard to categorize, most fall into two categories: generic or site specific. While the site-specific guidelines require a characterization of pH, organic matter, etc., at the site, generic guideline values are more independent of modifying factors and hence straightforward to legislate. Three major classes of tools for assessing ecological effects may be identified: standardized ecotoxicity experiments with single species exposed under controlled conditions to single chemicals spiked to soil; *ex situ* bioassays, here defined as simple laboratory assays where single species are exposed to historically contaminated soils collected in the field; and finally monitoring, analyzing, and mapping of population or community structures in the field. Furthermore, mesocosm, lysometer, or terrestrial model ecosystems (TME) may be useful; these may be considered as large (multispecies) bioassays or ecotoxicity tests [76-79]. TMEs have the advantage that they operate with the (relatively) undisturbed intrinsic soil populations that make up a small food web. TME hence allow the assessment of effects of toxicants that are mediated through changes in food supply or competition and predation.

One of the keystones in deriving environmental quality criteria is the use of standardized terrestrial test procedures. The emphasis of these prognostic tests is on reproducibility, standardization, international acceptance, and site independence. Although increasing in numbers, relatively few terrestrial tests are still approved by the International Standardisation Organisation (ISO) or Organization for Economic Cooperation & Development (OECD). However, other tests have shown promising results and are likely to be prepared for standardization in the future [80].

However, the major problem in using simple laboratory tests to extrapolate to contaminated land may not be the limitations of test species and the natural variation in species sensitivity. The problems associated with extrapolating from one or a few species, exposed under controlled and typically optimal conditions, to the complex interaction of species and chemicals found in most contaminated ecosystems should also cause concern. Although single-species laboratory tests with spiked materials have their obvious benefits, e.g., they measure direct toxicity of chemicals and interpretation is therefore simple, supplementary tools are often needed. Bioassays, as defined in this context are one of the more frequently used higher-tier alternatives. Basically the same test species may be used in bioassays for assessing the risk of a specific contaminated soil as in standard laboratory tests. However, bioassays have the advantage, compared to the use of spiked soil samples, that the exact toxicity of a specific soil may be accessed directly: this includes the combined and site-specific toxicological effect of the mixture of contaminants and their metabolites. Furthermore, the *in situ* bioavailability of that specific soil is (at least almost) maintained in the laboratory during the exposure period. Several studies have shown a reduction in bioavailability and/or toxicity of soils with an old history of contamination [80-85].

Bioassays are therefore often considered a more realistic tool than generic soil screening levels based on spiked laboratory soils. However, a number of uncertainties or problems may be associated with the use of bioassays and the interpretation of their results. First, the test species are still exposed to the contaminants in a relatively short period compared to the permanent exposure condition found at contaminated sites. Furthermore, they are exposed under more or less optimal conditions, in that stressors such as predation inter- and interspecies competition, drought, frost, and food depletion are eliminated during exposure. Finally, typically only a few species are tested individually.

To compensate for some of the limitations just described, contaminated soil may be assessed using multispecies mesocosms, lysometers, or TME. In these, species interactions may be evaluated by manually introducing several species to the systems or monitoring the intrinsic populations of the soil. Natural climatic conditions may be included if the test system is kept outdoors. However, if we want to get a more realistic and large-scale picture of the impact caused by, for example, pesticide use or sewage sludge application, or to assess the environmental health at waste sites, industrial areas, or gas works, it is often necessary to conduct some kind of field observations. Several case studies exist in which field studies have successfully elucidated the ecological risk of specific activities or the ecological impact at specific sites [85-87]. The small single-species bioassay, large multispecies TME, and field surveys have some drawbacks in common. First of all, it may be difficult to actually link the observed effect to a specific toxic component in the soil. Which of the many substances is actually causing the majority of the observed effects, or is it perhaps a combination of effects? For a hazard classification of soils or a ranking of soils this may not be so important.

However, to evaluate potential risk-reduction measures or risk management procedures it may be important to identify the most problematic substances. A comparison of soil screening values with measured concentrations for each chemical present at a site may be helpful to identify the most likely group of substances causing the observed effect. Other possible tools may include a toxicity identification evaluation (TIE) approach [88]. The TIE approach is a relatively new method, which aims to identify groups of toxicants in soils with mixed pollution. Potentially toxic components present in the soil are fractionated and determined, and the toxicity of each individual fraction is determined by a *Lux* bacteria-based bioassay or the Microtox bioassay. Although perhaps promising, TIE is a time-consuming and hence costly procedure not yet used routinely. Another crucial issue when analyzing the result of bioassays, TME, and field studies is the presence or absence of a proper reference site or soil. The control soil should in principle resemble the contaminated soil in all relevant parameters, e.g., texture, pH, organic matter, waterholding capacity, and nutrient content, a practical problem that very often is difficult to solve. The lack of adequate control or reference sites may, however, be conquered at least partially by the use of multivariate techniques [89], which relate the species composition and abundance to gradients of pollutants. It is not the intention of this chapter to present a review of statistical tools for ecological risk assessment, and hence a detailed discussion about the use of these is not given.

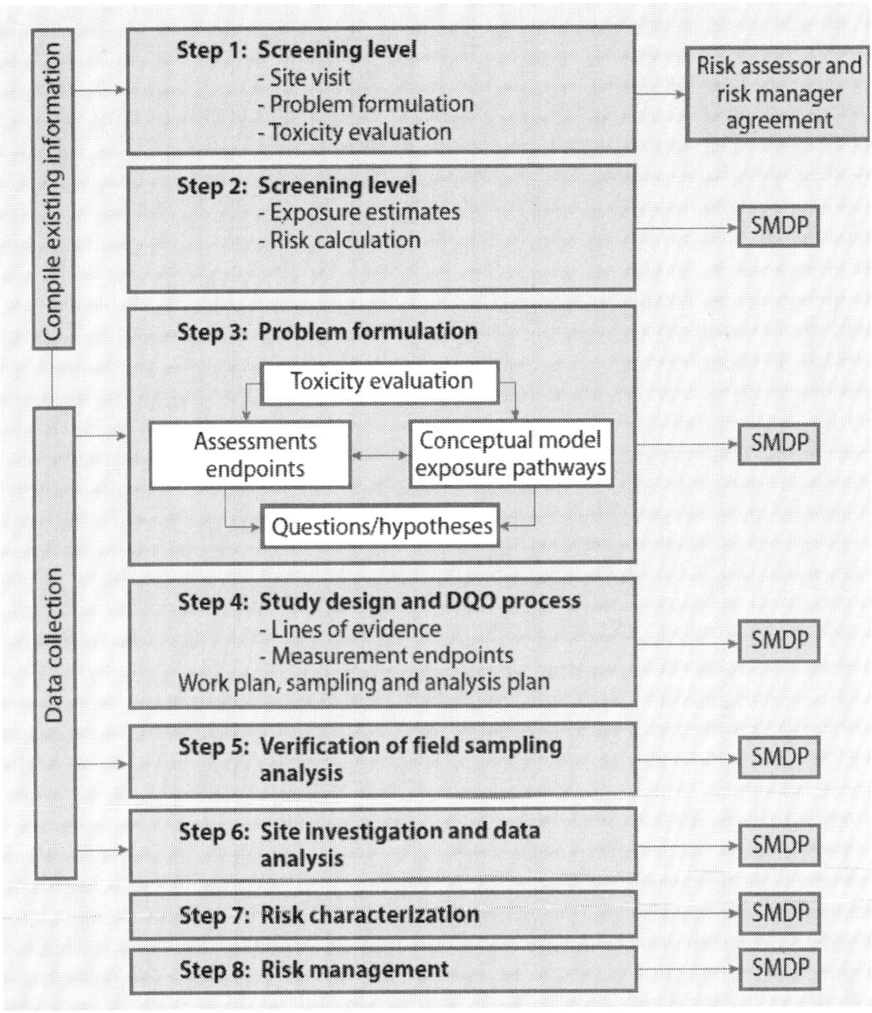

Figure 2. The eight steps in the US-EPA framework for risk assessment of contaminated Superfund sites. DQO = data quality objectives [89]

However, it is obvious that increased computer power and the presence of new easy-to-use software tools have increased the possibility to move away from more conventional univariate statistics such as analysis of variance (ANOVA) to more powerful multivariate statistics that use all collected data to evaluate effects at a higher level of organization. Statistical methods such as the power analysis may also be very useful in planning and designing large-scale ecotoxicity studies such as mesocosms,TME, or field surveys.

The US-EPA has published an Ecological Risk Assessment Guidance, which should be followed when assessing risks at Superfund sites. As all sites are considered unique this should always be done in a site-specific manner. The ERA process suggested by the US-EPA for Superfund sites follows an eight step process, which can be broken down into four categories, i.e. 1) planning and scoping, 2) problem formulation, 3) stressor response and exposure analysis and 4) risk characterisation. Essential for all steps are a negotiation and agreement of the need for further action between the risk assessor, the risk manager and other stakeholders, the so-called scientific-management decision points (SMDP).

SMDP made at the end of the screening-level assessment will not set an initial cleanup goal. Instead, hazard quotients, derived in this step, are used to help determine potential risk. Thus, requiring a cleanup based solely on those values would not be very likely, although it is technically feasible. There are three possible decisions at the SMDP:

1. There is enough information to conclude that ecological risks are very low or non-existent, and therefore there is no need to clean up the site on the basis of ecological risk.

2. The information is not adequate to make a decision at this point, and the ecological risk assessment process will proceed.

3. The information indicates a potential for adverse ecological effects, and a more thorough study is necessary.

In the Netherlands contaminated sites are first determined using a set of soil screening levels called target and intervention values, which take both human and ecological risks into account. At seriously contaminated sites remediation or other soil management decisions are required if the risks cannot be neglected based on a site-specific ecological and human risk assessment, and the chance for dispersion of the contaminants. Until now, the ecological risk assessment has been based on chemical analysis, including a Decision Table harbouring critical dimensions of the impacted area. The United Kingdom and Canada have also developed framework for ecological risk assessment of contamination land. A cornerstone in the UK framework of ERA is the connection to the statutory regime for identification and control of land potentially affected by contamination. The UK framework is based on schemes found in e.g. USA, Canada and the Netherlands. Like these it is a based on a tiered approach where the initial Tier 0 aims to determine whether a site falls under the Part IIA of the legislation. It involves the development of a Conceptual Site Model (CSM), which described what is already (historically) known about the site, e.g. whether there is a likely source-pathway-receptor linkage. The conceptual site model is followed by an initial screening phase (Tier 1) and an actual site-specific characterisation (Tier 2). Tier 1 is a simple deterministic comparison of chemical residue data and the soil quality guideline values supplemented with simple soil-specific toxicity testing. The final step (Tier 3) involves more detailed *in-situ* studies and for example ecological modelling based on a more advanced ecological theory. Tier 3 is not likely to be conducted at many sites.

3.1. Descision support system for ecological risk assessment

Ecological Risk Assessment is often a complex process with many variables to take into account. ERA involves many stakeholders and all have to be dealt with in a clear and consistent way. A stepwise or tiered approach is therefore useful to overcome the complexity of an ERA. In order to structure all the information collected, a Decision Support System (DSS) can be used. Each tier will lead to a decision to proceed or to stop. A number of decisions supporting systems or frameworks have already been developed in other countries, e.g. UK, the Netherlands and the USA. The DSS presented here is based on basic principles also common in the methodologies used in the USA and UK. However, in the present DSS measures of bioavailability and the use of the Triad approach may be built into the system more systematically. This chapter introduces the overall framework of a novel DSS including the Triad approach and the challenge to weight and scale results used in that process.

3.2. Framework for ecological risk assessment

Rutgers et al. (2000) developed a basic flowchart for Ecological Risk Assessment [90], which is used as the backbone of the decision support system (DSS) presented in the Figure 5.

The DSS is separated in three different stages, i.e.

- Stage I. Site characterization and description of land-use.
- Stage II. Determination of ecological aspects.
- Stage III. Site-specific tiered assessment (the Triad):

– Tier 1. Simple screening

– Tier 2. Refined screening

– Tier 3. Detailed assessment

– Tier 4. Final assessment.

Each of these four tiers is based on a weight of evidence (WoE) approach combining three lines of evidence (Chemistry, (eco)Toxicology and Ecology).

3.2.1. Boundaries of the DSS

The DSS in this chapter is not a full and comprehensive document for managing risk of contaminated land. It focuses strongly on supporting decisions made when considering risk to the terrestrial environment. Therefore it addresses only indirectly the risk to ground water and associated (connected) fresh water systems. Nevertheless information about e.g. reduced bioavailability may be useful when assessing potential risk for leaching of contaminants to ground water or fresh water. Furthermore, it is important to realise that the management of a contaminated site is more than assessing ecological risk. Issues like for example risk for humans, availability and cost of remediation solutions, development plans for the vicinity or the region are equally important.

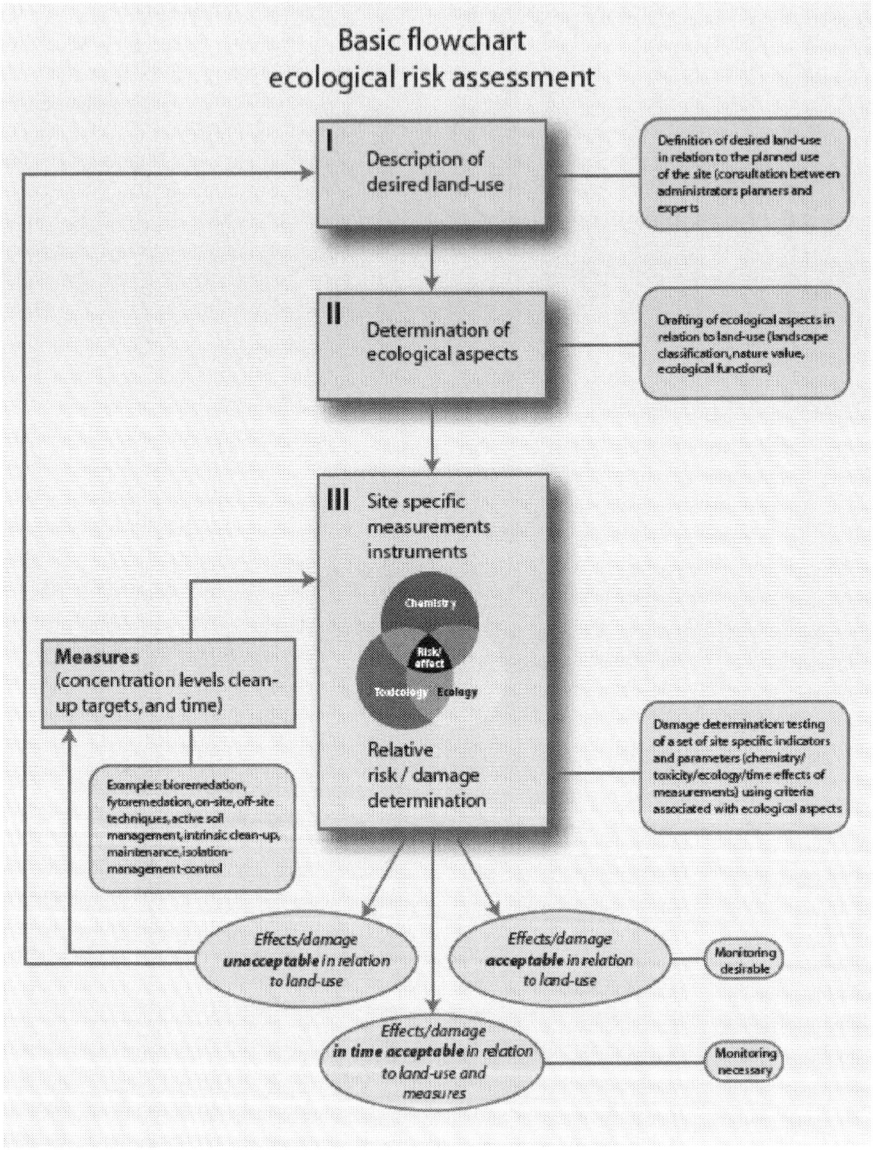

Figure 3. Basic flowchart for ecological risk assessment [90]

3.2.2. *Stage I — Site characterisation and land-use definition*

The first step in the DSS is to establish what is often referred to as a Conceptual Site Model. It aims at involving as many stakeholders as possible in order to describe site characteristics and to review all available information from the site, e.g. historical information about land-use, investigation of whether the site may be regulated under specific directives, obvious data gaps and urgency for reaction and data collection. The spatial borders of the site should be defined and the current and the future landuse have to be defined. Consultation between administrators, planners and experts therefore has to take place as early as possible in the process.

a. Initial requirements in the DSS

An inquiry among all stakeholders should be conducted as one of the first initiatives. The aim should be to collect as much information about soil characteristics as possible.

b. Defining land-use

One of the first actions to be taken among all stakeholders is to decide which landuse is required for the site, as this will determine the required data collection and testing. Many land-uses may be defined, but generally the four following overall categories of land-use classes are used:

- industrial area (including infrastructure and pavement).
- urban/residential area (including recreational and green areas).
- agricultural area.
- nature area.

c. When is an ecological risk assessment needed?

Most often a site specific ERA will be initiated only when soil concentrations exceed soil screening levels. However, this may not in itself be a sufficient criterion to go through the entire ERA procedure. Some boundary conditions, based on the present and future type of land-use, the level of contamination and various ecological considerations have to be met in order to rationalize an ERA. The experts and the rest of the stakeholders should answer a number of simple questions in order to conclude whether the required boundary conditions are fulfilled.

3.2.3. *Stage II — Determination of ecological aspects*

At stage II, site-specific ecological features and receptors relating to the land-use defined in Stage I need to be outlined. This includes aspects like key species and life support functions. The potential ecological receptors should be identified in order to determine whether potential source-pathway-receptor linkages can be established. This includes not only ecological receptors directly linked to the site but also those linked indirectly e.g. through leaching of contaminants to connected fresh water systems or (migrating) birds or mammals feeding in the area. In Table 9 some examples are given of land-use and related ecological aspects. This table can be used as a starting point for the selection of ecological aspect. Experts from ecotoxicology and ecology should be involved in the selection of ecological aspects.

3.2.4. *Stage III — Site specific instruments (the Triad)*

If after finalising Stage I and Stage II it is still considered that there is a need for a site specific evaluation of ecological risk the process continues to Stage III using the weight of evidence approach described below.

3.2.5. *Weight of evidence approaches*

In order to deal with conceptual uncertainties in a pragmatic way, it has been proposed to use weight of evidence (WoE) approaches for ERA [90-93]. The rationale is, like in justice, that many independent ways to arrive at one conclusion will provide a stronger evidence for ecological effects, making ERA less uncertain.

In the sediment research area the application of WoE started at an early stage and was called the Sediment Quality Triad. For terrestrial ecosystems WoE approaches and the Triad are still in a developing stage. The Triad approach is based on the simultaneous and integrated deployment of site-specific chemical, toxicological and ecological information in the risk assessment as given in Figure 6. The major assumption is that WoE in three independent disciplines will lead to a more precise answer than an approach, which is solely based on, for example, the concentrations of pollutants at the site. A multidisciplinary approach will help to minimise the number of false positive and false negative conclusions in ERA. It also gives acknowledgement to the fact that ecosystems are too complex to analyse in one-factorial approaches.

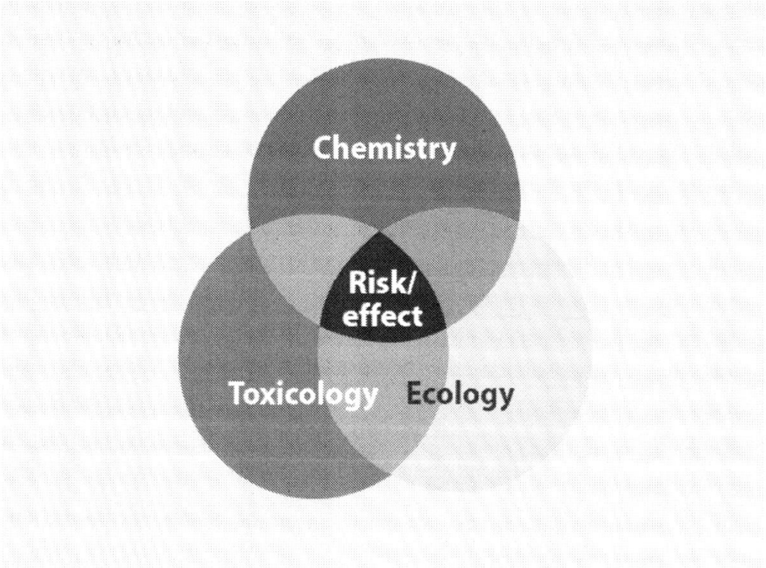

Figure 4. Schematic presentation of the integration of three fields of research according to a Triad [92]

http://dx.doi.org/10.5772/57287

- **Chemistry:** The concentration of contaminants in the environment (totals, bioavailable), accumulated in biota, or modelled via food-chains is used for calculation of risks on the basis of toxicity data from the literature.
- **Toxicology:** Bioassays with species across genera are carried out in order to measure the actual toxicity present in environmental samples from the site.
- **Ecology:** Field ecological observations at the contaminated site are compared to the reference site. Deviations from the reference site, which can be plausibly attributed to the contamination levels, are funnelled into the Triad.

3.3. Using the TRIAD in site specific assessment of contaminated soil

Triad is a powerful weight of evidence approach originally developed in order to evaluate sediment quality. In the terrestrial compartment less experience is available on the practical use of the Triad. This chapter describes the use of Triad in more detail and gives an insight into some of the important decisions risk assessors have to make when conducting the Triad in practise, e.g. how to scale, weight and integrate the outcome of the various investigations.

The Triad approach exists of three lines of evidence (LoE), the so-called Triad "legs", i.e. chemistry, (eco) toxicology and ecology. The Triad approach includes a tiered system in which each consecutive tier is increasingly fine-tuned to the site-specific situation. In the first tier the research is simple, broad and generic. In later tiers more specific and complex tests and analyses may be used. For each of the LoE in the Triad there are a variety of analyses or tests that can be chosen. Some examples are:

- **Chemistry:** Measurement of total concentrations, bioavailable concentrations, bioaccumulation, etc.
- **Toxicology:** Bioassays (in field and/or in lab), biomarkers etc.
- **Ecology:** Field observations of vegetation, soil fauna, micro-organisms, etc. In Chapter 6, a number of tests or tools that are for suitable for use in each tier are presented for the chemistry, toxicology and ecology LoE.

3.4. Decision charts in ecological risk assessment of contaminated sites

3.4.1. Flowcharts

This chapter is an attempt to present a decision support system, which can guide risk assessors in their assessment of site-specific ecological risk. A number of site-specific questions need to be answered before a final decision on performing an ecological risk assessment can be made. This chapter introduces a flow chart for ecological risk assessment of contaminated sites. The flowchart is presented as decision trees as shown in Figure 8 together with a more in-depth introduction to the relevant questions that needs to be addressed and answered when performing a site-specific ecological risk assessment.

3.4.2. Decision making in ERA

The assessment of ecological risk is performed stepwise in tiers. Higher tiers represent gradually more and more complex studies, but also more expensive and laborious studies. The full site-specific risk assessment covers four tiers, i.e.

- Simple screening: Tier 1.
- Refined screening: Tier 2.
- Detailed assessment: Tier 3.
- Final assessment: Tier 4.

The main principle in going from a simple screening over a more refined screening to a detailed assessment of the contaminated site is to minimize time and effort. The actual performance of the risk assessment and use of the various tiers may be very site-specific.

a. Tier 1 – Simple screening

After deciding in the two first stages of the ERA that ecological concern needs special consideration, the risk assessment starts typically with a simple evaluation at the screening level. This is done in order to minimize costs until new information indicates the need for further assessment and more sophisticated studies. Therefore, the tools used in the first screening need not only to be reasonably quick and easy, but also relatively cheap. The tools for use in Tier I are described in more detail in the toolboxes C1, T1 and E1. On the basis of the results of instruments used in Tier 1 it is decided to either stop further assessment or continue to a higher tier.

b. Tier 2 – Refined screening

Tier 2, still considered being at the screening level, aims at refining the measurement of exposure and at the same time to provide further insight into the toxicological and ecological properties of the contaminated soil. Tier 2 deviate from the conservatism normally associated with the use of total concentration in the risk assessment by taking (rough) estimations of bioavailability into consideration in the chemical LoE. A better screening of the toxicological and ecological properties of the soil compensates for the reduced conservatism in the Chemistry LoE of the Triad. The tools for use in Tier 2 are described in more details in the toolboxes C2, T2 and E2. On the basis of the results in Tier 2 a decision should be made to either stop further assessment or continue to a higher Tier.

c. Tier 3 – Detailed assessment

The tools in Tier 3 differ from the ones used in Tier 1 and Tier 2 in that they are more laborious, costly and may take longer. On the other hand they are (often) more realistic and/or ecological relevant in order to give a more comprehensive assessment of the ecological risk at the specific site. The stakeholders should beforehand negotiate a minimum set of tests. Is it for example necessary to consider all trophic levels in the toxicological and ecological LoE? Or does the land-use suggest otherwise? Is it necessary (or possible) to estimate the bioavailability of all the substances exceeding their SSL? If not, how are the non-investigated substances dealt with?

The tools described for use in Tier 3 are described in more details in the toolboxes C3, T3 and E3. Depending on the results from Tier 3 a decision should be made to either stop further assessment or continue with an even more detailed assessment in Tier 4

d. Tier 4 — Final assessment

In Tier 4, the aim of the studies is to answer any remaining questions and to decrease existing uncertainties and this may often require more in-depth research. Tools in Tier 4 can be similar to tools of Tier 3, but more focus has to be on site-specific circumstances. For example bioassays should be done with organisms, which normally occur at the site. Furthermore, it may be more relevant to consider ecological effects outside the contaminated area on e.g. predators or herbivores feeding in the area or effects in adjacent fresh water systems. This Tier requires specialised knowledge and experience with ERA, which implies that costs can be high and only a limited number of people may be able to perform the tests. Generally only on a very limited number of site evaluations will include investigations at this level. If the results of Tier 4 still indicate risk there are basically two possible solutions. Accept the risk and leave the contamination or remove (parts of) the contamination.

3.4.3. Screening tools

3.4.3.1. Triad based selection of methods

For each of the three Lines of Evidence (LoE) in the Triad various methods or tools are available. In order to facilitate the selection of appropriate tools in the right context, the tools have been compiled in subclasses or toolboxes. Each of these is a collection of tools considered to be potentially useful in the designated tiers and LoE of the Triad, i.e. chemistry, toxicology and ecology. Furthermore, the tools are arranged according to their complexity, price and practicability or in other words depending on whether they are most useful for screening or detailed assessment, i.e.

- Toolbox C1. Chemistry tools for simple screening.
- Toolbox T1. Toxicology tools for simple screening.
- Toolbox E1. Ecology tools for simple screening.
- Toolbox C2. Chemistry tools for refined screening.
- Toolbox T2. Toxicology tools for refined screening.
- Toolbox E2. Ecology tools for refined screening.
- Toolbox C3. Chemistry tools for detailed assessment.
- Toolbox T3. Toxicology tools for detailed assessment.
- Toolbox E3. Ecology tools for detailed assessment.
- Toolbox IV. Various tools for the final (Tier 4) assessments.

a. Toolbox C1 — Chemistry tools for simple screening

At the very first stage of the ERA process, total concentrations of all relevant chemicals are individually compared to soil screening levels (SSL) in order to evaluate whether there is a need for a site specific assessment of ecological risk. In the current Stage III of the ERA, this first generic evaluation of risk is followed by a more site-specific screening of risk including information from all three lines of evidence in the Triad. In the Chemistry part of the Triad more site-specific information is collected by:

Refining and targeting the comparison of soil concentrations with soil related benchmarks for site-specific purposes.

Incorporation of the accumulative risk of a mixture of contaminants by calculating the toxic pressure (TP) of a mixture and by doing so generating more site-specific insight to the potential ecological impact of a contaminated site. Each of these steps can be done separately or in combination, e.g. the TP can be calculated using existing SSL or using new developed benchmarks based on either NOEC or EC50 values or site-specific benchmarks can be compared to soil concentrations individually. The approach entirely depends on the strategy taken by the stakeholder group and the availability of data.

b. Toolbox T1 – Toxicology tools for simple screening

The main objective of the selected toxicity tests or bioassay at Tier 1 should be to screen the soil for presence of toxic compounds. This includes toxic degradation products or compounds, which are not routinely included in various national analytical programs for contaminated sites. This Tier is the first screening level of the ERA and the cost in form of manpower and money should hence be relatively low.

c. Toolbox E1 – Ecology tools for simple screening

Ecological surveys or monitoring studies are generally considered a time consuming effort performed by experts. This is in most cases true, wherefore detailed surveys normally take place in higher tier assessment. However, in order to ensure that also ecological information is collected and used in the Triad already in the screening phase, it is recommended to perform a limited examination of the site. A survey of the area with special focus on visible changes in e.g. plant cover or presence or absence of specific plants, trees or scrubs may indicate ecological damage, which can be associated to contaminants present at the site.

If any aerial pictures areavailable from the area these may give valuable information about the plant cover also historically, which may be helpful in identifying parts of the site where the impact may be highest (hot spots). At this stage the conclusion can in most cases only be indicative. Therefore if the results from the other line of evidence may cause any doubt or the survey indicated potential impact, it is recommended to either continue with a more refined screening in Tier 2 or go directly to the detailed assessment in Tier 3.

d. Toolbox C2 – Chemistry tools for refined screening

Selective solvent extraction

It may be considered useful to adjust the estimate of exposure by taking bioavailability into consideration and hereby deviating from the conservatism normally associated to the use of

http://dx.doi.org/10.5772/57287

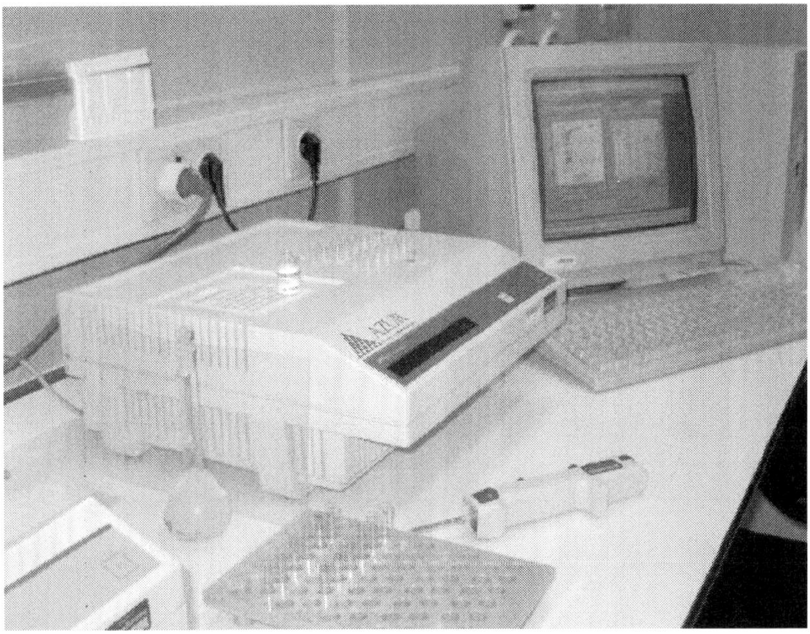

Figure 5. Equipment for measuring luminescence of Vibrio fischeri [94]

Figure 6. Simple survey of the site [95]

total concentration in the risk assessment. The principle in this refinement of the ecological risk assessment is to extract a more ecotoxicologically relevant fraction of the contamination than the total concentration. The latter generally tends to overestimate the risk of historically contaminated soils. In this screening phase no attempt is made to estimate the freely dissolved or readily bioavailable concentration of contaminants. Table 10 explains principal studies that employed chemical extractants to evaluate bioavailability.

Contaminant (Ref)	Solvent	Bioassay	Operation	Comments
Atrazine Phenanthrene (1)	Methanol/water, n-Butanol	Earthworm uptake Degradation	25 ml extractant, 10 g solid. Shaking for 2 h	Methanol/water best predictor for atrazine, n-butanol for phenanthrene.
DDT, DDE, DDD PAH (mixture) (2)	THF Ethanol	Earthworm uptake	15-20 ml extractant, 1 g soil, 10 sec mixing	Good correlation with earthworm accumulation.
Anthracene, Fluoranthene, Pyrene (3)	n-Butanol Propanol Ethyl acetate	Plant retention Earthworm uptake Degradation	25 ml extractant, 1-2 g soil, 5 sec mixing	Reasonable correlation with bioassays.
Phenanthrene Pyrene Chrysene (4)	n-Butanol	Earthworm uptake Degradation	15 ml extractant, 5-10 g soil, Mixing: 5 sec (worm) or 120 sec (degradation)	Applicable for bioavailability prediction.

Table 5. Outline of principal studies that employed chemical extractants to evaluate bioavailability

Instead the fraction of the contaminants is extracted, which can be directly compared to the existing soil screening levels. This is considered to be a relatively simple and quick method to screen for potential risk of contaminants in a more realistic way than using total concentrations. The extracted concentration (mg kg-1) is compared to the SSL and the result used in the Triad. It is therefore a prerequisite of this comparison that the extractability in the tests (with spiked soils) used for deriving SSL is close to 100% by the methods used. In most short-term tests (< four weeks) it will be reasonable to assume that only little "true" ageing or strong sequestering occurs and hence a majority of the spiked chemicals are still extractable with mild organic solvents. However, for most methods this still has to be fully validated.

Organic solvents most frequently used include methanol/water in different ratios, nbutanol, ethanol, propanpol, ethyl acetate and tetrahydrofuran (THF) given in Table 10. The method establishes preferential partitioning of hydrophobic contaminants to the extractant by increasing their solubility in the aqueous phase whilst removing pollutant compounds from soil surfaces establishing equilibrium conditions. No standard protocol has been adopted for mild chemical extractions in relation to bioavailability testing. Common methodology in literature primarily includes a soil sample to which a volume of chemical extractant is added (generally 1 – 10 g soil,15 – 25 ml extractant). This is followed by a period for mixing, e.g. vigorous mixing for10 – 120 seconds or shaking by orbital shakers for up to 2 hours. The extraction studies have mostly involved PAH and insecticides (including DDT, DDE, DDD

http://dx.doi.org/10.5772/57287

and atrazine) [93-95]. Studies that haverelated extractability with results from bioassays have generally focused on uptake and accumulation (% taken up by earthworms or plants) and bacterial degradation (%removed). Therefore, since convincing relationships between the chemical and biological tests were found it may indicate a potential for such extraction methods to predict bioavailability.

e. Toolbox T2 — Toxicology tools for refined screening

In the first simple screening of Tier I focus was on marine bacteria and aquatic/sediment living species. In Tier 2 relatively simple tests with soil dwelling species are used for a more refined screening of the soil samples, i.e. the earthworm survival tests and avoidance tests using soil invertebrates.

Figure 7. Organisms screening in soil [98]

The habitat function of soils is often assessed using the reproduction test with *Eisenia fetida*. The avoidance test with *Eisenia fetida* is a suitable screening test, which is less cost-intensive in terms of duration and workload than the reproduction test, and at the same time (normally) more sensitive than the acute test with the same species.

f. Toolbox E2 — Ecology tools for refined screening

In Tier 2 the observations from the survey may be expanded by simple on-site assessment of the overall soil functioning or biological activity of the soils. Recommended tools include bait-lamina sticks and simple microbial tests using general endpoints like soil respiration or C/N mineralization rates.

Figure 8. Bait-lamina sticks [99]

The main principle for tests at this level is to be relatively simple and cheap but at the same time to give valuable information whether or not the soil has lost some of its main services. Bait-lamina sticks for example have been demonstrated useful for describing biological activity of the soils in a general matter.

g. Toolbox C3 — Chemistry tools for detailed assessment

The objective of the tools found in this toolbox is to assess the bioavailable and freely dissolved fraction of pollutants found in pore water of soils from contaminated sites. The methods should (in principle) be able to mimic the fraction of organic pollutants available for uptake in biota. The collection of methods includes various non-depleting and depleting pore water extractions. Very few terrestrial ecotoxicity data are yet expressed as e.g. pore water concentrations. Instead, the outcome of the methodologies in this toolbox is compared with water quality standards.

h. Toolbox T3 — Toxicology tools for detailed assessment

The objective of the tools found in this toolbox is to evaluate the potential impact of contaminated soils to fauna and plants and hereby the entire ecosystems. Some of the methods use introduced, and not intrinsic, species. The benefit of this is a higher degree of standardisation, as the species used in these bioassays is easy to maintain in laboratory cultures compared to naturally occurring species. The drawback may be that their ecological relevance is less obvious. For example the compost worm *Eisenia fetida* is used as a surrogate to evaluate risk to soil dwelling earthworms. Two sets of bioassays are presented. One for directly assessing potential risk for soil dwelling species, including micro-organisms, plants and soil invertebrates, and one for assessing indirectly risk to aquatic species through e.g. leaching of contaminants. It is often anticipated that soil organisms are exposed to pollutants mainly through uptake from pore water. Therefore it may also be possible to evaluate, or at least to compare or rank, the risk of contaminated soil samples to soil dwelling organisms on the basis of the outcome of the aquatic test using elutriate or pore water. The choice of bioassays depends on a number of variables, e.g.:

- The current and future land-use, i.e. targets of protection.
- The size of the contaminated area.
- The potential for ground water or surface water contamination.
- The need of many simple tests or fewer more complicated tests.

Figure 9. Simple plant tests [91]

i. Toolbox E3 - Ecology tools for detailed assessment

In this late tier of the Triad, the objective of the activities is community or population response analysis, typically by conducting field surveys. As these studies (most often) are time consuming, costly and dependent on ecologically, taxonomically and statistical expertise they are most frequently done on large-scale sites with a long-term- remediation perspective. In fresh water ecosystem community surveys have been widely used with relative success. The absence of species from places where they would be expected to occur could be a strong identification of unacceptable levels of contaminants. However, this type of studies has only seldom been used for the terrestrial environment. The reasons for this are many. One of the dominants may be the lack of a concentration gradient and obvious "upstream" reference sites at most contaminated areas. No world-wide accepted guideline on how to plan and perform a terrestrial field survey is available and hence no straight-forward and easy-to-follow description can be given. The decision on when, where and how to conduct field surveys depends on a number of issues, e.g. the size of the area, the land-use, the type of contaminants present, time of the year and last but not least the time and money available to perform the study. Nevertheless, a number of general considerations have to be made in the planning phase of a successful field survey. These include (but are not limited to):

- Identify the targets of concern and the species to monitor.
- Elucidate the natural temporal and spatial variation before initiating a field study.
- Use statistical (power) analyses to determine the minimum number of samples or replicates needed to emonstrate the decided difference, e.g. 25% change.
- In order to establish a cause-effect relationship, a number of confounding parameters need to be characterized both at the reference and the test site, e.g. soil type, pH, salinity, hydrology, nutrient- and organic matter content and the presence of other contaminants.

As no single description on how to perform ecological surveys for contaminated sites can be given, some general considerations and useful references for this tier of the ecological risk assessment are given below for:

- Assessing impact in the overall biological activity and organic matter breakdown.
- Assessing impact on the microbial community.
- Assessing impact on the plant community.
- Assessing impact on the invertebrate community.

Reference data from reference sites, reference samples and literature

A crucial factor in a risk assessment is the quality of reference data, because the results of the site-specific ecological measurements or calculations are compared against these data. This is true for as well chemical information (i.e. background levels in that region), toxicological data from bioassays (i.e. site relevant reference soil and well characterized control soil in order to verify the test performance) and ecological field surveys.

The reference soil should in principle resemble the contaminated soil in all relevant parameters, e.g. texture, pH, organic matter, water-holding capacity, nutrient content. In practice, these ideal spots are difficult to find. If there is no or inadequate reference information, effects can only be determined in relative terms by comparison with other sites. This is usually adequate for determining the degree of urgency and/or the need for remediation. Reference data can be

Figure 10. Litter bags [94]

obtained by including reference sites (preferably more than one) in the sampling scheme, including reference measurements in the experimental set-up, or by obtaining reference data from the literature or by expert based judgment.

Higher tier assessment of the impact on biological activity and organic matter breakdown

In addition to the general information about biological activity in soils generated in Tier 2 from the bait-lamina test, other, slightly more laborious, tests may give additional information about the overall biological activity in soil, e.g. wheat straw decomposition (litter bag test) and cotton strip degradation. A review paper from Van Gestel et al. (2003) concluded that while the bait-lamina gave the best reflection of the biological activity of soil animals, e.g. earthworms, springtails and enchytraeids, the litter bag test and the cotton strip test are more indicative of the microbial activity in the soil [96]. Knacker et al. (2003) reviewed the usefulness of five different litter decomposition tests and concluded that the litter-bag test had distinct advantages over the others [97]. All of these simple tests only give insight into the overall activity in soils and the breakdown of organic material. They are hence most suitable on their own in cases of land-use with low sensitivity, e.g. industrial land. For land-uses where structural endpoints, e.g. biodiversity or specific species, are the target of protection other endpoint(s) should be monitored as well.

Higher tier assessment of the impact on the microbial community

The number of microorganisms, especially bacteria, in soil is extremely large. They differ widely in their function and sensitivity to chemicals. Besides more classical (and simple) measures of the microbial community like total bacterial biomass, the number of colony forming units and substance induced respiration rate (SIR), more advanced methods for assessing the impact of contamination on soil microorganisms have recently been made available.

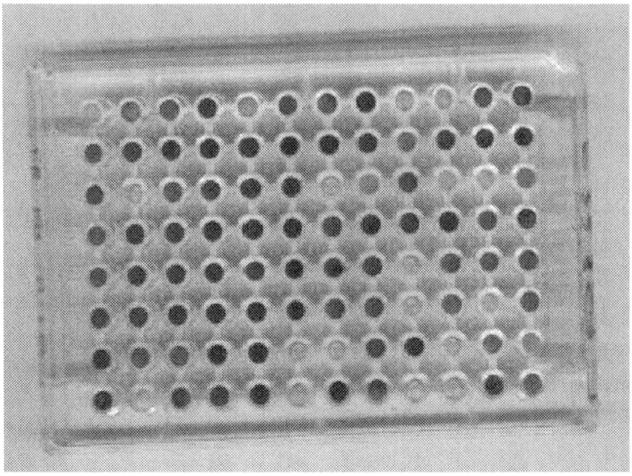

Figure 11. Microbial community [96]

This include microbial "fingerprinting" like phospholipid fatty acid analysis (PLFA) and community-level physiological-profiling (CLPP) based on metabolic response using BIOLOG plate systems, and the use of pollution induced community tolerance (PICT).

Higher tier assessment of the impact on the plant community

Plants interact dynamically with the physical and chemical characteristics of soils. Soil types and site characteristics, therefore, greatly influence the occurrence of plants and their total aboveground biomass (also called, Net Primary Production, i.e. NPP) within given climatic conditions and human management. Vegetation cover is an important indicator of soil quality and a diverse plant community is normally a good indication of essential soil functions such as the decomposition process, the mineralisation rate, and the occurrence of soil dwelling animals associated to (fresh) organic matter. Vascular plants are easy to sample. They are immobile and hence associated to soil contamination (and airborne pollution). Plant community parameters like plant cover, aboveground plant biomass, plant shoot/ root ratio, species diversity and the binary occurrence (presence/absence) of specific indicator species like metal-tolerant species may be used successfully in ERA.

Figure 12. Plant survey [95]

Higher tier assessment of the impact on the soil invertebrate community

Survey of soil biota in order to evaluate the effect of various sources of pollution on soil communities on historically contaminated sites have not yet been used on a larger scale by e.g. consultants. However, numerous (monitoring) studies by various research groups can be found in the open literature. Methods of surveying include:

- Collection of soil samples followed by extraction in the laboratory.
- Extraction or collection of organisms in the field, e.g. by hand-sorting or by the application of mustard or formalin.
- Trapping (surface dwelling) animals by the use of e.g. pit-falls. Monitoring species includes earthworms, snails, oribatid mites, nematodes, springtails, ants, ground-living beetles and spiders. Most of the studies have been done on metal contaminated sites (see references below). A substantial amount of work has been put into the challenge of developing a soil invertebrate system for evaluating risk of pollutants. The only soil invertebrate system that is used on a regularly basis in the context of ecological risk assessment of contaminated soils is most likely the nematode Maturity Index (MI). The system is based on the evidence, that rapid colonising species dominate nematode communities in disturbed ecosystems. In the

Netherlands experience with surveys of soil invertebrates from the monitoring programme Biological Indicator for Soil Quality (BISQ) has also been used in ERA..

Figure 13. Soil fauna sampling [98]

j. Toolbox for tests in Tier IV

The final assessment in the ERA process is not likely to be initiated for many contaminated sites. The choice of additional tests or monitoring at this level of the ERA is bound to be very site-specific and hence an issue for negotiation between stakeholders and experts. Accumulation in biota is included in this toolbox as the internal concentration in biota is believed, at least to some extend, to reflect uptake and then bioavailability. An alternative in this final tier could also be to model uptake in biota provided sufficient data is available [98].

4. Soil management program (Remediation)

4.1. Reporting norms and standards for contaminated land

The aim is to investigate whether the land has been contaminated, and if contamination has occurred whether the contamination presents a significant risk of harm. The commonly encountered international practice consists of three distinct reporting phases. It progresses

http://dx.doi.org/10.5772/57287

Figure 14. Detailed field survey [93]

from Phase 1 desktop and site walkover assessments with limited investigation and testing to a Phase 2 detailed invasive investigation and testing for site characterisation to a comprehensive Phase 3 report with an evaluation of remediation objectives and a proposed remediation plan, supported by control and monitoring measures for the activities.

The reporting system requires norms and standards of practice to be strictly applied, but also must retain flexibility to allow for decisions on the contaminated status of sites to be made in the most beneficial manner (considering ecological, social and economic aspects) also taking into account timeframes. In some cases, urgent priority works may require that the phased approach to reporting has to move forward in a concurrent single report.

4.2. Requirements for preliminary site assessment phase 1

A preliminary site assessment must consider the following elements

1. Site description-location and size

2. Nature and extent of contamination, contaminants of concern and historic activities that may be sources of contamination. List all present and past activities at the site that involves

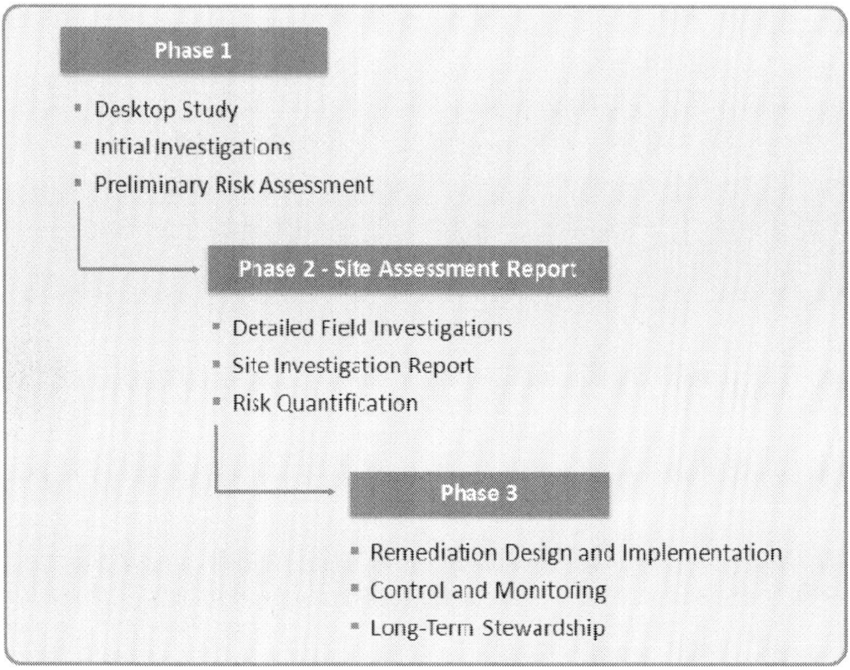

Figure 15. A phase approach for the assessment and remediation of contaminated land

storage and production, use, treatment or disposal of hazardous material that could contaminate the site.

3. Describe current condition of the site and contents and the results of any previous assessment report
4. Local topography and geology, drainage, surface cover, vegetation.
5. Status of ground water, approximate depth to water table
6. Proximity to surface water
7. Proximity to drinking water supplies
8. Annual rainfall and flood potential
9. Land and water use for the nearby areas and site
10. Any other regulations as Regulated by the ministry

All data may not be available, or data may vary in terms of uncertainty, it is thus important to recognize gaps in the knowledge base and to decide whether additional data must be obtained on the site characterisation. This may trigger the commencement of Phase 2 Investigations. The Phase 1 report must make clear recommendations on the status of the contamination risk posed by the site. If a complete site history clearly demonstrates that the site activities do not pose a contamination threat then no further investigation is warranted and the site should be recommended as suitable for reuse. In most cases it is likely that some level of preliminary investigation will be required to provide the level of certainty required to enable property re-development or transfer. A limited investigation of certain subsurface activities, for example, underground storage tanks, would be necessary to obtain a waiver on the contamination status of a site at a Phase 1 level of reporting.

4.3. Approval requirements

If soil contaminants are found at concentrations that exceed the applicable standards specified in the approval, the approval holder is required to implement a soil management program.The program must first address source control to stop on-going contaminant releases. After the sources of contamination have been stopped, further assessment and delineation of the contaminated area may be necessary. When the extent of contamination is understood, remediation objectives, as described below, must be adopted for the area. Once remediation objectives have been agreed to, appropriate treatment or containment technologies can be chosen and the management plan finalized.

4.4. Developing remediation objectives

Remediation objectives may be developed in a variety of ways ranging from generic guidelines to site-specific risk assessment. Generic guidelines are numerical concentration limits that are applicable under a variety of site conditions. When neither Environmental Protection nor the agency has a guideline for a particular substance, four options are available. First, the remediation objective may be based on the ambient background concentration for the site. Second, the guideline development protocol may be applied. Third, a remediation objective may be adopted from another jurisdiction if the proponent can show that the remediation objective is consistent with the environmental protection goals of the approval. Finally, a remediation objective may be developed by the proponent using site-specific risk assessment procedures. A risk assessment for an approved facility will focus on human health concerns but fundamental ecological concerns must also be addressed. Site-specific risk assessment is a means of quantifying the likelihood that soil contamination will have a harmful effect under conditions found at a specific site. The essential components of human health and ecological risk assessments are similar; however, ecological risk assessments tend to be more complex than those for human health because a wide variety of receptors may have to be considered. Very briefly, a site-specific risk assessment may be described as consisting of the following steps or components as shown in Figure 20;

Problem formulation involves developing a conceptual model of the possible contaminant effects on receptors at the site. The conceptual model describes contaminant distribution and concentration in relation to the receptors and their patterns of activity on the site. The *exposure assessment* describes the pathways by which soil contaminants may be taken up by the receptor.

This information is combined with receptor characteristics in order to estimate the contaminant uptake. The *toxicity assessment* describes the adverse effects that the contaminants may cause and the dose at which these effects occur.

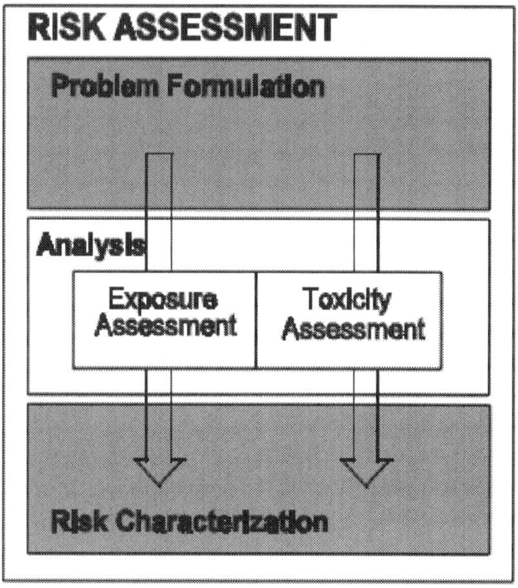

Figure 16. Risk Assessment Framework

The final step, *risk characterization,* compares this dose with the uptake rate (estimated during the exposure assessment) and determines whether or not an adverse effect is likely to occur. Risk assessment procedures can also be used to back calculate a contaminant concentration in soil at which no adverse effects are expected. At times, it may not be possible to remediate to a level compatible with industrial land-use objectives. In such cases, the approval holder must ensure that the contaminants are contained and receptor exposure does not occur. Formal risk assessment procedures are necessary to meet these requirements. Typically, an engineered containment system is required, the performance of which must be confirmed through periodic inspections and monitoring. The approval holder bears responsibility for the design, construction, assessment and maintenance of the risk management system, and any necessary emergency response should the system fail.

4.5. Enforcement policy

Enforcement is generally the last step in a regulatory process that aims to first prevent potential problems before they arise and to resolve existing problems in a cooperative manner. When discussions between the approval holder and the Department fail to prevent or resolve a contravention of an approval or the Act, a number of abatement and enforcement tools are available to the Department including:

- environmental protection orders;
- warning letters;
- tickets;
- administrative penalties;
- enforcement orders;
- prosecutions;
- court orders; and
- cancellation of approvals or certificates.

Due to various factors, including the high cost involved in remediation interventions, it is essential that a holistic and tiered, risk-based approach be adopted that is founded on international best practice, to address remediation in a uniform manner across the country. This is relevant irrespective of the sector of occurrence to safeguard both human health and the natural environment. The Framework is based on a review of international practice in the developed countries of the world and the emergence of remediation policy from developing countries, and an assessment of alternative approaches and methodologies that may find application in the development of a remediation framework.

5. Conclusion

Soil pollution is a result of many activities and experiments done by mankind which end up contaminating the soil. Industrial wastes such as harmful gases and chemicals, agricultural pesticides, fertilizers and insecticides are the most common causes of soil pollution. The others are ignorance towards soil management and related systems, unfavourable and harmful irrigation practices, improper septic system and management and maintenance of the same, leakages from sanitary sewage. There is urgent need for a tiered approach in ecological risk assessment of contaminated soils. Generic soil screening levels are needed as a first tier. Higher tiers of ecological risk assessment should, however, contain some kind of site-specific assessment. It is furthermore important to organize the various studies in a framework or decision support system that is transparent and useful for all stakeholders. A weight of evidence approach may be an obvious choice to deal with these uncertainties. The TRIAD approach, which incorporates and categorizes information in a triangle – chemistry, toxicology, and ecology – is an appropriate tool for handling conceptual uncertainties. Several remedies to these shortcomings have been proposed. Regarding ecotoxicity direct testing would allow for a major improvement in risk estimates. As to human health risks: including biological availability in risk estimates, more use of up to date knowledge about exposure routes, dose-effect relations and combination effects, and biomonitoring of effects are options for improvement.

Acknowledgements

Authors greatly acknowledge the University Malaya Research Grant (RG257-13AFR). Moreover, the comments of two anonymous reviewers are gratefully acknowledged.

Author details

Muhammad Aqeel Ashraf[1], Mohd. Jamil Maah[2] and Ismail Yusoff[1]

1 Department of Geology, Faculty of Science, University of Malaya, Kuala Lumpur, Malaysia

2 Department of Chemistry, Faculty of Science, University of Malaya, Kuala Lumpur, Malaysia

References

[1] Okrent D. 1999. On intergenerational equity and its clash with intragenerational equity and on the need for policies to guide the regulation of disposal of wastes and other activities posing very long time risks. Risk Analysis 19: 877-901.

[2] Belluck, D.A., Benjamin, S.L., Baveye, P., Sampson, J., Johnson, B. 2003. Widespread arsenic contamination of soils in residential areas and public spaces: an emerging regulatory or medical crisis? International Journal of Toxicology 22: 109-128.

[3] Richardson, G.M., Bright, D.A., Dodd, M. 2006. Do current standards of practice in Canada measure what is relevant to human exposure at contaminated sites? II: oral bioaccessibility of contaminants in soil. Human and Ecological Risk Assessment 12: 606-618.

[4] Van Zorge, J. A. 1996. Exposure to mixtures of chemical substances: is there a need for regulations? Food and Chemical Toxicology 34, 1033-1036.

[5] Swartjes, F.A. 1999. Risk-based assessment of soil and groundwater quality in the Netherlands: standards and remediation urgency. Risk Analysis 19:1235-1248.

[6] Eikelboom, R.T., Ruwiel E., Goumans, J.J.M. 2001. The building materials decree: an example of a Dutch regulation based on the potential impact of materials on the environment. Waste Management 21: 295-302.

[7] Crane, M. and Giddings, J.M. 2004. 'Ecologically acceptable concentrations' when assessing the environmental risks of pesticides under European Directive 91 414/EEC. Human and Ecological Risk Assessment 10: 733-747.

[8] Nathanail, P., McCaffrey, C., Earl, N., Forster, N.D., Gillett, A.G., Ogden, R. 2005. A deterministic method for deriving site-specific human health assessment criteria for contaminants in soil. Human and Ecological Risk Assessment 11: 389-410.

[9] Tarazona, J.V., Fernandez, M.D., Vega, M.M. 2005. Regulation of contaminated soils in Spain. Journal of Soil and Sediments 5:121-124.

[10] Evans, J., Wood, G., Miller, A. 2006. The risk assessment-policy gap: An example from the UK contaminated land regime. Environment International 32: 1066-1071.

[11] Huinink, J.T.M. 1998. Soil quality requirements of use in urban environments. Soil and Tillage Research 47: 157-162.

[12] Urzelai, A., Vega, M., Angulo, E. 2000. Deriving ecological risk-based soil quality values in the Basque Country. Science of the Total Environment 247: 279-284.

[13] Apitz, S.E. 2008. Is risk based, sustainable sediment management consistent with European policy. Journal of Soils and Sediments 8: 461-466.

[14] Provoost, J., Cornelis, C., Swartjes, F. 2006. Comparison of soil clean-up standards fort race elements between countries: why do they differ? Journal of Soil and Sediments 6: 173-181.

[15] Provoost, J., Reijnders, L., Swartjes, F., Bronders, J., Carlon, C., D'Allessandro, M., Cornelis, C. 2008. Parameters causing variation between soil screening values and the effect of harmonization. Journal of Soils and Sediments 8: 298-311.

[16] Toccalino, P.L. and Norman, J.E. 2006. Health-based screening levels to evaluate US Geological Survey groundwater quality data. Risk Analysis 26: 1339-1348.

[17] Patterson, M.M., Cohen, E., Prommer, H., Thomas, D.G., Rhodes, S., McKinley, A.I. 2007. Origin of mixed brominated ethane groundwater plume: contaminant degradation pathways and reactions. Environmental Science & Technology 41: 1352-138.

[18] Leon Paumen, M. 2008. Invertebrate life cycle responses to PAC exposure. PhD thesis. Amsterdam: University of Amsterdam.

[19] Nawrot, T., Plusquin, M., Hogervorst, J., Roels, H.A., Celis, H., Thijs, L., Vangronsveld, J., Van Hecke, E., Staessen, J. 2006. Environmental exposure to cadmium and risk of cancer: a prospective population-based study. The Lancet Oncology 7: 119-126.

[20] Simcox, N.J., Fenske, R.A., Wolz, S.A., Lee, I.W., Kalman, D.A. 1995. Pesticides in household dust and soil: exposure pathways for children of agricultural families. Environmental Health Perspectives 103: 1126-1134.

[21] Laidlaw, M.A.S., Mielke, H.W., Filippelli, G.M., Johnson, D.L., Gonzales, C.R. 2005. Seasonality and children's blood lead levels: developing a predictive model using

climate variables and blood data for Indianapolis, Indiana, Syracuse, New York and New Orleans Louisiana (USA). Environmental Health Perspectives 113: 793-800.

[22] Laidlaw, M.A.S., Mielke, H.W., Filippelli, G.M., Johnson, D.L. 2006. Blood lead in children: Laidlaw et al. respond. Environmental Health Perspectives 114: A 19.

[23] Rothenberg, S.J. and Rothenberg, J.C. 2005. Testing the dose-response specification in epidemiology: public health and policy consequences for lead. Environmental Health Perspectives 113: 1190-1195.

[24] Akesson, A. Lundh, T., Vahter, M., Bjellerup, P., Lidfeldt, J., Nerbrand, C., Samisoe, G., Strömberg, U., Skerfving, S. 2005. Tubular and glomerular kidney effects in Swedish women with low environmental cadmium exposure. Environmental Health Perspectives 113: 1627-1631.

[25] Nawrot, T.S., van Hecke, E., Thijs, L., Richart, T., Kuznetsova, T., Jin, Y., Vangronsveld, J., Roels, H.A., Staessen, J.A. 2008. Cadmium-related mortality and long-term secular trends in the cadmium body burden of an environmentally exposed population. Environmental Health Perspectives 116: 1620-1628.

[26] Lamphear, R.P., Dietrich, K., Auinger, P., Cox, C. 2000. Cognitive defects associated with blood lead concentrations, 10 microgram/dl in United States children and adolescents. Public Health Reports 115: 521-529.

[27] Wang, C.L., Chuang, H.J., Ho, C.K., Vang, C.Y. 2002. Relationship between blood lead concentrations and learning achievement among primary schoolchildren in Taiwan. Environmental Research A 89: 167-181.

[28] Wu, T., Buck, G.M., Mendola, P. 2003. Blood levels and sexual maturation in US girls. Environmental Health Perspectives 111: 737-741.

[29] Von Storch, H., Costa-Cabral, M., Hagner, C., Freser, F., Pacyna, Y., Pacyna, E., Kolb, S. 2003. Four decades of gasoline lead emissions and control policies in Europe: a retroactive assessment. Science of the Total Environment 311: 151-176.

[30] Mielke, H.W., Gonzales, C.R., Powell, E., Jartun, M., Mielke Jr., P.W.2007. Nonlinear association between soil lead and blood lead of children in Metropolitan New Orleans, Louisiana: 2000-2005. Science of the Total Environment 388: 43-53.

[31] Provoost, J., Cornelis, C., Swartjes, F. 2006. Comparison of soil clean-up standards for trace elements between countries: why do they differ? Journal of Soils and Sediments 6: 173–181.

[32] Miranda, M.L., Kim, D., Overstreet-Galeano, M.A., Paul, C.J., Hull, A.P., Morgan, S.P. 2007. The relationship between early childhood blood lead levels and performance on end-of-grade tests. Environmental Health Perspectives 115: 1242-1247.

[33] Patterson, B. M., Cohen, E., Prommer, H., Thomas, D. G., Rhodes, S., McKinley, A. J. 2007. Origin of a mixed brominated ethene groundwater plume: contaminant degradation pathways and reactions. Environmental Science & Technology 41: 1352–1358.

[34] Huel, G., Sahuquillo, J., Debotte, G., Oury, J., Takser, L. 2008. Hair mercury negatively correlates with calcium pump activity in human term newborns and their mothers at delivery. Environmental Health Perspectives 116: 263-267.

[35] Nawrot, T., Plusquin, M., Hogervorst, J., et al., 2006. Environmental exposure to cadmium and risk of cancer: a prospective population-based study. The Lancet Oncology 7: 119–126.

[36] Laidlaw, M. A. S., Mielke, H. W., Filippelli, G. M., Johnson, D. L. 2006. Blood lead in children: Laidlaw et al. respond. Environmental Health Perspectives 114: A19-A23.

[37] Akesson, A., Lundh, T., Vahter, M., et al., 2005. Tubular and glomerular kidney effects in Swedish women with low environmental cadmium exposure. Environmental Health Perspectives 113: 1627–1631.

[38] Mergler, D., Anderson, H.A., Chan, L.H.M., Mahaffey, K.R., Murray, M., Sakamoto, M., Stern, H.A. 2007. Methylmercury exposure and health effects in humans: a world-wide concern. Ambio 36: 3-11.

[39] ten Tusscher, G.W. 2002. Later childhood effects of perinatal exposure to background levels of dioxins in the Netherlands. PhD thesis. Amsterdam: University of Amsterdam.

[40] J. A. Van Zorge, "Exposure to mixtures of chemical substances: is there a need for regulations?"Food and Chemical Toxicology, vol. 34, no. 11-12, pp. 1033–1036, 1996.

[41] Filser, J., Koehler, H., Ruf, A., Rombke, J., Prinzing, A., Schaefer, M. 2008. Ecological theory meets soil ecotoxicology: challenge and chance. Basic and Applied Ecology 9: 346-355.

[42] Van Gestel, C.A.M. 2008. Physico-chemical; and biological parameters determine metal bioavailability in soils. Science of the Total Environment 406: 385-395.

[43] Sokolik, G.A., Ovsiannikova, S.V., Ivanova, T.G., Leinova, S.L. 2004. Soil-plant transfer of plutonium and americium in contaminated regions of Belarus after the Chernobyl catastrophe. Environment International 30: 939-947.

[44] Zhao, C., Ren, J., Xue, C., Lin, E., 2005. Study on the relationship between soil selenium and plant selenium uptake. Plant and Soil 277: 197-206.

[45] Liste, H., Prutz, I. 2006. Plant performance, dioxygenase-expressing rhizosphere bacteria, and biodegradation of weathered hydrocarbons in contaminated soil. Chemosphere 62: 1411-1420.

[46] van der Geest, H, Leon Paumen, M. 2008. Dynamics of metal availability and toxicity in historically polluted floodplain sediments. Science of the Total Environment 406: 419-425.

[47] Brouwer, A., Ahlborg, U.G., van den Berg, M., Birnbaum, L.S. 1995. Functional aspects of developmental toxicity of polyhalogenated hydrocarbons in experimental animals and human infants. European Journal of Pharmacology 293: 1-40.

[48] Rutgers, M., Tuinstra, J., Spijker, J., Mesman, M., Wintersen, A., Posthuma, L. 2008. Risico's voor het ecosysteem in stap twee van het saneringscriterium [Ecosystem risks in step 2 of the remediation criterion]. Bilthoven (the Netherlands): RIVM.

[49] Li, J., Zhou, B., Liu, Y., Yang, Q., Cai, W. 2008. Influence of coexisting contaminants on bisphenol A sorption and desorption in soil. Journal of Hazardous Materials 151: 389-393.

[50] Gennings, C., Carter Jr, W.H., Casey, M., Moser, V., Carchman, R., Simmons, J.E. 2004. Analysis of functional effects of five pesticides using a ray design. Environmental Toxicology and Pharmacology 14: 115-126.

[51] Hayes, T.B, Case, P., Chui, S., Chung, D., Haeffele, C., Haston, K., Lee, M., Mai, V.P., Marjuoa, Y., Parker, J., Tsui, M. 2006. Pesticide mixtures, endocrine disruption and amphibian declines: are we underestimating the impact. Environmental Health Perspectives 114, Supplement 1: 40-50.

[52] Perry, M.J., Venners, S.A., Barr, D.B., Xu, X. 2007. Environmental pyrethoid and organophosphorous insecticide exposures and sperm concentration. Reproductive Toxicology 23: 113-118.

[53] Laetz, C.A., Baldwin, D.H., Clllier, T.K., Hebert, V., Stark, J.D., Scholz, N.L. 2009. The synergistic toxicity of pesticide mixtures: implications for risk assessment and the conservation of the endangered Pacific salmon. Environmental Health Perspectives 117: 348-353.

[54] Roos, P. H., Tschirbs, S., Pfeifer, F., et al., 2004. Risk potentials for humans of original and remediated PAH-contaminated soils: application of biomarkers of effect. Toxicology 205: 181–194.

[55] Xiao, R. Y., Wang, Z., Wang, C. X., Yu, G., Zhu, Y. G. 2006. Genotoxic risk identification of soil contamination at a major industrialized city in northeast China by a combination of in vitro and in vivo bioassays. Environmental Science & Technology 40: 6170–6175.

[56] Lee, D. H., Lim, J. S., Song, K., Boo, Y., Jacobs Jr., D. R. 2006. Graded associations of blood lead and urinary cadmium concentrations with oxidative-stress-related markers in the U.S. population: results from the third National Health and Nutrition Examination Survey. Environmental Health Perspectives 114: 350–354.

[57] De Zwart, D. and Posthuma L. 2005. Complex mixture toxicity for single and multiple species: proposed methodologies. Environmental Toxicology and Chemistry 24: 2665-2676.

[58] Altenburger, R., Walter, H., Grote, M. 2004. What contributes to the combined effect of a complex mixture? Environmental Science & Technology 38: 6353-6362.

[59] Filser, J., Wittmann, R., Lang A. 2000. Response types in Collembola towards copper in the microenvironment. Environmental Pollution 107: 71-78.

[60] Drexler, J.W. and Brattin, J.W. 2007. An in vitro procedure for estimation of lead relative bioavailability: with validation. Human and Ecological Risk Assessment 13: 383-401.

[61] O'Halloran, K. 2006. Toxicological considerations of contaminants in the terrestrial environment for ecological risk assessment. Human and Ecological Risk Assessment 12: 74-83.

[62] Römbke, J. 2006. Tools and techniques for the assessment of ecotoxicological impacts of contaminants in the terrestrial environment. Human and Ecological Risk Assessment 12: 84-101.

[63] Roos, P.H., Tschirbs, S., Pfeifer, F., Welge, P., Hack, A., Wilhelm, M., Bolt, H.M. 2004. Risk potentials for humans of original and remediated PAH- contaminated soils: application of biomarkers of effect. Toxicology 205: 181-194.

[64] Xiao, R., Wang, Z., Wang, C., Yu, G., Zhu, Y. 2006. Genotoxic risk identification of soil contamination at a major industrialized city in north east China by a combination of in vitro and in vivo bioassays. Environmental Science & Technology 40. 6170-6175.

[65] Lee, D., Lim, J., Song, K., Boo, Y., Jacobs, D.R. Jr. 2006. Graded associations of blood lead and urinary cadmium concentrations with oxidative- stress-related markers in the US population: Results of the Third National Health and Nutrition Examination Survey. Environmental Health Perspectives 114: 350-354.

[66] Neumann, H.G. 1996. Toxic equivalence factors, problems and limitations. Food and Chemical Toxicology 34: 1045-1051.

[67] Dobrev, I.D., Andersen, M.E., Yang, R.S.H. 2002. In silico toxicology: simulating interaction thresholds for human exposure to mixtures of trichloroethylene, tetra-chloroethylene and 1,1,1 trichloroethane. Environmental Health Perspectives 110: 1031-1039.

[68] Silva, E., Rajapakse, N., Kortenkamp, A. 2002. Something from 'nothing'-eight weak estrogenic chemicals combined at concentrations below NOECs produce significant mixture effects. Environmental Science & Technology 36: 1751-1756.

[69] Richter, M. and Escher, B. 2005. Mixture toxicity of reactive chemicals by using two bacterial growth assays as indicators of protein and DNA damage. Environmental Science & Technology 39: 8753-8761.

[70] Caeiro, S., Costa, M. H., Ramos, T. B. 2005. Assessing heavy metal contamination in Sado Estuary sediment: An index analysis approach. Ecological Indicators 5: 151–169.

[71] Håkanson, L. 1980. An ecological risk index for aquatic pollution control: A sedimen-tological approach. Water Research 14: 975–1001.

[72] Long, E. R., MacDonald, D. D. 1998. Perspective: Recommended uses of empirically derived, sediment quality guidelines for marine and estuarine ecosystems. Human and Ecological Risk Assessment 4: 1019–1039.

[73] Pekey, H., Karakaş, D., Ayberk, S., et al., 2004. Ecological risk assessment using trace elements from surface sediments of İzmit Bay (Northeastern Marmara Sea) Turkey. Marine Pollution Bulletin 48: 946–953.

[74] Roach, A. C. 2005. Assessment of metals in sediments from Lake Macquarie, New South Wales, Australia, using normalization models and sediment quality guidelines

[75] Sutherland, R. A. 2000. Bed sediment-associated trace metals in an urban stream, Oahu, Hawaii. Environmental Geology 39: 611–627

[76] Bhattacharya, A., Routh, J., Jacks, G., et al., 2006. Environmental assessment of abandoned mine tailings in Adak, Västerbotten District (Northern Sweden). Applied Geochemistry 21: 1760–1780

[77] Usero, J., González-Regalado, E., Gracia, I. 1996. Trace metals in the Bivalve Mollusc Chamelea Gallina from the Atlantic Coast of Southern Spain. Marine Pollution Bulletin 32: 305–310.

[78] Cheng, J. L., Shi, Z., Zhu, Y. W. 2007. Assessment and mapping of environmental quality in agricultural soils of Zhejiang Province, China. Journal of Environmental Sciences, 19: 50–54

[79] Reeves, W.R., Barhoumi, R., Burghardt, R.C., Lemke, S.J., Mayura, K., McDonald, T.J., Phillips, T.D., Donelly, K.C.2001. Evaluation of methods for predicting the toxicity of polyaromatic hydrocarbon mixtures. Environmental Science & Technology 35: 1630-1636.

[80] El Masri, H., Mumtaz, M.M., Yushak, M.L. 2004. Application of physiologically-based pharmacokinetic modeling to investigate the toxicological interaction between chlorpyrifos and parathion in rat. Environmental Toxicology and Pharmacology 16: 57 71.

[81] Kortenkamp, A. and Altenburger, R. 1999. Approaches to assessing combination effects of estrogenic environmental pollutants. Science of the Total Environment 333: 131-140.

[82] Sumpter, J.P. and Johnson, A.C. 2005. Lessons from endocrine disruption and their application to other issues concerning trace organics in the aquatic environment. Environmental Science & Technology 39: 4321-4332.

[83] Brian, J.V, Harris, C.A., Scholze, M., Backhaus, T., Booy, P., Lamoree, M., Pojana, G., Jonkers, N., Runnalis, T., Bonfä, A., Marcomini, A., Sumpter, J. 2005. Accurate prediction of the response of freshwater fish to a mixture of estrogenic chemical. Environmental Health Perspectives 113: 721-728.

[84] Wölfe, D. 1997. Interactions between 2,3,7,8 TCDD and PCs as tumor promoters: limitations of TEFs. Teratogenesis, Carcinogenesis and Mutagenesis 17: 217-224.

[85] Marchant, C.A. 1996. Prediction of rodent carcinogenicity using the DEREK system for 30 chemicals currently being tested by the National Toxicology Program. Environmental Health Perspectives 104: 1065-1073.

[86] Verhaar, H.J.M., Morroni, J.R., Reardon, K.F., Hays, S,M., Gaver, D.P., Carpenter, R.L., Yang, R.S.H. 1997. A proposed approach to study the toxicology of complex mixtures of petroleum products. Environmental Health Perspectives 195 (Supplement 1): 179-195.

[87] Stevenson, C.N., Macmanus-Spencer, L.A., Luckenbach, T., Luthy, R.C., Epel, D. 2006. New perspectives on perfluorochemical ecotoxicology: inhibition and induction of an efflux transporter in the marine mussel Mytilus californianus. Environmental Science & Technology 40: 5580-5585.

[88] Bailer, A.J., Hughes, M.R., See, K., Noble, R., Schaefer, R. 2002. A pooled response strategy for combining multiple lines of evidence to quantitatively estimate impact. Human and Ecological Risk Assessment 8:1597-1612..

[89] Chapman, P.M., McDonald, B.G., Lawrence, G.S. 2002. Weight-of-evidence issues and frameworks for sediment quality (and other) assessments. Human and Ecological Risk Assessment 8:1489-1515.

[90] Rutgers, M., den Besten, P. 2005. Approaches to legislation in a global context.The Netherlands perspective – soil and sediment. In:Toxicity Testing Environmental. Thompson C, Wahhia K, Loibner AP (eds). Blackwell, CRC Press, Oxford, pp 269–289.

[91] Alexander, M. 1995. How toxic are toxic chemicals in soil? Environmental Science and Technology. 29:2713–2717.

[92] Alexander, M. 2000. Aging, bioavailability, and overestimation of risk from environmental pollutants. Environmental Science and Technology. 34:4259–4265.

[93] Callahan, C. A., Menzie, C. A., Burmaster, D. E.,Wilborn, D. C., Ernst, T. 1991 On-site methods for assssesing chemical impact on the soil environment using earthworms: a case study at the Baird and McGuire superfund site, Holbrook, Massachusetts. Environmental Toxicology and Chemistry. 10:817–826.

[94] Attrill, M. J., Depledge, M. H., 1997. Community and population indicators of ecosystem health: Targeting links between levels of biological organisation. Aquatic Toxicology 38:183–197.

[95] Keddy, C., Greene, J. C., Bonnell, M. A. 1994. A review of whole organism bioassays for assessing the quality of soil, freshwater sediment and freshwater in Canada. Scientific series no. 198. Environment Canada, Ottawa, Ontario, Canada.

[96] Van Gestel, C. A. M., Kruidenier, M., Berg, M. P. 2003. Suitability of wheat straw decomposition, cotton strip degradation and bait-lamina feeding tests to determine soil invertebrate activity. Biology and Fertility of Soils 37:115-123.

[97] Knacker, T., Förster, B., Römbke, J., Frampton, G.K. 2003. Assessing the effects of plant protection products on organic matter breakdown in arable fields – litter decomposition test systems. Soil Biology and Biochemistry 35:1269-1287.

[98] Jager, T., Fleuren, R.H.L.J., Hogendoorn, E.A., De Korte, G. 2003. Elucidating the routes of exposure for organic chemicals in the earthworm, Eisenia andrei (Oligochaeta). Environmental Science and Technology 37:3399-3404.

Chapter 2

Sustainability Aspects of In-Situ Bioremediation of Polluted Soil in Developing Countries and Remote Regions

Anders Jonsson and Henrik Haller

Additional information is available at the end of the chapter

http://dx.doi.org/10.5772/57315

1. Introduction

Xenobiotic and other toxic compounds from industrial activities are presently being accumulated in the ecosystems at an unsustainable rate. From an anthropocentric point of view such contamination is unsustainable because it threatens basic ecosystem functions and services that are necessary to maintain food security and provision of potable ground water. Furthermore natural habitats for plants and animals are threatened. Thousands of chemical compounds in varying concentrations and compositions have contaminated a wide range of habitats [1]. These contaminated sites demand diverse strategies and development of new technologies that are efficient and cheap enough to make remediation feasible even in remote areas and where financial support is limited.

Soil remediation has been a developing field for several decades and a common assumption is that it is a sustainable industry but very often clean-ups demand vast amounts of energy and lead to severe physical damage to the landscape. Traditional remediation technologies include soil excavation and landfill disposal. Sustainable remediation technology has been defined as a *"remedy or combination of remedies whose net benefit on human health and the environment is maximized through the careful use of limited resources"* [1]. Such technologies are only modestly implemented at present but new landfill legislations against "dig and dump" technologies can favour green solutions over chemical and energy-intensive techniques in the future [2].

Microorganisms have colonized some of the most extreme environments on the earth and some of them are efficient degraders of the pollutants that our industrial society produces. Bioremediation is a technology that uses living organisms and their enzymes to degrade, remove or immobilize toxic compounds [3]. Bioremediation have been gaining terrain recently [1] and some authors predict a further increase as technological advances surmount the limitations that exist today.

In this chapter we briefly discuss how principles of ecological engineering in concurrence with the application of basic thermodynamic principles and kinetic modelling can provide useful tools for the development of energy conserving and economically feasible bioremediation projects. We further discuss the potential of organic waste materials and by-products in locally adapted soil bioremediation. Finally we present some illustrative cases of novel research on sustainable bioremediation for tropical developing countries and remote locations and discuss some promising fields of future research and possible future applications.

2. Bioremediation and ecological engineering

Ecological engineering has been proposed as a theoretical framework to design *"sustainable ecosystems that integrate human society with its natural environment for the benefit of both"* [4]. Bioremediation is generally considered an ecological engineering practice but even if it addresses one of the core goals of ecological engineering, i.e. *restoration of damaged ecosystems* [5], bioremediation can be energy-intensive and have a serious impact on the remediated ecosystems particularly if excavation and ex situ methods are employed.

Mitsch and Jørgensen [5] identify 5 basic concepts that have been developed to distinguish ecological engineering from other approaches such as industrial ecology, biotechnology or environmental engineering. Ecological engineering:

1. is based on the self-designing capacity of ecosystems
2. can be the acid test of ecological theories
3. relies on systems approaches
4. conserves non-renewable energy sources
5. supports ecosystem conservation

If used collectively, these 5 concepts can provide guidelines for design of sustainable bioremediation projects. In a sustainable society the ultimate goal is to limit the rate of pollution to what ecosystems can assimilate and break down without long term negative effects on ecosystem functions or human health. Meanwhile that goal is attained we will have to give the ecosystems a hand to boost their intrinsic biodegradation capacity.

2.1. Self design

Self-design is the property of a system to reorganise in an unstable non-homogeneous environment [5]. The inclusion of self design in a project can facilitate the implementation of flexible strategies that adapt to new conditions. The powerful capacity of ecosystems to reorganise after change can be used in bioremediation projects as a means to reduce costs and energy use. Consortiums of soil microorganisms naturally restructure to adapt to new conditions including contamination and are therefore good candidates for self-design strategies. A typical microbial community in soil consist of less than 1 % hydrocarbon degraders but that fraction can increase to 10% after exposure to oil pollutants [6]. An example of how engineers can accelerate the ecosystems self-designing capacity is the introduction of a microbial strain with capacity to initiate degradation of high molecular-weight pollutants that later provide food (energy) for other microorganisms that take care of the metabolic by-products as the introduced strain dies off. In conventional engineering, items are often designed to behave in an as predictable way as possible since reliability and robustness are desired criteria of safety and quality. The outcomes of a technology that depend on self-design, on the contrary, are often a lot more difficult to predict. Bioremediation projects that include self-designing strategies thus require thorough monitoring to make sure that the degradation of the pollutants is satisfactory.

2.2. The acid test

Just as strong acids were traditionally used to distinguish gold from base metals, a full scale bioremediation project can be used to confirm or reject ecological theories. Ecological theories provide a useful framework that establishes restrictions and opportunities for bioremediation strategies. The complex nature of full scale in-situ bioremediation however might lead to outcomes somewhat different from what was expected. Since intense use of monitoring is required for full scale implementations of unproven technologies, the result from these can be used to refine the theories on which the project was based.

2.3. A systems approach

Reductionist analysis of the parts of a system in isolation from each other can provide important data and basic understanding of significant mechanisms but should be combined with a holistic view of the entire system in order to achieve a sustainable bioremediation project. The site specific conditions of the ecosystem that is subject to remediation must be quantified and analysed with appropriate models. That particular ecosystem however is part of a greater context whose characteristics must also be born in mind. When bioremediation strategies are chosen, socioeconomic and even cultural aspects must be taken into account. Questions like "what is the economic impact of the project?", "can polluted land be reclaimed for food production or construction?", "what resources are available in the human environment around the site?" might supply useful information. The use of locally available agricultural waste products, that are potential polluters by themselves, as amendments (see section about waste recycling) is an example of advantageous design that requires a systems approach. The source of the pollution must also be addressed to make sure that the remediation project isn't an encouragement to further pollution.

2.4. Non-renewable resources conservation

Energy use is one of the major sustainability issues for conventional remediation projects. Ex-situ remediation is typically too energy-intensive to be considered ecological engineering.

Furthermore it resembles what Jørgensen & Mitsch [5] call a *shell game,* in which pollutants are moved from one location to another. In sustainable bioremediation external energy input is ideally used only in the initiation phase to start a process that is subsequently powered by solar energy and the embodied chemical energy of the pollutant itself. The engineer's role is to facilitate the proper conditions in which such a process can take place.

2.5. Ecosystem conservation

The ability to cycle nutrients is a fundamental function of all ecosystems. Within the matter-cycling function lies the catabolic capacity to degrade and mineralise molecules of varying complexity, including pollutants. This function alone gives a nonmarket value to ecosystems, not always recognised by society. The role of ecological engineers is to identify ecosystem functions that are adoptable to human needs and apply them in ways that doesn't create any further degradation of nature. Since bioremediation projects aims at eliminating pollution that exerts stress on the ecosystem, it is by definition an ecosystem conservation approach. However, if large amounts of soil are physically removed from the site in ex-situ operations the remediation itself might be a threat to the ecosystem. Bioremediation projects are typically involved in the restoration of disturbed ecosystems but measures to prevent toxic compounds from entering sound ecosystems are likewise part of the bioremediation agenda.

3. Thermodynamic and kinetic considerations on bioremediation

The fact that most organic pollutants are composed of molecules with high embodied energies is an interesting starting point for any consideration about the best practice for sustainable soil remediation, especially in remote locations and developing countries where energy-intensive project are cost-prohibitive. Economic incentives are generally small for any kind of remediation practice in such regions and, furthermore, time constraints may not be as limiting as in highly developed urban environments. Technologies that manage to optimize the use of ecosystem services provided by indigenous organisms, in combination with a minimum input of extra energy and material resources, are potentially more sustainable in such conditions.

Figure 1 illustrates some basic principles including thermodynamic driving forces (from molecules with high to low embodied energy) and available transformation pathways in air and soil. In the atmosphere, abiotic photochemical reactions rapidly oxidize organic vapours. The rate determining step in these complex photochemical reactions is usually the initial attack on the primary pollutant by hydroxyl radical. Typical turnover times of i.e. reactive hydrocarbons are only a few hours in the sunlit atmosphere while lack of photochemistry in the subsurface of soils leads to significantly lower reaction rates and hence to much longer turnover times of pollutants.

In soils where conditions for microbial activity are unfavourable, degradation relies on abiotic chemical reactions. In such cases organic pollutants may persist for years to decades or longer and secondary, incompletely degraded products may accumulate. On the other hand if conditions are favourable (temperature, electron acceptors, nutrients, moisture, available

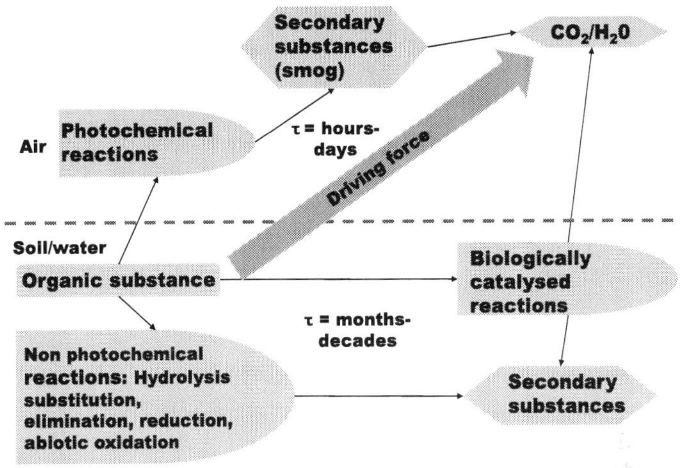

Figure 1. Chemical and biochemical transformations and mineralisation of organic substances in polluted soil and in air. The broad arrow indicates the thermodynamic driving force for the complete degradation of the pollutant in an oxidising environment. Typical average residence times of organic pollutants (τ) in air and soil are indicated.

carbon sources, low toxicity to microorganisms, pH, to mention some of the most determinant variables), microorganisms such as bacteria and fungi may use the organic pollutant as a source of energy and carbon. Enzymatic degradation is generally much faster than abiotic oxidation reactions and the organic pollutant that enters the metabolic pathways of soil microorganisms may be partly or completely mineralized. The driving force for such an enzymatic process is again the free energy potential that can be used by the organism. For typical hydrocarbon and oxygenate pollutants the free energy that is available by complete oxidation with oxygen as electron acceptor, is in the range of 3-12 x 10^3 KJ mol^{-1} or 37-46 MJ kg^{-1}, as can be seen in table 1. This energy potential may be used by a wide range of heterotrophic microorganisms.

In areas of hydrocarbon contamination the most commonly found heterotrophic bacteria are of the genera Pseudomonas, Achromobacter, Arthrobacter, Micrococcus, Vibrio, Acinetobacter, Brevibacterium, Corynebacterium, Flavobacterium, Mycobacterium and Nocardia [7]. Some soil microorganisms that are incapable of utilizing the full embodied energy potential of the pollutant may still be able to initiate catabolic processes that ultimately lead to complete mineralisation if oxygen is present at sufficient concentration. Filamentous fungi, such as e.g., white-rot fungi, can oxidize recalcitrant high molecular weight polycyclic aromatic hydrocarbons (PAH:s) with their extracellular enzymes to more water soluble substances. Both lignolytic and non lignolytic fungi has been shown to initiate degradation of high molecular-weight PAH and the oxidation products formed are thereby made available for further degradation by bacteria [8]. Such interactions between soil microorganisms is an important factor to consider in sustainable soil bioremediation and remediation practices that favour a high soil microbial biodiversity are likely to be more sustainable.

Category	Substance	Formula	Phase	M	S°	ΔH°$_f$	ΔS°$_f$	ΔG°$_f$	b	c	ΔG°(mineralisation)	
				g mol^{-1}	J mol^{-1}K^{-1}	KJ mol^{-1}	KJ mol^{-1}	KJ mol^{-1}			KJ mol^{-1}	MJ kg^{-1}
Gasoline	Benzene	C_6H_6	Liquid	78.11	173.26#	49.0#	-253.78	124.66	6	3	-3202	-41.00
Gasoline	Toluene	C_7H_8	Liquid	92.14	220.96#	12.0#	-342.59	114.14	7	4	-3823	-41.50
Gasoline	m-xylene	C_8H_{10}	Liquid	106.16	253.80#	-25.4#	-446.26	107.65	8	5	-4448	-41.90
Gasoline additive	MTBE	$C_5H_{12}O$	Liquid	88.15	265.3#	-315.4#	-650.52	-121.45	5	6	-3273	-37.14
Diesel	Hexadecane	$C_{16}H_{34}$	Liquid	226.44	586.18#	-456.3#	-1728.71	59.11	16	17	-10401	-45.93
Diesel	Octadecane	$C_{18}H_{38}$	Solid	254.49	480.2#	-457.4#	-2107.71	171.01	18	19	-11776	-46.27
Diesel	Phenanthrene	$C_{14}H_{10}$	Solid	178.23	215.06#	110.1#	-520.00	265.14	14	5	-6972	-39.12
Diesel	Perylene	$C_{20}H_{12}$	Solid	252.31	264.6#	182.4#	-636.14	372.07	20	6	-9682	-38.37
Mineralisation product	Carbon dioxide	CO_2	Gas	44.01	213.79#	-393.5#	2.81	-394.36				
Mineralisation product	Water	H_2O	Liquid	18.02	69.95#	-285.8#	-163.30	-237.19				
Element	Carbon	C	Solid (graphite)	12.01	5.83#	0	0	0				
Element	Hydrogen	H_2	Gas	2.02	130.68#	0	0	0				
Element	Oxygen	O_2	Gas	32.00	205.15#	0	0	0				

#NIST, National Institute of Standards and Technology website. Available at: http://webbook.nist.gov/chemistry/name-ser.html

Table 1. Thermodynamic properties and free energy of mineralisation of some representative "high energy" organic substances commonly found in polluted soil. Standard entropy of formation (ΔS°$_f$) was calculated from standard entropy values (S°) of the substances and their elements (phase indicated in table). Standard free energy of formation (ΔG°$_f$) was calculated from standard enthalpy of formation (ΔH°$_f$) and standard entropy of formation. Standard free energy of mineralisation (ΔG°) was calculated for the complete oxidative degradation of the organic compounds to the thermodynamically most stable end products according to the reaction: 1 mol of substance + a O_2 = b CO_2 + c H_2O.

In oxygen depleted soils microorganisms are unable to mineralise organic compounds to inorganic molecules only. Complete degradation leaves carbon dioxide and methane as end products. Intermediate fermentation products, that may be toxic to microorganisms and plants, may accumulate if they are not converted to methane [7]. Since methane and fermentation products has retained a substantial part of the embodied energy, the energy gain by microorganisms by anaerobic degradation is less than under aerobic conditions and the thermodynamic driving force for biodegradation is thus lower. Methane is also a highly potent greenhouse gas [9] and its unintended production should, from a sustainability point of view, be avoided. It is therefore in many cases preferable to maintain aerobic conditions in the soil during in-situ bioremediation projects. There are however some important exceptions. Xenobiotic and toxic chlorinated organic compounds such as e.g. highly chlorinated PCB:s, pentachlorophenol, tri- and perchloroethylene, gamma hexachlorocyclohexane (lindane) and many others are highly persistent to biodegradation under aerobic conditions. The reason for this may be lack of metabolic pathways available to common soil microorganisms that can generate energy gain for the organism. Persistence is also caused by high lipophilicity of many of these substances which reduce their bioavailability.

Chlorinated substances may be efficiently degraded by microorganisms capable of halorespiration under anaerobic conditions [10]. The degradation mechanism is characterized as reductive dechlorination. In contrast to aerobic co-metabolism which does not provide any energetic benefit for the degrading organism, reductive dechlorination may provide a substantial free energy yield. The thermodynamic driving force of such reactions is demonstrated by the fact that hexachloroethane for example, may be a stronger electron acceptor than oxygen [11].

3.1. Kinetic considerations

Biodegradation kinetics tells us not only about the rate of degradation of target compounds and of rate dependence on ambient factors, but by studying the shape of the degradation curves we may also get valuable information about degradation mechanisms. It is preferable to use models that have a theoretical basis rather than just empirical mathematical modelling that have limited predictive ability.

Natural attenuation of organic pollutants are typically mediated by slow growing organisms and if the concentration of pollutants is much lower than the concentration required for maximum growth of degrading organism (which is often the case in many polluted soils) then first order kinetics may apply, i.e. degradation rate is directly dependent on substrate concentration (Equation 1).

$$-dC/dt = kC \tag{1}$$

C is the substrate concentration and k is the first order rate constant. First order kinetics is often applied in degradation studies because of its simplicity but have little predictive power when conditions for growth change.

In situations with rapid growth of degrading organisms or when other factors than the substrate concentration controls the rate of degradation, first order kinetics will not describe the degradation process accurately. According to the earlier mentioned criteria for sustainable soil bioremediation, efforts should be made to stimulate growth of indigenous soil microorganisms that have or can achieve the ability to degrade target pollutants. The addition of organic waste products from the dairy industry that promotes growth of degrading organisms has been proved to be an efficient means to increase both rate and total degradation of some representative diesel range hydrocarbons in soil [12-14]. Although the addition of readily available carbon sources, such as lactose, lactate and amino acids (components of whey) to contaminated soil would be expected to trigger exponential growth of organisms, a common phenomenon in liquid culture media, it seems that the soil matrix itself does not readily promote exponential growth of degrading organisms. Growth of soil microorganisms is limited by a number of factors caused by the properties of the soil matrix. Among the limiting factors is diffusion of substrate and nutrients but also the specific area available for bacterial sorption on to soil particles. When a large fraction of this area is occupied by bacteria, further growth might be increasingly restricted by diffusion of substrate and nutrients into multilayers of bacteria. Kinetic models that take into consideration slow or restricted (linear growth) may more accurately describe degradation patterns in real life situations. Figure 2 illustrates the application of a three-half-order kinetic model (II) originally proposed by Brunner and Focht [15]. The model assumes linear biomass growth rate and simulates the accumulated $^{14}CO_2$-production from phenanthrene-9-^{14}C mineralisation in soil microcosms.

$$P(t) = S_0\left(1 - \exp\left(-k_1 t - k_2 t^2/2\right)\right) + k_0 t \tag{2}$$

P[t] is the cumulative $^{14}CO_2$-production (%) where t is the time (days), S_0 is the amount of substrate (%) that is "highly bioavailable", k_1 is the first-order rate constant (day$^{-1)}$, k_2 is a linear growth term (day$^{-2)}$, and k_0 is the zero order rate constant describing the mineralisation of "less bioavailable" substrate and indigenous mineralisation (% day^{-1}).The lag time was defined as the initial time until 1% of the added ^{14}C was detected as $^{14}CO_2$ [16]. From the end of the lag phase, iterative non-linear least squares regression was performed to fit data to the model by using the SOLVER function in Microsoft Excel 2000 [17].

The effects of applications of different amendments on phenanthrene biodegradation are clearly seen in figure 2. Kinetic parameters such as the linear growth term (k_2) and S_0 was significantly affected by the addition of the different amendments (Table 2). It should be noted that the stimulating effect of complex organic amendments such as whey and fermented whey may not only be an effect of the addition of a readily available carbon source (that promotes growth) but also of available micronutrients, such as B-vitamins, that is required to maintain efficient catabolic pathways for degradation of i.e. polycyclic aromatic hydrocarbons.

	Lag time (days)	S_0 (%)	k_1 (day^{-1})	k_2 (day^{-2})	k_0 (% day^{-1})	R	Mineralisation after 391 days (%)
Control	21	51.3^a	0	4.35^a x 10^{-5}	0	0.973	46.3^{ab}
FW low	24	53.8^a	0	5.83^b x 10^{-5}	0	0.997	52.1^a
FW high	87	-	-	-	-	-	14.4^c
SW low	21	48.6^a	0	4.09^a x 10^{-5}	0	0.974	43.4^b
SW high	18	63.2^b	0	20.8^c x 10^{-5}	0.0046^a	0.983	66.3^d
FW multi	15	65.0^b	0	63.6^d x 10^{-5}	0.014^a	0.992	70.8^e

Table 2. Tree-half-order kinetic constants for phenanthrene mineralisation in soil, at 25 °C. Lag time is expressed in units of days, S_0 and mineralisation in %, k_1 in day^{-1}, k_2 in day^{-2}, k_0 in % day^{-1} and R is the correlation coefficient. Values having the same superscript letters are not significantly different at a 95 % level. (From [18]. Reprinted with permission from Elsevier).

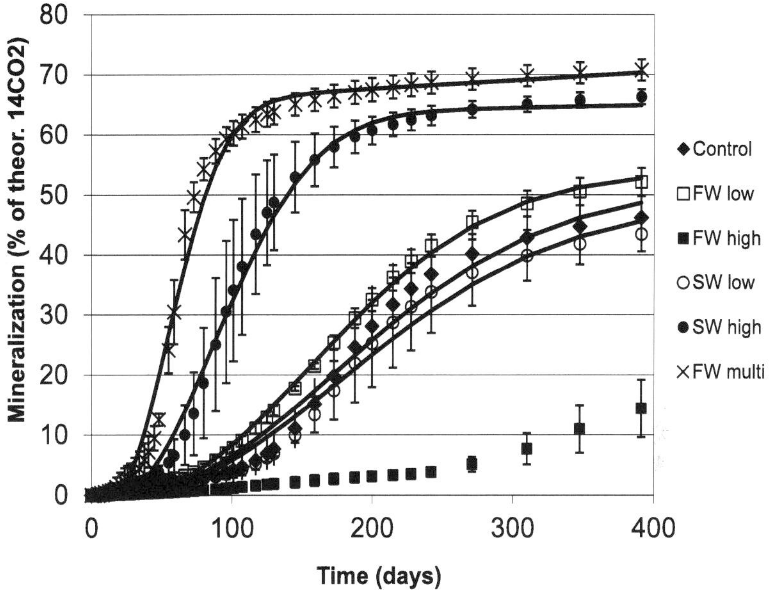

Figure 2. Effect of added fermented whey and sweet whey on ^{14}C-phenanthrene mineralisation to $^{14}CO_2$ in soil at 25 °C. FW low = Fermented whey 210 mg dw kg^{-1} soil dw; FW high = Fermented whey 2100 mg dw kg^{-1} soil dw; SW low = Sweet whey 210 mg dw kg^{-1} soil dw; SW high = Sweet whey 2100 mg dw kg^{-1} soil dw, all added at the beginning of the experiment. FW multi = Fermented whey 210 mg dw kg^{-1} soil dw added at the beginning of the experiment and then every 14th day. The curves are fitted to the data by use of the three-half-order kinetic model. Kinetic parameters are given in table 2 (From [18]. Reprinted with permission from Elsevier).

4. Waste recycling

Organic by-products are produced globally on a massive scale. Economic operations such as agriculture, livestock, fishing and forestry [19-21] generate a number of by-products that can be seen in Figure 3. Frequently unexploited organic waste products create serious environmental problems when disposed of [20]. Agriculture is the economic base of most tropical developing countries, accounting for 50 % of the employment [21]. Organic by-products from agriculture, forestry, fishery and livestock operations are thus available in abundance and are typically of limited economical value [19]. Animal feed has been proposed as an outlet for some of these by-products but due to low levels of protein, high levels of moisture and some anti-nutritional factors i.e. presence of tannins and polyphenols, this practice is limited [22]. However, being inexpensive and easily available sources of carbon, nutrients and bioactive compounds, many of these by-products can potentially be used as sources of readily available carbon for bioremediation purposes [23].

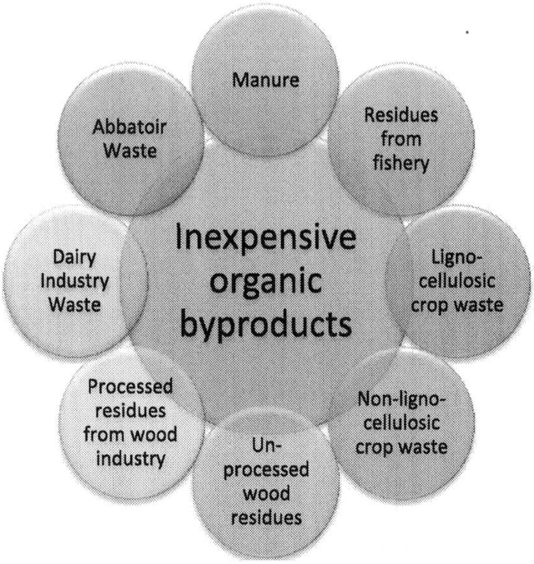

Figure 3. Some organic by-products that potentially can be used as amendments for bioremediation.

A number of organic by-products have the ability to enhance the degradation of different toxic compounds including lignocellulosic wastes such as sugarcane, bagasse and sawdust [26], crop residues such as coffee pulp [27] and molasses [28] and whey, a by-product of the dairy industry [12-14]. Three examples of inexpensive by-products with potential or confirmed bioremediation capacity are described in more detail below: pyroligneous acid, compost teas and whey.

4.1. Pyroligneous acid

Pyroligneous acid (PA) is a by-product of charcoal production [29] with a limited economic value that is also referred to as wood vinegar, pyrolysis oil, smoke extract, bio-oil etc. [30, 31]. PA is typically disposed of as waste [31] but in Japan it has been used for centuries to increase crop production and to combat agricultural pests [29, 32]. At present, the majority of the charcoal consumed worldwide is produced in traditional kilns in developing countries [33]. Such charcoal production lacks appropriate control of the gas and vapours produced in the process and causes land and air pollution [34]. Production systems exist however that recover these condensate co-products or use them as fuel in the pyrolysis process, reducing the emission to a minimum [30]. Several low tech kilns that recover the PA from the charcoal production have been developed including *the advanced Brazilian beehive kiln* and the *Casamance kiln* [35]. In recent years the research and technology concerning PA have advanced significantly [30], and the current trend in charcoal production is to improve the environmental performance while maintaining or improving charcoal yield and quality. The chemical composition of PA varies depending on factors such as type of feedstock, pyrolysis temperature and duration etc. [25, 30]. The most frequently identified compounds in PA include methanol, acetic acid, methyl acetone, acetaldehyde, furan, furfural and volatile organic acids. PA contains phenolic compounds that are known to have antimicrobial properties [36] and PA has successfully been used to control fungal growth on wood [30] and bacterial decay of foods [36, 37]. At low concentrations however, PA has been shown to stimulate germination, growth and yield in a wide range of plants [37]. Although the growth promoting effect of PA is not yet clearly understood it has been attributed to nutritious components such as nitrates of Ca, Mg, K, magnesium sulphate etc. but also to its stimulatory effect on symbiotic fungi and bacteria [31]. The formation of fruit bodies of hydrocarbon-degrading fungi such as *Lentinus edodes* and *Pleurotus ostreatus* was significantly increased after applications of PA to the growth medium [38]. Furthermore PA has been shown to significantly increase basal respiration and microbial biomass in highly weathered tropical soils suggesting that microbes used PA for their metabolism [29].

4.2. Compost teas

Compost teas (CT) are liquid compost extracts obtained when compost is soaked in water for a determined number of days. Optionally additives (derived from by-products) such as molasses, fish hydrolysate, rock dust, yeast extract, humic acids etc. are used to add nutrients and stimulate microbial activity. To date virtually all research on CT has focused on its potential to control plant diseases in agriculture. However, the fact that CT have proved to significantly increase soil microbial respiration and dehydrogenase activity [39] makes it an interesting candidate for bioremediation. A number of production parameters (e.g. aeration, compost source, nutrient additives etc.) have been manipulated in order to optimize plant disease suppression [40]. To date, consensus on optimum production parameters for disease suppression hasn't been reached and some studies report inconsistency in the performance [41]. However a number of reports have shown that aeration and additions of nutrients lead to significant increases of the cell mass of active bacteria [39, 40, 42] and the addition of humic acids and yeast extracts significantly increased the fungal populations [43]. Plate counts and microscopic examination studies have revealed that aerated CT has bacterial population dominated by *Bacillus sp., Lactobacillus sp., Micrococcus lutues, Staphylococcus sciuri, Pseudomonas putida, Burkholderia glumae* and *Clavibacter agropyri* while species of *Aspergillus, Penicillium* and *Trichoderma* dominated the fungal communities [43]. Many of these organisms have the capacity to degrade a number of toxic substances including PAH:s and organochlorine compounds [44, 45].

In a recent study, applications of compost teas made from vermicompost caused significantly greater dehydrogenase activity and microbial respiration [39] which makes it an interesting candidate for bioremediation. Vermicompost generally has higher concentrations of available nutrients (NO_3^-, exchangeable Ca, P and soluble K) and significantly larger and more diverse microbial populations than common compost. Microorganisms found in worm castings of *Eisenina fetida* have caused increased catabolic activity and enhanced degradation of PAHs in a number of experiments [46, 47].

4.3. Whey

Whey is produced when milk casein is removed from the milk in dairy operations to make cheese and other products. This liquid by-product constitutes between 85-95% of the milk volume and 55% of milk nutrients remain in the whey. About half of the yearly global production of 145 million tonnes is utilized for animal feed etc. The remaining large volumes end up as potential contaminants in the environment [48]. The chemical content of whey is characterised by lactose, a number of essential and non-essential amino acids in different proportions, vitamin B 1,2,6,7,12, folic acid and lactic acid [12, 13]. Relatively few studies have been performed on the effect of whey on biodegradation of organic pollutants [12]. However, research at Mid Sweden University has showed that the degradation of aliphatic and aromatic hydrocarbons can be significantly enhanced by the addition of whey and fermented whey [12, 13]. The biodegradation-enhancing effect of whey was primarily attributed to an increased microbial biomass stimulated by the readily available carbon source [12]. The degradation studies also showed a more complex dependence of carbon sources and growth factors, such as B-vitamins, on the degradation of an aromatic compound (phenanthrene) compared to an aliphatic compound (hexadecane) [18]. These results indicate that the presence of co-factors such as vitamins and micronutrients may be important to consider when evaluating the suitability of organic amendments for bioremediation of soil contaminated by various types of organic pollutants.

5. Case studies

Different technologies and strategies are necessary to remediate the present and future soil pollution. As our understanding of the degradation mechanisms of different pollutants evolves, so do our chances to develop alternative and complementary strategies that can take the concentrations of toxic compounds to a sustainable level. Since 2006, a research group at Mid Sweden University has undertaken research dedicated to the development of bioremediation methods for low priority remediation sites including, remote locations and tropical developing countries. Ecological engineering provide an appealing theoretical framework in such locations where energy and material costs are limiting factors. The research has implied laboratory experiments as well as pilot scale and in-situ projects in tropical and temperate climate. It includes pilot scale experiments in an experimental station that was developed in Nicaragua (se figure 4) and field trials in remote parts of northern Sweden where transport distances are long and the cold climate constitutes an additional challenge to in-situ bioremediation projects.

5.1. Field station in Nicaragua, designed for bioremediation experiments in a tropical climate

Waste management programs in developing countries are often deficient and the inappropriate disposal of hazardous waste creates pollution problems. Furthermore the use of pesticides and other toxic products is minimally restricted in many developing countries which leads to extensive contamination with persistent substances such as polycyclic aromatic hydrocarbons and organochlorine pesticides [49, 50]. The experimental station in Nicaragua (figure 5) is

http://dx.doi.org/10.5772/57315

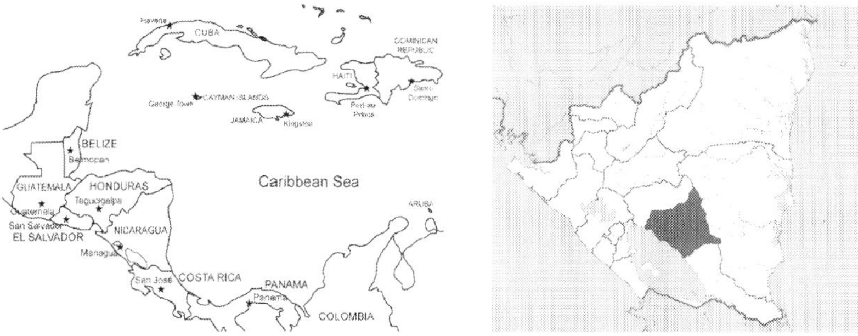

Figure 4. Left picture: Map of Central America and Nicaragua. Right picture: The red marked area is Chontales, the region where Mid Sweden University's pilot scale experimental station is located. (Map of Central America by Виктор В licensed under Creative Commons Attribution-Share Alike 2.0 Generic, Mapa de Nicaragua mostrando al departamento de Chontales by Vrysxy licensed under Creative Commons Attribution-Share Alike 3.0 Unported.)

Figure 5. Mid Sweden University´s experimental station at Casa Montesano in Nicaragua.

located on the agro ecological farm *Casa Montesano,* (11°59′70″ N 84°53′06″ E), where CIPP, a local NGO, dedicated to sustainable rural development projects has its headquarters. The region has a humid tropical climate with an average precipitation of 2000 mm per year and an annual average temperature between 25 and 28 °C. The soil on the experimental site is classified as ultisol, a common soil in the humid tropics occupying about 8 % of the global ice-free land area and supporting 18 % of the world's population.

Figure 6 shows the 24 compartments that were built in concrete with a smooth cement plaster in two sets of twelve compartments each. The compartments were built to fit 150 L each of agriculture soil. The purpose of the closed compartments is to test sustainable remediation methods in conditions as similar to in-situ remediation as possible without risks for pollutants to escape into nature. All technologies assessed in the experimental station are designed to be low cost/low technology methods, appropriate to economically marginalised people in developing countries. Notable is the use of locally available organic waste or by-products such as whey, pyroligneous acid and compost teas that can be achieved at a very low cost. The pilot scale research aim to find practical solutions to the obstacles that emerge when positive results from the laboratory are to be reproduced in the field. This implies dealing with issues like chemotaxis limitations, i.e. the importance of biodegrading microorganisms to be in close proximity to the pollutant. That can be achieved by the use of surfactants and solvents and by avoiding clogging of the soil pores.

The first experiment conducted in the station was on soil spiked with diesel at a concentration of 5.00 g diesel dw^{-1} kg^{-1} soil dw. Whey, pyroligneous acid and compost teas were applied at a high and a low dose on day 3, 18, 32, 46, 60, 74 and 88 after diesel contamination. The doses were adjusted to reflect the expected ideal dose (high) and an economically more viable dose (low). Soil samples were taken on day 2, 7, 12, 17, 31, 45, 59, 73, 87 and 101 after diesel contamination. Leachate water samples were collected by the end of the experiment in order to detect presence of metabolic by-products.

When this chapter was written the soil samples were not yet analyzed. The soil samples will be analysed by GC/FID to determine the effect on degradation from the different amendments. The amendments' effect on the microbial composition will subsequently be analysed by molecular methods. Whey has engendered significant enhancement of the degradation of aliphatic and aromatic hydrocarbons in previous laboratory experiments within our research group but compost tea and pyroligneous acid have not been previously tested.

In an upcoming project the research group envisages to use the experimental station for novel research on relevant pesticides. Many of the least biodegradable pesticides are relatively persistent to aerobic degradation [27] which poses additional challenges for in-situ remediation. These challenges will be addressed constructively in trials with redox barriers and other strategies.

5.2. Experiences from in-situ remediation trials in remote areas of Northern Sweden

5.2.1. Bioremediation field trial using whey treatment of diesel contaminated soil at a former gas station in Northern Sweden

A field experiment was carried out at a diesel contaminated site in Gäddede, in the county of Jämtland, north-western Sweden (65°30′N, 14°09′E) [51], see figure 7. The contaminated site is an old fuel station, established in the 1960:s. The contaminated site is about 10 m^2 in area

Figure 6. The compartments filled with soil at the experimental station in Nicaragua.

and about 0.5 m in depth. Previous investigations had pointed out an area with relatively high loads of aliphatic and aromatic hydrocarbons C_5-C_{35} (approximately 2200 mg kg^{-1} dw soil) dominated by diesel range aliphatics.

This site is representative of the many small contaminated areas that are spread out in rural areas, and are typically considered low priority sites. According to a study by the Swedish EPA in 2006 excavation and transport was the dominating remediation technology in Sweden [52] which is also the case in the county of Jämtland. In this case the distance from Gäddede to Gräfsåsen, the nearest approved treatment plant for contaminated soil, is 230 km. The alternative of excavation and transport will therefore be expensive and contribute to air pollution and greenhouse gas emissions. For such areas, a low cost, easy to apply in-situ remediation technique would be appropriate. In-situ whey treatment might be such a method. Addition of whey to increase biological activity in soil for remediation purposes has been previously demonstrated in laboratory studies [12, 13, 18].

5.2.1.1. Life cycle assessment of in-situ bioremediation with whey treatment in Gäddede in comparison to traditional excavation and composting

In order to compare the two alternative treatment strategies, a screening life cycle assessment model was set up. The purpose was to investigate the environmental performance of the whey method, to benchmark the whey method towards the excavation and composting practice and

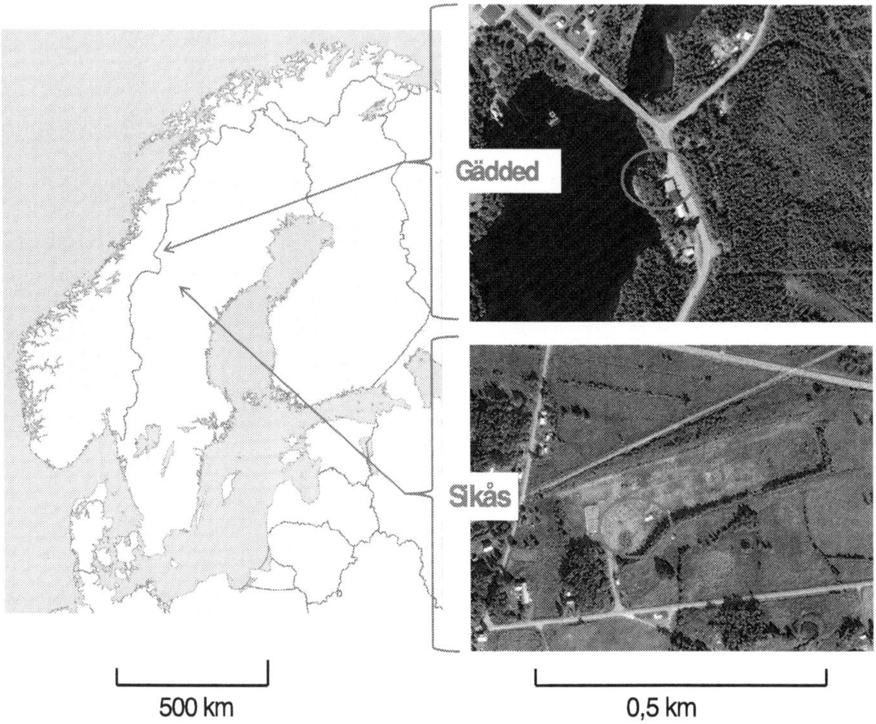

Figure 7. The test spots for in-situ bioremediation trials with whey in northern Sweden. (Location map of Scandinavia by NordNordWest licensed under the Creative Commons Attribution-Share Alike 3.0 Unported) Top picture: Diesel contaminated site at former gasoline station in Gäddede (Village). Bottom picture: Oil spot at lumberyard in Sikås (Village). © Lantmäteriet 2012/1067.

to identify environmental hot spots in the whey treatment life cycle [53]. The functional unit of this study was the treatment of 5 m^3 of contaminated soil reaching a diesel concentration of 500 mg kg^{-1} dw (upper limit in generic guidelines for "non sensitive areas"). For impact assessment the results are characterized as global warming potential (GWP), acidification potential (AP), eutrophication potential (EP) and photo oxidant creation potential (POCP). It was assumed that whey is added together with commercial fertilizer to avoid macronutrient deficit. Since the biological activity only takes place when it is warm enough (this mountainous region has a cold climate with at least 6 months of winter and a yearly average temperature of around 0°C) the total treatment time was estimated to two summers. The alternative treatment assumed in this study is to excavate the 5 m^3 of contaminated soil at the site in Gäddede, transport it to the waste treatment site in Gräfsåsen, Östersund, Sweden, and compost the soil in windrows after mixing the soil with manure. The windrows are turned using an excavator and when the concentrations of contaminants are low enough the compost is used as top cover of a landfill. Figure 8 illustrates a simplified model of the life-cycle process flow diagram for the whey treatment scenario.

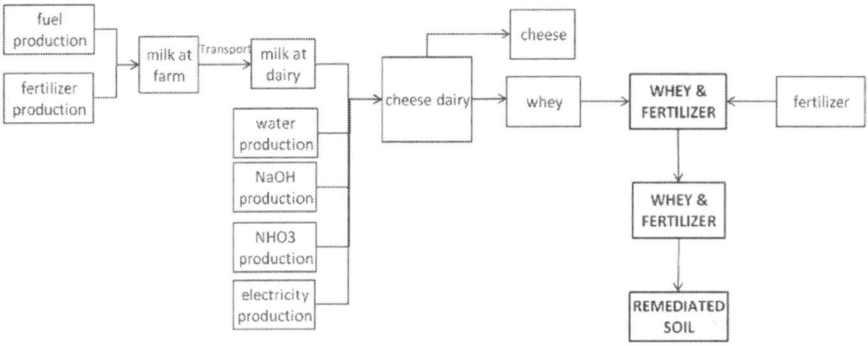

Figure 8. The life cycle inventory model of the whey scenario (From [53]).

The hydrocarbon pollution in the 5 m^3 soil to be treated (over an area of $10m^2$) was assumed to be the same in the whole mass of soil i.e. 2200 mg kg^{-1} dw. Comparisons with laboratory results [12] indicate that an addition of 210 mg dw whey kg^{-1} diesel polluted soil might give a sufficient reduction in diesel oil contamination to reach the higher target value of 500 mg kg^{-1} dw [54]. This gives a calculated need of about 2 kg dw whey, equivalent to 367 L of whey (5.5 g dw L^{-1}) for the area to be treated. The laboratory tests also indicate that a number of successive additions of whey will give a better degradation compared to one large addition, and eight additions (46 L of wet whey per addition) spaced out with 14 days in between was assumed to be a good practice. The total time needed for the degradation is from the laboratory tests estimated to 167 days. Since on-site biodegradation will only take place when it is warm enough [13] the degradation was assumed to take place during two summer months each year for two years. To ensure that biodegradation is not limited by macronutrient deficiencies addition of commercial fertilizer was assumed. It was estimated that a one-time addition of 1.5 kg YaraMila 18-4-14 would provide the necessary nutrients. During the remediation process whey and fertilizer was assumed to be manually added to the contaminated area. Whey additions of 46 L every 14 day over a period of two subsequent summers were assumed.

For the intended bioremediation a total of 2 kg dw whey is needed. Data on cheese and whey production was obtained from Milko dairy in Östersund. To produce 0.9 ton of Edam cheese and 1 ton of dry whey the cheese dairy on average requires 11000 L of milk. Inventory data was gathered to describe the use of average Swedish electricity, water and auxiliary chemicals [55]. Inventory data for milk production, including farming activities, transports and dairy operations was based on a LCI by the Swedish Diary Association.

Since whey is a by-product of cheese production, economic allocation was used giving an allocation of 80 % on cheese and 20 % on whey. In a sensitivity analysis the results were investigated, assuming that all production environmental impact is allocated to the whey. The fertilizer production is described using emissions data on the production of 1 kg of fertilizer. Whey and fertilizer will have to be transported to the contaminated site. Whey is produced in Sweden by Milko AB is transported over a distance of 230 km from Östersund to Gäddede. Fertilizer is obtained from commercial fertilizer store and transported the same distance. Environmental impacts from the transports were calculated using emissions factors assuming a light distribution truck, Euro 2, vehicle manufactured during 1996-2000.

Figure 9 illustrates a simplified model of the life-cycle process flow diagram for the excavation and composting treatment scenario. Assuming a soil density of 1922 kg m^{-3} for the 5 m^3 soil to be treated (over an area of 10 m^2) the calculated mass of contaminated soil was 9.6 ton.

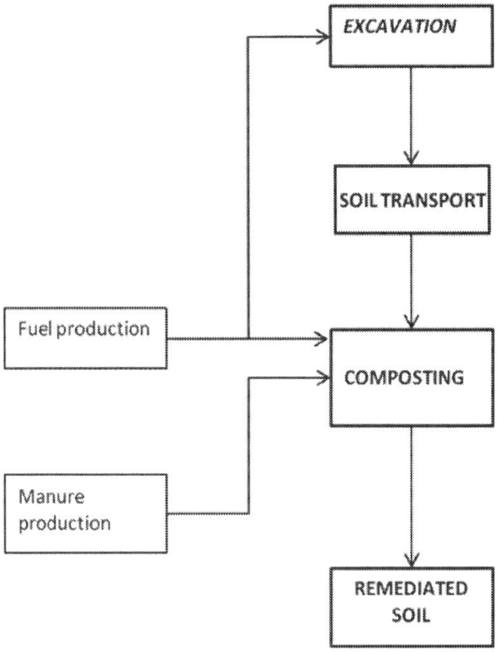

Figure 9. The life cycle inventory model of the excavation and composting scenario (From [53]).

It was assumed that the Gäddede case would need a 14 ton excavator for 15 minutes and use 2 L of diesel. Production of the diesel oil was described by inventory data from [56]. The excavated soil will then have to be transported 230 km from Gäddede to Östersund to a composting plant. Impacts from the transport is described using emission factors for a medium sized distribution truck with a payload of 8.5-14 ton, Euro class 2, produced between 1996 – 2000.

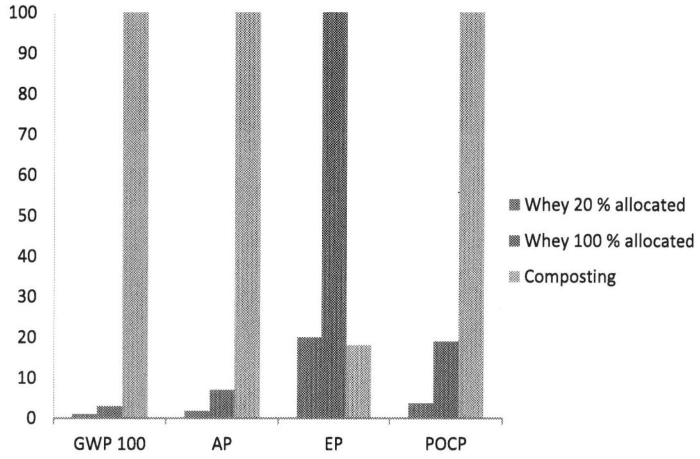

Figure 10. Results from the LCA models for the in-situ whey treatment scenario – with two levels of allocation of the whey production – and the excavation and off-site composting scenario. GWP 100, Global Warming Potential, 100 year time horizon; AP, Acidification Potential; EP, Eutrophication Potential; POCP, Photochemical Ozone Creation Potential. The allocation scenarios mean that respectively 20 and 100 % of the environmental effect of the cheese production was allocated to the whey. Normalized data, the highest result in each category is set to 100% (From [53]).

At Gräfsåsen, the waste treatment and landfill in Östersund, the contaminated soil is mixed with horse manure and placed in windrows in a string-like arrangement. Over the period 2005 to 2009 some 2320 tons of contaminated soil and 1560 tons of manure have been received at the composting plant in Gräfsåsen, giving an average manure use of 0.67 ton/ton. For the manure transport we assume the same emission factors as for the soil transport and an average transport distance of 50 km. The manure generations are totally allocated to the horse keeping. Every one month during the summer the windrows are mixed using a mixer mounted on a wheel loader. For this process half the environmental impact from the excavation in Gäddede was assumed. Use of the treated compost as cover of the land fill is not assumed to be connected to any environmental impacts. The results from the whey treatment scenario and the composting scenario for the four impact categories GWP, AP, EP and POCP is shown in Figure 10, together with the results from the sensitivity analysis where all environmental impacts from milk production and the cheese dairy are allocated to the whey.

For the GWP, AP and POCP impact categories the whey scenarios give lower results than the excavation and off-site composting scenario, whereas the EP category shows higher results. The EP result for the whey treatment is totally dependent on the whey production chain and originates at the farm (such as nutrient runoff, ruminating cattle, vaporization of ammonia from manure) or by the production of materials used at the farm (such as fertilizers). The milk farm activity is the most contributing in the whey scenario for all four impact categories shown.

When comparing the allocated and the unallocated whey treatment scenarios the milk-cheese chain has a rather large impact, especially for eutrophication (EP in diagram). It should be noted that the transport of whey and fertilizer for the whey scenario is described with emission factors of a truck transport, that means it is assumed to have the necessary whey and fertilizer as part of the load of the truck and the off-loading will be part of a delivery route where other parts of the load will be off-loaded in other places. Arranging good logistics is thus important both for the whey treatment and for the excavation and transport to off-site composting.

5.2.1.2. Results of bioremediation field trials and microcosm experiments using whey treatment of the contaminated soil in Gäddede

The result from whey treatment of the polluted site during a 3 ½ months field experiment was not promising. The experiment was initiated 10 May when the snow had melted and the soil temperature at 18 cm depth was 5 °C. The polluted area was given an initial dose of 545 g m^{-2} NPK fertilizer (18 % N; 4 % P; 10 % K) followed by bi-weekly doses of 3.2 L m^{-2} milk whey [51]. During the course of the summer the soil temperature rose to a maximum of 15 °C in the beginning of August and dropped thereafter. When the experiment was ended in mid August no detectable reduction of soil hydrocarbons was observed.

In parallel to the field experiment, microcosm studies were carried out on soil samples, from the vicinity of the contaminated site that had been artificially contaminated with diesel fuel. In the microcosms the soil was subjected to the same treatment as in the field study with the only difference that the soil was kept at 22 °C. During these conditions the diesel range aliphatics was almost completely degraded within 60 days. This is in accordance with previous studies on hexadecane that showed significant difference in degradation rates at 7 and 22 °C respectively [13]. Soil temperature is probably the most limiting factor for in-situ bioremediation in cold regions and the short period of warmth during 3 summer months is rarely enough to bring the soil temperature to levels that give sufficient rates of biodegradation. A possible way to circumvent this problem would be to heat up the soil with heat injection. Preliminary calculations carried out at our department reveal that the electric power needed to heat up a 100 x 100 x 2 m contaminated volume of polluted soil from 8 to 25 °C by e.g. a sufficient number of heat rods, would be 166 kW during a one month period corresponding to an input heating energy of 98 x 10^3 kWh. The calculation assumed that the soil surface was insulated and that there were no vertical and downwards heat losses during the heat-up period. The energy needed to maintain the temperature at 25 °C for another 11 month period assuming horizontal and vertical (downwards) heat losses would amount to 235 x 10^3 kWh. Total energy need for a one year period is then 333 x 10^3 kWh. The following input data was used: A linear temperature drop distance of 9.1 m to ambient temperatures, heat conductivity 1.5 W m^{-1} K^{-1}, specific heat capacity 800 J kg^{-1} K^{-1} and soil density 1300 kg m^{-3}. These are rough calculations based on some simplified assumptions about heat transfer in soil but may still give an estimate of the magnitude of the input energy needed for a temperature enhanced in-situ soil bioremediation project in cold climate. Downscaled to the 10 m^2 by 0.5 m of the Gäddede diesel oil plot the corresponding figures would be 27 and 387 kWh respectively for the heating and 11 months maintenance input energy need respectively giving a total energy need for a 12 month period of roughly 414 kWh. Part of this energy need could easily be provided by solar panels. Furthermore, taking into consideration the very rapid increase in degradation rate even at a moderately elevated temperature such as 22 °C it is possible that a successful temperature enhanced remediation of the Gäddede plot could even be carried out within 6 months during a period when solar panels can provide the total need of input energy. Solar panels can easily be moved from one project to another so the technology described here should have a potential for application in in-situ soil bioremediation in remote places and in cold regions.

5.2.2. One-time application of whey to oil spot at a lumberyard in Sikås

An experiment was carried out to test a low-cost and low-tech solution for remediation of soil contaminated by waste motor oil in a cold region. A 10 m^2 spot that had been polluted by leakage from a 200 l barrel with used motor oil was chosen for the experiment. This spot would certainly come up low in the priority list for remediation projects due to its location in a remote region in the northern parts of the county of Jämtland and also due to the scale of the pollution in proportions to the costs of traditional remediation practices. Such spots are however far from uncommon and the total impact from the highly toxic substances found in waste oil, such as carcinogenic PAH, may be significant. The polluted area was within a timber terminal and the soil was covered with highly compacted sand and gravel. The amount of waste oil that had leaked from the corroding barrel was estimated to 100 L. The permeability of the soil was very low and the oil had penetrated the soil to a depth of approximately10 mm or less. Samples from the plot showed that the oil mainly consisted of high molecular weight aliphatics and aromatics C_{16}-C_{35}.

The oil spot was divided in two parts separated by a shallow cut made by the shovel that can be seen in the top picture in Figure 11. 200 L of fresh milk whey was applied to the left side of the spot. The plot had a slight slope to the left and to the front in the picture. Approximately 50 % of the whey was lost by runoff in these directions due to the low permeability of the soil but some whey crossed the notch made by the shovel and penetrated the right part of the plot. The dashed line in the figure separates the treated left part of the plot from the untreated, right part of the plot. To make sure that lack of macronutrients would not limit the biodegradation a total of 6 kg commercial granulated fertilizer (N/P/K 18/4/10) was evenly applied to the whole spot (treated and control).

The photo from April 2007 was taken just before treatment. The snow had just melted and the average air temperature during April is around 2 °C at this latitude and rose to 15 °C during July, which is also normal for these latitudes. It is plausible that the soil surface could reach higher temperatures during a few hot summer days and that the dark colour of the oil would enhance heat absorption in such cases. Nevertheless, one year later, in May 2008, no difference whatsoever could be observed between treated and untreated spot, see middle picture in Figure 11 (note snow remaining in May).

The bottom picture in Figure 11 shows the plot two years and two months after application of whey to the left part. There is a remarkable difference between the darkness of the sectors and clearly, degradation of the oil in the treated part had commenced. The person in the picture indicating the divine between the light and the dark side of the plot had no a priori knowledge of the treatment strategy. Soil samples taken from the sectors indicate that the TPH concentration in the left part is half the concentration in the untreated part.

The long term effect of a one-dose of the liquid amendment (whey) is remarkable. It is worth to note that the summer 2008 was hotter than normal in the region and the monthly average temperature anomaly in July was + 1.5 °C according to the Swedish Meteorological and Hydrological Institute, www.smhi.se/klimatdata/meteorologi. Clearly the low soil temperature in this, just as in the above presented case, severely limits the rate of biodegradation, but when conditions becomes thermodynamically more favourable, the microorganisms are able to benefit from the extra carbon and micronutrients provided by the amendment, even after a one year incubation time.

6. Opportunities and future predictions of soil remediation

Remediation of contaminated soils is presently a costly venture but as new technologies emerge to treat soils in-situ, expenses can diminish considerably. Technology optimists like Ray Kurzweil forecast that pollution will discontinue to be produced in the future and predicts a set of technologies (including biotechnologies) that will be used to clean up the previously generated contamination. Kurzweil expresses a strong faith in nanotechnology that he believes will provide useful features for remediation of contaminants including; improved catalysis, chemical and atomic bonding, sensing, and mechanical manipulation. Currently, extensive research is undertaken to develop nanoproduced crystalline materials for catalysts that have the potential to improve chemical yields, reduce toxic by-products, and remove contaminants [59].

To overcome some of the difficulties to degrade recalcitrant compounds research on combined biological/abiological degradation technologies might provide useful technologies. Pretreatments by strong oxidizing agents such as ozone, Fenton´s reagent, potassium permanganate, ferrate etc. have shown promising results and Singh and Ward predict that combined chemical/biological treatment of pollutants will play an important role in the future [45]. Strong chemical oxidants may however severely affect soil microbial populations and furthermore requires input of energy and material resources.

The use of biocatalysts is another interesting method whose practical applications have been predicted to increase in the future. This kind of enzymatic remediation can potentially overcome difficulties such as low survival rate of whole cell organisms and generation of toxic by-products that are associated with bioaugmentation. Research on enzymatic bioremediation is at its infancy and few field studies exist. The major limitation for large-scale applications is currently the prohibitive cost for the production of the enzymes. Molecular tools are increasingly explored to overcome that obstacle which could decrease the costs significantly in the future [60]. As we have demonstrated in this chapter the use of readily available organic waste products as amendments and the appropriate application of principles of ecological engineering is potentially a way to make bioremediation more sustainable.

http://dx.doi.org/10.5772/57315

Figure 11. One time treatment of left part of oil spot with 200 l fresh milk whey.

7. Conclusion

It is a common belief that soil remediation is a sustainable industry but many remediation projects use vast amounts of energy and often physically damage the remediation site by removing its topsoil. Ecological Engineering can potentially make remediation projects more sustainable. The large embodied energy in many organic pollutants may be used as a driver of a self-designing in-situ remediation process by indigenous microorganisms in the soil.

The intrinsic capacity of soil microbial communities to initiate and accelerate degradation of soil pollutants rely on the fundamental thermodynamic driving force towards complete degradation of target organic pollutants to low molecular-weight inorganic compounds such as carbon dioxide and water. The degradation process releases quantities of useful "free" energy that can be used by soil microorganisms in their metabolism. Figure 1 illustrates this together with important abiotic and biotic degradation pathways and intermediate storages. Table 1 gives some numbers to illustrate the theoretical "energy gain" by degrading organisms as they catabolise representative soil pollutants. Turnover times for organic pollutants are typically short in the atmosphere, e.g. hours-days [61] but considerably longer in soil, e.g. months-decades [7, 44] mainly because of lack of photochemistry in the subsurface of the soil. The slow biodegradation kinetics in soil can however be considerably improved by the use of locally available waste materials, such as whey and fermented whey that stimulate microbial activity, figure 2 and table 2. There are a number of other waste materials that have been successfully tested by us [12-14, 18] or others [26-28] or remain to be tested (figure 3). Remediation strategies based on amendments from locally available by-products can be tailored to suit site-specific conditions and be adapted to a wide range of climates and contexts as illustrated in the case studies presented in this chapter.

Concepts of Ecological engineering, i.e. measures to stimulate the self-organizing capacity of ecosystems by the use of organic amendments, conservation of non-renewable energy resources and ecosystem conservation by in-situ treatment with locally available waste resources have the potential of making soil remediation projects in low priority sites in developing countries and in remote regions more sustainable than conventional methods that rely on costly and less environmentally benign technology. Life cycle assessment (LCA) was used to evaluate the different remediation technologies from a systems perspective. Results of the screening LCA study of the whey treatment method versus a conventional dig-transport-compotation scenario at a remote diesel oil polluted site in Northern Sweden (See map, Gäddede, figure 7) are shown in figure 10. Clearly, the in-situ whey treatment has an overall lower environmental impact compared to the transport-composting method. The most contributing activity for the composting scenario in all of the environmental impact categories studied was the transport of the excavated soil to the composting site. Thus the transport distance is of large importance. For sites where the transport distance is much shorter, the composting scenario would come out much better. This outcome is in accordance with what has been earlier reported by Suèr et al [57] and Toffoletto et al [58] that transportation and transport distance is a key element when choosing between in-situ and ex-situ remediation technology.

In this chapter we have demonstrated that Ecological engineering in concurrence with the application of basic thermodynamic principles and kinetic modelling of data from laboratory experiments can provide useful guidelines for the development of energy conserving and economically feasible bioremediation technologies. Such technologies have the potential of making in-situ bioremediation of polluted soil in low priority sites in developing countries and in remote regions more sustainable.

Acknowledgements

We gratefully acknowledge the financial support of the European Union Regional Development Fund. We are indebted to Morgan Fröling and Joseph Akambih Tajam for conducting the LCA screening analysis of the whey method and to Anja Wedermeyer and Sylvia Vespermann for calculating energy need for temperature-enhanced bioremediation in cold regions.

Author details

Anders Jonsson* and Henrik Haller

*Address all correspondence to: anders.jonsson@miun.se

Department of Ecotechnology and Sustainable Building Engineering, Mid Sweden University, Sweden

References

[1] Mudhoo A, Mohee R. Elements of Sustainability and Bioremediation. In: Bioremediation and Sustainability. John Wiley & Sons, Inc.; 2012..s. 1-41.

[2] Lynch JM, Moffat AJ, Lynch JM, Moffat AJ. Bioremediation - Prospects for the Future Application of Innovative Applied Biological Research. Ann. Appl. Biol.; 146(2:217-21.

[3] Vidali M. Bioremediation, An Overview. Pure and Applied Chemistry 2001;73(7): 1163-72.

[4] Mitsch WJ. What is Ecological Engineering? Ecological Engineering 2012;45(0):5-12.

[5] Mitsch WJ, Jørgensen SE. Ecological Engineering and Ecosystem Restoration. Wiley; 2004.

[6] Green AE, Voordouw G. Microbial Community Dynamics During Bioremediation of Hydrocarbons. In: Singh A, Ward OP. (eds) Biodegradation and Bioremediation. Berlin: Springer-Verlag; 2004.

[7] Riser-Roberts E. Remediation of Petroleum Contaminated Soils. Boca Raton: Lewis Publishers; 1998.

[8] Peng RH, Xiong AS, Xue Y, Fu XY, Gao F, Zhao W, et al. Microbial Biodegradation of Polyaromatic Hydrocarbons. FEMS Microbiology Reviews 2008;32(6):927-55.

[9] Solomons SD, Qin D, Manning M, Chen Z, Marquis M, Averyt KB, et al. Technical Summary. In: Climate Change 2007: The Physical Science Basis. Contribution of Working Group I to the Fourth Assessment Report of the Intergovernmental Panel on Climate Change. Cambridge, United Kingdom and New York, USA: Cambridge University Press; 2007.

[10] Mohn WW. Biodegradation and Bioremediation of Halogenated Organic Compounds. In: Sing A, Ward AP. (eds) Biodegradation and Bioremediation. Berlin, Heidelberg, New York: Springer 2004..p. 125-48.

[11] Adriano DC, Bollag JM, Frankenberger WT, Sims RCJ. (eds) Bioremediation of Contaminated Soil. Madison: American Society of Agronomy; 1999.

[12] Östberg TL, Jonsson AP, Bylund D, Lundström US. The Effects of Carbon Sources and Micronutrients in Fermented Whey on the Biodegradation of n-Hexadecane in Diesel Fuel Contaminated Soil. International Biodeterioration & Biodegradation 2007;60(4):334-41.

[13] Östberg TL, Jonsson AP, Lundström US. Enhanced Degradation of n-Hexadecane in Diesel Fuel Contaminated Soil by the Addition of Fermented Whey. Soil and Sediment Contamination: An International Journal 2007;16(2):221-32.

[14] Östberg TL, Jonsson AP, Lundstrom US. Accelerated Biodegradation of n-Alkanes in Aqueous Solution by the Addition of Fermented Whey. International Biodeterioration & Biodegradation 2006;57(3):190-4.

[15] Brunner W, Focht DD. Deterministic Three-Half-Order Kinetic Model for Microbial Degradation of Added Carbon Substrates in Soil. Applied and Environmental Microbiology 1984;47(1):167-72.

[16] Mohn WW, Stewart GR. Limiting Factors for Hydrocarbon Biodegradation at Low Temperature in Arctic Soils. Soil Biology & Biochemistry 2000;32:1161-72.

[17] Brown AM. A Step-by-step Guide to Non-linear Regression Analysis of Experimental Data Using a Microsoft Excel Spreadsheet. Computer Methods and Programs in Biomedicine 2001;65(3):191-200.

[18] Jonsson AP, Östberg TL. The Effects of Carbon Sources and Micronutrients in Whey and Fermented Whey on the Kinetics of Phenanthrene Biodegradation in Diesel Contaminated Soil. Journal of Hazardous Materials 2011;192(3):1171-7.

[19] Reddy N, Yang Y. Biofibers From Agricultural Byproducts for Industrial Applications. Trends in Biotechnology 2005;23(1):22-7.

[20] Ulloa JB, van Weerd JH, Huisman EA, Verreth JAJ. Tropical Agricultural Residues and their Potential Uses in Fish Feeds: the Costa Rican Situation. Waste Management 2004;24(1):87-97.

[21] Ruane J, Sonnino A. Agricultural Biotechnologies in Developing Countries and their Possible Contribution to Food Security. Journal of Biotechnology 2011;156(4):356-63.

[22] Aregheore EM. Chemical Composition and Nutritive Value of Some Tropical By-product Feedstuffs for Small Ruminants — In Vivo and In Vitro Digestibility. Animal Feed Science and Technology 2000;85(1–2):99-109.

[23] Singh H. Mycoremediation: Fungal Bioremediation. Wiley; 2006.

[24] Eggen T. Application of Fungal Substrate from Commercial Mushroom Production - Pleuorotus Ostreatus - for Bioremediation of Creosote Contaminated Soil. International Biodeterioration & Biodegradation 1999;44(2-3):117-26.

[25] Jayasinghe P, Hawboldt K. A Review of Bio-Oils from Waste Biomass: Focus on Fish Processing Waste. Renewable and Sustainable Energy Reviews 2012;16(1):798-821.

[26] Dzul-Puc JD, Esparza-García F, Barajas-Aceves M, Rodríguez-Vázquez R. Benzo(a)pyrene Removal from Soil by Phanerochaete Chrysosporium Grown on Sugarcane Bagasse and Pine Sawdust. Chemosphere 2005;58(1):1-7.

[27] Velázquez-Fernández JB, Ramírez-Sandoval M, Domínguez-Ojeda D, Martínez-Rizo AB. Biodegradation and Bioremediation of Organic Pesticides. Intech: 2012..http://www.intechopen.com/books/pesticides-recent-trends-in-pesticide-residue-assay/biodegradation-and-bioremediation-of-organic-pesticides (acessed 22 October 2013)

[28] Al-Hadhrami MN, Lappin-Scott HM, Fisher PJ. Studies on the Biodegradation of Three Groups of Pure n-Alkanes in the Presence of Molasses and Mineral Fertilizer by Pseudomonas Aeruginosa. Marine Pollution Bulletin 1997;34(11):969-74.

[29] Steiner C, Das KC, Garcia M, Förster B, Zech W. Charcoal and Smoke Extract Stimulate the Soil Microbial Community in a Highly Weathered Xanthic Ferralsol. Pedobiologia 2008;51(5–6):359-66.

[30] Mohan D, Pittman CU, Steele PH. Pyrolysis of Wood/Biomass for Bio-Oil: A Critical Review. Energy & Fuels 2006;20(3):848-89.

[31] Zulkarami B, Ashrafuzzaman M, Husni MO, Ismail MR, Zulkarami B, Ashrafuzzaman M, et al. Effect of Pyroligneous Acid on Growth, Yield and Quality Improve-

ment of Rockmelon in Soilless culture.(Report). Australian Journal of Crop Science; 5(12):1508.

[32] Spokas KA, Novak JM, Stewart CE, Cantrell KB, Uchimiya M, DuSaire MG, et al. Qualitative Analysis of Volatile Organic Compounds on Biochar. Chemosphere; 85(5):869-82.

[33] Chidumayo EN, Gumbo DJ, Chidumayo EN, Gumbo DJ. The Environmental Impacts of Charcoal Production in Tropical Ecosystems of the World: A synthesis. Energy for Sustainable Development.

[34] Zandersons J, Gravitis J, Kokorevics A, Zhurinsh A, Bikovens O, Tardenaka A, et al. Studies of the Brazilian Sugarcane Bagasse Carbonisation Process and Products Properties. Biomass and Bioenergy;17(3):209-19.

[35] FAO. Simple Technologies for Charcoal Making. Food and Agriculture Organization of the United Nations (Mechanical Wood Products Branch); 1983.

[36] Loo AY, Jain K, Darah I, Loo AY, Jain K, Darah I. Antioxidant Activity of Compounds Isolated from the Pyroligneous Acid, Rhizophora Apiculata. Food Chemistry;107(3):1151-60.

[37] Kadota M, Niimi Y, Kadota M, Niimi Y. Effects of Charcoal with Pyroligneous Acid and Barnyard Manure on Bedding Plants. Scientia Horticulturae;101(3):327-32.

[38] Yoshimura H, Washio H, Yoshida S, Seino T, Otaka M, Matsubara K, et al. Promoting Effect of Wood Vinegar Compounds on Fruit-Body Formation of Pleurotus Ostreatus. Mycoscience;36(2):173-7.

[39] Pant A, Radovich TJK, Hue NV, Arancon NQ. Effects of Vermicompost Tea (Aqueous Extract) on Pak Choi Yield, Quality, and on Soil Biological Properties. Compost Science & Utilization 2011;19(4):279-92.

[40] St. Martin CCG, Brathwaite RAI. Compost and Compost Tea: Principles and Prospects as Substrates and Soil-Borne Disease Management Strategies in Soil-less Vegetable Production. Biological Agriculture & Horticulture 2012;28(1):1-33.

[41] Ghorbani R, Koocheki A, Jahan M, Asadi GA. Impact of Organic Amendments and Compost Extracts on Tomato Production and Storability in Agroecological Systems. Agron. Sustain. Dev. 2008;28(2):307-11.

[42] Scheuerell SJ, Mahaffee WF, Scheuerell SJ. Variability Associated with Suppression of Gray Mold (Botrytis cinerea) on Geranium by Foliar Applications of Nonaerated and Aerated Compost Teas. Plant Disease;90(9):1201-8.

[43] Naidu YM, Sariah Kadir, Jugah; Siddiqui, Yasmeen. Microbial Starter for the Enhancement of Biological Activity of Compost Tea. International Journal of Agriculture and Biology 2010;12(1):51-6.

[44] Alexander M. Biodegradation and Bioremediation. 2 ed. San Diego: Academic Press; 1999.

[45] Singh A, Ward OP. Biodegradation and Bioremediation. Berlin: Springer; 2004.

[46] Tognetti C, Laos F, Mazzarino MJ, Hernandez MT, Tognetti C, Laos F, et al. Composting vs. Vermicomposting: A Comparison of End Product Quality. Compost Sci. Util.;13(1):6-13.

[47] Sinha RK, Agarwal S, Chauhan K, Chandran V, Soni BK, Sinha RK, et al. Vermiculture Technology: Reviving the Dreams of Sir Charles Darwin for Scientific Use of Earthworms in Sustainable Development Programs. Technology and Investment; 1(3):155.

[48] González-Siso MI. The Biotechnological Utilization of Cheese Whey: A Review. Bioresource Technology 1996;57(1):1-11.

[49] Crawford R. Bioremediation. In: Dworkin M, Falkow S, Rosenberg E, Schleifer K-H, Stackebrandt E. (eds.) The Prokaryotes. Springer New York; 2006..p. 850-63.

[50] Eddleston M, Karalliedde L, Buckley N, Fernando R, Hutchinson G, Isbister G, et al. Pesticide Poisoning in the Developing World—A Minimum Pesticides List. The Lancet 2002;360(9340):1163-7.

[51] Vilches AP, Bylund D, Jonsson AP. Enhanced Natural Biodegradation of Diesel Fuel Contaminants in Soil by Addition of Whey and Nutrients. In: Proceedings of Linnaeus ECO-TECH ′10; November 22-24..2010; Kalmar, Sweden; 2010.

[52] Naturvårdsverket. Åtgärdslösningar -Erfarenheter och Tillgängliga Metoder. Sweden: Naturårdsverket; 2006.

[53] Tajam JA, Fröling M, Jonsson AP. Small Scale In-situ Bioremediation of Diesel Contaminated Soil - Screening Life Cycle Assessment of Environmental Performance. In: Proceedings of Linnaeus ECO TECH ′10; November 22-24..2010; Kalmar, Sweden; 2010.

[54] Naturvårdsverket. Remediation of Contaminated Areas. Sweden: Naturvårsverket; 2008.

[55] Berlin J. Environmental Life Cycle Assessment (LCA) of Swedish Semi-Hard Cheese. International Dairy Journal 2002;12(11):939-53.

[56] CPM. SPINE LCI dataset: Diesel Productions. http://cpmdatabase.cpm.chalmers.se (accessed June 2013)

[57] Suèr P, Nilsson-Påledal S, Norrman J. LCA for Site Remediation: A Literature Review. Soil and Sediment Contamination: An International Journal 2004;13(4):415-25.

[58] Toffoletto L, Deschênes L, Samson R. LCA of Ex-Situ Bioremediation of Diesel-Contaminated Soil (11 pp). Int J Life Cycle Assess 2005;10(6):406-16.

[59] Kurzweil R. The Singularity Is Near: When Humans Transcend Biology. Penguin (Non-Classics); 2006.

[60] Peixoto RS, Vermelho AB, Rosado AS, Peixoto RS, Vermelho AB, Rosado AS. Petroleum-Degrading Enzymes: Bioremediation and New Prospects. Enzyme Research 2011, Article ID:475193..7 pp. doi 10.4061/2011/475193.

[61] Altshuller AP. Lifetimes of Organic Molecules in the Troposphere and Lower Stratosphere. In: Pitts JN Jr., Metcalf RL., Grosjean D. (ed.) Advances in Environmental Science and Technology. New York: John Wiley & Sons; 1980..p181-219.

Chapter 3

Chronic Toxicity of Weathered Oil-Contaminated Soil

Dinora Vázquez-Luna

Additional information is available at the end of the chapter

http://dx.doi.org/10.5772/57253

1. Introduction

Currently, the activities derived from the oil industry, such as extraction, transportation and processing of oil, have affected natural resources [1-3]. For decades, tropical lands have been contaminated by chronic oil spills, causing significant changes in physical and chemical characteristics of the soil, affecting plant development and reducing the growth of microorganisms [4]. Moreover, oil weathering and adaptation of some plants may hide the toxicity of high molecular weight hydrocarbons to other organisms. For this reason, it is necessary to incorporate aspects of *chronic toxicity of weathered oil-contaminated soil* in the study of *"Soil Pollution"*. This chapter aims to examine the chronic effects of old spills on soil, plants and beneficial microbes, in order to support the creation of new remediation technologies, focused on face the new challenges of soil contamination.

Oil pollution is a global problem of increasing importance [5], is estimated that every year numerous spills affecting natural resources of Southeast Mexico, in 2011 were contaminated more than 2,063 hectares as a result of 217 oil spills that affected soil and sea; of which 85 were caused by uncontrolled illegal connections, number which increased 204% from 2009-2011 [6]. With respect to the damages to the ground, in 2012 reported an increase of 87% of leaks and spills, its main causes were corrosion damage and failure of materials in pipelines, in that year 30.07 hectares were contaminated in the Southern Region, joined this, there is lag in the care of cases from previous years. This year, the oil industry closed its operations with a total of 163.63 hectares waiting to be remedied, of which, 39.5% are located in the southeast [7].

The environmental problems caused by oil spills, is not limited to visible pollution, because there are chronic effects that silently endanger ecosystems, biodiversity and environmental balance, due to bioaccumulation, leaching and extension of contaminants into groundwater with potential effects on all living organisms [8]. In Mexico, current environmental regulations is the NOM-138-SEMARNAT/SS-2003 which establishes the maximum permissible limits of hydrocarbons in soil and specifications for its characterization and remediation [9], although in this standard are omitted criteria for assessing chronic effects on soil microorganisms and plants as a result of weathered oil contamination. However, the contributions of several studies show that pollution and waste generation caused by oil activities have deteriorated soil quality [10,11]; ie, the social responsibility of the industry and the government should not only attend acute environmental emergencies [12].

Crude oil is a complex mixture of thousands of compounds that when released to the environment, is subjected to physical, chemical and biological processes called weatherability [13]. This process includes adsorption, volatilization, dissolution, biodegradation, photolysis, oxidation, and hydrolysis. The effects of weathering are difficult to predict because it depends on many biotic and abiotic factors [14]. Therefore, the mobility of hydrocarbons is also influenced by natural factors [15], which involve: a) chemical processes (hydrolysis, oxidation, reduction, photolysis), b) transport and physical processes (adsorption, advection, dispersion, diffusion, volatilization and dissolution), and c) biological processes (biodegradation, metabolism and toxicity) [16, 17].

When an oil spill occurs, it covers the soil surface, but initially at high viscosity prevents penetration towards the subsoil. The oil is retained in the topsoil, during this phase, the light fraction is photo-oxidized and volatilized through the soil pore space and transported to the atmosphere, in this process involves the first n-alkanes (methane, ethane, propane and butane), which are evaporated in less than 24 hours in tropical climates [18]. The hydrocarbons which are not evaporated are incorporated into the soil to form a waterproofing layer that prevents the normal flow of water. This layer or film affects the structure, porosity, absorption and water penetration into the soil [19].

Subsequently, the oil soluble fraction diffuses into the soil solution through infiltration. The behavior of this fraction in the soil depends on the type of texture. The presence of fine texture allows the volatilization of some compounds (C5-C7), but when the texture is coarse, it can leach out and transport themselves to the groundwater, affecting other organisms [20], including the human [21-23]. Otherwise, clay soils rich in organic matter immobilize some compounds, reducing their toxicity and decreasing its spread and leaching through the soil profile [24, 25].

The most stable fraction of crude oil is composed with more than 18 carbons (Polycyclic Aromatic Hydrocarbons [PAHs] and polar compounds), which are adhered on the soil matrix decreasing the solubility and volatility [26], and increasing the capacity of adsorption on the mineral and organic fractions, owing to the high content of active surface of the soil (clay 2:1) and to the high molecular weight hydrocarbons [27]. At this stage, the development of bacteria and fungi have influence on mineralization as part of natural attenuation process, but it is not sufficient to remove all hydrocarbons weathered [28].

Recent studies indicate that the major impediments to the biodegradation of hydrocarbons are the physical and chemical properties of the soil, the degree of contamination and the molecular weight of the compounds (C10-C40), but with Enhanced Natural Attenuation (ENA) may be observed a biodegradation of 26.4% [29] up to 60% with enriched amendments, although the n-alkanes are not removed completely [30]. Therefore, it is important to study the toxic potential, because there are reports indicating that can be bioaccumulated PAHs in vegetables such as *C. pepo sp.* [31].

2. Materials and methods

In southeastern Mexico, the land has been affected as a result of extraction; handling and transportation of oil, there are also zones of oil discharges, which are deposited in the open, without any environmental protection measure [32]. The bad condition of the pipelines and the dispersion through surface runoff of rainwater, resulting from weather conditions become more complicated to calculate the chronic effects of soil contamination by hydrocarbons [10]. Accordingly, this study was conducted in three stages, which were analyzed the chemical and physical properties of the soil, the toxic effects of weathered hydrocarbons on the growth and development of seedlings and after 150 and 240 days of exposure, and finally was evaluated the behavior of beneficial soil microorganisms in rhizospheric soil.

2.1. Effect on the physical and chemical properties of the soil

The study was undertaken at the facilities the *Colegio de Postgraduados*, located in Tabasco, Mexico. The soil with weathered oil was collected within 2 km of the Petrochemical "La Venta" (18 ° 04 '54 "N and 94 ° 02' 31" W), Figure 1. The pollution-free soil according to NOM-138-SEMARNAT/SS-2003 was located in the community of Santa Teresa Arroyo Hondo. The objective was to identify the types of soils and their level of similarity.

For both soils were determined the content of organic matter (Walkley and Black), pH (potentiometry), P and K exchangeable (by extraction with 1N ammonium acetate pH 7, quantification by atomic absorption and emission respectively), CEC (extraction 1N ammonium acetate pH 7, quantification by distillation and titration) and texture (Bouyoucos). The analytical methods used were those indicated in NOM-021-2000-RECNAT [33]. Total Petroleum Hydrocarbons (TPH) were determined by the technique reported in the NOM-138-SEMARNAT/SS-2003 [9].

2.2. Effects on plants

Bioassays to determine the phytotoxicity of weathered oil on two species of legumes (*Crotalaria incana* and *Leucaena leucocephala*) were carried out in the facilities of the Laboratory of Soil Microbiology, Colegio de Postgraduados, Tabasco, Mexico. Two states were studied phenological (plant and seedling), in two treatments: a) soil with 150 mg.kg^{-1} TPH (control treatment) and b) soil with 79.457 mg.kg^{-1} TPH weathered (this has been contaminated by over

25 years). Both soils were characterized as Gleysols, with the same pedogenetic origin, as described in the previous section.

Bioassays were established under a Completely Randomized Design (CRD) with three replications and two legumes, these were selected to be species that grow wild in oiled areas, but the former has tolerance, while the second shows sensitivity to high concentrations of crude oil [34]. In each bioassay was used 208 Protocol of the Organization for Economic Cooperation and Development (OECD) modified according to [35], which allows easily identify the symptoms of stress in the plant.

Seedling bioassays: 50 seeds were sown *C. incana* and 25 *L. leucocephala* by repetition, respectively. Glass containers were used (32 x 22 x 5.5 cm). The number of seeds sown per plant species was calculated according to the size of the seed [36], the viability of the species [37] and the area of the container. Seeds were previously scarified to remove impermeable integument, which constitute a barrier for germination [38]. Scarification consisted of immersing the seeds in sulfuric acid for 15 minutes and washed with tap water subsequent to remove all acid residues [39]. Germination tests were performed to determine the initial seed viability, finding viability standard values in both blocks [40]. The test lasted 30 days and the variables evaluated were mortality, height, root length and dry matter accumulation aerial and root.

Plant bioassays: Bioassays were established plants seedlings 30 days from uncontaminated soil. Subsequently, two of these were transplanted into containers, 15 days after a plant was removed from each container. Exposure of plants to pollutant lasted 150 days to *C. incana* and 240 days for *L. leucocephala*, because of its tolerance respective. The physiological variables were evaluated: height, root length, biomass (leaves and stems), root biomass and number of nodes, leaves and fruits. During both assays was provided with water to field capacity and were not supplied nutrients (N, P and K) to avoid interference on the growth of the specimens. The material was weighed on an analytical balance to obtain the values of DM. The multiple comparison of means was performed by Tukey test (a = 0.05). The numerical results were analyzed with SAS software version 9.1, using PROC GLM.

2.3. Effect on soil microorganisms

Quantification of microorganisms was determined by viable count method for serial dilutions [41] for *Rhizobium* extracts in nodules and Free Living Nitrogen Fixing Bacteria (FLNFB) in rhizospheric soil. Culture media were used combined carbon and yeast-mannitol agar [42].

3. Results and discussion

3.1. Effect on the physical and chemical properties of the soil

The soil profile and laboratory results confirmed that both soils are Gleysols, with the same pedogenetic origin [43] (Table 1), but the contamination had abnormalities in their chemical characteristics (Table 2), the which have been reported in other studies undertaken in the South East (Table 3).

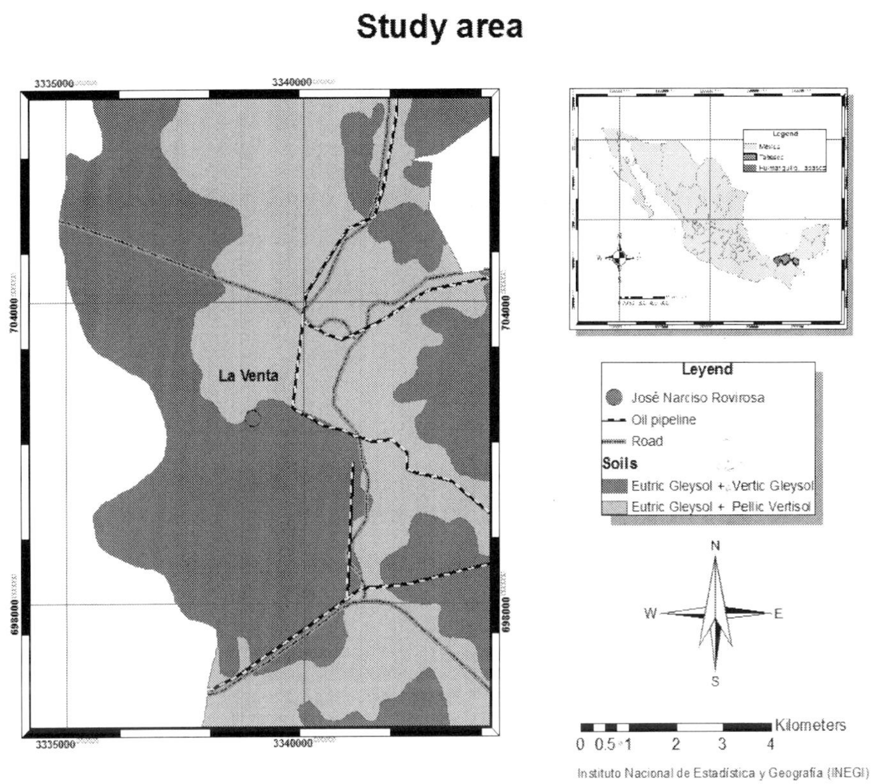

Figure 1. Study zone

Sampling 0-30 cm	pH (H_2O)	OM (%)	P mg.kg^{-1}	K Cmol(+)kg^{-1}	CEC	Clay	Silt (%)	Sand	Textural classification	TPH (mg.kg^{-1})
Soil †	6.3	10.2	23.1	0.35	45.2	61	21	18	Clayey	150
Soil ††	4.2	25.8	3.58	0.4	43.5	48	33	19	Clayey	79,457

* pH 1:2 (potentiometry), organic matter (Walkley and Black), P and K (extraction with 1N ammonium acetate pH 7 and quantification by atomic absorption and emission), CEC (Cation Exchange Capacity) (extraction with 1N ammonium acetate and quantification by distillation and titration), texture (Bouyoucos).

† Control soil

†† Soil with weathered oil (OM without previous removal of TPH)

Table 1. Chemical properties and concentration of TPH in the soils studied.

Sampling 0-30cm	pH	OM	P	K	CEC
		(%)	mg.kg^{-1}	C mol(+) kg^{-1}	
Control soil	Moderately acid	Medium	High	Medium	Very High
Polluted soil ††	Strongly acidic	Very high	Low	Medium	Very High

†† Soil with weathered oil (OM without previous removal of TPH)

Table 2. Interpretation of the characteristics of the soils studied. The interpretation was based on the ranges indicated in NOM-021-2000-RECNAT.

Alterations	Effects
High ratios of C / N and C / P	Unfavorable microbial growth [44]
Retaining TPH fractions in organic matter	Alter the solubility of phosphorus [45]
Increasing Na	Limitations in the production of plants [11, 46]
Decreasing pH	Decrease microbionas populations [47]
The electrical conductivity can be increased up to 5.6 times	Salinity [48]
Interruption in the interaction between cations Ca and K	Reduction in capacity of soil to retain Ca and K [32]

Table 3. Alterations reported in soils contaminated with TPH, in southeastern Mexico.

Also, many physical properties of soil are altered, such as water retention capacity; this is because when an oil spill occurs, hydrocarbons bit compete with water filling the pores [49]. On the other hand, increasing the moisture content in the soil reduces the adsorption of liquid hydrocarbons in the organic matter and clay and the filling of the pores and capillaries (Figure 2) [50].

Figure 2. Alteration in the retaining moisture of soil.

The oil can form macroaggregates and macropores that increase water flow (Figure 3), and these changes inhibit the water retention in the soil, which is moistened only after a long period of contact [51], so that the plants may suffer water stress and die.

Figure 3. Formation of macroaggregates in soil with 12,155 mg.kg^{-1} TPH weathered.

3.2. Effects on plants

The results indicated that the lower height and root length were found in soil with 79,457 mg.kg^{-1} of weathered oil, this can be explained because of the limited development of these. Both legumes formed less biomass in soils with higher concentrations of oil, which was associated with lower production of biomass in leaves and stems, as a result of the presence of the contaminant in the soil (Table 4 and 5), which limits the entry of water to the plant. However, the hydrophobic effect also affects the early stages of germination as evidenced with a delay of up to five days in the emergency [52], even in high concentrations of TPH, germination may be inhibited completely [53].

Some authors found that exposure to concentrations of 2 791, 9 025 and 79 457 mg.kg^{-1} of petroleum hydrocarbons in soil inhibited the vegetative growth and reduced plant biomass in seedlings of *Echinochloa polystachya, Brachiaria mutica* and *Cyperus sp* [10]. Others establish increased toxicity on the dry weight of rice seedlings (*Oryza sativa*) after 25 days of exposure to 90,000 mg.kg^{-1} weathered oil [54]. Biomass reduction is possibly due to widespread damage, which begins in the root system, hindering vegetative growth and therefore the accumulation of plant biomass. However, the first studies mentions that at low concentrations, oil could stimulate vegetative growth [55].

Crotalaria incana seedlings did not form nodules in soil contaminated with 79,457 mg.kg^{-1} of weathered oil after 30 days of exposure to oil. These results were similar to those obtained according to [34], who found that nodulation was completely inhibited in *Crotalaria sp.* and *Mimosa pigra* by concentrations above 50,000 mg.kg^{-1}.

In both plant species, growth decreased with higher concentration of hydrocarbons in soil (Table 5 and 6). This response may be related to decreased water absorption through the roots for the presence of the hydrophobic film formed by the oil added to the soil [56]. The lack of water absorbed decreases cell turgor, reduces or inhibits the processes of incorporation of nutrients and also affects vegetative growth [57, 58]. Water stress is related to the water potential gradient (which depends on the conditions present in the soil) and the membrane permeability to water (which depends on the species [59] therefore, *Leucaena leucocephala* shown to be more tolerant to water deficit).

Species	Concentration $mg.kg^{-1}$	Height	Root length	Mortality	Biomass				
					Leaves	Stems	Aerial	Root	Total
		cm		(%)			gr		
Leucaena	150	7.5a	19.0a	0.0a	3.3a	1.3a	4.6a	1.1a	5.7a
	79,457	5.1b	5.1b	4.2a	1.5b	0.5b	2.0b	0.5b	2.5b

Values with different letter are statistically different (Tukey, $p = 0.05$)

Table 4. Response of *Crotalaria incana* seedlings to 30 days of weathered oil exposure.

Species	Concentration $mg.kg^{-1}$	Height	Root length	Mortality	Biomass				
					Leaves	Stems	Aerial	Root	Total
		cm		(%)			gr		
Crotalaria	150	9.8a	9.8a	2.6a	1.6a	1.0a	2.6a	0.6a	3.2a
	79,457	2.9b	2.8b	28.6a	1.0a	0.5b	1.5b	0.4b	1.9b

Values with different letter are statistically different (Tukey, $p = 0.05$)

Table 5. Response of *Leucaena leucocephala* seedlings to 30 days of weathered oil exposure.

Crotaria incana plants showed a greater effect on biomass in response with exposure to the weathered oil (Figure 4), which may be related to increased toxicity of recalcitrants compounds. There are also some soil properties which allow the adsorption of pollutants [60]. Clay soils with high organic content and low pH may favor the persistence of toxic substances in the soil for a long time after the oil spill occurred [61, 34], due to the adhesiveness of organic matter [62].

Furthermore, fine texture allows the oil form a coarse structure on the outside and around the conglomerate making it waterproof [63], this has effects on root development, growth, and as a result will cause decrease in the accumulation biomass [59]. On the other hand, soil contamination by hydrocarbons can also modify some characteristics such as texture, bulk density, ratio of the particle size of the soil, reducing aeration and affecting the productive development of plants [64, 65].

Figure 4. Toxicity on *Leucaena leucocephala* seedlings with: a) 150 mg.kg-1 y b) 79,457 mg.kg-1 TPH weathered, and plants with c) 150 mg.kg-1 y d) 79,457 mg.kg-1 TPH weathered.

Species	Concentration	Height	Root length	Knot	Leaves	Fruits	Biomass						
							Aerial	Root	Leaves	Stems	Inflorescences	Seeds	Pods
	mg.kg-1	cm		Number			gr						
Crotalaria incana	150	91.7a	17.3a	16.0a	112.7a	54.0a	47.5a	5.6a	11.0a	14.7a	1.5a	9.6a	10.8a
	79,457	32.3b	7.0b	15.0a	2.7b	0.0b	0.6b	0.2b	0.1b	0.6b	0.0b	0.0b	0.0b

Values with different letter are statistically different (Tukey, p = 0.05)

Table 6. Response of *Crotalaria incana* to 150 days of weathered oil exposure.

Species	Concentration	Height	Root length	Knot	Leaves	Fruits	Biomass						
							Aerial	Root	Leaves	Stems	Inflorescences	Seeds	Pods
	mg.kg-1	cm		Number			gr						
Leucaena leucocephala	150	149.0a	107.7a	43.7a	19.1a	--	88.6a	34.0a	19.1a	69.5a	--	--	--
	79,457	66.7b	7.3b	31.0b	4.6b	--	15.9b	6.9b	4.6b	11.3b	--	--	--

Values with different letter are statistically different (Tukey, p = 0.05)

Table 7. Response of *Leucaena leucocephala* to 240 days of weathered oil exposure.

3.3. Effect on soil microorganisms

In plants *Leucaena leucocephala* high concentration of weathered oil did not affect populations of *Rhizobium* into nodules, as happened in the case of *C. incana*, in which both populations were significantly lower in a shorter time exposure (Figure 5). This is because the oil alters the physical and chemical characteristics of the soil, causing the blockage of gas exchange with the atmosphere and affecting microbial populations. Furthermore, the weathered oil is adsorbed in the ground, being less accessible and more difficult to degrade by microorganisms [66, 67]. This may bring a direct impact on rhizobia, because they are aerobic bacteria that remain in the soil as saprophytes, until they infect a radical hair.

Some authors mention that soil conditions have a marked effect on rhizobia, because they can impact the survival and the infectivity of root hairs [68, 69]. However, there are many other factors that influence on effectiveness symbiosis such as specificity and virulence of the bacterium *Rhizobium*, nutrimental factors, soil temperature and pH [70], the latter is of utmost importance because has been reported to decrease significantly in contaminated soil [38]. Added to this there are other factors such as the accumulation of heavy metals and salts that affect soil microbiota [11, 4].

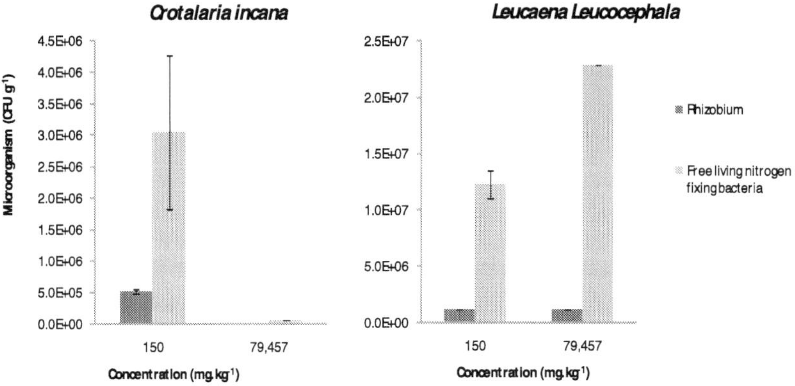

Figure 5. Quantification of populations of *Rhizobium* extracts in nodules and Free Living Nitrogen Fixing Bacteria in rhizospheric soil.

L. leucocephala almost doubled FLNFB populations with 79,457 mg.kg^{-1} of weathered oil. This can be explained because in tolerant species, some microorganisms can increase their populations in the presence of hydrocarbons [71], allowing support microbial growth. As well as [72] argue that tolerant plants are promising tools to accelerate the removal of PAH in long term polluted soils, due to their ability to thrive in a contaminated site, and its success is probably influenced by the relative amount of exudates and other compounds within the root, that stimulate microbial growth [73]. On the other hand, several authors argue that changing the C: N: P, the water content and the water retention capacity of clay soils are crucial to obtain the highest rate of degradation of TPH [74, 75].

4. Conclusion

Soil is one of the most valuable resources that humanity has due to the variety of services offered and which depend for food. Currently, there is tremendous competition for land use, either for use as human settlements, commercial, industrial or produce raw materials (wood) and food, so weathered oil pollution is one of the most important challenges for scientists, because not only it is decontaminated, it comes to restoring the quality and safety of the soil, its properties and microbial component vital for production and proper plant development.

Acknowledgements

The author wants to especially thank Dr. Macario Vázquez Rivera, Irma Luna Capetillo, Mayra Vázquez Luna and Daniel Lara Rodríguez for their valuable support and the Technical Council of COLPOS for their scientific consultancy. Finally, the author wants to thank the Universidad Veracruzana for financing the publication of the present article.

Author details

Dinora Vázquez-Luna*

Address all correspondence to: divazquez@uv.mx

Universidad Veracruzana, México

References

[1] García-Cuellar JA., Arreguín-Sánchez F., Vázquez SH., Lluch-Cota D. B. Impacto ecológico de la industria petrolera en la sonda de Campeche, México, tras tres décadas de actividad: una revisión. Interciencia 2004 29: 311–319.

[2] García-López E., Zavala-Cruz J., Palma-López DJ. Caracterización de las comunidades vegetales en un área afectada por derrames de hidrocarburos. Terra Latinoamericana 2006 (24), 17-26.

[3] Olguín EJ., Hernández ME., Sánchez-Galván G. Contaminación de manglares por hidrocarburos y estrategias de biorremediación, fitorremediación y restauración. Rev. Int. Contam. Ambient. 2007 (23), 139-154.

[4] Vázquez-Luna D. Environmental bases on the exploitation of crude oil in Mexico. In: Younes M. Crude Oil / Book 2. Rijeka: In Tech; 2012 p89-134.

[5] Hall C., Tharakan P., Hallock J., Cleveland C., Jefferson M. Hydrocarbons and the evolution of human culture. Nature 2003 (426), 318-322.

[6] PEMEX. PEMEX, 2011. Informe de responsabilidad social. D.F.: PEMEX; 2012

[7] PEMEX. Petróleos Mexicanos. Informe anual 2012. D.F.: PEMEX; 2013

[8] Ortínez BO., Ize LI., Gavilán, G. A. La restauración de los suelos contaminados con hidrocarburos en México. Gaceta ecológica 2003 (69), 83-92.

[9] SEMARNAT-SS. NOM-138-SEMARNAT/SS-2003. Norma Oficial Mexicana que establece los límites máximos permisibles de hidrocarburos en suelos y las especificaciones para su caracterización y remediación. México: Diario Oficial de la Federación; 2003

[10] Rivera-Cruz MC., Trujillo-Narcia A. Estudio de toxicidad vegetal en suelos contaminados con petróleos nuevo e intemperizado. Interciencia 2004 (29), 369-376.

[11] Adams RH., Zavala-Cruz J., Morales-García F. Concentración residual de hidrocarburos en el suelo del trópico II: Afectación a la fertilidad y su recuperación. Interciencia 2008 (33), 483-489.

[12] Rodríguez SM., Pereira ME., Ferreira DA., Silva E., Hursthouse AS., Duarte A. C. A review of regulatory decisions for environmental protection: Part I Challenges in the implementation of national soil policies. Environment International 2009 (35), 202-213.

[13] Hernández-Acosta E., Rubiños-Panta JE., Alvarado-López J. Restauración de los suelos contaminados con hidrocarburos: Conceptos básicos. Montecillo: Colegio de Postgraduados, 2004

[14] Barakat A. O., Qian Y., Kim M., Kennicutt, MC. Chemical characterization of naturally weathered oil residues in arid terrestrial environment in Al-Alamein, Egypt. Environment International, 2001 27, 291-310.

[15] Kostecki P., Morrison R., Dragun J. Hydrocarbons. In Hillel D. (ed) Encyclopedia of Soils in the Environment. Oxford: Elsevier; 2005. p217- 226

[16] Eweis J., Ergas S., Chang D., Schroeder E. Bioremediation principles. Series in Water Resources and Environmental Engineering. Boston: McGraw Hill; 1998

[17] Riser-Roberts E. Remediation of petroleum contaminated soils. Biological, Physical and Chemical Processes. Boca Raton: Lewis Publishers; 1998

[18] Botello, V. A. (1996). Características, composición y propiedades fisicoquímicas del petróleo. In: Botello VA., Rojas G., Benítez JA., Zárate-Lomelí D. (eds). Golfo de México. Contaminación e impacto ambiental: diagnóstico y tendencias. EPOMEX; 1996 400-666.

[19] King RF., Royle A., Putwain PD., Dickinson NM. Changing contaminant mobility in a dredged canal sediment during a three-year phytoremediation trial. Environmental Pollution 2006 (143), 318-326.

[20] González-Mille DJ., Ilizaliturri-Hernández CA., Espinosa-Reyes G., Costilla-Salazar R., Díaz-Barriga F., Ize-Lema I., Mejía-Saavedra J. Exposure to persistent organic pollutants (POPs) and DNA damage as an indicator of environmental stress in fish of different feeding habits of Coatzacoalcos, Veracruz, Mexico. Ecotoxicology 2010 (19), 1238-1248.

[21] Chichizola, C. Disruptores Endocrinos. Efectos en la Reproducción. Revista Argentina de Endocrinología y Metabolismo 2003 3 (40), 172-188.

[22] EA. Principles for Evaluating the Human Health Risks from Petroleum Hydrocarbons in Soils: A Consultation Paper. USA:EPA; 2003

[23] López E., Schuhmacher M., Domingo JL. Human health risks of petroleum-contaminated groundwater. Env Sci Pollut Res 2008 3 (15), 278-288.

[24] Totsche KU., Danzer J., Kogel-Knabner I. Dissolved organic matter-enhanced retention of polycyclic aromatic hydrocarbons soil miscible displacement experiment. J. Environ. Qual. 1997 (26), 1090-1100.

[25] Charman PE., Murphy B. W. Soils their properties and management. Hong Kong: Oxford University Press; 2007.

[26] Srogi K. Monitoring of environmental exposure to polycyclic aromatic hydrocarbons: a review. Environ Chem Lett 2007 (5), 169–195.

[27] Gutiérrez MC., Zavala CJ. Rasgos hidromórficos de suelos tropicales contaminados con hidrocarburos. Terra Latinoamericana 2002 (20), 101-111.

[28] Adetutu E M., Ball AS., Weber J., Aleer S., Dandie CE., Juhasz, AL. Impact of bacterial and fungal processes on 14C-hexadecane mineralisation in weathered hydrocarbon contaminated soil. Science of the Total Environment 2012, 414, 585-591.

[29] Adetutu E., Weber J., Aleer S., Dandie CE., Aburto-Medina A., Ball AS., Juhasz, AL. (2013). Assessing impediments to hydrocarbon biodegradation in weathered contaminated soils. Journal of Hazardous Materials 2013, http://dx.doi.org/10.1016/j.jhazmat.2013.01.052 (accessed 12 June 2013).

[30] Chaillan F., Chaineau C., Point V., Saliot A., Oudot, J. Factors inhibiting bioremediation of soil contaminated with weathered oils and drill cuttings. Environmental Pollution 2006 (144), 255-265.

[31] Parrish ZD., White JC., IM., Gent MP., Iannucci-Berger W., Eitzer BD., Kelsey JW., Mattina MI. Accumulation of weathered polycyclic aromatic hydrocarbons (PAHs) by plant and earthworm species. Chemosphere 2006 (64), 609–618.

[32] Hernández-Acosta E., Gutiérrez-Castorena MC., Rubiños-Plata JE., Alvarado-López J. Caracterización del suelo y plantas de un sitio contaminado con hidrocarburos. Terra Latinoamericana 2006 (24), 463-470.

[33] SEMARNAT. NOM-021-SEMARNAT-2000. Norma Oficial Mexicana que establece las especificaciones de fertilidad, salinidad y clasificación de suelos, estudio, muestreo y análisis. México: Diario Oficial de la Federación; 2000

[34] Rivera-Cruz MC., Trujillo-Narcía A., Miranda CM., Maldonado CE. Evaluación toxicológica de suelos contaminados con petróleos nuevo e intemperizado mediante ensayos con leguminosas. Interciencia 2005 (30), 326-331.

[35] Vázquez-Luna D., Castelán-Estrada M., Rivera-Cruz MC., Ortiz-Ceballos AI., Izquierdo RF. Crotalaria incana L. y Leucaena leucocephala Lam. (Leguminosae): Especies indicadoras de toxicidad por hidrocarburos del petróleo en suelo. Revista Internacional del Medio Ambiente 2010 (26), 183-191.

[36] Leishman M., Wright I., Moles A., Westoby M. The evolutionary ecology of seed size. In: Fenner M. (ed) Seeds: the ecology of regeneration in plant communities. Wallingford: CAB International; 2000 p31-57.

[37] Carreras M., Pascualides A., Planchuelo A. Comportamiento germinativo de las semillas de Crotalaria incana L. (Leguminosae) en relación a la permeabilidad de la cubierta seminal. Agriscientia 2001 (18), 45-50.

[38] Antoniolli Z., Bellé R., Giracca E.,Thomas, P. Dormancy break of Crotalaria seeds. Ciência Rural 1993 (23), 165-168.

[39] Huachin C., Carvajal J. Análisis de la semilla de doce leguminosas forrajeras tropicales. Memorias. XI Reunión Científica Tecnológica Forestal y Agropecuaria: INIFAP; 1998

[40] Bosco, O. Influência de tratamentos pré-germinativos, temperatura e luminosidade na germinação de sementes de Leucena (Leucaena leucocephala (Lam.) de Wit.), cv. Cunningham. Caatinga 2009 (22), 132-138.

[41] Madigan M., Martinko J., Parker J. Brock Biología de los microorganismos. Madrid, Prentice Hall Iberia; 1999

[42] CIAT. 1Simbiosis leguminosa-rizobio; manual de métodos de evaluación, selección y manejo agronómico. Cali: CIAT; 1988

[43] Palma-López D. J., Cisneros D. J., Moreno C. E., Rincón-Ramírez, JA. Suelos de Tabasco: su uso y manejo sustentable. Villahermosa: Instituto del Trópico Húmedo, Colegio de Postgraduados & Fundación Produce Tabasco A. C.; 2007

[44] Pardo CJ., Perdomo RC., López de Mesa JL. Efecto de la adición de fertilizantes inorgánicos en la degradación de hidrocarburos en suelos contaminados con petróleo. Nova 2004 2 (2), 40-49.

[45] Zavala-Cruz J., Gavi-Reyes F., Adams-Schroeder RH., Ferrera-Cerrato R., Palma-López DJ., Vaquera-Huerta H., Domínguez-Ezquivel JM. Derrames de petróleo en suelos y adaptación de pastos tropicales en el Activo Cinco Presidentes, Tabasco, México. Terra Latinoamericana 2005 (23), 293-302.

[46] Hernández NL. del C. Estudio de la contaminación con petróleo de los suelos agrícolas de la cuenca baja del río Tonalá mediante espectroscopía de infrarojo cercano. Tesis de Maestría en Ciencias, Programa en Producción Agroalimentaria en el Trópico. Colegio de Postgraduados Cárdenas; 2010

[47] Martínez EM., López FS. Efecto de hidrocarburos en las propiedades físicas y químicas de suelo arcilloso. Terra 2001 19, 9-17.

[48] De la Garza ER., Ortiz YP., Macias BA., García C., Coll D. Actividad biótica del suelo y la contaminación por hidrocarburos. Revista Latinoamericana de Recursos Naturales 2008 (4), 49-54.

[49] Cole, G. Assessment and remediation of petroleum contaminated sites. Boca Raton: Lewis Publisher; 1994

[50] Fine P., Graber E., Yaron B. Soil interaction with petroleum hydrocarbons: Abiotic processes. Soil Tech 1997 (10), 133-153.

[51] Romaniuk R., Brandt JF., Ríos PR., Giuffré L. Atenuación natural y remediación inducida en suelos contaminados con hidrocarburos. Ci. Suelo 2007 (25), 139-149.

[52] Quiñones AE., Ferrera-Cerrato R., Gavi RF., Fernández LL., Rodríguez VR., Alarcón A. Emergencia y crecimiento de maíz en un suelo contaminado con petróleo crudo. Agrociencia 2003 37, 585-594.

[53] Banks M., Schultz, K. Comparison of plants for germination toxicity tests in petroleum-contaminated soils. Water, Air, & Soil Pollution, 2005 167 (1-4), 211-219.

[54] García, R. D. E. Estudio de la toxicidad de los petróleos nuevo e intemperizado en el cultivo de arroz (Oryza sativa L.). Tesis de Ingeniería Ambiental. Universidad Juárez Autónoma de Tabasco Villaermosa; 2005

[55] Plice MJ. Some effects of crude petroleum on soil fertility. Soil Sci. Soc. Am. Proc. 1948 (13), 413-416.

[56] Reynoso-Cuevas L., Gallegos-Martínez M., Cruz-Sosa F., Gutiérrez-Rojas M. In vitro evaluation of germination and growth of five plant species on medium supplemented with hydrocarbons associated with contaminated soils. Bioresource Technol 2008 99, 6379-6385.

[57] Kisic I., Mesic S., Basic F., Brkic V., Mesic M., Durn G., Zgorelec Z., Bertovic L. The effect of drilling fluids and crude oil on some chemical characteristics of soil and crops. Geoderma 2009 (149), 209-216.

[58] Tang J., Wang M., Wang F., Sun Q., Zhou Q. Eco-toxicity of petroleum hydrocarbon contaminated soil. Journal of Environmental Sciences 2011 5 (23), 845-851.

[59] Salisbury F., Ross C. Fisiología de las plantas 3: desarrollo de las plantas y fisiología ambiental. Madrid: Thomson Paraninfo; 2000

[60] Rao NC., Afzal M., Malallah G., Kurian M., Gulshan S. Hydrocarbon uptake by roots of Vicia faba (Fabaceae). Environ Monit Assess 2007 (132), 439-443.

[61] Labud V., Garcia C., Hernández T. Effect of hydrocarbon pollution on the microbial properties of a sandy and a clay soil. Chemosphere 2007 (66), 1863-1871.

[62] Rivera-Cruz MC. Clasificación de suelos tropicales influenciados por derrames de petróleo en Tabasco. Tecnociencia Universitaria 2004 (7), 6-25.

[63] Li X., Feng Y., Sawatsky N. Importance of soil-water relations in assessing the end-point of bioremediated soils. Plant Soil 1997 (192), 219-226.

[64] Issoufi I., Rhykerd R., Smiciklas K. Seedling growth of agronomic crops in crude oil contaminated soil. J. Agron. Crop Sci. 2006 (192), 310-317.

[65] Inckot RC., Santos GD., De Souza LA., Bona C. Germination and development of Mimosa pilulifera in petroleum-contaminated soil and bioremediated soil. Flora - Morphology, Distribution, Functional Ecology of Plants 2011 3 (206), 261-266.

[66] Freedman, B. Enviromental Ecology. The Ecological effects of pollution, disturbance and other stresses. San Diego: Academic Press; 1995

[67] Rivera-Cruz MC., Ferrera-Cerrato R., Volke-Haller V., Fernández-Linares L., Rodríguez-Vázquez R. Poblaciones microbianas en perfiles de suelos afectados por hidrocarburos del petróleo en el estado de Tabasco, México. Agrociencia 2002 (36), 149-160.

[68] Alexander, M. Biodegradation and bioremediation. San Diego: Academic Press; 1999.

[69] Langer I., Syafruddin S., Steinkellner S., Puschenreiter M., Wenzel WW. Plant growth and root morphology of Phaseolus vulgaris L. grown in a split-root system is affected by heterogeneity of crude oil pollution and mycorrhizal colonization. Plant Soil 2010 (332), 339-355.

[70] Hubbell D. Legume infection by Rhizobium: a conceptual approach. Bioscience 1988 (31), 832-837.

[71] Salanitro JP. Bioremediation of petroleum hydrocarbons in soil. Advances in Agronomy 2000 (72), 53-105.

[72] Liste HH., Prutz I. Plant performance, dioxygenase-expressing rhizosphere bacteria, and biodegradation of weathered hydrocarbons in contaminated soil. Chemosphere 2006 (62), 1411–1420.

[73] Phillips LA., Greer CW., Farrell RE., Germida JJ. Plant root exudates impact the hydrocarbon degradation potential of a weathered-hydrocarbon contaminated soil. Applied Soil Ecology 2012 (52), 56– 64.

[74] Roldán-Martín A., Calva-Calva G., Rojas-Avelizapa N., Díaz-Cervantes MD., Rodríguez-Vázquez R. Solid culture amended with small amounts of raw coffee beans for the removal of petroleum hydrocarbon from weathered contaminated soil. International Biodeterioration & Biodegradation 2007 (60), 35–39.

[75] Ramírez M., Zapién B., Zegarra H., Rojas N., Fernández L. Assessment of hydrocarbon biodegradability in clayed and weathered polluted soils. International Biodeterioration & Biodegradation 2009 (63), 347-353.

Chapter 4

Heavy Metal Contamination in Brazilian Agricultural Soils due to Application of Fertilizers

Affonso Celso Gonçalves Jr., Herbert Nacke,
Daniel Schwantes and Gustavo Ferreira Coelho

Additional information is available at the end of the chapter

http://dx.doi.org/10.5772/57268

1. Introduction

The Brazilian grain production reaches record productivities each year and among the major crops, stand out the maize (*Zea mays* L.) and soybean (*Glycine max* L.). For these two cultures, the obtaining of the maximum yield has a fundamental importance in the current global scenario [1]. In this relentless pursuit of higher productivities, farmers use technologies related to various areas, such as new forms of fertilizer and pesticide application, different fertilizers, crop breeding, equipment and techniques of planting and harvesting. In order to increase the efficiency in this activity, due to the increased requirement for competitiveness of economic globalization, one of the tools found by farmers is the use of fertilizers containing micronutrients in their crops [2]. Among the various factors of production, the need of balanced and sustainable fertilizers occupies a prominent place, which should include not only the primary and secondary macronutrients, but also micronutrients, which are not considered in routine fertilization by most farmers. Currently, with the advancement of the industries of fertilizers, micronutrients are added directly in the formulation of basic fertilizers ($N:P_2O_5:K_2O$), which can be found in the market of domestic and imported agricultural inputs [3]. Among the several micronutrients used in Brazilian and Global agriculture, zinc (Zn) is a major one, having its need scientifically established over 70 years being its main functions related to photosynthesis, respiration, protein synthesis and plant membrane permeability [2].

Nutrient uptake by crops is influenced by various factors, including climatic conditions such as rain and temperature, genetic differences between varieties, type of soil, nutrient content of soil and various cultural practices.

In the strategy of reducing costs, both in manufacture of fertilizers, as well in croplands, the use of industrial waste for the obtaining of micronutrients has been widely used. These wastes, when used appropriately and rationally present themselves as a good solution for agriculture, but their improper use can cause serious damage to soil, plants and human beings. Several researches are touting in scientific works, facts that generate severe criticism to the fertilizer industry, mainly because the involvement of these industries in operations that endanger the environment, where the main factor is the use of raw materials containing toxic metals such as cadmium (Cd), lead (Pb) and chromium (Cr) in the manufacture of fertilizers [4]. In that way, the fertilizers used to supply micronutrients have often, in their composition, in addition to the desirable elements, toxic metals such as Cd, Pb and Cr [5,6].

The purpose of this chapter is to assist in developing a more adequate environmental legislation, that ensures a sustainable environment and a production of quality food. In that way, this chapter highlight the lack of criteria, in the current Brazilian environmental legislation, regarding the allowed limits of toxic heavy metals in agricultural soils and in the use of fertilizers contaminated with heavy metals in agriculture, as well, demonstrate that the use of these contaminated inputs can cause degradation of agricultural soils, making food production unviable at these locations. Furthermore, it was made the study of three cases about the existence of fertilizers contaminated with toxic metals in Brazilian agriculture and the impact caused by its use. The chapter also presents some strategies for recuperation and remediation of soils contaminated with toxic heavy metals.

The presence of heavy metals in fertilizers is a extremely important subject, since investigations in Brazil, coordinated by prosecutors in São Paulo, show evidence of entry loads containing toxic waste through irregular importation and smuggling, several times in Brazilian ports. These investigations shown that companies and manufacturers of agricultural inputs, are importing these toxic chemicals from United States, Canada, Mexico, Spain, Holland and England to be used as raw material in the manufacture of fertilizers to supply micronutrients.

Brazil presents itself as one of most promising nations of contemporary and future world, however, for the effective consolidation of its presence among the great powers is extremely necessary maintain the sustainability of its agroecosystem. Therefore, the Brazilian agricultural activity should be developed without damaging the environment and thus continue to grow in a rational and appropriate way, avoiding harm to future generations. In the world of agribusiness, which needs to be highly competitive, concepts such as traceability, environment and risk assessment are fundamental starting points for those who want to establish and grow.

2. Potential contamination of soil by agricultural activities

Due to its continental dimensions, Brazil has a great diversity of soil types in different regions of the country. Most of these soils have limitations for agriculture, characteristic of tropical regions such as acid, deep and highly weathered soils, with low nutrient availability and high exchangeable acidity (Al^{3+}), resulting in a low cation exchange capacity and low availability of P [7].

These factors highlight the need for adopting management strategies that aim to increase the chances of good crop of the culture when grown in soils of low fertility, giving them the ability to meet the nutritional requirements of these plants. The practices of fertilization and soil amendment aim to make soil suitable for the development of major crops, providing the necessary nutrients for the plants and decreasing the acidity of the soil. These practices are adopted throughout the world, and are based on the prerogative of the need to produce food to sustain the world population that grows rapidly each year.

According to the Food and Agriculture Organization of the United Nations (FAO) [8], in 1980 there were 4.4 billion people on the planet, 5.7 billion in 1995 and 6.9 billion in 2010. Studies estimate that by 2050 there will be 9.4 billion people and in 2200 about 11 billion [9]. Based on this scenario, agricultural activities have expanded greatly since 1945, currently holding approximately 38% of the land surface. Between 1985 and 2005, the acreage increased about 7%, generating large increases in food production [10].

This pressure on agricultural activities promotes an increasing fertilizer consumption per each year that goes around the globe (Figure 1). Studies show that in the early 60's used to be about 146 million tons of fertilizer, increasing in 2010 to 176 million of tons, demonstrating a significant increase in the consumption of these inputs [11]. Also according to the International Fertilizer Association (IFA) [11], Brazil is the fourth largest consumer of mineral fertilizers ($N:P_2O_5:K_2O$) in the world, only behind China, India and the United States. In recent years the consumption of nutrients by Brazilian agriculture has increased considerably being that in 1961 Brazil was only the 25th world consumer, however in the 1990, the country was already appearing as the seventh in the rankings. Regarding to micronutrients, there is an increase in consumption of 13.3 times in the period 1990-2003 [12].

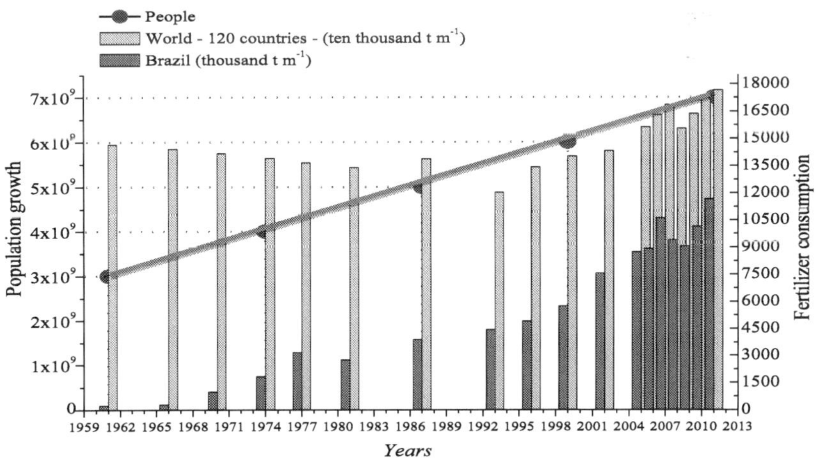

Figure 1. Population growth and consumption of fertilizers in Brazil and in the world [11]

However the use of these products is often not done correctly. The plan for fertilizer recommendation involves planning for the collection of soil samples, soil chemical analysis and recommendation by an appropriate professional, something that many producers do not perform, often using pre-determined recommendations without professional assistance. Another problem is the lack of crop rotation and the use of highly productive cultivar which, according to reference [13], require intense macronutrient application, resulting in micronutrient deficiencies in the soil.

Overuse of fertilizers can also provide soil contamination, it increases the concentration of a particular element or substance of environmental interest above its naturally occurring in the soil [7]. This contamination is related to eutrophication (excess nutrients), in this case the major nutrients to be considered are nitrogen (N) and phosphorus (P). Another concern is the availability of harmful elements such as toxic metals. These elements can express their pollutant potential directly on soil organisms, due to the availability to plants in phytotoxic levels, plus the ability to transfer to the food chain through the own plants, or by contamination of soil and water resources [14].

The presence of harmful elements in fertilizers is worrying because until the 70's, most fertilizers were obtained directly from ores found in nature, and although they also contained toxic metals, the beneficiation process of this material was less offensive from the environmental standpoint, because the contaminant levels were very low [6]. Recently, the shortage of mineral resources and modern extraction methods required large investments in equipment and technology, creating a technical difficulty and the pursuit of lower costs of production, which eventually encouraged companies to seek a cheaper solution such as using industrial waste, that may be feasible since it does not contaminate the environment. However, according to the Environmental Sanitation Technology Company of São Paulo State (CETESB) [15], in São Paulo, until the end of 2012, 4572 contaminated areas were found, and approximately 4% of these areas were contaminated with industrial waste, and among the contaminants, toxic metals are in 4th place, behind Polycyclic aromatic hydro carbons contaminants (PAHs), aromatic solvents and liquid fuels.

Between the sources of soil contamination from waste, there is the contamination of sewage sludge, that even though has a variable composition, with macro, micro nutrients and organic matter, their use may be limited by the presence of pathogens and toxic metals [16,17]. Studies evaluating the use of sewage sludge in sugarcane found that the doses used increased levels of chromium, nickel (Ni), Pb and zinc (Zn) in the soil [18].

Reference [19] studying the availability of Cd in dystrophic soil submitted to the application of sewage sludge, observed that at the first hour of contact between the sewage sludge and soil occurred transference of Cd to the solution. After this interval, the Cd concentration in the solution decreases, occurring stabilization in the following hours, due to adsorption and ion exchange which act providing less metal. According to the authors, in all observed time intervals, the levels of Cd were higher than reference values for CETESB (<0.5 mg kg^{-1}) [20], which characterize the contamination of the soil by sewage sludge.

The use of pig manure as fertilizer is also a practical often carried out mainly in south of Brazil, however, some studies have shown risks as to its overuse. Reference [21] evaluated the accumulation of metals in the soil under the systematic application of swine manure, and found an increase in the availability of copper (Cu), zinc and manganese (Mn). The authors also claim that in these cases there is a need to monitor these elements over time, so that the total contents do not exceed the critical values. In study conducted in China [22], the application of manure in soil for 16 years resulted in higher accumulation of Cd. One of the major sources of metals such as Pb, Ni, Cr and Cd in pig manure is mineral supplied by commercial feed [21].

The metals are also components of many pesticides, mainly Cu, Zn and Pb, which cause high soil contamination by these elements [23], another case is the use of Cu, that has been used since 1917 for the Bordeaux mixture [24]. In a study conducted in Caetés watershed in the county of Paty's Ensign, Rio de Janeiro, Brazil, it was aimed to evaluate the soil contamination by toxic metals, from the intensive use of agrochemicals, in which the authors found an increase in the total levels of metals in soil, stream water and dams. The soil metal concentration not reached critical established levels, however the result was worrying about the water stream and dam, which showed total contents of Cd, Mn and Pb above maximum standards established by the Brazilian legislation [23].

In England and Wales, reference [25] developed an inventory on the input of metals in agricultural soils. Forms of contamination are the most diverse, the main sources are generated by atmospheric deposition, sewage sludge, manure, inorganic fertilizers, lime, agrochemicals, irrigation water and industrial waste.

All these reports point to industrialization as the main cause for the large amount of metal dispersed in the environment, and human as the protagonist of all this pollution, because usually mankind seeks the easiest way supply their needs while generating countless volumes of waste. These actions put us in the position of predator. And thus there is a need of legislations that seek to limit the presence of these contaminants in the soil. In Table 1, the international maximum levels of metals allowed in soils of various places show what would be, in theory at least, a concern of many countries in the world about metal contamination.

Countries	**Toxic Metals (mg kg^{-1})**				
	As	**Cd**	**Pb**	**Cr**	**Hg**
(*) European Comunities	-	1-3	50-300	-	1-1,5
(**) United States	14	1.6	50-300	1500	0.5
(**) Poland (1977 – 1993)	30	1-3	70-150	50-80	5
(**) Germany (1984 – 1992)	20	3 – 1.5	100	100	2
(**) United Kingdon	10	3-15	500-2000	-	-
(**) Austria	50	5	100	100	5

Source: * [26], ** [27]

Table 1. Maximum permitted levels of toxic metals in the soil in some countries

Currently in Brazil, the legislation that deals about soil contamination is Resolution n°. 420 of 2009 of the National Council of Environment (CONAMA) [28], which aims to establish the criteria and values that guide of soil quality as the presence of chemical substances, among them the toxic metals. The resolution establishes three categories of guiding values which are: The Reference Value Quality (RVQ) which is the concentration of a substance that will define the natural quality of the soil, based on previous studies for each Brazilian state, the Prevention Value (PV), which refers to the concentration limit of a substance in the soil, and the Investigation Value (IV) which is the concentration of a substance in the soil of which there are potential risks, direct or indirect, to human health. Table 2 shows the values set for preventing and detecting toxic metals in soil.

However, the Resolution n°. 420 is being severely questioned by authorities and researchers, since it present several misconceptions that violate the Federal Constitution regarding the "preservation, enhancement and restoration of environmental quality propitious to life" (this issue will be addressed with greater emphasis on section 5 of this chapter).

Metal	PV	IV		
		Agricultural Area	Residential	Industrial
As (mg kg^{-1})	15.0	35.0	55.0	150.0
Cd (mg kg^{-1})	1.3	3.0	8.0	20.0
Pb (mg kg^{-1})	72.0	180.0	300.0	900.0
Cr (mg kg^{-1})	75.0	150.0	300.0	400.0
Hg (mg kg^{-1})	0.5	12.0	36.0	70.0

PV: Prevention Value; IV: Investigation value

Table 2. Values guiding to the presence of toxic metals in Brazilian soils in accordance with Resolution n° 420 of 2009 of CONAMA [28]

3. Fertilizers *versus* contaminants

There are three different categories of agricultural fertilizer: macronutrient fertilizer to supply major nutrients (N, P and K); macronutrient fertilizer to supply secondary nutrients [calcium (Ca), magnesium (Mg) and sulfur (S)] and micronutrients [Cu, Zn, Ni, Mn, iron (Fe), molybdenum (Mo), boron (B), chlorine (Cl), sodium (Na), cobalt (Co) and silicon (Si)]. All of these nutrients can be applied through simple fertilizer (only one nutrient) or compound form (formulations with more than one nutrient).There is a further classification based on the essentiality of nutrients, where metals may be classified as essentials, such as B, Cu, Fe, Mn, Mo and Zn; beneficial, such as Co, Ni and V; and toxic, such as Cd, Cr, Hg and Pb. Thus, it should be noted that the essential and beneficial elements can become toxic at high concentrations [30].

According to reports developed by the United States Environmental Protection Agency (USEPA) [31] and The Weinberg Group (requested by The Fertilizer Institute) [32], the main metal contaminants that are found in agricultural fertilizers are Cd, Cr, Co, Cu, Pb, Mo, Ni, Zn, arsenic (As), mercury (Hg) selenium (Se), vanadium (V) and radio 266 (Ra). It can be observed in this list, that among the metals commonly found as contaminants in fertilizers, occur the presence of micronutrients, required for plant development.

According to USEPA [31], among the fertilizers used to supply macronutrients, the phosphate shown the highest levels of toxic metals (especially Cd), followed by nitrogen and potassium fertilizers. With respect to micronutrients, the sources used to obtain it are varied, resulting in the occurrence of fertilizers from low to high levels of metals. In general, micronutrient sources have higher level of contaminant in comparison to sources of macronutrients, on the other hand, are applied larger amounts of fertilizer to supply the macronutrients, therefore, the possibility of soil and environment pollution with the use of any type of contaminated fertilizer is real and should be seriously studied.

The presence of contaminants in fertilizers occurs most of the time with the use of industrial waste or low cost sources for raw materials. These sources, when treated and used properly and rationally, present themselves as a good alternative to agriculture, however its improper use can cause serious damage to the soil, plants and the human beings.

This fact becomes even more important in the case of fertilizers to supply micronutrients, which often are obtained from raw materials originated from industrial wastes, which may show, in addition of the desirable elements, toxic metals. In this situation, the benefits are highly attractive for the fertilizer industries, once these companies acquire low quality raw material at low costs, without generating any type of waste, incorporating everything in their products.

One of the main cases regarding the presence of contaminants in fertilizers happened in the 90s in the United States, where many farmers and social actors reported the possibility of the presence of toxic metals in these inputs. The impact generated and population pressure led to the development of several research regarding the contamination of fertilizers and soil pollution, culminating with the publication by the U.S. government, of the first law regulating the content of toxic elements in the United States in 1998. From these studies and based on Canadian law, in 2002, the USEPA published the federal law that regulates the production of fertilizers to supply Zn from industrial waste [33].

Another case reported in the literature is the presence of contaminants in raw materials from China. In a publication from the Washington Department of Ecology (WDE) [34] it was reported the presence of Cd in raw materials for the manufacture of fertilizers, with samples showing more than 20% of Cd, i.e., more than 200,000 ppm. It is also reported that the final products produced with the use of these contaminated sources presented 10,600 ppm of Cd. The report also warns that Cd present in the analyzed material does not naturally occur in natural mineral sources, leading to the conclusion that the incorporation of Cd or the use of wastes contaminated with this metal was occurred.

In Brazil, there are frequent reports of companies selling fertilizer, mineral and organic, outside the standard quality. The most common cases are related to the presence of nutrients amounts below the specified labels, since this factor is more investigated. However, there are few studies evaluating the presence of toxic metals in these inputs, although the few that were conducted demonstrated the presence of these contaminants in fertilizers [6, 35, 36].

Currently, most grain-producing countries and other plant materials have laws regarding the limits of metals in fertilizer, which are presented in Table 3, where it can be observed a large variation between the allowed values. In general, the stipulated limits are for metals in phosphate fertilizer and micronutrient fertilizers, but there are some normatives that consider fertilizers in general, regardless of its composition or purpose. The limits of toxic metals can also be determined based on the amounts of nutrients or total amount in fertilizers.

Country	Fertilizer type	As	Cd	Pb	Cr
		mg kg^{-1}			
Switzerland[1]	Phosphate	NA	50.0	NA	NA
Finland[1]	Phosphate	NA	50.0	NA	NA
Sweden[1]	Phosphate	NA	100.0	NA	NA
Norway[1]	Phosphate	NA	100.0	NA	NA
Denmark[1]	Phosphate	NA	110.0	NA	NA
Belgium[1]	Phosphate	NA	210.0	NA	NA
Germany[1]	Phosphate	NA	210.0	NA	NA
Austria[1]	Phosphate	NA	275.0	NA	NA
U.S. (California)[2]	Phosphate	2.0*	4.0*	20.0*	NA
U.S. (Washington)[3]	Phosphate	13.0	165.0	61.0	1.0
Australia[1]	Phosphate	NA	300.0	NA	NA
Japan[1]	Phosphate	50.0	343.0	100.0	5.0
New Zealand[1]	Phosphate	NA	280.0	NA	NA
U.S. (Texas)[4]	Anyone	41.0	39.0	300.0	17.0
Canada[1]	Anyone	75.0	20.0	500.0	NA
China[4]	Anyone	50.0	8.0	100.0	NA
U.S. (California)	Micronutrient	13.0**	12.0**	140.0**	NA
U.S. (Washington)	Micronutrient	112.0**	83.0**	463.0**	6.0**
Australia	Micronutrient	NA	50.0	2000.0	5.0

1 [38], 2 [39], 3 [40], 4 [41]; * - Maximum allowed value per percentage point of P_2O_5 in the fertilizer; ** - Maximum allowed value per percentage point of the micronutrients sum in the fertilizer ; NA – not available

Table 3. International maximum limits of contaminants allowable in fertilizers

The Brazilian fertilizers law is based on the Normative Instruction (NI) 27/06 [37], which states like contaminating toxic metals, the elements As, Cd, Pb, Cr and Hg, determining the maximum allowable limits of these metals in four different categories of fertilizers (Table 4).

Fertilizer type	Maximum permitted quantity of the metal contaminant (mg kg^{-1})				
	As	Cd	Pb	Cr	Hg
Simple mineral fertilizer to suply P_2O_5	2.0*	4.0*	20.0*	40.0*	0.05*
Compound mineral fertilizer with micronutrient	500.0**	15.0**	750.0**	500.0**	10.0**
Compound mineral fertilizer with macronutrient and micronutrient	250.0	57.0	1000.0	NA	NA
Mineral fertilizer to suply micronutrient	4000.0	450.0	10000.0	NA	NA

* - Maximum allowed value per percentage point of P_2O_5 in the fertilizer; ** - Maximum allowed value per percentage point of the micronutrients sum in the fertilizer; NA – Not Available.

Table 4. Maximum allowable limits for metal contaminants in Brazilian fertilizers [37]

There are some doubts in relation to NI 27/06, which consider the legislation incipient, since that it does not determines which raw materials can be used as sources of nutrients in the fertilizers. Nutrient sources for macronutrients generally present good quality, however, for micronutrients, the use of low quality materials is frequently observed in Brazil, as mentioned previously, resulting in a possible contamination of the environment.

The legislation also does not tell how it will be the inspection of fertilizers industries, moreover tolerating the presence of contaminants in amounts up to 30% higher than the accepted values (Table 4). Another worth mentioning fact, is that the limits should be revised, since Article 5 affirms the need of a revision of the values predicted by NI 27/06 four years after its publication.

Nowadays, still does not exist an adequate legislation in Brazil regarding the use of toxic waste in the production of micronutrients. However, there are technical studies which classify these elements as hazardous materials, an example is the NBR 10004 of the Brazilian Association of Technical Standards (ABNT) [42], which stablish that the use of this wastes may present a risk to public health, causing or contributing to an higher incidence of diseases and mortality, also presenting adverse effects on the environment when handled or disposed inappropriately.

Another issue that is being discussed by the scientific community is that the determination of contaminants limits in fertilizers provides for the industries the possibility to include, as a source of nutrients, contaminated materials or with low quality, i.e., by the laws, the government provided the industries, exactly what they wanted, the possibility to incorporate low-quality raw materials and contaminated to supply nutrients [43]. It is observed so that there is a strong influence of the fertilizer industry in the studies and laws that limit the contaminants in these materials, showing that this problem is not only related to the environmental field, but also with the political and economic sector.

4. Metals: Toxicity in the foodchain

Due to its mobility in aquatic systems and to its toxicity to life forms, the toxic metals or heavy metals, once in surface or in groundwater, are mentioned as one of the most problematic inorganic contaminants in the environment.

The toxic meals are strongly pollutants elements and present many harmful effects to ecosystems, causing physical-chemical alterations to water, what promote the decrease of water quality and mortality of the fauna and flora, prejudicing consequently, the human health [44].

The term heavy metal, is many times used with connotations of pollution and toxicity. However, "heavy" is conventionally used to imply the high density of the metal, referring to the pure element or to an alloy of metallic elements. The expression heavy metal apply to elements which have specific mass higher than 5 g cm^{-3} or which have atomic number higher than 20 [6].

Some metals, as the case of Cd, Pb, Ni and others, even when its presence is diluted in waters, many times in undetectable quantities, have recalcitrant effect, i.e., its persistence in the environment imply in process as biomagnification, which regards to the accumulation of these harmful substances in the many trophic levels of the ecological chains of ecosystems.

The contamination by toxic metals in animals can occur due to inhalation, skin penetration or ingestion of contaminated waters and foods, this last caused mainly by the capacity of plants in bioccumulate these elements in high proportions and later offer them to animals and human beings [45].

In animals, the toxic metals act in highly specific reactions, enzymatic in the most part and in alterations in systems that evolves these reactions, resulting in negative effects. One classic example is the metalloenzymes, in which dimensional structure is changed if one "intruder metal" binds to its active site, what prevents the enzymatic action [46].

Metal contaminations in the environment results from a sum of natural processes, between which also make part the natural sources of contamination, however, it is a common sense that the sources so called anthropogenic are more frequent when compared to the natural sources of contamination [47].

Significant increases of the metals concentrations are found in areas of intense industrial activity, where the accumulation can be many times higher than the average levels from non-contaminated areas. However, areas distant from the industrial centers in many cases also show metals contamination, due to the atmospheric transportation of these compounds [48] and in the case of agricultural soils, due to the use of fertilizes contaminated with these metals [6].

The contamination by metals is arising from the continuous disposal of industrial, domestic or livestock wastes in soils, rivers, lakes and seas. These pollutants include the toxic metals, which cause great concern to public health, because they are incorporated very easily in the tropic chain. These harmful elements cannot be destroyed by biological degradation, as in the case of the most part of organic pollutants. Incidences of the accumulation of metals in fish, oysters, mussels, sediment and other components from the aquatic systems have been reported in many countries around the world [49, 50].

Once the metals are present in cultivated soils, they are likely of radicular absorption by plants. It is important to point out that some of these elements present biological function whereas others are much known as toxic even in low concentrations. This way, the plants can be considered as constituents of the foundation of the trophic chain, causing great concern due to the possibility of the occurrence of toxic concentrations of certain metals which are transported from plants to higher levels of the food chain [51].

The toxic metals are not only subjected to radicular absorption by plants when presents in the soil solution, because may also occur the fixation of the metals in the soil particles, seepage/leaching to lower levels of soil or even the surface runoff, reaching the groundwater, rivers, lakes and seas. In general, there are two types of soils in Brazil, sandy and clayey, so that in sandy soils, due to the low ion exchange capacity, the metals are easily moved in the environment by leaching or surface runoff, reaching easily the hydric resources. In the other hand, in clayey soils, due to the higher ion exchange capacity, the metals tend to stand immobilized in the soil particle, becoming available for the plants in more than one cultivation season.

The toxic metals resulting from the anthropogenic activities, when disposed incorrectly in the environment, directly or indirectly reaches the top of the chain, i.e., humans will have contact with these elements by the consumption of food from vegetal or animal origin, contaminated by water, soil or air. Once in the human organism, these metals present cumulative character, i.e., are accumulated in the live tissues, i.e., hardly are eliminated by the organism. Down below are related some of the characteristics of some toxic metals and its respective collateral effects caused to human organisms.

4.1. Arsenic (As) – CAS ID 7440-38-2

In humans, the ingestion of high levels of arsenic can result in death. Exposure to lower levels can cause nausea and vomiting, decreased production of red and white blood cells, abnormal heart rhythm, damage to blood vessels, and a sensation of "pins and needles" in hands and feet. The ingestion of inorganic arsenic can increase the risk of skin cancer and cancer in the liver, bladder, and lungs. Inhalation of inorganic arsenic can cause increased risk of lung cancer. Exposure to higher than average levels of arsenic occur mostly in the workplace, near hazardous waste sites, or in areas with high natural levels. At high levels, inorganic arsenic can cause death [52].

Aquatic organisms accumulate arsenic mainly as inorganic forms, and some of the organisms such as phytoplankton, bacteria, etc. transform them into methylated and organic forms [53].

As is associated with ores containing metals, such as copper and lead. Arsenic may enter the environment during the mining and smelting of these ores. Small amounts of arsenic also may be released into the atmosphere from coal-fired power plants and incinerators because coal and waste products often contain some arsenic. It cannot be destroyed in the environment. It can only change its form, or become attached to or separated from particles. It may change its form by reacting with oxygen or other molecules present in air, water, or soil, or by the action of bacteria that live in soil or sediment [52].

Researching the effects of arsenic accumulation from plants cultivated near pressure-treated wood, Cao and Ma [54] found out that elevated As concentrations were observed in the soils adjacent to fences and poles, because As was frequently used for wood treatment, in order to increase the wood life time. Eating vegetables from these contaminated soils may pose a risk of As exposure since vegetables would readily take up As from these soils.

As, according to the Agency for Toxic Substances and Disease Registry (ATSDR) [55] is the first substance on the Priority List of Hazardous Substances.

4.2. Cadmium (Cd) – CAS ID 7440-43-9

At room temperature (25 °C/73.4 °F) the Cd is presented at solid state, with atomic number of 48 and atomic mass of 112.41 g. This element belong to the group of the metal of gray silver metallic color and it is available in many forms including sheet, granule, pellet, powder, rod, wire, among other. It is a soft metal, highly toxic which can be easily cut with a knife, and mostly produced as a byproduct of the extraction of Zn, Cu and Pb [56].

It is considered one of the most toxic metals, presenting important harmful effects for biological activity on soil, plant metabolism, human and animal health. The abundance of Cd in magmatic and sedimentary rocks, generally not more than 0.3 mg kg^{-1}, being that this metal tends to be concentrated in clayey deposits. The Cd is strongly associated to Zn in the geochemistry, however, have a strong affinity with S, presenting a higher mobility than Zn in acid environments [27].

Generally ore (CdS) is associated as a contaminant (0.25 to 0.5%) to ore Zn and Pb. According to the International Cadmium Association [57], Cd has been mainly used in manufacturing NI-Cd batteries, though, also has other applications as pigments, coatings, stabilizers for PVC (polyvinyl chloride) and alloys.

One of the main problems associated with Cd is its final destination, because it can reach the soil or the air through the burning of urban waste or the burning of fossil fuels, in this way polluting the environment and causing damage to the ecosystem [58].

In humans, Cd affects several organ systems, as the Cardiovascular (Heart and Blood Vessels), Developmental (effects during periods when organs are developing), Gastrointestinal (Digestive), Neurological (Nervous System), Renal (Urinary System or Kidneys), Reproductive and Respiratory (From the Nose to the Lungs), Cd is also known as a carcinogen agent [52].

According to the Agency for Toxic Substances and Disease Registry [55] Cd is the seventh substance on the Priority List of Hazardous Substances.

4.3. Lead (Pb) – CAS ID 7439-92-1

The Pb is rarely found in its natural state, but in combination with other elements, being the most important ores which present this metal are: Galena, cerussite, anglesite, perite and others.

Among the toxic metals, Pb is present in larger amounts in the earth's crust, with an average concentration between 10 and 20 mg kg^{-1} [59], therefore, is among the most used metals worldwide scale of Fe> Al> Cu> Zn> Pb [60].

To be quite versatile, Pb is much used in industrial products, and can be hardened by the addition of other metals and metal alloy being used as producing components, welds and other materials, in addition to being able to form various compounds employed in chemical industry [61].

According to unofficial information from the National Department of Mineral Production (DNPM), the main consuming sectors Pb are: accumulators (batteries) 80%, representing approximately 88 tons of metal, oxides, 12%, equivalent to 13 tons; and electronic (alloys, solders and various), 8%, with 9000 tons.

The Pb in soil is extremely stable and highly toxic to humans and animals, is currently ranked as the second most dangerous element in the Priority List of Hazardous Substances [55]. Most of the lead enters the human body through the respiratory and gastrointestinal routes, and after absorption, it can be found in the blood, soft and mineralized tissues [62].

In humans, the Pb poisoning may initially cause lack of appetite, metallic taste in mouth, muscle discomfort, malaise, headache and abdominal cramps strong. However, in childhood, symptoms often linked to the deposition of this metal in the brain are predominant. In its inorganic Pb^{2+} is an inhibitor of enzymes, further affecting the nervous system [63].

The organ systems affected by Pb are the Cardiovascular (Heart and Blood Vessels), Developmental (effects during periods when organs are developing), Gastrointestinal (Digestive), Hematological (Blood Forming), Musculoskeletal (Muscles and Skeleton), Neurological (Nervous System), Ocular (Eyes), Renal (Urinary System or Kidneys) and Reproductive [52].

4.4. Chromium (Cr) – CAS ID 7440-47-3

With regard to heavy metal Cr, this has turned into a serious pollutant of air, soil and water. About 70% of the production of Cr is used in alloys, including stainless steel and 15% in industrial chemical processes, mainly in leather tanning, pigments, electrochemical and cooling towers [64].

The ions of Cr are found in different oxidation states, however, the trivalent form (Cr^{3+}) is the most common, while the hexavalent form (Cr^{6+}) depends on the more specific conditions of pH and redox potential [65].

The toxicity of Cr depends on its oxidation state. While the Cr^{3+} is relatively stable and immobile, the Cr^{6+} moves easily through the soil and aquatic environments, with a strong oxidizing agent capable of being absorbed even by the skin in this way, the Cr presents itself as a carcinogenic species [66, 67].

In humans, the Cr affects the immune, renal (kidney or urinary system) and respiratory tract (the nose to the lungs), and is known as a powerful carcinogen [52].

According to reference [68], a lethal dose of Cr^{6+} is between 50 and 100 mg kg^{-1}, a value much lower when compared to Cr^{3+}, the values are between 1900 and 3300 mg kg^{-1}. Since the Cr^{6+} occupies the 17th position in the Priority List of Hazardous Substances [55].

4.5. Nickel (Ni) – CAS ID 7440-02-0

The Ni occurs in the earth's crust in the order of 0.01%, mainly as sulfide, oxide and silicate minerals. Geological activities as natural weathering and volcanoes led the natural distribution of this metal in natural environments at modest levels. The Ni metal and its compounds are used in multiple applications as stainless steel and other alloys, castings, catalysts, batteries, electronics, ceramics, pigments and even coins [69].

Among the main uses of this metal can still cite it use in coating parts, in a process called electroplating, widely used around the world by Electroplating industries [70].

Although not an element extensively released to the environment, this can present a risk to human health. Epidemiological investigations and experimental studies have shown that certain compounds of Ni are extremely carcinogenic after inhalation, however, the same studies show that these risks are limited by conditions specific occupational exposure to this metal [71].

Results show that this element can cross the placenta and present embryological and teratogenic properties. The main danger of Ni to the man beside the carcinogenicity is related to the ability to cause sensitivity reactions [71].

The most common harmful health effect in the general population is allergic contact dermatitis elicited by prolonged skin contact of sensitized individuals with Ni. Therefore, the majority of studies related to health effects of Ni to humans concern airborne Ni particulate matter. Strongest evidence for high cancer risk was found for sulfidic nickel species (NiS, NiS_2 and Ni_2S_3) in Ni refinery dust. On the molecular level the toxic Ni species responsible for severe health effects, as allergic contact dermatitis and respiratory tract cancer, has been suggested to be caused by Ni^{2+} [69].

Ni affects the following organ systems: Cardiovascular (Heart and Blood Vessels), Dermal (Skin), Immunological (Immune System), Respiratory (From the Nose to the Lungs) [52]. Actually, Ni ranks the 57th position on the Priority List of Hazardous Substances [55].

4.6. Copper (Cu)- CAS ID 7440-50-8

The Cu, despite being a heavy metal, meets the criteria of essentiality for plants and microorganisms, being classified as micronutrient [72]. This metal is essential to human health, and is present in all human tissues and many fluids and with amino acids, fatty acids and vitamins, is required for the metabolic processes [73].

Excess copper may be toxic to humans because of the affinity metal with S-H groups of many proteins and enzymes associated with diseases such as epilepsy, melanoma and rheumatoid arthritis, as well as the loss of taste [72]. Toxic levels of Cu occur naturally in some soils, whereas others may contain high levels of Cu as a result of the anthropogenic release of contaminated materials into the environment through mining, smelting, manufacturing, agriculture and waste disposal technologies [72].

In plants, either deficient or in excess, Cu can cause disorders in plant growth and development by adversely affecting important physiological process. In particular photosynthetic electron transport is altered under deficiency and excess of Cu. Thus, for healthy plant growth and development Cu must be acquired from the soil, transported throughout the plant, distributed and compartmentalized within different tissues and its content carefully regulated within different cells and organelles. For this purpose, plants, like all other organisms, have homeostatic mechanisms to maintain the correct concentrations of essential metal ions. Membrane transport systems are likely to play a central role in these processes [73].

Copper ranks the 125th position on the Priority List of Hazardous Substances [55].

4.7. Manganese (Mn)- CAS ID 007439-96-5

According to the International Manganese Institute (IMnI) [74], this metal is the fourth most used in terms of tonnage, being ranked behind Fe, Al and Cu, being the production of Mn alloys about 17.7 million metric tons at 2011. Mn the second major abundant metal and 12th most abundant element in the earth's crust, being its ores deposits usually of sedimentary origin, with oxide ore layers inter-bedded with iron-rich formations [75].

Steel is basically an alloy of iron and carbon, consisting of an iron phase and iron carbides. Crude steel produced from iron contains an undesirable amount of oxygen and some sulphur. Mn plays a key role because of two important properties: its ability to combine with sulphur and its powerful deoxidation capacity [76].

High grade Mn ores (> 40%) are typically processed into suitable metallic alloy forms by pyrometallurgical processes. Low grade manganese ores (< 40%) are conventionally processed by pyrometallurgical reductive roasting or melting followed by hydrometallurgical processing for production of chemical manganese dioxide (CMD), electrolytic manganese (EM) or electrolytic manganese dioxide (EMD) [77].

In the human body, Mn displays a somewhat unique behavior with regard to its toxicity. It is relatively non-toxic to the adult organism except to the brain where it causes Parkinson-like symptoms when inhaled even at moderate amounts over longer periods of time [78].

This element, when in large amounts, affect fertility in mammals and are toxic to the embryo and fetus, what implicates that pregnant women should not be exposed to Mn anytime [78].

Due to the substitution of Pb by Mn in gasoline, reference [79], researching the cultivation of beans and oats in soil next to roads, found not only Mn concentrations significantly higher in organic soil but also higher accumulation found in fruits and stems, suggesting that the addition of MMT to gasoline can be the cause for this increase in exchangeable Mn in organic soils.

According to the Agency for Toxic Substances and Disease Registry [55], Mn is the 140th substance on the Priority List of Hazardous Substances.

5. Case studies: Contaminated fertilizers in Brazilian crops

Based on information obtained from seizures made in Brazilian seaports, related to the entry of loads of industrial waste with the presence of toxic metals for use as source of micronutrients in fertilizers [6], three studies were conducted to evaluate the possibility of soil and plants contamination using different sources and Zn levels, since this is main micronutrient applied in the Brazilian agricultural crops. All studies were conducted at the State University of West Paraná - Brazil, being one for soybean [2], one for the wheat crop [4] and the last one for the corn crop [5].

In the work done with the soybean crop [2], an experiment was conducted in 2007, in Palotina-PR, with coordinates 24° 18' S and 53° 55' W (Figure 2) in a soil classified as Hapludox (RED). It were used four different fertilizers for the Zn supply sold in Brazil, in three doses (0.0; 0.9 and 1.8 kg ha^{-1}). The results of this study demonstrated the presence of toxic metals (Cd, Pb and Cr) in the soybean leaf tissue (Table 5) at levels above of the tolerable in agricultural crops [27]. We also observed the occurrence of these three metals in the soil after harvest (Table 6), which compared to the Brazilian legislation, that establishes the maximum concentrations of contaminants in soils [28], classify the obtained values as above the maximum allowed for the soil be agriculturable (IV - Agricultural area) in the case of Cd. For Pb and Cr the values obtained in the study were below of the prevention value (PV), which is the maximum allowable value of the contaminant in soil to that it is able to sustain its biological processes.

The work done with the wheat crop [4] was conducted on the residual fertilization of the work with the soybean crop, since the concentrations of the metals Cd, Pb and Cr in the soil of the experiment were significant, especially in the case of Cd. The results of this study showed no accumulation of Cd in the leaves of wheat at the level of detection by the analytical method (flame atomic absorption spectrometry - FAAS), however, we found concentrations of Pb and Cr. For Pb the maximum concentration found was under the tolerable limit, and for the Cr, the maximum concentration found was above the tolerable level for crops [27]. One of the main conclusions of this work was the occurrence of availability of Cr and Pb to wheat plants with fertilization performed in previous crop (soybean), demonstrating the persistence of contaminants in soil and their occurrence in more than one crop agriculture.

The work conducted with maize crop was conducted in Mercedes, Paraná State-Brazil, with geographic coordinates of 24° 25' S and 54° 15' W (Figure 3) in a Rhodic Eutrudox. The fertilization of this experiment was performed with eight different sources of Zn in four different doses (0.0; 2.0; 4.0 and 6.0 kg ha^{-1}). In the chemical analysis performed in the fertilizers

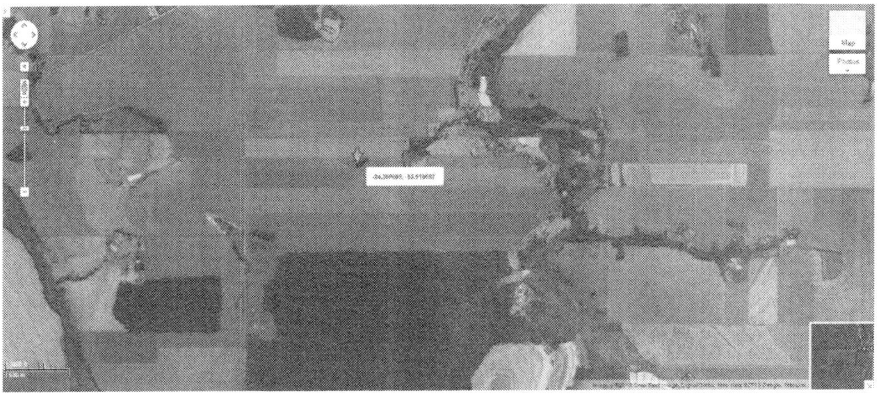

Figure 2. Local where the experiment involving fertilization with Zn in soybean and wheat culture were performed [80]

used in the work, it was observed that one of the sources presented 11039.46 mg of Pb kg^{-1}, being the maximum limit permitted by Brazilian law (NI 27/06) [37] 10000.00 mg of Pb kg^{-1} (Table 3). However, as previously mentioned in the section 3, the NI 27/06 tolerates the presence of contaminants up to 30% of the maximum permitted value, in this way, the fertilizer would not be considered as irregular, and can be freely traded.

Figure 3. Local where the experiment involving fertilization with Zn in corn crop was performed [80]

Regarding the leaf concentration of Cd, Pb and Cr (Table 5), it was observed that only the Pb concentrations was found at detectable levels by the used method (FAAS), where the leaf content of Pb in corn plants can be considered inside the tolerable limits [27]. For the soil, it was observed the occurrence of Cd, Pb and Cr (Table 6), so that the concentrations of Cd were below the maximum allowable limit so that the soil is agriculturable (IV - Agricultural area), and the Pb and Cr were below the PV, which is the maximum allowable value of the contaminant in the soil, to that it be able to sustain its biological processes.

Metal	Agricultural crop			Tolerable foliar level****
	Soybean*	Wheat**	Corn***	
Cd (mg kg^{-1})	0.73	ND	ND	0.05 a 0.50
Pb (mg kg^{-1})	25.53	1.95	8.67	0.50 a 10.00
Cr (mg kg^{-1})	3.36	9.28	ND	2.00

ND – not detected by the method used (FAAS) with a quantification limit of 0.01 mg L^{-1}; * - results from reference [2]; ** - results from reference [4]; *** - results from reference [5]; **** - Tolerable foliar level according to reference [27].

Table 5. Foliar levels of Cd, Pb and Cr in experiments conducted with mineral fertilizers to supply Zn in different crops in Brazil

Metal	Agricultural crop		Classification of the metal value in the soil*	
	Soybean*	Corn**	PV	IV – Agricultural area
Cd (mg dm^{-3})	5.52	2.52	1.30	3.00
Pb (mg dm^{-3})	24.25	51.48	72.00	180.00
Cr (mg dm^{-3})	22.19	18.43	75.00	150.00

* - results from reference [2]; ** - results from reference [5]; *** - Classification of the metal value in the soil according to reference [28]; PV–Prevention Value; IV – Investigation Value for agricultural area.

Table 6. Levels of toxic metals Cd, Pb and Cr in soil after cultivation of various crops fertilized with Zn in Brazil

Based on the results obtained in the three presented studies it can be inferred that the presence of metal contaminants in fertilizers is real and should be studied more carefully by regulatory authorities in Brazil. The foliar and soil levels of the metals Cd, Pb and Cr found cause concern from the environmental point of view, due the possibility of entry of these contaminants in the food chain, either by consumption of contaminated materials or pollution of water resources from contaminants in the soil. It should be noted that the values presented in these works were obtained with only one fertilization, so that with successive fertilizations may occur a greater availability of these contaminants to the environment.

It is worth affirm also that the laws related to the presence of contaminants [37], and levels of toxic heavy metals allowed in Brazilian soils [28], are being severely questioned by Brazilian researchers and government agencies. Among the key questions are the lack of criteria for determining and monitoring the presence of contaminants in fertilizers, passivity with the addition of toxic metals in brazilians soils by the minimum quality and lack of discussion with researchers and entities responsible for the drafting of legislation.

It should furthermore be highlighted that before approval of Resolution No. 420 of CONAMA [28], there was serious questioning on the part of Brazilian authorities and researchers in relation to the criteria used for its formulation by means of Proceeding 02000.002955/2004-69 [81-83], thus confirming the need for more societal discussion (especially among the specialized technical and scientific segments in a multidisciplinary manner) before this is submitted for approval of possible resolution. The resolution, moreover, violates the Brazilian Federal Constitution with respect to the "preservation, improvement and recovery of the environmental quality favorable to life" because it permits the introduction of contaminants in the soil up to the maximum limit permitted by the resolution, thus allowing the use of contaminated fertilizers and soil conditioners in agriculture until they reach and exceed these values and thereby degrade environmental quality. Therefore, the need for more in-depth field studies in relation to environmental legislation before it is drafted and approved becomes clear. These studies must be undertaken together with environmental regulatory bodies and research entities in the country. Based on the results obtained in this study, it is evident that some of the Zn fertilizer sold by sources in Brazil significantly affect the quantity of Cd, Pb, and Cr in soil. If their use is continued, there may be contamination of soil, surface and underground waters, plants, animals, and humans, leading to severe damage to health as well as the environment. This fact raises concern because according to references [84-86], even low levels of heavy metals in soil may be available to plants, an effect that is difficult to ascertain because the symptoms they case occur only over the long term, thus increasing even more the hazards associated with the use of contaminated materials in agricultural activities.

In a review about the environmental risks of adding phosphate fertilizers contaminated with trace elements in long-term, reference [87] warn about the possibility of increasing concentrations of As, Cd and Pb in agricultural soils using contaminated fertilizers. The authors also emphasize that to prevent the entry of potentially toxic elements in agriculture, it is necessary a strict control on the sources of macro and micronutrients used in agriculture. This demonstrates the importance and need for further studies related to the use of fertilizers contaminated with toxic metals in crops and the environmental impacts that these sources of nutrients can cause.

6. Sustainable strategies for soil remediation

The introduction of toxic chemicals by man in various environmental compartments (soil, water and atmosphere), the of controlled or uncontrolled elimination of wastes (accidental spills, mining processes and smelting of metallic ores, fertilizer application, sewage sludge and pig manure in agricultural soils) puts the ability to self-cleaning of contaminated ecosystems in doubt. Consequently, the pollutants that accumulate toxic metals are of concern in relation to human exposure of ecosystems and the potential impact [88].

Today there are several techniques for soil decontamination. It is worth mentioning that the choice of one or other technique, among other factors, also depends on the characteristics of the contaminant in question, between the different techniques is possible enumerate:

1. Thermal treatment: Use of heat to remove, stabilize or destroy the contaminants;

2. Physical treatment: Physical processes used to separate toxic substances from the environment;

3. Chemical treatment: Uses chemical reactions to remove, destroy or modify toxic substances;

4. Biological treatment: Makes use of natural metabolic means, such a microorganisms and other biological agents to remove, destroy or modify the contaminants;

5. Stabilization / solidification: This step the contaminants are stabilized chemically and / or modified to reduce the potential for contamination.

These technologies can in many cases be applied "in situ", i.e., at the site of origin of the contamination, or "ex situ" outside the place of contamination, which results in the removal and transport of the contaminated soil.

Among the physical-chemical treatments, these mostly are based on soil washing. These methods are based on the technological principle of the transfer of a contaminant from the soil for a liquid or gaseous phase. The main products are treated soil and the contaminants concentrated. The specific process of treatment depends on the type of contaminant, in particular as regards the type of connection established with the soil particles.

It is observed by the description above, that such treatments become process more expensive the decontamination because are many cases in which the soil treatment requires the removal of several layers of the same and specific treatment in laboratories.

Thus, many approaches have been developed, evaluated and performed to deal with soil pollution. Current technologies for decontamination soil are often expensive and of high energy consumption, and in many cases, the soil cannot be reused after treatment [89]. Thus, several studies have been developed to promote proper techniques and low cost of remediation in order to prevent the spread of these contaminants in the food chain [88, 90].

As an alternative, an ecological approach technology that involves the use of plants to clean up or remediate soils contaminated with toxic metals, called phytoremediation, has been developed and encouraged by the fact of being a simple and inexpensive option for remediation of contaminated soils [90-92].

Phytoremediation can be divided into phytoextraction (hiperacumulation) phytostabilization, rhizofiltration, fitotransformation and fitovolatilization (Figure 4).

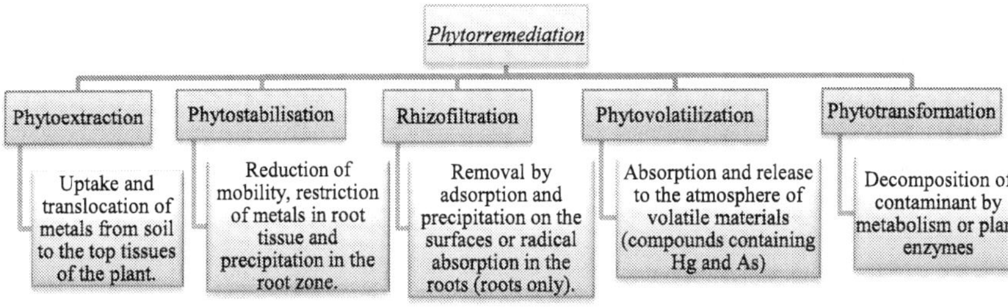

Figure 4. Types and characteristics of phytoremediation [92,93]

However for the phytoremediation to be effective it depends on the selection of plant species, preferably hyperaccumulators, which are able to grow in soils of low fertility, and degraded by the presence of contaminants [94].

In studies carried out in Iran using *Euphorbia macroclada* and *Centaurea virgata*, demonstrate that the sampled species not only were able to grow in soils highly contaminated by metals, as well as were capable of accumulating high concentrations of Zn, Mn, Cu, Pb and Fe, in this way these plants can be classified as hyperaccumulators, in other words, have adequate potential for phytoremediation of contaminated soils [95].

However, the metal accumulation by the plants is only efficient after the contaminant be removed from the soil, for example, by harvesting the plant matter. If most of the toxic metals are located on the shoots, the harvest can be performed using traditional farming methods. In general, it is necessary to harvest the plants before the fall of the leaves, or before his death and decomposition, so that contaminants do not disperse or return to the soil.

After harvesting, the biomass must be processed for extraction and recollection of the majority of metals. Alternatively, the volume and the weight of the biomass can be reduced by thermal processes, physical, chemical or microbial. In the case of burning plant, for example, the energy produced represents an economic value of the process and the ash may be treated as an ore, which may still be extracted the metallic contaminant (especially if the ashes are enriched in one or two metals).

Even Brazil showing great potential species for bioremediation and phytoremediation, the recovery of contaminated areas, due to the great biodiversity and climate, favoring biological processes in pollution treatment, knowledge about these species of plants and microbial communities in phytoremediation potential are still scarce [96].

In research carried out by reference [97], it was reported that some plants were able to remediate soils contaminated with metals such as *Brassica juncea* for phytoextraction of Cd and Cr^{6+}, Cu, Ni and Pb, *Avena sativa* for Zn, *Pisum sativum* L. and *Zea mays* L. for removal of Pb.

However, from an economic point of view, the use of these species of accumulator plants generally do not generate financial returns to the farmers, because these plants only remove the contaminant from the soil, being later discarded after the remediation. Thus if the farmer in question require to decontaminate a particular area, the phytoremediation only would generate costs, especially with seeds, agricultural machinery, irrigation, etc.

Nevertheless, currently there is a widespread idea that the species used for phytoremediation of soil can also generate economic gains beyond decontamination. The use of oil crops, for example, *Crambe abyssinica H., Moringa oleifera L.* and *Jatropha curcas,* in which the product marketing is not consumed as food, but as biofuel (biodiesel), is a possibility to decrease the cost of recovery of the contaminated area. Thus the fact that the plant has accumulated large amounts of heavy metal will have little influence on the end product of the culture, generating profit for the property.

Moreover, oil crops like *C. abyssinica, M. oleifera* and *J. curcas,* have been the subject of research in the removal of toxic metals such as Cd, Pb, Cr and water through the adsorption process, in which the pie their seeds are used as adsorbent [98-101]. The results of these studies demonstrated great potential for removal of these metals, and that these adsorbents are obtained at minimal charges, because they are wastes generated by the extraction of vegetable oil.

Thus, the use of cultures for vegetable oil production and recovery of contaminated areas can be an interesting alternative, which will generate the money from oil production and recovery / stabilization of the contaminated area with the development of these cultures, preventing them be lost through leaching or soil surface runoff and reach the groundwater and surface water and groundwater. However, more researches are still needed In order to evaluate this possibility, focusing on the development of oil-producing plants and cycles of addition of the adsorbent material and the biodegradability thereof. Between this and other issues to which scientific research is driven, many studies still need to be developed so that, in one way or another, the quest for sustainability will one day be achieved.

7. Conclusion

Based on information existing in the Brazilian and international literature, we can make some important observations:

The fertilization of plants is one of the major reasons of the increase in food production across the globe. Consequently, the use of fertilizers in agricultural crops is extremely high and growing, however, for the maintenance of this production capacity, we should be very careful with the fertilizers sources that are being used in agriculture.

The presence of toxic metals in fertilizers has been reported in the major food producing countries, like United States, China and Brazil. This demonstrates that this issue deserves further detailed studies of long-term about the successive addition and the effects of these contaminants to crops, soil, supply chain and the wider environment. Thus, the institutions of research and supervision, have to evaluate the products that are used for the manufacture of fertilizers on the market, as this is an activity that generate large amounts of money and that will always be a target of people with questionable character.

The current legislation, not only in Brazil, but internationally, with respect to the use of wastes as sources of nutrients, presence of contaminants in fertilizers and presence of toxic metals in soil, need higher accompaniments and answers for some of the community questions, because the legislation must ensure that the agricultural soils are not being degraded and becoming unproductive. In Brazil, it is strongly recommended that the laws must be harder and restrictive, until we have reliable and conclusive results related to the addition of contaminated fertilizer in long-term. In this case, it should be emphasized that these laws should be developed based on results from institutional researches and reliable groups, without the influence of political and industrial groups. Meanwhile, the damage caused by man in air, soil and waters, must be repaired, by remediation methods (such as phytoremediation) aiming not only the decontamination, but also ensuring a small profit, as an incentive to the farmer.

Brazil is considered as one of the countries with largest biodiversity and most promising in food production for the globe, however, for that prediction become a reality, it is very important that the productive areas are not degraded. The technologies developed by our civilization provide ways to avoid the degradation of our environment, but often they are not used because they prevent higher industrial profits, this should not occur, because the balance of the ecosystem and the quality of human life should be above of any monetary profit.

Author details

Affonso Celso Gonçalves Jr.[1], Herbert Nacke[2], Daniel Schwantes[3] and Gustavo Ferreira Coelho[4]

*Address all correspondence to: affonso133@hotmail.com

1 Western Paraná State University, Center of Agricultural Sciences, Marechal Cândido Rondon, Brazil

2 University Center Dynamic of Cataracts, Technological Center, Foz do Iguaçu, Brazil

3 Pontifical Catholic University of Paraná, Polytechnic School, Toledo, Brazil

4 Federal University of Paraná, Palotina, Brazil

References

[1] Nacke H, Gonçalves Jr AC, Stangarlin JR,Schwantes D, Strey L, Nava I. Productivity and yield components of maize fertilized with different sources and doses of zinc. Spanish Journal of Rural Development 2011;2(1) 47-55.

[2] Nava I, Gonçalves Jr AC, Nacke H, Guerini VL, Schwantes D. Availability of cadmium, lead and chromium toxic heavy metals in soil and soybean leaf tissue fertilized with different sources of NPK+Zn. Ciência e Agrotecnologia 2011;35(1), 884-892.

[3] Gonçalves Jr AC, Nacke H, Marengoni, NG, Carvalho, EA, Coelho GF. Yield and production components of soybean fertilized with different doses of phosphorus, potassium and zinc. Ciência e Agrotecnologia 2010;34(3) 660-666.

[4] Gonçalves Jr AC, Nacke H, Schwantes D, Nava IA, Strey L. Phytoavailability of toxic heavy metals and productivity in wheat cultivated under residual effect of fertilization in soybean culture. Water, Air and Soil Pollution 2011;220(1) 205-211.

[5] Nacke H, GonçalvesJr AC, Schwantes D, Nava I, Strey L, Coelho GF. Availability of heavy metals (Cd, Pb and Cr) in agriculture from commercial fertilizers. Archives of Environmental Contamination and Toxicology 2013;64(1), 371-379.

[6] Gonçalves Jr AC, Luchese EB, Lenzi E. Evaluation of phytoavailability of the cadmium, lead and chromium in soybean cultivated in the Latossolo vermelho escuro, treated with commercial fertilizers. Química Nova 2000;23(2) 173-177.

[7] Manzatto CV, Freitas Junior E, Peres JRR. Agricultural use of Brazilian soils. Rio de Janeiro: Embrapa Solos; 2002.

[8] FAO - Food and Agriculture Organization of the United Nations. The state of food and agriculture: Women in agriculture closing the gender gap for development. Rome: FAO; 2011.

[9] Nelson GC, Rosegrant MW, Palazzo A, Gray I, Ingersoll C, Robertson R, Tokgoz S, Zhu T, Sulser TB, Ringler C, Msangi S, You L. Food Security, Farming, and Climate Change to 2050: scenarios, results, policy options. Washington: International Food Policy Research Institute; 2010.

[10] Foley JA, Ramankutty N, Brauman KA, Cassidy ES, Gerber JS, Johnston M, Mueller ND, O'Connell C, Ray DK, West PC, Balzer C, Bennett EM, Carpenter SR, Hill J, Monfreda C, Polasky S, Rockstrom J, Sheehan J, Siebert S, Tilman D, Zaks DPM. Londons: Solutions for a cultivated planet. Nature 2011;478(1) 337-342.

[11] International Fertilizer Industry Association. http://www.fertilizer.org/ifa/ifadata/search (accessed 15 June 2013).

[12] Yamada T. Deficiências de micronutrientes, ocorrência, detecção e correção: o sucesso da experiência brasileira. Piracicaba: Potafos; 2004.

[13] Cakmark I. Plant nutrition research: Priorities to meet human needs for food in sustainable ways. Plant and Soil 2002;247(1) 03-24.

[14] Gonçalves Jr AC, Pessoa ACS. Phytoavailability of the cadmium, lead and chromium in soybean cultivated in argissolo vermelho eutrófico, treated with commercial fertilizers. Scientia Agrária2002;3(1-2) 19-23.

[15] CETESB - Environmental Sanitation Technology Company of São Paulo State. Relation of contamined areas in São Paulo 2012. http://www.cetesb.sp.gov.br/userfiles/file/areas-contaminadas/2012/texto-explicativo.pdf (accessed 15 June 2013).

[16] Lopes JC, Ribeiro LG, Araújo MG, Beraldo MRBS. Lettuce production using doses of sewage sludge. HorticulturaBrasileira 2005;23(1), 143-147

[17] Pegorini ES, Andreoli CV. Introduction. In: Andreoli, CV. (ed.) Alternative use of sanitary waste. São Paulo:ABES; 2006. p01-06.

[18] Marques MO, Nogueira TAR, Fonseca IM, Marques TA. Cr, Ni, Pb e Zn contents in sewage sludge added Red Argisol and sugarcane cropped. Revista de Biologia e Ciência da Terra2007;27(1) 133-143.

[19] Alamino RCJ, Polivanov H, Campos TMP, Silva VHG, Santos LV, Mendes JC. Bioavailability of Cadmium on a Latosol Increased of Sewage Sludge. Anuário do Instituto de Geociência 2007;30(2) 45-54.

[20] CETESB - Environmental Sanitation Technology Company of São Paulo State. Report establishment of guiding values for soils and groundwater in the State of São Paulo. São Paulo: CETESB; 2001.

[21] Mattias JL, Ceretta CA, Nesi CN, Girotto E, Trentin EE, Lourenzi CR, Vieira RCB. Copper, zinc and manganese in soils of two watersheds in Santa Catarina with intensive use of pig slurry. Revista Brasileira de Ciência do Solo 2010;34(4) 1445-1454.

[22] Wu L, Tan C, Liu L, Zhu P, Peng C, Luo Y, Christie P. Cadmium bioavailability in surface soils receiving long-term applications of inorganic fertilizers and pig manure. Geoderma 2012;173-174(1) 224–230.

[23] Ramalho JFGP, Sobrinho NMB, Velloso ACX. Heavy metals contamination of a watershed in Caetés by the use of agrochemicals. Pesquisa Agropecuária Brasileira 2000;35(7) 1289-1303.

[24] Baker DE. Copper. In: Alloway, BJ. (ed.) Heavy metals in soils. Glasgow: Blackie and Son; 1990. p.51-74.

[25] Nicholson FA, Smith SR, Alloway BJ, Carlton-Smith C, Chambers BJ. An inventory of heavy metals inputs to agricultural soils in England and Wales. The Science of the Total Environment 2003;311(1) 205–219.

[26] Council Directive of 12 June 1986 on the protection of the environment, and in particular of the soil, when sewage sludge is used in agriculture (86/278/EEC). http://eur-lex.europa.eu/LexUriServ/LexUriServ.do?uri=CELEX:31986L0278:EN:HTML (accessed 15 July 2013).

[27] Kabata-Pendias A, Pendias H. Trace elements in soils and plants. 3rd ed. Boca Raton: CRC Press; 2001.

[28] Brazil. Resolution 420. Brasília: Ministry of the Environment; 2009.

[29] Novais RF, Alvarez VVH, Barros NF, Fontes RLF, Cantarutti RB, Neves JCL. editors. Soil Fertilty: Viçosa: SBCS; 2007.

[30] USEPA – United States Environmental Protection Agency. Background Report on Fertilizer Use, Contaminants and Regulations. Columbus: USEPA; 1999.

[31] The Weinberg Group. Scientific Basis for Risk-Based Acceptable Concentrations of Metals in Fertilizers and Their Applicability as Standards. Washington: The Weinberg Group; 2001.

[32] USEPA – United States Environmental Protection Agency. Zinc fertilizers made from recycled hazardous secondary materials. Washington: USEPA; 2002.

[33] WDE – Washington Department of Ecology. Focus: Imported Cadmium-Contaminated Zinc Sulfate Used in Fertilizer and Other Products. Washington: WDA; 2000.

[34] Gabe U, Rodella AA. Trace elements in Brazilian agricultural limestones and mineral fertilizers. Communications in Soil Science and Plant Analysis 1999;30(5-6) 605-620.

[35] Martins ALC, bataglia OC, Camargo AO, Cantarella H. Corn yield and uptake of Cu, Fe, Mn and Zn from sewage sludge-amended soil with and without liming.Revista Brasileira de Ciência do Solo 2003;27(3) 563-574.

[36] Brazil. NormativeInstruction 27. Brasília: Ministry of Agriculture, LivestockandSupply; 2006.

[37] Hooda SP. Trace Elements in Soils. London: Wiley; 2010.

[38] CCR – California Code of Regulations. Standards and labeling. California: CCR; 2007.

[39] AAPFCO – Association of American Plant Food Control. AAPFCO's Statement of Uniform Interpretation and Policy (SUIP) #25: the heavy metal rule. http://www.aapfco.org/rules.html (accessed 15 June 2013).

[40] AFPC – Association of Fertilizer and Phosphate Chemists. Latest Metal Regulations in Fertilizers. http://afpc.net/Metal%20Regulations.html (accessed 15 June 2013).

[41] ABNT – Brazilian National Standards Organization. NBR 10004. Rio de Janeiro: ABNT; 2004.

[42] Rodella AA. Regulation of contaminant contents in fertilizers: A Case study. Revista Brasileira de ciência do solo 2005; 29(1) 797-801.

[43] Pino GAH. Biosorption of heavy metals using *Cocos nucifera*. Master's Dissertation. Pontifical Catholic University; 2005.

[44] Morillo J, Usero J, Gracia I. Partitioning of metals in sediments from the Odiel river (Spain). Environment International 2002;28(4) 263-271.

[45] Bueno BYM. Removal of Pb, Cr and Cu by a Combined Biosorption / Bioflotation process using a *Rhodococcusopacus* straint. Doctorate Thesis. Pontifical Catholic University; 2007.

[46] Nriagu JO, Pacyna JM. Quantitative assessment of worldwide contamination of air, water and soils by trace metals. Nature 1988;333(1) 134–139.

[47] Loska K, Wiechula D, Korus I. Metal contamination of farming soils affected by industry. Environmental International 2004;30(1) 159-165.

[48] Naimo TJ. A review of the effects of heavy metals on freshwater mussels. Ecotoxicology 1995;4(6) 341-362.

[49] Sayler GS, Nelson Jr JD, Colwell RR. Role of bacteria in bioaccumulation of mercury in the oyster *Crassostrea virginica*. Applied Microbiology 1975;30(1) 91–96.

[50] Peralta-Videa JR, Lopez ML, Narayan M, Saupe G, Gardea-Torresdey J. The biochemistry of environmental heavy metal uptake by plants: Implications for the food chain. The international Journal of biochemistry & Cell Biology 2009;41(8-9) 1665-1677.

[51] ATSDR – Agency for Toxic Substances and Disease Registry. Toxic Substances Index. http://www.atsdr.cdc.gov/az/a.html (accessed 15 June 2013).

[52] Rahman MA, Hasegawa H, Lim PR. Bioaccumulation, biotransformation and trophic transfer of arsenic in the aquatic food chain. Environmental Research 2012;116(1) 118-135.

[53] Cao X, MA LQ. Effects of compost and phosphate on plant arsenic accumulation from soils near pressure-treated wood. Environmental Pollution 2004;132(3) 435-442.

[54] ATSDR – Agency for Toxic Substances and Disease Registry. The Priority List of Hazardous Substances. http://www.atsdr.cdc.gov/SPL/index.html (accessed 15 June 2013).

[55] Brady JE, Humiston GE. General Chemistry. 2th ed. Rio de Janeiro: Livros Técnicos e Científicos; 1986.

[56] ICdA. INTERNACIONAL Cadmium ASSOCIATION, Revision of the Battery Directive: Public Consultation, 2001.

[57] PINO G.A.H., Biossorption of Heavy metals on green coconut powder. Master Dissertation, PUC-RJ, March, 113p. 2005.

[58] Paoliello MMB, Chasin AAM. Ecotoxicology of lead and its compounds. Salvador: CRA; 2001.

[59] ILZSG – International Lead and Zinc Study Group. Statistic of Lead and Zinc. 2010. http://www.ilzsg.org/static/home.aspx (accessed 15 June 2013).

[60] Kreush MA. Assessment with proposals for improving the industrial process of recycling lead and indicating applicability to slag generated. Master's Dissertation. Federal University of Parana; 2005.

[61] ATSDR – Agency for Toxic Substances and Disease Registry. Case studies in environmental medicine: lead toxicity. Atlanta: ATSDR; 1999.

[62] Capitani EM. Diagnosis and treatment of lead poisoning in children and adults. Medicina 2009;42(3) 319-329.

[63] Cervantes C, Campos-García J, Devars S, Gutiérrez-Corona F, Loza-Tavera H, Torres'Guzmán JC, Moreno-Sánchez R. Interactions of chromium with microorganisms and plants. FEMS Microbiology Reviews 2001;25(3) 335-347.

[64] WHO - World Health organization. Environmental health criteria 61 - Chromium. Geneva: WHO; 1988.

[65] Smith WLE, Gadd GM. Reduction and precipitation of chromate by mixed culture sulphate-reducing bacterial biofilms. Journal of Applied Microbiology 2000;88(6) 983-991.

[66] Kumral E. Speciation of chromium inwaters via sol–gel preconcentration prior to atomic spectrometric determination. Doctorate Thesis. The Graduate School of Engineering and Sciences of Izmir Institute of Technology; 2007.

[67] Frois SR, Grassi MT, Fernandes TC, Barreto RAS, Abate G. Preconcentration of Cr(III) and speciation analysis of chromium employing montmorillonite saturated with potassium ions. Química Nova 2011;34(3) 462-467.

[68] Schaumlöffel D. Nickel species: analysis and toxic effects. Journal of Trace Elements in Medicine and Biology 2012;26(1) 1-6.

[69] Veit MT, Silva EA, Fagundes-Klen MR, Tavares CRG, Gonçalves GC. Biosorption of nickel and chromium from a galvanization effluent using seaweed pre-treated on a fixed-bed column. Acta Scientiarum. Technology 2009; 31(2) 175-183.

[70] Leonard A, Gerber P, Jacquet P. Carcinogenicity, mutagenicity and teratogenicity of nickel.Mutation. Research/Reviews in Genetic Toxicology 1981;87(1) 1-15.

[71] Sodré FF, Lenzi E, Costa ACS. Applicability of adsorption models to the study of copper behaviour in clayey soils. Química Nova 2001;24(3) 324-330.

[72] Lima AJB, Cardoso, MG, Guerreiro MC, Pimentel FA. Using activated carbon to remove copper from sugar cane spirit. Química Nova 2006;29(2) 247-250.

[73] Yruela I. Copper in plants. Brazilian Journal of Plant Physiology 2205;17(1) 145-156.

[74] International Manganese Institute. Applications of Mn. http://www.manganese.org/about_mn/applications (accessed 15 June 1013).

[75] Das AP, Sukla LB, Pradhan N, Nayak S. Manganese biomining: A review. Bioresource Technology 2011;102(16) 7381-7387.

[76] Zhang W, Cheng, CY. Manganese metallurgy review. Part I: Leaching of ores/secondary materials and recovery of electrolytic/chemical manganese dioxide. Hydrometallurgy 2007;89(3-4) 137-159.

[77] Gerber GB, Leonard A, Hantson PH. Carcinogenicity, mutagenicity and teratogenicity of manganese compounds. Critical Reviews in Oncology/Hematology 2002;42(1) 25-43.

[78] Brault N, Loranger S, Courchesne F, Kennedy G, Zayed J. Bioaccumulation of manganese by plants: influence of MMT as a gasoline additive. Science of the Total Environment 1994;153(1-2) 77-84.

[79] Google. Google Maps. https://maps.google.com (accessed 15 June 2013).

[80] Paula Jr D. Office 2147/09-MP-PJCS-UMA. http://www.mma.gov.br/port/conama/processos/FE4582B1/Of2147-09-MP-PJCS-U MA_7maio09.pdf. 2009. (accessed 15 June 2013).

[81] Varjabedian R, Leite CBB. Office 1029/2009 – PGJ/CAO. http://www.mma.gov.br/port/conama/processos/FE4582B1/ConsideracoesMinisPublSaoPaulo.pdf. 2009 (accessed 15 june 2013).

[82] Brazilian Ministry of the Environment. Process 02000.002955/2004-69. 2011. http://www.mma.gov.br/port/conama/processo.cfm?processo=02000.002955/2004-69 (accessed 15 June 2013).

[83] Gimeno-García E, Andreu V, Boluda R. Heavy metals incidence in the application of inorganic fertilizers and pesticides to rice farming soils. Environmental Pollution 1996;92(1) 19–25.

[84] Ryser P, Sauder WR. Effects of heavy-metal-contaminated soil on growth, phenology and biomass turnover of *Hieraciumpiloselloides*. Environmental Pollution 2006;140(1) 52–61.

[85] Jan FA, Ishaq M, Khan S, Ihsanullah I, Ahmad I, Shakirullah M. A comparative study of human health risks via consumption of food crops grown on wastewater irrigated soil (Peshawar) and relatively clean water irrigated soil. Journal of Hazardous Materials 2010;179(1-3) 612–621.

[86] Jiao W, Chen W, Chang AC, Page AL. Environmental risks of trace elements associated with long-term phosphate fertilizers applications: A review. Environmental Pollution 2012;168(1) 44-53.

[87] Ghosh M, Singh SP. A Review on Phytiremediation of heavy metals and utilization of its byproducts. Applied ecology and environmental research 2005;3(1) 1-18.

[88] Lorestani B, Cheraghi M, Yousefi N. Phytoremediation Potencial of Native Plants Growing on a Heavy Metals Contaminated soil of Copper mine in Iran. World Academy of Science, Engineering and Technology 2011;53(1) 377-382.

[89] Susarla S, Medina VF, McCutcheon SC. Phytoremediation: An ecological solution to organic chemical contamination Ecological Engineering 2002;18(1) 647–658.

[90] Gratão PL, Prasad MNV, Cardoso PF, Lea PJ, Azevedo RA. Phytoremediation: Green technology for the clean up of toxic metals in the environment. Brazilian Journal of Plant Physiology 2005;17(1) 53-64.

[91] Lorestani B, Cheraghi M, Yousefi N. Accumulation of Pb, Fe, Mn, Cu and Zn in plantsand choice of hyperaccumulator plant in the industrial town of viean, Iran. Archives of Biological Science Belgrade 2011; 63(3) 739-745.

[92] Jing Y, He Z, Yang X. Role of soil rhizobacteria in phytoremediation of heavy metal contaminated soils Journal of Zhejiang University Science B 2007;8(3) 192-207.

[93] Mkumbo S, Mwegoha W, Renman G. Assessment of the phytoremediation potential for Pb, Zn and Cu of indigenous plants growing in a gold mining area in Tanzania. International Journal of Environmental Sciences 2012;2(4) 2425-2434.

[94] Cheraghi, M,Lorestani, B,Yousefi, N. Introduction of Hyperaccumulator Plants with Phytoremediation Potential of a Lead- Zinc Mine in Iran. World Academy of Science, Engineering and Technology 2011, 53(1) 163-168.

[95] Marques M, Aguiar CRC, Silva JJLS. Technical challenges and social, economic and regulatory barriers to phytoremediation of contaminated soils. Revista Brasileira de Ciência do Solo 2011;35(1) 1-11.

[96] Prasad MNV, Freitas HMO. Metal hyperaccumulation in plants – Biodiversity prospecting for phytoremediation technology.Eletronic Journal of Biotechnology 2003;6(3) 285-321.

[97] Gonçalves Jr. AC, Rubio F, Meneghel AP, Coelho GF, Dragunski DC, Strey L. The use of *Crambe abyssinica* seeds as adsorbent in the removal of metals from waters. Revista Brasileira de Engenharia Agrícola e Ambiental 2013;17(2) 306-311.

[98] Rubio F, Gonçalves Jr. AC, Meneghel AP, Tarley CRT, Schwantes D, Coelho GF. Removal of cadmium from water using by-product *Crambe abyssinca* Hochst seeds as biosorbent material. Water Science and Technology 2013;68(1) 227-233.

[99] Rubio F, Gonçalves Jr. AC, Strey L, Meneghel AP, Coelho, GF, Nacke, H. Applicability of *Crambe abyssinica* Hochst byproduct as biosorbent in the removal of chromium from water. Spanish Journal of Rural Development. 2013;4(1) 25-40.

[100] Gonçalves Jr. AC, Meneghel AP, Rubio F, Strey L, Dragunski DC, Coelho GF. Applicability of *Moringa oleifera* Lam. pie as an adsorbent for removal of heavy metals from waters. Revista Brasileira de Engenharia Agrícola e Ambiental 2013;17(1): 94-99.

[101] Nacke H, Gonçalves Jr. AC, Coelho GF, Strey L, Schwantes D, Laufer A. Removal of cadmium from aqueous solutions by adsorption on *Jatropha biomass* In: Bartolo H. (ed.) Green Design, Materials and Manufacturing Processes. Boca Raton: CRC Press Taylor & Francis Group; 2013. p367-372.

Chapter 5

Beneficial and Negative Impacts on Soil by the Reuse of Treated/Untreated Municipal Wastewater for Agricultural Irrigation – A Review of the Current Knowledge and Future Perspectives

Juan C. Durán–Álvarez and Blanca Jiménez–Cisneros

Additional information is available at the end of the chapter

http://dx.doi.org/10.5772/57226

1. Introduction

The scarcity of water for human use, such as food and energy production, manufacturing, drinking water and ecosystem conservation is a global problem for which the solution goes beyond merely the preservation of freshwater sources [1–2]. Although three quarters of the Earth's surface is covered by water, most of this water is either contained in oceans or confined in glaciers [3]. The volume of freshwater available for human activities (less than 1%) is unequally distributed throughout the globe; in some cases this water is confined to the deep sub–soil or is polluted [4]. Furthermore, the desertification of large areas caused by climate change has intensified the lack of water sources in cities and rural areas throughout the world [5]. Water scarcity results in food scarcity, since 70% of the water withdrawn for human activities goes to agriculture [6]. In zones where rain–fed agriculture is practiced, decay in crop yields is observed when droughts occur, which results not only in the scarcity of food but also the decrease in incomes due to falling crop sales [7]. The use of freshwater for agricultural irrigation limits the volume of freshwater available for human consumption; therefore, recycling of water becomes necessary for agricultural irrigation in dry zones. The idea of reusing wastewater to irrigate is not new; it actually originated around 3000 B.C. People in these ancient civilizations knew that wastewater contained both water and compounds that benefited the soil and thus they used it in a planned way to increase crop yields [8].

Commonly, reusing wastewater in agriculture is considered a deleterious practice since it may introduce pollutants to the environment, spread waterborne diseases, generate odor problems and result in aversion to the crops. Nevertheless, this kind of reuse may result in some benefits for soils, crops and farmers. Nowadays, the reuse of wastewater in agriculture is seen in some countries as a convenient environmental strategy [9–10]; municipal wastewater is therefore considered an appropriate option for reuse. This kind of wastewater contains a significant load of biodegradable organic material (carbon and nitrogen) as well as most of the mineral macronutrients (e.g. phosphorous, potassium, magnesium and boron) and micronutrients (e.g. molybdenum, selenium and copper) which are necessary for the growth of crops. Accumulation of organic matter in soil by irrigation with wastewater can be beneficial as it may result in the enhancement of the physical structure of the soil, the increase in the soil microbial activity and the improvement of soil performance as a filter and degrading media for pollutants. Conversely, a fraction of the organic matter contained in wastewater is due to the occurrence of organic pollutants (e.g. polyaromatic hydrocarbons and polychlorinated biphenyls) and pathogenic microbial agents [11–12]. Because of the presence of organic, inorganic and microbial pollutants in wastewater, a prior step of depuration is necessary before reuse in irrigation in order to avoid the pollution of soil, crops and the nearby water sources, and thus the dissemination of waterborne diseases or the degradation of soil. The extent at which wastewater has to be treated prior to irrigation depends on the restrictions established in local or international water quality criteria for irrigation [13]. Primary treatment schemes (coagulation–flocculation with sedimentation or aerobic/anaerobic stabilization pounds) are used for treating wastewater to irrigate crops that are not intended for human consumption (e.g. fodder), while secondary treatment of wastewater (biological treatment followed by disinfection) is recommended when unrestricted crops are irrigated [14–15]. In developing countries, most or the whole volume of wastewater produced in cities is treated prior to irrigation, while in low income countries wastewater treatment is not a priority, and thus untreated or partially treated wastewater or a mixture of treated and untreated wastewater is commonly used for agricultural purposes [12, 16]. In Mexico, China, India and Pakistan, for instance, large areas exist where untreated wastewater has been reused in agricultural irrigation for a considerable time [17]. The World Health Organization estimates that nearly 20 million hectares throughout the world are irrigated using untreated wastewater [18]. It is also reported that in some cities up to 80% of the vegetables locally consumed are produced using wastewater for irrigation [19]. The application of wastewater to soil, particularly untreated wastewater, followed by its infiltration poses a significant risk of pollution, not only to soil and crops but also to the surface and subterranean water sources surrounding the irrigated area [20–21].

Pollution by pathogenic agents is the main cause of concern regarding the application of treated/untreated wastewater to soil. Due to the variety of microorganisms entering the soil via the wastewater there is a high risk of enteric disease outbreaks for farmers and consumers [22–23]. This chapter addresses the contamination of wastewater irrigated soils by helminths (intestinal worms) and pathogenic bacteria common in developing countries (where untreated wastewater is used to a greater extent), as well as the risk of outbreaks of parasitic diseases for both farmers and consumers in agricultural areas where untreated wastewater is reused. The occurrence of antibiotic resistance in indigenous organisms of soil and pathogens reaching soil via wastewater is gaining the attention of scientists and health organizations around the world [24–25], thus a review of what it is known and the research opportunities in this field are presented in the text. With regard to organic pollution, a current topic of interest is the entry to the soil and potential risks within crops of so–called "contaminants of emerging concern". These pollutants are substances that have not previously been considered as pollutants since they are part of everyday products; however, due to the subtle but harmful effects that these substances may cause in a variety of aquatic and terrestrial organisms, concerns have risen due to their continuous entry into the environment via wastewater [26]. A review on the presence of some organic contaminants of emerging concern, such as pharmaceutical substances, personal care products and industrial additives, in wastewater–irrigated agricultural soils is presented in this chapter along with some of the known potential effects caused to soil organisms, plants and consumers. Such effects have just begun to be elucidated, and only for some groups of contaminants of emerging concern [27–28], even though it is now known that up to 7 million commercially available chemicals are routinely disposed of in sewage after use [29]. In this regard, this chapter makes some suggestions regarding the next steps in the toxicity studies for this class of pollutants, such as testing the synergistic effects of mixtures of contaminants of emerging concern in soil organisms.

In spite of the variety and quantity of contaminants that soil regularly receives through wastewater irrigation, this ecosystem possesses self–purification processes that maintain homeostasis within the system. Such self–purification processes may either inactivate or reduce the population of pathogenic microorganisms reaching the soil via wastewater through predation by the indigenous microbiota within the soil [30–31], the production of antibiotics by some organisms in the rhizosphere [32] and by retention of microorganisms in the surface layers of the soil profile through physical and chemical processes. For organic pollutants, mechanisms such as photolysis and biodegradation promote the dissipation of contaminants in the soil, while adsorption onto the soil particles lead to the retention –and the potential confinement– of organics within the solid matrix [33]. In this chapter, current knowledge concerning the environmental fate of pathogen and organic contaminants of emerging concern in wastewater irrigated soils is discussed, highlighting the laboratory approaches that show the best results in simulation of the conditions in the field. Knowledge of the environmental fate of contaminants in irrigated soils is important in order to perform more accurate risk assessment studies on contamination of water sources, soil and crops in wastewater irrigated areas; furthermore, it provides information to policy makers to make proper legislation aimed at promoting environmentally responsible management of treated/untreated wastewater in agricultural irrigation.

Depuration of wastewater prior to its reuse is the most plausible option to prevent soil pollution by wastewater reuse. However, since wastewater represents a cheap source of water and fertilizer for farmers [34], it is necessary to consider the needs of users before planning schemes of wastewater treatment. The use of wastewater treatment systems aimed at removing carbon, nitrogen, phosphorous and minerals in wastewater leads to the reduction in quality of effluents as fertilizers, impacting crop yields and thus in the livelihood of farmers. In this sense, the use of advanced primary treatment systems could be a feasible option to: a) remove suspended solids, pathogens and heavy metals in wastewater without significantly impacting the content of nutrients in effluent; b) preserve the quality of agricultural soils to properly perform ecosystem services such as the production of food; and, c) fulfill the needs of farmers that use wastewater as a source of water and nutrients. Treating wastewater by these kinds of systems may be an opportunity to couple sanitation with reuse within a program of comprehensive management of wastewater, the recycling of nutrients and the use of soil as a food producer and purification system.

This chapter aims to describe what it is known and what it is unknown regarding the positive and negative impacts of the reuse of treated/untreated wastewater in agricultural irrigation. It will be shown in detail how this practice can benefit soil and farmers, while at the same time posing a risk of contamination to the ecosystem. Emphasis is given to the purification processes occurring in the soil and how soil manages the continuous entrance of pollutants via wastewater. Lastly, some perspectives for further studies on the presence and environmental fate of pollutants in wastewater irrigated soils are proposed.

2. Impacts of wastewater reuse in agriculture

The reuse of wastewater results in both beneficial and negative impacts on soil, some of which are explained in this section. The aim is to identify both and to understand their origins in order to assist scientists and policy makers to balance them and even to greater advantage of the benefits compared to the drawbacks in certain situations.

2.1. Benefits of wastewater reuse in agriculture

Figure 1 summarizes the positive impacts of reusing wastewater in agricultural irrigation in all of its forms. The extent of the positive impacts depends on local conditions of the specific project.

2.1.1. Benefits in crops

Since wastewater is produced constantly and thus is always available, it is possible to select a wider range of crops to be sown year–round, specifically those of high profitability which normally have higher and more stringent water demands in terms of quantity and timing. The consistent use of wastewater in irrigation may stabilize the content of nutrients in the soil, even when growing crops with high nutritional requirements; this is because the continuous withdrawal of nutrients by plants is compensated by the constant input of organic and mineral components into the soil via wastewater. Examples of how the reuse of wastewater has led to increases in crop yields in arid zones can be found worldwide. Studies conducted in Hubli–Dharwad, India, showed that irrigation with treated and untreated wastewater made it possible to produce vegetables during the dry season; yields and selling prices increased by 3–5 times compared to the kharif (monsoon) season [35]. In Pakistan, Ghana and Senegal the reliability and flexibility of wastewater supply allows rural and urban farmers to cultivate profitable crops in a shorter time, resulting in several harvests per year (3 to 6) [36–37]. Treated/untreated wastewater is a source of organic matter and the same large diversity of nutrients contained in any formulated fertilizer. It is estimated that 1,000 m^3 of municipal wastewater applied to one hectare can contribute 16–62 kg of organic nitrogen, 4–24 kg of phosphorus, 269 kg of potassium, 18–208 kg of calcium and 9–110 kg of magnesium each year [16]. Table 1 shows the contribution of water and nutrients that untreated wastewater make to several crops.

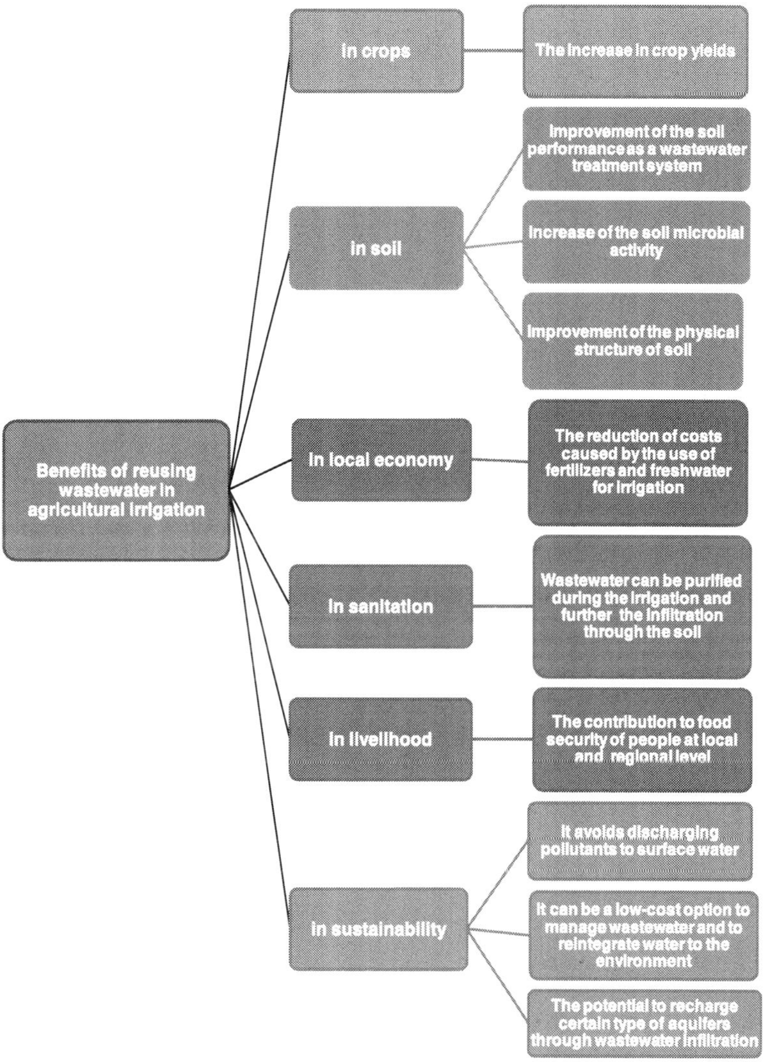

With information of references [14, 18, 21, 36, 37]

Figure 1. Beneficial impacts of reusing wastewater for agricultural irrigation

Nitrogen is a plant macronutrient which can be found in the form of nitrate ions ($N–NO_3$), mostly in treated wastewater, or as ammoniacal nitrogen ($N–NH_4^+$) and organic nitrogen in untreated wastewater. The sum of all these forms is known as total nitrogen (TN). Most crops absorb nitrates to the greatest extent (85% of the nitrate contained in wastewater); whereas 50% of ammoniacal and less than 30% of organic nitrogen contained in wastewater can be assimilated as it is by plants. The remaining nitrogen is taken up by soil microorganisms and transformed into nitrates or volatilized as N_2. In wastewater irrigated soils, organic nitrogen is transformed into nitrates by soil microorganisms to a greater extent than that observed in non–irrigated agricultural soils [38]. Problems related to high inputs of nitrate ions are due to their high solubility in water, and thus their rapid percolation through the soil to the aquifer.

Crop	Water requirements (mm/year)	Nutrients and sodium contribution by WW (kg ha/year)					
		N_{total}	P_{total}	K	Ca	Mg	Na
Maize	673	108–418	27–162	13–465	121–1401	61–741	182–1226
Green tomatoes	653	104–405	26–157	13–451	118–1358	59–718	176–1188
Chili	601	96–373	24–144	12–415	108–1250	54–661	162–1094
Beans	370	59–229	7–89	7–255	67–770	33–407	100–673
Wheat	520	83–322	21–125	10–359	94–1082	47–572	140–946
Barley	516	83–320	21–124	10–356	93–1074	46–568	139–939
Alfalfa	1360	218–843	54–326	27–938	245–2829	122–843	367–2475
Marrow	364	58–226	15–87	7–251	66–757	33–400	98–662
Oats	353.6	57–219	14–85	7–244	64–735	32–389	95–644

Source: reference [16]

Table 1. Contribution of nutrients and sodium from untreated wastewater and water requirements of demandant crops

A significant quantity of nitrate leaching through soil subsequently becomes unavailable for plants; this does not necessarily represent a problem, as nitrate is continuously supplied to soil via wastewater. More important, the presence of nitrates in subterranean water is related to occurrence of methemoglobinemia disease in infants ingesting nitrate at levels higher than 45 mg/L via drinking water [39]. The quantity of nitrogen washed out from soil depends on the irrigation rate, the frequency of rain events, the type of crops sown and the characteristics of the soil [40]. The amount of nitrogen that can be applied to soil to produce minimal nitrate leaching rates depends on the demand of crops, which usually varies between 50 and 350 kg of nitrogen per hectare [40]. Such demand is within or slightly above the amount of nitrate supplied by treated wastewater. In this sense, the limited removal of nitrogen by wastewater treatment would not significantly affect the input of this macronutrient to agricultural soils.

Phosphorous is another plant macronutrient, which is very scarce in soil, at the point it needs to be added through the application of fertilizers. Due to its stability and low solubility, this nutrient can be accumulated in soil. Wastewater normally contains small amounts of phosphorous, so its use for irrigation is beneficial to plants and it does not impact negatively upon the environment, even if applied consistently for long periods of time [40–41]. The recycling of phosphorous and nitrogen in wastewater–irrigated soils is important because it allows closure the P cycle rather than its breakage. Breakage of the cycle occurs when phosphorous is removed from wastewater during treatment, becoming trapped in sludge and dumped to confinement sites or landfills. An advantage of the availability of phosphorus in wastewater is that it is partly bound to organic components and thus it cannot form complexes with iron or aluminum ions upon its entry to soils [16]. In contrast to phosphorous, potassium is contained in soil at high concentrations (around 3% of the lithosphere) but in chemical forms that impede its bioavailability. As a result it is necessary to add potassium to soils via fertilizers. Approximately 185 kg of potassium per hectare are required to cultivate some crops [16]. Sewage contains low concentrations of potassium, insufficient to cover the theoretical demand in most cases. Meeting the demand for potassium in irrigated soils will depend on the amount of wastewater supplied at each irrigation event, the wastewater quality and the frequency of irrigation. Fertilization with potassium has not resulted in adverse impacts to the environment [42]. Recycling nutrients by the reuse of wastewater promotes savings in energy, which would otherwise be consumed in the production of fertilizers [43]. In particular, the recycling of phosphorus is important since the world's phosphorus reserves are becoming scarce [44]. Fertilizing agricultural soils by the reuse of wastewater invariably leads to the increase of crop yields. An example of this can be found in Mezquital Valley, Mexico [45]; in this respect, Table 2 shows the differences in the agricultural production in croplands of Mezquital Valley when either untreated wastewater or groundwater is used for agricultural irrigation.

Crop	Crop yield (tons/ha)		Increment (%)
	Untreated wastewater	Groundwater	
Corn	5.0	2.0	150
Barley	4.0	2.0	100
Tomato	35.0	18.0	94
Oats for forage	22.0	12.0	83
Alfalfa	120.0	70.0	71
Chili	12.0	7.0	70
Wheat	3.0	1.8	67

Source: references [16, 52]

Table 2. Comparison of crop yields for some vegetables in plots where wastewater and groundwater are used for agricultural irrigation (Mezquital Valley, central Mexico)

The use of wastewater in Mezquital Valley has also contributed to changing the landscape of the zone, transforming barren soils into productive and green vibrant soils, as shown in Figure 2.

Figure 2. Comparison of untreated wastewater irrigated (right side) and rain–fed (left side) croplands from Mezquital Valley, central Mexico

2.1.2. Benefits in soil quality

In order to define the improvements in soil quality produced by the application of treated/untreated wastewater it is necessary to establish the use of the irrigated soil. It is known that soil complies with five ecological functions: a) a medium for plant growth (including agriculture); b) a biodiversity pool and habitat for plants and (micro and micro) fauna; c) a carbon sink; d) a storage, filter and transforming medium for nutrients, pollutants and water; and, e) a landscaping and engineering medium. [46]. This chapter focuses on the functions of soil as a medium for plant growth as well as in its role as a transforming medium for nutrients and pollutants.

In addition to the continuous supply of nutrients to the soil, irrigation with treated/untreated wastewater confers significant improvements in soil quality. Favorable changes reported in irrigated soils comprise: a) an improvement in the physical structure of soil; b) an increase in soil microbial activity; and, c) the improvement of the soil performance as a wastewater treatment system.

Improvement of the physical structure of soil. The physical structure of soil is defined as the arrangement of the solid particles and the size, shape and interconnection of pores and voids. Soil structure is closely related to its capacity to store and transport gases and water (and thus dissolved substances) [47]. Gas exchange between the soil and the atmosphere determines whether aerobic, anoxic or anaerobic conditions prevail within the soil. This in turn regulates the metabolism of soil microorganisms and impacts, inter alia, upon the nitrogen fixation, the transformation of soil organic matter and the degradation of pollutants. Additionally, the physical structure of soil affects the plant growth by influencing root distribution and thus the ability to take up water and nutrients [48]. Improvements in the physical structure of soil are related to the increase in both the stability of the soil aggregates and soil porosity. The enhancement of the physical structure of soil results in a rise in agronomic productivity, the augmentation of water infiltration through soil to the aquifer and a decrease in erodibility [49]. The hierarchical theory of aggregation proposes that microaggregates (particle size below 250 μm) in the soil are formed initially by the attachment of organic material to some inorganic components of soils (e.g. clay and hydroxides); in turn these microaggregates join together to form macroaggregates (particle size above 250 μm). Alternatively, macroaggregates can form around the particulate organic matter, while exudates produced by soil microorganisms serve as cementing agents, making micro and macroaggregates more stable [50]. Microaggregates can be also formed from bacterial colony clusters which use bacterial polysaccharide exudates to bind with clay particles. The clay particles act as a protective shell for clusters and macro-aggregate formation continues as described above [51].

Since the formation of aggregates in the soil is related to the presence of organic matter, and in some cases microorganisms, it might be expected that the continuous supply of these two elements via wastewater would result in the increased formation and stability of soil aggregates and thus an improvement in the physical structure of soil. For example, the study referred in [52] establishes that increased soil microbial activity due to the augmentation of organic carbon content by the application of wastewater impacts positively upon the stability of soil aggregates. Furthermore, there are substances contained in wastewater other than organic matter and microorganisms that may contribute to the formation and stability of soil aggregates. Calcium and magnesium cations, which are abundant in wastewater, increase the formation of microaggregates through cationic bridging between clay and organic matter, resulting in aggregation. In arid soils and soils with low organic matter contents, insoluble calcium and magnesium carbonates can trigger the formation of soil micro and macroaggregates [51]. Additionally, calcium can inhibit clay dispersion, and thus the breakup of aggregates, when sodium concentration increases in soil [53]. Dissolved organic matter in wastewater can form complexes with iron and aluminum in soil forming mobile organo-metallic compounds which can further precipitate and act as cores for microaggregates formation. Particulate organic matter (i.e. suspended solids in wastewater) may enhance the binding of microaggregates to subsequently form macroaggregates; for instance, extracellular polysaccharides of microorganisms in the surface of suspended solids can act as binding agents in the formation of macroaggregates [54]. In the case of phosphorous, the formation of insoluble aluminum and calcium phosphates in the soil can induce the formation of microaggregates and additionally it may act as a macroaggregate binding agent [55]. The entry of certain chemicals to the soil via wastewater increases the stability of soil aggregates. For example, hydrophobic substances (e.g. surfactants, lipids and hydrocarbons) decrease the wettability of aggregates by inducing water repellency, which in turn leads to increased cohesiveness and low decomposition rates of soil aggregates [51]. Agricultural activities in wastewater irrigated soils may also contribute to the improvement of the physical structure of soil. Previous studies have found that some crops (i.e. maize, alfalfa and leguminous plants) have beneficial effects on the conservation of the physical structure of soil. Aggregation of soil particles tends to increase when planting crops characterized by high density and long length of roots; this is because chemicals released by roots (i.e. mucilage) enhance the stability of soil aggregates in the rhizosphere by increasing the bond strength and decreasing the wetting rate [56]. According to the study reported in reference [57], roots of leguminous crops increase the aggregation of soil particles. Corn (*Zea mays*) residues (leaves and shoots) also increase aggregation of soil particles compared with other crops; this is attributable to the liberation of phenolic compounds from plant tissues, since phenols favor the agglutination of particles and prevent wetting [57–58]. Municipal and industrial wastewater may also be a source of phenolic compounds to soil through irrigation, producing similar effects to those of corn wastes [59]. The study referred to in [58] demonstrated that the stability of soil aggregates is high for continuous cultivation of alfalfa (*Medicago sativa*), while the opposite effect was observed for soybean. This is attributable to the low concentration of phenols in the latter [60]. Some studies have addressed the changes in the physical structure of agricultural soils caused by long–term irrigation with wastewater. The results of these studies show a decrease in soil porosity caused either by occlusion of pores by the suspended solids contained in wastewater or by the augmentation of micropores (radius < 0.01 μm) in the soil matrix [61–62]. Depending on the method of water application during irrigation, an increase in the compaction of soil may be observed in the plot after an irrigation event [63]. Soils irrigated by flooding exhibit high compaction while water dropping effects (erosion) may be observed in soils irrigated by spraying. In any of both cases, wastewater irrigated soils exhibit large populations of earth-worms which may assist in the formation and connection of pores within the soil matrix. Undoubtedly wastewater contains agents that improve the physical structure of soil. However, studies performed so far show contrasting results, either an increase in the soil microporosity or soil compaction. It is therefore necessary to carry out studies aimed at measuring changes in the physical structure of soil throughout several irrigation cycles and for longer periods (months or years); additionally, it is of interest to assess changes in the physical structure of soil at landscape level (piedmont or catena), as it may be useful for evaluating the horizontal displacement of soil particles and nutrients.

Increase of soil microbial activity. Either due to the extra supply of organic carbon or because of the addition of microorganisms via wastewater, microbial activity in wastewater irrigated soils tends to be higher than that found in non–irrigated soils [64–65]. This increase in the microbial activity of the soil brings benefits to both agriculture and the development of flora and fauna in the soil ecosystem. According to the study reported in reference [66], the C/N ratio in soils irrigated with wastewater for long periods tends to decrease by up to 45%, which implies an improvement in the nutritional conditions for soil microorganisms. The authors report an increase in the population of copinotrofic and oligotrophic bacteria (234 and 217%, respectively), as well as in the populations of actinomycetes (234%) and fungi (206%) in soils irrigated with wastewater for 100 years compared with those populations found in non–irrigated soils. Rises in the metabolic activity of soil, measured as the production of ATP and enzymatic activity have been also reported [65-66]. According to reference [66], soil enzymatic activity remained unchanged 20 years after wastewater irrigation ceased. In contrast, the study referred to in [67] shows that elevated microbial activity in soils irrigated with treated wastewater decreases after few days without irrigation. Due to the augmentation of the populations of bacteria, actinomycetes and fungi in the irrigated soil, a rise in the rhizospheric activity is experienced, resulting in: a) the increase in the growth and development of plants;

b) high rates in stabilization of organic matter entering the soil through wastewater; c) higher performance of the depuration of wastewater and degradation of the pollutants fixed in the soil in comparison with non–irrigated soils; and, d) the improvement in the formation and stability of soil aggregates. The latter may be explained by the role of polysaccharides exuded by bacteria as transient binding agents, which initialize aggregation of soil microaggregates [67]. The transformation of carbon and nitrogen by soil microorganisms supports the proliferation of soil (micro and macro) fauna which is essential for soil formation as well as for the development of plants. According to the work referred to in [68], the use of treated wastewater to irrigate an agricultural soil over 20 years has resulted in the improvement of the metabolic efficiency of soil microflora to transform carbonaceous and phosphorous substances into nutrients readily available to plants and macrofauna.

Soil biomass has proven to be capable of adsorbing a certain proportion of heavy metals contained in the wastewater. For instance, the study referred in [69] found biosorption rates for cadmium and nickel within the range of 5 to 55 mg/g of biomass in a soil that had been irrigated with wastewater for two decades. In that soil, the predominant bacteria after irrigation were *Enterobacteriaceae* and *Pseudomonas*.

The effect of wastewater irrigation on soil nitrogen fixing organisms has been little studied. An increase in soil nitrifying activity accompanied by a low rate of denitrification has been observed in wastewater irrigated forest soils [70], while in the study referred to in [71] a peak in N_2O production in a soil irrigated with treated wastewater was reported, followed by an immediate drop in gas production. So far, the metabolic processes performed by different soil microbial species in wastewater irrigated soils have been little explored. However, it is important to keep in mind the important role that soil microorganisms play in both the development of the soil and plants as well as in the purification of wastewater when planning agricultural systems based on the reuse of wastewater. Even when soil microbial populations show some kind of resilience to a wide variety of contaminants, some other chemicals can cause not only toxic effects to soil microorganisms but the proliferation of pathogenic organisms and the occurrence of antibiotic resistance within the agricultural soils.

Improvement of soil performance as a wastewater treatment system. As it is known, the application and infiltration of wastewater through soil results in its purification. In practice, specific wastewater treatment systems are based on soil infiltration, which have been demonstrated to improve water quality to levels obtained using tertiary treatment systems [72–73]. Purification of wastewater is one of the ecological functions of soil; through this mechanism, soil maintains, at least partially, the quality of surface and groundwater bodies. The extent at which this natural system works is highly variable, from almost nonexistent to very high, depending on local conditions and types of pollutants. Table 3 shows the extent to which pollutants in wastewater are removed by infiltration through the soil. The application of wastewater to soil reduces the content of pathogenic microorganisms by 6–7 log units for bacteria and 100% for helminths and other protozoa. Total organic carbon can be reduced by up to 90%, while levels of recalcitrant compounds in wastewater, such as phosphorus (20–90%), nitrogen (20–70%), and metals (70–95%) are also reduced dramatically. In sewage, organic phosphorus (5–50 mg/L) is biologically converted to phosphate; subsequently, in alkaline or calcareous soils, phosphate precipitates with calcium to form calcium phosphate and remains available for plants. In contrast, in acidic soils phosphate reacts with iron and aluminum oxides to form insoluble compounds, which are unavailable to plants. Sometimes soluble phosphate is initially immobilized by adsorption onto soil particles and then slowly returns to insoluble forms, allowing for further adsorption of mobile phosphate. This process is generally known as phosphate aging [72].

Variable	Effect
Organic matter	Biodegradable material is reduced by more than 90%, while less readily biodegradable material is adsorbed and later biodegraded or volatilized.
Nitrogen	Nitrogen is removed from water at a level similar to tertiary treatment systems by transformation in soil as well as by assimilation by soil microorganisms and plants.
Phosphorus	Phosphorous is reduced to levels of 1 mg/L or less by assimilation by plants.
Microorganisms	Helminth eggs and protozoa are easily removed by straining in the soil surface; bacteria and viruses can also be adsorbed onto the soil particles and then desiccated or killed by indigenous soil microorganisms. The performance of these processes depends on the texture, physical structure and organic matter content of soil.
Heavy Metals	Heavy metals can be removed by the formation of complexes with soil organic matter, precipitation or methylation at efficiencies of 70–95%.
Toxic organic compounds	Most are retained in soil and then biodegraded at different rates.

Table 3. Processes in soil that improve the quality of the wastewater, relative to selected parameters

Most of the organic compounds (natural and synthetic) in sewage are rapidly transformed in soil to stable, and in some cases non–toxic, organic compounds (e.g. humic and fulvic acids). Actually, soil biodegrades a greater amount and variety of organic pollutants than that reported for water streams. Wastewater application to soil under controlled conditions (e.g. limited irrigation rate and intermittent flooding) permits the biodegradation of hundreds of kilograms of carbonaceous substances per hectare per day, with no impact on the environment [72]. Total organic carbon levels in wastewater are dramatically reduced from levels of 80–200 mg/L to 1–5 mg/L in the infiltrated water [74]. Heavy metals can be removed from wastewater during soil infiltration and confined within the organic domain of the soil for several hundred years. Metals are retained in the surface layer of the soil either by complexation with soil organic matter or by precipitation at high pH values. Only a small fraction of metals infiltrates to lower layers of the soil profile and even less can be assimilated by crops. For instance, around 80–94% of cadmium, copper, nickel, and zinc can be removed in the first 5–15 cm of the soil profile, 5–15% is leached to lower layers and only 1–8% can be absorbed by grass [75]. A similar process occurs with fluorine [76]. This phytoremediation process is used to treat wastewater in planned natural treatment systems such as wetlands. However, it is necessary to be aware that some edible crops are able to take up heavy metals to a greater degree than grasses [77].

The capability of soil to act as a filter and transforming medium for wastewater pollutants can be observed in both long–term and newly wastewater irrigated soils [72, 78]. The operation of this natural purification system is closely related to the physical and chemical properties of the soil and thus modifications in soil characteristics caused by irrigation with wastewater may either improve or worsen the performance of this natural wastewater treatment system. The increase in the soil organic matter content is the main factor resulting in an improvement in the removal of biological, organic and inorganic pollutants as wastewater leaches through the soil. This is because soil organic matter promotes the immobilization of pollutants either by adsorption or formation of complexes, while at the same time stimulating the proliferation of degrading microorganisms [78–79]. Regularly, heavy metals are fixed in the upper layers of the soil profile by complexation with organic matter [65], thus organic matter enrichment in wastewater irrigated soils results in greater retention of heavy metals by the solid matrix. Heavy metals cannot be biodegraded but they may be modified by soil microorganisms. Biological methylation of metals and metalloids, such as selenium, arsenic and mercury, has been reported in wastewater irrigated soils. It is expected that this process is elevated in wastewater irrigated soils, where microbial biomass occurs at higher levels than in non–irrigated soils. Methylation of heavy metals leads either to reduced toxicity or increased loss of metals in soil through volatilization [80–81]. Another process observed in long–term wastewater irrigated soils, related to those aforementioned, is the potential of soil microorganisms to develop resistance to the harmful effects caused by the presence of heavy metals in the solid matrix [69, 82]. Such resistance is similar to that developed to antibiotics and has been reported for cadmium, chromium, zinc and nickel in soils irrigated with wastewater over the long term [69, 83]. It is plausible that the expression of these resistances results in an increase of heavy metal methylation in the soil, which allows soil microorganisms to survive and to continue with those metabolic functions that increase agricultural productivity and purify wastewater. With regard to organic contaminants, the increase in the soil organic matter content produces, in most cases, an incremental boost in the adsorption of solutes onto soil particles. Studies referred to in [84] found increased adsorption of organic contaminants (i.e. pesticides, pharmaceuticals and estrogenic hormones) in long–term wastewater irrigated soils compared to rain–fed soils from the same agricultural area. Organic compounds displaying high hydrophobicity are adsorbed by soil not only faster and to a greater extent but with greater strength than is observed for semi–polar and polar compounds [85]. The increase in the hydrophobicity of soil due to the application of wastewater increases the capacity of such soils to strongly retain non–polar organic contaminants within the solid matrix. The increase in the adsorption of organic pollutants by soil results in an extended retention time in the solid matrix, encouraging biodegradation processes. Similar to the results reported for adsorption, higher rates of biodegradation of organic pollutants have been observed in treated/untreated wastewater irrigated soils compared to non–irrigated ones [86]. This may be caused, on the one hand, by the continuous supply of organic matter to the soil via wastewater, which can be used by soil microorganisms as co–substrate in the biodegradation of target organic pollutants, and on the other hand, by the prolonged exposure of soil organisms to pollutants. The latter case can be understood as the acclimation of the degrading organisms to the occurrence of organic pollutants in the soil followed in the short term by the acquisition of the capability for using organic contaminants as a carbon source. The increase in the soil organic matter content caused by wastewater irrigation has a positive impact not only in the adsorption of organic compounds but also on the retention by soil of wastewater–

borne pathogens. This is due to the high affinity of the cell membranes to the organic domain of soil. The study referred to in [87] reports a higher adsorption of enteric bacteria *Escherichia coli* and the enteric protozoa *Giardia lamblia* in long–term wastewater irrigated soils compared with long–term groundwater irrigated soils from the same agricultural zone.

In general terms, an increase in pH values has been observed in agricultural soils irrigated with treated/untreated wastewater; although in less cases soil pH tended to decrease following the application of wastewater [81]. The first phenomenon is attributed to the continuous addition of salts (carbonates, calcium, magnesium, sodium) in wastewater. The second case is explained by the high mineralization rate of organic matter in the irrigated soil, which is highly dependent on the soil type, the climatic conditions of the site, and the quality of wastewater, among other reasons. The increase in soil pH, in combination with the continuous supply of organic matter, results in the buffering of soil pH, which prevents the drop of soil pH values during rain events (including acid rain). Stabilization of soil pH values also contributes to the retention of heavy metals in the surface layers of the soil by the formation of insoluble basic salts. Furthermore, basic values of soil pH can facilitate the adsorption of neutral and basic organic contaminants; as these compounds tends to be better adsorbed to neutral and basic soils than to acidic ones.

Since wastewater irrigation improves the physical structure of soil (i.e. increased formation and greater stability of aggregates), aerobic conditions may be maintained within the soil matrix; which in turns contributes to an increase in the aerobic biodegradation rate of organic pollutants. Additionally, an increase in the adsorption of pollutants can be achieved in better structured soils due to the increase in the specific surface area of soil particles. Moreover, higher biodegradation of the adsorbed contaminants can be expected as long as they remain available to microorganisms after adsorption. In irrigated soils where occlusion of the pores by the suspended solids in wastewater occurs, anoxic conditions may be achieved. Under such conditions, toxic species of heavy metals are chemically reduced into non–toxic species (e.g. Cr^{+6} into C^{+3} and As^{+5} into As^{+3}), then they may be immobilized by the formation of insoluble hydroxides. The extent to which wastewater irrigation contributes to the function of the soil as filter and degradation medium for pollutants is just beginning to be studied. The potential of soil to act as an efficient wastewater depuration system is a powerful argument to convince policy makers that agricultural irrigation with treated/untreated wastewater can be an appropriate strategy to simultaneously solve problems of water stress and low agricultural productivity with no negative impacts in the quality of water sources surrounding the irrigation site. This, of course, is achieved when all of the appropriate precautions to avoid contamination are taken at each site.

2.2. Negative impacts of wastewater reuse in agriculture

The main drawback of reusing treated/untreated wastewater in agriculture is the pollution of soil, the potential contamination of crops and water sources, and the inherent risk of harmful effects that contamination poses to the exposed organisms. Even when soil acts as an efficient living filter to remove, inactivate and transform the pollutants contained in wastewater, it is not fully effective at eliminating some of them. Moreover, as a result of the increasing industrial development, wastewater irrigated soils continuously receive newly synthesized substances, which may negatively impact the effectiveness of soil as a treatment system by poisoning the degrading microorganisms, destroying the physical structure of soil or damaging the natural cycles occurring within soil. The pollutants received by soil via wastewater may be different in developing and developed countries. Examples of this include pathogenic microbial agents. In developed countries most wastewater is treated prior to reuse and thus pathogens are not present in irrigation water, while in developing countries untreated wastewater is used in most of cases. Pathogens vary for different zones; for instance, the enteric protozoa *Giardia* is commonly found in wastewater of developing countries (Latin American and African countries), while the parasitic protozoa *Cryptosporidium* occurs in developed countries (United States and western European countries). Similar to microorganisms, some organic pollutants can be found in wastewater from developing countries and not in developed countries. Examples include some herbicides (e.g. DDT and atrazine) whose use is restricted in developed countries; on the other hand, nanomaterials and new–generation antibiotics, all of which are much more likely to occur in wastewater of developed countries. The determination of pollutants in soil initially requires specific sampling methods which take into consideration the heterogeneity of the soil matrix. In addition, specialized extraction techniques able to efficiently isolate analytes (or microorganisms) from soil are necessary prior to analysis. Specialized analytical methods have been developed and validated for the determination of trace contaminants and microorganisms in soil. However, in most cases, these methods are time–consuming, expensive and require the use of specialized reagents and personnel. It is therefore necessary to continue research towards the development of simpler and environmentally–friendly analytical techniques. Determining the occurrence and concentration of contaminants in soil is a task that requires a significant effort; however, this is only a part of the job. The study and understanding of the environmental fate of contaminants in soil is also a priority task to accomplish truly useful environmental risk assessment studies comprising soil, water sources, crops, farmers and consumers. Knowing the environmental fate of contaminants in the soil is necessary to understand the potentialities and limitations of each soil as a natural purification system of wastewater and an effective tool to define the capacity of each site to support wastewater irrigation in agriculture. Since soil is a complex and heterogeneous matrix, the fate of contaminants can vary significantly from one site to another. In this sense, it is worth defining which parameters are determinant in the fate of contaminants within soil and, on the basis of this knowledge, elucidating the fate of contaminants in other sites using mathematical tools to achieve such extrapolations. In this section, attention will be focused on pathogenic microbial agents, heavy metals and organic pollutants contained in municipal wastewater. The occurrence of such pollutants in wastewater–irrigated soil as well as their environmental fate in soil is addressed; additionally the most significant effects of these contaminants will be treated in some detail. Lastly, perspectives for further studies on the occurrence and fate of the studied pollutants in soil are presented.

2.2.1. Soil pollution by pathogenic microbial agents

Contamination of soil and crops by pathogenic agents is the effect of wastewater reuse in agriculture that receives most attention from environmentalists and scientists. Municipal wastewater contains a huge quantity and variety of bacteria, protozoa and viruses passed from human and animal feces and urine; therefore this water is a vector for intestinal infections (although some other diseases can spread from the environment via wastewater). Exposure may be direct through contact or ingestion of wastewater and soil, or indirect through contact with sick people or by ingestion of polluted crops, meat or milk. There are four groups at risk: a) farmers and their families, b) crop handlers, c) product consumers and d) people living nearby to irrigated fields. For any of these groups children and elderly are the most vulnerable, especially when they are undernourished. The most affected group is agricultural workers due to their high exposure to wastewater and contaminated soils [18]. Table 4 shows the risk of infection of water–borne diseases for vulnerable groups in irrigated areas using treated/untreated wastewater.

Effects caused by microbial pollution in soil. Several diarrheal outbreaks have been associated with the use of wastewater to irrigate [18, 88]; however, since this occurs in places where sanitation, hygiene practices and drinking water are of low quality it is always difficult to define their specific contribution to the total diseases burden. Cholera, caused by the bacterium *Vibrio cholera*, is one infection closely linked to wastewater irrigation in poor countries. Other intestinal diseases related to the use of wastewater to irrigate are traveler's diarrhea caused by *Escherichia coli*, shigellosis caused by *Shigella* spp., gastric ulcers caused by *Helicobacter pylori*, giardiasis caused by the parasitic protozoan *Giardia intestinalis* and amebiasis caused by *Entamoeba histolytica*. Additionally, viral enteritis (caused by rotaviruses) and Hepatitis A are the most reported viral infections caused by consumption of polluted vegetables [89]. Some studies [90] report skin diseases, such as dermatitis (eczema), in farmers that come into contact with untreated wastewater and wastewater irrigated soil. Nail problems in farmers, such as *koilonychias* (spoon–formed nails), have also been reported as related to the presence of fungi in wastewater irrigated soils [91]. Health and growth problems have been observed in cattle that consume forage produced by wastewater irrigation. Furthermore, in low income areas where water is scarce, cattle are not only fed with fodder grown using wastewater but also they are allowed to drink the wastewater used for irrigation. Some protozoa can survive in the surface layers of soil or even in aerial parts of crops; animals can be infected after eating these crops, although this is a remote way of transmission. There is some evidence indicating that beef tapeworm (*Taenia saginata*) can be transmitted from livestock fed with wastewater–irrigated forage to meat consumers. Furthermore strong evidence indicates that cattle grazing on fields freshly irrigated with raw wastewater or drinking from raw wastewater canals or ponds can become heavily infected by *Cysticercus bovis*, the early stage of the *Taenia saginata* life cycle [88].

Microbial agents in wastewater irrigated soils. The study of microbial contamination by the use of treated/untreated wastewater in agricultural irrigation is focused in the pollution of crops rather than the soils receiving wastewater. This is because, on the one hand, a greater number of people are exposed to pathogenic microorganisms through consumption of contaminated crops, meat and milk than by direct contact with irrigated soils, and on the other hand, the difficulties in the analysis of microorganisms in soil; for instance, the inherent problems of extracting microorganisms from such a complex matrix as the soil. Studies in the

Group exposed	Helminth infections	Bacterial/viral infections	Protozoan infections
Consumers	Significant risk of *Ascaris* infection for both adults and children consuming vegetables contaminated with helminth ova.	Cholera, typhoid and shigellosisoutbreaks reported due to the consumption of polluted crops. Seropositive responses for *Helicobacter pylori* in crop consumers. Increase in risks of suffering non–specific diarrhea when concentration of thermotolerant bacteria in wastewater used for irrigation exceeds 10^4 CFU/100 mL.	Evidence of parasitic protozoa found on the surface of wastewater–irrigated vegetables, but no direct evidence of disease transmission.
Farm workers and their families	Significant risks of *Ascaris* infection for both adults and children in contact with untreated wastewater and irrigated soils. Risk remains, especially for children, when wastewater presents more than 1 nematode egg per litre. Increased risk of hookworm infection in farmers.	Increased risk of diarrheal diseases for children in contact with wastewater when it exceeds 10^4 CFU/100 mL for thermotolerant coliforms. Elevated risk of *Salmonella* infection in children exposed to untreated wastewater and wastewater irrigated soil. Elevated seropositive responses to norovirus in adults exposed to partially treated wastewater and wastewater irrigated soil.	Risk of *Giardia intestinalis* infection insignificant for contact with both treated/ untreated wastewater and soil. Increased risk of amoebiasis observed due to contact with untreated wastewater and wastewater irrigated soil.
Nearby communities	High risk of infections when flood and furrow irrigation is used. *Ascaris* transmission not studied for sprinkler irrigation.	Sprinkler irrigation with untreated wastewater and high aerosol exposure associated with Increased rates of bacterial infections due to the use of partially treated wastewater (10^4–10^5 CFU/100 mL or less). No risks of viral infection associated with sprinkler irrigation.	No data of protozoan infections transmission during irrigation with wastewater.

Table 4. Summary of health risk associated with the use of wastewater in agriculture

Mezquital Valley, Central Mexico, found the occurrence of fecal contamination indicators (*Escherichia coli*). *Giardia lamblia* cysts and helminth eggs (*Ascaris lumbricoides*) at different depths of long–term wastewater irrigated soils. Results shown in Figure 3 evidence the accumulation of the three microorganisms in the first few centimeters of the soil profile, indicating that infectious agents are removed from wastewater at the beginning of percolation through soil; such removal can be achieved by several physical and chemical phenomena. In this study, the content of pathogenic microorganisms in soils with different time under irrigation was also evaluated. Results showed that the accumulation of microorganisms in the tested soils is not related to the time under irrigation, suggesting that soils have mechanisms to inactivate and/or destroy these microorganisms after irrigation.

As mentioned above, different types of microorganisms can be found in wastewater irrigated soils depending on the zone where reuse is taking place. For example, the study referred in [92] showed a higher prevalence of *Cryptosparidium* spp. compared with *Giardia* spp. in wastewater irrigated and manure amended soils of dairy farms in southeastern New York. *Cryptosporidium* is a protozoan commonly found in developed countries, while different species of *Giardia* are widespread in developing countries. In this respect, the study referred in [93] found the occurrence of *Ascaris lumbricoides*, hookworm and *Trichiuris trichiura* in 69% of the soil samples taken in an untreated wastewater irrigated area in West Bengal, India.

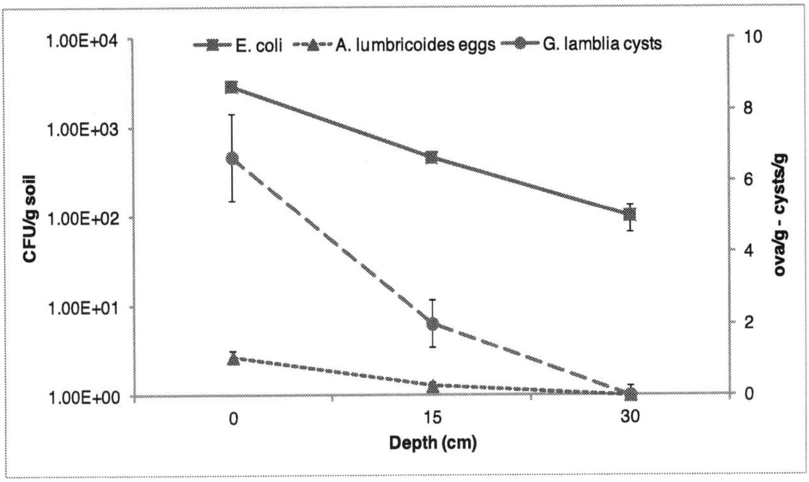

Figure 3. Abundance of three pathogenic microorganisms in a long–term wastewater irrigated soil at different depths

The entry of antibiotic–resistant pathogens (ARPs) and antibiotic resistance genes (ARGs) into the soil via wastewater is an emerging issue. Since municipal wastewater contains both sub–therapeutic amounts of antibiotics, ARPs and ARGs –which occur to a greater extent when sewer systems combine municipal and hospital wastewater– [94], these substances can reach the soil, modifying the dynamic of soil microbial populations. Antibiotic resistance may occur naturally in the soil, and to a greater extent in the rhizosphere, which functions as a hotspot for both antibiotic–resistant bacteria and ARGs [32]. Previous studies have found the presence of opportunistic pathogens (*Stenotrophomonas maltophilia,* responsible for respiratory tract infections and endocarditis) in the rhizosphere of *Brassicacaea* type plants [95]. The transfer of ARGs from these opportunistic bacteria to human pathogens reaching the soil through wastewater has not yet been demonstrated. The ARPs reaching the soil through wastewater may survive on the soil surface and, if conditions are appropriate, reproduce or migrate to surface and groundwater sources. ARGs may be mobilized into aquifers by infiltration of wastewater or into surface water sources by runoff. So far a relationship between the presence of traces of antibiotics in wastewater and the occurrence of antibiotic resistance in the irrigated soil has not been categorically established. In previous studies the incidence of two sulfonamide resistance genes (*sul1* and *sul2*) was determined in the Mexico City wastewater, agricultural soils irrigated with such wastewater over different time periods and rain–fed soils [96]. The authors found the presence of ARGs in the three analyzed matrices; the concentration of resistance genes was 150 to 1500 times higher in irrigated soils than in non–irrigated ones. The occurrence of ARGs was positively related to the time under irrigation, with a higher content of resistance genes occurring in *Enterococci* bacteria living in soils irrigated for longer periods of time [96]. Such behavior may indicate that prolonged irrigation with wastewater promotes both the proliferation of indigenous ARPs in soil, due to the high and constant supply of nutrients via wastewater, and the increase in the assimilation of resistance genes due to the higher biomass content in old wastewater irrigated soils. Conversely, studies reported in [97] found that the abundance of isolates resistant to tetracycline, ciprofloxacin, sulfonamides and erythromycin were identical in wastewater irrigated soils and freshwater irrigated soils despite the high load of ARGs and ARPs in the wastewater used for irrigation. In this regard, the study in reference [98] found, by comparing the resistome of soils irrigated either with wastewater or groundwater, that *Entetococci* bacteria in freshwater irrigated soil were highly resistant to a greater number of antibiotics (erythromycin, tylosin, tetracycline, and ciprofloxacin) than long–term wastewater irrigated soil, which showed resistance to lincomycin and daptomycin. Furthermore, no differences were found in the content of ARPs when wastewater and freshwater irrigated soils were compared, suggesting that ARPs rarely survive after they enter soil via wastewater. Even though it seems unlikely that development of antibiotic resistance to human pathogens in wastewater irrigated soil is related to the input of antibiotics and resistant organisms via wastewater, it is worth, as a next step, studying the exchangeable genetic material (e.g. plasmids), since such material can be assimilated by soil microorganisms, inducing antibiotic resistance. Many questions remain about the mechanisms leading to the transference of this type of genetic material [99].

Microbial pollution in crops. Crops are polluted by direct contact with wastewater during irrigation. Pollution of the edible parts of plants depends not only on the quality of water, but also on the quantity applied to soil, the irrigation method and the type of crop. For example, zucchini when spray–irrigated with wastewater accumulate higher levels of pathogens on their surface than other crops. Zucchini have a hairy and sticky cover and grows close to the ground, which favors the attachment of pathogens. Microbial contamination of crops can occur not only as a result of wastewater irrigation but also during washing, packing, transportation and marketing. These problems are frequently not addressed, giving the impression that irrigation is the only source of microbial pollution [100]. In a previous study, referred in [101], it was found that less microbial pollution of crops is caused if irrigation is performed by subsurface dripping than through sprinklers, furrows or flooding. Moreover, the study reported in [102] showed that subsurface irrigation does not pollute crops even when using wastewater with 6–7 x 10^5 CFU/100 mL of fecal coliforms and 225 helminth ova/L. Microbial pollution of crops also depends on the type of crop. Fruits from trees are rarely polluted when irrigation is not provided using sprinklers (this is not a common procedure used to apply wastewater since sprinkler heads tend to become clogged). Fruits grow far from the watering sites when furrow and flood methods are used. The microbial contamination of crops in wastewater irrigation systems is closely related to the survival of microorganisms. Table 5 shows the survival times of some pathogens in agricultural soils and crops irrigated with wastewater.

Pathogen	Survival time (days)	
	Soil	Crops
Ascaris lumbricodes eggs	180	30
Salmonella spp.	80	25
Fecal coliforms	<70, but usually <20	<30, but usually <15
Vibrio cholera	<20, but usually <10	<5, but usually <2
Entamoeba histolytica	<20	<10
Trichuris trichiura eggs	>180	<60, but usually <30
Taenia saginata eggs	>180	<60, but usually <30
Enterovirus	<40	<20

Source: references [105–106]

Table 5. Survival of selected pathogens in soil and crops irrigated with wastewater

Both pathogenic and non–pathogenic microorganisms display differences in their survival in soil and crops. For instance, the non–pathogenic fecal coliform indicator *E. coli* can survive in soil for nearly a month, while the pathogenic strain of *E. coli* O157:H7 survives at most for 14 days in spinach leaves [103]. It is known in some detail that survival of pathogenic bacteria can increase by internalization within the plant tissues [104]. Previous studies indicate that *E. coli* can translocate from soil to leaves of lettuce through the root system [107]. In contrast, the results reported in reference [108] indicate that translocation of pathogenic bacteria to the edible parts of crops via the root system is quite unlikely. It is more likely that pathogens enter to the edible parts of crops through wounds in vegetal tissues [109]. Wounded tissues have been demonstrated allow the entrance of *Salmonella* and *E. coli* to lettuce and tomato plants [110–111]. Similarly, it is reported that *E. coli* can use the stomatal cavities in leaves to enter the internal structure of lettuce [115]. The pathway of this kind of entry is still unknown. Once inside the plant tissues, pathogen survival rates improve since they can use cellulose as their main source of carbon. Protozoa are larger in size than bacteria and thus they cannot access the internal parts of the plants; however, these pathogenic organisms can adhere to the surface of edible plants and remain there by the excretion of polymers which facilitate adhesion. Table 6 shows some examples of the occurrence of protozoa in crops irrigated with treated/untreated wastewater.

http://dx.doi.org/10.5772/57226

Pathogen	Crop	Occurrence	Reference
Giardia lamblia	Potatoes	5.1 cysts/kg	Crops irrigated using untreated wastewater in Marrakesh. [112]
	Coriander	254 cysts/kg	
	Mint	96 cysts/kg	
	Carrots	155 cysts/kg	
	Radish	59.1 cysts/kg	
Ascaris lumbricoides	Potatoes	0.18 eggs/kg	
	Turnip	0.27 eggs/kg	
	Coriander	2.7 eggs/kg	
	Mint	4.63 eggs/kg	
	Carrots	0.7 eggs/kg	
	Radish	1.64 eggs/kg	
Enterobius vermicularis	Lettuce	10–40 cysts/kg	Crops irrigated using treated and untreated wastewater in Kahramanmaras, Turkey. [113]
	Parsley	10–60 cysts/kg	
	Cress	10–20 cysts/kg	
	Spinach	1–3 cysts/kg	
Entamoeba hystolitica	Lettuce	10–50 cysts/kg	
	Parsley	10–50 cysts/kg	
Giardia lamblia	Lettuce	10–20 cysts/kg	
Ascaris lumbricoides	Lettuce	10–30 eggs/kg	
	Parsley	10–30 eggs/kg	
Trichuris trichiura	Spinach	3.3% of the analyzed samples.	Crops grown in soils irrigated with raw wastewater in West Bengal, India. [93]
	Pudina	3.1% of the analyzed samples.	
	Coriander	5% of the analyzed samples.	
Hookworm	Lettuce	9.4% of the analyzed samples.	
	Parsley	3.3% of the analyzed samples.	
	Spinach	6.7% of the analyzed samples.	
	Pudina	9.4% of the analyzed samples.	
	Celery	3.6% of the analyzed samples.	
	Coriander	5% of the analyzed samples.	
Ascaris lumbricoides	Lettuce	43.8% of the analyzed samples.	
	Parsley	23.3% of the analyzed samples.	
	Spinach	36.7% of the analyzed samples.	
	Pudina	50% of the analyzed samples.	
	Celery	25% of the analyzed samples.	
	Coriander	35% of the analyzed samples.	
Helminth eggs	Leafy vegetables	100 eggs/kg	Vegetables irrigated with untreated wastewater in Faisalabad, Pakistan. [114]
	Cauliflower		

Table 6. Occurrence of some pathogen protozoa on the surface of crops irrigated using treated/untreated wastewater

Fate of pathogenic microorganisms in soil. Upon their arrival to irrigated soils, microorganisms can either survive or be inactivated/killed by the physical and chemical processes naturally occurring in soil as well as by predation by indigenous soil organisms. Given the case that these microorganisms can survive in the soil, they may subsequently colonize soil particles, infiltrate the soil to the aquifer or migrate through across the landscape by runoff. Processes affecting the environmental fate of the pathogenic microorganisms in soil are shown in Figure 4. Previous experiments have demonstrated that some microorganisms can vertically and/or horizontally mobilize through the soil, travelling long distances from the initial point of contamination [116]. Bacterial migration in soil has been reported up to 830 meters, while for viruses such displacement is significantly lower, i.e. up to 408 m [117–118]. Survival of pathogens is related with their environmental fate since the longer the lifetime of the microorganisms the larger the distance they can travel. As indicated in Table 5, bacteria can survive for long periods compared to viruses, and thus bacteria can be transported farther. Climatic conditions also impact upon pathogen transportation; for instance, in frozen soils pathogens can survive longer and thus they can be transported farther than in tropical and desert soils [119]. Microorganisms can be more easily displaced through coarse textured soils than fine textured ones. The study referred to in [118] found greater mobilization of coliforms in sand–gravel soil than in fine sand. In fact, in coarse sandy soils, the vertical movement of microorganisms can be as rapid as that observed for inorganic tracers. In this regard, the results reported in reference [120] evidence that infiltration of streptomycin–resistant *E. coli* can be compared with that of the chloride tracer in undisturbed soil columns, even when different soil textures are compared. Since the transportation of microorganisms is similar to that observed for tracers, the physical structure of soil is the determinant factor in them reaching the aquifer; therefore, a greater occurrence and interconnection of pores within the solid matrix may result in efficient infiltration of water and thus bacteria. Studies on the movement of pathogens in the field confirm the rapid movement of pathogenic bacteria observed in laboratory tests. These studies also found a high concentration of bacteria and viruses in groundwater [121]. In addition to the higher quantity and interconnection of pores, the increased transport of bacteria and viruses through the soil can be explained by the presence of preferential paths within the soil matrix. Such preferential paths are referred to cracks, fractures, worm holes and channels formed by plant roots or fauna in the soil. Studies reported in reference [122] show that larger microorganisms (*E. coli*) can mobilize deeper into soil than smaller coliphages. Moreover, the study referred to in [123] confirms that bacterial cells smaller than 1 μm in diameter are more rapidly transported through soil than larger organisms.

The chemical properties of soil can also impact upon the vertical and horizontal transport of microorganisms. The mineral composition of soil can favor adhesion of microbial cells, eggs or cysts onto soil particles. Several types of bacterial cells have been shown to strongly adhere to the mineral domain of soil and aquifer material [124]; once adhered, bacteria can replicate and form biofilms on the surface of soil particles. In wastewater irrigated soils, the accumulated organic matter as well as the continuous input of dissolved organic matter via wastewater may enhance the proliferation of bacteria. With regard to parasites, the study referenced in [87] found that *Ascairs lumbricoides* eggs and *Giardia lamblia* cysts adhere to the mineral fraction of wastewater irrigated soils more rapidly and more strongly than to the organic domain. In the

http://dx.doi.org/10.5772/57226

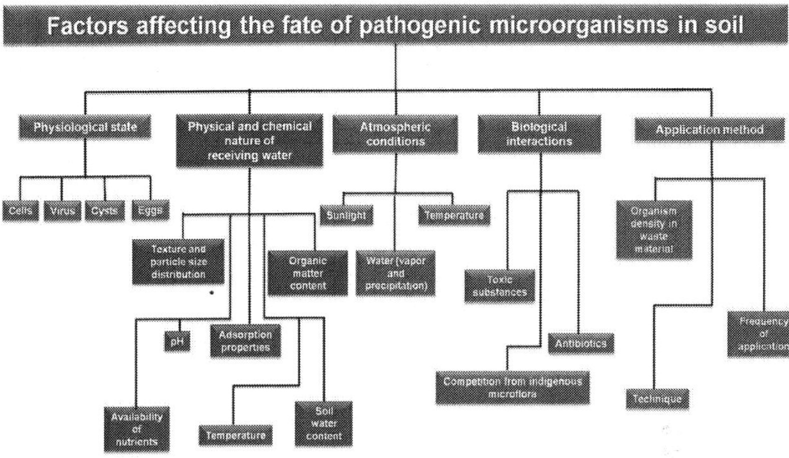

Source: modified from [121]

Figure 4. Factors affecting the environmental fate of pathogenic microorganisms in wastewater irrigated agricultural soils

case of *Ascaris* eggs, adhesion occurs with the silica in sand particles. In contrast to protozoan eggs, studies referred to in [125] suggest that adhesion or adsorption of protozoan cysts may be related to soil organic matter rather than the mineral fraction of soil. This has been attributed to the hydrophobic nature of the cysts walls. According to the findings reported in [126] detachment of bacteria from soil particles is effected by the composition of the irrigation water. In that study, *Pseudomonas* sp. showed enhanced transport when distilled water was used for detachment in column experiments, compared with 0.01 M NaCl. Such results suggest that clean water can efficiently wash off the polysaccharides excreted by bacterial cells which act as an adhesive between soil particles and bacteria. The opposite effect has been observed for Ascaris eggs. When soil is washed with NaOCl, eggs are effectively detached from soil particles; this is because sodium hypochlorite can destroy the albuminose layer that coats the surface of helminth eggs and which anchors with the soil particles [127]. According to the established in [87], the environmental relevance of studying the impact of this salt on the detachment of eggs from soil relies on the fact that NaOCl can be found in reclaimed water, as it is commonly used for disinfection of effluents.

Once microorganisms are retained by soil, either by adsorption/adhesion or straining, they can be inactivated or eliminated by desiccation. This phenomenon is particularly important in arid areas where high levels of solar radiation are reported. The environmental fate of microorganisms in soil also depends on the native microorganisms living in the solid matrix. Predators of wastewater-borne pathogenic bacteria in soil include Streptomycetes, Myxobacteria, *Bdellovibrio* and nematodes [121]. The presence of plants may affect the persistence and movement of microorganisms in soils. On the one hand, pathogen can found favorable conditions for survival in the rhizosphere due to the high content of nutrients in this zone; and on the other hand, native bacteria in rhizosphere can be natural predator of those pathogens, while roots may excrete antibiotics that inhibit or kill pathogenic microorganisms.

2.2.2. *Soil pollution by heavy metals*

Given that most agricultural wastewater irrigation is performed using municipal wastewater, which contains negligible amounts of heavy metals [11], the occurrence of these elements in wastewater irrigated soils is usually significantly lower than the maximum permissible concentrations established by international regulations. However, there are some cases where care should be taken when reusing wastewater in irrigation, e.g. close to tanneries, metal processing or mining areas [91]. Different levels of risk are perceived for the different heavy metals. While some of them are nutrients for plants at trace concentrations, others have been shown to produce harmful effects on exposed organisms, or are absorbed by plants and accumulated through the food web. Table 7 presents the risks that are incurred by the presence of some heavy metals in soil.

Risks characteristics	Metal
Low risk	Mn, Fe, Zn, Cu, Se, Sb
High risk	Cr, As, Pb, Hg, Ni, Al, Cd
Essential micronutrient to plants	Cu, Fe, Mn, Mo, Zn, Ni
Beneficial for some crops	Co, Na, Si
Can accumulate in crops to levels that are toxic for consumers	Cd, Cu, Mo
No human toxicological threshold established for wastewater intended for irrigation	Hg
Relatively high threshold for wastewater used in irrigation	Cu, Fe, Mn, Zn
Low absorption by plants	Co, Cu, Mn, Zn

Source: with information from [75]

Table 7. Heavy metal risk characteristics during irrigation

Cadmium is the metal with the highest associated risk. It is toxic to humans and animals in doses much lower than those that visibly affect plants; furthermore crop uptake (which is notably high in acidic soils) can increase the dose consumed by organisms and in turn accumulation in animal tissue. Absorbed cadmium in animals is stored in kidney and liver, although meat and milk products have shown to be little affected by cadmium accumulation [75]. There is a relatively good knowledge to allow the setting of limits regarding the acceptable amount of heavy metals contained in wastewater used to irrigate. In the study referred to in [128], numerical calculation of the limits for the maximum tolerable pollutant concentration in wastewater irrigated soils was carried out (health–based targets). This was based on the acceptable daily human intake (ADI) for selected heavy metals and the amount that can be "permitted" to accumulate in soil before harmful effects occur in consumers of crops (Table 8). This analysis assumed: a) only two exposure routes (wastewater → soil → plant → human; and, wastewater → crop → human); b) a global diet in which the daily intake of grains/cereals, vegetables, root/tuber crops and fruit accounts for ~75% of daily adult food consumption; c) a body mass for adults of 60 kg; d) all of the food grain, vegetables, root/tuber crops and fruits are obtained from land irrigated with wastewater; and, e) a total daily intake of pollutants by this consumption path of 50% of the ADI (the remaining 50% of the ADI was attributed to background exposure). Table 8 shows the inputs of heavy metals by wastewater to irrigated soils, assuming an application of treated wastewater of approximately 1.2 m/year, which is roughly the amount of water required to produce a crop cycle in an arid zone.

Element	**Maximum input by wastewater (kg/ha/year)**	**Maximum tolerable concentration (mg/kg)**
Arsenic	0.6–12	9
Cadmium	0.06–0.24	7
Chromium	1.2–60	3200
Lead	1.2–60	150
Mercury	0.12–0.12	5
Nickel	0.24–12	850
Selenium	0.24–0.6	140
Silver	1.2	3

Source: reference [128]

Table 8. Maximum tolerable concentration of heavy metals in wastewater irrigated soils

Health effects associated with the use of water heavily contaminated with industrial discharges for agricultural irrigation have been reported. In Japan, itai–itai disease, a bone and kidney disorder associated with chronic cadmium poisoning, occurred in areas where rice paddies were irrigated with water from the contaminated Jinzu River [129]. In some parts of China, the use of industrial wastewater for irrigation was associated with a 36% increase in hepatomegaly (enlarged liver) and 100% increase in both cancer and congenital malformation rates [130].

With regard to the occurrence of heavy metals in agricultural soils irrigated using wastewater, the study referred in [131] presents an inventory of sources of some heavy metals (zinc, copper, nickel, lead, chromium and cadmium) in agricultural soils of England and Wales. Results showed that the greatest contribution of heavy metals in those soils comes from the application of sludge from wastewater treatment plants, while irrigation appeared to be of little importance as a source of heavy metals in soils. According to this investigation, which followed the rates of deposition of heavy metals in the studied soils, the time required for metal concentrations to reach maximum values permitted by international regulations is 80 years for zinc and at least 1256 years for cadmium. In this respect, study referred to in [132] showed that concentrations of heavy metals in long–term untreated–wastewater irrigated soils in central Mexico were 10 times lower than the limits set by the Danish regulations; moreover, the authors estimated that another century of irrigation is necessary to exceed these values. In most cases, metals have little impact on aquifers. According to the results reported in [133] the most toxic metals to humans –cadmium, lead, and mercury– were absent in groundwater at five sites in the United States after 30–40 years of applying secondary and primary effluents at rates between 0.8 m/year and 8.6 m/year to different crops. The reason given was that the pH values greater than 6.5 in soil and wastewater resulted in the precipitation of the entire amount of metals. Metals are normally bonded into the organic matter through the formation of organo–metallic complexes, which are not bioavailable to plants. The addition of lime and wastewater to soil assists the precipitation of metals, while the addition of chemical fertilizers has the opposite effect, since over the long term they tend to lower the soil pH and thus solubilize metals.

In contrast, agricultural soils have been reported in which the concentration of heavy metals, such as cadmium and zinc, are close to reaching the maximum levels set out in international regulations. In these cases, the factors leading to an exacerbated soil contamination and thus increased risk of groundwater and crop pollution are: a) sandy soil texture; b) acidic to neutral soil pH; c) low organic matter content; and/or, d) the use of industrial wastewater for agricultural irrigation [134–135]. In such cases, the cessation of agricultural irrigation with wastewater is recommended, together with allowing the recovery of soil through remediation techniques such as phytoremediation.

2.2.3. *Soil pollution by organic compounds*

Pollution of soil by organic substances has been a matter of concern to scientists and organizations regulating the quality of soil, water sources and food for several decades. An extensive body of work exists addressing the degradation of soil by conventional organic pollutants (e.g. pesticides, polyaromatic hydrocarbons, organochlorides, paraffin, organic solvents, etc.). However, in sites where treated/untreated wastewater is disposed of by agricultural irrigation one can find organic substances different to those commonly studied and reported in literature treating oil spills, mining zones or soil polluted by industrial wastewater. Most of the dissolved and particulate organic matter contained in municipal wastewater is produced by the degradation of human and animal excreta, hence organic matter in wastewater is composed mainly by saccharides, lipids, amino acids and proteins; however, a tiny fraction of the organic material in wastewater originates from chemicals contained in everyday consumer products used and disposed of via sewage by people in urban and rural areas. According to [24], thousands of organic compounds are contained in municipal wastewater at trace levels and there is a lack of knowledge regarding the effects that such substances may cause to the exposed organisms, either by themselves or in combination with other compounds or groups of organic compounds. This group of chemicals is referred as "organic pollutants of emerging concern" (OPECs) [136]; though they should actually be listed as priority pollutants in cases where wastewater is used to irrigate crops, since these contaminants are in contact with soils, crops and water sources near the irrigation area [137]. Over the last three decades, significant work in the field of analytical chemistry has been carried out in order to extract, isolate and quantify some of these pollutants in wastewater and soils. Frequently found OPECs in such complex matrices are pharmaceutically active compounds (PACs) and their metabolites, personal care products (e.g. disinfectants, fragrances, insect repellents, sunscreens, etc.), sweeteners, stimulants (e.g. caffeine and psychoactive drugs), detergents and their metabolites, plasticizers and industrial additives (e.g. additives in gasoline) [137]. Almost all of the studies addressing the removal of OPECs in wastewater treatment plants report that most of these substances are partially degraded/removed in primary and secondary treatment systems –and some pollutants are only partially removed even in tertiary treatment systems– [138]. Because of this, OPECs occur in irrigated soils if either treated or untreated wastewater is used in irrigation. Effluents of wastewater treatment plants contain a small fraction of the parent substance as well as the by–products generated during treatment. However, some of the compounds may be retained and concentrated in the sludge produced during wastewater treatment and reach the environment via the use of sludge (or biosolids) as soil amendments in agriculture. Due to continuous industrial development, the number of organic substances contained in wastewater is constantly increasing; in fact, most of these substances are not tested before they are released onto the market, and therefore their potential risks or the side effects they cause in non–target organisms in soils or water bodies is yet unknown.

Effects caused by domestic wastewater–related organic pollutants in irrigated soils. As mentioned above, due to the ever growing pool of organic compounds discharged to the soil via wastewater, there is a general lack of knowledge regarding the effects that such substances cause to exposed organisms. In general terms, municipal wastewater is the main vector of OPECs to reach the environment, so that these substances are ubiquitous at sites where wastewater streams occur. Pharmaceutically active compounds (PACs) are designed to cause a defined effect on target organisms; however, when trace amounts of these substances are transported by wastewater into environment, they can interact with non–target organisms. One effect that has captured the attention of the scientific community in recent years is the development of antibiotic resistance by pathogenic microorganisms due to the occurrence of antibiotics in wastewater, surface water bodies and soils receiving wastewater [139–140]. However, a large number of studies on this subject report that proliferation of antibiotic–resistant pathogens is quite unlikely in wastewater irrigated soils [97–99]. Conversely, the study referenced in [96] attempts to relate the occurrence of sulfonamide and fluoroquinolone antibiotics with the emergence of antibiotic resistances in wastewater and long–term wastewater irrigated soils. The authors reported a relationship between time under irrigation and the frequency of detection of antibiotic resistance genes in soils. In the case of non–antibiotic PACs, the most studied compounds –because they are the most used worldwide– are the analgesic and anti–inflammatory drugs [141]. Compounds such as ibuprofen, naproxen, diclofenac, paracetamol and ketoprofen have been shown to cause systemic damages in aquatic species; damages in liver, gills and kidney are commonly reported [142]. The non–steroidal anti–inflammatory drug diclofenac has been demonstrated to cause visceral gout in vultures; in fact, the presence of diclofenac in livestock was the cause of the mass death of three species of vulture in India and Africa [143]. Other studies show that chronic exposure to traces of anti–inflammatory drugs leads to a lessening in the development of human embryo cells [144]. The occurrence of psychotropic agents at trace levels in water bodies polluted by wastewater discharges has been shown to alter the behavior of some fish species, suppressing their survival instincts against predators [145]. With regard to OPECs that are not pharmaceutically active compounds, there is significant concern that they may alter hormone homeostasis in organisms. These substances, known as endocrine disruptors, can mimic or compete with natural hormones by binding with active sites on hormone receptors, causing reduced or disproportionate hormonal responses in the affected organisms [146]. The most potent endocrine disruptors found so far in municipal wastewater are the natural and artificial estrogenic hormones –the latter are used as birth control agents– and the regulators of thyroidal function, followed by plasticizers (e.g. phthalates and bisphenols), surfactants and their metabolites and some industrial additives [147]. Endocrine disruptors are suspected of causing the feminization or masculinization of fish and reptile populations as well as the occurrence of breast cancer, imbalances in thyroidal function, teratogenic effects (e.g. cryptorchid) in mammals, and even obesity in mammals (obesogens) [148–150]. There is a serious lack of knowledge regarding the effects caused by OPECs in soil organisms. Studies on this field have been little developed compared to those for water bodies. Table 9 shows some examples of effects caused to soil organism by the occurrence of OPECs.

The effects caused by this class of pollutants are not limited to soil organisms and impacts can be observed in the soil matrix. For example, surfactants can, on the one hand, decrease the capillarity and penetrability of soil as well as increase the solid–liquid contact angle, the shape factor and the sorptivity of soil particles. On the other hand, the input of these substances can increase the desorption of previously sorbed organic molecules on the soil particles, which in turn increases the bioavailability and mobility of the desorbed compounds [163]. To evaluate the toxic effects caused by the occurrence of OPECs to soil organisms two approaches are commonly used, i.e. acute and chronic toxicity studies. For the former, high concentrations of target pollutants are supplied to studied organisms under controlled conditions for a short period; chronic toxicity tests, on the other hand, are based on prolonged exposure of organisms to low (i.e. environmentally representative) doses of the studied pollutants. So far, most toxicity studies dealing with OPECs have been carried out using the acute toxicity approach. Even though these studies do not fully represent the conditions observed in the field, they provide valuable information on the subject of impacts caused by this kind of contaminants to soil organisms. Studies evaluating chronic toxic effects of pollutants are more representative of field conditions, i.e. toxic substances enter to soil in small doses over long periods. In this regard, conducting long–term toxicity studies that evaluate the chronic effects caused by OPECs in soil organisms are a priority. Several toxicity studies report that the effects of organic pollutants on soil organisms (i.e. reduction in soil respiration, enzymatic activity and nitrification/denitrification rates) are observed in the early days of exposition; then, after a short period (4 to 10 days) soil recovers to its basal conditions [153–154, 159, 161]. The next step in toxicity studies for these emerging pollutants is to determine the dynamics of the toxic effects on soil organisms after tens or hundreds of growing cycles in which target contaminants are continuously supplied; i.e. under conditions similar to what occurs in long–term irrigated areas.

Compound	Effect in soil organisms	Reference
Estrone, 17β estradiol (hormones)	Negative impacts on the vegetative cycle of alfalfa (*Medicago sativa).*	[151]
Sex hormones	Shift in sex ratio of free life nematode communities in soil.	[152]
Triclosan (antibacterial agent)	Inhibition in plant growth (rice and cucumber). Effect concentrations 50 (EC50, i.e. 50% of exposed population was affected) were 57 and 108 mg/kg for rice and cucumber respectively. Inhibition of soil respiration and phosphatase activity at concentration levels higher than 10 mg/kg.	[153]
	Reduction in soil respiration 4 days after supplying the compound. The observed effects were dependent on the adsorption of the compound onto the soil.	[154]
Bisphenol A (plasticizer)	Shift in sex ratio to female individuals in isopod (soil arthropod) communities.	[155]
Abamectin (anthelminthic)	Negative impacts on reproduction of *Folsomia fimetaria* and *Folsomia candida* (soil arthropods) at concentrations of 0.25 and 0.5 mg/kg of soil (dry mass), respectively. Negative impacts on reproduction of soil earthworms at concentration levels of 0.06 mg/kg of soil (dry mass).	[156]
Fenbendazole and cypermethrin (antiparasitic)	Negative impacts on degrading microorganisms of dung.	[157]
Sulfonamide and tetracycline antibiotics	Inhibition of the soil microbial activity by 10% (ED10) at concentrations of 0.003–7.35 μg/g of soil (dry mass). Shifts in fungi:bacteria ratio.	[158]
Sulfadiazine (antibiotic)	Decrease in denitrification rates when the input of antibiotic was 100 mg/kg of soil (dry mass).	[159]
	Significant decrease in the bacteria:fungi ratio	[160]
Chlortetracycline, tetracycline, tylosin, sulfamethoxazole, sulfamethazine and trimethoprim (antibiotics)	Decrease in crop growth (sweet oat, rice and cucumber). Inhibition of the microbial activity of soil (soil respiration and phosphatase enzyme activity).	[161]
Human and veterinary pharmaceutically active substances	Decrease in growth and development of *Phaseolus vulgaris L., Glycine max, Medicago sativa, Zea mays,* and several other crops.	[162]

Table 9. Summary of negative effects on soil organisms caused by the occurrence of pollutants of emerging concern at trace levels

Occurrence of domestic wastewater–related organic pollutants in irrigated soils. In spite of the fact that wastewater is the main vehicle allowing OPECs to reach soil, very few studies reporting the presence of these pollutants in wastewater irrigated soils have been carried out.

This finds an explanation, on the one hand, in the inherent difficulty of extracting and isolating organic compounds at trace levels from the soil matrix and, on the other hand, in the fact that analyzing this type of pollutants is relatively expensive. Figure 5 shows the sites where monitoring studies aimed at determining the occurrence of OPECs in wastewater irrigated soils have been performed. In this figure the number of sites monitored is contrasted with the 20 countries with the highest use of untreated wastewater for agricultural irrigation. Most of the monitoring studies are concentrated in China, the country using the highest volume of untreated wastewater in agriculture [17], followed by the United States and Mexico –the latter is the second placed country in terms of reuse of untreated wastewater for irrigation–.

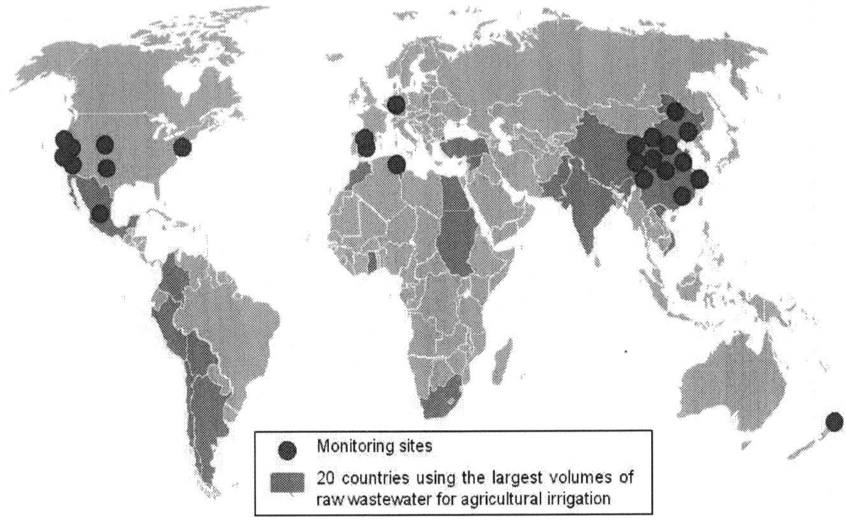

Figure 5. Monitoring studies for pollutants of emerging concern in wastewater irrigated soils throughout the world and comparison with the 20 countries using the largest volumes of raw wastewater for agricultural irrigation

Efforts in monitoring emerging pollutants in developing countries where the use of raw wastewater is widespread are of value; this requires cooperation with research centers where analytical techniques are currently validated to perform soil analyses or by sharing "know how" and technology with developing countries in order to perform analysis on site. Determination of OPECs in soil requires an exhaustive extraction step, which in most cases has to be carried out at a moderately high temperature, particularly in the case of analysis of thermolabile compounds (e.g. sulfonamide antibiotics). Extraction methods such as pressurized fluids extraction, microwave assisted extraction and ultrasonic assisted extraction are preferred over traditional Soxhlet extraction techniques, since they guarantee greater contact between the solvent and the soil particles, resulting in higher recoveries of analytes. Analysis of OPECs is commonly accomplished using either liquid or gas chromatography techniques; although liquid chromatography is preferred as it is more suitable for the analysis of polar compounds, i.e. most PACs [164]. Monetary costs of these analyses are relatively high and analysis entails the use of potentially dangerous chemicals, which is in part the reason why monitoring studies for OPECs in soils are not carried out in poor countries. So far, the most reported emerging pollutants in wastewater irrigated soils are the pharmaceutically active substances (e.g. antibiotics, non–steroidal anti–inflammatory agents, anticonvulsants, anticoagulants and sex hormones), followed by plasticizers (e.g. phthalic acid esters and bisphenol A), metabolites of surfactants (e.g. nonylphenol, octylphenol) and antibacterial and antimycotic agents (e.g. triclosan and triclocarban). Table 10 shows the concentrations of some OPECs reported for wastewater irrigated areas.

Overall, higher concentrations of OPECs are found in the first 30 cm of the soil profile. Such behavior suggests that these compounds are retained by the organic matter accumulated over time of irrigation; which is consistent with the organic nature of these contaminants, although several of them display some polarity. Concentration levels reported for the monitored PACs range from below the detection limits of the analytical techniques to tens of μg/kg of soil (dry mass). Monitoring studies referred to in [165] report concentration levels of the pharmaceuticals ibuprofen, naproxen and carbamazepine in the range of 0.25 to 6.48 μg/kg of soil (dry weight) in Phaeozem and Leptosol soils that have been irrigated using untreated wastewater for eight decades. Other studies [168] found average concentrations of 1.8 μg/kg for triclosan and 2.5 μg/kg for estrone. In contrast to PACs, concentrations reported for plasticizers and surfactants are of the order hundreds of μg/kg of soil (dry mass). For example, in the study referenced in [170] concentrations of 14–80 μg/kg are reported for nonylphenols, while concentrations of 140–2610 μg/kg were observed for some plasticizers. Concentrations of up to 7110 μg/kg of the plasticizer di–2(ethylhexyl)phthalate, have been reported elsewhere [175]. High concentrations of plasticizers in soils are explained by the ubiquity of these compounds in environment. Phthalic acid esters are contained in almost all plastic products and can easily leach from the solid matrix (i.e. the plastic articles). Once phthalates are released from the solid matrix, they can get into environment not only via wastewater but by aerial deposition, using dust particles as carriers [176]. Nonylphenols, the major by–products of the anaerobic biodegradation of surfactants [177], are commonly found in wastewater irrigated soils due to the significant presence of detergents in municipal wastewater in combination with the anaerobic conditions prevailing in sewerage systems. In contrast to PACs, plasticizers and surfactant metabolites are non–polar in nature and for this reason, higher adsorption can occur for these compounds in soil, causing not only their build up in the surface layer of soil but the potential decrease in their bioavailability to soil microorganisms. Most monitoring studies of OPECs in environmental solid matrices are focused on determining these contaminants in biosolids amended soils rather than in wastewater irrigated soils. This is necessary since: a) biosolids in wastewater treatment plants concentrate organic pollutants during water depuration, hence a greater concentrations of contaminants are expected in biosolids amended soils than in treated/untreated wastewater irrigated soils; b) the use of biosolids as agricultural soil amendment is a more socially acceptable practice than reusing wastewater, thus it tends to be more practiced (or at least more reported) than wastewater irrigation, and it therefore becomes necessary to determine the pollutant load reaching the soil in this manner; c) since analysis of

Compound	Concentration (μg/kg)	Comments	Reference
Carbamazepine	0.28–0.94	Concentration range observed in the surface layer (0–10 cm depth) of a treated wastewater irrigated soil during an irrigation cycle (May to October). The lowest concentration was observed before irrigation started while the highest concentration was determined in soil at the end of the irrigation cycle. Irrigation at the site has been occurring for the last 30 years.	[165–167]
	5.14 and 6.48	Concentrations found in the surface layer (0–10 cm depth) of Leptosol and Phaeozem soils, respectively that has been irrigated using untreated wastewater for 85 years.	
	4.92, 2.9 and 1.92	Concentrations found in forested, grass–covered and cultivated soil irrigated with treated wastewater for more than 25 years. Carbamazepine was found mainly in the first 30 cm of the soil profile.	
Ibuprofen	<LOD–3; <LOD–3	Concentration ranges observed in loamy sand and sandy loam turf soils (0–30 cm depth) irrigated with treated wastewater at an irrigation rate of 1.1–1.2 and 1.5–1.6–fold the evapotranspiration rate, respectively. Wastewater irrigation has been occurring at the site for almost 20 years.	[168]
Naproxen	<LOD–12.5; <LOD–9.5		
Triclosan	<LOD–6; <LOD–2.8		
Bisphenol A	<LOD–1.25; <LOD–1		
Estrone	<LOD; <LOD–5.3		
Ibuprofen	<LOD and 0.25	Concentrations found in the surface layer (0–10 cm depth) of Leptosol and Phaeozem soils that have been irrigated using untreated wastewater for 85 years.	[165]
Naproxen	0.73 and 0.55		
Nonylphenols	123 and 41		
Triclosan	18.6 and 4.4		
Bisphenol A	14.8 and <LOD		
Di-*n*-butylphthalate	552 and 244		
Butylbenzylphthalate	346 and 171		
Di-2-(etylhexyl)phthalate	2079 and 820		
Clofibric acid	<LOD–9	Concentration range observed in soil from a golf course irrigated with reclaimed wastewater	[169]
Triclocarban	<LOD –105	Concentration ranges for pharmaceuticals and endocrine disrupting chemicals in agricultural soils of Hebei province, north China, which have been irrigated using treated wastewater for more than 50 years.	[170]
4-nonylphenol	14.2–60.3		
Salicylic acid	1.4–10.7		
Tetracycline	<LOQ –19.9		
Oxytetracycline	1.1 –16, maximum 212		
Trimethoprim	<LOQ –2.6		
Primidone	<LOQ –3.3		
Omeprazole	6.5–24.3	Ranges of concentration found in two agricultural soils irrigated with treated wastewater in Spain. Pollutants were found at higher concentrations in the surface layer of the soils.	[171]
Spironolactone	0.6		
Diazepam	4.65	Concentration found in an agricultural soil irrigated with treated wastewater. Pollutants were accumulated in the surface layer of the studied soil (0–30 cm).	[172]
Carbamazepine	5.77		
Butylbenzylphthalate	59–1580	Ranges of concentration of phthalate esters in agricultural soils irrigated with untreated wastewater in the peri–urban area of Guangzhou city	[173]
Di-2-(etylhexyl)phthalate	107–29370		
Di-*n*-butylphthalate	9–2740		
Di-*n*-amylphthalate	1–80		
Caffeine	14	Concentrations reported for volcanic soils (Vitric, Orthic, Allophanic soils) irrigated using treated wastewater for more than 15 years (at rates of 70 mm/weak) in Rotorua, New Zealand.	[174]
Amitriptyline	<5		
Carbamazepine	217		
Chlorpromazine	<5		
17α ethynilestradiol	<5		
Diltiazem	<248		
Thioridazine	<259		

<LOD: concentrations below the limit of detection of the analytical method used for determination.
<LOQ: below the limit of quantification of the analytical technique.

Table 10. Concentrations of organic pollutants of emerging concern in treated/untreated wastewater irrigated soils

OPECs in soil is expensive, these types of monitoring studies are conducted mainly in developed countries, where the use of biosolids as soil amendment is practiced more intensively than the reuse of treated/untreated wastewater in agricultural irrigation. At present there are no regulations that establish maximum permissible concentrations for organic pollutants of emerging concern in soils. The development of such regulations relies on the results obtained in both acute/chronic toxicity tests and in health risk assessments.

Environmental fate of organic pollutants of emerging concern (OPECs) in wastewater irrigated soils. The environmental fate of OPECs in the soil is governed by the physical and chemical properties of both the compounds and the soil as well as by the climatic conditions of the site where reuse is taking place. The chemical properties of the organic pollutants significantly impacting the environmental fate of OPECs are polarity, hydrophobicity and volatility. Table 11 shows some OPECs found in municipal wastewater which serve as examples of the differences in the chemical properties affecting the environmental fate of OPECs in soil. Due to the organic nature of OPECs, soil organic matter, mainly its non–polar fraction, plays a determinant role in the retention of these pollutants in soil [178]. However, sorption onto soil organic matter does not occur equally for all contaminants, since polar molecules tend to remain soluble in water rather than be retained in the soil organic matter; conversely, non–polar molecules are instantaneously adsorbed by soil organic matter [179]. The polarity of organic compounds is determined by the presence of ionizable radicals within the molecules; carboxyl, phenol, amine and amide moieties may gain or lose protons, depending on soil pH values, acquiring a positive or negative charge, respectively. The compounds for which functional groups lose protons may be poorly retained by soil due to repulsion forces between the deprotonated radical and the negatively charged soil particles (i.e. organic matter and clay); this results in the facilitated leaching of organic pollutants into the aquifer [180]. However, when functional groups within organic molecules gain positive charge, they may be retained onto the soil particles by cation exchange –as occurs for some tetracycline antibiotics– [181]. In both cases, the organic moiety within OPEC molecules may be held in the soil organic domain by hydrophobic affinity. In general, the pH of wastewater irrigated soils tends to be neutral to basic [81], which results in low retention of negatively charged compounds compared to neutral or positively charged organic compounds [180]. Studies referenced in [182] found that non–steroidal anti–inflammatory drugs (NSAIDs) such as naproxen, which can produce negatively charged molecules after the ionization of the carboxyl functional group, are adsorbed to a lower extent than other compounds displaying higher hydrophobicity, such as carbamazepine or triclosan, in organic soils with high clay content. Organic compounds lacking of ionizable functional groups or displaying non–ionizable functional groups express their hydrophobicity by spontaneously migrating from water to the soil organic domain [183]. In wastewater irrigation systems, dissolved and particulate organic matter contained in wastewater tends to accumulate in the surface soil horizons, significantly favoring the build up of these compounds in topsoil. In studies referred to in [184] a greater accumulation of hydrophobic compounds, such as carbamazepine and esters of phthalic acid, was found in surface horizons of the irrigated soils, whereas hydrophilic compounds, namely ibuprofen, naproxen and diclofenac, were found in subsurface horizons. This behavior is explained, on the one hand, because hydrophilic compounds remain dissolved in water rather

Polarity (ionization state at commonly found soil pH values)

Positive	Negative	Positive/Negative (zwitterions)
Erythromycin (antibiotic)	Naproxen (non–steroidal analgesic drug)	Ofloxacin (antibiotic)
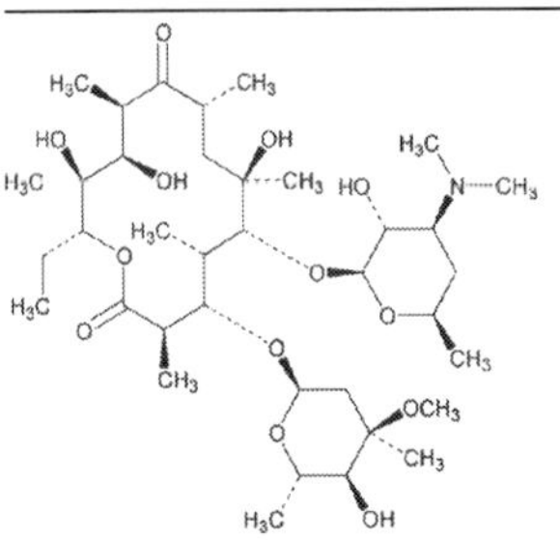	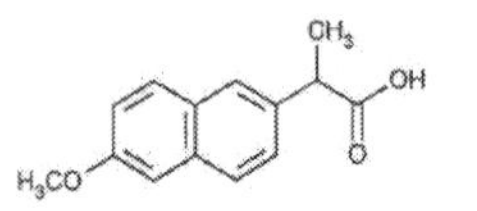	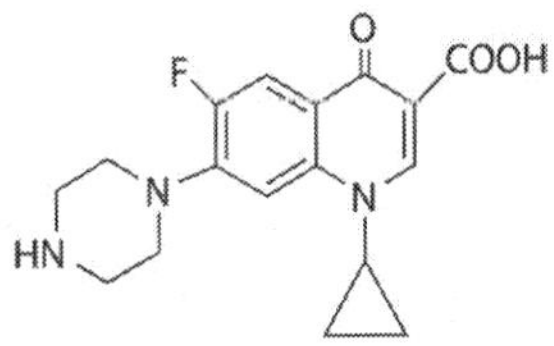
pK_a 8.91	pK_a 4.15	pK_a 6.27 (COOH); pK_a 8.87 (NH_2^+)

Hydrophobicity

Hydrophobic	Hydrophilic
Di–2–(ethylhexyl)phthalate (plasticizer)	Ciprofloxacin hydrochloride (antibiotic)
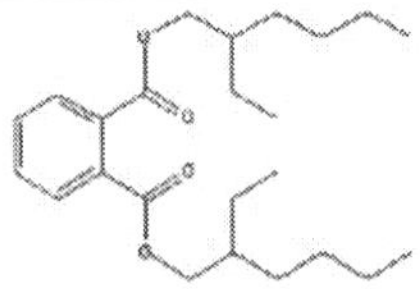	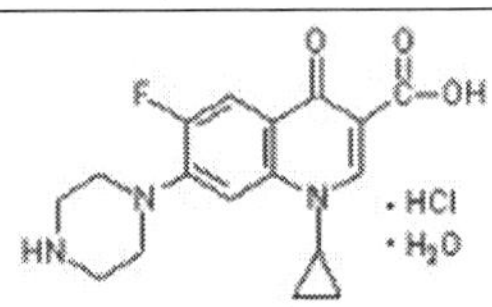
pK_{ow} 7.5	pK_{ow} – 0.82
Solubility in water at 25°C: 4.1x10^{-2} g/L	Solubility in water at 25°C: 30 g/L

Volatility

Volatile	Non–volatile
Galaxolide (fragrance used in detergents)	Bisphenol A (plasticizer precursor)
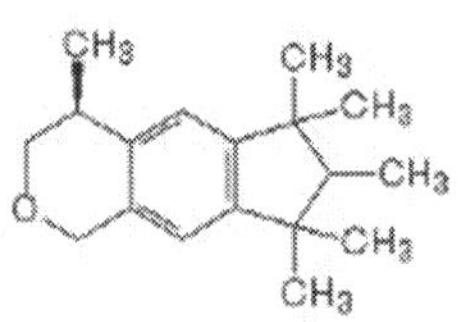	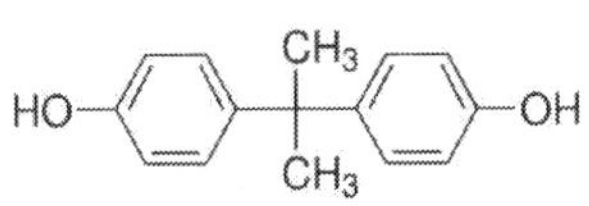
Vapor pressure: 7.27x10^{-2} Pa	Vapor pressure: 9.33x10^{-6} Pa

Table 11. Relevant physical and chemical properties in terms of the environmental fate of emerging pollutants in soil

than being retained in soil and, on the other hand, because of hydrophilic compounds are more susceptible to desorption from soil either during further irrigation or heavy rain events, and thus tend to rapidly reach subsoil and the aquifer [180, 182].

The chemical structure also affects the environmental fate of OPECs in soil. Molecules displaying aromatic moieties, such as carbamazepine and naproxen, have been shown to be

http://dx.doi.org/10.5772/57226

strongly retained by soil organic matter –both to the aliphatic and aromatic fractions of soil organic matter– compared with compounds that have no resonance structures [182, 185–186]. This behavior is explained by the formation of bonds between aromatic rings within the solute molecules and the soil organic matter [185]. Nonylphenols and octylphenol compounds have surfactant properties as they possess an aliphatic chain and a phenol moiety at the edge of the molecule [177]. Due to this structure, these compounds can promote resolubilization of organic contaminants retained in soil, although the estimated risk of this occurrence is considerably low [187]. The presence of heteroatoms in organic molecules can impact upon their environmental fate in soil. For example, oxygen atoms within the ciprofloxacin molecule can form covalent bonds with aluminum and iron oxides in soil, resulting in irreversible adsorption of the compound onto the solid matrix [188]. With respect to volatile OPECs, artificial fragrances represent the best example of this feature; these compounds are contained both in personal care products and detergents. Typically, the more volatile compounds are also hydrophobic, so they can be spontaneously retained in topsoil and then volatilize when temperature increases [189]. Since irrigation using untreated wastewater, which contains large amounts of fragrances, is carried out in arid areas, it is expected that a significant fraction or all of the fragrance molecules are rapidly volatilized upon their input to soil via wastewater. Volatilization of OPECs in wastewater irrigated soils is still an unexplored issue; studies aimed at determining the fraction of organic contaminants that can be volatilized in soil enriched with organic matter via wastewater irrigation are still needed.

Natural attenuation processes leading to the removal and dissipation of OPECs in soil are shown in Figure 6. Contaminants may either dissipate in soil by photodegradation, biodegradation or chemical degradation (hydrolysis, oxidation or reduction) mechanisms; they may be accumulated in soil by adsorption or removed from soil by volatilization. There is a significant lack of information in the literature with regard to the natural photodegradation (i.e. photolysis of compounds by sunlight) of emerging pollutants in soil. The information available on the photodegradation of pesticides in agricultural soils is useful in elucidating, to some extent, the potential for photodegradation of OPECs in soil. Studies on natural photodegradation of the pesticide quinalphos showed that photodegradation takes place only in the first 2–5 mm of soil (photic layer); this photolysis takes place in two stages, each one at a different depth [190]. In the uppermost soil layer (the first 2 mm) direct photodegradation of the organic compounds (i.e. the transformation of compounds due to the direct incidence of photons) occurs; in this same layer, the production of free radicals (e.g. hydroxyl radicals and excited dissolved organic matter) occurs due to the breakdown of the soil organic matter. In the second stage, the free radicals migrate to the lower photic layer through facilitated transport by soil moisture; subsequently, photolytic transformation occurs below by the action of free radicals generated in the upper layer of soil (indirect photodegradation) [191]. As a result of the aforementioned aspects, soil moisture content as well as organic matter are determinant factors in the photodegradation of organic contaminants retained in the soil surface [191]. The physical structure of the soil can also significantly impact upon the photodegradation of organic pollutants, as it defines the depth to which solar radiation can penetrate the soil. In the study referenced in [192], sunlight photolysis of 4–nonylphenol in biosolids amended soils was studied. Photolysis resulted in 40% conversion of the compound within 30 days, with photodegradation observed in the first 5 mm of the soil. Since natural photodegradation occurs only in the soil surface layers, the organic compounds retained in the topsoil will be the most exposed to direct sunlight, although this does not necessarily imply increased rates of photodegradation. An example of this is the anticonvulsant agent carbamazepine, which is prominently retained in topsoil but has demonstrated poor photodegradation in water studies. Conversely, the anti–inflammatory drug diclofenac has been shown to present significant photoactivity [193], but it is less well retained in the topsoil. Due to this, photodegradation is unlikely to occur in soil by direct or indirect means. In spite of almost all of the studies evaluating the natural photodegradation of OPECs have been carried out in aqueous matrices [194], the results obtained in these experiments provide valuable information concerning the photoactivity of such compounds; which can be useful for studying the photolysis of organic pollutants in soil. For example, it is known that the NSAID ibuprofen and the anticonvulsants drugs carbamazepine and primidone are poorly photodegraded in water whereas the antibacterial agent triclosan, the antibiotic drug sulfamethoxazole and the NSAIDs diclofenac and naproxen are readily photodegraded [194]. These results can be the basis to establish experiments aimed at determining or modeling the photodegradation of organics in soil. In general terms, the natural sunlight reaching the troposphere (i.e. the surface of earth) does not possess enough energy to mineralize most photodegradable compounds [195]; therefore a wide variety of by–products occurs when organic contaminants in water and soil are photodegraded. It is known that, in some cases, more harmful compounds can be produced by photodegradation of some organic pollutants. For example, 2,8–dichlorodibenzo–*p*–dioxin is produced by the natural photodegradation of the antibacterial agent triclosan [196]. Differently to triclosan, its breakdown product has the potential to cause cancer in mammals. Another example is the antiepileptic drug carbamazepine, which photodegrades to acridone [197], a compound related to the occurrence of cancer in aquatic species.

Most photodegradation studies of OPECs (in water matrices) have been carried out in developed countries at latitudes higher than 30°N [194]. It is therefore necessary to investigate the intensity of photodegradation processes occurring at lower latitudes, in zones where higher incidence of sunlight occurs and treated/ untreated wastewater irrigation is more intensively practiced.

Biodegradation of OPECs in soil has been studied in greater detail than photodegradation. Laboratory studies have found that biodegradation of emerging pollutants occurs optimally under aerobic conditions, while negligible transformations have been observed under anaerobic conditions [198]. This implies that biodegradation of this kind of contaminants is more likely to occur in well–structured soils, where tillage activities are frequently carried out, which allows better gas exchange through the soil matrix. The opposite behavior may be observed in anoxic/anaerobic soils, for instance in paddy fields. The antiepileptic drug carbamazepine is reported as one of the most refractory organic pollutants in soil, which has led researchers to consider this antiepileptic agent as a marker for anthropogenic contamination of surface and groundwater bodies [199]. In the study referenced in [200], mineralization of carbamazepine in soil was found to be less than 2%, after 120 days of incubation under aerobic conditions, while the reported in [201] show half–life times of 472 days in aerobic

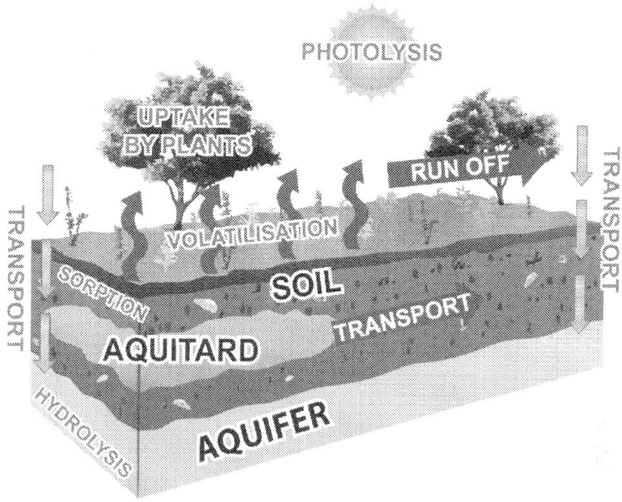

Figure 6. Processes involved in the environmental fate of emerging pollutants in soil

biosolids amended soils. Other compounds listed as recalcitrant in soil are the X–ray contrast media iopaminol, iomeprol and iohexol, whose biodegradation kinetic rate constants range from 0.29 to 0.46 μM/day [202]. Pharmaceuticals, such as the antiepileptic drug primidone and the psychoactive diazepam have shown recalcitrance in water [203]; further studies are necessary in order to elucidate whether such behavior may also occur in soil. Substances designed to exert an effect on microorganisms have been shown to be rapidly biodegraded in soil. Examples of these are antibacterial agents such as triclosan, triclocarban and antibiotic substances [204–205]. Triclosan and triclocarban have been shown to be biodegraded in aerobic soils after 18 and 108 days, respectively [198], whereas the antibiotic compounds erythromycin, oleandomycin, tylosin, tiamulin and salinomycin displayed half–lives in aerobic soil of 20, 27, 8, 16 and 5 days, respectively [205]. Endocrine disrupting compounds, such as phthalate esters have been shown to efficiently biodegrade in agricultural soils, displaying half–lives of 7.8 to 8.3 days for di-butyl phthalate and 26–30 days for di–2–(ethylhexyl)phthalate [206]. Currently, few soil microorganisms have been identified as degraders of emerging pollutants. For example, the fungi *Trametes versicolor* has been demonstrated to degrade naproxen [207], while *Rhodococcus rhodochorus* bacteria [208] have been shown to degrade carbamazepine down to levels of 15% of its initial concentration in soil. In the case of phthalates (plasticizers) bacteria belonging to groups of *Corynebacterium*, *Mycobacterium* and *Nocardia* were demonstrated to degrade up to 90% of di–butyl phthalate within 48 hours in biodegradation experiments using isolated bacteria cultivated in saline solution [209]. Knowing the species of microorganisms that perform the biodegradation of OPECs in soil is useful in order to design engineered systems to treat wastewater and polluted soils based on the increased ability of degraders to degrade specific compounds by acclimatization and bioaugmentation. Such systems were tested in [210] using the fungus *Trametes versicolor* to degrade up to 94% of carbamazepine in wastewater after 6 days in an air pulsed bed bioreactor. Biodegradation of OPECs in soil is influenced by the sorption phenomenon, therefore soil characteristics such as the content of organic matter, soil texture and soil pH are crucial for this process to occur. Adsorption of the organic contaminants onto the surface of the soil particles may favor biodegradation when the sorbed compounds are still bioavailable; conversely, when strong adsorption occurs (chemisorption on soil organic matter, clay or soil micropores) it can result in decreased bioavailability of the compounds and thus in the confinement of the pollutants within the soil matrix. Other properties of soil involved in the biodegradation of OPECs are: a) the climatic conditions of the site; b) the physical structure of soil; c) the soil moisture; and, d) the adaptation of soil organisms to biodegrade the target pollutants. It is possible that microorganisms in long–term wastewater irrigated soils more efficiently biodegrade OPECs than those living in non–irrigated soils or soils irrigated for a shorter time. This is due to the ability of soil microorganisms to adapt to using emerging pollutants as a carbon source. In this sense, studies comparing the degradation efficiency of OPECs in long–term wastewater irrigated soils with that observed in non–irrigated soils or newly irrigated soils are needed in order to establish appropriate strategies to prevent contamination of groundwater. Very few efforts have been made to determine the nature and quantity of by–products generated in soil by the biodegradation of OPECs. As shown in [200], biodegradation of emerging pollutants can generate by–products that can be more harmful than the original substance [196–197], thus the presence of these by–products as well as their environmental fate should be priority for further research.

Those emerging pollutants that are not degraded by soil microorganisms may either accumulate in soil, be assimilated by plants (if they are bioavailable) or be degraded by other mechanisms (e.g. photodegradation or hydrolysis). In the case of carbamazepine, studies referred to in [184] explain that this compound is one of the most highly accumulated in wastewater irrigated soils. Moreover, carbamazepine can be assimilated by plants in wastewater irrigation systems at environmentally relevant concentrations (i.e. within the range 1–3 μg/L). The study referenced in [211] shows that cucumber (*Cucumis sativus L.*) can accumulate carbamazepine in different parts of the plant: 4.5 μg/kg in roots, 1.9 μg/kg in stem, 39.9 μg/kg in leaves and 2.1 μg/kg in fruit. According to the authors, phytotoxic effects were observed when carbamazepine was supplied to soil by irrigation at concentrations as high as 10,000 μg/L. Results of this study show that consumption of carbamazepine polluted cucumber results in doses of 1 ng of carbamazepine per gram of fruit. Other studies show that soybean (*Glycine max L.*) can take up carbamazepine, triclosan and triclocarban in roots, stems and leaves at concentrations of 1.3–3.4 μg/kg for carbamazepine and 2.4–13.7 μg/kg for the antibacterial agents triclosan and triclocarban. Concentrations of antibacterial agents in plants at a second harvesting were found to be higher than those obtained in the first one; this may be due to the accumulation of contaminants in the soil, as a bioavailable pool, between each irrigation events [212]. To date, the study of the assimilation of OPECs by plants in wastewater irrigated soils is still limited; moreover, priority should be given to develop health risk assessment studies related to the consumption of contaminated crops.

Adsorption (i.e. retention of solutes on the surface of the soil particles) of OPECs in the soil is a decisive process in their environmental fate, since through this process contaminants may either be retained or migrate into the aquifer. In cases where organic pollutants are retained in topsoil, photodegradation or volatilization phenomena can easily take place. The strength of the bonds that pollutants establish with soil particles determines the bioavailability of molecules to plants and soil microorganisms. Adsorption of pollutants onto soil is measured by the distribution coefficient (K_d) which relates the amount of compound retained by soil to the mass remaining in the liquid phase [213]. Several models to determine the distribution coefficient of organic compounds have been developed; such models vary in complexity and the accuracy with which they represent the field conditions; yet simple adsorption models such as linear, Langmuir and Freundlich are the most used [213]. Due to their organic nature, OPECs tend to be rapidly and strongly adsorbed by soil organic matter; due to this effect, non–polar emerging pollutants, such as phthalates, have been shown to instantly adsorb onto organic soils [214]. On the other hand, OPECs displaying negative charge at the soil pH values, as occurs for NSAIDs, exhibit less adsorption by soil due to the repulsive forces between the negatively charged moiety within the molecule and the soil particles displaying negative charges (i.e. organic matter and clay) [182]. Accumulation of organic matter in wastewater irrigated soils increases the soil's ability to adsorb organic compounds. The proof of this can be found in the study referenced in [96], which reports greater adsorption of the antibiotics sulfamethoxazole and ciprofloxacin in long–term wastewater irrigated soils compared to non–irrigated ones from the same area. In addition to soil organic matter, OPECs may be retained by the inorganic domain of soil; for instance, ciprofloxacin showed strong and instantaneous adsorption by iron oxides and clay in agricultural soils, which was achieved by the formation of covalent bonds between metals in the soil and the oxygen atoms within ciprofloxacin molecules [188]. Furthermore, adsorption of carbamazepine by smectite type clays has been reported by [215]. According to studies referred to in [182, 185, 215], the adsorption of OPECs with multiple aromatic rings is more efficient in soils displaying a high content of humified organic matter –which displays higher aromaticity than labile organic matter–. Polyaromatic compounds can establish π–π bonds between the aromatic rings within the pollutant molecules and aromatic compounds contained in soil organic matter. The formation of such bonds should be studied in future research in order to determine the optimum chemical characteristics of soil organic matter which enable better retention of contaminants, hence preventing their mobilization into the aquifer and/or making them available for uptake by plants. OPECs may be adsorbed by dissolved organic matter to soil via wastewater. Adsorption of organic pollutants to dissolved organic matter increases the solubility of the compounds and hence facilitates the lixiviation through soil. Studies referenced in [216–217] report that compounds such as naproxen, carbamazepine and sex hormones can be adsorbed onto dissolved organic matter, notably to the hydrophobic and neutral hydrophilic fractions of dissolved organic matter. The speed of formation and strength of bonds between organic compounds and the dissolved organic matter varies depending on the quality of both wastewater and dissolved organic matter in soil [217]. The continuous occurrence of OPECs in wastewater irrigated soils can impact upon the adsorption of other organic pollutants; this is because at the time emerging pollutants enter to soil via wastewater, some of the active adsorption sites in soil are still occupied by previously adsorbed pollutants. In the study referred to in [182], the distribution coefficients of three OPECs, namely naproxen, carbamazepine and triclosan, were determined by an adsorption model which takes into account the previous presence of organic pollutants in the soil (the initial mass model [218]). The authors found modest differences between the values obtained in their study and those reported in the literature. However, it was observed that compounds previously adsorbed onto soil, i.e. naproxen and carbamazepine, were released from the solid matrix each time wastewater "washes" the soil in each irrigation event, resulting in a risk of contamination of the aquifer.

The transportation of OPECs through soil is closely related to their adsorption onto the solid matrix. Transport studies can be performed using different approaches, either packed soil columns or undisturbed soil columns tests. Transport of OPECs and pathogens is better described using the undisturbed soil column approach; through this approach, it is possible to evaluate the impact of both physical and chemical properties of soil on the transport of pollutants. In transport assays using undisturbed soil columns it is possible to assess the impact of preferential paths on the transport of solutes and particles, at the same time determining the effect of chemical properties of soil in the retention of solutes under dynamic flow conditions. The type of clays in soil significantly impacts on the transport of organic pollutants. The presence of expansive clays in soil results in the disappearance of preferential paths in the porous network of soil once clay becomes wet, which in turn provokes the decay in transport of contaminants contained in water. However, in such cases, soil conditions become anaerobic and thus organic pollutants are biodegraded with difficulty. The understanding of the environmental fate of OPECs in wastewater irrigated soil still has many gaps. It is therefore important to carry out studies on the laboratory scale and then in the field (plot level or landscape level) in order to determine the fate of these substances under real conditions. Results of these studies are of great importance, on the one hand, to allow more accurate and useful risk assessment studies and, on the other hand, to determine the characteristics of the sites suitable for irrigation with treated/untreated wastewater without posing a risk to the health of organisms and to the quality of crops and water sources. Lastly, regulations for OPECs in soil should be established in order to set maximum concentration limits for the accumulation of these compounds in terms not only of the effects caused to soil organisms, but also their potential to reach groundwater.

3. Perspectives for further studies

Reuse of wastewater in agricultural irrigation is a complex issue that requires the development of numerous studies in different disciplines; in this section some perspectives for further studies are presented.

1. Long–term studies aimed at determining the improvement of soil properties to produce food. Such studies should compare the rate of entry and conversion of carbon, nitrogen and phosphorus in irrigated soils in order to obtain a mass balance showing either the sustainability or the accumulation of organic matter in wastewater irrigated soils.

Moreover, studies demonstrating the long–term increase in the soil's ability to treat wastewater used for irrigation should be carried out for each of the properties addressed in this chapter, as well as those considered appropriate in each system.

2. The determination of OPECs and pathogens in soils irrigated with wastewater. Such monitoring studies can be used to establish an inventory of contaminated sites that reflects the level of pollution in developed and developing countries. This can help in proposing *ad hoc* solutions for each site.

3. Determining feasibility and the mechanisms that can lead to horizontal propagation of antibiotic resistance genes in soil microorganisms (either innocuous microorganisms or opportunistic pathogens).

4. Chronic toxicity studies of OPECs in wastewater irrigated soils covering either several crop cycles, several generations of organisms or several years. Toxicity studies should address the effects of the presence of mixed contaminants at trace levels (environmentally relevant concentrations) on soil organisms. Such studies should be conducted including new emerging contaminants, e.g. nanoparticles.

5. The study of the environmental fate of emerging contaminants using different model molecules in soil. Such environmental fate studies should be carried out at laboratory and field scale. In the case of environmental fate studies at laboratory scale, conditions used should be those that best emulate field conditions, e.g. sunlight lamp intensities similar to those observed in the field for testing photodegradation or undisturbed soil columns in transport assays through soil.

6. The determination and quantification of the by–products appearing in soil upon dissipation of OPEC. Harmful compounds such as dioxins, chlorophenols and polyaromatics may be produced in soil from substances such as triclosan and carbamazepine. Discerning the occurrence and fate of these substances in soil should be addressed in future studies

7. Determination of the environmental fate of organic, inorganic and microbial contaminants in agricultural soil remaining after irrigation with wastewater has ceased. Worldwide, notably in developed countries, there are several sites where irrigation with wastewater has been stopped after a considerable time; in such cases, it is necessary to know the fate of the pool of pollutants that accumulated in soil during continuous input via wastewater. Phenomena such as the release of heavy metals confined in soil organic matter can occur when soil organisms start to mineralize organic carbon accumulated in the soil. In addition, the soil microorganisms can lose the capacity to treat pollutants in wastewater, leaving the soil vulnerable in cases where wastewater irrigation is restarted.

8. Studies elucidating the conditioning methods for agricultural soils newly irrigated with wastewater. Since in arid regions a considerable increase in the area under irrigation is being observed, it is necessary to use current knowledge to implement regulations establishing the optimal conditions for soils candidate to receive treated/untreated wastewater. These are necessary to prevent soil degradation and contamination of water sources in the irrigated area.

9. Studies on the migration of contaminants in soil due to extreme events caused by climate change. Extreme rainfall events can cause an incremental increase in the mobilization of organic contaminants retained in the surface layers of soil into aquifers or to non–irrigated soils affected by runoff. However, increases in temperature can decrease the biodegradation of organic pollutants in the soil due to excessive drying of the solid matrix.

10. The development and implementation of wastewater treatment systems to remove organic, inorganic and biological pollution without reducing the content of organic matter in the water. These systems must be inexpensive for dissemination in developing countries. Advanced primary treatment systems may represent a plausible strategy in such cases.

11. The development and validation of environmentally–friendly analytical techniques for the determination of OPECs in soils..

4. Conclusions

The reuse of treated/untreated municipal wastewater for agricultural irrigation definitely has positive impacts on soil as a medium for the development of plants and animals; additionally, this practice results in positive impacts on the welfare of farmers due to the monetary savings and profits that they obtain by the use of wastewater as a fertilizer and water source for crops. Similarly, the soil's ability to self–cleanse and treat the wastewater supplied at each irrigation event increases with the reuse of wastewater. The accumulation of organic matter in the soil surface results in changes in soil pH to neutral and basic values, an improvement of soil physical structure and an increase in the soil microbial activity. Together with this, soil organisms become acclimatized to the presence of contaminants and thus their resilience to the harmful effects caused by pollutants increase. These phenomena lead to an improvement in the ability of the soil to act as a filter and transforming medium for contaminants and thereby to an increase in its capacity to treat wastewater. Such an improvement in soil functions can be capitalized by the State and the conventional treatment regime can be changed to a cheaper one driven by natural attenuation mechanisms. This in turn improves the quality of life of people living in the area by increasing food production and the possibility of obtaining profit by sales of produce. The responsible reuse of municipal wastewater for agricultural irrigation can help to mitigate three problems which are a priority in developing countries: a) water stress in arid areas where rain–fed agriculture makes development uncertain. In such areas freshwater sources are used for agriculture rather than human consumption, and therefore the reuse of municipal wastewater not only results in savings of freshwater but also in the recharge of the aquifer in the irrigated area. Recharge is with good quality water produced by infiltration of wastewater through the soil; b) the food crisis and the lack of jobs in rural and peri–urban areas in developing countries. Reuse of wastewater represents a way of producing food for consumption and sale; and, c) the treatment of municipal wastewater generated in urban and rural areas through a low cost natural treatment systems which in turn generate profits for population.

http://dx.doi.org/10.5772/57226

In order to reuse wastewater responsibly and exploit its inherent benefits for soil and people living in the irrigated area, the occurrence of contaminants in wastewater –especially untreated wastewater– must be kept in mind. The presence of pathogenic microorganisms and the potential for antibiotic resistance dissipation via wastewater should be priority concerns in designing wastewater reuse schemes in agricultural areas, notably when using raw wastewater. Attention should be paid to the fate of emerging contaminants in wastewater irrigation schemes including its transportation through irrigation canals, storage in dams and deposition in agricultural soils and transport to aquifers. Another priority is the elucidation of the chronic toxic effects caused by the continuous presence of traces of emerging contaminants in irrigated soils. Since the group of OPECs is quite broad, model compounds should be selected to determine the rate at which they are dissipated or retained/transported through soil, as well the risk of these compounds reaching the aquifer or being assimilated by plants. Despite the spread of antibiotic resistance in the environment it has not been conclusively shown the role that irrigation with treated/untreated wastewater plays in this. To date, the concentrations of OPECs found in soil irrigated with wastewater are lower than the toxicity thresholds reported in literature. The precautionary principle states that wastewater must be minimally treated before irrigation in order to remove pathogenic microorganisms and trace of heavy metals, as well as to reduce as much as possible the concentration of emerging pollutants. Other areas of opportunity to be developed in order to reduce the risk of soil degradation and effects on soil organisms are: a) the development of environmentally friendly everyday–consumer products, containing organic compounds that have been proven to have no harmful effects on living organisms even at trace concentrations. Consumer products must follow strict risk assessments before release to the market; b) an improvement in health systems in cities in order to reduce the incidence of infectious diseases that ultimately generate biological contamination of soil, especially in irrigation systems using raw wastewater; c) the maintenance of wastewater irrigation schemes fed with municipal wastewater in order to avoid a high input of heavy metals and refractory organic compounds to soil and crops through irrigation; and, d) the *ad hoc* treatment of municipal wastewater to allow its reuse in agricultural activities. Low cost treatment systems aimed at removing microorganisms, suspended solids and trace heavy metals are recommended to treat wastewater without affecting its properties as a fertilizer and source of organic matter to improve physical, chemical and microbiological soil properties. Such an approach allows soil to fulfill its ecological functions as a generator of food and livelihoods and as a protective barrier to the aquifer.

Author details

Juan C. Durán–Álvarez[1] and Blanca Jiménez–Cisneros[2]

1 Centre of Applied Science and Technological Development, National Autonomous University of Mexico, Mexico D.F., Mexico

2 International Hydrological Programme (IHP), UNESCO, Paris, France

References

[1] Elimelech M. The Global Challenge for Adequate and Safe Water. Aqua – Journal of Water Supply: Research and Technology 2006;55(1) 3–10.

[2] Rijsberman F.R. Water Scarcity: Fact or Fiction. Agricultural Water Management 2006;80(1–3) 5–22.

[3] United States Geological Survey. USGS: The USGS Water Science School. The World's Water – Distribution of Earth's Water. http://ga.water.usgs.gov/edu/earthwherewater.html (accessed on July 13 2013).

[4] Shiklomanov I.A. World Water Resources – A New Appraisal and Assessment for the 21st Century. St Petersburg: UNESCO; 1998.

[5] Food and Agriculture Organization of the United Nations, FAO. Coping with Water Scarcity: An Action Framework for Agriculture and Food Security. Rome: FAO; 2012.

[6] Earth Trends: Environmental Information. http://www.wri.org/project/earthtrends/ (accessed on August 4th 2013).

[7] Pereira L.S., Cordery I., Iacovides I. Coping with Water Scarcity: Addressing the Challenges. Paris: UNESCO; 2009.

[8] DeFeo G., Mays L.W., Angelakis A.N. Water and Wastewater Management Technologies in the Ancient Greece and Roman Civilizations. In: Wilder P. (ed.) Treatise in Water Science. Oxford: Elsevier; 2011. p3–22.

[9] Lazarova V., Levine B., Sack J., Cirelli G., Jeffrey P., Muntau H., Salgot M., Brissaud F. Role of Water Reuse for Enhancing Integrated Water Management in Europe and Mediterranean Countries. Water Science and Technology 2001;43(10) 25–33.

[10] Anderson J. The Environmental Benefits of Water Recycling and Reuse. Water Science and Technology: Water Supply 2003;3(4) 1–10.

[11] Asano T. Wastewater Reclamation and Reuse: Water Quality Management Library. Pennsylvania: Technomic Publishing Company; 1998.

[12] Hussain I., Raschid–Sally L., Hanjra M.A., Marikar F., van der Hoek W. Wastewater Use in Agriculture: Review of Impacts and Methodological Issues in Valuing Impacts. Colombo: International Water Management Institute, IWMI; 2002.

[13] Kretschmer N., Ribbe L., Gaese H. Wastewater Reuse for Agriculture. Technology Resource Management and Development – Water Management 2002;2(1) 35–61.

[14] World Health Organization, WHO. Health Guidelines for the Use of Wastewater and Excreta in Agriculture. Geneve: WHO; 1989.

[15] Shelef G., Azov Y. The Coming Era of Intensive Wastewater Reuse in the Mediterranean Region. Water Science and Technology 1996;33(10–11) 115–125.

[16] Jiménez B. Wastewater Reuse to Increase Soil Productivity. Water Science and Technology 1995;32(12) 173–180.

[17] Jimenez B., Asano T. Water Reclamation and Reuse Around the World. In: Jimenez B., Asano T. (eds.) Water Reuse – An international Survey of Current Practice, Issues and Needs. London: IWA Publishing; 2008. p3–27.

[18] World Health Organization, WHO. Guidelines for the Safe Use of Wastewater, Excreta and Greywater. Geneva: WHO; 2006.

[19] Ensink J.H.J., Simmons R.W., van der Hoek W. Wastewater Use in Pakistan: The Cases of Haroonabad and Faisalabad. In: Scott C.A., Faruqui N.I., Raschid–Sally L. (eds.) Wastewater Use in Irrigated Agriculture: Confronting the Livelihood and Environmental Realities. Wallingford: CAB International; 2004. p91–99.

[20] Toze S. Reuse of Effluent Water – Benefits and Risks. Agricultural Water Management 2006;80(1–3) 147–159.

[21] Keraita B., Jiménez B., Drechsel P. Extent and Implications of Agricultural Reuse of Untreated, Partially Treated and Diluted Wastewater in Developing Countries. CAB reviews: Perspectives in Agriculture, Veterinary Science, Nutrition and Natural Resources 2008;3(58) 1–15.

[22] Cifuentes E., Gomez M., Blumenthal U., Tellez–Rojo M.M., Romieu I., Ruiz–Palacios G., Ruiz–Velazco S. Risk Factors for Giardia Intestinalis Infection in Agricultural Villages Practicing Wastewater Irrigation in Mexico. American Journal of Tropical Medicine and Hygiene 2000;62(3) 388–392.

[23] Mara D.D., Sleigh P.A., Blumenthal U.J., Carr R.M. Health Risks in Wastewater Irrigation: Comparing Estimates from Quantitative Microbial Risk Analyses and Epidemiological Studies. Journal of Water and Health 2007;5(1) 39–50.

[24] Baquero F., Martínez J.L., Cantón R. Antibiotics and Antibiotics Resistance in Water Environments. Current Opinion in Biotechnology 2008;19(3) 260–265.

[25] Knapp C.W., Dolfing J., Ehlert P.A.I., Graham D.W. Evidence of Increasing Antibiotic Resistance Gene Abundances in Archived Soils Since 1940. Environmental Science and Technology 2010;44(2) 580–57.

[26] Beausse J. Selected Drugs in Solid Matrices: A Review of Environmental Determination, Occurrence and Properties of Principal Substances. Trends in Analytical Chemistry 2004;23(10–11) 753–761.

[27] Schmitt H., Römbke J. The Ecotoxicological Effects of Pharmaceuticals (Antibiotics and Antiparasiticides) in the Terrestrial Environment – A Review. In: Kúmmerer K. (ed.) Pharmaceuticals in the Environment: Sources, Fate, Effects and Risks. Heidelberg: Springer; 2008. p285–303.

[28] Colborn T., vom Saal F.S., Soto A.M. Developmental Effects of Endocrine–Disrupting Chemicals in Wildlife and Humans. Environmental Health Perspectives 1993;101(5) 378–384.

[29] Daughton C.G. Non–Regulated Water Contaminants: Emerging Research. Environmental Impact Assessment Review 2004;24(7–8) 711–732.

[30] Foppen J.W., Schijven J.F. Evaluation Data from the Literature on the Transport and Survival of Escherichia Coli and Thermotolerant Coliforms in Aquifers Under Saturated Conditions. Water Research 2006;40(3) 401–426.

[31] Read D.S., Sheppard S.K., Bruford M.W., Glen D.M., Symondson W.O.C. Molecular Detection and Predation by Soil–Arthropods on Nematodes. Molecular Ecology 2006;15(7) 1963–1972.

[32] Berg G., Eberl L., Hartmann A. The Rhizosphere as a Reservoir of Opportunistic Human Pathogen Bacteria. Environmental Microbiology 2005;7(11) 1673–1685.

[33] Yamamoto H., Nakamura Y., Morigushi S., Nakamura Y., Honda Yuta., Tamura I., Hirata Y., Hayashi A., Sekizawa J. Persistence and Partitioning of Eight Selected Pharmaceuticals in the Aquatic Environment: Laboratory Photolysis, Biodegradation and Sorption Experiments. Water Research 2009;43(2) 351–362.

[34] Fausto Cereti C., Rossini F., Federici F., Quaratino D., Vassilev N., Fenice M. Reuse of Microbially Treated Olive Mill Wastewater as Fertiliser for Wheat (Triticum Durum). Bioresource Technology 2004;91(2) 135–140.

[35] Bradford A., Brook R., Hunshal C. Wastewater irrigation: Hubli–Dharwad, India. International Symposium on Water, Poverty and Productive uses of Water at the Household Level, 21–23 January 2003, Muldersdrift, South Africa.

[36] Van der Hoek W., Ul–Hassan M., Ensink J.H.J., Feenstra S., Raschid–Sally L., Munir S., Aslam R., Ali N., Hussain R., Matsuno Y. Urban Wastewater: A Valuable Resource for Agriculture. Colombo: International Water Management Institute Research; 2002.

[37] Gaye M., Niang S. Epuration des eaux usées et l'agriculture urbaine – Etudes et Recherches. Dakar: ENDA–TM; 2002.

[38] Ramírez–Fuentes E., Lucho–Constantino C., Escamilla–Silva E., Dendooven L. Characteristics and Carbon and Nitrogen Dynamics in Soil Irrigated with Wastewater for Different Lengths of Time. Bioresource Technology 2002;85(2) 179–187.

[39] Fan A.M., Steinberg V.E. Health Implications of Nitrate and Nitrite in Drinking Water: an Update on Methemoglobinemia Occurrence and Reproductive and Developmental Toxicity. Regulatory toxicology and pharmacology 1996;23(1) 35–43.

[40] Girovich M. Biosolids Treatment and Management: Processes for Beneficial Use. New York: Marcel Dekker Inc; 1996.

[41] Degens B., Schipper L., Claydon J., Russell J., Yeates G. Irrigation of an Allophanic Soil with Dairy Factory Effluent for 22 Years: Responses of Nutrient Storage and Soil Biota. Australian Journal of Soil Research. 2000;38(1) 25–35.

[42] Mikkelsen R., Camberato J. Potassium, Sulfur, Lime and Micronutrient Fertilizers. In: Rechcigl J. E. (ed.) Soil Amendments and Environmental Quality. Boca Raton: CRC Press; 1995. p.109–137.

[43] Sala L., Serra M. Towards Sustainability in Water Recycling. Water Science and Technology 2004;50(2) 1–8.

[44] Steen I. Phosphorus Availability in the 21st Century: Management of a Non–Renewable Resource. Phosphorus and Potassium 1998;217(1) 25–31.

[45] Jiménez B., Chávez A. Treatment of Mexico City Wastewater for Irrigation Purposes. Environmental Technology 1997;18(7) 721–729.

[46] Coleman D.C., Crossley J.D.A., Hendrix P.F. Fundamentals of soil ecology. London: Academic Press; 2004.

[47] Lal R. Soil Structure and Sustainability. Journal of Sustainable Agriculture 1991;1(4) 67–92.

[48] Pardo A., Amato M., Quaglietta Chiaranda F. Relationships between Soil Structure, Root Distribution and Water Uptake of Chickpea (Cicer arietinum L.). Plant Growth and Water Distribution. European Journal of Agronomy 2000;13(1) 39–45.

[49] Boix–Fayos C., Calvo–Cases A., Imeson A.C., Soriano–Soto M.D., Tiemessen I.R. Spatial and Short Term Variations in Runoff, Soil Aggregation and Other Soil Properties Along a Mediterranean Climatological Gradient. Catena 1998;33(2) 123–138.

[50] Edwards A.P. Bremner J.M. Microaggregates in Soil. European Journal of Soil Science 1967;18(1) 64–73.

[51] Bronick C.J., Lal R. Soil Structure and Management. Geoderma 2005;124(1–2) 3–22.

[52] Martens D.A., Frankenberger W.T. Modification of Infiltration Rates in an Organic–Amended Irrigated. Agronomy Journal 1992;84(4) 707–717.

[53] Armstrong A.S.B., Tanton T.W. Gypsum Application to Aggregated Saline–Sodic Clay Topsoils. Journal of Soil Science 1992;43(2) 249–260.

[54] Jastrow J.D. Soil Aggregate Formation and the Accrual of Particulate and Mineral–Associated Organic Matter. Soil Biology and Biochemistry 1996;28(4–5) 665–676.

[55] Haynes R.J., Naidu R. Influence of Lime, Fertilizer and Manure Applications on Soil Organic Matter Content and Soil Physical Structure: A Review. Nutrient Cycling in Agroecosystems 1998;51(2) 123–137.

[56] Czarnes S., Hallett, P.D., Bengough A.G., Young I.M. Root– and Microbial–Derived Mucilages Affect Soil Structure and Water Transport. European Journal of Soil Science 2000;51(3) 435–443.

[57] Haynes R.J., Beare M.H. Influence of Six Crops Species on Aggregate Stability and Some Labile Organic Matter Fractions. Soil Biology and Biochemistry 1997;29(11–12) 1647–1653.

[58] Raimbault B.A., Vyn T.J. Crop Rotation and Tillage Effects on Corn Growth and Soil Structural Stability. Agronomy Journal 1991;83(6) 979–985.

[59] Brunetti G., Senesi N., Plaza C. Effects on Amendments with Treated and Untreated Olive Oil Mill Wastewaters on Soil Properties, Soil Humic Substances and Wheat Yield. Geoderma 2007;138(1–2) 144–152.

[60] Martens D.A. Plant Residue Biochemistry Regulates Soil Carbon Cycling and Carbon Sequestration. Soil Biology and Biochemistry 2000;32(3) 361–369.

[61] Cox L., Celis R., Hermosin M.C., Becker A., Cornejo J. Porosity and Herbicide Leaching in Soils Amended with Olive–Mill Wastewater. Agriculture, Ecosystem and Environment 1997;65(2) 151–161.

[62] Coppola A., Santini A., Botti P., Vacca S., Comegna V., Severino G. Methodological Approach for Evaluating the Response of Soil Hydrological Behavior to Irrigation with Treated Municipal Wastewater. Journal of Hydrology 2004;292(1–4) 114–134.

[63] Wang Z., Chang A.C., Wu L., Crowley D. Assessing the Soil Quality of Long–Term Reclaimed Wastewater–Irrigated Cropland. Geoderma 2003;114(3–4) 261–278.

[64] Tam N.F.Y. Effects of Wastewater Discharge on Microbial Populations and Enzyme Activities in Mangrove Soils. Environmental Pollution 1998;102(2–3) 233.242.

[65] Friedel J.K., Langer T., Siebe C., Stahr K. Effects of Long–Term Waste Water Irrigation on Soil Organic Matter, Soil Microbial Biomass and Its Activities in Central Mexico. Biology and Fertility of Soils 2000;31(5) 414–421.

[66] Filip Z., Kanazawa S., Berthelin J. Characterization of Effects of a Long–Term Wastewater Irrigation on Soil Quality by Microbiological and Biochemical Parameters. Journal of Plant Nutrition and Soil Science 1999;162(4) 409–413.

[67] Kay B.D. Soil Structure and Organic Carbon: A Review. In: Lal R., Kimble J.M., Follett R.F., Stewart B.A. (eds.) Soil Processes and the Carbon Cycle. Boca Raton: CRC Press; 1998. p169–197.

[68] Androver M., Farrus E., Moya G., Vadell J. Chemical Properties and Biological Activity in Soils of Mallorca Following Twenty Years of Treated Wastewater Irrigation. Journal of Environmental Management 2012;95 188–192.

[69] Ansari M.I., Malik A. Biosoption of Nickel and Cadmium by Metal Resistant Bacterial Isolates from Agricultural Soil Irrigated with Industrial Wastewater. Bioresource Technology 2007;98(16) 3149–3153.

[70] Kim D.Y., Burger J.A. Nitrogen Transformations and Soil Processes in a Wastewater–Irrigated, Mature Appalachian Hardwood Forest. Forest Ecology and Management 1997;90(1) 1–11.

[71] Russell J.M., Cooper R.N., Lindsey S.B. Soil Denitrification Rates at Wastewater Irrigation Sites Receiving Primary–Treated and Anaerobically Treated Meat–Processing Effluent. Bioresource Technology 1993;43(1) 41–46.

[72] Bouwer H. Groundwater Recharge with Sewage Effluent. Water Science and Technology 1991; 23(10–12) 2099–2108.

[73] Bouwer H. Soil–Aquifer Treatment of Sewage. Rome: Food and Agriculture Organization of the United Nations; 1987.

[74] Quanrud D.M., Hafer J., Karpiscak M.M., Zhang J., Lansey K.E., Arnold R.G. Fate of Organics During Soil–Aquifer Treatment: Sustainability of Removals in the Field. Water Research 2003;37(14) 3401–3411.

[75] Pescod M. Wastewater Treatment and Use in Agriculture, FAO irrigation and drainage Paper 47. Rome: Food and Agriculture Organization; 1992.

[76] Ayres D., Westtcott W. Water Quality for Agriculture, FAO Paper 29. Rome: Food and Agricultural Organization; 1985.

[77] Arora M., Kiran B., Rani S., Rani A., Kaur B., Mittal N. Heavy Metal Accumulation in Vegetables Irrigated with Water from Different Sources. Food Chemistry 2008;111(4) 811–815.

[78] Idelovitch E., Michail M. Soil–Aquifer Treatment: A New Approach to an Old Method for Wastewater Reuse. Journal (Water Pollution Control Federation) 1984;56(8) 936–943.

[79] Lin C., Shacahr Y., Banin A. Lin, C. Heavy Metal Retention and Partitioning in a Large–Scale Soil–Aquifer Treatment (SAT) System Used for Wastewater Reclamation. Chemosphere 2004;57(9) 1047–1058.

[80] Gadd G.M. Microbial Influence on Metal Mobility and Application for Bioremediation. Geoderma 2004;122(2–4) 109–119.

[81] Kunhikrishnan A., Bolan N.S., Müller K., Laurenson S., Naidu R., Kim W.I. The Influence of Wastewater Irrigation on the Transformation and Bioavailability of Heavy Metal(loid)s in Soil. In: Sparks D.L. (ed.) Advances in Agronomy, Vol. 115. London: Academic Press; 2013. p.219–273.

[82] Aleem A., Isar J., Malik A. Impact of Long–Term Application of Industrial Wastewater on the Emergence of Resistance Traits in Azotobacter chroococcum Isolated from Rhizospheric Soil. Bioresource Technology 2003;86(1) 7–13.

[83] Altaf M.M., Masood F., Malik A. Impact of Long–Term Application of Treated Tannery Effluents on the Emergence of Resistance Traits in Rhizobium sp. Isolated from Trifolium Alexandrinum. Turkish Journal of Biology 2008;32(1) 1–8.

[84] Müller K., Duwig C., Prado B., Siebe C., Hidalgo C., Etchevers J. Impact of Long–Term Wastewater Irrigation on Sorption and Transport of Atrazine in Mexican Agricultural Soils. Journal of Environmental Science and Health, Part B 2012;47(1) 30–41.

[85] von Oepen B., Kördel W., Klein W. Sorption of Nonpolar and Polar Compounds to Soils: Processes, Measurements and Experience with the Applicability of the Modified OECD–Guideline 106. Chemosphere 1991;22(3) 285–304.

[86] Müller K., Magesan G.N., Bolan N.S. A Critical Review of the Influence of Effluent Irrigation on the Fate of Pesticides in Soil. Agriculture, Ecosystems and Environment 2007;120(2) 93–116.

[87] Landa–Cansigno O., Durán–Álvarez J.C., Jiménez B. Retention of Escherichia coli, Giardia lamblia Cysts and Ascaris lumbricoides Eggs in Agricultural Soils Irrigated by Untreated Wastewater. Journal of Environmental Management 2013;128 22–29.

[88] Shuval H., Adin A., Fattal B., Rawutz E., Yekutiel P. Wastewater Irrigation in Developing Countries: Health Effects and Technical Solutions – Technical Paper No. 51. Washington: The World Bank; 1986.

[89] Seymour I.J., Appleton H. Foodborne Viruses and Fresh Produce. Journal of Applied Microbiology 2001;91(5) 759–773.

[90] [90] Bos R., Carr R., Keraita B. Assessing and Mitigating Wastewater–Related Health Risks in Low–Income Countries: An Introduction. In: Drechsel Pay, Scott C.A., Raschid–Sally L., Redwood M., Bahri A. (eds) Wastewater Irrigation and Health: Assessing and Mitigating Risk in Low–Income Countries. London/Ottawa/Colombo: Earthscan/IDRC/IWMI; 2010. p.29–51.

[91] Jiménez B., Drechsel P., Koné D., Bahri A., Raschid–Sally L., Qadir M. General Wastewater, Sludge and Excreta Use Situation. In: Drechsel Pay, Scott C.A., Raschid–Sally L., Redwood M., Bahri A. (eds) Wastewater Irrigation and Health: Assessing and Mitigating Risk in Low–Income Countries. London/Ottawa/Colombo: Earthscan/IDRC/IWMI; 2010. p.3–29.

[92] Barwick R.S., Mohammed H.O., White M.E., Bryant R.B. Prevalence of Giardia spp. and Cryptosporidium spp. on Dairy Farms in Southeastern New York State. Preventive Veterinary Medicine 2003;59(1–2) 1–11.

[93] Gupta N., Khan D.K., Santra S.C. Prevalence of Intestinal Helminth Eggs on Vegetables Grown in Wastewater–Irrigated Areas of Titagarh, West Bengal, India. Food control 2009;20(10) 942–945.

[94] [94] Rizzo L., Manaia C., Merlin C., Schwartz T., Dagot C., Ploy M.C., Michael I., Fatta–Kassinos D. Urban Wastewater Treatment Plants as Hotspots for Antibiotic Resistant Bacteria and Genes Spread into the Environment: A Review. Science of the Total Environment 2013;447(1) 345–360.

[95] Berg G., Marten P., Ballin G. Stenotrophomonas maltophilia in the Rhizosphere of Oilseed Rape – Occurrence, Characterization and Interaction with Phytopathogenic Fungi. Microbiology Research 1996;151(1) 19–27.

[96] Dalkmann P., Broszat M., Siebe C., Willaschek E., Sakinc T., Huebner J., Amelung W., Grohmann E., Siemens J. Accumulation of Pharmaceuticals, Enterococcus, and Resistance Genes in Soils Irrigated with Wastewater for Zero to 100 Years in Central Mexico. PLoS ONE 2012;7(9) e45397.

[97] Negreanu Y., Pasternak Z., Jurkevitch E., Cytryn E. Impact of Treated Wastewater Irrigation on Antibiotic Resistance in Agricultural Soils. Environmental Science and Technology 2012;46(9) 4800–4808.

[98] McLain J.E.T., Williams C.F. Development of antibiotic resistance in bacteria of soils irrigated with reclaimed wastewater. In: proceedings of the 5th National Decennial Irrigation Conference, 5–8 December 2010, Phoenix, USA. 2010.

[99] Gatica J., Cytryn E. Impact of Treated Wastewater Irrigation on Antibiotic Resistance in the Soil Microbiome. Environmental Science and Pollution Research 2013;20(6) 3529–3538.

[100] Jiménez B. Wastewater Use in Agriculture: Public Health Considerations. In: Trimble T.W., Trimble S.W. (eds.) Encyclopedia of Water Science. London: CRC Press; 2007. p.1303–1306.

[101] Oron G., DeMalach Y., Hoffman Z., Manor Y. Effect of Effluent Quality and Application Method on Agricultural Productivity and Environmental Control. Water Science and Technology 1992;26(7–8) 1593–1601.

[102] Najafi P., Mousavi S., Feizi M. Effects of Using Treated Municipal Wastewater in Irrigation of Tomato. Journal of Agricultural Science and Technology 2003;15 (1) 65–72.

[103] Patel J., Millner P., Nou X., Sharma M. Persistence of Enterohaemorrhagic and Nonpathogenic E. coli on Spinach Leaves and in Rhizosphere Soil. Journal of Applied Microbiology 2010;108(5) 1789–1796.

[104] Heaton J.C., Jones K. Microbial Contamination of Fruit and Vegetables and the Behaviour of Enteropathogens in the Phyllosphere: A Review. Journal of Applied Microbiology 2008;104(3) 613–626.

[105] Feachem R.G., Bradley D.J., Garelick H., Mara D.D. Sanitation and Disease – Health Aspects of Excreta and Wastewater Management. Chichester: The World Bank; 1983.

[106] Strauss M. Human Waste (Excreta and Wastewater). Swiss Federal Institute of Aquatic Science and Technology (EAWAG). http://www.eawag.ch/organisation/abteilungen/sandec/publikationen/publications_wra/downloads_wra/human_waste_use_ETC_SIDA_UA.pdf (accessed on August 20 2013).

[107] Solomon E.B., Yaron S., Matthews K.R. Transmission of Escherichia coli O157:H7 from Contaminated Manure and Irrigation Water to Lettuce Plant Tissue. Applied and Environmental Microbiology 2002;68(1) 397–400.

[108] Zhang G., Ma L., Beuchat L.R., Erickson M.C., Phelan V.H., Doyle, M.P. Lack of Internalization of Escherichia coli O157:H7 in Lettuce (Lactuca sativa L.) After Leaf Surface and Soil Inoculation. Journal of Food Protection 2009;72(10) 2028–2037.

[109] Mitra R., Cuesta–Alonso E., Wayadande A., Talley J., Gilliland S., Fletcher J. Effect of Route of Introduction and Host Cultivar on the Colonization, Internalization, and Movement of the Human Pathogen Escherichia coli O157:H7 in Spinach. Journal of Food Protection 2009;72(7) 1521–1530.

[110] Guo X., Chen J., Brackett R.E., Beuchat L.R. Survival of Salmonellae on and in Tomato Plants from the Time of Inoculation at Flowering and Early Stages of Fruit Development Through Fruit Ripening. Applied and Environmental Microbiology 2001;67(10) 4760–4764.

[111] Barker–Reid F., Harapas D., Engleitner S., Kreidl S., Holmes R., Faggian R. Persistence of Escherichia coli on Injured Iceberg Lettuce in the Field, Overhead Irrigated with Contaminated Water. Journal of Food Protection 2009;72(3) 458–464.

[112] Amahmid O., Asmama S., Bouhoum K. The Effect of Waste Water Reuse in Irrigation on the Contamination Level of Food Crops by Giardia Cysts and Ascaris Eggs. International Journal of Food Microbiology 1999;49(1) 19–26.

[113] Erdogrul Ö., Şener H. The Contamination of Various Fruit and Vegetable with Enterobius vermicularis, Ascaris Eggs, Entamoeba histolyca Cysts and Giardia Cysts. Food Control 2005;16(6) 557–560.

[114] Ensink J.H., Mahmood T., Dalsgaard A. Wastewater-Irrigated Vegetables: Market Handling Versus Irrigation Water Quality. Tropical Medicine and International Health 2007;12(s2) 2–7.

[115] Gomes C., Da Silva P., Moreira R.G., Castell–Perez E., Ellis E., Pendleton M. Understanding E. coli Internalization in Lettuce Leaves for Optimization of Irradiation Treatment. International Journal of Food Microbiology 2009;135(3) 238–247.

[116] Chen M. Pollution of Ground Water by Nutrients and Fecal Coliforms from Lakeshore Septic Tank Systems. Water, Air, and Soil Pollution 1988;37(3–4) 407–417.

[117] Stewart L.W., Reneau R.B. Spatial and Temporal Variation of Fecal Coliform Movement Surrounding Septic Tank–Soil Absorption Systems in Two Atlantic Coastal Plain Soils. Journal of Environmetal Quality 1981;10(4) 528–531.

[118] Keswick B.H., Gerba, C.P. Viruses in Groundwater. Environmental Science and Technology 1980;14(11) 1290–1297.

[119] Gerba C.P., Melnick J.L., Wallis C. Fate of Wastewater Bacteria and Viruses in Soil. Journal of the irrigation and drainage division 1975;101(3) 157–174.

[120] Smith M.S., Thomas G.W., White R.E., Ritonga D. Transport of Escherichia coli Through Intact and Disturbed Soil Columns. Journal of Environmental Quality 1985;14(1) 87–91.

[121] Abu–Ashour J., Joy D.M., Lee H., Whiteley H.R., Zelin S. Transport of Microorganisms Through Soil. Water, Air, and Soil Pollution 1994;75(1–2) 141–158.

[122] Gerba C.P., Bitton G. Microbial Pollutants: Their Survival and Transport Pattern to Groundwater. In: Bitton G., Gerba C.P. (eds.) Groundwater Pollution Microbiology. New York: John Wiley & Sons, Inc.; 1984. p.65–88.

[123] Gannon J.T., Manilal V.B., Alexander M. Relationship Between Cell Surface Properties and Transport of Bacteria Through Soil. Applied and environmental microbiology 1991;57(1) 190–193.

[124] Scholl M.A., Mills A.L., Herman, J.S., Hornberger, G.M. The Influence of Mineralogy and Solution Chemistry on the Attachment of Bacteria to Representative Aquifer Materials. Journal of Contaminant Hydrology 1990;6(4) 321–336.

[125] Hsu B.M., Huang C. Influence of Ionic Strength and pH on Hydrophobicity and Zeta Potential of Giardia and Cryptosporidium. Colloids and Surfaces A: Physicochemical and Engineering Aspects 2002;201(1) 201–206.

[126] Gannon J., Tan Y.H., Baveye P., Alexander, M. Effect of Sodium Chloride on Transport of Bacteria in a Saturated Aquifer Material. Applied and Environmental Microbiology 1991;57(9) 2497–2501.

[127] Gaspard P.G., Wiart J., Schwartzbrod J. Étude Expérimentale de l'Adhésion des Ceufs d'Helminthes Ascaris suum: Consequences pour l'Environnent. Revue des Sciences de L'Eau 7, 367–376.

[128] Chang A.C., Page A.L., Asano T. Developing Human Health–Related Chemical Guidelines for Reclaimed Wastewater and Sewage Sludge Applications in Agriculture. Geneva: World Health Organization; 1995.

[129] World Health Organization, WHO Environmental Health Criteria 135: Cadmium–Environmental Aspects. Geneva: World Health Organization; 1992.

[130] Yuan Y. Etiological Study of High Stomach Cancer Incidence Among Residents in Wastewater Irrigated Areas. Environmental Protection Science 1993;19(1) 70–73.

[131] Nicholson F.A., Smith S.R., Alloway B.J., Carlton–Smith C., Chambers B.J An Inventory of Heavy Metals Inputs to Agricultural Soils in England and Wales. Science of the Total Environment 2003;311(1) 205–219.

[132] Siebe C., Cifuentes E. Environmental Impact of Wastewater Irrigation in Central Mexico: An Overview. International Journal of Environmental Health Research 1995;5(2) 161–173.

[133] Leach L., Enfield C., Harlin C. Summary of Long–Term Rapid Infiltration System Studies. EPA Report EPA–600/2–80–165. Oklahoma: U.S. EPA; 1980.

[134] Mapanda F., Mangwayana E.N., Nyamangara J., Giller K.E. The Effect of Long–Term Irrigation Using Wastewater on Heavy Metal Contents of Soils Under Vegetables in Harare, Zimbabwe. Agriculture, Ecosystems & Environment 2005;107(2) 151–165..

[135] Muchuweti M., Birkett J.W., Chinyanga E., Zvauya R., Scrimshaw M.D., Lester J.N. Heavy Metal Content of Vegetables Irrigated with Mixtures of Wastewater and Sewage Sludge in Zimbabwe: implications for Human Health. Agriculture, Ecosystems & Environment 2006;112(1) 41–48.

[136] Barceló D. Emerging Pollutants in Water Analysis. Trends in Analytical Chemistry 2003;22(10) xiv–xvi.

[137] Kümmerer K. Pharmaceuticals in the Environment: Sources, Fate, Effects and Risks. Berlin: Springer–Verlag; 2009.

[138] Bolong N., Ismail A.F., Salim M.R., Matsuura T. A Review of the Effects of Emerging Contaminants in Wastewater and Options for Their Removal. Desalination 2009;239(1) 229–246.

[139] Kümmerer K. Significance of Antibiotics in the Environment. Journal of Antimicrobial Chemotherapy, 2003;52(1), 5–7.

[140] Thiele-Bruhn S. Pharmaceutical Antibiotic Compounds in Soils – A Review. Journal of Plant Nutrition and Soil Science 2003;166(2) 145–167.

[141] Santos L.H., Araújo A.N., Fachini A., Pena A., Delerue–Matos C., Montenegro M.C.B.S.M. Ecotoxicological Aspects Related to the Presence of Pharmaceuticals in the Aquatic Environment. Journal of Hazardous Materials 2010;175(1) 45–95.

[142] Fent K., Weston A.A., Caminada D. Ecotoxicology of Human Pharmaceuticals. Aquatic Toxicology 2006;76(2) 122–159.

[143] Swan G.E., Cuthbert R., Quevedo M., Green R.E., Pain D.J., Bartels P., Cunningham A.A., Duncan N., Meharg A.A., Oaks J.L., Parry–Jones J., Shultz S., Taggart M.A., Verdoorn G., Wolter K. Toxicity of Diclofenac to Gyps Vultures. Biology Letters 2006;2(2) 279–282.

[144] Pomati F., Castiglioni S., Zuccato E., Fanelli R., Vigetti D., Rossetti C., Calamari D. Effects of a Complex Mixture of Therapeutic Drugs at Environmental Levels on Human Embryonic Cells. Environmental Science and Technology 2006;40(7) 2442–2447.

[145] Mennigen J.A., Stroud P., Zamora J.M., Moon T.W., Trudeau V.L. Pharmaceuticals as Neuroendocrine Disruptors: Lessons Learned from Fish on Prozac. Journal of Toxicology and Environmental Health, Part B 2011;14(5–7) 387–412.

[146] Welshons W.V., Thayer K.A., Judy B.M., Taylor J.A., Curran E.M., Vom Saal F.S. (2003). Large Effects from Small Exposures. I. Mechanisms for Endocrine–Disrupting Chemicals with Estrogenic Activity. Environmental Health Perspectives 2003;111(8) 994–1006.

[147] Petrovic M., Eljarrat E., De Alda M.L., Barceló D. Endocrine Disrupting Compounds and Other Emerging Contaminants in the Environment: A Survey on New Monitoring Strategies and Occurrence Data. Analytical and Bioanalytical Chemistry 2004;378(3) 549–562.

[148] Bigsby R., Chapin R.E., Daston G.P., Davis B.J., Gorski J., Gray L.E., Howdeshell K.L., Zoeller R.T., vom Saal, F.S. Evaluating the Effects of Endocrine Disruptors on Endocrine Function During Development. Environmental Health Perspectives 1999;107(4) 613–618.

[149] Grün F., Blumberg B. Endocrine Disrupters as Obesogens. Molecular and Cellular Endocrinology 2009;304(1) 19–29.

[150] Birnbaum L.S., Fenton, S.E. Cancer and Developmental Exposure to Endocrine Disruptors. Environmental health perspectives 2003;111(4) 389–394.

[151] Shore L.S., Kapulnik Y., Ben-Dor B., Fridman Y., Wininger S., Shemesh M. Effects of Eestrone and 17 β-Estradiol on Vegetative Growth of Medicago sativa. Physiologia Plantarum 1992;84(2) 217–222.

[152] Hu C., Hermann G., Pen–Mouratov S., Shore L., Steinberger Y. Mammalian Steroid Hormones Can Reduce Abundance and Affect the Sex Ratio in a Soil Nematode Community. Agriculture, Ecosystems and Environment 2011;142(3) 275–279.

[153] Liu F., Ying G.G., Yang L.H., Zhou Q.X. Terrestrial Ecotoxicological Effects of the Antimicrobial Agent Triclosan. Ecotoxicology and Environmental Safety 2009;72(1) 86–92.

[154] Butler E., Whelan M.J., Sakrabani R., van Egmond R. Fate of Triclosan in Field Soils Receiving Sewage Sludge. Environmental Pollution 2012;167 101–109.

[155] Lemos M.F., van Gestel C.A., Soares A.M. Endocrine Disruption in a Terrestrial Isopod Under Exposure to Bisphenol A and Vinclozolin. Journal of Soils and Sediments 2009;9(5) 492–500.

[156] Diao X., Jensen J., Hansen A.D. Toxicity of the Anthelmintic Abamectin to Four Species of Soil Invertebrates. Environmental Pollution 2007;148(2) 514–519.

[157] Sommer C., Bibby B.M. The Influence of Veterinary Medicines on the Decomposition of Dung Organic Matter in Soil. European Journal of Soil Biology 2002;38(2) 155–159.

[158] Thiele–Bruhn S., Beck I.C. Effects of Sulfonamide and Tetracycline Antibiotics on Soil Microbial Activity and Microbial Biomass. Chemosphere 2005;59(4) 457–465.

[159] Kotzerke A., Sharma S., Schauss K., Heuer H., Thiele–Bruhn S., Smalla K., Wilke B.M., Schloter M. Alterations in Soil Microbial Activity and N–Transformation Processes Due to Sulfadiazine Loads in Pig–Manure. Environmental Pollution 2008;153(2), 315–322.

[160] Hammesfahr U., Heuer H., Manzke B., Smalla K., Thiele–Bruhn S. Impact of the Antibiotic Sulfadiazine and Pig Manure on the Microbial Community Structure in Agricultural Soils. Soil Biology and Biochemistry 2008;40(7) 1583–1591.

[161] Liu F., Ying G.G., Tao R., Zhao J.L., Yang J.F., Zhao L.F. Effects of Six Selected Antibiotics on Plant Growth and Soil Microbial and Enzymatic Activities. Environmental Pollution 2009;157(5) 1636–1642.

[162] Jjemba P.K. The Potential Impact of Veterinary and Human Therapeutic Agents in Manure and Biosolids on Plants Grown on Arable Land: A Review. Agriculture, Ecosystems and Environment 2002;93(1) 267–278.

[163] Abu–Zreig M., Rudra R.P., Dickinson W.T. Effect of Application of Surfactants on Hydraulic Properties of Soils. Biosystems Engineering 2003;84(3) 363–372.

[164] Beausse J. Selected Drugs in Solid Matrices: A Review of Environmental Determination, Occurrence and Properties of Principal Substances. Trends in Analytical Chemistry 2003;23(10) 753–761.

[165] Durán–Álvarez J.C., Becerril–Bravo E., Castro V.S., Jiménez B., Gibson R. The Analysis of a Group of Acidic Pharmaceuticals, Carbamazepine, and Potential Endocrine Disrupting Compounds in Wastewater Irrigated Soils by Gas Chromatography–Mass Spectrometry. Talanta 2009;78(3) 1159–1166.

[166] Fenet H., Mathieu O., Mahjoub O., Li Z., Hillaire–Buys D., Casellas C., Gomez E. Carbamazepine, Carbamazepine Epoxide and Dihydroxycarbamazepine Sorption to Soil and Occurrence in a Wastewater Reuse Site in Tunisia. Chemosphere 2009;88(1) 49–54.

[167] Walker C.W., Watson J.E., Williams C. Occurrence of Carbamazepine in Soils Under Different Land Uses Receiving Wastewater. Journal of Environmental Quality 2012;41(4) 1263–1267.

[168] Xu J., Chen W., Wu L., Green R., Chang A.C. Leachability of Some Emerging Contaminants in Reclaimed Municipal Wastewater-Irrigated Turf Grass Fields. Environmental Toxicology and Chemistry 2009;28(9) 1842–1850.

[169] Xu J., Wu L., Chen W., Chang A.C. Simultaneous Determination of Pharmaceuticals, Endocrine Disrupting Compounds and Hormone in Soils by Gas Chromatography–Mass Spectrometry. Journal of Chromatography A 2008;1202(2) 189–195.

[170] Chen F., Ying G.G., Kong L.X., Wang L., Zhao J.L., Zhou L.J., Zhang L.J. Distribution and Accumulation of Endocrine–Disrupting Chemicals and Pharmaceuticals in Wastewater Irrigated Soils in Hebei, China. Environmental Pollution, 2011;159(6) 1490–1498.

[171] Pérez–Carrera E., Hansen M., León V.M., Björklund E., Krogh K.A., Halling–Sørensen B., González–Mazo E. Multiresidue Method for the Determination of 32 Human and Veterinary Pharmaceuticals in Soil and Sediment by Pressurized–Liquid Extraction and LC–MS/MS. Analytical and Bioanalytical Chemistry 2010;398(3) 1173–1184.

[172] Vazquez–Roig P., Segarra R., Blasco C., Andreu V., Picó Y. Determination of Pharmaceuticals in Soils and Sediments by Pressurized Liquid Extraction and Liquid Chromatography Tandem Mass Spectrometry. Journal of Chromatography A 2010;1217(16) 2471–2483.

[173] Zeng F., Cui K., Xie Z., Wu L., Liu M., Sun G., Lin Y., Lou D., Zeng, Z. Phthalate Esters (PAEs): Emerging Organic Contaminants in Agricultural Soils in Peri–Urban Areas Around Guangzhou, China. Environmental Pollution 2008;156(2) 425–434.

[174] Gielen G.J.H.P. The fate and effects of sewage–derived pharmaceuticals in soil. PhD thesis. University of Canterbury; 2007.

[175] Hu X.Y., Wen B., Shan X.Q. Survey of Phthalate Pollution in Arable Soils in China. Journal of Environmental Monitoring 2003;5(4) 649–653.

[176] Halden R.U. Plastics and Health Risks. Annual Review of Public Health 2010;31 179–194.

[177] Ahel M., Giger W., Koch M. Behaviour of Alkylphenol Polyethoxylate Surfactants in the Aquatic Environment—I. Occurrence and Transformation in Sewage Treatment. Water Research 1994;28(5) 1131–1142.

[178] Ahmad R., Kookana R.S., Alston A.M., Skjemstad J.O. The Nature of Soil Organic Matter Affects Sorption of Pesticides. 1. Relationships With Carbon Chemistry as Determined by 13C CPMAS NMR Spectroscopy. Environmental Science and Technology 2001;35(5) 878–884.

[179] Chiou C.T., Kile D.E. Deviations from Sorption Linearity on Soils of Polar and Nonpolar Organic Compounds at Low Relative Concentrations. Environmental Science and Technology 1998;32(3) 338–343.

[180] Hyland K.C., Dickenson E.R., Drewes J.E., Higgins C.P. Sorption of Ionized and Neutral Emerging Trace Organic Compounds onto Activated Sludge from Different Wastewater Treatment Configurations. Water Research 2012;46(6) 1958–1968.

[181] Sassman S.A., Lee L.S. Sorption of Three Tetracyclines by Several Soils: Assessing the Role of pH and Cation Exchange. Environmental Science and Technology 2005;39(19) 7452–7459.

[182] Durán–Álvarez J.C., Prado–Pano B., Jiménez–Cisneros B. Sorption and Desorption of Carbamazepine, Naproxen and Triclosan in a Soil Irrigated with Raw Wastewater: Estimation of the Sorption Parameters by Considering the Initial Mass of the Compounds in the Soil. Chemosphere 2012;88(1) 84–90.

[183] Karickhoff S.W., Brown D.S., Scott T.A. Sorption of Hydrophobic Pollutants on Natural Sediments. Water Research 1979;13(3) 241–248.

[184] Gibson R., Durán–Álvarez J.C., Estrada K.L., Chávez A., Jiménez–Cisneros B. Accumulation and Leaching Potential of Some Pharmaceuticals and Potential Endocrine Disruptors in Soils Irrigated with Wastewater in the Tula Valley, Mexico. Chemosphere, 2010;81(11) 1437–1445.

[185] Chefetz B., Mualem T., Ben–Ari J. Sorption and Mobility of Pharmaceutical Compounds in Soil Irrigated with Reclaimed Wastewater. Chemosphere 2008;73(8) 1335–1343.

[186] Chefetz B., Xing B. Relative Role of Aliphatic and Aromatic Moieties as Sorption Domains for Organic Compounds: A Review. Environmental Science and Technology 2009;43(6) 1680–1688.

[187] Fox K.K., Chapman L., Solbe J., Brennand V. Effect of Environmentally Relevant Concentrations of Surfactants on the Desorption or Biodegradation of Model Contaminants in Soil. Tenside, Surfactants, Detergents 1997;34(6) 436–441.

[188] Gu C., Karthikeyan K.G. Sorption of the Antimicrobial Ciprofloxacin to Aluminum and Iron Hydrous Oxides. Environmental Science and Technology 2005;39(23) 9166–9173.

[189] Peck A.M., Hornbuckle K.C. Synthetic Musk Fragrances in Lake Michigan. Environmental Science and Technology, 2004;38(2) 367–372.

[190] Goncalves C., Dimou A., Sakkas V., Alpendurada M. F., Albanis T.A. Photolytic Degradation of Quinalphos in Natural Waters and on Soil Matrices Under Simulated Solar Irradiation. Chemosphere 2006;64(8) 1375–1382.

[191] Frank M.P., Graebing P., Chib J.S. Effect of Soil Moisture and Sample Depth on Pesticide Photolysis. Journal of Agricultural and Food Chemistry 2002;50(9) 2607–2614.

[192] Xia K., Jeong C.Y. Photodegradation of the Endocrine–Disrupting Chemical 4–Nonylphenol in Biosolids Applied to Soil. Journal of Environmental Quality 2004;33(4) 1568–1574.

[193] Tixier C., Singer H.P., Oellers S., Müller S.R. Occurrence and Fate of Carbamazepine, Clofibric Acid, Diclofenac, Ibuprofen, Ketoprofen, and Naproxen in Surface Waters. Environmental Science and Technology 2003;37(6) 1061–1068.

[194] Boreen A.L., Arnold W.A., McNeill K. Photodegradation of Pharmaceuticals in the Aquatic Environment: A Review. Aquatic Sciences 2003;65(4) 320–341.

[195] Zepp R.G., Cline D.M. Rates of Direct Photolysis in Aquatic Environment. Environmental Science and Technology 1977;11(4) 359–366.

[196] Latch D.E., Packer J.L., Arnold W.A., and McNeill K. Photochemical Conversion of Triclosan to 2, 8–Dichlorodibenzo–p–dioxin in Aqueous Solution. Journal of Photochemistry and Photobiology A: Chemistry, 2003;158(1) 63–66.

[197] Calisto V., Domingues M.R.M., Erny G.L., Esteves V.I. Direct Photodegradation of Carbamazepine Followed by Micellar Electrokinetic Chromatography and Mass Spectrometry. Water Research 2011;45(3) 1095–1104.

[198] Ying G.G., Yu X.Y., Kookana R.S. Biological Degradation of Triclocarban and Triclosan in a Soil Under Aerobic and Anaerobic Conditions and Comparison with Environmental Fate Modeling. Environmental Pollution 2007;150(3) 300–305.

[199] Gasser G., Rona M., Voloshenko A., Shelkov R., Tal N., Pankratov I., Elhanany S Lev O. Quantitative Evaluation of Tracers for Quantification of Wastewater Contamination of Potable Water Sources. Environmental Science and Technology, 2010;44(10) 3919–3925.

[200] Li J., Dodgen L., Ye Q., Gan J. Degradation Kinetics and Metabolites of Carbamazepine in Soil. Environmental Science and Technology 2013;47(8) 3678–3684.

[201] Walters E., McClellan K., Halden R.U. Occurrence and Loss Over Three Years of 72 Pharmaceuticals and Personal Care Products from Biosolids–Soil Mixtures in Outdoor Mesocosms. Water Research 2010;44(20) 6011–6020.

[202] Kormos J.L., Schulz M., Kohler H.P.E., Ternes T.A. Biotransformation of Selected Iodinated X–ray Contrast Media and Characterization of Microbial Transformation Pathways. Environmental Science and Technology 2010;44(13), 4998–5007.

[203] Pomiès M., Choubert J.M., Wisniewski C., Coquery M. Modelling of Micropollutant Removal in Biological Wastewater Treatments: A Review. Science of the Total Environment 2013;443 733–748.

[204] Kookana R.S., Ying G.G., Waller N.L. Triclosan: its Occurrence, Fate and Effects in the Australian Environment. Water Science and Technology 2011;63 (4): 598–604.

[205] Schlüsener M.P., Bester K. Persistence of Antibiotics such as Macrolides, Tiamulin and Salinomycin in Soil. Environmental Pollution, 2006;143(3) 565–571.

[206] Xu G., Li F., Wang Q. Occurrence and Degradation Characteristics of Dibutyl Phthalate (DBP) and Di–(2–ethylhexyl) Phthalate (DEHP) in Typical Agricultural Soils of China. Science of the Total Environment 2008;393(2) 333–340.

[207] Marco–Urrea E., Pérez–Trujillo M., Blánquez P., Vicent T., Caminal G. Biodegradation of the Analgesic Naproxen by Trametes versicolor and Identification of Intermediates Using HPLC–DAD–MS and NMR. Bioresource Technology 2010;101(7) 2159–2166.

[208] Gauthier H., Yargeau V., Cooper D.G. Biodegradation of Pharmaceuticals by Rhodococcus rhodochrous and Aspergillus niger by co–Metabolism. Science of the Total Environment 2010;408(7) 1701–1706.

[209] Chao W.L., Lin C.M., Shiung I.I., Kuo Y.L. Degradation of Di–butyl–phthalate by Soil Bacteria. Chemosphere 2006;63(8) 1377–1383.

[210] Jelic A., Cruz–Morató C., Marco–Urrea E., Sarrà M., Perez S., Vicent T., Petrovic M., Barcelo D. Degradation of Carbamazepine by Trametes versicolor in an Air Pulsed Fluidized Bed Bioreactor and Identification of Intermediates. Water Research 2012;46(4) 955–964.

[211] Shenker M., Harush D., Ben–Ari J., Chefetz B. (2011). Uptake of Carbamazepine by Cucumber Plants–A Case Study Related to Irrigation with Reclaimed Wastewater. Chemosphere, 2011;2(6) 905–910.

[212] Wu C., Spongberg A.L., Witter J.D., Fang M., Czajkowski K.P. Uptake of Pharmaceutical and Personal Care Products by Soybean Plants from Soils Applied with Biosolids and Irrigated with Contaminated Water. Environmental Science and Technology 2010;44(16) 6157–6161.

[213] Limousin G., Gaudet J.P., Charlet L., Szenknect S., Barthes V., Krimissa M. Sorption Isotherms: a Review on Physical Bases, Modeling and Measurement. Applied Geochemistry 2007;22(2) 249–275.

[214] Murillo–Torres R., Durán–Álvarez J.C., Prado B., Jiménez–Cisneros B.E. Sorption and Mobility of Two Micropollutants in Three Agricultural Soils: A Comparative Analysis of Their Behavior in Batch and Column Experiments. Geoderma 2012;189–190 462–468.

[215] Bi E., Schmidt T.C., Haderlein S.B. Environmental Factors Influencing Sorption of Heterocyclic Aromatic Compounds to Soil. Environmental Science and Technology 2007;41(9) 3172–3178.

[216] Maoz A., Chefetz B. Sorption of the Pharmaceuticals Carbamazepine and Naproxen to Dissolved Organic Matter: Role of Structural Fractions. Water Research 2010;44(3) 981–989.

[217] Stumpe B., Marschner B. Long–Term Sewage Sludge Application and Wastewater Irrigation on the Mineralization and Sorption of 17β–Estradiol and Testosterone in Soils. Science of the Total Environment 2007;374(2) 282–291.

[218] Kaiser K., Guggenberger G., Zech W. Sorption of DOM and DOM Fractions to Forest Soils. Geoderma 1996;74(3) 281–303.

Chapter 6

The Complex Nature of Pollution in the Capping Soils of Closed Landfills: Case Study in a Mediterranean Setting

Jesús Pastor, María Jesús Gutiérrez-Ginés,
Carmen Bartolomé and Ana Jesús Hernández

Additional information is available at the end of the chapter

http://dx.doi.org/10.5772/57223

1. Introduction

1.1. Waste landfills capped with soil

In most developed countries there is serious concern about the state of waste landfills that were closed towards the end of the past century. Economic growth and urban development during this period generated vast amounts of domestic and industrial waste, and this waste was deposited in landfills without its separation or prior treatment. Today, countries with emerging economies or countries in settings of poverty are facing a similar situation, whereby the uncontrolled disposal of waste has led to regions with worryingly high levels of pollutants that affect the atmosphere, soil and water resources.

In the Mediterranean setting, most landfills have been sealed simply by capping with soil from the surroundings. This soil has given rise to a plant cover emerging from the existing seed bank. Besides recovering the visual impacts on the landscape of mountains of rubbish, a plant cover will avoid the spread of pollutants to other ecosystems once the landfill has been closed. However, the implantation of such a cover is conditioned by interactions among several factors, which are responsible for the complex nature of soil pollution affecting closed landfills.

In the literature, there is a lack of work related to pollution of the capping soils of closed landfills. Bibliography concerning landfills is mostly focused on leachates and their effect on water. In case of soil-related works, they were not accomplished in Mediterranean environment. Specially, quantitative data of landfill soil pollution and its possible effects on colonizing plant population is not available.

For all these reasons, the purposes of this chapter are: i) to describe the profile of solid waste landfills sealed with soil in the Mediterranean setting; ii) to focus on the study of a given case in order to present a research methodology that can be used in other scenarios with a similar problem. In the sections detailing this case study, we describe the methods and techniques employed for studying landfill's remediation and discuss the data obtained to give an overview of the topic examined.

2. Profile of solid waste landfills in the Mediterranean region

Numerous waste tips in the central Iberian Peninsula capped with soil over the 1980s and 90s have been widely described from an interdisciplinary perspective in [1]. Research efforts have focused on the pollutants present in the soils used to seal 20 of these landfills and on factors inducing the spread of pollutants. These studies have been aimed at designing measures to remediate the visual impacts of solid waste landfills (Figure 1).

Figure 1. Based on [7]

The figure shows the main impacts (not only visual) produced by a landfill sealed with soil. The most important impact is pollution produced by surface and deep leachates of polluting

http://dx.doi.org/10.5772/57223

substances generated by surface run off and rainwater infiltration [2-6]. This type of pollution especially affects ecosystems in the main areas of leachates discharge i.e., the foots of landfill slopes grazed by domestic and wild animals [5]. Effects are nevertheless also produced on crops, particularly cereals such as barley, which are sometimes grown on the landfill itself (usually on its platforms). We have noted that this cereal accumulates heavy metals (unpublished data). In addition, soil pollution may spread to nearby rivers, on which these mountains of waste seem to hang, or to streams, which transport pollutants to areas beyond the landfill.

Table 1 summarizes the main plant communities detected at the 20 landfill sites examined, along with a summary of their main characteristics according to [8]. Despite the 20 years passed since the landfills were closed, plant cover generally lacks a bush stratum. Existing communities are those classified as ruderals and nitrophiles with a dominance of annual species whose life cycle is typical of the Mediterranean region. Many of these landfills still show large expanses of soil unable to sustain plant growth while other areas boast good plant cover, though with a low diversity. In general, all landfill sites are grazed by itinerant herds of sheep.

Phytosociological class	**Main characteristics**
9. ISOETO-NANOJUNCETEA Br.-Bl. & Tüxen ex Westhoff, Dijk & Passchier 1946	Pionner annual and dwarf perennial ephemeral isoetid communities on periodically flooded bare soils
12. PHRAGMITO-MAGNOCARICETAE Klika in Klika & V. Novák 1941	Swampy, fenny, lacustrine and riverine helophyte communities dominated by perennial graminoids, sedges, forbs and herbs of fresh and brackish waters
20. JUNCETEA MARITIMI Br.-Bl. in Br.-Bl. Roussine & Nègre 1952	Perennial grasslands growing on coastal and inland temporary wet or inundated salt marshes
34. ARTEMISIETEA VULGARIS Lohmeyer, Preising & Tuxen ex von Rochow 1951	Perennial and tall biennial forbs, grasses and thistle pioneer ruderal and nitrophilous sunny communities growing on rich soils.
37. PEGANO-SALSOLETEA Br.-Bl. & O. Bolós 1958	Nitrophilous or halo-nitrophilous dwarf scrub communities, including anthropogenic alloctonous shrubby vegetation
38. POLYGONO-POETEA ANNUAE Rivas Martinez 1975	Annual pioneer ephemeral and exceptionally small creeping perennial nitrophilous anthropozoogenic heavy trodden communities of urban and rural paths
39. STELLARIETEA MEDIAE Tüxen, Lohmeyer & Preising ex von Rochow 1951.	Annual ephemeral weeds, ruderal, nitrophilous and semi-nitrophilous communities
39A. Stellarienea mediae	-Cultivated field weed communities
39B. Chenopodio-Stellarienea Rivas Goday 1956	-Ruderal, nitrophilous and seminitrophilous communities.
39e. Thero-Brometalia (Rivas Goday&Rivas Martinez ex Esteve 1973) O. Bolós 1975	-Subnitrophilous Mediterranean annual ephemeral grassland-like spring blooming communities.
39f. Sisymbrietalia officinalis J. Tüxen in Lohmeyer & al. 1962 em. Rivas-Martinez, Fernándz-González & Loidi 1991	- Nitrophilous and temperate annual ephemeral grassland-like. Path, roadside and rural often trampled communities

Phytosociological class	Main characteristics
40. GALIO-URTICETEA Passarge ex Kopeck�	Perennial hemycriptophyte and climbing tall herbs of nitrified wood fringes and other semi-shaded anthropogenic biotope communities.
41. CARDAMINO HIRSUTAE-GERANIETEA PURPUREI (Rivas-Martínez, Fernández-González & Loidi 1999) Rivas-Martínez, Fernández-González & Loidi classis nova, stat. Nov.	Annual spring and summer ephemeral internal and external shrub fringes slightly nitrified semi-shaded communities, growing on rich organic nutrient soils.
43. TRIFOLIO-GERANIETEA Müller 1962	Semi-shaded perennial herb communities of scarce moisture external fringe woodlands. Calcareous or mesoeutrophic rich soils in temperate submediterranean central Iberian territories.
50. TUBERARIETEA GUTTATAE (Br.-Bl. in Br.-Bl., Roussine & Nègre 1952) Rivas Goday & Rivas Martínez 1963 nom. mut. Propos.	Therophytic grasslands. Pioneer spring and early summer ephemeral plant acidophilous or calcifugous communities, dominated by non nitrophilous annual short herbs and grasses, but localized only in dry or initial soils, mostly in submediterranean or step territories.
51. FESTUCO-BROMETEA Br.-Bl. & Tüxen ex Br.-Bl. 1949	Perennial xerophytic and mesophytic grasslands. Anthropogenic grazed baso-neutrophilous or slightly acidophilous mesophytic or slightly xerophytic nutrient rich-pastures largely covered by perennial grasses.
54. POETEA BULBOSAE Rivas Goday & Rivas Martínez in Rivas-Martinez 1978	Western Mediterranean oceanic thermo- to supramediterranean upper semiarid to humid pastures, grazed and manured, dominated by dwarf perennial grasses and other nutritious prostrate chamaephytes…..
56. LYGEO-STIPETEA Rivas-Martinez 1978 nom. Conserv. Propos.	Mediterranean perennial basophilous xerophytic tall bunchy dense or short open grasslands.
57. STIPO GIGANTAE-AGROSTIETEA CASTELLANAE Rivas-Martínez, Fernández-González & Loidi 1999	Silicicolous perennial grasslands rich in endemics, serial of *Quercus rotundifolia* and other *Quercus* natural potential forest communities.
59. MOLINIO-ARRHENATHERETEA Tüxen 1937	Mesophile to wet often manured meadows and pasture communities on deep and moist soils, widely spread by grazed and anthropic activities

Table 1. Phytosociological classes and mean characteristics of the main species found at the landfills.

Most of the capped landfills are mixed dumps containing both domestic and industrial waste. Besides mitigating the visual impacts of a landfill, the plant cover prevents its collapse and the pollution of other ecosystems by deposited waste materials.

However, in such scenarios the stability of plant communities that become established from the seed bank of the capping soil layer is threatened. Among others, the factors that give rise to this situation are continued waste disposal after the landfill's initial sealing, the scarce volume of capping soil present and land use projects implemented without a priori planning.

3. Case study: The Getafe landfill (Madrid)

3.1. Geomorphological characterization

Here we examine the case of a closed landfill in the Madrid Autonomous Community. This site can be described as one of the most complex scenarios observed among the soil-capped solid waste landfills of the central Iberian Peninsula despite its many features common to all the landfills examined in this region [1]. Located in the municipal district of Getafe (Madrid), this landfill was first described by [9], when it occupied an area of around 70,000 m^3. Fifteen years later (in 2009), the site covered some 95,000 m^2 of land.

Continuous waste dumping and subsequent capping with soil from the surroundings has determined the complex morphology of this landfill. In the photo in Figure 2, the landfill appears as a flattened hill rising out of a plain.

Figure 2. Picture of the whole landfill in spring of 2009

The landfill site has three main zones: a zone (western) mostly containing solid domestic waste, and two zones (central and eastern) mainly accommodating industrial waste and some inert compounds. We have designated these latter zones "rubble tips" to distinguish them from the landfill proper (Figure 3).

The flatted tops of the landfill correspond to platforms, yet more outstanding are its 12 slopes showing a high variety of exposures (across their 360º). Slope heights are 10-20 m and gradients are 50%. Their profiles are straight and many slopes overlap one another. Many slopes show signs of erosion, especially in troughs, often exposing their waste materials. Leachate surface runoff may be observed in three main discharge areas. The westernmost discharge area occupies a wetland. The other two areas, south of the rubble tips form shallow water sheets in the wettest months and quickly dry when rain ceases at the end of spring. In all these discharge zones and at the foots of the slopes, sheep herds may be found grazing. What is more, these and other animals drink any water that accumulates in these areas in 5 to 6 months of the year.

3.2. Composition of the capping soil layer: Factors linked to fertility, salinity, metal toxicity, organic compounds and erosion

To identify the soil factors that mainly determine the landfill's vegetation, mostly arising from the seed bank of the capping soil, we used a stratified sampling procedure (platforms, adjacent rubble tips and main surface leachate discharge zones). At each site, samples were collected using a hoe from the top soil layer (0-10 cm) to give an average soil sample. 57 of such samples were transported to the laboratory, where they were air-dried and sieved (< 2 mm). These samples were then subjected to each of the techniques mentioned in the following sections. Sampling sites 27 to 31 correspond to piles of waste deposited directly in the easternmost discharge zone with no type of cover at all. Although these samples do not correspond to the capping soil, they were collected to assess the possible effect of these waste materials on the soils of the discharge zone in future studies. Corresponding results do not appear in the tables provided below.

Figure 3. Main areas of the landfill and sites where capping soil samples were collected. 03/08/2009 Google Earth Image. UTM coordinates: X=442599 Y=4459459, 30T.

a. Soil fertility indicators

In the 57 soil samples, we examined several variables related to soil fertility with consequent impacts on the vegetation. The procedures described in [10] were used to determine: pH in water and in a saturated soil paste, percentage organic matter by potassium dichromate reduction, Kjedahl total nitrogen, pseudototal (by extraction with nitric and perchloric acids at 4:1) and exchangeable (by extraction with ammonium acetate, pH 7) concentrations of Ca, K, Na and Mg, and pseudototal and bioavailable P and P_2O_5 concentrations, analyzed by plasma emission spectroscopy (ICP-OES).

Sampling point	pH	OM.	N	Ca		K		Na		Mg		P		P_2O_5
				T	E	T	E	T	E	T	E	T	E	
Platform														
G-23	8.5	2.9	0.09	61566	6	1729	5	923	2	7111	3	205	0	130
G-24	8.0	1.2	0.06	5483	66	8855	4	244	5	4869	6	598	7	240
G-45	8.1	1.3	0.08	7345	30	6892	4	223	4	7085	4	553	3	100
G-46	7.8	1.4	0.08	15253	28	11362	7	276	6	22144	4	542	8	260
G-47	8.3	1.8	0.10	14890	22	10051	3	257	5	11553	4	569	4	110
G-48	7.9	1.5	0.10	9818	34	9400	5	218	7	11036	5	462	7	150
Slope														
G-6	7.5	4.0	0.05	6627	27	2495	5	86	16	1316	7	70	0	40
G-8	8.1	0.9	0.05	42871	9	3600	7	355	6	10842	2	32	0	50
G-11	8.2	2.5	0.12	42156	12	6790	12	337	16	7992	4	542	4	300
G-16	7.8	2.2	0.12	46571	14	4478	14	502	12	5795	5	317	0	110
G-17	8.0	2.3	0.06	57028	8	2524	10	604	6	6123	6	132	0	45
G-18	8.2	2.8	0.18	38558	9	3262	9	760	15	6124	7	113	0	15
G-19	7.6	1.2	0.19	11643	27	8270	4	241	15	6617	8	441	0	60
G-20	7.8	1.4	0.09	4051	59	11010	5	243	6	6765	9	547	9	240
G-21	7.9	0.6	0.05	8366	41	7671	4	222	7	5506	8	494	8	200
G-22	8.1	0.9	0.06	7296	53	9209	4	254	5	5810	5	590	6	190
G-39	8.1	1.1	0.05	25150	8	4800	4	350	6	8250	3	50	0	20
Foot of slope														
G- 3 A	7.9	0.9	0.06	3969	53	6902	7	221	27	6222	7	258	10	140
G- 3 F	7.6	2.1	0.09	5922	46	7492	6	214	18	10126	5	258	14	200
G-9	8.2	1.2	0.06	42176	10	3206	7	467	5	4391	4	106	0	60
G-10	8.2	0.9	0.03	29893	9	4759	4	339	5	5086	5	137	0	20
G-37	7.8	3.7	0.17	10122	41	8360	8	224	9	4598	7	429	19	320
G-38	7.8	2.3	0.11	14063	35	5774	5	215	32	4861	11	444	9	200
Discharge zone														
G- 2 A	7.7	6.2	0.29	13490	39	7245	10	324	52	9071	6	279	9	140
G- 2 B	7.8	7.8	0.37	28560	21	6702	9	252	34	8336	7	366	12	220
G- 4 A	7.5	3.8	0.13	12337	20	3570	12	162	26	3884	5	10	0	1050
G- 4 B	7.8	3.8	0.10	11135	19	3468	8	177	32	3577	8	8	0	70
G-7	7.5	5.6	0.25	37831	33	7442	3	308	32	7953	10	276	0	115

Table 2. pH, organic matter (OM, %), nitrogen (N, %), pseudo-total concentration (T) of nutrient elements (mg kg^{-1}) and percentage of exchangeable fraction (E) in soil samples collected from landfill proper.

The results of these determinations in each soil sample are provided in Tables 2 and 3. All results are provided to highlight the huge variation existing for each factor. pH varied from 7.0 to 8.5, given the alkaline nature of the surrounding soils used to cap the landfill. The distributions of all variables failed to vary significantly between the landfill proper and rubble tips.

b. Heavy metals and trace elements toxic for plants

Although some preliminary results regarding soil pollution due to heavy metals, organic compounds and salinity have been already described [1, 7, 11, 12] here we examine this issue in detail. All 57 soil samples were subjected to inductively coupled plasma optical emission spectometry (ICP-OES) to determine pseudototal (after prior extraction with nitric and perchloric acids, 4:1 [13]) and bioavailable (after prior extraction with ammonium acetate + EDTA using the method [14]) concentrations of Al, Mn, Zn, Cu, Pb, Cr and Ni. In addition, total As concentrations were determined by X-ray fluorescence in 48 samples. Total Hg levels were determined using an Advanced Mercury Analyser (AMA-254, LECO Company, Czeck Republic) according to the procedure described by [15] in 34 selected samples of the 57 soil samples collected [16].

Tables 4 and 5 provide the metal and trace element concentrations detected in the capping soil and discharge area samples. We also examined total Al and Mn levels: Al concentrations in the landfill ranged from 8123 to 50747 mg kg^{-1}, and Mn concentrations from 205 to 7432 mg kg^{-1}. In the rubble tips, concentration ranges were higher for Al and lower for Mn. Given the alkaline nature of the soils, these elements are not considered hazardous for plant populations and are therefore not included in the tables.

The sites showing the highest levels of all elements occurred on the landfill's slopes and these showed an uneven spatial distribution. However, the most contaminated sites were simultaneously polluted by all elements. The percentage of a metal found in its bioavailable form was also highly variable. Despite being poorly mobile, Pb showed high bioavailability percentages. Cd was also highly bioavailable. Most variation was shown by Zn and Cu. The metals appearing in lowest concentrations were Cr and Ni.

Apart from the trace element bioavailability study conducted according to the method of [14], we performed a more exhaustive analysis of metal bioavailability in the soil samples. To this end, we undertook sequential extraction by the BCR method optimized by [17]. Sequential extraction serves to indicate the fractions of each metal that are bioavailable (F1: exchangeable), reducible (F2: bound to oxyhydroxides), oxidizable (F3: bound to organic matter) and residual (F4). Given that it is the landfill proper that shows the higher concentrations of these metal pollutants, 5 sites were selected representing platform, slope and discharge zones showing variable concentrations of these types of pollutant. The sites were selected according to their known distributions of metals; we have preserved the numbers assigned to their collection sites. Table 6 provides total concentrations of each metal in each sample calculated as the sum of all fractions. The reader may find the percentages of each metal found in each fraction in Figure 4.

Sampling point	pH	OM.	N	Ca		K		Na		Mg		P		P_2O_5
				T	E	T	E	T	E	T	E	T	E	
Platform														
G- 49	8.1	1.7	0.11	13588	24	7139	7	102	15	10111	5	613	3	18
G- 50	7.6	1.6	0.12	10910	38	7464	7	120	11	9279	4	297	11	1171
G- 51	7.8	2.2	0.10	14428	23	9093	5	134	8	10778	2	873	6	66
G- 52	7.8	0.8	0.04	5342	42	6711	3	106	11	6411	4	283	4	12
G- 53	7.7	2.3	0.10	15831	21	7671	8	97	12	7853	3	580	12	54
Slope														
G- 13	8.3	1.5	0.08	26245	16	2653	16	187	11	3343	6	106	21	30
G- 15	7.7	1.2	0.09	16589	50	7331	5	198	8	4009	7	249	4	8603
G- 32	7.8	0.7	0.05	63966	14	6989	33	238	12	7583	5	261	3	4008
G- 40	8.1	1.0	0.04	14200	28	7550	3	250	6	10350	2	350	4	264
G- 42	8.2	0.2	0.03	9617	35	3704	4	158	16	4643	4	289	12	26
Foot of slope														
G- 12	7.9	1.4	0.07	10790	37	7968	4	214	7	7278	9	423	4	144
G- 14	8.0	1.0	0.06	17651	27	3658	7	127	11	3126	9	68	0	159
G- 33	7.9	1.4	0.09	17519	27	7125	6	190	7	5067	6	536	6	72
G- 34	7.0	2.5	0.13	31469	17	7942	5	248	22	6187	7	469	5	790
G- 35	7.9	1.2	0.08	25913	19	8208	5	173	12	7176	7	428	5	44
G- 41	7.9	1.1	0.10	24281	20	8442	4	246	6	8838	1	593	19	25903
G- 43	8.0	0.8	0.05	72211	6	7503	3	246	6	10099	3	362	8	993
G- 44	7.9	1.3	0.07	17827	19	7959	3	315	13	5894	5	253	4	287
Discharge zone														
G- 1 A	7.7	4.2	0.20	72005	34	4234	5	1281	74	8466	22	177	10	5918
G- 1 B	7.7	2.0	0.11	43501	21	7456	4	292	19	7316	3	265	6	3830
G- 5	7.3	5.8	0.24	25041	30	8150	8	412	40	11338	3	449	6	3305
G- 25	8.0	2.6	0.12	15164	28	8552	6	310	27	5753	10	407	5	74
G- 26	7.8	1.1	0.07	64650	9	6624	3	198	15	6081	5	240	4	819
G- 36	7.5	5.6	0.30	34429	25	9961	5	306	27	8024	8	498	3	4245

Table 3. pH, organic matter (OM, %), nitrogen (N, %), pseudo-total concentration (T) of nutrient elements (mg kg^{-1}) and percentage of exchangeable fraction (E) in soil samples collected from rubble tips.

Sampling point	Zn		Cu		Pb		Cd		Cr		Ni		As	Hg
	T	B	T	B	T	B	T	B	T	B	T	B		
Platform														
G-23	9491	23	4593	3	4421	50	93	35	531	0.9	231	1.1	271	11
G-24	137	20	14	18	19	70	0.0		3.2	2.2	8.7	1.5	n.d	0.0
G-45	148	2.1	8.9	13	28	8.2	0.0		7.3	0.6	5.6	4.1	14	-
G-46	147	2.1	38	17	35	48	0.0		2.2	2.3	12	4.0	22	-
G-47	126	3.8	12	14	11	43	0.0		4.7	1.1	9.9	3.4	n.d	-
G-48	175	5.3	43	16	22	40	0.0		18	0.6	21	3.4	n.d	-
Slope														
G-6	640	22	1916	21	132	29	0.0		256	0.1	35	2.5	119	0.3
G-8	13029	27	1055	8.1	12689	23	185	34	269	1.7	86	1.9	282	4.0
G-11	3247	20	240	11	579	51	10	67	154	0.6	43	2.0	n.d	0.5
G-16	10777	25	748	9.3	5734	38	142	41	150	3.0	42	2.6	n.d	3.0
G-17	17416	26	1313	8.5	12612	28	308	37	242	1.5	60	2.2	n.d	4.9
G-18	22992	28	1804	9.4	18136	30	306	36	298	1.8	80	3.7	685	4.2
G-19	5190	25	125	14	1198	47	29	52	40	3.3	15	2.6	n.d	0.2
G-20	1085	2.6	18	9.4	106	12	0.3	64	9.8	0.5	9.0	0.9	12	0.0
G-21	129	9.0	11	21	18	37	0.0		3.5	1.4	7.0	1.7	13	0.0
G-22	168	9.3	14	20	25	53	0.0		9.1	0.8	8.8	1.8	23	0.0
G-39	17830	76	1445	20	12555	68	190	60	587	2.2	210	2.5	492	-
Foot of slope														
G- 3 A	61	2.2	9.1	14	18	59	0.0		2.3	12	5.1	9.2	-	-
G- 3 F	69	4.6	31	6.2	8.1	40	0.0		9.2	3	5.2	7.0	-	-
G-9	12632	25	1181	10	7179	37	257	48	236	2.4	90	2.1	306	2.4
G-10	15184	33	1493	13	10085	48	155	42	504	2.2	206	2.2	n.d	3.3
G-37	559	66	20	25	28	61	0.0		3.8	4.7	8.5	5.9	15	0.0
G-38	108	18	15	24	23	65	0.0		15	1.0	7.4	6.7	19	0.0
Discharge zones														
G- 2 A	528	17	36	27	117	36	0.0		6.6	6.1	8.6	5.1	-	-
G- 2 F	366	14	35	18	64	28	0.0		5.7	4.9	7.3	5.5	-	-
G- 4 A	577	19	882	27	149	29	0.0		110	0.1	156	1.5	-	-
G- 4 F	477	19	1260	26	139	19	0.0		153	0.2	213	1.5	-	-
G-7	2417	33	151	27	290	48	2.5	86	9.5	3.4	34	8.9	54	0.3
Ref. (pH"/>7)	450		210		300		3		150		112		29*	1.5

Table 4. Pseudo-total concentration (T) of trace elements (mg kg^{-1}) and percentage of bioavailable fraction (B) in soil samples collected from landfill proper. nd: not detected; -: not analyzed. Reference levels for alkaline soils according to Spanish law (RD1310/1990), *As Dutch reference level.

Sampling point	Zn		Cu		Pb		Cd		Cr		Ni		As	Hg
	T	B	T	B	T	B	T	B	T	B	T	B		
Platform														
G-49	10	0.0	13	21	0.0		0.0		1.8	0.0	5.4	0.0	n.d	-
G-50	17	0.0	21	23	22	54	0.0		3.4	0.0	7.8	0.0	n.d	-
G-51	13	0.0	17	19	0.0		0.0		2.4	0.0	6.5	0.0	n.d	-
G-52	9	0.0	14	18	0.0		0.0		3.6	0.0	8.0	0.0	12	-
G-53	13	0.0	14	29	0.0		0.0		6.8	0.0	7.7	0.0	15	-
Slope														
G-13	1261	18	783	7.0	164	35	0.0		378	0.1	223	1.2	42	0.8
G-15	159	15	34	14	28	48	0.0		17	0.7	12	3.3	29	0.0
G-32	166	11	40	54	24	42	0.0		14	3.8	11	3.4	26	0.1
G-40	120	35	15	14	15	86	0.0		3.5	3.4	5.0	6.6	14	-
G-42	72	12	10	9.2	10	43	0.0		6.7	1.5	4.1	7.1	n.d	-
Foot of slope														
G-12	1123	2	83	14	129	10	0.0		10	1.2	25	2.6	25	0.0
G-14	341	16	266	11	64	32	0.0		94	0.1	75	1.6	22	0.2
G-33	146	19	29	14	26	56	1.2	90	7	2.5	8.8	3.5	22	0.0
G-34	224	18	28	22	48	40	1.5	3.6	10	1.9	12	4.4	37	0.1
G-35	76	4.8	14	18	11	33	0.9	24	13	1.2	9.1	3.2	31	0.0
G-41	1259	1.3	38	13	148	7.6	1.4	26	9.1	1.2	11	4.1	21	-
G-43	226	2.8	22	5.4	24	12	0.0		1.1	9.1	10	2.2	100	-
G-44	1869	10	73	21	215	54	3.5	67	18	0.9	22	3.6	30	-
Discharge zone														
G- 1 A	238	23	18	32	57	43	0.0		2.9	12	4.6	9.0	-	-
G- 1 F	74	4.0	13	19	17	22	0.0		1.5	18	6.4	5.2	-	-
G-5	252	65	56	31	66	41	0.0		10	0.1	8.3	5.2	31	0.2
G-25	171	13	25	19	44	45	0.0		5.7	3.3	10	4.0	27	0.4
G-26	126	8.1	63	8.5	83	32	0.0		24	1.0	16	2.0	36	0.2
G-36	904	35	68	41	152	62	1.7	11	9.2	2.8	14	7.5	40	0.6
Ref. (pH"/>7)	450		210		300		3		150		112		29*	1.5

Table 5. Pseudo-total concentration (T) of trace elements (mg kg^{-1}) and percentage of bioavailable fraction (B) in soil samples collected from rubble tips. nd: not detected; -: not analyzed. Reference levels for alkaline soils according to Spanish law [18], *As Dutch reference level.

The results of these tests prompt the following conclusions. Cd and Zn showed highest percentages in the bioavailable fraction. In the case of Cd, this finding is of major concern given its high concentration in the soils examined and this situation has been also described by [19]. Arsenic shows the highest residual percentage and thus its available levels are low. The bioavailable fraction of Cu is fairly low, while remaining fractions vary according to the soils. The behavior of Pb was more irregular among the different soils. In general, the residual fraction was low. The organic matter fraction was variable being greatest at site 18. Its bioavailable fraction was very high at site 39, which is worrying given the high concentration of this heavy metal at this site.

Sample		Zn	Cu	Pb	Cd	As
G-7	wetland	1425	102	278	3	16
G-18	slope	127529	1441	19500	364	32
G-23	platform	33446	2042	11057	106	10
G-39	slope	76120	1272	20032	223	14
G-46	platform	610	41	182	2	0

Table 6. Total concentration of trace elements (mg kg^{-1}) in soils selected for conducting the sequential fractionation

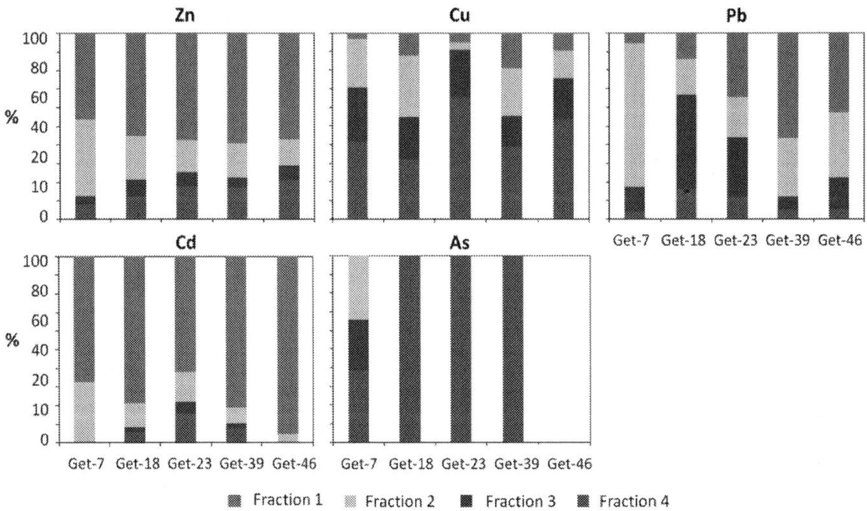

Figure 4. Percentage of metal content that is in each fraction

Clearly the presence of very high concentrations of trace elements and heavy metals is a problem for the establishment of plant populations, worsened by the fact that the sites of highest concentrations coincide with zones of intense slope. In effect, zones corresponding to samples 17 and 18, along with 10 and 39, are naked slopes with practically no plant cover in comparison with surrounding zones.

c. Salinity

Salinity has been described as one of the main impacts on the plant populations and animals of the closed landfills of the Iberian Peninsula's central region [5, 12, 20, 21] as well as landfills in other environments [22-25]. This problem is therefore closely examined in the case of the Getafe landfill. Electrical conductivity was determined in all the soil samples collected, along with F^-, Cl^-, NO_2^-, NO_3^-, PO_4^{3} and SO_4^{2-} anion concentrations by ion chromatography. These results may be found in Tables 7 and 8. As for the other chemical properties of the soil, points showing greatest salinity were unevenly distributed throughout the capping soil.

By comparing soils from the landfill and rubble tips by principal components analysis (PCA) (Statgraphics 15), we were able to observe that electrical conductivity was related to the Cl^- and SO_4^{2-} contents of the landfill cap and also to F^-, Na, NO_3^- concentrations in the case of the rubble tips. The distribution of sites appearing on the new coordinate axes seems to indicate higher salinity in discharge areas. To confirm this observation, we conducted an analysis of variance (ANOVA) of the factors electrical conductivity and Cl^- level in the different landfill areas. Our findings indicate that both variables were significantly higher in the soils of the wetlands where the landfill's runoff is deposited.

d. Organic compounds

Pollution by organic compounds is also a concern emerging from studies designed to address the topic of sealed landfills, as many recently banned compounds, dumped in landfills and numerous affected ecosystems, have been detected [26]. The organic compounds determined in the soil samples and the techniques used for this purpose were: total hydrocarbons by infrared spectrometry (UNE 77307); organochlorine insecticides and polychlorinated biphenyls (PCBs) by gas chromatography (ISO 10382); and polycyclic aromatic hydrocarbons (PAHs) (ISO 18287) and phenols (U.S. E.P.A 3550B, U.S. E.P.A 3650B and U.S. E.P.A 8401) by gas chromatography. The reader is referred to [1] for descriptions of these techniques and their modifications for the present purposes.

Table 9 shows the great variety of organic pollutants that may be found in the Getafe landfill. Those detected at concentrations higher than permitted levels and widely distributed at the site were total hydrocarbons, PCBs, the PAHs with a greater number of rings and some organochlorine insecticides. In general, the sites showing most pollution of this type were those also showing most heavy metal pollution.

Given that total hydrocarbons were detected at all the sites in which these were examined (N=43), Table 10 presents the differences detected.

Landfill samples	EC	F^-	Cl^-	NO_2^-	NO_3^-	PO_4^{3-}	SO_4^{2-}
Platform							
G- 23	132	5.5	4.6	1.4	5.4	0.0	26
G- 24	107	1.8	8.1	1.6	52	2.1	11
G- 45	114	2.6	12	1.5	3.3	2.1	8.5
G- 46	161	7.4	8.5	1.1	5.3	2.3	31
G- 47	116	2.4	8.9	1.7	7.6	1.6	10
G- 48	113	2.6	13	1.6	3.7	3.1	8.0
Slope							
G- 6	175	2.8	6.4	0.7	3.0	0.0	147
G- 8	317	19	6.0	1.4	17	0.0	398
G- 11	159	7.6	7.7	1.0	3.6	0.0	41
G- 16	1853	10	38	1.3	70	0.0	4063
G- 17	587	14	10	1.4	22	0.0	592
G- 18	985	31	121	2.6	193	0.0	635
G- 19	1329	13	32	2.0	1152	0.0	691
G- 20	154	1.6	5.2	1.3	38	2.2	15
G- 21	171	2.5	6.3	0.9	94	0.9	91
G- 22	157	1.7	4.9	1.1	15	1.1	123
G- 39	366	18	8.6	3.0	41	0.0	214
Foot of slope							
G- 3 A	270	3.2	33	1.2	28	0.0	96
G- 3 F	340	1.6	30	1.5	110	5.4	67
G- 9	212	18	8.7	1.3	13	0.0	139
G- 10	237	17	5.3	1.1	12	0.0	194
G- 37	391	1.5	27	1.0	112	0.0	65
G- 38	369	4.2	60	1.3	7.3	0.0	656
Discharge zones							
G- 2 A	1490	3.8	135	2.9	93	2.0	1938
G- 2 F	1960	1.6	77	1.5	0.0	0.0	3322
G- 4 A	1490	3.7	125	1.9	148	2.9	1647
G- 4 F	1500	2.9	43	1.0	34	0.0	1729
G- 7	1878	10	85	1.3	20	1.0	4866

Table 7. Electrical conductivity (EC, μS cm^{-1}) and anion concentration (mg kg^{-1}) in soil samples collected from landfill proper.

http://dx.doi.org/10.5772/57223

Tip samples	EC	F^-	Cl^-	NO_2^-	NO_3^-	PO_4^{3-}	SO_4^{2-}
Platform							
G-49	135	1.5	6.6	0.0	0.0	5.2	18
G-50	826	1.1	3.2	2.4	0.0	0.0	1171
G-51	169	1.7	8.3	0.0	0.0	5.0	66
G-52	60	2.4	2.9	2.6	0.0	2.8	12
G-53	206	1.2	5.1	2.6	0.6	8.0	54
Slope							
G- 13	170	7.3	7.2	0.8	18	0.0	30
G- 15	1719	3.7	4.4	0.8	12	0.0	8603
G- 32	2360	9.4	6.9	1.0	67	0.0	4008
G- 40	299	2.8	5.1	1.0	3.8	0.0	264
G- 42	99	3.0	7.9	1.5	4.3	1.8	26
Foot of slope							
G- 12	90	0.0	1.1	0.0	0.0	0.0	144
G- 14	193	4.2	15	1.9	13	0.0	159
G- 33	327	2.5	13	1.0	143	0.0	72
G- 34	744	3.8	56	1.5	21	1.5	790
G- 35	198	2.7	13	2.1	37	2.8	44
G- 41	1716	1.9	4.8	1.2	19	0.0	25903
G- 43	877	2.1	3.6	1.0	11	0.0	993
G- 44	384	4.4	46	1.4	5.0	0.0	287
Discharge zones							
G- 1 A	8220	8.6	7570	0.0	495	0.0	5918
G- 1 F	2350	2.9	199	2.3	29	0.0	3830
G- 5	2180	3.8	260	2.0	201	0.0	3305
G- 25	280	3.3	35	1.9	24	0.0	74
G- 26	709	6.0	12	1.5	60	0.0	819
G- 36	2500	5.2	69	3.6	43	0.0	4245

Table 8. Electrical conductivity (EC, $\mu S\ cm^{-1}$) and anion concentration ($mg\ kg^{-1}$) in soil samples collected from rubble tips.

Pollutant	Max conc.	Ref*	Pollutant	Max conc.	Ref*
PAHs			*Insecticides*		
Naphthalene	0.23	1	Alfa-HCH	0.01	0.01
Acenaphthene	0.04	6	Beta-HCH	0.27	0.01
Fluorene	0.09	5	Gamma-HCH	0.48	0.01
Anthracene	0.46	45	Hexachlorobenzene	0.04	0.01
Fluoranthene	2.59	8	Endosulfan	0.07	0.6
Pyrene	2.03	6	p.p'-DDE	0.02	0.6
1,2-benzanthracene	0.99	0.2	*Hydrocarbons*		
Chrysene	1.11	20	Total concentration	3408	50
Benzo(b)fluoranthene	1.85	0.2	Conc. of aromatics	335	
Benzo(k)fluoranthene	0.89	2	Conc. of aliphatics	3073	
Benzo(a)pyrene	1.56	0.02	*Phenols*		
Indene-1,2,3-(cd)pyrene	1.63	0.3	Phenol	0.05	7
Dibenzo(a,h)anthracene	0.14	0.03	Cresols	0.02	4
			2,4,6-trichlorophenol	0.01	0.9
PCBs	3.05	0.01	Pentachlorophenol	0.01	0.01

Table 9. Maximum concentration of organic pollutants found in soils of Getafe landfill (mg kg^{-1}) and maximum allowed values according to Spanish law (*Ref, [27])

e. Factors linked to soil erosion

Signs of soil erosion observed on the landfill's slopes prompted us to address this matter, given the significant effect that soil particle size and the loss of certain fractions can have on the ability of plant species to take root.

The traditional method of Bouyoucos to determine sand, mud and clay fractions was used on all 57 soil samples. In addition, the Mastersizer-S was used to assess particle size by the dispersion and diffraction of a laser light beam as it crosses a suspension of the sample. This technique and the sample preparation method are described in [1]. Particle size was determined in 43 of the samples to establish the type of particle that may be lost through erosion. Significant differences in this variable were detected in several fractions of fine sand between soil from the landfill cap and soil from the rubble tips. These results are provided in Figure 5 and table 11. The high standard deviation of the data determined that only differences in the sand fraction of the rubble tip soil were significant.

Although the results obtained using both granulometric techniques are not comparable since the first method gives a percentage weight while the second procedure provides percentage volumes, both revealed that the most marked differences among the higher zones, slopes and lower zones occur in the rubble tips adjacent to the landfill. Table 11 shows the different

Landfill		Landfill		Tips	
Sampling point	**HC**	**Sampling point**	**HC**	**Sampling point**	**HC**
Platform		*Foot of slope*		*Slope*	
G-23	901	G- 3 A	13	G-13	215
G-24	78	G- 3 F	13	G-15	52
Slope		G-9	92	G-32	33
G-6	2423	G-10	154	*Foot of slope*	
G-8	93	G-37	33	G-12	174
G-11	230	G-38	26	G-14	169
G-16	62	*Discharge zone*		G-33	52
G-17	63	G- 2 A	5.1	G-34	19
G-18	3408	G- 2 F	5.1	G-35	22
G-19	95	G- 4 A	854	*Discharge zone*	
G-20	33	G- 4 F	854	G- 1 A	7.5
G-21	42	G-7	123	G- 1 F	7.5
G-22	78			G-5	67
G-39	87			G-25	24
				G-26	18
				G-36	36
Ref.	50	Ref.	50	Ref.	50

Table 10. Total concentration of hydrocarbons (HC, mg kg-1) in points of landfill proper and rubble tips and maximum allowed values according to Spanish law (Ref, [27])

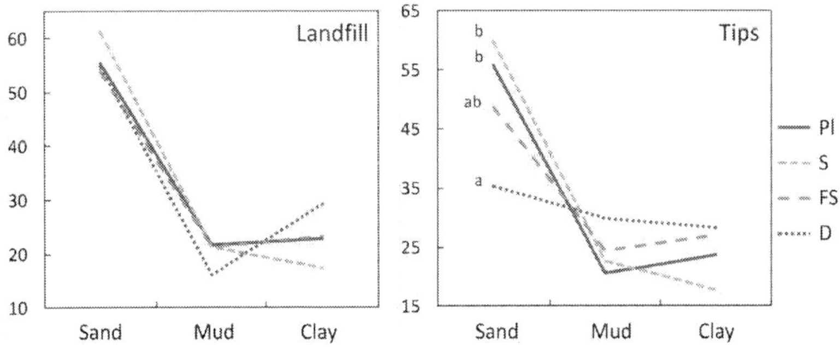

Figure 5. Mean percentage of each textural fraction determined by Bouyoucos technique in samples from platforms (P), slopes (S), foots of slopes (FS) and discharge zones (D) in landfill proper and tips. Different letters mean significant differences between means in the same area (Bonferroni, 95%)

granulometric fractions analyzed. For the rubble tips, results indicate the dragging of fine sands from slopes towards the lower zones accompanied by the consequent build-up of coarse sands. Although with a lack of significance, differences were also observed in the remaining fractions.

These data do not seem to clearly indicate the signs produced by the in situ transport of particles from the higher to the lower zones of slopes and discharge areas. No distinguishing factors were revealed in a discriminatory analysis (figure 6). The findings of such a study also indicate the heterogeneity of the situations arising on even a single slope and increase the complexity of understanding the plant colonization pattern, which may vary as small patches depending on these variations produced on a small scale.

Area		Range of particle size (mm)							
		Clay	Mud	Fine sand A	Fine sand B	Fine sand C	Medium sand	Coarse sand	Very coarse sand
		<0.002	0.002-0.02	0.02-0.05	0.05-0.1	0.1-0.2	0.2-0.5	0.5-1	1-3.2
Landfill									
Platform	M	0.34	1.93	2.69	7.64	13.7	16.4	23.8	33.4
	SD	*0.23*	*0.65*	*1.39*	*4.28*	*6.72*	*3.10*	*9.05*	*14.5*
Slope	M	0.46	3.12	3.32	8.85	14.0	14.8	16.7	38.6
	SD	*0.50*	*2.28*	*0.64*	*2.06*	*3.4*	*3.51*	*7.65*	*13.9*
Foot of Slope	M	0.76	3.87	3.59	9.15	14.3	16.2	13.8	38.4
	SD	*0.65*	*2.81*	*0.64*	*1.87*	*1.81*	*3.95*	*1.55*	*3.84*
Discharge zone		0.35	2.30	4.43	11.9	17.6	15.6	13.0	34.8
Tips									
Slope	M	0.20	1.72	3.10	8.89 a	14.5 a	13.3	14.2 b	44.1
	SD	*0.25*	*0.62*	*1.37*	*4.14*	*6.21*	*3.8*	*2.92*	*15.8*
Foot of Slope	M	0.34	2.30	4.45	12.9 ab	20.1 ab	16.7	12.9 ab	30.4
	SD	*0.23*	*0.45*	*1.02*	*3.27*	*5.03*	*5.4*	*1.79*	*14.5*
Discharge zone	M	0.26	2.76	5.78	16.0 b	22.9 b	17.8	9.73 a	24.8
	SD	*0.21*	*1.08*	*2.30*	*4.95*	*3.42*	*1.30*	*2.20*	*10.2*

Table 11. Mean (M) and standard deviation (SD) of percentages of each granulometric fraction in different areas of landfill and tips. Different letters in the same range of particle size mean significant differences between means (Bonferroni, 95%)

3.3. Heterogeneous distribution of pollutants

Through PCA, we tried to gain insight into the structure of the soil cap used to seal the landfill. In Figure 7A, it may be seen that the first axis, or component, is closely and positively linked to heavy metal and organic compound pollution although Na and F also appeared in this group of variables, and negatively related to soil fertility due to the presence of K and P. The second component was more related to soil salinity, represented by electrical conductivity, chlorides, sulfates, nitrates and nitrites. When organic components and the trace elements Hg and As were excluded, results failed to vary significantly and the first component continued to be positively and closely linked to the presence of heavy metals and negatively linked to that of K (Figure 7B). The second component, more related to salinity or electrical conductivity, this time was linked more to chlorides than the other anions.

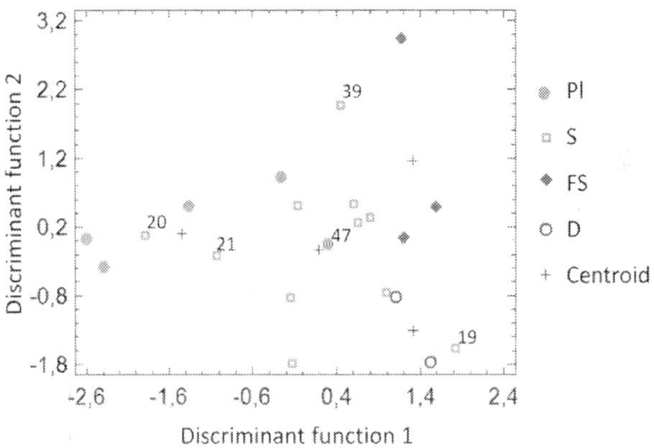

Figure 6. Representation of discriminant functions calculated with Mastersizer results of landfill soil samples, grouped in platforms (P), slopes (S), foots of slopes (FS) and discharge zones (D).

These findings confirm our previous results indicating that despite the uneven distribution of pollutants, at the most polluted sites all pollutants contribute to this contamination. The PCA plot of points on the new axes (Figures 7C and 7D) serves to visually identify the sites showing highest heavy metal pollution as the landfill slopes and those with the greatest salinity as the rubble tips. The platforms emerged as the least polluted sites both in terms of heavy metals and salts contents.

The chemical analysis results reveal great heterogeneity in both the distributions and concentrations of pollutants. As an example of the complexity of the problem addressed, Zn concentrations range from 9 mg kg^{-1} to 23000 mg kg^{-1}; maximal Cd concentrations are 308 mg kg^{-1} (of which 85% represents the easily soluble fraction) and the maximal concentration of total hydrocarbons is 3408 mg kg^{-1}.

The spatial distributions of these factors determined using a Geographical Information System (ArcMap™ software, v. 9.3.1., ESRI) are depicted in Figure 8.

4. Understanding the complex nature of landfill soil caps with the view of restoring the impacts of pollution

The mountains of waste and rubble we have created are new landscape features that most often emerge in areas around cities. These scenarios can be viewed as laboratories for research into the environmental impacts of landfills that were capped without prior treatment of the deposited waste. Even considering that the restoration of degraded ecosystems is a systemic topic, the functionality of this epistemological approach arises from the fact that ecosystems are dynamic systems that evolve and co-evolve with human activity.

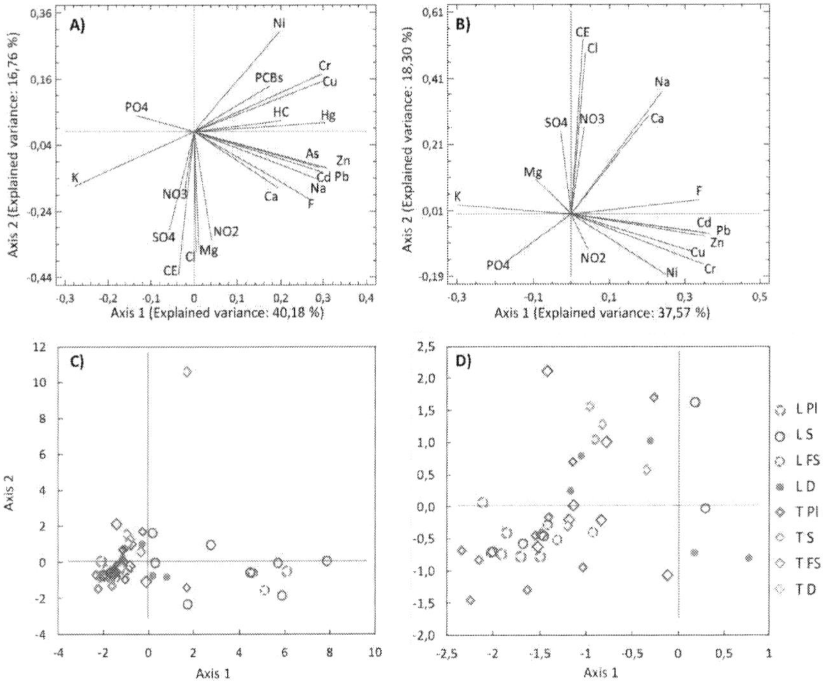

Figure 7. Principle components analysis of the soil chemical variables showing points appearing on the new coordinate axes. A) PCA of the whole set of variables, N = 29, B) PCA excluding As, Hg, hydrocarbons and PCBs increasing the number of cases to N = 52, C) representing the 52 points on the new axes created by the PCA in Figure B), D) expansion of plot C) from −2.5 to 1 abscissa and -2 to 2.5 ordinate. L, landfill proper; T, rubble tips; PI, platforms; S, slopes; FS, foot of slopes; D, discharge zone.

The complexity of the problem faced arises from questions related to the secondary ecological succession (from the capping soil's seed bank), which interacts with the primary succession that is possible in this new ecosystem in the landscape. Besides restoring its impacts, efforts need to also focus on revegetating the landfill system itself.

Hence, these landfills may be considered a new type of ecosystem in which primary and secondary successions coincide. They are thus of great interest for ecological science since they provide a real scenario for investigating the measures we should install to restore a degraded and polluted ecosystem and help us identify the plant species related to their varied forms of pollution. This will enable researchers to select the most appropriate plant species for revegetation efforts rather than simply establishing a green cover once a landfill has been capped.

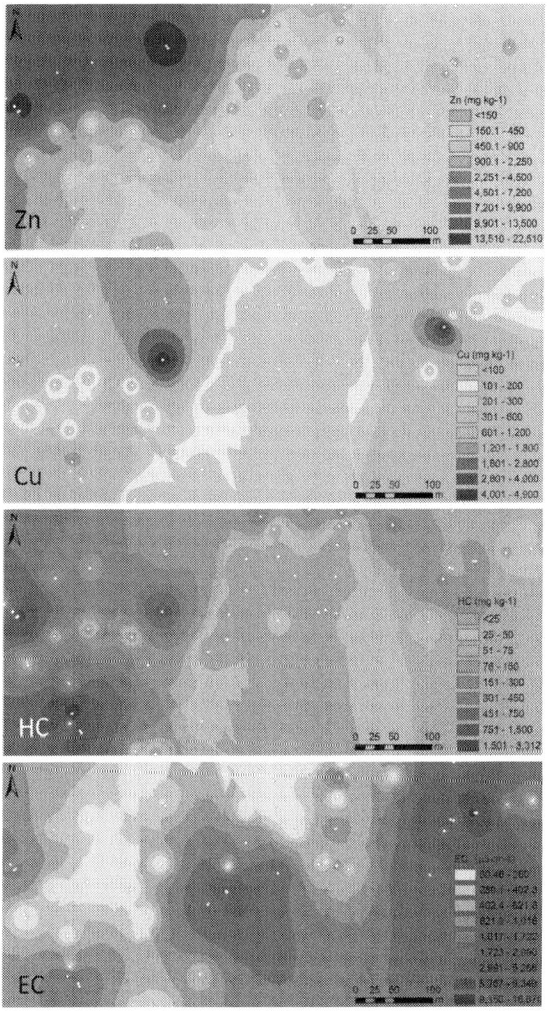

Figure 8. Spatial distribution of Zn, Cd, total hydrocarbons and electrical conductivity

The ecological theory that is most applicable to the restoration of the environmental impacts of capped landfills addresses the stress and ecological strategies of herbaceous species.

The classification of plant life cycle strategies described by Grime combines the stress intensity with the perturbation intensity [7, 28]. Thus, "competing species" are more appropriate for landfills with a low intensity of perturbation and stress, "ruderal" species adapt better to conditions of low stress and intense perturbation, and "stress-tolerants" are ideal for settings of intense stress and scarce perturbation. When both these factors are excessive, this approach is ineffective.

It should not be forgotten, however, that different types of ecosystem respond differently to a given perturbation, and vice-versa, that a given ecosystem can respond in many different ways to different perturbations. We also need to be aware of the vast environmental variability and randomness that exists along with other associated forms of uncertainty [29].

5. Conclusions

If a sealed landfill needs to be revegetated, it will be necessary to study the fertility of its soil cover, heavy metals and trace elements that can cause plant toxicity, salinity and organic compounds in the capping soil layer. The research methodology used in the landfill case study can be followed in other scenarios with a similar problem.

The analysis of all considered parameters and the heterogeneous distribution of pollutants indicate that a single-species cover should be avoided. It will be necessary to create a multi-species cover that will adapt to the heterogeneous distribution of the organic and inorganic pollutants present in capping soils and to the morphological features of the landfill's slopes.

From a scientific viewpoint, the scenario of the closed waste landfill has enabled the in depth study of what we have called the erosion-pollution binomial. This is the complex situation found in the capping soils of closed landfills in the Mediterranean setting. The plant species used for their revegetation should have the capacity to show an adequate response to this biome. To find such species, there is an urgent need for autecological studies and studies designed to assess native and commercial plant species that are able to adapt to these particular conditions. This is the reason why these results should not be extrapolated to other non-Mediterranean settings.

Acknowledgements

Authors acknowledge program P2009/AMB-1478 Community of Madrid Program (EIADES). MJGG was funded by the FPU fellowship (AP2008-02934) of Spain's Ministry of Education.

Author details

Jesús Pastor[1], María Jesús Gutiérrez-Ginés[1*], Carmen Bartolomé[2] and Ana Jesús Hernández[2]

*Address all correspondence to: mjesus.gutierrezg@uah.es

1 Department of Environmental Biology, MNCN, CSIC, Madrid, Spain

2 Department of Life Sciences, Alcalá University, Alcalá de Henares, Spain

References

[1] Hernández AJ, Bartolomé C, editors. Estudio multidisciplinar de vertederos sellados. Diagnóstico y pautas de recuperación. Alcalá de Henares: Servicio de Publicaciones de la Universidad de Alcalá; 2010.

[2] Adarve MJ, Hernández AJ, Gil A, Pastor J. B, Zn, Fe and Mn content in four grassland species exposed to landfill leachates. Journal of Environmental Quality 1998;27: 1286-1293.

[3] Hernández AJ, Adarve MJ, Pastor J. Some impacts of urban waste landfills on Mediterranean soils. Land Degradation & Development 1998;9: 21-33.

[4] Pastor J, Urcelay A, Oliver S, Hernández AJ. Impact of Municipal Waste on Mediterranean Dry Environments. Geomicrobiology Journal 1993;11: 247-260.

[5] Pastor J, Alía M, Hernández AJ, Adarve MJ, Urcelay A, Antón FA. Ecotoxicological studies on effects of landfill leachates on plants and animals in Central Spain. The Science of the Total Environment, 1993;140: 127-134.

[6] Pastor J, Urcelay A, Adarve MJ, Hernández AJ, Sánchez A. Aspects of contamination produced by domestic waste landfills on receiving waters in Madrid province. In: Nath B. et al. (eds.) Environmental Pollution. Science, Policy, Engineering. London: European Centre for Pollution Research; 1993. p254-261.

[7] Hernández AJ, Pastor J. Validated Approaches to Restoring the Health of Ecosystems Affected by Soil Pollution. In: Domínguez JB (ed.) Soil Contamination Research Trends. New York: Nova Science Publishers; 2008. p51-72.

[8] Rivas-Martínez S, Diaz TE, Fernández-Gonzalez F, Izco J, Loidi J, Lousa M, Penas A. Vascular Plant Communities of Spain and Portugal. Addenda to the syntaxonomical checklist of 2001. Itinera Geobotanica 2002;15: 433-992.

[9] Adarve MJ. Análisis de la incidencia ambiental de vertederos de residuos sólidos urbanos en aguas, suelos y especies vegetales de zonas de descarga. PhD thesis. Universidad de Alcalá; 1993.

[10] Hernández AJ, Pastor J. Técnicas analíticas para el estudio de las interacciones suelo-planta. Henares, Revista de Geología 1989;3: 67-102.

[11] Hernández AJ, Pérez-Leblic MI, Bartolomé C, Rodríguez J, Álvarez J, Pastor J. Ecotoxicological diagnosis of a sealed municipal landfill. Journal of Environmental Management 2012;95: S50-S54.

[12] Pastor J, Hernández AJ. La restauración en sistemas con suelos degradados: estudios de casos en vertederos, escombreras y emplazamientos de minas abandonadas. In: Millán R, Lobo C (eds.) Contaminación de Suelos: Tecnologías para su recuperación. Madrid: CIEMAT; 2008. p539-560.

[13] Walsh LM. Soil Society of America. Instrumental Methods for analysis of soils and plant tissue, vol VII. Soil Science Society of America, Wisconsin;1971.

[14] Lakanen E, Ervio R. A comparison of eight extractants for the determination of plant available micronutrients in soils. Acta Agricultura Fennica 1971;123: 223-232.

[15] Millán R, Gamarra R, Schmid T, Sierra MJ, Quejido AJ, Sánchez DM, Cardona AI, Fernández M, Vera R. Mercury content in vegetation and soils of the Almadén mining area (Spain). Science of The Total Environment 2006;368(1): 79-87.

[16] Gutiérrez-Ginés MJ. AJ Hernández. R Millán. J Pastor. Study of Hg in soil cover of sealed urban landfills and its toxic impact in Lupinus albus L. grown in landfill soils with high levels of this trace element. In: 9th Iberian and 6th Iberoamerican Congress on Environmental Contamination and Toxicology, CICTA2013, 1-4 July 2013, Valencia, Spain; 2013.

[17] Mossop KF, Davidson CM. Comparison of original and modified BCR sequential extraction procedures for the fractionation of copper, iron, lead, manganese and zinc in soils and sediments. Analytica Chimica Acta 2003;478: 111–118.

[18] RD 1310/1990. Real Decreto 1310/1990 de 29 de octubre, por el que se regula la utilización de los lodos de depuración en el sector agrario. Boletín Oficial del Estado del 1 de noviembre de 1990.

[19] Liphadzi MS, Kirkham MB. Physiological Effects of Heavy Metals on Plant Growth and Function. In: Huang B (ed.,) Plant-environment interactions. New York: Taylor & Francis; 2006. p243-269.

[20] Hernández AJ, Adarve MJ, Gil A, Pastor J. Soil salination from landfill leachates: effects on the macronutrient content and plant growth of four grassland species. Chemosphere 1999;38: 1693-1711.

[21] Pastor J, Hernández AJ. Heavy metals, salts and organic residues in old solid urban waste landfills and surface waters in their discharge areas: Determinants for restoring their impact. Journal of Environmental Management 2012;95: S42-S49.

[22] Cheng CY, Chu LM. Phytotoxicity data safeguard the performance of the recipient plants in leachate irrigation. Environmental Pollution 2007;145(1):195-202.

[23] Fatta D, Papadopoulos A, Loizidou MA. Study of the landfill leachate and its impact on the groundwater quality of the greater area. Environmental Geochemistry and Health 1999;21: 175-190.

[24] Tatsi AA, Zouboulis AI. A field investigation of the quantity and quality of leachate from a municipal solid waste landfill in a Mediterranean climate (Thessaloniki, Greece). Advances Environmental Research 2002;6: 207-219.

[25] Kalčíková G, Zagorc-Končan J, Zupančič M, Žgajnar Gotvajn A. Variation of landfill leachate phytotoxicity due to landfill ageing, Journal of Chemical Technology and Biotechnology 2012;87(9): 1349-1353.

[26] Weber R, Watson A, Forter M, Oliaei F. Persistent organic pollutants and landfills. A review of past experiences and future challenges. Waste Management Research 2011;29: 107-121.

[27] RD 9/2005. Real Decreto 9/2005 de 14 de enero, por el que se establece la relación de actividades potencialmente contaminantes del suelo y los criterios y estándares para la declaración de suelos contaminados. Boletín Oficial del Estado del 18 enero 2005.

[28] Ursic KA, Kenkel NC, Larson DW. Revegetation dynamics of cliff faces in abandoned limestone quarries. Journal of Applied Ecology 1997;34: 289-303.

[29] Urcelai A. Estructura de sistemas herbáceos mediterráneos sometidos a la acción antrópica y posibles mecanismos de resiliencia. PhD thesis. Universidad de Alcalá; 1997.

Chapter 7

Assessment of Soil Contamination of a Cattle Market around River Ogun Basin, Isheri, Nigeria

A.A. Adeyi, O.M. Omidiran and O. Osibanjo

Additional information is available at the end of the chapter

http://dx.doi.org/10.5772/57302

1. Introduction

Continuous increase in environmental pollution caused by the expansion and amplification of domestic, industrial, and agricultural waste generation has resulted in steadily growing number of contaminants release into the environment [1]. The uncontrolled discharge of pollutants into natural sinks such as aquatic ecosystems in developing countries like Nigeria has prompted a need for periodic assessment of the level of pollutants released into the environment. Contamination of water resources and soil with toxic metals and priority organic pollutants represents a major environmental challenge being a source of human exposure to these toxicants with potential deleterious human health effects. Soils in several parts of Nigeria have accumulated considerable amount of toxic elements arising largely from unsustainable consumption and production activities such as illegal mining of solid minerals, uncontrolled land filling of post-consumer hazardous wastes including electronic waste or e-waste [2, 3], uncontrolled industrial emissions, and vehicular emissions; as well as environmentally unsound petroleum exploration, production and exploitation activities, and accidental crude and petroleum products oil spills etc., [3].

Soil is the foundation of terrestrial ecosystems and a non- renewable fundamental agricultural resource, inextricably linked to productivity, land development and environmental quality [4]. Soils may contain many organic and inorganic compounds of natural and anthropogenic origin. The concentrations and toxicity of these compounds present in environmental media range very widely and depend on possible interactions (synergies) among chemicals. The continuous introduction of these contaminants into the environment has resulted in their accumulation. Soil pollution with heavy metals occurs not only near mines and smelters, but also on agricultural land because of the application of sewage sludge as fertiliser and metal-based pesticides [5]. Extensive soil contamination with hazardous pollutants including inorganic (i.e. Cd, Pb, Cu, Hg, As etc) and organic (i.e. pesticides, dioxins, polyaromatic hydrocarbons etc) pollutants has increasingly been of global concern over the last decades, especially in Asian [6] and other developing countries, where urbanization and industrialization have occured rapidly and haphazardly.

Heavy metals continue to receive increasing attention due to a better understanding of their fate as well as toxicological relevance in ecosystems and human health [7]. Pollution of soils by heavy metals represents one of the most important ecological problems today [8]. Anthropogenic emission of heavy metals in the environment during the last century has led to increasing accumulation of metals in soils and natural waters in both urban and rural areas. Heavy metals are ecologically harmful because they tend to bio-accumulate over time in soils and plants with ability to have a negative influence on the physiological activities of plants (e.g. photosynthesis, gaseous exchange and nutrient absorption), influencing reductions in plant growth, dry matter accumulation and yield [9]. Heavy metal pollution exerts toxic effects on soil microbial biomass [10] and microbial processes such as soil respiration, nitrogen mineralization, and enzyme activities [11]. Soil contamination by heavy metals has also been shown to inhibit soil microbial activities [12], in turn reducing soil fertility and inhibiting the germination of certain seed plants [13]. Heavy metals contamination may also produce nutrient imbalance in plants with adverse effects on the synthesis and functioning of many biologically active compounds [14]. Subsequent accumulation of metals in the food chain [15] is detrimental to human health.

Heavy metals transport in soil profile is a major environmental concern because even slow transport through the soil may eventually lead to deterioration of groundwater quality. Preferential flow can accelerate the movement of water and solutes through soil profile [16]. Soil pollution can lead to water pollution if toxic chemicals leach into groundwater, or if contaminated runoff reaches streams, lakes, or oceans. The decomposition of organic materials in soil can release sulphur dioxide and other sulphur compounds, causing acid rain which can leach heavy metals from soil. Thus, there is need to reduce the introduction of heavy metals into the environment from anthropogenic sources since metals are persistent, toxic and non-biodegradable [3]. In addition, chemicals that are not water-soluble contaminate plants that grow on polluted soils, and they also tend to accumulate increasingly toward the top of the food chain. Both organic and inorganic contaminants are important in soil management. The most prominent chemical groups of organic contaminants are petroleum hydrocarbons, Polycyclic Aromatic Hydrocarbons (PAHs), Polychlorinated Biphenyls (PCBs), chlorinated aromatic compounds, detergents, and pesticides. Inorganic species include nitrates, phosphates, and heavy metals such as cadmium, chromium and lead.

Organochlorine pesticides not only accumulate in animal tissues; many are extremely stable and persist in soil and plants, and if soluble enough can reach groundwater or surface waters. They can therefore enter the food chain not only via their target and non-target organisms, but also imbibed water and via plants eaten by herbivores, including cattle [17]. Their persistence in soils depends both on the nature of the soil and their own physico-chemical properties. Light soils facilitate the water-borne transport of soluble pesticides [18]. Soils with high clay and organic matter contents tend to retain both the more soluble pesticides (because of their high water storage capacity) and the more hydrophobic pesticides (because of their high specific surface area and other sorption-favouring properties) [19,20]. Because of the dangers they pose to non-target organisms, DDT and a number of other organochlorine pesticides were banned in most countries in the 1970s and 1980s, at least for agricultural use. However, because of their resistance to degradation processes and immobility, many are still found in high concentrations in soils to which they were formerly applied, where they constitute a reservoir and secondary source of ongoing contamination of waters, wildlife, and crops [21,22,23].

1.1. Health and environmental effects of heavy metals

Metals are particularly toxic to the sensitive, rapidly developing systems of foetuses, infants, and young children, otherwise known as "vulnerable group". Some metals, such as lead and mercury, easily cross the placenta of mothers and damage the brain of foetuses. Childhood exposure to some metals such as lead and mercury can result in learning difficulties, memory impairment, damage to the nervous system, and behavioural problems such as aggressiveness and hyperactivity. At higher doses, heavy metals can cause irreversible brain damage. Children may receive higher doses of metals from food than adults, since they consume more food for their body weight than adults [24].

Exposure to manganese (Mn) is usually via inhalation, which results in the main cause of its toxicity. Mn toxicity has been reported through occupational (e.g. welder, miner) and dietary overexposure and is evidenced primarily in the central nervous system, although lung, cardiac, liver, reproductive and fetal toxicity have been noted [24]. Mn neurotoxicity results from an accumulation of the metal in brain tissue [25]. Brain permeability to manganese is higher than that to iron and zinc. Manganese is easily concentrated in the brain, especially in the basal ganglia, and can cause an irreversible neurological syndrome similar to Parkinson's disease [24].

Following long-term exposure to cadmium, the main health concerns are its toxicity to the kidney and bones, arising via ingestion and inhalation, and its lung carcinogenicity seen in exposed workers following inhalation [26]. Over a period of time, cadmium accumulates in the kidney and, if the organ concentration exceeds a critical threshold, the tubule cells become damaged and renal function impaired. This cadmium build-up also affects vitamin D metabolism, disturbing the calcium balance within the body, which may lead to a decrease in the mineral content within the bones, resulting in osteoporosis and osteomalacia [26,27]. Its toxicity is linked with reproduction problem because it affects sperm and reduces birth weight. It is a potential carcinogen and seems to be a causal factor in cardiovascular diseases and hypertension. Large concentrations of Cd in the soil are associated with parent material (black slates) and most are manmade (burning of fossil fuels, application of fertilizers, sewage sludge, and plastic waste) [28,29]. In humans, long-term exposure is associated with renal dysfunction. The average daily intake for humans is estimated as 0.15 μg from air and 1 μg from water. Smoking a packet of cigarettes can lead to the inhalation of around 2-4 μg of cadmium, but levels may vary widely [27].

Mercury is a toxic substance which has no known function in human biochemistry or physiology and does not occur naturally in living organisms [27]. Mercury is a toxic heavy metal and a persistent environmental pollutant. Known sources of Hg in soil e.g. mercury-in-bulb thermometers used in hospitals; alkyl mercury compounds especially methyl mercury used as fungicide in agriculture, present in coal and hydrocarbon gas etc. Exposure to mercury is associated with serious adverse health and developmental effects, especially in pregnant women, developing foetuses, and young children [24]. This heavy metal is toxic even at low concentration to a wide range of organisms including humans. The organic form of mercury can be particularly toxic, and the methyl-and ethyl-forms have been the cause of several major epidemics of poisoning in humans resulting from the ingestion of contaminated food, e.g. fish. Two major epidemics in Japan were caused by the release of methyl and other mercury compounds from an industrial site followed by accumulation of the chemicals in edible fish. The poisoning became well-known as Minamata disease [28,29]. Inorganic mercury poisoning is associated with tremors, gingivitis and/or minor psychological changes, together with spontaneous abortion and congenital malformation. Monomethylmercury causes damage to the brain and the central nervous system, while foetal and postnatal exposure have given rise to abortion, congenital malformation and developmental changes in young children [27]. The negotiations under UNEP of a new Minamata Mercury Convention is under way to underscore international concerns about the environmental and health impact of mercury and the need for coordinated global action to address the issue effectively. Average concentrations of mercury in soil are 6.6 mg/kg, 6.6 mg/kg, 24 mg/kg and 50 mg/kg in agricultural, residential/ parkland, commercial and industrial soils, respectively [30].

Lead has been known to be toxic since the 2nd century BC in Greece. It is a widespread contaminant in soils. Lead poisoning is one of the most prevalent public health problems in many parts of the world. It was the first metal to be linked with failure in reproduction. It can cross the placenta easily. It also affects the brain, causing hyperactivity and deficiency in the fine motor functions, thus, it results in damage to the brain. The nervous systems of children are especially sensitive to Pb leading to retardation. It is also cardiotoxic and contributes to cardiomyopathy (disease of the heart muscle leading to the enlargement of the heart) [28,29]. Lead affects almost every organ system in the human body. The central nervous system is particularly vulnerable in infants and children under age six. The effects are the same whether it is breathed or swallowed. Large amounts of lead exposure may lead to blood anaemia, severe stomach ache, muscle weakness, and brain damage. Lower levels of exposure, may affect a child's mental and physical growth leading to learning disabilities and seizures [24]. Major sources of human exposure to lead include old lead water pipes, tetraethyl lead from gasoline, lead in paint and improper disposal of used lead acid battery (ULAB) in dump sites. In humans, exposure to lead can result in a wide range of biological effects depending on the level and duration of exposure. Various effects occur over a broad range of doses, with the developing foetus and infant being more sensitive than the adult. High levels of exposure may result in toxic biochemical effects in humans, which in turn cause problems in the synthesis of haemoglobin, effects on the kidneys, gastrointestinal tract, joints and reproductive system, and acute or chronic damage to the nervous system. Lead poisoning, which is so severe as to cause evident illness, is now less common. This could be due to international efforts at phasing out lead in gasoline and paints in many parts of the world including developing countries [31] indeed. At intermediate concentrations, however, there is persuasive evidence that lead can have small, subtle, subclinical effects, particularly on neuropsychological developments in children. Some studies suggest that there may be a loss of up to 2 IQ points for a rise in blood lead levels from 10 to 20 μg/dl in young children [26]. Uncontaminated soil contains lead concentrations less than 50 ppm but soil lead levels in many urban areas exceed 200 ppm [32]. The EPA's standard for lead in bare soil in play areas is 400 ppm by weight and 1200 ppm for non-play areas [33].

Arsenic is well-known as a poison and a carcinogen. It has an average concentration in the soil of 5 to 6 mg/kg [28]. Its amount in the soil is related to rock type and industrial activity [34]. It is used in combination with other materials in pigments, poison gases and insecticides (such as Paris green, calcium arsenate and lead arsenate) and is well known from former use as a rat poison. Arsenic has a long history of medical applications; before penicillin was developed an arsenic compound was used to treat syphilis and yaws. It is used in ammunition manufacturing, semi-conductor manufacturing, as a preservative in tanning and taxidermy, as well as on the exterior of wood such as deck and playground materials as well as by-product of copper smelting [34]. Acute (short-term) arsenic poisoning may cause nausea, vomiting, diarrhea, and weakness, loss of appetite, shaking, cough and headache while chronic (long-term) exposure may lead to a variety of symptoms including skin pigmentation, numbness, cardiovascular disease, diabetes, and vascular disease. Arsenic is also known to cause a variety of cancers including skin cancer (non-melanoma type), kidney, bladder, and lung, prostate and liver cancer.

Chromium is required for carbohydrate and lipid metabolism and the utilization of amino acids. Its biological function is also closely associated with that of insulin and most Cr-stimulated reactions depends on insulin. However, excessive amount of the metal can cause toxicity. Toxic levels are common in soils applied with sewage sludge [28,29]. Chromium is used in metal alloys and pigments for paints, cement, paper, rubber, and other materials. Low-level exposure can irritate the skin and cause ulceration. Long-term exposure can cause kidney and liver damage, and damage to circulatory and nerve tissue. Chromium often accumulates in aquatic life, adding to the danger of eating fish that may have been exposed to high levels of chromium [27].

Nickel occurs in the environment only at very low levels. Humans use nickel for many applications like the use of nickel as an ingredient of steel and other metal products [27], in the metallurgical, chemical and food processing industries, especially as catalysts and pigments [35]. Foodstuffs have low natural content of nickel but high amounts can occur in food crops growing in polluted soils. Humans may also be exposed to nickel by inhalation, drinking water, smoking, and eating contaminated food [27]. Uptake of high quantities of nickel can cause death, systemic effects, respiratory effects, cardiovascular effects, gastrointestinal, haematological musculoskeletal, hepatic, renal, endocrine, metabolic, immunological and lymphoreticular, neurological, reproductive, developmental effects and cancer [36]. Nickel occurs naturally in the Earth's crust with an average concentration of 0.0086% (86 ppm). The nickel content of soil may vary depending on local geology [37].

Copper is an essential substance to human life, but in high doses it can cause anaemia, liver and kidney damage, and stomach and intestinal irritation. People with Wilson's disease are at greater risk for health effects from overexposure to copper. Copper normally occurs in drinking water from copper pipes, as well as from additives designed to control algal growth [28]. The major sources of environmental copper releases include the mining, smelting and refining of copper, industries producing products from copper such as wire, pipes and sheet metal, and fossil fuel combustion [38].

1.2. Health and environmental effects of some common Persistent Organic Pollutants (POPs)

Persistent organic Pollutants (POPs) are associated with serious human health problems, including cancer, neurological damage, birth defects, sterility, and immune system defects. US Environmental Protection Agency (USEPA) and the Stockholm Convention have classified certain POPs as probable human carcinogens, including aldrin, dieldrin, chlordane, Dichlorodiphenyltrichloroethane (DDT), heptachlor, Hexachlorobenzene (HBC), toxaphene, and Polychlorinatedbiphenyls (PCBs) [39,40,41]. Laboratory studies have shown that low doses of POPs may affect organ systems. Chronic exposure to low doses of certain POPs may affect the immune and reproductive systems. Exposure to high levels of certain POPs can cause serious health effects or death. The primary potential human health effects associated with POPs are cancer; immune system suppression; nervous system disorder; reproductive damage; altered sex ratio; reduced fertility; birth defects; liver, thyroid, kidney, blood, and immune system damage; endocrine disruption; developmental disorders; shortened lactation in nursing women; and chloracne and other skin disorders.

Studies have linked POPs exposure to declines, diseases, or abnormalities in a number of wildlife species, including certain kinds of fish, birds, and mammals [39,42]. Wildlife also can act as sentinels for human health: abnormalities or declines detected in wildlife populations can sound an early warning bell for people. For example, the behavioural abnormalities and birth defects in fish, birds, and mammals in and around the Great Lakes. People are mainly exposed to POPs through contaminated foods. Less common exposure routes include drinking contaminated water and direct contact with the chemicals. In people and other mammals alike, POPs can be transferred through the placenta and breast milk to developing offspring. In addition, sensitive populations, such as children, the elderly, and those with suppressed immune systems, are typically more susceptible to many kinds of pollutants, including POPs. Because POPs have been linked to reproductive impairments due to their endocrine disrupting properties, men and women of child-bearing age may also be at risk [43-47].

Organochlorine pesticides are a large class of multipurpose chlorinated hydrocarbon chemicals, some of which are POPs. They break down slowly in the environment and accumulate in the fatty tissues of animals. Thus, they stay in the environment and food web long after being applied [48]. Many organochlorine pesticides are endocrine disrupting chemicals, meaning they have subtle toxic effects on the body's hormonal systems [27]. Endocrine disrupting chemicals often mimic the body's natural hormones, disrupting normal functions and contributing to adverse health effects. Organochlorine pesticide is ubiquitous environmental contaminants because they break down very slowly. The effects that some of the most common organochlorine pesticides have on humans as well as animal's health are as follows:

Acute exposure to chlordane can possibly cause neurological effects while long-term exposure can damage the liver, kidney, and other internal organs. Significant immune system changes have also been reported in exposed workers. There is some evidence of carcinogenicity [49]. Research has also indicated that chlordane is an endocrine disruptor. The half-life of chlordane in soil has been reported to be from one to three years. Chlordane binds to aquatic sediments and bioconcentrates in the fat of organisms.

DDT is a possible human carcinogen; some of its metabolites Dichlorodiphenyldichloroethylene (DDE) and Dichlorodiphenyldichloroethane (DDD) are probable human carcinogens [49]. DDT and its breakdown products are thought to be endocrine disruptors. DDT is lipophilic and will both bio-concentrate and bio-magnify. It is present almost everywhere in the environment, and residue has been detected in the arctic. DDT breaks down into the related compounds; DDE and DDD. In the environment, DDT and its metabolites have a half life of 10 to 15 years in soil. DDT is highly toxic to birds and fish, DDE and DDD are also toxic and persistent [50]. Some DDT may evaporate from soil and enter the air, and some may be broken down by the sun or microorganisms.

Heptachlor is metabolised in animals to heptachlor epoxide, whose toxicity is similar to that of heptachlor [49]. Both binds to animal fat. Heptachlor and heptachlor epoxide are possible human carcinogens and are thought to be endocrine disruptors. At high levels or from chronic exposure, they can cause central nervous system and liver damage. The half life of heptachlor in temperate soil is up to two years. Heptachlor has been found in treated wastewater from industrial processes including coal mining, foundries, and nonferrous metals manufacturing. It has been detected in the blood of cattle in the U.S. and heptachlor has been strongly implicated in the decline of several wild bird populations [50].

Mirex is probably a carcinogen and is also thought to be an endocrine disruptor. It may cause damage to skin, liver, nervous, and reproductive systems at high levels [49]. Crustaceans may be the most sensitive organisms to mirex. It can bio-accumulate and bio-magnify and is very persistent, with a half life of up to 10 years in soil. It has been detected in arctic freshwater.

Lindane, which belongs to the class of new POPs under the Stockholm Convention may be a carcinogen and is thought to be an endocrine disruptor. It can cause nervous system effects and pulmonary oedema from short, and acute exposures. Long-term exposure can cause liver and kidney damage, as well as various blood disorders [49]. Lindane bio-concentrates slightly in fish and crustaceans, and can remain in the air for up to 17 weeks and travel long distances. It is broken down quickly in water [50]. Lindane exposure has been associated with recurrent miscarriage. Lindane also disrupts natural levels of estrogens, androgens and thyroid hormones in rodents [51]. Lindane produces behavioural and neurochemical changes in developing rats at doses that do not produce symptoms in adults. This is significant because it suggests that lindane may have the greatest effect during development [52]. Both acute and sub-chronic exposures to lindane were found to reduce dopamine concentrations in the brains of rats by about 46%. Dopamine is an important neurotransmitter with effects on mood and behaviour [53]. Lindane exposure during postnatal development in rats was associated with alterations in levels of detoxification enzymes in the brain that lasted until adulthood. It is not known how this alteration might affect the outcome of future exposures [54]. Exposure to lindane decreased both thyroid hormones and progesterone in developing lambs [55S].

Dieldrin can cause poisoning in humans and mammals following exposure via the skin, oral intake or inhalation. It acts as a stimulant to the central nervous system and accumulates in fatty tissue causing severe damage to the liver and kidneys. Animal experiments have revealed a carcinogenic effect, but as yet no teratogenic action. Residue levels of aldrin (mostly present as dieldrin) do not accumulate indefinitely, as the result of yearly applications at a constant dosage rate, but will reach a maximum level which is, in practice, of the order of that resulting from a single annual dose [56,57]. There is virtually no leaching of aldrin into deeper soil layers and thus no contamination of groundwater, which is eventually used as drinking or irrigation water [58,59].

Endosulfan is highly toxic to the nervous system and can cause circulatory problems, headache, vomiting and diarrhea. Thus, it is one of the new POPs under the Stockholm Convention. It is also a suspected hormone disruptor [49]. Endosulfan has shown no potential to accumulate over time in animals. It is more water soluble than other organochlorine pesticides, such as DDT, and is less persistent in the body because it metabolises quickly [60].

1.3. Health and environmental effects of Polycyclic Aromatic Hydrocarbons (PAHs)

Because of combustion of fossil fuels and organic waste, PAHs are ubiquitous in the environment. Studies show that certain PAHs metabolites interact with DNA and are genotoxic, causing malignancies and heritable genetic damage in humans. Many of these compounds have carcinogenic and mutagenic activities and present a hazard for human health [61]. In humans, heavy occupational exposure to mixtures of PAHs entails a substantial risk of lung, skin, or bladder cancer. PAHs generally have a low degree of acute toxicity to humans. The most significant endpoint of PAHs toxicity is cancer. Increased incidences of lung, skin, and bladder cancers are associated with occupational exposure to PAHs. It is difficult to ascribe observed health effects in epidemiological studies to specific PAHs because most exposures are to PAHs mixtures. Animal studies show that certain PAHs affect the hematopoietic, immune, reproductive, and neurologic systems and cause developmental effects [62].

1.4. Soil contamination by cattle wastes

Cattle manure is a major waste product of agricultural practice involving the breeding and rearing of cattle. Inappropriate disposal of manure can create environmental problems such as odours and leaching of nitrate ion and other pollutants into groundwater. Repeated annual application of manure with high salt content caused a build up of soluble salts in soils, sufficient to lower their productivity [63]. During the last three decades, dairy farm facilities have received attention from the public and regulations due to increased environmental concern. These facilities concentrate in certain regions of the USA (e.g. California, New York, Minnesota, and Pennsylvania) and generate considerable amount of manure, which can harm soil and water quality. Manure varies in mineral composition depending on the type of animal and the ration fed. It was reported that cattle (Bos Taurus) manure contains 2.7 to 9.5% N, 0.5 to 0.8%

O, 1.7 to 2.9% K, 1.4 to 20% Ca, and 0.61 to 0.76% Mg (% dry waste basis) [64]. In addition, varying amounts of Al, Fe, Mn, S, Cl, Cu, Zn, and B and trace concentrations of Co, Cd, Cr, Ni, As and Se are present. Large fractions of these elements are present in a water-soluble form. With high precipitation, almost all regions where dairy farm facilities concentrate, great qualities of these elements are lost through surface runoff and leaching into groundwater. Manure contains a large amount of soluble and insoluble organic substances. Organic matter content ranges between 80% and 90% (dry weight basis) [64]. Pyrolysis field ionization mass spectroscopy (Py-FIMS) was also used to analyse cow manure samples [65]. The organic components that were identified include lignins, dimeric lignins, pesticides, lipids, monoesters, fatty acid, sterols, and heterocyclic nitrogen compounds.

The aim of this study was to determine the distribution, concentration and profiles of some ubiquitous environmental pollutants such as heavy metals, polycyclic aromatic hydrocarbons (PAHs) and organochlorine pesticides in soil of a cattle market around River Ogun Basin, Isheri, Nigeria so as to ascertain the level of contamination of the soil resulting from the anthropogenic activities taking place in the area.

2. Experimental

2.1. General description of ogun river basin

The River Ogun basin is situated in south western part of Nigeria covering a total area of 22.44 cubic kilometres. It arises in Oyo state, on the southern side of Yoruba plateau, and runs south into Ogun state, passing through Abeokuta before entering Lagos state and finally discharging into the Lagos Lagoon. The elevation of the northern, eastern, southern, and western boundaries are 457 m, 366 m, 426 m, and 230 m, respectively above the sea level [66]. Various organizations are active in the development of socio-economic conditions of the Ogun river basin. The Ogun-Osun River Basin Development Authority (OORBDA) is responsible for the management of the water resources. The Ogun river basin is among others contributing a diversion at Mokoloki for irrigated farming of the upland. Another is the Ogun State Agricultural Development Project, which has incorporated a fishery component to increase fish production and availability. Ogun River is used as a source of water supply for domestic, industrial, agricultural, and recreational purposes along the states it traverses. This site has become a place of interest considering its constant and continuous pollution owing to the fact that it serves as a focal point of some commercial activities in the ever growing cattle market around the basin [67]. Figure 1 shows the map of the study area.

2.2. Sampling design

The soil samples were collected from the cattle market situated around the Ogun River, Isheri along Lagos-Ibadan Express Road. The soil samples were collected at two different depths of 0-15 cm and 15-30 cm with non-metallic sampler. The sampling site was divided into three zones as follows:

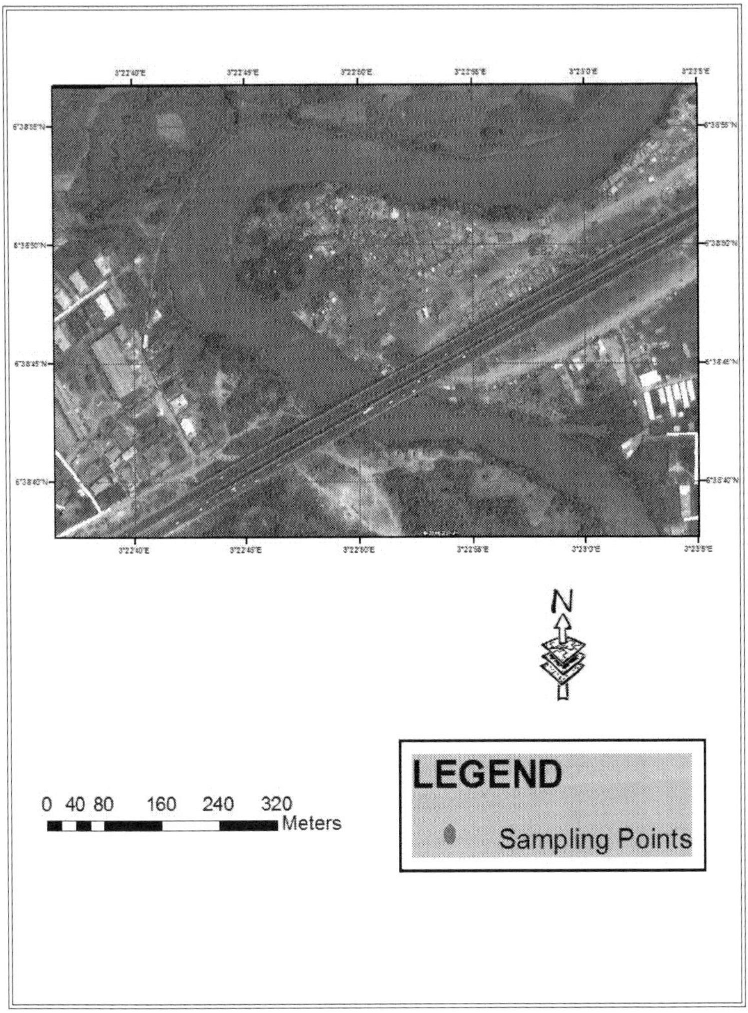

Figure 1. Map of river Ogun showing the Sampling Points

1. Samples were collected at five different spots (SA1-SA5) upstream the market across the river at a distance of about 100 m to the market. The major human activities on this side of the river were sand mining and fishing. The fishing activity was simple and involved the use of paddled canoe and boat with fishing net and hooks. The natural physical feature of these areas was shrub growth. The samples from this area were pooled together to form a representative sample, A1S1 (representing composite sample of the random samples

http://dx.doi.org/10.5772/57302

collected from 0-15 cm depth), and A2S2 (representing composite sample of the random samples collected from 15-30 cm depth).

2. Random samples were also collected at five different locations including the market centre and the areas surrounding it. The activities and the features observed at the different points were;

i. SB1: Old car scraps, marshy area located about 50 m to the market centre

ii. SB2: Auto-mechanic workshop and residential tents built with planks and polypropylene materials

iii. SB3: Cattle rearing and selling with other petty trading

iv. SB4: Slaughter houses, wastewater discharge point, animal hair and horn burning with tyres, animal blood cooking and processing, and solid waste dumpsite

v. SB5: Children playground at the river bank

The samples from this area were pooled together and a representative sample, B1S1 (representing composite sample of the random samples collected from 0-15 cm depth), and B2S2 (representing composite sample of the random samples collected from 15-30 cm depth) were obtained for the analysis.

3. Two samples were collected downstream the market from two points to form a composite samples C1S1 and C2S2 for the 0-15 cm and 15-30 cm depths, respectively. The activities around this point are

i. SC1: Petty commercial activities around a police post in a slum with most houses built with plank, rust iron sheet and polypropylene material;

ii. SC2: A dumpsite that was almost swept off by erosion into the river situated at the river bank.

2.3. Sample collection, treatment and preservation

Two samples were collected from each area of sampling around the cattle market to form a composite. Control sample was also collected in a rural community, Lalupon in Ibadan. The sample portion that is to be analysed for PAHs and organochlorine pesticides were collected in a glass bottle wrapped with aluminium foil to prevent exposure to sunlight, as organochlorine pesticides are known to be sensitive to sunlight, while the sample to be analysed for heavy metals and physico-chemical parameters were collected in a polythene bag. The samples were preserved in an ice chest for onward transportation to the laboratory, where they were kept in a refrigerator. The samples collected for heavy metals and physico-chemical parameters were air-dried at room temperature until they were properly dried and large objects (sticks, stones, wood, and e.t.c) were manually removed. The samples were then gently ground with a porcelain mortar and pestle, and sieved through a 2 mm sieve and stored in polythene bags until sample digestion. The samples collected for PAHs and organochlorine pesticides analyses were not air-dried to minimise loss of those components which can vaporise easily [68].

2.4. Analytical procedure

2.4.1. Determination of soil pH

Approximately 20.0 g each of the air-dried and sieved soil samples (< 2 mm) were weighed, 50 ml distilled water was added and then mixed and allowed to stand for 30 minutes. Suspension was stirred every 10 minutes during this period. The suspension was allowed to settle for another 30 minutes and a pH meter electrode (Jenway 3510) was placed in the suspension and the reading was taken after some seconds. The electrode was removed from the suspension and rinsed thoroughly with distilled water and excess water was carefully dried.

2.4.2. Determination of soil organic matter

About 0.70 g each of the air-dried and sieved soil samples were weighed and 10.0 ml of 1.0 N K_2CrO_7 solution and 20.0 ml concentrated H_2SO_4 were added immediately [69]. The solution was mixed for 1 minute by swirling until the soil and reagents were thoroughly mixed. The mixture was allowed to stand for 30 minutes after which 100 ml distilled water was added and then allowed to cool under running water. 3-4 drops of Ferroin indicator was added and then titrated with 0.5 N Ferrous Ammonium Sulphate. Near the end-point, the solution took a greenish colour and later changed to dark green. At this point, the 0.5 N Ferrous Ammonium Sulphate was added drop by drop until the colour changed from green to reddish-brown at the end point. The Ferrous Ammonium Sulphate solution was standardized using 1 N potassium dichromate before use. Blank titration was carried out in the same manner, but without the soil sample.

The % organic Carbon was determined as follows:

$$\% \; Organic \; Carbon = \frac{(meq. \; FAS \;\; for \; blank \; - \; meq \; FAS \;\; for \; sample)\,(0.003 \; x \; 100f)}{Weight \; in \; g \; of \; air-dried \; soil}$$

Where, f = 1.334

% Organic matter = 1.729 x % Total Organic Carbon

meq- Milligram equivalent = normality of solution x ml of solution used.

2.4.3. Determination of soil particle size by the Bouyoucos hydrometer method

This is a measure of the size distribution of individual particles in a soil sample. Particle size distribution analysis is often used in soil science to evaluate soil texture. The soil texture is based on different combination of sand, silt, and clay separately, that makes up the particle size distribution. The size range of sand is 2.0 mm-0.05 mm, silt is 0.05 mm-0.002 mm and clay is less than 0.002 mm (< 0.002 mm). Particle size distribution analysis result can be used to predict the water retention capacity and unsaturated hydraulic conductivity of soils [70]. About 50.0 g of air-dried soil sample sieved to < 2 mm was weighed into 250 ml beaker. 20 ml of 5.0 % Sodium Hexametaphosphate was added along with 350 ml distilled water. The mixture was stirred mechanically on an end-to-end shaker at 100 revolutions per minutes for 5 minutes. The water-soil slurry was transferred to a sedimentation cylinder and sufficient water was added to bring the level to the 1000 ml mark. The top of the cylinder was covered and inverted several times until all the soil particles are in suspension. The cylinder was placed on the flat surface and the time was recorded. The soil hydrometer was placed on the suspension carefully and slided slowly into the suspension until the hydrometer was floating. The first reading was taken on the hydrometer at 60 s after the cylinder was set down. The hydrometer was removed and the temperature of the suspension was measured. After the first hydrometer reading, the suspension was left standing for 3 hours and the second reading was taken. The temperature of the suspension was also measured at this point. The first reading measured the percentage of silt and clay in suspension while the second reading indicated the percentage of clay in the suspension. When taking the hydrometer reading, the hydrometer was carefully lowered into the cylinder about 25 s before the reading was taken to ensure that it came to rest before the appointed reading time. The readings were taken at the top of the meniscus formed by the suspension around the stem of the hydrometer. As soon as the reading was taken, the hydrometer was carefully removed and rinsed with distilled water. The results were corrected to a temperature of 20 °C and 2.0 were subtracted to compensate for the added dispersing agent.

Correction factor for Temperature = 0.3 [Observed Temperature (T °C) – 20 °C]

$\% \text{ Clay} + \% \text{ Silt} = [H_1 + 0.3\ (T_1 - 20) - 2.0]\ 2$

$\% \text{ Clay} = [H_2 + 0.3\ (T_2 - 20) - 2.0]\ 2$

$\text{Sand} = 100 - [H_1 + 0.3\ (T_1 - 20) - 2.0]\ 2$

$\text{Silt} = 100.0 - (\% \text{ Sand} + \% \text{ Clay})$

Where,

H_1 = Hydrometer reading at 60 s

T_1 = observed Temperature at 60 s (= 28 °C)

H_2 = Hydrometer reading at 3 hours

T_2 = observed Temperature at 3 hours (= 28 °C)

2.4.4. Metal analysis

Acid digestion procedure was employed for the determination of the concentrations of the environmentally available metals in the soil samples. Prior to the determination of the metal concentration in soil, sample digestion is a necessary pre-treatment step. The most common method for determining the concentrations of metal contained in a soil sample is the total elemental analysis. The review on total metal concentration determination in soil samples revealed that heavy metals bound to non-silicates and silicate structures were determined with concentrated acids such as HCl, $HClO_4$, and HF [71]. However, in recent times, strong acid extractant such as concentrated nitric acid, aqua regia (a mixture of concentrated HCl and concentrated HNO_3 in ratio 3:1) and dilute form of the acids are used to determine total metals in contaminated soils [72]. The acids dissolve almost all elements that could become environmentally available especially metal oxides and carbonates [73].

However, heavy metals bound to silicate structures are not normally dissolved since they are not usually mobile in the environment [74] but the extractant normally give a reliable measure of metals added to soils as non-silicates from industrial sources that has potential for natural leaching and biological processes. The use of perchloric acid and hydrofluoric acid is generally fading away since complete dissolution of the soil is no longer required for total metal analysis in environmental work [75].

Most current environmental analysis work that involved total metal analysis has employed the use of 2 M HNO_3 [3,76,77]. The levels of heavy metals extracted with 2 M HNO_3 have been reported to represent maximum contents of potentially available metals for plants [77]. In this study, approximately 1.0 g each of the composite soil samples were digested with 20 ml of 2 M HNO_3 by heating the vessel in a water bath between 90-100°C for 2 hours with shaking every 20 minutes. The vessel was allowed to cool to room temperature. The digestate was filtered and made up to mark with distilled water. Two replicate samples were digested in the same way together with a reagent blank and all the samples were analyzed using Buck 200A Atomic Absorption Spectrophotometer using air-acetylene flame.

2.5. Organochlorine pesticides and PAHs analysis

2.5.1. Soil sample extraction

A solvent mix of pesticide residue (PR) grade acetone and methylene chloride (50:50) was prepared and 50 ml of the solvent mix was added to about 10 g of the sample, spiked with 1 ml of surrogate standard [78,79]. The sample was placed in the ultrasonic bath (Grant Instruments) and sonicated for about 10-15 minutes at 70°C. Then 10 g of anhydrous sodium sulphate (Na_2SO_4) was added to the sample to remove the water present initially in the sample and shaken gently until a clear extract was developed. The procedure was repeated once more with an additional 50 ml of solvent mix. The solvent was then concentrated on a rotary evaporator (Buchi equipment) and exchanged with 5 ml of n-hexane then re-concentrated to 1 ml. The concentration of the sample extract was necessary to remove the acetone and methylene chloride used in the extraction. In the course of the concentration of the solvent on the evaporator, the acetone and methylene chloride was removed leaving n-hexane with the extract. This is necessary to prepare the extract for clean-up and fractionation in a column containing n-hexane-mixed silica slurry. The extracted sample was then fractionated into the aliphatic and aromatic fractions (PAHs) using silica gel column [80]. The same procedure was repeated for the extraction of organochlorine pesticide in the sample using acetone and dichloromethane (DCM) solvent mixture (50:50).

2.5.2. Column packing for fractionation and clean-up of extract

Glass column was packed with 10 g of 100-200 mesh silica gel preconditioned (baked) at 105°C overnight. The silica was mixed with n-hexane to form slurry. The column was then eluted with about 15 ml of PR grade n-hexane and the solvent was collected to waste. Caution was taken not to allow the column to dry up. Using 1 ml pipette, 1 ml of the extract in n-hexane was added onto the column and then eluted with 60 ml of n-hexane. The aliphatic fraction was collected and the column was then eluted with 40 ml of PR grade dichloromethane (DCM) and another fraction was collected, which is the Polycyclic Aromatic Hydrocarbons (PAHs) fraction. Each of these fractions was concentrated to 1 ml using a rotary evaporator. After concentrating the PAH fraction to 1 ml, 5 ml of PR grade n-hexane was added and was further concentrated to remove the DCM using the rotary evaporator. The concentrates were later transferred into 2 ml sample vials using the graduated 2 ml pipettes. The final volumes of the extracts were noted and the sample extracts were subjected to GC-MS analysis. The same clean-up procedure was carried out on the organochlorine pesticide extract by eluting the column with PR grade n-hexane.

2.5.3. *Instrumental analysis*

The extracts were analysed for PAHs and organochlorine pesticides using a GC equipped with MS detector. Separation was achieved by injecting about 1 μL of each extract into the GC system (in a splitless mode) through a capillary column (3.0 m length, 0.25 mm internal diameter, 0.25 μm film thickness). Helium gas was used as the carrier gas. Table 1 summarize the GC/MS condition. The quantification limit of the PAHs in the standard and the samples was 0.001 ppm, while that of the organochlorine pesticide residue was 0.002 ppm. The average response factor for the weight ranges were calculated and used for sample quantification.

GC/MS condition	PAHs	Organochlorine pesticide
Instrument	Shimadzu GC-MS QP2010	Shimadzu GC-MS QP2010
Column	HP-1MS (Cross linked PH ME siloxane) 19091S-933 Film thickness: 0.25 μm, Length: 3.0 m, Column ID: 0.25 mm	HP-1MS (Cross linked PH ME siloxane) 19091S-933 Film thickness: 0.25 μm, Length: 3.0 m, Column ID: 0.25 mm
Injection method	Splitless mode	Splitless mode
Injection volume	1.0 μL	1.0 μL
Carrier gas	Helium (1.2 ml/min)	Helium (1.18 ml/min)
Injection temperature	250 °C	250 °C
Oven temperature programme	GC oven temperature was kept 60 °C for 1 min First ramp at 5 °C/ min to 180 °C Second ramp at 10 °C/min to 280 °C Final temperature at 10 °C/min to 300 °C	GC oven temperature was kept 80 °C for 1 min First ramp at 5 °C/ min to 200 °C Final temperature at 10 °C/min to 270 °C
MS mode	Total ion current (TIC) mode	Selected Ion Monitoring Mode (SIM)
Ion source temperature	200 °C	200 °C
Interface temperature	250 °C	250 °C

Table 1. Operational conditions of GC/MS for PAHs and organochlorine pesticide

The concentration of each analyte was determined by calculating the amount of analyte or hydrocarbon range injected from the peak response in area ratio. The contribution from the solvent front and the surrogate compound were excluded from the total area of the sample.

$$C_f = \frac{A(p)\ x\ R_f\ x\ V_f\ x\ D_f\ x\ 1000}{W_i}$$

Where,

C_f = Final Sample concentration (μg/L)

A(p) = Measured area of peak (peaks)

W_i = Initial weight extracted (g dry weight)

V_f = Final extract volume (ml).

D_f = Dilution factor of sample or extract if diluted.

R_f= Response factor from the calibration standard calculation

$$R_f = \frac{Concentration\ (P)}{Area\ (P)}$$

Concentration (p) = Total concentration of range

3. Results and discussion

3.1. Soil pH

Soil pH is one of the most important physico-chemical properties which control many other soil physical, chemical, and biological properties. The pH of a soil is affected by the concentration of CO_2 in the soil air, salt concentration (salt effect), and the presence of colloidal particles (suspension effect). The higher the CO_2 concentration in the soil solution, the lower the pH, and the pH of a neutral or calcareous soil is very sensitive to small changes in CO_2 concentration. pH greatly affects the solubility of minerals in soils. Most minerals are more soluble in acid soils than in neutral or slightly basic solutions [81]. The pH of a soil is related to the bioavailability of metals in the soil to plants. At low pH, metals are easily bioavailable because there is increase in the solubility of metals. However, at high pH i.e. slightly alkaline to highly alkaline conditions, metals are not easily mobilized, because they are not easily soluble at high pH, hence not easily bio available [82]. The pH of the composite soils collected at the five different points (SA1-SA5) opposite the market at a distance of about 100 m to the market was slightly acidic with the pH of the topsoil (0-15cm) slightly lower than that of the subsoil (15-30cm) except at the downstream of the river (sampling area C) (Table 2).

3.2. Soil Organic Matter

Soil organic matter is any material produced originally by living organisms (plant or animal) that is returned to the soil and goes through the decomposition process. At any given time, it

Physical Parameter	Sampling Area A		Sampling Area B		Sampling Area C	
	A_1S_1	A_2S_2	B_1S_1	B_2S_2	C_1S_1	C_2S_2
	(0-15cm)	(15-30cm)	(0-15cm)	(15-30cm)	(0-15cm)	(15-30cm)
pH	5.43	5.58	6.69	7.19	6.82	6.11
% Organic matter content	0.83	1.05	0.97	1.80	0.88	1.31
% Clay	24.8	20.8	22.8	26.8	24.8	28.8
% Silt	20	24	18	28	24	22
% Sand	55.2	55.2	59.2	45.2	51.2	49.2

Table 2. Summary of Soil physicochemical parameters

consists of a range of materials from the intact original tissues of plants and animals to the substantially decomposed mixture of materials known as humus. Most soil organic matter originates from plant tissue. Plant residues contain 60-90 percent moisture. The remaining dry matter consists of carbon (C), oxygen, hydrogen (H) and small amounts of sulphur (S), nitrogen (N), phosphorus (P), potassium (K), calcium (Ca) and magnesium (Mg). Although present in small amounts, these nutrients are very important from the viewpoint of soil fertility management [83]. Soil organic matter (SOM) serves as a soil conditioner, nutrient reservoir, substrate for microbial activity, preserver of the environment, and major determinant for sustaining and increasing agricultural productivity. The OM content of soils ranges from less than1% in desert soils to close to 100% in organic soils. A typical agricultural soil may contain between 1 and 5% OM in the top 15 cm [84]. The percentage organic matter of the soil samples collected around the site is shown in Table 2. Generally, the subsoil has more organic matter content than their respective topsoil. The top soil contains less than 1% OM while the subsoil contains higher values. Soil organic matter performs very important functions in the soil such as: acts as a binding agent for mineral particles, this is responsible for producing friable (easily crumbled) surface soils; increases the amount of water that a soil may hold; and provides food for organisms that inhabit the soil. Humus is an integral component of organic matter because it is fairly stable and resistant to further decomposition. Humus is brown or black and gives soils its dark colour. Like clay particles, humus is an important source of plant nutrients.

3.3. Particle size distribution (soil texture)

Particle size is a fundamental property of any sediment, soil or dust deposit that can provide important clues to nature and provenance and influences a variety of other properties [85]. Soil particle size distribution is of greta importance to soil water movement, soil erosion and soil solute migration [86]. It is the determination of the proportion of each soil fraction in the samples and includes the % clay, % silt and % sand. Generally, in all the soil samples collected, sand has the highest percentage. The topsoil of the site has the highest percentage of sand in all the sampling areas. Figure 2 shows the variation in the physico-chemical parameters in the soil.

3.4. Concentration of heavy metals in soil

The results of heavy metals analyzed in the soil samples are shown in Table 3 while Figure 3 shows the distribution around the study site. There is variation in the distribution of the metals around the market with some of the metals such as lead (23.8 mg/kg) and copper (10.4 mg/kg) having the highest concentrations in the subsoil (15-30 cm). The concentrations of all the metals around the market is the highest when compared to the downstream and upstream of the river. This may be attributed to the anthropogenic activities in this area. In most cases, the subsoil has higher concentration of the metals than their respective topsoil, except Zn, Cd, and Ni. However, Cadmium compounds being of very low solubility in water will have little downward movement and accumulate in the topsoil [87]. This may be due to strong adsorption of Cd by the surface soils, which have a higher organic matter content and higher pH; factors which are known to increase Cd adsorption in soils [88]. Also, it was reported that Cd accumulated in the soil surface layers and had low vertical movement when mobility of heavy metals contained in the sludge and the wastewater used for the irrigation was studied in the soil samples collected from the soil profiles from the surface to 100 cm depth of farmland [89]. The mean concentrations and standard deviation of Pb, Zn, Cu, Cr, Cd and Ni in the topsoil (0 – 15cm) and subsoil (15 – 30cm) around the river basin at Isheri are 14.6 ± 4.5 mg/kg and 18.1 ± 8.1 mg/kg; 38.0 ± 27 mg/kg and 44.6 ± 28 mg/kg; 6.41 ± 3.3 mg/kg and 7.77 ± 3.7 mg/kg; 7.35 ± 0.1 mg/kg and 10.5 ± 0.4 mg/kg; 0.74 ± 0.3 mg/kg and 0.63 ± 0.03 mg/kg ; and 3.81 ± 0.5 mg/kg and 5.70 ± 0.3 mg/kg, respectively.

Comparing the results of heavy metal concentrations obtained in this study with the control sample collected at a rural community, Lalupon in Ibadan and some typical values in rural and urban soil around the world (Table 4). The average metal concentrations in both the top and subsoil were higher than the values obtained from the control sample. There are no background and baseline data on toxic metals concentrations in Nigeria. Metal concentrations in the study site when compared to the control sample suggested that anthropogenic inputs have occurred over the years. The concentrations of all the metals were below the values reported [90] except Cd which appeared higher in both the top and the subsoil of the study area.

Metals	Sampling Area A		Sampling Area B		Sampling Area C		Mean Conc.	
	A_1S_1	A_2S_2	B_1S_1	B_2S_2	C_1S_1	C_2S_2	Topsoil	Subsoil
	(0-15cm)	(15-30cm)	(0-15cm)	(15-30cm)	(0-15cm)	(15-30cm)	(mg/kg)	(mg/kg)
Pb	12.9	13.9	17.7	23.8	11.4	12.4	14.6±4.5	18.1±8.1
Zn	20.8	20.7	57.0	64.7	19.0	24.5	38.0±4.5	44.6±28
Cu	6.12	6.19	8.73	10.4	4.09	5.14	6.41±3.3	7.77±3.7
Cr	10.4	10.4	7.3	10.7	7.4	10.2	7.35±0.1	10.5±0.4
Cd	0.73	0.55	0.93	0.61	0.55	0.65	0.74±0.3	0.63±0.03
Ni	5.27	4.72	4.19	5.91	3.43	5.48	3.81±0.5	5.70±0.3

Table 3. Heavy metal concentrations (mg/kg) in the soil around the cattle market in Isheri, Nigeria

There is positive and significant correlation between Zn, Pb and Cu and Ni with Cr. Atmospheric fallout of the metals such as lead from the motor vehicle exhaust from the nearby busy Lagos-Ibadan express road as well as movement of goods around the market may have contributed to the high concentrations of heavy metals in the soil. Also, there is positive and significant correlation between organic matter and Ni, Pb, and Cu.

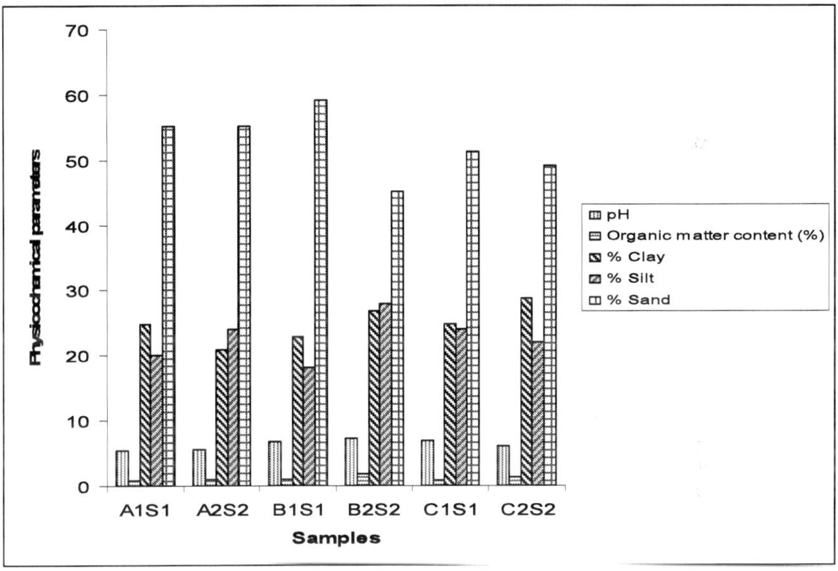

Figure 2. Variation in the physicochemical parameters in the soil around the cattle market in Isheri, Nigeria

	Pb	Zn	Cd	Cu	Cr	Ni
Topsoil (this study)	14.6	38	0.74	6.41	7.35	3.81
Subsoil (this study)	18.1	44.6	0.63	7.77	10.5	5.70
Control site (0-15cm)	9.12	ND	ND	4.57	4.83	0.57
Rural–urban Soils average (Bowen, 1979)	35	90	0.35	30	70	-

Table 4. Comparison of mean concentrations (mg/kg) of metals in this study with the control sample and typical soil values around the world

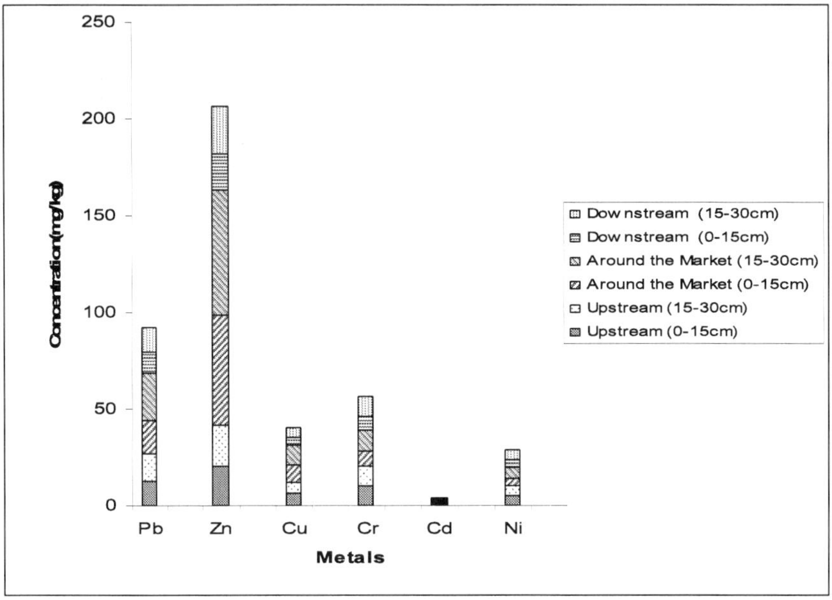

Figure 3. Heavy metal distribution in the soil around the cattle market in Isheri, Nigeria

3.5. Concentration of Polynuclear Aromatic Hydrocarbons (PAHs)

The results of the PAHs in the soil samples collected along the River Ogun basin at Isheri, along Lagos-Ibadan Express road are shown in the Table 5. Regular burning of all kinds of waste, tyres etc. was observed in this area and this might be responsible for the presence of PAHs around the study site. Some of the PAHs such as naphthalene, phenanthrene, fluoranthene, chrysene benzo(a)anthracene and perylene, were found at significantly higher concentrations around the area. Out of the seven species regarded as potential carcinogenic compounds, namely: chrysene, benzo(a)anthracene, benzo(k)flouranthene, benzo(g,h,i)perylene, dibenzo(a,h)anthracene, indeno(1,2,3-cd)pyrene, and benzo(a)pyrene, chrysene has the highest concentration in the samples collected at all the sampling points. Its concentration was relatively higher in the subsoil (15-30cm) than the topsoil (0-15cm) except at sampling point C (downstream), where there is little variation in the concentrations in both layers. Figure 4 shows the variation in the concentration of PAHs with depth.

PAHs Component	Sampling area A		Sampling area B		Sampling area C	
	A_1S_1	A_2S_2	B_1S_1	B_2S_2	C_1S_1	C_2S_2
	(0-15 cm)	(15-30 cm)	(0-15 cm)	(15-30 cm)	(0- 15 cm)	(15-30 cm)
Naphthalene	419	244	159	663	668	609
2-methyl Naphthalene	158	103	64.3	234	245	227
Acenaphthylene	8.1	4.0	8.4	24.2	20.2	25.8
Acenaphthene	3.7	19.7	3.8	9.3	9.1	69.2
Flourene	17.9	9.3	18.3	18.1	29.6	18.4
Phenanthrene	60.4	47.5	91.7	76.4	129	85.7
Anthracene	14.8	11.9	17.2	8.3	22.9	6.4
Flouranthene	41.2	121	69.1	31.8	69.8	38.9
Pyrene	14.8	69.1	57.5	18.7	43.7	19.6
Chrysene	93.4	273	81.4	239	193	192
Benzo(a)anthracene	105	115	ND	53.2	50.3	ND
Benzo(k)flouranthene	19.8	59.3	25.2	7.1	12.0	9.3
Benzo(b)flouranthene	5.8	18.0	4.1	18.5	51.1	49.5
Perylene	10.6	30.3	181	ND	641	399
Benzo(g,h,i)perylene	ND	12.5	8.5	ND	ND	ND
Dibenzo(a,h)anthracene	ND	12.6	ND	ND	ND	ND
Indeno(1,2,3-cd)pyrene	3.6	ND	ND	ND	32.0	ND
Total PAHs Concentration (μg/kg)	976	1150	790	1400	2220	1750

Table 5. PAHs concentration (μg/kg) in the soil around the cattle market in Isheri, Nigeria

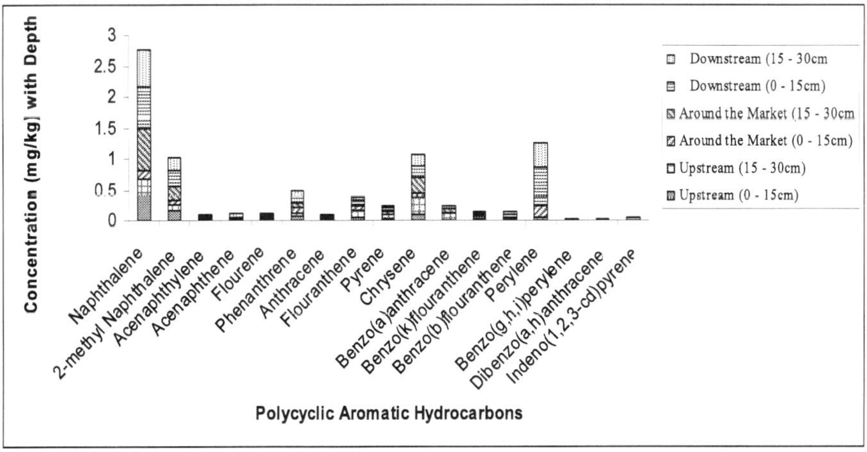

Figure 4. Variation in the concentration of PAHs with respect to depth in the soil around the cattle market in Isheri, Nigeria

Comparing the values obtained in this study with background soil concentrations of PAHs (Table 5) [91], the concentrations obtained in this study was higher than the concentrations in rural and agricultural soils in most cases except benzo(g,h,i) perylene and indeno(1,2,3-cd)pyrene. There is a significant and positive correlation between phenanthrene versus flourene and perylene; flouranthene versus pyrene, benzo(k)flouranthene, dibenzo(a,h)anthracene and benzo(g,h,i)perylene; and benzo(k) flouranthene versus benzo(g,h,i)perylene and dibenzo(a,h)anthracene.

Compounds	Concentrations (x 10^{-3} mg/kg)		
	Rural Soil	Agricultural Soil	Urban Soil
Acenaphthene	1.7	6	NE
Acenaphthylene	NE	5	NE
Anthracene	NE	11-13	NE
Benzo(a)anthracene	5-20	56-110	169-59,000
Benzo(a)pyrene	2-1,300	4.6-900	165-220
Benzo(b)fluoranthene	20-30	58-220	15,000-62,000
Benzo(c)perylene	NE	53-130	60-14,000
Benzo(g,h,i)perylene	10-70	66	900-47,000
Benzo(k)fluoranthene	10-110	58-250	300-26,000
Chrysene	38.3	78-120	251-640
Fluoranthene	0.3-40	120-210	200-166,000
Fluorine	NE	9.7	NE
Indeno(1,2,3-cd)pyrene	10-15	63-100	8,000-61,000
Phenanthrene	30.0	48-140	NE
Pyrene	1-19.7	99-150	145-147,000

Note: NE-Not established

Table 6. Background soil concentrations of PAHs

3.6. Organochlorine pesticides

There is variation in the concentrations of the organochlorine pesticides (OCP) in the soil samples collected around the cattle market (Table 6). The major OCP residue identified in the area with their percentage concentrations were p,p'-DDT, 344 μg/kg (24.5%), δ-BHC, 290 μg/kg (20.7%), Endosulfan sulphate, 186 μg/kg (13.2%), lindane (γ-BHC), 129 μg/kg (9.1%), Endrin, 110 μg/kg (7.8%), Aldrin, 93.9μg/kg (6.7%), Dieldrin, 93.8 μg/kg (6.7%), o,p-DDE, 45.0 μg/kg (3.2%), β-BHC, 42.1 μg/kg (3.0%), Endosulfan II, 34.5 μg/kg (2.5%), and Heptachlor, 22.8 μg/kg (1.6%). The variation in the concentrations of the organochlorine pesticides with respect to depth around the cattle market is presented in Figure 4. Dieldrin, a metabolite of aldrin is present at higher concentration in most of the sampling points when compared to the parent aldrin. The presence of o,p' DDE revealed that there is historical use as well as recent us of the parent DDT in the vicinity of the cattle market. Though, DDT has been banned for over 20

http://dx.doi.org/10.5772/57302

years, it's presence around the cattle market indicate that it is illegally imported into the country as it is not produced in Nigeria. Also, lindane, one of the nine new POPs was found at high concentration around the market. The presence of some of the banned chemical residue around the cattle market revealed that there is weak regulation and low enforcement on banned chemicals in the country.

In most cases, the concentrations of organochlorine pesticides is highest at the sampling point C which could be attributed to the use of pesticides to control household pests and insects in the area; as the area is a slum with most houses built of planks, rust iron sheets and polypropylene material and a dumpsite at the bank of the river. There is very strong and significant correlation at the 0.01 levels (2-tailed) between DDT versus δ-BHC, heptachlor, aldrin and dieldrin; dieldrin versus δ-BHC, heptachlor and aldrin; aldrin versus heptachlor as well as endrin versus δ-BHC.

	Sampling area A		**Sampling area B**		**Sampling area C**	
	A_1S_1	**A_2S_2**	**B_1S_1**	**B_2S_2**	**C_1S_1**	**C_2S_2**
Organochlorine Pesticide	(0-15cm)	(15-30cm)	(0-15cm)	(15-30cm)	(0-15cm)	(15-30cm)
α-BHC	13.2	ND	ND	ND	ND	ND
β-BHC	34.5	7.6	ND	ND	ND	ND
Lindane	108.5	12.0	7.5	ND	ND	ND
δ-BHC	47.4	ND	53.7	60.8	64.4	63.5
Heptachlor	3.3	ND	3.7	5.3	6.0	4.5
Aldrin	13.6	ND	14.2	16.4	34	14.7
Heptachlor-epoxide	ND	ND	ND	ND	ND	ND
cis-Chlordane	ND	ND	ND	ND	ND	ND
Endosulfan I	ND	ND	ND	ND	ND	ND
trans-Chlordane	ND	ND	ND	ND	ND	ND
DDE	ND	ND	14.0	14.3	16.7	ND
Dieldrin	15.2	ND	15.1	16.4	22.2	24.9
Endrin	23.4	ND	28.0	23.7	13.5	20.9
Endosulfan II	ND	ND	ND	ND	15.8	18.7
Endosulfan sulphate	ND	ND	ND	ND	186	ND
p,p' DDT	45.6	ND	59.4	66.0	97.8	75.2
Methoxychlor	ND	ND	ND	ND	ND	ND
Total organochlorine pesticides concentration (μg/kg)	305	19.6	196	203	456	222

Table 7. Organochlorine pesticides concentrations (μg/kg) in the soil around the cattle market in Isheri, Nigeria

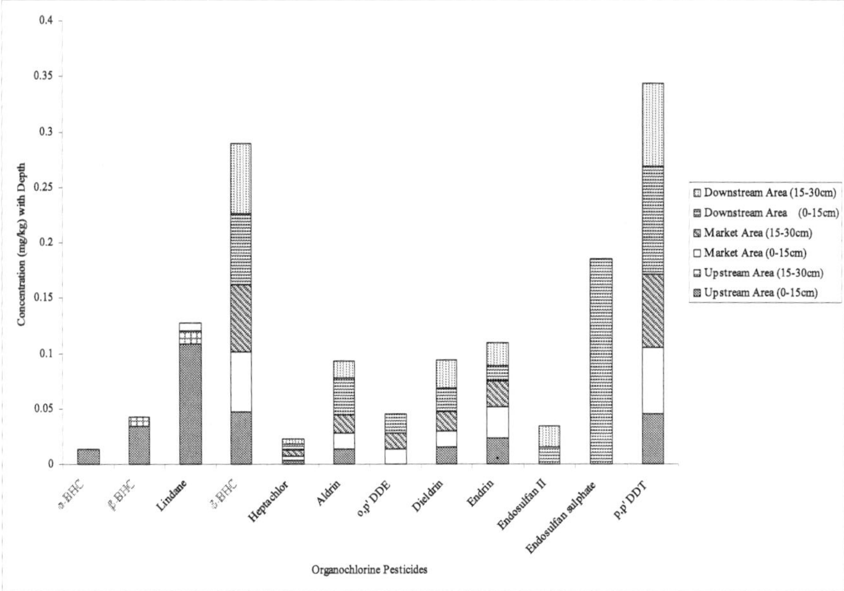

Figure 5. Variation in organochlorine pesticides residue concentrations with respect to depth in the soil around the cattle market in Isheri, Nigeria

4. Conclusion

In this study, the distribution, concentration and profiles of some ubiquitous environmental pollutants such as heavy metals (Pb, Ni, Cu, Cr, Zn and Cd), polycyclic aromatic hydrocarbons (PAHs) and organochlorine pesticides in soil of a cattle market around River Ogun Basin, Isheri, Nigeria were assessed as well as some soil physico-chemical characteristics. The pH of the top soil was observed to be lower than the corresponding subsoil in most of the areas sampled except the areas downstream of the river. The pH of the topsoil being lower than the subsoil may be responsible for the high metal concentrations in the subsoil. Low pH of top soil means more heavy metals will be dissolved in top soil and the texture of the soil being sandy and loose also might favour the migration/leaching of heavy metals to lower fractions. The solubilising heavy metals in the top soil may be migrating into the subsoil, except in cadmium, which was concentrated in the top soil. The soil of the entire sample area was sandy loamy soil with low concentration of organic matter. The cattle wastes released in the area might have relatively increased the organic matter content of the soil. Though, the topography of the area and the location of the cattle market, being sloppy and close to the river basin favoured the washing away of the top soil by run-off into the river and this might be responsible for the high organic matter in the subsoil. The pH of the soil of the area was slightly acidic to neutral.

The soil sample with high organic matter content has high metal concentrations in most cases except Cd, Pb, Cu and Cr. Generally, there is positive and significant correlation between the organic matter content versus Pb, Zn, Cu, Ni and Cr. There is variation in the concentrations of PAHs and organochlorine pesticides residues with depth around the river basin. The detection of banned organochlorine pesticides in the soil of the area could be attributed to both the historical use as well as recent use of these chemicals in the area. Thus, there is need for stricter regulations on banned chemicals to safe our environment from ubiquitous and persistent environmental pollutants.

Author details

A.A. Adeyi, O.M. Omidiran and O. Osibanjo

Analytical and Environmental Chemistry Unit, Department of Chemistry, University of Ibadan, Ibadan, Nigeria

References

[1] Oketola AA, Fagbemigun KT. Determination of Nonylphenol, Octylphenol and Bisphenol-A in Water and Sediments of Two Major Rivers in Lagos, Nigeria. Journal of Environmental Protection 2013;4(7A) 38-45.

[2] Ogundiran MB, Osibanjo O. Mobility and Speciation of Heavy Metals in Soils Impacted by Hazardous Waste. Chemical Speciation and Bioavailability 2009; 21(2) 59-69.

[3] Oketola AA, Adebisi AA, Morakinyo O. Distribution and Bioavailability of Metals in Gasoline Contaminated Sites in Lagos, Nigeria. Journal of Solid Waste Technology and Management 2013; 39 (3) 161-172.

[4] Zhu WX. Consideration of Soil Ecological Processes in Restoration and Succession Phytoecological Sinica 2005;29(3) 479-486.

[5] Aoyama M, Tanaka R. Effects of Heavy Metal Pollution of Apple Orchard Surface Soils Associated with Past Use of Metal-Based Pesticides on Soil Microbial Biomass and Microbial Communities. Journal of Environmental Protection 2013;4(4A) 27-36.

[6] Oh K, Li T, Cheng H, Hu X, He C, Yan L, Shinichi Y. Development of Profitable Phytoremediation of Contaminated Soils with Biofuel Crops. Journal of Environmental Protection 2013;4(4A) 58-64.

[7] Wang XS, Qin Y. Some Characteristics of the Distribution of Heavy Metals in Urban Topsoil of Xuzhou, China. Environmental Geochemistry and Health 2007; 29 11-19.

[8] Mikanova O. Effects of Heavy Metals on Some Soil Biological Parameters. Journal of Geochemical Exploration 2006;88(1-3) 220-223.

[9] Devkota B, Schmidt GH. Accumulation of Heavy Metals in Food Plants and Grasshoppers from the Taigetos Mountains, Greece. Agriculture, Ecosystems and Environment 2000;78(1) 85-91.

[10] Chander K, Brookes PC, Harding SA. Microbial Biomass Dynamics Following Addition of Metal-Enriched Sewage Sludges to a Sandy Loam. Soil Biology and Biochemistry 1995;27(11) 1409-1421.

[11] Giller KE, Witter E, McGrath SP. Toxicity of Heavy Metals to Microorganisms and Microbial Processes in Agricultural Soils: A Review. Soil Biology and Biochemistry 1998;30(10-11) 1389-1414.

[12] Nosir S, Stanislav P, Yosef S. The Impact of the Almalyk Industrial Complex on Soil Chemical and Biological Properties. Environmental Pollution 2005;136(2) 331-340.

[13] Chugh LK, Sawhney SK. Effect of Cadmium on Germination, Amylase and Rate of Respiration of Germination Pea Seed. Environmental Toxicology 1996;134(6) 45-61.

[14] Cherngkh NA. Alteration of the Concentration of Certain Elements in Plants by Heavy Metal in the Soil. Soviet Soil Science 1991; 23(6) 45-53.

[15] Johnson FM. The Genetic Effects of Environmental Lead. Journal of Environmental Toxicology 2004; 410(2) 123-140.

[16] Van E. Long Term Heavy Metal Leaching from Soils, Sewage Sludge and Soil Sewage Mixtures. Environmental Science Technology 1991; 25 706-711.

[17] WHO. Public Health Impact of Pesticides used in Agriculture. World Health Organization; 1990.

[18] Walker A, Turner IJ, Cullington JE, Welch SJ. Aspects of the Adsorption and Degradation of Isoproturon in a Heavy Clay Soil. Soil Use and Management 1999;15(1) 9-13.

[19] Weed SB, Weber JB. Pesticide-Organic matter interactions. In: Guelzi WD. (ed.) Pesticides in Soil and Water. SSSA: Madison, WI; 1974. p39-44.

[20] Harner T, Wideman, JL, Jantunen LM, Bidleman TF, Parkhurst WJ. Residues of Organochlorine Pesticides in Alabama Soils. Enviromental Pollution 1999;106(3) 323-332.

[21] Gong ZM, Tao S, Xu FL, Dawsob R, Liu WX, Cui Y H, Cao J, Wang XJ, Shen WR, Zhang WJ, Quing BP, Sun R. Level and Distribution of DDT in Surface Soils from Tianjin, China. Chemosphere 2004;54(8) 1247-1253.

[22] Andrade ML, Reyzabal ML, Covelo EF, Vega FA. Organochlorine Pesticides Content in Soils from Bahia Blanca Horticultural Belt (Argentina). Canadian Journal of Soil Science 2005;85(2) 273-282.

[23] Zhu YH, Liu H, Xi Z, Cheng X, Xu X. Organochlorine Pesticides (DDTs and HCHs) in Soils from the Outskirts of Beijing, China. Chemosphere 2005;60(6) 770-778.

[24] Nádaská G, Lesn◉ J, Michalík I. Environmental Aspect of Manganese Chemistry. Hej: Env 100702-A.

[25] Crossgrove J, Zheng GW. Manganese Toxicity upon Overexposure. NMR Biomedicine 2004;17(18) 544-553.

[26] Environment Agency. Contaminants in Soil: Updated Collation of Toxicological Data and Intake Values for Humans; Cadmium. Science Report SC050021/SR TOX7. Bristol: Environment Agency; 2009.

[27] Lenntech. Nickel. http://www.Lenntech.com/periodic-chart-elements/Ni-en.htm (accessed 18 June 2013).

[28] Oliver MA. Soil and Human Health: A Review. European Journal of Soil Science 1997;48(4) 573-592.

[29] Puschenreiter M, Horak O, Friesel W, Hartl W. Low-cost Agricultural Measures to Reduce Heavy Metal Transfer into the Food Chain- A Review. Plant Soil and Environment 2005;51(1) 1-11.

[30] Canadian Council of Ministers of the Environment. Canadian Soil Quality Guidelines for the Protection of Environmental and Human Health: Mercury (1norganic) Winnipeg. Publication No. 1299.

[31] Gilbert SG. A Small Dose of Toxicology. The Health Effects of Common Chemicals. 2nd Edition. Heathy World Press. USA; 2012.

[32] American Academics of Pediatrics. Lead Poisoning: from Screening to Primary Prevention. Pediatrics 1993; 92; 176-183.

[33] Agency for Toxic Substances and Disease Registry. Case Studies in Environmental Medicine (CSEM). Lead Toxicity; 2010.

[34] http://www.kingcounty.gov/healthservices/health/ehs/toxic/ArsenicFacts.aspx (accessed 21 June 2013).

[35] Cempel M, Nikel G. Nickel: A Review of its Sources and Environmental Toxicology. Polish Journal of Environ. Studies 2006; 15(3) 375-382.

[36] Agency for Toxic Substances and Disease Registry. Toxicological Effects of Nickel; 2005.

[37] Duke JM. Nickel in rocks and ores. In: Nriagu JO, ed. Nickel in the Environment. New York, NY: John Wiley and Sons, Inc. 1980; 27-50.

[38] New Hampshire Department of Environmental Services. Environmental Fact sheet. ARD-ehp-9; 2013.

[39] UNEP Chemicals. Persistent Organic Pollutants (POPs). United Nations Environment Programme. http://www.chem.unep.ch/pops/ (accessed 20 March 2009).

[40] Intergovernmental Forum on Chemical Safety. Framework for the Management of PCBs; 2001.

[41] www.epa.gov/oia/toxics/pop.htm (accessed 21 June 2013).

[42] Marta Venier. Legacy and New POPs: What the North Americans Great Lakes can Tell Us. 6th Network Conference on Persistent Organic Pollutants 17-18 April 2012.

[43] Burken JG, Schnoor JL. Uptake and Fate of Organic Contaminants by Hybrid Poplar Trees: conference proceedings, April 13-17, 1997, 213th ACS National Meeting, American Chemical Society Environmental Division Symposia, San Francisco.

[44] Charles MJ, Schell MJ, Willman E, Gross HB, Lin Y, Sonnenberg S, Graham ML. Organochlorines and 8-Hydroxy-2-deoxyguanosine in Cancerous and Noncancerous Breast Tissue: Do The Data Support the Hypothesis that Oxidative DNA Damage Caused by Organochlorines Affects Breast Cancer? Archives of Environmental Contamination and Toxicology 2001; 41(3) 386-95.

[45] Charlier C, Pitance F, Piomteux G. PCB Residues in a Breast Cancer Patient Population. Bulletin of Environmental Contamination and Toxicology 2003;71(5) 887-891.

[46] Osibanjo O. Organochlorine in Nigeria and Africa. In: Fiedler H. (ed.) The Hand book of Environmental Chemistry. Springer-Verlag: Berlin Heidelberg: 2002. 3 p321-354.

[47] Sala M, Ribas-Fito N, Cardo E, de Muga ME, Marco E, Mazon C, Verdú A, Grimalt JO, Sunyer J. Levels of Hexachlorobenzene and Other Organochlorine Compounds in Cord Blood: Exposure Across Placenta. Chemosphere 2001;43 (4-7) 895-901.

[48] Swackhamer D, Hites RA. Occurrence and Bioaccumulation of organochlorine Compounds in Fish from Siskiwit Lake, Isle Royale, Lake Superior. Environmental Science and Technology 1988; 22(5) 543–548.

[49] Schafer K, Kegley S, Patton S. Nowhere to Hide, Pesticide Action Network http://www.panna.org/resources/documents/nowhereToHideAvail.dv.html (accessed 15 December 2008).

[50] Kegley SE, Hill BR, Orme S, Choi AH. PAN Pesticide Database, Pesticide Action Network. http://www.pesticideinfo.org (accessed 5 December 2008).

[51] Schantz SL, Widholm JJ. Cognitive Effects of Endocrine-Disrupting Chemicals in Animals. Environmental Health Perspectives 2001;109(12) 1197-1206.

[52] Rivera S, Rosa R, Martınez E, Sunol C, Serrano MT, Vendrell M, Rodrıguez-Farre E, Sanfeliu C. Behavioral and Monoaminergic Changes after Lindane Exposure in Developing Rats. Neurotoxicololy and Teratology 1998;20(2) 155–160.

[53] Martinez AO, Martinez-Conde E. The Neurotoxic Effects of Lindane at Acute and Subchronic Dosages. Ecotoxicology and Environmental Safety 1995;30(2) 101-105.

[54] Ashu Johri A, Dhawan A, Singh RL, Parma D. Persistence in Alterations in the Ontogeny of Cerebral and Hepatic Cytochrome P450s following Prenatal Exposure to Low Doses of Lindane. Toxicological Sciences 2008;101(2) 331–340.

[55] Beard AP, Rawlings NC. Thyroid Function and Effects on Reproduction in Ewes Exposed to the Organochlorine pesticides lindane or Pentachlorophenol (PCP) from Conception. Journal of Toxicology and Environmental Health Part A 1999;58 (8)509–530.

[56] Decker GC, Bruce WN, Bigger JH. The Accumulation and Dissipation of Residues Resulting from the Use of Aldrin in Soils. Journal of Economic Entomology 1965;58(2) 266-271.

[57] Elgar KR. Analysis of Crops and Soils for Residues of the Soil Insecticides Aldrin and Telodrin. Journal of the Science of Food and Agriculture 1966;17(12) 541-545.

[58] Lichtenstein EP. Persistence and Behaviour of Insecticidal Residues in Soils and their Translocation into Crops. Archives of Environmental Health 1965; 10(6) 825-826.

[59] Edwards CA. Insecticide Residues in Soils. Residue Review 1966; 13 83-132.

[60] New Zealand Food Safety Authority. Endosulfan. Wellington, New Zealand. http://www.nzfsa.govt.nz/consumers/chemicals-nutrients-additives-and toxins/endosulfan/index.htm (accessed 25 March 2009).

[61] Tsibart AS, Gennadiev AN. Polycyclic Aromatic Hydrocarbons in Soils: Sources, Behavior, and Indication Significance (A Review). Eurasian Soil Science 2013; 46(7) 728-741.

[62] Agency for Toxic Substances and Disease Registry (ATSDR) Case Studies in Environmental Medicine. Toxicity of Polycyclic Aromatic Hydrocarbons (PAHs). US Department of Health and Human Services. Public Health Service. 2009.

[63] Wallingford GW, Powers WL, Murphy LS. Present knowledge of the effect of land application of animal wastes. 1975 International Symposium on Livestock Waste April 21-24 1975. ASAE, St Joseph, MI.

[64] Eck H.V., Stewart B.A. Manure. In: Rechcial J.E. (ed.) Soil Amendment and Environmental Quality. Lewis Publisher, New York: 1995. p169-198.

[65] Liang BL, Gregorick EG, Schnitzer M, Schulten HR. Characterization of Water Extracts of Two Manures and their Adsorption on Soils. Soil Science Society of America 1996; 60(6) 1758-1763.

[66] Teckon and Associate. Federal Republic of Nigeria, Federal Department of Water Resources. Improvement of River Channels along River Ogun (Lagos Lagoon to Olokemeji). Final Report. 1987. p163.

[67] Oketola AA, Adekolurejo SM, Osibanjo O. Water Quality of Rivr Ogun using Multivariate Statistical Techniques. Journal of Environmental Protection 2013; 4(5) 466-478.

[68] United State Environmental Protection Agency. Polycyclic aromatic Hydrocarbons (PAHs). Office of Solid Waste, Washinton DC 20460, 2008.

[69] Walkey A, Black IA. An Examination of Degtjareff Method for Determining Soil Organic Matter and a Proposed Modification of the Chromic Acid Titration Method. Soil Science 1934; 37(1) 29-38.

[70] Brady N.C. 1984. The Nature and Properties of Soils. Macmillan Publishing Co., New York; 1984. 9th ed. p366.

[71] Abdul-Saheb I, Schwab AR, Banks MK, Hetrick BA. Chemical Characterization of Heavy-metal Contaminated Soil in Southeast Kansas. Water, Air and Soil Pollution 1994;78(1-2) 73-82.

[72] Myung CJ, Thornton I. Heavy Metal Contamination of Soils and Plants in the Vicinity of a Lead-Zinc Mine, Korea. Applied Geochemistry 1996;11(1-2) 53-59.

[73] Kennedy JH. Analytical Chemistry: Principles. Sanders Company: United States of America; 1990. 2nd Ed.

[74] USEPA Report. Recent Development for In-situ Treatment of Metals Contaminated Soils, U.S. Environmental Protection Agency, Office of Solid Waste and Emergency Response; 1996.

[75] Ogundiran MA. Assessment and Chemical Remediation of Soil Contaminated by Lead-Acid Battery Waste in Lalupon, Oyo State, Nigeria. Ph.D thesis. University of Ibadan, Ibadan, Nigeria; 2007.

[76] Smejkalova M, Mikanova L, Boruvka L. Effects of Heavy Metal Concentrations on Biological Activity of Soil Microorganisms. Plant Soil Environment 2003; 49(7) 321-326.

[77] Onianwa PC, Fakayode SO. Lead Contamination of Topsoil and Vegetation in the Vicinity of a Battery Factory in Nigeria. Environmental Geochemistry and Health 2000; 22(3) 211-218.

[78] American Society for Testing and Materials. Methodology for Comparison of Petroleum Oil by Gas Chromatography. ASTM Method D 3328-78, 1982.

[79] United State Environmental Protection Agency. Polynuclear Aromatic Hydrocarbons. Washington DC, 1986.

[80] Massachusetts Department of Environmental Protection (DEP). Methods for the Determination of Extractable Petroleum Hydrocarbons (EPH). Massachusetts DEP, Wall Experiment Station; 1995.

[81] Bolan N.S., Kandaswamy K. pH. In: Daniel Hillel (ed.) Encyclopedia of Soils in the Environment. Elsevier Ltd; 2005. p196-202.

[82] Chuan MC, Shu GY, Liu JC. Solubility of Heavy Metals in a Contaminated Soil: Effects of Redox Potential and pH. Water, Air and Soil Pollution 1996;90(3-4) 543-556.

[83] Bot A., Benites J. The Importance of Soil Organic Matter. FAO Soil Bulletin; 2005.

[84] Schnitzer M. Organic Matter, Principle and Processes. In: Daniel Hillel (ed.) Encyclopedia of Soils in the Environment. Elsevier Ltd: 2005. p86-93.

[85] Ward A. The Importance of Particle Size Distributions to the Characterization of Soils. Caribbean Institute for Meteorology and Hydrology; 2012.

[86] Hu HC, Tian FQ, Hu HP. Soil Particle Size Distribution and its Relationship with Soil Water and salt under Mulched Drip Irrigation in Xinjiang of China. Science China Technological Sciences 2011;54(6) 1568-1574.

[87] Sidle RC, Kardos LT. Van Genuchten M.Th. Transport of Heavy Metals in Sludge-treated Forested Area. Journal of Environmental Quality 1977;6(4) 438-443.

[88] Loganathan P, Hedley MJ. Downward movement of Cadmium and Phosphorus from Phosphate Fertilisers in a Pasture Soil in New Zealand. Environmental Pollution 1997; 95(3) 319-324.

[89] Behbahaninia A, Mirbagheri AS, Javid AHT. Heavy Metals Transport in the Soil Profiles under the Application of Sludge and Wastewater. World Academy of Science, Engineering and Technology 2008;9 53-55.

[90] Bowen HJM. Environmental Chemistry of the Elements. Academic Press, London. 1979.

[91] Jones KC, Stratford JA, Waterhouse k. Polynuclear Aromatic Hydrocarbons in U.K Soils: Long Term Temporal Trends and Current Levels. Trace Subst. Environmental Health 1987 2: 140-148.

Chapter 8

Assessment of Historical Heavy Metal Pollution of Land in the Proximity of Industrial Area of Targoviste, Romania

Carmen Cristina Elekes

Additional information is available at the end of the chapter

http://dx.doi.org/10.5772/58304

1. Introduction

Contamination of soil and water with organic and inorganic pollutants is a subject of interest to European policy, looking for new ways of preventing pollution and remediating polluted sites. Source and type of contamination influences the nature of pollution and methods of remediation. Thus, accidental discharge can be isolated more easily from a small area of soil, while the contamination of water is more difficult to control. Deposition of particulate matter can contaminate wide surfaces with low concentrations of contaminants. However, particulate matter is the main source of contamination of soils with heavy metals.

Many international researches have studied the concentrations of metals in soil, in correlation with texture, structure and pH of soil. In order to reduce the concentration of heavy metal to a level allowing development of all crop plants under a secure intake level of heavy metals, pollution mitigation measures should be applied on those soils. Depending on land use, the legislation of each country establishes the normal and maximal limits of heavy metals in soil.

The development of metallurgical activities in an irresponsible manner and without taking into account the environmental damage, lead to a historical accumulation of heavy metals in soil. The concentration of heavy metals in soil varies significantly depending on the type of soil and geographic region. This indicates that the parental material, climatic conditions and human activities have a predominant impact on the chemical forms (speciation) and on the mobility of metals in soil. In addition to the methods of prevention and mitigation of pollutants emissions, there are a number of methods of remediation, based on soil stripping and replacement or methods of bioremediation.

Across Europe, an extensive study was conducted concerning the concentration of heavy metals in soil [1], involving the collaboration of several organizations: EuroGeo Survey, Geological Survey of Finland (GTK) and the Forum of European Geological Survey Directors (FOREGS). The project was conducted between 1996 and 2003, but unfortunately did not cover the Romanian territory. At the end of the project were drawn distribution maps of all metals in soils and sediments found along rivers and a Geochemical Atlas of Europe was designed.

Another significant study is Alina Kabata-Pendias book "Trace Elements in Soils and Plants" published so far in four editions [2]. This book provides a concise but comprehensive overview of the biogeochemistry of trace elements found in the soil-plant system. Includes over 400 references to recent studies that have been conducted to determine the metal content of the soil-plant system and highlights the significance of anthropogenic factors leading to the change of state elements in soil and plants. Subjects are bioindicators behaviour in the environment, soil remediation, and hyperaccumulation and hyperextraction of heavy metals from soil.

Organizations such as the Food and Agriculture Organization (FAO) and World Health Organization (WHO) have established very comprehensive reports related to the concentration of metals in food and doses considered daily necessary, maximum intake for different age groups, maximum limits in food and soils (Table 1). In Romania, the reference values for trace metals in soils are governed by Order 756 / 3rd of November 1997 [3]. It regulates normal values, alert thresholds and action levels for different trace elements by use of soils.

Heavy metal	Canadian Standard			Dutch standard		
	Arable land	Inhabited area	Industrial area	Arable land	Inhabited area	Industrial area
Cu	150	100	150	36	100	500
Zn	600	500	600	140	500	3000
Cd	3	5	3	0, 8	5	20
Pb	375	500	375	85	150	600

Table 1 Maximum limits of heavy metals in soil, according with Canadian and Dutch standards (ppm) [4]

The present study completes FOREGS project and aims to establish the concentrations of heavy metals in an industrial area, near the city of Targoviste. Depending on the climatic characteristics, on topography and on pollution rose, we have established representative points of soil sampling for the area. Samples were collected from both the surface of soil to determine the horizontal distribution of heavy metals in the industrial area, as well as on profile (0-40 cm) to determine the vertical distribution of the metals and to assess the extent of historical pollution.

2. Heavy metals occurrence in soil and potential impact on life quality

The problem of soil contamination with heavy metals, fuel and other toxic materials is a reality worldwide. Following the accidents which occurred with discharges of toxic materials on the ground, the affected area increased by infiltration of substances into groundwater. The groundwater is carrying the pollutants to residential areas, endangering the health of residents.

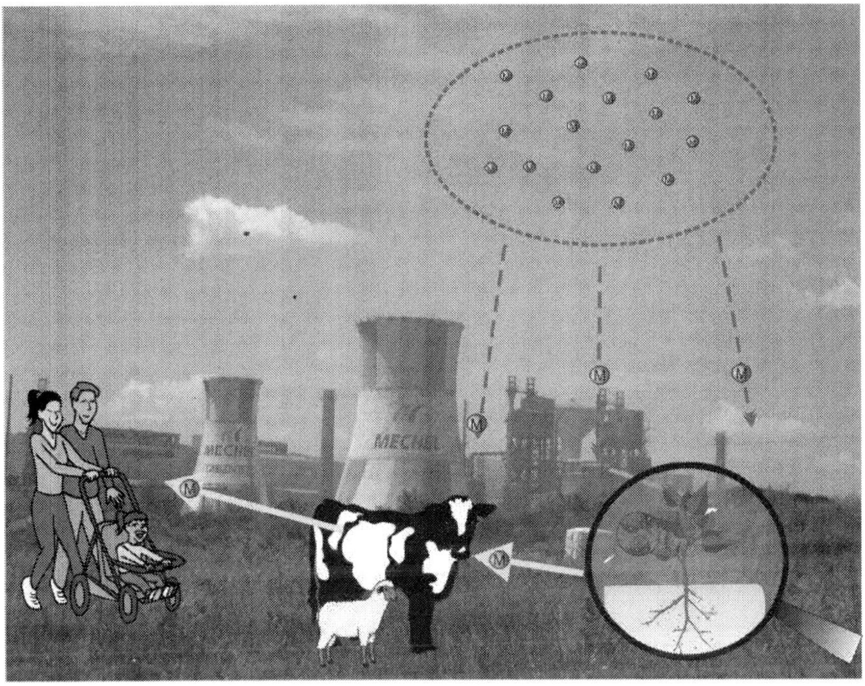

Figure 1. Transfer of metallic particulate to human body

Heavy metals are natural components, which occurred in high concentrations under natural conditions. In the twentieth century, metalliferous uploading of air, water, soil and therefore of plants and the human body, has become an important concern of international researchers. Heavy metals are considered a risk for living organisms because they tend to bioaccumulate. Bioaccumulation is an increase of concentration of a chemical in living organisms, as compared to the concentration of the element in the environment. Compounds are accumulated in living organisms by uptake from environment and storage at a higher rate than that of metabolism or excretion.

Emissions from metallurgical plants are transported by air masses and then deposited on the ground, leading to an increase of metal concentration in the upper layer of soil. Plants, perennial grasses especially have a high storage capacity of such metals in their shoots. The plants loaded with large amounts of heavy metals are consumed by animals that are grazing of this land (Figure 1).

Metal concentration in soil varies significantly depending on the soil type, but also by region [5, 6]. This indicates that the parent material and climatic conditions have a predominant impact on the chemical state of metals in soil. Kabata-Pendias and Krakowiak [7] set a factor of soil parameters (RDI - Relative explanation index) based on the calculation of correlation coefficients matrix for metals and some soil parameters: pH, clay content, cation exchange capacity, substance the organic content of the soil and iron. For approx. 1000 samples, the strongest positive linear correlation was obtained for metals and fine fraction of the soil. This relationship varies for the metals studied and is well illustrated by the content of heavy metals in soil which increases with increasing clay content. The highest value of the RDI when correlated with clay content (60-75%) was calculated on Zn, Fe, Ni and Cr, while the lower value (10-30%) was calculated for Cd, Pb, Cu, and Mn (Figure 2).

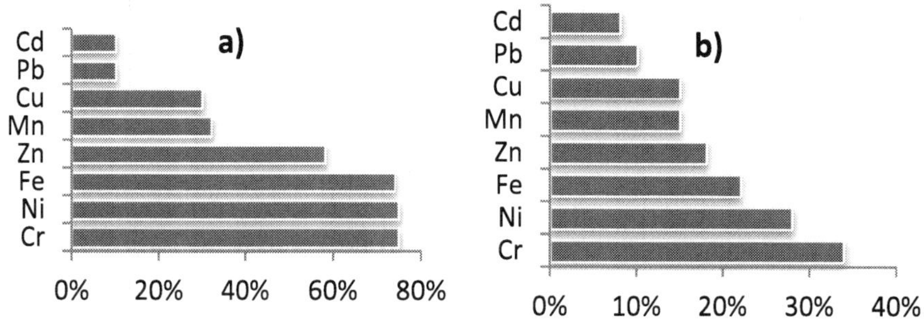

Figure 2. Relative explanation index (RDI), of statistically significant relationship at 99% confidence level, between heavy metals in soil and clay content < 0.02 mm (a) and cation exchange capacity in soil (b) [2]

Copper (Cu) is an important element for all life forms, but can be toxic in high concentrations. The average Cu content in the lithosphere is 70 ppm. In natural soils, the average concentration is 2-40 ppm. As described for Ni, Cu has no similarities with any other metal regarding his chemical behaviour in soil. Significant quantities of Cu in the soil are connected in the minerals, therefore, this metal is supplied only by a very slow decay processes. Cu may occur in the form of readily soluble salts (copper nitrate, copper sulphate), and as an oxide and hydroxide. It binds to organic matter, ferric oxide and Al. Intake of Cu in plants can be increased by low pH and organic fertilizers. Cu concentration can increase significantly under the effect of anthropogenic activities (non-ferrous metal processing, the use of substances for plant protection). In humans, the contamination has not yet been notified by the dietary intake of Cu in the body, but its high concentration may cause liver damage.

Zinc (Zn) is widespread in nature and the average content of Zn in the lithosphere is about 80 ppm. Unpolluted soil contains an average of 15-100 ppm of Zn. Zinc binds to organic matter and ferric and Mn oxides. It occurs in large amounts in the layers of the clay minerals. Under natural conditions, in A horizon of soil from wet areas, with slightly acid pH, more than half of the Zn is bound to organic material [4]. Because of the extensive use of Zn in industry, the Zn content in soil surrounding the industrial areas can reach even 5000 ppm [2]. The effect of

Zn is particularly harmful because its accumulation leads to accumulation of other heavy metals, such as Pb, Cu and cadmium (Cd). Concomitantly, in the lime-rich soils, the plants show Zn deficiency symptoms. Also, Cd is strongly chemical bond to the Zn, as the proportion Zn/Cd in soil is constant. The availability to plants can be influenced by the concentration of cadmium in soil, pH conditions, temperature, amount of organic matter, and presence of other metals. Cadmium is irreversibly bound by ferric and manganese oxides in soil, and by clay minerals that influence the Cd mobility.

Lead (Pb) in soil is largely associated with colloidal organic matter, which results in a high proportion of Pb accumulated in the top 5-15 cm of contaminated soils. The Pb concentration decreases with depth in a soil profile [2]. In the geosphere, the average Pb concentration is about 16 ppm. The increasing of Pb concentration may be caused by the accumulation of fuel combustion residues from the transportation, by application of sewage sludge and by the use of some pesticides in gardens or orchards. Increasing soil pH may decrease the absorption of Pb in soil. Plants are able to accumulate significant amounts of Pb (300-400 ppm) in pollution conditions without noticeable symptoms [2]. High concentration of Pb particularly affects the neurovegetative functions, hampers blood and cause chronic emphysema in humans.

Tin (Sn) concentrations in soil are generally low, with values of 2-3 mg/kg in unpolluted areas and can reach 200 – 1000 mg/kg in areas of high tin deposits [8] or in areas influenced by anthropogenic activities including smelters of ferrous and non-ferrous metals and coal-fired power plant [9].

Cobalt (Co) is widely distributed in rocks and soils and always occurs in nature in association with nickel and usually with arsenic [10]. The common Co minerals are smaltite ($CoAs_2$) and cobaltite (CoAsS) and the most important sources of cobalt are residues from the smelting of arsenical ores of nickel, cobalt and lead [10]. Cobalt in environment may represent a hazard to human health and is considered a metal with marked allergenic potential.

Chromium (Cr) is used in alloying metals, in the industry of paints, cement, paper, rubber and other materials. Exposure to low concentrations of Cr produce skin irritation and ulceration, and long-term exposure can cause kidney and liver diseases, and diseases of the circulatory system and nervous tissue. Chromium accumulates especially in aquatic fish and the consumption increases the risk of a high intake of this metal.

Nickel (Ni) is necessary in small quantities in the human body to produce red blood cells [4], but greater amounts (>1.0 mg/d) may become toxic. Short-term exposure to Ni does not cause health problems, but over a long period leads to weight loss, heart and liver diseases, and skin irritation. Nickel can accumulate in aquatic organisms, but its presence increases for higher levels on the food chain.

Manganese (Mn) is a metal naturally ubiquitous in the environment, found in many types of rocks and soil, essential for normal physiologic functioning in humans and animal. Exposure to low levels of Mn is considered to be nutritional for humans. Long-term exposure to high levels of Mn by inhalation in humans may result in central nervous system effects. The metallic Mn is used in steel production to improve hardness, stiffness and strength. Mn is also used in carbon steel, stainless steel and high-temperature steel, along with cast iron and superalloys [11]. The average Mn levels in soil range from 40 to 900 ppm [11].

Molybdenum (Mo) is a valuable alloying agent which contributes to the hardness and toughness of quenched and tempered steels. Mo also improves the strength of steel at high temperature. In the environment, Mo differs from the other micronutrients in soils because it is less soluble in acid and more soluble in alkaline soils. Mo availability to plants is sensitive to pH and drainage conditions. Some plants can have up to 500 ppm of the metal when they grow on alkaline soils [12].

2.1. Heavy metal pollution of soil

The presence of heavy metals in natural and contaminated soils shows a great variability in both the horizontal and vertical dimensions [2]. Chemical pollution of soils in Romania is affecting approx. 0.9 million ha of soil, of which 0.2 million ha are affected by excessive pollution. Adverse effects are particularly strong to pollution by heavy metals (Cu, Pb, Zn, Cd) and sulphur dioxide, identified especially in Baia Mare, Zlatna and Copsa Mica. Although in last years a number of industrial units have been closed and others have reduced activity, the soil pollution is quite high in some areas: Targu Mures, Turnu Magurele, Tulcea and Slatina. Oil pollution and salt water from oil wells and transport affects approximately 50, 000 ha. Soil damaged by excavation comprises 15, 000 ha and constitutes the most serious form of damage to soil, encountered in the mining industry, for example in the mining basin of Oltenia. Suitability of land affected by this type of pollution decreased by 1-3 classes, and some of these areas has become unproductive. Soil cover with solid waste and residues caused sealing of approximately 18 000 ha of farmland and meadows [13]. Direct economic damage on agricultural production due to these restrictions is estimated by reducing it by about 20% per year.

The study of Lăcătuşu and Ghelase [14] aimed to assess the abundance of anthropogenic heavy metals in soils at various distances from Romfosfochim SA Valea Călugărească. The researchers compared the specific data of metal concentrations in polluted soils with those of similar soils, not subject to pollution effects. The results showed a decrease in the percentage of geogenic abundance with proximity to the source of pollution, although the concentrations of Cu, Pb, Zn and Cd are significantly higher at distances between 0-500 m compared to the distance of 6 km [14]. The depth of penetration into the soil of heavy metals from industrial emissions is shallow (up to 15 cm) in forest soils and up to 30-40 cm in arable soils [15].

Vrînceanu et al. [16] have published the results of research on polluted soils from Copsa Mica, showing metal concentrations for Cu, Zn, Pb and Cd in soil, with values between 69-136 mg/kg, 962-2191 mg/kg, 1182 -1978 mg/kg and 30-42 mg/kg respectively. In 2002 a study showed that in the soil from Baia Mare the Cu concentration exceeded 9.5 times the maximum limits and 4.8 times the alert and action limits. The concentration of Pb exceeded 132 times the alert threshold and 66 times the limit of intervention and the concentration of Zn exceeded 11 and 6 times these limits respectively [17]. Recent researches in Baia Mare showed some decreasing of heavy metal concentration, but the average values of these concentrations exceed 6 times the maximum level of lead. In the case of Cu, Zn and Cd the average values exceed the maximum level by 10, 3 and 7 times. Multiple pollution average index for the four heavy metals

http://dx.doi.org/10.5772/58304

determines the classification of this area as excessive pollution class (values greater than 16) in the layer of 0-10 cm and as very strong pollution class in the layer of 10-20 cm. The maximum values of this index reached 78.2 in some excessively polluted areas. In 1994, a land of 21, 875 ha (3, 245 ha of forest and 18, 630 ha of agricultural land) have been severely affected by heavy metal pollution by exceeding the maximum limits for Pb, Cu, Zn and Cd [16].

Maximum levels of metal pollution were detected in Baia Mare for Cd, in Copşa Mica for Pb and Zn, and in Valea Călugărească for Cu [15]. Besides these considerations we can add the ecological accident occurred at Baia Mare in 2000 that led to the contamination of water and soil with cyanide from extraction plants. Specific conditions of those soils (moderate to severe acidity) are favouring the translocation of pollutants from soil to plants, animals and humans, leading to an increase of metal toxicity and a reduction of soil and water quality. Research conducted in 2000 in several areas in the south of Romania have shown the persistence of severe soil pollution with heavy metals in the vicinity of industrial plants (S.C. Neferal and S.C. Acumulatorul-Bucureşti-Pantelimon, S.C. Turnu S.A. – Turnu Măgurele) and pollution with fluoride near ALRO-Slatina.

All these environmental assessments showed the necessity of soil quality improving policy, by changing the land use and by replacing the food crops with industrial crops. Also these areas need measures to reduce the pollution and its toxic effects. Exploit of Romanian peat deposits for the purpose of complete the organic matter of soil, could be a source for improving soil quality and reducing pollution [18].

2.2. Availability of toxic metal compounds in the soil for plants

A global statistical evaluation of the substances exchange between soil and plants, led to the conclusion that the percentage itself is influenced by the following parameters:

- Soil texture;
- Carbon content of organic substances;
- Cation exchange capacity of the soil;
- Calcium carbonate equivalent;
- Oxides and hydroxides, mainly Fe, Mn and Al;
- pH values - dominant factor.

Soil organic fraction plays an extremely important role because they can delay both the accumulation and transfer of metals and their movement into the soil. Metal toxicity in soil can be increased or reduced by soil organic fraction. Soil pH directly influences the availability of metals as soil acidity determines solubility of element and its ability to move in the soil solution. Regarding the content of phosphorus in soil, the presence of high doses of P_2O_5 can increase or decrease metal uptake. In addition, the accumulation of metals is directly influenced by the plant physiology. For example, Cd uptake in grain has been described to be either antagonistic or synergistic with high concentrations of Pb in the soil [19].

In discussions about soil protection and remediation, pollutant limits for various elements have been established only under certain conditions and soil parameters. It was not taken into account the specific conditions such as the fact that on low-carbon light soils there is strong influence of rainfall leading to a strong acid mobilization and uptake into plants of toxic heavy metals. This does not happen on heavier soils rich in limestone.

The solubility of Zn in soil was studied by Herms and Brummer [20], which demonstrated the extent to which this element is dissolved by increasing acidity of the soil and became available for plants uptake. A pH value of 5 of low-Zn soil could lead to lasting effect of uptake large amounts of Zn, with all the negative consequences that result. The balance of Zn in the soil solution is carried out according to the pH of soil: at 1200 mg/kg of Zn and a pH of 7, at 100 mg/kg of Zn and pH of 6, and at only 40 mg/kg Zn and a pH of 5. This indicates that also the low-Zn soil can store dangerous amounts of available Zn.

3. Studied area

Metallurgy activities produce gas, wastewater and waste containing pollutants that can be sources of risk under normal handling, and especially for irresponsible handling of equipment. The main resulted pollutants are: CO_2, CO, NO, SO, VOCs (e.g. BaP, PAHs, dioxins, Freon), particulate heavy metals (Pb, Cu, Zn, As, Cr), cyanides, phenols, heavy metals and sometimes toxic organic compounds in waste material.

Annual average concentrations exceed the maximum permitted levels in many localities (Baia Mare, Copsa Mica, Medias, Targoviste, Arad, Deva, etc.), for both particulate matter and for sediment, coming mainly from industry of steel.

The main polluter from industrial zone of Targoviste city was SC Mechel Targoviste SA, located in the south of the city. By the metallurgical activity, emissions were resulting, with significant concentrations of pollutants, including heavy metals. In the process of obtaining steels at SC Mechel SA, technological flow was served by Electric steelmaking 2, EBT electric oven with a capacity of 70 tons/hour and Continuous casting plant (in billets) with a maximum capacity of production of 2.5 tons/hour. The activity SC Mechel Targoviste SA was assisted by filtration systems so that the environmental impact to be reduced to a lower limit.

During the elaboration of steel in Electric steelmaking no. 2, total dust are emitted to air, metals (Cd 0.05 mg/Nm^3, Pb 0.3 mg/Nm^3 and Cr + Cu + Mn + Ni + Zn 5.0 mg/Nm^3), organic compounds, nitrogen oxides (NOx), carbon monoxide (CO) and sulphur oxides (SOx) [21]. In addition to these emissions of heavy metal particulates, SC Mechel SA polluted the soil by the waste and slag dumps generated during the metallurgical processes and stored in an open deposit.

Dispersion of pollutants in the atmosphere largely depends on emission characteristics, meteorological factors, topography, soil roughness, the height of buildings, especially the stack height. Among the meteorological factors, wind direction, intensity, and thereof frequency, are the dominant factors on which we can determine the wind rose indicating the direction of predominant movement of air masses. Based on the wind rose [22] can be designed the

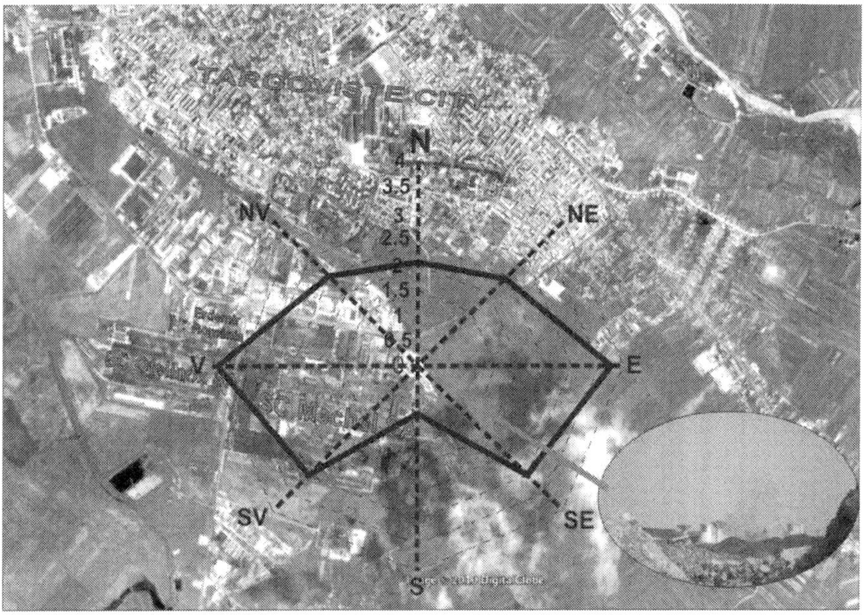

Figure 3. Dispersion and transport of pollutants particles according to the pollution direction compass, relative to the most significant pollution source, the furnace of the Electric steelmaking no. 2 belonging to SC Mechel SA

pollution direction compass which indicates the direction of dispersion and transport of pollutants particles (Figure 3). In the middle of pollution direction compass is the most significant pollution source, the furnace of the Electric steelmaking no. 2 belonging to SC Mechel SA. East was one of the predominant directions of pollutants transport and the particles have deposited on the ground (agricultural fields) during the process of steel elaboration.

In addition to the pollution produced by ground deposition of particles emitted by the SC Mechel SA chimney, these soils are affected by deposit of slag and skim from the same plant. The dumps that are within walking distance (less than 10 m) of agricultural land of residents from Colanu, Romania, are subject to a continuous process of erosion and transport by wind, smaller particles from the surface of stockpiles being transported up to 10 km. Dispersion of the slag particles is due to transport activities to and from the place of storage of metal waste.

Characteristic types of soils for studied area are gray luvisol and gray brown luvisol (Figure 4) according SRTS-2003 (reddish brown after SRCS 1980), soils of a reddish hue, quite evident in the upper horizon and really evident in the middle horizon. Water and air permeability of the soil was moderate. Humus content was about 3%, nutrient supply was moderate, the soil reaction was low-acid with pH values in the range of 6.0 to 6.4, and the degree of base saturation was 80% to 85%.

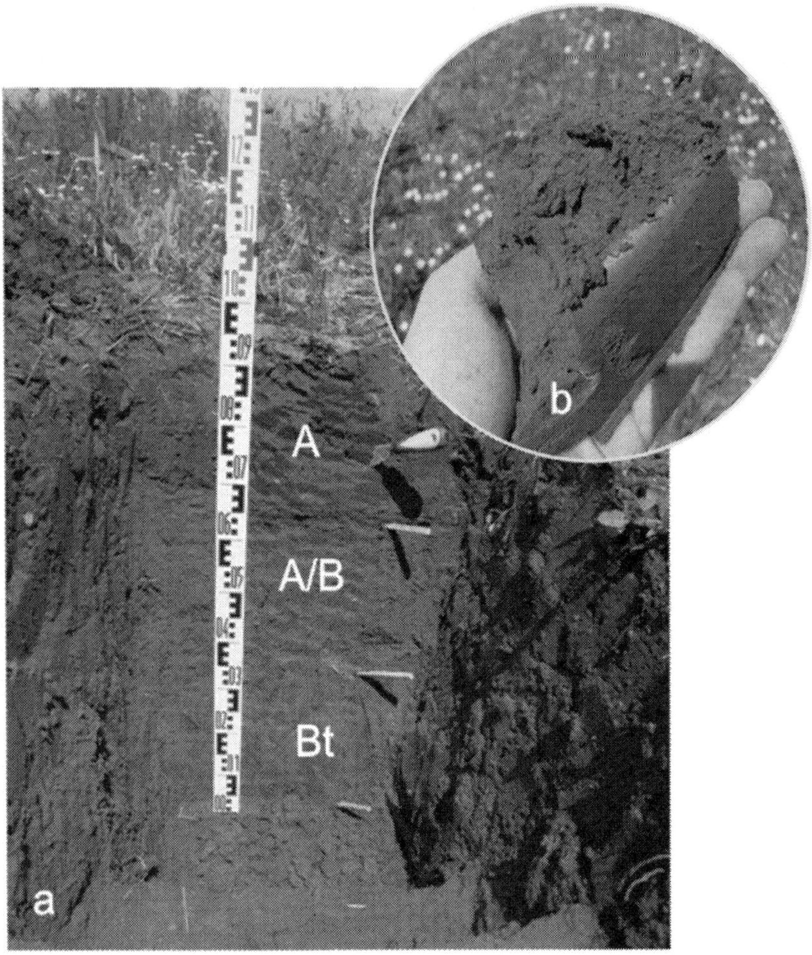

Figure 4. Gray brown luvisol profile (a) and brown-yellowish-rusty upper horizon (b)

Bioaccumulation in upper and middle horizon was low; plant debris was mostly decomposed by the action of fungi, and could be observed the formation of small amounts of humus with predominating fulvic acids. Due to intense alteration of mineral component occurs removal of clay colloid from the surface, with accumulation in the Bt horizon, where the profile shows a textural differentiation.

In soils with alluvial B horizon, such as soil in the industrial area of the Targoviste city, colloidal order mineral fraction of this horizon contains large amounts of colloidal hydroxides (e.g. $Fe(OH)_3$ and $Al(OH)_3$) and various hydrated iron and aluminium sesquioxides free-form

Figure 5. Iron hydroxides

($mR_2O_3\,nH_2O$) [23]. Large amounts of iron hydroxides in the humus horizon could be observed for the brown and reddish-brown soils from the industrial zone of Targoviste (Figure 5). The presence of these hydroxides is manifested by brown-yellowish, brown, reddish-brown, yellowish-rusty or rusty lit of the horizon where they are deposited.

4. Matcrial and method

4.1. Experimental design

Sampling points were chosen to reflect a snapshot of the impact of metallurgical activities in this area by particles emissions. Based on weather conditions and pollution direction compass of Dâmbovița County (Figure 6), were determined that areas found at SW and W towards the emission source are the most affected. In that location have been chosen the harvest area Zone I – industrial fields, with two subzones, for SW and W directions respectively. Land at NE, E and SE from the source of pollution, are grouped according with the use category in Zone II – agricultural fields sensitive to high concentrations of heavy metals.

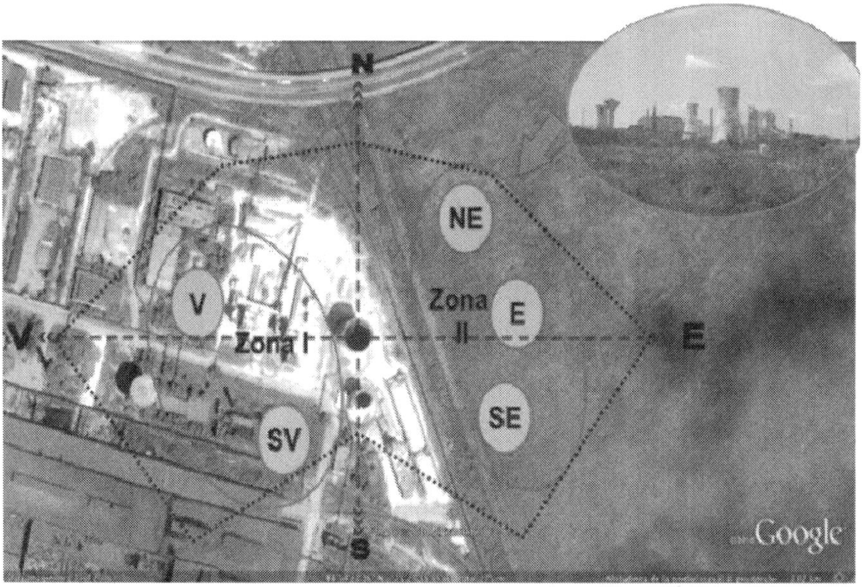

Figure 6. Sampling directions from industrial area of Targoviste city according to the pollution rose

Sampling of soil was done at distances between 50 and 1000 meters from the source of pollution, from five different points, chosen according to triangle method. The results of metal concentration represent the average of these five samples. From each sampling point, samples were taken from three layers: the upper layer (0-5 cm depth), middle layer (5-20 cm depth) and lower layer (20-40 cm depth). These layers were chosen according with the depth to which the roots of culture plant normally develop.

The soil samples were processed in the laboratory for elemental analysis by ICP-AES. The soil samples were dried at 40 °C for 24 hours, ground to a fine powder, sieved at 250 μm (according to SR ISO 11464).

4.2. Analytical methods

Determination of heavy metal concentration in soil was done on replicates samples by Inductively Coupled Plasma - Atomic Emission Spectrometry method (ICP-AES). The soils samples were mineralized in Berghof microwave digester, using a mixture 1:1 with nitric acid (according with Berghof method) prior to ICP-AES analyses. The advantage of ICP-AES is the multielemental detection (Cu, Zn, Sn, Pb, Co, Ni, Mn, Cr and Mo) [24]. For this research, analyses were conducted with Liberty 110 spectrometer of Varian brand. The minimal detection limits of device range according to the analysed element and is 0.4 mg/kg for Zn, Mn

http://dx.doi.org/10.5772/58304

and Cu; 0.5 mg/kg for Cr and Co; 0.6 mg/kg for Sn, Ni and Pb. The concentrations values for analysed metals were expressed in milligrams of metal per kilogram of dry soil (mg/kg).

The soil pH was determined with a portable pH-meter, WTW 3110 SET 2, with precision of 0.01 units. For pH analyses, 5 g of each soil sample were mixed with 50 ml KCl 0.1N, F 1000, Tt 0.0056 g/ml and homogenized for 15 minutes with a magnetic stirrer.

5. Results

5.1. Statistical results of heavy metal concentration

In the industrial area of Targoviste city, the concentration of heavy metals in soil was highly dependent on the metal species, the position of sampling point towards the source of pollution, the stack, and the depth of sampling. The general statistical results of heavy metal concentration in studied area are presented in Table 2.

The variation of Cu concentration in the 0-5 cm layer of soil was higher than the variation on the profile. Cu concentration ranged from 578.4 mg/kg in the surface layer, to 170.4 mg/kg (5 -20 cm) and 152.4 mg/kg (20-40 cm). The Cu concentration variation showed that 75% of the samples had values of concentration up to 135.9 mg/kg, and only 25% of the samples had higher concentrations than this value. The same pattern of variation of Cu concentration was followed for the deeper layers. For the layer of 5-20 cm depth, most samples (75%) had concentrations within a narrow range, from 28.7 to 43.6 mg/kg and for the layer of 20-40 cm depth 75% of samples range between 22.0 to 55.2 mg/kg.

The variation of Zn concentration was very high at the surface of soil profile (827.7 mg/kg) and lower in the deeper layers of soil (128.8 mg/kg and 186.9 mg/kg respectively. In all the three studied layers, the concentration of Zn in most of the samples was placed in a very small range as compared to the magnitude of total concentration range. In the surface layer, 50% of the samples showed a concentration of Zn from 42.6 to 86.1 mg/kg and 75% were found between 42.64 to 225.03 mg/kg. The same pattern of distribution of Zn concentration was followed in the deeper layers, 75% of the samples being in a relatively narrow range of concentration from 44.4 to 72.9 mg/kg for the depth of 5-20 cm and 33.1 to 81.8 mg/kg for the layer of 20-40 cm depth.

The values of Sn concentration showed a higher homogeneity in soil when compared to Cu and Zn concentration. Sn concentration ranged from 92.7 mg/kg at the surface of profile to about 40 mg/kg in deeper layers (5-40 cm). The four quadrants of Sn concentration distribution showed narrow values, though there was a concentration of Sn values (50%) in the range 32.7 to 53.9 mg/kg for 0-5 cm layer. In the middle layer (5-20 cm), about 75% of the sample ranged from <LD to 20.9 mg/kg. At 20-40 cm depth, quadrants II and III of the concentration distribution were distributed on the interval from 15.39 to 22.8 mg/kg, the remaining 50% of the samples having values lower or higher than this range.

Metal	Depth	Mean±SD	Min-Max	Median	Q1	Q3	Skewness	Kurtosis
Cu	0-5 cm	125.5±162.6	22.0-600.4	36.5	30.9	135.9	2.1185	4.6664
	5-20 cm	61.1±55.0	28.7-199.1	35.1	32.5	43.6	1.8284	1.9995
	20-40 cm	63.3±53.1	22.0-174.5	38.1	35.4	55.2	1.5680	0.7874
Zn	0-5 cm	223.9±269.0	42.6-870.3	86.1	73.3	225.0	1.6278	1.3552
	5-20 cm	74.3±42.9	44.4-173.2	54.1	48.4	72.9	1.5888	1.0601
	20-40 cm	75.6±50.61	33.1-220.1	53.1	49.4	81.8	1.9591	4.0515
Sn	0-5 cm	65.1±30.2	32.7-125.4	53.9	41.9	84.8	0.7305	-0.7436
	5-20 cm	12.9±12.9	<LD-44.4	14.9	<LD	20.9	0.8094	0.8786
	20-40 cm	17.8±10.9	<LD-38.2	19.4	15.4	22.8	-0.3501	0.0826
Pb	0-5 cm	76.4±98.9	0.6-294.3	43.3	0.66	92.1	1.3987	0.8529
	5-20 cm	47.9±42.6	<LD-121.0	42.5	20.86	64.1	0.7683	-0.5585
	20-40 cm	34.0±43.6	<LD-145.6	26.3	<LD	35.6	1.5800	2.0421
Co	0-5 cm	16.2±4.8	7.1-23.5	17.3	12.9	19.1	-0.5281	-0.4808
	5-20 cm	14.7±5.1	6.7-21.5	13.6	10.3	19.4	-0.0702	-1.6268
	20-40 cm	13.1±4.0	6.7-19.9	14.9	9.5	15.6	-0.3423	-0.9854
Ni	0-5 cm	65.1±58.5	13.8-185.4	37.6	22.0	84.8	1.0999	-0.1291
	5-20 cm	20.2±15.9	8.7-52.7	13.1	12.3	14.8	1.6371	0.8719
	20-40 cm	23.2±23.9	4.4-72.4	14.2	8.3	17.7	1.5576	0.7595
Mn	0-5 cm	1579.7±352.3	1159.9-2348.0	1419.1	1373.2	1731.3	0.9420	0.0714
	5-20 cm	1348.9±243.6	758.9-1677.3	1384.1	1322.7	1504.3	-1.2771	1.5204
	20-40 cm	1367.5±327.9	720.4-1763.1	1486.3	1258.4	1601.1	-1.0239	-0.1494
Cr	0-5 cm	114.6±125.5	13.7-315.6	25.1	21.6	205.7	0.7343	-1.3288
	5-20 cm	46.0±51.7	16.2-168.3	22.7	17.81	30.4	1.7522	1.5205
	20-40 cm	41.0±52.9	8.2-159.9	18.3	13.8	21.5	1.6981	1.1754
Mo	0-5 cm	7.2±7.8	0.6-23.4	4.5	0.6	10.5	1.0334	-0.1858
	5-20 cm	3.0±3.4	<LD-10.8	1.2	0.7	5.1	1.1289	0.2488
	20-40 cm	2.4±3.3	<LD-9.7	1.0	0.6	1.6	1.6104	0.9883

LD – limit of detection

Table 2 Statistical results of heavy metal concentration (mg/kg) in soil surrounding the industrial area of Targoviste city

The variability of Pb concentration was very high, most of the samples showing very low levels, even below the detection limit of the analytical method, in the entire soil profile studied, while some samples presented significant concentrations of Pb up to 294.3 mg/kg (0-5 cm), 121.0 mg/

kg (5 to 20 cm deep) and 145.6 mg/kg (20-40 cm). The distribution of the samples in terms of the concentration was varying, 75% of the samples from the surface of soil profile were in the range of concentrations from 0 to 92.1 mg/kg, and the remaining 25% were in the range of 92.1 to 294.3 mg/kg. The Pb distribution was less variable in the deeper layers, the four quadrants of the concentration being distributed at relatively equal intervals, especially for the depth 5-20 cm. Heterogeneity of the sample in terms of the concentration of lead in the ground was indicated by the high value of the standard deviation, higher than the average concentration throughout the studied soil profile.

The concentrations of Co in the samples were distributed homogeneously throughout the range of concentrations for both the upper layer and deeper layers of 5-20 cm and 20-40 cm respectively. The homogeneity of the samples was indicated by the low values of standard deviation, between 29% and 34% of the average, and by the median value that was very close to the average value, 17.3 mg/kg, 13.6 mg/kg, respectively 14.9 mg/kg for the three depths. On the surface of the soil profile, 50% of the analysed samples were in the range of concentration from 12.9 to 19.1 mg/kg.

In surface layer of soil, the range of Ni concentration varied widely, up to 185.4 mg/kg. The soil sample from surface layer presented heterogeneous distribution of Ni, with 75% of the samples from 0-5 cm depth showing a concentration ranging 13.8 to 84.8 mg/kg, and only 25% of the samples ranging 84.8 to 185.4 mg/kg. The Ni concentration variation was lower in deeper layers; most of the samples (75%) were covered by a much narrower range of concentration from 8.7 to 14.8 mg/kg for 5-20 cm depth, and 4.4 to 17.7 mg/kg for 20-40 cm depth.

Distribution of samples within the range of Mn concentration was uniform and 50% of the samples were within the range of 1373.2 to 1731.3 mg/kg, while the remaining 50% of the samples from the surface of the soil profile were higher or lower than this range. The same pattern of samples distribution was followed in the deeper layers, where 50% of the samples were concentrated in the middle of the range of Mn concentration. The standard deviation of the Mn concentration indicates that the samples had similar values of concentration for each layer, accounting 22%, 18% and 24% respectively for the three layers of 0-5 cm, 5-20 cm and 20-40 cm.

The wide variation of Cr concentration in the samples was reflected by the high value of standard deviation, greater than the average concentration for all three depths. Half of the samples from 0-5 cm layer of soil showed Cr concentrations in a very small range between 13.7 and 25.1 mg/kg. The remaining 50% of the samples were distributed in a wider concentration range between 25.1 and 315.6 mg/kg. At depths greater than 5 cm, 75% of the investigated samples had concentrations of Cr in a narrow range of 14.2 mg/kg for the depth of 5-20 cm and 13.3 mg/kg for the depth of 20-40 cm, while the remaining 25% had concentrations in the range of 137.9 mg/kg and 138.4 mg/kg respectively for the two layers of soil.

The range of Mo concentration was higher in the surface of soil profile (22.8 mg/kg) compared to the range of concentration of deeper layers of soil (10.8 mg/kg and 9.7 mg/kg respectively). The majority of samples (75%) showed low levels of Mo, up to 10.5 mg/kg on the surface and 5.1 mg/kg and 1.6 mg/kg in the deeper layers of 5-20 cm and 20-40 cm respectively. The concentration varied greatly, with values of standard deviation higher than the mean concentration.

5.2. Horizontal distribution of heavy metals

The horizontal distribution of heavy metals and level of metal pollution in the industrial area of Targoviste city was established by comparing to the Romanian legislation [3], which regulates normal values, alert thresholds and action levels for different trace elements by use of soils, agricultural and industrial land (Table 3).

Limit	Cu	Zn	Sn	Pb	Co	Ni	Mn	Cr	Mo
NV	20	100	20	20	15	20	900	30	2
ATA	100	300	35	50	30	75	1500	100	5
ATI	250	700	100	250	100	200	2000	300	15

NV – normal values; ATA – alert threshold for agricultural soil; ATI – alert threshold for industrial soil

Table 3 Normal values and alert thresholds of heavy metal concentration (mg/kg) for agricultural and industrial soil in Romania [3]

The concentration of Cu on the surface layer of soil differed greatly between the two studied zones (I and II) (Table 4). Thus, in Zone I, on the SW and W directions towards the emission source, the concentration of Cu reached 401.44 mg/kg and 134.58 mg/kg, values of 7 to 20 times higher than normal levels in this category of soils [25]. In Zone II, the concentrations of Cu on the NE, E and SE directions were slightly greater than the normal value of concentration, ranging between 22.37 and 36.61 mg/kg.

In the case of Zn, the maximum values allowed for industrial soil, 700 mg/kg were overcome on W direction to the source, while the value on SW direction was below that limit, but exceeded the normal value of Zn in soil. The soils in Zone II showed a deficiency of Zn at surface layer and in the deeper layers. In the upper horizon, the Zn concentration was between 46.03 and 86.07 mg/kg, values that are below the normal value of this element in soil [26].

Sn concentrations in the analysed soils were above normal values (Table 4) in both the Zone I and Zone II. A concentration of about 6 times the normal limit was found in soils under the SW direction. For other samples, the concentration of Sn in soil ranged from 33.77 to 82.90 mg/kg, the lowest value being on E direction and the highest on SE direction.

Although the normal value of Pb in soil is 20 mg/kg, the concentration of this element in the studied samples of soils did not exceed the threshold. The soil in Zone II showed the lowest concentration. In the SE direction, Pb concentration in soil was the highest, 43.35 mg/kg. The soil in the Zone I, showed values of Pb concentration that exceeded 4 to 12 times the normal values: 85.38 mg/kg in the SW direction and 252.00 mg/kg on the W direction.

The concentrations of Co in soils were not much higher than the normal values of metal concentration in soil. Higher values were found in the soil from Zone I, 19.93 mg/kg in the W

Heavy metal	Direction towards the emission source				
	SV	V	NE	E	SE
Cu	401.44	134.58	32.41	22.37	36.61
Zn	194.28	719.13	74.45	46.03	86.07
Sn	112.31	52.79	43.78	33.77	82.90
Pb	85.38	252.00	0.60	0.76	43.25
Co	20.42	19.93	14.11	8.94	17.48
Ni	166.64	83.84	15.01	22.42	37.39
Mn	1264.26	2156.48	1726.44	1362.45	1388.85
Cr	200.31	312.33	13.88	24.98	21.68
Mo	9.79	0.6	20.57	0.65	4.45
	Under normal limit	Exceed normal limit	Exceed the alert threshold for agricultural soil	Exceed the alert threshold for industrial soil	

Table 4 Horizontal distribution of heavy metal concentrations (mg/kg) in the industrial area of Târgovişte city

direction and 20.42 mg/kg in the SW direction. The soils in the Zone II showed values of Co concentrations that varied between 8.94 and 17.48 mg/kg. The lowest value was on E direction, and the highest value on SE direction.

The concentration of Ni in soils varied widely in the two studied zones. In Zone I, the concentration of Ni was between 4 and 8 times higher than the normal value of metal concentration in soil, 83.84 mg/kg in the W direction and 166.64 mg/kg on the SW direction. The soil in Zone II showed Ni concentration close to normal values, ranging from 15.01 to 37.39 mg/kg. The highest value was for the Ni concentration on SE direction.

Concentration of Mn in the soil was two times higher than the normal value and range between 1264.26 mg/kg and 2156.48 mg/kg. The lowest value of Mn concentration was on SW direction and the higher on the W direction.

Concentrations of Cr in soils from the industrial area were very different between the two studied zones. In Zone I, the concentrations were extremely high compared to the normal limit (30 mg/kg), reaching values over 10 times higher: 200.33 mg/kg and 312.33 mg/kg on SW and W directions respectively. In Zone II, Cr concentration showed values below the normal concentration of Cr in this type of soil, and varied between 13.88 and 24.98 mg/kg [27].

Molybdenum concentration in the surface layer of soil varied within wide limits, irrespective of the position towards the source of pollution. In Zone I, concentration of Mo has the minimum value in the W direction (0.6 mg/kg), and the maximum value in the SW direction (9.79 mg/kg). This value was about 5 times higher than the normal value of Mo concentration in soil. In Zone II, the concentration of this metal varied also within wide limits, from 0.65 mg/kg in the

E direction to 20.57 mg/kg in the NE direction. The recorded values were more than 10 times higher than normal values.

5.3. Vertical distribution of heavy metals

The vertical distribution of heavy metals in the soil profile (0 – 40 cm) is shown in figures 7 – 14, which indicates the level of heavy metals in the three soil layers: 0 – 5 cm, 5 – 20 cm and 20 – 40 cm. For each layer, linear regression was calculated to indicate correlations of heavy metal concentrations with the pH.

The high concentrations of Cu in the SW direction were maintained at very high level in the depth of soil profile (Figure 7). In the W direction, Cu concentration decrease to depth of 40 cm to levels of 27 mg/kg. The soil in Zone II showed similar values of Cu concentration on the entire profile of soil. Between the Cu concentration and the pH of soil was observed a moderate correlation in the surface layer of soil (0.3) and very low correlations for the two deeper layers. All the correlations were positive.

Extremely high Zn concentration from the surface layer of soil did not maintained in depth. Below 5 cm depth the concentration of Zn decreased to values lower than 35 mg/kg (Figure 8). In the SW direction, the concentration of Zn in the deeper layers of soil remained at levels comparable to the surface layer, ranging between 154.38 and 194.28 mg/kg. The lowest values were in the middle of the soil profile, and the highest values on surface layer. The design of the Zn distribution in soil was maintained in Zone II. The highest concentrations were on the surface, decreased in the middle of soil profile, and increased in the lower part of the profile. The concentrations of Zn to this area were between 46.03 and 86.07 mg/kg. Correlation between the Zn concentrations in the soil with its pH is strong in the upper layer (0.48), and very low to depth.

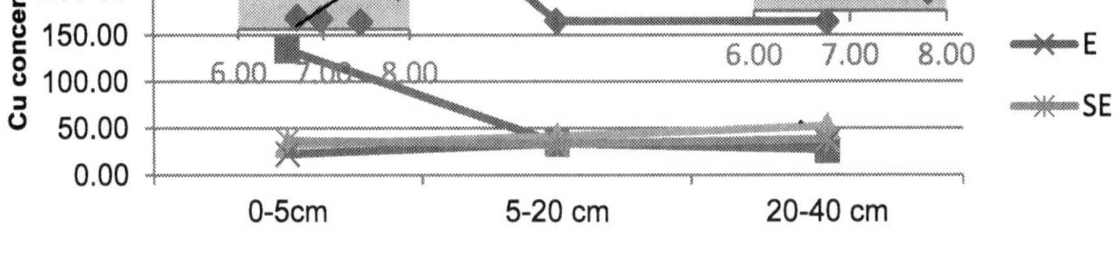

Figure 7. Copper vertical distribution in soil and correlation with pH

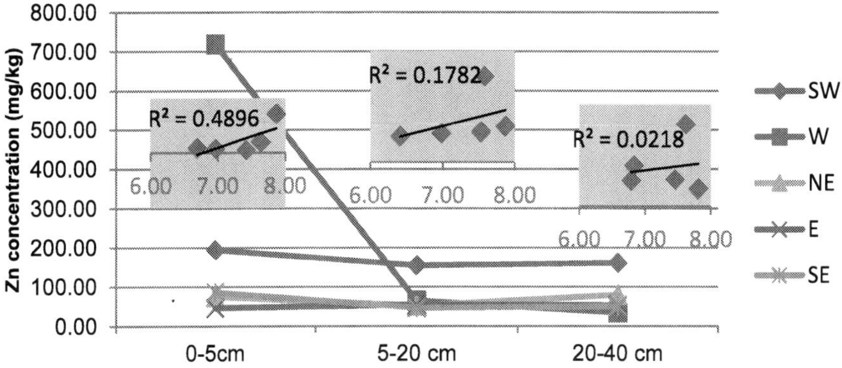

Figure 8. Zinc vertical distribution in soil and correlation with pH

In contrast to Cu and Zn, Sn distribution in the soil profile showed a very wide variance (Figure 9). The surface layer showed Sn concentration much higher than in depth. For all studied direction, was observed that the concentration of Sn decreased towards the centre, and grow on the bottom of the profile. Correlation between the Sn concentrations and the pH of soil is positive, but to a different extent, depending on the depth. Thus, the surface layer had a very low correlation, in the 5-20 cm layer the correlation was low, and the bottom of soil profile the correlation was moderate (> 0.4).

Distribution of Pb in the soil did not follow a pattern, but varied from one area to another. Thus, the high Pb concentration from the surface, in the W direction, decreased sharply towards the middle of soil profile, and continues to decrease to the bottom of the profile (Figure 10). In the SW and E directions, the Pb concentration was higher in the middle and lower soil profile than in the upper horizon, far exceeding the normal value of Pb concentration in soil. In the NE and SE directions, Pb concentrations were similar throughout the entire soil profile. The correlation between the Pb concentrations in soil and the pH decreased from the surface to the depth of soil profile. The correlation was moderate positive in the upper layer (0.47), and low but positive to the bottom of soil profile.

The distribution of Co on the soil profile varied from one point to another (Figure 11). In the SW, SE and E directions, the Co concentration in the soil increased slightly in the middle of soil profile and decreased sharply towards the bottom of it. In the W direction, the Co concentration decreased sharply towards the centre of soil profile and continues to decrease to the bottom thereof. In the NE direction, the Co concentration is higher in the lower part of the profile as compared to the upper layers. The correlation between the Co concentrations in soil and pH varied on the soil profile as intensity. In the upper layers, the correlation is positive, but very low (<0.2), while the lower layers showed weak negative correlation (0.2 - 0.4).

Nickel concentration varied along the profile of soil. Much higher values were observed in the surface layer of soil compared to the values representing the concentration of Ni in the middle of the profile. At the bottom of the profile, the concentrations increased slightly (Figure 12).

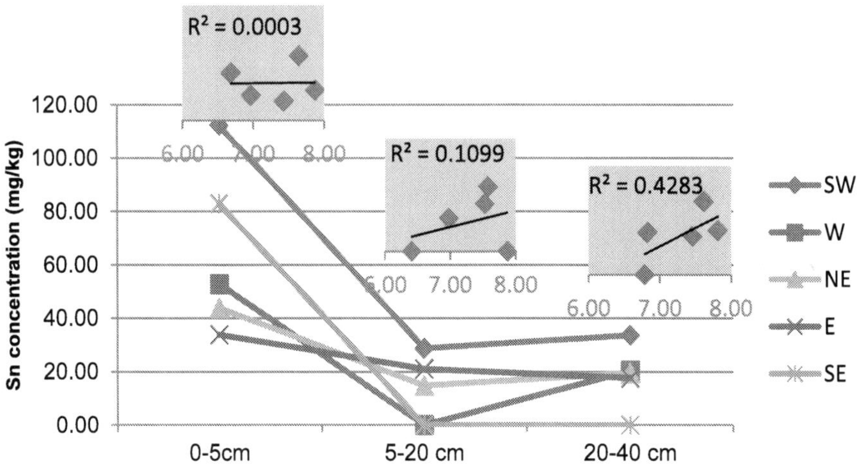

Figure 9. Tin vertical distribution in soil and correlation with pH

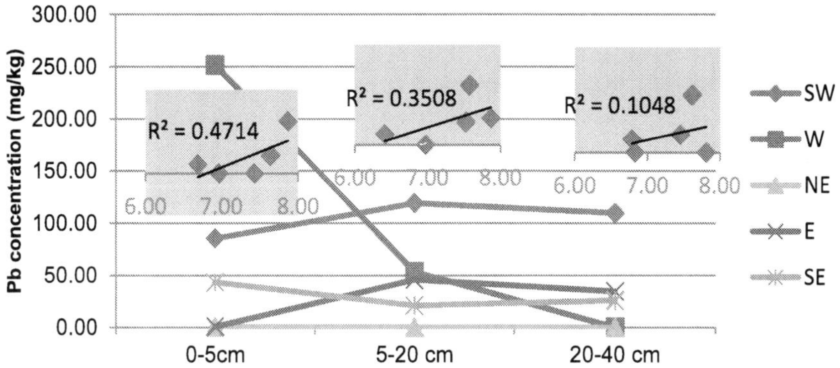

Figure 10. Lead vertical distribution in soil and correlation with pH

For soil samples collected in the SW direction, the metal concentrations maintained at high value on the entire soil profile compared to other directions. In the W direction, the surface layer showed much higher concentrations of Ni, and in the deeper layers the value of Ni concentration were still high. The correlation between the Ni concentrations in soil and the pH was low, positive and decreased in intensity with the depth profile.

Except for samples collected west from the pollution source, distribution of Mn was approximately uniform throughout the soil profile, the concentration ranging between 942.32 mg/kg and 1578.64 mg/kg at 5-20 cm depth, and between 796.68 and 1631.13 mg/kg at 20-40 cm depth (Figure 13). The surface layer showed a low correlation between the Mn concentrations and

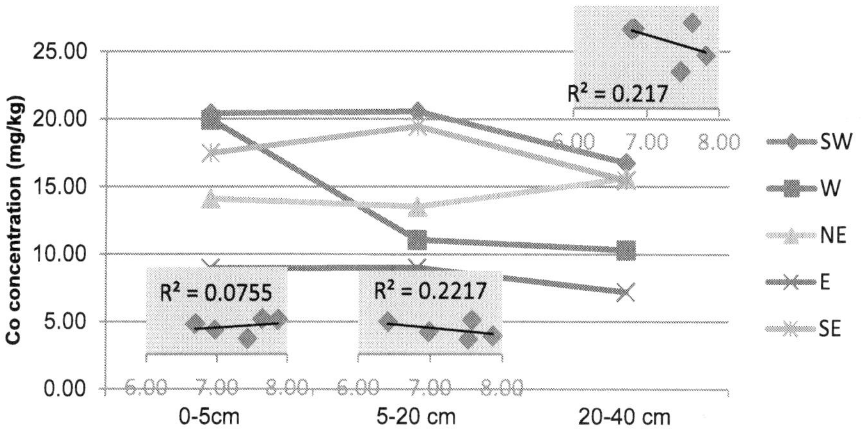

Figure 11. Cobalt vertical distribution in soil and correlation with pH

pH, in the 5-20 cm layer the correlation was strong (> 0.6) and in the lower part of the soil profile correlation was moderate.

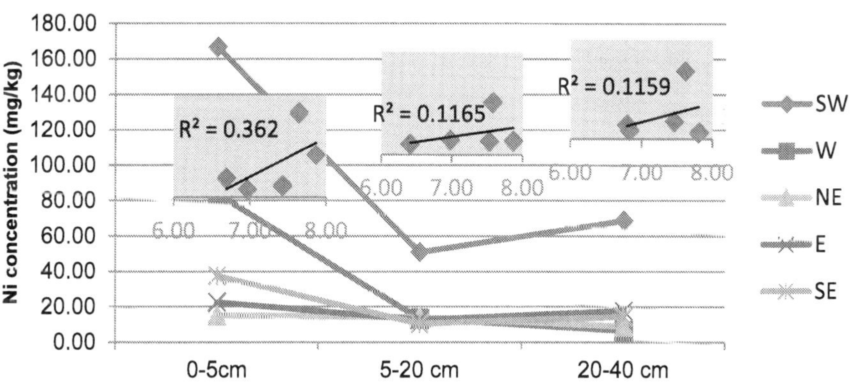

Figure 12. Nickel vertical distribution in soil and correlation with pH

The soils in Zone II showed similar values of Cr concentration on the entire soil profile, observing only a slight increase in the median layer (Figure 14). In the SW and W directions, Cr distribution differed on the soil profile. In the surface layer the concentration had high values, but they decreased with the depth. Only in the surface layer the concentration of Cr in the soil had a strong correlation with pH of soil (> 0.6).

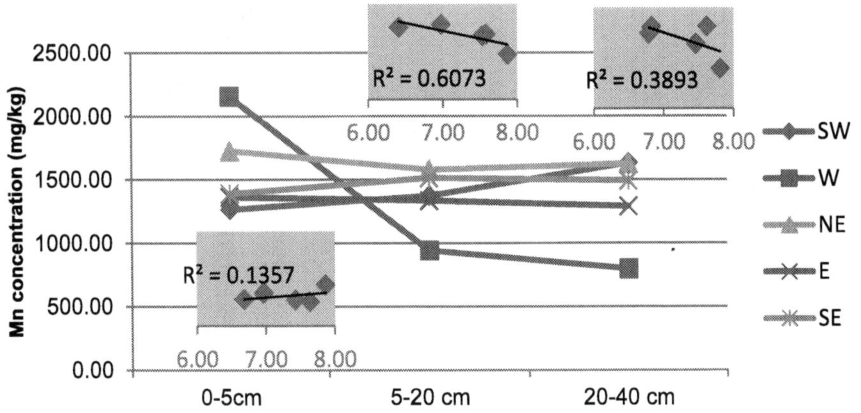

Figure 13. Manganese vertical distribution in soil and correlation with pH

The Mo concentration varied widely in depth of soil profile (Figure 15). In the SW, NE and SE directions the concentrations decreased significantly in the middle of the profile and remain at the same value at the bottom of it. In the W direction, a higher concentration of Mo (4.94 mg/kg) was found at a depth of 5-20 cm. The best correlation between the Mo concentration in soil and pH was at depth of 5-20 cm, with values > 0.4. The upper and lower layers had a lower correlation between Mo concentration and pH.

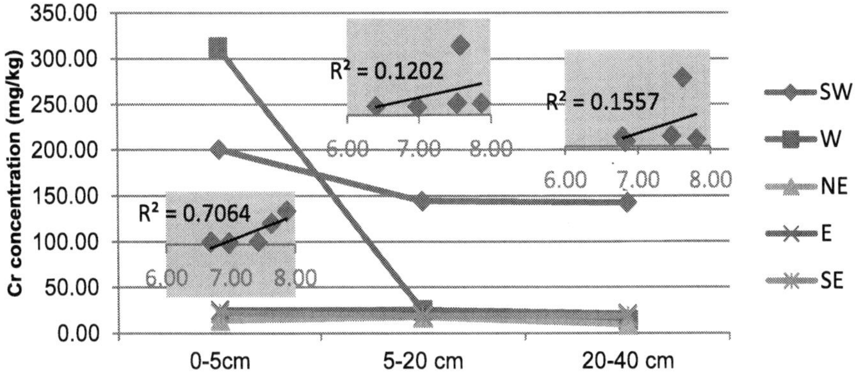

Figure 14. Chromium vertical distribution in soil and correlation with pH

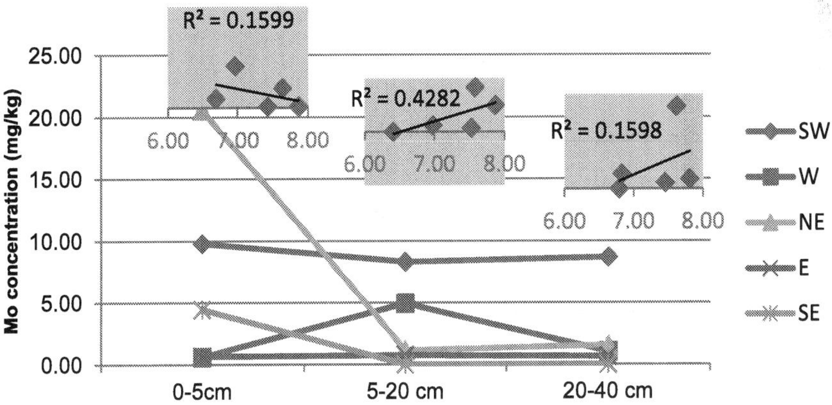

Figure 15. Molybdenum vertical distribution in soil and correlation with pH

The statistical analysis indicated that the heavy metal concentration in soil is negatively correlated with the depth, as the metal concentration decreased with the increasing of depth (Table 5). The correlation was statistically significant at level lower than 5%. Heavy metal concentration in soil was positively correlated with the pH of soil, with low to moderate intensity, except the Mn concentration which showed a very low negative concentration.

	Cu	Zn	Sn	Pb	Co	Ni	Mn	Cr	Mo
Depth	-0.2454 [a]	-0.3469 [a]	-0.5965 [a]	-0.2588 [a]	-0.2907 [a]	-0.3879 [a]	-0.2639 [c]	-0.3277 [a]	-0.3469 [c]
pH	0.3796 [a]	0.4171 [a]	0.1554 [b]	0.5066 [a]	0.5675 [c]	0.3991 [a]	-0.0995 [c]	0.5007 [a]	0.0932 [a]
[a] - $p < 0.05$; [b] - $p < 0.01$; [c] - $p < 0.001$									

Table 5 Pearson coefficient of correlation between heavy metal concentrations in soil and depth and pH of soil

5.4. Vertical mobility index of heavy metals in soil

The vertical mobility index (VMI) represents the relative explanation of heavy metal concentration between two underlying layer of soil: between 0-5 cm and 5-20 cm; between 5-20 cm and 20-40 cm (Figure 16). The metal mobility between layers was statistically significant at 95% confidence level. VMI will indicate a very weak mobility of metal at values lower than 20%, weak mobility of metal at values between 20 and 40%, moderate mobility of metal at values between 40 and 60%, strong mobility at values between 60 and 80%, and very strong mobility at values higher than 80%.

Distribution of Cu in the soil profile was given by the very strong correlation of the concentration of this element in the three layers of soil profile. High values of vertical mobility index were observed between the concentrations of Cu in the two underlying layers, which demonstrated the very high mobility of Cu in the soil, mobility influenced also by the pH. Values of correlation between concentrations at different depths indicated weak Zn mobility on the soil profile to depth of 20 cm and a very strong mobility between 20 and 40 cm.

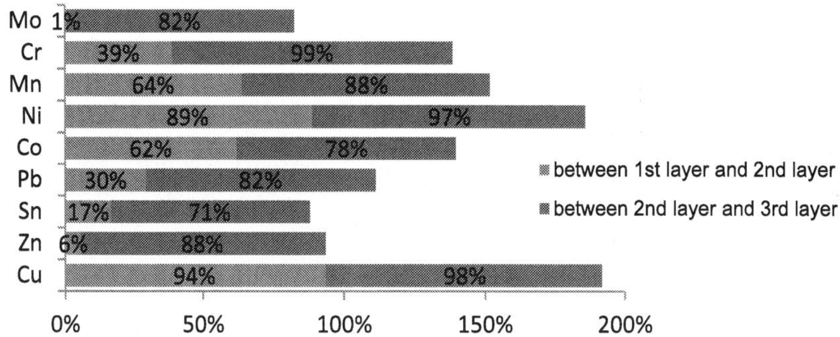

Figure 16. Vertical mobility index (VMI) of heavy metals between the layers of the soil profile, statistically significant at 95% confidence level

The mobility index of Sn was different depending on the depth. In the surface layer, this metal had a weak mobility; the correlation between the two layers (0-5cm and 5-20 cm) was very low. At greater depths, Sn mobility index was higher, indicated by the strong correlation of the concentrations at depths greater than 20 cm. Correlation of Pb concentration between layer of 0-5 cm and the other two layers was very low (<0.3). A strong correlation exists only between 5-20 cm and 20-40 cm depth which indicate a very strong mobility of Pb. The correlation between the concentrations of Co in the soil profile indicated a strong mobility of metal between the surface layer and underlying layers. The correlation between the Ni concentrations of different depths of soil profile is very strong, indicating a very strong mobility of Ni in soil which increased with depth.

The correlation between the concentrations of Mn in different depths of the soil profile is very strong, which indicated that a significant increase in the concentration of Mn in the upper layer will lead to an increase of Mn concentration also in the depth due to very strong mobility of this metal. The correlation between the concentrations of Cr in different depths of soil profile indicated a weak mobility of Cr between the surface layer and the middle layer of soil profile and a very strong mobility in the lower part of the soil profile. Correlations between the Mo concentrations were different on the soil profile: very weak in its upper layer and strong in the lower layers of soil profile.

6. Discussion

Because of the metallurgical activities carried out in the vicinity of Targoviste, which produced in time significant amounts of particulate matter with a high content of Pb, Cr, Cu, Mn, Ni and

http://dx.doi.org/10.5772/58304

Zn, the quality of agricultural soils is negatively influenced by the concentrations of heavy metals, which represent a risk of toxicity to humans. In order to evaluate the degree of loading of soils with heavy metals, the present study shows: (i) the horizontal distribution of heavy metals, depending on the prevailing direction of the air masses movement (wind rose) and (ii) vertical distribution to highlight the mobility of heavy metals into the soil profile, depending on the depth to which the plant roots can be found [15].

Previous studies have shown Targoviste area as an area of potential risk of heavy metal toxicity, along with other sources of pollution in Romania, as Copşa Mică, Deva and Baia Mare. This study presents the results of the concentration of heavy metals in soils near Targoviste, and are considered the high risk toxic metals such as Cu, Zn, Sn, Pb, Co, Ni, Mn, Cr, Mo. The analysis of horizontal distribution of heavy metals (Table 4) indicated that the agricultural soil in the vicinity of metallurgical unit were loaded with heavy metals concentrations exceeding several times the normal values and even alert threshold. Some values indicated concentrations of Cu and Zn even higher than in Copşa Mică [16] and Baia Mare [17].

Soils most affected by heavy metal pollution were those located on SV and V directions against the source of pollution, which are the predominant direction of air masses movement in the studied area. Also the other studied directions are characterized by concentration of heavy metals higher than normal limits, constituting a potential risk in the use of those soils in agriculture.

Analysis of vertical distribution of heavy metals indicated that high levels of heavy metals are caused primarily by industrial activity and less by the concentration of metals in the parent material. The concentration of all metals was higher in surface layer of soil (0-5 cm) because of deposition of particulate matter [21] and lower in the deeper layers of soil, possibly due to binding of metals to soil organic fractions (in the surface layer) that have slowed the metal mobility on the profile and decreased the transfer of metals to the underlying layers. Statistical analysis of correlations between the metal concentrations in soil and depth of the soil layer showed a negative correlation, which confirmed the hypothesis that the high concentration of heavy metals in soils from the industrial area of Targoviste, comes from a source of pollution and not from the parent material.

The metal mobility from the surface layers to the deeper layers can be explained on the basis of vertical distribution pattern of heavy metal. In order to measure the metal mobility on the profile of soil, a vertical mobility index (VMI) was calculated. This index indicated the extent to which that metal from a top layer of soil was transferred to the deeper layer by leaching and adsorption to soil particles. Depending on the type of soil, the mobility varies, promoting or inhibiting the rate of metal transfer from surface soil to groundwater, where the toxicity hazard is more difficult to control and the remediation actions are trickier. The most mobile heavy metals in investigated soils were Cu and Ni, with a mobility of 94 % and 89 % respectively, between the two layers from the surface of soil profile and a mobility of 98 % and 97 % respectively, between the two deeper layers. The hazard risk of these two metals to get into the groundwater was expressed by the high values of VMI. Also Mn and Co showed quite high values of VMI, which placed them as potential hazardous metals. Analysis of heavy metal mobility is important because once they reached the groundwater they present a greater risk to humans. Long-term exposure to high concentrations of these metals can cause serious illnesses in humans [4, 10, 11], and the intervention for remediation of soil is urgent, before the heavy metals will reach the groundwater.

7. Conclusions

Following chemical analysis by ICP-AES, we found the heavy metal concentration levels above the normal value of an industrial soil, and sometimes even exceeding the alert limit (according to Order 756/1997).

Distribution of heavy metals on soil profile showed that accumulation of metal in the soil decreases with the depth. The concentrations were higher on the surface of the soil profile. This indicates a continuous impact of metallurgical activities on soil by accumulating new quantities of metals.

The concentration of heavy metals in soil was also considered to be in connection with the pH and chemical composition of soil: the presence or absence of other metallic elements.

The vertical mobility index indicated moderate to strong mobility of majority of studied metals, especially in the lower layer of the soil profile, 20-40 cm.

Acknowledgements

The research is part of the PhD thesis "Studies and research concerning the remediation of heavy metal polluted soil by eco-technological procedure", in Romanian, realized by the author in 2011. Is a good opportunity to thank my colleagues Sorin Ciulei for his support in soil assessment and evaluation, Irina Fierascu and Radu Claudiu Fierascu for their support in the sample analyses by ICP-AES. Also I want to thank my thesis coordinator, Prof. Georghe Ionita for guidance in preparing the thesis and to Prof. Dumitru Ion for the financial support for the publication.

Author details

Carmen Cristina Elekes*

Address all correspondence to: cristina_elekesh@yahoo.com

"Valahia" University of Târgovişte, Dâmboviţa, Department of Environmental Engineering, Romania

References

[1] FOREGS Geochemical Baseline Mapping Programme Geochemical Atlas of Europe, 2005, ISBN 951-690-913-2, http://www.gsf.fi/publ/foregsatlas/.

[2] Kabata-Pendias A. Trace elements in soils and plants, fourth edition. Boca Raton, U.S.A.: CRC Press, Taylor and Francis Group; 2010.

[3] Order no. 756 from 3rd November 1997 for the approval of the Regulation of environmental pollution assessment.

[4] Géczi A, Bódis K. Környezeti monitoring Verespatak-Aabrudbánya vidékén. Kriterion, Cluj Napoca 2003, http://www.rosiamontana.ro/img_upload/c77c3453789af5de5049783 baaa35f31/Anexa_2_verespatak _kutjel_pdf_RO.pdf.

[5] Sumner ME. Handbook of Soil Science. Boca Raton, Florida: Taylor & Francis; 2000.

[6] Salmon SU. Geochimical Modelling of Acid Mine Drainage in Mill Tailings, KTH, Land and Water Resources Engineering, Stockholm, Sweden: Doctoral Thesis; 2003, ISBN 91-7283-607-5.

[7] Kabata-Pendias A, Krakowiak A. Useful phytoindicator (dandelion) for trace metal pollution, Proc. 5th Int. Conf. Transport, Fate and Effects of Silver in the Environ., Andren A.W. and Bober T.W., eds., Argentum, Madison, WI 1998, 145.

[8] Schafer SG, Femfert U. Tin – a toxic heavy metal? A review of the literature, Regulatory and Environmental Health 1985; 4 57 – 69.

[9] Kaminski MD, Landsberger S. Heavy metals in urban soils of East St. Louis, IL. Part I: Total concentration of heavy metals in soils, Journal of the Air & Waste Management Association 2000; 50(9) 1667-1679.

[10] Domingo JL. Cobalt in the Environment and Its Toxicological Implications, Rev Environ Contam Toxicol 1989; 108 105-132.

[11] Agency for Toxic Substances and Disease Registry (ATSDR). Toxicological Profile for Manganese (Update). Draft for Public Comment. U.S. Public Health Service, U.S. Department of Health and Human Services, Atlanta, GA. 1997.

[12] Lenntech, Chemical properties of molibdenum – Health effects of molybdenum – Environmental effects of molybdenum, http://www.lenntech.com/periodic/elements/mo.htm [accessed 02.12.2013].

[13] Dumitru M, Răuţă C, Toti M, Gamenţ E, Motelică DM. Impact of emissions loaded with heavy metals on soil land plant, Contaminated soil. Netherlands: Kluwer Academic Publishers; 1995, 607-607.

[14] Lăcătuşu R, Ghelase I. Heavy metal abundance in soil, in Romanian, Mediul înconjurător 1992; III (4) 44-52.

[15] Răuţă C, Lăcătuşu R, Cârstea S. (1992) Heavy metal pollution of soils in Romania, Mediul înconjurător 1992; III (4) 33-44.

[16] Vrînceanu NO, Motelică DM, Dumitru M, Toti M, Gamenţ E, Tănase V. Aspects concerning soil pollution with heavy metals in Copşa Mică area, Proceedings of International Conference "Soil under Global Change", Constanţa, România 2002, 357-365.

[17] Lăcătuşu R, Cârstea S, Lungu M, Kovacsovics B, Lazăr R. Soil pollution with cyanides and heavy metals in the Baia Mare area; ecological reconstruction, Soil Science 2002; 36(1) 77-87.

[18] Dumitrescu C et al. Methods and technics of evaluation and neutralization of pollutants, in Romanian, Bucharest; 2002, ISBN 973-652-643-7, 30-110.

[19] Müller G. Soil biology, in Romanian, Bucharest: Ed. Agro Silvică; 1965.

[20] Herms U, Brümmer GW. Einflussgrössen der Schwermetallöslichkeit und –binding in Boden. Z. Pflanzenernaehr, Bodenkd 1984; 147, 400.

[21] Regional Environmental Protection Agency Piteşti, Integrated Environmental Authorization, no. 159/30.10.2007, requested by S.C. Mechel Târgovişte S.A.

[22] MEMDUR: Sustainable resource management system for environmental monitoring and risk assessment in order to prevent negative effects of crisis management - Scientific and Technical Report (RST) no. D11-037/2007.

[23] Teodorescu-Soare E. Pedology, in Romanian, Iaşi: Ed. „Ion Ionescu de la Brad"; 2006.

[24] Costea T, Istudor V, Fierascu RC, Fierascu I, Botez A. Researches upon indigenous herbal products for therapeutic valorification in metabolic diseases Note I. Betulae folium and Rubi idaei folium, sources of micro- and macroelements. Farmacia 2013; 61: 1.

[25] Elekes CC, Ioniţă Gh, Busuioc G. The bioconversion factor of some green plants growth in the metallurgic industrial area. Annals of Food Science and Technology, Târgovişte 2009; 10(2) 580-585.

[26] Elekes CC, Dumitriu I. The impact of metallurgic activities on the soil quality in Dâmbovita County. Scientific Papers. Agronomy Series, Ed. "Ion Ionescu de la Brad" 2009; 52(1) 345-350.

[27] Ioniţă Gh, Elekes CC. Studies and research concerning the polluted soils decontamination by biosystems. Conference Proceedings of the Academy of Romanian Scientists, Bucharest 2009; 1(1) 201-210.

Chapter 9

Assessment of Soil Pollution with Heavy Metals in Romania

Anca Maria Moldoveanu

Additional information is available at the end of the chapter

http://dx.doi.org/10.5772/57595

1. Introduction

Human activities, industry, erosion and the continuous spread of urbanization can induce soil pollution in various ways. The pollution of soil can be dangerous for human health because the toxic substances can enter the crops and the ground water. The soil, due to its properties and structure, plays the role of a filter that can retain and be a deposit for toxic substances. The most frequent contaminants of soil in Europe are heavy metals and mineral oil. The group 'heavy metals' for the purpose of discussing health risks or impacts generally includes: Arsenic (As), Lead (Pb), Cadmium (Cd), Chromium (Cr) (although only the form Cr(VI) is toxic), Copper (Cu), Mercury (Hg), Nickel (Ni), Zinc (Zn).

Several of these elements are necessary for human health and are beneficial when taken into the body in foods or as supplements at appropriate, low levels. Conversely, cadmium, lead and mercury have no known biological function and are toxic to humans.

The sources of heavy metals that pollute the soil can be human activities like metalliferous mining (As, Cd, Cu, Ni, Pb, Zn), smelters (As, Cd, Pb), metallurgy, electronic industry (where metals are used in batteries, semiconductors, circuits), rolling (Ni, Cd, Pb, Hg, Se), dyes and paints industry (Pb, Cr, As, Se, Mo, Cd, Co, Ba, Zn), plastics industry (Cd, Zn, Pb, Sn are used as polymer stabilizers), chemical industry (using Pb, Ni, Nb, Hg, Pt, Ru as electrode catalysts), wood industry (As, Cr and Cu). In the vicinity of furniture factories and wood-processing, these elements were often identified as soil and water pollutants [1, 2, 3, 4, 5, 6, 7, 8, 9].

Also, by storing municipal waste, special and hazardous waste, the soil may be contaminated with various heavy metals.

The combustion of fossil fuels leads to the presence of Cd, Zn, As, Se, Cu, Mn, V, in the ashes and particulates of combustion. Some heavy metals (Se, Te, Pb, Mo, Li) are added to fuels and lubricants to improve their properties. [1, 2, 3, 4, 5, 6, 7, 8, 9].

Corrosion of metals contained in various materials (Cu, Pb pipes and roofs, Cr, Ni, Co in stainless steel, Cr, Pb painting) can cause soil pollution. Agro-livestock activities can also cause soil pollution (As, Cu, Zn can be added to feed pigs and poultry, Cd, As, Pb, Mn, Cu, Zn can be present in phosphate fertilizers and pesticides). [1, 2, 3, 4, 5, 6, 7, 8, 9].

Pollution by heavy metals and many organic contaminants is practically irreversible (European Commission, 2012).

Excess heavy metal accumulation in soils is toxic to humans and other animals.

Exposure to heavy metals is normally chronic (exposure over a longer period of time), due to food chain transfer. Acute (immediate) poisoning from heavy metals is rare, but possible, through ingestion or dermal contact. Chronic problems associated with long-term heavy metal exposures are [10, 11, 12, 13, 14, 15, 16, 17, 18, 19] for:

1. Lead [20-60] – is well known to be toxic and its effects have been more extensively reviewed than the effects of other trace metals. Lead can cause serious injury to the brain, nervous system, red blood cells, and kidneys. Exposure to lead can result in a wide range of biological effects depending on the level and duration of exposure. Various effects occur over a broad range of doses, with the developing young and infants being more sensitive than adults. Lead poisoning severe enough to cause evident illness is now very rare. Lead can be harmful after uptake from food, air, water or contaminated dust. Lead is a particularly dangerous chemical, as it can accumulate in individual organisms, as well as entire food chains.

2. Cadmium – affects kidney, liver, and GI tract [28, 30, 61, 62, 63].

3. Arsenic – is associated with skin damage, increased risk of cancer and problems with circulatory circulatory system, affects kidneys and the central nervous system [64, 65, 66, 67, 68, 69].

4. Chromium is associated with allergic dermatitis in humans [70, 71].

5. Mercury is associated with kidney damage [72, 73, 74, 75].

6. The most common problem caused by cationic metals (metallic elements whose forms in soil are positively charged cations e.g., Pb2+) are mercury, cadmium, lead, nickel, copper, zinc, chromium and manganese. The most common anionic compounds (elements whose forms in soil are combined with oxygen and are negatively charged e.g., $MoO4^{2-}$) are arsenic, molybdenum, selenium, and boron.

1.1. The aim of the study

This study tried to assess the level of soil contamination with heavy metals (lead, cadmium, mercury, manganese, nickel, total chromium, zinc, cobalt, copper and arsenic) in 34 counties in Romania. No previous data exists as a reference level of soil pollution.

2. Material and methodology

Sampling was done in 2012, in the fall (September – October). Soil samples were taken from max 20 cm deep, from areas around children's units like nurseries, kindergartens, schools or playgrounds for children. The samples were taken in county capital cities. Also, soil samples were taken from areas of land used for agricultural purposes like farmlands and gardens.

For metal analysis, 0.2 g of ground soil were digested with aqua regia using microwave digestion (EN 13346 :02) followed by filtration and adjusting the volume in a volumetric flask. Concentrations of Pb, Cd, Cu, Cr were measured using the electrothermal atomic absorption spectrometry (ISO 11464:2006, SR ISO 11047/1999, EPA 6020). Concentrations of Zn, Mn, Co, Ni, Cr were measured using flame AAS (atomic absorption spectrometry) and concentrations of Hg and As were measured using atomic absorption spectrometry with cold vapour (HVG-AAS) and SR EN 1485/2007. [76, 77]

Despite the importance of soil pollution for our society, there is no EU legislation specifically targeting the protection of soil [78, 79]. A regulation regarding the soil pollution exists in Romania (Order nr. 756, November, the 3th, 1997). In this Order [80], the following definitions are of importance: environmental impact, types of land uses that require a certain soil quality, soil characterized by a maximum acceptable level of pollutants - sensitive and less sensitive soil, remediation targets, potentially significant pollution, significant pollution, threshold alert – (concentrations of pollutants in air, water, soil or emissions / discharges, which are designed to alert the competent authorities of a potential impact on the environment and determines the triggering of additional monitoring and / or reduction of concentrations of pollutants in emissions / evacuations), intervention threshold, reference sample [80].

Heavy Metals	Reference threshold for sensitive soil	Alert thresholds - sensitive soil	Alert thresholds - soil less sensitive	Action levels - soil sensitive	Action levels – soil less sensitive
Arsenic (As)	5	15	25	25	50
Cadmium (Cd)	1	3	5	5	10
Cobalt (Co)	15	30	100	50	250
Chromium (Cr) : Total Chromium total Hexavalent Chromium	30 1	100 4	300 10	300 10	600 20
Copper (Cu)	20	100	250	200	500
Manganese (Mn)	900	1500	2000	2500	4000
Mercury (Hg)	0.1	1	4	2	10
Nickel (Ni)	20	75	200	150	500
Lead(Pb)	20	50	250	100	1000
Zinc (Zn)	100	300	700	600	1500

Table 1. Reference values for traces of heavy metals in soil - Heavy metals (mg/kg dry substance). Values are stipulated in Order 756/1997 for the approval of the Regulation on the assessment of environmental pollution [80].

3. Results and discussions

Cadmium in the soil registered levels that ranged from 0 and 0.86 mg/kg, in the urban areas and in the rural areas. Compared to the threshold value it can be seen that the levels of cadmium in soil, in both urban and rural areas, have not exceeded the threshold value (1mg/kg). It can be seen that the level of cadmium in soil is slightly higher in rural areas compared to urban areas.

Lead in soil, registered levels that ranged from 0 and 96.11 mg/kg in the urban areas and from 0 to 147.67 mg/kg in rural areas. It can be seen that, both in the urban and rural areas, the lead levels in soil exceeded the threshold value (20 mg/kg). The alert threshold for sensitive soil (50 mg/kg) was exceeded in both the urban and rural areas.

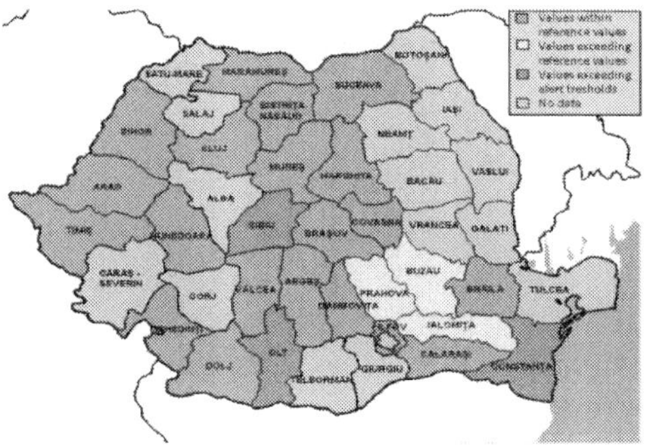

Figure 1. Lead levels in soil in urban areas

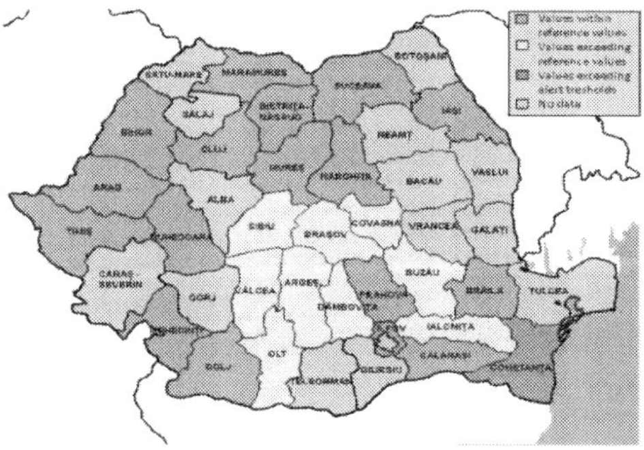

Figure 2. Lead levels in soil in rural areas

Comparing the two charts it can be seen that, in the same area (Constanța county), the level of lead in soil is over the alert threshold in the urban and in the rural area. In other two counties

(Buzău and Ialomiţa) the levels of lead in soil are exceeding the threshold limit both in the urban area and in the rural area. There are other two counties beside Constanta, where the levels of lead are exceeding the alert threshold in urban areas (Olt and Sibiu). In rural areas it can be seen that the number of counties where lead levels in soil exceed the threshold is higher than in urban areas.

Copper in soil registered levels that ranged from 0 and 128.5 mg/kg in the urban areas and from 0 to 80.43 mg/kg in rural areas. It can be seen that the levels identified in soil exceeded, both in the urban and rural areas, the threshold value (20 mg/kg). Also, the alert threshold for sensitive soil (100 mg/kg) was exceeded in a single urban area. In rural areas a slightly higher level of copper in the soil was observed.

Figure 3. Copper levels in soil in urban areas

Figure 4. Copper levels in soil in rural areas

The alert threshold was exceeded in a single county in urban areas. The number of counties where was found levels of copper in soil, in rural areas, that exceeded the threshold was double in comparison with urban areas.

Nickel in soil registered levels that ranged from 0 and 44 mg/kg in the urban areas and from 0 to 41.6 mg/kg in rural areas. The levels of nickel found in soil samples exceeded the threshold value (20 mg/kg), in the urban and rural areas. Nickel recorded slightly higher levels in the soil of rural areas, but, also, were found peaks in urban soil. The alert threshold for sensitive soil (75 mg/kg) was not exceeded in either urban or rural area.

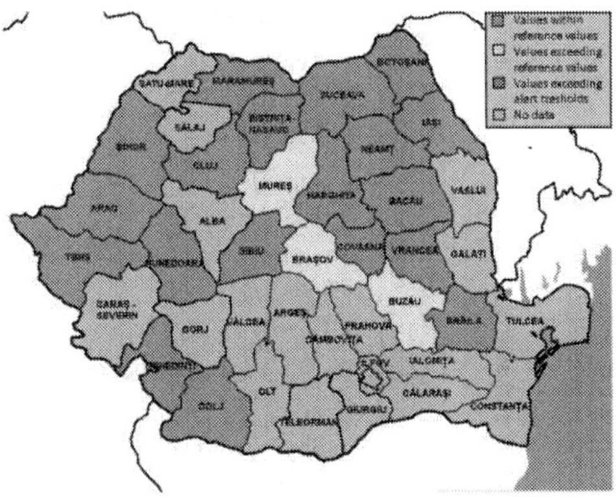

Figure 5. Nickel levels in soil in urban areas

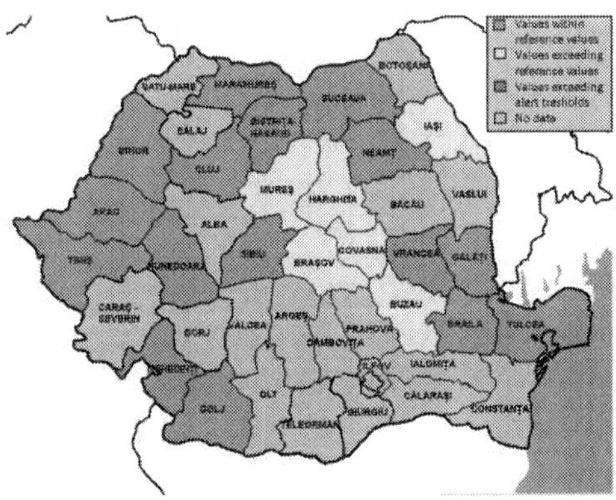

Figure 6. Nickel levels in soil in rural areas

As it can be seen in the charts above, the soil pollution with nickel is similar in the urban and rural areas, regarding the number of counties affected.

Manganese in soil registered levels that ranged from 0 and 1256.57 mg/kg in the urban areas and in rural areas the levels ranged from 0 to 1072. 62 mg/kg. Compared to the threshold value, the levels of manganese found in soil exceeded the threshold, both in the urban and rural areas (900 mg/kg). Alert threshold (1500 mg/kg) was not exceeded in neither rural areas nor urban areas.

Figure 7. Manganese levels in soil in urban areas

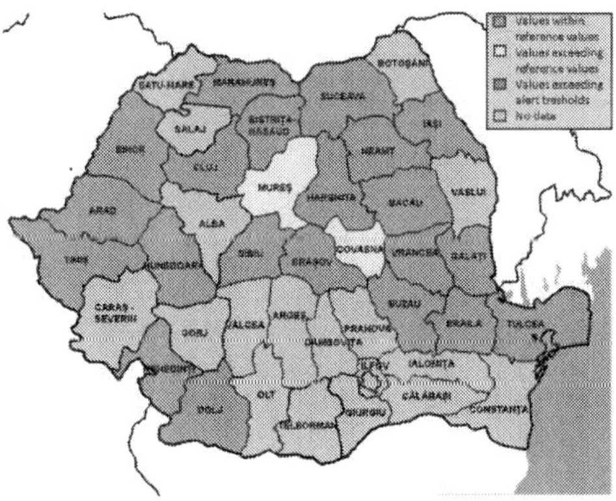

Figure 8. Manganese levels in soil in rural areas

The threshold was exceeded, both in urban and rural areas, in only one county in the middle of the country (Mureş).

Concerning the levels of **total chromium** in soil, values are similar, in urban and rural areas. The levels found in soil were between 0 and 40.81 mg/kg in urban areas and between 0 and

57.23 mg/kg in rural areas. Only the threshold (30 mg/kg) was exceeded in the urban and rural areas and not the alert threshold (100 mg/kg).

Figure 9. Total chromium levels in soil in urban areas

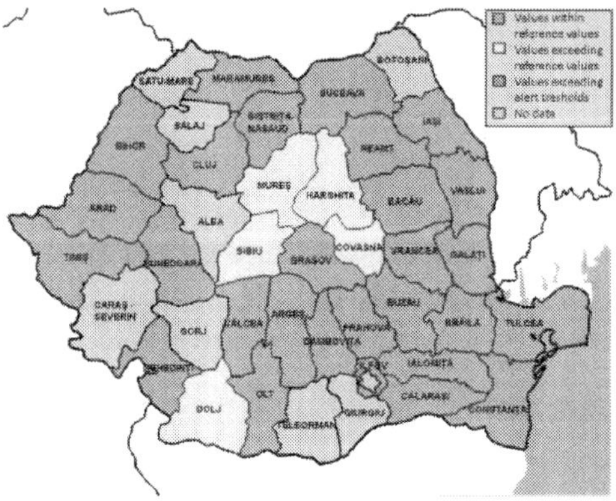

Figure 10. Total chromium levels in soil in rural areas

The total chromium levels registered in soil exceeded the threshold in a small number of counties. Only in one county (Mureș), levels of chromium exceeding the threshold were found in both the urban and rural areas.

Cobalt levels were higher in rural areas compared to urban areas. The highest value was found in the soil of a farmland in Covasna county, where the level of cobalt was equal to 20.23 mg/kg, exceeding the threshold value (15 mg/kg) but not the alert threshold (30 mg/kg).

http://dx.doi.org/10.5772/57595

Figure 11. Total cobalt levels in soil in urban areas

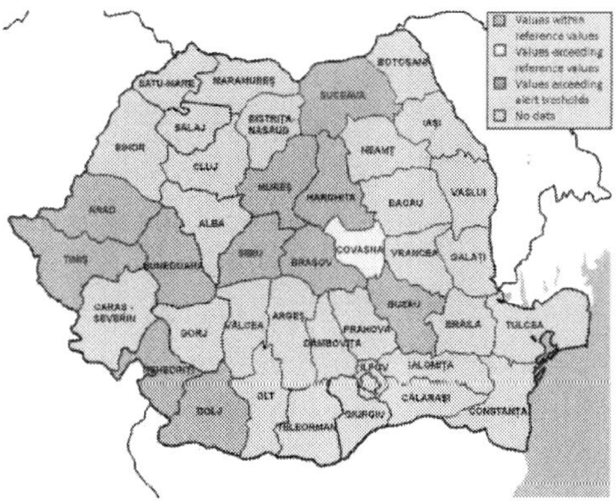

Figure 12. Total cobalt levels in soil in rural areas

From the charts, it can be seen that it is only one hot spot where the levels of cobalt in soil exceeded the threshold.

Zinc levels ranged between 0 and 275.35 mg/kg in the soil of urban areas and between 0 – 131.79 mg/kg in the soil of rural areas. The levels recorded were similar in urban and rural areas. One single higher value was found in Sibiu county.

Figure 13. Zinc levels in soil in urban areas

Figure 14. Zinc levels in soil in rural areas

A similar situation with zinc pollution of the soil was found in the urban and in the rural areas of Romania as number of counties affected.

Mercury levels in the soil ranged between 0 and 0.265 mg/kg in the soil of urban areas and between 0 – 0.305 mg/kg in the soil of rural areas. The threshold value is equal to 0.1 mg/kg and was exceeded in urban and rural areas, but the alert threshold is equal to 1 mg/kg and was not exceeded in neither urban nor in rural areas. Special mention must be made for district Ferneziu of the town Baia Mare where there is a historical pollution of the environment. The level of mercury found in this area was equal to 0.265 mg/kg.

http://dx.doi.org/10.5772/57595

Figure 15. Mercury levels in soil in urban areas

Figure 16. Mercury levels in soil in rural areas

For the situation of soil pollution with mercury it must be stressed that, in the area of Maramureş there is a historical pollution due to a non-ferrous industry existing for a long period of time.

Arsenic levels in soil were between 0 and 8.22 mg/kg in urban areas and 0 and 9.59 mg/kg soil. Arsenic concentrations were higher in rural areas compared to urban soil. The reference threshold is equal to 5 mg/kg and the alert threshold is equal to 15 mg/kg. The threshold value was exceeded but the not the alert threshold.

Figure 17. Arsenic levels in soil in urban areas

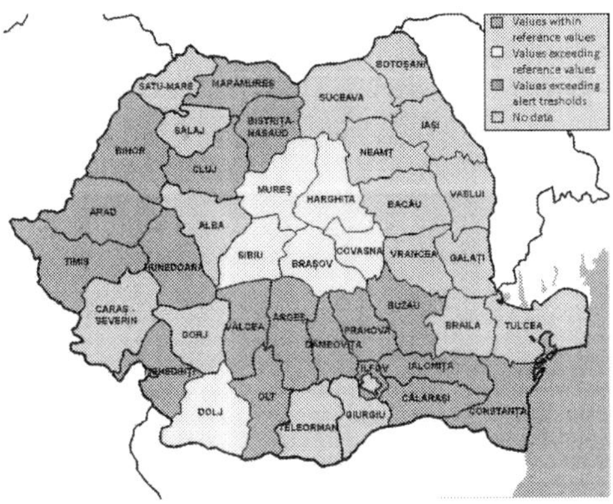

Figure 18. Arsenic levels in soil in rural areas

Comparing the two maps, it can be seen that, levels of pollution exceeding the threshold were found in the Sibiu and Mureș County, in the middle of the country, in both urban and rural areas.

http://dx.doi.org/10.5772/57595

4. Conclusions

This study is the first study that tried to assess the situation of soil pollution with heavy metals in Romania.

After assessment of the soil contamination with heavy metals in 34 counties in Romania, it can be concluded that for all investigated metals except cadmium the threshold values were exceeded.

Alert thresholds were exceeded for lead and copper, both in urban and rural areas, especially in two counties: Maramureş county and Sibiu county. Areas like Maramureş county or Sibiu county have a historic pollution due to the industry present since 1844 in Baia Mare, Maramures county and 1936 în Copşa Mică, Sibiu county.

A national screening program was performed in the years 2008-2010 in Maramureş county (Baia Mare city and especially Ferneziu district) to see the impact on human health in connection with lead pollution. In the years 2008 and 2009 a group of 91 children was investigated [81, 82].

The results showed the following situation:

Year of study	Lead in blood up to 10 μ_g/dl	Lead in blood up to 20 μ_g/dl	Lead in blood up to 30 μ_g/dl	Lead in blood up to 35 μ_g/dl	Lead in blood > 35 μ_g/dl
2008 - % of investigated children	95.4	97.8	97.8	97.8	2.2
2009 - % of investigated children	58.70	71.74	84.06	89.13	10.87

Table 2. Lead levels in blood in city Baia Mare, Maramureş county in 2008 and 2009.

A special situation had the children living near the factory that processes lead in the district Ferneziu in Baia Mare county.

During the year 2010, the study group was made of 94 pregnant women and 94 new born babies. The blood was taken from the new born cord in the case of children [83].

Level of lead in blood of pregnant women	0-10 μ_g/dl	10-20 μ_g/dl	20-30 μ_g/dl	> 35 μ_g/dl
% of the study group	94.7	1.06	2.12	2.12

Table 3. Lead levels in blood in the group of pregnant women, in city Baia Mare, Maramureş county, in 2010.

Level of lead in blood of cord of new born babies	0-10 μ_g/dl	10-20 μ_g/dl	20-30 μ_g/dl	> 35 μ_g/dl
% of the study group	93.6	1.06	-	5.34

Table 4. Lead levels in blood in the cord of new born babies, in city Baia Mare, Maramureş county, in 2010

For copper, alert thresholds was exceeded, both in urban and rural areas. In Suceava county, copper mining was performed since 1922. In this area, there were taken rehabilitation measures in the period between1999-2005. Still, some of the effects of the ancient pollution situation can be seen today.

All the other investigated metals in the soil, both in residential areas and agricultural areas (which both can be classified as sensitive lands), do not exceed the alert threshold for sensitive lands.

For cadmium, nickel, manganese, cobalt and arsenic higher levels were recorded in the soil of rural areas than in the soil of urban areas.

For total chromium, zinc and mercury similar levels were recorded in the soil of urban and rural areas.

It must be stressed that in the areas where the alert thresholds were exceeded, pollution with impact on the environment can be present. The situation can be labelled as "potentially significant pollution". In this case, the competent authorities can take action to reduce the pollutants emissions in the air or to take measures to diminish the possibilities to contaminate the soil directly or through water.

Author details

Anca Maria Moldoveanu[1,2*]

Address all correspondence to: anca.moldoveanu@gmail.com

1 University of Medicine and Pharmacy, "Carol Davila" Bucharest, Romania

2 National Institute of Public Health, Bucharest, Romania

References

[1] Raymond A.Wuana and Felix E. Okieimen, HeavyMetals in Contaminated Soils: A Review of Sources, Chemistry, Risks and Best Available Strategies for Remediation;

International Scholarly Research Network ISRN Ecology, Volume 2011, Article ID 402647, 20 pages, doi:10.5402/2011/402647,

[2] Jintao Liang , Cuicui Chen, Xiuli Song, Yulan Han, Zhenhai Liang , Assessment of Heavy Metal Pollution in Soil and Plants from Dunhua Sewage Irrigation Area, Int. J. Electrochem. Sci., 6 (2011) 5314 – 5324, www.electrochemsci.org,

[3] Heavy Metal Soil Contamination, Soil Quality – Urban Technical Note No. 3, United States Department of Agriculture, September, 2000,

[4] EPA. Soil Screening Guidance: User's Guide. Publication 9355.4, July 23th, 1996

[5] European Commission, Science for Environment Report, In-Depth Report, Soil Contamination: Impacts on Human Health, September 2013, Issue 5. http://ec.europa.eu/environment.

[6] EUGRIS *Portal for land and soil management in Europe* [online].Available at: www.eugris.info

[7] European Commission (2012) *The State of Soil in Europe – A contribution of the JRC to the European Environment Agency's State and Outlook Report – SOER 2010.* Luxembourg: Publications Office of the European Union. Available at: http://ec.europa.eu/dgs/jrc/downloads/ jrc_reference_report_2012_02_soil.pdf

[8] Rodriguez Lado, L., Hengl, T., Reuter, H.I., (2008) Heavy metals in European soils: a geostatistical analysis of the FOREGS Geochemical database. *Geoderma* 148: 189-199.

[9] Huber & Prokop (2012), Progress in the management of contaminated sites. EIONET Workshop on Soil, Ispra, 10- 12 December 2012.

[10] D. R. Baldwin and W. J. Marshall, "Heavy metal poisoning and its laboratory investigation, " *Annals of Clinical Biochemistry*, vol. 36, no. 3, pp. 267–300, 1999.

[11] WHO/UNECE (2006): *Health risks of heavy metals from long-rangetransboundary air pollution.* Draft of May 2006. Copenhagen: WorldHealth Organization Regional Office for Europe and Geneva: UnitedNations Economic Commission for Europe (UNECE). Available at:www.euro.who.int/__data/assets/pdf_file/0007/78649/E91044.pdfWHO (2013) *Ten chemicals of major health concern.* [online]. Retrievedfrom: www.who.int/ipcs/assessment/public_health/chemicals_phc/en/index.html,

[12] Brevik, E.C. & Burgess, L.C. eds (2013) *Soils and Human Health.* Boca Raton: CRC Press.

[13] Kibble, A. and Russell, D. (2010) *Contaminated Land and Health.* In Maynard, R.L. *et al.* (eds) Environmental Medicine. Boca Raton, Florida: CRC Press. Pp: 565–573

[14] Meerschman, E., Cockx, L., Islam, M.M., Meeuws, F., and Van Meirvenne, M. (2011) Geostatistical Assessment of the Impact of World War I on the Spatial Occurrence of Soil Heavy Metals. *Ambio.*40(4): 417–424.

[15] Navarro, M.C., Perez-Sirvent, C., Martinez-Sanchez, M.J., Vidal, J., Tovar, P.J., Bech, J. (2008) Abandoned mine sites as a source of contamination by heavy metals: A case study in a semi-arid zone. *Journal of Geochemical Exploration* 96: 183–193

[16] Environment Agency (England) (2009) *Human health toxicological assessment of contaminants in soil.* Science report - Final SC050021/ SR2. Bristol: Environment Agency. Available at: www.environmentagency. gov.uk/static/documents/Research/ TOX_guidance_report_-_ final.pdf

[17] Oliver, M.A. (1997) Soil and human health: a review. *European Journal of Soil Science.* 48: 573-592

[18] Morgan, R. (2013) *Soil, Heavy Metals, and Human Health.* In Brevik, E.C. & Burgess, L.C. (2013) Soils and Human Health. Boca Raton. FL: CRC Press, pp. 59-80.

[19] Gil, F., Capitan-Vallvey, L.F., De Santiago, E., Ballesta, J., Pla, A., Hernandez, A.F., Gutierrez-Bedmar, M., Fernandez-Crehuet, J., Gomez, J., Lopez-Guarnido, O., Rodrigo, L., Villanueva, E. (2006) Heavy metal concentrations in the general population of Andalusia, South of Spain A comparison with the population within the area of influence of Aznalcollar mine spill (SW Spain). *Science of the Total Environment.* 372 (2006): 49-57.

[20] WHO Europe (2009) *Levels of lead in children's blood fact sheet 4.5 (RPG4_Chem_Ex1)* Copenhagen: World Health OrganizationRegional Office for Europe. Available at: www.euro.who.int/__data/ assets/pdf_file/0003/97050/4.5.-Levels-of-lead-in-childrens-blood- EDITING_layouted.pdf

[21] Canfield, R.L., Henderson, C.R. Jr, Cory-Slechta, D.A., Cox, C., Jusko, T.A., Lanphear, B.P. (2003) Intellectual impairment in children with blood concentrations below 10 μg per deciliter. *New England Journal of Medicine.* 348:1517–1526.

[22] Chen, A., Dietrich, K.N., Ware, J.H., Radcliffe, J., Rogan, W.J. (2005) IQ and blood lead from 2 to 7 years of age: are the effects in older children the residual of high blood lead concentrations in 2-year-olds? *Environmental Health Perspectives.* 113:597–601.

[23] Jakubowski, M., Trzcinka-Ochocka. M., Raźniewska, G., Christensen, J.M., Starek, A. (1996) Blood lead in the general population in Poland. *International Archives of Occupational and Environmental Health.* 68:193–198.

[24] Mielke, H.W. & Reagan, P.L. (1998) Soil is an important pathway of human lead exposure. *Environmental Health Perspectives.* 106 (Suppl 1): 217–229.

[25] Tong, S., von Schirnding, Y.E., Prapamontol, T. (2000) Environmental lead exposure: a public health problem of global dimensions. *Bulletin of the World Health Organization.* 78:1068–1077.

[26] Ducoffre, G. et al. Lowering time trend of blood lead levels in Belgium since 1978. *Environmental research,* 51: 25–34 (1990).

[27] *Inorganic lead*. Geneva, World Health Organization, 1995 (Environmental Health Criteria, No. 165).

[28] Elinder, C.G. et al. Lead and cadmium levels in blood samples from the general population of Sweden. *Environmental research*, 30: 233–253 (1983).

[29] Brunekreef, B. The relationship between air lead and blood lead in children: a critical review. *Science of the total environment*, 38: 79–123 (1984).

[30] Ewers, U. et al. Contribution of lead and cadmium in dust fall to blood lead and blood cadmium in children and adults living in two nonferrous smelter areas of West Germany.*In*: Lekkas, T.D., ed. *International Conference on Heavy Metals in the Environment, Athens, September 1985*. Edinburgh, CEP Consultants, 1985.

[31] *Guidance manual for the integrated exposure uptake biokinetic model for lead in children*. Washington, DC, US Environmental Protection Agency, 1994 (EPA 540-R-93-081).

[32] Ziegler, E.E. ET AL. Absorption and retention of lead by infants. *Pediatric research*, 12:29–34 (1978).

[33] Brunekreef, B. *The relationship between environmental lead and blood lead in children: a study in environmental epidemiology*. Wageningen, Department of Environmental and Tropical Health, Agricultural University of Wageningen, 1985(Report 1985-211).

[34] *Lead*. Geneva, World Health Organization, 1977 (Environmental Health Criteria, No. 3).

[35] Barry, P.S.I. Concentrations of lead in the tissues of children. *British journal of industrial medicine*, 38: 61–71 (1981).

[36] Buchet, J.P. et al. Mobilization of lead during pregnancy in rats. *International archives of occupational and environmental health*, 40: 33–36 (1977).

[37] Silbergeld, E.K. Lead in bone: Implications for toxicology during pregnancy and lactation. *Environmental health perspectives*, 91: 63–70 (1991).

[38] Silbergeld, E.K. et al. Lead and osteoporosis: mobilization of lead from bone in postmenopausal women. *Environmental research*, 47: 79–94 (1989).

[39] Grandjean, P. et al. Reference intervals for trace elements in blood: significance of risk factors. *Scandinavian journal of clinical and laboratory investigation*, 52: 321–337 (1992).

[40] Smith, M. et al. The effects of lead exposure on urban children: the Institute of Child Health/Southampton Study. *Developmental medicine and child neurology*,. 47 (Suppl):1–54 (1983).

[41] Fergusson, D.M. et al. A longitudinal study of dentine lead levels, intelligence, school performance and behaviour. Part I Dentine lead levels and exposure to environmental risk factors. *Journal of child psychology and psychiatry*, 29: 781–792 (1988).

[42] Baghurst, P. et al. Lifelong exposure to environmental lead and children's intelligence at age seven: the Port Pirie cohort study. *New England journal of medicine*, 327: 1269–1284 (1992).

[43] Winneke, G. et al. Follow-up studies in lead-exposed children. *In*: Smith, M.A. et al., ed. *Lead exposure and child development, an international assessment*. Dordrecht, Kluwer Academic Publishers, 1989, pp. 260–270.

[44] Altmann, L. et al. Impairment of long-term potentiation and learning following chronic lead exposure. *Toxicology letters*, 66: 105–112 (1993).

[45] *Air quality criteria for lead*. Research Triangle Park, NC, US Environmental Protection Agency, 1986.Chapter 6.7 Lead Air Quality Guidelines - Second Edition,

[46] *Air quality criteria for lead*. Research Triangle Park, NC, US Environmental Protection Agency, 2000.Chapter 6.7 Lead Air Quality Guidelines

[47] *Toxicological profile for lead – draft for public comment*. Atlanta, GA, US Department of Health and Public Services, 1992.

[48] Hernberg, S. & Nikkanen, J. Enzyme inhibition by lead under normal urbanconditions. *Lancet*, 1: 63–64 (1970)

[49] Piomelli, S. et al. Threshold for lead damage to heme synthesis in urban children.*Proceedings of the National Academy of Sciences of the USA*, 79: 3335–3339 (1982).

[50] Grandjean, P. et al. Delayed blood regeneration in lead exposure: an effect on reserve capacity. *American journal of public health*, 79: 1385–1388 (1989).

[51] Piotrowski, J.K. & O'brien, B.J. *Analysis of the effects of lead in tissue upon human health, using dose–response relationships*. London, University of London Monitoring and Assessment Research Centre, 1980 (MARC Report No. 17).

[52] Schwartz, J. et al. Lead-induced anemia: dose–response relationships and evidence for a threshold. *American journal of public health*, 80: 165–168 (1990).

[53] Seppäläinen, A.M. et al. Subclinical neuropathy at "safe" levels of lead exposure.

[54] *Archives of environmental health*, 30: 180–183 (1983).

[55] Davis, J.M. and Svendsgaard, D.J. Nerve conduction velocity and lead: a critical review and meta-analysis. *In*: Johnson, B.L., ed. *Advances in neurobehavioural toxicology*. Chelsea, Lewis Publishers, 1990, pp. 353–376.

[56] Stollery, B.T. et al. Short term prospective study of cognitive functioning in lead. *British journal of industrial medicine*, 48: 739–749 (1991).

[57] Schwartz, J. Low-level lead exposure and children's IQ: A meta-analysis and search for a threshold. *Environmental research*, 65: 42–55 (1994).

[58] Needleman, H. et al. Lead-associated intellectual deficit (letter). *Journal of the American Medical Association*, 306:367 (1982)

[59] Staessen, J.A. et al. Impairment of renal function with increasing blood lead concentrations in the general population. *New England journal of medicine*, 327: 151–156 (1992).

[60] Schwartz, J. & Otto, D.A. Blood lead, hearing thresholds and neurobehavioral development in children and youth. *Archives of environmental health*, 42: 153–160 (1987).

[61] Schwartz, J. & Otto, D.A. Lead and minor hearing impairment. *Archives of environmental health*, 46: 300–305 (1991).

[62] Pan, J., Plant, J.A. Voulvoulis, N., Oates, C.J. Ihlenfeld, C. (2010). Cadmium levels in Europe: implications for human health. *Environmental Geochemistry and Health.* 32 (1): 1-12

[63] Schoeters, G., Den Hond, E., Zuurbier, M., Naginiene, R., Van Den Hazel, P., Stilianakis, N., Ronchetti, R., Koppe, J.G. (2006): Cadmium and children: Exposure and health effects. *Acta Paediatrica*, 95(Suppl.): 50-54.

[64] UNEP (2010) *Final review of scientific information on cadmium.*Nairobi: United Nations Environment Programme. Available at: www.unep.org/hazardoussubstances/Portals/9/Lead_Cadmium/docs/ Interim_reviews/ UNEP_GC26_INF_11_Add_2_Final_UNEP_ Cadmium_review_and_apppendix_Dec_2010.pdf,

[65] WHO (2010a) *Exposure to Arsenic: A major public health concern.* Geneva: World Health Organization. Available at: www.who.int/ipcs/ features/arsenic.pdf,

[66] Xu, J. and Thornton, I. (1985) Arsenic in garden soils and vegetable crops in Cornwall, England: Implications for human health. *Environmental Geochemistry and Health.* 7(4): 131-133.

[67] Klinck, B., Palumbo, B., Cave, M. and Wragg, J. (2005) *Arsenic dispersal and bioaccessibility in mine contaminated soils: a case study from an abandoned arsenic mine in Devon, UK.* British Geological Survey Research Report RR/04/003 52pp. (ISBN 0 85272 483 7). Available at: http://nora.nerc.ac.uk/3681/1/RR04003.pdf

[68] Martinez-Sanchez, M.J., Martinez-Lopez, S., Martinez-Martinez, L.B., Perez-Sirvent, C. (2013) Importance of the oral arsenic bioaccessibility factor for characterising the risk associated with soil ingestion in a mining-influenced zone. *Journal of Environmental Management.* 116: 10-17.

[69] Mitchell, P. and Barre, D. (1995) The nature and significance of public exposure to arsenic: a review of its relevance to South West England. *Environmental Geochemistry and Health.* 17(2):57-82.

[70] Philipp, R., Hughes, A.O., Robertson, M.C. and Mitchell, T.F. (1984) *Soil levels of arsenic and malignant melanoma incidence.* In: Environmental Contamination, pp. 432–437. CEP Consultants, Edinburgh.

[71] Chromium, Safety 101: Health Effects of Hexavalent Chromium, Minnesota Safety Council Minnesota OSHA at http://www.dli.mn.gov/OSHA/PDF/fact_hexchrom.pdf., 2006,

[72] ATSDR – Chromium - ToxFAQsTM Internet address is http://www.atsdr.cdc.gov/toxfaqs/index.asp, October 2012,

[73] UNEP (2013b) Global Mercury Assessment 2013: Sources, Emissions, Releases and Environmental Transport. Geneva: United Nations Environment Programme. Available at : www.unep.org/PDF/ PressReleases/GlobalMercuryAssessment2013.pdf

[74] UNEP (2013a) *Mercury: time to act.* Nairobi: United Nations Environment Programme Available at: www.unep.org/PDF/PressReleases/Mercury_TimeToAct.pdf

[75] Bellanger, M., Pichery, C., Aerts, D., Berglund, M., Castano, A., Cejchanova, M., Crettaz, P., Davidson, F., Esteban, M., Fischer, M.E., Gurzau, A.E., Halzlova, K., Katsonouri, A., Knudsen, L.E., Kolossa-Gehring, M., Koppen, G., Ligocka, D., Miklavcic, A., Reis, M.F., Rudnai, P., Tratnik, J.S., Weihe, P., Budtz-Jorgensen, E., Grandjean, P. (2013) Economic benefits of methylmercury exposure control in Europe: Monetary value of neurotoxicity prevention. *Environmental Health.*12: 3.

[76] Greenwood, M, R. (1985) Methylmercury poisoning in Iraq. An epidemiological study of the 1971-1972 outbreak. *Journal of Applied Toxicology.* 5(3):148-59.

[77] Iustina Popescu, Rodica Stănescu, Mattia Biasioli, Franco Ajmone Marsan, Ionel Constantinescu, Assessing Human Risks Through *CSOIL* Exposure Model For A Soil Contamination Associated To Heavy Metals, U.P.B. Sci. Bull., Series B, Vol. 75, Iss. 1, 2013 ISSN 1454-2331,

[78] Vern Grubinger and Don Ross, University of Vermont, Interpreting The Results Of Soil Tests For Heavy Metals, Sept. 15, 2011,

[79] Directive 2004/35/CE– regarding the soil protection,

[80] Directive proposal establishing a framework for the protection of soil - COM (2006) 232,

[81] Romanian law - Order 756/1997 for the approval of the Regulation on the assessment of environmental pollution,

[82] Health monitoring of the population in Baia Mare city in relation with exposure to lead, Public Health Department, Baia Mare, Maramureș 2008,

[83] Health monitoring of the population in Baia Mare city in relation with exposure to lead, Public Health Department, Baia Mare, Maramureș, 2009,

[84] Anca Maria Moldoveanu, Andra Neamțu, Daniela Rusea, Sima Carmen, Mariana Vlad, Biological screening of the population lead exposure, www.insp.gov.ro, 2012.

Section 2

Emerging Approaches for Soil Quality Assessment

Chapter 10

Eco-Technological Solutions for the Remediation of Polluted Soil and Heavy Metal Recovery

Carmen Cristina Elekes

Additional information is available at the end of the chapter

http://dx.doi.org/10.5772/57314

1. Introduction

Increasing industrial activities and the lack of appropriate measures to counteract its effects are causing a progressive pollution of air, water and soil with heavy metal emissions. Studies have shown that after the downturn of the industrial activity of metallurgical plant, as is in case of the industrial platform of Targoviste (Romania), heavy metals do not persist in air or water, but tend to concentrate especially in soil and sediment. The heavy metal concentration exceeding the threshold in soil can be considered as risk for human health and remediation technics should be applied to decrease the metal content in soil. The classical methods of soil remediation are expensive and some of them involve the removal of huge volume of soil. An alternative of such methods are the bioremediation methods which involve only eco-friendly materials and procedures, lead to metal recovery with minimal impact on the environment and are cost-effective.

Phytoremediation is a process which uses green plants to remediate the soil polluted with heavy metals or other contaminants. The use of different species of plants in the bioremediation process of polluted soils is an adequate option, with minimal influence over the environment, without destroying the soil, which also provides the opportunity to recover the heavy metals. Phytoremediation is a cheaper method, by 50-80% compared to other methods of bioremediation [1]. The disadvantage of this method is that it can be a much more slowly process of remediation, requiring several seasons of plant growth. The contaminants may reduce the growth of plants and the resulted biomass, enriched with heavy metals, is potentially harmful in the food chain.

Through the application of phytoremediation on heavy metals polluted soils, the resulting plant biomass will have a high content of toxic metals [2]. This biomass is considered waste and requires controlled and responsible disposal because of the risk of toxicity for environment, and transfer into the food chain. With the decomposition of plant biomass, metals can be washed by rain and transported back into the soil. In order that the phytoremediation process to result in effective outcomes and the level of heavy metals from the environment to decrease, not only to move those metals from one area to another, the remediation of polluted soils should end with quantitative recovery of metals [3]. The recovery of heavy metals has the advantage of increasing the economic value of the phytoremediation process by transforming this method in a financial self-supporting approach of environmental remediation.

There have been numerous studies on the phytoremediation process, having examined the species of plants that have greatest ability to accumulate heavy metals, factors affecting results of phytoremediation and the areas to be covered with plants for remediation purpose, but studies on treatment, storage or use of resulting biomass are insufficient. Some studies presented the possibilities of heavy metal recovery from different waste, even from agricultural waste. The present research aims to put in one sentence the phytoremediation process and the recovery of heavy metals from the phytoremediation by-products.

The research focused on identifying methods of heavy metal recovery from ash, resulted from the incineration of biomass. The phytoremediation process needs to end up with the heavy metal recovery to obtain (a) de-polluted soil, (b) ash with low content of heavy metals, that can be used as fertilizer in agriculture and (c) amounts of heavy metals that can be recovered in the industry to obtain an economic advantage by financially self-supporting of the phytoremediation process. Because of the lack of researches in this domain, this research was conducted based on the results of those studies that aim to recover metals from different kind of waste (from agriculture, sewage sludge, or woods).

2. Heavy metals in soil

Heavy metals are harmful for the human health, because they tend to accumulate in the living organisms. This bioaccumulation is caused by the high rate of absorption comparing to the rate of metabolism or excretion of the harmful compounds. The emissions from the metallurgical plants are transported by air masses up to 10 km from the pollution source and are deposited on the ground leading to an increase of heavy metal concentration in the upper layer of soil.

In soil, the soluble metals go into the soil solution and can be absorbed or immobilized by plants or can be leached to the deeper layer of soil and to the ground water (Figure 1). Some metals are chemical or physical adsorbed to soil particles. The fate of metals in soil depends on the depth of soil layer, on the erosion processes and on the pH. The heavy metals adsorbed to the soil particles from the upper layer can be subject to the erosion processes and transported by surface waters or by wind. The metals absorbed to deeper soil particles can be subject to microbiological and chemical degradation, can be stabilized by the plant root, or can cross from stable to available forms according with pH. The biological activity influences the speed and rate of pollutants degradation and the clay-humus complex represents an efficient buffer in neuter or alkaline soil reaction [4].

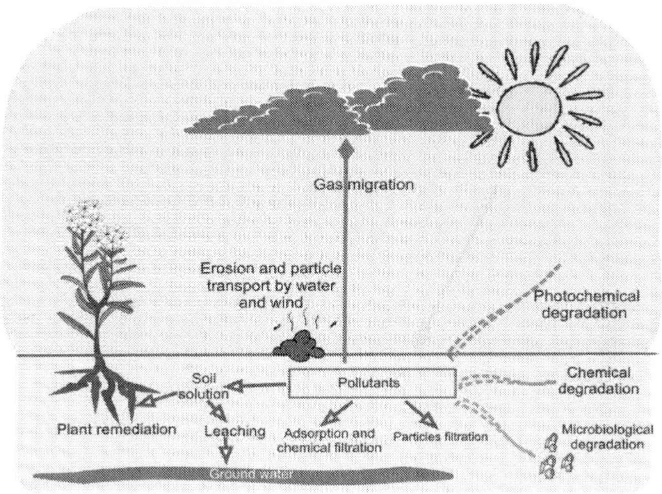

Figure 1. The fate of pollutants in soil [2]

The availability of metallic compounds in soil for plants depends on the soil texture, organic matter content of soil, cation exchange capacity, calcium carbonate equivalent and pH. Soil organic substances play an extremely important role because they can delay both the accumulation and transfer of metals and their movement into the soil. Metal toxicity in soil can be increased or reduced by the organic substances. Soil pH directly influences the availability of metals as soil acidity determines the metal solubility and its ability to move in the soil solution. Concerning the content of phosphorus in soil there are areas where uptake of metals is accelerated or rather diminished due to the presence of high doses of P_2O_5. In addition, the physiology of the plant species influences accumulation of metals. For example, in the case of cadmium uptake by grain was noted either a competition or a synergism in case of high concentrations of lead in the soil [5].

For soil protection, the limits for various pollutants have been established only under certain conditions and soil parameters. It was not taken into account the fact that on light soils, low-carbon, there is a strong influence of acid rainfall leading to a strong mobilization and uptake into plants of toxic heavy metals. This does not happen on heavier soils rich in limestone.

The solubility of zinc in soil was studied by Herms and Brummer [6], who demonstrated the extent to which this element is dissolved by increasing acidity of the soil and became available to plants absorption. In a low-zinc soil, a pH value of 5 could lead to a lasting effect of uptake large amounts of zinc, with all the negative consequences that result. Zinc equilibrium in the soil solution is realized at the level of 1200 mg/kg soil and a pH of 7, at level of 100 mg/kg soil and pH of 6 and at a level of only 40 mg/kg at a pH of 5. These levels of equilibrium make the low-zinc soil to release in the soil solution dangerous amounts of this element.

3. Phytoremediation process

Phytoremediation is defined as the process which uses green plants for the relief, transfer, stabilization or degradation of pollutants from soil, sediments, surface waters and groundwater.

In order to be used in phytoremediation, the selected plant species must be tolerant for the pollutant to be extracted, to quickly develop high biomass, to accumulate metals in harvestable parts, to have a well-developed root system and have a high bioaccumulation factor. This factor must be 20 or more for the phytoremediation to reduce the contamination of soil by 50% over a period of 10 crops [7]. The level of metal bioaccumulation and recovery is directly proportional to the quantity of biomass.

Plants that accumulate high levels of metals are known as hyperaccumulators and can accumulate 50-100 times more metal than a normal plant. There are about 400 hyperaccumulator species and the level of concentration is 10000 mg/kg for Zn and Mn, 1000 mg/kg for Cu, Co, Ni and As, and 100 mg/kg for Cd [8].

Thlaspi caerulescens species is the most studied and known as a hyperaccumulator plant with tolerance capacity for high concentrations of heavy metals in soil (e.g. Cu, Zn, Ni, Cd). Other tolerant species for heavy metals is *Berkheya coddii*, which was studied in South Africa, on ultramafic soils enriched with Ni [9]. The authors found, in leaves, values of Ni concentration about 18000 mg/kg, exceeding several times the metal content in soil (1300 mg/kg) without presenting symptoms of toxicity.

Uptake by the root is the most important way to get trace elements in plants, but have been observed absorption and adsorption processes of metals at level of other tissues. Metal uptake in plants is influenced by the species-specific ability, pedological factors, of which the most important are pH, Eh, fluid regime, clay content, cation exchange capacity, nutrient balance and concentration of other heavy metals. Also, the weather demonstrated some indirect effects on metals absorption rate in plants, mainly due to the influence on the amount of water in the environment [10]. In general, a high-temperature environment positively influences the absorption of micro-nutrients by plants [11].

A disadvantage of phytoremediation is that many hyperaccumulator plants produce a small quantity of biomass. For example, *Thlaspi caerulescens* produces only 2-5 tons/ha, but are plants that produce a larger quantity of biomass 9 t/ha for *Alyssum bertolonii*, or even 22 t/ha for *Berkhya coddii* [12].

Hyperaccumulation involves the absorption, transport and translocation of metals in tissues, where can be stored large amounts of these elements. One of the most studied mechanisms for the metal isolation use metallothionein-derived peptides and phytochelatins. Metal binds to organic sulfite in cysteine, which form the majority of metallothionein-derived peptides. It has been shown that metallothioneins and phytochelatins are stimulated by exposure to metals [13].

Phytoremediation of heavy metal polluted soils involves the following processes (Figure 2):

a. Phytoextraction represents the process which uses plants for the absorption, translocation and accumulation of pollutants from soil, to root systems and shoots. Herbaceous plants are suitable for phytoextraction because they grow quickly, forming a large amount of biomass and can remediate different types of soils. Four species of grass, *Vetiveria zizoniodes, Paspalum notatum, Pennisetum glaucum, Stenotaphrum secundatum* were used to decontaminate open mines. The first species is the most effective one, and the addition of fertilizers is not necessary.

b. Phytodegradation represents the process that uses plants for degradation of organic compounds. The organic compounds are degraded by plants or isolated in vacuoles for later degradation. In general, organic compounds can be degraded on several levels: partial conversion to less toxic compounds, partial degradation and subsequent isolation and complete degradation.

c. Phytovolatilization represents the process of absorption of pollutants by plants and volatilization into the atmosphere by the foliar system. Some plants are able to convert the metal ions in the more volatile forms by the phytovolatilization, which may reduce the toxicity and can translocate the metals to the stomata. Poplar can volatilize trichlorethylene, eucalyptus can volatilize methyltertiarybutylether, selenium can be transformed into Indian mustard and methyl mercury associated with plant roots can be converted from Hg^{2+} to volatile Hg.

d. Phytostabilization represents the process that use the plants which showed immobilization capacity for some metals or capacity for create binding condition for metals which will be adsorbed to soil particles and will be less available. These plants can also transform some toxic molecules in less toxic forms (e.g. Cr^{6+} to Cr^{3+}). To stabilize heavy metals in soil have to be used plants with low capacity of accumulation to reduce the dispersion of metals by grazing or by plant death. Among the remediation techniques of mining areas, phytostabilization is a method with good results in prevention of acid mine discharges and metal stabilization [14].

Heavy metals behave differently and have a different mobility depending on plant species. Therefore, Pb, Cr and Cu tend to be stabilized and retained in the root, Cd, Ni and Zn are more easily translocated to aerial tissues, and Cd is transported even to the harvestable tissues of plants [15].

For Pb there are some plant species, such as *Brassica juncea, Vetiveria ziziniioides, Cardaminopsis halleri, Cynodon dactylon* and *Sorghum halepense* presenting hyperaccumulative capacity. To improve the ability of plants to accumulate heavy metals, the polluted soil can be amended with chelates that increase the metal bioavailability. This method has disadvantages of the high risk of metals to be more easily leached to the groundwater, in addition to the higher cost of remediation.

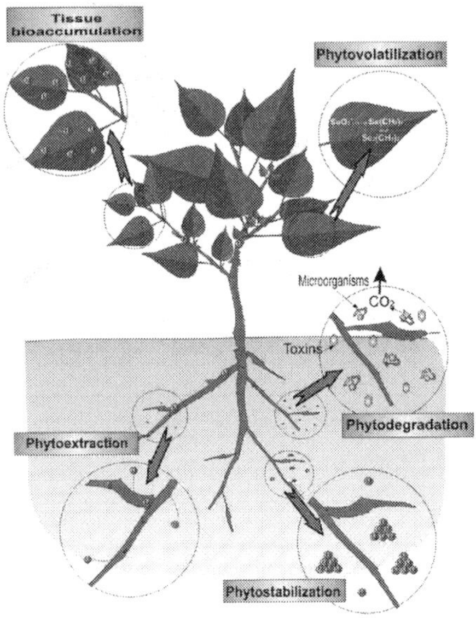

Figure 2. Phytoremediation of heavy metals polluted soil [2]

Among the plants that can be used in phytoremediation of mine tailings, studies have focused on *Eriophorum angustifolium,* a plant resistant to substrates with a wide range of pH from 10.9 to 2.7. Other species of plants that can be grown in a low pH environment are *Carex rostrata, Eriophorum scheuchzeri, Phragmites australis, Typha angustifolia, Typha latifolia,* which grows to a pH value of 2.1, 4.4, 2.1, 3.0 and 2.5 respectively [14].

The studies made since 1977 by the American biologist Dr. Robert Brooks [16], have shown that metals can be extracted from plants (e.g. Ni, Zn, Pb and Au), but the facility of this process depends on the density and solubility of the elements. From the first experiments was obtained 0.01 g of Ni from few kilograms of plant biomass and, more recently, 10 g of Au were obtained from a two hectares of rape culture, established in the vicinity of abandoned mines in California.

4. Methods for heavy metal recovery

So far there have been numerous studies on the phytoremediation process, having examined the species of plants that have a greater ability to accumulate heavy metals, factors affecting the results of phytoremediation and areas that should be remediated with plants. In terms of treatment, storage or heavy metal recovery from the biomass resulted from the phytoremediation process the reference studies are scant.

Delplanque et al. [17] conducted a study on the behavior of metals during the combustion of leaves and shoots of *Salix* (grown for phytoremediation), describing the type of ash obtained from the use of biomass as an energy source and the level of heavy metal concentration. At the end of their study, the authors concluded that the combustion of biomass obtained from phytoremediation reduces the waste volume, but ash cannot be used as fertilizer in agriculture due to high levels of heavy metals.

The international researches in the last decade deal with developing several techniques and methods of incineration and recovery of heavy metals from industrial waste [18,19], garbage [20] and vegetable waste resulting from agriculture [21,22,23]. Unlike industrial waste and garbage, the biomass obtained from the phytoremediation process has a higher content of organic matter and low ash content, only 5% of the dry matter [24]. Compared with agricultural waste, collected from the fields containing heavy metals within normal levels, biomass resulting from phytoremediation process contains much higher concentrations of heavy metals. Considering these aspects and morphological differences between the species used in phytoremediation, heavy metal recovery methods must be adapted for each type of biomass used (grass species, tree species), according to the concentration and metal mixture.

4.1. Thermal treatment

During the thermal treatment (incineration or pyrolysis), the organic matter from biomass is destroyed and metals remain in ash as oxides [25] which can be recovered by hydrometallurgical processes [22,23,26] and bio-hydrometallurgical processes [27], ion exchange [28,29,30,31,32], flotation [19,33], magnetic field [34,35] or electrolysis [3,18,21] and bio-electrochemical procedures [36].

Due to the high content of oxygen, plant biomass can be easily incinerated, requiring only small volume of air during the combustion. In addition, relatively low sulfur content of the biomass is an advantage because there are no harmful gases released in the atmosphere during the combustion [20]. A negative aspect of incineration at very high temperatures (over 600 °C) is the fact that some heavy metals, including those with the greatest risk of toxicity, are volatile: Pb, Zn, Cd, Se [37]. Volatilization of metals can be exploited as an advantage in the recovery of metals, because these elements condense in the cooler areas of the incineration oven and are adsorbed on fine particles of ash retained in the cyclone or on filters [37,38]. The metal recovery is more efficient due to higher metal concentrations by mass of ash [20,37]. The mechanism which determines the behavior of metals during incineration process is characterized by three aspects: (a) evaporation in the combustion areas and condensation of metals in the lower temperature zones of the furnace, (b) physical adsorption on specific surface area of ash particles, and (c) chemo-absorption [38]. The surface area of ash particle is determined primarily by unburned carbon and assessed by electron microscopy (SEM) and is a determining factor for the adsorption of volatilized metals [18,38].

To minimize the risk of volatile metals reaching the atmosphere, the incineration plant shall be fitted with ash particle filtering and retrieval (filters and cyclone). Mercury, selenium and arsenic are metals with the highest percentage of vaporization into incineration system.

Mercury is highly volatile and can be delivered almost entirely as vapor in the form of HgO and $HgCl_2$. Oxidized form of mercury is easily collected from air pollution control system [37].

Keller et al. [3] demonstrated by their study that, better results of the thermal treatment of the biomass are obtained by pyrolysis, under reducing conditions, compared to incineration. The researchers aimed only the recovery of volatile metals (especially copper and zinc), from shoots of Salix use in phytoremediation. These statements are subject to the heavy metal content of ash, which have to lie under the maximum permissible level if they are used as amendment.

4.2. Hydrometallurgical processes

Extraction of heavy metals by ash leaching is a complex chemical process which offers the possibility of obtaining quantitative precipitated metal. The solutions used for leaching of ash must be environmentally friendly, efficient, cheap and with a high capacity for regeneration.

In the leaching process, the extraction of heavy metals is subject to such factors as the solubility and availability of the metal. The solubility can be influenced by pH, the chemical form of inorganic species, organic matter, and the reducing properties. Most metals from waste indicated a higher solubility in acid solutions [39]. Singer et al. [40] tested the extraction of aluminum with citric acid at different temperatures and the results showed that the extraction of metal was considerably influenced by the concentration of the acid used and by temperature. Based on this study, Machado et al. [23] have studied the recovery of Ni and Zn in a multi-component solution by precipitation in the form of alkali metal hydroxide. As a result of the experiment the researchers obtained a recovery rate of over 99% and a purity of 92% for Ni and 99.4% for Zn. The main factor which influences the precipitation of metals from NaOH solution was the pH, which defined the precipitation of Ni at pH 14 and the precipitation of Zn at pH 10. Recovery of Cd, Cu and Pb by leaching with $NaHCO_3$ was investigated by Lezcano et al. [22] who obtained different amounts of metal in relation to pH.

To increase the solubility of metals from ores, Hoque & Philip [27] proposed the introduction of a microbial population, to convert insoluble metal sulfides into soluble metal sulfates. For example, for the extraction of copper from copper sulfide, it was oxidized by microorganisms to copper sulfate. The metal ions have been concentrated in the aqueous phase and the solid residues were removed [27]. A similar technology is used for the conversion of solid metal in water soluble form, in the presence of microorganisms. The technology is called biooxidation and is used for the microbiological oxidation of metal-containing minerals to be extracted.

Bio-hydrometallurgical processes are used in copper metallurgy in the presence of the bacterium *Thiobacillus ferooxidans*, carrying bivalent iron to trivalent iron oxidation [41]:

$$CuFeS_2 + 4\,O_2 \rightarrow CuSO_4 + FeSO_4$$

$$FeSO_4 + H_2SO_4 + \tfrac{1}{2}\,O_2 + \text{bacteria} \rightarrow Fe_2(SO_4)_3 + H_2O$$

$$2\,Fe_2(SO_4)_3 + CuFeS_2 + 3\,O_2 + 2\,H_2O \rightarrow CuSO_4 + 5\,FeSO_4 + H_2O$$

The reaction takes place in aqueous solution, and the last two reactions are cyclic ensuring the continuous development of the leaching process of the chalcopyrite, while copper passes in the solution as sulfate. Another type of bacteria is used to oxidize the sulfur (*Thiobacillus sulfooxidans*). The solutions obtained by leaching, after purification and concentration, can be processed to extract metal ions which we are interested in.

4.3. Electrochemical processes

Electrochemical processes for the extraction of heavy metals have the advantage of selective recovery of metals, depending on the metals reduction potential of the metal to be extracted, but in order to obtain a high purity for every metal, tests must be performed to optimize the factors which influence the metal deposition on the electrode (pH and the electrolyte concentration, the temperature of the electrolytic bath and the metal species). The metal extraction by electrolysis is common use in the metallurgy of zinc, copper, nickel, etc.

Each metal has a specific ion discharge potential, which corresponds to the minimum potential at which an ion electrode begins to discharge continuously and visible (substance discharge). When a substance has more ions, they are discharged successively as they achieve the potential of each download. On this basis is realized the separation and selectively deposition of a number of metal ions from the same solution, if their discharge potentials are differing by at least 0.2 V, otherwise they are deposited at the same time. The ions can be electro-gravimetric separated with a determination error of less than 0.1%.

Fukuta et al. [42] have obtained the selective recovery of Cu, Ni and Zn with sodium sulfide, but in 2011, Machado et al. [43] conclude that, because of the difference in deposition potential of only 0.25 V, the separate recovery of Ni and Zn by electrolysis may be compromised, with the risk of co-deposition of the two metals. For this reason, the research was continued and the electrolyte was used to test the removal of the two metals by precipitation. Ni extraction by electrochemical processes was tested also by Lee [18]. The subject solution of this experiment was a spent electro-less nickel plating solution, the electrode used as the anode was made of platinum, and the cathode was made of stainless steel. Just as in the previous studies, pH played a key role in the extraction of Ni, and, at the end of the process, the metal was obtained in the form of nickel hydroxide and nickel fine particles.

The study conducted by Kirkelund et al. [21] to remove the Cd from the residual plant material by electrochemical methods is based on the principle of electro-migration of ions in solution, in an electric field. The researchers used membranes for the anions and cations exchange to optimize the process. Optimization of results for metal extraction was pursued also by Modin et al. [36] who applied a bio-electrochemical process for recovering Cu, Pb, Cd and Zn in dilute solutions. The anode was inoculated with micro-organisms from the sewage sludge and in the anode chamber a nutrient solution was circulated. The advantage of this method is the less energy consumption for the metal discharge and the selectivity of metal extraction [44].

5. Material and method

The present study was conducted on seven perennial grass species from *Juncaceae* and *Poaceae* family, to find the best solutions for the phytoremediation of soils in the vicinity of metallurgical plant of Targoviste (Figure 3). The aim of research was to evaluate the capacity of these plants species to accumulate heavy metals which were found in high concentrations (above the normal range in agricultural soils) on the studied site. The best accumulative species, *Lolium perenne* was used in an experiment of heavy metal extraction from plant biomass to test the efficiency of metal recovery methods in case of phytoremediation procedure.

Figure 3. Studied area in the vicinity of metallurgical plant of Targoviste

5.1. Research course

Sampling points of plants and soil were chosen so that the results to reflect a snapshot of the impact of metallurgical activities in this area by particles emissions. Sampling was done at distances between 500 and 1000 meters from the source of pollution, from three different points, chosen according to triangle method. The results of metal concentration represent the average of these three samples. The depth of sampling was chosen according with the depth to which the roots of plant culture normally develop. Was formed a mean sample from the column of soil between 0 – 20 cm depth.

Plants were harvested in two seasons, summer and autumn 2008. Was harvested the entire plant, including the root system. For each plant sample, the soil underneath was collected, down to the horizon where the plant developed its root system. The soil was used to establish the bioaccumulation factor of each plant species by comparing the metal concentration in plants with the metal concentration in soil.

The harvested plants were wild growing species which already were adapted to high pollution level of the soil. They were perennial grasses, which usually are used as forage for animals: *Lolium perenne, Festuca pratensis, Stipa capillata, Agrostis alba, Cynodon dactylon, Agrostis tenuis* and *Luzula campestris*.

The plants and soil samples were processed in the laboratory for elemental analysis by ICP-AES (see section 5.2). After harvesting, the fresh plants sample were cleaned with deionized water to remove the soil particles, dried at 60 °C for few hours, ground to a fine powder and

http://dx.doi.org/10.5772/57314

analyzed to establish the metal concentrations. The soil samples were dried at 40 °C for 24 hours, ground to a fine powder, sieved at 250 µm (according to SR ISO 11464).

Based on the results of bioaccumulation for studied plant species, *Lolium perenne* was chosen for the subsequent experiment for heavy metal recovery. In the vicinity of metallurgical plant, a 10x10 m experimental plot was cultured with this species. During the growing period no amendments were added. After one growing season, the plants were mowed and used for the experiment of heavy metal extraction by hydrometallurgical and electrochemical processes.

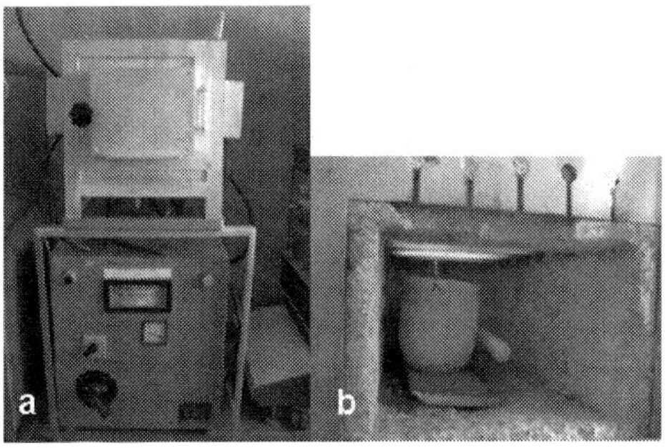

Figure 4. Plant biomass incineration: (a) heat treatment oven with forced rod, (b) incineration process

First the plants were dried for few hours, at 105 °C to remove the water and to decrease the biomass volume [45]. After drying stage, the plants were incinerated in alumina crucibles using a heat treatment oven with forced rod (Figure 4), nominal voltage of 220 V, nominal amperage of 30 A, and maximum temperature 1300 °C. The incineration was conducted at temperature between 400 °C and 600 °C in two stages. The first stage was the heating up to 500 °C for 30 minutes, followed by a 90 minutes of burning at temperature of 500 - 600 °C. Weighing was made for the raw biomass, for the biomass after drying and after incineration to establish the percentage of dry matter and percentage of ash.

Figure 5. Electrolyses cell

Leaching of ash was made by mixing 10 g of ash with 200 ml of nitric acid (HNO_3), concentrated to 65%. The mixture was heated to boiling point for 60 minutes and then cooled down to room temperature. After complete cooling, in the mixture were added, on ice, 200 ml of concentrated sulfuric acid (H_2SO_4). After half an hour of rest, the mixture was filtered.

For the extraction of heavy metals, an electrolytic cell was designed (Figure 5), with stainless steel electrodes and 200 ml of the filtrate obtained from the leaching of the ash as electrolyte. The electrolysis was carried out for 90 minutes with an electric intensity of 1.5 A, and a voltage of 11.4 V. The microscopic and gravimetric methods were used to assess the metal deposition on electrodes.

5.2. Analytical methods

Determination of heavy metal concentration, in both plants and soil underneath, was done by Inductively Coupled Plasma - Atomic Emission Spectrometry method (ICP-AES). For analyze, the samples were mineralized in Berghof microwave digester, plants by mixture with 10 ml of nitric acid concentrated 65% and 2 ml of hydrogen peroxide, and soil in mixture 1:1 with nitric acid (according with Berghof method). The advantage of this method is the multielemental detection, which give the possibility, in one shot, to read a wide range of elements [46]. For this research, analyzes were conducted with Liberty 110 spectrometer of Varian brand. The minimal detection limits of device range according to the analyzed element and is 0.4 mg/kg for Zn, Mn and Cu; 0.5 mg/kg for Cr and Co; 0.6 mg/kg for Sn, Ni and Pb. The concentrations values for analyzed metals were expressed in milligrams of metal per kilogram of dry soil or plants (mg/kg).

The soil pH was determined with a portable pH-meter, WTW 3110 SET 2, with precision of 0.01 units. For pH analyzes, 5 g of each soil sample were mixed with 50 ml KCl 0.1N, F 1000, Tt 0.0056 g/ml and homogenized for 15 minutes with a magnetic stirrer.

The deposition on electrodes was evaluated by microscopy and quantitative assessed by Energy Dispersive X-Ray Fluorescence method (EDXRF) [47], using a PW4025 – MiniPal – Panalytical type EDXRF Spectrometer. The XRF determinations were conducted in Helium atmosphere, excited for 300 s, without any filter, at 16 kV voltage. The current intensity was automatically adjusted by the use of a 3.6 μm Mylar tissue [48]. The surface of electrodes was evaluated for heavy metal concentration before and after electrolysis.

5.3. Data analysis

The bioaccumulation factor (BF) for studied plants was calculated as the ratio between metal concentration in plants and metal concentration in soil:

$$BF = \frac{C_{plant}}{C_{soil}} \tag{1}$$

where: C_{plant} represents the metal concentration in plants and C_{soil} represents the metal concentration in soil.

To calculate the percentage of metal concentration by drying, the metal content of dry matter was reported to the element content of the fresh biomass and to calculate the percentage of metal concentration by incineration, the metal content of ash was reported to the element content of the dry matter.

Percentage of metal extraction into solution (%) was calculated as ration between the metal content of 400 ml solution (filtered leachate) and the metal content of 10 g of ash:

$$\text{Percentage of metal extraction into solution}\left(\%\right)=\frac{\text{solution concentration}*400/1000}{\text{ash concentration}*10/1000} \quad (2)$$

The quantity of metal recovered by electrolysis was calculated according with the mass of metal deposition on cathode (0.7 mg), percentage of ash from the fresh biomass (5.1%) and the mass of fresh plants that can be harvested from one hectare (40 t):

$$\text{Recoverd metal}\left(\mu g/5\,g\,\text{ash}\right)=\frac{\text{metal concentration in cathode deposit}\left(mg/kg\right)}{1000}*0.7 \quad (3)$$

From 40 t of fresh biomass of *Lolium perenne* that can be harvested from one hectare, with a percentage of ash of 5.1%, can be obtained 2.04 t of ash.

$$\text{Recoverd metal}\left(g\text{ per ha}\right)=\frac{\text{recoverd metal}\left(mg/5\,g\,\text{ash}\right)}{5}*2.04 \quad (4)$$

6. Results

6.1. Heavy metal concentration in soil

The soil samples consisted in the upper layer of soil, 0 – 20 cm, where the most of the roots can be found. The content of soil in macronutrients was about 13 g/kg for Ca, 3 g/kg for Mg, and 1 g/kg for P and K. The soil reaction had the value of 7.30±0.42. Heavy metal concentration in soil (Table 1) was compared to the normal values of agricultural soils and alert thresholds for industrial soils according with the Romanian regulations [49]. The average content of Cu, Sn and Pb in soil exceeded the alert threshold for agricultural soils, 100, 35, 50 mg/kg respectively, but in some sampling points the concentrations exceeded even the alert threshold for industrial soils, 250, 100, 250 mg/kg. For Zn, the mean concentration in soil was in normal limit for agricultural soil (300 mg/kg), but in some sampling points exceeded the alert threshold of 700 mg/kg. The Co concentration in soil had low values with a uniform distribution between the sampling points. None of the samples showed values of Co concentration higher than 30 mg/kg, the alert threshold for agricultural soils. The mean value of Ni concentration did not exceed the alert threshold for agricultural soils (75 mg/kg), but the concentration was varying in a wide range from one sampling point to another and some results showed values of concentration close to the alert threshold for industrial soil (200mg/kg). The mean concentration of Mn in studied sample of soil was higher than alert threshold for agricultural soil (1500 mg/kg), and some of results showed values of Mn concentration that exceeded the alert threshold for industrial soils (2000 mg/kg). The mean concentration of Cr was just under the alert threshold for agricultural soils (100 mg/kg), but some sampling points showed values for Cr concentration higher than the alert threshold for industrial soils (300 mg/kg).

Metal	Soil concentration (mg/kg)		Metal	Soil concentration (mg/kg)	
	Mean ±SD	Range		Mean ±SD	Range
Cu	152.4±177.7	21.9-600.4	Co	15.9±4.3	7.1-23.54
Zn	194.3±231.7	42.6-870.3	Ni	53.4±48.6	11.9-185.4
Sn	65.7±30.7	24.6-125.41	Mn	1545.7±334.07	1159.9-2348.76
Pb	65.2±87.2	<LD-294.3	Cr	98.0±112.0	12.9-315.6

<LD – below limit of detection

Table 1. Heavy metal concentration in soil underneath plants (mg/kg dry matter)

The wide variability of metal concentration in soil was according with the orientation and the distance against the pollution source. Some heavy metals showed a uniformity in metal distribution (e.g. Co and Sn), which is probably because of the geological origin of these metals, more than from the pollution source.

6.2. Heavy metal bioaccumulation in plants

Perennial grasses develop a large plant biomass in a relatively short time and are known as heavy metal tolerant biosystems, accumulating high levels of these elements. Also, perennial grasses have a high content of dry matter: *Lolium perenne* 36%, *Festuca pratensis* 33%, *Stipa capillata* 43%, *Agrostis alba* 42%, *Cynodon dactylon* 40%, *Luzula campestris* 50% and *Agrostis tenuis* 46%, compared to other species such as *Papaver rhoeas, Cirsium arvense* or *Artemisia vulgaris* which have only 12 – 36 % dry matter.

According to the perennial grasses tolerance for heavy metals and because of environmental and weather conditions favorable to their development, analyzed plant species have adapted to the toxic heavy metal concentration in soil and accumulated these elements to high levels [50]. In addition to these aspects, the concentration of metals in plants was influenced by plant age, topography and synergistic and antagonistic effects of the elements found in soil.

http://dx.doi.org/10.5772/57314

The heavy metal concentration in perennial grasses was widely different between the species for all studied metals (Table 2). Copper concentration range between 1.76 and 113.83 mg/kg, with the highest value for *F. pratensis*. In the same species was found the highest value for the tin concentration, 379 mg/kg, while the lower value of tin concentration was for *L. campestris*, 8 mg/kg. Zinc concentration range for studied species between 62 – 922 mg/kg, and lead concentration varies between not detectable level of concentration in most of studied species and 201 mg/kg. The maximum values of zinc and lead concentration were found for *L. perenne* species, 922 mg/kg and 201 mg/kg respectively.

The cobalt concentration was below limit of detection for three of studied species. The highest values of this element concentration were found for species *S. capillata* and *A. tenuis*. The accumulation of Co was influenced by the metal concentration in soil and by the soil moisture which lead to the leaching of some cobalt compound and increasing of cobalt availability for plants. The mean values of Ni concentration in studied species of plants varied widely from one species to another even inside of the same genus. The lowest value of Ni concentration was 3.88 mg/kg for *A. tenuis* and the highest was for *A. alba*, 60.23 mg/kg. The range of Mn concentration was between 165.9 mg/kg and 703.92, with the highest values of concentration for *L. perenne* species. The mean concentration of Cr in studied plants ranged between 10.04 mg/kg for *L. campestris* and 191.99 mg/kg for *S. capillata*.

Species	Cu	Zn	Sn	Pb
L. perenne	61.95±15.7	921.67±136.2	217.83±40.1	201.23±14.9
F. pratensis	113.83±22.8	130.62±48.2	379.23±103.2	<LD
S. capillata	10.04±3.9	88.22±2.8	265.25±17.4	<LD
A. alba	31.83±5.5	85.34±7.12	250.04±74.2	5.21±3.2
A. tenuis	0.99±0.6	72.33±3.8	235.94±12.1	2.72±0.1
C. dactylon	25.11±2.4	62.09±2.0	186.13±31.4	<LD
L. campestris	1.76±0.4	71.69±2.9	8.38±0.9	7.38±0.5
Species	**Co**	**Ni**	**Mn**	**Cr**
L. perenne	<LD	12.55±4.6	703.92±156.2	114.19±35.4
F. pratensis	<LD	33.58±11.6	341.90±41.5	173.99±28.6
S. capillata	4.49±0.4	27.72±2.5	342.13±16.6	191.99±12.8
A. alba	3.20±2.0	60.23±20.5	362.07±79.2	56.75±15.1
A. tenuis	4.60±0.2	3.88±0.4	269.39±16.5	21.96±2.53
C. dactylon	2.61±1.1	28.60±10.6	296.33±48.7	28.95±3.9
L. campestris	<LD	17.18±1.41	165.90±0.8	10.04±1.4

<LD – below limit of detection

Table 2. Mean concentration of heavy metals in perennial grasses (mg/kg dry matter)

For phytoremediation process to be effective it is better to use those biosystems species adapted to the climatic and soil conditions of the area to be de-polluted. For this reason, the species used in the studies were chosen from those plants that normally grow in the industrial area of the city of Targoviste, perennial grass which are effective to mowing and rebuild their vegetative mass. In addition, the losses caused by death of leaves are greatly reduced.

The bioaccumulation capacity of plants was estimated as the ratio of metal content in soil and the metal concentration in plant. This ratio is called bioaccumulation factor (BF) [7] and we evaluated as weak accumulators the species which have a BF value between 0.8 - 1.2, as good accumulators the species with a value of BF between 1.5 - 5.0 and hyperacumulators those species with higher BF than 5.0 (Table 3).

Absorption and accumulation of metals in perennial grasses was influenced by both species and the soil underneath, pH, moisture and metal content in soil. The bioaccumulation of the studied metals was differently influenced by pH of soil and metal content.

Even *F. pratensis* and *L. perenne* showed the highest values of Cu, Zn, Sn concentration, they did not show the highest accumulation capacity for those metals. The best accumulator for Cu, Zn and Sn were the plants of *C. dactylon* species which showed BF values of 1.12, 1.37 and 6.06 respectively for Cu, Zn and Sn. Lead was very well accumulated by *L. campestris* which showed a very high level of metal bioaccumulation, 12.3. Tin was the metal with best bioaccumulation in perennial grasses.

Metal	BF	Accumulation gradient	Metal	BF	Accumulation gradient
Cu	0.88±0.2	*A. alba* – weak	**Pb**	1.04±0.1	*L. perenne* – weak
	1.12±0.1	*C. dactylon* – weak		4.54±0.2	*A. tenuis* – good
Zn	0.92±0.1	*L. campestris* – weak		12.3±0.9	*L. campestris* – hyper
	0.98±0.1	*A. tenuis* – weak	**Co**	<0.8	Not accumulative
	1.00±0.1	*A. alba* – weak	**Ni**	1.26±0.05	*L. campestris* – good
	1.31±0.3	*L. perenne* – good		1.27±0.42	*C. dactylon* – good
	1.37±0.1	*C. dactylon* – good		1.63±0.63	*A. alba* – good
Sn	2.43±0.1	*S. capillata* – good	**Mn**	<0.8	Not accumulative
	3.00±0.8	*A. alba* – good	**Cr**	0.83±0.08	*F. pratensis* – weak
	4.10±0.8	*F. pratensis* – good		1.16±0.23	*C. dactylon* – weak
	4.11±0.6	*L. perenne* – good		1.51±0.11	*A. tenuis* – good
	5.85±0.1	*A. tenuis* – hyper		2.11±0.10	*S. capillata* – good
	6.06±0.3	*C. dactylon* – hyper		2.68±0.75	*A. alba* – good

Table 3. Bioaccumulation factor (BF) of heavy metals in plant species – metal accumulation capacity of plants (not accumulative, weak accumulative, good accumulative or hyper accumulative)

None of the studied species of perennial plants showed accumulative capacities for either Co or Mn. This was probably because of the exclusion mechanism of these plants for the two elements. For Ni, three of studied species showed accumulative capacity: *L. campestris, C. dactylon* and *A. alba,* while for Mn five studied species showed accumulative capacity. For the phytoremediation of soils polluted with Ni and Cr, the best species to use is *A. alba,* because it showed the highest values of BF, 1.63 and 2.68 respectively.

6.3. Heavy metal extraction from plant biomass

Following the phytoremediation of soil polluted with heavy metals, a crop of 400 kg of plants was obtained which contain heavy metals, but the concentrations of these elements in the biomass is very low for the use of these plants for heavy metal extraction process. In order to concentrate the metal, plants were subjected to a drying process in which the mass of substance decreased by 64%, registering a percentage of metal concentration of 243.9%. The percentage of ash resulted from the incineration of dry biomass was 14.3% and the percentage of metal concentration in the plant ash increased by values between 544.5 - 2282.1% (Table 4) The differences of metal concentration by incineration was because of volatilization of some elements or because of the adsorption on flying ash which was lost during the incineration. Because of that, future researches will be needed for the construction of an integrated system of plant biomass incineration, designed with filters and cyclone for the recovery of all particles of ash.

In this experiment, the ash was used for metal extraction by leaching. The results showed low effectiveness of this method, because the metals from ash were not fully extracted in the leachate. The weakest extraction in the leachate was for tin, only 16.3% from the total quantity of tin in the ash. A fair extraction was only for Ni and Cr which were extracted 92.6% and 87.8% respectively from the ash (Table 5).

Future research should be conducted to establish methods of metal extraction by leaching of ash with better results for the majority of heavy metals to be recovered.

Metal	Concentration in plants (mg/kg)			Percentage of concentration %	
	fresh biomass	dry matter	ash	by drying	by incineration
Cu	81.35	198.41	1454.08	243.9%	732.8%
Zn	172.87	421.63	2295.92	243.9%	544.5%
Sn	106.77	260.42	3367.35	243.9%	1293.0%
Pb	3.05	7.44	76.53	243.9%	1028.6%
Co	< LD	< LD	306.12	-	-
Ni	< LD	< LD	2142.86	-	-
Mn	179.99	438.99	3214.29	243.9%	732.2%
Cr	65.08	158.73	3622.45	243.9%	2282.1%

Table 4. Metal concentration in plant biomass by drying and incineration (%)

Metal	Ash concentration (mg/kg)	Solution concentration (mg/l)	Percentage of metal extraction into solution* (%)
Cu	1454.08	11.20	30.8%
Zn	2295.92	30.45	53.1%
Sn	3367.35	13.70	16.3%
Pb	76.53	1.25	65.3%
Co	306.12	< LD	-
Ni	2142.86	49.60	92.6%
Mn	3214.29	35.5	44.2%
Cr	3622.45	79.55	87.8%

* 10 g of ash were used to prepare 400 ml solution

Table 5. Heavy metal extraction by leaching of ash

In the process of electrolysis, the cathode layout has changed on the surface. At macroscopic level could be observed the metal deposition after the electrolysis - oxides spots and at the microscopic level could be observed a smoothing of the surface (Figure 6).

There was a difference in mass of the cathode of 0.7 mg, from 4.6242 g before to 4.6249 g after electrolysis.

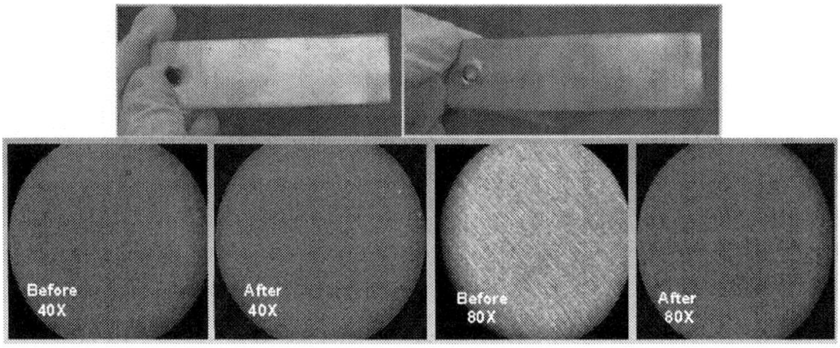

Figure 6. Metal depositions on electrode at 1x, 40x and 80x magnification [2]

The quantitative evaluation of metal deposition on cathode was made by EDXRF and different metal concentrations were observed. The most effective deposition was for Ni, Mn and Cr, which showed concentration of 3.07, 2.2 and 7.3 g/kg respectively.

6.4. Balance of phytoremediation and heavy metal extraction

The experiment of phytoremediation of heavy metal polluted soil in the vicinity of Targoviste city showed the results for one growth season of perennial grass *Lolium perenne* (Table 6). The metal concentration in soil as evaluated before and after the plant culture and after mowing the results showed a decreasing of metal concentration in soil with 0.3 – 5.9%. The concentration of cobalt and nickel had the lowest decreasing. The best accumulated heavy metals by plants were Zn and Pb. The accumulation of Pb was because of the high concentration of this metal in soil and the Zn accumulation was because of the synergic effect of the Pb concentration in soil on the accumulation of Zn [2].

Metal	Metal concentration in soil (mg/kg dry matter)		Percentage of metal extraction by plants (%)	Metal content in plant (mg/kg dry matter)	BF
	before	after			
Cu	313.54	303.62	3.2%	198.41	0.653
Zn	141.04	132.72	5.9%	421.63	3.177
Sn	107.36	102.78	4.3%	260.42	2.642
Pb	71.99	68.23	5.2%	7.44	0.109
Co	16.77	16.72	0.3%	< LD	-
Ni	52.93	52.63	0.6%	< LD	-
Mn	1176.02	1152.63	2.0%	438.99	0.381
Cr	90.66	87.56	3.4%	158.73	1.813

<LD – below limit of detection

Table 6. Results of the phytoremediation of heavy metal polluted soil near Targoviste

The synergic effect of the Pb concentration in soil can be observed also in the value of Zn bioaccumulation factor, 3.177. During the experiment of phytoremediation, *Lolium perenne* showed good accumulative capacity for chromium also.

By reporting the results to the quantity of ash used in the experiment, small quantities of heavy metals were obtained (Table 7).

According with the biomass quantity that can be harvested from one hectare cultivated with *Lolium perenne* species (40 tons) and ash content of raw material (5.1%), in a growing season of this species, can be extracted from soil about 0.13 g of Cu, 0.27 g of Zn, 0.07 g of Sn, 0.04 g of Co, 0.88 g of Ni, 0.63 g of Mn, and 2.09 g of Cr. The efficiency of heavy metals extraction by electrolysis can be improved by increasing the leaching efficiency.

Even if these amounts are very small and the economic value of process is almost nonexistent, the immense advantage of metal recovery is the extraction of these elements from soil and the decreasing of toxicity risk caused by the presence of heavy metals in the environment.

Metal	Metal concentration in leaching solution (mg/l)		Metal concentration in cathode deposit (mg/kg)	Recovered metal	
	before	after		μg/5 g ash	g per ha
Cu	11.20	7.40	445.92	0.31	0.13
Zn	30.45	22.40	946.39	0.66	0.27
Sn	13.70	11.45	260.91	0.18	0.07
Pb	1.25	0.55	< LD	-	-
Co	< LD	< LD	142.31	0.10	0.04
Ni	49.60	23.35	3076.37	2.15	0.88
Mn	35.5	16.935	2222.486	1.56*	0.63
Cr	79.55	17.70	7312.619	5.12**	<2.09

* 2% of Mn deposit is probably because of anodic dissolution

** The results were probably contaminated because of high content of Cr in the stainless steel (14%)

Weight of metal deposition – 0.7 mg; fresh biomass per hectare – 40 t; percentage of ash – 5.1%

Table 7. Results of heavy metal extraction from *Lolium perenne,* by electrolysis

7. Discussion

The present study concerned a novel approach of the phytoremediation process: phytoremediation followed by heavy metal extraction. The researches concerned first the analysis of soil to reveal the necessity of remediation, then researches concerned the species used for phytoremediation and methods of processing the resulted biomass. The treatment of biomass enriched with heavy metals led to the recovery of these elements as alloy, which can be economically exploited. Also, the resulted ash was characterized as material with low heavy metal concentration, which can be used in agriculture or can be disposed without the risk of toxicity.

Analysis of soil samples from the vicinity of metallurgical plant of Targoviste indicated an exceeding of alert threshold for Cu, Sn, Pb, Mn and Cr, with metal concentration levels that could be harmful for both the agricultural and industrial soil [49]. According to these aspects and because the land was used as pasture, the soil needed interventions for decreasing the heavy metal concentration. The best way to remediate this land was the use of wild growing perennial plants to extract heavy metals from soil and the resulted biomass to be mowed and disposed as hazardous material. The wild growing plants have the advantage of ecological adaptability and low costs for starting and maintaining the crop [2].

Metal concentration was evaluated for seven perennial grasses, regarding the bioaccumulation capacity of these species, to establish the most effective species to be used in remediation of studied area. Results indicated heavy metal concentrations for *Lolium perenne* similar to results from previous articles: chromium and copper concentration showed values similar to the results presented by Arienzo et al., 35 and 70 mg/kg for Cr and Cu respectively [50]. Instead, the results of present study indicated 2 and 6 times higher concentrations for Pb and Zn when compared with plants from unpolluted sites and similar results when compared with plants from metallurgical site [50]. The same values of Pb and Zn concentrations were obtained also by Bidar et al. [51] when have researched the behavior of *Trifolium repens* and *Lolium perenne* growing in a heavy metal contaminated field. This study indicated that the two metals, Pb and Zn, are accumulated predominantly in roots (269.98 mg/kg for Pb and 1511.18 mg/kg for Zn), and in lower concentrations in shoots (45.65 mg/kg for Pb and 218.15 mg/kg for Zn) [51]. The results of soil phytoremediation can be improved by the addition of chelates such as EDGA which lead to a 2 times increasing of heavy metal uptake by plant [52].

For a correct evaluation of the most efficient species to be used in phytoremediation, the bioaccumulation factor should be investigated because it indicated the metal accumulation ability of each species by comparing the metal concentration in plant with the metal content of the soil [7]. The results of present study indicated that, despite some species showed a high metal concentration in shoots, the metal uptake was caused by the high metal content of soil, not by the accumulation capacity of that species [2]. Only some of the studied species accumulated heavy metals in higher concentration than the metal content of soil (BF > 1) (Table 3).

For effective results of the phytoremediation, the process should be implemented for a long period of time, with many growing seasons followed by mowing of plants [7]. Therefore, from a phytoremediation process implemented at commercial scale will result huge quantities of plant biomass which should be treated as hazardous material because of the contamination with heavy metals [53]. The leaching test showed that the composting of this biomass is not efficient, because by composting process are formed soluble organic compounds that enhanced metal solubility and availability for plants [53]. Based on these requirements, the present study aimed the development of a novel method of thermal treatment of the biomass from phytoremediation to reduce the initial volume to 5.1%, by drying and incineration. A similar percent (2 – 5%) of mass reduction was indicated by Ghosh and Singh [53], after biomass combustion. Thereby, the much smaller quantities of ash can be properly disposed, but still with the risk of heavy metal toxicity. Also, the resulted ash can be used in phytomining process to recover the saleable heavy metals [53].

In 1994, the development of novel methods for metal recovery from ash resulted from the plant incineration have been referred as future research, known as extraction from bio-ore [54]. The recovered metals could reveal the secondary potential of hyperaccumulating species, the economic value besides the environmental benefits.

In the incineration process, the organic matter is destroyed, releasing metals as oxides [53]. At higher temperature, over 600ºC, the most toxic metals became volatile [37], and because of this reason the thermal treatment should not exceed this temperature [2]. Thus, by the incineration of plant biomass, the metal concentration is increasing by ash mass and the volatilization should be avoided. A very important factor that influenced the metal recovery efficiency was the leaching solution, which determined the metal solubility according with the pH of leachate. The use of nitric acid and sulfuric acid as leaching reagent had a positive influence on the nickel and chromium solubility. The lowest solubility showed tin and copper. The experimental results showed that the methods used for thermal treatment of plant biomass and the leaching method were efficient and a quantity of 0.7 mg of alloyed (Ni, Cr and Mn was majority) metals was obtained. For a selective recovery of heavy metals is necessary that the electrochemical process should be conducted according to the differences of deposition potential of each metal [42, 43].

The balance of phytoremediation and heavy metal extraction indicated that the thermal treatment followed by ash leaching and electrolysis was an efficient method of metal extraction from the phytoremediation by-products – the plant biomass. By this process could be recovered saleable heavy metals and the waste resulted from phytoremediation was a heavy metal low-concentration material, without toxicity risk.

8. Conclusions

For the studied area, in the vicinity of metallurgical plant of Targoviste city, the heavy metal concentration in soil for Cu, Sn, Pb and Mn exceeded the alert threshold for agricultural soil. For Zn, Ni and Cr, some of the results showed values that exceeded also the alert threshold for industrial soils.

The metal concentration in soil was according with the position against the pollution source.

The heavy metal concentration was widely different in the seven studied species of perennial grasses, and the bioaccumulation capacity was different according with the species and metal concentration in soil. Even a species accumulate high concentrations of metals, this could be because of the high content of metal in soil, not because the species showed accumulative capacity – BF higher than 1 (Table 8).

Metal	Maximal concentration in studied species (mg/kg)		Maximal value of bioaccumulation factor	
Cu	113,83	*Festuca pratensis*	1,12	*Cynodon dactylon*
Zn	921,67	*Lolium perenne*	1,36	*Cynodon dactylon*
Sn	379,23	*Festuca pratensis*	6,06	*Cynodon dactylon*
Pb	66,30	*Lolium perenne*	12,29	*Luzula campestris*
Co	4,6	*Agrostis tenuis*	0,40	*Agrostis tenuis*
Ni	60,23	*Agrostis alba*	1,63	*Agrostis alba*
Mn	703,93	*Lolium perenne*	0,33	*Lolium perenne*
Cr	191,99	*Stipa capillata*	2,67	*Agrostis alba*
Mo	25,63	*Festuca pratensis*	15,34	*Cynodon dactylon*

Table 8. The maximal values of heavy metal concentration and bioaccumulation factor for studied plant species

The thermal treatment of plant biomass was an effective method for metal concentration in the material which will be used in metal recovery.

The leaching method should be improved for the extraction of a higher percentage of metals from ash. A quantity of 0.7 mg of metals deposition was obtained after the run of an electrochemical procedure for metal recovery.

By thermal treatment, leaching and electrolysis, small quantities of heavy metals can be recovered from plants used in phytoremediation of polluted soils. The extraction of these elements has advantage for environmental protection, by decreasing the risk of toxicity of heavy metals in soil.

Acknowledgements

The research is part of the PhD thesis "Studies and research concerning the remediation of heavy metal polluted soil by eco-technological procedure", in Romanian, realized by the author in 2011. Is a good opportunity to thank my colleagues Adrian Catangiu for his involvement in the experiment of metal recovery by electrolysis and Irina Fierascu and Radu Claudiu Fierascu for their involvement in the sample analyses by ICP-AES and EDXRF. Also I want to thank my thesis coordinator, Prof. Georghe Ionita for his support.

Author details

Carmen Cristina Elekes*

Address all correspondence to: cristina_elekesh@yahoo.com

Department of Environmental Engineering, "Valahia" University of Târgovişte, Dâmboviţa, Romania

References

[1] Salt DE, Blaylock M, Kumar NPBA, Dushenkov V, Ensley BD, Chet I, Raskin I. Phytoremediation: A novel strategy for the removal of toxic metals from the environment using plants. Bio-Technologies 1995; 13: 468-474.

[2] Elekes CC. Studies and research concerning the remediation of heavy metal polluted soil by eco-technological procedure, in Romanian, PhD thesis, "Valahia" University of Târgovişte, 2011.

[3] Keller C, Ludwig C, Davoli F, Wochele J. Thermal Treatment of Metal-Enriched Biomass Produced from Heavy Metal Phytoextraction, Environ. Sci. Technol. 2005; 39(9): 3359–3367.

[4] Scheffer F, Schachtschabel P. Lehrbuch der Bodenkunde, Heidelberg, Berlin: Spektrum Akademischer Verlang; 2002.

[5] Müller G. Soil biology, (in Romanian: Biologia solului). Bucureşti: Ed. Agro Silvică; 1965.

[6] Herms U, Brümmer GW Einflussgrössen der Schwermetallöslichkeit und –binding in Boden, Z. Pflanzenernaehr, Bodenkd; 1984.

[7] Scragg A. Environmental Biotechnology. New York: Oxford University Press; 2005.

[8] Schnoor JL, Licht LA, McCutcheon SC, Wolfe NL, Carriera LH. Phytoremediation: an emerging technology for contaminated sites, 2005. http://www.engg.ksu.edu/HRSC/Abstracts/schnoor.html (2007 01 12).

[9] Mesjasz-Prybylowicz J, Nakonieczny M, Migula P, at al. Uptake of cadmium, lead, nikel and zinc from soil and water solutions by the nickel hyperaccumulator *Berkheya coddi*. Acta biol. Cracoviensia Ser. Bot. 2004; 46: 75-85.

[10] Kabata-Pendias A. Trace elements in soils and plants, fourth edition. Boca Raton, U.S.A.: CRC Press, Taylor and Francis Group; 2010.

[11] Caggiano R, D'Emilio M, Macchiato M, Ragosta M. Heavy metals in ryegrass species versus metal concentrations in atmospheric particulate measured in an industrial area southern Italy. Environmental Biomonitoring and Assessment 2005; 102, 67-84.

[12] Cunningham SD and Berti WR. Remediation of contaminated soil with green plants, Vitro Cell. Dev. Biol. 1993; 29: 207-212.

[13] Marchand L, Mench M, Jacob DL, Otte ML. Metal and metalloid removal in constructed wetlands, with emphasis on the importance of plants and standardized measurements: A review. Environ. Pollut. 2010; 158: 3447-3461.

[14] Stoltz E. Phytostabilisation – use of wetland plants to treat mine tailings, Departament of Botany, Stockholm, Doctoral Thesis; 2004, ISBN 91-7265-972-6.

[15] Nixdorf B, Fyson A, Krumbeck H. Review: plant life in extremely acidic waters. Environ. Exp. Bot. 2001; 46: 203-211.

[16] Brooks RR. Plants that hyperaccumulate heavy metals, their role in phytoremediation, microbiology, archeology, mineral exploration and phytomining. New York: CAB International; 1998.

[17] Delplanque M, Collet S, Del Gratta F, Schnuriger B, Gaucher R, Robinson B, Bert V. Combustion of Salix used for phytoextraction: The fate of metals and viability of the processes. Biomass and Bioenergy 2013; 49: 160 – 170.

[18] Lee HY. Separation and Recovery of Nickel From Spent Electroless Nickel-Plating Solutions with Hydrometallurgical Processes. Separation Science and Technology 2013; DOI:10.1080/01496395.2012.756523.

[19] Min X, Yuan C, Liang Y, Chai L, Ke Y. Metal recovery from sludge through the combination of hydrothermal sulfidation and flotation. Procedia Environmental Sciences 2012; 16: 401 – 408.

[20] Tan Z, Lagerkvist A. Phosphorus recovery from the biomass ash: A review. Renewable and Sustainable Energy Reviews 2011; 15: 3588–3602.

[21] Kirkelund GM, Damoe AJ, Ottosen LM. Electrodialytic removal of Cd from biomass combustion fly ash suspensions. Journal of Hazardous Materials 2013; 250– 251: 212–219.

[22] Lezcano JM, González F, Ballester A, Blázquez ML, Muñoz JA, García-Balboa C. Sorption and desorption of Cd, Cu and Pb using biomass from an eutrophized habitat in monometallic and bimetallic systems. Journal of Environmental Management 2011; 92: 2666-2674.

[23] Machado MD, Soares EV, Soares HMVM. Selective recovery of copper, nickel and zinc from ashes produced from Saccharomyces cerevisiae contaminated biomass used in the treatment of real electroplating effluents. Journal of Hazardous Materials 2010; 184: 357–363.

[24] Tarhon P. Plant Physiology, (in Romanian: Fiziologia plantelor), Chişinău: Ed. Lumina; 1992.

[25] Kikuchi T, Tanaka S. Biological Removal and Recovery of Toxic Heavy Metals in Water Environment. Critical Reviews in Environmental Science and Technology 2012; 42:1007–1057.

[26] Kumar D, Gaur JP. Metal biosorption by two cyanobacterial mats in relation to pH, biomass concentration, pretreatment and reuse. Bioresource Technology 2011; 102: 2529–2535.

[27] Hoque MdE, Philip OJ. Biotechnological recovery of heavy metals from secondary sources - An overview. Materials Science and Engineering C 2011; 31: 57–66.

[28] Banerjee D, Rao MA, Wattal PK. Separation and Recovery of Cs from High Active Waste Simulant using Resorcinol Formaldehyde Polycondensate Resin: Batch and Column Studies. Separation Science and Technology 2013; 48(1): 133-139.

[29] Guo F, Nishihama S, Yoshizuka K. Selective Recovery of Copper, Cobalt, and Nickel from Aqueous Chloride Media using Solvent Impregnated Resins. Solvent Extraction and Ion Exchange 2012; 30(6): 579-592.

[30] Sivrikaya S, Altundag H, Zengin M, Imamoglu M. Separation, Preconcentration, and Recovery of Pd(II) Ions using Newly Modified Silica Gel with Bis(3-Aminopropyl)Amine. Separation Science and Technology 2011; 46: 2032–2040.

[31] Mizoguchi K, Ida J, Matsuyama T, Yamamoto H. Straight-chained thermo-responsive polymer with high chelating group content for heavy metal ion recovery. Separation and Purification Technology 2010; 75: 69–75.

[32] Kır E, Çengeloğlu Y, Ersoz M. Influence of Chelating Agents on the Recovery of Al(III), Fe(III), Ti(IV) and Na(I) from Red Mud by Cation Exchange Membranes. Separation Science and Technology 2006; 41(5): 961-977.

[33] Stoica L, Oproiu GC. Cu(II) Recovery from Aqueous Systems by Flotation. Separation Science and Technology 2004; 39(4): 893-909.

[34] Tsai TH. Iron Removal during Recovery of Silicon from Sawing Waste by Applying Magnetic Field. Separation Science and Technology 2011; 46(5): 702-707.

[35] Jainae K, Sanuwong K, Nuangjamnong J, Sukpirom N, Unob F. Extraction and recovery of precious metal ions in wastewater by polystyrene-coated magnetic particles functionalized with 2-(3-(2-aminoethylthio)propylthio)ethanamine. Chemical Engineering Journal 2010; 160: 586–593.

[36] Modin O, Wang X, Xue Wu, Rauch S, Fedje KK. Bioelectrochemical recovery of Cu, Pb, Cd, and Zn from dilute solutions. Journal of Hazardous Materials 2012; 235– 236: 291– 297.

[37] Nzihou A, Stanmore B. The fate of heavy metals during combustion and gasification of contaminated biomass–A brief review. Journal of Hazardous Materials 2013; 256-257: 56-66.

[38] Li C, Xiea F, Mab Y, Caia T, Li H, Huanga Z, Yuana G. Multiple heavy metals extraction and recovery from hazardous electroplating sludge waste via ultrasonically enhanced two-stage acid leaching. Journal of Hazardous Materials 2010; 178: 823–833.

[39] Meawad Amr S, Bojinova DY, Pelovski YG. An overview of metals recovery from thermal power plant solid wastes. Waste Management 2010; 30: 2548–2559.

[40] Singer A, Navrot J, Shapira R. Extraction of aluminum from fly ash by commercial and microbiologically produced citric acid. Applied Microbiology and Biotechnology 1982; 16: 228–230.

[41] Constantin I. Heavy metals – processes and technologies, (in Romanian: Metale grele – procese si tehnologii). Bucureşti: Ed. Printech; 2009, ISBN 978-606-521-270-1.

[42] Fukuta T, Ito T, Sawada K, Kojima Y, Matsuda H, Seto F. Separation of Cu, Zn and Ni from plating solution by precipitation of metal sulfides. Kag. Kog. Ronbunshu 2004; 30: 227–232.

[43] Machado MD, Soares EV, Soares HMVM. Selective recovery of chromium, copper, nickel, and zinc from an acid solution using an environmentally friendly process. Environ Sci Pollut Res 2011; 18: 1279–1285.

[44] Won SW, Lim A, Yun YS. Recover y of high-puri ty metalli c Pd from Pd(II)-sorbed biosorbents by incinę ration. Bioresour. Technol. 2013; 137: 400-403.

[45] Dumitrescu C. Methods and technics of assessment and neutralization of pollutants, (in Romanian: Metode şi tehnici de evaluare şi neutralizare a poluanţilor). Bucureşti: Polytechnic University; 2002, ISBN 973-652-643-7, 30-110;

[46] Costea T, Istudor V, Fierascu RC, Fierascu I, Botez A. Researches upon indigenous herbal products for therapeutic valorification in metabolic diseases Note I. Betulae folium and Rubi idaei folium, sources of micro- and macroelements. Farmacia 2013; 61: 1.

[47] Fierascu RC, Fierascu I. Archaeometallurgical investigation of historical coins. Metalurgia International 2013; 18(7): 277-278.

[48] Fierăscu RC, Dumitriu I, Ion ML, Catangiu A, Ion RM. Surface and analytical techniques study of Romanian coins. European Journal of Science and Theology 2009; 5(1): 17-28.

[49] Order no. 756 from 3rd November 1997 for the approval of the Regulation of environmental pollution assessment.

[50] Arienzo M, Adamo P, Cozzolino V. The potential of *Lolium perenne* for revegetation of contaminated soil from a metallurgical site. The Science of the Total Environment 2004; 319: 13-25.

[51] Bidar G, Garçon G, Pruvot C, Dewaele D, Cazier F, Douay F, Shirali P. Behaviour of *Trifolium repens* and *Lolium perenne* growing in a heavy metal contaminated field: Plant metal concentration and phytotoxicity. Environmental Pollution 2007; 147: 546-553.

[52] Bouwman LA, Bloem J, Römkens PFAM, Japenga J. EDGA amendment of slightly heavy metal loaded soil affects heavy metal solubility, crop growth and microbivorous nematodes but not bacteria and herbivorous nematodes. Soil Biology & Biochemistry 2005; 37: 271-278.

[53] Ghosh M, Singh SP. A review on phytoremediation of heavy metals and utiliyation of its byproducts. Applied Ecology and Environmental Research 2005; 3(1): 1-18.

[54] Nicks L, Chambers MF. Nickel farm. Discover 1994; September: 19.

Chapter 11

Selected Bioremediation Techniques in Polluted Tropical Soils

Emmanuel Uzoma Onweremadu

Additional information is available at the end of the chapter

http://dx.doi.org/10.5772/58381

1. Introduction

Ecosystems are regularly confronted with natural environmental variations and disturbances over time and geographic space. A disturbance is any process that removes biomass from a community, such as fire, flood, drought, or predation. Disturbances occur over vastly different ranges in terms of magnitudes as well as distances and time periods (Levin, 1992) and are both the cause and product of natural fluctuations in death rates, species assemblages, and biomass densities within an ecological community. These disturbances create places of renewal where new directions emerge out of the patchwork of natural experimentation and opportunity implying a good measure of ecological resilience is a cornerstone theory in ecosystem (Folke, *et al., 2004*).One of such disturbances is pollution which alters ecological balance.

Intense industrial activity and urbanization in recent times, especially in developing countries, have led to serious environmental pollution, resulting in a large number and variety of contaminated sites which became a threat to the local ecosystems. In all these, natural resources such as soils, water, air and vegetation are adversely affected.

Industrial revolution gave birth to environmental pollution which continued till today. It was a revolution that led to the emergence of great factories and consumption of immense quantities of fossil fuels which was associated with an unprecedented rise in air pollution and large volume of industrial chemical discharges. This was added to the growing population with a load of untreated human waste. The Second World War made pollution to become a popular issue due to radioactive fallout from atomic warfare and testing. Pollution began to draw major public attention with the emergence of cities and megacities associated with a stockpile of refuse and characterized by substantial output of sewage and particulate matter.

Pollution defines the introduction of harmful substances often referred to as contaminants into the natural environment that cause adverse change. The term contamination is in some cases used interchangeably with pollution in environmental chemistry, where the main interest is the harm done on a large scale to humans or to organisms or environments that are important to human beings. Common soil contaminants include chlorinated hydrocarbons, heavy metals such as chromium, cadmium–found in rechargeable batteries, and lead–found in lead paint, aviation fuel and still in some countries, gasoline, zinc, arsenic and benzene. Recycling industrial byproducts into fertilizer may result in the contamination of soils with various metals. Ordinary municipal landfills are the source of many chemical substances entering the soil environment and often reaching groundwater, emanating from the wide variety of refuse.

In the case of the term contamination, it is the presence of a minor and unwanted constituent in a material, in a physical body or in the natural environment. In chemistry, contamination usually refers to a single constituent, but in specialized fields the term can also mean chemical mixtures, even up to the level of cellular materials.

Pollution may take various forms including discharge of deleterious chemical substances on natural substances. Pollution can be point source or nonpoint source pollution.

Sometimes pollution takes the form of harmful energy such as noise, heat or light. Generally speaking, foreign substances and energies which contaminate natural resources are referred to as pollutants. Substances contain some level of impurity; and this may become an issue if the impure chemical is mixed with other chemicals or mixtures and causes additional chemical reactions. Sometimes, the additional chemical reactions are beneficial, in which case the label 'contaminant' is replaced with reactant or catalyst. When additional reactions are detrimental, other terms such as toxin or poison depending on the chemistry involved are used. However, if no remedial action is undertaken, the availability of arable land for cultivation will decrease, because of stricter environmental laws limiting food production on contaminated lands. Inorganic and organic contaminants typically found in urban areas are heavy metals and petroleum-derived products. The presence of both types of contaminants on the same site presents technical and economic challenges for decontamination strategies. There have also been some unusual releases of polychlorinated dibenzodioxins, commonly called dioxins for simplicity.

In Nigeria, there is paucity of soil information leading to several forms of soil degradations. Except in recent times environmental impact assessments (EIAs) are rarely conducted on natural resources before embarking on major projects. The EIAs are often not backed up with necessary implementation legislations. Mineral exploration and exploitation as well as various construction activities are known to have negative impact on surface and subsurface soils, surface and groundwater, rocks and rocklike minerals, atmospheric resources, vegetation and wildlife.

Available soil data are not problem-solving (Lal and Ragland,1993).Non-use of soil survey data and information has led to soil and soil-related environmental problems such as nutrient depletion, nutrient imbalances, multiple nutrient deficiencies, nutrient toxicity, general decline in soil quality and yield decline. The situation is often aggravated by socioeconomic pressures mainly resulting from poverty and inability to afford relevant inputs of agricultural production. Sound characterization and classification of soils based on quality and proper presentation of such information in user-friendly form is a necessary adjunct in sustained use of soils. Again, soil quality data will go a long way in promoting bio-safety of farm products for both local consumption and their internationalization.

Primarily, this paper is aimed at reviewing crude oil and non-crude oil polluted soils of tropical soils with particular emphasis on Nigeria. Specifically, some biotechnological methods are suggested for the amelioration of contaminated soils. A good knowledge of status and distribution of polluted soils will go a long way in assisting in the production of land use maps which will facilitate policy and legislations on soil and soil-related natural resources. Land use maps derived from soil survey and land evaluation are useful in soil management as well as in vulnerability and risk assessments. This is true as soil quality problems vary requiring different remediation strategies to overcome.

Remediation deals with the removal of pollutants or contaminants from natural resources. The affected natural resources may include soil, groundwater, surface water sediment, vegetation, rock minerals, wildlife and air. A major aim of remediation is the recovery and general protection of human health and the environment. Sometimes, remediation is done in places intended for redevelopment. Remediation goes with an array of regulatory requirements, and its assessments are based on human health and ecological risks.

Several approaches are used in the remediation of polluted soils, ranging from biological, chemical and engineering techniques. Sometimes, it may require a combination of organic and inorganic strategies. For instance the Neapolitan yellow tuff (NYT) was utilized as a component of an organo-mineral sorbent/exchanger soil conditioner with pellet manure (NYT/PM) to reduce the mobility of Cd and Pb and recover plant performance in heavily polluted soils from illegal dumps near Santa Maria La Fossa (Lower Volturno river basin, Campania Region, southern Italy). Pot experiments were performed by adding the NYT/PM mixture (1:1, w/w) to polluted soil at the rates of 0%, 25%, 50% or 75% (w/w). Wheat (*Triticum aestivum*) was used as the test plant. The addition of organo-zeolite NYT/PM mixture significantly reduced the DTPA (diethylene-triamine-pentaacetic acid)-extractable Cd and Pb from 1.01 and 97.5 mg kg^{-1} in the polluted soil, to 0.14 and 11.6 mg kg^{-1}, respectively, in the soil amended with 75% NYT/PM. The best plant response was observed in amended soil systems treated with 25% NYT/PM, whereas larger additions induced plant toxicities due to increased soil salinity.

When a soil on site is found to be contaminated to a depth of several metres and construction work needs to get started in a few months' time, soil replacement is the fastest remedy. However, some of the contaminated areas can be restored by combining modern and age-old methods. This is where plants and their microbial partners may enter the picture now and in the future. This because heavy metals in soils with residence times of thousands of years present numerous health dangers to higher organisms (Garbisu and Alkorta,2001). They are also known to decrease plant growth, ground cover and have a negative impact on soil microflora (McGrath *et al.*,2001). There is increasing and widespread interest in the maintenance of soil quality and remediation strategies for management of soils contaminated with trace metals, metalloids or organic pollutants. Heavy metals are deposited in soils by atmospheric input and the use of mineral fertilizers or compost, and sewage sludge disposal. Conventional remediation methods usually involve excavation and removal of contaminated soil layer, physical stabilization and washing of contaminated soils with strong acids or HM chelators (Steele and Pichtel,1998). Bioremediation, that is. the use of living organisms to manage or remediate polluted soils, is an emerging technology. It is defined as the elimination, attenuation or transformation of polluting or contaminating substances by the use of biological processes.

It is no new discovery that many plant species can grow in soils contaminated by various pollutants. Some species can even sequester or decompose contaminants. Soil and plant microbes help plants survive in harsh conditions.

Bioremediation includes the productive use of biodegradative processes in the elimination or detoxification of pollutants that have found their way into the environment, especially where such pollutants are capable of threatening public health. Some of the methods are *ex situ* while others are *in situ*. The *ex situ* bioremediation techniques involve the excavation or removal of soil from ground. A good number of *in situ* bioremediation techniques are generally the most desirable options due to cheapness and less disturbances since they provide the services in place avoiding excavation and transport of contaminants. Processes include phytoremediation, phytostabilization,phytotransformation,phytoextraction,rhizofiltration and phytoscreening.

Phytoremediation involves the treatment of polluted natural resource through the use of plants that mitigate the problem without the need to excavate the contaminant material and dispose of it elsewhere. The use of plants in remediation has been growing rapidly in popularity worldwide for the last twenty years or so. Phytoremediation may be defined as use of vegetation to contain, sequester, remove, or degrade organic and inorganic contaminants in soils, sediments, surface water and groundwater. Phytoremediation is a technology that uses plants to remove contaminants from soil and water (Raskin and Ensley,2000). The basic idea that plant can be used for environmental remediation is very old and cannot be traced to any particular source. However, a series of fascinating scientific discoveries combined with an interdisciplinary research approach have allowed the development of this idea into a promising, cost-effective, and environmental friendly technology.

Certain plants and microorganisms are able to precipitate metal compounds in the rhizosphere. Efficacy was shown by the use of lead pyromorphite (Cotter-Howells *et al.*,1999), as phytoremediation may provide an effective means to reduce metal toxicity as well as metal mobility (Cotter-Howells and Caporn,1996). This is referred to as phytoimmobilisation. Although the application of microbial biotechnology has been successful with petroleum-based constituents, microbial digestion has met limited success for widespread residual organic and metals pollutants. Vegetation-based remediation shows potential for accumulating, immobilizing, and transforming a low level of persistent contaminants. We can find five types of phytoremediation techniques, classified based on the contaminant fate: phytoextraction, phytotransformation, phytostabilization,phytodegradation, rhizofiltration, even if a combination of these can be found in nature.

Phytoremediation consists of reducing or eliminating pollutant concentrations in contaminated soils, water, or air, with plants. Selected plant species are able to contain, degrade, or eliminate metals, pesticides, solvents, explosives, crude oil and its derivatives, and various other contaminants from the media that contain them. Boyd and Javre (2001) reported phytoenrichement of soils by *Sebertia acuminata* in New Caledonia. In phytoremediation, the assumption is that certain plants called hyperaccumulators are able to bioaccumulate, degrade,or render harmless contaminants found in natural resources such as soils, water, and air. The maize plant *(Zea mays)* showed high tolerance towards Cr with negligible concentration in leaves (Lasat *et al.*,1998). A plant is said to be a hyperaccumulator if it can concentrate the pollutants in a minimum percentage which varies according to the pollutant involved. More than 1000 mg/kg of dry weight for nickel, copper, cobalt, chromium or lead; or more than 10,000 mg/kg for zinc or manganese are recommended (Baker and Brooks,1989). In addition to this, it is assumed that hyperaccumulating plants can be found thriving under very harsh conditions or under situations that are not ideal for plant growth.

Some plants are able to translocate and accumulate particular types of contaminants. Plants can be used as biosensors of subsurface contamination, thereby allowing investigators to quickly delineate contaminant plumes (Burken et al.,2011). Chlorinated solvents have been observed in tree trunks at concentrations related to groundwater concentrations (Vroblesky et al.,1998). Phytoscreening often leads to more optimized site investigations and reduce contaminated site cleanup costs. Phytoremediation has become increasingly popular and has been employed at sites with soils contaminated with lead, uranium, and arsenic and it has the advantage that environmental concerns may be treated *in situ*.

The technology of phytoremediation has been successfully used in the restoration of abandoned metal-mine sites, reducing the impact of sites where polychlorinated biphenyls have been dumped during manufacture and mitigation of on-going coal mine discharges.

There are a range of processes mediated by plants which are useful in soil and soil-related environmental problems. Processes include phytostabilization, phytotransformation,phytoextraction,rhizofiltration and phytoscreening.

Phytostabilization entails the reduction of the mobility of substances in the environment. This could be done by limiting the leaching of substances from the soil. Its main focus is on long-term stabilization and containment of the pollutant. Plants can reduce wind erosion; or their roots can prevent water erosion, immobilize the pollutants by adsorption or accumulation, and provide a zone around the roots where the pollutant can precipitate and stabilize. Phytostabilization focuses mainly on sequestering pollutants in soil near the roots but not in plant tissues. By this, pollutants become less bioavailable to livestock and wildlife, and human exposure is drastically reduced.

Phytoextraction is the uptake and concentration of substances from the environment into the plant biomass. The use of plants to mine toxicants is called phytomining. Phytoextraction employs metal hyperaccumulator plant species to transport high quantities of metals from soils into the harvestable parts of roots and aboveground shoots (Chaney *et al.*,1997). Phytoextraction is an innovation using higher plants for *in situ* decontamination of metal-polluted soils, sludges and sediments (Wenzel and Jockwer,1999). Large biomass production and high rates of metal uptake and translocation into the shoot system are critical in achieving reasonable metal extraction rates. Effective phytoextraction requires both plant genetic ability and the development of optimal agronomic management practices (Gupta and Sinha,2007). Hyper accumulators are defined as plants that contain in their tissue more than 1,000 mg kg-1 dry weight of Ni, Co, Cu, Cr, Pb, or more than 10,000 mg kg-1 dry weight of Zn, or Mn (Steele and Pichte,1998). Hyper accumulation is thought to benefit the plant by means of allelopathy, defence against herbivores, or general pathogen resistance in addition to metal tolerance (David *et al.*,2001). In-situ phytoextraction of Ni by a native population of *Alyssum murale* on an ultramafic site (Albania) have been reported (Bani *et al.*,2007). In the case of phytomining, the use of native flora (including local populations of hyperaccumulators) with limited agronomic practices (extensive phytoextraction) could be an alternative to intensively managed crops. The use of plants in remediation has been growing rapidly in popularity worldwide for the last twenty years or so. In general, this process has been tried more often for extracting heavy metals than for organics The technique of phytoextraction uses plants to remove contaminants from soils, sediments or water into harvestable plant biomass. Such organisms that absorb larger-than-normal amounts of contaminants from the soil are referred to as hyperaccumulators. Examples of hyperaccumulators are *Athyrium yokoscense* (Japanese false spleenwort), *Avena strigosa* (Brittle oat), *Crotalaria juncea* (Sunn hemp), *Eichhornia crassipes* (water hyacinth), *Pistia stratiotes* (water lettuce). *Helianthus annuus* (Sunflower), *Salix viminalis* (Basket willow), *Lemna minor* (Duckweed), *Amaranthus retroflexus* (Redroot Amaranth), *Glomus intradices* (Mycorrhizal fungus), Eragrostis bahiensis (Bahia lovegrass), *Cynodon dacvtylon* (Bermuda grass), *Festuca arundinacea* (Tall fescue), Lolium perenne (Perennial ryegrass), *Panicum virgatum*, (Switchgrass), *Phaseolus acutifolius* (Tepary beans), *Cocos nucifera* (Coconut tree), *Spirodela polyrhiza* (Giant duckweed), *Tagetes erecta* (African-tall) and *Zea mays* (Maize)

In phytoremediation, plants absorb contaminants through the root system and store them in the root biomass and/or transport them up into the stems and/or leaves. A living plant may continue to absorb contaminants until it is harvested. Thereafter the process, the cleaned soil can support other vegetation with significant healthfulness.

Some transgenic plants containing genes for bacterial enzymes have been found to be effective hyperaccumulators (Meagher, 2000). Salt-tolerant plants like sugar beets are commonly used for the extraction of sodium chloride in reclaiming soils previously flooded by salt water. Sunflower (*Helianthus annuus*) is an effective hyperaccumulator in cleaning soils contaminated with arsenic. In general, plants with non-invasive and moisture-tolerant root systems can be planted on the embankments. Crops most commonly planted in decontamination systems in Colombia are plantain (*Musa paradisiaca*), papaya (*Carica papaya*), bore (*Alocasia macrorrhiza*), sugar cane (*Saccharum officinarum*) and nacedero tree (*Trichanthera gigantea*). They are commonly used for forage production in Colombia. Under local conditions it produces about 10 tons of dry matter ha/year with 18 per cent of protein in the foliage dry matter. A good number of them grow very well in the sub-Saharan Africa, therefore are suggested for phytoremediation in that region.

There are two major forms of phytoextraction, namely assisted or natural phytoextraction. In induced or assisted phytoextraction, hyper-accumulators are cultivated for the purpose of remediation. It is associated with the use of chelators in soils to increase metal solubility or mobilization so that the plants can absorb them more easily. In natural phytoextraction, plants naturally take up the contaminants in soil unassisted. Many natural hyperaccumulators are metallophyte plants that can tolerate and incorporate high levels of toxic metals.

An advantage of phytoextraction is friendly moderate impact in the soil ecosystem. Most traditional methods commonly used for cleaning up heavy metal-contaminated soil disrupt soil structure and reduce soil productivity, but phytoextraction has the ability of cleaning up the soil without causing any kind of harm to soil quality and soil structural integrity. In addition to this, phytoextraction is cost-effective when compared with other soil remediation techniques, although it is frequently argued argued that significant effects are only achieved in the long term.

Phytotransformation describes chemical modification of environmental substances as a direct result of plant catabolic and anabolic activities. These activities lead to inactivation, degradation or immobilization. The degradation as caused by plants is referred to as phytodegradation, On the other hand, immobilization is known as phytostabilization which is a process of reducing the mobility of substances in the environment, for example, by limiting the leaching of substances from the soil.

Certain plants render organic pollutants, such as pesticides, explosives, solvents, industrial chemicals, and other xenobiotic substances non-toxic by their metabolism. Sometimes, microorganisms living in association with plant roots may metabolize these substances in soil or water. These complex and recalcitrant compounds cannot be broken down to basic molecules (water, carbon-dioxide, etc.) by plant molecules, and, hence, the term phytotransformation represents a change in chemical structure without complete breakdown of the compound. The term "Green Liver Model" is used to describe phytotransformation, as plants behave analogously to the human liver when dealing with these xenobiotic compounds or foreign compounds (Burken et al., 2004). After uptake of the xenobiotics, plant enzymes increase the polarity of the xenobiotics by adding functional groups such as hydroxyl groups (-OH).

This is known as Phase I metabolism, similar to the way that the human liver increases the polarity of drugs and foreign compounds. Whereas in the human liver enzymes such as Cytochrome P450s are responsible for the initial reactions, in plants enzymes such as nitroreductases carry out the same role.

In the second stage of phytotransformation, known as Phase II metabolism, plant biomolecules such as glucose and amino acids are added to the polarized xenobiotic to further increase the polarity (known as conjugation). This is again similar to the processes occurring in the human liver where glucuronidation (addition of glucose molecules by the UGT (e.g. UGT1A1) class of enzymes) and glutathione addition reactions occur on reactive centres of the xenobiotic.

Phase I and II reactions serve to increase the polarity and reduce the toxicity of the compounds, although many exceptions to the rule are seen. The increased polarity also allows for easy transport of the xenobiotic along aqueous channels.

In the final stage of phytotransformation (Phase III metabolism), a sequestration of the xenobiotic occurs within the plant. The xenobiotics polymerize in a lignin-like manner and develop a complex structure that is sequestered in the plant. This ensures that the xenobiotic is safely stored, and does not affect the functioning of the plant. However, preliminary studies have shown that these plants can be toxic to small animals (such as snails), and, hence, plants involved in phytotransformation may need to be maintained in a closed enclosure.Hence, the plants reduce toxicity (with exceptions) and sequester the xenobiotics in phytotransformation. Trinitrotoluene phytotransformation has been extensively researched and a transformation pathway has been proposed (Subramanian *et al.*,2006).

In the case of organic pollutants, such as pesticides, explosives, solvents, industrial chemicals, and other xenobiotic substances, certain plants, such as Cannas, render these substances non-toxic by their metabolism. In other cases, microorganisms living in association with plant roots may metabolize these substances in soil or water. These complex and recalcitrant compounds cannot be broken down to basic molecules (water, carbon-dioxide, etc.) by plant molecules, and, hence, the term *phytotransformation* represents a change in chemical structure without complete breakdown of the compound. The mechanism is likened to the Green Liver Model which is used to describe phytotransformation, as plants behave analogously to the human liver when dealing with these foreign compound/pollutant (Burken, 2004), After uptake of the xenobiotics, plant enzymes increase the polarity of the xenobiotics by adding functional groups such as hydroxyl groups (-OH).

This is known as Phase I metabolism, similar to the way that the human liver increases the polarity of drugs and foreign compounds. Whereas in the human liver enzymes such as Cytochrome P450s are responsible for the initial reactions, in plants enzymes such as nitroreductases carry out the same role.In the Phase II metabolism, plant biomolecules such as glucose and amino acids are added to the polarized foreign compound pollutants to further increase the polarity. This is known as conjugation and is again similar to the processes occurring in the human liver where glucuronidation and glutathione addition reactions occur on reactive centres of the xenobiotic.

Phase I and II reactions serve to increase the polarity and reduce the toxicity of the compounds, although many exceptions to the rule are seen. The increased polarity also allows for easy transport of the xenobiotic along aqueous channels.In the Phase III metabolism, the foreign pollutant compounds are a sequestered within the plant. The xenobiotics polymerize in a lignin-like manner and develop a complex structure that is sequestered in the plant where they are safely stored. However, such plants can be toxic to small animals like snails, and, hence, plants involved in phytotransformation may need to be maintained in a closed enclosure. Plants therefore reduce toxicity and sequester the xenobiotics through phytotransformation. Trinitrotoluene phytotransformation has been extensively researched and a transformation pathway has been proposed (Subramanian *et al.*,2006).

A significant number of organic chemicals and many inorganic ones are subject to enzymatic attack through the activities of living organisms. Efficacy of microbes in decontamination depends on some edaphic properties such as soil pH soil aeration, soil nutrient status, soil moisture, soil temperature, soil texture and type of heavy metal (Vidali,2001). According to

Thapa *et al.* (2012,) most of modern society's environmental pollutants are included among these chemicals, and the actions of enzymes on them are usually lumped under the term *biodegradation*. The productive use of biodegradative processes eliminate or detoxify pollutants that have found their way into the environment and threaten public health, usually as contaminants of soil, water, or sediments is *bioremediation* (Thapa *et al.*,2012).

Some microbes can reduce activity of different types of heavy metals. Agricultural wastewater treatment can be effectively undertaken through biological processes involving the activity of microorganisms such as bacteria, algae, fungi, plants and animals (Chara *et al.*,1999). This they can do by their ability to convert active forms of toxic metals to inactive forms. However, choice of microbes depends on the availability of energy sources of the organisms in question. Other environmental conditions like temperatures, oxygen, moisture and the presence of hazardous contaminant contribute immensely in influencing efficacy of microbes in remediation programmes. The aerobic bacteria recognized for their degradative abilities are *Pseudomonas, Alcaligenes, Sphingomonas.* These microbes have often been reported to degrade pesticides and hydrocarbons, both alkanes and polyaromatic compounds. Many of these bacteria use the contaminant as the sole source of carbon and energy. The contact between the bacteria and contaminant is a precondition for degradation. Some bacteria are mobile and exhibit a chemotactic response, sensing the contaminant and moving toward it (Burken *et al.*, 2011).

Soil fungi are very helpful in cleaning the pedosphere. The use of fungi in remediation is mycoremediation. Mycoremediation is a form of bioremediation in which fungi are used to decontaminate the area. The term *mycoremediation* refers specifically to the use of fungal mycelia in bioremediation. One of the primary roles of fungi in the ecosystem is decomposition, which is performed by the mycelium. The mycelium secretes extracellular enzymes and acids that break down lignin and cellulose, the two main building blocks of plant fiber. These are organic compounds composed of long chains of carbon and hydrogen, structurally similar to many organic pollutants. The key to mycoremediation is determining the right fungal species to target a specific pollutant. Certain strains have been reported to successfully degrade the nerve gases VX and sarin.

In one conducted experiment, a plot of soil contaminated with diesel oil was inoculated with mycelia of oyster mushrooms ; traditional bioremediation techniques (bacteria) were used on control plots. After four weeks, more than 95% of many of the PAH (polycyclic aromatic hydrocarbons) had been reduced to non-toxic components in the mycelial-inoculated plots. It appears that the natural microbial community participates with the fungi to break down contaminants, eventually into carbon dioxide and water. Wood-degrading fungi are particularly effective in breaking down aromatic pollutants (toxic components of petroleum), as well as chlorinated compounds.

Rhizofiltration is the uptake of metals into plant roots. Mycofiltration is a similar process, using fungal mycelia to filter toxic waste and microorganisms from water in soil. Soils Arbuscular mycorrhizae (AM) are ubiquitous symbiotic associations between higher plants and soil fungi (Brown and Wilkins,1985) and their extra-radical mycelium form bridges between plant roots and soil, and mediate the transfer of various elements into plants. There is also a growing body of evidence that arbuscular mycorrhizal fungi can exert protective effects on host plants under conditions of soil metal contamination. Binding of metals in mycorrhizal structures and immobilization of metals in the mycorrhizosphere may contribute to the direct effects. Indirect effects may include the mycorrhizal contribution to balanced plant mineral nutrition, especially P nutrition, leading to increased plant growth and enhanced metal tolerance. It has been widely reported that ectomycorrhizal and ericoid mycorrhizal fungi can increase the tolerance of their host plants to heavy metals when the metals are present at toxic levels. The underlying mechanism is thought to be the binding capacity of fungal hyphae to metals in the roots or in the rhizosphere which immobilizes the metals in or near the roots and thus depresses their translocation to the shoots (Smith and Read, 1997). Arbuscular mycorrhizal plants may exhibit much lower shoot concentrations of Zn and higher plant yields than non-mycorrhizal controls, indicating a protective effect of mycorrhizas on the host plants against potential Zn toxicity (Diaz *et al.*,1996). It has been demonstrated that at high soil heavy metal concentrations, arbuscular mycorrhizal infection reduced the concentrations of Zn, Cd and Mn in plant leaves (Heggo *et al.*,1990). Field investigations have indicated that mycorrhizal fungi can colonize plant roots extensively even in metal contaminated sites (Sambandan *et al.*,1992).

Phytodegradation is commonly applied as a phytoremediation measure. Phytodegradation (also rhizodegradation) is the breakdown of contaminants through the activity existing in the rhizosphere. Rhizobacteria are effective in nickel extraction (Abou-Shanab *et al.*,2003). It is facilitated by the presence of proteins and enzymes produced by the plants or by soil organisms such as bacteria, yeast, and fungi. Rhizodegradation is a symbiotic relationship where the plants provide nutrients necessary for the microbes to thrive, while microbes provide a healthier soil environment.

Rhizofiltration is a water remediation technique that involves the uptake of contaminants by plant roots. Rhizofiltration is used to reduce contamination in natural wetlands and estuarine areas.

Phytodegradation or rhizodegradation is the breakdown of contaminants through the activity existing in the rhizosphere due to the presence of proteins and enzymes produced by the plants or by soil organisms such as bacteria, yeast, and fungi. Rhizodegradation is a symbiotic relationship where the plants provide nutrients necessary for the microbes to thrive, while microbes provide a healthier soil environment.

Soils that have been contaminated for a long time may undergo prolonged remediation (Olson *et al.*,2007) and are less responsive to rhizodegradation than their freshly contaminated counterparts (Gunderson *et al.*,2007). There is therefore a need for enhancement of bioavailability as a key for successful biodegradation. Often times, selection and engineering of plants and microbial strains that modify solubility and transport of organic pollutants through exudation of biosurfactants become necessary and promising (Wang et al.,2007). In enhancing rhizodegradation, gene cloning of plants containing bacterial enzymes for the degradation of organic pollutants such as PCBs will be helpful in this regard. Other practices include the use of of root-colonising bacteria like *Pseudomonas fluorescens* expressing degradative enzymes such as ortho-monooxygenase for toluene degradation (Yee *et al.*,1998).In Nigeria, soils and sediments polluted with crude oil hydrocarbons are of major environmental concern on various contaminated sites. Hydrocarbon-degrading microorganisms are ubiquitously distributed in soils and constitute less than 1% of the total microbial communities but may increase to 10% in the presence of crude oil (Atlas,1995). However, use of fertilizers in hydrocarbon-contaminated soils act as biostimulants in such conditions. Some microbes are able to use HC as a carbon and energy source (van Hamme *et al.*,2003) preferentially in the absence of a readily available carbon source like labile natural organic matter. Read e*t al.* (2003) observed increased phosphorus mobilisation due to exudation of biosurfactants by lupine (*Lupinus angustifolius)*

Rhizofiltration is a water remediation technique that involves the uptake of contaminants by plant roots. Rhizofiltration is used to reduce contamination in natural wetlands and estuary areas.

Bioremediation can be classified as *ex situ and in situ* bioremediation. The former techniques involve the excavation or removal of soil from ground. Important *ex situ* treatments are composting, biopiles land farming, and bioreactors. *In situ* is a simple technique in which contaminated soil is excavated and spread over a prepared bed and periodically tilled until pollutants are degraded. The goal is to stimulate indigenous biodegradative microorganisms and facilitate the aerobic degradation of contaminants. The practice is limited to the treatment of superficial 10–35 cm of soil. Since land farming has the potential to reduce monitoring and maintenance costs, as well as clean-up abilities, it has received much attention as a disposal alternative. In land farming, contaminated soils are combined with nonhazardous organic amendments such as manure or agricultural wastes. Organic materials in land farming supports the development of a rich microbial population and elevated temperature Composting is a process of piling contaminated soil organic substances such as manure or agricultural wastes. The added organic material supports the development of a rich microbial population and elevates temperature of the pile. Stimulation of microbial growth by added nutrients results in effective biodegradation in a relatively short period of time characteristic of composting. Sometimes, biopiles are used in bioremediation. A biopile is a hybrid of land farming and composting; and is used for treatment of surfaces contaminated with petroleum hydrocarbons. Biopiles are improved forms of land farming that tend to control physical losses of the contaminants through leaching and volatilization. Land farming is a method in which contaminated soil is spread over a prepared bed along with some fertilizers and occasionally rotated. It stimulates the activity of bacteria and enhances the degradation of oil. But, the use of biopiles provides a favourable environment for autochthonous aerobic and anaerobic microorganisms.

Composting is a process of piling contaminated soil organic substances such as manure or agricultural wastes. The added organic material supports the development of a rich microbial population and elevates temperature of the pile. Stimulation of microbial growth by added nutrients results in effective biodegradation in a relatively short period of time (Thapa *et al.*, 2012).

Most *in situ* bioremediation techniques are generally the most desirable options due to cheapness and less disturbances since they provide the services in place avoiding excavation and transport of contaminants. This could useful in pro-poor communities common in sub-Saharan Africa. However, *in situ* remediation is among other factors governed by depth of soils for its efficacy. In many soils effective oxygen is also a prerequisite. Examples of important in situ bioremediation include are biosparging, bioventing, in situ biodegradation, and bioaugmentation. The *Deinococcus radiodurans* is used for metal remediation in radioactively polluted environments (Brim *et al.*,2000).

Crude oil is a mixture of thousands of varying chemical compounds. Given that composition of each type of oil is unique, there are different ways to bioremediate them using microbes and flora. Bioremediation can occur naturally or can be encourage with addition of microbes and fertilizers.

The microbes present in the soil at early stage recognize the oil and its constituents by biosurfactants and bio emulsifiers. After this, they attach themselves and use the hydrocarbon present in the petroleum as a source of energy. However, low solubility and adsorption of high molecular weight hydrocarbons can pose as a limiting factor to their availability to microorganisms. But, addition of biosurfactants enhances the solubility and removal of these contaminants. Again, rates of oil biodegradations increases with addition of biosurfactants.

Volatility, volubility, and susceptibility to biodegradation differ distinctly among constituents of crude oil. Some compounds are easily degraded, some resist degradation and some are non-biodegradable (Mukred *et al.*,2008). Yet, biodegradation of different petroleum compounds occurs simultaneously but at different rates because different species of microbes preferentially attack different compounds. This scenario leads to progressive and successive disappearance of constituents of crude oil over time.

Microbes produce enzymes in the presence of carbon sources, and these enzymes are responsible for the break down of hydrocarbon molecules. Many different enzymes and metabolic pathways are involved in the degradation of hydrocarbons contained in crude oil polluted soils. It implies that complete hydrocarbon degradation requires an appropriate enzyme, unavailability of which either prevents or minimizes its breakdown.

Bioremediation has various benefits of outstanding environmental and agricultural implications.

People perceive bioremediation as an acceptable strategy for the transformation of a wide variety of pollutants, often involving recycling (Polprasert, 1989).

Byproducts from bioremediation treatment are usually harmless products. Such residues include carbon dioxide, water, and cellular biomass, implying that most hazardous contaminants can be transformed to harmless products thereby eliminating the chance of future liability associated with treatment and disposal of contaminated material.

Processes involved in bioremediation can be conducted on-site without causing a major disruption of normal activities of the ecosystem. But, this, they need to transport quantities of waste off site and the potential threats to human health and the environment that can arise during transportation are eliminated.

Bioremediation is cheap when compared with other technologies that are used for clean-up of toxic waste. Some of the contaminants are sources of energy to the soil microbes thereby sustaining microbial biodiversity. Certain bacteria are mobile and exhibit a chemotactic response, sensing the contaminant and moving toward it.

Bioremediation was described as a strategy for integrated and sustainable development (Preston and Murgueitio, 1992). More possibilities of recycling wastes within farming systems become available as wastes from one process become inputs for another (Preston and Murgueitio 1992).

2. Limitations of bioremediation

1. Bioremediation is limited to those compounds that can be degraded biologically. However, not all compounds are susceptible to rapid and complete degradation. Some substrates such as straw, saw dust and maize cobs can be used to facilitate contact between soil microbes and toxicants.

2. Biological processes are often highly specific. For instance, anaerobic bacteria used for bioremediation of polychlorinated biphenyls in river sediments, dechlorination of the solvent trichloroethylene. The white rot fungus Phanaerochaete chrysosporium have the ability to degrade an extremely diverse range of persistent or toxic environmental pollutants

3. Research is needed to develop and engineer bioremediation technologies that are appropriate for sites with complex mixtures of contaminants that are not evenly dispersed in the environment.

4. Bioremediation often takes longer than other treatment options, such as excavation and removal of soil or incineration.

5. Regulatory uncertainty remains regarding acceptable performance criteria for bioremediation and there are no acceptable endpoints for bioremediation treatments. 6) There are some concerns that the products of biodegradation may be more persistent or toxic than the parent compound.

3. Justification

Soil pollution is widespread in Nigeria leading to varying forms of degradation. Soil pollution in Nigeria is associated with loss of bioresources especially plant materials. In reaction to this, it becomes imperative to use biological techniques in restoring and resisting further degradation.

4. Materials and methods

Nigeria, the most populous country in Africa, is situated on the Gulf of Guinea in West Africa with a land area of 351,649 sq mi (910,771 sq km); and total area of 356,667 sq mi (923,768 sq km). Nigerian population is estimated to be152,217,341 with a growth rate: of 1.9%; Its neighbors are Benin, Niger, Cameroon, and Chad. The lower course of the Niger River flows south through the western part of the country into the Gulf of Guinea. Swamps and mangrove forests border the southern coast; while inland areas are hardwood forests.The vegetation of Nigeria northwards include derived savannah,guinea savannah,sudan savannah and sahel savannah. Seven major soil groups in Nigeria include Alfisols, Ultisols, Inceptisols, Entisols, Vertisols, Oxisols and Histosols. Arable farming is a major socioeconomic activity while crude oil prospecting is a chief source of national revenue. Oil and oil-related activities plus urban wastes constitute significant mechanism of pollution.

Figure 1. Location map of the study area

Soil samples were collected by random sampling in polluted soils studied. Soil samples were collected from epipedons (0-15 cm depth) only. The soil samples were sieved using 2-mm sieve before they were subjected to various laboratory analyses.

Cation exchange capacity was measured by ammonium acetate method at a pH 7 (Soil Survey Staff, 2003). Soil pH was determined using a 1:1 soil :water ratio (Soil Survey Staff, 2003). Total carbon content of soils was estimated by loss on ignition using LECO equipment (Leco Corp, St. Joseph, M. I.). Exchangeable Ca and Mg in ammonium acetate extracted solution (leachate) were estimated with atomic absorption spectrophotometer (Perkin Elmer Model) while exchangeable K and Na were measured by flame photometry. Total nitrogen was estimated by microkjeldahl (Bremner, 1996). Available phosphorus was obtained using Bray P No.2 method (Olsen and Sommers, 1982).

Soil heavy metals were extracted with 1 M ammonium acetate (NH_4OAc) [1:4 ratio of soil / 1 M NH_4OAc extraction solution : 60 minutes reaction time (Reed and Martens, 1996]. After extraction, the suspension was centrifuged at 7500 x g relative centrifuge force for 30 minutes and the supernatant was passed through a Whatman No.42 paper. The filter paper was allowed to drip dry. The metal concentrations in the supernatant after centrifugation were determined using Inductively Coupled Plasma Optical Emission Spectroscopy, IRIS N701776 (Thermo Jarrell Ash Corporation). The total petroleum hydrocarbon (TPH) was determined using Fourier Transform Infra-red spectrometry (FTIR) (QAL/AM/S 16) at wavelengths ranging from 2800-3200 cm. The digestion and analytical procedures were checked by analysis of DOLT-3 Matrix Certified Reference Material with known concentration for heavy metals (Cantillo and Calder, 1990).

5. Results

Table 1 shows chemical composition of cassava sludge as opposed to compost derived from it. The result showed that cassava sludge had higher values of total carbon and total nitrogen than the compost manure formed from it.Resultantly, the C:N ratio was 17 when compared to 10 got from the compost. Values of exchangeable calcium and potassium were 9.8 and 18.3 cmol/kg, respectively in the compost manure as opposed to 9.3 and 7.4 cmol/kg in cassava sludge. Composting of cassava sludge reduced total values of heavy metals in it (Table 1). The pH_{KCl} value of composted cassava sludge increased to 6.3 (Table 1).

Properties	Sludge	Final Compost
Total carbon(g/kg)	342	138
Total nitrogen(g/kg)	20	13
Carbon-nitrogen ratio	17	10
Exchangeable calcium(cmol/kg)	9.3	9.8
Exchangeable magnesium(cmol/kg)	7.8	5.3
Exchangeable potassium(cmol/kg)	7.4	18.3
Available phosphorus(mg/kg)	2.2	1.5
Total chromium(mg/kg)	2.6	0.9
Total cadmium(mg/kg)	3.3	2.4
Total vanadium(mg/kg)	2.8	1.2
Total nickel(mg/kg)	7.	2,3
Total petroleum hydrocarbon(mg/kg)	58.6	12.4
pH_{KCl}	4.9	6.3

(Source: Onweremadu, 2008)

Table 1. Typical chemical characteristics of the cassava sludge and the produced compost (dry weight basis)

Certain factors influence ability of microbes to cause degradation (Table 2).Soil moisture requirement for optimum degradation of oil is high to very high (30-90 %) while 25- 28 % soil

Environmental conditions affecting degradation. Parameters	Condition required for microbial activity	Optimum value for an oil degradation
Soil moisture	25–28% of water holding capacity	30–90%
Soil pH	5.5–8.8	6.5–8.0
Oxygen content	Aerobic, minimum air-filled pore space of 10%	10–40%
Nutrient content	N and P for microbial growth	C:N:P = 100:10:1
Temperature (°C)	15–45	20–30
Contaminants	Not too toxic	Hydrocarbon 5–10% of dry weight of soil
Heavy metals	Total content 2000 ppm	700 ppm
Type of soil	Low clay or silt content	
Sources:Vidali,2001; Thapa *et al.*2012,		

Table 2. Factors Influencing Microbial Activities

moisture is required since a good number of these organisms are aerobes. Again, optimal soil pH range for oil degradation is 6.5 to 8.0 while microorganisms require 5.5 to 8.8.

Pollutant	Time in Days					
((mg/kg))	30	60	90	120	180	$LSD_{0.05}$
Chromium	15	25	35	40	50	1.25
Cadmium	35	60	65	75	90	0.92
Vanadium	20	25	30	35	40	1.08
Nickel	25	40	60	80	85	0.96
TPH	1.8	2.5	6.5	7.5	9.0	0.09
TPH= total petroleum hydrocarbon						
(Source: Onweremadu, 2008)						

Table 3. Effect of composted sludge on removability of contaminants at room temperature with time (days)

There was significant reduction (p=0.05) in the concentration of heavy metals and total petroleum hydrocarbon with time when treated with compost manure derived from cassava sludge (Table 3).However, removability rate varied among soil pollutants over time. Higher of values of Nickel were removed between 60 and 120 days while Vanadium was steadily degraded in the soil. Generally, more values of these soil toxicants were removed in cumulative terms from soils on a long-term. But, heavy metal concentration differed between rainy and dry seasons prevalent in the area as well as between land use types (Table 4). There were greater variations among heavy metals in rainy season in soils affected by automobile services when compared with values in dry season in the same land use. The variation was highest in cadmium (CV=79%), followed by Nickel (CV=48 %) and least in mercury (CV=0 %).Similar trend was observed in arable soils of the area (Table 4).

	Automobile wastes			
	Heavy metal	**Arable land**		**Automobile**
	RS	DS	RS	DS
Cd	34	35	79	17
Cr	35	31	37	25
Ni	25	37	48	18
Hg	54	1	0	49
Pb	20	20	36	32

CV=coefficient of variation in percentage, DS=dry season, RS=rainy season, Cd=cadmium, Cr=chromium, Ni=nickel, Hg=mercury, Pb=lead

(Onweremadu *et al.*, 2007).

Table 4. Seasonal variability (CV) of heavy metal concentration (mg/kg) in soils affected by

6. Discussion

Exchangeable basic cations are significantly deficient in Nigerian soils particularly polluted soils. Application of composted cassava sludge will enrich these soils with basic cations thereby reducing predominance of acidic radicals. Composted cassava sludge has high content of these neutralizing cations plus appreciable values of total nitrogen which is easily leached in these tropical soils (Table 1).Again, the pH value of the compost (6.3) (Table1) is optimal for microbial activity in soils (Table 2) as most tropical crops perform optimally at ph range of 5.5 to 6.5 (Ahn,1979). Composted cassava sludge progressively removed heavy metals and total petroleum hydrocarbons in crude oil polluted soils of Nigeria (Table 3). In a similar study, Wen *et al. (2002)* reported a depression in the phytoavailability of copper using sludge compost but Sims and Kline (1991) observed variations in characteristics of different composts and extractants, implying compost type influences uptake and availability.in different media..It is possible sorption ability of organic materials my depend on the specific surface area of each type. However,the concentration of these toxicants vary with season as indicated in Table 4, implying that their application strategies will follow the temporal variability. It was observed that impact of composted sewage waste became significant after thirteen years in an experiment conducted by Nogueirol *et al.* (2013) using three rates of sludges with maize and sugar cane as test crops.

Knowledge on the fate of heavy metals in soil–plant system is of great importance as that helps in predicting food quality especially in popular crops. Highest concentrations of heavy metals were recorded in the topmost layer of soils (Yedilar *et al.*,1994) and they reported that concentrations in rice leaves and grains were lower than values observed in rice roots in China.

Composted sludge has high content of organic carbon which has the capacity of absorbing organic pollutants in soil thereby reducing their solubility as demonstrated by several authors (Hernandez-Soriano et al.,2007; Stevens-Garmons et al.,2012). In addition, organic carbon inputs might enhance the soil buffer capacity. The application of slightly acidic sewage (6.3), will also affect the mobility and bioavailability of metals present in soils. The evolution of heavy metal distributions and bioavailability depended not only on total metal concentration but also on other properties such as pH, organic matter decomposition and dissolved organic carbon (Miaomiao *et al.*,2009) and they reported that composting systems decreased soil pH, soil organic matter and dissolved carbon. Their results varied between sewage sludge and swine waste. In some studies conducted in China, soil pH, soil organic matter and clay content were used to predict phytoavailability of Cadmium in paddy soils,stating that pH negatively correlated with cadmium cont of soils (Krebs et al.,1998). Martinez and Motto, (2000) reported that solubility availability, mobility and toxicity of heavy metals to plants increases as pH decreases. In a study at Suwon, Korea. Kim *et al.* (2009) remarked that translocation of heavy metals depended more on the variety of crop rather on soil physicochemical properties.

7. Conclusion

Polluted soils and soil-related natural resources can be decontaminated using living organisms including plant species classified as hyperaccumulators. Besides several microbial species can be used in bioremediation process since they can perform and efficient biodegradation activities. Efficacy in microbial degradation as well as microbial growth and activity are readily affected by edaphic characteristics like aeration, pH, temperature, and moisture. Bioremediation ensures a minimum impact on the ecosystem, minimizing the presence of potentially toxic by-products after the process.. There are *ex situ* and *in situ* methods of bioremediation; but, most *in situ* bioremediation techniques are generally the most desirable options due to cost-effectiveness and limited impact since they can be performed in situ; avoiding excavation and transport of contaminants. The *in situ*technique proves more affordable in rural communities with higher prevalence of poverty. The main drawback is associated with long-term character of bioremediation. However, most bioremediation techniques are constrained by high specificity of operation, long period of degradation and lack of acceptable endpoints of treatment. Further research is needed to overcome the time constraint of bioremediation processes as well and to increase the spectrum of activity of microbial species to address a variety of wastes that are currently generated and entering the soil..

http://dx.doi.org/10.5772/58381

Author details

Emmanuel Uzoma Onweremadu*

Address all correspondence to: uzomaonweremadu@yahoo.com

Department Of Soil Science and Technology, Federal University Of Technology, Owerri, Nigeria

References

[1] Abou-Shanab, R. A. , J. S. Angle, T. A. Delorme, R. L.Chaney, P. Berkum, H. Moawad, K. Ghanem and H. A. Ghozlan.2003.Rhizobacterial effects on Nickel extraction from soil and uptake by Alyssum Murale. New Phytologist,158 (1): 219-224.

[2] Ahn,P.M.1979.West African soils.Oxford University Press London.

[3] Atlas, R. M .1995.Bioremediation of petroleum pollutants.Intern. Biodeteriorat. and Biodegrad.,35 (1-3):317-327.

[4] Baker, A. J. M.; Brooks, R. R. .1989.Terrestrial higher plants which hyperaccumulate metallic elements : A review of their distribution, ecology and phytochemistry.Biorecovery 1 (2): 81–126

[5] Bani,A., G. Echevarria, S. Sulçe, J. L. Morel, and A.Mullai.2007.In-situ Phytoextraction of Ni by a native population of Alyssum murale on an Ultramafic site (Albania).Plant and Soil, 293 (1) 79-89.

[6] Boyd, R. S. and T. Javre (2001). Phytoenrichement of soil content by Sebertia acuminata in New Caledonia and the concept of elemental alelopathy, South African Journ.of Sci, 97 (2): 535-538.

[7] Bremner, J. M. 1996. Nitrogen – total. In: methods of soil analysis, chemical methods (Ed. D. L. Sparks). ASA and SSSA, Madison, Wisconsin, USA pp. 1085 – 1121.

[8] Brim H, S.C. McFarlan, J.K. Fredrickson, K.W. Minton, M. Zhai, L.P. Wackett, M.J. and Daly .2000. Engineering Deinococcus radiodurans for metal remediation in radioactive mixed waste environments. Nature Biotechnol. 18 (1): 85–90.

[9] Brown M. T. and D. A. Wilkins.1985.Zinc tolerance of mycorrhizal Betula. New Phytologist, 99:101-106.

[10] Burken, J.G. 2004. Uptake and metabolism of organic compounds: Green-Liver model .In: McCutcheon, S.C.; Schnoor, J. L., Phytoremediation: Transformation and Control of Contaminants, A Wiley-Interscience Series of Texts and Monographs, Hoboken, NJ: John Wiley, p. 59,

[11] Burken, J.; Vroblesky, D.; Balouet, J.C. 2011.Phytoforensics, Dendrochemistry, and hytoscreening: New green tools for delineating contaminants from past and present, Environ. Sci. and Technol. 45 (15): 6218–6226.

[12] Cantillo, A, and J. Calder, 1990. Reference materials for marine science. Fresewnius J. Anal. Chem. 338: 380 – 382.

[13] Chaney, R. L, M. Malik, Y. M. Li, S. L. Brown, J. S. Angle and A. J. M. Baker.1997.Phytoremediation of soil metals: Current opinion. Biotechnol., 8 (3): 279-284.

[14] Chará, J.G. Pedraza and N. Conde. 1999.The productive water decontamination system: A tool for protecting water resources in the tropics. Livestock Research for Rural Development11 (1):4-18

[15] Cotter-Howells, J. D, and S. Caporn.1996.Remediation of contaminated land by formation of heavy metal phosphates. Appl. Geochem., 11 (2):335-342.

[16] Cotter-Howells, J. D. P. E. Champness and J. M. Charnock,1999.Mineralogy of lead phosphorus grains in the roots of Agrostis capillaris L by ATEM and EXAFS.Min. Mag. (Lond), . 63 (6): 777-789.

[17] Davis, M. A., R. S. Boyd and J. H. Cane..2001.Host switching does not circumvent the Ni-based Defense of the Ni hyperaccumulator Streptanthus polygaloides (Brassicaceae), South African Journ. Sci. 97, (2) 554-557.

[18] Folke, C., S.Carpenter.B. Walker, M.Scheffer, T.Elmqvist, L.Gunderson, C.S. Holling. 2004.Regime shifts, resilience, and biodiversity in ecosystem management. Ann. Rev. Ecol. Syst, 35: 557–581.

[19] Garbisu,C. and I. Alkorta, 2001.Phytoextraction: A cost effective plant-based technology for the removal of metals from the environment," Biores. Technol., 77 (3): 229-236.

[20] Gunderson, J. J., J. D. Knight and K. C. J. van Rees.2007.Impact of ectomycorrhizal colonization of hybrid poplar on the remediation of diesel-contaminated soil. Journ. Environ. Qual. 36 (4) 927-934.

[21] Gupta, A. K. and S. Sinha, 2007.Phytoextraction capacity of the Chenopodium album L. grown on soil amended with tannery sludge, Bioresour. Technol., 98 (2): 442-446

[22] Heggo, A., J. S. Angle and R. L. Chaney.1990.Effects of vesicular arbuscular Mycorrhizal fungi on heavy metal uptake by soybeans, Soil Biol. and Biochem.,22 (6): 865-869

[23] Hernandez-Soriano M.C., Peña A. and Mingorance M.D. 2007. Retention of organophosphorous insecticides on a calcareous Mediterranean soil modified with carbon-rich amendments and a surfactant. Sci. Total Environ.378: 109-113

[24] Hendershot, W.H., H.Lalande and M.Duquette. 1993. Soil reaction and exchangeable acidity. In: Carter, M.R. (Ed.). Soil Sampling and mettods of soil analysis. Can. Soc. Soil Sci. Lewis Publishers, London. Pp. 141- 145.

[25] Kim,W.J.E.Yang. G.Jung,B.Park,S.Park.J.Kim, J.Kwon and G.Ryu.2009.Bioavailability and safety issues of heavy metals in paddy soil-rice soils continuum in Korea. National Institute of Agricultural Science and Technology,Suwon,Korea..S.K.Gupta,and R.Schulin.1998.Solubility and uptake of metals with and without liming of sludge amended soils.J.Environ.Qual.27:18-23.

[26] Lasat, M. M., M. Fuhrman, S. D. Ebbs, J. E. Cornish and L. V. Kochian 1998.Phytoremediation of radiocesium-contaminated soil: Evaluation of cesium-137 bioaccumulation in the shoots of three plant species," Journ. Environ. Qual.27 (1): 165-169.

[27] Levin, S. A. 1992. The problem of pattern and scale in ecology: The Robert H. MacArthur Award. Ecology 73 (6): 1943–1967.

[28] Martinez,C.E.and H.L.Motto.2000.Solubility of lead, zinc and copper added to mineral soils. Environ. Pollut.,107:153-158.

[29] McGrath, S. P. , F. J. Zhao and E. Lombi. 2001.Plant and rhizosphere processes involved in phytoremediation of metal contaminated soils. Plant and Soil, 232, (1): 207-214.

[30] Meagher, R.B .2000.Phytoremediation of toxic elemental and organic pollutants, Current Opinion in Plant Biology 3 (2): 153–162.

[31] Miaomiao, H,L.Wenghong, L Xinqiang , W.Donglei and T.Guangming.2009. Effect of composting process on phytotoxicity and speciation of copper ,zinc and lead in sewage sludge and swine manure.Waste Manage.,29 (2): 590=597

[32] Mukred,A.M., A. A. Hamid, A. Hamzah and W.M.W, Yusoff.2008. Development of Three Bacteria Consortium for the Bioremediation of Crude Petroleum-oil in Contaminated Water, OnLine Journ. Biol. Sciences, 8 (4) 73-81.

[33] Nelson, D.W. and L.E. Sommers. 1982. Total carbon and organic matter. In: Page, A.L, R.H.

[34] Miller and D.R. Keeney (Eds.). Methods of soil analysis, part 2. Agronomy Monogr. No.9, SSSA Madison W.I. pp.539 – 579.

[35] Nogueirol,R.C.,W.J.de Melo, E.I.Bertoncini, L.Reynaldo and F Alleoni.2013.Concentrations of Cu, Fe, Mn and Zn in tropical soils amended with sewage sludge and composted sewage sludge.Environ. Monit. And Assess.185 (4):2929-2938.

[36] Olsen, S.R. and Sommers, L.E. 1982. In: Page, A.L; Miller, R.H. and Keeney, D.R. (eds).

[37] Methods of soil analysis, 2nd ed. Agron. Monogr. ASA and SSSA, Madison, WI. pp. 403-430.

[38] Olson, P. E., A. Castro, M. Joern, N. M. DuTeau, E. A. H. Pilon-Smits and K. F. Reardon.2007.Comparison of plant families in a greenhouse phytoremediation study on an aged polycyclic aromatic hydrocarbon-contaminated soil. Journ. Environ. Qual., 36:1461-1469.

[39] Onweremadu.E.U,,E.T.Eshett and G.E.Osuji.2007.Temporal variability of selected heavy metals in automobile soils.Int. Journ.Environ.Sci.Technol.,4 (1):35-41.

[40] Onweremadu.E.U.2008.Characterization and removability of priority pollutants in oil-spilled site using composted cassava sludge.Life Sci.Journ.5 (3):62-66.

[41] Polprasert, .C. 1989. Organic waste recycling, Asian Institute of Technology. Bangkok. John Wiley and Sons London. 357p.

[42] Preston. T. R. and Murgueitio .E. 1992. Strategy for sustainable livestock production in the tropics. Cali. SAREC-CIPAV. 89p.

[43] Raskin,I. and B. D. Ensley.2000.Phytoremediation of toxic Metals using plants to clean-up the environment. Wiley, New York, .

[44] Reed, S. T. and D. C. Martens. 1996. Copper and Zinc p. 703 – 722. In D. L. Sparks (ed.) methods of soils analysis. Part 3, SSSA Book Series No. 5. SSSA and ASA, Madison, WI

[45] Sambandan, K., K. Kannan and N. Raman,1992.Distribution of vesicular-arbuscular mycorrhizal fungi in heavy metal polluted soils of Tamil Nadu, India. Journ. Environ. Biol.13 : 159-167.

[46] Sims,J.S. and J.T.Kline.1991.Chemical fractionation and uptake of heavy metals in soils amended with co-composted sewage sludge.J.Environ.Qual. 20:387-395.

[47] Smith, S. E. and D. J. Read1997.Mycorrhizal symbiosis. 2nd Edition, Academic Press, London.

[48] Soil Survey Staff. 2003. Keys to soil taxonomy. Ninth edition. United States Department of Agriculture. 332 pp.

[49] Steele,M.C. and J. Pichte.1998. Ex-situ remediation of metal contaminated superfund soil using selective extractants, Journ. Environ. Eng., 124 (7): 639-645.

[50] Stevens-Garmon 2012. Sorption of emerging trace organic compounds onto wastewater sludge solids. Water Res. 45 (11):3417-3426

[51] Subramanian, M.; O. J.David, . and J. V .Shanks,. 2006.. TNT Phytotransformation pathway characteristics in Arabidopsis: Role of aromatic hydroxylamines, Biotechnol. Prog. 22 (1): 208–216.

[52] Thapa,B.,A.K.C. Kumar and A. Ghimire (2012). A review on bioremediation of petroleum hydrocarbon contaminants in soil . Kathmandu Univ. Journ. Sci. Eng.Technol., 8 (I): 164-170.

[53] Van Hamme, J. D., A. Singh and O. P. Ward.2003.Recent advances in petroleum microbiology. Microbiol. And Molecular Biol. Rev., 67 (4): 503-549.

[54] Vidali.M.2001. Bioremediation.:An overview. Pure Appl. Chem., 73 (7) :1163-1169.

[55] Vroblesky, D.; Nietch, C.; Morris, J. 1998.Chlorinated ethenes from groundwater in tree trunks, Environ. Sci.and Technol. 33 (3): 510–515.

[56] Wang, Q., X. Fang, B. Bai, X. Liang, P. J. Shuler and W. A. Goddard.2007.Engineering bacteria for production of rhamnolipid as an agent for enhanced oil recovery. Biotechnol. Bioeng. 8 (4): 842-853.

[57] Wen,G., T.E..Bates, S.Inanaga, R.P.Voroney,K. Hamamura and D.Curtin.2002. A yield control approach to assess phytoavailability of Zn and Cu in irradiated, composted sewage sludges and composted manures in field experiments II:Copper.Plant and Soil, 246 (2):241-248..

[58] Wenzel, W. W. and F. Jockwer,.1999.Accumulation of heavy metals grown on mineralized soils of the Austrian Alps. Environ. Pollut. 104, (1):145-155.

[59] Yedilar,A.,P. Grill.,T.Sun and.A.Kettrup.1994.Fate of heavy metals in a land treatment system irrigated with municipal waste water.Chemosphere,28 (2):375-381.

[60] Yee, D. C.,J. A. Maynard and T. K. Wood.1998.Rhizoremediation of trichloroethylene by a recombinant, rootcolonizing Pseudomonas fluorescens strain expressing toluene ortho-monooxygenese constitutively. Appl. Environ. Microbiol. 64 (1):112-118.

Chapter 12

Thermal Removal of Polycyclic Aromatic Hydrocarbons from Gasification Biochars

Sebastian Meyer, Bruno Glaser, Daniel Fischer,
Peter Quicker, Yves Noel and Georg Kuffer

Additional information is available at the end of the chapter

http://dx.doi.org/10.5772/57269

1. Introduction

The number of small-scale (< 200 kW electrical capacity) wood gasifiers used for electricity and heat provision in Central Europe is increasing. After the wood gasification process, about 10% of the dry-mass of the wood feedstock input are left in form of wood gasification residues [1] consisting of mixture of char and ash. Inspired by the Terra preta phenomenon and the intention to generate own humus rich black earths [2], especially farmers are interested in the opportunity to use the carbonaceous and nutrient-rich gasification residues produced by their own gasifiers for farm fertilizer production, soil amelioration and for carbon sequestration [3].

However, gasification residues are known for their high content in polycyclic aromatic hydrocarbons (PAH) [4]. PAH are carcinogenic, persistent, accumulate in organisms and partly inhibit reproduction [5]. Due to that, gasification residues with very high PAH contents need to be classified as hazardous waste. As a moderate increase in temperature substantially increases the vapor pressure of PAH, thermal processes can be used to volatize and subsequently remove PAH from background matrixes [6]. To reduce the PAH content in gasification residues, a thermal process – the so called PAH volatilization unit has been developed by a German wood gasifier manufacturer. In the following, a critical assessment of the functional efficiency of this patented process is described.

The research work was carried out with the objective to evaluate the effectiveness of the PAH volatilization unit in reducing the PAH content in gasification residues. This objective was chosen to better assess the environmental suitability of using gasification residues as soil amendment. Our working hypothesis was that it is possible to technically reduce the PAH content in gasification residues to a level which allows for an agricultural use of the gasification residues in line with the soil protection regulations in Germany. As a precondition for this evaluation, the suitability of the analysis methods DIN 13877:A and DIN 13877:B for the determination of the PAH content in wood gasification residues had to be checked. Based on the results of [7], our working hypothesis in this regard was that analysis method DIN 13877:B would be more suitable for the analysis task as compared to analysis method DIN 13877:A.

2. Materials and methods

2.1. Comparison of PAH analysis methods for gasification residues (1st experiment)

Hilber et al. 2012 [7] demonstrated that the selection of an appropriate solvent is crucial to determine the PAH content in biochars. They recommend a Soxhlet-extraction with toluene for the PAH analysis of biochars. To complement and to reassess this work, the PAH content (sum of the 16 PAH defined by the US Environmental Protection Agency EPA) of three samples of the same production batch of gasification residues was determined using two different extraction methods (cold extraction with acetone according to DIN 13877:A and Soxhlet extraction with toluene for 5 hours according to DIN 13877:B). The gasification residues had been produced from poplar wood chips in a commercially operated fixed-bed Joos-Spanner biomass gasifier (30 kW electrical capacity) of the German manufacturer Spanner Re² GmbH. The PAH analyses of all experiments described on the following pages were carried out by the commercial laboratory Görtler Analytical Services GmbH in Vaterstetten, Germany.

2.2. Testing the functional efficiency of the PAH volatilization unit (2nd and 3rd experiment)

In a next step, the functional efficiency of a PAH volatilization unit developed by a cooperation of the RWTH Aachen and the German manufacturer Spanner Re² GmbH was assessed. Gasification residues produced under identical process conditions as described in the first experiment were treated in the PAH volatilization unit as illustrated in Figure 1:

This experimental setting was slightly modified for a 3rd experiment: The sealing air stream was reduced. In addition to that, air from the external source (17) was intermittently (within the standstill periods of the conveyor screws) blown into the gasification residue stream within the heating chamber to supply oxygen to the hot gasification residues. In this technical design version of the process, about 4% of the flue gas from the PAH volatilization unit was recirculated back into the air stream which supplied both the combustion chamber and the sealing air stream.

The PAH content of three samples of the processed batches of gasification residues was analyzed according to DIN 13877:B in both experiments.

2.3. Redesign of the PAH volatilization unit (4th experiment)

In a redesigned version of the PAH volatilization unit, the manufacturer prevented the recirculation of PAH into the PAH volatilization unit by blowing clean air from an external

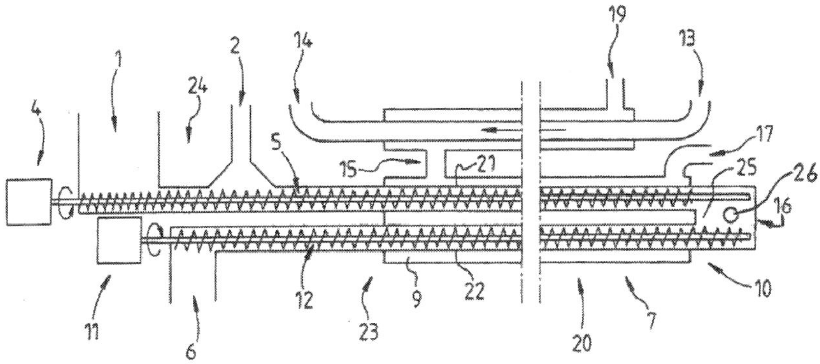

Figure 1. PAH volatilization unit (Source: Patent Specification) The gasification residues entered the unit on the left side (1) and were being transported by a motor- (4) driven long conveyor screw (5) in an upper tube (24) through a heat exchanger (23) until the heating chamber (16) on the far right side of the unit. From this point, they were transported back through the heat exchanger by a motor- (11) driven second long conveyor screw (12) in a second tube (22) on the bottom side of the unit until the exit (6) on the bottom left site of the unit. Both conveyor screws were operated according to the following time schedule: 15 seconds operation, 30 seconds stop, 15 seconds operation and so forth... Within the heating chamber, the gasification residues got in contact with the outer side of a hot (surface temperature: about 600 °C) tube (26) transporting hot wood gas produced in the wood gasifier (not shown in the figure). It is technically possible to supply air from an external source (16) to the gasification residue stream inside the heating chamber. However, this option was not used for the 2nd experiment. A hot external air stream used for the air supply of the wood gasifier entered (13) the heat exchanger at a temperature of about 300 °C and left (14) the heat exchanger at a temperature of about 375 °C, thereby effectively cooling the wood gasification residues on their way through the heat exchanger. Sealing air from the same external source as the heating chamber air stream was blown into the left side (6) of the bottom tube of the PAH volatilization unit in counter flow principle and left the unit on the top left side (1) of the unit. This counter current air flow was introduced to prevent volatized PAH from leaving the PAH volatilization unit through the exit for the processed gasification residues. In the 2nd experiment, about 5 % of the air leaving the unit was recirculated back into the sealing air stream.

source in the sealing air nozzle and by discharging the airstream with the volatized PAH in the gas engine where the wood gas of the gasifier was burned. Further modifications of the redesigned PAH volatilization unit included larger conveyor screw diameters and conveyor screw tube diameters. Gasification residues from a 45 kW_{el} wood gasifier fed by a mixture of chips from different sort of woods were treated in this modified PAH volatilization unit in a 4th experiment. The sealing air stream and the heating chamber air stream volumes were increased as compared to the 2nd and 3rd experiment. The PAH content of one sample of the processed batch of gasification residues was analyzed according to DIN 13877:B.

2.4. PAH volatilization in a laboratory scale experiment (5th experiment)

To evaluate the general capability of thermal processes to volatize and subsequently remove PAH from gasification residues, the following experiment was carried out:

20 gram samples of gasification residues from a 45 kW_{el} wood gasifier fed by a mixture of chips from pine trees and spruce were filled in an open steel container (20mm width x 20mm depth x 100mm height). This container was placed in an electrical box furnace (Nabertherm, Model

LH 30/14) which had been heated to specified temperature levels (550° C, 650° C and 700° C in three consecutive trials). During the experiments, the box furnace was purged by inert gas of type Argon 4.6. After 30 minutes, the container was removed from the furnace and immediately cooled down in a water quench. During the cooling process, the container was purged from above with Argon 4.6. Due to that, the complete heating and cooling procedure took place in an oxygen-free environment.

The PAH content of the three samples (one for each temperature level) was analyzed according to DIN 13877:B.

2.5. Statistical analysis

All statistical tests were conducted with the open source software R 3.0.1 (R CORE TEAM 2012). Due to the low number of samples per treatment (n=3), particular care and attention was paid to the statistical requirements and assumptions. In this respect, our treatments didn't show neither a normal distribution nor homogenous variances among groups so that the Student's t-test couldn't be applied. Therefore, we applied a permutation version of ANOVA according to [8] for the statistical analysis of the obtained results from the experiments 1, 2 and 3. For this purpose we used the package "ImPerm" [9].

3. Results

3.1. Comparison of PAH analysis methods for gasification residues (1st experiment)

Table 1 summarizes the Σ EPA16 PAH content in the three gasification residue samples in mg/kg dry mass (DM) as determined by the two different analytical methods. On average, the analysis according to DIN 13877:B resulted in PAH contents more than 4 times higher compared to the analysis according to DIN 13877:A. The standard errors of the mean (SEM) are indicated after the average values. The difference between the sample means (n=3) of the two analysis methods is significant ($p < 0.01$) according to the applied permutation version of ANOVA.

Figure 2 depicts the results from Table 1 in two box-and-whisker plots.

3.2. Functional efficiency of the PAH volatilization unit (2nd and 3rd experiment)

Table 3 summarizes the Σ EPA16 PAH content in the gasification residue samples of the production batch after the treatment in the PAH volatilization unit (PAH-VU) in the 2nd experiment. On average, the PAH content of the processed gasification residues was 58% lower as compared to the unprocessed residues from the 1st experiment. The difference between the sample means (n=3) of the two analysis methods is significant ($p < 0.01$) according to permutation version of ANOVA applied. The standard errors of the mean are indicated after each average value.

Figure 3 depicts the results from Table 3 in two box-and-whisker plots.

Gasification Residues	**Extraction with acetone** according to DIN 13877:A	**Extraction with toluene** according to DIN 13877:B
(untreated)	Σ EPA16 PAH content	Σ EPA16 PAH content
	mg/kg DM	mg/kg DM
sample 1	542	3,056
sample 2	571	1,009
sample 3	504	2,702
Average	**539 ± 16 (SEM)**	**2,255 ± 516 (SEM)**

Table 1. Comparison of the two PAH analysis methods using either acetone or toluene for extraction (1st experiment). SEM indicates standard error of the mean.

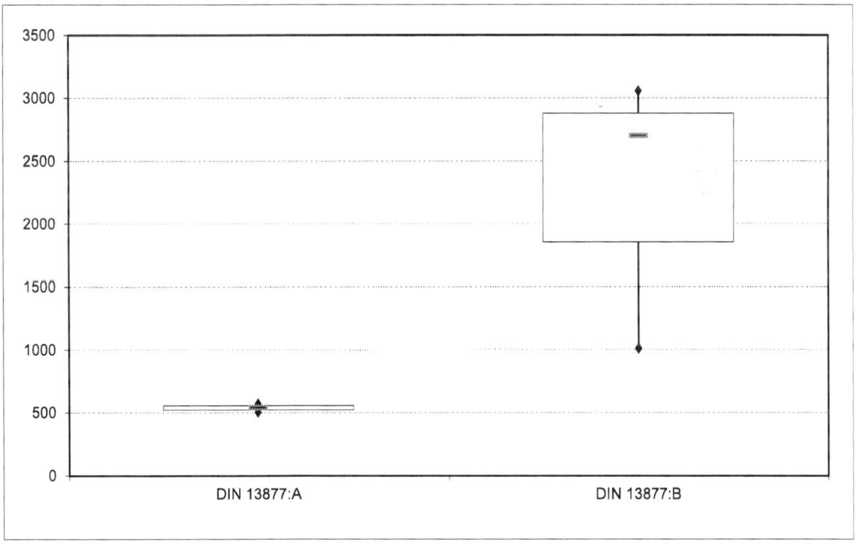

Figure 2. Boxplots indicating the median (central black bar), the minimum and maximum (lower and upper whisker) and the lower and upper quartile (lower end upper end of the box) of the Σ EPA16 PAH contents in mg/kg DM in three gasification residue samples after application of the analysis method DIN 13877:A (left) and DIN 13877:B (right).

Gasification Residues	Extraction with acetone according to DIN 13877:A			Extraction with toluene according to DIN 13877:B		
(untreated)	Σ EPA16 PAH content			Σ EPA16 PAH content		
	Sample 1	Sample 2	Sample 3	Sample 1	Sample 2	Sample 3
	mg/kg DM			mg/kg DM		
Naphthalene	460	480	420	1,200	570	1,200
2-Methylnapthalene	63	66	61	490	130	330
1-Methylnapthalene	63	67	62	470	100	260
Acenaphtylene	43	47	43	580	120	340
Acenapthene	14	15	14	110	15	54
Flourene	8,5	8,4	8,9	180	38	120
Phenanthrene	12	15	13	690	170	580
Anthracene	1,8	2,4	2	120	32	120
Flouranthen	1,5	1,9	1,7	94	31	140
Pyrene	1,4	1,6	1,5	70	27	110
Benzo(a)anthracene	< 0,01	< 0,01	< 0,01	4,5	2,4	12
Chrysene	< 0,01	< 0,01	< 0,01	4,8	2,4	18
Benzo(b)flouranthene	< 0,01	< 0,01	< 0,01	1,5	0,43	3,4
Benzo(k)flouranthene	< 0,01	< 0,01	< 0,01	0,24	0,05	0,64
Benzo(a)pyrene	< 0,01	< 0,01	< 0,01	0,65	0,19	1,9
Indeno(1,2,3.cd)pyren	< 0,01	< 0,01	< 0,01	0,25	0,11	0,9
Dibenz(a,h)anthracene	< 0,01	< 0,01	< 0,01	0,05	< 0,01	0,31
Benzo(g,h,i)perylene	< 0,01	< 0,01	< 0,01	0,16	0,16	0,4
Σ EPA16 PAH	**542**	**571**	**504**	**3,056**	**1,009**	**2,702**

Table 2. Comparison of the two PAH analysis methods using either acetone or toluene for extraction (1st experiment). Detailed results.

Gasification Residues	Production batch without treatment from 1st experiment	Production batch with treatment in the PAH-VU from 2nd experiment
	Σ EPA16 PAH content (DIN 13877:B, extraction with toluene)	Σ EPA16 PAH content (DIN 13877:B, extraction with toluene)
	mg/kg DM	mg/kg DM
sample 1	3,056	1,291
sample 2	1,009	731
sample 3	2,702	806
Average	**2,255 ± 516 (SEM)**	**943 ± 143 (SEM)**

Table 3. PAH reduction in PAH volatilization unit (2nd experiment). SEM indicates standard error of the mean.

Figure 3. Boxplots indicating the median (central black bar), the minimum and maximum (lower and upper whisker) and the lower and upper quartile (lower end upper end of the box) of the Σ EPA16 PAH contents in mg/kg DM in three gasification residue samples from the 1st experiment (left) and from the 2nd experiment (right).

Gasification Residues (processed)	Σ EPA16 PAH content (DIN 13877:B)		
	Sample 1	Sample 2	Sample 3
		mg/kg DM	
Naphthalene	850	370	510
2-Methylnapthalene	41	15	17
1-Methylnapthalene	32	14	14
Acenaphtylene	14	4.3	7.9
Acenapthene	3.4	1.4	1.1
Flourene	3.6	4	0.94
Phenanthrene	330	250	200
Anthracene	40	37	32

Gasification Residues (processed)	Σ EPA16 PAH content (DIN 13877:B)		
	Sample 1	Sample 2	Sample 3
	mg/kg DM		
Flouranthen	26	32	33
Pyrene	19	26	18
Benzo(a)anthracene	1.5	2	1.2
Chrysene	2.4	3.4	1.9
Benzo(b)flouranthene	0.27	0.54	0.23
Benzo(k)flouranthene	0.05	0.04	0.05
Benzo(a)pyrene	0.06	0.1	0.06
Indeno(1,2,3.cd)pyren	0.18	<0.01	0.03
Dibenz(a,h)anthracene	<0.01	<0.01	<0.01
Benzo(g,h,i)perylene	0.04	0.07	0.05
Σ EPA16 PAH	**1,291**	**731**	**806**

Table 4. PAH reduction in PAH volatilization unit (2nd experiment). Detailed results.

Table 5 summarizes the Σ EPA16 PAH content in the gasification residue samples of the production batch after the treatment in the PAH volatilization unit in the 3rd experiment. On average, the PAH content of the processed gasification residues was 36% lower as compared to the unprocessed residues from the first experiment. The difference between the sample means (n=3) of the two analysis methods is not significant ($p < 0.05$) according to permutation version of ANOVA applied. The standard errors of the mean are indicated after each average value.

Gasification Residues	Production batch without treatment from 1st experiment	Production batch with treatment in the PAH-VU from 3rd experiment
	Σ EPA16 PAH content (DIN 13877:B, extraction with toluene)	Σ EPA16 PAH content (DIN 13877:B, extraction with toluene)
	mg/kg DM	mg/kg DM
sample 1	3,056	1,713
sample 2	1,009	1,292
sample 3	2,702	1,298
Average	**2,255 ± 516 (SEM)**	**1,434 ± 113 (SEM)**

Table 5. PAH reduction in PAH volatilization unit (3rd experiment). SEM indicates standard error of the mean.

http://dx.doi.org/10.5772/57269

Figure 4 depicts the results from Table 5 in two box-and-whisker plots:

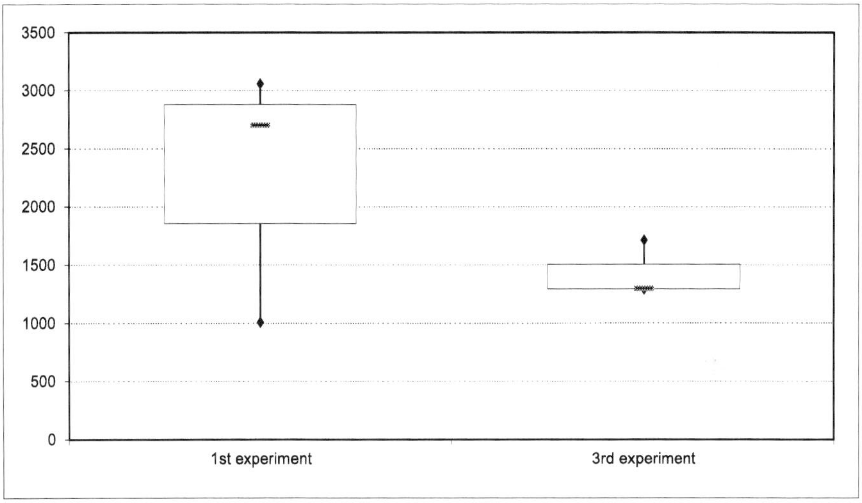

Figure 4. Boxplots indicating the median (central black bar), the minimum and maximum (lower and upper whisker) and the lower and upper quartile (lower end upper end of the box) of the Σ EPA16 PAH contents in mg/kg DM in three gasification residue samples from the 1st experiment (left) and from the 3rd experiment (right).

Gasification Residues (processed)	Σ EPA16 PAH content (DIN 13877:B)		
	Sample 1	Sample 2	Sample 3
		mg/kg DM	
Naphthalene	1,000	860	980
Acenaphtylene	1.1	0.47	0.52
Acenapthene	0.06	0.05	0.03
Flourene	<0.01	<0.01	<0.01
Phenanthrene	480	390	240
Anthracene	86	23	34
Flouranthen	87	12	26
Pyrene	59	6.8	17
Benzo(a)anthracene	<0.01	<0.01	<0.01
Chrysene	<0.01	<0.01	<0.01

Gasification Residues (processed)	Σ EPA16 PAH content (DIN 13877:B)		
	Sample 1	Sample 2	Sample 3
		mg/kg DM	
Benzo(b)flouranthene	<0.01	<0.01	<0.01
Benzo(k)flouranthene	<0.01	<0.01	<0.01
Benzo(a)pyrene	<0.01	<0.01	<0.01
Indeno(1,2,3.cd)pyren	<0.01	<0.01	<0.01
Dibenz(a,h)anthracene	<0.01	<0.01	<0.01
Benzo(g,h,i)perylene	<0.01	<0.01	<0.01
Σ EPA16 PAH	**1,713**	**1,292**	**1,298**

Table 6. PAH reduction in PAH volatilization unit (3rd experiment). Detailed results.

3.3. Redesign of the PAH volatilization unit (4th experiment)

Table 7 summarizes the Σ EPA16 PAH content in the gasification residue sample of the production batch after the redesign of the PAH volatilization unit (PAH-VU) in the 4th experiment. The PAH content of the processed gasification residues was 82% lower as compared to the average PAH content in the unprocessed residues from the 1st experiment.

Gasification Residues	Production batch without treatment from 1st experiment	Production batch with treatment in the PAH-VU from 4th experiment
	Σ EPA16 PAH content (DIN 13877:B, extraction with toluene)	Σ EPA16 PAH content (DIN 13877:B, extraction with toluene)
	mg/kg DM	mg/kg DM
sample 1	3,056	396
sample 2	1,009	
sample 3	2,702	
Average	**2,255 ± 516 (SEM)**	

Table 7. PAH reduction in PAH volatilization (4th experiment). SEM indicates standard error of the mean.

3.4. PAH volatilization in a laboratory scale experiment (5th experiment)

Table 9 summarizes the Σ EPA16 PAH content in the gasification residue samples treated in the laboratory scale experiment in an oxygen-free environment. Following a temperature treatment of at least 650 °C, the PAH contents in the gasification residues decreased drastically.

http://dx.doi.org/10.5772/57269

Gasification Residues (processed)	Σ EPA16 PAH content (DIN 13877:B)
	Sample 1
	mg/kg DM
Naphthalene	160
Acenaphtylene	58
Acenapthene	2.8
Flourene	12
Phenanthrene	74
Anthracene	14
Flouranthen	35
Pyrene	32
Benzo(a)anthracene	2.5
Chrysene	3.2
Benzo(b)flouranthene	0.96
Benzo(k)flouranthene	0.32
Benzo(a)pyrene	0.59
Indeno(1,2,3.cd)pyren	0.49
Dibenz(a,h)anthracene	0.1
Benzo(g,h,i)perylene	0.5
Σ EPA16 PAH	**396**

Table 8. PAH reduction in PAH volatilization (4th experiment). Detailed results.

Gasification Residues	Σ EPA16 PAH content (DIN 13877:B, extraction with toluene)
	mg/kg DM
550 °C	1,000
650 °C	1.3
700 °C	0.28

Table 9. PAH content in gasification residues from the laboratory scale experiment (5th experiment)

4. Discussion

The comparison of the PAH analysis methods described in section 2.1 confirms the recommendation of Hilber et al. 2012 [7]: Soxhlet extraction with toluene resulted in a much higher extraction of PAH from the gasification residues as compared to the cold extraction with

Gasification Residues (processed)	Σ EPA16 PAH content (DIN 13877:B)		
	550 °C	650 °C	700 °C
		mg/kg DM	
Naphthalene	590	0.77	0.18
Acenaphtylene	0.3	0.19	< 0.01
Acenapthene	1.3	< 0.01	< 0.01
Flourene	0.2	< 0.01	< 0.01
Phenanthrene	230	0.23	< 0.01
Anthracene	43	0.04	< 0.01
Flouranthen	57	0.02	< 0.01
Pyrene	57	0.03	< 0.01
Benzo(a)anthracene	6.6	< 0.01	< 0.01
Chrysene	13	< 0.01	< 0.01
Benzo(b)flouranthene	3,6	< 0.01	< 0.01
Benzo(k)flouranthene	1,3	< 0.01	< 0.01
Benzo(a)pyrene	0.95	< 0.01	< 0.01
Indeno(1,2,3.cd)pyren	2.5	< 0.01	< 0.01
Dibenz(a,h)anthracene	0.1	< 0.01	< 0.01
Benzo(g,h,i)perylene	0.56	< 0.01	< 0.01
Σ EPA16 PAH	**1,000**	**1.3**	**0.28**

Table 10. PAH content in gasification residues from the laboratory scale experiment (5th experiment). Detailed results.

acetone (see Table 1). Thus, analysis method DIN 13877:B was applied for the PAH analysis in the subsequent experiments 2-5.

[10] observed a positive correlation between the pyrolysis temperature (between 200 °C and 500 °C in the experiment carried out) applied for the production of biochars from lake sediments and the sorption of Phenanthrene to these biochars. As gasification temperatures are higher [1] than the pyrolysis temperatures applied in this experiment, we explain our analysis results by a strong sorption of PAH to the carbonaceous matrix of the gasification residues.

[11] recently compared recovery rates after reflux extraction with toluene and a 1:1 acetone/cyclohexane mixture of the three surrogate PAH Acenaphtene-d_{10}, Phenanthrene-d_{10} and Chrysene-d_{10} which had been previously added to a pyrolysis char made form orchard pruning. The experiment resulted in higher recovery rates with the solvent toluene as compared to the solvent mixture acetone/cyclohexane for the high molecular weight PAH Phe-

nanthrene-d_{10} (68% compared to 41%) and Chrysene-d_{10} (58% compared to 7%). The recovery rate with the solvent toluene was though lower as compared to the acetone/cyclohexane mixture for the low molecular weight PAH Acenaphtene-d_{10} (68% compared to 80%).

The differences in the recovery rates stated above are yet small if compared to our analysis results (see Table 1). Reflux extractions are usually carried out at temperatures close to the boiling point of the solvent applied. We thus presume that the strong difference in our analysis results after applying the analysis methods DIN 13877:A (low temperature extraction) and 13877:B (high temperature extraction) can be mainly explained by the temperature difference between the two extraction methods. We recommend to directly compare Soxhlet extractions with the solvents toluene and a 1:1 acetone/cyclohexane mixture for the determination of the Σ EPA16 PAH content of gasification residues for future research.

Based on the results described above, the effectiveness of the legal provisions of the German Federal Soil Protection Act (Bundesbodenschutzgesetz [BBodSchG]) [12] and the German Federal Soil Protection and Contaminated Sites Ordinance (Bundesbodenschutzverordnung [BBodSchV]) [13] for the protection of agricultural soils against PAH pollution in Germany was analyzed. Special attention was paid to the suitability of the PAH analysis methods specified in these regulations. The German Federal Soil Protection and Contaminated Sites Ordinance defines precautionary values for the PAH content of soils. These precautionary values are set at 10 mg/kg DM for the Σ EPA16 PAH content in the finely granulated part of soils with humus contents larger than 8% and at 3 mg/kg DM for the Σ EPA16 PAH content of soils with humus contents of less or equal 8%. The German Federal Soil Protection and Contaminated Sites Ordinance specifies a number of analysis methods which may be used to determine the PAH content of soils in accordance with this ordinance. Among them are the analysis methods DIN 13877:A (cold extraction with acetone) and DIN 13877:B (Soxhlet extraction with toluene). However, our analysis results in section 3.1 have clearly shown that DIN 13877:A is not suitable to determine the PAH content in gasification residues. Thus, the regulations of the German Federal Soil Protection Act and the German Federal Soil Protection and Contaminated Sites Ordinance currently cannot prevent the application of gasification residues with high PAH contents to agricultural soils, if the cold extraction with acetone is applied for their characterization. It is recommended to close this legislative loophole. Apart from that, it should be noted that there is no boundary value for PAH loads added to soil (however for the PAH content in the soil) in the German soil legislation.

While thermal processes are already used to remove PAH from contaminated soils [6], this principle was applied for the PAH removal from wood gasification residues for the first time to our knowledge. The results displayed in Table 3, Table 5 and Table 7 indicate a reduction of the Σ EPA16 PAH content of the gasification residues by 36% to 82% after the treatment in the PAH volatilization unit. Still, the residual PAH contents in the gasification residues are too high to allow for an agricultural use.

The residual PAH content in the gasification residues was higher in the third experiment as compared to the second experiment. This result might be explained by a difference in the PAH content of the untreated gasification residues (although these were produced under the same production conditions) or by a change in the airflow conditions within the PAH volatilization unit caused by modifications of the sealing air stream and the heating chamber air stream volumes. As already mentioned, gasification residues which had already passed through the hot heating chamber got in contact with the volatilized PAH contained in the sealing air which was blown into the lower conveyor screw. We assume that this is one of the reasons for the high residual PAH content of the gasification residues in the 2nd and 3rd experiment.

This hypothesis is supported by the comparably low PAH content of 396 mg/kg DM which was reached after the redesign of the PAH volatilization unit which prevented the recirculation of volatized PAH. Independent from the avoided recirculation effect, the increase in the air supply to the heating chamber in the 4th experiment has with high probability promoted the oxidation (combustion) of gasification residues in the heating chamber. The resulting temperature increase might have supported a more complete volatilization of the PAH from the gasification residues as compared to the 2nd and 3rd experiment.

The 5th experiment proves that thermal volatilization processes are capable in reducing the PAH content of gasification chars to levels which are acceptable for agricultural applications. It can be derived from the data presented in Table 9, that minimum process temperatures of about 650 °C are necessary for an effective removal of PAH from gasification residues.

Based on the laboratory scale experiment, the following additional technical design modifications are suggested to further improve the functional efficiency of the PAH volatilization unit:

- The gasification residues need to be sufficiently hot to allow for a complete volatilization of the PAH sorbed to the residue surfaces. In a modified heat exchanger, the hot wood gas should be used to heat the gasification residues in counter-flow principle. It should be noted in this context that the boiling temperatures of the analyzed PAH are in the range of 218 °C (naphthalene) to 536 °C (Indeno(1,2,3-cd)pyrene) [6],[14].

- The aim of the PAH volatilization unit is to reduce the PAH content in the gasification residues. If the gasification residues can be sufficiently heated in the heat exchanger, it is not necessary anymore to supply oxygen into the PAH volatilization unit to burn a part of the residues. In this case, cheating chamber air (and perhaps sealing air as well) would not be needed anymore.

- The volatilized PAH need to be efficiently removed from the gasification residues before condensing again to the surface of the residues. If the sealing air stream is removed, there is a need for another transport method to remove the volatized PAH. It is suggested to vacuum the volatized PAH through a heat-resistant filter installed at the hottest point of the heat exchanger and thus separate them from the hot gasification residues.

Since the use of biochars in agriculture promises beneficial effects for soil amelioration and climate change mitigation (see [15] and [16]), it is recommended to continue the technical development to obtain gasifier residues with low PAH contents which might subsequently be used as soil amendment. The joint composting of gasification residues and organic feedstock sources could help to further reduce any remaining PAH in the gasification residues via biodegradation [14] and to prevent nutrient losses in the composting process. In case further experiments would affirm the viability of the preparation of gasification residues with very low PAH contents which also meet all other applicable environmental standards for soil amendments, a disposal problem could be turned into the valuable resource supply option for the agricultural sector.

5. Conclusion

The results endorse the suitability of PAH analysis method DIN 13877:B for the determination of the PAH content in gasification residues. Related to that, our findings give evidence that DIN 13877:A is not suitable to determine the PAH content in gasification residues. Thus, the regulations of the German Federal Soil Protection Act and the German Federal Soil Protection and Contaminated Sites Ordinance currently cannot prevent the application of gasification residue with high PAH contents to agricultural soils. It is recommended to close this legislative loophole.

A reduction of the Σ EPA16 PAH content in gasification residues by 36% to 82% was achieved by the tested PAH volatilization unit. However, the residual Σ EPA16 PAH content in the gasification residues (between 396 and 1,713 mg/kg DM) is still far too high to allow for an agricultural use of the residues. There are though promising technical redesign options available to further improve the effectiveness of the examined thermal process. In a laboratory scale experiment, the Σ EPA16 PAH content of gasification residues was reduced to values below 2 mg/kg DM following thermal processing at temperatures of 650 °C and more.

Acknowledgements

We thank Matthias Wanderwitz for his support to the sample preparation and Spanner Re² GmbH for financial support to the research work.

Author details

Sebastian Meyer[1*], Bruno Glaser[1], Daniel Fischer[1], Peter Quicker[2], Yves Noel[2] and Georg Kuffer[3]

*Address all correspondence to: seppmeyer@gmx.net

1 Soil Biogeochemistry, Martin-Luther-University Halle-Wittenberg, Germany

2 Unit of Technology of Fuels, RWTH Aachen University, Aachen, Germany

3 Spanner Re² GmbH, Germany

References

[1] Meyer, S.; Glaser, B.; Quicker, P. Technical, Economical, and Climate-Related Aspects of Biochar Production Technologies: A Literature Review. *Environmental Science & Technology* 2011, *45* (22), 9473–9483.

[2] Glaser, B.; Haumaier, L.; Guggenberger, G.; Zech, W. The'Terra Preta'phenomenon: a model for sustainable agriculture in the humid tropics. *Naturwissenschaften* 2001, *88* (1), 37-41.

[3] Kuzyakov, Y.; Subbotina, I.; Chen, H.; Bogomolova, I.; Xu, X. Black carbon decomposition and incorporation into soil microbial biomass estimated by 14C labeling. *Soil Biology and Biochemistry [Online]* 2009, *41* (2), 210-219.

[4] Schimmelpfennig, S.; Glaser, B. One Step Forward toward Characterization: Some Important Material Properties to Distinguish Biochars. *Journal of Environmental Quality* 2012, *41*, 1001–1013.

[5] Bruno Veyrand; Véronique Sirot; Sophie Durand; Charles Pollono; Philippe Marchand; Gaud Dervilly-Pinel; Alexandra Tard; Jean-Charles Leblanc; Bruno Le Bizec. Human dietary exposure to polycyclic aromatic hydrocarbons: Results of the second French Total Diet Study. *Environment International [Online]* 2013, *54* (0), 11–17.

[6] Harmon, T. C.; Burks, G. A.; Aycaguer, A.-C.; Jackson, K. Thermally enhanced vapor extraction for removing PAHs from lampblack-contaminated soils. *Journal of Environmental Engineering [Online]* 2001, *127* (11), 986.

[7] Hilber, I.; Blum, F.; Leifeld, J.; Schmidt, H.-P.; Bucheli, T. D. Quantitative Determination of PAHs in Biochar: A Prerequisite To Ensure Its Quality and Safe Application. *Journal of Agricultural and Food Chemistry* 2012, *60* (12), 3042–3050.

[8] Kabacoff, R. I. R in Action. Data analysis and graphics with R; Manning Publications Co.: Shelter Island, NY 11964, 2011.

[9] Wheeler, B. lmPerm. Permutation tests for linear models. http://cran.r-project.org/web/packages/lmPerm/index.html (accessed August 02, 2013).

[10] Min Wu; Bo Pan; Di Zhang; Di Xiao; Hao Li; Chi Wang; Ping Ning. The sorption of organic contaminants on biochars derived from sediments with high organic carbon content. *Chemosphere [Online]* 2013, *90* (2), 782–788.

[11] Fabbri, D.; Rombola, A. G.; Torri, C.; Spokas, K. A. Determination of polycyclic aromatic hydrocarbons in biochar and biochar amended soil. *Journal of Analytical and Applied Pyrolysis* 2013, *103* (SI), 60–67.

[12] Bundesministeriums der Justiz. Gesetz zum Schutz vor schädlichen Bodenveränderungen und zur Sanierung von Altlasten (Bundes-Bodenschutzgesetz -BBodSchG).

www.gesetze-im-internet.de/bundesrecht/bbodschg/gesamt.pdf (accessed March 15, 2013).

[13] Bundesministeriums der Justiz. Bundes-Bodenschutz- und Altlastenverordnung (BBodSchV). http://www.gesetze-im-internet.de/bundesrecht/bbodschv/gesamt.pdf (accessed March 15, 2013).

[14] A.K. Haritash; C.P. Kaushik. Biodegradation aspects of Polycyclic Aromatic Hydrocarbons (PAHs): A review. *Journal of Hazardous Materials [Online]* 2009, *169* (1–3), 1–15.

[15] Fischer, Daniel, Glaser, Bruno. Synergisms between Compost and Biochar for Sustainable Soil Amelioration. http://www.intechopen.com/books/management-of-organic-waste/synergism-between-biochar-and-compost-for-sustainable-soil-amelioration (accessed April 20, 2012).

[16] Meyer, S.; Bright, R. M.; Fischer, D.; Schulz, H.; Glaser, B. Albedo Impact on the Suitability of Biochar Systems To Mitigate Global Warming. *Environmental Science & Technology* 2012, *46* (22), 12726–12734.

Chapter 13

Electrochemical Detection of Mercury Removal from Polluted Bentonite and Quartz using Different Removing Agents

I. Robles, Luis A. Godínez, J. Manríquez,
F. Rodríguez, A. Rodríguez and E. Bustos

Additional information is available at the end of the chapter

http://dx.doi.org/10.5772/57446

1. Introduction

Heavy metals (also known as the trace metals) are a large group of elements which are industrially and biologically important; in consequence they are defined as the group of elements with an atomic density greater than 6 g cm^{-3}. Some of these heavy metals are toxic to living organisms in high concentrations. Heavy metals of greatest concern in terms of human health, agriculture and ecotoxicology are arsenic (As), cadmium (Cd), mercury (Hg), lead (Pb), tallium (Tl), and uranium (U). However, some other heavy metals, such as cobalt (Co), chrome (Cr), manganese (Mn), and zinc (Zn), are necessary for the healthy growth of plants and / or animals but only in small concentrations. In addition, deficiencies of "essential" trace elements or heavy metals (often called "micronutrients") have limited agricultural productivity in many parts of the world, which include the lack of Zn, Cu and Mn in crops and Co, Mn, Cu and Zn in livestock (Alloway, 1990).

Mercury (Hg) has a high surface tension, meaning it easily forms small, compact, spherical drops. Although mercury droplets are stable compared to those of other metals, they can evaporate at high vapor pressures. High mercury concentration in the environment can be dangerous and increases as elemental mercury evaporates and enters the atmosphere. In addition, high concentrations of indoor mercury may pose an inhalation risk.

Mercury, which is commonly found in the earth's crust as salts such as mercuric sulfide, is released into the environment by volcanic eruptions. In addition to volcanic eruptions, erosion of rocks and soils can also release mercury. Anthropogenic sources of mercury include mercury produced as a by-product of the mining and refinement of others metals like copper, gold, lead and zinc. Mercury can also be released through recycling operations and is sometimes obtained from natural gas or other fossil fuels.

Approximately a third of the mercury circulating in the environment is produced naturally, while the remaining two-thirds are from industrial activity or other human activities. Due to the increase in industrial activity, the quantity of mercury circulating in the environment (atmosphere, soils, lakes, streams and oceans) has increased two to fourfold since the beginning of the industrial era. As a result, mercury levels in our environment are dangerously high.

Most of the mercury in the environment, released from both natural and anthropogenic sources, is elemental mercury (Hg^0), including 99% of total atmospheric mercury. However, biogeochemical transformations of mercury can create Hg^{1+} and Hg^{2+}. Most inorganic mercury compounds (Hg^{1+}) are mildly water soluble and can be found in soils and sediments. In contrast, inorganic Hg^{2+} compounds vary depending on the local chemical atmosphere. Hg^{2+} has a short atmospheric lifetime (minutes) and is quickly removed through deposition due to its high surface activity and water solubility. Soil microorganisms oxidize mercury to its 2+ form (Leopold et al, 2010; Godarzi et al, 2012; Slowey et al, 2005; Mikac, 1999, Shi et al, 2005).

For the determination of metal in low concentrations a number of techniques can be applied (Table 1), in particular colorimetry and atomic absorption spectrometry. Colorimetric determinations do not need expensive instrumentation, but there are some disadvantages such as the need for destruction of the organic material and a separation step to avoid metallic interferences. Atomic absorption is very useful for the sensitive detection and quantitative determination of many metals, especially when using the flameless technique, but the instrumental requirements are rather expensive, another disadvantage of this techniques is that they require pretreatment of the sample differs often from metal to metal so that they do not lend themselves for systematic analyses in which it is not known which metal (s) may, or not be presented (Bodle et al, 1980; Bigham, 1990; Sarzaniniet al, 1994).

Techniques for total mercury determination include hydride generation atomic absorption (Chapple, 1990) and cold vapor atomic absorption (US-EPA Method 7471). For a general determination of mercury species, high-performance liquid chromatography with inductively coupled plasma mass spectrometry can be used (Fodör, 2000, 2005, 2007; Franke and Zeeuw, 1976). In consequence, there is a need for the relatively simple and rapid analytical procedure capable to carry out systematic screening for heavy metals (Franke and Zeeuw, 1976). Some electrochemical techniques can determine all the different mercury species and oxidation states.

Electrochemistry provides analytical techniques characterized by instrumental simplicity, moderate cost and portability. Some as stripping methods use a variety of electrochemical procedures which all share a characteristic initial stage. First, the analyte is deposited on an electrode, usually starting from a stirring solution. After an exact period of time, electrolysis is interrupted, stirring is stopped and the quantity of analyte deposited is measured using voltammetric procedures. During the second stage of the analysis, the analyte is redissolved or liberated from electrode (Woolever et al, 2001).

Technique	Hg / ng mL^{-1}
X – rayFluorescense.	10
Neutron Activation.	2
Gold Film.	0.5
Differential Pulse Voltammetry.	0.04
Cold Vapour Atomic Absorption (CVAAS).	0.01
Cold Vapour Atomic Fluorescense (CVAFS).	0.0001
Inductively Coupled Plasma – Mass Spectrometry (ICP – MS).	0.001
Inductively Coupled Plasma – Atomic Emission Spectrometry (ICP – AES).	50

Table 1. Analytical techniques to determine mercury.

In Anodic Stripping Voltammetry (ASV, Figure 1), the electrode behaves as a cathode during deposition and as an anode during redissolution, where it is oxidized by the analyte again and returns to its original form (Skoog, 1997; Dogan-Topal et al, 2010; Pineda et al, 2009; Anguiano et al, 2012; Bustos, 2012). ASV has been used to detection of trace metals ions in solution at ~10^{-11} M (or sub-ppb) concentrations have been reported (Hubert, 2001). This technique is advantageous to other analytical techniques is its simplicity of use, low cost of instrumentation, and being monodestructive (Wooleveret al, 2001).

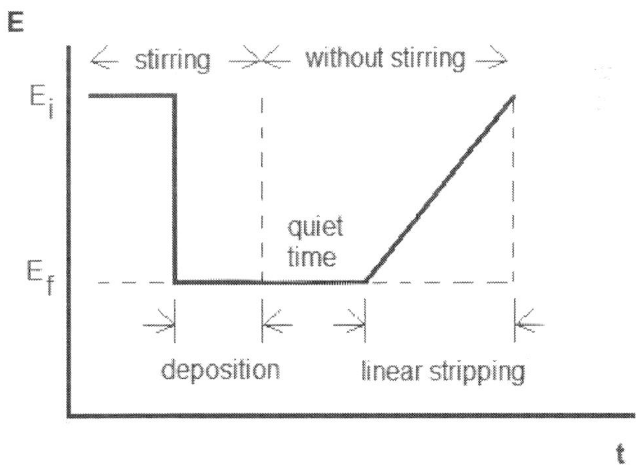

Figure 1. Potential sequence used as perturbation in an ASV experiment.

As shown in Figure 1, there are three sequential steps in a typical ASV experiment: deposition or preconcentration, quiet time and linear stripping. During the deposition step, the electro-active species are reduced up to their zero-valence form under stirred electrolytic solutions. In this sense, E_f must to be a more negative potential than E_i. Later, the stirring is stopped during the quiet time in order to avoid convection transport of ions through the solutions. Finally, the species previously reduced are selectively oxidized during an anodic linear sweep and the current is recorded as function of the potential applied without stirring of the electrolytic solutions (Scholz, 2002).

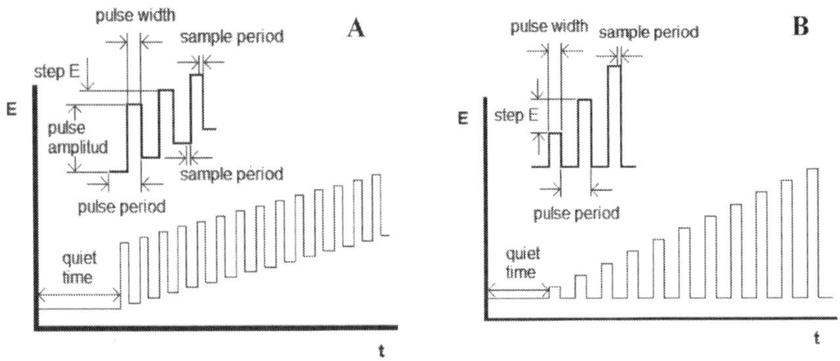

Figure 2. Potential wave form used as perturbation in a DSPV (A) and NPV (B) experiment.

Differential Pulse Stripping Voltammetry (DSPV, Figure 2A) is comparable to Normal Pulse Voltammetry (NPV, Figure 2B) in that the potential is also scanned with a series of pulses. However, it differs from NPV because each potential pulse is fixed, of small amplitude (10 to 100 mV), and is superimposed on a slowly changing base potential. Current is measured at two points for each pulse, first point just before the application of the pulse and second one at the end of the pulse (Ari et al, 1990; Brett and Oliveira, 1998).These sampling points are selected to allow for the decay of the nonfaradaic (charging) current. The difference between current measurements at these points for each pulse is determined and plotted against the base potential (Woolever et al, 2001; Franke and Zeeuw, 1976).

Square Wave Voltammetric (SWV) technique is among the most sensitive means, for the direct evaluation of concentrations; it can be widely used for the trace analysis. The perturbation consists of a square wave having constant amplitude, superimposed at the same time upon a staircase wave form. The current is measured at the end of each forward half-cycle (I_f) and at end of each reverse half-cycle (I_r). The difference between both current values (I_f-I_r) is displayed as a function of the applied potential E as shown in Figure 3. Therefore, the effect of the charging current is notably decreased in SWV since any residual charging current is completely removed (Bard and Rubinstein, 1999; Ari et al, 1990; Scholz, 2002; Zbigniew et al, 1994; Brett and Oliveira, 1998).

SWV has several advantages, has an excellent sensitivity and the rejection of background currents, this speed coupled with computer control and signal averaging allows for experi-

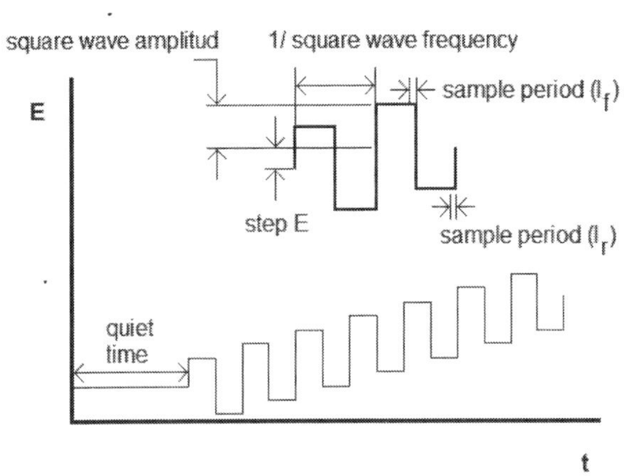

Figure 3. Potential wave form used as perturbation in a SWV experiment.

ments to be performed repetitively and increases the signalto- noise ratio. Applications of SWV include the study of electrode kinetics with regard to preceding, following, or catalytic homogeneous chemical reactions (Franke and Zeeuw, 1976; Dogan – Topal et al, 2010).

Using ASV, the concentration of mercury in liquid samples can be determined. For the determination of mercury in soil, ASV is used after an acid digestion which removes mercury from soil samples (Lamble and Hill, 1998). Some alternative methods for the removal of mercury in soil samples also exist, such as the use of removing agents as called in Table 2, in these studies removing agents were used to remove some heavy metals from different samples (Reddy and Camesselle, 2009; Reddy, 2005; Haheb, 2012; Tandy et al, 2004; Rhazi, 2002; Cox, 1996; Robles et al, 2012). In this paper, we compared the efficiency of eight removing agents showed in Table 2 in the ability to remove mercury from polluted bentonite and quartz comparing DSPV, SWV and ASV.

Removing agent	Concentration	Sample	Removed Metal	η / %
Ethylenediamine tetraacetic acid (EDTA)	0.2 M	Low permeability soil polluted with heavy metals	Hg	20
Diethylenetriamin pentaaceticacid (DTPA)	0.2 M			50
Potassium iodide (KI)	0.2 M			80
Hydroxypropyl-β-cyclodextrine (HPCD)	10%			15
Ethyleneediaminetetraacetic acid + cysteine+ sodium chloride	275 mgL^{-1}, 1.25%,	Rawfish	Hg	90

Removing agent	Concentration	Sample	Removed Metal	η / %
(EDTA +Cys+NaCl)	0.5 % Respectively			
Ethylenediamine tetraacetic acid (EDTA)	0.4 and 4 mM	Polluted soil	Cu, Pb	(Cu)- EDTA (84), NTA (66)
Ethylenediamine disuccinicacid (EDDS)				EDDS (67)
Iminodisuccinicacid (IDSA)				(Pb)- EDTA (94), NTA (65),
Ácidometilglicindi acetic (MGDA)				EDDS (67)
Nitrilotriacetic acid (NTA)				
Chitosan	0.02 mMg^{-1} (Co^{2+} and Ca^{2+})	Polluted water	Co, Ca, Cr,	Cu (53),
	1.2 mMg^{-1} (Cr^{3+})		Cu	Co (11)
	2 mMg^{-1} (Cr^{3+})			
Potassium iodide (KI)	0.1 M	Polluted soil	Hg	99
Potassium iodide (KI)	0.1 M	Polluted soil	Hg	62
Ethylenediamine tetraacetic acid (EDTA)	0.1 M			75

Table 2. Scientific publications about metal extractions facilitated by removing agents.

Table 2 contains references about different removing agents reported to remove different metals. These studies analyzed the removal of metals from different samples, obtaining the highest efficiencies using removing and complexing agents by exchanging charges of cations to remove mercury in the first case, or coordinating cations in the second case (Wypych, 2004; Montuenga, 1979; Přibil, 1982; Buffle, 1990; Spencer et al. 2000; Malone, 1999).

2. Experimental procedure

Stripping voltammetric measurements were carried out using the potensiostat / galvanostat Epsilon modular electrochemical analysis system. Three-electrode electrochemical cell (Figure 4) was used with glassy carbon electrode as working electrode, a platinum wire as auxiliary electrode and an Ag/AgCl as reference electrode (Pineda et al, 2009; Anguiano et al, 2012; Bustos, 2012). Glassy carbon electrode was polished in cloth with an alumina suspension of 1, 0.3 and 0.05 μm. Between each polish it was rinsed with deionized water, and then it was sonicated during 5 min in deionized water to eliminate any residual alumina.

http://dx.doi.org/10.5772/57446

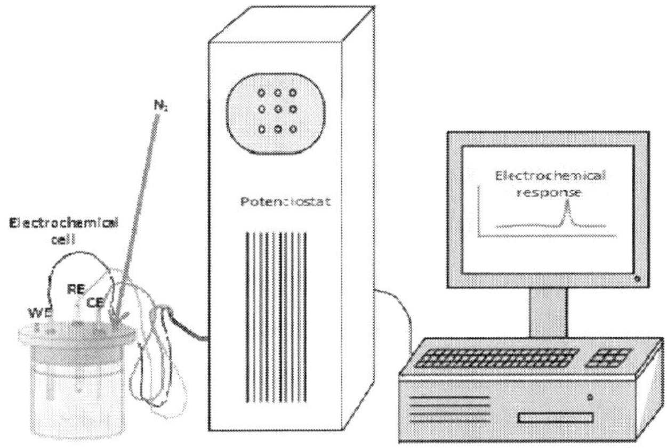

Figure 4. Scheme of a 3-electrode cell used in voltammetry techniques, where gas inlet is used for bubbling electrolyte solution with an inert gas and with controlled temperature.

2.1. Reagents and solutions

Reagents used in this study were: potassium chloride, potassium iodide, potassium hydroxide, sodium hydroxide, etilendiamintetracetic acid, and hydrochloric acid obtained from J. T. Baker; sodium chloride from Sigma Ultra; chitosan from crab shells practical grade, β-ciclodextrin hydrate, L-cysteine were obtained from Aldrich. For mercury pollution were used mercury chloride (II) from Merk, and mercury oxide (II) obtained from Hach. Calcium bentonite from Lodbent Bentonite, and sand white quartz from Sigma Aldrich.

Removing agents tested were 0.1 M potassium iodide (KI), 0.1 M potassium chloride (KCl), 0.1 M potassium hydroxide (KOH), 0.1 M hydrochloric acid (HCl), 0.1 M ethylenediaminetetraacetic acid (EDTA), 10% hydroxypropyl-β-cyclodextrine (HPCD) in deionized water, 0.01 M chitosan and a mixture of 275 mgL^{-1} EDTA, 1.15% cysteine and 0.5% sodium chloride (NaCl). All of the agents, with the exception of chitosan, had previously been tested for the ability to remove metals in soil samples. Control experiments were carried out with water. Solutions were prepared using water type I, according to ASTM-D1193-99. Chitosan were dissolved in acetic acid. Samples of quartz and calcium bentonite were polluted with mercuric chloride ($HgCl_2$) and mercuric oxide (HgO) at concentrations of 10 and 25 mgL^{-1}. Eight removing agents were tested to find the most effective. The percentage of mercury removed was quantified by ASV after extracting the liquid from the bentonite/quartz samples.

2.2. Techniques and procedures

Electrochemical techniques as Anodic Stripping Voltammetry (ASV), Differential Stripping Pulse Voltammetry (DSPV) and Square Wave Stripping Voltammetry (SWSV) were tested to select adequate technique to quantify mercury removal efficiency. After select adequate technique, calibration curves were created for all different removing agents for the addition of both $HgCl_2$ and HgO.

Experimental conditions for ASV were as follows: pre-concentration potential –0.6 V vs. Ag/AgCl, deposition time 6 min, quiet time 30 s, scan rate 20 mV s^{-1}. An increase in signal due to increasing mercury was monitored and recorded along with the increment in current associated with the concentration addition. For SWSV were used an initial potential of -0.2 mV, a deposition potential of -0.6 V for a deposition time of 10 s; a quiet time of 5 s, a SW frequency of 50 Hz, a potential step of 0.005 V. For DSPV were used an initial potential of -0.2 mV, a deposition potential of -0.6 V for a deposition time of 10 s; a quiet time of 5 s, a potential step of 4 mV, a pulse width of 50 ms, a pulse period of 200 ms, pulse amplitude of 50 mV. All experiments were carried out at room temperature (25±1°C) (Anastasiadou et al, 2010). Calibration curves for mercuric quantification were done using electrochemical techniques to select the best.

3. Results and discussion

3.1. Selection of detection technique

To select the technique to quantify mercury was realized calibration curves with both mercury compounds by ASV, DSPV and SWSV. Figure 5 shows the comparison of mercury detection using different electrochemical techniques. Mercury chloride was not showed but resembled same behavior. Table 3 shows comparison of electrochemical techniques of both mercury compounds, where it shows that any technique can be used to quantify mercury by its low detection and quantification limits, but the use of ASV shows the best fit with the lowest DL and QL. In consequence, ASV was selected to quantify mercury in solution, which was removed from polluted bentonite and quartz.

Mercury Compound	Technique	Linear Equation	R^2	m (μA mM^{-1})	DL (pM)	QL (nM)
$HgCl_2$	ASV	2549.30x - 0.0355	0.993	2549.30	112.043	0.373
	DPSV	299.43x - 0.5035	0.978	299.43	422.226	1.407
	SWSV	467.50x - 1.3763	0.981	467.50	479.046	1.597
HgO	ASV	6793.30x - 5.4037	0.998	6793.30	42.046	14.015
	DPSV	1313.70x - 1.0952	0.968	1313.70	96.237	32.079
	SWSV	2347.90x - 1.6533	0.971	2347.90	95.385	31.795

Table 3. Parameters comparison of stripping voltammetry to quantify mercury.R^2 is the correlation coefficient, m represents the slope of linear regression; DL means Detection Limits, QL represents Quantification Limits.

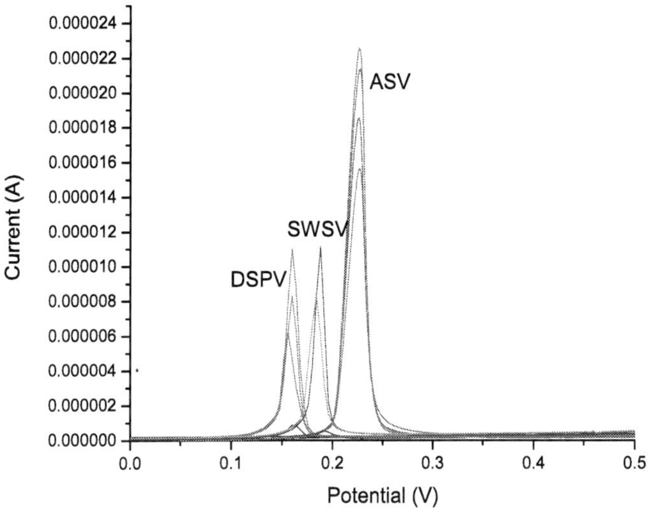

Figure 5. ASV-DSPV-SWSV comparison to select technique to quantify mercury removed.

3.2. Electrochemical responses of removing and complexing agents

After obtaining encouraging calibration results for mercury detection using ASV with sequential addition on removing agents proposed. Electrochemical answer of removing and complexing agents in the presence of mercury compounds was obtained (Figure 6).

Table 4 shows corresponding equations of different removing agents, including fitting (R^2), sensibility (obtained from the slope m), detection and quantification limits (DL and QL respectively) for both mercury compounds. DL and QL represent fundamental performance characteristics of measurement processes, where DL or Limit of Detection (LOD) is defined as an indicator of the minimum detectable analyte net signal, amount or concentration. His term is widely understood and quoted by most chemists as a measure of the inherent detection capability. In general, the LOD is taken as the lowest concentration of an analyte in a sample that can be detected, but not necessarily quantified, under the stated conditions of the test. In another hand, QL or Limit of Quantification (LOQ) is the lowest concentration of an analyte in a sample that can be determined with acceptable precision and accuracy under the stated conditions of test. The general equation to determine detection and quantification limit is:

$$\frac{LOD}{LOQ} = \frac{F(SD)}{b}$$

Where F is a factor of 33 and 10 for LOD and LOQ, respectively. SD represent the standard deviation of the ordinate intercept, or residual standard deviation of the linear regression; and b the slope of the regression line. For a linear calibration curve, it is assumed that the instrument response y is linearly related to the standard concentration x for a limited range of concentra-

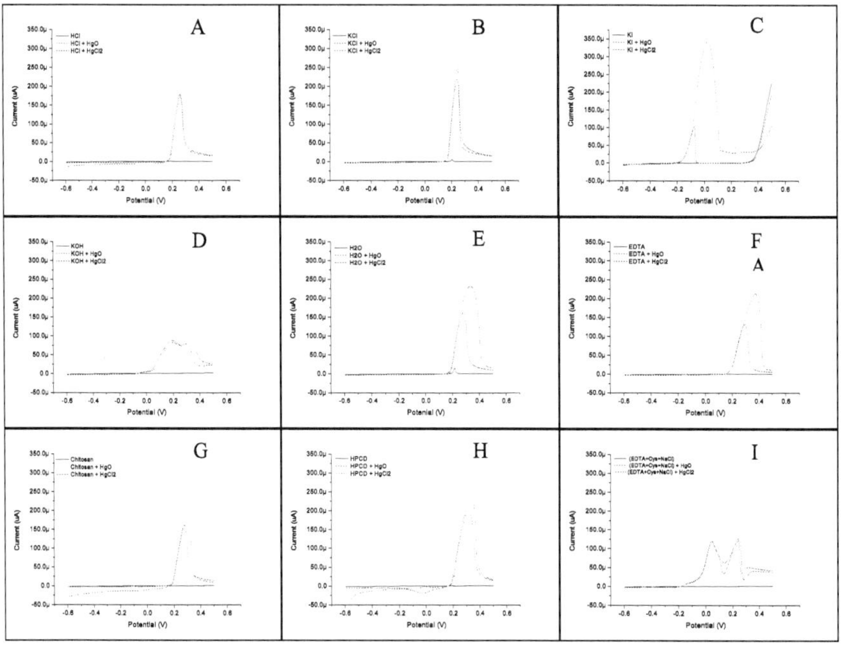

Figure 6. Electrochemical behavior of mercury compounds to removing agents: (A) HCl, (B)KCl, (C)KI, (D) KOH, (E) H_2O, (F) EDTA, (G) chitosan, (H) HPCD and (I) EDTA+Cys+NaCl.White the ASV showed before, calibration curves of each of one mercury compounds added to removing agents was obtained as Figure 7 shows with the ASV response for chitosan.

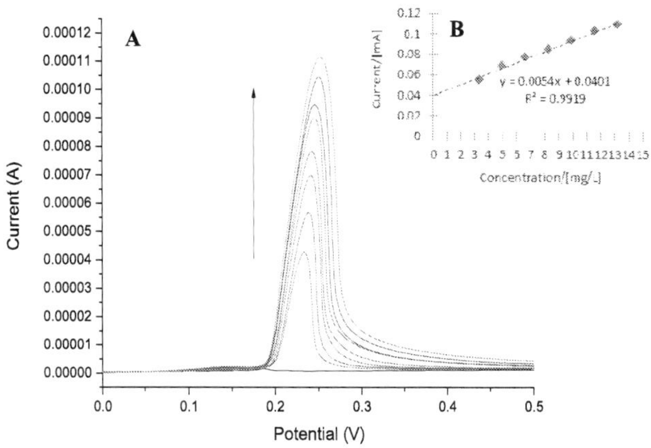

Figure 7. ASV detection of HgO addition to chitosan in 0.1 M HCl using vitreous carbon, platinum wire and Ag|AgCl as work, counter and reference electrode with a scan speed of 70 mV s^{-1} (A), and linear fit of HgO addition to chitosan (B).

http://dx.doi.org/10.5772/57446

Removing Agent	HgO				
	Equation	R^2	m (mA mg^{-1})	DL (μg L^{-1})	QL (μg L^{-1})
HCl	y=0.0077x+0.0980	0.993	0.0077	0.2000	0.6680
KCl	y=0.0093x+0.0421	0.999	0.0093	0.0676	0.2250
KI	y=0.0141x+0.1194	0.982	0.014	0.0292	0.0974
KOH	y=0.0035x+0.0280	0.994	0.0035	0.2750	0.9160
EDTA	y=0.0039x+0.0369	0.988	0.0039	1.2400	4.1400
HPCD	y=0.0069x+0.0730	0.982	0.0069	0.1100	0.3660
Cys+EDTA+NaCl	y=0.0083x+0.0162	0.999	0.0083	0.0454	0.1510
Chitosan	y= 0.0054x+0.0401	0.992	0.0054	0.1480	0.4930
Complexing Agent			$HgCl_2$		
HCl	y=0.009x+0.0576	0.995	0.009	0.1710	0.5710
KCl	y=0.0092x+0.0538	0.993	0.0092	0.0684	0.2280
KI	y=0.0106x+0.1018	0.980	0.0106	0.0386	0.1290
KOH	y=0.0005x+0.0484	0.970	0.0005	1.9200	6.4100
EDTA	y=0.0033x+0.0421	0.975	0.0033	1.4700	4.8900
HPCD	y=0.0056x0.0487	0.995	0.0056	0.1350	0.4510
Cys+EDTA+NaCl	y=0.0163x-0.1353	0.987	0.0163	0.0231	0.0770
Chitosan	y=0.0046x+0.0406	0.986	0.0046	0.1740	0.5790

Table 4. Calibration curves corresponding to each removing agents.

tion. This model is used to compute the sensitivity b and the LOD and LOQ. Therefore, the LOD and LOQ can be expressed as

$$LOD = \frac{3S_a}{b};\ LOD = \frac{10S_a}{b}$$

Where S_a is the standard deviation of the response and b is the slope of the calibration curve. The standard deviation of the response can be estimated by the standard deviation of either y-residuals, or y-intercepts, of regression lines. This method can be applied in all cases, and it is most applicable when the analysis method does not involve background noise. It uses a range of low values close to zero for calibration curve, and with a more homogeneous distribution will result in a more relevant assessment (Currie 1995, 1999; Guidance 2000).

As can be seen in Table 4, we obtained a good QL and DL of mercury compounds in the presence of removing and complexing agents. So ASV is a good technique to determine mercury concentration in presence of removing agent too. In this way, removal efficiencies of mercury were analyzed in the next section.

3.3. Mercury removal efficiencies

Removal percentages were calculated in base of fitting equations for all the different removing agents, they are show in Figures 8 and 9. Due to structural differences in bentonite and quartz (Figure 10), mercury removal behaved differently for the two.

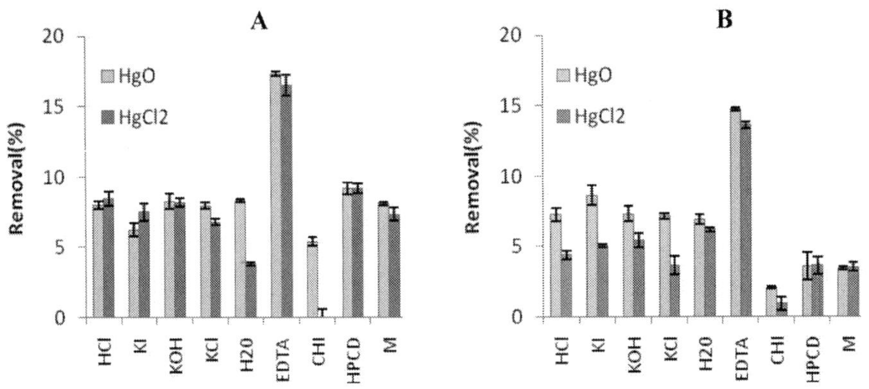

Figure 8. Removal percentages of mercury ($HgCl_2$ and HgO) in bentonite using different removing agents with 10 mg L^{-1} (A) and 25 mg L^{-1} (B) mercury concentration.

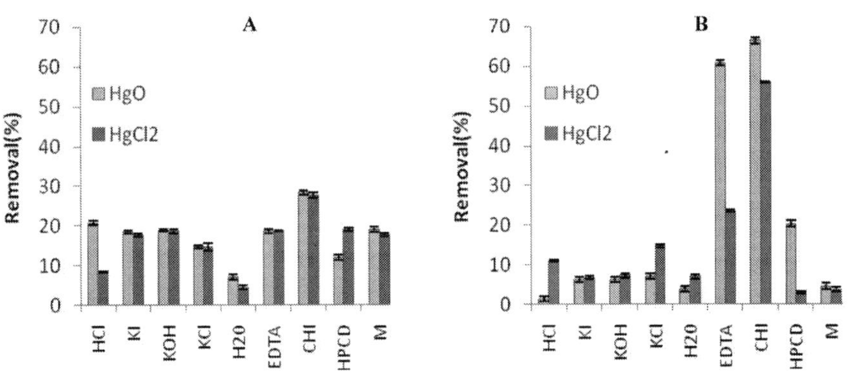

Figure 9. Removal percentages of mercury ($HgCl_2$ and HgO) in quartz using different removing agents with 10 mg L^{-1} (A) and 25 mg L^{-1} (B) mercury concentration.

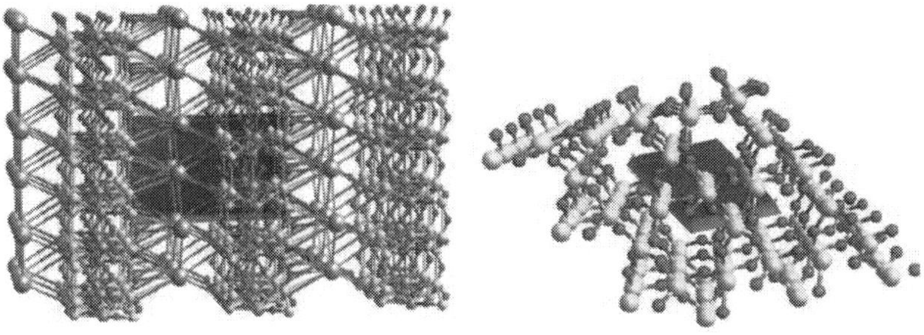

A B

Figure 10. Representation of the structure of bentonite (A) and quartz (B) (Bruker AXS Advanced X-Ray Solutions Software).

Quartz is ordered as strong matrix of oxygen-silicon bonds, which prevent metal ions from penetrating its structure (Figure 10B). On the other hand, metal ions can easily slip into bentonite, which is formed of stacked layers, due to the gaps between the layers (Figure 10A). Because of the strong organized bonds in quartz, mercury is more easily removed from quartz than bentonite. Moreover, the particle size of bentonite is lower than quartz, and therefore has a bigger surface area and as a synthetic colloid, in addition, it has negative charges which attract positive ions. Once bentonite and quartz samples were contaminated with HgO or $HgCl_2$, removing agents were added. After 24 h the sample was collected and analyzed by ASV.

In order of the results obtained, bentonite adsorbed chitosan easily, swelling up dramatically which made it difficult to remove the supernatant. Therefore, the removal of mercury from bentonite using chitosan was inefficient (Figure 8, less than 5 %). This effect was consequence of the affinity of chitosan to the layers of bentonite and it was retained inside of this clay. In contrast, when the chitosan was used to remove Hg^{2+} from quartz (Figure 9), this got a removal close to 30 % when it was 10 mg L^{-1}, and the double of removal was when quartz had 25 mg L^{-1} mercury; this increase was proportional to the concentration of mercury ions over quartz, where chitosan took off the pollutant from particle surface to the solution. Anyway, all the different extracting agents used to remove mercury remove metals by electrostatic forces, forming ionic bonds. Thus, magnitude depends on ionic charge.

On the other hand, complexing agents act differently with metallic ions. Metals tend to lose electrons during chemical reactions, creating metallic ions. The positive charge of these cations attracts negative ions to form complexes held together by covalent bonds. Donating species (ligands) needs to have a lone pair of electrons which can be donated to form a bond. Water, ammonia and halides are common inorganic ligands (Buffle, 1990; Malone, 1999; Montuenga, 1979).

Stronger compounds as EDTA and chitosan tend to remove more mercury contamination present in the bentonite and quartz samples, than 0.1 M KI, 0.1 M KCl, 0.1 M KOH, 0.1 M HCl, 0.1 M EDTA, 10 % HPCD, 0.01 M chitosan and a mixture of 275 mg L^{-1} EDTA, 1.15 % cysteine and 0.5 % NaCl by the different arrangement of mercury ions in these chemical solutions, with less than 10 % removal from bentonite (Figure 8), and less than 20 % from quartz (Figure 9) in both concentrations of Hg^{2+}: 10 and 25 mg L^{-1}.

With ASV has been possible study the thermodynamic and kinetic of adsorption of Hg (II) on Ca-Bentonite, detecting the pollutant in solution to confirm that Ca-Bentonite has a good adsorption capacity of Hg^{2+} adjusting the results to Freundlich isotherm as a mathematical model, with a pseudo second order of reaction. Using the standard potential (E°) obtained after the ASV, we obtained different thermodynamic parameters as equilibrium constant (K_{eq}), free standard energy ($\Delta G°$) and entropy ($\Delta S°$) of Hg^{2+} – Ca-Bentonite. These results indicate us that the process of adsorption is spontaneous, endothermic and irreversible by a possible inclusion and interchange of Hg^{2+} with Ca^{2+} between the Bentonite slides.

In addition, we have used the ASV to evaluate the electrorremediation of mercury polluted soil using complexing agents like EDTA removing up to 75 % of metal contaminants in mercury polluted soil samples by wetting them with 0.1M EDTA, placing them in an experimental cell equipped with Ti electrodes, and then applying a 5 V electric field for 6 hours in a batch reactor; Hg^{2+} was removed around 87 % in a time of 9 hours close to the anode side by the presence this complexing agent (Robles et al, 2012).

4. Conclusions

ASV was selected as an electrochemical technique to quantify the percentage of mercury removed in liquid samples, which were derived from bentonite and quartz samples previously polluted with two mercury compounds (HgO and $HgCl_2$). Eight removing agents were analyzed: 0.1 M KI, 0.1 M KCl, 0.1 M KOH, 0.1 M HCl, 0.1 M EDTA, 10 % HPCD, 0.01 M chitosan and a mixture of 275 mg L^{-1} EDTA, 1.15% cysteine and 0.5 % NaCl, using both extracting (ionic bond complexes) and complexing agents (covalent bond complexes).

The best result of the removal of Hg^{2+} from bentonite and quartz were obtained with complexing agents: EDTA and chitosan. For bentonite, the most effective removing agent was EDTA which removed 17 % of both Hg compounds. In the case of quartz, chitosan was the best removing agent, removing 62 % of HgO and 53 % of $HgCl_2$.

This is the first report of the use of chitosan to remove mercury ion from bentonite and quartz, which is a very good new alternative of complexing agent because it is biodegradable, cheap and easy to obtain (derived from shrimp exoskeleton), with potential application to quantify mercury in polluted soil during some remediation strategy to verify its cleaning without a pre-treatment of sample.

http://dx.doi.org/10.5772/57446

Acknowledgements

The authors would like to thank to Consejo Nacional de Ciencia y Tecnología de los Estados Unidos Mexicanos (CONACyT) with the project CONACYT 193883-UANL, L'Oreal, United Nations Educational, Scientific and Cultural Organization (UNESCO), Academia Mexicana de Ciencias (AMC), and Fundación México – Estados Unidos para la Ciencia (FUMEC) for the funding of this research. I. Robles is grateful to CONACyT for her scholarship.

Author details

I. Robles, Luis A. Godínez, J. Manríquez, F. Rodríguez, A. Rodríguez and E. Bustos*

*Address all correspondence to: ebustos@cideteq.mx

Centro de Investigación y Desarrollo Tecnológico en Electroquímica S. C. Parque Tecnológico Querétaro Sanfandila, Pedro Escobedo, Mexico

References

[1] Alloway, B. J. (1990). Heavy metals in soils, John Wiley and Sons, Inc. New York.

[2] Anastasiadou, Z. D., Jannakoudakis, P. D., Girousi, S. T. (2010). Square wave anodic stripping voltammetry determination of eco-toxic metals in samples of biological and environmental importance. *Central EuropeanJournal of Chemistry*, 8 (5), 999 - 1008.

[3] Anguiano, D. I.,García, M. G., Ruíz, C., Torres, J., Alonso – Lemus, I., Alvarez – Contreras, L., Verde – Gómez, Y., Bustos, E. (2012). Electrochemical Detection of Iron in a Lixiviant Solution of Polluted Soil Using a Modified Glassy Carbon Electrode. *International Journal of Electrochemistry*, 2090 - 3537.

[4] Ari, I., Andrzej, L., Rolf, S. (1990). Contemporary Electroanalytical Chemistry. Plenum Press. New York and London, 419 - 429.

[5] Bard, A. J., Rubinstein, I. (1999). Electroanalytical Chemistry: A Series of Advances. Marcel Dekker, Inc. USA, Vol. 14.

[6] Bigham, M. D. (1990). Field Detection and Implications of Mercury in Natural Gas, SPE Prod. Eng., 120 - 124.

[7] Bodle, W. W., Attari, A., Serauskas, R. (1980). Considerations for Mercury in LNG Operations", Inst. of Gas Tech., 6th. Int. Conference on LNG, Kyoto Japan, Paper 1.

[8] Brett, Ch. M. A., Oliveira, A. M. (1998). Electroanalysis. Oxford University Press Inc. New York.

[9] Buffle, J. (1990). Complexation Reactions in Aquatic systems, an Analytical Approach, Editorial Ellis Horwood, Great Britain.

[10] Bustos, E. (2012). Design, Construction and Characterization of an Electrochemical Sensor of Mercury in Soil.Environmental Influences of Mercury Ore Processing: Case Studies Selected at Slovenian, Mexican, Hungarian Group Meeting in Idrija, Editors: Mateja Gosar, Tatjana Dizdarevič, Miloš Miler, UNESCO, Slovenian National Committee of the International Geoscience Programme, Idrija, Slovenia.

[11] Chapple, G. (1990). The determination of arsenic, selenium and mercury levels in U.S. EPA quality control samples using the GBC HG3000 continuous-flow hydride generator, GBC AA Aplications, N° 17, Australia.

[12] Cox, C. D. (1996).Electrokinetic remediation of mercury contaminated soils using iodide / iodine lixiviant. *Environmental Science and Technology*, 30, 1933 - 1938.

[13] Currie, L. A. (1999). Detection and quantification limits: origins and historical overview. *Analytica Chimica Acta* 391, 127 - 134.

[14] Currie, L. A. (1995).Nomenclature in evaluation of analytical methods including detection and quantification capabilities (IUPAC recommendations). *Pure & Applied Chemistry*, 67 (10), 1699 - 1723.

[15] Fodör, P. (2000). Development of analytical systems for the simultaneous determination of the speciation of arsenic and Selenium, *Analytical Chimica Acta*, 413, 13 – 23.

[16] Fodör, P. (2005). Arsenic speciation in farmed hungarian freshwater fish. *Journal of Agricultural and Food Chemistry*, 53, 9238 - 9243.

[17] Fodör, P. (2007). Selenium speciation studies in Se- enriched chives by HPLC-ICP-MS. *Food Chemistry* 101, 1398 – 1406.

[18] Franke, J. P., Zeeuw, R. A. (1976). Differential Pulse Anodic Stripping Voltammetry as a Rapid Screening Technique for Heavy Metal Intoxications. *Archives of Toxicology*, 37, 47 - 55.

[19] Godarzi, N. M., Shahbazi, B., Grigoryan, K. (2012). The study of mercury pollution distribution around a chlor-alkali petrochemical complex, Bandar Imam, southern Iran. *Environmental Earth Science*, 67, 1485 – 1492.

[20] Guidance for industry, Analytical Procedures and Methods Validation: Chemistry, Manufacturing, and Controls, Federal Register (2000) 65, 776 - 777.

[21] Haheb, P. (2012). Reduction of mercury from mackerel filled using combined solution of cysteine, EDTA, and sodium chloride. *Journal of Agricultural and Food Chemistry*, 60, 6069 - 6076.

[22] Hubert H. G. (2001). Electrochimie Physique etAnalytique. Presses Polytechniqueset Universitaires Romandes, 345.

[23] Lamble, K., J., Hill S. J. (1998). Microwave digestion procedures for environmental matrices. *Analyst*, 123, 103R – 133R.

[24] Leopold K., Foulkes, M., Worsfold, P. (2010).Methods for the determination and speciation of mercury in natural waters- a review. *AnalyticaChimicaActa*, 663, 127 – 138.

[25] Malone, L. J. (1999) Introducción a la química, Segunda Edición, Editorial Limusa, México D. F.

[26] Mikac, N. (1999). Speciation of mercury in sediments of the Seine estuary (France). *Applied Organometalic Chemistry*, 13, 715 – 725.

[27] Montuenga, C. (1979). Formación de complejos en química analítica, editorial Alhambra, Madrid.

[28] Pineda, K. A., Bustos, E., García, M. G. (2009). Determination of Heavy Metals in Samples with Caolite of Guanajuato by Voltammetry with Anodic Redisolution. *ECS Transactions*, 20 (1), 21 – 29.

[29] Přibil R. (1982). Applied complexometry, Volume 5 of Pergamon series in analytical chemistry.

[30] Reddy K. (2005).Electrokinetic remediation of metal-contaminated field soil. *Separation Science and Technology*, 40, 1701-1720.

[31] Reddy, K., Camesselle C. (2009).Electrochemical remediation technologies for polluted soils, sediments and grounwater, Willey, New Jersey.

[32] Rhazi, M. (2002). Influence of the nature of the metal ions on the complexation with chitosan. Aplication to the treatment of liquid waste. *European Polimer Journal*, 38, 1523 - 1530.

[33] Robles, I., García, M. G., Solís, S., Hernández, G., Bandala, Y., Juaristi, E. and Bustos E. (2012). Electroremediation of mercury polluted soil facilitated by complexing agents. *International Journal of Electrochemistry Science*, 7, 2276 – 2287.

[34] Sarzanini, C., Sacchero, G., Aceto, M., Abollino, O., Mestasti, E. (1994). Determination of mercury in naphtha. *Analytical Chimica Acta*, 284, 661 - 667.

[35] Scholz, F. (2002). Electroanalytical Methods: Guide to Experiments and Applications. Springer-Verlag, Berlin, Heidelberg, 191 - 208.

[36] Shi, J., Liang, L., Jiang, G., Jin, X. (2005). The speciation and bioavailability of mercury in sediments of Haihe River, China. *Environment International*, 31 (3), 357 – 365.

[37] Skoog, D. A. (1997). Fundamentos de química analítica, Editorial Reverté, Barcelona.

[38] Slowey, A. J., Rytuba, J. J., Brown, G. E. (2005). Speciation of mercury and mode of transport from placer gold mine tailings. *Environmental Science & Technology*, 39, 1547-1554.

[39] Spencer, J. N., et al (2000). Química, Estructura y Dinámica, Editorial CECSA, México.

[40] Tandy Tandy, S., Bossart, K., Mueller, R., Ritschel, J., Hauser, L., Schulin, R., Nowack, B. (2004). Extraction of heavy metals from soils using biodegradable chelating agents, *Environmental Science and Technoly*, 38, 937 - 944.

[41] US - EPA, Method 7471 (http://www.epa.gov/osw/hazard/testmethods/sw846/pdfs/7471b.pdf).

[42] Woolever, C. A., Starkey, D. E., Dewald, H. D. (2001). Differential pulse anodic stripping voltammetry of barium and lead in gunshot residues. *Forensic Science International*, 117, 185 - 190.

[43] Wypych, F. (2004). Clay Surfaces: Fundamentals and Applications, Elsevier Academic Press.

[44] Zbigniew, G., Chalmers, T. A., Bryce, W. A. J. (1994). Fundamentals of Electrochemical Analysis. 2nd edition. Polish Scientific Publishers, 136 – 140.

Chapter 14

Treatment Methods for Radioactive Wastes and Its Electrochemical Applications

V. Valdovinos, F. Monroy-Guzman and E. Bustos

Additional information is available at the end of the chapter

http://dx.doi.org/10.5772/57445

1. Introduction

There are many uses of radioactive materials which improve or facilitate human activities or quality of life of people. These uses are given in different fields of technology, ranging from power generation to supply entire cities or areas, to medical and industrial uses, even the smoke detectors in buildings. All these applications generate radioactive waste that may represent risks to the environment or to human beings, but it is necessary to have special attention to the management of radioactive waste.

In this chapter there are information about the generalities of radioactive wastes, such as its definition, origin, classification and stages of radioactive waste management. In addition, there are information about the current state of research and technologies which have been proposed for the treatment of radioactive waste, with their advantages and disadvantages, in special case of the electrochemical techniques to treat radioactive waste with theoretical considerations and cases of study. At the end of this chapter, there is information about the risk assessment and development of future strategies.

2. Radioactive waste

2.1. Origin of radioactive waste

Radioactive waste are created from all activities that radioactive materials are used, either as part of the process or the use of such materials as a constituent of equipment or instruments that allow the realization of a practice. Due to large differences in the characteristics of the waste generated in the different areas and to a better understanding of these origins, is convenient to differentiate the activities coming from nuclear fuel cycle of applications coming from medical, research activities and industrial uses (Radiation Safety General Regulations, 1988; Méndez de Vigo, 2000).

2.2. Classification of radioactive wastes

Classification of radioactive waste is in order of any stage from its origin just to their collection, segregation, treatment, conditioning, storage, transportation and final disposal. In this case, in the Figure 1 is the physical classification of radioactive waste (IAEA, 1970 and 1994).

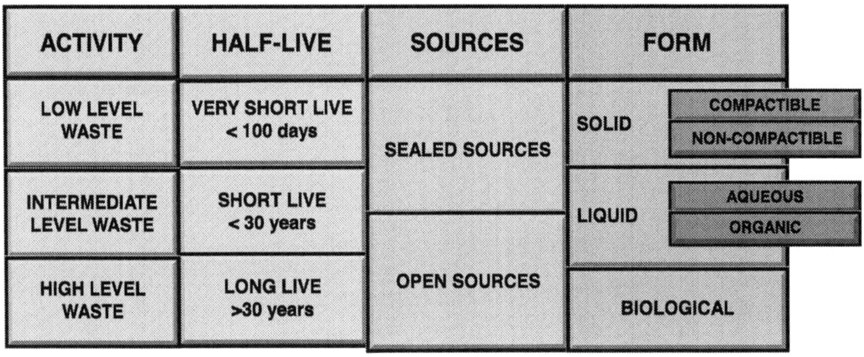

Figure 1. Classification of radioactive waste (own creation).

Quantitative values of allowable activity content for each significant radionuclide will be specified on the basis of safety assessments for individual disposal sites, and the radiological classification of radioactive waste is (IAEA, 2009):

- **Exempt waste (EW):** Waste that meets the criteria for clearance, exemption or exclusion from regulatory control for radiation protection purposes (IAEA, 2004).
- **Very short lived waste (VSLW):** waste that can be stored for decay over a limited period of up to a few years and subsequently cleared from regulatory control according to arrangements approved by the regulatory body, for uncontrolled disposal, use or discharge. This class includes waste containing primary radionuclides with very short half-lives often used for research and medical purposes.
- **Very low level waste (VLLW):** Waste that does not necessary meet the criteria of EW, but that does not need a high level of containment and isolation and, therefore, is suitable for disposal in near surface landfill type facilities with limited regulatory control. Such landfill type facilities may also contain other hazardous waste. Typical waste in this class includes

soil and rubble with low levels of activity concentration. Concentrations of longer lived radionuclides in VLLW are generally very limited.

- **Low level waste (LLW):** Waste that is above clearance levels, but with limited amounts of long lived radionuclides. Such waste requires robust isolation and containment for periods of up to a few hundred years and is suitable for disposal in engineered near surface facilities. This class covers a very broad range of waste. LLW may include short lived radionuclides at higher levels of activity concentration, and also long lived radionuclides, but only at relatively low levels of activity concentration.
- **Intermediate level waste (ILW):** Waste that because of its content, particularly of long lived radionuclides, requires a greater degree of containment and isolation that provided by near surface disposal. However, ILW needs no provision, or only limited provision, for heat dissipation during its storage and disposal. ILW may contain long lived radionuclides, in particular, alpha emitting radionuclides that will not decay to a level of activity concentration acceptable for near surface disposal during the time for which institutional controls can be relied upon. Therefore, waste in this class requires disposal at greater depths, of the order of tens of meters to a few hundred meters.
- **High level waste (HLW):** Waste with levels of activity concentration high enough to generate significant quantities of heat by the radioactive decay process or waste with large amounts of long lived radionuclides that need to be considered in the design of a disposal facility for such waste. Disposal in deep, stable geological formations usually several hundred meters or more below the surface is generally recognized option for disposal of HLW.

2.3. Management of radioactive waste

The ultimate goal of waste management lies in its restraint and seclusion of the human environment, for a period of time and under conditions such that any release of radionuclides does not pose unacceptable radiological risk to people or the environment. Management should ensure that all charges are minimal for future generations.

A responsible management of radioactive waste requires the implementation of measures aimed at protecting human health and the environment. The basic steps for effective management of radioactive waste are part of a global system, ranging from waste generation to final disposal are: minimization of radioactive waste, pretreatment, characterization, treatment, conditioning, transport, storage and disposal (IAEA, 1970; Figure 2).

A management system should to be applied in all steps of radioactive waste management, in order to ensure that activities, facilities, equipment and waste products in meeting the overall safety, health, environmental, security, quality and economic requirements, with safety and environmental protection being of primary importance (IAEA-2008, IAEA 2006a, 2006b).

Management system has to contain a description of the processes and supporting information that explain how work is to be prepared, reviewed, carried out, recorded, assessed and improved. In the design of work processes, a detailed sequence of steps in the activities for

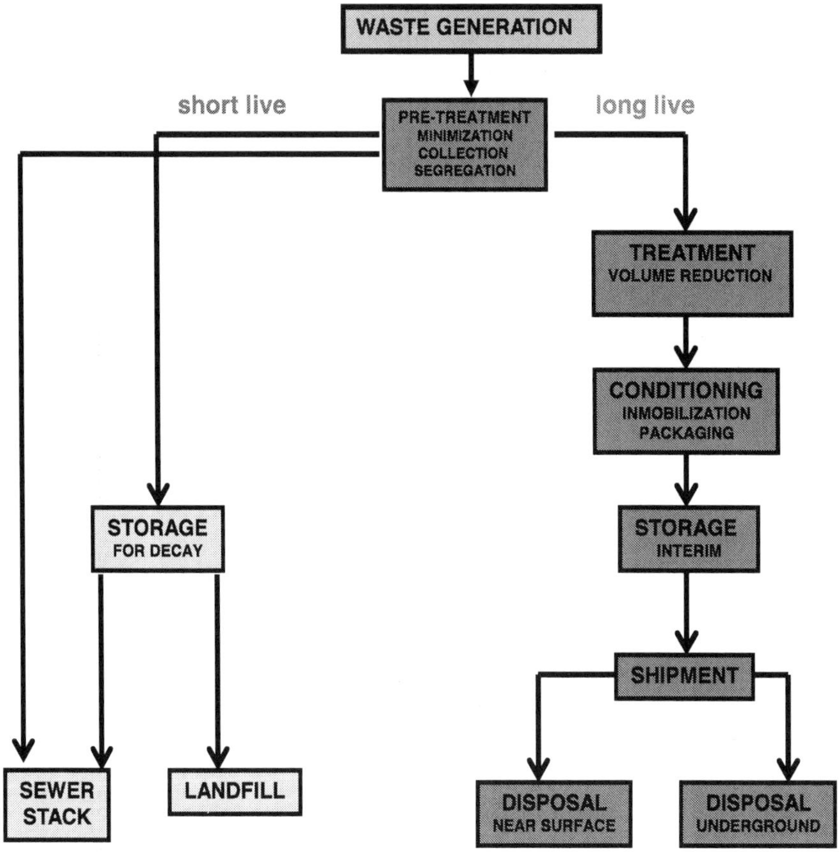

Figure 2. Steps for managing radioactive wastes (IAEA, 2001).

pretreatment, treatment, conditioning and disposal wastes should be considered, such as: (i) characterization of waste at each step in the overall waste management program, (i) analytical methods such as sampling protocols for waste characterization or process control; (iii) monitoring of discharges; (iv) monitoring for clearance purposes; (v) non-destructive examination and testing; (v) heat treatment, (vi) use of special handling tools and techniques, protective clothing or facilities for radiation protection, etc.

Identification of items should be established and documented on the basis of the importance to safety and environmental protection or waste isolation. Records shall specify: (a) the origin of the waste and the processes that generated it; (b) pretreatment of the waste; (c) clearance of the waste; (d) discharge of the waste; (e) characterization of the waste; (f) treatment of the waste; (g) design of the containers and/or packages and of equipment, structures, systems and components for the pretreatment, treatment of the waste.

Additionally, assessments must perform on work processes used in the waste management. Inspections and measurements have to being performed and the associated records maintained. Consequently, controls to activities, facilities, equipment and waste products will have to be designed, considering factors such as: (a) the quantities and potential hazards (radiological and non-radiological, for example chemical) of the waste, and the necessary degree of isolation; (b) the dispersibility and mobility of the waste forms involved and the necessary degree of containment; (c) the interval before disposal; (d) experience with, and maturity of, the technology and the potential for future advances; (e) the reliability of equipment and its function in relation to safety and environmental protection; (f) the complexity and degree of standardization of the activities; (g) the novelty and maturity of the activities; (h) ease of operation, maintenance of equipment and eventual decommissioning of the facility, etc.

Quality assurance requirements should be established for all phases of the waste management process to ensure that each waste is correctly processed and the final waste form has the required properties. This requires quality assurance parameters to be specified. Performance data must be documented for each parameter in order to demonstrate compliance. The quality assurance program should cover the following aspects: (1) waste characterization, (2) waste management process specification, (3) processing conditions, (4) product specification and (5) storage or disposal (HSEEA 2007; IAEA 2008, IAEA 2006).

2.4. Safety guide

Safety is a top priority in radioactive waste management, because of this, the purpose of this section is to present a brief guideline of recommended procedures for working with radioactive wastes. The safety aspects and environmental protection which need to be considered in the handling and processing of radioactive wastes are mainly associated with: flammability, toxicity, explosion and radioactivity.

Some of the most important aspects to consider regarding safety and environmental protection in the radioactive waste management are (IAEA 2008, IAEA 2006a,2006b):

1. The handling, processing and storage radioactive wastes require a permit which is authorized by a local regulatory agency.

2. The wastes must be adequately characterized; chemically, physically, and radiological as a precursor to waste management. The composition of waste should be known with sufficient accuracy that nuclear and conventional safety and environmental protection are not compromised. Toxic or hazardous constituents should be characterized by analytical means or from knowledge of the processes, so that hazards associated with treatment methods of waste can be identified. The reachability of radionuclides, toxic materials and the generation rates for volatile organic compounds or powders and other hazardous gases should be determined. It is important to know the chemical stability of radioactive waste: flammability, corrosively, reactivity, pyrophoricity, rapid oxidation promotion, biodegradability and the chemically incompatible waste forms should be carefully controlled. The amount of mobilizing agents such as chelating compounds, particularly stable ones, should be kept to a minimum. Waste containing hazardous constituents that

are mobile in the environment, or constituents that enhance the mobility of radionuclides should avoid.

3. Several possible process options have to be identified for treating radioactive wastes and before selecting it should include a safety analyses. To prefer processes the cheapest and simplest to procure and operate. The best practical environmental options should be selected, with radiation doses and discharges as low as reasonably achievable (ALARA).

4. Radioactive liquid and gaseous discharges should be as low as reasonably achievable (ALARA) and meet regulatory and environmental restrictions.

5. It is suggested that cheap simple structures will be adequate for installation.

6. Personnel protection should be used in installation: protective clothing, rubber gloves, face mask, visors and personal dosimeters.

7. It is necessary to assess the radiation exposure of the operators on the site and to estimate the frequency and consequences of possible accidents related to the facility and the materials which it handles;

8. The waste processing area must to have suitable fire protection systems, adequate ventilation, and for the waste contains liquids a collection sump.

9. To ensure that radiation exposure to the workforce is as low as reasonably achievable (ALARA).

10. It is recommended having a central waste management facility.

11. Details of the safety methodology will be given in a planned technical document.

3. Treatment of radioactive waste

The aim of the radioactive waste treatment is to minimize the volume of waste requiring management. Treatment process selection for waste depends upon its radiological and physicochemical properties and the quantity (IAEA, 2001; Adenot et al, 2005; Chang, 2001). The objective of this section is to provide a brief overview of usually techniques applied for the treatment of radioactive waste as Figure 3 shows.

3.1. Treatment of aqueous waste

The processes available for treating aqueous radioactive waste are mainly: ion exchange/ sorption, chemical precipitation, evaporation or ultrafiltration/reverse osmosis. However, liquid containing suspended matter must be treated to remove the particulates before primary treatment or after it. Sedimentation, decantation, filtration or centrifugation are treatments used commonly to clear the effluent wastes or to remove miscellaneous debris or insoluble particles (IAEA, 2001; Adenot et al, 2005; Abdel et al, 2011).

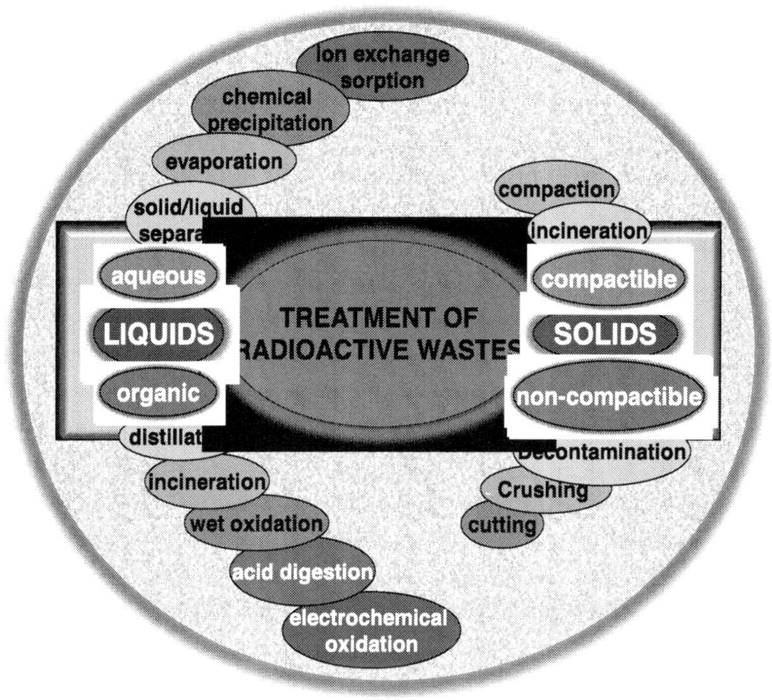

Figure 3. Classification of the treatment process applied to radioactive wastes (authors own creation).

3.1.1. Chemical precipitation

Chemical precipitation processes are regularly used for removing radioactivity from low and intermediate level aqueous wastes at fuel reprocessing facilities, research laboratories and power stations. Precipitation processes are greatly versatile, relatively low investment and operational costs; and may treat from large volumes of liquid effluents containing relatively low concentrations of active species to those containing large amounts of particulates or high concentrations of inactive salts. However, in some cases, a pretreatment stage, such as oxidization of organic contaminants, decomposition of complexed species, pH adjustment, change of the valency state or adjust the ionic species, should be applied prior to the formation of precipitate in order to improve the process. Radionuclides can be removed by precipitation, co-precipitation with a carrier or sorption on to particulates present in the waste (IAEA, 2001; Adenot et al, 2005).

3.1.2. Ion exchange/sorption

Ion exchange methods have extensive applications to remove soluble radionuclides from liquid waste produced in nuclear fuel cycle operations, radioisotope production and research facilities. It is very effective at transferring the radioactive content of a large volume of liquid into a small volume of solid.

Ion exchange process involves the replacement of cations or anions between an insoluble solid matrix containing ionizable polar groups and a liquid solution. When the ionic groups are negatives the exchange will involve cations and when they are positively charged they involve anions. The process is selective, stoichiometric and, as a rule, reversible; therefore ion exchangers can be "regenerated" and radioactive liquid waste recovered with high activity content or if the exchangers become "exhausted" they are removed and treated as radioactive wastes.

A wide range of materials is available for the ion exchange treatment of radioactive liquids: (a) natural ion exchangers (clays, zeolites, cellulose, charcoals, collagen) and (b) synthetic materials such as zeolites, hydrous oxide gels of metals or organic resins formed by highly polymerized cross-linked hydrocarbons containing ionic groups (sulfonic acid, carboxylic acid, amino groups, etc.).

Ion exchange processes can be operated in batch or continuous modes and if the wastes contain high concentrations of salts, suspended solids, organic contaminants or the radionuclide ionic form not suitable, the liquid wastes will have to be pre-treated before exchange process (IAEA, 2001 and 2002; Adenot et al, 2005).

3.1.3. Evaporation

Evaporation process is effective for concentrating or removing salts, heavy metals and a variety of hazardous materials from waste effluent, reducing large volumes of liquid wastes with high factor decontaminations. The process is commonly used for the treatment of high, intermediate and low level waste effluents; in particular for the treatment of small volumes of highly active effluents and may be carried out through the use of commercially available evaporation equipment. However, evaporation has some important limitations: unsuitable for waste effluents containing large concentrations of inactive salts, expensive because its large energy requirement and the presence of some organic compounds can produce explosions during evaporation (IAEA, 2001; Adenot et al, 2005).

3.2. Treatment of radioactive organic liquid

Liquid scintillation, solvents, oils and diverse biological fluids, generated in nuclear research centers, medical centers or industries are considered as radioactive organic liquid wastes. These wastes may present radioactive and chemical or biochemical hazards requiring treatments to remove or destroy chemically or biochemically hazardous components. The objective is to reduce the volume of radioactive waste which requires storage, transport, conditioning and land disposal, eliminating the organic components to enhance compatibility of the treated waste with secondary conditioning processes.

Processes such as incineration, wet oxidation, acid digestion, electrochemical oxidation and distillation, can be applied for treating radioactive organic liquid (IAEA, 1992 and 2001).

3.2.1. *Incineration*

Incineration is used for reduction of solid and liquid radioactive waste volume, downscaling land requirements for disposal. Incineration combusts or oxidizes wastes at high temperatures, generating as end products of the complete incineration: CO_2, H_2O, SO_2, NO, and HCl gases. Emission control equipments for particulates, SO_2, NO_x and products of incomplete oxidation are needed to control emissions of regulated air pollutants.

The disadvantages of radioactive waste treatment with incineration are: off-gas filtering system are required to control radioactive discharges, thickening and dewatering wastes pretreatment may be required, is not economical for small solid waste plants and secondary waste volumes may be large (Chang, 2001; IAEA, 1992, 2001 and 2006; NEA, 1999).

3.2.2. *Wet oxidation*

The organic components of radioactive wastes such as ion exchange resins, foams, cellulosic waste and liquid scintillation can be transformed, degraded, or immobilized using wet oxidation.

Wet oxidation remedial process involves injecting an oxidizing agent, such as hydrogen peroxide, activated sodium persulfate, ozone, Fenton's Reagent (hydrogen peroxide with an iron catalyst) or other oxidant into the subsurface to destroy organic compounds.

The complete mineralization of carbon-based compound wastes by most chemical oxidizers produces carbon dioxide, water, and oxygen as well as minor concentrations of nontoxic ions, salts, and acids. Wet oxidation is thus a process analogous to incineration, with the advantage of using low temperatures (Twissel and Holt, 1996; IAEA, 1992; Chang, 2001).

3.2.3. *Acid digestion*

Acid digestion is an oxidative destruction technology for some liquid organic wastes (hexane, TBP) and organic constituents of mixed waste such as cellulose (paper), polyethylene, latex rubber, Tyvek™, neoprene, polyvinylchloride, polystyrene ion exchange resins, filters, plastics, and/or chlorinated cutting oils organic, that may reduce the waste volume of 20 to 100 times. Acid digestion process uses a mixed of nitric acid in a phosphoric acid carrier solution at temperatures below 200°C and at atmospheric or moderate pressures (< 20 psig). The principal organic portion of the waste is broken down and mineralized by the acid solution producing: inorganic constituents in solution, which can be immobilized easily in a glass or ceramic, and gases (CO_2, CO, O_2 and NO_x) that can be treated in an off-gas scrubbing system, to convert NO_x to reusable nitric acid (Report OST, 1999; IAEA, 1992 and 2001; Adenot et al, 2005).

3.2.4. *Distillation*

Distillation is a radioactive waste volume reduction technique used for pretreating liquid scintillation and miscellaneous solvent waste in conventional equipment. The process is simple, known, and cost effective if the valuable solvent is recycled or reused. The active residue could be either immobilized or destroyed by incineration (IAEA, 2001).

3.3. Treatment of solid waste

Solid wastes are produced by all applications and uses of radioactive materials, in normal operations and maintenance activities. Solid, low and intermediate level wastes are generally segregated into combustible, compactible and non-compactible forms.

Treatments for solid waste are used to reduce the waste volume and/or convert the waste into a form suitable for handling, storage and disposal (IAEA, 2001 and 2002; Chang, 2001; Adenot et al, 2005; NEA, 1999).

3.3.1. Decontamination

Decontamination is defined as the removal of contamination from areas or surfaces of facilities or equipment by washing, heating, chemical or electrochemical action, mechanical cleaning or by other means. The decontamination objectives are mainly: to reduce the volume of equipment and materials requiring storage and disposal in licensed disposal facilities, to remove contamination from components or systems, to reduce dose levels in the installations and to restore sites and facilities to an unconditional-use condition. Decontamination processes may divide into chemical, electrochemical and mechanical processes:

- **Chemical decontamination.** In the chemical decontamination are used concentrated or dilute chemical reagents in contact with the contaminated item, to dissolve the contamination layer, covering the base metal and eventually a part of the base metal.
- **Decontamination by melting** presents the particular advantage of homogenising a number of radionuclides in the ingots and concentrating other radionuclides in the slag and filter dust resulting from the melting process, thus decontaminating the primary material. The problem with inaccessible surfaces or complex geometries is eliminated and the remaining radioactivity content is homogenised over the total mass of the ingot.
- **Mechanical and manual decontamination** included wet or dry abrasive blasting, grinding of surfaces and removal of concrete by spalling or scarifying, washing, swabbing, foaming agents, and latex-peelable coatings. These techniques are most applicable to the decontamination of structural surfaces which may be cleaned by sweeping, wiping, scrubbing or removed by grit blasting, scarifying, drilling and spalling.

A wet abrasive-blasting system uses a combination of water, abrasive media and compressed air, and is normally applied in 24 a self-contained, leaktight, stainless steel enclosure. The dry abrasive-blasting technique, commonly called sandblasting or abrasive jetting, uses abrasive materials suspended in a medium that is projected onto the surface being treated, resulting in a uniform removal of surface contamination. The scarification process removes the top layers of a contaminated surface down to the depth of the sound, uncontaminated surface.

There are two basic disadvantages with the mechanical methods: the surface of the workpiece has to be accessible and many methods may produce airborne dust (IAEA, 2001 and 2002; Chang, 2001; Adenot et al, 2005; NEA, 1999).

3.3.2. Compaction

Compaction is performed in order to reduce the waste volume and concentrates the radionuclides. Plastics, paper, absorbent material, and cloth are compatible in conventional compactors. Metal pipe, valves, conduit, wood, and other like items are compatible in super compactors. Compactors can range from low-force compaction systems (~5 tons or more) through to presses with a compaction force over 1000 tons (super compactors). Volume reduction factors are typically between 3 and 10, depending on the waste material being treated (IAEA, 2001 and 2002).

3.3.3. Cutting

Cutting and sawing operations are carried out mainly on large items which consist usually of metals or plastics. This waste has to be reduced in size to make it fit into packaging containers or to submit it to treatment such as incineration. The cutting is carried out either in the dry state in cells, using remote control when necessary and with conventional tools, or underwater. The cutting may also be done with plasma-jets, laser torches, or explosive fuses.

3.3.4. Crushing

Crushing techniques may be used for size reduction of friable solids (glass, concrete, ceramics). In principle, all types of mill, grinder, and crushing machines of conventional technology can be used.

3.3.5. Shredding

Shredding reduces void space and is particularly effective when plastics are compacted. Air, which is trapped between the folds of bulk plastic and in plastic bags and sleeving, takes up storage space. When the plastic is shredded, better use is made of the waste container space.

3.3.6. Incineration

The size reduction, mixing and blending of the solid wastes is necessary for successful combustion operation.

4. Electrochemical treatment of radioactive waste

4.1. Electroremediation

The electrochemical treatment, electroremediation, also known as electrokinetic remediation (EKR) process is classified as a physicochemical technology by the electrochemical transformation or destruction of organic and inorganic wastes, which offers many advantages such as the capacity to remove organic and inorganic pollutants by applying direct electric current into the soil. The EKR is easy to operate and involves the installation of electrodes into the organic or inorganic waste and the application of a low voltage gradient or direct current through them (Vazquez et al, 2007). This process is capable of mineralizing the organics into carbon dioxide and water completely, without emission of any toxic materials like dioxins. Several metal oxidizing agents like Ag(II)/Ag(I), Ce(IV)/Ce(III), Co(III)/Co(II), etc., have been tested with this process in pilot and commercial scale systems (Adenot et al, 2005; Chang, 2001; Prabhakaran et al, 2009; Farmer et al, 1991).

This technology requires having humid waste into which electrodes are inserted, the electrodes' terminals are connected to a power supply, and an electric current or potential gradient is applied to generate an electrical field. During this process, electrode reactions take place on its surface, generating protons (H^+) and hydroxyl (^-OH) at the anode and the cathode, respectively. The concentration of these ions near the electrodes creates an acid front that moves from anode to cathode and a basic front that moves from cathode to anode. These species interact with the soil to carry pollutants out into the pore solution. Some transport phenomena occur in the liquid phase of soil when direct current or voltage gradient is passed through the electrodes, such as ion migration (electromigration), electroosmosis and electrophoresis (Murillo – Rivera et al, 2009; Alcántara et al, 2008), inducing complex and coupled electrochemical and properties of matrix. In this case, the transport phenomenon depends on the chemistry of the pollutant (Pamuku and White, 1992; Reddy and Chinthamreddy, 2003; Braud et al, 1998).

If the pollutant is inorganic, as metals, these can be removed by electromigration, because the compounds are dissociated in ions. If the ions have negative charge (anions) they will move toward the anode, and if they have positive charge (cations) they will move toward the cathode, an important characteristic which can determine where the metal, in ionic form, can be recovered (Virkutytea et al, 2002; Figure 4). Non-polar pollutants or organic pollutants can be removed by electroosmosis, attributed to the excess charges on the soil surface. There occurs the net ionic migration that represents the bulk movement along pore fluid through the electrical double layer of charge at the solid–liquid interface (Al-Shahrani and Roberts, 2005). And finally, electrophoresis is the movement of charged solid particles, including clay particles and bacterial cells with size less than 20 m, in response to the electrostatic potential gradient. As in electromigration, positively charged particles migrate towards the cathode, and negatively charged particles migrate towards the anode, or they both simply move by a mechanical transport (Hamed et al, 1991; Méndez et al, 2012).

On other hand, many different soil remediation technologies are available to clean polluted waste, and most of them remove pollutants from the soil–water complex for further treatment or disposal in a more concentrated form. However, when the pollutants are persistent, toxic or simply have low solubility and a strong adsorption to soil surfaces and organic matter, the traditional remediation technologies are used, such as washing, and land-farming, amongst others. In these cases, electroremediation, also known as electrokinetic remediation (EKR) process is classified as a physicochemical technology, which offers many advantages such as the capacity to remove organic and inorganic pollutants by applying direct electric current into the soil, even in clays (Virkutytea et al, 2002). The EKR is easy to operate and involves the installation of electrodes into the soil or waste and the application of a low voltage gradient or direct current through them. This technology has the capacity to removed heavy metals (Pb°, Hg°, Cd°, Ni°, Cu°, Zn°, Cr°); toxic anions (NO_3^-, SO_4^{2-}); mixtures of organic and ionic pollutants; cyanide; explosive compounds; hydrocarbons (gasoline, diesel, oil, BTEX: benzene, toluene, ethyl benzene and xylene); polychlorinated biphenyls (PCBs) and radioactive species (^{137}Cs, ^{90}Sr, ^{60}Co, ^{238}Ur); from both saturated and unsaturated soils, sludges and sediments (Virkutytea et al, 2002; Hamed et al, 1991).

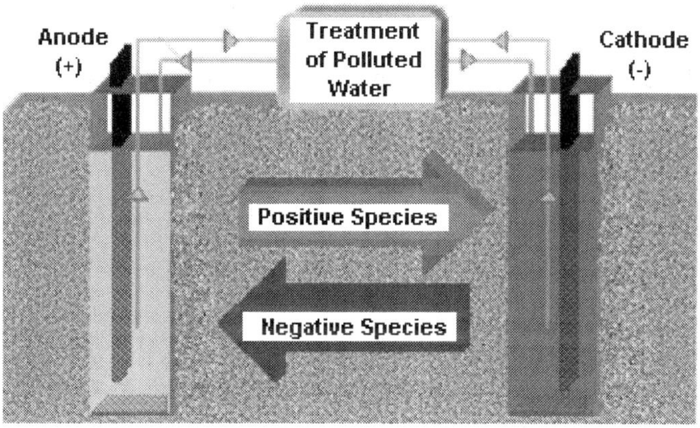

Figure 4. Electromigration of ions in an electrochemical arrangement to remove pollutant from the matrix (authors own creation).

Some researchers have published (Pamuku et al, 1990 and 1992; Reddy et al, 2003; Hamed and Bhadra, 1997; Reed et al, 1999; Khan and Alam, 1994) that the EKR application was better in clayed soils that had low concentration of heavy metal than even expensive regular mining procedures. However, a disadvantage was the time required to achieve over 90 % metal removal.

In order to improve the EKR process and diminish the removal time, some efforts have been focused on changing some operational parameters. For example, some researcher have modified the pH and the current density, introducing chemical compounds on electrolyte chambers (Yeung et al, 1996; Cox et al, 1996), or have added complexing and chelating agents to desorb the pollutant from soils (Cundy and Hopkinson, 2005; Ruiz et al, 2011; Colleta et al, 1997), or have modified the system, placing reactive barriers into the soil matrix (Ribeiro et al, 2005; Gómez et al, 2009) or simply they have tested different electrode material.

Thus, the efficiency of pollutant removal using EKR depends of many factors. However, it is important to select a good electrode material, especially when electrochemical technologies are used. Their selection can be according the thermodynamic and kinetic reactions carried out on their surfaces or by their mechanical, thermal and corrosion resistance.

4.2. Electrodes during the electrokinetic remediation of wastes

In that sense, some materials, as the case of titanium (Vázquez et al, 2004), platinum, gold, silver, stainless steel, among others used in EKR, suffer a kind of passivation, generating an oxide film on their surfaces which cover the active sites. This behavior occurs during the experimental conditions, which increases the electrical resistances in the system. For that reason, it is necessary to pre-treat or pre-activate the material before using, to increase roughness or surface active sites. Also, carbon electrodes have been used in EKR processes because of their low cost and accessibility (Saichek and Reddy, 2003; Hu et al, 2002) and because they are inert. However, these kinds of materials commonly form bonds with the species in solution or form oxide film. As well, they can adsorb some species on their surface.

In order to increase the active sites, eliminate the passivation phenomenon, increase electrode life and improve the oxidant activity, it is necessary to modify electrode surfaces to obtain high overpotentials. Consequently, some electrode materials have been modified with metallic oxide, forming a thin layer on a base metal (usually titanium), i.e. $Ti|SnO_2–Sb_2O_3$, $Ti|IrO_2–Ta_2O_5$, $Ti|IrO_2$, $Ti|RuO_2$, and carbon$|TiO_2$, among others (Hu et al, 2002; Méndez et al, 2012). These kinds of electrodes can be used as anodes in order to promote electrochemical oxidation. For that reason, the name "Dimensionally Stable Anodes" (DSA) was proposed by Comninellis and Pulgarin (Comninellis and Pulgarin, 1991) who demonstrated the high reactive surface of iridium DSA. These are able to degrade organic compounds by the presence of hydroxyl radicals on their surface, while being mechanically resistant to this reactive species and to pH changes (Comninellis, 1994).

4.3. Configuration of electrodes during an electrokinetic remediation of wastes

Furthermore, most conventional *in situ* methods have trouble managing the remediation of fine-grained soils with one or more of the following conditions: high water content, high organic matter content, high metal concentration; whereas electroremediation is suitable for wastes (Saichek and Reddy, 2005). Although, there are some aspects of electroremediation that require attention before the technology can be successfully implemented in the field, such as: soil characterization, type and concentration of pollutant and electrode material, as well as electrode array configuration and spacing.

In the last three decades, different research groups have been working in the electrokinetic treatment at different levels: laboratory, pilot system and directly in the site pursuing to remove organic pollutants (Gilbon, 2001), inorganic pollutants and radioactive wastes, among others (Hamed et al, 1991). Also, there are reports about the successful implementation of electrokinetic treatment *in situ*; one example is reported by Monsanto, DuPont and General Electric, who used the Lasagna™ remediation treatment *in situ* to remove trichloroethylene with 98 % of efficiency (USEPA, 1997). Another practical example was developed by Sandia National Laboratories, who removed chromium *in situ* using Iridium/Titanium electrodes protected with a porous ceramic coating (Fernández, 2012).

Previous research has been focused on maximize electrokinetic extraction efficiency. Most laboratory or pilot electroremediation studies have been carried out in one dimension (1D) array, having only one anode (+) and one cathode (-), separated by the polluted soil. For field applications, the system must be effective and efficient; therefore it is required an appropriate electrode configuration, in order to accomplish this goal EKR treatment should be implemented in a bidimensional system (2D), this means that electrode array must be set up in respect to the *x-y* axis; however, there are few reports evaluating the performance of electrode configurations in the *x-y* plane, and there is no experimental data about performance of electrokinetic treatment with different electrode arrays (Méndez et al, 2012; Pérez – Corona et al, 2013; Reddy et al, 2001).

In addition, several flushing solutions have been investigated, such as water, acids, bases, chelating agents, alcohol, and other additives (Chaiyaraksa and Sriwiriyanuphap, 2004). In practice, acid washing and chelator soil washing are the two most prevalent removal methods (Giannis et al, 2007; Rampley and Ogden, 1988). Also, most of the electrokinetic equipment has been manufactured as a horizontal type. Laboratory-scale electrokinetic remediation has performed for TRIGA soil during the past 5 years, but in this time, a study related to pilot-scale electrokinetic remediation was performed (Kim et al, 2002 and 2003).

4.4. Destruction of radioactive organic wastes

The process developed for the removal of organic contaminants from bulk water using graphite based adsorbents with electrochemical regeneration at the University of Manchester (Brown and Roberts, 2007), was adjusted for the destruction of radioactive organic wastes, specifically oils contaminated with alpha radioactivity produced at Magnox Ltd nuclear decommissioning site in UK (Wickenden, 2001). This approach comprises four stages:

1. *Emulsification:* the oils contaminated are emulsified in water using CLAX 200S as organic emulsifying agent to give a stable emulsion.

2. *Adsorption: the* emulsion is vigorous mixing with the graphitic adsorbent (NyexTM 1000 supplied by Arvia Technology Ltd) by fluidizing the adsorbent using air sparging. A quick adsorption is produce by the non-porous nature of the NyexTM.

3. *Sedimentation:* when the fluidising air is turn off, NyexTM particles precipitate to form a bed in the anode compartment of the electrochemical cell.

4. *Electrochemical destruction:* two electrodes are placed either side of the bed and a direct electric current is passed through the bed which destroys the pollutant through anodic oxidation of the organic matter. Oxidation of the organic matter may produce soluble breakdown products or off gases (CO, CO_2) and small amounts of H_2 and Cl_2 at the electrodes. After electrochemical treatment, the regenerated adsorbent is ready for immediate reuse and the whole cycle is repeated (Brown et al, 2013).

The treatment of radioactive oils by adsorption and electrochemical regeneration systems has been achieved at pilot scale 200 L. The latter consisted of three bipolar stacks of six electrochemical cells of each with an electrode area of 2 500 cm^2. Graphite plate bipolar electrodes were used and a micro-porous polyethylene membrane (Daramic, Grace GMBH) separated the adsorbent bed from the cathode. The catholyte solution, 0.3 wt % NaCl solution acidified to a pH of less than 2, was stored in a small tank and pumped through the cathode compartments of the six cells. The distribution of the radioactivity after the process was examined in the samples of the adsorbent and electrolyte after each regeneration cycle or after each destruction cycle for 1.5 L of contaminated oil using high-resolution gamma spectrometry for ^{241}Am, ^{154}Eu, ^{155}Eu, ^{137}Cs and ^{60}Co, liquid beta scintillation counting for ^{3}H,^{14}C, ^{55}Fe, ^{63}Ni, ^{90}Sr, ^{241}Pu and ^{36}Cl, and alpha spectrometry for plutonium isotopes ($^{239/240}Pu$, Pu) and ^{241}Am (Brown et al, 2013).

The process was achieved with a specific treatment rate of 0.63 $\mu L\ cm^{-2}\ h^{-1}$ using a current density of 10 mA cm^{-2}. Regeneration was carried out a current of 1 A (20 mA cm^{-2}) for 25 h L $^{-1}$ of oil with a regeneration energy of 48 kWh L^{-1} of oil. And the oil loading on the adsorbent was less than 25 wt % on the adsorbent to avoid excessive cell voltage.

The process of adsorption coupled with electrochemical regeneration can remove and destroy around 95 % radioactive oils in the first cycle, and over 99 % of the emulsified oil. Around 80 – 90 % of the majority of the radioactive species are transferred to the aqueous phase with a negligible discharges of tritium gaseous; hence no off-gas treatment before direct discharge to atmosphere is necessary (Brown et al, 2013).

Electrochemical oxidation using a boron-doped diamond (BDD) anode coupled with sonication is a promising method to oxidise (Low Level Waste) LLW or Intermediate Level Wastes (ILW) oils to carbon dioxide and water. The boron-doped diamond (BDD) electrodes contain non-aggressive and non-corrosive chemicals, are ease of disposal of the spent electrolyte and allow simple electrochemical cell configuration. Electro-oxidation tests were performed into an electrochemical cell which comprised a 250 mL beaker fitted with a rubber bung that held a BDD (DIAFILM PE TM) anode and a stainless steel cathode applied 0.1 A, a cell potential of 5 – 15 V and sonication. The electrolyte contained oil, sodium sulphate and sodium hydroxide added at the start of the test to maintain an alkaline pH in order to trap the carbon dioxide as carbonate. The method was effective for unused hydraulic oil, vacuum pump oil and a waste used machine tool oil (Taylor et al, 2009).

Mediated Electrochemical Oxidation (MEO) process has been used by destroying the organic components of combustible mixed wastes and for dissolving radioactive materials, such as transuranic oxides (PuO_2). The radioactive components of the wastes dissolved in the electrolyte, can be recovered or immobilized for disposal (Chiba et al, 1995). The destroying organic process is accomplished via a mediator (oxidizing agent).

Several metal oxidizing agents like Ag(II)/Ag(I), Ce(IV)/Ce(III), Co(III)/Co(II),Fe(IV)/Fe(III) etc., in nitric or sulfuric acids have been tested (Chiba et al, 1995; Farmer et al, 1995). In particular, the Ag(II)/Ag(I) based MEO system is capable of destroying cutting oil, cellulose (paper and cloth), rubber (latex), plastics (Tyvek, polyethylene and polyvinyl chloride), biomass (bacteria) and ion exchange resins, attaining high destruction efficiencies at reasonable coulombic efficiencies (Chiba et al, 1995).

Ag(II) in a nitric acid solution produces highly reactive OH and NO_3 radicals which attack organic compounds converting most of them to CO_2, water, and inorganic ions. The resulting Ag(I) is recycled to Ag(II) at the anode of an electrochemical cell to maintain a supply of oxidant and minimize consumption of Ag. The Ag(II) migrates back into the bulk electrolyte to continue the oxidation process. A microporous membrane is usually placed between the electrodes to prevent the oxidizer produced at the anode from being reduced at the cathode. Ag(II) is a very effective oxidizing agent for the destruction of nonhalogenates organic compounds. Unfortunately, halide ions liberated during the destruction of halogenated organics react with Ag(II) to form insoluble precipitates. Therefore systems tolerant to halide anions such as Ce(IV)/Ce(III), Co(III)/Co(II) and Fe(IV)/Fe(III) have been tested. Tests at Pacific Northwest Laboratory with the Cerium/HNO_3 system has shown that cerium provides the same or better oxidation rates for cutting oils that silver system (Schwinkendorf and Hart, 1995).

4.5. Examples of electrochemical treatments of radioactive waste

4.5.1. ^{137}Cs and ^{134}Cs removal from radioactive ash

Combustible waste contaminated with ^{137}Cs and ^{134}Cs, generated by the tsunami within Fukushima, was incinerated, producing a large quantity of radioactive ash. A washing–electrokinetic decontamination method was developed by Kim and collaborators to decontaminate these radioactive ashes (Kim et al, 2002 y 2003). The process consists in a washing of contaminated ash with HNO_3 in three cycles using pH of 0 and ash (1g) – nitric acid (1 mL) mixture; the waste solution generated is precipitated with CaO and reused only those produced in the third washing. If the residual radioactivity of the washed ash is higher than the clearance concentration level, the washed ash is treated by electrokinetic equipment for decontamination. The removal efficiency of ^{137}Cs from radioactive ash is inversely proportional to an increase in the mixing ratio of ash weight (g) / solution volume (mL) and directly proportional to the increase in the concentration of HNO_3.

A pilot-sized washing–electrokinetic equipment was manufactured to remove cesium from lots of radioactive ash. The equipment consisted of 200 L washing equipment, 50 L electrokinetic equipment, and 150 L precipitation equipment. The electrokinetic equipment consisted of a couple of anode rooms, electrokinetic ash cells, cathode rooms and metal oxide separators. Cesium from radioactive ashes moves to the cathode room through electro-migration and electro-osmosis. 87 – 89 % of ^{134}Cs and ^{137}Cs in the radioactive ashes were removed in three washings and around 98 – 100 % from the waste solution after precipitation with CaO. If the radioactive ashes contain more than 30 – 40 Bq g^{-1} of ^{134}Cs and ^{137}Cs the electrokinetic process requires between 8 – 10 days after three washings to reach below 2.0 Bq g^{-1} (Kim et al, 2013).

4.5.2. Sodium recovery from alkaline nuclear waste

An electrochemical salt-splitting process has been developed to recover and recycle NaOH from radioactive wastes containing large amounts of sodium salts. Sodium separation process can save costs by reducing the disposal volume of wastes and by producing NaOH for recycle into waste treatment processes such as sludge leaching, regenerating ion exchange resins, inhibiting corrosion in carbon-steel tanks, or retrieving tank wastes (Fontain et al, 2009). The process is based on a two-compartment cell separated by a cation selective membrane. The waste enters into the anolyte electrodes, sodium ions migrate across membrane into the catholyte, under the influence of an applied electrical potential. Hydroxides form at the cathode by the reduction of water producing thus a sodium hydroxide solution (Kurath et al, 1997; Hobbs, 1999; Ambashta and Sillanpää, 2012; Fountain et al, 2009).

Platinized titanium (Pt/Ti) anodes and cathodes, and a Ceramatec® NASD membrane were tested by Hobb, applying a current density of 300 A m^{-2} and a voltage of 5 - 5.6 V during a run time of 70 - 104 h. This approach allows retain anionic species such as nitrate, aluminate or sulfate in the compartment anolyte of the electrochemical cell, and can produce caustic from radioactive wastes with low levels of gamma radioactivity, which could be released for off-site use without further treatment (Hobbs, 1999).

Pacific Northwest National Laboratory (PNNL) and Ceramatec Inc. tested a NaSelect® (sodium super ionic conductor) ceramic membrane to separate sodium from Low Activity Waste (LAW) streams at Hanford site (Fountain et al, 2009; Pendleton et al, 2011). A pilot scale system was designed, built and operated to process 7 L h^{-1} of LAW. The process selectively removed up to 80 % of sodium hydroxide from LAW and produced up to 50 % concentrated caustic for reuse in removal aluminum during sludge washing as a pretreatment step in the vitrification of radioactive waste; reducing about 39 % the waste volume. NaSelect® ceramic membrane prevents migration of mono valent and multi valent elements and other radionuclides to the sodium hydroxide stream (Pendletonet al, 2011; Figure 5).

Electrolytic decontamination is accomplished by applying a low dc voltage through an electrolyte to induce a chemical reaction. Contamination is removed at the anode the working electrode and goes into solution. The cathode (counter electrode) can be constructed from stainless steel. Electrolyte solutions such as phosphoric or sulfuric acid, used normally in electropolished, have to be changed after a few articles decontaminated, resulting in large quantities of radioactive toxic waste that must be treated and/or disposed. Due to this, electrolytes with a very low toxicity from which the radioactive materials can be easily separated and recycled, are investigated and applied. Sodium nitrate has been chosen as electrolyte to decontaminate metals contained Pu and Am (Wedman et al, 1996). In this medium, both actinides can be precipitated or entrained in the ferric hydroxide formed as surface metal is removed, resulting a clean surface, free of contamination, and the separation of the radioactive waste from the solution. Thus, the electrolyte solution can be recycled, greatly reducing waste.

For this process, the appropriate current densities appear to be in the range of 0.1 to 0.2 A cm^{-2}. Higher current densities result in higher metal removal rates, but adversely affect the

http://dx.doi.org/10.5772/57445

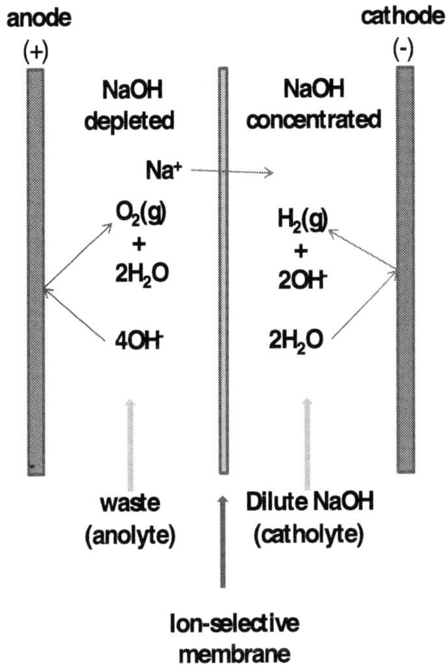

Figure 5. Electrochemical cell for the recovery of sodium hydroxide from alkaline salt solutions (Hobbs, 1999).

surface morphology by causing roughening, pitting, or burning. Optimum situation for the electrolytic decontamination process is the treatment of metal surfaces that have been electro-polished before contamination (Wedman et al, 1996).

NaOH solutions have been used as electrolytes to decontaminate metal surface contaminated by tritium. In this approach, the metal to be decontaminated is submerse in the electrolyte, connected to the negative pole, and the anode (polytetrafluoroethylene wax-impregnated graphite), to positive pole; applying a current densities in the range of 10 - 50 mA cm^{-2}. The tritium adsorbed on the cathode surface is replaced by the hydrogen and ejected to the electrolytic solution. This process is proper to small metallic parts of complex geometry or large pieces with simple geometry; and does not modify the surface metal, thus the objet can be reused (Bellanger, 1992).

Other decontamination process for contaminated stainless steel equipment is accomplished by dissolution of the metal surface that is being decontaminated when it is in contact with a solution composed by an inorganic reduction-oxidation systems such as Ce(IV)/Ce(III), Cr(VI)-Cr(III, Ag(II)-Ag(l) or Co(III)-Co(II) in nitric acid. This solution is subsequently reoxidized by electrolysis (Lerch et al, 1980).

4.5.3. *Remotion of ^{137}Cs from soil*

The thick-target PIXE analysis was successfully applied at least to the analysis of the electrokinetic behavior of cesium mixed with the soil as CsBr. By using cesium as a test pollutant, we could clearly observe the effect of the electrolysis without taking into account complicated interactions between the contaminant and the soil, such as dissolution of soil particles and adsorption phenomena.

The water supply to the anode well was effective to enhance the removal rate. From the observed fast migration of the pore water, as well as from the result of a simple calculation on the electrophoretic flow velocity, it was concluded that the migration of cesium observed was due mainly to the electroosmotic flow. Remediation performance by increasing the electric conductivity of the soil by mixing NaCl was possible. This result was consistent with the reduction of the electroosmotic flow velocity due to the elevated ionic strength. It was found that the addition of NaCl makes no sense also from the viewpoint of the potential hazard due to toxic gas emission as well as the cost of electricity. Owing to the interference by major metallic elements in the soil, the detection sensitivity of the present method based on the simple LX-ray measurement was not enough to investigate behavior of the trace level Cs contaminants. In order to attenuate only Ti-Kα but to allow high transmission of Cs-Lα radiation, a thin foil of a single element having an absorption edge at an appropriate energy.

4.5.4. *Remotion of ^{60}Co and ^{137}Cs from soil*

The pilot-scale electrokinetic remediation equipment suitable for the geological characteristics of a South Korean nuclear facility site was developed for the remediation of radioactive soil. The optimal experimental conditions were obtained with 50 L electrokinetic remediation equipment and the results are as follows: the removal efficiencies Co^{2+} and Cs^{+} from the artificially contaminated soil after 15 days were 98.4 % and 94.9 % respectively, and the generated effluent volume was 3.4 mL g^{-1}. The removal efficiencies of ^{60}Co and ^{137}Cs by nitric acid were increased by 3.1 % and 2.0 % more than those by acetic acid. The removal efficiencies of ^{60}Co and ^{137}Cs from the soil of high concentrations (1 320 Bq kg^{-1} and 1 720 Bq kg^{-1}) were increased by 14.4 % and 3.8 % more than those from the soil of low concentrations (110 Bq kg^{-1} and 120 Bq kg^{-1}).

The removal efficiency of ^{137}Cs from the soil of average 1.4 mm particle size was increased by about 2.3 % more than that from the soil of an average 0.7 mm particle size. The removal efficiencies of ^{60}Co and ^{137}Cs by the application of an electric current of 15 mA cm^{-2} were increased by 1.4 % and 4.4 % more than those by the application of 10 mA cm^{-2}. A total removal efficiency of ^{60}Co and ^{137}Cs from the radioactive soil of about 2 000 Bq kg^{-1} was 95.8 % during electrokinetic remediation by the application of an electric current of 15 mA cm^{-2} for 55 days (Gye – Nam et al, 2013).

On other hand, a soil washing system with a hydro-cyclone was very effective for decontaminating the radionuclides in the TRIGA (Training, Research, Isotope, General Atomic) soil in Korea. The size of the contaminated TRIGA soil ranged from very fine to more than 1.0 mm. The volume of soil particles larger than 1.0 mm was 28 - 52 %, and the volume of soil particles smaller than 0.06 mm was less than 4 - 10.5 %, but the volume of medium-sized soil particle was 43 - 61 %. The radioactive concentration was strongest in the soil particle smaller than 0.063 mm, as predicted. Oxalic acid was the best chemical agent for washing, especially for cobalt. A scrubbing time of 4 h was the optimum time to obtain a removal efficiency of more than 75 % for ^{137}Cs and ^{60}Co. A mixing ratio of the soil weight to the volume of the oxalic acid solution of 1:10 was the best for washing; two scrubbing cycles with 1.0 M oxalic acid avoided the generation of a considerable amount of waste solution. The removal efficiency with a hydro-cyclone was 30 % higher than that without. Vertical plates and alum had important roles in reducing the sedimentation time. The waste solution could be reused up to five times after passage through a column of a strong acid resin (Gye – Nam et al, 2007).

4.5.5. Remotion of ^{238}U, ^{137}Cs and ^{85}Sr from soil

Experimental study seems to demonstrate that the feasibility of electrokinetic soil remediation for the removal of radionuclide contaminants from soils. The experimental results indicate that the technique is effective in radionuclide contaminants from soils with a relatively small amount of energy. Uranium and strontium were efficiently removed from kaolinite by electrokinetic remediation. In the case of cesium, the removal rate may be significantly slower than those of uranium and strontium. This is due to the lower ionic mobility and the affinity of cesium onto kaolinite. Acetic acid was effective as enhancing agent for buffering hydroxide ions produced by the cathode reaction, and prevented the precipitation of uranium ions in the cathode region.

Accordingly, the acetic acid increased the removal efficiency and decreased energy consumption. The use of citric acid was not efficient in removing uranium from kaolinite, because the direction of electromigration was opposite to that of electroosmosis. Since most metal–citrate chelates were negatively charged, they were transported toward the anode by electromigration while electroosmosis flowed toward the cathode. Therefore, removal efficiency significantly decreased. This result indicates that the selection of enhancement agent should be considered with respect to contaminant type and site characteristics. The electrokinetic removal of uranium from the soil weathered from uraniferous black shale was not efficient. This was due to the low proportion of the mobile fraction, since most uranium exists as residual fractions derived from enriched uraniferous parent rocks (Kyeong-I lee et al, 2003).

5. Risk assessment and development of future strategies

Nuclear site operations and successful site restoration depend on the availability of suitable waste management routes and facilities. Effective management of both radioactive and non-radioactive waste is essential to the delivery is a significant part of the process.

Strategic decisions about waste management are informed by the following key principles: risk reduction is a priority, centralised and multi-site approaches should be considered where it may be advantageous, waste should be minimised and the waste hierarchy should be used as a framework for waste management decision making and enables an effective balance of priorities including value for money, affordability, technical maturity and the protection of health, safety, security and the environment.

For Low Level Waste, disposal will be in fit for purpose facilities that reflect the nature of the wastes to be managed. Within this overall framework our priority is to achieve risk reduction by dealing with waste in ageing storage facilities and placing it into safer modern storage conditions. Diverse radioactive waste management and disposal solutions will be pursued where these offer benefits over previous arrangements. New waste management approaches will often require different transport arrangements and will be a matter of great interest to planning authorities and people living close to the sites involved.

Some general ideas about the underpinning strategies are showed in Figure 2, the basic steps for effective management of radioactive waste are part of a global system, ranging from waste generation to final disposal are: minimization of radioactive waste, pretreatment, characterization, treatment, conditioning, transport, storage and disposal (IAEA, 1970). All of these ideas with the intention to reduce the volume of radioactive wastes.

6. Conclusions

Radioactive materials are extensively used in industrial and research activities into medical, agricultural and environmental applications, and in various other areas. During the production and use of these materials, radioactive waste will inevitably arise; this must be managed with particular care owing to its inherent radiological, biological, chemical and physical hazards.

Producers and users of radioactive materials must be sure that a waste management strategy exists prior to the start of waste generation. A well-developed waste management strategy should consider the entire sequence of waste management operations, from the waste's production until its final disposal, including the various regulatory, sociopolitical and economic issues. The identified goal of radioactive waste management can be met with reasonable cost and resource use by implementing a carefully planned waste management strategy using appropriate technologies.

Waste containing long lived radionuclides must be treated, conditioned, stored and disposed of at a repository specifically designed for this purpose. Sample storage capacity is needed for the decay of short lived radionuclides and for storing long lived waste prior to, and after, the treatment and conditioning steps. Decay is the only natural way of reducing radioactivity (the process of transmutation of some long lived radionuclides is not considered viable at this time).

Since radionuclides have decay rates ranging from days to thousands of years, proper segregation of wastes depending on their half-lives, and separate treatment and conditioning of these wastes, is an important factor in the overall scheme of radioactive waste management. Treatment of waste involves operations intended to benefit safety and/or economy by changing the characteristics of the waste. There are different procedures to the treatment of radioactive waste, like:

1. *Treatment of aqueous waste:* ion exchange/sorption, chemical precipitation, evaporation, ultrafiltration / reverse osmosis (UF/RO), incineration, solid / liquid separation (sedimentation, filtration), centrifugation, hydrocyclone techniques, chemical precipitation (co-precipitation or isomorphous precipitation with the carrier, adsorption on to the floc or on added absorbers, flocculation, sedimentation, pH adjustment, chemical oxidation and reduction, ion exchange / sorption, evaporation and reverse osmosis.

2. *Treatment of radioactive organic liquid:* incineration, wet oxidation, electrochemical oxidation, acid digestion, distillation, phase separation by adduct formation and biological digestion.

3. *Treatment of solid waste:* decontamination, compaction, incineration and electroremediation.

In the case of electroremediation treatment of radioactive wastes offers many advantages such as the capacity to remove organic and inorganic pollutants by applying direct electric current into the soil, even in clays, it is easy to operate and involves the installation of electrodes into the soil or waste and the application of a low voltage gradient or direct current through them. In consequence, the EKR is a very good technological alternative to remove radioactive species as ^{22}Na, ^{137}Cs, ^{134}Cs, ^{85}Sr, ^{90}Sr, ^{60}Co and ^{238}Ur from both saturated and unsaturated radioactive soils, sludges, ash and sediments, with less time and high efficiencies of removal in comparison of the other alternatives to remove radioactive wastes.

Acknowledgements

The authors would like to thank to Consejo Nacional de Ciencia y Tecnología de los Estados Unidos Mexicanos (CONACyT) with the project 193883 of projects and thematic networks strengthening research, L'Oreal, United Nations Educational, Scientific and Cultural Organization (UNESCO), Academia Mexicana de Ciencias (AMC), Fundación México – Estados Unidos para la Ciencia (FUMEC). V. Valdovinos is grateful to CONACyT for her scholarship.

Author details

V. Valdovinos[1], F. Monroy-Guzman[2] and E. Bustos[1*]

*Address all correspondence to: ebustos@cideteq.mx

1 Centro de Investigación y Desarrollo Tecnológico en Electroquímica, S. C., Parque Tecnológico, Querétaro, Sanfandila, Pedro Escobedo, Querétaro, México

2 Centro Nuclear "Dr. Nabor Carrillo Flores", La Marquesa, Ocoyoacac, Carretera México-Toluca, Edo, México

References

[1] Abdel, R. O., Rahman, H. A., Hung, Y. –T. (2011). Liquid Radioactive Wastes Treatment: A Review. *Water*, 3, 551-565.

[2] Adenot, F., Advocat, T., Bisel, I., Broudic, J. C., Brun, P., Cau-Dit-Coumes, C., Delage, F., Dussossoy. J. L., Faure, S., Fournel, B., Girold, C., Grandjean, A., Hollebecque, J. Fl., Lacombe, J., Ladirat, C., Lemort, F., Peuget, S., Pinet, O., Poitou, S., Rouppert, F., Vistoli, P. P. (2005). Le traitement et le conditionnement des desechts. Rapport technique DTCD/SCD/2005/001, CEA, France.

[3] Alcántara,M. T., Gómez, J., Pazos, M., Sanromán, M. A. (2008). Combined treatment of PAHs cotaminated soils using the sequence extraction with surfactant – electrochemical degradation. *Chemosphere*, 70, 1438 - 1444.

[4] Al-Shahrani, S. S., Roberts, E. P. L. (2005). Electrokinetic removal of caesium from kaolin, *Journal of Hazardous Materials*, B122, 91 - 101.

[5] Ambashta, R. D., Sillanpää, M. E. T. (2012). Membrane purification in radioactive waste management: a short review. *Journal of Environmental Radioactivity*. 105, 76 – 84.

[6] Bellanger, G. (1992). Process for decontaminating the surface of a metallic component contaminated by tritium, and device for using said process, European Patent EP0274329.

[7] Braud, F., Tellier, S., Astruc, M. (1998). Modelling of decontamination rate in an electrokinetic soil processing, *International Journal of Environmental Analytical Chemistry*, 68, 105 – 121.

[8] Brown, N. W., Roberts, E. P. L. (2007). Electrochemical pre-treatment of effluents containing chlorinated compounds using an adsorbent. *Journal of Applied Electrochemistry*. 37, 1329 – 1335.

[9] Brown, N. W., Campen, A. K., Wickenden, D. A., Roberts, E. P. L. (2013). On-site destruction of radioactive oily wastes using adsorption coupled with electrochemical regeneration. *Chemical Engineering Research and Design*. 91 (4), 713 – 721.

[10] Colleta, T. F., Bruell, C. J., Ryan, D. K., Inyang, H. I. (1997). Cation-Enhanced Removal of Lead from Kaolinite by Electrokinetics. *Journal of Environment Engineering*, 123 (12), 1227 - 1233.

[11] Comninellis, Ch. (1994). Electrocatalysis in the electrochemical conversion/combustion of organic pollutants for waste water treatment. *Electrochimica Acta*, 39, 1857 - 1862.

[12] Comninellis, Ch., Pulgarin, C. (1991). Anodic oxidation of phenol for wastewater treatment. *Journal of Applied Electrochemistry*, 21 (8), 703 - 708.

[13] Cox, C. D., Shoesmith, M. A., Ghosh, M. M. (1996). Electrokinetic remediation of mercurycontaminated soil. *Environmental Science and Technology*, 30 (6), 1933 - 1039.

[14] Cundy, A. B., Hopkinson, L. (2005). Electrokinetic and land remediation and soil engineering, *Applied Geochemistry*, 20, 841 - 848.

[15] Chaiyaraksa, C., Sriwiriyanuphap, N. (2004). Batch washing of cadmium from soil and sludge by a mixture of $Na_2S_2O_5$ and Na_2EDTA. *Chemosphere*, 56 (11), 1129 – 1135.

[16] Chang, H. O. (2001). Hazardous and Radioactive Waste Treatment Technologies Handbook.CRP Press, USA.

[17] Chiba, Z., Schumacher, B., Lewis, P., Murguia, L. (1995). Mediated electrochemical oxidation as an alternative to incineration for mixed wastes, UCRL-JC-119133, Lawrence Livermore National Laboratory, USA.

[18] Farmer, J. C., Hickman, R. G., Wang, F. T., Lewis, P. R., Summers, L. J. (1991). Initial study of the complete mediated electrochemical oxidation of ethylene glycol, Report No. UCRL-LR-106479, Lawrence Livermore National Laboratory, Livermore, CA.

[19] Farmer, J. C., Wang, F. T., Hawley - Fedder, R., Lewis, P. R., Leslie, J. (1995). Summers, and Linda Foiles. Initial Study of Halide-Tolerant Mediators for the Electrochemical Treatment of Mixed and Hazardous Wastes. Lawrence Livermore National Laboratory, USA.

[20] Fernández, C. D. (2012). Descontaminación de Suelos.Departament of Pedology, University of Granada, Granada, Spain.

[21] Fountain, M. S., Kurath, D. E., Sevigny, G. J., Poloski, A. P., Pendleton, J., Balagopal, S., Quist, M., Clay, D. (2009). Caustic recycle from handford tank waste using NaSICON ceramic membrane salt splitting process. PNNL-18216. Pacific Northwest National Labortory, USA.

[22] Giannis, A., Gidarakos, E., Skouta, A. (2007). Application of sodium dodecyl sulfate and humic acid as surfactants on electrokinetic remediation of cadmium-contaminated soil. *Desalination*, 211, 249 - 260.

[23] Gilbón, A. J. M. (2001). Electroremediación de Suelos Contaminados Con Hidrocarburos. Universidad Autónoma Metropolitana, Mexico City, Mexico.

[24] Gómez, J., Alcántara, M. T., Pazos, M., Sanromán, M. A. (2009). A two – stage process using electrokinetic remediation and electrochemical degradation for treating benzoapyrene spiked kaolin. *Chemosphere*, 74 (11), 1516 - 1521.

[25] Gye-Nam, K., Wang-Kyu, Ch., Chong-Hun, J., Jei-Kwon, M. (2007). Development of a Washing System for Soil Contaminated with. Radionuclides Around TRIGA Reactors. *Journal of Industrial and Engineering Chemistry*, 13 (3), 406 - 413.

[26] Gye–Nam, K., Seung–soo, K., Hye-Min, P., Wan-Suk, K., Uk-Ryang, P., Jei-Kwon M. (2013). Cs-137 and Cs-134 removal from radioactive ash using washing–electrokinetic equipment. *Annals of Nuclear Energy*. 57, 311 – 317.

[27] Hamed, J., Acar, Y. B., Gale, R. J. (1991). Pb(II) Removal from Kaolinite by Electrokinetics, *Journal of Geotechnical Engennering*. ASCE, 117 (2), 241 - 271.

[28] Hamed, J. T., Bhadra, A. (1997). Influence of current density and pH on Electrokinetics, *Journal of Hazardous Materials*, 55, 279 - 294.

[29] Hobbs, D. T. (1999). Caustic recovery from alkaline nuclear waste by an electrochemical separation process. *Separation and Purification Technology*, 15, 239 - 253.

[30] HSEEA (2007) Fundamentals of the management of radioactive waste. Health and Safety Executive, the Environment Agency and the Scottish Environment Protection Agency. December 2007.

[31] Hu, J. M., Meng, H. M., Zhang, J. Q., Cao, C. N. (2002). Degradation mechanism of long service life Ti/IrO_2-Ta_2O_5 oxide anodes in sulphuric acid, *Corrosion Science* 44 (8), 1655 - 1668.

[32] IAEA (1970). Standardization of Radioactive Waste Categories, Technical Reports Series No. 101, International Atomic Energy Agency (IAEA), Vienna.

[33] IAEA (1992). Treatment and conditioning of radioactive organic liquids. IAEA-TEC-DOC-656 International Atomic Energy Agency (IAEA), Vienna.

[34] IAEA (1994). Classification of Radioactive Waste, Safety Series No. 111-G-1.1, International Atomic Energy Agency (IAEA), Vienna.

[35] IAEA (2001). Handling and Processing of Radiactive Waste from Nuclear Applications. Technical Reports Series No. 402. International Atomic Energy Agency (IAEA), Vienna.

[36] IAEA (2001). Decontamination and Decommissioning of Nuclear Facilities. Technical Reports Series No. 401, . International Atomic Energy Agency (IAEA), Vienna.

[37] IAEA (2002). Application of Ion Exchange Processes for the Treatment of Radioactive Waste and Management of Spent Ion Exchanger. Technical Reports Series No. 408. International Atomic Energy Agency (IAEA), Vienna.

[38] IAEA (2004). Application of the Concepts of Exclusions, Exemption and Clearance, IAEA Safety standards Series No. RS-G-1.7, International Atomic Energy Agency (IAEA), Vienna.

[39] IAEA (2006). Application of Thermal Technologies for Processing of Radioactive Waste, IAEA-TECDOC-1527, International Atomic Energy Agency (IAEA), Vienna.

[40] IAEA(2006) Fundamental Safety Principales. IAEA SAFETY STANDARDS SERIES. SF-1, IAEA, Vienna.

[41] IAEA (2006b)The Management System for Facilities and Activities. Safety Requirements No. GS-R-3.IAEA 2006.

[42] IAEA(2008) The Management system for the processing, handling and storage of radioactive waste. Safety Guide. No. GS-G-3.3, International Atomic Energy Agency (IAEA), Vienna.

[43] IAEA (2009). Classification of Radioactive Waste, Safety standards Series No. GSG-1, International Atomic Energy Agency (IAEA), Vienna.

[44] Khan, L. I., Alam, S. (1994). Heavy metal removal from soil by coupled electro-hydraulic gradient. *Journal of Environment Engineering* 120 (6), 1524 - 1543.

[45] Kim, S. O., Kim, K. W., Stüben, D. (2002). Evaluation of electrokinetic removal of heavy metals from tailing soils, *Journal of Environment Engineering*, 128, 705 - 715.

[46] Kim, G. N., Oh, W. Z., Won, H. J., Choi, W. K. (2003). Removal of cesium and cobalt from soil around TRIGA reactor using electrokinetic method, *Journal of Industrial and Enginnering Chemistry*, 9 (3), 306 - 313.

[47] Kim, G. –N., Kim, S., Park, H. –M., Kim, W. –S, Park, U. –R, Moon, J. –K. (2013). Cs-137 and Cs-134 removal from radioactive ash using washing – electrokinetic equipment. *Annals of Nuclear Energy*, 57, 311 – 317.

[48] Kurath, D. E., Brooks, K. P., Hollenberg, G. W., Sutija, D. P., Landro, T., Balagopal, S. (1997). Caustic recycle from high-salt nuclear wastes using a ceramic-membrane salt-splitting process. *Separation and Purification Technology*, 11, 185 - 198.

[49] Kyeong-Hee, K., Kim, K., Soon-Oh, L., Chang-Woo, L., Myung-Ho, K., Kyoung-Woong, (2003). Electrokinetic processing for the removal of radionuclides in soils. *Separation Science and Technology*, 38 (10), 2137 – 2163.

[50] Lerch, R. E., Kennewick, J. A., Partridge, R. (1980). Decontamination of metals using chemical etching. US Patent Aug. 12, 4,217,192.

[51] Méndez de Vigo, G. E. (2000). Origin and Radioactive Waste Management, Official College of Physicians Illustrious. The 3rd. Edition, 105, 188.

[52] Méndez, E., Pérez, M., Romero, O., Beltrán, E. D., Castro, S., Corona, J. L., Corona, A., Cuevas, M. C., Bustos, E. (2012). Effects of electrode material on the efficiency of hydrocarbon removal by an electrokinetic remediation process. *Electrochimica Actu*, 86, 148 – 156.

[53] Murillo-Rivera, B., Labastida, I., Barrón, J., Oropeza - Guzman, M. T., González, I.,Teutli - Leon, M. M. M. (2009). Influence of anolyte and catholyte composition on TPHs removal from low permeability soil by electrokinetic reclamation. *Electrochimica Acta*, 54 (7), 2119 – 2124.

[54] NEA (1999). Decontamination techniques used in decommissioning activities. A Report Nuclear Energy Agency (NEA), Task Group on Decontamination.

[55] Pamucku, S., Khan, L. I., Fang, H. –Y. (1990). Zinc detoxification of soils by electro-osmosis. Transportation Research record 1288. *Geotechnical Engineering*, 41 - 51.

[56] Pamucku, S., White J. K. (1992). Electrokinectic removal of selected heavy metals from soil. Environmental Progress 11 (3), 241 – 250.

[57] Pamukcu, S., Wittle, J. K. (1992). Electrokinetic Removal of Selected Heavy Metals from. *Soil, Environmental Progress.* 11 (3) (1992) 241 - 250.

[58] Pendleton, J., Dayton, T., Duffey, K., Bhavaraju, S., Balagopal, Sh. (2011). Pilot-scale system for recycling caustic from LLW simulant. WM2011 Conference, Phoenix, AZ

[59] Pérez – Corona, M., Ochoa, B., Cárdenas, J., Hernández, G., Solís, S., Fernández, R., Teutli, M., Bustos, E. (2013). Comparison of different arrangements of electrodes during the electrokinetic treatment of polluted soil with hydrocarbons and its final application *in situ.* Recent Research Development in Electrochemistry, Transworld Research Network, 37/661 (2), Kerala, India, 59 - 80.

[60] Prabhakaran, D., Kannadasan, T., Basha, C. A. (2009). Mediated electrochemical oxidation process for destruction of TOC in a batch recirculation reactor. *International Journal of Chemical Technology Research*, 1 (4), 962 – 969.

[61] Radiation Safety General Regulations (1988). Official Journal of the Federation, November 22.

[62] Rampley, C. G., Ogden, K. L. (1988). Removal of Lead from Surrogate and Real Soils Using a. Water Soluble Chelator, Adsorption and Batch Extraction. *Environ. Sci. Technol.* 32 (7), 987 - 995.

[63] Reed, B. E., Berg, T. E., Thompson, J. C., Hatfield, J. H. (1995). Chemical conditioning of electrode reservoir during electrokinetc soil flushing of Pb-contaminated silt loam. *Journal of Environment Engineering*, 121, 805 - 815.

[64] Reddy, K. R., Donahue, M., Saicheck, R. E., Sasaoka, R. (1999). Preliminary Assessment of Electrokinetic Remediation of Soil and Sludge Contaminated with Mixed Waste. *Journal of the Air & Waste Management Association*, 49 (7), 823 - 830.

[65] Reddy, K. R., Xu, C. Y., Chinthamreddy, S. (2001). Assessment of electrokinetic removal of heavy metals from soils by sequential extraction analysis. *Journal of Hazardous Materials.* B84 (2 – 3), 279 - 296.

[66] Reddy, K. R., Chinthamreddy, S. (2003). Volumetric strains of clean sands subject to cyclic loads. *Journal of Geotechnical and. Geoenvironmental Engineering*, 263 – 272.

[67] Reddy, K. R., Chaparro, C., Saichek, R. E. (2003). Removal of Mercury from Clayey Soils Using Electrokinetics. *Journal of Environmental Science and Health*, Part A – Toxic / Hazardous Substances & Environmental Engineering, A38 (2), 307 - 338.

[68] Reddy, K. R., Chaparro, C., Saichek, R. E. (2003). Iodide – enhanced electrokinetic remediation of mercury – contaminated soils. *Journal of Environmental Engineering,* ASCE, 129 (12), 1137 – 1148.

[69] Report OST (1999). Acid digestion of organic waste. Mixed Waste Focus Area, Reference #1827, U.S. Department of Energy, USA.

[70] Ribeiro, A. B., Rodríguez - Maroto, J. M., Mateus, E. P., Gomes, H. (2005). Removal of organic contaminants from soils by an electrokinetic process: the case of atrazine: Experimental and modeling. *Chemosphere,* 59 (9), 1229 - 1239.

[71] Ruiz, C., Anaya, J. M., Ramírez, V., Alba, G. I., García, M. G., Carrillo - Chávez, A, Teutli, M. M. M., Bustos, E. (2011). Soil arsenic removal by a permeable reactive barrier of iron coupled to an electrochemical process. *International Journal of the Electrochemical Science,* 6, 548 - 560.

[72] Saichek, R. E., Reddy, K. R. (2003). Effect of pH control at the anode for the electrokinetic remoal of phenanthrene from kaolin soil. *Chemosphere,* 51 (4), 273 - 287.

[73] Saichek, R. E., Reddy, K. R. (2005). Electrokinetically enhanced remediation of hydrophobic organic compounds in soils: a review. *Critical Review of Environmental Science Technolog* y, 35, 115 - 192.

[74] Schwinkendorf, W. E., Hart. P. W. (1995). Low - temperature alternatives to incineration for doe mixed waste. WM'95 Conference Proceedings, Tucson, Az. USA, 1296 - 1303.

[75] Taylor, G. T., Newey, A. W. E., Bates, C. J., King, C. R., Dawes, K. (2009). Demonstration of electrochemical oxidation of oils using boron - doped diamond electrodes and its potential role in the disposal of radioactively contaminated waste lubricants. WM2009 Conference, March 1-5, Phoenix, USA.

[76] Twissel, M. A., Holt, N. S. (1996). Wet oxidation of organic radioactive waste. European Commision Nuclear Science and Technology. EUR 16776, Luxembourg. European Communmunities.

[77] USEPA, (1997). Resource guide for electrokinetics laboratory and field processes applicable to radioactive and hazardous mixed wastes in soil and groundwater from 1992 to 1997. US Environmental Protection Agency, Washington, DC.

[78] Vázquez, M. V., Hernández, F., Grandoso, D., Arbelo, C. D. (2004). Study of the electrical resistance of andisols subject of electro-remediation treatment. *Portugaliae Electrochimica Acta,* 22, 399 - 410.

[79] Vázquez, M. V., Hernández, F., Benjumea, D., Grandoso, D., Lemus, M., Arbelo, C. D. (2007). Electrokinetic determination of the buffer capacity of Andisols. *Science of the Total Environment,* 378 (1-2), 214 - 217.

[80] Virkutytea, J., Sillanpää, M., Latostenmaa, P. (2002). Electrokinetic soil remediation – critical overiew. *Science of the Total Environment,* 289 (1-3), 97 - 121.

[81] Yeung, A. T., Hsu, Ch., Menon, R. M. (1996). EDTA-enhanced electrokinetic extraction of lead. *Journal of Geotechnical and Geoenvironmental Engineering,* 122 (8), 666 - 673.

[82] Wedman, D. E., Martinez, H. E., Nelson, T. O. (1996). Electrolytic decontamination of conductive materials for hazardous waste management. LA-UR-96-3795, USA.

[83] Wickenden, D. (2011). Innovative treatment of radioactive oil waste – the Arvia® process. In: Proceedings of the 11th Tritum User Group Conference, Chaplecross, UK.

Chapter 15

Optical Fibers to Detect Heavy Metals in Environment: Generalities and Case Studies

J. A. García, D. Monzón, A. Martínez, S. Pamukcu,
R. García and E. Bustos

Additional information is available at the end of the chapter

http://dx.doi.org/10.5772/57285

1. Introduction

The fiber optic sensors can be used to create a truly distributed chemical sensing capability for selectively detecting metal compounds by spatial and temporal acquisition over large distances in the subsurface. In addition the fiber optic sensors have several advantages such as small size, light weight, immunity to electromagnetic interference (EMI), high temperature performance, large bandwidth, high sensitivity, and environmental ruggedness (Krohn, 1988). Most current technologies capable of detecting contaminants use strategically placed sensing or monitoring devices. This works reasonably well if plausible event location is known, hence settle recording vast amounts of benign data over time until the appearance of the suspected event. This approach remains limited for application in large spatial scales in the geo-environment and subsurface. A simple approach is to suppress all the benign data by triggered transmittal of the signals only at the spatial and temporal vicinity of the event. This, in essence the "truly distributed" sensing capable of delivering the event signal "*wherever*" and "*whenever*" it might occur, as opposed to only at strategic places where the sensing devices are pre-located. The revolutionary advances in flexible sensing and distributed data processing permits us sensing in this truly distributed manner.

Sensors based on fiber optic cable functions make use of the following important features of the cable to sense the environment: (1) optical loss: intrinsic and extrinsic energy loss properties, (2) refractive index: index profile in radial direction and the reduction of index fluctuation along the axial direction; (3) shape: cross sectional shape and size, the surface finish and the fluctuation of the size along axial direction. Present fiber optic sensors mostly use energy loss principles (i.e., changes in optical power in linearly positioned wave-guides) for chemical detection. These can be limited for distributed applications if energy depletes over a short stretch of the fiber sensor, or frequent sensor points are needed at a prohibitively expensive cost. Other sensors use the changes in refractive index and/or cross sectional size of the fiber cable that change the light scattering property in optical fibers, known as Brillouin scattering (Horiguchi et al, 1995; Kee et al, 2000).Fiber optic sensing based on Brillouin scattering has been used successfully in civil infrastructure for health monitoring (Bao et al, 2001; Ohno et al, 2001). In this chapter a background on use of optical fibers for chemical sensing and new developments and proposed advancements are discussed.

2. Background

2.1. Overview of fiber optic sensors

In an optical fiber sensor a physical, chemical or biological variable can interact with the light and produce a change in one of their parameters. It is desirable to produce an optical signal related uniquely to the parameter of interest. These sensors use the optical fiber either as the sensing element (intrinsic sensors), or as a means of relaying signals from remote sensing area to the signal processor (extrinsic sensor), or both. Optical fiber sensors take advantage of the inherent fiber optic characteristics which include their lightweight, of very small size, passivity, low-power requirement, resistance to electromagnetic interference, environmental impact and corrosion, their bandwidth, and flexibility. They can be installed in areas normally inaccessible by conventional sensors, they can be interfaced with data communication systems and pose no risk of electric shock in live measurements. These attributes have allowed optical fiber sensors to displace traditional sensors for measurement and monitoring of rotation, acceleration, electric and magnetic field, temperature, pressure, acoustics, vibration, linear and angular position, strain, humidity, viscosity, pH, gas and chemical content among many others.

Use of Fiber Optic Sensors is a viable real-time data gathering approach by surface-adhering or embedding the fiber to a specimen under evaluation. The concept of embedding fiber-optic sensors into structures has generated a great deal of interests in aerospace engineering initially and more recently in civil engineering. There are several types of chemical sensing techniques based on optical waveguides (Ho et al, 2001). Among those are fiber Bragg gratings (FBG), which is marking of a fiber with a laser to create a local narrow band pass filter sensitive to environmental parameters (Guemes et al, 1998; Schulz et al., 1998). Optical time domain reflectometry (OTDR) consists of sending a powerful light pulse and observe modification in the reflected light due to local in homogeneties along the fiber. The pulse losses correspond to specific environmental interaction. The evanescent pulse technique is also based on OTDR, in which the fiber cladding is modified to interact with the environment and the pulse travels partially through the cladding. These sensors demand large optical power, due to the cumulative energy loss at the points of contact with the chemicals.

Over the last decade, there has been rapid development in the area of smart sensor technologies, in particular using structurally integrated optical fiber to form the basis for smart structure technology. A variety of configurations have been developed for measurement of strains and deformations in structures, including localized-type such as fiber Bragg gratings and multiplexed long gauge interferometric sensors, and distributed sensing schemes including Stimulated Brillouin Scattering (SBS) or Brillouin Optical Time Domain Analysis (BOTDA) (Bao et al, 2001) and Brillouin Optical Time Domain Reflectometry (BOTDR) (Pamukcu et al, 2006; Anastasio et al, 2007).

Between different types of optical sensors reported, there are those based on sensitive coatings onto the fiber surface, Fabry-Perot interferometers, long-period fiber gratings (LPFG), LPFG with sensitive films, hetero-core devices, fiber Bragg gratings on doped fibers (i.e, Germanium doped). Fiber gratings are structures consisting of a periodic perturbation of the optical and/or geometrical properties of an optical fiber. Depending on the pitch of the perturbation, fiber gratings fall into two distinct categories: short period gratings, known as fiber Bragg gratings (FBGs) and, long period gratings (LPFGs). Stretching the fiber gratings causes a change in grating period, hence the wavelength of the reflected light. This makes the FBGs ideal for localized temperature and strain measurements. Unlike FBGs in which counter directional coupling occurs in the core, co-directional coupling occur in LPFGs between the core and cladding. This feature renders LPFGs sensitive not only to temperature and strain, but also to bending causing a curvature, to hydrostatic pressure, to torsion and to ambient refractive index changes. The closer the ambient refractive index to that of the cladding the stronger the sensitivity to refractive index changes. It is this high sensitivity that has piqued the interest in development of various types of refractive index-based LPFG sensors which constitute most of the chemical sensing applications (Orellana and Haigh, 2008; Kasik et al, 2010).

Point detection fiber optic sensors have been developed successfully for measurement of liquid levels, chemical species, drugs, environmental agents (such as pollutants and pesticides), biochemical reactions, and to monitor a wide variety of chemical processes (Wolfbeis, 2000). A fiber optic laser induced breakdown spectroscopy method was demonstrated in the field using a push-cone device, which is a single point, single time measurement technique. The most common configuration for optical pH sensors, and other environmental parameters, employs a fluorescence indicator (Lee et al, 2000). Among the different types of optical fiber devices used in pH sensing are, hetero-core fibers, U-bend fibers, fiber Bragg and long-period gratings, fibers and fiber tips with active doped cladding, among others (Kocincovaet al, 2007). Some of the substances that can be detected or identified using optical fiber sensors are volatile organic compounds (alcohols, formaldehydes, methane, ketones, CO_x, O_2, and H_2), some metallic ions like Ca, Al, Cu, Zn, Hg, V and Pb (Jeronimoet al, 2007; Wolfbeis, 2008).

Wide application of advanced chemical sensing in the environment may suffer from scaling issues. The real-world conditions often require self-referencing, spatially distributed, temporally continuous, and chemically selective sensors for monitoring regions spanning over long lengths or wide areas. When large area monitoring for chemical agent intrusion is required, use of currently available point sensors can be cost prohibitive. Other non-point, distributed detection methods based on energy loss principles (Buerck et al, 2001) may also be inadequate when scaled to wide area monitoring due to extensive energy input requirements.

One of the unique features of the optical fiber technology is the possibility to construct distributed sensors, in which the measuring can be determined along a line of space with a given spatial resolution (Galindez-Jamioy et al, 2012) by, for example, Brillouin optical time domain analysis (BOTDA) (Cui et al., 2009, 2010 and 2011); an hetero-core LPFG sensors. In here, we examine current and proposed application of these techniques to spatially distributed, temporally continuous, and chemically selective sensing applications in soil and water environment. The premise of Brillouin technique goes back to 1920 when physicist Leon Brillouin first studied the diffusion of light by acoustic waves. The phenomenon he observed was a frequency change of scattered light. The first major papers related to distributed fiber optic sensor based on Brillouin were generated in mid-nineties (Bao, et al. 1995; Fellay et al, 1997). Current research on Brillouin sensing may be divided in three categories: photonics (the physics of Brillouin); data processing and post processing to improve signal to noise ratios, and applications of distributed sensing to civil infrastructure and environment.

2.2. Fundamentals of optical techniques

The interest on optical techniques to measure or detect chemical agents have been continuously extending and growing over the last forty years. Special attention has been focused on the development of optical sensor to detect heavy metals, due to the hazardous effects of these ions on the health of human beings and ecosystems. Optical methods have the advantage of being fast, simple, compact, portable, low-cost, with sensitivities and resolutions improved to detect in the picomolar range.

Combined with other technologies, like microfluidics systems, optical waveguides, or MEMs, optical methods are suitable for application where conventional electrodes cannot be used because of their large size or because of the risk of electrode shock during in vivo measurements. Due to their minute size; these optical microsystems are capable of gathering diverse data with a small amount of analyte. The diversification of optical techniques have made possible to construct novel sensing platforms to detect heavy metals in air, water or soil, food and beverages, or biological samples.

Optical sensors to detect heavy metals employ an optical transduction technique, i. e. an element that "translate" the chemical variable into an optical signal (intensity, wavelength, polarization or phase), to yield analyte information (McDonagh et al, 2008, Grattan and Meggitt, 1999). Optical chemical sensors can be categorized, according to the transduction technique, in direct sensors and reagent-mediated sensing systems. In direct sensors the element of interest is detected directly via an optical property of the sample such as scattering or florescence, for example.

However, most heavy metals optical sensor uses an intermediate agent. Most of the optical chemical techniques to detect heavy metals are based on optical absorption, fluorescence, Raman spectroscopy, or surface plasmon resonance, whereby the perturbed signal is related to the reaction of the intermediate agent under the presence of a specific heavy metal. In general, all these techniques involves the interaction of an incident beam over an analyte or indicator element yielding transmitted, reflected or fluorescent signal. A schematic representation of the spectroscopic principle, the working mechanism of an optical sensor is shown in Figure 1.

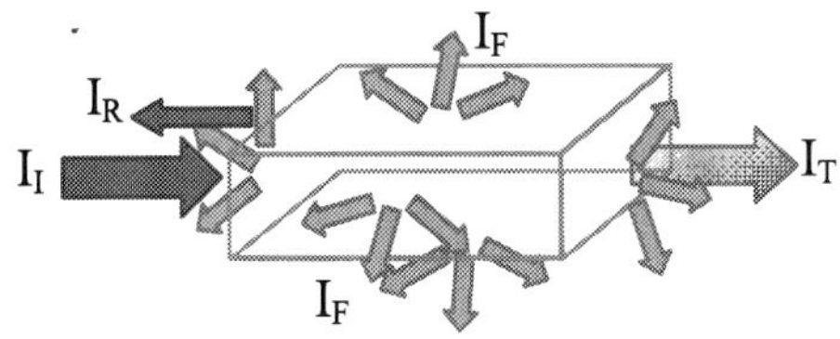

Figure 1. Representation of the optical signals in an optical chemical sensor. The incident beam I_I interacts with the heavy metal sensitive layer and depending on the optical properties of the sample one of the three signals (transmitted I_T, reflected I_R, and generated by fluorescence or scattering I_F) will be produced and will give information about the type and concentration of heavy metal present on the sample.

2.2.1. *Absorbance-based techniques*

Among the optical chemical techniques, the simplest to implement is that based on the measurement of light absorbed by a sensitive heavy metal layer. Absorption in a gas or liquid, where it is assumed that each single molecules equally contributes to the total light absorbed, may be characterized by a Beer-Lambert law, or simply the Beer law,

$$I_T = I_I 10^{-\varepsilon C d} \qquad (1)$$

where I_T and I_I represents the intensity of the transmitted and incident beam, ε is the molar absorptivity ($Lmol^{-1}$ cm), and C is the concentration (mol L^{-1}) of the absorbing species and d is the absorption path length (cm). In the case of a solid, absorbing and homogeneous medium, the transmitted signal is calculated using the Lambert Bouguer law, expressed as $I_T = I_I e^{-\alpha x}$, where x is the thickness of the medium and α is the extinction coefficient. The Beer law can also be expressed in terms of the absorbance (or optical density) A:

$$A = \log_{10}\left(\frac{I_I}{I_T}\right) = \varepsilon C d \qquad (2)$$

There exists a linear relation between the absorbance and the concentration of the element to be measured. However, in order to observe the linear dependence of absorbance on concentration, the incident beam should be ideally monochromatic. In the case that a wide broadband light source is used, the contribution of all wavelengths must be considered, in such cases the equation (2) becomes:

$$\bar{A} = \log_{10}\left(\frac{\int I_I(\lambda)d\lambda}{\int I_T(\lambda)d\lambda}\right) \tag{3}$$

Also the presence of highly absorbing or highly scattering media should produce a deviation from perfect Beer law behavior. In the case that more than one absorbing material is present the absorbance contribution of each species must be considered. In most of the absorbance-based heavy metals sensors an intermediate agent, an optical film that changes its absorbance according to the concentration of a specific heavy metal, is used (Antico et al, 1999;Guo et al, 2006).

A special case of the absorption-based sensors are those schemes where materials that change their color under the presence of a specific heavy metal are used (Balaji et al, 2006; Prabhakaran et al, 2007). The reaction of the sensitive components to the concentration of a specific ion produces a photochromic reaction that can be observed with a naked eye. Such materials are often in solution, but for sensing the most attractive are those that can be deposited as thin films over a substrate. The instrumentation of absorption-based sensors is the simpler of the optical heavy metal techniques, since it can be implemented with a monochromatic light source and a photodetector. This also makes this technique very susceptible to be implemented in microscopic opto-fluidic configurations that could diversify the technique.

2.2.2. Reflectance-based techniques

It is well-known that chemical reactions could lead to changes in the complex refractive index of a substance; this fact has been impulse researchers to design and fabricate materials that react with heavy metals that can be used as transducers. When these materials, commonly in the form of a thin layer, are illuminated with an appropriated light the signal will be partially or totally reflected. However, this reflectance will change when the layer is in contact with a specific metal that it reacts with. If the refractive index of the layer is purely real, the changes in the reflected signal can be estimated by using the Fresnel formulae.

However, in most cases the optical response of these materials under the presence of heavy metals are more complex and involve a change in the real and imaginary parts of the refractive index, that produce changes in reflectivity and absorbance. Also, there is a contribution of scattered light. So, the reflected signal is composed of light from different sources, however, also in this complex response the signal reflected is used to deduce, directly or indirectly, the concentration C of the heavy metals.

The reflected-based techniques are specially used in optical fiber schemes since the set-up is very simple to implement (Yusofand Ahmad, 2003, Guillemain et al, 2009). The material sensitive to the heavy metals are directly deposited over the fiber tip or in a substrate that will be illuminated by an optical fiber. The reflected signal is usually collected by the same fiber, but frequently another fiber or fibers are used to collect it. The reflected signal propagates along the fiber to the detector, where it is analyzed in order to determine the heavy metal present and their concentration (Figure 2).

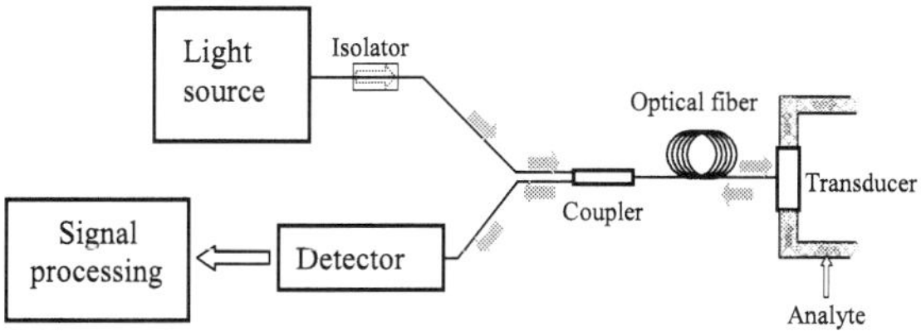

Figure 2. Diagram of a reflectance-based optical chemical sensor based on optical fibers.

2.2.3. *Fluorescence-based techniques*

Some materials have the property of being fluorescent when they are illuminated with a light source of appropriated wavelength. The fluorescence is the optical radiation generated when electrons of an atom or molecule return from the excited to the ground state after absorption of a photon from an excitation light source. In general the energy of the excited photon is lower than the absorbed one so the wavelength of the fluorescence signal is longer than that of the excitation.

The intensity of the fluorescent signal (I_F) is proportional to the intensity of light absorbed by the sample (I_I-I_T), therefore it is possible to establish a direct relation between the intensity of the fluorescent signal and the concentration of an absorbing material. This feature is very important for sensing since intensity of the fluorescence increases as the concentration of the absorbing species augments. Although, we have just made reference to the fluorescence intensity, for sensing, the decay time of the fluorescence signal is more frequently used because this parameter is less sensitive to source fluctuations, interference from ambient light or drift due to aging of detector. It is possible to design and fabricate a fluorescent material sensitive to a specific heavy metal. Thus, the intensity, wavelength and life time of the fluorescent signal will change under the presence of this metal. Fluorescence-based techniques are the most used to detect the presence of heavy metals due to its extraordinary sensibility (Mayra et al., 2008;Achatz et al, 2011, Aksuner, 2011).

2.2.4. *Surface plasmon resonance-based techniques*

The most popular label-free refractometric technique is the Surface Plasmon Resonance (SPR), since it allows the direct observation of chemical reactions in real time without the use of markers or labels. SPR is a quantum optical-electrical phenomenon produced by the interaction of light with a metal surface. Actually, the surface plasmon is a charge density oscillation that exists at a metal-dielectric interface. The plasmon propagates in a direction parallel to the metal-dielectric interface in the boundary of the metal and the external medium (Figure 3).

These oscillations are very sensitive to any change in the optical refractive index of the material at the boundary. The optical excitation of plasmon can be achieved in a three-layer system consisting of a thin metal film sandwiched between two isolators of different dielectric constant

(Maier, 2007), where the phase-matching condition between the optical and plasmon wave vector is fulfilled. In the optical domain, the surface plasmon excitation will be observed as an intensity transmission loss at a specific wavelength. The wavelength of the dip depends on the refractive index of the two dielectrics and the thin metal film, and the propagation constant of the optical waveguide. There are three common method to excite surface plasmon, using a prism coupler and the attenuated total reflection, a periodic grating, and an optical waveguide planar (Figure 3A) or cylindrical. The prism coupler technique is the most popular since exhibits a good sensitivity, stability, and reproducibility for the measurement of heavy metals (Forzani et al, 2007; Lin et al, 2009; Abdi et al, 2011; Fen et al, 2012 and 2013; Fen and Yunus, 2013). For heavy metal detection a sensitive thin film layer is deposited over the thin metal film, so when the target heavy metal interacts with the layer a refractive index change is produced. The surface plasmon conditions changes and the peak wavelength shifts as can be seen in Figure 3B. SPR is the most sensitive refractometric method, with a theoretical resolution of $1x10^{-7}$, so it is possible to detect very small traces of heavy metals.

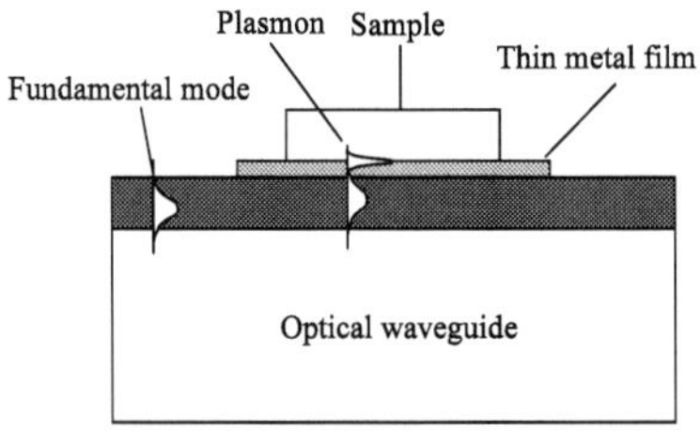

A

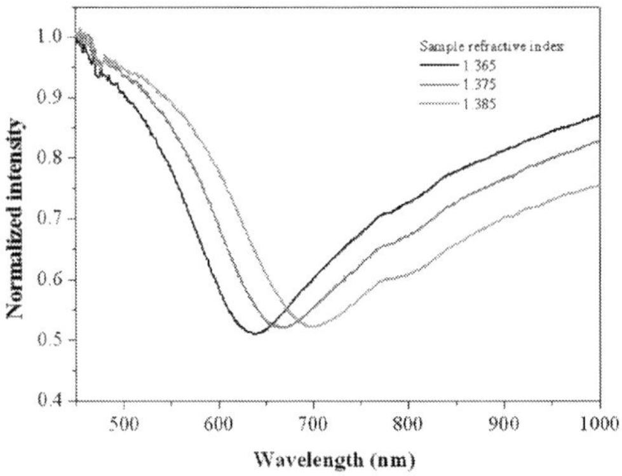

B

Figure 3. (A) A schematic representation of the surface plasmon excitation in a dielectric planar waveguide coupler configuration. (B) Optical transmission spectra of an hetero-core optical fiber coated with a 20 nm thin gold film when the fiber is immersed in a liquid with a refractive index of 1.365, 1.375, and 1.385, black, red and green line respectively.

2.3. Fundamentals of BOTDA and LPFG based sensing

Figure 4, 5 and 6 show the stimulated Brillouin scattering based BOTDA photonics configuration and the principle of measurement used at Lehigh University Geo-sensing laboratory, respectively (Texier et al, 2005; Pamukcu et al, 2006; Turel and Pamukcu, 2006; Anastasio et al, 2007). Brillouin is a nonlinear effect, in which light is scattered at well-defined points along the fiber where the acoustic properties of the fiber are locally modified by the environment. The stimulated Brillouin scattering (SBS) it is an acoustic – optical process which is useful for distributed measurements of a probe beam by the SBS interaction with a counter-propagating nanosecond pump pulse. In the SBS technique, as in a null detector, the pump and probe are initially de-tuned by a (frequency) that is slightly greater than the Brillouin frequency. Therefore, in unstressed fiber, the base line remains flat resulting in a self-referenced sensor eliminating the need for duplication with another reference fiber.

The Brillouin line being intrinsically narrow (~20 - 50 MHz), the initial de-tuning can be quite small so that the amount of strain required to generate a signal is also quite small (0.001 %), allowing for higher resolution and sensitivity of the sensor compared to other fiber-based measurement techniques. When SBS based sensors are used for environmental sensing the fiber is hitched or bonded with selective polymer transducers that are mass detectors in direct contact with the surrounding medium. The polymer reacts to the surrounding (i.e. moisture, pH, target chemical) by selectively absorbing the target compound and it swells. Localized swelling of the bonded or hitched polymer produces tangential, axial or radial stresses on the fiber depending on the physical coupling. These stresses result in axial straining of the fiber and a measurable change in its local acoustic properties, hence a Brillouin scatter of the transmitted light. The location of the generated signal is determined by time domain reflectometry.

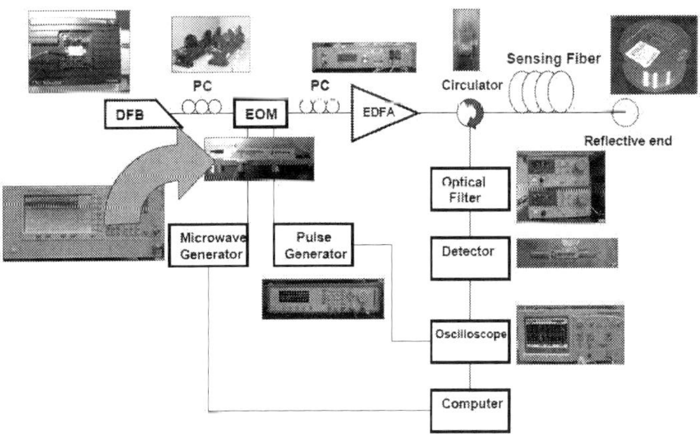

Figure 4. SBS based BOTDR photonics set up at Lehigh Geo-Sensing Laboratory.

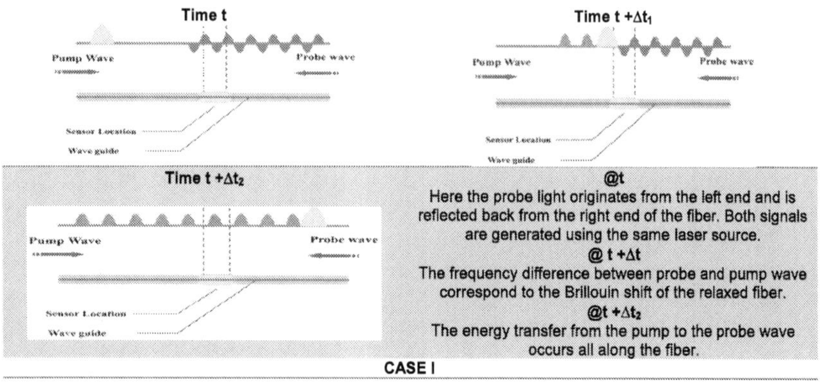

Figure 5. Measurement principles of SBS based sensing: case I.

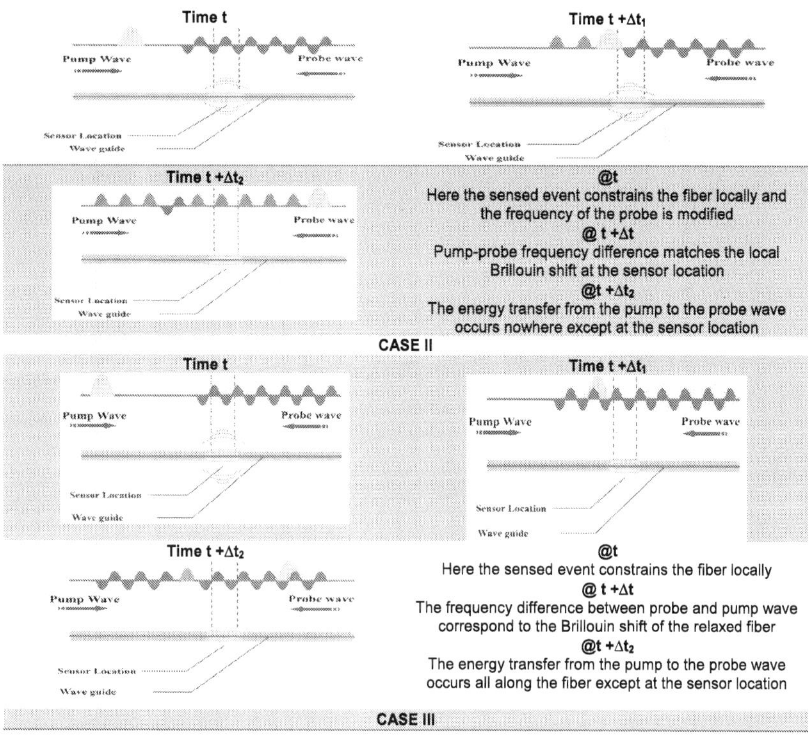

Figure 6. Measurement principles of SBS based sensing: case II and III.

In the case of LPFG`s the environmental changes produce a spectral shifts, that required a sophisticate or a complicate equipment. These devices are very sensitive to temperature changes so to measure another different parameter it is necessary to make the corresponding compensation. One alternative to avoid these difficulties are hetero-core fibers. These devices are constructed by changing the diameter of the core in a small length (mm) in a transmission line (Figure 5 and 6), which causes the optical wave to expand within the cladding in the single-mode region of the hetero-core, thus the evanescent field can easily interact with the external medium.

Owing to the core diameter mismatch, some of the light is guided by the cladding of the SM fiber (Figure 7). This makes the transmission of the device dependent on the refractive index of the external medium. The sensor exhibits maximum transmission changes when the index of the sample medium approaches that of the SM fiber cladding. The device can operate at different wavelengths as well as when coated with thick films made of variable index materials. Moreover, standard emitters, fibers, detectors, etc., are needed to fabricate the sensor, which makes it attractive for diverse applications (Villatoro and Monzon-Hernandez, 2006).

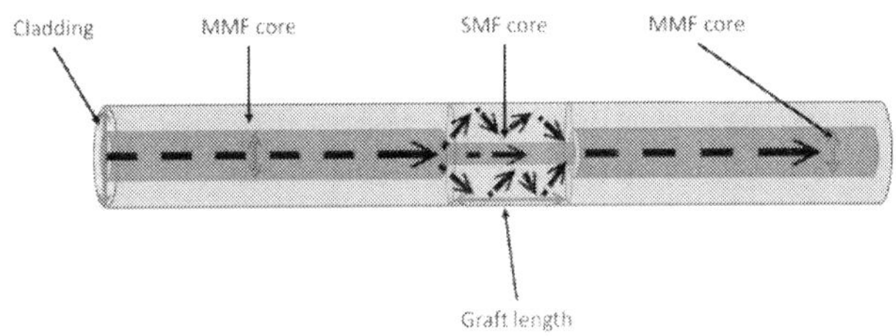

Figure 7. Schematic representation of the hetero-core fiber structure. A piece of single-mode fiber (SMF) of length L is inserted between two multimode fibers (MMF). Cladding diameter of both fibers is 125 μm but the core diameters are different.

3. Investigation

3.1. BOTDA/BOTDR applications and proposed sensor configurations

The usefulness of SBS for sensing is general: any change in external conditions, that affects the acoustic properties of the optical fiber, can in principle be detected. This is true of direct temperature and pressure changes, but can also be true of changes in chemical environment that can be made to result in temperature and pressure changes. An SBS based BOTDR sensing system was used to detect water content changes in soil. Water transducers (hydrophilic polymers) were tested to correlate Brillouin strain response to the water content of the surrounding soil environment. In these experiments, the optical fiber was wound and secured about discretely placed discs (2 cm length x 5 cm diameter) of AEP60 hydrophilic polymer (Figure 8A), stringed along 100-m fiber continuous optical fiber. The diameter of the polymer disc was selected to accommodate the minimum curvature of bending of the fiber, as shown in Figure 8B.

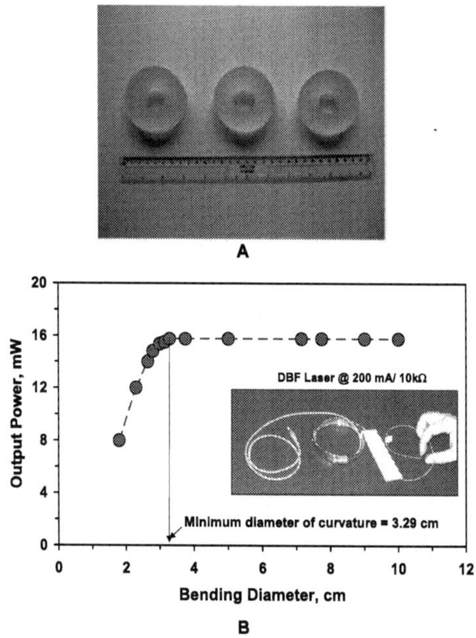

Figure 8. A photograph of the AEP60 hydrogel polymer discs used in sensor (A); assembly and test results for minimum fiber bending radius assessment (B).

The optical fibre at the inlet and outlet of the string of four transducers were spliced to spools of fibre on each end, and connected to the photonic set-up. Each water transducer was then embedded in a wet clay sample of predetermined water content (5, 10, 20 and 30 % by dry weight of clay), as shown in the inset sketch of Figure 9B.

The clay samples were packed in equal volume, watertight, cylindrical cells of 14-cm diameter and 28-cm height. The experiments were conducted in a temperature-controlled environment, at 25°C so that Brillouin scattering measurements were not influenced by thermal expansion or contraction of the fibre. The Brillouin shift was measured with 5 to 20 minute intervals up to the maximum observable swelling. When no significant change in Brillouin shift was recorded for three consecutive measurements, the transducers were removed from the soil chambers and left for open air-drying.

The Brillouin shift measurements were recorded with 5 to 20 minute intervals until the drying phase was completed. Since Brillouin readings could be recorded for strains as low as 10, very small volume change due to water absorption could be detected in a few minutes. Figure 9B shows the time rate of Brillouin signal changes during the hydrogel swelling and drying cycles of the 4-different water content clay hosts.

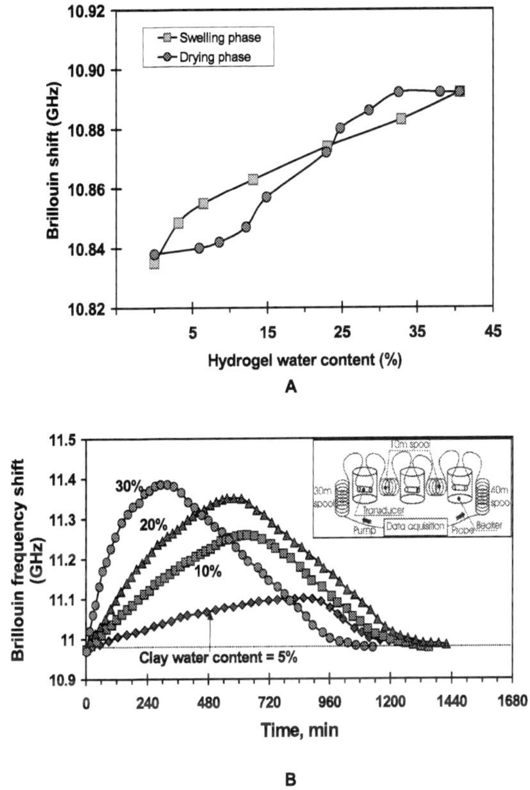

Figure 9. (A) Calibration of Brillouin Shift of fiber versus the water content by dry weight of the hydrogel disc. (B) Brillouin signal shift as a function of swelling/shrinkage time and clay water content (*Inset* – sketch of experimental set up).

In the experiments described above, the AEP60 polymer used would typically expand from 38 % to 400 % over dry volume when exposed to water. They are non-toxic and are manufactured in medical grades, approved for use in human wound care applications. This group of polymers does not swell in hydrocarbons and chlorinated solvents and has high thermal resistance. They are cross-linked to give them mechanical stability and accurate expansion characteristics. The water absorption and expansion factor can be accurately pre-defined at the formulation stage. Full expansion is reproducible over many wetting and drying cycles and is consistent over a wide range of pH and dissolved solid concentrations. A different integration of optical fiber and polymer transducer was used to improve the polymer response kinetics as shown in Figure 10. In this design, the polymer was reduced to smaller size discrete sleeves (1cm length x 0.3 cm diameter) that were bonded over the optical fiber. The bonding adhesive used was Locktite 414, a super bonding, quick drying adhesive containing cyanoacrylate and is intended for plastics and vinyl. The Locktite 414 was applied to each end of the polymer sleeve also.

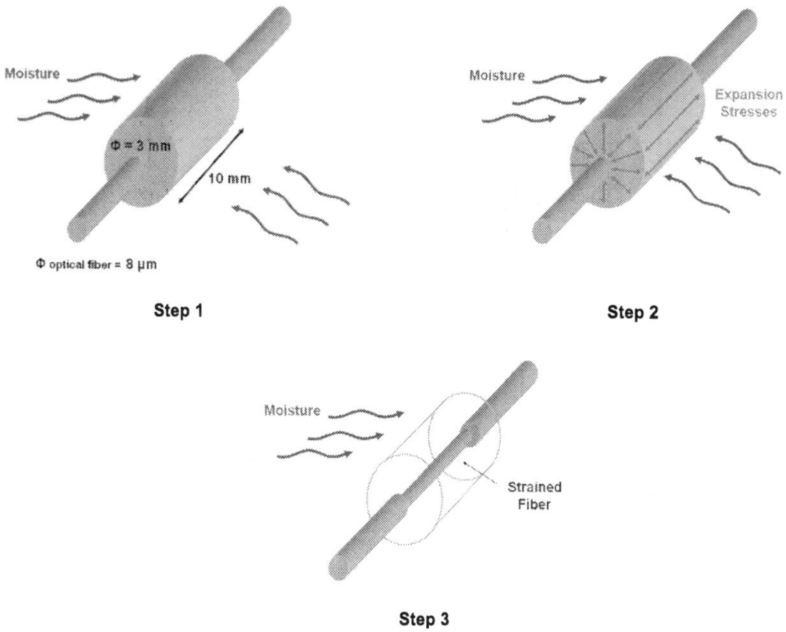

Figure 10. The working principle of the integrated sensor. Step 1 - Solution sorbed by the chemical selective polymer sleeve bonded onto the fiberoptic cable. Step 2 - The selective polymer swell upon encountering the target compound or ion in the solution. Step 3 - The swelling of the bonded polymer sleeve induces a "tangential pull" or "axial strain" on to the fiber locally.

The reduced size was anticipated to improve the swelling kinetics and alleviate the hysteresis affects observed in the previous configuration. The working principle of the integrated sensor is also depicted in Figure 10, where first the influx of the target substance (e.g. water) into the polymer transducer causes swelling of the bonded polymer. The swelling causes the bonding interface to strain and cause the fiber elongate in tangential pull. The fiber strain can then be recorded with location and amplitude, as shown in Figure 11, indicating where along the fiber line the influx of the target substance had occurred, and also the calibrated quantity of the substance based on the degree of swelling of the polymer, respectively.

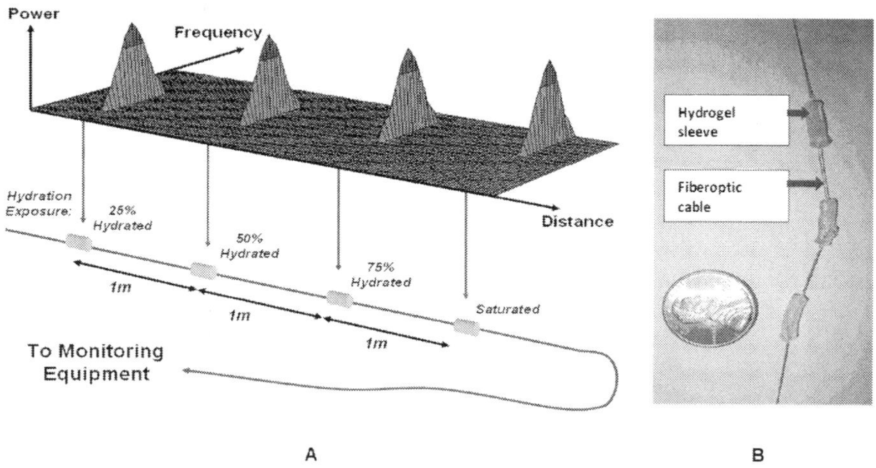

Figure 11. The operation schematic and the photograph of a spent water sensor: (A) Conceptual schematic of the integrated sensor assembly and its working principle as a distributed water sensor. (B) A laboratory prototype of the integrated water sensor.

Repeated swelling and shrinkage tests of the polymer sleeve component of the integrated water sensor showed hysteresis of length and diameter change. Both the length and diameter of the of the polymer sleeve expanded by 35 % (~ 0.35 cm and ~0.1 cm, respectively) after being soaked in water for three days. The majority of this expansion occurred during the first 12 hours of soaking. Increasing with every cycle, the final dry length of the polymer was greater than the original by ~ 0.025 cm (~2.5 %). As more cycles were completed, the diameter increased to its maximum faster, but the value of this maximum decreased. The final diameter of the polymer sleeve was fairly constant for each cycle, slightly less than the original by ~ 0.0005 cm (~0.2 %).

The magnitude and rate of swelling correlated directly with the initial water content when the polymer sleeves were embedded in test clay specimens of different water contents. Once again the full swelling occurred between 8 to 12 h when the sleeves were embedded in wet clay. Figure 12A shows that the polymer linear extension and clay water content relation was fairly linear. The linear relation is desirable for robust calibration. Figure 12B shows the dimensionless frequency shift response of the integrated water sensor to clay water content increase. The figure plots two spectrums, the shift spectrum at the location of the expanded polymer and a spectrum near the polymer location that does not undergo the swelling stress.

The actual Brilloin frequency shift was measured 0.0432 GHz corresponding to axial strain of 0.098 % for the fiber. The actual elongation of the polymer rod was 3 mm, which fully transferred to the fiber would have corresponded to an average strain of 0.14 % for the fiber over the 12.6 ns pulse. This meant that about 70 percent of the axial tug generated by the swelling of polymer was transferred to the fiber. The other 30 percent can be attributed to slippage

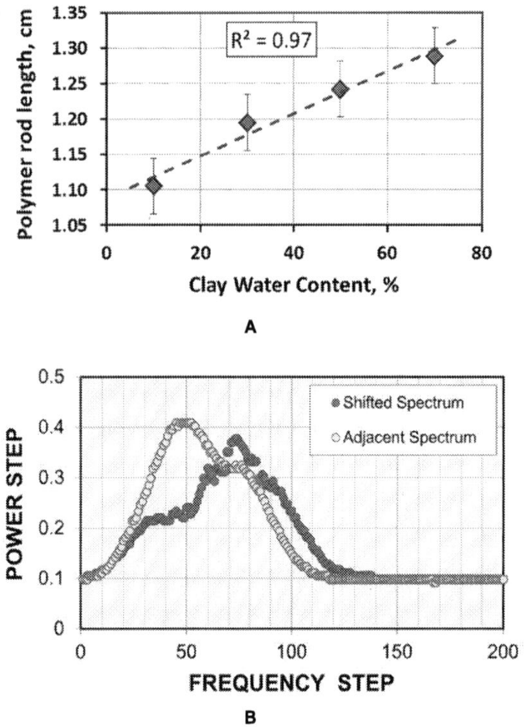

Figure 12. (A) Variation of embedded AEP60 polymer sleeve length vs clay w/c in swelling behavior. (B) Shifted frequency spectra for integrated sensor under swelling stresses with increased water.

across the length of the bond between the fiber and the polymer sleeve or other process related inefficiencies.

These test results demonstrated the viability of integration of optical fiber with reactive polymer as a BOTDR based sensor. Yet, the slow kinetics of the response (~12 hr for full swelling in case of water) rendered the prototype assembly limited for quick detection and measurement purposes. Hence a new polymer and fiber-optic cable configuration is proposed where a thin layer (e.g. on the order of few hundred μm) of the reactive polymer is brushed and bonded onto the fiber-optic cable. Figure 13 presents variations of the conceptual sensor where the reactive polymer coat is continuous. The working principle of this new configuration is similar to the previous ones described, all based on BOTDR, with the exception that the continuous coating of a thin layer reactive polymer is anticipated to provide a truly distributed and fast detection mechanism.

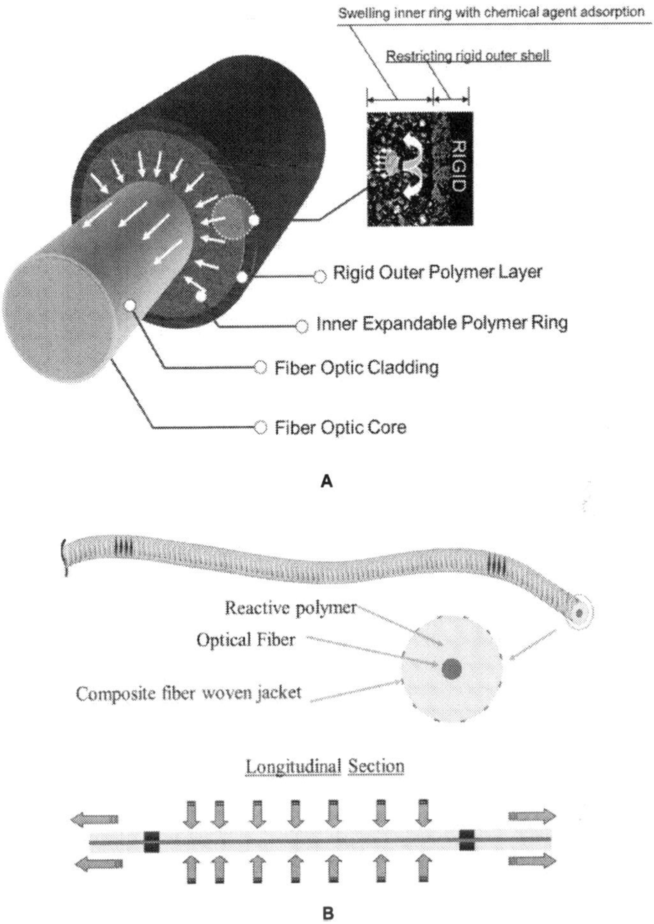

Figure 13. Advanced integrated chemical sensors based on BOTDR/BOTDA: (A) Cross-section of integrated sensor with core/shell polymer layers; (B) Working principal of an integrated sensor with polymer/jacket combination.

3.2. Reactive polymers principles used as sensors

Point detection fiber optic sensors have been developed for measurement of liquid levels, chemical species (inorganic and organic), drugs, environmental agents (such as pollutants and pesticides), biochemical reactions, and to monitor a wide variety of various chemical processes (Wolfbeis, 2000). One of the major components of a sensor system is the sensing or recognition element. Polymers have often been utilized as a chemical sensing material. The interaction of an analyte with the polymer coating is of prime importance. A signal needs to be generated once this interaction occurs, and much work has been carried out to understand the interaction of the polymer coating with the analyte molecules, especially in terms of the diffusion behavior of the analyte through the polymer coating to the actual sensor itself and its subsequent enrichment within the polymeric coating layer (Philips, et al, 2003).

The integrated chemical sensor based on BOTDR discussed in here is a single SBS (Stimulated Brillouin Scattering) sensing optical fiber bonded with such a polymer, which swells selectively in the presence of a target chemical. The polymer coat is cross-linked to swell in a preferential direction. During swelling, the bonded polymer coat exerts a radial or tangential force at the polymer-fiber interface, hence an axial strain on the fiber. The SBS signals are generated along the fiber at the discrete points of chemical contact triggered by polymer swelling, hence the local axial strains in the fiber. Consequently, the location of the target chemical can be detected simply by linear positioning of the SBS signals over the entire length of the cable. Because the detection is based on local physical changes in the fiber and not on loss of transmitted light, widely distributed sensing is possible without high power requirements. The specific detection and measurement components of the integrated sensor described above include, a core/shell type (multi-layer) polymer coating, and an inner fiber optic cable core, or a rigid woven jacket conceptualized in Figure 13.

In this conceptual design, the outer cross-linked polymer coating (shell) serves as a rigid, high permeability filter and confinement to the inner polymer (core). The core is a flexible, chemically selective polymer, preferably with rapid mass sorption kinetics. While the highly networked rigid polymer shell confines and directs the cores welling toward the fiber optic cable, the selective detection of chemicals is based on the thermodynamics and kinetics of chemical sorption and swelling of the core polymer layer. A volume change or "swelling" occurs in the inner flexible polymer layer as a result of mass sorption of the surrounding analyte. Solute/solvent sorption interrupts the intermolecular forces between individual chains of a lightly cross-linked or a linear polymer adjacent to the glass fiber, resulting in swelling forces. The outer, permeable but rigid polymer layer comprised of highly networked cross-linked chains help confine the volume change tendency and direct part of the swelling forces inward (radially and/or longitudinally) thus creating a hoop or a tangential stress on the fiber at the polymer interface. This "pinch" or tug" of the fiber cause changes in the elastic and refractive properties of the fiber locally, generating a shift in its original Brillouin frequency at that local. Comparing the new frequency to the original, it is then possible to quantify the change in terms of the swelling forces and the quantity of the absorbed substance.

A multi-chemical sensor can be developed by bundling polymer-coated fibers of different functions and sensitivities. The entire sensor assembly can be several tens of kilometers of optical fiber hosting several 100 measurement locations on the same line, each at a minimum spatial resolution of 1-meter. The photonics assembly connected to one end of the fiber allows fast detection of discrete sensing locations rendering the entire assembly a multiplexed network of many point sensors on the same transmittal line. Hence, the fiber line coupled with the stringed transducers can be embedded linearly or laced into a host medium (i.e. water pipeline, paved surfaces, porous media such as soil or concrete) to detect target chemical(s) online over large distances, areas or spaces by linear positioning of the fiber.

http://dx.doi.org/10.5772/57285

3.2.1. *Reactive polymers*

The molecularly imprinted polymers are often used to improve selectivity (Wolfbeis, 2000; Philips et al, 2003). The incorporation of desired functional monomers into the polymer structure further enhances the selectivity to a given analyte. These polymers are cross-linked and prepared by free radical polymerization processes such as solution or dispersion polymerization with acrylic or vinyl monomers. Polyelectrolyte gels are charged cross-linked three-dimensional networks of monomers that possess high swelling capability due to solvent sorption. The amount of swelling is known to be a string function of pressure, temperature, ion concentrations and pH changes (Siegel, 1993; Siegel et al, 1998; Matsuo and Tanaka, 1988). Their swelling and kinetics depend on parameters such as the degree of cross-linking (Skouri et al, 1995), external salt molarity (Yin et. al, 1992), and the degree of gel ionization rule (Katchalsky and Michaeli, 1995; Yin et al, 1992).

An emulsion or miniemulsion polymerization approach can be utilized to prepare film-forming polymer latexes with desired functional moieties to be used to coat optical fibers. The base latex polymer may be based on acrylic (e.g., *n*-butyl acrylate, n-butyl methacrylate) or styrene/acrylic film-forming compositions (i.e., with glass transition temperatures (T_g) of room temperature or lower). These latexes are prepared by conventional emulsion polymerization or by a miniemulsion polymerization process in the case where monomers with very low water solubility are used. In the miniemulsion polymerization process, the monomer would be emulsified in the presence of an aqueous surfactant (such as sodium lauryl sulfate) / costabilizer (e.g., hexadecane dissolved in the (co)monomer mixture) combination using a high shear device to form miniemulsion droplets which could then be polymerized in the presence of a free radical such as potassium persulfate.

The polymer would also be crosslinked to varying extents using crosslinking monomers such as ethylene glycol dimethacrylate, divinylbenzene, or bisacrylamide. In addition, functional monomers can be copolymerized along with the base acrylic or styrene/acrylic monomers. One type of monomer is a carboxylic acid such as methacrylic acid (MAA), which copolymerizes well with the base monomers. At high concentrations, this latex could function as an alkali-swellable latex whereby the latex particle size, and coating swellability, would increase dramatically upon neutralization in aqueous solutions of high pH (e.g. > 10) which would trigger a sensor response. N-methylol acrylamide (NMA) may also be incorporated into the base copolymer composition to obtain a crosslinked polymer, which can also act as a hydrogel, which could also swell when exposed to water. In addition, the monomer, N-(isopropylacrylamide) (NIPAM) will also be utilized for forming hydrogel particles which can swell when exposed to water. Incorporating NIPAM into a polymer composition would also lead to the formation of a thermosensitive polymer coating since poly(NIPAM) exhibits a strong phase transition above 32°C. It is also possible to copolymerize a alkoxysilane-containing monomer with the acrylic or styrene-acrylic monomers via miniemulsion polymerization to enhance the compatibility of the polymer coating with the glass optical fiber.

3.2.2. *Bonding/lamination*

The extent of bonding of the polymer coating to the glass optical fibers is critical. The optical fibers can either be coated with the manufacturer's cladding removed or in place. Preliminary experiments have shown that it is difficult to coat the uncoated glass fibers. These fibers are brittle without the manufacturer's cladding in place; the composition of which is unknown. The fiber can be passed through a coagulant bath prior to its immersion in the latex bath. Similar to dip coating, which has been utilized in preliminary coating experiments, the latex will coagulate onto the glass fiber. The surface of the fiber needs to be made hydrophilic for this process. This can be achieved by either physical adsorption of nonionic water-soluble polymer such as poly(vinyl alcohol) (PVOH) or by corona treatment of the fiber surface. In addition, the polymer processing techniques used in wire coating applications can also be applied to the case of the optical fibers. Important coating parameters would include the solids content of the latex (a high solids content is needed to control the rheology of the dispersion to be coated; a reasonable viscosity is needed for effective coating). Latexes can also be made self-thickening by the incorporation of carboxyl groups into the latex particles. A thickener can also be added to a latex composition to adjust the coating viscosity.

In addition, the surface tension of the latex would need to be controlled to give good wetting onto the glass fiber. Contact angle measurements on glass substrates can be used to determine the optimum wetting behavior before moving on to the glass fiber itself. The thickness of the polymer coating would also need to be varied to determine the necessary thickness needed to give a good, measurable response when exposed to solvent or water containing the heavy metal ions. If the coating is not thick enough, the response to the analyte may be too weak. If the coating is not uniform on the fiber, there would be unexposed regions of the fiber which would affect the detection limit and sensitivity. In addition, there needs to be good adhesion of the coating to the fiber, otherwise delamination could occur. Silane adhesion promoters can be explored to enhance adhesion of the polymer coating onto the glass fiber substrate. The drying temperature and drying conditions (e.g., time and temperature that the latex-coated fibers are dried in an oven to ensure good film integrity or the use of forced heated air flow over the fibers) are also critical coating variables to be investigated.

3.2.3. *Kinetics*

The kinetics of swelling of the polymer coatings when exposed to aqueous or organic media needs to be evaluated by monitoring the changes in the dimensions of the coating or the gravimetric uptake of the media by the polymer. The time-dependent changes can be analyzed to give an idea of the best polymer architecture to obtain an optimum sensor response when exposed to a given chemical. The time constant for the sensor response needs to be determined and correlated with the swelling kinetics of the polymer coating to achieve the best sensor performance.

3.3. Development of a prototype ph sensor with reactive polymer coating

The development of an optical fiber pH sensor based on hetero-core fiber structure coated with an acrylic polymer doped with Prussian blue is discussed here. In this design, the pH changes of the surrounding medium affects the Prussian blue present in the layer and produce a change in the refractive index of the layer. The pH changes are then observed as an increment in the hetero-core transmission signal.

3.3.1. Building of hetero-core optical fibers with reactive layer

The hetero-core fibers were constructed using two different length and two different types of optical fiber. In this case two types of single-mode fibers (SMA and SMB) and two of multimode fibers (MMA and MMB) were used. First, two pieces of MM fiber, stripped of its coating polymer (3 cm section) were spliced to a stripped SM fiber on each side. The hetero-core fibers were treated with Prussian Blue 0.1 mM (PB), polyvinyl alcohol(PVOH) at 4 %,acrylic polymer emulsion (APE) at 50 % plus and their combinations, like PVOH + PB and APE + PB to develop a reactive coat over the stripped surfaces. A small U-shape container made of a glass capillary was fixed to a mechanical mount and was filled with the mixture of polymer support and Prussian blue sensitive material. Then the single-mode section of the hetero-core fiber was immersed for 5 minutes into the solution after which the fiber was removed and dried at room temperature. In this manner the sensitive material was adhered to the single-mode section of the hetero-core fiber.

One end of the hetero-core fiber was connected to a white light source Yokogawa AQ4305 and the other to the spectrum analyzer Ando AQ6315A (Figure 14A). The set-up was used to measure the transmission light during the modification process of the fiber and later to measure the response of the modified fiber to pH changes. In order to test the sensitivity of the device to changes in pH, a test was designed which consisted of immersing the optical fiber section modified with PVOH/PB or APE/PB in a Petri dish where the pH was varied by adding 0.1 M NaOH or 0.1 M HCl, recording each transmission spectrum changes in the wavelength range from 350 nm to 1700 nm (Figure 14B).

The transmission spectra of two hetero-core fibers with PVOH/A (5 mm and 10 mm sections) were measured in different pH solutions are shown in Figure 14A and 14B, respectively. As seen in figure 15 the device has good sensitivity (-1.5 dB and -2 dB approximately), however the signal is erratic and not repeatable for different pH changes. This was attributed to solubility of PVOH in acidic conditions, and checked visually and with the transmission spectrum analysis. A new polymer, acrylic polymers emulsion (APE) was selected to replace PVOH. This polymer has similar characteristics as PVOH. It is water soluble, inexpensive, and colorless when dried, and has been reported as a good support in manufacturing of modified electrodes for pH determination.

The transmission spectra of two hetero-core fibers with APE/A (5 mm and 10 mm sections) were measured in different pH solutions are shown in Figure 16A and 16B, respectively. As seen in Figure 16A, the device showed very obvious changes when subjected to acidic and basic conditions. The presence of three peaks in visible region, 400, 500 and 700 nm wave-

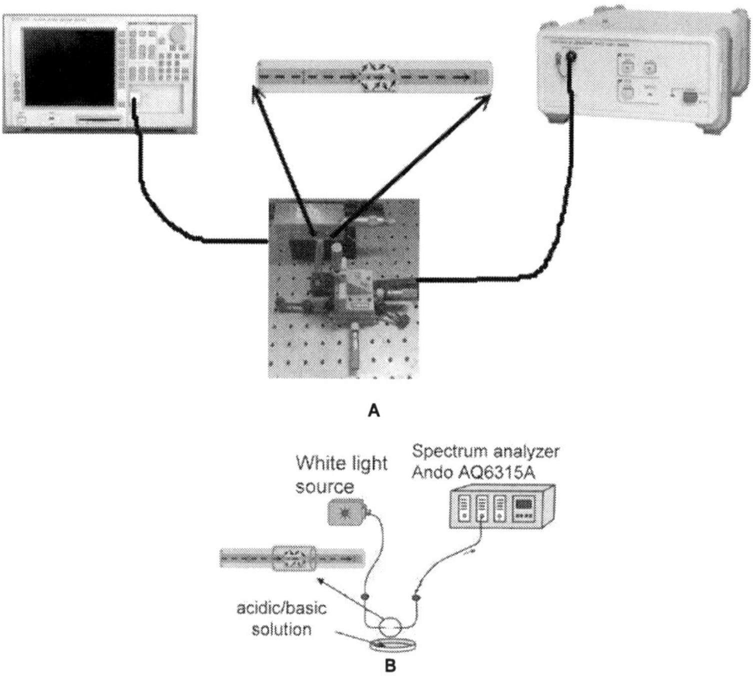

Figure 14. Set up to test prototype hetero-core fibers for chemical detection: (A) Diagram of the set-up to measure the transmission signal of the hetero-core fibers; (B) Diagram of the test system to determine the sensitivity of hetero-core fiber to pH changes.

lengths were noted when optical fiber was in air. These signals were attributed to light absorption and loss of light by index refraction changes by the composite material. In acidic pH values the light losses were in the range of 1.5 to 3 dB while for basic pH levels transmission near to 0.5 dB.

In order to identify the origin of signals found, the concentration of Prussian blue was increased to 2 mM. As we can see in Figure 16B, the intensity of transmission peaks at 400, 500 and 700 nm was increased, suggesting that they are due to increased concentration of PB in the composite. It also shows that the device sensitivity increased from 1.5 to 3 dB with 1mM concentration of PB until 4 to 6 dB with 2 mM concentration of PB for acidic pH solutions, but behaved same as previous in basic solutions. Finally there was a good return to initial conditions after each change of interface (Figure 17).

Subsequently we performed a sensitivity analysis for pH changes by taking the APE/PB modified fiber signal in air to use as a normalizing reference. As shown in Figure 18, there is good sensitivity to pH values lower than 7 with gains up to 6 dB at 400 nm (absorption or loss peak), whereas above pH 7 the peak is inverted, turning in a gains peak which may be due to

http://dx.doi.org/10.5772/57285

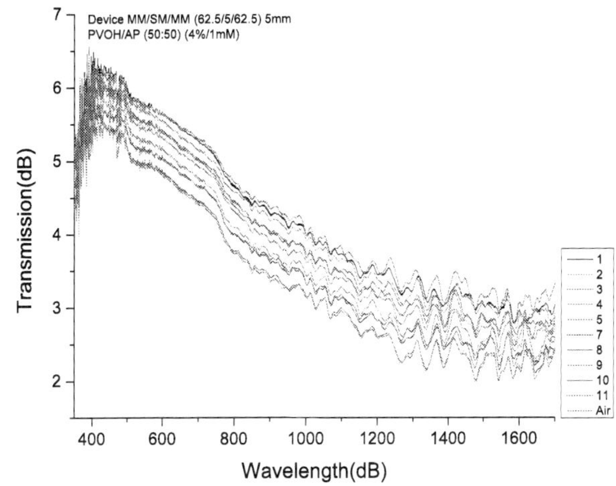

A

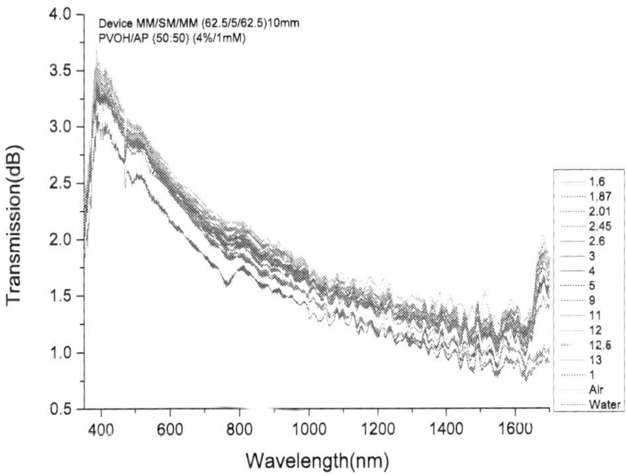

B

Figure 15. Transmission spectra of hetero-core fiber with PVOH/PB and its sensitivity to pH change with 5 mm (A) and 10 mm (B) length.

the hydration process of the polymer and breaking of complex of PB by hydration. To identify the changes in transmission spectrums, the most characteristic signals (400, 700 and 800 nm) and the response to 1500 nm (common wavelength in telecommunication systems) were plotted independently.

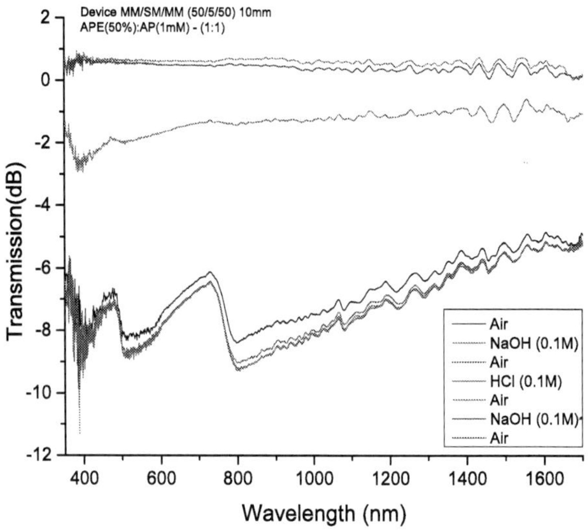

A

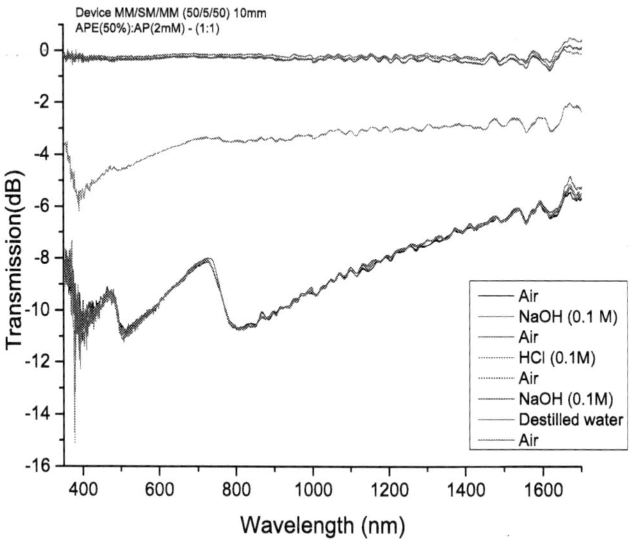

B

Figure 16. Transmission spectra of hetero-core fiber with APE/PB and its sensitivity to pH change with 1 mM (A) and 2 mM (B) PB.

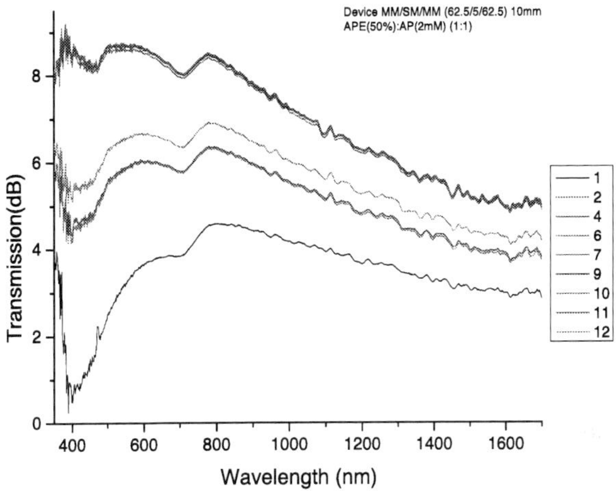

Figure 17. Transmission spectrum normalized to air of the modified hetero-core device with APE/PB (2 mM) and its sensibility to pH changes.

Analyzing the charts of Figure 18 shows that the transmission intervals are decreasing with increasing wavelength, which demonstrates good sensitivity of the device. As previously mentioned, the pH changes were more evident at pH values less than 7, due to that pH values higher than 7, it promotes the process of hydration of Prussian blue complex (Equation 4) and the signal grows weak on each pH change (García – Jareño et al, 1996).

$$Fe_4\left[Fe(CN)_6\right]_3 + 12OH^- \Leftrightarrow 4Fe(OH)_3 + 3\left[Fe(CN)_6\right]^{4-} \quad (4)$$

Based on the results obtained so far, the 10 mm rather than the 5 mm length hetero-core device is recommended since their sensitivity and the evanescent wave field is bigger than 5 mm length devices. Also the 10 mm length device provides a gain of about 2 dB at throughout the analysis spectrum.

Actually, all devices showed in this paper are preparing to their application in real scenarios, with the intention to quantify physicochemical properties directly to polluted soil without extraction from the field and pre-treatment of sample, which could reduce time and costs of analytical determination, increasing the sensibility, detection and quantification limits in comparison with spectroscopic and spectrometric techniques, to take the best professional decision to remediate in the better technical conditions the polluted soil.

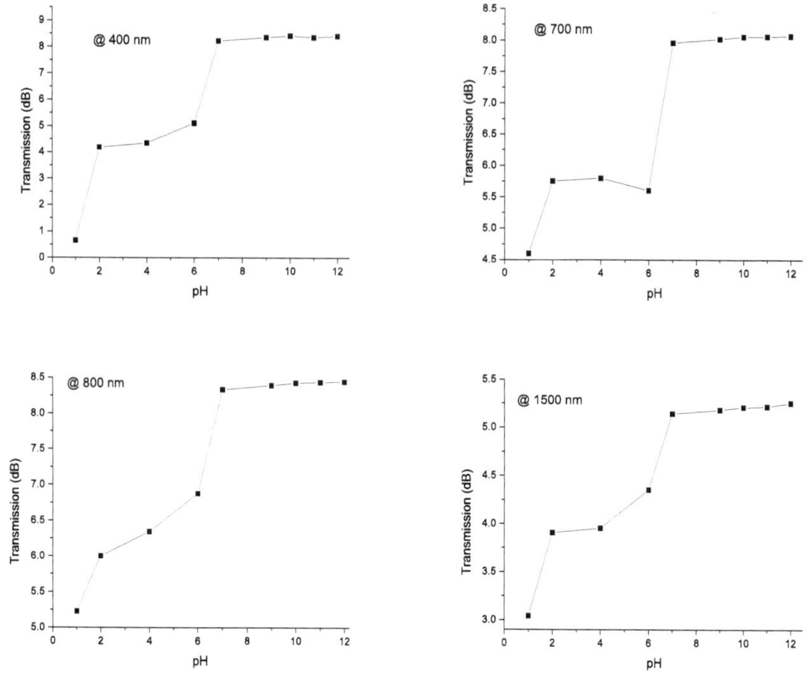

Figure 18. Transmission spectrum of hetero-core device modified with APE/PB and its sensibility to pH changes at 400, 700, 800 and 1500 nm wavelength.

4. Conclusion

For pollution detection and soil remediation purposes it is essential to have relevant and reliable information on the soil structure, the hydrogeological circumstances and accumulation zones of the detected pollutants. Combined application of geological, hydrogeological and geophysical investigations prior the placement of the optical fiber in the field may increase the efficiency of the monitoring technique.

Spatially resolved mapping of chemical constituents is an important need in a variety of environmental and geo-environmental applications. For example, spatially resolved analyte monitoring can simultaneously indicate and locate when an accepted level of exposure to toxic or explosive species has been exceeded, and can track its source.

The capability of long-range distributed sensing is unique to optical – fiber technology. A distributed fiber optic sensor returns a value of a target measurement as a function of the linear position along the fiber length. The only contact between the point to be measured and the observation area is the optical fiber.

Acknowledgements

The authors would like to thank the Consejo Nacional de Ciencia y Tecnología de los Estados Unidos Mexicanos (CONACyT), L'Oreal, Academia Mexicana de Ciencias (AMC) and Fundación México – Estados Unidos para la Ciencia (FUMEC).J. A. García is grateful to CONACyT for his scholarship.

Author details

J. A. García[1], D. Monzón[2], A. Martínez[2], S. Pamukcu[3], R. García[4] and E. Bustos[1*]

*Address all correspondence to: ebustos@cideteq.mx

1 Centro de Investigación y Desarrollo Tecnológico en Electroquímica S.C., Querétaro, México

2 Centro de Investigaciones en Óptica A. C., León, México

3 Fritz Engineering Laboratory, Lehigh University, Bethlehem, USA

4 Laboratorio de Química Atmosférica, Centro de Ciencias de la Atmósfera, Universidad Nacional Autónoma de México, Ciudad Universitaria, Coyoacán, Mexico

References

[1] Abdi, M. M., Abdullah, L. Ch., Sadrolhosseini, A. M., Yunus, W. M. M., Moksin, M. M., Tahir, P. M. (2011). Surface plasmon resonance sensing detection of mercury and lead ions based on conducting polymer composite, *Plos One*, 6, e24578.

[2] Achatz, D. E., Ali, R., Wolfbeis, O. S. (2011). Luminescent chemical sensing, biosensing, and screening using upconverting nanoparticles, *Top Current Chemistry*, 300, 29 – 50.

[3] Aksuner, N. (2011). Development of a new fluorescent sensor based on a triazolo-thiadiazin, derivative immobilized in polyvinyl chloride membrane for sensitive detection of lead (II) ions, *Sensors and Actuators B*, 157, 162 – 168.

[4] Anastasio, S., Pamukcu, S., Pervizpour, M. (2007). BOTDR Detection of Chemical & Liquid Content, Proc. of the 7th FMGM, GSP 175 Int. Sym. on Field Measurements in Geomechanics, ASCE, Boston, MA, 1 - 12.

[5] Anastasio. S, Pamukcu, S., Pervizpour, M. (2007). Chemical Selective BOTDR Sensing for Corrosion Detection on Structural Systems, Chang, ed., Proc. of the 7th Int. Workshop on Structural Health Monitoring (IWSHM 2007), Stanford, CA, 1701 - 1708.

[6] Antico, E., Lerchi, M., Rusterholz, B., Achermann, N., Badertscher, M., Valiente, M., Pretsch, E. (1999). Monitoring Pb^{2+} with optical sensing films, *AnalyticalChimicaActa*, 388, 327 - 338.

[7] Balaji, T., Sasidharan, M., Matsunaga, H. (2006).Naked eye detection of cadmium using inorganic–organic hybrid mesoporous material, Analytical Bioanalytical Chemistry, 384, 488-494.

[8] Bao, X., DeMerchant, M., Brown, A., and Bremner, T. (2001). Tensile and compressive strain measurement in the lab and field with the distributed Brillouin scattering sensor.*Journal of Lightwave Technology*, 19, 1698.

[9] Bao, X., Dhliwayo, J., Heron, N., Webb, D. J., Jackson, D. A. (1995). Experimental and theoretical studies on a distributed temperature sensor based on Brillouin scattering, *Journal of Light Technology*, 13, 1340 - 1348.

[10] Buerck, J., Roth, S., Kraemer, K., Mathieu, H. (2001). OTDR distributed sensing of liquid hydrocarbons using polymer-clad optical fibers, Proceedings of The Second International Symposium and Workshop on Time Domain Reflectometry for Innovative Geotechnical Applications, C. H. Dowding, ed. Academic, Evanston, Il., 496 - 509.

[11] Cui, Q., Pamukcu, S., Xiao, W., Guintrand, C., Toulouse, J., Pervizpour, M. (2009). Distributed fiber sensor based on modulated pulse base reflection and Brillouin gain spectrum analysis, *Applied Optics*, 48 (30), 5823 - 5828.

[12] Cui, Q., Pamukcu, S., Lin, A., Xiao, W., Toulouse, J. (2010). Performance of double side band modulated probe wave in BOTDA distributed fiber sensor, *Microwave and Optical Technology Letters*, 52, 2713 - 2717.

[13] Cui, Q., Pamukcu, S., Lin, A., Xiao, W., Herr, D., Toulouse, J., Pervizpour, M. (2011). Distributed temperature sensing system based on Rayleigh scattering BOTDA, *IEEE Sensors Journal*, 11(2), 399 - 403.

[14] Cui, Q., Pamukcu, S., Xiao, W., Pervizpour, M. (2011). Truly distributed fiber vibration sensor using pulse base BOTDA with wide dynamic range, *IEEE Photonics Technology Letters*, 3 (24), 1887 - 1889.

[15] Fellay, A., Thévenaz, L., Facchini, M., Niklès, M., Robert, P. (1997). Distributed sensing using stimulated Brillouin scattering: towards ultimate resolution, Optical Fiber Sensors, OSA Technical Digest Series, Optical Society of America, Washington, D.C. 16, 324 – 327.

[16] Fen, Y. W., Mahmood, W., Yunus, M., Yusof, N. A. (2012). Surface plasmon resonance optical sensor for detection of Pb^{2+} based on immobilized p-tert-butylca-

lix[4]arene-tetrakis in chitosan thin film as an active layer, *Sensors and Actuators B*, 171 – 172, 287 – 293.

[17] Fen, Y. W., Yunus, W. M. M. (2013). Utilization of chitosan - based sensor thin films for the detection of lead ion by surface plasmon resonance optical sensor, *IEEE Sensors Journal*, 13, 1413 - 1418.

[18] Fen, Y. W., Yunus, W. M. M., Talib, Z. A. (2013). Analysis of Pb(II) ion sensing by crosslinked chitosan thin film using surface plasmon resonance spectroscopy, *Optik*, 124, 126 – 133.

[19] Forzani, E. S., Foley, K., Westerhoff, P.,Tao, N. (2007). Detection of arsenic in groundwater using a surface plasmon resonance sensor, *Sensors and Actuators B*, 123, 82 – 88.

[20] García-Jareño, J. J., Navarro-Laboulais, J., Vicente, F. (1996). Electrochemical Study of Nafion Membranes / Prussian Blue Films on ITO Electrodes. *Electrochimica Acta*, 41, 17, 2675 – 2682.

[21] Galindez-Jamioy, C. A., López-Higuera, J. M. (2012). Brillouin Distributed Fiber Sensors: An Overview and Applications. *Journal of Sensors, 204121, pp 17.*

[22] Grattan, K. T. V., Meggitt, B. T. (1999). Optical Fiber Sensor Technology, Chemical and Environmental Sensing, KlumerAcademinc Publishers, Vol. 4.

[23] Guillemain, H., Rajarajan, M., Sun, T., Grattan, K. T. V. (2009). A self-referenced reflectance sensor for the detection of lead and other heavy metal ions using optical fibres, *Measurement Science Technology*, 20, 045207.

[24] Guo, L., Zhang, W., Xie, Z., Lin, X., Chen, G. (2006). An organically modified sol–gel membrane for detection of mercury ions by using 5,10,15,20-tetraphenylporphyrin as a fluorescence indicador, *Sensors & Actuators B*, 119, 209 - 214.

[25] Horiguchi, T., Shimizu, K., Kurashima, T., Tateda, M., Koyamada, Y. (1995). Development of a distributed sensing technique using Brillouin scattering, *Journal of Lightwave Technology*, 13, 1296 – 1302.

[26] Jerónimo, P., Araújo, A., Conceição, B. S. M., Montenegro, M. (2007). Optical sensors and biosensors based on sol-gel films, *Talanta*, 72, 13 – 27.

[27] Kasik, I., Mrazek, J., Martan, T., Pospisilova, M., Podrazky, O., Matejec, V., Hoyerova, K., Kaminek, M. (2010) Fiber-optic pH detection in small volumes of biosamples, *Analytical and Bioanalytical Chemistry*, 398, 1883 – 1889.

[28] Katchalsky, A., Michaeli, I. (1995). Polyelectrolyte gels in salt solutions, *Journal of Polymer Science*, 15 (69).

[29] Kee, H. H., Lees, G. P., Newson, T. P. (2000). All-fiber system for simultaneous interrogation of distributed strain and temperature sensing by spontaneous Brillouin scattering.*Optics Letters*, 25, 695.

[30] Kocincova, A., Borisov, S., Krause, C., Wolfbeis, O. (2007).Fiber-optic microsensors for simultaneous sensing of oxygen and pH, and of oxygen and temperature.*Analytical chemistry*, 79, 8486 – 8493.

[31] Krohn, D. A. (1988).Fiber Optic sensors: Fundamental and applications, Instrument Society of America.

[32] Lee, S. -H., Kumar, J., Tripathy, S. K. (2000). Thin film optical sensors employing polyelectrolyte assembly. *Langmuir*, 16.

[33] Lin, T. –J., Chung, M. –F. (2009). Detection of cadmium by a fiber-optic biosensor based on localized surface plasmon resonance, *Biosensors and Bioelectronics*, 24, 1213 – 1218.

[34] Maier, S. A. (2007). Plasmonics Fundamental and applications, Springer.

[35] Matsuo, E. S., Tanaka, T. (1988). Kinetics of discontinuous volume phase transition of gels, *Journal of Chemical Physics*, 89 (3): 1695.

[36] Mayra, T., Klimant, I., Wolfbeis, O. S., Werner, T. (2008). Dual lifetime referenced optical sensor membrane for the determination of copper (II) ions, *Analytical ChimicaActa*, 462, 1 - 10.

[37] McDonagh, C., Burke, C. S., MacCraith, B. D. (2008). Optical Chemical Sensors, *Chemical Review*, 108, 400 - 422.

[38] Ohno, H., Naruse, H., Kihara, M., Shimada, A. (2001) Industrial applications of the BOTDR optical fiber strain sensor, *Optical Fiber Technology*, 7,45 – 64.

[39] Orellana, G., Haigh, D. (2008) New trends in fiber-optic chemical and biological sensors. *Current Analytical Chemistry*, 4, 273 – 295.

[40] Pamukcu, S., Cetisli, F., Texier, S., Naito, C., Toulouse, J. (2006).Dynamic strains with Brillouin scattering distributed fiber optic sensor, GeoCongress 2006, 187, ASCE pp. 31 - 36.

[41] Pamukcu, S., Texier, S., Toulouse, J. (2006).Advances in water content measurement with distributed fiber optic sensor, GeoCongress 2006a, 187, ASCE, pp. 7 - 12.

[42] Phillips, C., Jakusch, M., Steiner, H., Mizaikoff, B., Fedorov, A. G. (2003).Model-based optimal design of polymer-coated chemical sensors, *Analytical Chemistry*, 75, 1106 - 1115.

[43] Prabhakaran, D., Nanjo, H., Matsunaga, H. (2007). Naked eye sensor on polyvinyl chloride platform of chromo-ionophore molecular assemblies: A smart way for the colorimetric sensing of toxic metal ions, *Analytical ChimicaActa*, 601, 108 - 117.

[44] Siegel, R. A. (1993). Hydrophobic weak polyelectrolyte gels: studies of swelling equilibria and kinteics, *Advanced Polymer Science*, 109, 233.

[45] Siegel, R. A., Falamarzian, M., Firestone, B. A., Moxley, B. C. (1988).pH-controlled release from hydrophobic/polyelectrolyte copolymer hydrogels, *Journal of Controlled Release*, 8, 179.

[46] Skouri, R., Schosseler, F., Munch, J. P., Candau, S. J. (1995). Swelling and elastic properties of polyelectrolyte gels, *Macromolecules*, 28, 197.

[47] Texier, S., Pamukcu, S., Toulouse, J. (2005). Advances in subsurface water-content measurement with a distributed Brillouin scattering fibre-optic sensor, Proc.of SPIE 5855, 17th Int. Confer. On Optical Fibre Sensors, OFS-17, Bruggs, Belgium, pp. 555 - 558.

[48] Texier, S., Pamukcu, S., Toulouse, J., Ricles, J. (2005).Brillouin scattering fiber optic strain sensor for distributed applications in civil infrastructure, Chang, ed., 5th Int. Workshop on Structural Health Monitoring (IWSHM 2007), Stanford, CA, pp. 1395 - 1402.

[49] Turel, M., Pamukcu, S. (2006) Brillouin scattering fiber optic sensor for distributed measurement of liquid content and geosynthetic strains in subsurface, Geoshanghai GSP: Site and Geomaterial Characterization, ASCE, Shanghai, PRC, pp. 72 - 79.

[50] Villatoro, J., Monzón-Hernández, D. (2006). Low-cost optical fiber refractive-index sensor based on core diameter mismatch.*Journal of Light-wave Technology*, 24, 1409 - 1413.

[51] Wolfbeis, O. (2008). Fiber-optic chemical sensors and biosensors.*Analytical chemistry* 80, 4269 – 4283.

[52] Wolfbeis, O. S. (2000). Fiber-Optic Chemical Sensors and Biosensors, *Analytical Chemistry*, 72, 81R.

[53] Yin, Y. L., Prud'homme, R. K., Stanley, F. (1992). Chapter 6: Relationship between poly(acrylic acid) gel structure and synthesis. In A.J Harland and R.K. Prud'homme, editors, Polyelectrolyte Gels. ACS Symp. Series 480, ACS Washington D.C.

[54] Yusof, N. A., Ahmad, M. (2003). A flow-through optical fibre reflectance sensor for the detection of lead ion based on immobilized gallocynine, *Sensors and Actuators B*, 94, 201 – 209.

Chapter 16

Metabolomics for Soil Contamination Assessment

Maria C. Hernandez-Soriano and
Jose C. Jimenez-Lopez

Additional information is available at the end of the chapter

http://dx.doi.org/10.5772/58294

1. Introduction

The evaluation of biological responses to assess and predict the impact of environmental changes in ecosystems functioning is receiving increasing attention, and current research focuses on overcoming concerns about the specificity of biomarkers (Amiard-Triquet et al., 2012). Generally, biomarkers are chemicals, metabolites, susceptibility characteristics, or physiological changes that relate to the exposure of an organism to a chemical. Accordingly, a selected biomarker, i.e. biological response, can be linked to a specific environmental exposure, being representative of the health status of the ecosystem studied. The identification of biomarker profiles has been possible upon the development of metabolomics. Those profiles allow the genuine identification of the relevant biological response/s associated to a particular exposure while the assessment of a single biomarker could only estimate the potential response of the ecosystem to a particular pollutant.

Metabolomics is the generic name assigned to a scientific field that addresses the characterization of low molecular weight organic metabolites released by living organisms in response to environmental stimuli. Morrison et al. (2007) provided an extended definition "the application of metabolomics to the investigation of both free-living organisms obtained directly from the natural environment (whether studied in that environment or transferred to a laboratory for further experimentation) and of organisms reared under laboratory conditions (whether studied in the laboratory or transferred to the environment for further experimentation), where any laboratory experiments specifically serve to mimic scenarios encountered in the natural environment".

The methodological approach of metabolomics relies on a comprehensive analysis of the set of metabolites or "metabolome" produced in response to particular environmental stimuli. Accordingly, the metabolome is the pool of metabolites, small molecules, within a cell, tissue, organ, biological fluid, or entire organism (Miller, 2007). Exposure of an organism to an external stressor will result in changes at the level of the metabolome (Ankley et al., 2006; van Ravenzwaay et al., 2007), and such changes may constitute a highly sensitive indicator of an external stress. Therefore, metabolomics has potential as a sensitive and rapid technique that can elucidate the relationships between metabolite levels and an external stressor, such as contaminant exposure, nutritional deficit or a disease.

The main advantage of metabolomics over traditional research is to overcome the bias associated to the assessment of predefined metabolites (Singh, 2006). Among the diverse applications of omic profiling methods to the environmental sciences, ecotoxigenomics addresses the response of organisms to pollutants based on the different sensitivity of species to toxicants (Spurgeon et al., 2008). Currently, the implementation of metabolomics in ecological risk assessment is still at an early stage, mostly applied as a screening tool to assess the potential toxic effects of pollutants or to determine the mode of action (MOA) of a toxicant. Otherwise, application of metabolomics for environmental monitoring allows the study of a large variety of species from relatively uncontrolled environments.

Bundy et al. (2009) highlight the challenge of identifying a large number of metabolites and the necessity of creating metabolite databases specifically dedicated to environmental issues. Multivariate statistical analysis has proved highly effective for metabolite identification. Thus, principal component analysis (PCA) is used to identify differences between metabolic profiles of organisms exposed to organic or inorganic pollutants (Jones et al., 2014; Kwon et al., 2012; Lankadurai et al., 2011). Besides, association between the metabolic profile and biological factors evaluated as markers for exposure to pollutants can be modelled by partial least squares (PLS) regression analysis (Ellis et al., 2012).

The implementation of metabolomics for the assessment of soil contamination is nevertheless at an early stage (Viant, 2009). A basic screening of published research in the web of science returns circa 100 items for the search "soil pollution-metabolomics", with a significant launch in 2011 (Figure 1), reduced to 21 records when the search is narrowed with the term "biomarkers", published in the period 2007-2013. However, emerging regulatory challenges demand the advance of toxicity testing. Toxicogenomics tools have been presented as an advanced from the current methodologies used for regulatory decision making in ecotoxicology, which entirely rely on whole animal exposures and adverse effects on survival, growth, and reproduction (Ankley et al., 2006). From the acquisition of reproducible metabolic profiles as response to the presence of specific pollutants in soil (Jones et al., 2008) to the application of metabolomics techniques to the study of the response of the entire community of a soil to factors such as pollution and climate change (Jones et al., 2014), the implementation of metabolomics in ecotoxicology is a sound answer to the current needs of society and the environment (van Ravenzwaay et al., 2012).

During the last decade a number of general revisions about the application of metabolomics in environmental health assessment have been published (Bundy et al., 2009; Miller, 2007; Snape et al., 2004; van Ravenzwaay et al., 2007; Viant et al., 2003). The present review specifically summarizes the most significant research concerning implementation of

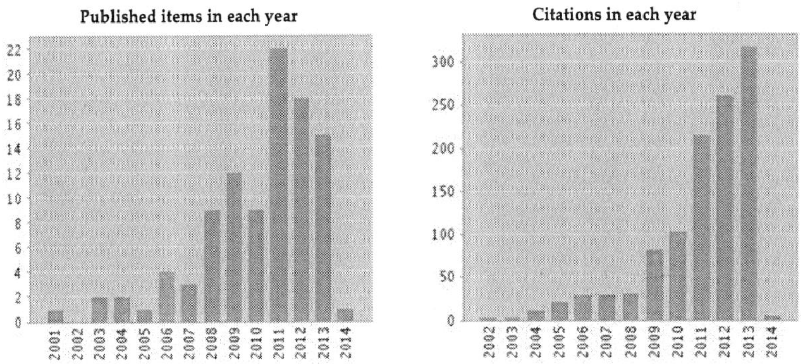

Figure 1. Citation report for the search "soil pollution-metabolomics" as obtained from Thomson Reuters (January 2014).

metabolomics in soil contamination assessment. The main objectives of this revision are i) to provide a systematized outline for the application of metabolomics in risk assessment of soil contamination, ii) to provide a rapid guide to the methodological approaches currently optimized and iii) to unify and simplify the knowledge currently available in the topic to provide an accessible tool for further advance in the implementation of metabolomics in risk assessment.

2. Methodological approaches

2.1. Metabolites isolation

Generally, metabolites are extracted from intact organisms (occasionally from selected relevant tissues) that have been exposed to the studied toxicant by moderate chemical extraction (Baylay et al., 2012; Yuk et al., 2010). The organisms commonly selected for toxicity testing are earthworms (Table 1), particularly the genus *Eisenia*, which are a classic model organism for toxicity assays (Sanchez-Hernandez, 2006; Van Gestel et al., 1992; van Gestel et al., 1989) and have been since long included in official guidelines (OECD, 1984, 2004). Earthworms ingest large amounts of soil and uptake a significant amount of contaminant through the skin. Therefore they are continuously exposed to contaminants. Extractions performed with methanol-chloroform (Baylay et al., 2012) or phosphate buffer solution (Yuk et al., 2010) on pulverized or lyophilised organisms are described to extract the maximum number of metabolites while allowing the performance of reliable analyses.

The isolated extracts usually might not require further sample treatment prior to analysis, which minimizes the introduction of artefacts but also facilitates the development of low cost, rapid methodologies.

2.2. Metabolites determination: chromatography, spectroscopy and spectrometry

The leading analytical techniques in metabolomics for soil contamination assessment are proton nuclear magnetic resonance spectroscopy (^{1}H NMR) and gas chromatography–mass spectrometry (GC-MS), as thoroughly reported in Table 1, both allowing the identification of compounds at molecular level in the analysed substances. Several authors have also implemented high pressure liquid chromatography (HPLC) and ultra high pressure liquid chromatography (UPLC) coupled with mass spectrometry detector (MS) for the assessment of biological responses to soil contamination with heavy metals (Hédiji et al., 2010; Hughes et al., 2009). Overall, these analytical techniques allow the determination and identification of the metabolites that foremost represent the metabolic alterations related to the toxic effects of organic or inorganic contaminants in soil.

Technique	**Organic toxicant**	**Animal model tested**	**Biomarkers of contaminant exposure**	**Reference**
^{1}H NMR GC/MS	2-fluro-4-methylaniline	*E. veneta*	^{D}[2-hexyl-5-ethyl-3-furansulfonate, maltose], ^{I}inosine monophosphate	(Bundy et al., 2002)
^{1}H NMR GC/MS	3-trifluromethyaniline	*E. veneta*	^{I}lactate	(Lenz et al., 2005)
^{1}H NMR GC/MS	3-trifluromethylaniline	*E. veneta*	^{I}[Ala, Gly, Asn, glucose, citrate, succinate]	(Warne et al., 2000)
^{1}H NMR GC/MS	3-fluro-4-nitrophenol	*E. veneta*	^{D}[acetate, malonate], ^{I}[succinate, trimethylamine-N-oxide]	(Bundy et al., 2001)
^{1}H NMR GC/MS	3,5-difluroaniline	*E. veneta*	^{D}2-hexyl-5-ethyl-3-furansulfonate, ^{I}inosine monophosphate	(Bundy et al., 2002)
^{1}H NMR GC/MS	4-fluroaniline	*E. veneta*	^{D}[Maltose, hexyl-5-ethyl-3-furansulfonate]	(Bundy et al., 2002)
^{1}H NMR	Aroclor 1254	*E. fetida*	No significant changes	(Fitzpatrick et al., 1992)
^{1}H NMR	Caffeine	*E. fetida*	^{I}fumarate	(McKelvie et al., 2011)
^{1}H NMR	Carbamazephine	*E. fetida*	^{D}[Fumarate, Glu, Val, ^{D}Leu]	McKelvie et al. (2011)
^{1}H NMR	Carbaryl	*E. fetida*	^{D}[Phe, Tyr, Lys, Ala, Val, Leu]	(Heimbach, 1988) (Edwards and Bater, 1992)
^{1}H NMR	Chlorpyrifos	*E. fetida*	No significant changes	(Yu et al., 2006)
^{1}H NMR	Chlorpyrifos	*L. rubellus*	^{I}fumarate	(Baylay et al., 2012)

http://dx.doi.org/10.5772/58294

Technique	Organic toxicant	Animal model tested	Biomarkers of contaminant exposure	Reference
^{1}H NMR spectroscopy, GC–MS	Chlorpyrifos	*C. elegans*	^{I}[Ala, betaine, ornithine], ^{D}[choline, Glu, Gly, Ile, lactate, n-buterate, taurine]	(Baylay et al., 2012)
^{1}H NMR	Dimethyl phthalate	*E. fetida*	^{D}[Phenylalanine, Alanine, Leucine, Valine]	(Drewes and Vining, 1984)
^{1}H NMR GC/MS	DTT and Endosulfan	*E. fetida*	^{I}[maltose, Ala, Leu]	(McKelvie et al., 2009)
1-D & 2-D ^{1}H NMR spectroscopy	Endosulfan	*E. fetida*	Significant fluctuations in glutamine/GABAeglutamate cycle metabolites and spermidine	(Yuk et al., 2013)
1-D & 2-D ^{1}H NMR spectroscopy	Endosulfan Sulfate	*E. fetida*	Significant fluctuations in glutamine/GABAeglutamate cycle metabolites and spermidine	(Yuk et al., 2013)
^{1}H NMR	Estrone	*E. fetida*	^{D}[Adenine, Glu]	McKelvie et al. (2011)
^{1}H NMR GC-MS analysis	Imidacloprid/ Thiacloprid	*L. rubellus*	^{D}Toxicological endpoints (survival, weight loss, and reproduction)	Baylay et al. (2012)
^{1}H NMR	Naphthalene	*E. fetida*	^{D}[Ala, Leu, Val, Lys]	(Brown et al., 2009)
^{1}H NMR	Nonylphenol	*E. fetida*	^{D} [adenine, Glu]	McKelvie et al 2011
^{1}H NMR	Polybrominated diphenyl ethers (PBDE) 209	*E. fetida*	^{D}Maltose, ^{I}[Lys, Glu]	McKelvie et al. (2011)
^{1}H NMR	Perfluorooctanoic acid	*E. fetida*	^{I}[succinate, HEFS, Glu], ^{D}[leu, Val, LyS, Phe, Arg, maltose, ATP]	(Lankadurai et al., 2012)
^{1}H NMR	Perfluorooctane sulfonate	*E. fetida*	^{I}[succinate, HEFS, Glu], ^{D}[leu, Val, Lys, Phe, Arg, maltose, ATP]	(Lankadurai et al., 2012)
^{1}H NMR	PHA	*E. fetida*	Leu, Val, Ala, Lys and maltose changed in response to PAH exposure.	(Brown et al., 2009)
^{1}H NMR	Phenanthrene	*E. fetida*	^{I}[Ala, betaine, Scyllo- and myo-inositol, cholesterol, phosphatidylcholine], ^{D}Glu	(Lankadurai et al., 2011) Brown et al. (2010)

Technique	Organic toxicant	Animal model tested	Biomarkers of contaminant exposure	Reference
^{1}H NMR	Polychlorinated biphenyl	*E. fetida*	^{I}ATP	McKelvie et al. (2011) (Whitfield Åslund et al., 2011)
^{1}H NMR	Poly brominated diphenyl ethers	*E. fetida*	^{I}[lys, Glu], ^{D}maltose	McKelvie et al. (2011)
^{1}H NMR GC/MS	Pyrene	*L. rubellus*	^{D}[lactate, tetradecanoic acid, hexadecanoic acid, octadecanoic acid], ^{I}[Ala, Leu, Val, Ile, Lys, Tyr, methionine]	(Jones et al., 2008)
1-D & 2-D ^{1}H NMR spectroscopy	Rifluralin	*E. fetida*	^{I}[Ala, Gly, maltose, ATP]	Yuk et al. (2011)
^{1}H NMR	Thiacloprid	*L. rubellus*	No significant changes were reported	Baylay et al. (2012)
1-D & 2-D ^{1}H NMR spectroscopy	Trifluralin	*E. fetida*	^{I}[Ala, Gly, ATP], ^{D}maltose	Yuk et al. (2011)
^{1}H NMR spectroscopy and UPLC-MS	Cd	*C. elegans*	^{I}phytochelatins, ^{D}cystathionine	(Hughes et al., 2009)
^{1}H NMR, HPLC-PDA	Cd	*S. lycopersicum*	^{I}choline, ^{D}[glucose, citrate, malate, glutamine, asparagine, Phe, Tyr, Val, Ile, trigonelline]	(Hédiji et al., 2010)
^{1}H NMR, GC-MS analysis	Chlorpyrifos + Ni	*L. rubellus*	^{I}[Phe, glucose, malate, arachidonic acid, fumarate, Lys, Tyr, monosaccharides, monophosphorylated form of inositol, succinate, uracil, ethanolamine, Pro, putrescine], ^{D}myo-inositol,	(Baylay et al., 2012)
^{1}H NMR spectroscopy, GC–MS	Chlorpyrifos + Ni	*C. elegans*	^{I}[Ala, creatine, His, lactate, betaine, carnosine], ^{D}[choline, Gly, Ile, leu, lys, Val]	(Jones et al., 2012)
^{1}H NMR	Cu (II)	*L. rubellus*	^{I}His	(Gibb et al., 1997)

Technique	Organic toxicant	Animal model tested	Biomarkers of contaminant exposure	Reference
^{1}H NMR spectroscopy, GC–MS	Ni	*C. elegans*	^{I}[Asn, choline, Gln, lactate, succinate], ^{D}[Ala, Ile, Gly, Val]	Jones et al. (2012)
^{1}H NMR	Pb, Zn	*S. salsa*	i) Exposure to Pb - ^{D}Tyr. ii) Exposure to Zn - ^{I}[Val, Ile, Leu, Thr, Ala, Asn, Phe, fosfocoline], ^{D}[acetate, Gly, glucose, fumarate and ferulate]. iii) Exposure to Pb & Zn - ^{I}[Ala, Asn, Tyr, Phe], ^{D}[acetate, succinate, Asp, malonate, fructose, glucose, fumarate and ferulate]	(Wu et al., 2013)
^{1}H NMR	Mixed metals (Cadmium, copper, lead and zinc)	*L. rubellus*	^{I} [maltose, His]	(Bundy et al., 2004)
^{1}H NMR	Tellurite	*P. pseudoalcaligenes*	^{I}[Thr, Leu, Tyr, betaine, Ser, Lys, Ile, Ala, Arg, Val, glutathione, adenosine], ^{D}[Lactate, Gly]	(Tremaroli et al., 2009)
^{1}H NMR	Titanium dioxide	*E. fetida*	^{I} [Phe, Tyr, Lys, Glu, Ala, lactate, Val, Leu], ^{D}maltose	(Whitfield Aslund et al., 2012)

D = decrease; I = increase

Table 1 Metabolic responses of test organism following exposure to selected environmental contaminants in contact tests.

2.3. Metabolomics data analysis

The general approach to data analysis in metabolomics can be summarized in three main stages: explorative, supervised and biological interpretation (Smilde et al., 2010). The explorative phase aims to find groups, clusters and outliers in metabolites and samples studied while the supervised discriminates two or more groups to make predictive models and to find biomarkers (Amiard-Triquet et al., 2012; Dallinger-Marianne, 2000; van Ravenzwaay et al., 2007). Multivariate methods are currently preferred, although univariate and semi-univariate methods have been commonly used for selecting biomarkers. For instance, the lysosomal system was identified as a particular target for the toxic effects of pollutants in soil organisms. However, it is nonspecific as a marker and only included in a suite of biomarkers among diverse soil invertebrate species can provide the necessary specificity for risk assessment purposes (Kammenga et al., 2000). Finally, the biological interpretation seeks the links between metabolome data and underlying metabolic networks through metabolite set enrichment, pathway analysis and metabolic network inference (Trygg et al., 2006). Thus, finding metabolite relationships is essential to determine comprehensive and meaningful metabolic changes as biological response to environmental stimuli (Ellis et al., 2012; Morrison et al., 2007). Accordingly, such extensive evaluation of the impact of pollutants in the metabolism of target organisms is the approach that can add value to the assessment of soil health and viability of soil organisms undergoing stress from pollution.

3. Metabolomics bioinformatics

Information processing by bioinformatics tools and computational biology methods has become essential for solving complex biological problems in genomics, proteomics, and metabolomics. Understanding "omics" data requires both common statistical and computational based methods due to the multi-dimensional and complexity level of the data.

Data-analytical methods for the study of biological systems as developed in the field of computational biology provide a suit of indispensable tools to survey the outcome of metabolomics studies. First, computational biology allows a fast screening of the large biological and chemical data sets generated (Shulaev, 2006), and therefore the identification of the most relevant metabolites, i.e. compounds specifically representative of the metabolic changes in the model system following exposure to different concentrations of organic and inorganic toxicants. As a result of the large number of variables (metabolites) studied, metabolomics studies encompass a significant statistical power for the systematic detection of biological responses to environmental changes (van Ravenzwaay et al., 2012). Second, the mathematical models developed in computational biology allow the identification of relationships between the external stimuli and the metabolic response (Zhang et al., 2010). Third, the implementation of computational algorithms to structural biology makes possible to discover the structure-function of new macromolecular compounds, the functional enzymatic conversion and changes in their activity, as well as their molecular interaction and relationship with others compounds in the pathways where they are involved (Jimenez-Lopez et al., 2013). Moreover, it is possible to detect patterns in such biological responses and establish significant dose-response relationships. Besides, pattern recognition reduces the metabolomics data from hundreds of variables to two or three components that are orthogonal to each other. Overall, this advance of computational biology has been possible due to three significant technological breakthroughs: high-information-content data streams, novel bio-statistical methods, and the computational power to analyse these data.

Data processing and statistical analyses are commonly performed using multivariate (typically a principal component analysis (PCA) and (or) partial least squares (PLS) regression analysis) and univariate (t-test) analyses (Brown et al., 2010; Jones et al., 2014; McKelvie et al., 2011; Yuk et al., 2013). These analyses are performed in combination with the quantification and identification of the metabolites. Subsequently, biological interpretation of the data is necessary for understanding the link between the external stimulus and the metabolic response of the organisms.

Principal component analysis is the most widely used multivariate statistical approach in metabolomics, used to explain the overall variability in a data set via a a set of uncorrelated variables called principal components (PCs), which are linear combinations of the original variables (Trygg et al., 2006). The organization of samples in PCA scores plots is based on the similarities between their metabolic profiles. Thus, PCA allows for dimensional reduction of the data into a low dimensional plane, such as PC1 versus PC2. The scores plot (e.g., PC1 versus PC2) allows for a visual examination of the relationship between the samples based on their metabolic profiles. In a 1-D PCA loadings plot, the contribution (or weight) of each metabolite to the discrimination of the sample classes along one component is represented by the intensity of the metabolite peak. In the 2-D PCA loadings plot discrimination is performed by selecting the points that are scattered further away from the tight cluster of points found near the origin.

Other widely used multivariate statistical tools in metabolomics are PLS regression analysis and PLS discriminant analysis (PLS-DA). Both PLS-regression and PLS-DA are methods for samples classification, with pre-defined variables added to maximize the separation between the sample classes and to construct predictive models. The predefined variables for PLS-regression are measurable quantities such as the contaminant exposure concentration. Validation methods such as the leave-one-out cross validation are used to test the robustness of the models generated by PLS-regression, PLS-DA, OPLS, and OPLS-DA (Whitfield Åslund et al., 2011).

Although metabolomics studies mostly use multivariate statistics, univariate statistical analyses can contribute to the information gained from a study. Thus, t-tests can be used to assess the significance of the separation between the controls and stressed organisms in PCA and PLS-DA scores plots. Also, t tests can be used to determine which metabolites in the ^{1}H NMR spectra of the treatment class increased or decreased significantly relative to the controls.

4. Biomarkers

The somewhat secondary significance of biological responses for soil contamination assessment was customarily associated to the limitation of biomarkers as measurable responses to contaminants, which classically could only provide an indication of exposure to contaminants in soil (Sanchez-Hernandez, 2006). The development of metabolomics, considered an "emerging field" as late as mid-2010, has provided the tools for the determination of multiple biomarkers across different levels of biological organization, and therefore a better assessment of the ecological consequences of contamination. Since the creation of the first metabolomics web database, METLIN (Smith et al., 2005), 60,000 metabolites has been incorporated, a rapid development closely related to the evolution of mass spectrometry instrumentation and data analysis tools. Currently, the number of databases and metabolites registered is continuously increasing. Table 2 summarizes some of the most relevant databases operative and the corresponding website is also indicated. Further information on metabolomics databases can be obtained from the metabolomics society (http://www.metabolomicssociety.org). For instance, ChemSpider is an aggregated database of organic molecules containing more than 20 million compounds from many different providers. At present the database contains information from such diverse sources as a marine natural products database, ACD-Labs chemical databases, the EPA's DSSTox databases and from a series of chemical vendors. It has extensive search utilities and most compounds have a large number of calculated physico-chemical property values.

One of the goals in bioinformatics is to establish automated and efficient ways to integrate large, biological datasets from multiple sources. This objective is challenging because data sources are heterogeneous in terms of their functions, structures, data access methods and dissemination formats. In addition, the enormous quantity of information produced by "omics" is handled via computers that systematically analyze and store the accumulating sequence, structure and function data. Databases are essential in metabolomics because they provide a rapid and specific tool to identify the compounds isolated from an organism exposed to a particular environmental challenge. Thus, the KNApSAcK package provides tool for analysing datasets of mass spectra as well as for retrieving information on metabolites by entering the name of a metabolite, the name of an organism, molecular weight or molecular formula. A list of metabolites that are associated to a taxonomic class can be obtained by search with the taxonomic name, from which information of individual metabolites can be retrieved. The NIST Chemistry WebBook provides access to chemical and physical property data for chemical species. The data provided in the site are from collections maintained by the NIST Standard Reference Data Program and outside contributors. Data in the NIST Chemistry WebBook can be found by direct searches for chemical species or indirect searches based on related data. Specific databases are also being developed, such as LIPID MAPS, currently the largest database of lipid molecular structures. Otherwise, SetupX combines mass spectrometric and biological metadata, which is a step forward in the organization of information generated by metabolomics analysis.

METLIN	http://metlin.scripps.edu/index.php
LIPID MAPS	http://www.lipidmaps.org/
KEGG	http://www.genome.jp/kegg/pathway.html
ChemSpider	http://www.chemspider.com/
SetupX	http://fiehnlab.ucdavis.edu/projects/binbase_setupx
KNApSAcK	http://kanaya.naist.jp/KNApSAcK/
NIST	http://webbook.nist.gov/chemistry/
MassBank	http://www.massbank.jp/
HMP	http://www.hmdb.ca/
IIMDB	http://metabolomics.pharm.uconn.edu/iimdb/

Table 2 Selected metabolomic databases.

Metabolomic databases are thus accompanied by accurate description of the biological study design and accompanying metadata reporting on the laboratory workflow from sample preparation to data processing.

Currently, standard analyses focus on the determination of amino acids, mono- and disaccharides, lipids/fatty acids, short chain fatty acids and small phenolics. Accordingly, it is possible to already launch the standardization of metabolomics analysis. For instance, the Northwest Metabolomics Research Center (University of Washington) has established a relevant list of target compounds to evaluate biological responses to changes in the environment. The list of compounds is summarized in Table 3.

Metabolic Pathways	Number of Metabolites
Alanine, aspartate and glutamate metabolism	15
Arginine and proline metabolism	23
Butanoate metabolism	18
Citrate cycle (TCA cycle)	11
Cysteine and methionine metabolism	14
Fatty acid metabolism	3
Glutathione metabolism	14
Glycine, serine and threonine metabolism	21
Glycolysis / Gluconeogenesis	16
Histidine metabolism	13
Lysine biosynthesis	7
Lysine degradation	6
Nitrogen metabolism	9
Oxidative phosphorylation	6
Pentose phosphate pathway	10
Phenylalanine metabolism	10
Phenylalanine, tyrosine and tryptophan biosynthesis	8
Purine metabolism	30
Pyrimidine metabolism	30
Pyruvate metabolism	10
Synthesis and degradation of ketone bodies	4
Tryptophan metabolism	15
Tyrosine metabolism	18
Valine, leucine and isoleucine biosynthesis	11
Valine, leucine and isoleucine degradation	5

Table 3 Summary of metabolites and metabolic pathways representative of biological responses to environmental stimuli.

The information of metabolites and metabolic pathways has been obtained from the website of Kyoto Encyclopedia of Genes and Genomes (Kegg, http://www.genome.jp/kegg/). According to the research results summarized in Table 1, the implementation of metabolomics in the assessment of soil contamination indicates that contaminants in soil affect several of the major metabolic pathways in living organisms (Table 3), including glycolysis, trycarboxylic acids cycle and amino acids metabolism. Moreover, data analysis indicates an overall reduction in the production of the associated metabolites. For instance, the interference in amino acids specialized pathways results in a decreased synthesis of purine and pyrimidine nucleotides (Brown et al., 2010; McKelvie et al., 2011). These nucleotides are essential for the production of the energy (ATP molecules) that drive most of the enzymatic reactions in living organisms, but also protein synthesis is consequently hampered, which explain the negative effect in processes such as antioxidant activity.

Another emerging group of biomarkers, as highlighted in several studies, are lipids (Rochfort et al., 2009; Sanchez-Hernandez, 2006). Rochfort et al., (2009) indicate that lipophilic extracts can be used in field based metabolomics experiments to investigate different treatment effects on earthworms. Lipid metabolism is highly sensitive to environmental contaminants (Vega-López et al., 2013), with increasing production of lipoprotein vesicle and lipid peroxidation rate during early stages of the biological response to the presence of a toxicant (Lankadurai et al., 2011). Relatedly, earthworm esterases has been proposed as biomarkers for pesticide contamination in soil (Sanchez-Hernandez, 2010). Esterases are directly involved in the natural tolerance of earthworms to pesticides, and can therefore be used as specific biomarkers, but furthermore, their characterization by metabolomics approach might help to select the appropriate earthworm species for regulatory toxicity testing. Overall, the increasing specificity of the research performed in ecotoxigenomics will allow a realistic and meaningful incorporation of biological responses in ecological risk assessment.

5. Oxidative stress in contaminated soil

The induction of the oxidative stress response by the presence of toxic compounds in the environment is a primary mechanisms of defence, although prolonged exposure to contaminants is likely to overwhelm this short-term defence (Regoli et al., 2002).

Metabolites such as proline possibly detoxify the ROS under stress in vivo (Smirnoff, 1993). Exposure of plants to both redox active, for example, Cu and Hg, and other metals, for example, Cd and Zn, induces the generation of free radicals that leads to oxidative stress. This represents one of the major causes of toxicity particularly due to redox metals. The cells are equipped with an elaborate network of antioxidative enzymes and low molecular weight metabolites which mitigate the oxidative stress. Proline scavenges different free radicals in certain in vitro generation and detection systems.

Proline quenches ROS and reactive nitrogen species (RNS), which relieves the oxidative burden from the glutathione system. Moreover, polyamines also have an antioxidative role by quenching the accumulation of $O_2^{\cdot-}$ probably through inhibition of NADPH oxidase (Paschalidis and Roubelakis-Angelakis, 2005). This may facilitate phytochelatin synthesis and enhance metal tolerance (Siripornadulsil et al., 2002).

Overall, oxidative defence response to toxicity or other environmental stress involves the generation of oxygenated metabolites from exposed organisms and activation/inhibition of the production of antioxidants enzymes and metabolites such as glutathione. The depletion of antioxidants for prolonged exposures might result in the decrease of the response effectiveness and eventual imbalance between generation and elimination of reactive oxygen species. Depletion of glutathione appears to be a major mechanism in short-term heavy metal toxicity (Schutzendubel and Polle, 2002). In accordance with this hypothesis, a good correlation between glutathione contents and tolerance index was observed with 10 pea genotypes differing in Cd sensitivity (Metwally et al., 2005). High GSH concentrations in hyperaccumulator *T. Goesingense* coincided with high constitutive activity of serine acetyl transferase (SAT); SAT catalyses the acetylation of L-Ser to OAS which in turn provides the carbon skeleton for Cys biosynthesis. Elevated GSH levels in *T. Goesingense* also coincided with the ability both to hyperaccumulate Ni and to resist its damaging oxidation effects.

The significance of glutathione and the metal-induced phytochelatins (PCs) in heavy metal tolerance has been studied intensely (Rauser, 1995). However, PCs are important for detoxification of only a limited set of metals such as Cd^{2+}, Cu^{2+} and AsO_2^{2-} while Zn^{2+} and Ni^{2+} are poor inducers of PCs and exhibit low binding affinity. Most other metals lack significant binding.

Evaluation of metabolites related to oxidative response constitutes a relevant group of target compounds for risk assessment. Although oxidative response to soil contamination has been classically addressed in plants, the study of this response in soil microorganisms is already being introduced in ecotoxicology as a fundamental part of the biological response of soil microorganisms to soil contamination (Boer et al., 2013; Tremaroli et al., 2009). Accordingly, Boer et al. (2013) describe the attenuation of the oxidative response for springtails in laboratory tests, which constitutes and early detection of soil pollution, and standardized test have been developed.

6. Metabolites related to soil contamination with organic compounds

The importance of the identification of biomarkers and metabolic pathways specifically related to soils contamination with a particular pollutant or group of pollutants has been already highlighted through this chapter. From the information summarized in Table 1 and Table 3 it is possible to infer that soil contamination with organic compounds, namely pesticides o polycyclic aromatic hydrocarbons, abates essential metabolic pathways such as the trycarboxylic acid cycle and the oxidative stress response, while lipid metabolism appears to be enhanced. However, the advance in the application of bioinformatics is providing further progress in terms of identification of specific biomarkers for risk assessment of individual target compounds. Thus, toxicity of endosulfan has been directly related with alterations of the GABA-glutamine cycle (Yuk et al., 2013), while chlorpyrifos depresses the Cori cycle and reduces the production of phospholipids, as indicated by lower levels of choline (Jones et al., 2012). Baylay et al. (2012) specifically relates chlorpyrifos toxicity to increased levels of fumarate, an intermediate of the trycarboxylic acid cycle. Research conducted with the same earthworm (*E. fetida*) and other families of organic compounds revealed a different metabolic response (Brown et al., 2010; Lankadurai et al., 2012), confirming the capability of metabolomics to discriminate the metabolic pathways involved in the response to a particular toxic compound. Moreover, the results strongly suggest that sets of biomarkers might be soon sufficiently reliable as for their implantation in in toxicity standardized test.

The relevance of these and future studies on the development of risk assessment strategies is aggravated by the inherent risk of soil contamination for human health. Soil contaminants may be responsible for health effects costing millions of euros. Health problems range from cancer (arsenic, asbestos, dioxins), to neurological damage and lower IQ (lead, arsenic), kidney disease (lead, mercury, cadmium), and skeletal and bone diseases (lead, fluoride, cadmium).

Overall, few studies have been conducted on the toxicity of complex chemical mixtures in soils. The effects of the soil and organisms within it upon organic pollutants are unknown. The data currently available correspond mostly to short-term studies and high level exposure of these chemicals, which is less relevant to the potential low-level, long term health impacts on living organisms near to contaminated soil.

7. Metabolites related to soil contamination with heavy metals

The uptake of excess metal ions is toxic to most organisms, and the biochemical impact of metal ions on the cells varies with the chemistry of the element as their chemical nature. In plants, phytotoxicity of heavy metals in most parts can be attributed to symplastic accumulation of heavy metals, such as the cytosol and chloroplast stroma. Metal-induced changes in development are the result of either a direct and immediate impairment of metabolism or signaling processes that initiate adaptive or toxicity responses that need to be considered as active processes of the organism. Transport processes have been recognized as a central mechanism of metal detoxification and tolerance (Hall, 2002; Hall and Williams, 2003).

Some metals, for example, Zn and Cu, are essential for normal plant growth and development as they serve as structural and functional components of specific proteins. Other metals, for example, Cd and Pb, have no known function in plants although a Cd requirement for carbonic anhydrase from marine diatoms has been reported (Lane and Morel, 2000).

Upon exposure to metals, organisms often synthesize a set of diverse metabolites that accumulate to concentrations in the millimolar range, particularly specific amino acids, such as proline and histidine, peptides such as glutathione and phytochelatins (PC), and the amines spermine, spermidine, putrescine, nicotianamine, and mugineic acids that can be detected as response to these metals exposure. The advance of toxicogenomics in relation to organic contaminants is significantly ahead of the equivalent research in metal contaminated soil (Table 1). Nevertheless, research conducted up to date has yielded a number of biomarkers representative of the biological response of soil microorganisms to metals toxicity. Thus, soil contamination with Pb has been related with an enhancement of lipid metabolism (Sanchez-Hernandez, 2006) and more directly with reduction of tyrosine levels (Wu et al., 2013).

Otherwise, Cd toxicity promotes the secretion of phytochelatins in *C. elegans*, likely at the expenses of the sulphur metabolism, as suggested by the reduction in cystathionine (Hughes et al., 2009), while the response of tomato plants to Cd involves several biochemical pathways (Hédiji et al., 2010). These examples illustrate the genuine specificity of biological reactions to different metals but also the variation in representative biomarkers among different organisms. Accordingly, exposure of *C. elegans* to Ni (Jones et al., 2012) yields a different metabolome than Cd since different biochemical pathways are affected.

In plants, data currently available demonstrate the significance of nitrogen-containing metabolites beyond phytochelatins and glutathione in plant response and acclimation to heavy metals. The various metal ions have specific chemical properties and induce distinct responses of adaptation and damage development. Thus, accumulating N-metabolites display a variety of functions, i.e. metal ion chelation, antioxidant defence, protection of macromolecules, and possibly signalling.

Proline is an extensively studied molecule in the context of plant responses to abiotic stresses. Up-regulation of proline is often encountered in plants under heavy metal stress, comparable to what occur under other abiotic stresses. When compared at equal toxic strength, proline accumulation decreased in the order Cd > Zn > Cu (Schat et al., 1997). In addition, it has been suggested different functions of proline under metal-stress, being involved in osmoregulation, metal chelation, antioxidant, and regulator of specific functions in plant morphogenesis.

Furthermore, Ni-hyperaccumulation has been specifically linked to histidine production (Krämer, 2005), particularly for *Saccharomyces cerevisiae* (Pearce and Sherman, 1999). The beneficial role of high histidine levels has been shown in transgenic *Arabidopsis thaliana* which accumulated about 2-fold higher histidine levels than wild- type plants and showed more than 10-fold increased biomass production in the presence of toxic Ni in the growth medium (Wycisk et al., 2004).. Moreover, cell surface-engineered yeast displaying a histidine oligopeptide (hexa-His) has been shown to adsorb 3–8 times more copper ions than the parent strain, being more resistant to Cu than the parent (Kuroda et al., 2002).

Otherwise, polyamine contents are altered in response to the exposure to heavy metals. Weinstein et al. (1986) showed an increment in putrescine content in Cd-treated oat seedlings and detached oat leaves with a marginal rise in spermidine and spermine content. They influence a variety of growth and development processes in plants and have been suggested to be a class of plant growth regulators and to act as second messengers (Kakkar and Sawhney, 2002). It has been suggested that they could stabilize and protect the membrane systems against the toxic effects of metal ions, particularly the redox active metals.

Overall, the number of studies remains rather scarce, and the preliminary results available in the literature merely constitute a launching platform for this promising research field.

8. Future perspectives

The main objective of metabolomics implementation in soil risk assessment is to meet the continuously increasing demand of safety data from human and ecological risk assessments.

Accordingly, regulatory programs worldwide are currently incorporating tests with end-points that involve the effects of chemicals and the impact in specific metabolic pathways (Ankley et al., 2006). Toxicological end-points can be general biological responses such as survival or weight loss (Baylay et al., 2012), but specific biomarkers provide the accuracy that was classically elusive for test with living organisms

Several issues immediately arise from the summary here presented, such as the need to perform field toxicological test, with natural soils rather than use artificial soils, as was the case with some of the studies listed in Table 1. Ecotoxigenomics can also benefit from the incorporation of further analytical techniques. Techniques based on mass spectrometry are certainly required to understand the mechanisms involved in the alteration of metabolic pathways as response to toxicants. However, for screenings which merely require the detection of differences between metabolic phenotypes, optical methods such as FT-IR would be suitable, particularly if extremely high sample throughput is required (Bundy et al., 2009). Although no data was available in the existing literature, Figure 2 illustrates the change in the fingerprint of organic compounds in a soil amended with different sources of carbon collected 10 after the application. While some of the groups of compounds might be merely related to the sources of carbon added, the variations in the signal associated to polysaccharides (600-1000 cm^{-1}) can be associated to changes in the metabolic fingerprint of the soil system and therefore linked to microbiological activity in soil. Overall, the introduction of these results seeks to encourage further characterization of families of compounds in intact soil (or functional pools such as aggregates) in relation with soil processes, an approach that can find immediate application in the assessment of biological responses to toxic compounds in soil.

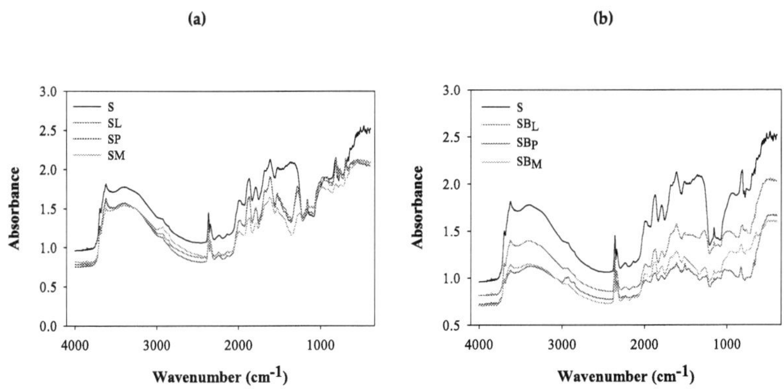

Figure 2. Absorption spectra obtained by Fourier transform infrared spectroscopy (FTIR) for an agricultural soil (S), soil amended with fresh residues (a): dry leaf litter (SL), peanut shell (SP), maize residue (SM), and soil amended with bio-char (b) derived from those feedstocks (BL, BP or BM). Spectra presented (after 10 d incubation) are the average of 5 spectra obtained for different samples of each treatment. Hernandez-Soriano et al., unpublished data.

The variability of biological responses has been one of the main obstacles for their implementation in standardized risk assessment. However, the examination of changes in biological

http://dx.doi.org/10.5772/58294

processes by accurate analytical techniques and powerful statistical tools has launched a new era in our understanding of the soil processes. The possibility of identifying the most sensitive metabolites for a certain toxicant and develop a tailored standardized test is the ultimate goal pursued.

Acknowledgements

MCH-S thanks The University of Queensland for a postdoctoral research fellowship. JCJ-L thanks the European research program Marie Curie (FP7-PEOPLE-2011-IOF) for his PIOF-GA-2011-301550 grant.

The funders had no role in study design, data collection and analysis, decision to publish, or preparation of the manuscript.

Author details

Maria C. Hernandez-Soriano[1] and Jose C. Jimenez-Lopez[2]

*Address all correspondence to: m.hernandezsoriano@uq.edu.au

1 School of Agriculture and Food Sciences, The University of Queensland, St Lucia QLD, Australia

2 The UWA Institute of Agriculture, The University of Western Australia, Crawley, Perth WA, Australia

References

[1] Amiard-Triquet, C., Amiard, J.C., Rainbow, P.S., 2012. Ecological Biomarkers: Indicators of Ecotoxicological Effects. Taylor & Francis.

[2] Ankley, G.T., Daston, G.P., Degitz, S.J., Denslow, N.D., Hoke, R.A., Kennedy, S.W., Miracle, A.L., Perkins, E.J., Snape, J., Tillitt, D.E., Tyler, C.R., Versteeg, D., 2006. Toxicogenomics in Regulatory Ecotoxicology. Environmental Science & Technology 40, 4055-4065.

[3] Baylay, A.J., Spurgeon, D.J., Svendsen, C., Griffin, J.L., Swain, S.C., Sturzenbaum, S.R., Jones, O.A.H., 2012. A metabolomics based test of independent action and concentration addition using the earthworm Lumbricus rubellus. Ecotoxicology 21, 1436-1447.

[4] Boer, M., Ellers, J., Gestel, C.M., Dunnen, J., Straalen, N., Roelofs, D., 2013. Transcriptional responses indicate attenuated oxidative stress in the springtail Folsomia candida exposed to mixtures of cadmium and phenanthrene. Ecotoxicology 22, 619-631.

[5] Brown, S.A.E., McKelvie, J.R., Simpson, A.J., Simpson, M.J., 2010. 1H NMR metabolomics of earthworm exposure to sub-lethal concentrations of phenanthrene in soil. Environmental Pollution 158, 2117-2123.

[6] Brown, S.A.E., Simpson, A.J., Simpson, M.J., 2009. 1H NMR metabolomics of earthworm responses to sub-lethal PAH exposure. Environmental Chemistry 6, 432-440.

[7] Bundy, J., Davey, M., Viant, M., 2009. Environmental metabolomics: a critical review and future perspectives. Metabolomics 5, 3-21.

[8] Bundy, J., Spurgeon, D., Svendsen, C., Hankard, P., Weeks, J., Osborn, D., Lindon, J., Nicholson, J., 2004. Environmental Metabonomics: Applying Combination Biomarker Analysis in Earthworms at a Metal Contaminated Site. Ecotoxicology 13, 797-806.

[9] Bundy, J.G., Lenz, E.M., Bailey, N.J., Gavaghan, C.L., Svendsen, C., Spurgeon, D., Hankard, P.K., Osborn, D., Weeks, J.M., Trauger, S.A., Speir, P., Sanders, I., Lindon, J.C., Nicholson, J.K., Tang, H., 2002. Metabonomic assessment of toxicity of 4-fluoroaniline, 3,5-difluoroaniline and 2-fluoro-4-methylaniline to the earthworm Eisenia veneta (rosa): Identification of new endogenous biomarkers. Environmental Toxicology and Chemistry 21, 1966-1972.

[10] Bundy, J.G., Osborn, D., Weeks, J.M., Lindon, J.C., Nicholson, J.K., 2001. An NMR-based metabonomic approach to the investigation of coelomic fluid biochemistry in earthworms under toxic stress. FEBS Letters 500, 31-35.

[11] Dallinger-Marianne, J., 2000. Biomarkers in terrestrial invertebrates for ecotoxicological soil risk assessment. Reviews of Environmental Contamination and Toxicology 164, 93.

[12] Drewes, C.D., Vining, E.P., 1984. In vivo neurotoxic effects of dieldrin on giant nerve fibers and escape reflex function in the earthworm, Eisenia foetida. Pesticide Biochemistry and Physiology 22, 93-103.

[13] Edwards, C.A., Bater, J.E., 1992. The use of earthworms in environmental management. Soil Biology and Biochemistry 24, 1683-1689.

[14] Ellis, J.K., Athersuch, T.J., Thomas, L.D.K., Teichert, F., Perez-Trujillo, M., Svendsen, C., Spurgeon, D.J., Singh, R., Jaerup, L., Bundy, J.G., Keun, H.C., 2012. Metabolic profiling detects early effects of environmental and lifestyle exposure to cadmium in a human population. Bmc Medicine 10.

[15] Fitzpatrick, L.C., Sassani, R., Venables, B.J., Goven, A.J., 1992. Comparative toxicity of polychlorinated biphenyls to earthworms Eisenia foetida and Lumbricus terrestris. Environ Pollut 77, 65-69.

[16] Gibb, J.O.T., Holmes, E., Nicholson, J.K., Weeks, J.M., 1997. Proton NMR spectroscopic studies on tissue extracts of invertebrate species with pollution indicator potential. Comparative Biochemistry and Physiology Part B: Biochemistry and Molecular Biology 118, 587-598.

[17] Hall, J.L., 2002. Cellular mechanisms for heavy metal detoxification and tolerance. J Exp Bot 53, 1-11.

[18] Hall, J.L., Williams, L.E., 2003. Transition metal transporters in plants. J Exp Bot 54, 2601-2613.

[19] Hédiji, H., Djebali, W., Cabasson, C., Maucourt, M., Baldet, P., Bertrand, A., Boulila Zoghlami, L., Deborde, C., Moing, A., Brouquisse, R., Chaïbi, W., Gallusci, P., 2010. Effects of long-term cadmium exposure on growth and metabolomic profile of tomato plants. Ecotoxicology and Environmental Safety 73, 1965-1974.

[20] Heimbach, F., 1988. A comparison of laboratory methods for toxicity testing with earthworms, In: Edwards, C.A., Neuhauser, E.F. (Eds.), Earthworms in waste and environmental management. SPB Academic Publishing, The Hague, The Netherlands, pp. 329-335.

[21] Hughes, S.L., Bundy, J.G., Want, E.J., Kille, P., Stu®rzenbaum, S.R., 2009. The Metabolomic Responses of Caenorhabditis elegans to Cadmium Are Largely Independent of Metallothionein Status, but Dominated by Changes in Cystathionine and Phytochelatins. Journal of Proteome Research 8, 3512-3519.

[22] Jimenez-Lopez, J., Kotchoni, S., Hernandez-Soriano, M., Gachomo, E., Alché, J., 2013. Structural functionality, catalytic mechanism modeling and molecular allergenicity of phenylcoumaran benzylic ether reductase, an olive pollen (Ole e 12) allergen. Journal of Computer-Aided Molecular Design 27, 873-895.

[23] Jones, O.A.H., Sdepanian, S., Lofts, S., Svendsen, C., Spurgeon, D.J., Maguire, M.L., Griffin, J.L., 2014. Metabolomic analysis of soil communities can be used for pollution assessment. Environmental Toxicology and Chemistry 33, 61-64.

[24] Jones, O.A.H., Spurgeon, D.J., Svendsen, C., Griffin, J.L., 2008. A metabolomics based approach to assessing the toxicity of the polyaromatic hydrocarbon pyrene to the earthworm Lumbricus rubellus. Chemosphere 71, 601-609.

[25] Jones, O.A.H., Swain, S.C., Svendsen, C., Griffin, J.L., Sturzenbaum, S.R., Spurgeon, D.J., 2012. Potential New Method of Mixture Effects Testing Using Metabolomics and Caenorhabditis elegans. Journal of Proteome Research 11, 1446-1453.

[26] Kakkar, R.K., Sawhney, V.K., 2002. Polyamine research in plants – a changing perspective. Physiologia Plantarum 116, 281-292.

[27] Kammenga, J.E., Dallinger, R., Donker, M.H., Kohler, H.R., Simonsen, V., Triebskorn, R., Weeks, J.M., 2000. Biomarkers in terrestrial invertebrates for ecotoxicological soil risk assessment. Rev Environ Contam Toxicol 164, 93-147.

[28] Krämer, U., 2005. Phytoremediation: novel approaches to cleaning up polluted soils. Current Opinion in Biotechnology 16, 133-141.

[29] Kuroda, K., Ueda, M., Shibasaki, S., Tanaka, A., 2002. Cell surface-engineered yeast with ability to bind, and self-aggregate in response to, copper ion. Applied Microbiology and Biotechnology 59, 259-264.

[30] Kwon, Y.-K., Jung, Y.-S., Park, J.-C., Seo, J., Choi, M.-S., Hwang, G.-S., 2012. Characterizing the effect of heavy metal contamination on marine mussels using metabolomics. Marine Pollution Bulletin 64, 1874-1879.

[31] Lane, T.W., Morel, F.M.M., 2000. A biological function for cadmium in marine diatoms. Proceedings of the National Academy of Sciences 97, 4627-4631.

[32] Lankadurai, B.P., Simpson, A.J., Simpson, M.J., 2012. 1H NMR metabolomics of Eisenia fetida responses after sub-lethal exposure to perfluorooctanoic acid and perfluorooctane sulfonate. Environmental Chemistry 9, 502-511.

[33] Lankadurai, B.P., Wolfe, D.M., Simpson, A.J., Simpson, M.J., 2011. 1H NMR-based metabolomics of time-dependent responses of Eisenia fetida to sub-lethal phenanthrene exposure. Environmental Pollution 159, 2845-2851.

[34] Lenz, E.M., Weeks, J.M., Lindon, J.C., Osborn, D., Nicholson, J.K., 2005. Qualitative high field 1H-NMR spectroscopy for the characterization of endogenous metabolites in earthworms with biochemical biomarker potential. Metabolomics 1, 123-136.

[35] McKelvie, J.R., Wolfe, D.M., Celejewski, M.A., Alaee, M., Simpson, A.J., Simpson, M.J., 2011. Metabolic responses of Eisenia fetida after sub-lethal exposure to organic contaminants with different toxic modes of action. Environmental Pollution 159, 3620-3626.

[36] McKelvie, J.R., Yuk, J., Xu, Y., Simpson, A.J., Simpson, M.J., 2009. H-1 NMR and GC/MS metabolomics of earthworm responses to sub-lethal DDT and endosulfan exposure. Metabolomics 5, 84-94.

[37] Metwally, A., Safronova, V.I., Belimov, A.A., Dietz, K.J., 2005. Genotypic variation of the response to cadmium toxicity in Pisum sativum L. J Exp Bot 56, 167-178.

[38] Miller, M.G., 2007. Environmental Metabolomics: A SWOT Analysis (Strengths, Weaknesses, Opportunities, and Threats). Journal of Proteome Research 6, 540-545.

[39] Morrison, N., Bearden, D., Bundy, J., Collette, T., Currie, F., Davey, M., Haigh, N., Hancock, D., Jones, O.H., Rochfort, S., Sansone, S.-A., Štys, D., Teng, Q., Field, D., Viant, M., 2007. Standard reporting requirements for biological samples in metabolomics experiments: environmental context. Metabolomics 3, 203-210.

[40] OECD, 1984. Earthworm, acute toxicity tests. Guideline for testing chemicals. No. 207. OECD, Paris, France.

[41] OECD, 2004. Earthworm reproduction test. Guideline for testing chemicals. No. 222. OECD, Paris, France.

[42] Paschalidis, K.A., Roubelakis-Angelakis, K.A., 2005. Sites and regulation of polyamine catabolism in the tobacco plant. Correlations with cell division/expansion, cell cycle progression, and vascular development. Plant Physiol 138, 2174-2184.

[43] Pearce, D.A., Sherman, F., 1999. Toxicity of copper, cobalt, and nickel salts is dependent on histidine metabolism in the yeast Saccharomyces cerevisiae. J Bacteriol 181, 4774-4779.

[44] Rauser, W., 1995. Phytochelatins and related peptides. Structure, biosynthesis, and function. Plant Physiology 109, 1141-1149.

[45] Regoli, F., Gorbi, S., Frenzilli, G., Nigro, M., Corsi, I., Focardi, S., Winston, G.W., 2002. Oxidative stress in ecotoxicology: from the analysis of individual antioxidants to a more integrated approach. Marine Environmental Research 54, 419-423.

[46] Rochfort, S.J., Ezernieks, V., Yen, A.L., 2009. NMR-based metabolomics using earthworms as potential indicators for soil health. Metabolomics 5, 95-107.

[47] Sanchez-Hernandez, J.C., 2006. Earthworm Biomarkers in Ecological Risk Assessment, In: Ware, G., Whitacre, D., Albert, L., Voogt, P., Gerba, C., Hutzinger, O., Knaak, J., Mayer, F., Morgan, D.P., Park, D., Tjeerdema, R., Yang, R.H., Gunther, F. (Eds.), Reviews of Environmental Contamination and Toxicology. Springer New York, pp. 85-126.

[48] Sanchez-Hernandez, J.C., 2010. Environmental applications of earthworm esterases in the agroecosystem. Journal of Pesticide Science 35, 290-301.

[49] Schat, H., Sharma, S., Vooijs, R., 1997. Heavy metal-induced accumulation of free proline in metal-tolerant and a nontolerant ecotype of Silene vulgaris. Physiologia Plantarum 101, 477-482.

[50] Schutzendubel, A., Polle, A., 2002. Plant responses to abiotic stresses: heavy metal-induced oxidative stress and protection by mycorrhization. J Exp Bot 53, 1351-1365.

[51] Shulacv, V., 2006. Metabolomics technology and bioinformatics. Briefings in Bioinformatics 7, 128-139.

[52] Singh, O.V., 2006. Proteomics and metabolomics: The molecular make-up of toxic aromatic pollutant bioremediation. PROTEOMICS 6, 5481-5492.

[53] Siripornadulsil, S., Traina, S., Verma, D.P., Sayre, R.T., 2002. Molecular mechanisms of proline-mediated tolerance to toxic heavy metals in transgenic microalgae. Plant Cell 14, 2837-2847.

[54] Smilde, A.K., Westerhuis, J.A., Hoefsloot, H.C.J., Bijlsma, S., Rubingh, C.M., Vis, D.J., Jellema, R.H., Pijl, H., Roelfsema, F., Greef, J., 2010. Dynamic metabolomic data analysis: a tutorial review. Metabolomics 6, 3-17.

[55] Smirnoff, N., 1993. The role of active oxygen in the response of plants to water deficit and desiccation. New Phytologist 125, 27-58.

[56] Smith, C.A., O'Maille, G., Want, E.J., Qin, C., Trauger, S.A., Brandon, T.R., Custodio, D.E., Abagyan, R., Siuzdak, G., 2005. METLIN: a metabolite mass spectral database. Ther Drug Monit 27, 747-751.

[57] Snape, J.R., Maund, S.J., Pickford, D.B., Hutchinson, T.H., 2004. Ecotoxicogenomics: the challenge of integrating genomics into aquatic and terrestrial ecotoxicology. Aquatic Toxicology 67, 143-154.

[58] Spurgeon, D.J., Morgan, A.J., Kille, P., 2008. Current research in soil invertebrate ecotoxicogenomics, In: Hogstrand, C., Kille, P. (Eds.), Comparative Toxicogenomics, pp. 133-163.

[59] Tremaroli, V., Workentine, M.L., Weljie, A.M., Vogel, H.J., Ceri, H., Viti, C., Tatti, E., Zhang, P., Hynes, A.P., Turner, R.J., Zannoni, D., 2009. Metabolomic investigation of the bacterial response to a metal challenge. Appl Environ Microbiol 75, 719-728.

[60] Trygg, J., Holmes, E., Lundstedt, T., 2006. Chemometrics in Metabonomics. Journal of Proteome Research 6, 469-479.

[61] Van Gestel, C.A.M., Dirven-Van Breemen, E.M., Baerselman, R., Emans, H.J.B., Janssen, J.A.M., Postuma, R., Van Vliet, P.J.M., 1992. Comparison of sublethal and lethal criteria for nine different chemicals in standardized toxicity tests using the earthworm Eisenia andrei. Ecotoxicology and Environmental Safety 23, 206-220.

[62] van Gestel, C.A.M., van Dis, W.A., van Breemen, E.M., Sparenburg, P.M., 1989. Development of a standardized reproduction toxicity test with the earthworm species Eisenia fetida andrei using copper, pentachlorophenol, and 2,4-dichloroaniline. Ecotoxicology and Environmental Safety 18, 305-312.

[63] van Ravenzwaay, B., Cunha, G.C.-P., Leibold, E., Looser, R., Mellert, W., Prokoudine, A., Walk, T., Wiemer, J., 2007. The use of metabolomics for the discovery of new biomarkers of effect. Toxicology Letters 172, 21-28.

[64] van Ravenzwaay, B., Herold, M., Kamp, H., Kapp, M.D., Fabian, E., Looser, R., Krennrich, G., Mellert, W., Prokoudine, A., Strauss, V., Walk, T., Wiemer, J., 2012. Metabolomics: A tool for early detection of toxicological effects and an opportunity for biology based grouping of chemicals—From QSAR to QBAR. Mutation Research/Genetic Toxicology and Environmental Mutagenesis 746, 144-150.

[65] Vega-López, A., Ayala-López, G., Posadas-Espadas, B.P., Olivares-Rubio, H.F., Dzul-Caamal, R., 2013. Relations of oxidative stress in freshwater phytoplankton with heavy metals and polycyclic aromatic hydrocarbons. Comparative Biochemistry and Physiology Part A: Molecular & Integrative Physiology 165, 498-507.

[66] Viant, M.R., 2009. Applications of metabolomics to the environmental sciences. Metabolomics 5, 1-2.

[67] Viant, M.R., Rosenblum, E.S., Tjeerdema, R.S., 2003. NMR-Based Metabolomics: A Powerful Approach for Characterizing the Effects of Environmental Stressors on Organism Health. Environmental Science & Technology 37, 4982-4989.

[68] Warne, M., Lenz, E.M., Osborn, D., Weeks, J., Nicholson, J., 2000. An NMR-based metabonomic investigation of the toxic effects of 3-trifluoromethyl-aniline on the earthworm Eisenia veneta. Biomarkers 5, 56-72.

[69] Weinstein, L.H., Kaur-Sawhney, R., Rajam, M.V., Wettlaufer, S.H., Galston, A.W., 1986. Cadmium-induced accumulation of putrescine in oat and bean leaves. Plant Physiol 82, 641-645.

[70] Whitfield Åslund, M., Simpson, A., Simpson, M., 2011. 1H NMR metabolomics of earthworm responses to polychlorinated biphenyl (PCB) exposure in soil. Ecotoxicology 20, 836-846.

[71] Whitfield Aslund, M.L., McShane, H., Simpson, M.J., Simpson, A.J., Whalen, J.K., Hendershot, W.H., Sunahara, G.I., 2012. Earthworm sublethal responses to titanium dioxide nanomaterial in soil detected by (1)H NMR metabolomics. Environ Sci Technol 46, 1111-1118.

[72] Wu, H., Liu, X., Zhao, J., Yu, J., 2013. Regulation of Metabolites, Gene Expression, and Antioxidant Enzymes to Environmentally Relevant Lead and Zinc in the Halophyte Suaeda salsa. Journal of Plant Growth Regulation 32, 353-361.

[73] Wycisk, K., Kim, E.J., Schroeder, J.I., Krämer, U., 2004. Enhancing the first enzymatic step in the histidine biosynthesis pathway increases the free histidine pool and nickel tolerance in Arabidopsis thaliana. FEBS Letters 578, 128-134.

[74] Yu, Y.L., Wu, X.M., Li, S.N., Fang, H., Zhan, H.Y., Yu, J.Q., 2006. An exploration of the relationship between adsorption and bioavailability of pesticides in soil to earthworm. Environmental Pollution 141, 428-433.

[75] Yuk, J., McKelvie, J.R., Simpson, M.J., Spraul, M., Simpson, A.J., 2010. Comparison of 1-D and 2-D NMR techniques for screening earthworm responses to sub-lethal endosulfan exposure. Environmental Chemistry 7, 524-536.

[76] Yuk, J., Simpson, M.J., Simpson, A.J., 2013. 1-D and 2-D NMR-based metabolomics of earthworms exposed to endosulfan and endosulfan sulfate in soil. Environmental Pollution 175, 35-44.

[77] Zhang, Q., Bhattacharya, S., Andersen, M.E., Conolly, R.B., 2010. Computational Systems Biology and Dose-Response Modeling in Relation to New Directions in Toxicity Testing. Journal of Toxicology and Environmental Health, Part B 13, 253-276.

Section 3

Current Trends in Phytoremediation

Chapter 17

Phytoremediation of Soils Contaminated with Metals and Metalloids at Mining Areas: Potential of Native Flora

Paulo J.C. Favas, João Pratas, Mayank Varun, Rohan D'Souza and Manoj S. Paul

Additional information is available at the end of the chapter

http://dx.doi.org/10.5772/57469

1. Introduction

Contaminated soils and residues can be remediated by various methods, such as: removal, isolation, incineration, solidification/stabilization, vitrification, thermal treatment, solvent extraction, chemical oxidation, etc. These methods have the disadvantage of being very expensive and in some cases, they involve the movement of contaminated materials to treatment sites thus, adding risks of secondary contamination [1-3]. Therefore, currently preference is being given to *in situ* methods that are less environmentally disruptive and more economical. In this context, biotechnology offers phytoremediation techniques as a suitable alternative.

Phytoremediation can be understood as the use of plants (trees, shrubs, grasses and aquatic plants) and their associated microorganisms in order to remove, degrade or isolate toxic substances from the environment [3-8]. The word "phytoremediation" derives from the Greek «phyton», meaning "plant", and Latin «remedium», which means "to remedy" or "to correct".

Substances that may be subjected to phytoremediation include metals (Pb, Zn, Cd, Cu, Ni, Hg), metalloids (As, Sb), inorganic compounds (NO^{3-} NH^{4+}, PO_4^{3-}), radioactive chemical elements (U, Cs, Sr), petroleum hydrocarbons (BTEX), pesticides and herbicides (atrazine, bentazone, chlorinated and nitroaromatic compounds), explosives (TNT, DNT), chlorinated solvents (TCE, PCE) and industrial organic wastes (PCPs, PAHs), and others [5].

2. Phytoremediation strategies

Phytoremediation techniques include different modalities, depending on the chemical nature and properties of the contaminant (if it is inert, volatile or subject to degradation in the plant or in the soil) and the plant characteristics (Figure 1). Thus, phytoremediation essentially comprise six different strategies, though more than one may be used by the plant simultaneously.

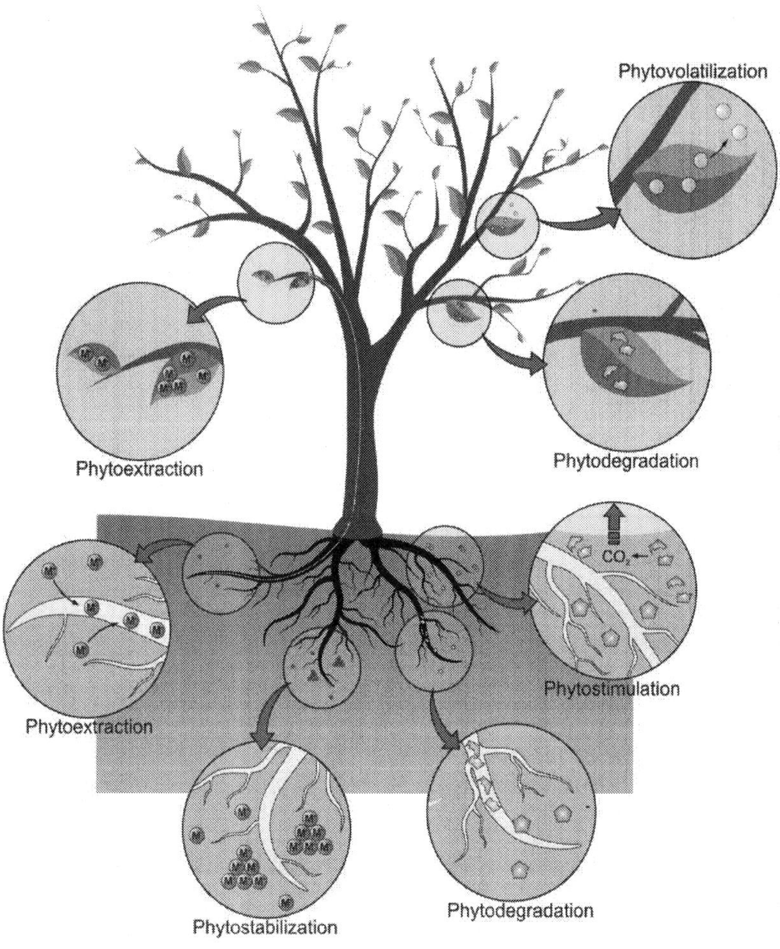

Figure 1. Schematic representation of phytoremediation strategies.

1. **Phytodegradation** (Phytotransformation): organic contaminants are degraded (metabolized) or mineralized inside plant cells by specific enzymes that include nitroreductases (degradation of nitroaromatic compounds), dehalogenases (degradation of chlorinated solvents and pesticides) and laccases (degradation of anilines). *Populus* species and *Myriophyllium spicatum* are examples of plants that have these enzymatic systems [9,10].

2. **Phytostabilization** (Phytoimmobilization): contaminants, organic or inorganic, are incorporated into the lignin of the cell wall of roots cells or into humus. Metals are precipitated as insoluble forms by direct action of root exudates and subsequently trapped in the soil matrix. The main objective is to avoid mobilization of contaminants and limit their diffusion in the soil [3,11-13]. Species of genera *Haumaniastrum, Eragrostis, Ascolepis, Gladiolus* and *Alyssum* are examples of plants cultivated for this purpose.

3. **Phytovolatilization:** this technique relies on the ability of some plants to absorb and volatilize certain metals/metalloids. Some element ions of the groups IIB, VA and VIA of the periodic table (specifically Hg, Se and As) are absorbed by the roots, converted into non-toxic forms, and then released into the atmosphere. As examples the species *Astragalus bisulcatus* and *Stanleya pinnata* for Se or transgenic plants (with bacterial genes) of *Arabidopsis thaliana, Nicotiana tabacum, Liriodendron tulipifera* or *Brassica napus* for Hg can be mentioned [13-18]. This technique can also be used for organic compounds.

4. **Phytoextraction** (Phytoaccumulation, Phytoabsorption or Phytosequestration): this involves the absorption of contaminants by roots followed by translocation and accumulation in the aerial parts. It is mainly applied to metals (Cd, Ni, Cu, Zn, Pb) but can also be used for other elements (Se, As) and organic compounds. This technique preferentially uses hyperaccumulator plants, that have the ability to store high concentrations of specific metals in their aerial parts (0.01% to 1% dry weight, depending on the metal). *Elsholtzia splendens, Alyssum bertolonii, Thlaspi caerulescens* and *Pteris vittata* are known examples of hyperaccumulator plants for Cu, Ni, Zn/Cd and As, respectively [3,19-26].

5. **Phytofiltration:** this uses plants to absorb, concentrate and/or precipitate contaminants, particularly heavy metals or radioactive elements, from an aqueous medium through their root system or other submerged organs. The plants are kept in a hydroponic system, whereby the effluents pass and are "filtered" by the roots (Rhizofiltration), or other organs that absorb and concentrate contaminants [13,27,28]. Plants with high root biomass, or high absorption surface, with more accumulation capacity (aquatic hyperaccumulators) and tolerance to contaminants achieve the best results. Promising examples include *Helianthus annus, Brassica juncea, Phragmites australis, Fontinalis antipyretica* and several species of *Salix, Populus, Lemna* and *Callitriche* [3,16,29-31].

6. **Rhizodegradation** (Phytostimulation): growing roots promote the proliferation of degrading rhizosphere microorganisms which utilize exudates and metabolites of plants as a source of carbon and energy. In addition, plants may exude biodegrading enzymes themselves. The application of phytostimulation is limited to organic contaminants [3,27]. The microbial community in the rhizosphere is heterogeneous due to variable spatial

distribution of nutrients, however species of the genus *Pseudomonas* are the predominant organisms associated with roots [13,32,33].

There are other strategies, which are considered categories of phytoremediation by some authors, but actually, they are mixed techniques or variations of the above mentioned strategies. These include:

a. **Hydraulic barriers:** some large trees, particularly those with deep roots (e.g., *Populus* sp.), remove large quantities of groundwater during transpiration. Contaminants in this water are metabolized by plant enzymes, and vaporized together with water or simply sequestered in plant tissues [3,34].

b. **Vegetation covers:** Herbs (usually grasses), eventually shrubs or trees, establish on landfills or tailings, are used to minimize the infiltration of rain water, and contain the spread of pollutants. The roots increase soil aeration thus, promoting biodegradation, evaporation and transpiration [7,35-37]. The difficulty of this technique is that tailings generally are not suitable for the development of plant roots. However, various investigations have been undertaken with the aim of developing processes of cultivation in tailings. For example a technique in which an organic soil composed of sawdust, plant remains, and some NPK-fertilizers is deposited on the surface was utilized by Hungarian agronomists (*Biological Reclamation Process, BRP*), [38]. The workers were able to obtain, at the end of a single biological cycle, 76 different plant species including cereals, shrubs, fruit trees and even large trees like oaks and pines.

c. **Constructed wetlands:** these are ecosystems consisting of organic soils, microorganisms, algae and vascular aquatic plants in areas where the water level is at/near the surface, at least part of the year. All the components work together in the treatment of effluents, through the combined actions of filtration, ion exchange, adsorption and precipitation [27,39,40]. It is the oldest method of wastewater treatment and is not regarded as proper phytoremediation, since it is based on the contributions of the entire system [3,41]. Good cleaning efficiency, low cost of construction along with easy operation and maintenance are the main advantages. It is widely applied in the treatment of domestic, agricultural and industrial waste water, but has proved to be suitable also for treating acid mine drainages [42-45].

d. **Phytodesalination:** it is a recently reported [13,46] emerging technique that utilizes halophytes to remove excess salts from saline soils. The potential of *Suaeda maritima* and *Sesuvium portulacastrum* in removal and accumulation of NaCl, from highly saline soils, has been demonstrated [47]. Although it has its peculiarities, this technique is a modality of phytoextraction.

3. Advantages and limitations of phytoremediation

Phytoremediation offers several advantages, but also some disadvantages, which should be considered when seeking to apply this technology (Table 1). If low cost is an advantage, the time necessary to observe the results can be long. The pollutant concentration and the presence of other toxins should be within the tolerance limits of the plant to be used. Selecting plants with the efficiency for remediating varied contaminants simultaneously is not easy. These limitations and the possibility of these plants entering in the food chains, should be taken into account when applying this technology.

Advantages	**Limitations**
In situ and passive technique	Limited to shallow soils or where contamination is localized to the surface (< 5 m)
Uses solar energy and is low cost	Still under development and therefore not accepted by many regulatory agencies
Has reduced environmental impact and contributes to the landscape improvement	There is little knowledge of farming, genetics, reproduction and diseases of phytoremediating plants
High acceptance by the public	Metal concentrations in the soil can be toxic and lethal to plants
Provides habitat for animal life	Generally, plants are selective in metal remediation
Reduction in dispersal of dust and contaminants by wind	Treatment slower than the traditional physico-chemical techniques
Reduction of surface runoff	Contamination may spread through the food chain if accumulator plants are ingested by animals
Reduction of leaching and mobilization of contaminants in soil	Efficient phytoremediating plants may not adapt to climatic and environmental conditions at contaminated sites
Harvesting of the plants or organs that have accumulated metals is easy to accomplish with existing technology	If the plants release compounds to increase the mobility of the metals, these can be leached into groundwater
The harvested biomass can be economically valuable	The area to be decontaminated must be large enough to allow application of cultivation techniques
Plant process more easily controlled than those of microorganisms	Toxicity and bioavailability of degradation products remain largely unknown

Table 1. Advantages and limitations of phytoremediation [3,7,8,48,49].

4. Phytoextraction and phytostabilization

Phytoextraction and phytostabilization are the two techniques most useful for phytoremediation of metal and metalloid contaminated soils. Phytoextraction has been widely studied, mainly due to the potential for high efficiency and possible economic value (in metal recovery, energy production) [3,23,24,48,50,51]. Preferably, plants used in phytoextraction should present, among others, the following characteristics [13,23,52,53]:

1. tolerance to high concentrations of metals;
2. accumulate high concentrations in their aerial tissues;
3. rapid growth;
4. high biomass production;
5. profuse root system;
6. easy to cultivate and harvest.

Phytoextraction can only be considered effective if the accumulated contaminant is subsequently removed through harvesting (Figure 2). If most of the captured heavy metals are translocated to shoots, traditional farming methods can be used for harvesting. It is important to harvest the plants before leaf-fall or death and decomposition to ensure that contaminants do not disperse or return to the soil [20].

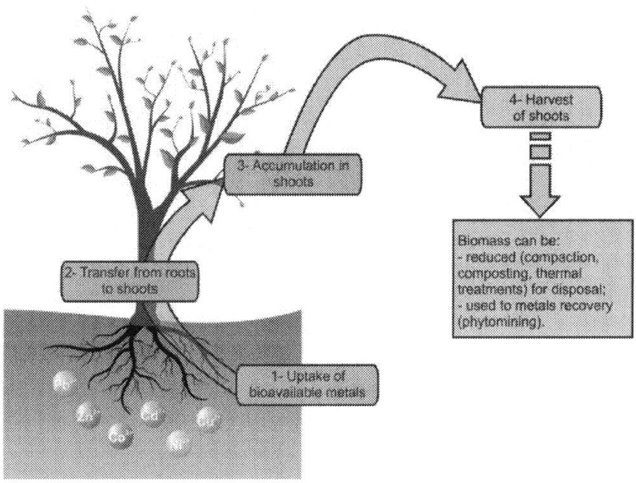

Figure 2. Schematic representation of phytoextraction of metals from soil.

After harvesting, biomass may be processed for extraction and recovery of metals (phytomining). The commercial value of metals such as Ni, Zn, Cu or Co may encourage the phytoremediation process. Alternatively, thermal, physical, chemical or microbiological processes can be used to reduce the volume/ weight of biomass. In the case of incineration of plants the energy produced represents an economic opportunity, and the ash can be further processed for extraction of metals. However, this process must be very careful, given the possible chemical elements accumulated, to prevent any dispersion mechanisms of contaminants.

According to McGrath and Zhao [22], phytoextraction efficiency is determined by two key factors: the ability to hyperaccumulate metals and the biomass production. Therefore, if these factors influence the phytoextraction, they can be optimized to improve the phytoremediation process. One possibility is the addition of chemical agents into the soil in order to increase the bioavailability of metals and their root uptake [54,55]. This form of assisted phytoremediation (or induced phytoremediation) has shown great potential and has been widely studied (Figure 3).

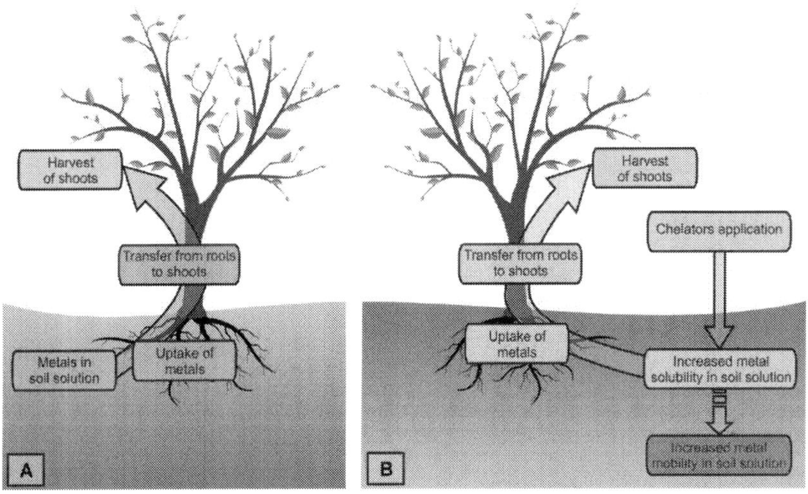

Figure 3. Schematic representation of the processes of natural (A) and assisted (B) phytoextraction.

Although hyperaccumulators are phytoextractors par excellence, usually they are low biomass producers. Thus, it is generally accepted that plants with a significant biomass production capacity can compensate their relatively lower metal accumulation capacity, to an extent where the amount of metal removed can be higher [51].

Phytoextraction potential can be estimated by calculation of bioconcentration factor (or biological absorption coefficient) and translocation factor [51,56]. The bioconcentration factor (BCF), which is defined as the ratio of the total concentration of element in the harvested plant tissue (C_{plant}) to its concentration in the soil in which the plant was growing (C_{soil}), is calculated as follows:

$$BCF = \frac{C_{plant}}{C_{soil}} \tag{1}$$

Translocation factor (TF), defined as the ratio of the total concentration of elements in the aerial parts of the plant (C_{shoot}) to the concentration in the root (C_{root}), is calculated as follows:

$$TF = \frac{C_{shoot}}{C_{root}} \quad (2)$$

The commercial efficiency of phytoextraction can be estimated by the rate of metal accumulation and biomass production. Multiplying the rate of accumulation (metal (g)/plant tissue (kg)) by the growth rate (plant tissue (kg)/hectare/year), gives the metal removal value (g/kg of metal per hectare and per year) [3,19,54,57]. This rate of removal or extraction should reach several hundred, or at least 1 kg/ha/year, for the species to be commercially useful, and even then, the remediation process may take from 15 to 20 years [3].

Some soils are so heavily contaminated that removal of metals using plants would take an unrealistic amount of time. The normal practice is to choose drought-resistant fast-growing crops or fodder which can grow in metal-contaminated and nutrient-deficient soils.

In contrast to phytoextraction, phytostabilization aims at reducing the mobility of contaminants in the soil. In this technique, contaminated soil is covered by vegetation tolerant to high concentrations of toxic elements, limiting the soil erosion and leaching of contaminants in to groundwater. Mobility of contaminants can be reduced by surface adsorption/accumulation in roots as well as their precipitation in rhizosphere by induced changes in pH or by oxidation of the root environment [3,12,58]. For example, the immobilization of arsenic in iron plates in the rhizosphere of salt marsh plants [58,59]. Phytostabilization can also be promoted by plant species with the capacity to exude high amounts of chelating substances. These substances lead to immobilization of contaminants by preventing their absorption, while simultaneously reducing their mobility in soil. Thus, plants with phytostabilization potential can be of great value for the revegetation of mine tailings and contaminated areas [58,60].

5. Phytoremediation potential of native flora of contaminated soils

It is possible to find a wide variety of plant species that can colonize areas highly polluted with heavy metals and metalloids, such as mine tailings or soils degraded and contaminated by mining/industrial activities. These are referred to as metallophyte and pseudometallophyte species.

Metallophytes are endemic plant species of natural mineralized soils and, therefore, have developed physiological mechanisms of resistance and tolerance to survive on substrates with high metal levels [61,62]. Since metallophytes, in general, and hyperaccumulators, in particular, are relatively rare and usually produce reduced biomass, the study of pseudometallophytes, indigenous species of contaminated soils, is of great value. Pseudometallophyte species (or facultative metallophytes) aren't specialized in metalliferous soils and have a more extensive distribution, but, due to selective pressure, are capable to survive in metalliferous soils [63-65]. Thus, the high pressure of metalliferous soils (natural or contaminated by human action) allows the selection of populations of common species, with higher tolerance than other populations of the same species. Therefore, their capacity of adaptation to these environments and, eventually, of accumulation of metals and metalloids, can be very interesting with a view to their use, for example, in ecological restoration, phytoremediation and bioindication actions.

In recent decades many studies have been conducted in contaminated mining and industrial areas and in natural metalliferous soils [35,51,58,66-78] in order to inventory and screen the indigenous species and evaluate their potential for phytoremediation of contaminated soils.

5.1. Phytoremediation potencial of native flora of contaminated soils in Portugal

Several studies to survey the indigenous plant species of diverse contaminated areas and evaluate their potential for phytoremediation have been performed in Portugal [79-93]. The authors have undertaken studies to evaluate the phytotechnological potential (phytoremediation, phytomining, bioindication, biogeochemical prospecting) of native flora of soils enriched with metals and metalloids, in distinct abandoned mining areas of tin/tungsten (Sn/W), copper (Cu), lead (Pb), uranium (U), and chromium (Cr) and the results are presented in this chapter.

5.1.1. Native flora of old mining areas

In the old mining areas studied, several line transects were made in mineralized and non-mineralized zones as well as tailings. Soils and plants were collected at 20 m intervals along the line transects (0, 20, 40 m, etc.) in circle of ≅2 m radius. At each location four random partial soil samples weighing 0.5 kg each were collected from 0 to 20 cm depth and mixed to obtain one composite sample to save time and costs. These were oven-dried at a constant temperature, manually homogenized and quartered. Two equivalent fractions were obtained from each quartered sample. One was used for the determination of pH, and the other for chemical analysis. The samples for chemical analysis were sieved using a 2 mm mesh sieve to remove plant matter and subsequently screened to pass through a 250 μm screen. Samples were also obtained from all species of plants whenever found growing within the 2 m radius of each sampling point. The plant sample focused on the aerial parts, taking into consideration similar maturity of the plants and the proportionality of the different types of tissues, or the separation of different types of tissues (leaves and stems) in some species. In the laboratory, the vegetal material was washed thoroughly, first in running water followed by distilled water, and then dried in a glasshouse. When dry, the material was milled into a homogenous powder. Soil pH was determined in water extract (1:2.5 v/v). The soil and plant samples were acid-digested for elemental analysis. Analytical methods included colorimetry for W, atomic absorption spectrophotometry (AAS, Perkin-Elmer, 2380) for Ag, Co, Cr, Cu, Fe, Mn, Ni, Pb and Zn and hydride generation system (HGS) for As and Sb. Fluorometry (Fluorat-02-2M analyzer, Lumex) was the methodology that was adopted for the determination of the U content in the plant and soil samples. Data quality control was performed by inserting triplicate samples into each batch. Certified references materials were also used.

5.1.1.1. Tin/tungsten mines

The studied areas included several abandoned Sn/W mines (Sarzedas mine, Fragas do Cavalo mine, Tarouca mine, Vale das Gatas mine, Adoria mine, Ervedosa mine, Regoufe mine, and Rio de Frades mine). Results obtained from Sarzedas (Central Portugal) and Vale das Gatas mines (Northern Portugal) are presented.

A summary of trace element data in soil from the Sarzedas mine is shown in Table 2. Among the elements present in the soils, Ag, As, Pb, Sb and W show the most relevant anomalies. Soil pH was negatively correlated to mineralization. Low pH values observed near the mineralized area can be explained by the presence of sulfides in the mineralization [84]. High levels of sulfides, in particular pyrite and arsenopyrite that are easily weathered, favors the dissolution of toxic elements, allowing higher dispersion and bioavailability.

	Range	Mean	Median	Standard deviation
pH	3.3 – 5.2	4.7	4.8	0.5
Ag	0.69 – 1.91	0.98	0.92	0.32
As	11.1 – 651	76.3	19.9	181
Co	5.40 – 14.9	8.80	8.41	2.60
Cr	50.7 – 129	96.5	100	26.3
Cu	15.5 – 78.2	40.7	35.1	21.1
Fe	21,881 – 58,644	39,981	37,356	12,883
Mn	22.0 – 92.0	50.0	47.0	22.0
Ni	11.2 – 52.5	21.8	19.7	10.2
Pb	35.7 – 417	85.4	53.9	106
Sb	30.5 – 5,986	663	87.8	1,689
W	0.80 – 684	663	2.90	52.3
Zn	29.0 – 127	58.8	53.3	24.3

Table 2. Trace elements content (mg/kg) and pH of soil samples (N=24, Sarzedas mine).

In the flora of Sarzedas mine area, As was accumulated in aerial tissues of *Pinus pinaster* and *Digitalis purpurea*. Therefore, these species are suited for recognizing the anomaly. High accumulation of As was present in leaves (Figure 4), and it increased in the older tissues. This translocation is a common mechanism in plants to avoid toxicity in young leaves as their metabolic activity is higher [84]. *Digitalis purpurea* also accumulated substantial amount of Sb (Figure 5), indicating its tolerance to this element, although the assimilation occured at low concentrations in the soil. Species that are capable of accumulating W are *D. purpurea, Cistus ladanifer, P. pinaster, Calluna vulgaris* and *Helichrysum stoechas* (Figure 6).

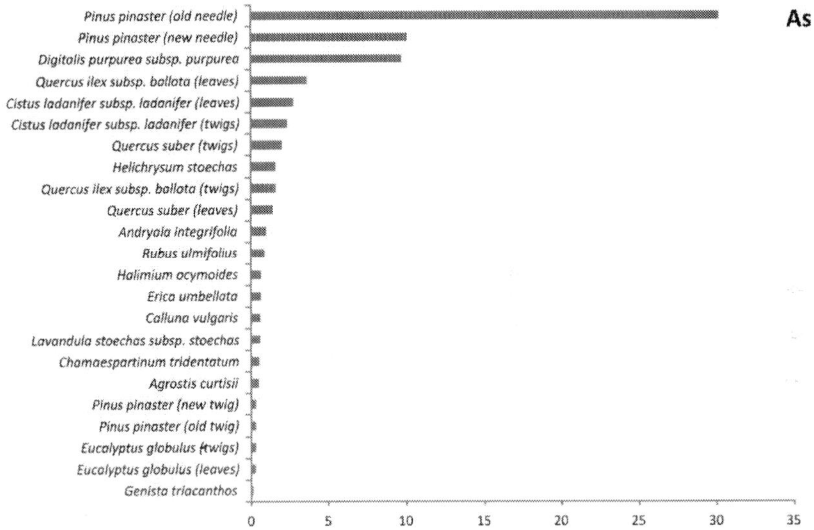

Figure 4. Accumulation of As (mg/kg DW) in plant species of the Sarzedas mining area.

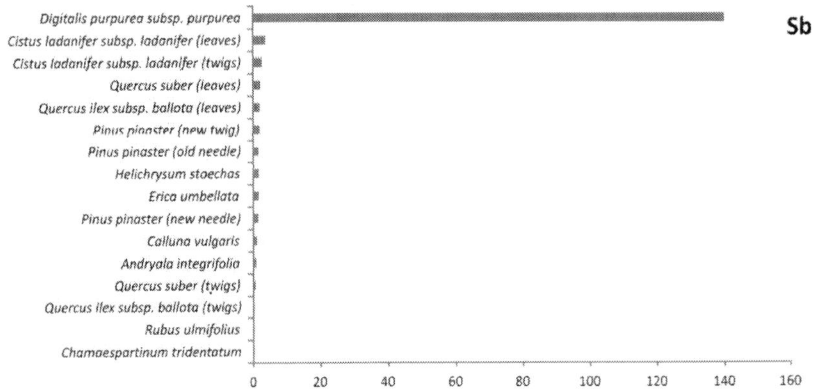

Figure 5. Accumulation of Sb (mg/kg DW) in plant species of the Sarzedas mining area.

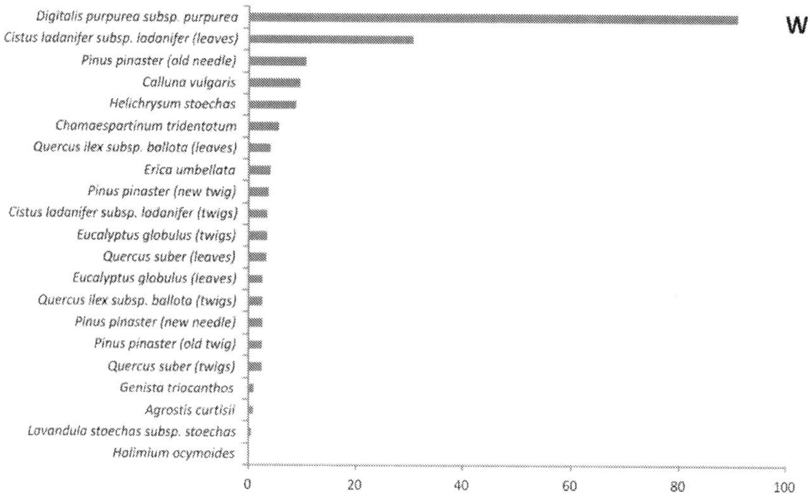

Figure 6. Accumulation of W (mg/kg DW) in plant species of the Sarzedas mining area.

It was concluded that the species and organs best suited for biogeochemical prospecting and/or with potential for mine restoration in the Sarzedas mine area are by order of importance: 1) As: old needles of *P. pinaster*, aerial tissues of *C. vulgaris*, *Chamaespartium tridentatum*, leaves of *C. ladanifer*, *Erica umbellata* and *Quercus ilex* subsp. *ballota*; 2) Sb: *D. purpurea*, *E. umbellata*, stems of *C. ladanifer*, *C. vulgaris*, *C. tridentatum* and stems of *P. pinaster*; 3) W: *D. purpurea*, *C. tridentatum*, old stems and needles of *P. pinaster*, stem and leaves of *C. ladanifer*, *E. umbellata* and stems and leaves of *Q. ilex* [84].

	Range	Mean	Median	Standard deviation
pH	3.5 – 6.3	5.0	5.0	0.8
As	26.7 – 5,770	446	56.7	1,178
Cu	11.7 – 352	88.0	29.0	101
Fe	18,482 – 60,100	33,039	29,443	12,463
Mn	103 – 898	336	167	248
Ni	11.6 – 61.2	30.6	23.6	15.1
Pb	55.4 – 6,299	499	102	1,285
Zn	63.1 – 469	180	125	112
W	2.00 – 636	73.8	10.6	162

Table 3. Trace elements content (mg/kg) and pH of soil samples (N=69, V. Gatas mine).

Very high maximum values for Pb (6,299 mg/kg), As (5,770 mg/kg) and W (636 mg/kg) were observed at the Vale das Gatas mine (Table 3). The Cu-Mn-W-As-Pb-Zn association, which reflects the presence of mineralised veins in the area, is inversely correlated with pH [93]. In general, the content variations in plant materials were strongly related to the content variations in soils. It has also been verified that in contaminated locations or tailings, the concentration of metals in plant tissues is high due to the high metal concentrations in the soil.

The leaves of *Agrostis castellana* and *Holcus lanatus* reflect the Cu, Pb and Ni pedogeochemical anomalies. The aerial parts of *Pteridium aquilinum* and *Juncus effusus* seem to be indicative of Zn anomalies in the soil [94]. *Holcus lanatus* and *A. castellana* were the main accumulators of As (Figure 7), Cu (Figure 8), Fe (Figure 9) and Pb (Figure 10) and good accumulators of Zn (Figure 11). *Pteridium aquilinum* was a good accumulator of As, Pb and Zn (Figures 7, 10, 11). *Juncus effusus* appeared to be a Zn accumulator (Figure 11).

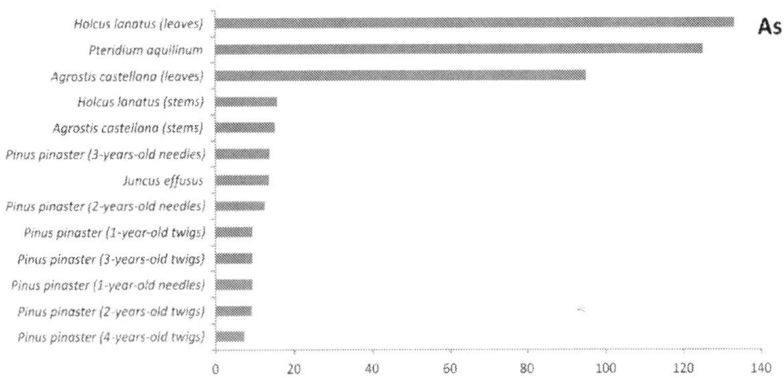

Figure 7. Accumulation of As (mg/kg DW) in plant species of the V. Gatas mining area.

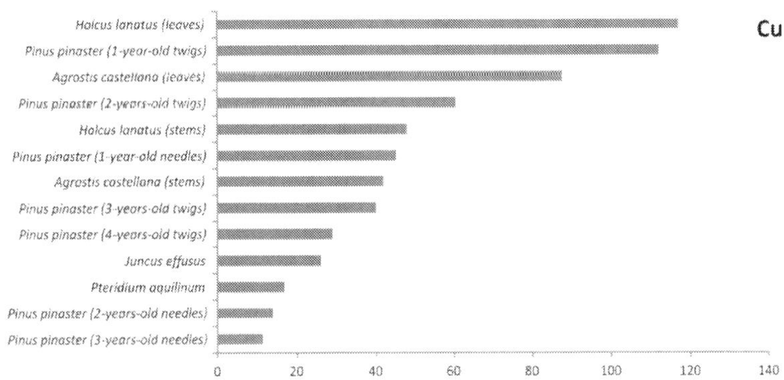

Figure 8. Accumulation of Cu (mg/kg DW) in plant species of the V. Gatas mining area.

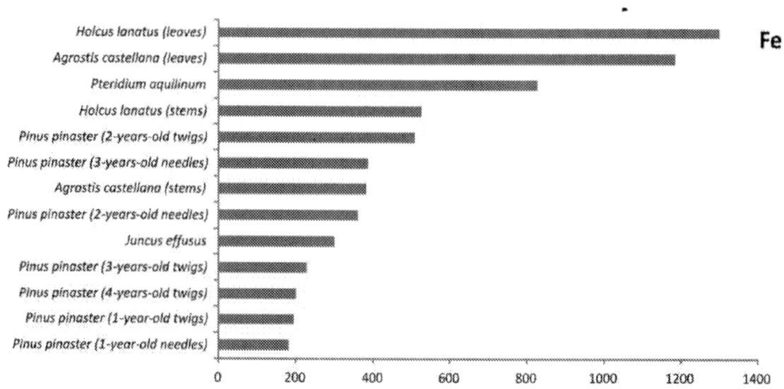

Figure 9. Accumulation of Fe (mg/kg DW) in plant species of the V. Gatas mining area.

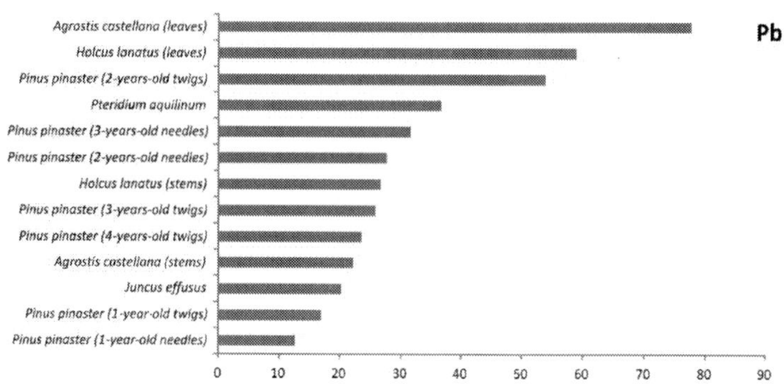

Figure 10. Accumulation of Pb (mg/kg DW) in plant species of the V. Gatas mining area.

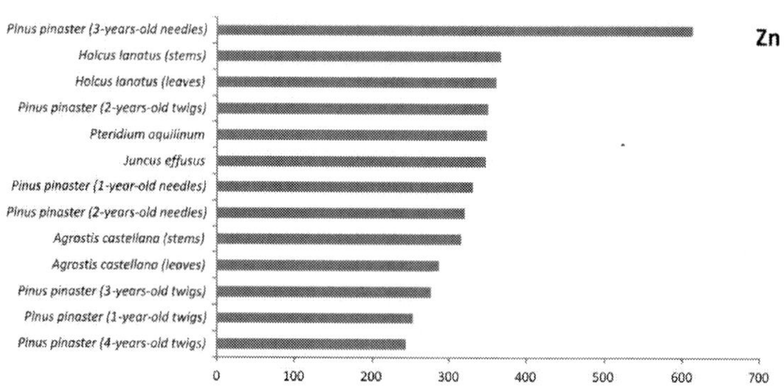

Figure 11. Accumulation of Zn (mg/kg DW) in plant species of the V. Gatas mining area.

The *P. pinaster* trees growing on the tailings and contaminated soils of Vale das Gatas mine accumulated the studied elements in quantities greater than observed in plants of the areas representative of the local geochemical background. These values were also higher than those typically observed in this species.

In the *P. pinaster* samples from tailings and contaminated soil locations, the older needles (2- and 3-years-old) show a tendency to accumulate higher concentrations of As, Fe, Zn, Pb and W while Ni and Cu were preferentially accumulated in young needles and stems (1-year-old) [93]. This allowed the authors to conclude that the metal/metalloid concentrations of elements in plants depend as much on the plant organ as on its age and in biogeochemical studies, it is important not to mix foliar and woody material in the same sample. The species showed a great variability in the accumulation behaviour of As, Fe, Mn, Cu, Zn, Pb, Ni, and W with the age of the organ. Thus, the 1-year-old needles and stems accumulated higher levels of Cu (Figure 8) and Ni (Figure 12). While the older needles accumulated higher levels of As, Fe, Pb, Zn and W (Figures 7, 9, 10, 11 and 13). The 2-years-old stems may also be appropriate samples to detect higher levels of Fe, Zn and Pb.

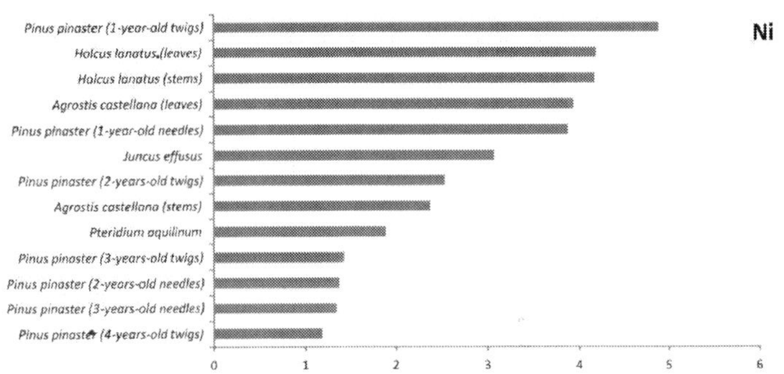

Figure 12. Accumulation of Ni (mg/kg DW) in plant species of the V. Gatas mining area.

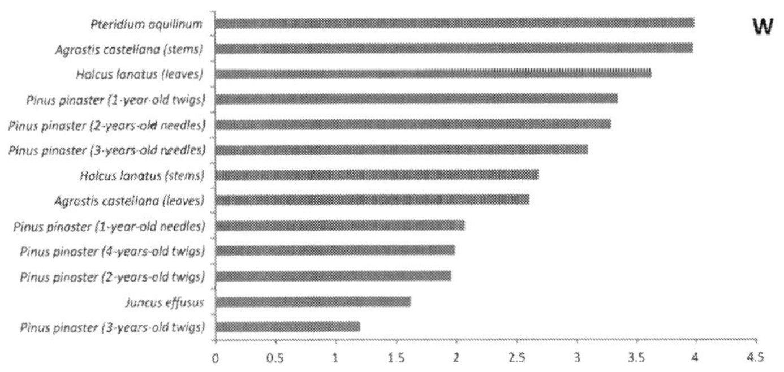

Figure 13. Accumulation of W (mg/kg DW) in plant species of the V. Gatas mining area.

5.1.1.2. Copper mines

The São Domingos mine (abandoned in 1966) located in south-east Portugal is also included in this study. This is one of the historical mining centres, known for its activity since pre-Roman times, with extraction of gold, silver and copper [95] though copper production was the highlight.

A summary of soil trace element data is presented in Table 4. High levels of As, Cu, Pb and Zn were recorded in the soils. Copper concentration in soils reached up to 1,829 mg/kg as a result of the former activities at the site (copper smelter). Maximum concentration of As in soils was very high, reaching 1,291 mg/kg. The concentration of Pb in the soil was also very high, 2,694 mg/kg as the average value registered. The average Zn concentration in soils was of 218 mg/kg but it could reach 714 mg/kg, a level that can be extremely toxic for plants. Cobalt and Cr concentrations in soils were normally low, ranging from 20.1 to 54.3 mg/kg for Co and 5.1 to 84.6 mg/kg for Cr. Nickel and Ag were also low, varying from 27.2–52.9 mg/kg and 2.5–16.6 mg/kg, respectively.

	Range	**Mean**	**Median**	**Standard deviation**
pH	4.0 – 6.7	5.1	5.1	0.6
Ag	2.50 – 16.6	7.50	7.00	3.60
As	37.2 – 1291	393	353	324
Co	20.1 – 54.3	31.0	29.4	8.40
Cr	5.10 – 84.6	26.5	8.30	31.7
Cu	87.3 – 1,829	553	444	443
Ni	27.2 – 52.9	42.2	43.9	6.60
Pb	234 – 12,218	2,694	2,355	2,345
Zn	104 – 714	218	163	145

Table 4. Trace elements content (mg/kg) and pH of soil samples (N=21, S. Domingos mine).

In plants, Pb concentration was rather high for some species, varying from 2.9 to 84.9 mg/kg dry weight (DW) (Figure 14). Semi-aquatic species sampled in the mining area, *Juncus conglomeratus* and *Scirpus holoschoenus,* showed high accumulation of Pb in plant tissues. Lead above 20 mg/kg DW was found in leaves of two species of *Cistus,* typical Mediterranean shrubs known for their tolerance to drought and low nutrients availability. Arsenic concentration in plant tissues ranged from 0.3 to 23.5 mg/kg DW. Maximum As was recorded in *J. conglomeratus, Thymus mastichina, J. effusus* and *S. holoschoenus* [82]. Semi-aquatic species from the Juncaceae family showed the highest content of both metals. Copper concentration in plant tissues ranged from 3.60 to 28.9 mg/kg DW (Figure 15). These Cu values are within the range considered normal for plants [96]. The species *Cistus monspeliensis* and *Daphne gnidium* showed the highest Zn concentrations [82]. A few trees, *Eucalyptus, Quercus* and *Pinus* species, were found in the contaminated area showing accumulation of different metals in the aboveground tissues. Due to their high biomass, they can be very effective for metals phytoextraction and

http://dx.doi.org/10.5772/57469

phytostabilization especially when established in the less contaminated soils on the peripheral zone of the study area [82].

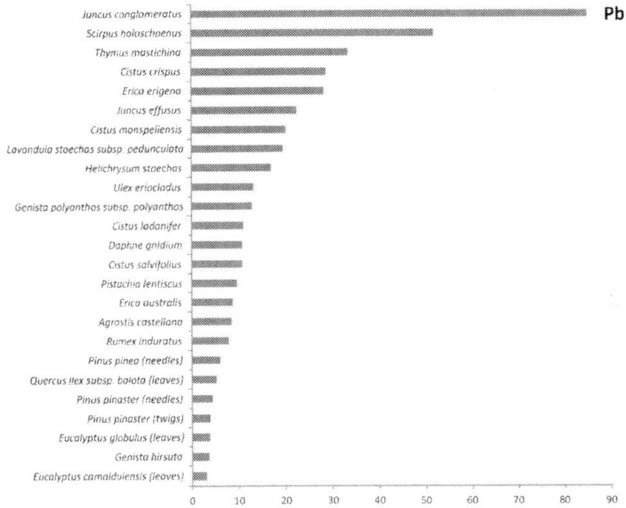

Figure 14. Accumulation of Pb (mg/kg DW) in plant species of the S. Domingos mining area.

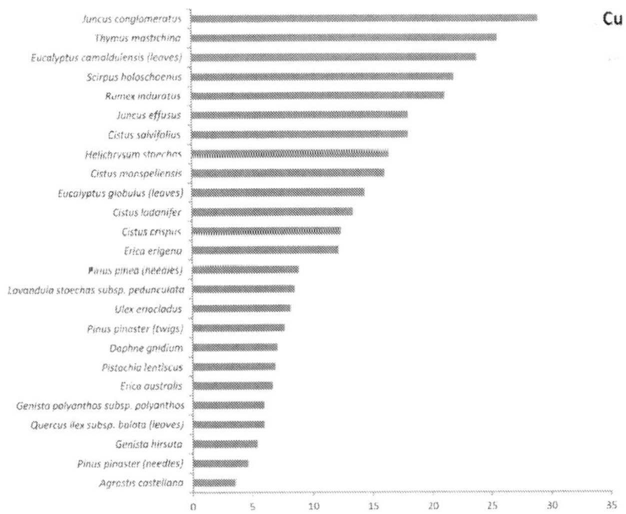

Figure 15. Accumulation of Cu (mg/kg DW) in plant species of the S. Domingos mining area.

5.1.1.3. Lead mines

The Barbadalhos mine is an abandoned Pb mine in Central Portugal. It was exploited for Pb by underground mining from 1887 till the 1940s. The concentrated ore was smelted on site. As per the usual practice at the time, tailings were deposited on the ground.

Metal concentrations in soil are shown in Table 5. Lead concentration in soils reached 9,331 mg/kg while the average value was 928 mg/kg; obviously due to mining of galena at the site. In soils from mineralized zone, the mean Pb concentration (2,380 mg/kg) was nearly 9 times the threshold for industrial soils suggested by Canadian Environmental Quality Guidelines [97].

	Range	Mean	Median	Standard deviation
pH	3,6 – 6.4	4.7	4.6	0.5
Ag	0.71 – 13.0	1.71	1.06	2.03
As	2.77 – 208	16.9	8.07	31.7
Co	3.74 – 50.5	20.1	16.7	12.4
Cr	61.3 – 196	89.0	85.7	22.1
Cu	21.4 – 193	41.7	34.5	27.9
Fe	24,145 – 98,510	40,283	38,497	13,751
Mn	44.4 – 2,224	596	381	588
Ni	7.68 – 87.0	30.5	28.1	12.2
Pb	24.4 – 9,331	928	68.8	2,119
Zn	30.4 – 517	134	90.3	109

Table 5. Trace elements content (mg/kg) and pH of soil samples (N=45, Barbadalhos mine).

Samples from 49 species of the native flora were investigated at this site. Individual elements and species displayed different trends of accumulation. All plants collected along mineralized zone accumulated eight metals (Ag, Co, Cr, Cu, Fe, Ni, Pb, and Zn) but many plants from non-mineralized zone accumulated only five metals (Ag, Cu, Fe, Pb, and Zn). A few however did accumulate the remaining three (Co, Cr, and Ni); bringing the count of metals accumulated at par with those of mineralized zone [92].

Most plants were seen to be tolerant of soil Pb concentrations. In mineralized zone, Pb concentrations in plants ranged from 1.11 to 548 mg/kg DW. This is far above the 100 – 400 mg Pb/kg content considered toxic for most plants [98]. Significant accumulation of Pb was seen in *Cistus salvifolius* (548 mg/kg), *Lonicera periclymenum* (318 mg/kg), *Anarrhinum bellidifolium, Phytolacca americana, Digitalis purpurea, Mentha suaveolens* (255 – 217 mg/kg) [listed in decreasing order] (Figure 16). Pteridophytes like *Polystichum setiferum, Pteridium aquilinum,* and

Asplenium onopteris also showed 117 – 251 mg/kg Pb in aerial parts. In plants from non-mineralized zone, Pb content was not significant ranging from 0.94 to 11.6 mg/kg.

Though at first glance maximum Pb content observed in trees like *Acacia dealbata* (84 mg/kg: leaves), *Olea europaea* (62 mg/kg: twigs), and *Quercus suber* (58 mg/kg: twigs) from mineralized zone is not very impressive compared to that of smaller plants mentioned above, nevertheless these trees can be very effective due to their higher biomass. When combined with the hardy nature, biomass and abundance of this species, the moderate accumulation indicates immense potential for phytoextraction of Pb in the area [92].

In mineralized zone, Zn concentrations in plants reached 1,020 mg/kg in *D. purpurea*. And ranged from 262 to 887 mg/kg in *L. periclymenum*, *P. americana*, *Solanum nigrum*, *P. setiferum*, *M. suaveolens*, *Viola riviniana*, and *A. bellidifolium* [listed in decreasing order] [92].

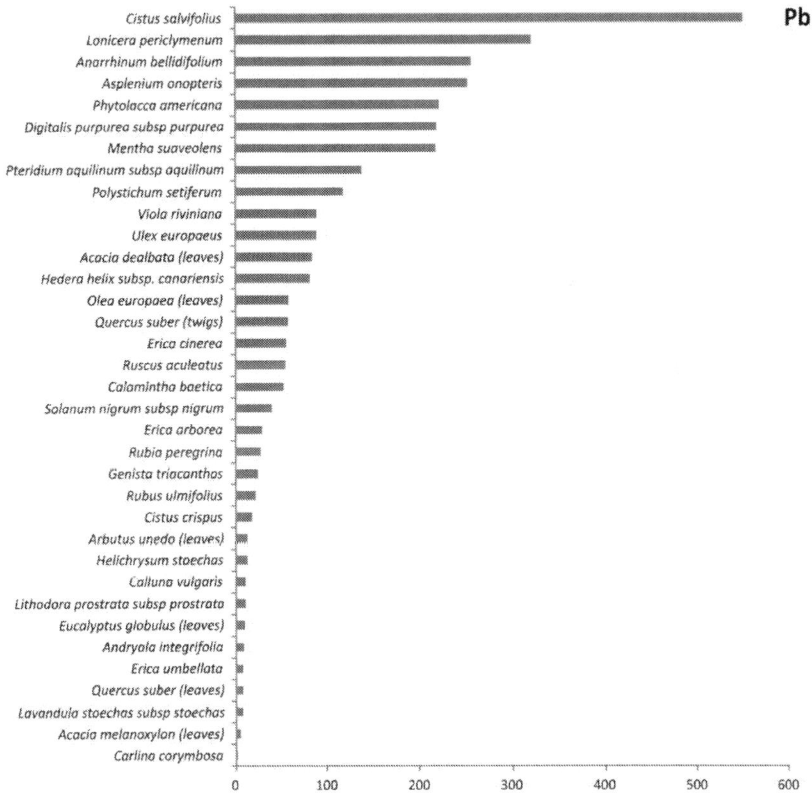

Figure 16. Accumulation of Pb (mg/kg DW) in plant species of the Barbadalhos mining area.

5.1.1.4. Uranium mines

The old U mine of Sevilha (Central Portugal) is one of several small mines exploited by ENU (Portuguese Uranium Company). After the removal of the main ore body, the site was filled with the mine wastes and a reclamation process was initiated. This action was somewhat unsuccessful because the selected allochthonous plant species (*Lupinus* sp.) did not survive.

Current U soil contamination on the Sevilha mine ranges from 8 to 560 mg/kg [99]. Species of Compositae and Ericaceae (among the most abundant families of terrestrial plants) accumulated highest U concentrations (Figure 17). Among Compositae members, an average of 4.91 mg/kg DW and a maximum of 13.1 mg/kg DW was found in *Helichrysum stoechas* and an average of 4.07 mg/kg DW and a maximum of 10.5 mg/kg DW was recorded in *Hypochaeris radicata* (Figure 17). In *Erica umbellata* an average of 1.70 mg/kg DW and a maximum of 7.50 mg/kg DW were obtained (Figure 17). Even though the concentrations obtained in the latter are not high, it is particularly interesting because it has a high bio-productivity. This accumulation potential might be intensified if uptake enhancement strategies, such as addition of citric acid, are adopted. A restoration program can be applied to the soils of Sevilha mine by adopting revegetation with endemic species allied to a process of continuous phytoremediation that avoids dispersion of U into the streamlets.

Even though the soils in this mine are not highly contaminated, the lixiviation of refilling materials has been responsible for the dispersion of U into ground and superficial water bodies. These waters are being used for subsistence agriculture and, therefore, the risk of contamination spreading to humans can be acute, due to food chain accumulation [99]. The plant survey revealed that some of the native plant species are well adapted to U contamination in soils, therefore, they are metallotolerants. Their phytoremediation potential has to be evaluated. Dispersion of U into the streamlets can be minimized by a strategic combination of terrestrial and aquatic plant phyto-systems. Revegetation with *Helichrysum stoechas, Hypochaeris radicata* and *Erica umbellata* will allow fixation of U in the plants and a consequent reduction in its dispersion. This site can be an excellent prototype for the restoration of other mines in Portugal where levels of contamination are a matter of concern.

5.1.1.5. Chromium mines (Serpentine soils)

The abandoned mining area of Pingarela in North-east Portugal has serpentine soils and associated flora. These soils are disproportionately rich in trace elements like Ni, Cr, Co and poor in Ca. Serpentine outcrops have been referred to as barrens because they are often sparsely vegetated and extremely poor in essential nutrients, hence not of much agricultural value. Serpentine ecosystems can generally be distinguished by their grey-green or reddish rocky soils (soils are very thin), and shrubby or stunted vegetation with plants having small leathery leaves.

Plant species found on serpentine soils can be divided into two groups: (a) serpentine-tolerant or serpentine-facultative plants, which are able to survive on serpentine but grow better elsewhere; (b) serpentinicolous, serpentine-endemic or serpentine-obligate plants, which grow exclusively on serpentine soils and are not found on other substrates [83,100]. Both these

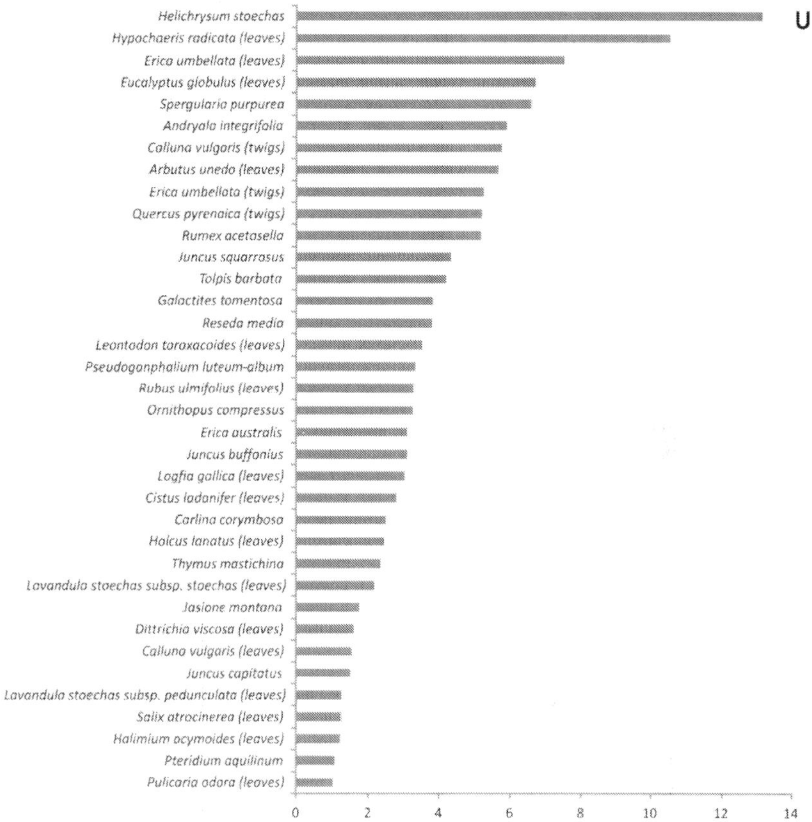

Figure 17. Accumulation of U (mg/kg DW) in plant species of the Sevilha mining area.

groups include species with different efficiencies to uptake or exclude a variety of metals. Serpentinophytes often experience drought, nutrient stress and excessive exposure to heavy metal and high light intensity. This means there is less substrate in which nutrients and water can be held and made available to plants.

An area of ~8,000 ha in north-east Portugal is serpentinized with characteristic geology and flora. The serpentine plant community and respective soils were analyzed [83] to examine the trace metal budget in different tissues of the plants exhibiting resistance to trace metals. 135 plant species belonging to 39 families and respective soils were analyzed for total Co, Cr, Cu, Fe, Mn, Ni, Pb and Zn. Heavy metal concentrations recorded in sampled soils are shown in Table 6. The high contents of Ni and Cr obtained were to be expected, considering the geochemistry of the sampling site. However, the high variability of Cr, Ni, Fe and Mn in the soils is noteworthy.

	Range	Mean	Median	Standard deviation
pH	4.9 – 9.3	6.7	6.7	0.6
Co	56.0 – 151	82.3	81.1	18.5
Cr	200 – 6,822	1,622	1,410	1,064
Cu	30.8 – 221	101	99.3	34.7
Fe	95.0 – 110,418	82,950	84,711	14,502
Mn	1,007 – 1,835	1,389	1,363	175
Ni	102 – 2,295	918	883	464
Pb	18.5 – 46.6	29.2	29.1	5.78
Zn	63.0 – 242	110	112	24.8

Table 6. Trace element content (mg/kg) and pH of serpentine soil samples (N=74, Pingarela mine).

The Ni hyperaccumulating endemic of this region is *Alyssum serpyllifolium* subsp. *lusitanicum,* which concentrated 38,105 mg Ni/kg DW in the aboveground tissues (Figure 18). *Bromus hordeaceus* with 1,467 mg Ni/kg DW and *Linaria spartea* with 492 mg Ni/kg DW in the aerial parts, also showed high concentration of Ni. Four other taxa *viz.—Plantago radicata, Ulmus procera, Lavandula stoechas* and *Cistus salvifolius* showed more than 100 mg Ni/kg DW (Figure 18).

Chromium has low solubility in the serpentine soil solution due to the relatively high pH values of these soils [83]. This is reflected in the low uptake of this element by plants, which in general did not exceed 40 mg/kg. However, concentrations of 707 mg Cr/kg DW were reported in the above ground parts of *L. spartea* (Figure 19). *Alyssum serpyllifolium* also presented high content of Cr, reaching a maximum of 130 mg Cr/kg DW. *Ulmus procera* showed a content of 173 mg Cr/kg DW in the twigs.

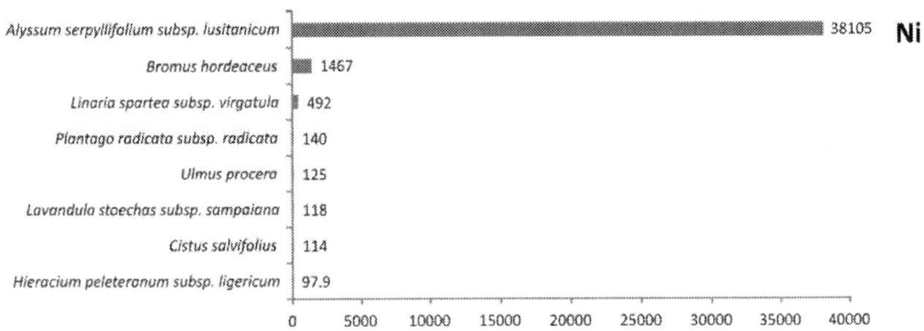

Figure 18. Accumulation of Ni (mg/kg DW) in serpentine plant species of the Pingarela mining area.

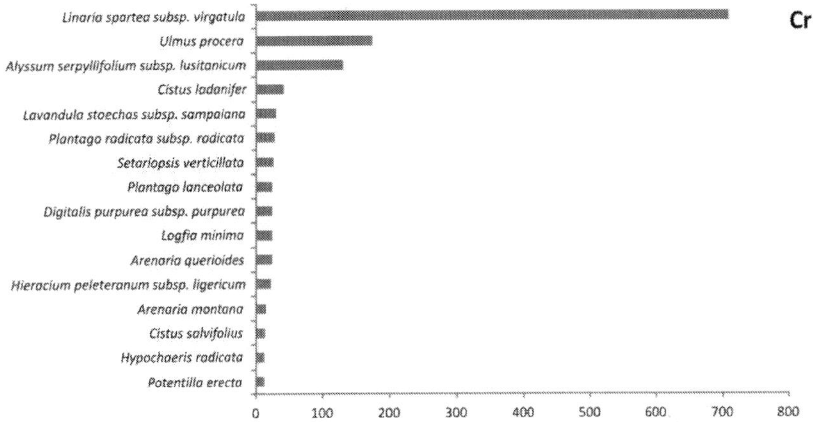

Figure 19. Accumulation of Cr (mg/kg DW) in serpentine plant species of the Pingarela mining area.

Although leaves of serpentine plants show 10 times higher Co levels than those of plants growing on non-serpentine soils, the absolute concentrations are only about 10 mg/kg DW in most species. In this study highest accumulation of Co was found in the aboveground tissues of *A. serpyllifolium* with 145 mg Co/kg DW and *L. spartea* with 63.2 mg Co/kg DW [83].

6. Conclusion

The physico-chemical properties of the metalliferous or metal-contaminated soils tend to inhibit soil-forming processes and plant growth. In addition to elevated metal/metalloid concentrations, other adverse factors included absence of topsoil, erosion, drought, compaction, wide temperature fluctuations, absence of soil-forming fine materials and shortage of essential nutrients [84,101]. Degraded soils of mines usually have low concentrations of important nutrients, like N, P and K [102]. Toxic metals can also adversely affect the number, diversity and activity of soil organisms, inhibiting soil organic matter decomposition and N mineralization processes. The chemical form of the potential toxic metal, the presence of other chemicals which may aggravate or attenuate metal toxicity, the prevailing pH and the poor nutrient status of contaminated soil affects the way in which plants respond to it. Substrate pH affects plant growth mainly through its effect on the solubility of chemicals, including toxic metals and nutrients.

Metal toxicity issues do not generally arise in the case of native flora, considering that native plants become adapted over time to the locally elevated metal levels [76,103]. Native plants may be better phytoremediators for contaminated lands than the known metal hyperaccumulators because these are generally slow growing with shallow root systems and low biomass. Plants tolerant to toxic metals and low nutrient status with a high rate of growth and biomass are the ideal species to remediate degraded soils and habitats like those around mines. The native flora displayed its ability to withstand high concentrations of heavy metals in the soil. Some species also displayed variable accumulation patterns for metals at different soil concentrations. This variation was also observed in different parts of the same plant suggesting that full consideration of plant–soil interactions should be taken into account when choosing plant species for developing and utilizing methods such as phytoremediation.

Indigenous plant species growing on tailings and contaminated soils show tolerance to imposed stress conditions (metal-contamination and nutrient deficiency) and can fulfill the objectives of stabilization, pollution attenuation and visual improvement. Besides, these species are drought-resistant and some even exhibit high biomass and bioproductivity. In fact, the constraints related to plant establishment and amendment of the physical–chemical properties of the metalliferous soils depends upon the choice of appropriate plant species. Hence, the plant community tolerant to toxic trace elements plays a major role in remediation of degraded mine soils.

The existing natural plant cover at abandoned mining sites can be increased manifold by wide-scale planting and maintenance of native species with higher metal accumulation potential for some years. Even dispersal of seeds obtained from plants on site is to be encouraged. Adding organic amendment is essential to facilitate the establishment and colonization of these "pioneer plants". They can eventually modify the man-made habitat and render it more suitable for subsequent plant communities. Allowing native species to remediate soils is an attractive proposition since native wild species do not require frequent irrigation, fertilization, and pesticide treatments, while simultaneously a plant community comparable to that existing in the vicinity can be established.

Therefore, mine restoration could benefit from a broader perspective including different groups of plant species as they can perform distinct functional roles in the remediation process. The use of leguminous plants, for example, may enrich the nutrient content and the combined used of perennials and annuals can provide substantial inputs in terms of organic matter and nutrient recycling, thus contributing in distinct ways to the development of the soil [82,104]. This approach requires more information about plant communities growing on metal-contaminated soils in order to accurately determine their potential for remediation of polluted soils at abandoned mines. Ideal phytoremedial candidates can be screened out from the native flora and after assessing their individual requirements, suitable conditions/amendments can be created to develop them as good competitors with enhanced growth and proliferation than their counterparts growing on the same metal contaminated nutrient depleted soils.

Significant accumulation of heavy metals and metalloids in both soils and native wild flora suggests that metal contamination is a matter of great concern in the studied mining areas. The native flora displayed its ability to withstand high concentrations of heavy metals/ metalloids in the soil. However, accumulation patterns of metals/metalloids in the plants tested differed. As metal concentrations in above ground parts were maintained at low levels, metal tolerance in most cases may mainly depend on their metal excluding ability. However, metal/metalloid concentrations higher than toxic level in some species like *Agrostis castellana* (for As, and Fe), *Cistus ladanifer* subsp. *ladanifer* (for Cr, and W), *Cistus*

http://dx.doi.org/10.5772/57469

salvifolius (for Ni, and Pb), *Digitalis purpurea* subsp. *purpurea* (for Sb, W, and Zn), *Helichrysum stoechas* and *Hypochaeris radicata* (for U), *Holcus lanatus* (for As, Cu, and Fe), *Lonicera periclymenum, Mentha suaveolens* and *Phytolacca americana* (for Pb, and Zn), *Pinus pinaster* (for As, W, and Zn), *Polystichum setiferum* and *Solanum nigrum* subsp *nigrum* (for Zn), *Pteridium aquilinum* (for As), as well as the serpentine plant species *Alyssum serpyllifolium* subsp. *lusitanicum, Lavandula stoechas* subsp. *sampaiana, Linaria spartea* subsp. *virgatula* and *Ulmus procera* (for Cr, and Ni) and *Bromus hordeaceus* and *Plantago radicata* subsp. *radicata* (for Ni) indicate that internal detoxification metal tolerance mechanisms might also exist; therefore, their utility for phytoremediation is possible. Furthermore, the plants could grow and propagate in substrata with low nutrient conditions which would be a great advantage in the revegetation of mine tailings. It was also observed that despite lower accumulation, trees of the studied regions can be very effective due to their higher biomass.

Some of the studied species also showed variable accumulation patterns for metals at different soil concentrations. This difference was also noted between parts of the same plant suggesting that full consideration of plant–soil interactions should be taken into account when choosing plant species for developing and utilizing methods such as phytoremediation.

Acknowledgements

This study was partially supported by the European Fund for Economic and Regional Development (FEDER) through the Program Operational Factors of Competitiveness (COMPETE) and National Funds through the Portuguese Foundation for Science and Technology (PEST-C/MAR/UI 0284/2011, FCOMP 01 0124 FEDER 022689).

Author details

Paulo J.C. Favas[1,3*], João Pratas[2,3], Mayank Varun[4], Rohan D'Souza[4] and Manoj S. Paul[4]

*Address all correspondence to: pjcf@utad.pt

1 University of Trás-os-Montes e Alto Douro, UTAD, School of Life Sciences and the Environment, Vila Real, Portugal

2 Department of Earth Sciences, Faculty of Sciences and Technology, University of Coimbra, Coimbra, Portugal

3 IMAR-CMA Marine and Environmental Research Centre, Faculty of Sciences and Technology, University of Coimbra, , Portugal

4 Department of Botany, St. John's College, Agra, India

References

[1] Evanko CR, Dzombak DA. Remediation of metals-contaminated soils and groundwater. Technology Evaluation Report. Pittsburgh: GWRTAC – Ground-Water Remediation Technologies Analysis Center; 1997.

[2] Gómez Orea D. Recuperación de espacios degradados. Madrid, Barcelona, México: Ediciones Mundi-Prensa; 2004.

[3] Prasad MNV. Phytoremediation of metals and radionuclides in the environment: the case for natural hyperaccumulators, metal transporters, soil-amending chelators and transgenic plants. In: Prasad MNV. (ed.) Heavy metal stress in plants: from biomolecules to ecosystems, Second Edition. Berlin: Springer; 2004. p345-391.

[4] Chaney RL, Malik M, Li YM, Brown SL, Angle JS, Baker AJM. Phytoremediation of soil metals. Current Opinion in Biotechnology 1997; 8 279-284.

[5] Ensley BD. Rationale for use of phytoremediation. In: Raskin I, Ensley BD. (ed.) Phytoremediation of toxic metals. Using plants to clean up the environment. New York: John Wiley & Sons, Inc.; 2000. p3-11.

[6] Prasad MNV, Freitas HMO. Metal hyperaccumulation in plants – Biodiversity prospecting for phytoremediation technology. Electronic Journal of Biotechnology 2003; 6(3) 285-321.

[7] Mendez MO, Maier RM. Phytoremediation of mine tailings in temperate and arid environments. Reviews in Environmental Science and Biotechnology 2008; 7 47-59.

[8] Dickinson NM, Baker AJM, Doronila A, Laidlaw S, Reeves RD. Phytoremediation of inorganics: realism and synergies. International Journal of Phytoremediation 2009; 11 97-114.

[9] Schnoor JL, Licht LA, McCutcheon SC, Wolfe NL, Carreira LH. Phytoremediation of organic and nutrient contaminants. Environmental Science & Technology 1995; 29 318A-323A.

[10] Rylott EL, Bruce NC. Plants disarm soil: engineering plants for the phytoremediation of explosives. Trends in Biotechnology 2008; 27(2) 73-81.

[11] Berti WR, Cunningham SD. Phytostabilization of metals. In: Raskin I, Ensley BD. (ed.) Phytoremediation of toxic metals. Using plants to clean up the environment. New York: John Wiley & Sons, Inc.; 2000. p71-88.

[12] Domínguez MT, Madrid F, Marañón T, Murillo JM. Cadmium availability in soil and retention in oak roots: potential for phytostabilization. Chemosphere 2009; 76 480-486.

[13] Ali H, Khan E, Sajad MA. Phytoremediation of heavy metals – Concepts and applications. Chemosphere 2013; 91 869-881.

[14] Brooks RR. Phytoremediation by volatilisation. In: Brooks RR. (ed.) Plants that hyperaccumulate heavy metals: their role in phytoremediation, microbiology, archaeology, mineral exploration and phytomining. New York: CAB International; 1998. p289-312.

[15] Pilon-Smits E, Pilon M. Breeding mercury-breathing plants for environmental cleanup. Trends in Plant Science 2000; 5(6) 235-236.

[16] Poschenrieder CH, Barceló J. Estrés por metales pesados. In: Reigosa MJ, Pedrol N, Sánchez A. (eds.) La ecofisiologia vegetal: Una ciencia de síntesis. Madrid: Thomson; 2004. p413-442.

[17] Ruiz ON, Daniell H. Genetic engineering to enhance mercury phytoremediation. Current Opinion in Biotechnology 2009; 20 213-219.

[18] Pilon-Smits EAH, LeDuc DL. Phytoremediation of selenium using transgenic plants. Current Opinion in Biotechnology 2009; 20 207-212.

[19] McGrath SP. Phytoextraction for soil remediation. In: Brooks RR. (ed.) Plants that hyperaccumulate heavy metals: their role in phytoremediation, microbiology, archaeology, mineral exploration and phytomining. New York: CAB International; 1998. p261-287.

[20] Blaylock MJ, Huang JW. Phytoextraction of metals. In: Raskin I, Ensley BD. (ed.) Phytoremediation of toxic metals. Using plants to clean up the environment. New York: John Wiley & Sons, Inc.; 2000. p53-70.

[21] Ma LQ, Komar KM, Tu C, Zhang W, Cai Y, Kennelley ED. A fern that hyperaccumulates arsenic. Nature 2001; 409 579.

[22] McGrath SP, Zhao FJ. Phytoextraction of metals and metalloids from contaminated soils. Current Opinion in Biotechnology 2003; 14 277-282.

[23] Hernández-Allica J, Becerril JM, Garbisu C. Assessment of the phytoextraction potential of high biomass crop plants. Environmental Pollution 2008; 152 32-40.

[24] Pedron F, Petruzzelli G, Barbafieri M, Tassi E. Strategies to use phytoextraction in very acidic soil contaminated by heavy metals. Chemosphere 2009; 75 808-814.

[25] Xie QE, Yan XL, Liao XY, Li X. The arsenic hyperaccumulator fern *Pteris vittata* L. Environmental Science & Technology 2009; 43(22) 8488-8495.

[26] Van der Ent A, Baker AJM, Reeves RD, Pollard AJ, Schat H. Hyperaccumulators of metal and metalloid trace elements: Facts and fiction. Plant Soil 2013; 362 319-334.

[27] Frers C. El uso de plantas acuáticas en el tratamiento de aguas residuales. Carmen de Areco, Argentina: El Planeta Azul; 2009.

[28] Dhote S, Dixit S. Water quality improvement through macrophytes – a review. Environmental Monitoring and Assessment 2009; 152 149-153.

[29] Dushenkov S, Kapulnik Y. Phytofiltration of metals. In: Raskin I, Ensley BD. (ed.) Phytoremediation of toxic metals. Using plants to clean up the environment. New York: John Wiley & Sons, Inc.; 2000. p89-106.

[30] Pratas J, Favas PJC, Paulo C, Rodrigues N, Prasad MNV. Uranium accumulation by aquatic plants from uranium-contaminated water in Central Portugal. International Journal of Phytoremediation 2012; 14 221-234.

[31] Favas PJC, Pratas J, Prasad MNV. Accumulation of arsenic by aquatic plants in large-scale field conditions: Opportunities for phytoremediation and bioindication. Science of the Total Environment 2012; 433 390-397.

[32] Crowley DE, Alvey S, Gilbert ES. Rhizosphere ecology of xenobiotic-degrading microorganisms. In: Kruger EL, Anderson TA, Coats JR. (eds.) Phytoremediation of Soil and Water Contaminants. Washington: ACS Symposium Series; 1997. p20-36.

[33] Khan MS, Zaidi A, Wani PA, Oves M. Role of plant growth promoting rhizobacteria in the remediation of metal contaminated soils. Environmental Chemistry Letters 2009; 7 1-19.

[34] Schnoor JL. Phytostabilization of metals using hybrid poplar trees. In: Raskin I, Ensley BD. (ed.) Phytoremediation of toxic metals. Using plants to clean up the environment. New York: John Wiley & Sons, Inc.; 2000. p133-150.

[35] Williamson NA, Johnson MS, Bradshaw AD. Mine Wastes Reclamation. The Establishment of Vegetation on Metal Mine Wastes. London: Mining Journal Books; 1982.

[36] Brooks RR, Chiarucci A, Jaffré T. Revegetation and stabilization of mine dumps and other degraded terrain. In: Brooks RR. (ed.) Plants that hyperaccumulate heavy metals: their role in phytoremediation, microbiology, archaeology, mineral exploration and phytomining. New York: CAB International; 1998. p227-247.

[37] Jorba M, Vallejo R. La restauración ecológica de canteras: un caso con aplicación de enmiendas orgánicas y riegos. Ecosistemas 2008; 17(3) 119-132.

[38] González V. A indústria extractiva e o ambiente. Boletim de Minas 1990; 27(3) 311-323.

[39] Vymazal J. The use constructed wetlands with horizontal sub-surface flow for various types of wastewater. Ecological Engineering 2009; 35 1-17.

[40] Fonder N, Headley T. The taxonomy of treatment wetlands: A proposed classification and nomenclature system. Ecological Engineering 2013; 51 203-211.

[41] Horne AJ. Phytoremediation by constructed wetlands. In: Terry N, Bañuelos G. (eds.) Phytoremediation of contaminated soil and water. New York: Lewis Publishers; 2000. p13-39.

[42] Sobolewski A. A review of processes responsible for metal removal in wetlands treating contaminated mine drainage. International Journal of Phytoremediation 1999; 1(1) 19-51.

[43] López Pamo E, Aduvire O, Barettino D. Tratamientos pasivos de drenajes ácidos de mina: estado actual y perspectivas de futuro. Boletín Geológico y Minero 2002; 113(1) 3-21.

[44] Nyquist J, Greger M. A field study of constructed wetlands for preventing and treating acid mine drainage. Ecological Engineering 2009; 35 630-642.

[45] Adams A, Raman A, Hodgkins D. How do the plants used in phytoremediation in constructed wetlands, a sustainable remediation strategy, perform in heavy-metal-contaminated mine sites? Water and Environment Journal 2013; 27(3) 373-386.

[46] Zorrig W, Rabhi M, Ferchichi S, Smaoui A, Abdelly C. Phytodesalination: a solution for salt-affected soils in arid and semi-arid regions. Journal of Arid Land Studies 2012; 22 299-302.

[47] Ravindran KC, Venkatesan K, Balakrishnan V, Chellappan KP, Balasubramanian T. Restoration of saline land by halophytes for Indian soils. Soil Biology & Biochemistry 2007; 39 2661-2664.

[48] Glass DJ. Economic potential of phytoremediation. In: Raskin I, Ensley BD. (ed.) Phytoremediation of toxic metals. Using plants to clean up the environment. New York: John Wiley & Sons, Inc.; 2000. p15-31.

[49] Van Nevel L, Mertens J, Oorts K, Verheyen K. Phytoextraction of metals from soils: How far from practice? Environmental Pollution 2007; 150 34-40.

[50] Nascimento CWA, Xing B. Phytoextraction: a review on enhanced metal availability and plant accumulation. Scientia Agricola 2006; 63(3) 299-311.

[51] Zhuang P, Yang QW, Wang HB, Shu WS. Phytoextraction of heavy metals by eight plant species in the field. Water, Air, and Soil Pollution 2007; 184 235-242.

[52] Sakakibara M, Ohmori Y, Ha NTH, Sano S, Sera K. Phytoremediation of heavy metal contaminated water and sediment by *Eleocharis acicularis*. Clean: Soil, Air, Water 2011; 39 735-741.

[53] Shabani N, Sayadi MH. Evaluation of heavy metals accumulation by two emergent macrophytes from the polluted soil: an experimental study. Environmentalist 2012; 32 91-98.

[54] Pierzynski G, Kulakow P, Erickson L, Jackson L. Plant system technologies for environmental management of metals in soils: Educational materials. Journal of Natural Resources and Life Science Education 2002; 31 31-37.

[55] Saifullah, Meers E, Qadir M, Caritat P, Tack FMG, Laing GD, Zia MH. EDTA–assisted Pb phytoextraction. Chemosphere 2009; 74(10) 1279-1291.

[56] Tu S, Ma LQ. Comparison of arsenic and phosphate uptake and distribution in arsenic hyperaccumulating and nonhyperaccumulating fern. Journal of Plant Nutrition 2004; 27(7) 1227-1242.

[57] Liang HM, Lin TH, Chiou JM, Yeh KC. Model evaluation of the phytoextraction potential of heavy metal hyperaccumulators and non-hyperaccumulators. Environmental Pollution 2009; 157 1945-1952.

[58] Antosiewicz DM, Escudě-Duran C, Wierzbowska E, Skłodowska A. Indigenous plant species with the potential for the phytoremediation of arsenic and metals contaminated soil. Water, Air, & Soil Pollution 2008; 193 197-210.

[59] Doyle MO, Otte ML. Organism-induced accumulation of iron, zinc and arsenic in wetland soil. Environmental Pollution 1997; 96 1-11.

[60] Mains D, Craw D, Rufaut C, Smith C. Phytostabilization of gold mine tailings from New Zealand. Part 2: Experimental evaluation of arsenic mobilization during revegetation. International Journal of Phytoremediation 2006; 8 163-183.

[61] Baker AJM. Accumulators and excluders – strategies in the response of plants to heavy metals. Journal of Plant Nutrition 1981; 3 643-654.

[62] Baker AJM, Walker PL. Ecophysiology of metal uptake by tolerant plants. In: Shaw AJ. (ed.) Heavy metal tolerance in plants: Evolutionary aspects. Boca Raton, Florida: CRC Press, Inc.; 1990. p155-193.

[63] Becerril JM, Barrutia JL, Garcia Plazaola A, Hernández A, Olano JM, Garbisu C. Especies nativas de suelos contaminados por metales: aspectos ecofisiológicos y su uso en fitorremediación. Ecosistemas 2007; 16(2) 50-55.

[64] Becerra-Castro C, Monterroso C, Prieto-Fernández A, Rodríguez-Lamas L, Loureiro-Viñas M, Acea MJ, Kidd PS. Pseudometallophytes colonising Pb/Zn mine tailings: A description of the plant–microorganism-rhizosphere soil system and isolation of metal-tolerant bacteria. Journal of Hazardous Materials 2012; 217/218 350-359.

[65] Quintela-Sabarís C, Ribeiro MM, Poncet B, Costa R, Castro-Fernández D, Fraga MI. AFLP analysis of the pseudometallophyte *Cistus ladanifer*: comparison with cpSSRs and exploratory genome scan to investigate loci associated to soil variables. Plant and Soil 2012; 359 397-413.

[66] Ernst W. Mine vegetation in Europe. In: Shaw AJ. (ed.) Heavy Metal Tolerance in Plants: Evolutionary Aspects. Boca Raton, Florida: CRC Press, Inc.; 1990. p21-37.

[67] Robinson BH, Chiarucci A, Brooks RR, Petit D, Kirkman JH, Gregg PEH, DeDominicis V. The nickel hyperaccumulator plant *Alyssum bertolonii* as a potential agent for phytoremediation and phytomining of nickel. Journal of Geochemical Exploration 1997; 59 75-86.

[68] Poschenrieder C, Bech J, Llugany M, Pace A, Fenes E, Barceló J. Copper in plant species in a copper gradient in Catalonia (North East Spain) and their potential for phytoremediation. Plant and Soil 2001; 230(2) 247-256.

[69] Madejón P, Murillo JM, Marañon T, Cabrera F, López R. Bioaccumulation of As, Cd, Cu, Fe and Pb in wild grasses affected by the Aznalcóllar mine spill (SW Spain). Science of the Total Environment 2002; 290 105-120.

[70] Kidd PS, Monterroso C. Metal extraction by *Alyssum serpyllifolium* ssp. *lusitanicum* on mine-spoil soils from Spain. Science of the Total Environment 2005; 336 1-11.

[71] Yanqun Z, Yuan L, Jianjun C, Haiyan C, Li Q, Schvartz C. Hyperaccumulation of Pb, Zn and Cd in herbaceous grown on lead-zinc mining area in Yunnan, China. Environment International 2005; 31 755-762.

[72] Rodríguez N, Amils R, Jiménez-Ballesta R, Rufo L, De la Fuente V. Heavy metal content in *Erica andevalensis*: an endemic plant from the extreme acidic environment of Tinto River and its soils. Arid Land Research and Management 2007; 21 51-65.

[73] Saraswat S, Rai JPN. Phytoextraction potential of six plant species grown in multimetal contaminated soil. Chemistry and Ecology 2009; 25(1) 1-11.

[74] D'Souza R, Varun M, Masih J, Paul MS. Identification of *Calotropis procera* L. as a potential phytoaccumulator of heavy metals from contaminated soils in Urban North Central India. Journal of Hazardous Materials 2010; 184 457-464.

[75] Lorestani B, Cheraghi M, Yousefi N. Phytoremediation potential of native plants growing on a heavy metals contaminated soil of copper mine in Iran. World Academy of Science, Engineering and Technology 2011; 53 377-382.

[76] Varun M, D'Souza R, Pratas J, Paul MS.. Metal contamination of soils and plants associated with the glass industry in North-central India: prospects of phytoremediation. Environmental Science and Pollution Research 2012; 19 269-281.

[77] Maric M, Antonijevic M, Alagic S. The investigation of the possibility for using some wild and cultivated plants as hyperaccumulators of heavy metals from contaminated soil. Environmental Science and Pollution Research 2013; 20 1181-1188.

[78] Zhang Z, Sugawara K, Hatayama M, Huang Y, Inoue C. Screening of As-accumulating plants using a foliar application and a native accumulation of As. International Journal of Phytoremediation 2013; doi:10.1080/15226514.2013.773277.

[79] De Koe T, Beek M, Haarsma M, Ernst W. Heavy metals and arsenic grasses and soils of mine spoils in North East Portugal, with particular reference to some Portuguese goldmines. In: Nath B. (ed.) Environmental Pollution 1. Proc. Int. Conf. ICEP-1; 1991. p373-380.

[80] Henriques FS, Fernandes JC. Metal uptake and distribution in rush (*Juncus conglomeratus* L.) plants growing in pyrite mine tailings at Lousal, Portugal. Science of the Total Environment 1991; 102 253-260.

[81] Alvarenga PM, Araújo MF, Silva JAL. Elemental uptake and root-leaves transfer in *Cistus ladanifer* L. growing in a contaminated pyrite mining area (Aljustrel – Portugal). Water, Air, and Soil Pollution 2004; 162 81-96.

[82] Freitas H, Prasad MNV, Pratas J. Plant community tolerant to trace elements growing on the degraded soils of São Domingos mine in the south east of Portugal: environmental implications. Environment International 2004; 30 65-72.

[83] Freitas H, Prasad MNV, Pratas J. Analysis of serpentinophytes from north-east of Portugal for trace metal accumulation – relevance to the management of mine environment. Chemosphere 2004; 54 1625-1642.

[84] Pratas J, Prasad MNV, Freitas H, Conde L. Plants growing in abandoned mines of Portugal are useful for biogeochemical exploration of arsenic, antimony, tungsten and mine reclamation. Journal of Geochemical Exploration, 2005; 85 99-107.

[85] Díez Lázaro J, Kidd PS, Monterroso Martínez C. A phytogeochemical study of the Trás-os-Montes region (NE Portugal): Possible species for plant-based soil remediation technologies. Science of the Total Environment 2006; 354 265-277.

[86] Branquinho C, Serrano HC, Pinto MJ, Martins-Loução MA. Revisiting the plant hyperaccumulation criteria to rare plants and earth abundant elements. Environmental Pollution 2007; 146 437-443.

[87] Turnau K, Henriques FS, Anielska T, Renker C, Buscot F. Metal uptake and detoxification mechanisms in *Erica andevalensis* growing in a pyrite mine tailing. Environmental and Experimental Botany 2007; 61 117-123.

[88] Abreu MM, Tavares MT, Batista MJ. Potential use of *Erica andevalensis* and *Erica australis* in phytoremediation of sulphide mine environments: São Domingos, Portugal. Journal of Geochemical Exploration 2008; 96 210-222.

[89] Marques APGC, Moreira H, Rangel AOSS, Castro PML. Arsenic, lead and nickel accumulation in *Rubus ulmifolius* growing in contaminated soil in Portugal. Journal of Hazardous Materials 2009; 165 174-179.

[90] Anawar HM, Freitas MC, Canha N, Santa Regina I. Arsenic, antimony, and other trace element contamination in a mine tailings affected area and uptake by tolerant plant species. Environmental Geochemistry and Health 2011; 33(4) 353-362.

[91] Abreu MM, Santos ES, Ferreira M, Magalhães MCF. *Cistus salviifolius* a promising species for mine wastes remediation. Journal of Geochemical Exploration 2012; 113 86-93.

[92] Pratas J, Favas PJC, D'Souza R, Varun M, Paul MS. Phytoremedial assessment of flora tolerant to heavy metals in the contaminated soils of an abandoned Pb mine in Central Portugal. Chemosphere 2013; 90 2216-2225.

[93] Favas PJC, Pratas J, Prasad MNV. Temporal variation in the arsenic and metal accumulation in the maritime pine tree grown on contaminated soils. International Journal of Environmental Science and Technology 2013; 10(4) 809-826.

[94] Favas PJC, Pratas JAMS, Conde LEN. Bioaccumulation of heavy metals in plants that colonize tailings (Vale das Gatas mine – Vila Real – North of Portugal). In: Prego R, Duarte A, Panteleitchouk A, Santos TR. (eds.) Studies on Environmental Contamination in Iberian Peninsula. Lisbon: Instituto Piaget; 2002. p259-272. (in Portuguese)

[95] Gaspar OC. História da mineração dos depósitos de sulfuretos maciços vulcanogénicos da Faixa Piritosa Portuguesa. Boletim de Minas 1998; 35(4) 401-414.

[96] Kabata-Pendias A. Trace Elements in Soils and Plants, 4th ed. Boca Raton: CRC Press; 2010.

[97] Canadian Council of Ministers for the Environment. Canadian soil quality guidelines for the protection of environmental and human health, update 6.02, Publication no. 1299, ISBN 1-896997-34-1. 2006.

[98] Alloway BJ. Heavy Metals in Soils. London: Blackie Academic & Professional; 1990.

[99] Favas P, Pratas J. Uranium in soils, waters and plants of the an abandoned uranium mine (Central Portugal). In: 12th International Multidisciplinary Scientific GeoConference (SGEM 2012) Conference Proceedings Vol. V, 17-23 June 2012, Albena, Bulgaria. Albena: STEF92 Technology Ltd.; 2012. p1023-1028.

[100] Reeves RD, Baker AJM, Borhidi A, Berazain R. Nickel-accumulating plants from the ancient serpentine soils of Cuba. New Phytologist 1996; 133 217-224.

[101] Wong MH. Ecological restoration of mine degraded soils with emphasis on metal contaminated soils. Chemosphere 2003; 50 775-780.

[102] Huenneke LF, Hamburg SP, Koide R, Mooney HA, Vitousek PM. Effects of soil resources on plant invasion and community structure in Californian serpentine grassland. Ecology 1990; 71(2) 478-491.

[103] D'Souza R, Varun M, Pratas J, Paul MS. Spatial distribution of heavy metals in soil and flora associated with the glass industry in North Central India: implications for phytoremediation. Soil and Sediment Contamination 2013; 22 1-20.

[104] Hooper DU, Vitousek PM. The effects of plant composition and diversity on ecosystem processes. Science 1997; 277 1302-1305.

Chapter 18

Contribution of Vegetation to Alleviate Slope's Erosion and Acidity

Normaniza Osman, Mohammed Saifuddin and
Aimee Halim

Additional information is available at the end of the chapter

http://dx.doi.org/10.5772/57228

1. Introduction

Most of the slope soils in the tropical region such as in Malaysia is arid and infertile due to lack of buffering capacity and low clay activities which resulted in soil acidity. In addition, high intensity rainfall and extreme conditions of slope such as transient drought and lack of nutrients have reduced the survival and growth of potential plants. Rain water percolation which leaches basic elements such as calcium, magnesium, potassium and sodium from the soil profile is also another factor which contributes to soil acidity.

Soil acidity has a huge negative impact on fertility, biological activities and plant productivity. Fortunately, the use of vegetation and their association with microbes have great potential to alleviate soil acidity and erosion problems. Moreover, heavy metal-tolerant plants also play an important role in phytoremediation. Therefore, proper plantation and management of plants in polluted and acidic soil may significantly contribute to restoring the natural environment.

The practice of using vegetation, so-called "Bio-engineering" technique, combines an ecological, mechanical and hydrological concept which has been successfully applied to alleviate soil erosion in Malaysia [1]. The combined right choices of plants and planting technique confer numerous advantages such as producing a high biodiversity, low cost-maintenance, self-sustainability as well as environmental friendly slope [1]. Nonetheless, the interaction of vegetation and soil are complex as it is involved with, *inter alia,* varying soil properties, soil pollutants and different type of plant coverage. Therefore, in light of fast developmental land transformation and climate change, it is of a great challenge to integrate all plant-soil properties and its interaction with the whole ecosystem and environment.

1.1. Objectives of the chapter

This book chapter will explore the attributions of vegetation to alleviate the most common slope problems in Malaysia; acidity and erosion. This chapter will also provide potential characteristics of plants to reinforce soil and the mechanism of Aluminium (Al) accumulator plant to alleviate soil acidity. The specific aims of this book chapter are to:

a. Describe the ecological approach towards slope stabilization;

b. Outline the vegetation effects on erosion rate and carbon sequestration;

c. Elucidate the mechanism of Al accumulator plant to alleviate soil acidity and

d. Exhibit some potential slope plants.

2. Ecological approaches to stabilize slope

There are many types of natural and manmade slopes exist in Malaysia. Due to the nature of topography e.g. slope angle, length, aspect, gradient and curvature, and the weather conditions, these slopes become unstable and make serious geologic hazards. Around the world, landslide depends on the geological characteristics, hydrological condition and rainfall distribution [2,3]. In Malaysia, significant numbers of slope failure are reported on manmade and residual soil slopes especially at the time of high intensity rainfall due to rapid change of soil properties, particularly physical properties, such as bulk density, cohesiveness and shear strength [4]. However, there are three common triggering factors for slope failure in respect to Malaysia which are intense rainfall, water level change and change of slope loading (Figure 1), indicating hydrological condition gives infavorable impact on the slope stability. Apart from that, landslides also occurred due to the human activities such as cultivation, excavation for housing, foot paths and deforestation or construction works in hillsides and lack of plantation in hilly area [5,6]. In addition, improper slope design might cause manmade slope failure which has been acknowledged as one of the most reasons for frequent disaster around the world [7]. To overcome this problem, the slopes need to be proper designed, considering the geological characteristics, structural model, local weather and soil characteristics [8]. From eco-physiological point of view, the plant availability, so-called bioengineering systems, has become an alternative solution to minimize the impacts and stabilize the slope [1,9].

Bioengineering systems are extremely beneficial as it incorporates with geotechnical engineering, ecological and biological aspects [7,11]. In bioengineering techniques, less heavy equipments and live plants are usually used which are cost effective and provided long-term soil stability [12]. Generally, newly cut and manmade slope soils are unstable and expose to soil loss. Hence, it is needed to be made more resistant to erosion. In our study, Nordin et. al. [13], have discovered that different composition of root distribution and branching between two potential slope plants, namely; *Acacia mangium* and *Leucaena leucocephala* (Table 1) caused the differences in pull-out values. In tandem with *L. leucocephala*'s ability to establish on slope's harsh condition and reinforced the soil [14], it exhibited the higher force of pull-out, thus,

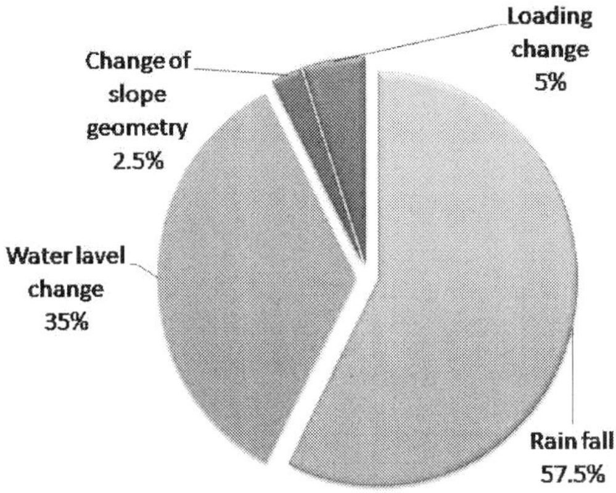

Figure 1. Common triggering factors which is related to slopes erosions [10].

showing its capability to resist the uprooting forces. This feature helps the species anchorage and resists the slope against forces such as wind. Interestingly, the study also found that the pull-out force is much affected by the tensile strength (R=0.88; Figure 2). The tensile strength decreased with increasing root diameter. Thus, this property of higher tensile strength in the finer roots would ultimately yield an increase in shear strength of the root-soil composite in the field and provide better ductility to the root-soil composite with a higher capacity to withstand surface erosion and runoff [15].

Species	Pull-out Force (KN)	Tensile strength (N/mm^2)
Acacia mangium	1.69 ± 0.34	54.37 ± 10.80
Leucaena leucocephala	2.25 ± 0.45	104.83 ± 18.72
Significant difference between species at $LSD_{p<0.05}$ = 0.73 (pull-out) and 18.9 (tensile strength)		

Table 1. Pull-out test and tensile strength of *Acacia mangium* and *Leucaena leucocephala* (data are means ± standard error)

Knowing the importance of vegetation, however, the problem of vegetation establishment on freshly cut slope is a succession problem, in which naturally, the process takes a longer time. Fortunately, with the human helps and input, we can accelerate this process. Normaniza and Barakbah [1] proposed a concept for establishing vegetation cover on slopes which enhanced the process of natural succession (Figure 3). The introduction of pioneer species on slope is to initiate the succession process and later to accelerate the ecological restoration as a whole. Initially, the proposed pioneer species is to improve the quality of soil, be it acidic or infertile soil.

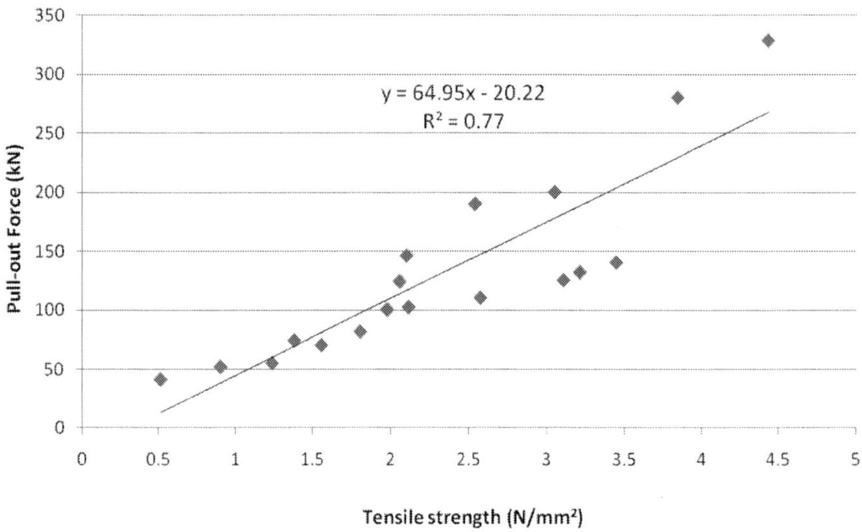

Figure 2. The relationship between pull-out force and root tensile strength [13].

In relation to the proposed concept, the selected pioneer must exhibit prominent characteristics which include high growth rate, good root profiles, water relations and high tolerance to a wide range of adverse factors with regard to soil quality, microclimate and mechanical stress [9,16,17]. It is anticipated that once the pioneer is established, the succession process would be enhanced through the changes of abiotic and biotic factors. Consequently, influx of other species will enrich the plant biodiversity of slopes. This plant community changes would not only hasten the process of natural succession but would also attract small animals such as insects and birds to the ecosystem. This fauna association would assist in promoting "flora-fauna" interaction *via* being the agents of seed dispersal, which would ultimately enhance the natural plant succession process.

This ecological approach towards slope stabilization has been proven on slope projects conducted by our team at several highways cut slopes in Malaysia [1,18] (Figure 4). Our study showed that legume trees which act as pioneers had increased the species diversity and slope stability parameters of the slope (Table 2). The legume tree plots exhibited higher root length density and resulted in a lower saturation level of the soil. In addition, penetration resistance was observed to increase with presence of legume trees. The increment of shear strength at 30 cm soil depth was prominent in legume tree plot. Overall increment of slope stability parameters indicated the positive effect of legume trees in reducing the probability slope failures, since most surface failures occur at a depth of 20-50 cm [1]. In contrast, the grasses plots had almost saturated, 83.9-93.2%, a characteristic of failing slope.

http://dx.doi.org/10.5772/57228

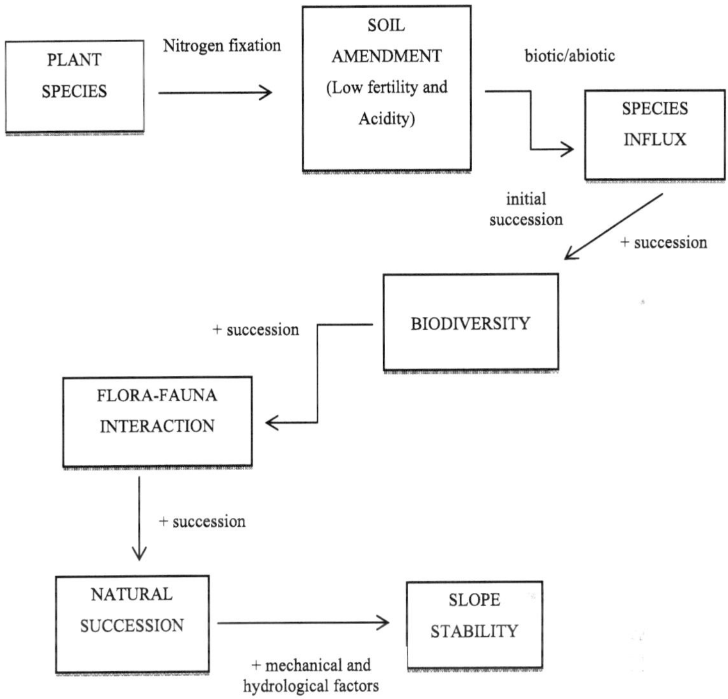

Figure 3. A proposed concept for establishing vegetation cover on slopes and enhancing the process of natural succession [1].

Parameters	Grasses	Legume trees
Saturation level (%)	83.9-93.2	73.6-81.2
Root length density (Km m^{-3})	0.86	9.4
Penetrability (24 months) (MPa)	0.97-1.5	1.44-2.01
Shear strength (at 30 cm soil depth at 24 months) (KPa)	100.5±5.5	104.6±2.4

Table 2. Slope stability parameters between the plots (grasses and legume trees)

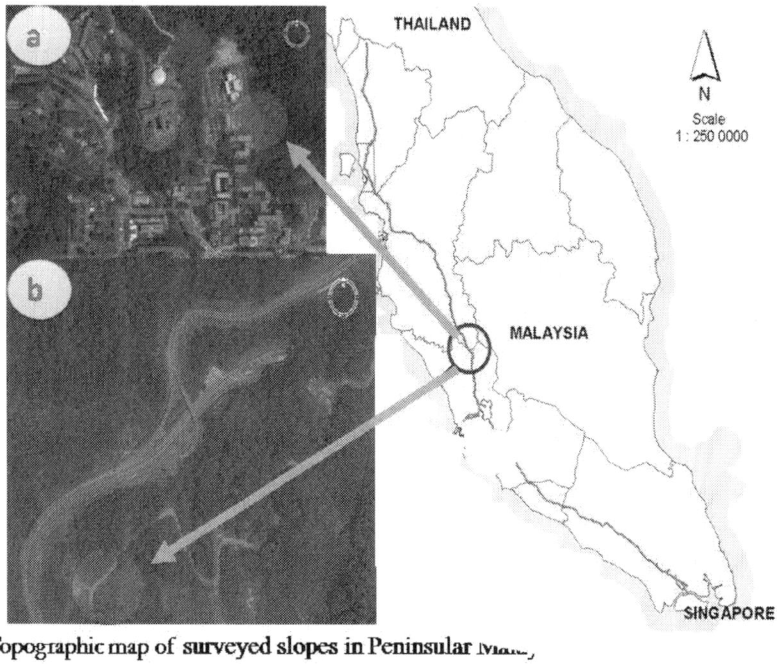

Figure 4. Topographic map of surveyed slopes in Peninsular Malaysia; a) Faculty of Science, University of Malaya; latitude 03⁰ 07′ 28.5′′ N, longitude 101⁰ 39′ 14.6′′ E and b) Batu 38, Pusat Pengajian Luar, University of Malaya, Ulu Gombak; latitude 03⁰ 20′ 45.27′′ N, longitude 101⁰ 46′ 26.52′′ E.

3. Impacts of vegetation on soil erosion and carbon sequestration

3.1. Soil erosion

The use of vegetation to control soil erosion has been practiced for many centuries, firstly introduced in China in 16th century to stabilize dam [19]. Nowadays, this practice of vegetation has been successfully applied to stabilize slope throughout the world. The vegetation and erosion process are interrelated by the ability of the plant life growing on soil and the interaction of root and soil [20]. But the interaction of vegetation and soil are complex as it involved with, inter alia, the combination of soil types, plant coverage and the steepness of slope. There are many factors also responsible for controlling soil erosion such as soil elements, soil density, slope length, existing plant species and plant position on slope, plant age, plant coverage and plant root distribution. Moreover, the revegetation process also influenced by the plant-soil interaction such as soil acidity, nutrient content, and drought conditions. Toriman and Shukor

[21] found that in a forest area of Malaysia, interception reduces 23.9% of the total rainfall and it is varying subjected to plant canopy, density and types of plants.

In our research findings, the plant density treatment (i.e. low, medium and high densities) of the potential slope plant species, *Melastoma malabathricum* provided the significant findings on the interception process at the sloping areas. A higher plant density increased the leaf area index (LAI) (Figure 5) as well as contributed to a higher plant growth. In addition, the highest plant density in the studied plots recorded the lowest erosion rate, indicating that soil erosion was lower at the area with a higher vegetation density by intercepting rainfall by plant canopy (LAI) (Figure 6).

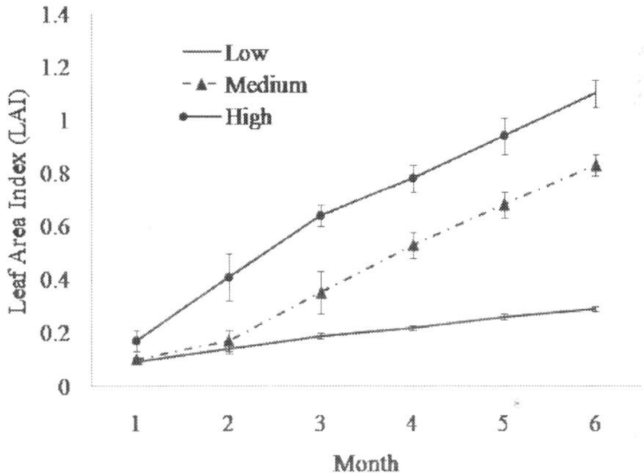

Figure 5. The Leaf Area Index (LAI) of *M. malabathricum* of three different density treatments.

Furthermore, the higher the plant density, the higher the soil carbon content is. This result also indicates that the increased of species density influenced in carbon cycle via storing the large amount of carbon in soil through photosynthesis and respiration. It can be explained by the large amount of litter fall on the soil surface, thus, enhanced the decomposition process which in turns, increased the organic matter and mineral content at the top layer of soil. Apart from that, the amount of soil carbon was directly related to the root length density (Figure 7). Higher soil carbon content was observed at the greater root length density (RLD) area, indicating the distribution of soil carbon was induced by the root distribution. The root system supplies decomposable organic matter in soil and supports a large microbial community in the rhizosphere [22,23], thus, help in distribution of soil carbon. In addition, a higher plant density produced more litter on the top layer of the soil surface, which in turns increased the decomposition process via carbon and nitrogen cycles [24], hence, increased the soil pH value as well. In aftermath, the soil pH was enhanced.

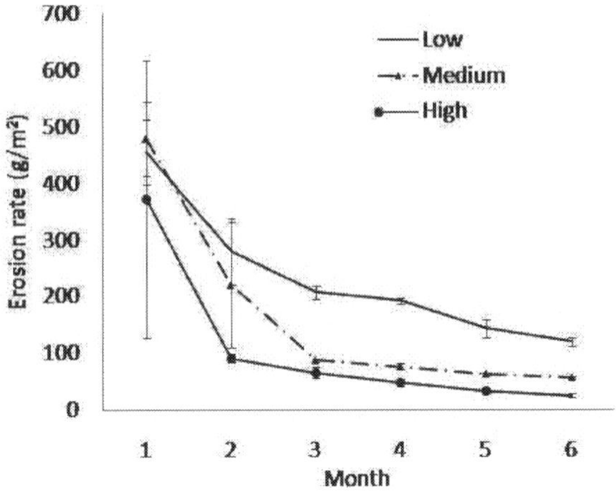

Figure 6. The erosion rate on slope at different plant density of *M. malabathricum*.

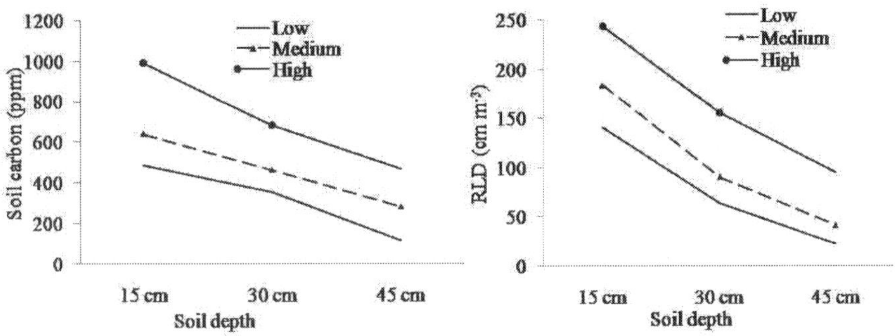

Figure 7. Soil carbon content and root length density (RLD) with depth of the species studied at different plant density.

3.2. Carbon sequestration and carbon sink potentiality

Carbon sequestration, a natural processes in ecosystems where CO_2 is absorbed from the atmosphere and stored it in plants and soil. During the photosynthesis process, plants absorb CO_2 and converted into carbohydrate or starch [25]. In this way, atmospheric carbon is stored in the leaves, stems, and roots for a long period of time (Figure 8). When a tree is utilized for wood, its ability to sequester carbon is extended, and the carbon is not released until the product burns or decomposes. Vegetation plays an important role in sequestrating carbon, as one way to alleviate global warming, a global issue discussed nowadays. Whereas, carbon sink potential is defined as a natural entity, process, activity or mechanism such as plants which can alleviate greenhouse gases from the atmosphere. It has been reported that forest, for example the tropical rainforest, is one of the largest carbon sinks in the world [26]. The higher the potential of the plant to absorb CO_2, the greater is the capacity of the plant to be a carbon sink potential. The quantitative measure on the CO_2 absorption by individual plant will assist to assess the carbon sink potentiality of plants. The plants which exhibited high photosynthetic components i.e. A_{max}, A_{400}, light and CO_2 saturation levels, are the good carbon sink plants.

A_{max} is an indicator of acclimatisation towards elevated CO_2 and can be used to determine the plant growth capacity in a future climatic situation. In our study, *Leucaena leucocephala, Peltophorum pterocarpum* and *Justicia betonica* exhibited higher Maximum Assimilation Rate (A_{max}) and Quantum Efficiency (QE) than *Lantana camara* and *Thunbergia erecta* (Table 3). From the simulated CO_2 experiments, *L. leucocephala, P. pterocarpum* and *J. betonica* seem to show the ability to utilize high concentrations of CO_2 in order to enhance photosynthetic rate. Furthermore, in the simulated light experiments, no photo-oxidation occurs. It has been reported that plants which can maintain the use of captured light energy for NADPH and ATP synthesis, may provide more sink capacity. This ability, then, would diminish the accumulation of excitation energy in the photosynthetic pigments, which is a major cause of photo-oxidative damage. Thus, the capacity to resist photo-oxidation is an indication that *L. leucocephala, P. pterocarpum* and *J. betonica* are the good potential carbon sink species. In addition, *L. leucocephala* and *P. pterocarpum* remain photosynthetically comparatively active at lower CO_2 concentrations, indicating low CO_2 is required to initiate the photosynthesis process of *L. leucocephala* and *P. pterocarpum*. Changes in CO_2 concentrations affected the photosynthesis of both species similarly; *L. leucocephala* and *P. pterocarpum* seemed to show higher ability to utilize high concentration of CO_2 in order to enhance photosynthetic rate as compared to other species studied. As inferred from these findings, *L. leucocephala* is a good carbon sink plant. Other species also showed considerably higher carbon sequestration capacity in which they can be regarded as supportive plants enhancing the carbon sink source when combine-grown on slope.

4. Rehabilitation of acidic slope

Malaysia's climate is described as typical tropical with warm, high rainfall intensity as well as humidity throughout the year. As its experiences high precipitation ranging from 2000 mm to 2500 mm per annum, the highland areas especially slopes are prone to erosion and landslides [10]. Due to the high rainfall and humid climate throughout the year in Malaysia, most of the slopes are facing the acidity problem. The acidic and infertile condition of slope does not encourage the vegetation establishment and consequently the slope turns barren. The barren slope is prone to erode when rainfall dislodge soil particles and carries them off the acidic slope. It would form rills and gullies which can trigger landslides. Rainfall hitting the soil surface can also seal the soil particles and make a crust that prevents infiltration and creates runoff.

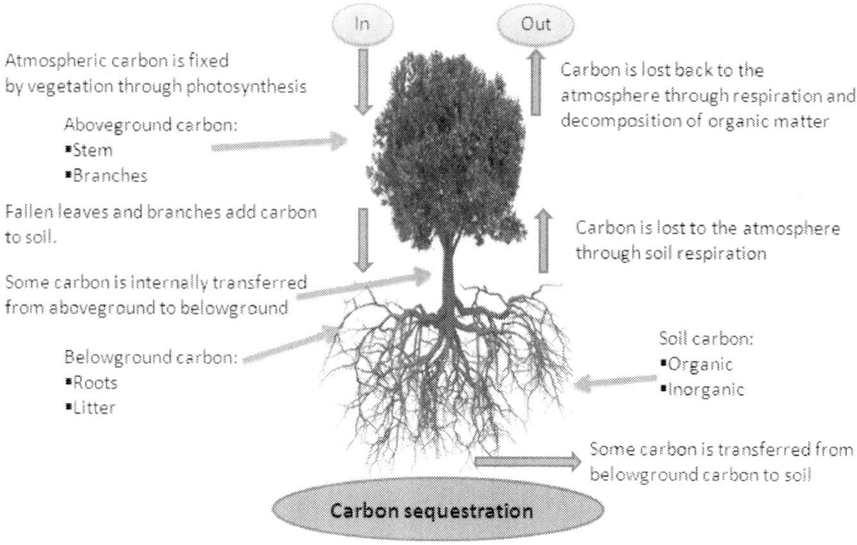

Figure 8. The process of carbon sequestration [27].

Plant species	Light response curves			Carbon response curves			
	Amax (μmol m^{-2} s^{-1})	Γ_1 (μE m^{-2} s^{-1})	QE (μE m^{-2} s^{-1})	Amax (μmol m^{-2} s^{-1})	A400 (μmol m^{-2} s^{-1})	gm (μmol m^{-2} s^{-1})	Γ_{CO2} (ppm)
LL	62	27	0.13	124	58	3	2
PP	36	85.5	0.1	80	34	2	9
JB	12.6	42.8	0.06	37.6	16.2	0.11	70
LC	9.3	55.9	0.06	16.7	10	0.08	48.5
TE	4.3	31.0	0.04	17.7	2.4	0.08	96.5

Maximum Assimilation Rate (A_{max}), Light Compensation point (Γ_1), Quantum Efficiency (QE), Photosynthesis at ambient CO_2 concentration (A_{400}), Mesophyll conductance (g_m), CO_2 compensation point (Γ_{CO2}), LL: *Leucaena leucocephala*, PP; *Peltophorum pterocarpum*, JB: *Justicia betonica*, LC: *Lantana camara*, TE: *Thunbergia erecta*

Table 3. Photosynthetic components of plants grown on slope [18]

Acid soil rehabilitation is an essential process of minimizing the acidity level of the soil and providing a better environment for plant growth. This process also will help offset greenhouse gas emissions, guarantee more food for an increasing population and contribute to the economic progress of future generations. There are several methods used in rehabilitating acidic soil. However, only three methods, which are liming, planting legumes and acid tolerant plants will be discussed in this chapter.

4.1. Liming

Lime is a material that contains calcium (Ca) or magnesium (Mg) and will neutralize soil acidity. Carbonates are the most available and widely used liming materials. Lime decreases the acidity level of soil (increases pH) by changing the hydrogen ions of soil into water and carbon dioxide (CO_2) molecule. One calcium ion (Ca^{++}) from the lime replaces two hydrogen ions (H^+) of soil complex. In addition, one carbonate ion (CO_3^{2-}) reacts with water molecule (H_2O) to form bicarbonate ion (HCO_3^-). These react with hydrogen ion (H^+) to form H_2O and CO_2. Thus, the pH of soil increases due to the concentration of hydrogen ions (H^+) has been reduced (Figure 9).

Figure 9. Reduction of soil acidity (or H^+ ions) by lime [28].

Regarding the mechanism of acid soil rehabilitation, liming changes the biological, structural and chemical components of soils. Organic matter decay is slow in acid soils due to the low activity levels of soil organisms. However, excessive liming rate would pollute the soil and cause environmental hazards.

4.2. Planting legumes species

Legumes foster production of a greater total biomass in the soil by providing additional nitrogen. Soil microbes use the increased nitrogen to break down carbon-rich residues of crops like wheat or corn. Normaniza and Barakbah [1] introduced a planting technique and documented that legumes plants showed high tolerance to acidic condition when planted with lime, rock phosphate and sphagnum moss (Figure 10). This planting technique known as "Microclimate Plant Propagation Technique" and the supplied chemicals ($CaCO_3$, K_2SO_4 and $MgSO_4$, and rock phosphate) assist as a suitable plant supplement to enhance the plant growth.

Another advantage of tree legumes is their deep root systems, a characteristic which confers persistence even on infertile soils [18]. Several legumes have aggressive taproots reaching six to eight feet deep and half inch in diameter that open pathways deep into the soil. Legumes contribute to an increased diversity of soil flora and fauna lending a greater stability to the total life of the soil [18]. In conclusion, nitrogen-fixing abilities of legumes plants are important for alleviating soil acidity, maintaining ecosystem fertility and long-term slope stabilization.

4.3. Planting Al-tolerant plants

Aluminium was the major factor for slope acidity and the presence of Aluminium was unavoidable because it was a part of most clay particles. The mechanism of Al accumulator plant to alleviate soil acidity has attracted the interest of plant ecologist and physiologist as well as evolutionary biologist. In the presence of Al, the tolerant cultivars have efficiently uptake and utilized Ca and P. The susceptible (Al-sensitive) and intermediate cultivars exhibited less Ca and P uptake and utilization. The effect of Al on roots indicated that the nutrient solution having Al at a concentration below 40 mM has stimulated root growth, increasing the size and number of central cap cells. Beyond 60 mM, root growth was strongly inhibited with cellular damage in peripheral root cap cells.

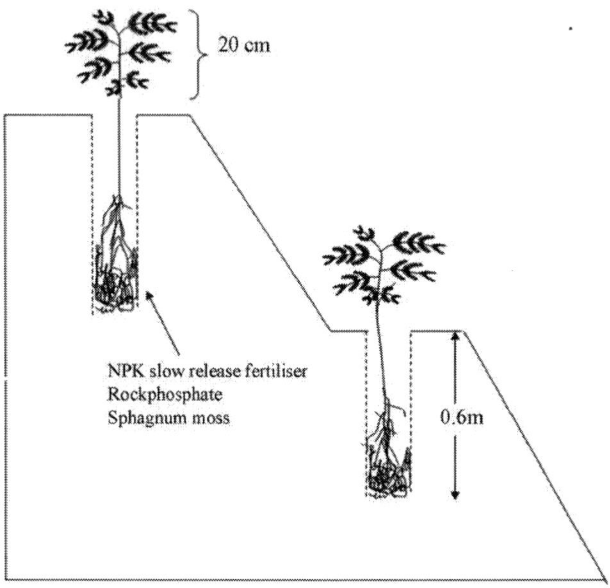

Figure 10. Plantation of legume seedling on slope [1].

Revegetation with Al-tolerant plants can be a valuable rehabilitation tool. Al-tolerant plants can tolerate and accumulate high concentrations of Al in the shoot whereby the growth of the plants was not affected by Al toxicity. Plants can deal with Al toxicity by setting up several aluminium tolerance mechanisms. Therefore, on such Al-contaminated soil, planting Al-tolerant plants plays increasingly important phytoremediation role. Proper management of these kind of plants in acidic soil may significantly contribute to restoring the natural environment. On the other hand, most of Al-tolerant plants are shrubs for example, *M. malabathricum*. These kind of shrubs have woody root systems (M type) that give mechanical support to slopes stability. Moreover, when *M. malabathricum* are planted with grass, they can help to prevent sloughing of the shallow sod layer. The woody top growth also helps to stabilize rehabilitated areas by reducing surface wind velocity. These shrubs also improve soil and forest floors by drying them out, adding organic matter, and fix some nitrogen.

4.4. Tolerance mechanism of plants in acidic slope

Some tolerance mechanisms to ensure its survival and growth vary amongst the species. In some cases, the plant produces small leaves as a response to acidic condition of the soil, for example smaller and thinner leaves in *L. leucocephala* as observed in our research (Figure 11).

Figure 11. Leaf necrosis to reduce leaf area - a visual symptom of Aluminium toxicity, a mechanism to resist acidic condition [18].

Additionally, this visual observation indicated that plants reduced leaf number by leaf wilting and necrosis. This necrosis symptom is the sign of acid tolerance mechanism of plants, possibly as a mechanism to reduce leaf area [18]. However, high leaf chlorophyll content was observed in adapted plants in acidic soil and the plant seems to have recovered from the toxicity effect by increasing the nodulation activity. In addition, leaf analysis of acidic and non-acidic treated *L. leucocephala* showed that Aluminium concentration was higher by 36% in acidic treated than those in non-acidic treated *L. leucocephala,* implying a high Al uptake of *L. leucocephala* in acidic slope. The excessive accumulation of Al in leaf may indicate the mechanism of acidity tolerance of this species. Moreover, the plant which exhibits Al concentration in leaf more than 1000 ppm is reportedly called an Al accumulator, a mechanism of Al adaptation of a plant to acidity (Figure 12). *M. malabathricum* accumulated the highest concentration of Aluminium, which was almost 1850 ppm, which makes both species an Al accumulator (more than 1000 ppm) and *A. mangium* is non-accumulator.

As similarly reported by Watanabe and Osaki [29], most of the plant samples (*Evodia latifolia* and *Justicia betonica*) exhibited general symptoms of Al toxicity which includes curly young leaves, reduced leaf number and necrosis, as mechanisms to reduce leaf area.

In relation to the Al concentration of the leaf, the soil pH grown with *M. malabathricum* increased up to 6.0, meanwhile 5.5 and 5.1 for *L. leucocephala* and *A. mangium,* respectively (Figure 13). The results imply a positive relationship between the concentration of Al in the leaf and soil pH; as the Aluminium accumulation in leaf increased, the value of soil pH increased as well. The experiment showed the importance role of plant as an Aluminium accumulator in rehabilitating the acidic slope.

Figure 12. Al concentration of leaves of three species studied

Other possible tolerance mechanisms that could be identified in this project were increasing root length, stomatal conductance and LAI (Figure 14-17). The interaction and compilation of all tolerance mechanism contribute to the rehabilitation of the acidic soil. The photosynthetic rate and chlorophyll content of Al-tolerant plants increased with the increasing of Al concentration. Therefore, as the Al concentration increased, the tolerance mechanism has also enhanced by increasing the transpiration rate of plant.

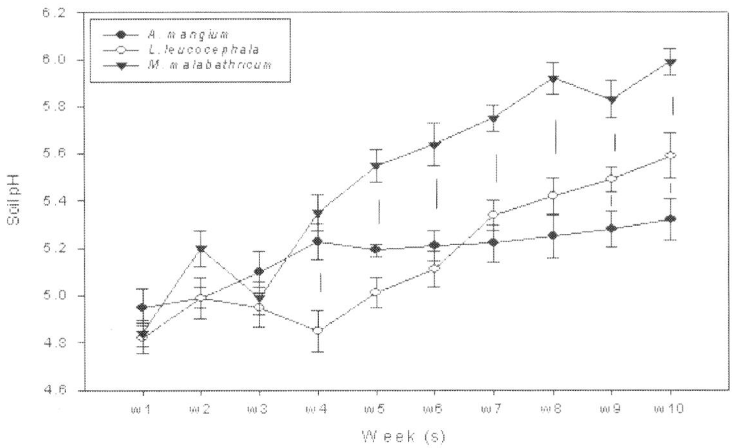

Figure 13. Soil pH changes grown by treated plants of the three species studied.

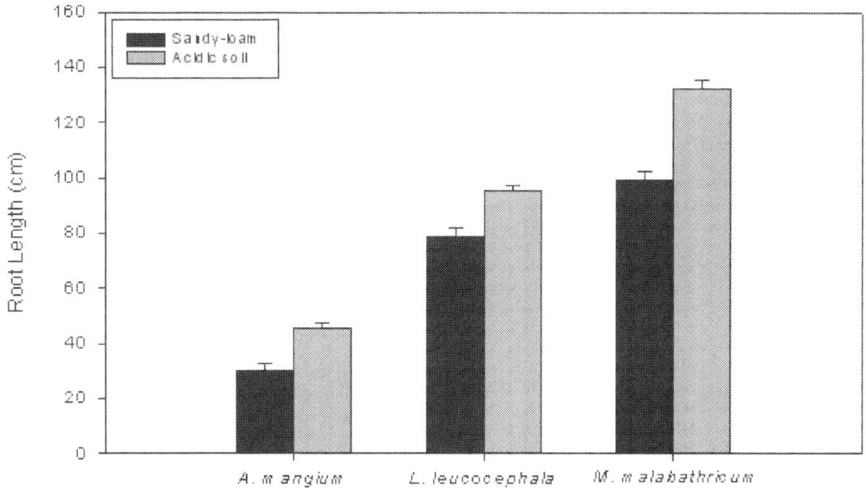

Figure 14. Root length in sandy loam and acidic soil of three species studied

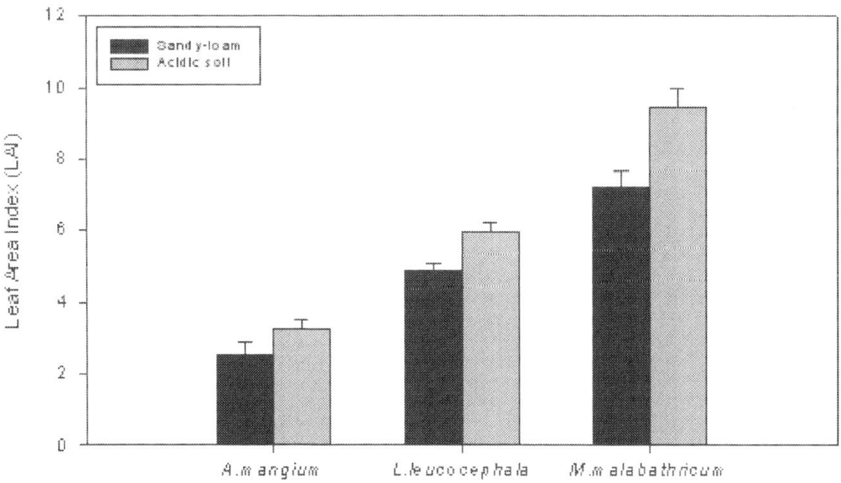

Figure 15. Leaf Area Index in sandy loam and acidic soil of three species studied

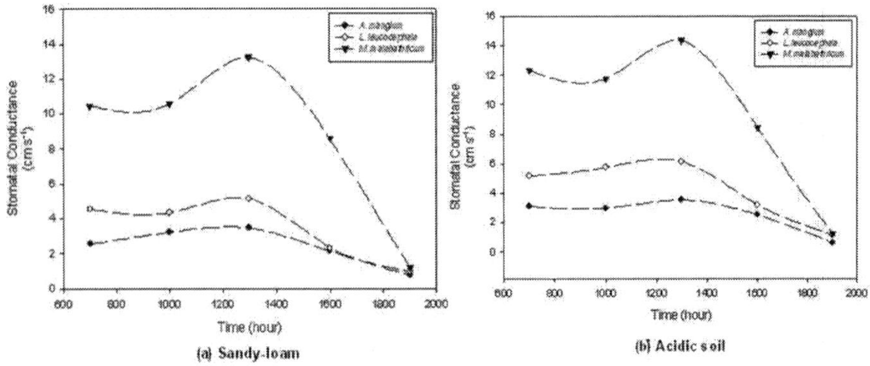

Figure 16. Stomatal conductance in (a) sandy loam and (b) acidic soil of three species studied

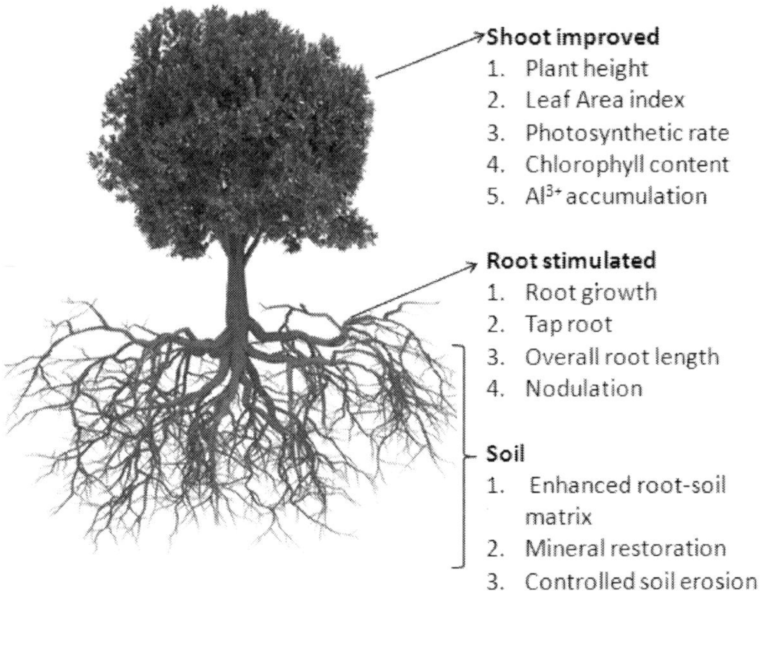

Figure 17. Tolerance mechanism of plants [30]

5. Potential slope plants

Numerous studies have been conducted to determine the plant species which are suitable as slope plants. Normaniza and Barakbah [1] and Stokes et al. [31] referred that native plants usually increased the success rate of the planting program and reduced the long-term maintenance requirements. Karim and Mallik [32] suggested that the selected plants should be adapted to local climate and be able to prevent landslides or erosion. Mafian et al. [12] showed that the reinforcement of soil by vegetation is highly promising solution and this approach would be more beneficial if the species acutely possessed the mechanical (through reinforcement of soils by plant root), hydrological (through reduction in runoff and by keeping the slope relatively dry) and environmental (through the increase in carbon sequestration to counter the rising carbon dioxide level in atmosphere) aspects.

Different plant species has different hydro-mechanical characteristics and can perform different roles on slope but certain types of plants are better than others in terms of soil reinforcement and surface protection [33]. Many problems may occur after planting any type of plants which does not fulfill the slope plant characteristics. Therefore, selection of plant species by observing the potential slope plant characteristics is crucial. A set of criteria was formulated to select potential species for plantation on slope [1,9,16,31]. Physiological characteristics such as the high photosynthetic rate, transpiration rate and growth rate and root profiles, such as high cellulose content in roots, fine roots, root biomass, root volume and root length are considered as major criteria [14,16]. Additionally, the selected plant should exhibit other prominent characteristics such as good plant-water relations and tolerance of wide range of adverse condition with regard to soil acidity and water stress [9].

A list of potential tree and shrubs species was presented in Table 4. Based on the observations, *L. leucocephala* and *P. pterocorpum* showed the higher bioengineering characteristics than *A. mangium and M. malabathricum*. It was discovered that root architecture of *L. leucocephala* and *P. pterocorpum* was VH and R type, respectively. The VH- and R type root architecture was considered to be the most effective root system for slope stabilization and soil reinforcement [34]. The H-types were found to be beneficial for wind resistance. The M-types are regarded to be beneficial for controlling soil erosion. The extensive root growth and tensile strength *L. leucocephala* and *P. pterocorpum* were claimed to be the cause of tremendous enhancement of mechanical impacts on soil. Thus, in terms of root properties, *L. leucocephala* and *P. pterocorpum* were more prominent to play a major mechanical role on soil and their high root tensile strength would ultimately improved soil shear strength as well. These introduced tropical plants will indeed assist eco-engineer to establish bioengineering technique on slope and provide long-term soil reinforcement.

Many legumes, especially woody trees are particularly planted for controlling soil erosion, slope stabilization and restoration in tropical countries [13]. *Leucaena leucoephala* and *Peltophorum pterocorpum*, have a potential to be slope plants. *L. leucoephala* is one of the most productive fast growing, semi ever green and nitrogen fixing tropical legume trees. In Malaysia, *L.*

leucocephala is used as a potential slope pioneer and wind protection. It has aggressive taproots reaching six to eight feet deep and half inch in diameter that open pathways deep into the soil (Figure 18). Nodules on the root of plant can fix atmospheric nitrogen and this is perhaps the most notable aspect that sets them apart from other plants. Additionally, *P. pterocarpum* is a woody ornamental plant and has a R type root system [16] (Figure 19). This tree usually is planted along roadsides, parks and slope. It has high atmospheric nitrogen-fixing potentiality.

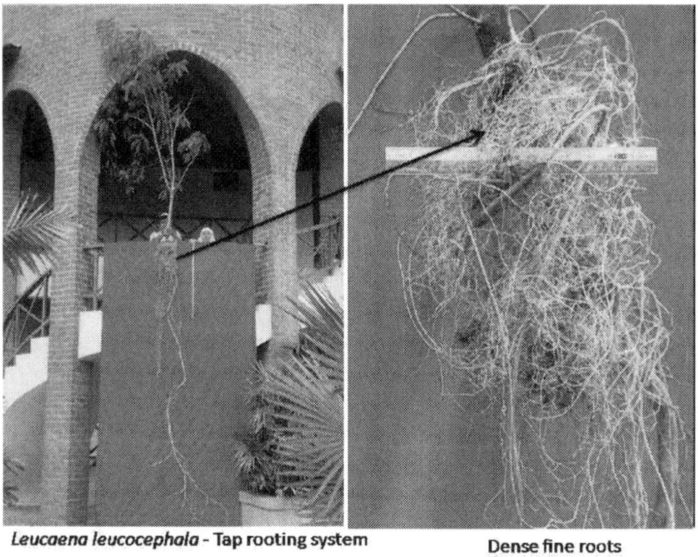

Figure 18. Root profile of a potential slope plant- *Leucaena leucocephala* [18].

M. malabathricum produced the M type root system that makes it suitable to grow at slope area (Figure 20). Acidic treated *M. malabathricum* showed a higher root length than non-acidic treated, implying high water absorption to perform a basic metabolic process such as photosynthesis. Plant released the absorbed water to the atmosphere by transpiring through pores on the leaves. As a result, the excessive water were removed and resulted in a drier and more stable slope. Moreover the flowering feature of *M. malabathricum* can help to enhance the flora-fauna interaction of the slopes by increasing the biodiversity. Different species have different mechanical characteristics and ranges acidic soil rehabilitation capacity. Potential slope plants and their mechanical characteristics were shown in Table 4. Additionally, a list of potential tree and shrubs species for planting in acidic slope was shown in Table 5.

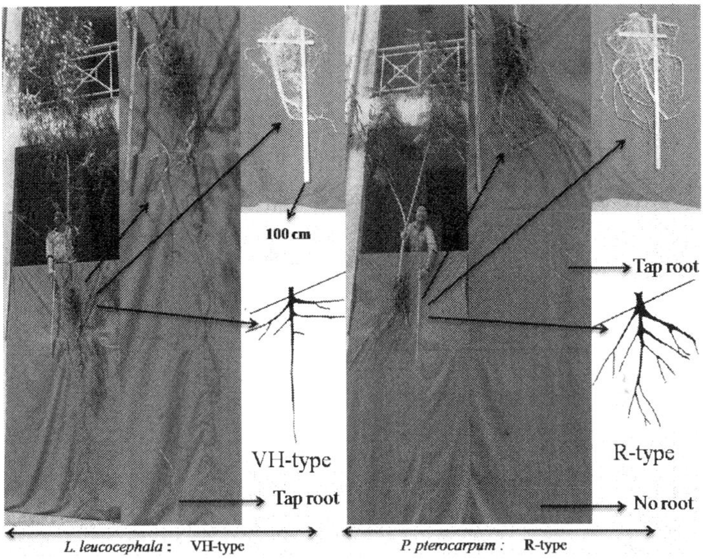

Figure 19. Potential slope plant- *Peltophorum pterocarpum* [16].

Figure 20. (a) Plant profile, (b & c) root profile and (d) flowering feature of *M. malabathricum*.

Species	Root pattern	Type of root system	Root tensile strength (MPa)
Leucaena leucocephala	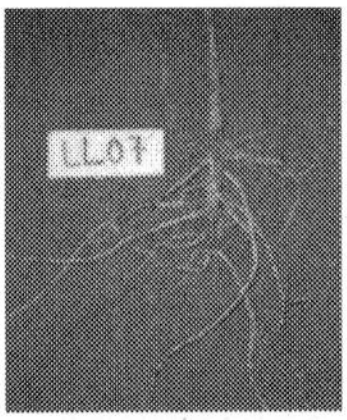 **VH type**	 **VH type**	104
Peltophorum pterocarpum	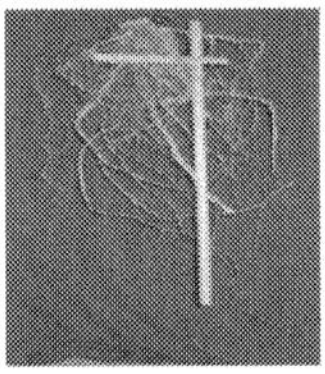 **R type**	 **R type**	61
Acacia mangium	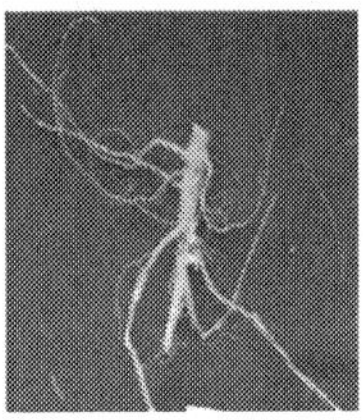 **H type**	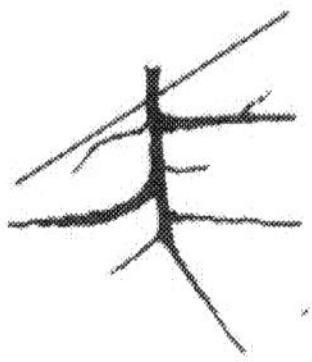 **H type**	54
Dillenia suffruticosa	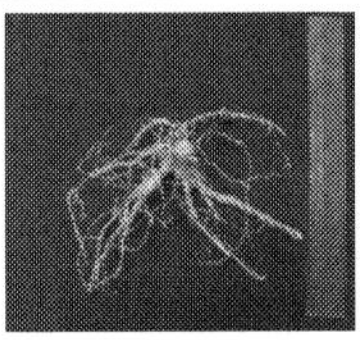	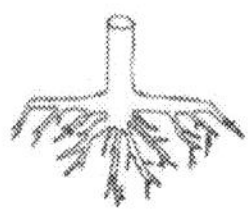 **Heart**	-
Melastoma malabathricum	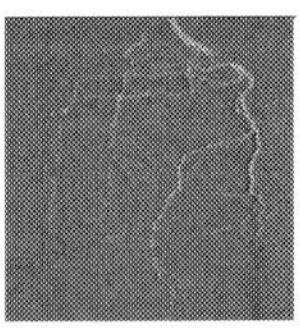 **M type**	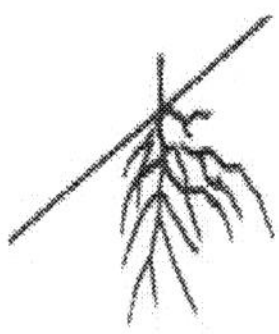 **M type**	29

Table 4. Potential slope plants and their mechanical characteristics [13,16,35].

Species	Genus	Family	Site (soil pH)
Bixa orellana	*Baxia*	Bixaceae	> 5
Leucaena leucocephala	*Leucaena*	Mimosoideae	> 4, Moderate acidic
Acacia mangium	*Acacia*	Fabaceae	> 4, Moderate acidic
Bauhinia purpurea	*Bauhinia*	Fabaceae	> 4, Moderate acidic
Melastoma malabathricum	*Melastoma*	Melastomataceae	< 3, Severe acidic
Thunbergia erecta	*Thunbergia*	Acanthaceae	> 5
Justicia betonica	*Justicia*	Acanthaceae	> 5
Lantana camara	*Lantana*	Verbenaceae	> 5
Hibiscus mutabilis	*Hibiscus*	Malvaceae	> 5
Vetiver zizanioides	*Chrysopogon*	Poaceae	> 5

Table 5. List of species for planting in acidic slope and classified by slope characteristic [13, 18].

6. General discussion

The observation of this study provides the key findings and contribution of tropical plants to alleviate soil acidity and soil erosion. The bioengineering characteristics of selected tropical plants have been intensively assessed to identify their potentiality towards slope stabilization. This observation also revealed the contribution of pioneer species to enhance the process of natural succession on slope. Amongst the tropical plants, few were suitable for reinforcing slope and rehabilitation of acidic slope, exhibiting tolerance mechanism in soil acidic condition. Based on the field studies, significant morphological and physiological changes were observed in *M. malabathricum* in response to severe acidic (< 3) condition. These changes include the improved photosynthetic rate, transpiration rate, LAI and root system. In addition, the highest concentration of Al (> 1200 ppm) in the *M. malabathricum* leaves, either Al-treated or not, has made this species the most suitable plant for severe acidic slope. This study also reveals an alternative approach to alleviate the acidity. Besides liming, legume trees were recommended for acidic slope rehabilitation due to their nitrogen-fixing abilities and extensive root systems that can penetrate a deeper soil depth. The deeper the soil depth, the higher the soil pH (less acidic), which makes this condition more conducive for the root establishment as well as the plant growth as a whole [18].

In terms of the alleviation of soil erosion, the soil reinforcement of slope stability is mainly depending on the properties of root systems of plant species. Root biomass, root architecture and tensile strength were remarked as prominent engineering properties of plants to assess its potentiality for soil reinforcement [13]. The extensive root growth and development of the taproot (VH-type) of *L. leucocephala* has resulted in a tremendous enhancement of mechanical effects on soil. As inferred from these findings, M-type root was suitable in protecting ground cover from surficial erosion. In relations to this mechanical aspect, plant density also showed an important contribution for controlling soil erosion. Research findings exhibited that higher plant density would reduce the soil erosion rate, indicating a high soil-root interaction and canopy interception. Apart from that, carbon sink potentiality of plants was also measured *via* photosynthetic components of plants grown on slope. It can be envisaged that the outstanding carbon sink potential of the slope plant community would confer numerous advantages, not only to improve the global warming and sustain the ecosystem but also to produce more stable and safe slopes.

7. Conclusion

The new information and findings presented in this chapter have mostly been achieved through the bioengineering research, carried out in Malaysia. Aluminium accumulating plants and new potential tropical slope plants are introduced and recommended for the revegetation programme of the slope to alleviate soil acidity and erosion. These potential tropical plants' attributions will be beneficial for enlisting database which can be further applied for ground bio-engineering techniques in other tropical countries in the South-East Asia. It is our earnest hope that the new discoveries and the outcomes of this knowledge would benefit the soil science as well as other related disciplines.

Acknowledgements

The authors would like to thank University of Malaya (UMRG-RG042-09SUS, RG120-11SUS, RG019-09AET, PV084-2012A, PV052-2011A), Slope Engineering Branch of Public Works Department Malaysia, Ministry of Science, Technology and Innovation (SF16-02-03-6013) and Ministry of Higher Education (FP070-2010A) for the research grants throughout this project.

Author details

Normaniza Osman*, Mohammed Saifuddin and Aimee Halim

*Address all correspondence to: normaniza@um.edu.my

Institute of Biological Sciences, Faculty of Science, University of Malaya, Kuala Lumpur, Malaysia

References

[1] Normaniza O. Barakbah SS. The effect of plant succession on slope stability. Ecological Engineering 2011;37 139-147.

[2] Chau KT. Sze YL. Fung MK. Wong WY. Fong EL. Chan LCP. Landslide hazard analysis for Hong Kong using landslide inventory and GIS. Computers and Geosciences 2004;30 429-443.

[3] Dong JJ. Tu CH. Lee WR. Jheng YJ. Effects of hydraulic conductivity/strength anisotropy on the stability of stratified, poorly cemented rock slopes. Computers and Geotechnics 2012;40 147-159.

[4] Mizal-Azzmi N. Mohd-Noor N. Jamaludin N. Geotechnical Approaches for Slope Stabilization in Residential Area. Procedia Engineering, 2011;20 474-482.

[5] Rajakumar P. Sanjeevi S. Jayaseelan S. Isakkipandian G. Edwin M. Balaji P. Ehanthalingam G. Landslide susceptibility mapping in a hilly terrain using remote sensing and GIS. Journal of the Indian Society of Remote Sensing 2007;35 31-42.

[6] Mugagga F. Kakembo V. Buyinza M. A characterisation of the physical properties of soil and the implications for landslide occurrence on the slopes of Mount Elgon, Eastern Uganda. Natural Hazards 2012;60 1113-1131.

[7] Song W. Yongguo Z. Xiaoxu G. Xiaoning L. Slope Landscape Classification and Application Security in the Special Section of Western. Procedia Environmental Sciences Part A 2012a;12 146-151.

[8] Song YS. Hong WP. Woo KS. Behavior and analysis of stabilizing piles installed in a cut slope during heavy rainfall. Engineering Geology 2012;129–130 56-67.

[9] Normaniza O. Barakbah SS. Parameters to predict slope stability—Soil water and root profiles. Ecological Engineering 2006;28 90-95.

[10] Huat BBK. Kazemian S. Study of Root Theories in green tropical slope stability. Electronic Journal Of Geotechnical Engineering 2010;15 1825-1834.

[11] Stokes A. Ghani MA. Salin F. Danjon F. Jeannin H. Berthier S. Kokutse AD. Frochot H. Root morphology and strain distribution during tree failure on mountain slopes Eco-and Ground Bio-Engineering: The Use of Vegetation to Improve Slope Stability. In: Stokes, A., Spanos, I., Norris, J. E. & Cammeraat, E. (eds.). Springer Netherlands; 2007.

[12] Mafian S. Huat BBK. Rahman NA. Sing H. Potential plant species for live pole application in tropical environment. American Journal of Environmental sciences 2009;5 759-764.

[13] Nordin MA. Osman N. Abdullah CH. Pull-out and tensile strength properties of two selected tropical trees. Sains Malaysiana 2011;40 577-585.

[14] Normaniza O. Faisal HA. Barakbah SS. Engineering properties of *Leucaena leucocephala* for prevention of slope failure. Ecological Engineering 2008;32 215-221.

[15] Fattet M. Fu Y. Ghestem M. Ma W. Foulonneau M. Nespoulous J. Le Bissonnais Y. Stokes A. Effects of vegetation type on soil resistance to erosion: Relationship between aggregate stability and shear strength. Catena 2011;87 60-69.

[16] Saifuddin M. Normaniza O. Physiological and root profile studies of four legume tree species. Life Science Journal 2012;9 1509-1518.

[17] Stokes A. Mattheck C. Variation of wood strength in tree roots. Journal of Experimental Botany 1996;47 693-699.

[18] Normaniza, O., Barakbah, S.S. and Faisal, H.A. 2009. The contribution of plant community to accelerate the process of natural succession. *American Journal of Environmental Sciences*. 5 (1): 7-15.

[19] Evette A. Labonne S. Rey F. Liebault F. Jancke O. Girel J. History of bioengineering techniques for erosion control in rivers in western Europe. Environmental Management 2009;43 972-984.

[20] Preti F. Giadrossich F. Root reinforcement and slope bioengineering stabilization by Spanish Broom (Spartium junceum L.). Hydrology and Earth System Sciences 2009;13 1713-1726.

[21] Toriman ME. Shukor MN. An analysis of rainfall interception on the selected experimental plot of Pangkor Hill Reserved Forest. Journal Wildlife and National Park 2006; 169-178.

[22] Mahdavi K. Choupanian A. Gheytouri M. Mahdavi M. Effect of physiographic factors on soil carbon sequestration in Kermanshah (Iran). International Journal of Forest Soil and Erosion 2012;2 159-162.

[23] Gyssels G. Poesen J. The importance of plant root characteristics in controlling concentrated flow erosion rates. Earth Surface Processes and Landforms 2003;28 371–384.

[24] Kidd PS. Proctor J. Effects of Aluminum on the growth and mineral composition of Betulapendula Roth. Journal of Experimental Botany 2000;51 1057–1066.

[25] Raven JA. Karley AJ. Carbon Sequestration: Photosynthesis and Subsequent Processes. Current Biology 2006;6 R165-R167.

[26] Malhi Y. Grace J. Tropical forests and atmospheric carbon dioxide. Trends in Ecology & Evolution 2000;15 332-337.

[27] Lal R. Soil carbon sequestration to mitigate climate change. Geoderma 2004;123 1-22.

[28] Meiwes KJ. Application of lime and wood ash to decrease acidification of forest soils. Water, Air, and Soil Pollution 1995;85 143-152.

[29] Watanabe T. Osaki M. Mechanisms of adaptation to high aluminum condition in native plant species growing in acid soils: A review. Communications in Soil Science and Plant Analysis 2002;33 1247–1260.

[30] Idris RM. Rehabilitation of acidic soil using potential slope plants. M.Sc Thesis, University of Malaya; 2011.

[31] Stokes A. Atger C. Bengough A. Fourcaud T. Sidle R. Desirable plant root traits for protecting natural and engineered slopes against landslides. Plant and Soil 2009;324 1-30.

[32] Karim MN. Mallik, AU. Roadside revegetation by native plants: I. Roadside microhabitats, floristic zonation and species traits. Ecological Engineering 2008;32 222-237.

[33] Coppin NJ. Richards IG. Use of Vegetation in Civil Engineering, Construction Industry Research & Information Association, 2007.

[34] Reubens B. Poesen J. Danjon F. Geudens G. Muys B. The role of fine and coarse roots in shallow slope stability and soil erosion control with a focus on root system architecture: a review. Trees - Structure and Function 2007;21 385-402.

[35] Yen CP. Tree root patterns and erosion control. In: Jantawat, S. (Ed.), Proceedings of the international workshop on soil erosion and its countermeasures. Soil and Water Conservation Society of Thailand, Bangkok, 1987; 92–111.

Chapter 19

Metal Contamination of Soils and Prospects of Phytoremediation in and Around River Yamuna: A Case Study from North-Central India

Manoj S. Paul, Mayank Varun, Rohan D'Souza,
Paulo J.C. Favas and João Pratas

Additional information is available at the end of the chapter

http://dx.doi.org/10.5772/57239

1. Introduction

The rapid industrialization and intensive agricultural activities over the last few decades have resulted in accumulation of various pollutants in the environment, which are distributed over wide areas by means of air and water. This has caused visible detrimental effects to the ecosystem and consequences to human health. Today, many soils throughout the world have undesirably high concentrations of heavy metals. These include lead (Pb), cadmium (Cd), zinc (Zn), mercury (Hg), arsenic (As), silver (Ag), chromium (Cr), copper (Cu), iron (Fe), and the platinum group elements. At low or background concentrations, heavy metals are not pollutants. They occur naturally in the environment due to their presence in bedrocks. Some heavy metals such as Zn and Cu are also essential micronutrients for living organisms. Therefore, the term *heavy metal pollution* refers to heavy metal levels that are abnormally high relative to normal background levels. All heavy metals at high concentration have strong toxic effects and are regarded as environmental pollutants.

Some heavy metals (like Fe, Zn, Ca and Mg) have been reported to be of bio-importance to man and their daily medicinal and dietary allowances have been recommended. However, some others (like As, Cd, Pb, and methylated forms of Hg) have been reported to have no known bio-importance in human biochemistry and physiology and consumption even at very low concentrations can be toxic [1]. Even for those that have bio-importance, dietary intakes have to be maintained at regulatory limits, as excesses result in poisoning or toxicity [2]. Although individual metals exhibit specific signs of their toxicity, the following have been reported as general signs associated with Cd, Pb, As, Hg, Zn,

Cu and Al poisoning: gastrointestinal disorders, diarrhoea, stomatitis, tremor, ataxia, paralysis, vomiting and convulsion, depression, and pneumonia when volatile vapours and fumes are inhaled [3]. The nature of effects could be toxic (acute, chronic or sub-chronic), neurotoxic, carcinogenic, mutagenic or teratogenic.

Pb, Zn, Cu, Co, Mn, Fe, Cr and Cd have been found in the streams and rivers of the Americas, Europe, Asia, Africa and Australia [4-9]. In India, presence of heavy metals has been reported in the Brahmaputra [10]; the Kali and Hindon [11]; and more recently, in the Gomti [12]; the Cauvery [13]; and the Ganga [14].

The Yamuna (also Jamuna or Jumna) is the largest tributary of the Ganga in northern India, having the total length of about 1376 km. The source of Yamuna is Yamunotri in the Uttarakhand Himalaya, which is north of Haridwar in the Himalayan mountains. Yamuna river flows through the states of Uttarakhand, Delhi, Haryana and Uttar Pradesh and finally merges with river Ganges at a sacred spot known as Triveni Sangam in Allahabad. A number of prominent cities such as Delhi, Mathura and Agra lie on the bank of river Yamuna. Over 57 million people depend on the Yamuna waters. Just like the Ganges, the Yamuna too is highly venerated in Hinduism and worshipped as goddess Yamuna, throughout its course.

Due to high density population growth, rapid industrialization, today Yamuna is one of the most polluted rivers in the world, especially around New Delhi, where 15 drains discharge waste water into the river. The city dumps ~58% of its waste into it. When the river enters the city, it is already contaminated with 7500 coliform content per 100 ml. when it leaves the city, it carries with a dangerously high coliform content of 24 million per 100 ml. Even the ground water has been affected by leachates that pass down from the dumping sites. According to the Central Pollution Control Board (CPCB), 70% of the pollution in river is from untreated sewage and the remaining 30% is from industrial sources, agricultural run-off, garbage etc. The water quality of Yamuna River falls under the category "E" which makes it fit only for recreation and industrial cooling, completely ruling out the possibility for underwater life. Almost every year mass death of fishes is reported. Biological Oxygen Demand (BOD) load increased by 2.5 times between 1980 and 2005: from 117 tonnes per day in 1980 to 276 in 2005.

Although the government of India has spent nearly $500 million to clean up the river, the river continues to be polluted with garbage while most sewage treatment facilities are underfunded or malfunctioning. The Ministry of Environment and Forests (MoEF) of the Government of India (GOI) took measures to curb pollution in 12 towns of Haryana, 8 towns of Uttar Pradesh, and Delhi under an action plan (Yamuna Action Plan-YAP) which is being implemented since 1993 [15]. However in 2009, the Union government admitted the failure of the Ganga Action Plan (GAP) and the Yamuna Action Plan (YAP), saying that "rivers Ganga and Yamuna are no cleaner now than two decades ago" despite spending over Rs 1, 700 crore to control pollution [16]. In August 2009, Delhi Jal Board (DJB) initiated its plan for resuscitating a 22 km stretch of the Yamuna in Delhi by constructing interceptor sewers, at the cost of about Rs 1, 800 crore [17].

There are three main sources of pollution in the river, namely household and municipal disposal sites, agricultural run-off, and industrial effluents and run-off. Urban runoff and agricultural runoff are mainly non–point sources. The major sources of pollution from agriculture are fertilizers containing superabundant nutrients such as nitrogen and phosphorus, and heavy metals such as Cd, Cu, Pb and Zn. Water quality may also be altered by other factors, such as livestock manure, human waste, and atmospheric deposition. Atmospheric pollutants are often the largest source of waterborne metals. It is estimated that 70% of lead in water and over 50% of many of the other trace metals in the Great Lakes (USA) are derived from atmospheric transfer. In general, freshwater ecosystems have low natural background metal levels and therefore tend to be sensitive to even small additions of most trace metals. Heavy metal contamination of soils and water from industrial and traffic sources in urban environments has been studied in North America and Europe [18-22]. Agencies like the World Health Organization (WHO) and the United states Environment Protection Agency (USEPA) have set stringent standards for maximum permissible limits of heavy metals, but there is a paucity of detailed studies on heavy metal pollution and its remediation within industrial zones in developing countries. Yamuna outnumbers any other river in the number of industries on its bank. This is because it passes through many major industrial cities. About 22, 42, and 17 large and medium industrial units in the states of Haryana, Delhi, and Uttar Pradesh have been identified as polluting the river in the action plan area. In addition, the water in this river remains stagnant for almost 9 months in a year aggravating the situation.

According to the Agra District Industrial Centre officials, there were 226 iron foundries and about 340 metal casting units functioning in Agra in the decade of 1990-2000. Before the revised pollution control directives put the Agra diesel generator manufacturing industry off its track, the foundry industry of this town ranked among the country's largest assemblies of metal casting industrial units, generating business of over Rs 6, 000 crores. The ban on coking coal in the blast furnaces utilized by the foundry and metal-casting industry was a serious setback and the number of industrial units reduced drastically. In August 1999, the Supreme Court ordered the closure of 53 iron foundries and 107 other factories in Agra. In September 2010, it again ordered the closure of 212 of the 1, 715 small industries that had failed to disclose their toxic emission levels to the Uttar Pradesh Pollution Control Board (UPPCB). Another 299 were required to install pollution controlling devices, failing which they too would face closure. However, the ground realities are still nowhere near the reduced pollution levels targeted in Yamuna and its adjacent areas whether Agra or elsewhere, after it leaves the Himalayan foothills. The status quo, thus, ultimately leaves much to be desired.

Phytoremediation is an emerging technology that employs the use of green plants for the clean up of contaminated environment. It takes the advantage of the fact that a living plant acts as a solar-driven pump, which can extract and concentrate certain metals from the environment [23]. This remediation method maintains the biological properties and physical structure of the soil. The technique is environmentally friendly, cost-effective, visually unobtrusive, and offers the possibility of bio-recovery of the metals. In the case of heavy metal contamination in soil, phytoremediation techniques are narrowed down to *Phytoextraction,* where plants remove metals from the soil by concentrating them in their harvestable parts [24], and *Phytostabilization,* where plants reduce the mobility and bioavailability of pollutants by immobilization [25].

Phytoremediation is becoming possible because of the successful basic and applied research much of it conducted with the productive interdisciplinary cooperation of plant biologists, soil chemists, microbiologists and environmental engineers. Extensive progress has been made in characterizing and modifying the soil chemistry of the contaminated site to accelerate phytoremediation. The greatest progress in phytoremediation has been made with metals [26, 27]. Phytoremediation leaves the topsoil in usable condition and it is aesthetically pleasing. It requires minimal equipment and less energy inputs as plants do most of the work using solar energy. Thus, it is an eco-friendly process. The plants used can later be harvested, processed and disposed off in an environmentally sound manner. This technology has been receiving attention lately as an innovative, cost-effective alternative to the otherwise tedious and expensive methods in use which are not only a burden on the exchequer but also require efforts on recurring basis.

Phytoremediation employing indigenous species can be an ecologically viable option for sustainable and cost-effective management. Native plants often become adapted to locally elevated levels of metals in soil at contaminated sites, e.g. mines and industrial zones [28-30] and metal toxicity issues do not generally arise. Many native, well adapted plants have been investigated and even used for heavy metal bioindicatoring and phytoremedial purposes including lemongrass and other wild grasses, vetiver, *Sesbania, Avena, Crotalaria, Crinum asiaticum, Typha latifolia* and *Calotropis procera* etc. [31-35, 28]. Native wild species are also important to remediate soils in context of the studied area due to a remark (April, 2006) of the Supreme Court prohibiting the cultivation of plants requiring fertilizers and pesticides along the Yamuna. In the light of this limitation, native wild species are a viable option since these do not require agronomic inputs.

Since the river Yamuna is the life line of Mathura and Agra, the existing pollution level has posed a serious threat not only to the environment but also to the human population. Adjacent areas are highly polluted and are a sink for a variety of chemicals including heavy metals. The present study was undertaken: (i) to get a comprehensive profile of eight metals in water and adjacent soils of the river Yamuna within Mathura, Agra and Bateshwar; (ii) to get a qualitative and quantitative estimate of the species present at test sites through phyto-sociological surveys; and (iii) to inventorize species with potential for phytoremediation present on sites by comparing with those previously reported by the authors as suitable in this context.

2. Case study

Agra (27°10′N, 78°05′E, 169 msl), on the banks of the river Yamuna, is located in Uttar Pradesh in the north central part of India. It is roughly 200 km south-east of the national capital, New Delhi. Bounded by the Thar desert of Rajasthan on its south-east, west and north-west peripheries, it is a semi-arid area. The world renowned Mughal monument, the Taj Mahal is situated here. It is world renowned for its leather industry and marble handicrafts but it also boasts a cast iron and engineering goods industry. Mathura (27.28°N 77.41°E) is located approximately 60 km north of Agra and 145 km south-east of Delhi. According to Hindu scriptures Mathura is the birthplace of Lord Krishna. It is a fast expanding city with about half a million residents. Mathura oil refinery is one of the biggest oil refineries of Asia. Textile printing, dyeing and silver ornament manufacturing are major industries. Apart from these there are units manufacturing taps, household items, and cotton materials. Bateshwar (26.93°N 78.54°E) is a village on the banks of Yamuna about 120 km downstream from Agra. It is an important spiritual and cultural centre for Hindus.

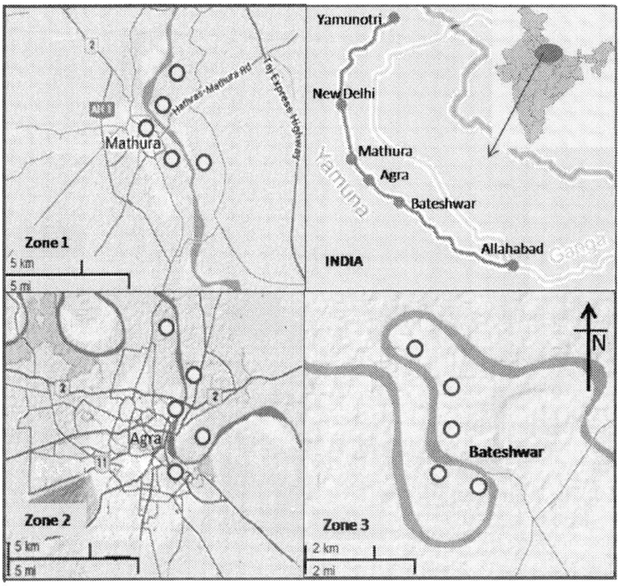

Figure 1. Map of the study area

The study area was divided into three zones (Figure 1); all three along the course of Yamuna and covering two cities viz. Mathura (zone 1) and Agra (zone 2) and a large village i.e. Bateshwar (zone 3). The distance between zones 1 and 2 is 80 km and zones 2 and 3 is 120 km downstream. In all, a total distance of 200 km was covered along the course of river. In each zone, 5 sites were selected ~1 km apart. Five random soil samples were taken from 0-15 cm depth at each site. A total of 75 soil samples (25 from each zone) were analyzed in order to obtain a complete profile. The same number of river water samples was collected from midstream at a depth of about 0.3 m. Soil from the botanical garden of St. John's college, Agra, was utilized as control.

The statistical significance of differences among mean metal content in water and soil was independently determined by one-way analysis of variance (ANOVA) followed by Fisher's LSD test. Pearson's coefficient for correlation of water and soil data was analyzed at a significance level of $P < 0.05$ and $P < 0.01$ with SPSS 16.0 statistics software.

3. Physico-chemical profile

3.1. Water

Physico-chemical properties of the water samples collected from the study zones are mentioned in Table 1. The pH values indicate neutral nature of river water acceptable as per BIS [36] and WHO [37] guidelines. A reading of 6.5 to 7.5 is considered neutral, suitable for general plant growth [38]. Conductance which reflects the status of major ions/inorganic pollution and is a measure of total dissolved solids and ionized species in the water, varies between 434 – 503 μmho/cm. Total dissolved solids were highest in zone 2. The hardness of water body is regulated largely by the levels of Ca and Mg salts. Other metals if present such as Fe, Al and Mn may also contribute to hardness. Most parameters were within their respective acceptable limits [36, 37]. Electrical conductivity was low. High COD, BOD and low DO in zones 1 and 2 are due to the discharge of huge amount of the untreated urban and industrial wastewater/ effluents indiscriminately. All three zones were faecally contaminated. Bacterial contamination ranged from19000—93000 coliform/100ml; the values are much higher than recommended values of 1coliform/100ml. Most of these coliforms were of faecal type due to gravity discharge of faecal wastes in adjacent areas along the river.

Parameters	**Zone 1**	**Zone 2**	**Zone 3**	**Acceptable Limits [36, 37]**
pH (1:2.5)	7.31	7.23	7.61	6.5-8.5
Total Dissolved Solids (mg/l)	266	314	245	500
Conductivity (μmho/cm)	462	503	434	
N-NO_3 (mg/l)	3.54	4.37	4.11	10
N-NH_3 (mg/l)	2.34	2.36	1.63	10
Total hardness (mg/l)	227	223	210	250
Total alkalinity (mg/l)	203	188	209	
Chemical Oxygen Demand (mg/l)	23.5	24.3	11.3	
Chloride (mg/l)	8.5	8	9	250
Fluoride (mg/l)	0.37	0.37	0.32	
Dissolved Oxygen (mg/l)	5.63	-	8.67	>5
Biological Oxygen Demand (mg/l)	9.63	10.7	5.34	
Sodium (mg/l)	6.81	6.62	7.15	
Potassium (mg/l)	0.38	0.4	0.38	
Ca hardness (mg/l)	124	131	115	
Mg hardness (mg/l)	83	96.4	91.6	
Faecal coliforms (MPN/100ml)	86000	93000	19000	
Streptococcus (MPN/100ml)	64000	71000	17000	

Table 1. Physico-chemical profile of water

3.2. Soil

The soil of the study area is characterized by alluvium, which is an admixture of gravel, sand, silt and clay in various proportions deposited during the quaternary period. The area is a part of Indo-Gangetic alluvium of quaternary age and is made up of recent unconsolidated fluviatile formations comprising sand, silt, clay and *kankar* with occasional beds of gravel. The topsoil is coarse and angular sand with small clay fraction. The sub-soil is sandy throughout. The stabilized topsoil is reddish brown with sand and clay mixed. The minimum depth of topsoil layer is 60 cm.

Physico-chemical properties of soil samples are given in Table 2. The topsoil in the study area is sandy loam (sand 60-80%, silt 10-24%, clay 8-16%). It has high exchangeable sodium percentage (ESP) values and moderate water retaining capacity. The sub-soil is sandy throughout. Soil pH ranged from neutral to alkaline. Zones 3, 2 and 1 were classified as very low, low and medium in organic matter, respectively.

Zone	pH (1:2.5)	Electrical Conductivity (dS/m) (1:2.5)	Organic Matter (%)	Avail. Phosphate (kg/ha)	Avail. Potash (kg/ha)	Avail. Nitrogen (kg/ha)
1	7.06-7.12	0.46-.50	1.44-1.53	108-115	236-298	53.4-60.8
2	6.24-6.81	0.33 0.38	0.8-1.2	131-140	65-94	50.1-55.2
3	7.43-7.6	0.44-0.47	0.5-0.72	50-65	143-178	75.2-87.8
Control	7.21	0.54	1.68	70.5	393	112.9

Table 2. Physico-chemical profile of soils

The electrical conductivity (EC) of soils ranged from 0.33-0.54 dS/m. Zone 1 and 2 soils fall in very high (>100 kg/ha), soils from zone 3 and control site in the high (50-100 kg/ha) phosphate availability bracket. Soils from zones 1, 3 and control displayed medium (130-330 kg/ha) potash levels while zone 2 was low (<130 kg/ha) in available potash. Nitrogen content in the soil samples ranged from 50.1 – 112.9 kg/ha.

4. Heavy metal profile

4.1. Water

Concentrations of heavy metals in the water samples collected from different location have been summarized in Table 3. It is clearly evident from the table that heavy metals were consistently higher in zone 2 compared to zones 1 and 3. Cr content was markedly higher among the metals in zone 2 followed by zone 1. The concentration of heavy metals in water samples ranged from 0.018 – 0.095 mg Pb l^{-1}, 0.025 – 0.341 mg Cd l^{-1}, 0.47 – 1.76 mg Zn l^{-1}, 0.27

– 1.58 mg Cu l^{-1}, 0.001 – 0.005 mg Co l^{-1}, 0.80 – 9.37 mg Cr l^{-1} and 0.078 – 0.32 mg Ni l^{-1}. As was not detected in any sample.

Zone		Pb	Cd	Zn	Cu	Co	Cr	Ni
1	Range	0.025-0.041	0.05-0.136	0.7-1.02	0.86-0.98	0.002-0.004	2.87-4.23	0.078-0.12
	Avg.	0.036a	0.088a	0.868a	0.916a	0.003a	3.55a	0.097a
	SD	0.007	0.037	0.128	0.046	0.0005	0.540	0.016
2	Range	0.066-0.095	0.159-0.341	1.37-1.76	1.27-1.58	0.004-0.005	6.42-9.37	0.17-0.32
	Avg.	0.082b	0.243b	1.56b	1.40b	0.005b	7.914b	0.256b
	SD	0.013	0.072	0.166	0.112	0.001	1.138	0.071
3	Range	0.018-0.028	0.025-0.03	0.47-0.61	0.27-0.33	0.001-0.002	0.8-0.97	0.009-0.015
	Avg.	0.023c	0.028c	0.54c	0.3c	0.001c	0.864c	0.01c
	SD	0.004	0.004	0.058	0.042	0.0003	0.066	0.003
	F value	*	*	*	*	*	*	*
Permissible limits	WHO [37]	0.01	0.003	5	2	-	-	-
	USEPA [39]	0.015	0.005	5	1.3	-	-	-
World Average		0.004	0.001	0.2	1.4	-	-	-

As content below detection limit.

F value :"*" statistically significant. Different letters in the same column denote significant statistical difference (P≤0.001) in mean metal contents in water samples from different zones

SD- Standard deviation.

WHO – World Health Organization.

USEPA – United States Environment Protection Agency.

Table 3. Heavy metal content of water samples (mg L^{-1})

All the metals in water samples were positively (P<0.01) correlated with each other (Table 4). In other words, metal concentration trends were identical and increased simultaneously for Pb, Cd, Zn, Cu, Co, Cr, and Ni.

	Cd	Zn	Cu	Co	Cr	Ni
Pb	0.970**	0.985**	0.875**	0.924**	0.960**	0.962**
Cd		0.977**	0.902**	0.944**	0.954**	0.963**
Zn			0.925**	0.953**	0.976**	0.966**
Cu				0.963**	0.940**	0.899**
Co					0.961**	0.930**
Cr						0.947**

** Correlation is significant at the 0.01 level (two-tailed) (two-tailed; n=75)

Table 4. Correlation coefficients: water heavy metal concentrations

Higher concentrations of metals in zone 2 (Figure 2) may be attributed to the discharge of industrial effluents from various sources including untreated sewage, municipal waste and agrochemical runoff from the nearby villages directly into the river. The concentrations of Co and Ni were found to be negligible at all sites. Due to the neutral to alkaline nature of river water, most of the heavy metals have precipitated and settled as carbonates, oxides, and hydroxide bearing sediments and elevated levels indicates higher exposure risks to the benthic biota of the river. Based on the WHO [37] and USEPA [39] drinking water standards (Table 3) the results in the present investigation show that Pb, Cd and Cr at all sites and Ni at most sites far exceeded the prescribed limits. Cu values from zone 2 were above the USEPA [39] threshold. One Way ANOVA and Fisher's LSD test indicate the difference in mean content of each metal among zones was highly significant statistically ($P \leq 0.001$).

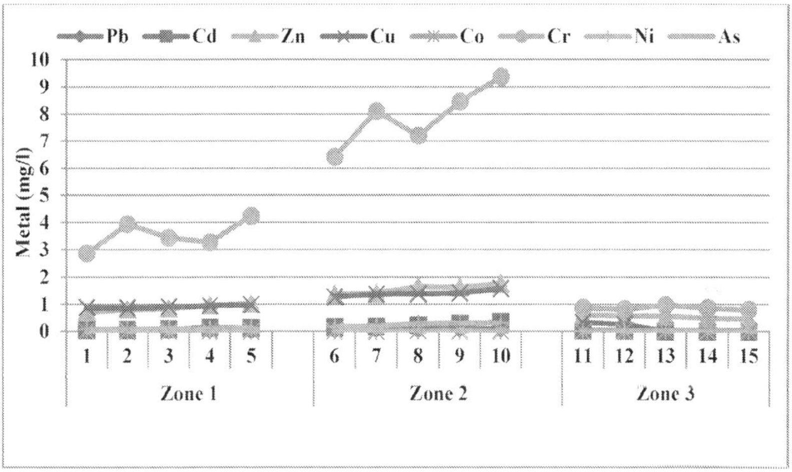

Figure 2. Average heavy metal content in water samples

When compared with the metal profile of the rivers around the world (Table 5) the situation does not seem that desperate here, at least as far as heavy metal contamination is concerned. The picture, however, is quite different when we consider the WHO guidelines for drinking water and World average of trace elements in unpolluted rivers [56, 57], the concentration ranges of Pb and Cd were well above the international guidelines and acceptable concentrations for drinking water (Table 3). When compared to the world average of trace elements for unpolluted rivers, the river was considered polluted by Pb, Cd, Zn and Cu.

Rivers	Pb	Cd	Zn	Cu	Co	Cr	Ni	References
Yamuna river (present study)	0.018-0.095	0.05-0.341	0.47-1.76	0.27-1.58	0.001-0.005	0.8-9.37	0.009-0.32	
Cauvery river, India	13.35	-	47.51	4.57	8.25	1.01	4.53	[13]
Brahmaputra river, India	-	-	916	108	168	222	179	[10]
Ganga river, India	76.36	11.5	332.5	48.39	-	5.36	4.88	[14]
Gomti river, India	3.058	-	63.022	-	-	0.064	0.013	[12]
Challawa river, Nigeria	0.44	-	1.2	0.22	-	0.47	-	[40]
Mghogha river, Morocco	48.25	0.36	299.5	56.7	-	86.4	46.83	[41]
Sava river, Croatia	34	0.5	91	24	-	-	-	[42]
Pasig river, Philippines	70	-	530	-	160	-	21.2	[43]
Rhine river, Netherland	188.2	7.1	684.3	62.5		6.4	33.7	[44]
Zhujiang, China	75.2	-	212	51	17.8	70.6	61.8	[45]
Almendares river, Cuba	93	2.5	262	158	-	90	-	[46]
Montevideo, Uruguay	44-128	1-1.6	174-491	58-135	-	79-253	-	[47]
Ribeira river, Brazil	767102	0.2-5.5	15-5090	60	-		-	[48]
Amazon river, Brazil	83	-	110	37.5	-	65	26.7	[49]
Danube river, Serbia and Montenegro	28.65	3.12	253.74	36.29	-	76.26	70.1	[50]
Msimbazi river, Tanzania	-	0.9	79	14	-	12	8.7	[51]
Brisbane River, Australia	20.1-81.9	1.9	40.8-144.0	31.1-30.2	-	14.2-54.3	-	[52]
Siahroud river, Iran	9.7	0.05	14.9	-	-	1.03	-	[53]
Gediz River, Turkey	1.3	-	2.6	-	1.6	-	4	[54]
Avg. shale value/ world avg.	20	0.3	95	45		90	68	[55]

Table 5. Average heavy metal concentrations of rivers around the world (mg L^{-1})

4.2. Soil

Concentrations of heavy metals in the soil samples have been summarized in Table 6. Quantitatively the metals were observed in the sequence Pb > Zn > Cr > Ni > Cu > As > Cd > Co (Figure 3), though their thresholds for concern, mobility in soil and toxicity are different so this trend does not necessarily reflect the threat of individual metals. Pb and Zn were found in fairly higher concentrations at all the sampling locations. Generally, an overall linear increasing trend of metal contamination was noted from site 1, before the Yamuna enters the city of Mathura, to site 10 where the river leaves Agra. Thus, maximum values for all metals were observed in the samples pertaining to Agra. In the third zone metal concentrations were seen to decrease gradually. One-way ANOVA and Fisher's LSD test indicate that mean Pb and Co content was different at all sites ($P \leq 0.001$); while mean Cr, Cd, Cu, Ni, and As in zone 2 differed significantly from zone 1 and 3 ($P \leq 0.001$). The latter did not differ significantly among themselves. Mean Zn content in zone 1 differed significantly from zone 2 and 3 ($P \leq 0.05$). The difference between the latter was not significant statistically.

Zone		**Pb**	**Cd**	**Zn**	**Cu**	**Co**	**Cr**	**Ni**	**As**
1	Range	157-230	8.6-20.6	87.3-136	22.4-41.5	1.84-4.8	26.3-53.2	23.1-41.2	14.2-20.4
	Avg.	200a	13.4a	116a	30.5a	3.71a	40.7a	33.6a	17.2a
	SD	29.6	5.36	21.4	8.14	1.25	9.77	6.78	2.97
2	Range	241-285	17.8-25.2	129-222	50.4-64.2	5.91-15.2	76.3-104	57.8-71.3	22.2-28.6
	Avg.	261b	20.3b	173b	57b	9.6b	86.7b	63.5b	25.2b
	SD	16.1	3.13	37.6	5.93	3.80	10.5	5.44	3.10
3	Range	111-136	6.02-14.6	115-167	22.4-30.1	2.91-6.4	14.7-45.6	17.8-32.4	7.9-17.6
	Avg.	125c	10.2a	144b	25.7a	4.95c	28.1a	23.3a	13.8a
	SD	10.4	3.71	21.9	3.43	1.29	11.8	5.57	4.01
Control	Range	13.6-18.4	1.23-1.87	39.6-54.3	12.8-24.3	1.68-2.53	10.2-14.3	7.3-9.7	3.03-5.7
	Avg.	15.6	1.6	47.4	18	2.1	12	8.6	4.2
	SD	1.88	0.25	5.3	4.53	0.31	1.55	1	0.99
	F value	*	*	*	*	*	*	*	*
Suggested thresholds in soil [58]	Industrial	600	22	360	91	-	87	50	12
	Residential	140	10	200	63	-	64	50	12
Suggested thresholds in soil [59]	Background	-	-	140	-	-	100	35	29
	Intervention		-	720	-	-	380	210	55
Threshold values [60]	Class I	35	0.2	100	35	-	90	40	15
	Class II	250	0.3	200	50	-	150	60	30
	Class III	500	1.0	500	400	-	300	200	40

F value :'*' statistically significant. Different letters in the same column denote significant statistical difference ($P \leq 0.05$) in mean metal contents in soil samples from different zones.

SD- Standard deviation.

Table 6. Heavy metal content of soil samples (mg kg^{-1})

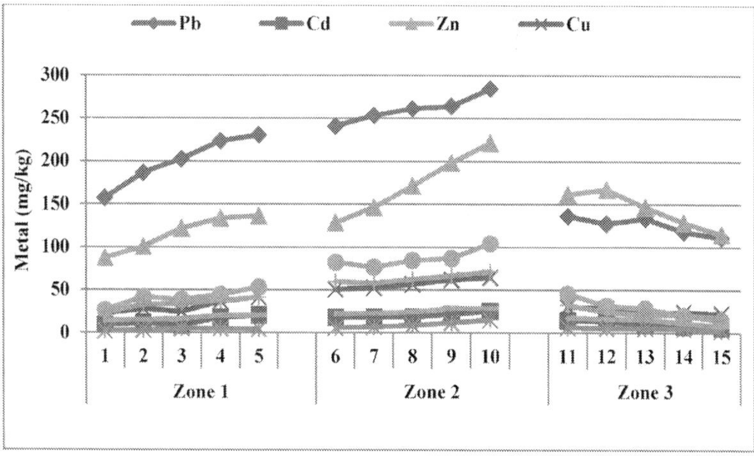

Figure 3. Average heavy metal content in soil samples

All the metals in soils were positively (P<0.01) correlated with each other (Table 7). Significant negative correlation was observed between metal concentrations and soil pH (P<0.01). The same was observed in the case of Zn and Co with Organic matter. Phosphate is able to increase water-soluble lead forms from contaminated soils by 56.8– 100% [61]. This is clearly shown by the phosphate values (Table 2) obtained for different samples with maximum in zone 2 followed by zone 1 which probably led to higher Pb values in zones 1 and 2 (Table 6). Fertilizers contain from trace to several ppm of Pb, Zn, Cu, Mg [62, 63]. High P_2O_5-blended fertilizers and the pure phosphates, contain significant concentrations of several elements of potential environmental or agronomic concern [62, 64].

	Cd	Zn	Cu	Co	Cr	Ni	As	OM	pH
Pb	0.821**	0.479**	0.889**	0.629**	0.909**	0.933**	0.894**	0.426**	-0.802**
Cd		0.701**	0.886**	0.724**	0.862**	0.848**	0.906**	0.0876	-0.621**
Zn			0.723**	0.915**	0.689**	0.653**	0.744**	-0.493**	-0.371**
Cu				0.805**	0.966**	0.972**	0.932**	0.0260	-0.773**
Co					0.809**	0.784**	0.806**	-0.304**	-0.529**
Cr						0.991**	0.944**	0.0816	-0.834**
Ni							0.936**	0.138	-0.837**
As								0.106	-0.739**
OM									-0.242*

* Correlation is significant at the 0.05 level (two-tailed; n=75)

** Correlation is significant at the 0.01 level (two-tailed)

Table 7. Correlation coefficients: soil heavy metal concentrations

Agra is the fourth most populated city in Uttar Pradesh, India. With a population of 1.7 million (2011 census) it generates about 700 tonnes of solid wastes every day. It is also a major cause for adding contamination to soil and groundwater. Solid waste is also discharged from 200 hospitals and nursing homes along with 168 foundries, 52 tanneries, 300 shoe industries, 200 petha (a local sweet) manufacturing units, 50 dairies, 56 electroplating units, 15 silver vibrators and 15 galvanizing units. Significantly higher amount of metal pollution in the samples from the city (sites 6-10) is obviously due to untreated domestic/wastewater, sewage and industrial effluent discharged at these sites throughout the year. The increasing contamination as one proceeds downstream mirrors the extent of damage caused to the pedosphere.

Mean concentrations of heavy metals in soils at the sites studied were compared with threshold values of soil suggested by the Canadian Environmental Quality Guidelines [58]. It was observed that As (sites 1-13) and Ni (sites 6-10) crossed their respective industrial thresholds while the other metals (Pb, Zn and Cu) are well within it. Mean concentrations of As at sites 4-10 were approximately twice the thresholds suggested. Cd and Cr levels were above their thresholds only at site 10. However, the situation is drastically different in the perspective of the residential limits where in addition to these, the thresholds are exceeded even by Pb, Cd (10 sites each), Cr (5 sites) and also Zn and Cu at one site.

On comparing metal concentrations with the values suggested for soil remediation by VROM, Netherlands [59], values of Zn (sites 7-13), Ni (sites 4-10) and Cr (site 10) were above the background values but below the intervention level. It is significant to note that in studies similar to the present one, the degree of contamination and the resulting 'hazard indices' for soils may vary when different thresholds, existing in only a few countries, are considered [65]. To increase the reliability of risk estimation due to contaminants, global consensus on such thresholds is urgently needed.

The concentrations of As are usually low, less than 6 ppm, for geological and soil environment [64]. It is estimated that about 60% As in the environment is from anthropogenic sources including As-based pesticides, fertilizers, and wastes from mines, smelter and tannery industries [66]. The relatively high values of As in the samples seem to be directly related to the discharge of domestic and industrial effluent as well as use of phosphate fertilizers, pesticides used in the agricultural activities in the region.

Highly significant positive correlation ($P<0.01$) was observed between soil and water content of Pb, Cd, Cu, Co, Cr and Ni. The results also indicate that metal concentrations in soil were higher than those in the water. This distribution pattern of heavy metals between the water phase and soil is expected as most heavy metal speciation studies have reported a similar pattern of distribution both in sea water as well as in lakes [67-69].

Several authors have pointed out the need for a better knowledge of urban soils [18, 70]. In the past few years, studies on urban soils in many cities have been carried out around the world. Some examples are Spanish [19, 71] and Italian cities [21, 72]. Other examples for European cities are Aberdeen [73], Athens [74], Oslo [22] and Belgrade [18]. The mean heavy metal contents for all zones are compared in Table 8 to those of some cities around the world. The differences concerning population, living habits, industrial activities, etc., cause significant differences in the metal contamination profile. Compared to average concentrations in urban soils in the world, the mean concentrations of Pb and Cu are up to 2—4 times higher in some cases but still less than London, Naples and Palermo. In the case of Cd, it is many times higher than Kattedan (India). Zn and Cr contents do not differ much; still they are less than those of Naples and Madrid. Ni content is more than almost all European cities, but less than Kattedan and Firozabad in India. Co values are less than those reported from other industrial regions of India. As content is less than that of Firozabad.

City		Pb	Cd	Zn	Cu	Co	Cr	Ni	As	Reference
London		294	-	183	73	-	-	-	-	[75]
Madrid		161	-	210	72	-	75	14	-	[76]
Rostock		83	-	100	35	-	48	30	-	[77]
Sevilla		161	-	107	64.6	-	42.8	23.5	-	[19]
Belgrade		53.2	-	129.1	29	-	33.2	67.4	-	[18]
Palermo		253	-	151	77	-	39	19.1	-	[72]
Naples		262	-	251	11	-	74	-	-	[21]
Nanjing		107.3	-	162.6	66.1	-	84.7	-	-	[78]
Hong Kong		93.4	-	168	24.8	-	n.a.	-	-	[79]
Kattedan		195-6241	0.08-0.16	130-3191	72-1450	12-36	77-586	63-494	0.10-0.21	[80]
Firozabad		35.5-781	3.64-107	76.4-1247	22.4-300	10.9-63.7	19.1-158	23-218	9.25-204	[29]
Present Study	Zone 1	200	13.4	116	30.4	3.70	40.7	33.6	17.2	
	Zone 2	261	20.3	173	57.0	9.60	86.7	63.5	25.2	
	Zone 3	125	10.2	144	25.7	4.96	28.1	23.3	13.8	

Table 8. Average heavy metal concentrations in urban soils from different cities across the world (mg kg^{-1})

It is encouraging to note that the mean concentrations of individual metals are below those reported from other industrial hubs within India i.e. Kattedan (Andhra Pradesh) [except Cd and As] and Firozabad (Uttar Pradesh). Kattedan Industrial Development Area (KIDA) is a major industrial area of Andhra Pradesh and houses 400–500 industries, including 150 large scale industries and 300 small-scale industries. Major sources of metals pollution are battery, electrode, oil refining, metal plating, textile, pharmaceutical, chemical paints, rubber, petrochemicals, glass, therapeutics, and Pb extraction facilities [81]. This is also one of the contaminated areas identified by the Central Pollution Control Board (CPCB) in New Delhi, and referred to as an ecological disaster area [81]. Firozabad is the hub of the Indian Glass industry.

5. Assessment of heavy metal contamination in soil

Assessment of soil contamination was performed by the contamination index (P_i) and integrated contamination index (P_c) as expressed by fuzzy functions [82, 29, 28]. Class I criteria [60] could be used as no-polluted threshold; Class II as lowly polluted threshold value; and while Class III as highly polluted threshold value. P_i values ≤ 1 indicate no contamination; $1 \leq P_i \leq 2$

indicates low contamination; $2 \le P_i \le 3$ indicates moderate contamination; while $P_i > 3$ indicates high contamination.

Individual elements displayed remarkably different patterns of accumulation in soils. Furthermore, observed differences in the magnitude of accumulation suggest that the relative contribution of the individual elements to total heavy metal contamination varies. Figure 4 shows the proportions of contamination levels (from P_i values) in the soil samples from all the sites studied. Except for 76% samples from zone 2, which showed moderate Pb contamination, the rest exhibited low contamination zone as did all samples from zones 1 and 3. In case of Cd, all samples were in the high contamination zone. For Zn, 24% samples from zone 2 were moderately contaminated while 72%, 76%, and 100% samples from zones 1, 2, and 3, respectively were in the low contamination range. For Cu, 88% samples from zone 2 were moderately contaminated while 36% and 12% samples, from zones 1 and 2, respectively were in the low contamination range. All samples from zone 3 indicated no contamination. Except for zone 2 (20% samples) in the low contamination zone, the remaining samples did not indicate Cr contamination. For Ni, 12% and 72% samples from zones 1 and 2, respectively were moderately contaminated while 16% and 22% samples from, respectively were in the low contamination range. All samples from zone 3 indicated no contamination. In the case of As, 64%, 100%, and 60% samples from zones 1, 2, and 3 were in the low contamination range.

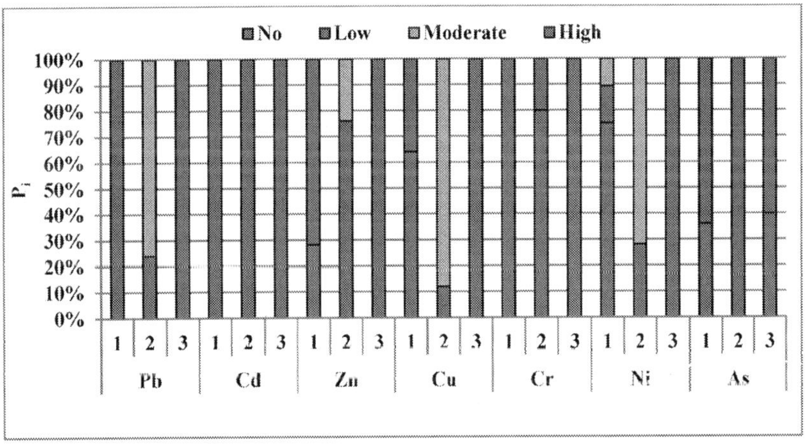

Figure 4. Contamination indices (Pi) of heavy metals in soil samples

Thus, zone 1 was found to be lowly contaminated with Pb, Zn, Cu, Ni and As but highly contaminated with Cd. Zone 2 exhibited low to moderate contamination of Pb, Zn, Cu, Ni; low Cr and As contamination; and high Cd contamination. Zone 3 was lowly polluted with Pb and Zn. As contamination ranged from none to low. No Cu, Cr and Ni contamination was observed. These results agree with the findings regarding metal contamination of soil due to the glass industry at Firozabad, India [29]. Of the nine elements studied, Zn, Cd, and As showed a greater accumulation in all soils, whereas, accumulation of Ni and Cu was high in limited samples.

Integrated Contamination Indices (P_c) were calculated for all soils to assess the extent of heavy metal contamination at the sites. P_c is defined as the summation of the difference between the contamination index for each metal and 1 (one). It is categorized under the following heads: $P_c \leq 0$ no contamination; $0 \leq P_c \leq 7$ low contamination; $7 \leq P_c \leq 21$ moderate contamination; $P_c > 21$ high contamination. Threshold values for Co could not be obtained hence this metals was excluded in the calculation. A clear ascending trend is visible in the P_c values for all sites (Figure 5). P_c values generally show a moderate to high contamination at studied sites. The P_c indices indicate that 45% sampling locations fall in the moderate contamination while 55% of the samples fall in the high contamination range. All the samples from zone 2 fell under the high contamination category. While in zone 1, 60% samples come under moderate and 40% under high contamination level category. In zone 3, 76% and 24% samples were in the moderate and high contamination range, respectively.

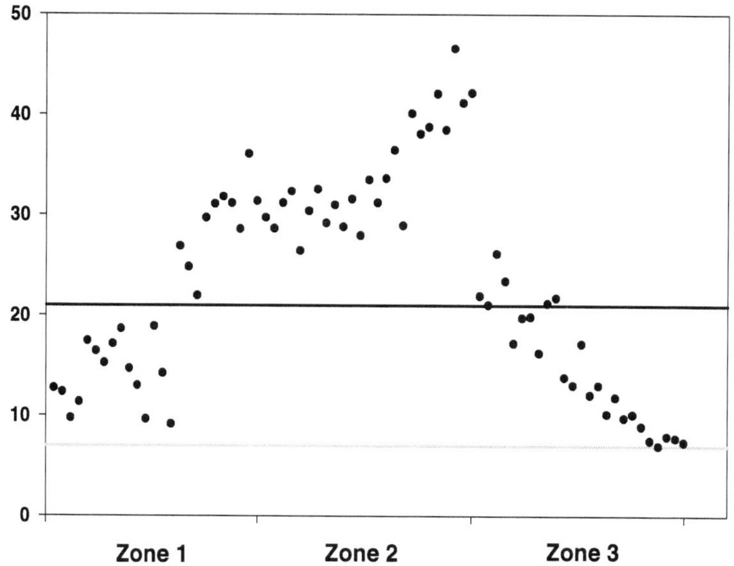

Figure 5. Integrated contamination indices (Pc) of soil samples. Black and gray lines are the upper threshold values of moderate and low contamination, respectively

The effect of the glass industry on urban soil metal characterization was assessed at 25 test sites at Firozabad, India [29]. The area is characterized by little or no monitoring of industrial processes, usage and disposal of hazardous chemicals. A comprehensive profile of Zn, Mn, Co, Cd, Pb, Cr, Ni, Cu and As contamination was obtained. Zn, Cd, and As showed a greater accumulation in all soils, whereas, accumulation of Ni and Cu was high in limited samples.

Integrated contamination indices (P_c) indicate that 60% of the sites were in the high contamination range and 28% were in the moderate contamination range with just 12% sites on the border of the moderate to low contamination range. [83] assessed the impact of both landuse and soil textures on Cd, Zn, Pb and Cu based on samples collected from the major landuse/ landcover pattern of Dutch forests and aerable soils drawn from six different sites. Metal content in agricultural and industrial soil is found to be higher than the forest soil.

The fact that no P_c value in the present investigation fell within the low contamination range was not surprising, given the fact that the study was being carried out in an area which has already been contaminated with metals, but moderate to high indices in zone 1 and 2 are alarming because these include heavily populated areas. The local populace is, thus, exposed to wide range of historically well established toxins and even carcinogens. The situation is surely compounded by vehicular pollution at urban sites (1-10). Vehicular emissions are a significant source of many pollutants [21, 84].

6. Phytosociological studies

Plants show differing morpho-physiological responses to soil metal contamination. Most are sensitive to very low concentrations; others have developed tolerance, and a reduced number accumulate metals. The latter capacity has practically opened up the way to phytoextraction. Hyperaccumulation is an unusual occurrence, seen in a narrow range of species which often grow in metal-rich soils. The following thresholds for metal hyperaccumulation in shoots, without evident symptoms of toxicity, have been suggested [85]: 100 mg kg^{-1} for Cd, 1, 000 mg kg^{-1} for Ni, Cu, Co and Pb, and 10, 000 mg kg^{-1} for Zn and Mn. Known hyperaccumulators are generally minor vegetation components in most European and North American habitats. Currently, more than 400 hyperaccumulator species are known, belonging to 45 different botanical families, among which the most frequent are Brassicaceae and Fabaceae [86].

Lack of information on the agricultural management of hyperaccumulators, together with slow-growing and poor shoot and root growth, increase the difficulties in the practical application of these species in remediation projects [87]. Hence, the potential for any plant species to remediate successfully heavy metal contaminated sites depends on all of the following prerequisite factors: a) the amount of metals that can be accumulated by the candidate plant, b) the growth rate of the plant in question, and c) the planting density [88]. The growth rate of a plant in a chemically contaminated soil is important from the perspective of biomass. Parameters like basal area, canopy, abundance, dominance of species can help obtain a more rounded picture in the case of mixed planting or natural flora at a contaminated site. The rate of metal removal from the soils can be calculated if information on the above mentioned parameters is available. In addition, versatility of the candidate plant to tolerate and at the same time accumulate multiple metal contaminants and/or metal-organic mixtures would be an asset for any phytoremediation system.

The choice of plant species is thus, an important task in any phytoremediation based technique. Decontaminating a site in a reasonable number of harvests requires plants that are both high yielders of biomass and good metal accumulators by dry weight. It has been demonstrated [89, 90] that, wild native plants may be better phytoremediators for waste lands than the known metal bioaccumulators like *Thlaspi caerulescens* and *Alyssum bertolonii* because the latter are slow growing with shallow root systems and low biomass. Also, the technology for their large-scale cultivation is not fully developed; therefore, their use is rather limited [91].

If soil at contaminated sites, e.g. mines, industrial zones is naturally high in a particular metal, native plants will often become adapted over time to the locally elevated levels [28-30] and metal toxicity issues do not generally arise. Successful establishment and colonization of several pioneer plant species tolerant to Pb/Zn mine spoils has also been demonstrated with tolerant plants including *Phragmites australis, Vetiveria zizanioides,* and *Sesbania rostrata* [31, 92]. Many native, well adapted plants have been investigated and even used for heavy metal bioindicatoring and phytoremedial purposes including lemongrass and other wild grasses, vetiver, *Sesbania, Avena, Crotalaria, Crinum asiaticum, Typha latifolia* and *Calotropis procera* etc. [31-35]. Phytoremediation employing indigenous species can be an ecologically viable option for sustainable and cost-effective management.

An important component of any ecosystem is the species it contains. Species also serve as good indicators of the ecological condition of a system [93]. Ecological surveys are necessary for an adequate characterization of a plant community and also to know the diversity and dispersion status of species in the area. Phytosociology aims to characterize and classify plant communities in terms of composition and structure.

At all sampling sites within a zone, ecological indices [relative frequency, relative density, relative dominance and importance value index (IVI)] were estimated, by using a $1m^2$ quadrat. Sampling was done randomly at 10 spots at each site within a zone. The data were compiled and analysed according to some workers [94-96].

Relative density is the proportion of density of a species (plants/unit area) to that of the stand as a whole. The dispersion of species in relation to that of all the species is termed as relative frequency of a species. Relative dominance is the proportion of the basal area of a species to the sum of the basal area of all species present. Basal area refers to area covered by the plant's stem and leaves one inch above the ground surface. The overall picture of ecological importance of a species in relation to the community structure can be obtained by adding the values of the above three parameters [97].

A total of 22 weed species were recorded from the sites (Table 9). Most of the weeds recorded are herbs except *Calotropis procera* and *Datura stramonium* which are shrubby in nature. Two grasses i.e. *P. annua* and *C. dactylon* were observed. The phytosociological parameters obtained from the sites clearly indicate that there are naturally occurring plant species which have the capacity to tolerate the heavy metal content of the soils. The floral composition of the three zones varied qualitatively and quantitatively. Most species were seen to grow vigorously. Relative frequency, relative density, relative dominace and IVI indicate that *Calotropis procera, Parthenium hysterophorus, Chenopodium murale, Croton bonplandianum, Rumex dentatus, Amaranthus spinosus, Datura stramonium* and *Withania somnifera* were the most abundant weeds. All of these species have been reported as potential phytoremediators in earlier studies. It is

http://dx.doi.org/10.5772/57239

important to note that floral diversity decreased with increasing contamination profile of the sites. Maximum species (20) were observed in zone 3, followed by zones 1 and 2.

Zone	Plants	Relative frequency	Relative density	Relative dominance	IVI
1	*Amaranthus spinosus*	8.42	10.55	0.15	19.12
	Rumex dentatus	6.32	6.00	0.07	12.38
	Calotropis procera	8.42	10.31	0.18	18.92
	Croton bonplandianum	6.32	6.00	0.23	12.54
	Chenopodium murale	6.32	6.24	0.01	12.56
	Datura stramonium	7.37	9.83	0.16	17.37
	Stellaria media	5.26	4.32	0.00	9.58
	Withania somnifera	8.42	9.59	0.13	18.15
	Heliotropium ellipticum	5.26	3.12	0.00	8.38
	Achyranthes aspera	6.32	5.76	0.01	12.08
	Parthenium hysterophorus	7.37	6.95	0.01	14.33
	Amaranthus alba	5.26	4.56	0.01	9.83
	Boerhaavia diffusa	3.16	2.64	0.01	5.80
	Euphorbia hirta	6.32	6.00	0.00	12.31
	Sida longifolia	4.21	4.56	0.00	8.77
	Gnaphalium luteo-album	5.26	3.60	0.02	8.88
2	*Parthenium hysterophorus*	11.59	16.32	0.22	28.13
	Abutilon indicum	7.25	4.75	0.06	12.06
	Calotropis procera	11.59	15.13	0.20	26.93
	Croton bonplandianum	7.25	8.90	0.14	16.28
	Amaranthus spinosus	4.35	2.08	0.00	6.43
	Rumex dentatus	11.59	13.65	0.15	25.39
	Withania somnifera	8.70	8.61	0.10	17.40
	Cynodon dactylon	7.25	3.56	0.00	10.81
	Chenopodium murale	7.25	9.50	0.11	16.85
	Achyranthes aspera	5.80	4.75	0.02	10.57
	Sida longifolia	7.25	2.97	0.00	10.21
	Gnaphalium luteo-album	5.80	4.15	0.00	9.95
	Boerhaavia diffusa	4.35	5.64	0.00	9.99

Zone	Plants	Relative frequency	Relative density	Relative dominance	IVI
3	*Parthenium hysterophorus*	7.14	9.65	0.17	16.97
	Abutilon indicum	4.76	3.46	0.01	8.23
	Calotropis procera	7.14	12.57	0.18	19.89
	Croton bonplandianum	5.56	5.65	0.17	11.37
	Cynodon dactylon	4.76	2.91	0.06	7.74
	Chenopodium murale	6.35	12.39	0.14	18.88
	Poa annua	4.76	3.28	0.01	8.05
	Rumex dentatus	3.97	1.64	0.01	5.62
	Barleria diffusa	4.76	4.55	0.00	9.32
	Achyranthes aspera	5.56	4.19	0.01	9.76
	Sida longifolia	4.76	3.64	0.00	8.41
	Withania somnifera	4.76	3.83	0.09	8.68
	Boerahvia diffusa	4.76	2.55	0.00	7.31
	Sida cordifolia	4.76	3.28	0.00	8.04
	Amaranthus spinosus	7.14	8.56	0.10	15.80
	Gnaphalium luteo-album	4.76	3.83	0.03	8.62
	Euphorbia hirta	3.17	3.46	0.00	6.64
	Ageratum conyzoides	3.17	2.37	0.00	5.55
	Datura stramonium	4.76	4.19	0.00	8.95
	Tridex procumbens	3.17	4.01	0.00	7.18

IVI- Importance Value Index

Table 9. Phytosociological parameters of flora at test sites

C. procera has been demonstrated as a potential phytoremediator species. The shrub showed good accumulation of metals and is a potential phytoextractor for As and Zn as well as a promising phytostabiliser for Pb, Cd, Cu and Mn [28, 29, 35]. *C. procera* was observed to have a high degree of sociability i.e. relative frequency, relative density, relative dominance and IVI. *P. hysterophorus* was also important in this context and was most dominant in zone 2. This species has been identified for As phytoextraction along with *A. spinosus, C. bonplandianum,* and *D. stramonium* [28]. The latter two have also been indicated for phytoextraction—*C. bonplandianum* for Mn and *D. stramonium* for Mn, Cr, and Cu—together with *R. dentatus* for Pb. Another species with high IVI, *C. murale* has been suggested for Zn, Cd, Pb and Cu phytoextraction [28, 29]. Among the less dominant species, *Tridex procumbens* and *Euphorbia hirrta* have also been reported as promising tools for phytoextraction of Mn and As, respectively [28]. *E. hirrta* and *D. stramonium* were not found in zone 2.

Poa annua has been identified as a phytostabilizer for Mn, Cd, and As and phytoextractor for Cu and Pb. Cu concentrations up to 742.06 mg kg^{-1} dry weight have been reported in *P. annua* shoots [28, 29]. *Poa annua* was observed only at sites in zone 1.

Other species found at the sites have also been indicated for further studies following initial field surveys. *Gnaphalium luteo-album* (Mn and As); *Withania somnifera* (Cu); and *Heliotropium ellipticum* (As) have shown promise as phytostabilisers for these metals and metal combinations [28, 29].

7. Discussion

The occurrence as well as concentrations of heavy metals like Pb, Zn, Cu, Co, Mn, Fe, Cr and Cd in streams and rivers all over the world is increasing. In the present case study, heavy metal contamination was consistently higher in city of Agra, which may be attributed to the heavy industrialization combined with agricultural and urban runoff. The situation is made worse by atmospheric deposition, again attributable to industrial and vehicular pollution. In general, freshwater ecosystems have low natural background metal levels and therefore tend to be sensitive to even small additions of most trace metals. The river water far exceeded the limits of metals prescribed by WHO and USEPA for drinking water standards and Pb, Cd, and Cr content at all sites and Ni at most sites exceeded the prescribed limits. In a heavily populated country like India where a sizeable portion of the population is illiterate and resides in slums/ poorly planned neighbourhoods without proper sanitation and drainage, day-to-day activities also contribute to the overall pollution load. Provision of suitable alternatives along with proper education and awareness is integral to the minimization of this problem at the source. Apart from taking measures like effluent treatment before it enters the river and subsequent treatment of river water at the most polluted sites, a steady flow of water is to be ensured throughout the year, by way of channelizing the river with canals at crucial points. Such measures can address this problem to a substantial extent. Expenditure of more than US$ 500 million without much success appears to be an unjustified proposition.

Phytoremediation has been receiving attention lately as an innovative, cost-effective alternative to the otherwise tedious and expensive methods in use, which are not only a burden on the exchequer but also, require efforts on a recurring basis. Lack of information on the agricultural management of hyperaccumulator species, together with their poor biomass and root proliferation, increases the difficulties in their practical application. It has been amply demonstrated that wild native plants may be better phytoremediating tools. These species can be an ecologically viable option for sustainable and cost-effective management especially in scenarios where expertise, technical expertise and/or funding is a limiting factor. Ecological surveys are necessary for adequate characterization of a plant community and subsequent identification of prospective candidates for phytoremedial strategies since metal toxicity issues generally do not arise in plants already established on contaminated soils. Allowing native species to remediate site is an attractive proposition since these species do not require frequent irrigation, fertilizers, and pesticide treatments, while simultaneously a plant community comparable to that existing in the vicinity can be established. The outcome is, thus site remediation, ecological restoration and addition in aesthetic value. This is also in concurrence with the ruling (2006) of the Hon'ble Supreme Court of India prohibiting cultivation of plants requiring fertilizers and pesticides along the river Yamuna. Using these perennial phytoremedial candidates without any special needs holds much promise in this context. In addition, versatility of the candidate plant to tolerate and at the same time accumulate multiple metal contaminants and/or metal-organic mixtures would be an asset for any phytoremediation strategy.

Acknowledgements

Financial support from University Grants Commission [F. no. 35-47/2008(SR)] is gratefully acknowledged. This study was partially supported by the European Fund for Economic and Regional Development (FEDER) through the Program Operational Factors of Competitiveness (COMPETE) and National Funds through the Portuguese Foundation for Science and Technology (PEST-C/MAR/UI 0284/2011, FCOMP 01 0124 FEDER 022689.

Author details

Manoj S. Paul[1*], Mayank Varun[1], Rohan D'Souza[1], Paulo J.C. Favas[2,4] and João Pratas[3,4]

*Address all correspondence to: mspaul07@gmail.com

1 Department of Botany, St. John's College, Agra, India

2 Department of Geology, School of Life Sciences and the Environment, University of Trás-os-Montes e Alto Douro, Vila Real, Portugal

3 Department of Earth Sciences, Faculty of Sciences and Technology, University of Coimbra, Coimbra, Portugal

4 IMAR-CMA Marine and Environmental Research Centre, Faculty of Sciences and Technology, University of Coimbra, Coimbra, Portugal

References

[1] Nolan K. Copper Toxicity Syndrome. Journal of Orthomolecular Psychiatry 2003;12(4) 270–282.

[2] Marschner H. Mineral Nutrition of Higher Plants. London: Academic Press Limited; 1995.

[3] McCluggage D. Heavy Metal Poisoning, NCS Magazine. Columbus, USA: The Bird Hospital; 1991. http://www.cockatiels.org/articles/ Diseases/metals.html.

[4] Katz BG, Bricker OP, Kennedy MM. Geochemical mass balance relationships for selected ions in precipitation and stream water, Catoctin Mountains, Maryland. American Journal of Science 1985;285 931–962.

[5] Houba C, Remade J, Dubois D, Thorez J. Factors affecting the concentrations of cadmium, zinc, copper and lead in the sediments of the Vesdre river. Water Research 1983;17 1281–1286.

[6] Zhang J, Huang WH. Dissolved trace metals in the Huanghe: The most turbid large river in the world. Water Research 1993;27 1–8.

[7] Pizarro J, Vila I, Manuel C. Chemical compositions of Chilean rivers. Verth-Int. Ver. Theory of predictive Limnology 1998;26 948–951.

[8] Alexandrine E, Jean-Marie M, Thevenot DR. Urban runoff impact on particulate metal concentrations in the river Seine. Water, Air, and Soil Pollution 1998;108 83–105.

[9] Pistelok F, Galas W. Zinc pollution of the Przemsza river and its tributaries. Polish Journal of Environmental Studies 1999;8 47–54.

[10] Subramanian V, Van't Dack L, Van Grieken R. Chemical composition of river sediments from the Indian sub-continent. Chemical Geology 1985;48 271–279.

[11] Kumar A, Kaur I, Mathur RP. Water quality and metal enrichment in bed sediment of the river Kali and Hindon, India. Environment and Geochemistry Health 1998;20 53–60.

[12] Singh VK, Singh KP, Mohan M. Status of heavy metals in water and bed sediments of river Gomti – a tributary of the Ganga river, India. Environmental Monitoring and Assessment 2005;105 43–67.

[13] Abida B, Ramaiah M, Khan I, Veena K. Heavy Metal Pollution and Chemical Profile of Cauvery River Water. E-Journal of Chemistry 2009;6(1) 47–52.

[14] Pandey J, Shubhashish K, Pandey R. Heavy metal contamination of Ganga river at Varanasi in relation to atmospheric deposition. Tropical Ecology 2010;51(2) 365–373.

[15] Daniel Pepper. India's rivers are drowning in pollution. Fortune (magazine); 2007.

[16] Karthikeyan, A. "Failure of Ganga, Yamuna projects." (http://articles.timesofindia.indiatimes.com/2009-09-04/chennai/28093527_1_cooum-treatment-plants-sewage-treatment). (Assessed 9 september 2009)

[17] "Inflow to Yamuna to be cleaned up at last" (http://www.expressindia.com/latest-news/inflow-to-yamuna-to-be-cleaned-upat-last/509240/). Indian Express. (Assessed 31 August 2009).

[18] Crnković D, Ristić M, Antonović D. Distribution of HeavyMetals and Arsenic in Soils of Belgrade (Serbia and Montenegro). Soil and Sediment Contamination 2006;15 581–589.

[19] Madrid L, Diaz-Barrientos E, Madrid F. Metals in urban soils of Sevilla: seasonal changes and relations with other soil components and plant contents. European Journal of Soil Sciences 2004;55 209–217.

[20] Chirenje T, Ma LQ, Reeves M, Szulczewski M. Lead distribution in near surface soils of two Florida cities: Gainesville and Miami. Geoderma 2004;119 113–120.

[21] Imperato M, Adamo P, Naimo D, Arienzo M, Stanzione D, Violante P. Spatial distribution of heavy metals in urban soils of Naples city (Italy). Environment Pollution 2003;124 247–256.

[22] Tijhuis L, Brattli B, Saether OM. A geochemical survey of topsoil in the city of Oslo, Norway. Environmental Geochemistry and Health 2002;24 67–94.

[23] Raskin I, Smith RD, Salt DE. Phytoremediation of metals: using plants to remove pollutants from the environment. Current Opinion in Biotechnology 1997;8 221–226.

[24] Salt DE, Smith RD, Raskin I. Phytoremediation. Annual Review of Plant Physiology and Plant Molecular Biology 1998; 49 643–668.

[25] Vangronsveld J, Cunningham SD. 1998. Introduction to the Concepts. In: Vangronsveld J, Cunningham SD. (ed.) Metal Contaminated Soils: In situ Inactivation and Phytorestoration. Georgetown, USA: RG Landes Co.; 1998. p1-15.

[26] Blaylock MJ, Huang JW. 2000. Phytoextraction of metals. In: Raskin I, Ensley BD (eds.) Phytoremediation of toxic metals: using plants to clean-up the environment. New York: John Wiley & Sons Inc; 2000. p53–70.

[27] Salt DE, Blaylock M, Kumar NPBA, Dushenkov V, Ensley D, Chet I, Raskin I. Phytoremediation: a novel strategy for the removal of toxic metals from the environment using plants. Biotechnology 1995;13 468–474.

[28] D'Souza R, Varun M, Pratas J, Paul MS. Paul Spatial distribution of heavy metals in soil and flora associated with the glass industry in North Central India: Implications for Phytoremediation. Soil and Sediment Contamination 2013;22 1–20.

[29] Varun M, D'Souza R, Pratas J, Paul MS. Metal contamination of soils and plants associated with the glass industry in North Central India: Prospects of phytoremediation. Environment Science and Pollution Research 2012;9(1) 269-281.

[30] Pratas J, Favas PJC, D'Souza R, Varun M, Paul MS. Phytoremedial assessment of flora tolerant to heavy metals in the contaminated soils of an abandoned Pb mine in Central Portugal. Chemosphere 2013;90 2216–2225.

[31] Yang B, Shu W, Ye Z, Lan C, Wong M. Growth and Metal Accumulation in Vetiver and two Sebania species on Lead/Zinc Mine Tailings. Chemosphere 2003;52 1593–1600.

[32] Uraguchi S, Watanabe I, Yoshitomi A, Kiyono M, Kuno K. Characteristics of cadmium accumulation and tolerance in novel Cd-accumulating crops, *Avena strigosa* and *Crotalaria juncea*. Journal of Experimental Botany 2006; 57(12) 2955-2965.

[33] Varun M, D'Souza R, Kumar D, Paul MS. Bioassay as monitoring system for lead phytoremediation through *Crinum asiaticum* L. Environment Monitoring and Assessment 2011a; 178(1-4) 373–381.

[34] Varun M, D'Souza RJ, Pratas J, Paul MS. Evaluation of phytostabilization, a green technology to remove heavy metals from industrial sludge using *Typha latifolia* L. Biotechnology Bioinformatics and Bioengineering 2011b; 1(1) 137-145.

[35] D'Souza R, Varun M, Masih J, Paul MS. Identification of *Calotropis procera* L. as a potential phytoaccumulator of heavy metals from contaminated soils in Urban North Central India. Journal of Hazardous Material 2010;184 457–464.

[36] BIS. Drinking water specification. IS:10500:1991. New Delhi, India: Bureau of Indian Standard; 1991.

[37] WHO (World Health Organisation). Guidelines for drinking water quality. Vol. 1, recommendations, seconded. World health organization, Geneva; 1993.

[38] Parkpain P, Sreesai S, Delaune RD. Bioavailability of heavy metals in sewage sludge-amended Thai soils. Water, Air, and Soil Pollution 2000;122 163–182.

[39] USEPA (United States Environmental Protection Agency). Current drinking water standard. http://www.epa.gov/safewater/mcl.html. (accessed 2002).

[40] Azumi, D.S, Bichi, M.H. Industrial pollution and heavy metals profile of Challawa river in kano, Nigeria. Journal of Applied Sciences and Environmental Sanitation 2010;5(1) 23–29.

[41] Rodríguez-Barroso MR, Benhamou M, El Moumni B, El Hatimi I, García-Morales JL. Evaluation of metal contamination in sediments from north of Morocco: geochemical and statistical approaches. Environmental Monitoring And Assessment 2009;159(1-4) 169–181.

[42] Halamić J, Galović I, Šparica M. Heavy Metal (As, Cd, Cu, Hg, Pb and Zn) distribution in topsoil developed on alluvial sediments of the Drava and Sava Rivers in NW Croatia. Geologia Croatica 2003;56(2) 215–232.

[43] Amore RR. Concentration, speciation, provenance and pollution levels of heavy metals in the bottom sediments of the pasig river. M.Sc. Thesis. University of the Philippines, Diliman, Quezon; 2000.

[44] Middelkoop FI. Heavy metal pollution of the river Rhine and flood plains in The Netherlands, Netherlands. Journal of Geosciences 2000;79 411–428.

[45] Chen HM, Zheng CR. Evaluation of purification ability of suspended substance with log ks for heavy metal in water bodies. Acta pedologica 1995;32 86–92.

[46] Olivares-Rieumont S, de La Rosa D, Lima L, Graham DW, Alessandro KD, Borroto J, Martínez F, Sánchez J. Assessment of heavy metal levels in Almendares River sediments - Havana City, Cuba. Water Research 2005;39(16) 3945–3953.

[47] Muniz P, Danulat E, Yanicelli B, García-Alonso J, Medina G, Bícego MC. Assessment of contamination by heavy metals and petroleum hydrocarbons in sediments of Montevideo harbor, Uruguay. Environment International 2004; 29 1019–1028.

[48] Eysink GGJ, Padua HB, Bertoletti SAEP, Martins MC, Pereira DN, Roberto S. Heavy metals in the Vale do Ribeira and in Iguape-Cananéia CETESB. Ambiente 1988;2(1) 6–13. (in Portuguese)

[49] Siqueira GW. Study of heavy metals levels and other elements in surface sediments at the Santos Estuarine System and Shelf Coastal of the Amazon. PhD thesis. São Paulo University, Instituto Oceanogr áfico, Brazil; 2003.

[50] Milenkovic N, Damjanovic M, Ristic M. Study of Heavy Metal Pollution in Sediments from the Iron Gate (Danube River), Serbia and Montenegro. Polish Journal of Environmental Studies 2005;14(6) 781–787.

[51] Akhabuhaya J, Lodenius M. Metal pollution of river msimbazi, Tanzania. Environment International 1988;14 511–514.

[52] Mackey AP, Hodgkionson M, Nardella R. Nutrient levels and heavy metals in mangrove sediments from the Brisbane River, Australia. Marine Pollution Bulletin 1992;24(8) 418–420.

[53] Charkhabi AH, Sakizadeh M, Bayat R. Land use effects on heavy metal pollution of river sediments in Guilan, southwest of the Caspian sea. Caspian Journal of Environmental Sciences 2008;6(2) 133–140.

[54] Kucuksezgin F, Uluturhan E, Batki H. Distribution of heavy metals in water, particulate matter and sediments of Gediz River (Eastern Aegean). Environmental Monitoring and Assessment 2008;141 213–225.

[55] Turekian K, Wedepohl KH. Distribution of the elements in some major units of the earth's crust. Geological Society of American Bulletin 1961;72 175–192.

[56] Meybeck M, Helmer R. The quality of rivers: From pristine stage to global pollution. Paleography, Paleoclimatology, and Paleoecology 1989;75 283–309.

[57] Schiller AM, Boyle EA. Variability of dissolved trace metals in the Missisipi river. Geochimistry and Cosmochim. Acta 1987;51 3273–327.

[58] CCME (Canadian environmental quality guidelines). The Canadian Council of Ministers of the Environment; 2003.

[59] VROM (Dutch Ministry of Housing, Spatial Planning and the Environment) Ministerial circular on target and intervention values for soil remediation. Annex A: target values, soil remediation intervention values and indicative levels for serious contamination, DBO/1999226863, 2004; February 4.

[60] SEPAC (State Environmental Protection Administration of China). Chinese Environmental Quality Standard for Soils; 1995; (GB15618-1995).

[61] Skowronski GA, Seide M, Abdel-Rahman MS. Oral bioaccessibility of trivalent and hexavalent chromium in soil by simulated gastric fluid. Journal of Toxicology and Environmental Health-Part A 2001;63(5) 351–362.

[62] Onianwa PC. Monitoring atmospheric metal pollution: a review of the use of mosses as indicators. Environmental Monitoring and Assessment 2001;71 13–5.

[63] Zhang M, He Z, Calvert DV, Stoffella PJ, Jiang X. Surface runoff losses of copper and zinc in sandy soils. Journal of Environmental Quality 2003;32 909–915.

[64] Adriano CD. Biogeochemistry of Trace Metals. United States: Lewis publishers; 1992.

[65] Modis K, Komnitsas K. Optimum sampling density for the prediction of acid mine drainage in an underground sulphide mine. Mine Water Environment 2007;26 237–242.

[66] Loska K, Wiechula D, Barska B, Cebula E, Chojnecka A. Assessment of arsenic enrichment of cultivated soils in Southern Poland. Polish Journal of Environmental Studies 2003;12(2) 187–192.

[67] Maina HM, Barminas JT, Nkafamiya II. Levels and distribution of some heavy metals in soils in the vicinity of ashaka cement factory, gombe state Nigeria. Journal of Chemical Society of Nigeria 2009;34 1–5.

[68] Lone MI, He ZH, Peter J, Stoffella I, Yang X. Phytoremediation of heavy metal polluted soils and water: Progresses and perspectives. Journal of Zhejiang University Science B 2008;9(3) 210–220.

[69] Almå AR, Kweyunga C, Manoko LK. Investigation of trace metal concentrations in soil, sediments and waters in the vicinity of "Geita Gold Mine" and "North Mara Gold Mine" in North West Tanzania (with Barrick Response). IPM-report; 2009.

[70] Nijkamp P, Rodenburg AC, Wagtendonk JA. Success factors for sustainable urban brownfield development: A comparative study approach to polluted sites. Ecology Economics 2002;40 232–252.

[71] Sánchez-Martin MJ, Garcia-Delgado M, Lorenzo LF, Rodriguez-Cruz MS, Arienzo M. Heavy metals in sewage sludge amended soils determined by sequential extractions as a function of incubation time of soils. Geoderma 2007; 142 262–273.

[72] Manta DS, Angelone M, Bellanca A, Neri R, Sprovieri M. Heavy metals in urban soils: a case study from the city of Palermo (Sicily), Italy. Science of the Total Environment 2002;300 229–243.

[73] Paterson E, Sanka M, Clark L. Urban soils as pollutant sinks-a case study from Aberdeen, Scotland. Applied Geochemistry 1996;11 129–131.

[74] Chronopoulos J, Haidouti C, Chronopoulou-Sereli A, Massas I. Variations in plant and soil lead and cadmium content in urban parks in Athens. Science of the Total Environment 1997;196 91–98.

[75] Thornton I. Metal contamination of soil in urban areas. In: Bullock P, Gregory PJ (ed) Soils in the Urban Environment. Blackwell. London: Oxford; 1991. p47–75.

[76] De Miguel E, Jiménez de Grado M, Llamas JF, Martín-Dorado A, Mazadiego LF. The overlooked contribution of compost application to the trace element load in the urban soil of Madrid (Spain). Science of the Total Environment 1998;215 113–122.

[77] Kahle P. Schwermetallstatus Rostocker Gartenböden. Journal of Plant Nutrition and Soil Science 2000;163 19–196.

[78] Li XD, Lee SL, Wong SC, Shi WZ, Thornton I. The study of metal contamination in urban soils of Hong Kong using a GIS-based approach. Environment Pollution 2004;129(1) 113–124.

[79] Lu Y, Gong ZT, Zhang GL, Burghardt W. Concentrations and chemical speciation of Cu, Zn, Pb and Cr of urban soils in Nanjing, China. Geoderma 2003;115 101–111.

[80] Sekhar KC, Chary NS, Kamala CT, Vairamani M, Anjaneyulu Y, Balaram V, Sorlie JE. Risk communications: around the world environmental risk assessment studies of heavy metal contamination in the industrial area of Kattedan, India—A case study. Human and Ecological Risk Assessment 2006;12 408–422.

[81] Prashanthi V, Rao J, Sreenivasa K, Raju A. Soil pollution due to a land disposal of industrial effluents. Journal of Industrial Pollution 2001;17 9–18.

[82] Huang R. Environmental Pedology. Beijing: Higher Education Press; 1987.

[83] Romkens PFAM, Salomons W. Cd, Cu, and Zn solubility in arable and forest soils: consequences of land use changes for metal mobility and risk assessment. Soil Science 1998;163(11) 859–871.

[84] Viard B, Pihan F, Promeyrat S, Pihan JC. Integrated assessment of heavy metal (Pb, Zn, Cd) highway pollution: bioaccumulation in soil, Graminaceae and land snails. Chemosphere 2004;55 1349–1359.

[85] Baker AJM, Brooks RR. Terrestrial higher plants which hyperaccumulate metallic elements- a review of their distribution, ecology and phytochemistry. Biorecovery 1989;1 81–126.

[86] Baker AJM, McGrath SP, Reeves RD, Smith JAC. Metal hyperaccumulator plants: a review of the ecology and physiology of a biological resource for phytoremediation of metal-polluted soils. In: Terry N, Bañuelos G. (eds.) Phytoremediation of Contaminated Soil and Water. Boca Raton, FL, USA:CRC Press; 2000. p85–108.

[87] Lasat MM. Phytoextraction of toxic metals: A review of biological mechanisms. Journal of Environmental Quality 2002; 31 109–120.

[88] Saxena PK, Raj SK, Dan T, Perras MR, Vettakkorumakankav NN. Phytoremediation of metal contaminated and polluted soils. In: Prasad MNV, Hagemeyer J. (eds.) Heavy metal stress in plants – From molecules to ecosystems. Heidelberg, Berlin, New York: Springer–Verlag; 1999. p305–329.

[89] Porębska G, Ostrowska A. Heavy metal accumulation in wild plants: implications for phytoremediation. Polish Journal of Environmental Studies 1999;8(6) 433–442.

[90] Saraswat S, Rai JPN. Phytoextraction potential of six plant species grown in multimetal contaminated soil. Chemical ecology 2009;25 1–11.

[91] Pulford ID, Watson C. Phytoremediation of heavy metal contaminated land by trees —a review. Environment International 2003;29(4) 529–540.

[92] Chiu KK, Ye ZH, Wong MH. Growth of Vetiveria zizanioides and Phragmities australis on Pb/Zn and Cu mine tailings amended with manure compost and sewage sludge: a greenhouse study. Bioresource Technology 2006; 97 158–170.

[93] Morgenthal TL, Cilliers SS, Kellner K, Hamburg HV, Michael MD. The vegetation of ash disposal sites at Hendrina Power Station II: Floristic composition. South African Journal of Botany 2001;67(4) 520-532.

[94] Oosting HJ. The study of Plant Communities. San Francisco: W.H. Freeman and Co; 1958.

[95] Phillips EA. Methods of vegetation study. New York: Henry Holt & Co. Inc.; 1959.

[96] Hanson HC, Churchill ED. The plant community. New York: Reinhold Publishing Corporation; 1961.

[97] Curtis JT, McIntosh RP. An upland forest continuum in the prairie forest border region of Wisconsin. Ecology 1951; 32 476-496.

Chapter 20

Continuous and Induced Phytoextraction — Plant-Based Methods to Remove Heavy Metals from Contaminated Soil

Maciej Bosiacki, Tomasz Kleiber and
Bartosz Markiewicz

Additional information is available at the end of the chapter

http://dx.doi.org/10.5772/57257

1. Introduction

As a result of the continuous civilization progress and the increasing human population we have been observing an accelerating process of environmental pollution, frequently leading to its complete degradation. The primary cause of environmental pollution is associated with the rapid development of motorization and industry (particularly power industry and mining), progressing urbanization, improved standards of living worldwide, intensive farming (application of high amounts of mineral fertilizers and herbicides), along with numerous other anthropogenic factors. These sources contribute to increased concentrations of many chemical elements and compounds in the atmosphere, soil, water and plants (including crops with edible parts for human consumption).

The cycle of chemical elements and compounds in nature is influenced not only by human activity, but also nature itself, in which progressing geological processes occur such as volcanic eruptions, shifts of tectonic plates or natural disasters.

In recent years the ecological awareness of the general public has increased, with decontamination of the polluted environment being perceived as an essential requirement. The aim of reclamation of polluted areas is to restore ecosystems polluted by human activity to the condition resembling their former natural state.

There are many methods applied in environment decontamination worldwide, including const-intensive, conventional physico-chemical methods. Scientists continue to search for novel, more effective and economically viable methods of pollutant inactivation. In recent years an increasing body of research has focused on engineering bioremediation, such as e.g. an *in situ* technology of phytoremediation, utilizing the capacity of plants to accumulate high amounts of pollutants from the environment. Research conducted for many years now has identified a relatively numerous group of plants potentially applicable in this new environment decontamination technology.

At present a major ecological problem is connected with the penetration to the environment of heavy metals, which at higher concentrations are strongly toxic to humans and animals (resulting in e.g. increased disease incidence), and have a negative effect on soil properties as well as quality and physiological activity of plants.

Literature sources present various definitions of an element to be considered a heavy metal. One of these hypotheses says that they are elements with specific gravity greater than 4.5, 5, 6 and 7 g cm^{-3}. It is a physical term, which is understood and used differently in various contexts. There are also many definitions based on atomic number. Some of them are considered to be nutrients for living organisms (essential nutrients e.g. Fe, Cu, Zn, Mn, Ni) and others are redundant or toxic (e.g. Cd, Pb, Hg, Al, As). Their common characteristic is connected with the fact that at excessive concentrations in the environment they have an adverse effect on plant growth and development, and when incorporated in the food chain may also pose a hazard for animals and humans.

2. Soil factors influencing availability of heavy metals to plants

Physico-chemical properties of soil influencing contents of phytotoxic forms of heavy metals include the type of soil, its grain size distribution, reaction, organic matter content, sorption properties and redox potential [43, 67, 68, 73].

Mother rock is a natural source of heavy metals in soils. The amounts of elements coming from mother rock constitute the geochemical background posing no threat to soil fertility [67]. Other sources of heavy metals include geochemical processes and anthropogenic factors. In soil heavy metals are subjected to processes affecting their concentrations and chemical forms [4]. In individual soil horizons their content depends on anthropogenic and climatic factors [67, 114].

Soil reaction is a major factor influencing the form and availability of heavy metals to plants [67]. Soil acidity may lead to their increased concentrations in soil [3, 13, 30, 48, 93, 112] and their excessive uptake by plants [42, 93]. According to Tyler and Olsson [118], concentrations of Cu and Pb increase also at pH 7.5 – 7.8 as a result of formation of stable complexes with ligands, which solubility is connected with solubility of organic substance. Liming results in a reduced content of available forms of heavy metals in soil [39]. Soil reaction is a factor determining the force with which heavy metals are bound by organic substance and mineral compounds [2, 60].

Organic substance found in soil contributes to the limitation of the amounts of heavy metals available to plants [38, 47], since it binds very strongly Cr^{3+}, Fe^{3+}, Pb^{2+}, Ni^{2+} and Co^{2+} ions and to a lesser extent also Mn^{2+} and Zn^{2+} [68]. Organic matter binds heavy metals into water insoluble forms or forms sparingly soluble in water [79], thus reducing the share of plant-available forms of heavy metals [19], and in this way it limits their toxicity to plants [48, 87].

Heavy metals differ in the force, with which they are bound by the sorption complex and they are connected mainly with the silt fraction [67]. In soil they undergo exchange and biological sorption. They may also be precipitated in the form of insoluble compounds [83]. Availability of heavy metals in soil is influenced by the cation sorption capacity. Introduction of compounds enhancing sorption capacity to soil causes a reduction of amounts of available metal forms in soil [106].

Availability of heavy metals is also dependent on the redox reactions taking place in soil [83]. Oxide forms of heavy metals become readily available to plants at a low redox potential [33].

The application of mineral fertilizers changes amounts of phytoavailable forms of heavy metals in soil. The effect of used fertilizers on physico-chemical and biological properties of soil causes a reduction or an increase of amounts of phytotoxic forms in the soil environment [105]. When applying mineral fertilizers we introduce heavy metals to soil, which contents in fertilizers are connected both with the raw material from which they were produced and the technological process of fertilizer production. The greatest contamination with heavy metals, particularly cadmium, is found in phosphorus fertilizers [52] and calcium fertilizers, mainly being by-products of various branches of industry [27, 69]. Heavy metal contents in phosphorus fertilizers depend on the fertilizer type [52] and solubility [57]. The application of phosphorus fertilizers leads to the transition of soluble phosphate forms into sparingly soluble zinc, copper, cadmium and lead phosphates [30], leading to the reduction of heavy metal contents in plants.

3. Admissible heavy metal contents in soil binding in Poland

In Poland respective boundary heavy metal contents are specified in the Ordinance of the Minister of the Environment of 9 September 2002 on soil quality standards and land quality standards. These standards were established taking into consideration the current and forecasted functions for the following categories of land types (Table 1):

1. Category A:

   a. Landed property incorporated into an area legally protected on the power of the regulations of the Water Act,

   b. Areas protected on the power of regulations on nature protection if maintenance of the current soil pollution levels does not pose a threat for human health or the environment – for these areas the concentrations meet the standards resulting from the actual status, subject to points 2 and 3;

2. Category B – land classified as agriculturally utilized area except for land covered by ponds and ditches, forested areas as well as areas covered by trees and shrubs, barren

land, as well as developed and urbanized areas except for industrial areas, surface mining land in use and municipal areas;

3. Category C – industrial areas, surface mining land in use, transportation areas.

Attachment to the Ordinance of the Minister of the Environment of 9 September 2002 (item 1359)

Pollutant	Category A	Category B					Category C		
		Depth [m below ground level]							
		0-0.3	0.3-15.0		>15		0-2	2-15	
		Water permeability of soil [m/s]							
			Up to	below	Up to	below		Up to	below
			$1 \cdot 10^{-7}$		$1 \cdot 10^{-7}$			$1 \cdot 10^{-7}$	
Arsenic	20	20	20	25	25	55	60	25	100
Barium	200	200	250	320	300	650	1000	300	3000
Chromium	50	150	150	190	150	380	500	150	800
Tin	20	20	30	50	40	300	350	40	300
Zinc	100	300	350	300	300	720	1000	300	3000
Cadmium	1	4	5	6	4	10	15	6	20
Cobalt	20	20	30	60	50	120	200	50	300
Copper	30	150	100	100	100	200	600	200	1000
Molibdenum	10	10	10	40	30	210	250	30	200
Nickel	35	100	50	100	70	210	300	70	500
Lead	50	100	100	200	100	200	600	200	1000
Mercury	0,5	2	3	5	4	10	30	4	50

Table 1. Admissible concentrations in soil (mg kg^{-1} dry matter)

Based on the multiannual analyses the Institute of Soil Science and Plant Cultivation in Puławy, Poland (IUNG) specified boundary heavy metal contents in soils to be utilized agriculturally (Table 2). As it was reported by Ociepa et al. [94], the mean content of heavy metals in agriculturally utilized soils in Poland is lower than in countries of Western Europe or the USA, which results mainly from a lesser share of farms with intensive agricultural production systems and a lesser intensity of industrial processes. The same authors reported that approx. 90% agriculturally utilized area have natural levels of toxic metals. Several percent of this area have elevated contents (I^0) of mainly cadmium and zinc, while approx. 3% are polluted (II-V^0) with metals. The greatest percentages of polluted soils are found in the Śląskie, MałoPolande and Wrocławskie provinces of Poland.

http://dx.doi.org/10.5772/57257

Metal	Soil class*	Soil pollution rates					
		0	I	II	III	IV	V
lead (Pb)	a	30	70	100	500	2500	>2500
	b	50	100	250	1000	5000	>5000
	c	70	200	500	2000	7000	>7000
zinc (Zn)	a	50	100	300	700	3000	>3000
	b	70	200	500	1500	5000	>5000
	c	100	300	1000	3000	8000	>8000
copper (Cu)	a	15	30	50	150	750	>750
	b	25	50	80	100	500	>500
	c	40	70	100	150	750	>750
nickel (Ni)	a	10	30	50	100	400	>400
	b	25	50	75	150	600	>600
	c	50	75	100	300	1000	>1000
cadmium (Cd)	a	0, 3	1, 0	2	3	5	>5
	b	0, 5	1, 5	3	5	10	>10
	c	1, 0	3, 0	5	10	20	>20

Pollution rate: 0 – natural content, I – elevated content, II – weak pollution, III – medium pollution, IV – strong pollution, V – very strong pollution

*soil classes:

a. Very light soils with low silt and clay contents (< 10%), irrespective of pH (in KCl)

- light soils (10 - 20% silt and clay), highly acid (pH < 4.5) and acid (pH 4.5 – 5.5)

b. Light soils (10 - 20% silt and clay), neutral reaction (pH > 6.5)

- medium soils (20 - 35% silt and clay), highly acid (pH < 4.5) and acid (pH 4.5 – 5.5)

c. heavy soils (>35% silt and clay), highly acid (pH < 4.5)

- mineral organic soils (organic substance at 6 - 10%)

d. Medium heavy soils (20-30% silt and clay) and heavy soils (> 35%).

Table 2. Boundary values in mg kg^{-1} heavy metals in the surface layer (0 – 20 cm) of soils with different degrees of contamination (IUNG 1992)

4. Contents of heavy metal soluble forms and reaction of surface horizon (0 - 20 cm) of soils in green belts adjacent to selected transportation routes in the city of Poznań (western Poland)

Anthropopressure affects physical, biological and chemical properties of soil. Soils in urban areas, located along transportation routes, are exposed to heavy metal pollution, originating from substances produced during combustion of fuels, abrasion of road surfaces and tires, granular materials falling onto the ground during transport, etc. [24, 36]. Platinum metals, which have been employed in production of car catalysts, reach environment and cause contamination of soil, plants and water [84]. Moreover, chemical substances used in winter to eliminate black ice (e.g. sodium or calcium chlorides) as well as deposition of dusts and water migration of elements contribute to soil degradation and deterioration of plant growth conditions. In urban areas strong alkalization of soil is frequently observed, which significantly reduces contents of soluble forms of metallic components [12, 27, 74].

According to the Ordinance of the Minister of the Environment of 9 September 2002 on soil quality standards and land quality standards currently binding in Poland (the Journal of Law Dziennik Ustaw no. 165, item 1359), soil category B comprises soils in urbanized areas, for which the admissible heavy metal level (mg kg^{-1} dry matter) in the upper 0 - 30 cm layer is Ni 100, Cd 4, Pb 100 and Cr 150, respectively.

Legal regulations pertain to total heavy metal contents in soils; however, many authors claim that it is not always a direct indicator of their bioavailability [44]. In the opinion of Gorlach and Gambuś [53], the most appropriate measure is to assess soil contents of soluble forms of trace elements, as they may be absorbed by plants.

The aim of the studies conducted by the authors was to determine what amounts of soluble forms of heavy metals are available for plants in the 0 - 20 cm layer of soil in green areas located in the vicinity of selected transportation routes in the city of Poznań, Poland. Collected soil samples were tested for soil reaction and contents of selected heavy metals (cadmium, lead, chromium and nickel, classified as metallic micronutrients, essential elements at the same time having a negative effect on plants when found in greater amounts).

Material and methods

Soil samples from green areas were collected in October 2012 in the vicinity of transportation routes in the city of Poznań, Poland. Forty five streets were selected for analyses and soil samples were collected using an Egner sampling stick from a 0 - 20 cm layer at a distance of 0.5 - 2.0 m from the roadway. Along each analyzed street 4 bulk samples were collected, comprising 15 individual samples (4 x 15 = 60 individual samples). Heavy metals (Cd, Pb, Cr and Ni) were extracted from soil using Lindsay's solution containing in 1 dm^3: 5 g EDTA (ethylenediaminetetraacetic acid), 9 cm^3 25% NH_4OH solution, 4 g citric acid and 2 g $Ca(CH_3COO)_2 \cdot 2H_2O$. Next they were assayed by flame atomic absorption spectroscopy (FAAS) with an AAS 3 Zeiss apparatus. Active acidity expressed in pH (H_2O) was determined by potentiometry (soil : water = 1:2), [50].

Results were analyzed determining their minimum, maximum and mean values (in the case of pH its logarithmic value was considered), standard deviation, coefficients of variation and empirical distribution for individual chemical parameters. Mean results of chemical analyses are presented in Table 3, while stem-and-leaf displays are given in Figs. 1 - 2.

In these analyses soil pH fell within the range of 4.32 – 8.26, while its coefficient of variation was as low as 9% (Table 3). For most plants optimal soil reaction is pH_{H2O} 6.0 – 6.5. Only 2.2% soils had a highly acid reaction (at pH 4.3 – 5.0), whereas a vast majority had an alkaline reaction pH>7.4 (46.7% soil samples) and a neutral reaction (37.8%), comprising jointly 84.5% samples

http://dx.doi.org/10.5772/57257

(Figure 1). Elevated soil pH is connected with an excessive content of alkaline ions, i.e. calcium, magnesium, sodium and potassium. Alkalization of urban soils may result e.g. from the use of salts such as sodium chloride to remove black ice from roadways, as well as strongly alkalizing dusts (containing e.g. CaO, MgO, Na_2O, K_2O), being by-products of carbon combustion to heat houses [27, 74]. In turn, Kleiber [74] to a much greater degree suggests strong alkalization of soil (pH in H_2O up to 10.95) as a factor having a potentially negative effect on plant growth and development.

In the opinion of Klimowicz and Melke [75], in urban areas traffic-related pollution is more dangerous than industrial pollution, since it is spread in relatively large amounts, at low heights, in the respiration zone of humans, animals and plants. Relief features, distance from a roadway and intensity of vehicle traffic have a decisive effect on the contents of heavy metals in soils adjacent to transportation routes in urban green areas. In the soil found at the car market in Słomczyn near Warsaw (one of the biggest car markets in Poland) contents of zinc, lead and copper are as follows (in mg kg^{-1} d.m.): the car sale point Zn at 612.3, Pb at 397.8 and Ni at 94.5, while at the spare part sale point and warehouse they are: Zn at 679.1, Pb at 420.4 and Ni at 114.0, respectively [115].

Soil pollution with heavy metals is typically assessed on the basis of total contents of elements [44]. Those authors claimed that this assessment should be supplemented with an analysis of heavy metal contents directly available to living organisms.

This study consisted in the determination of contents of soluble heavy metal forms (Table 3). Cadmium content in soils fell within the range of 0.16 – 0.42 mg Cd dm^{-3} and it was of relatively limited variability (CV=17.50%). As many as 55.1 % samples contained this heavy metal at <0.21 - 0.26 mg Cd dm^{-3} (Figure 2). Bach [9] in soils of green areas adjacent to transportation routes in the city of Krakow, Poland found contents of soluble cadmium forms to range from 0.21 to 1.54 mg kg^{-1} d.m. soil. In most tested soils Bach [9] recorded cadmium content from 0.4 to 0.8 mg kg^{-1} d.m. soil. According to Kabata-Pendias and Pendias [67], natural total cadmium content (the so-called background) in Polish soils is 0.3 mg Cd kg^{-1} d.m.

Lead content ranged from 0.79 to 42.96 mg Pb dm^{-3}, at the same time being highly variable depending on the location (CV=110.78%). As many as 84.4% tested soils had low contents of this heavy metal (up to 9.22 mg Pb dm^{-3}), while samples with 42.96 mg Pb dm^{-3} accounted for 2.2.% (Figure 2). A low lead content (max. up to 4.3 mg Pb dm^{-3}) was reported in his study by Kleiber [74]. Lead content (soluble forms) in soils adjacent to transportation routes in the city of Krakow ranged from 11.1 up to 142.8 mg kg^{-1} d.m. soil, of which the highest proportion (53%) comprised soils with its contents from 40 to 80 mg Pb kg^{-1} d.m. soil. Mean content of bioavailable Pb forms in soils of allotment gardens located in the right-bank Warsaw was 332.7 mg kg^{-1} d.m. [41].

The primary source of lead pollution in soils (adjacent to transportation routes) up to 2005 was connected with tetraethyl lead, commonly added to gasoline as an antiknock agent [62]. In turn, the authors of this chapter in their analyses recorded a low content of soluble lead content in the surface soil layer in areas adjacent to transportation routes in the city of Poznań, Poland.

No lead pollution in soils adjacent to exit routes leading from Poznań was detected by Hofman and Wachowski [62].

In turn, chromium content ranged from 0.26 to 0.67 mg Cr dm^{-3}. The coefficient of variation for chromium was similar to that calculated for nickel, amounting to CV 22.34%. The greatest proportion (44.4%) of soils was found within the range of chromium contents from < 0.36 to 0.46 mg Cr dm^{-3} (Figure 2). Results of analyses conducted by Bach [9] indicate that the content of chromium (soluble forms) in soils in areas adjacent to transportation routes in the city of Krakow ranged from trace values to 10.23 mg Cr kg^{-1} d.m.

The recorded nickel content ranged from 0.43 to 1.25 mg Ni dm^{-3}. The variability of levels for this metal was relatively medium-ranged (CV=23.19%). Most of soil samples (84.4%) were characterized by nickel contents falling within the range of values from < 0.43 to 0.63 mg Ni dm^{-3} (Figure 2). In soils found in green areas adjacent to transportation routes in Krakow, Poland the content of nickel soluble forms ranged from 1.07 to 6.38 mg kg^{-1} d.m. soil [9]. In contrast, in soils collected from allotment gardens located in right-bank Warsaw the mean content of bioavailable nickel forms was 28.4 mg kg^{-1} d.m. [41].

Environmental pollution with heavy metals constitutes a serious problem in some regions of Poland. According to Dmochowski et al. [41], high emissions of heavy metals originating from a dense network of transportation routes with high intensity vehicle traffic causes their accumulation in soils and crops produced in allotment gardens located in Praga Południe, the right-bank district of Warsaw, Poland.

In the opinion of Heck et al. [59], introduction of advanced catalytic systems, containing platinum, rhodium and palladium and constituting a source of environmental pollution also for soils in areas adjacent to transportation routes makes their monitoring a necessary practice. In the nearest future it will be required to create an effective system of environmental pollution monitoring.

No.	Street	pH (H_2O)	Contents (mg kg^{-1} d.m.)			
			Ni	Cd	Pb	Cr
1	28 czerwca 56 r.	7.67	0, 49	0, 23	14, 81	0, 31
2	Aleje Niepodległości	7.58	0, 51	0, 28	10, 74	0, 33
3	Aleje Solidarności	7.37	0, 58	0, 27	3, 36	0, 49
4	Arciszewskiego	7.30	0, 48	0, 20	6, 55	0, 41
5	Armii Poznań	7.72	0, 59	0, 21	42, 96	0, 55
6	Biskupińska	7.76	0, 52	0, 23	2, 18	0, 46
7	Dolna Wilda	8.26	0, 44	0, 18	4, 00	0, 38
8	Droga Dębińska	7.37	0, 48	0, 23	6, 76	0, 34

http://dx.doi.org/10.5772/57257

No.	Street	pH (H_2O)	Contents (mg kg^{-1} d.m.)			
			Ni	Cd	Pb	Cr
9	Fredry	7.30	0, 62	0, 33	6, 83	0, 60
10	Głogowska	6.86	0, 62	0, 31	3, 56	0, 57
11	Hercena	7.00	0, 58	0, 21	1, 91	0, 46
12	Hetmańska	7.45	0, 43	0, 19	3, 92	0, 40
13	Jarochowskiego	7.60	0, 55	0, 24	3, 24	0, 46
14	Kopanina	6.64	0, 73	0, 26	7, 45	0, 39
15	Królowej Jadwigi	6.55	0, 86	0, 22	8, 70	0, 35
16	Krzywoustego	6.98	0, 58	0, 28	4, 31	0, 57
17	Księcia Mieszka I	7.17	0, 50	0, 23	18, 89	0, 39
18	Kuźnicza	6.29	0, 70	0, 28	6, 17	0, 43
19	Lechicka	8.18	0, 55	0, 25	2, 89	0, 52
20	Leszczyńska	7.63	0, 57	0, 28	3, 46	0, 67
21	Lodowa	7.72	0, 51	0, 21	3, 10	0, 48
22	Maczka	7.48	0, 60	0, 30	6, 07	0, 56
23	Małopolska	7.47	0, 62	0, 29	2, 97	0, 64
24	Marcelińska	7.89	0, 55	0, 26	2, 78	0, 59
25	Niestachowska	5.84	0, 71	0, 25	10, 97	0, 38
26	Nowowiejskiego	7.14	0, 49	0, 19	3, 02	0, 36
27	Opieńskiego	7.32	0, 47	0, 22	0, 79	0, 39
28	Opolska	7.76	0, 62	0, 28	3, 65	0, 64
29	Ożarowska	7.50	0, 51	0, 24	3, 76	0, 41
30	Piątkowska	7.40	0, 50	0, 26	2, 64	0, 46
31	Piłsudskiego	7.50	0, 60	0, 26	2, 10	0, 57
32	Polanka	6.45	0, 60	0, 24	5, 73	0, 41
33	Poznańska	6.99	0, 44	0, 16	2, 49	0, 26
34	Przemysłowa	7.36	0, 60	0, 28	2, 54	0, 60
35	Serbska	7.79	0, 60	0, 24	2, 34	0, 63

No.	Street		pH (H_2O)	Contents (mg kg^{-1} d.m.)			
				Ni	Cd	Pb	Cr
36	Słowiańska		6.87	0, 63	0, 23	5, 31	0, 29
37	Stablewskiego		7.55	0, 57	0, 23	3, 28	0, 46
38	Starołęcka		7.12	0, 63	0, 29	6, 01	0, 53
39	Stróżyńskiego		7.53	0, 49	0, 23	2, 71	0, 46
40	Szpitalna		4.32	0, 82	0, 23	11, 84	0, 50
41	Umultowska		6.79	1, 25	0, 42	3, 03	0, 65
42	Widłakowa		7.41	0, 50	0, 22	2, 07	0, 45
43	Witosa		7.09	0, 48	0, 21	6, 07	0, 43
44	Wojska Polskiego		6.53	0, 69	0, 26	9, 48	0, 39
45	Zwierzyniecka		7.64	0, 52	0, 23	2, 49	0, 40
Content		mean	7.22	0.59	0.25	6.00	0, 47
		minimum	4.32	0.43	0.16	0.79	0, 26
		maximum	8.26	1.25	0.42	42.96	0, 67
Standard deviation			0, 65	0.14	0.04	6.64	0.10
Coefficient of variation (%)			9, 0	23.19	17.50	110.78	22.32

Table 3. Contents of soluble forms of heavy metals and soil reaction in the surface soil layer (0 - 20 cm) in green areas adjacent to selected transportation routes in the city of Poznań, Poland

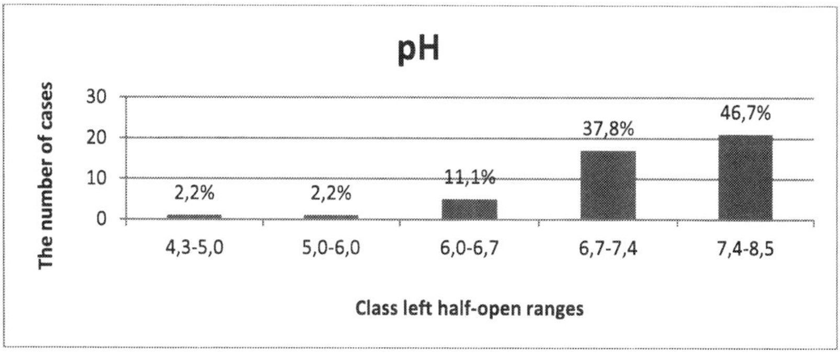

Figure 1. The stemplot for pH (H2O) of soil in green areas located near to routes in Poznan

http://dx.doi.org/10.5772/57257

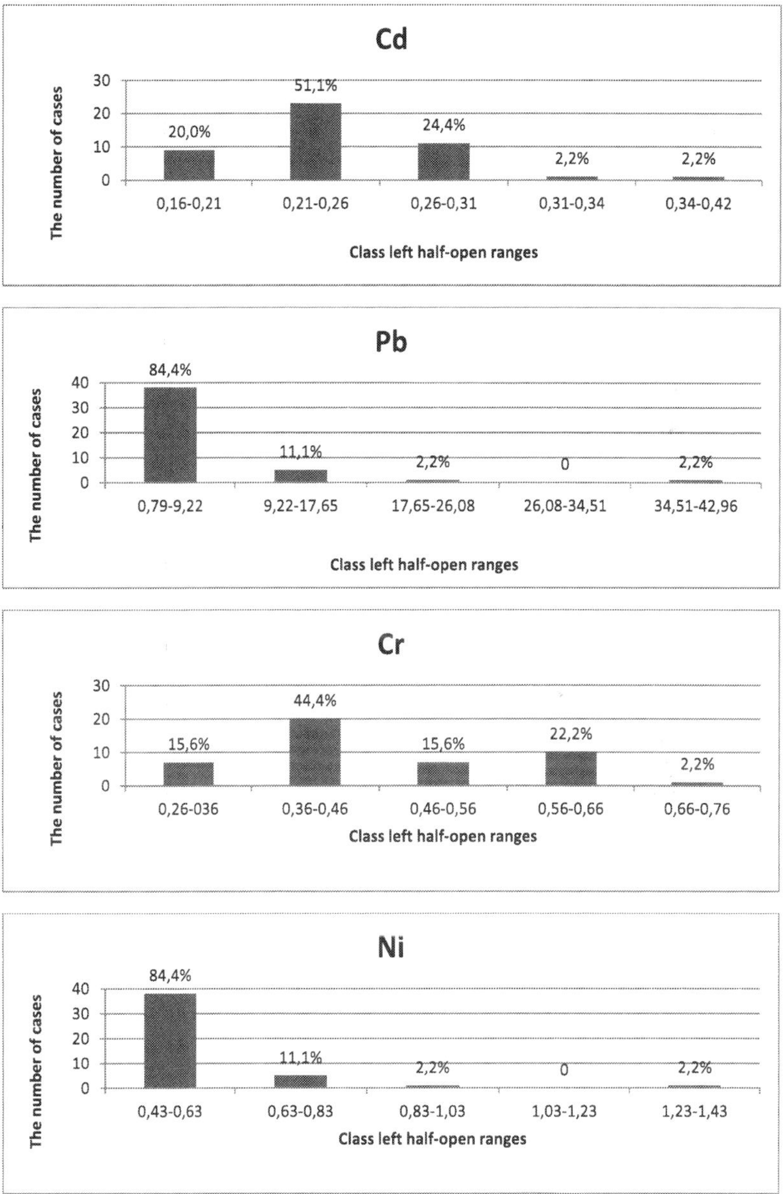

Figure 2. The stemplot for content of heavy metals in soil of green areas located near to routes in Poznan

4.1. Concluding remarks

Based on analyses of soils collected from green areas located in the vicinity of selected transportation routes in the city of Poznań, Poland in most examined locations an alkaline reaction pH >7.4 (46.7% soil samples) and a neutral reaction (37.8%) were found, which may significantly affect habitat conditions for plants.

Moreover, most tested soil samples contained low amounts of soluble forms of cadmium, lead, chromium and nickel.

5. Heavy metal contents in leaves of selected ornamental plant species growing in green areas adjacent to selected transportation routes in the city of Poznań (western Poland)

Effects of soil pollution with heavy metals may be identified by analyzing their content in plants. Plants may be considered good bioindicators of environmental pollution with heavy metals. Most frequently healthy leaves are indicator parts of plants; however, it depends on the species and occasionally even on the cultivar.

The aim of the analyses conducted by the authors in September 2012 was to determine contents of selected heavy metals in leaves of trees growing in the vicinity of transportation routes in Poznań (western Poland).

Material and methods

It was decided to conduct analyses on healthy plants with no symptoms of damage. Leaves of trees were collected from the central parts of long shoots, distributed at various sides around the circumference of the crown. A total of 15 - 20 leaves were collected from each tree. Leaves were collected from 10 trees of a given genus (1 bulk sample comprised approx. 150 - 200 leaves). The collected material was dried at a temperature of 45 - 50°C and then homogenized. In order to determine total forms of heavy metals they were mineralized in a mixture of nitric and perchloric acids (v:v=3:1; Σ 30 cm^3). Following mineralization Cd, Cr, Pb and Ni contents were determined by atomic absorption spectrometry (AAS) in a Carl Zeiss Jena apparatus.

High variation was observed in the contents of heavy metals in leaves of tested trees (*Acer sp., Betula sp., Robinia sp., Salix sp., Tilia sp.*), (Table 4). Mean content of cadmium amounted to 1.18 - 1.74 mg Cd kg^{-1} (for *Tilia sp.* and *Betula sp.*), for chromium it was 0.46 - 0.82 mg Cr kg^{-1} for *Tilia sp.* and *Acer sp.* (except for *Betula sp.* – trace content), that of lead ranged from 1.55 to 3.28 mg Pb kg^{-1} (for *Betula sp.* and *Salix sp.*, respectively). Nickel content ranged from 1.63 mg Ni kg^{-1} (for *Tilia* sp.) up to 2.23 mg Ni•kg^{-1} (for *Betula sp.)*

Table (5) present contents of heavy metals in leaves of selected genera of trees reported by Tomašević et al. [117], Lawal et al. [81] and Knezevic et al.[76]. In this study cadmium and nickel contents in leaves did not exceed the levels detected by those authors. In contrast, they found markedly lower chromium contents in leaves of *Tilia* sp., *Aesculus hippocastanum* and

Azadirachta indica. In this study the content of lead in leaves was markedly lower than that recorded by those researchers for *Tilia* sp. and *Aesculus hippocastanum,* while it was comparable to that in *Azadirachta indica* and *Paulownia elongata* (Tables 4 and 5).

Sample	Content	Heavy metals			
		Cd	Cr	Pb	Ni
Acer sp.	Minimum	0.25	Tr.	0.83	0.80
	Maximum	1.85	1.79	3.19	2.40
	Average	1.31	0.82	2.10	1.81
Betula sp.	Minimum	1.16	Tr.	0.55	1.90
	Maximum	2.87	Tr.	3.49	2.40
	Average	1.74	Tr.	1.55	2.23
Robinia sp.	Minimum	0.47	Tr.	1.06	0.50
	Maximum	2.54	2.13	5.73	2.90
	Average	1.37	0.48	2.78	2.07
Salix sp.	Minimum	1.12	Tr.	2.43	1.60
	Maximum	1.71	1.69	4.61	2.50
	Average	1.39	0.48	3.28	2.10
Tilia sp.	Minimum	0.29	Tr.	2.10	0.80
	Maximum	2.09	1.16	4.28	2.50
	Average	1.18	0.46	3.19	1.63

Table 4. Contents of heavy metals (mg kg^{-1} d.m.) in leaves of selected genera of trees growing in Poznań (tr. – trace amounts)

Sample	Content	Heavy metals			
		Cd	Cr	Pb	Ni
Tilia sp.[1]	Min	dl	dl	1.88	-
	Max	1.4	0.02	11.4	-
Aesculus hippocastanum 1	Min	0.4	dl	5.35	-
	Max	4.9	0.33	20.3	-
Azadirachta indica[2]	Min	-	0.17	0.77	1.14
	Max	-	0.37	2.25	1.92
Paulownia elongata S.Y. Hu[3]	Min	-	-	0.94	3.23
	Max	-	-	3.16	6.63

dl detection limit

[1] Tomašević et al. 2004; [2] Lawal et al. 2011, [3] Knezevic et al. 2009

Table 5. Contents of heavy metals (mg kg^{-1} d.m.) in leaves of selected genera of trees

6. The principle of phytoremediation

Among soil purification methods biological methods are increasingly often focused on as particularly promising [1, 99, 108]. One of these is phytoremediation, based on the activity of living organisms [40]. It is an alternative method, competitive in relation to other technologies extensively applied in pollutant removal from soil. There are methods facilitating deactivation or removal of toxic substances from the substrate. In most cases they are based on methods of physico-chemical extraction, but their application is connected with excessive costs and complete elimination of soil microorganisms. Reconstruction of semi-natural ecosystems in such cases is a lengthy process.

Obvious advantages of this biological method include its applicability at the contamination site, as well as relatively low investment outlays and low operating costs at the simultaneous high effectiveness of the process [95]. In the opinion of Salt et al. [107], other factors promoting its more common use are connected with the fact that it is an environmentally friendly process, which does not disturb soil structure, and that it may use many plant species.

The term phytoremediation originates from Greek *phyton* – a plant and Latin *remediare* – to purify. Salt et al. [108], Blaylock and Huang [14], Schnoor [109], Schwitzguebel et al. [111], McGrath and Zhao [86], Vassilev et al. [122], Vangronsveld et al. [121] and Larcher et al. [80] stated that phytoremediation is a technology using higher plants to stabilize and either remove or reduce amounts of soil pollutants, bottom deposits or surface and underground waters. Cunningham et al. [37] and Salt et al. [107] define a plant as a system of filters and pumps powered by solar energy and extracting from its environment and accumulating specific elements.

As it was reported by Pandolfini et al. [96], this biological method is based on the practical use of three types of physiological response to substances found in the environment, i.e. exclusion, accumulation and hyperaccumulation.

The term phytoremediation refers to the following methods using higher plants to purify environmental matrices [49, 98, 108, 119]:

- phytodegradation – the use of plants and microorganisms to degrade organic pollutants,
- phytostabilization – the use of plants to reduce bioavailability of pollutants in the environment,
- phytoextraction – the use of plants absorbing pollutants and accumulating them in organs removed from fields together with crops in order to purify soil from heavy metals and organic substances,
- phytovolatilization – the use of plants to volatilize pollutants and release them to the atmosphere,
- rhizofiltration – the use of plant roots to absorb pollutants from water and sewage,
- rhizodegradation – the use of plants to supplement the bioremediation process performed by microorganisms colonizing the rhizosphere.

In the opinion of Negri et al. [91], the above mentioned phytoremediation technologies act at three different detoxication levels on pollutants accumulated in the environment, i.e. the pollution-loaded soil matrix (phytostabilization, rhizofiltration and rhizodegradation), plants (phytodegradation, phytoextraction, rhizofiltration) and the atmosphere (phytovolatilization). Studies are also being conducted on the application of plants in the technology to excavate heavy metals, referred to as phytomining [5, 90, 97]. Soils with high heavy metal contents are planted with plants capable of growing under such adverse conditions and accumulating selected elements in their biomass. Such biomass is next treated as the so-called bio-ore. The biomining technlogy to extract heavy metals from soil, created by R. L. Chaney, J.S. Angle, A.J. Baker and J.M. Li, was patented in 1989 [45].

7. Continuous and induced phytoextraction

Phytoextraction, a phytoremediation method observed in 1885 by Bauman [11], is an excellent concept of soil purification, which is not yet commonly applied. In many centers worldwide research is being conducted on phytoextraction in the search for plant species capable of accumulating heavy metals in their aboveground parts. Within phytoextraction of heavy metals from soil we may distinguish the so-called continuous and induced phytoextraction.

Plants used in phytoextraction of heavy metals from soil should exhibit:

- good tolerance to high concentrations of heavy metals
- capacity to absorb and accumulate heavy metals in their aboveground parts
- rapid growth
- high increase in biomass
- resistance to diseases, pesticides and adverse atmospheric conditions
- low cultivation requirements
- easy harvesting and processing.

In continuous phytoextraction heavy metals are absorbed by plants and accumulated continuously with plant growth [14, 20, 46, 85, 110].

Apart from it being economically attractive, continuous phytoextraction is also environmentally friendly, leaving the site suitable for cultivation of other plants [37, 63].

In urbanized areas continuous phytoextraction may be used in two types of sites. One comprises degraded soils in post-industrial areas, while the other, highly promising as a future application of phytoextraction, is connected with soils in the vicinity of transportation routes and in urban green areas. The potential of ornamental plant species most frequently planted in urban locations is investigated in many research centers worldwide.

Such species include also *Tagetes erecta* L. It results from studies conducted by Bosiacki [16, 17] that it is a species exhibiting tolerance to high concentrations of cadmium and lead. When comparing several cultivars of this species the author found that cv. 'Hawaii' has the greatest contents of cadmium and lead in individual aboveground organs. The highest cadmium content (96.06 mg kg^{-1} d.m.) was found in leaves of *Tagetes erecta* L. cv. 'Hawaii', while that of lead (145.00 mg kg^{-1} d.m.) was recorded in stems. Accumulation of heavy metals in individual plant organs is not uniform. Bosiacki [15] observed variation in cadmium contents in individual organs of selected ornamental plant species. The highest content of this metal was detected in *Tagetes erecta* 'Inca Yellow' in its roots, in *Salvia splendens* 'Fuego' in leaves and stems, while in *Helianthus annuus* 'Pacino' it was in inflorescences. Liu et al. [82] stated that *Calendula officinalis* L. accumulated more Cd and Pb in roots than in shoots, while *Althaea rosea* L. collected greater amounts of these elements in shoots than in roots.

Zhou and Wang [124] when investigating the effect of cadmium on growth in three ornamental plant species stated that *Salvia splendens* L. is most sensitive to Cd, while *Abelmoschus manihot* Medik and *Tagetes erecta* L were most resistant.

Phytoremediation of heavy metals from contaminated areas, including urbanized areas, according to Porębska and Gworek [101] may be conducted using many species of ornamental plants and vegetables, e.g. *Salix viminalis* L. (accumuating Zn and Cd), *Alyssum bertolonii* Desv. (Ni), *Brassica pekinensis* Rupr.(Pb), *Thlaspi caerulescens* J. Presl & C. Presl. (Zn). The same authors stated that wild-growing plants, commonly considered weeds in cultivation of agricultural and horticultural crops, may also be used in bioremediation, e.g. *Atriplex nitens* Schkuhr, *Artemisia vulgaris* L., *Chenopodium album* L., (accumulating greatest amounts of Pb among the tested plants) and *Lacuca serriola* L. (accumulating greatest amounts of Zn).

Antonkiewicz and Jasiewicz [6] assessed suitability of *Helianthus tuberosus* L., *Zea mays* L., *Sida hermaphrodita* Rusby, *Amaranthus sp.* L. and *Cannabis sativa* L. in phytoextraction of heavy metals from soil. In their study the highest contents of cadmium and zinc were detected in amaranth, while that of lead and nickel in Virginia fanpetals. Jerusalem artichoke had the greatest content of copper.

Antonkiewicz et al. [7] when investigating phytoextraction of heavy metals (Cd, Pb, Ni, Zn and Cu) from soil using Virginia fanpetals (*Sida hermaphrodita* Rusby) found that this species absorbed the highest rates of nickel from soil and the lowest of copper.

Many authors for phytoextraction of soils polluted with heavy metals recommended sunflower, corn, rape, amaranth, willows, Miscanthus and strong growing cereals, while for phytoextraction in urban soils he recommended plants with high tolerance to pollution, e.g. London plane, northern red oak, Japanese larch, poplars, field maple, ashes and dogwoods, desert false indigo, false Spirea and forsythia.

Larcher et al. [80] conducted pilot-scale studies in the industrial area of Turin (Italy) using two plant species in phytoremediation of soil to remove heavy metals. They found *Helianthus annuus* L. "Holeko HO" and *Brassica juncea* Czern. "Red Giant" to be highly suitable for phytoremediation. Those authors stated that further studies in this respect are needed.

In urban green areas the predominant forms are lawns and turf-covered areas in escarpments, embankments, spoil tips, belts separating roadways, parking lots, gas stations, landfills and industrial waste dumps. Depending on the use of turf areas it is highly important to select appropriate species and cultivars of lawn grasses. It results from experiments conducted by Bosiacki and Zieleziński [23] on the potential of three grass species (*Poa pratensis* L. 'Evora', *Festuca arundinacea* Schleb. 'Asterix', *Festuca rubra* L. *sensu lato* 'Jasper') in phytoextraction of nickel that *Poa pratensis* L. 'Evora' and *Festuca arundinacea* Schleb. 'Asterix' are species showing the greatest capacity to accumulate this metal. In turn, *Festuca rubra* L. *sensu lato* 'Jasper' turned out to be the species showing the greatest capacity to accumulate cadmium and lead [21, 22].

Studies are also conducted on induced phytoextraction using plants producing large amounts of biomass, but additionally such substances as e.g. chelating carriers affecting mobility of individual elements and enhancing pollutant accumulation in plant organs (particularly aboveground parts) are introduced to the environment [35, 92, 123].

Chelating substances are to transform forms of heavy metals sparingly soluble or insoluble in water into forms available to plants. Studies are being conducted using different chelators, both natural and synthetic (e.g. humus substances, low molecular organic acids, citric acid, tartaric acid, amino acids as well as EDTA, EGTA, EDDS, HEDTA, CDTA, EDDHA, DTPA, NTA). Some of them are biodegraded in soil, while others in combination with a heavy metal may be leached to ground waters contaminating them. For this reason it is essential in this technology to determine an appropriate dose and date for the application of a given chelating agent.

It results from preliminary studies (unpublished data) conducted in 2012 by Bosiacki at the Department of Plant Nutrition, the Poznań University of Life Sciences, Poland that EDTA introduced to mineral soil contaminated individually with cadmium at 1.5 mg dm^{-3} in the form of cadmium sulfate ($3CdSO_4$ $8H_2O$), lead 100 mg dm^{-3} as lead acetate [$(CH_3COO)_2Pb$ $3H_2O$] and nickel 50 mg dm^{-3} as nickel sulfate ($NiSO_4$ $6H_2O$) caused an increase in the contents of these metals in leaves of *Tagetes erecta* 'Taishan Orange'. Doses of EDTA were introduced to soil in the solution form 35 days after plants had been planted to mineral soil contaminated with these metals. Samples of plant material (leaves) were collected 25 days after the application of EDTA to soil. Contents of individual heavy metals in leaves of *Tagetes erecta* 'Taishan Orange' are presented in Figs. 3, 4 and 5. The research found that acid (EDTA) has improved the efficiency of phytoextraction of cadmium, lead and nickel in soil by Tagetes erecta L. 'Taishan Orange'.

Identification of compounds complexing toxic heavy metals and at the same time biodegradable in the soil medium is crucial for induced phytoremediation.

At present studies are being conducted at the Department of Plant Nutrition, the Poznań University of Life Sciences, Poland on the application of a biodegradable compound in phytoextraction of heavy metals from contaminated soils.

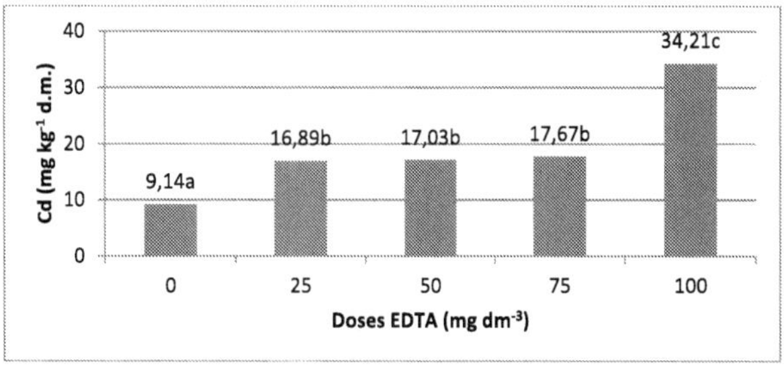

*homogeneous groups were identified using the Duncan test, p = 0.05 (values denoted with identical letters do not differ significantly)

Figure 3. The effect of increasing EDTA doses introduced to soil (contaminated with 1.5 mg Cd dm^{-3}) on cadmium content in leaves of *Tagetes erecta* 'Taishan Orange'

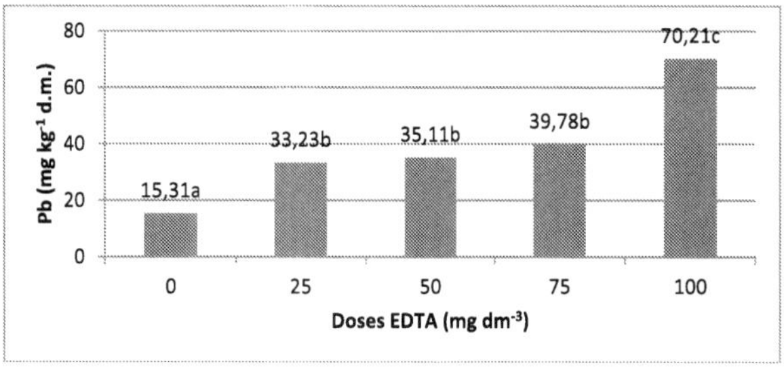

*homogeneous groups were identified using the Duncan test, p = 0.05 (values denoted with identical letters do not differ significantly)

Figure 4. The effect of increasing EDTA doses introduced to soil (contaminated with 100 mg Pb dm^{-3}) on lead content in leaves of *Tagetes erecta* 'Taishan Orange'

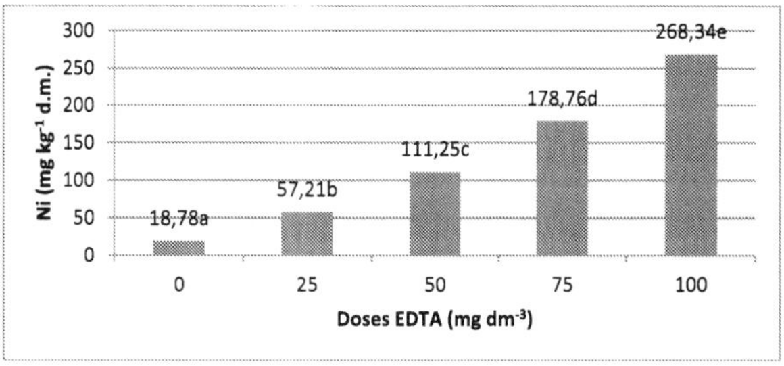

*homogeneous groups were identified using the Duncan test, p = 0.05 (values denoted with identical letters do not differ significantly)

Figure 5. The effect of increasing EDTA doses introduced to soil (contaminated with 50 mg Ni dm^{-3}) on nickel content in leaves of *Tagetes erecta* 'Taishan Orange'

8. The phenomenon of hyperaccumulation

As it was stated by Brooks [28] the discovery and description of a phenomenon termed hyperaccumulation has contributed to the practical use of plants to remove metallic pollutants from soil. According to Boyd and Martens [25] and Brown et al. [29], plant species referred to as hyperaccumulators are genetically and physiologically capable of accumulating large amounts of heavy metals with no symptoms of toxicity. Threshold values of metal concentrations have been used to define metal hyperaccumulation, including 100 mg kg^{-1} dry weight of shoots for Cd, 1000 mg kg^{-1} for Cu, Ni, Pb and 10 000 mg kg^{-1} for Zn [10, 28, 86]. Cline et al. [34] stated that concentrations of heavy metals in tissues of hyperaccumulator plants should be 1 - 2%. Van der Ent et al. [120] recommend the following concentration criteria for different metals and metalloids in dried foliage:100 μg g^{-1} for Cd, Se and Ti; 300 μg g^{-1} for Co, Cu and Cr; 1000 μg g^{-1} for Ni, Pb and As; 3000 μg g^{-1} for Zn; 10000 μg g^{-1} for Mn, with plants growing in their natural habitats. There are over 400 known plant species from 45 families classified as hyperaccumulators. Most species belong to the families *Brasicaceae* and *Fabaceae* [103]. Approximate numbers for various elements are as follows: Ni (450), Cu (32), Co (30), Se (20), Pb (14), Zn (12), Mn (12), As (5), Cd (2), Tl (2), [120]. Moreover, hyperaccumulators typically accumulate most efficiently one heavy metal and to date no plant has been identified which would be capable of accumulating in its tissues all toxic metals [37]. According to Vangronsveld et al. [121] successful phytoremediation depends first of all on the selection of an appropriate plant species. For this reason in the opinion of Baker et al. [10] it would be a practical solution to grow many different species at the same time.

Most natural hyperaccumulators are plants characterized by slow growth and production of low amounts of biomass. These traits result in a limited applicability of these plant species in phytoextraction of heavy metals from soil [37]. An example in this respect may be provided by *Thalspi caerulescens* L., producing only 5 ton dry matter per hectare [32], whereas it is cost-effective to use plants, which yield 20 ton per hectare and the accumulation of metal in aboveground parts is over 1% [63].

Studies are being conducted worldwide on the use of plants producing large amounts of biomass (for energy generation purposes) in the phytoextraction of heavy metals from polluted soils. They include annual plants such as e.g. cereals and rape, and perennials e.g. willows, which capacity for phytoextraction of heavy metals was confirmed by Greger and Landberg [56] and Boyter et al. [26]. Other perennial species of energy crops include *Spartina pectinata, Sida Hermaphrodita, Rosa multiflora, Helianthus tuberosus, Populus spp.,* and *Robinia pseudoacacia.* The most significant parameters of these plants include high annual increment of biomass and its high calorific value [116].

Ociepa et al. [94] stated that plants grown for energy purposes may play a considerable role in view of the assumptions of the Common Agricultural Policy and the environmental protection policy of the European Union. One of such plant species is *Miscanthus × giganteus* Greef and Deu.

In the years 2008 – 2011 at the Department of Plant Nutrition, the Poznan University of Life Sciences, Poland, studies were conducted to assess applicability of *Miscanthus × giganteus* Greef and Deu. in phytoextraction of heavy metals from soil. The aim of the conducted analyses was to determine what amounts of cadmium and lead are transported to aboveground parts of *Miscanthus* × *giganteus* and whether *Miscanthus* × *giganteus* would be suitable for rapid phytoextraction of cadmium, lead from polluted soils.

Material and methods

The vegetation experiment was conducted in an unheated plastic tunnel with suspended sides, of 6 x 30 m in size. at the Marcelin Experimental Station of the Poznan University of Life Sciences.

Seedlings of *Miscanthus* x *giganteus* were produced at a tissue culture laboratory of Vitroflora. Plants were planted in the beginning of May in drainless containers (of 7 dm^3) filled with previously prepared substrate. The experiment comprised 16 combinations (in each year of the study) and each combination consisted of six replications. A replication comprised one plant growing in pots.

Phytoremediation of cadmium and lead by *Miscanthus* x *giganteus* was investigated in two years of growth with the plants being grown in two substrates. at four levels of metal contents. Since light soils with low contents of organic matter predominate in Poland. such soil was selected for the experiment. Another substrate was a mixture of this mineral soil with highmoor peat. Highmoor peat was added to increase the amount of organic matter in the mineral soil.

Substrates: mineral soil (sand) and mineral soil with highmoor peat (1:1 v/v)

Doses of cadmium: control (native contents of cadmium), 3, 5 and 10 mg dm^{-3})

Doses of lead: control (native contents of cadmium), 250, 1000 and 5000 mg dm^{-3}).

In mineral soil the method according to Mocek and Drzymała [88] was used to determine particle density (which amounted to 2.65 g cm^{-3}) and bulk density (1.62 g cm^{-3}). Total porosity of mineral soil was 38.9%. Moreover. grain size distribution of mineral soil was determined by the densimetric method according to Prószyński [88]. On the basis of the percentages of fractions the grain size class of soil was identified (according to the guidelines of the Polish Society of Soil Science) - sand.

Prior to the establishment of the experiment, Corg content in mineral soil was determined according to the Tiurin method [51]. Content of organic carbon in sand (the Tiurin method) was 0.55% (0.95% humus). while the percentage of organic matter in the mixture of sand and highmoor peat (from loss on ignition) was 10.05%. In the substrate composed of a mixture of mineral soil with highmoor peat (1:1 v/v) the percentage of organic substance was determined by loss on ignition the substrate by the direct method at high temperature in the presence of oxygen, under the influence of which organic substance is decomposed (carbon is released in the form of CO_2. hydrogen in the form of H_2O and nitrogen as N_2, while the other elements remain in ash).

Experiments were conducted using highmoor peat by Hartmann (sphagnum peat. ground. fractional with acid reaction (pH 4.50). This peat has a high water capacity, at the same time retaining an elastic structure. The weight of 1 dm^3 peat was 490 grams.

In order to obtain an appropriate pH for growing of *Miscanthus* x *giganteus* a neutralization curve was plotted for the analyzed substrates. On its basis the dose of $CaCO_3$ required for the maintenance of pH within the range of 6.5-7.0 was established. The reaction of the substrate (mineral soil + highmoor peat) was regulated using 3 g dm^{-3} $CaCO_3$ (chemically pure reagent). The substrate composed of mineral soil did not require reaction control. Despite that fact 1 g dm^{-3} $CaCO_3$ was applied in order to maintain pH at 6.5-7.0. An adequate amount of calcium carbonate was introduced to each experimental container with the substrate. Cadmium and lead contents in analyzed substrates after liming amounted: in mineral soil Cd 0.27, Pb 27.32 mg dm^{-3} and mineral soil with highmoor peat Cd 0.18, Pb

12.03 mg dm^{-3}. Two weeks after liming nutrients cadmium and lead were introduced to the substrate. Cadmium and lead was introduced only in the first year of the experiment in the form of chemically pure reagents (C.P.): cadmium sulfate ($3CdSO_4 8H_2O$), lead acetate [$(CH_3COO)_2Pb\bullet3H_2O$]. Pre-vegetation fertilization (in the first year) with macro- and micronutrients was determined taking into consideration initial nutrient contents in substrates. after liming reaching the following levels (in mg dm^{-3}): N 200, P 120, K 250, Mg 100, Fe 50, Mn 20, B 1.5 and Mo 1.5. All macro- and micronutrients were introduced in the form of solutions using chemically pure reagents (potassium monophosphate, potassium nitrate, ammonium nitrate, magnesium saltpeter, magnesium sulfate, iron sulfate, copper sulfate, zinc sulfate, manganese sulfate, ammonium molybdate, borax). In the second year of the experiment an identical experimental design was used as in the first year. After plant cutting in the first year of the experiment containers with polluted substrates were stored in an unheated tunnel to the next vegetation year (the second year of the study). In the second year of the study in March - prior to the beginning of vegetation substrate samples were collected and chemical analyses were performed to determine nutrient contents. On this basis nutrient fertilization was established (leading to nutrient contents at the same levels. which were applied in the first year of the experiment). Nutrients in substrates were determined using the *"Universal"* method [78] in CH_3COOH solution at a concentration of 0.03 mol dm^{-3}, pH in water was determined by potentiometry (the substrate to water ratio of 1:2), while conductometry was applied to determine EC (mS cm^{-1}), (the substrate to water ratio of 1:2), [50].

The following nutrient determination techniques were applied: $N-NH_4$ and $N-NO_3$ by microdistillation (Bremner modified by Starck), P by colorimetry using the vanadium-molybdenum method, K, Ca and Na by flame photometry, Mg by atomic absorption (AAS), Cl and $S-SO_4$ by nephelometry [78].

In October in each year of the study prior to harvesting plant height was measured. Dry weight of plants was recorded and samples of plant material were collected for analyses.

Harvested plant material (entire aboveground mass) was dried in an extraction drier at a temperature of 105°C for 48 h. Next the material was ground and at 2.5 g from each sample it was digested in a mixture of concentrated HNO_3 (ultra pure) and $HClO_4$ (analytically pure) at a 3:1 ratio [18]. Content of cadmium and lead in the plant material were determined by flame atomic absorption spectrophotometry (FAAS), AAS-3 spectrophotometer by Zeiss. Moreover, content of metals in the reference material (*Pseudevernia furfuracea* BCR®-482/2009) was determined.

In the first and second year of the study samples of substrate were collected after harvest, from which metals were extracted using the Lindsey's solution containing in 1 dm^3: 5 g EDTA (ethylenediaminetetraacetic acid), 9 cm^3 25% NH_4OH solution, 4 g citric acid and 2 g $Ca(CH_3COO)_2 2H_2O$. Next this metal was assayed by flame atomic absorption spectrophotometry (FAAS), AAS-3 spectrophotometer by Zeiss.

Results of content of cadmium and lead in substrates and aboveground parts of *Miscanthus × giganteus* were elaborated statistically in the Statobl program applying a one-way analysis of variance for orthogonal factorial experiments, with differences between means determined at a significance level p=0.05.

Miscanthus × giganteus Greef and Deu. (Figure 6) is an interspecies hybrid of a diploid Chinese silver grass [*Miscanthus sinensis* (Thunb.) Anderss.] and a tetraploid Amus silver grass [*Miscanthus sacchariflorus* (Maxim.) Benth], belonging to the family *Poaceae* [54, 55, 61]. After three years of culture the yield of dry matter is 20 to 35 t and plant height is 3 - 4 m [102]. *Miscanthus × giganteus* originates from South-Eastern Asia and it has been grown in Europe, initially as an ornamental plant, for over 50 years. In Poland it still is not a popular crop. In the nearest future interest in growing of this plant in Poland will be increasing. This abundant grass forms new shoots stretching outwards from rhizomes (underground runners), forming an increasingly bigger rounded cluster [102].

Analyses conducted by physiologists classify *Miscanthus* to the group of plants of the C-4 pathway, which are characterised by a highly efficient photosynthesis process, ensuring a rapid and high increase in biomass, at a simultaneous lower transpiration coefficient, i.e. lower water consumption [58, 89].

Proposed methods to manage Miscanthus after phytoextraction of heavy metals from soil include combustion (ashes – hazardous waste), bio-ore, paper and pulping industry, production of particleboards as well as chemical industry (packaging plastics).

Figure 6. *Miscanthus x giganteus* Greef and Deu. in the third year of growth

The aim of the conducted analyses was to determine the effect of increasing doses of cadmium, lead introduced to mineral soil (sand) and to mineral soil with an addition of highmoor peat (at a 1:1 ratio, v/v), on the tolerance index of *Miscanthus x giganteus*. The tolerance index (T_i) was calculated, i.e. the ratio of the yield (dry biomass) obtained in metal-polluted soil to the yield (dry biomass) produced in unpolluted soil. This index is considered as the most reliable indicator of the toxic effect of heavy metals contained in soils and substrates exercised on plants.

$T_i < 1$ value lower than one - inhibition of growth or plant death

$T_i = 1$ value equal one - no effect of increased metal contents on yielding

$T_i > 1$ value greater than one - positive effect of metal on yielding.

http://dx.doi.org/10.5772/57257

In the first year of growth of *Miscanthus × giganteus* the tolerance index greater than one was recorded in plants in mineral soil polluted with 3 mg Cd dm^{-3} and 1000 mg Pb dm^{-3} (Table 6). In the second year of growth the tolerance index greater than one was found in plants in mineral soil contaminated with all the used cadmium pollution rates and in soil, to which 250 and 1000 mg Pb dm^{-3} were introduced. The tolerance index greater than one was also recorded in the substrate constituting a mixture of mineral soil with highmoor peat contaminated with 3 mg Cd dm^{-3} (in the first year of growth) and 5 mg Cd dm^{-3} (in the second year of growth).

Metal	Dose of metal (mg dm^{-3})	Substrates			
		Mineral soil		Mineral soil + highmoor peat	
		The first year of growth	The second year of growth	The first year of growth	The second year of growth
Cd	3	**1.02**	**1.20**	**1.02**	0.88
	5	0.99	**1.34**	0.83	**1.01**
	10	0.70	**1.24**	0.95	0.85
Pb	250	0.80	**1.30**	0.78	0.99
	1000	**1.01**	**1.28**	0.99	0.90
	5000	1.00	0.98	0.93	0.58

Table 6. The tolerance index (Ti) in *Miscanthus x giganteus*

In the opinion of Arduini et al. [8] and Kozak et al. [77], *Miscanthus* is a less tolerant species to high concentrations of heavy metals in comparison to osier and it is necessary to conduct further tests on the applicability of varieties of this species in the phytoextraction of heavy metals.

Kalembasa and Malinowska [72], when testing different clones of *Miscanthus* found differences in their capacity to accumulate individual heavy metals, particularly cadmium. Iżewska [65] in the third year of *Miscanthus* culture recorded a greater content of cadmium and a lower content of lead in its biomass. In the opinion of Kalembasa and Malinowska [71], contents of cadmium and lead in biomass are also dependent on the harvest date of Miscanthus. The highest cadmium concentration was recorded in *Miscanthus* biomass at the begining of June, lead at the beginning of September and October. Those authors stated that the triploid Miscanthus species was characterized by higher contents of cadmium and lead in its biomass in comparison to the diploid species. According to those authors, nitrogen, phosphorus and potassium fertilization has an effect on cadmium content in biomass of diploid Miscanthus genotypes.

Based on studies conducted by the authors of this chapter it was found that cadmium applied at 3 and 5 mg dm^{-3} mineral soil in the first year of growth of *Miscanthus × giganteus* had no significant effect on the content of this metal in aboveground parts of plants in comparison to those growing in soil with no addition of this metal (Table 7). In the second year of growth in mineral soil to which 5 mg Cd dm^{-3} were introduced a significantly greater content of this metal was detected in aboveground parts of plants in comparison to plants growing in unpolluted soil. In plants growing in mineral soil the greatest Cd content was found in the second years of growth of *Miscanthus* × *giganteus* growing in soil polluted with 10 mg Cd dm^{-3}.

In a mixture of mineral soil with highmoor peat, to which 3 mg Cd dm^{-3} were introduced both in the first and second year of growth of *Miscanthus* × *giganteus* no significant differences were observed in the contents of this metal in aboveground parts. The greatest cadmium content was recorded in plants growing in the substrate polluted with 10 mg Cd dm^{-3} in the first and second years of growth.

Substrate	**Dose of metal ($mg\ dm^{-3}$)**	**Year of culture**							
		1st year				**2nd year**			
		min.-max. range	**range R**	**SD**	**mean**	**min.-max. range**	**range R**	**SD**	**mean**
Mineral soil	Control	0.98-1.29	0.31	0.12	**1.17 ab**	1.29-1.52	0.23	0.10	**1.39 ab**
	Cd 3	1.31-1.99	0.68	0.24	**1.70 bc**	1.14-2.59	1.45	0.51	**2.06 a-d**
	Cd 5	1.33-3.69	2.36	0.78	**2.61 bcd**	1.78-4.69	2.91	0.97	**3.52 d**
	Cd 10	3.56-8.34	4.78	1.62	**5.51 e**	6.16-12.62	6.46	2.81	**9.30 g**
Mineral soil + highmoor peat	Control	0.64-1.33	0.69	0.25	**0.97 a**	0.95-1.46	0.51	0.19	**1.20 ab**
	Cd 3	1.34-1.93	0.59	0.22	**1.60 abc**	1.38-1.90	0.52	0.20	**1.66 abc**
	Cd 5	2.39-3.96	1.57	0.67	**2.97 cd**	2.68-4.78	2.10	0.97	**3.41 d**
	Cd 10	5.23-9.34	4.11	1.55	**7.46 f**	5.54-11.17	5.63	2.48	**8.66 fg**

*homogeneous groups were identified using the Duncan test, p=0.05 (values denoted with identical letters do not differ significantly)

Table 7. Contents of Cd (mg kg^{-1} dry weight) in aboveground parts of *Miscanthus* x *giganteus* growing in substrates polluted with cadmium

A significantly greater lead content in aboveground parts of *Miscanthus* × *giganteus* in comparison to the control was found in plants growing in all tested substrates contaminated with this metal (Table 8). In plants growing in mineral soil contaminated with 10 mg Pb dm^{-3} a higher content of this metal was detected in the first year of growth and it was the highest content recorded in the analyses. No differences were observed in lead contents in aboveground parts of *Miscanthus* × *giganteus* in the first and second year of growth in a mixture of mineral soil with highmoor peat.

Substrate	Dose of metal (mg dm^{-3})	Year of culture							
		1st year				2nd year			
		min.-max. range	range R	SD	mean	min.-max. range	range R	SD	mean
Mineral soil	Control	1.23-1.77	0.54	0.21	**1.54 a**	1.76-2.74	0.98	0.38	**2.31 a**
	Pb 250	29.87-48.25	18.38	7.44	**39.29 b**	27.37-46.78	19.41	7.71	**36.83 b**
	Pb 1000	56.89-75.43	18.54	6.92	**62.17 b**	53.45-73.56	20.11	7.58	**64.88 bc**
	Pb 5000	71.46-280.62	209.16	85.24	**210.20 e**	58.56-131.34	72.78	29.19	**104.49 d**
Mineral soil + highmoor peat	Control	1.14-1.62	0.48	0.18	**1.47 a**	1.12-1.67	0.55	0.21	**1.38 a**
	Pb 250	27.89-40.99	13.10	5.65	**34.29 b**	21.67-45.62	23.95	8.00	**31.92 b**
	Pb 1000	48.67-61.78	13.11	4.74	**53.48 b**	40.56-62.38	21.82	7.80	**49.30 b**
	Pb 5000	73.56-110.34	36.78	12.56	**92.98 cd**	53.56-137.89	84.33	31.89	**96.46 d**

*homogeneous groups were identified using the Duncan test, p=0.05 (values denoted with identical letters do not differ significantly)

Table 8. Contents of Pb (mg kg^{-1} dry weight) in aboveground parts of *Miscanthus* x *giganteus* growing in lead-polluted substrates

In the conducted analyses the cadmium and lead concentration indexes were calculated for aboveground parts of *Miscanthus* × *giganteus*. The metal concentration index was caluclated from the formula

$C = a : b$

a - content in a plant growing in polluted substrate

b - content in a plant growing in unpolluted substrate.

The greatest concentration index was recorded for lead in the first year of growth in the case of plants growing in mineral soil (Table 9). Plants growing in mineral soil were characterized by a greater concentration index for cadmium and lead in the first year of growth. An identical dependence was found in plants growing in a mixture of soil and peat, except for plants growing in a substrate contaminated with 5000 mg Pb dm^{-3}, in which a higher lead concentration index was found in the second year of growth.

Miscanthus sp. was tested on heavy metal contaminated arable soil in Southern Poland [110]. The authors concluded that this species accumulates high amounts of metals what may cause high emission of contaminants during biomass combustion.

According to Kalembas [70] in ash of *Miscanthus sinensis* Thumb. the content of individual heavy metals ranks in the following decreasing levels: Zn>Cd>Pb>Ni>Cu>Cr.

Metal	Dose of metal ($mg\ dm^{-3}$)	Substrates			
		Mineral soil		Mineral soil + highmoor peat	
		The first year of growth	The second year of growth	The first year of growth	The second year of growth
Cd	3	1.45	1.48	1.65	1.38
	5	3.06	2.53	3.06	2.84
	10	7.69	6.69	7.69	7.22
Pb	250	25.51	15.94	23.33	23.13
	1000	40.37	28.09	36.38	35.72
	5000	136.49	45.23	63.25	69.90

Table 9. Metal concentration indexes in aboveground parts of *Miscanthus x giganteus*

Both in soil and in a mixture of soil and peat a lower cadmium content was recorded after the second year of culture except for substrates contaminated with 10 mg Cd dm^{-3}, in which this dependence was not observed (Table 10).

When analyzing lead content in tested substrates after the completion of growth a lower Pb content was found also in the second year except for a mixture of mineral soil with peat, to which lead was not introduced (table 11).

In the substrate being a mixture of soil with peat lower contents of cadmium and lead were observed in comparison to those recoded in mineral soil in all the experimental variants (Tables 10 and 11).

Type of pollution	Substrate	Year of growth	
		1st year	2nd year
Control (native content of Cd $mg\ dm^{-3}$)	mineral soil	0.09 c	0.07 b
	soil + peat	0.07 b	0.05 a
Weak pollution (Cd 3 $mg\ dm^{-3}$)	mineral soil	2.23 d	1.09 b
	soil + peat	1.79 c	0.77 a
Medium pollution (Cd 5 $mg\ dm^{-3}$)	mineral soil	4.12 d	2.65 c
	soil + peat	2.51 b	1.32 a
Strong pollution (Cd 10 $mg\ dm^{-3}$)	mineral soil	6.37 b	6.39 b
	soil + peat	5.43 a	5.22 a

*homogeneous categories were identified with the Duncan test, p=0.05 (values denoted with identical letters do not differ significantly)

Table 10. Cadmium contents (extracted with Lindsey's solution) in substrates (in $mg\ dm^{-3}$) after the completion of plant growth in the first and the second year of analyses

http://dx.doi.org/10.5772/57257

Type of pollution	Substrate	Year of growth	
		1st year	2nd year
Control	mineral soil	24.84 c	15.18 b
(native content of Pb mg dm^{-3})	soil + peat	6.95 a	4.79 a
Weak pollution	mineral soil	227.71 d	192.68 c
(Pb 250 mg dm^{-3})	soil + peat	130.31 b	104.74 a
Medium pollution	mineral soil	837.70 d	575.34 b
(Pb 1000 mg dm^{-3})	soil + peat	665.30 c	236.88 a
Strong pollution	mineral soil	3755.06 d	3304.71 c
(Pb 5000 mg dm^{-3}).	soil + peat	2431.14 b	1401.27 a

*homogeneous groups were identified with the Duncan test, p=0.05 (values denoted with identical letters do not differ significantly)

Table 11. Contents of lead (extracted with Lindsey's solution) in substrates (in mg dm^{-3}) after the completion of plant growth in the first and the second years of analyses

9. The amount of ash (%) after incineration of aboveground parts of *Miscanthus × giganteus*

Aboveground parts of *Miscanthus × giganteus* were incinerated at a temperature of 450ºC in a Linn Elektro Therm furnace. Incineration was performed in two stages:

- stage I - preliminary carbonization at a temperature of 100°C for 1 h
- stage II – combustion at 450°C for 5 h.

Incineration was performed on 20 randomly selected samples of aboveground dry matter of *Miscanthus × giganteus.* Assuming dry matter used in the incineration process as 100%. after combustion a mean 8.20% ash was obtained (SD 0.55), (Table 12).

The amount of ash left after combustion of aboveground parts of *Miscanthus × giganteus* was analyzed in terms of empirical probability distribution. Mean observations of the amount of ash were ordered in the so-called stemplot. in which separate intervals called classes were identified. Five classes were determined for the amount of ash left after incineration and next the number of observations (the so-called class frequencies) belonging to them were specified for each class. In order to ensure uniqueness of classification of observations it was assumed that classes are left side open intervals and right side closed intervals. Moreover, frequencies were added which when multiplied by 100 determine the percentage of observations of the amount of ash left after combustion of plants as classified to individual classes.

Among all observations after the combustion process the highest percentage (40%) was found for the amount of ash within the range of 8.44 to 8.88%. The amount of ash within the range of 8.00 to 8.44% ranked second constituting 30% observations.

In a study conducted by Kalembasa [70] largest content of raw ash was obtained from the *Sida hermaphrodyta* Rusby. (59.5 kg t^{-1}), less *Miscanthus sinensis* Thumb. (52.5 kg t^{-1}) and at less *Salix sp.* (31.5 kg t^{-1}).

Miscanthus x giganteus	Classes	Number f_i	Frequencies f_i/n	Cumulative frequencies
Amount of ash after combustion in (%)	V = (6.68-7.12]	1	1/20	1/20
	IV = (7.12-7.56]	1	1/20	2/20
	III = (7.56-8.00]	4	4/20	6/20
	II = (8.00-8.44]	6	6/20	12/20
	I = (8.44-8.88]	8	8/20	20/20
Total		20	1	

Table 12. The stemplot for the amount of ash (%) obtained after combustion of aboveground pars of *Miscanthus* x *giganteus*

Concluding remarks

Miscanthus × *giganteus* may be used in phytoextraction of cadmium and lead from polluted soils; however, when assessing its potential for phytoextraction of these metals from soil it was found not to be considerable. This plant need to be further tested in terms of cadmium and lead phytoextraction capacity from contaminated soil in natural *environment.*

The study was financed from funds for science in the years 2008 - 2011 as a research project no. N N305 085535

Acknowledgements

The authors express their gratitude to the Management Board of the Municipal Green Areas in Poznań and the Management Board of the Municipal Road Network in Poznań for facilitating the soil monitoring project.

Author details

Maciej Bosiacki*, Tomasz Kleiber and Bartosz Markiewicz

*Address all correspondence to: mbos@up.poznan.pl

Department of Plant Nutrition, University of Life Sciences in Poznan, Poland

References

[1] Alizadeh SM., Zahedi-Amiri G., Savaghebi-Firoozabadi G., Etemad V., Shirvany A., Shirmardi M. Assisted phytoremediation of Cd-contaminated soil using poplar rooted cuttings. Int. Agrophys., 2012;26 219-224.

[2] Alloway BJ. Heavy metals i soils. Blackie Gong. and London. New York: John Willey & Sons; 1990; p319.

[3] Alloway BJ., Ayres DC. Chemical pollution base. Warszawa: PWN; 1999 *(in Polish).*

[4] Amir S., Hafidi M., Merilna G., Revel JC. Sequential extraction of heavy metals during composting of sewage sludge. Chemosphere 2004; 59 801 – 810.

[5] Anderson CWN., Brooks RR., Chiarucci A., Lacoste CJ., Leblanc M., Robinson BH., Simcock R., Stewart RB. Phytomining for nickel, thallium and gold. Journal of Geochemical Exploration 1999;67 407-415.

[6] Antonkiewicz J., Jasiewicz C. Estimation of usefulness of different plant species for phytoremediation of soils contaminated with heavy metals. Acta Sci. Pol. Formatio Circumectus 2002;1(1-2) 119-130 *(in Polish).*

[7] Antonkiewicz J., Jasiewicz C., Lošák T. Using *Sida hermaphrodita* Rusby for extraction of heavy metals from soil. Acta Sci. Pol. Formatio Circumectus 2006;5(1) 63-73 *(in Polish).*

[8] Arduini I., Masoni A., Ercoli L., Mariotti M. Growth and cadmium uptake of *Miscanthus sinensis* as affected by cadmium. Agric. Mediterran. 2003;133(3-4) 169-178.

[9] Bach A. Studies on soil pollution with heavy metals in green areas adjacent to transportation routes and assessment of their salinity rates with a determination of soil pH. Uniwersytet Rolniczy im. Hugona Kołłątaja w Krakowie. A study commissioned by the City Office of Krakow; 2011 Available on-line 14.06.2012 <http:www.bip.krkow.pl.> *(in Polish).*

[10] Baker AJM., McGrath SP., Reeves RD., Smith JAC. Metal hyperaccumulator plants: a review of the ecology and physiology of a biochemical resource for phytoremediation of metal-polluted soils. In: Terry N., Banuelos G. (ed.) In Phytoremediation of Contaminated Soil and Water. Lewis Publishers 2000; p85-107.

[11] Bauman A. Das verhalten von Zinksalzen gegen pflanzen und im boden. Landwirtisch. Verss 1885;31 1-53.

[12] Bielicka A., Ryłko E., Bojanowska I. Contents of metals in soils and vegetables from Gdansk and Straszyn allotments. Ochrona Środowiska i Zasobów Naturalnych 2009;40 209-216.

[13] Blake L., Goulding KWT. Effects of atmospheric deposition, soil pH and acidification on heavy metal content in soils and vegetation of semi-natural ecosystems AT Rothamsted Experimental Station. UK. Plant and Soil 2002;240 235 – 251.

[14] Blaylock M., Huang J. Phytoextraction of metals. In: Raskin I, Ensley B (ed.) Phytoremediation of toxic metals: using plants to clean up the environment. New York: Wiley; 2000. p53-69.

[15] Bosiacki M. Accumulation of cadmium in selected species of ornamental plants. Acta Sci. Pol. Hortorum Cultus 2008;7(2) 21-31.

[16] Bosiacki M. Phytoextraction of cadmium and lead by selected cultivars of *Tagetes erecta* L. Part I. Effect of Cd and Pb on yielding. Acta Sci. Pol. Hortorum Cultus 2009;8(2) 3-13.

[17] Bosiacki M. Phytoextraction of cadmium and lead by selected cultivars of *Tagetes erecta* L. Part II. Contents of Cd and Pb in plants. Acta Sci. Pol. Hortorum Cultus 2009;8(2) 15-26.

[18] Bosiacki M., Roszyk J. The comparing methods of mineralization of plant material on the content of heavy metals. Apar. Bad. Dydakty 2010;XIV(4) 37-41 *(in Polish).*

[19] Bosiacki M., Tyksiński W. Dependence between the content of organic carbon and the content of cadmium and lead in horticultural substrates. Acta Agrophysica 2006;7(3) 517-526 *(in Polish).*

[20] Bosiacki M., Wojciechowska E. Phytoextraction of nickel by selected ornamental plants. Ecological Chemistry and Engineering S 2012;19(3) 331-345.

[21] Bosiacki M., Wolf P. Evaluation of the usefulness of selected soecies of grasses to phytoremediation of cadmium and lead. Part I. Cadmium. Apar. Bad. Dydakty. 2008;13(3) 19-27 *(in Polish).*

[22] Bosiacki M., Wolf P. Evaluation of the usefulness of selected soecies of grasses to phytoremediation of cadmium and lead. Part II. Lead. Apar. Bad. Dydakty. 2008b; 13(3) 28-36 *(in Polish).*

[23] Bosiacki M., Zieleziński Ł. Phytoextraction of nickel by selected species of lawn grasses from substrates contaminated with heavy metals. Acta Sci. Pol., Hortorum Cultus 2011;10(3) 155-173.

[24] Botre C., Tosi M., Mazzei F., Bocca B., Petrucci F., Alimonti A. Automotive catalytic converters and environmental pollution: Role of the platinum group elements in the redox reactions and free radicals production. International Jurnal of Enviromnent and Health. 2007;1(1) 142-152.

[25] Boyd RS., Martens SN. Nickel hyperaccumulated by *Thlaspi montanum var. montanum* is acutely toxic to an insect herbivore. Oikos 1994;70 21-25.

[26] Boyter MJ., Brummer JE., Leininger WC. Growth and metal accumulation of Geyer and mountain willow grown in topsoil versus amended mine tailings. Water Air Soil Pollut. 2009;198 17-29.

[27] Breś W. Anthropopressure factors causing trees to die off in Urban landscape. Nauka Przyr. Technol. 2008;2(4) 31 *(in Polish).*

[28] Brooks RR. Plants that Hyperaccumulate Heavy Metals. Wallingford, UK: CAB International 1998.

[29] Brown SL., Chaney RL., Anle JS., Baker AJM. Zinc and cadmium uptake of *Thlaspi caerulescens* grown in nutrient solution. Soil Sci. Soc. Am. J. 1995;59125-133.

[30] Brümmer G., Gerth J., Herms U. Heavy metals species, mobility and availability in soils. Z. Pflanzenern Bonenk 1986;149 382 – 389 *(in Deutsch).*

[31] Brümmer G., Herms U. Influencing factors of heavy metals. Solubility, retention and availability in the soil. Bielefelder Ökol. Beitr. 1985;1 117 – 139 *(in Deutsch).*

[32] Chaney RL., Brown SL., Y-M Li. Potential use of metal hyperaccumulators. Mining Environm. Manag. 1995;9 9-11.

[33] Chłopecka A. The effect of various compounds of cadmium, copper, lead and zinc to form these metals in the soil and their content in plants. IUNG Seria R 1994 *(in Polish).*

[34] Cline RG., Powell PE., Szaniszko PJ., Reid PP. Comparison of the abilities of hydroxamic, synthetic and other natural organic acids to chelate ion and other ions in nutrient solution. J. Amer. Soc. Soil Sci. 1982;46 1158-1164.

[35] Cooper EM., Sims JT., Cunningham SD., Huang JW., Berti WR. Chelate-assisted phytoextraction of lead from contaminates soils. J. Environ. Qual. 1999;28 1709-1719.

[36] Crucq R., Frennet A. Catalysis and automotive pollution control. Elsevier Science Publishers B.V., Amsterdam 1987.

[37] Cunnigham SD., Berti WR., Huang JW. Phytoremetiation of contaminated soils. Trends Biotechnol. 1995;13 393-397.

[38] Curyło T., Jasiewicz C. Comparison of the impact of multi-organic fertilizer and mineral and mineral yield and intake of heavy metals by plants. Fol. Univ. Agr. Stet.–Agr. 1998;72 35-41 *(in Polish).*

[39] Czekała J., Jakubus M., Gładysiak S. The content of soluble forms of trace elements, depending on the pH of the soil and the extraction solution. Zesz. Probl. Post. Nauk. Rol. 1996;434(1) 371 – 376 *(in Polish).*

[40] Dhankher OP., Doty SL., Richard B. Meagher RB., Pilon-Smits E. Biotechnological approaches for phytoremediation. In: Arie Altman and Paul Michael Hasegawa (ed.) Plant Biotechnology and Agriculture. Oxford: Academic Press 2011; pp. 309-328.

[41] Dmochowski D., Prędecka A., Mazurek M., Pawlak A. Hazards related to the emission of heavy metals in view of ecological safety. Example of allotments in urban areas. The Polish Journal of Aviation Medicine and Psychology 2011;3(17) 257-265 (*in Polish).*

[42] Domska D., Bobrzecka D., Wojtkowiak K. Changes in the content of some nutrients in the soil depending on their acidity. Zesz. Probl. Post. Nauk. Rol. 1998;456 255 – 559 (*in Polish*).

[43] Fergusson JE. The heavy elements. Chemistry, environmental impact and health effects. Pergamon Press. Oxford 1990.

[44] Filipek-Mazur B., Tabak M. Soil, traffic pollution, copper, zinc, chromium, sequential extraction. Ecological Chemistry and Engineering A 2011;18(11) 1533 -1538.

[45] Gałuszka A., Migaszewski Z. Problems of sustainable use of mineral resources. Problems of Sustainable Development 2009;4(1) 123-130 (*in Polish*).

[46] Gangrong Shi., Qingsheng Cai. Cadmium tolerance and accumulation in eight potential energy crops. Biotechnology Advances, 2009;27 555-561.

[47] Gawęda M. The effect of organic matter in soil on the lead level in edible parts of lettuce and carrot. Acta Hort. 1995;379 221-228.

[48] Gębski M. Factors influencing soil and fertilizer on heavy metal uptake by plants. Post. Nauk. Roln. 1998;5 3 – 16 (*in Polish*).

[49] Gerhardt KE., Huang XD., Glick BR., Greenberg BM. Phytoremediation and rhizoremediation of organic soil contaminants: Potential and challenges. Plant Sci. 2009;176 20–30.

[50] Golcz A. Soil salinity and acidity. Research Methods In Plant Sciences vol. 3. Soil Sickness. In Narwal SS., Politycka B., Fengzhi Wu, Sampietro DA. (ed.) Studium Press LLC, Huston USA. 2011; p43-53.

[51] Golcz, A., Bosiacki, M. Soil Organic Matter, Research methods in plant sciences vol. 3. Soil Sickness. In Narwal SS.. Politycka B.. Fengzhi Wu. Sampietro DA. (ed.) Studium Press LLC. Huston US, 2011; p68-78.

[52] Gorlach E., Gambuś F. Phosphate and compound fertilizers as a source of soil contamination with heavy metals. Zesz. Prob. Post. Nauk Roln. 1997;448a 139-146.

[53] Gorlach E., Gambuś F. Potentially toxic trace elements in soils (excess, harmfulness and countermeasures). Zesz. Prob. Post. Nauk Roln. 2000;472 275-296 (*in Polish*).

[54] Greef JM., Deuter M. Syntaxonomy of *Miscanthus × giganteus* (GREEF et DEU), Angew. Bot. 1993;67 87–90.

[55] Greff J., Deuter M., Jung C., Schondelmaier J. Genetic diversity of European *Miscanthus* species revealed by AFLP fingerprinting. Genetic Resources and Crop Evolution 1997;44 185-195.

[56] Greger M., Landberg T. Use of willow in phytoextraction. Int. J. Phytoremed. 1999;1 115-123.

[57] He QB., Singh BR. Crop uptake of cadmium from phosphorus fertilizers. I. Yield and cadmium content. Water, Air, and soil Poll. 1994;74: 251 – 265.

[58] Heaton E., Voigt T., Long SP. A quantitative review compering the yields of two candidate C-4 perennial biomass crops in relation to nitrogen, temperature and water. Biomass & Bioenergy 2004;27 21-30.

[59] Heck RM., Farrauto RJ., Gulati S. Catalytic air pollution control. John Wilwy and Sons INC., New York 2002.

[60] Herms U., Brümmer G. Influencing factors of heavy metal solubility and retention in soil. Z. Pflanzenern Bonenk. 1991;147 400 – 424 (*in Deutsch*).

[61] Hodkinson TR., Chase MW., Takahashi C., Leitch IJ., Bennett MD., Renvoize SA. The use of DNA sequencing (ITS and *trn L-F*), AFLP, and fluorescent in situ hybridization to study allopolyploid *Miscanthus (Poaceae)*. American Journal of Botany 2002;89 279-286.

[62] Hofman M., Wachowski L. Platinum and lead along the main exit routes from the city of Poznan. Environmental pollution control, Journal of Polish Sanitary Engineers Association 2010;32(3) 43-47 *(in Polish)*.

[63] Huang JW., Chan J., Berti WR., Cunningham SD. Phitoremediation of lead+contaminated soils Role of synthetic chelate in lead phytoextraction. Environ. Sci. Technol. 1997;31 800-805.

[64] IUNG Report on the condition of soil and agriculturally utilized area in Poland 1980 -1990. IUNG 1992 *(in Polish)*.

[65] Iżewska A. Contents of heavy metals in *Miscanthus sacchariflorus* as an indicator of usefulness of sewage sludge and composted sludge]. Zesz. Probl. Post. Nuk Rol. 2006;512 165-171 *(in Polish)*.

[66] Jóźwiak M. Influence of cement industry on accumulation of heavy metals in bioindicators. Ecological Chemistry and Engineering. S 2009;16(3) 323-334.

[67] Kabata–Pendians A., Pendias H. Biochemistry trace elements. Warszawa: PWN 1999 *(in Polish)*.

[68] Kabata–Pendians A., Pendias H. Trace elements in soils and plants. USA: CRC Press. Boca Raton. FL. 2001.

[69] Kabata-Pendias A., Piotrowska M. 1987. Trace elements as a criterion for the suitability of agricultural wastes. IUNG Puławy: 1987; Seria P 39.

[70] Kalembasa D. The amount and chemical composition of ash obtained from biomass of energy crops. Acta Agrophysica 2006;7(4) 909-914 *(in Polish).*

[71] Kalembasa D., Malinowska E. Contents of cadmium, lead and nickel at different development stages of selected *Mscanthus* genotypes. Ecological Chemistry and Engineering. A 2009;16(4) 349-356.

[72] Kalembasa D., Malinowska E. Chemical composition and yield of biomass in selected clones of Miscanthus grasses. Element cycle in nature. Instytut Ochrony Środowiska. Warszawa: Monograph 2005;3 315-318 *(in Polish).*

[73] Kalembasa S., Godlewska A. Cao effect of the addition of sewage sludge to the content of heavy metals in different extracts. Zesz. Probl. Post. Nauk. Rol. 1998;456 193 – 196 *(in Polish).*

[74] Kleiber T. 2009. Nutritional resources of soil in the localities of monumental large-leaved linded (*Tilia platyphyllos f. aurea*) alleys. Ecological Chemistry and Engineering 2009;16(3) 277-286.

[75] Klimowicz Z., Melke J. The content of heavy metals in soils in the vicinity of traffic roads using chosen stretches of road as examples. Roczniki Gleboznawcze 2000;T. LI(3/4) 37–46.

[76] Knezevic M., Stankovic D., Krstic B., Sijacic Nikolic M., Vilotic D. Concentrations of heavy metals in soil and leaves of plant species *Paulownia elongata* S.Y.Hu and *Paulownia fortunei* Hemsl. African Journal of Biotechnology 2009;8(20) 422-5429, 19 October, 2009, DOI: 10.5897/AJB09.844.

[77] Kozak M., Kotecki A., Dobrzański Z. The *Miscanthus giganteus* response to chemical contamination of soil. Górecki H. (red) Chemistry and biochemistry in the agricultural production and environment protection. Czech-Pol_Trade, Prague: 2006; 520-524.

[78] Kozik E., Golcz A. Plant nutrients. Research Methods In Plant Sciences vol. 3. Soil Sickness. In Narwal S.S., Politycka B., Fengzhi Wu, Sampietro D.A. (ed.) Studium Press LLC, Huston USA 2011; 21-41.

[79] Łabętowicz J., Rutkowska B. Factors determining the concentration of trace elements in the soil solution. Post. Nauk Roln. 2001;6 75-85 *(in Polish).*

[80] Larcher F., Vigetti A., Merlo F., Ajmone-Marsan F., Devecchi M. New methods for the recovery of post industrial areas: choosing plants for phytoremediation. ISHS Acta Horticulturae 881: II International Conference on Landscape and Urban Horticulture, Bologna, Italy: 2010.

[81] Lawal AO., Batagarawa SM., Oyeyinka OD., Lawal MO. Estimation of heavy metals in neem tree leaves along Katsina – Dutsinma – Funtua Highway in Katsina State of Nigeria J. Appl. Sci. Environ. Manage. 2011;15(2) 327 – 330.

[82] Liu Jia-Nv, Zhou Qi-Xing, Sun T., Ma Lena Q, Wang S. Growth responses of three ornamental plants to Cd and Cd–Pb stress and their metal accumulation characteristics. Journal of Hazardous Materials 2008;151(1) 261-267.

[83] Majewska M., Kurek E. Microorganisms - factor that modifies the concentration of cadmium in the soil solution. Post. Nauk Roln. 2002;1 3-13 (*in Polish).*

[84] Matusiewicz H., Gała P. Optical emission spectrometry for determination of trace amounts of platinum metals (Pt, Pd, Ru, Rh, Ir) and Ca, Mg, Pb, Cr, Mn, Fe, Co, Ni, Cu, Zn, Cd in environmental samples. Ecological Chemistry and Engineering S 2009;16(4) 497-532.

[85] Maxted AP., Black CR., West HM. Crout NMJ., McGrath SP., Young SD. Phytoextraction of cadmium and zinc by *Salix* from soil historically amended with sewage sludge. Plant Soil 2007;290 157-172.

[86] McGrath SP., Zhao FJ. Phytoextraction of metals and metalloids from contaminated soils. Current Opinion in Biotechnology 2003;14 277-282.

[87] Mercik S., Kubik I. Chelation of heavy metals by humic acids and the effect of peat to download Zn, Pb, Cd by plants. Zesz. Prob. Post. Nauk Roln.1995;422 19-30 (*in Polish).*

[88] Mocek A., Drzymała S. The genesis, analysis, classification of soils. Poznan: Publishing company by Poznan Uniwersity of Life Sciences 2010 *(in Polish).*

[89] Naidu SL., Long SP. Potential mechanisms of low-temperature tolerance of C-4 photosynthesis in *Miscanthus* × *giganteus* an in vivo analysis. Planta 2004;220 145-155.

[90] Nedelkoska TV., Doran PM. Characteristics of heavy metal ustaje by plant species with potential for phytoremediation and phytomining. Minerals Engineering 2000;15(5) 549-561.

[91] Negri MC., Hinchman RR., Gatliff EG. Phytoremediation using green plants to clean up contaminated soil, groundwater and wastewater. International Tropical Meeting on Nuclear and Hayardous Waste Managemant, Spectrum 96, American Nuclear Societz, Seattle, WA: 1996.

[92] Neugschwandtner RW., Tlustos P., Komarek M., Szakova J. Phytoextraction of Pb and Cd from a contaminated Agricultural soil Rusing different EDTA application regimes: laboratory versus field scale measures of effciency. Geoderma 2008;144 446-454.

[93] Nowak W., Wojtasik A. The content of cadmium and nickel in carrots grown on two soil types using different fertilizers. Zesz. Prob. Post. Nauk Roln. 1997;448a 269-272 *(in Polish).*

[94] Ociepa A., Łach J., Gałczyński Ł. Benefits and limitations arising from the development of soils polluted by heavy metals under the crops of industrial-energetic plants. 2, 1: 231-235. Proceedings of ECOpole 2008;2(1) 231-235 *(in Polish).*

[95] Orban A. Compendium of soil Clen-Up technologies and Soil Remediation Companies. ECE/CHEM/115, United Nations, New York: 1997.

[96] Pandolfini T., Gremigni P., Gabbrielli R. Biomonitoring of soil health by plants. CAB International Wallingfford 1997; 325-347.

[97] Patra HK., Mohanty M. Phytomining: an innovative post phytoremrdiation management technology. International Quarterly Journal of environmental sciences 2013;3 15-20.

[98] Phytoremediation Work Team US EPA. Phutoremediation decision tree., ITRC Work Group, November 1999.

[99] Pilon-Smits EAH. Phytoremediation. Annual review of Plant Biology 2005;56 15–39.

[100] Pogrzeba M., Krzyżak J., Sas-Nowosielska A., Majtkowski W., Małkowski E., Kita A. A heavy metal environment al threat resulting from combustion of biofuels of plant origin. Environmental Heavy Metal Pollution and Effects on Child Mental Development. NATO Science for Peace and Security, Series C: Environmental Security 2011;1 213-225.

[101] Porębska G., Gworek B. Evaluation of the usefulness of plants to remediation of soils contaminated with heavy metals. Ochrona Środowiska i Zasobów Naturalnych 1999;17 81-89.

[102] Pyter R., Heaton E., Dohleman F., Voigt T., Long S. Agronomic Experiences with *Miscanthus × giganteus* in Illinois, USA. Jonathan R. Mielenz (ed.), Biofuels: Methods and Protocols, Methods in Molecular Biology. Humana Press, a part of Springer Science+Business Media, LLC 2009;581 41-52.

[103] Raskin I., Kumar NPBA., Dushenkov V., Salt DE. Bioconcentration of heavy metals by plants. Curn. Opin. Biotechnol. 1994;5 285-290.

[104] Sady W. Fertilization of field-grown vegetables. Kraków: 2000; p33 *(in Polish).*

[105] Sady W., Rożek S. The effect of physical and chemical soil properties on the accumulation of cadmium in carrot. Acta Hort. 2002;571 73-75.

[106] Sady W., Rożek S., Domagała-Świątkiewicz I. Bioaccumulation of cadmium in carrots, depending on the selected soil properties. Zesz. Nauk. AR Kraków 2000;364 171-173 *(in Polish).*

[107] Salt DE., Blaylock M., Kumar NPBA., Dushenkov V., Ensley BD., Chet I., Raskin I. Phytoremediation a novel strategy for the removal of tonic metal from the environment using plants. Biotechnology 1995;13 468-474.

[108] Salt DE., Smith RD., Raskin I. Phytoremediation. Ann. Rev. Plant Physiol. Plant Mol. Biol. 1998;49 643-649.

[109] Schonoor JL. Phytoremediation of soil land groundwater. GWARTAC Technology Report TE-02-01 (March 2002).

[110] Schwartz C., Echevarria G., Morel JL. Phytoextraction of cadmium with *Thlaspi caerulescens.* Plant Soil 2003;249 27-35.

[111] Schwitzguebel JP. van der Lelie D., Glass DJ., Vangonsveld J., Baker AJM. Phytoremediation: European and American trends, successes, obstacles and needs. J. Soils Sediments 2002;2 91-99.

[112] Smal H., Misztal M., Ligęza S., Stachyra J. Effect of soil acidification on the content of selected trace elements in the soil solution in laboratory test conditions. Zesz. Prob. Post. Nauk Roln. 1998;456 565-571.

[113] Sowa I., Wójciak-Kosior M., Kocjan R. The content of some trace elements in selected medicinal plants collected in the province of Lublin.. Acta Sci. Pol., Hortorum Cultus 2012;11(6) 15-22.

[114] Strączyńska S., Strączyński S. Cadmium in soils derived from different source rocks of the massif 'Śnieżnik'. Cadmium in the environment - environmental issues and methodological. Zesz. Nauk. Kom. PAN 2000;26 73 – 76 (*in Polish).*

[115] Szczepocka A. Criteria for estimating soil pollution from heavy metals. Zeszyty Naukowe Szkoły Głównej Służby Pożarniczej 2005;32 13-29 *(in Polish).*

[116] Szempliński W., Dubis B. Preliminary studies on yielding and energetical efficiency of selected crops grown for biogas generation. Fragm. Agron. 2011;28(1) 77–86 *(in Polish).*

[117] Tomašević M., Rajšić S., Đorđević D., Tasić M., Krstić J., Novaković V. Heavy metals accumulation in tree leaves from urban areas. Environ Chem Lett 2004;2 151–154 DOI 10.1007/s10311-004-0081-8.

[118] Tyler G., Olson T. Concentration of 60 elements in the soil solution as related to the soli acidity. Europ. J. Soli. Scie. 2001;52 151- 165.

[119] US EPA Technology Innovative Office. A citizen's guide to phytoremediation. Technology fact sheet, EPA 542-F-98-011, August 1998.

[120] Van der Ent A., Baker AJM., Reeves RD., Pollard AJ., Schat H. Hyperaccumulators of metal and metalloid trace elements: Facts and fiction. Plant Soil 2013;362 319-334.

[121] Vangronsveld J., Herzig R., Weyens N., Boulet J., Adriaensen K., Ruttens A., Thewys T., Vassilev A., Meers E., Nehnevajova E., van der Lelie D., Mench M. Phytoremedia-

tion of contaminated soils and groundwater: lessons from the field. Environ Sci Pollut Res. 2009;16 765-794.

[122] Vassilev A., Schwitzguebel JP., Thewys T., van der Lelie D., Vangronsveld J. The use of plants for remediation of metal contaminated soils. Scientific World J. 2004;4 9-34.

[123] Wenzel WW., Unterbruner R., Sommer P., Sacco P. Chelate-assisted phytoextraction using canola (*Brassica napus* L.) in outdoors pot and lysimeter experiments. Plant Soil 2003;249 83-96.

[124] Zhou Qi-Xing., Wang Xiao-Fei. Ecotoxicological effects of cadmium on three ornamental plants. Chemosphere 2005;60(1) 16-21.

Section 4

Risk Assessment of Organic Contaminants - Case Studies

Chapter 21

Diagnosis and Management of Field Pollution in the Case of an Organochlorine Pesticide, the Chlordecone

T. Woignier, F. Clostre, P. Cattan, J. Levillain,
Y.M. Cabidoche and M. Lesueur-Jannoyer

Additional information is available at the end of the chapter

http://dx.doi.org/10.5772/57263

1. Introduction

1.1. Background

The consequences of pesticide pollution of soil and water can be extremely damaging for both the environment and human health. Chlordecone (CLD), an organochlorine insecticide ($C_{10}Cl_{10}O$), was used to control the banana black weevil more than twenty years ago in the French West Indies but continues to contaminate the environment today [1, 2]. CLD pollution occurs in many parts of the world [3-5] and CLD is very persistent, sorptive and highly lipophilic [6]. Its persistence in soils is due to its low solubility in water [7], high affinity for organic matter with a K_{oc} (soil organic carbon content/water partitioning coefficient) of between 2.5 and 20 m^3kg^{-1} [1, 8], and to its chemical structure (Figure1), which makes it poorly biodegradable [9]. Similar characteristics are shared by all organochlorine pesticides and persistent organic pollutants.

CLD causes diffuse pollution in agricultural soils [10], which in turn become a continuous source of contamination for water resources, crops and animals [1, 3, 11, 12]. In addition, its long-term effects on human health and child development linked to the consumption of polluted food and water are now a serious concern [13-15].

1.2. Objective

This chapter explores the multifactorial nature of soil pollution and its evolution at field and regional scale. We focus on two key factors that determine CLD contamination and dispersion in ecosystems. One is physical and the other anthropogenic. The first is **the influence of clay microstructure** on (1) the concentration of CLD in the soil, (2) the bioavailability of the pollutant for crops and water resources, and (3) the efficiency of the remediation process. The second factor is past and present **farming practices** through their effects on (1) the stock of pollutant and its components in the soil and its distribution at the plot and regional scales, and (2) the diffusion of the pollutant in the different compartments of the environment. These two factors determine the contributing areas and their pollutant stock, the potential availability and fluxes of the pollutant, and hence its potential transfer to different environmental compartments.

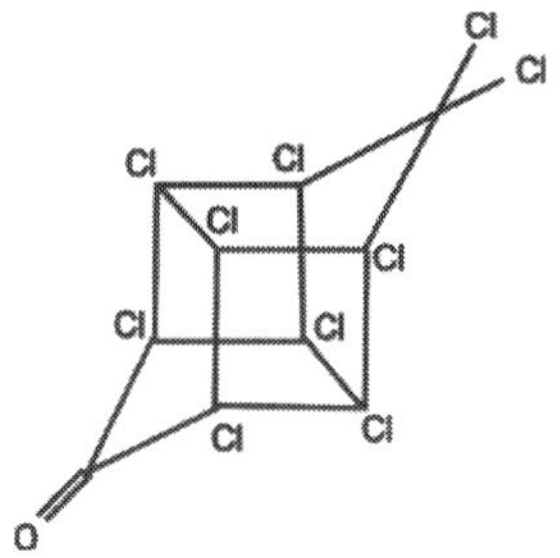

Figure 1. The chlordecone molecule ($C_{10}Cl_{10}O$)

2. Influence of clay microstructure

The studied soils come from Guadeloupe (16°15 N, 61°35 W) and Martinique (14°40 N, 61°00 W) in the French West Indies. These volcanic islands rise to 1467 m and 1388 m elevation respectively. Rainfall is high and ranges from 1000 to 10 000 mm/year depending on the elevation and geographical area. All primary minerals of andesitic rocks are weathered, so that soils have a high content of secondary minerals (clays): halloysite for nitisol, halloysite and Fe-oxihydroxides for ferralsol, and allophane for andosol, the three main soil types contaminated in French West Indies [1, 10, 16].

These volcanic soils have a high infiltration capacity (saturated hydraulic conductivity greater than 60 mm h^{-1}) [1, 10]. However in the "clay" matrix of andosols, there are pores smaller than 1 micron, where water and solute transfers are slow. All these soil types are acidic (pH = 4.5 -6), which prevents clay dispersion and sheet erosion [1]. Among these soils, the carbon content of andosol is particularly high, which may influence the retention of the compound (stock) and its availability.

Indeed all soils are not equivalent in terms of pesticide contamination and in their ability to transfer the pollutant to water and to plants [17, 18]. For example, although andosols are highly polluted [1, 10, 16], data show they release less pesticide to percolating water and crops than other soils [1, 19]. In the case of CLD, one explanation for the retention effect reported in the literature is the high organic content of these soils and the high affinity of the pesticide for soil organic carbon [1, 8].

However, these volcanic soils contain amorphous clay, allophane, whose structure and physical properties differ from those of the crystalline clays found in nitisols and ferralsols. Allophane clay is amorphous and has physical features that closely resemble those of nano porous materials: large pore volume and water content, a broad pore size distribution, a high specific surface area and a fractal structure [20-22]. One objective of this chapter is to show the influence of the clay microstructure on the accumulation and retention of chlordecone in soils.

2.1. Allophane microstructure

Figure 2 shows the pore volume and specific surface area as a function of the allophane content of a set of andosols. There was a clear increase in these two textural features with an increase in allophane content, showing that allophane clay favors larger porous features. The pore volume and the specific surface area were well correlated with allophane content (respectively $P<0.0001$ $r^2=0.87$and $P<0.0001$, $r^2=0.80$). The specific surface area was as high as 180 $m^2.g^{-1}$ and pore volume close to 2.5 $cm^3.g^{-1}$. This combination of high specific surface area and large pore volume suggests that the porous structure is made up of both micro- and mesopores.

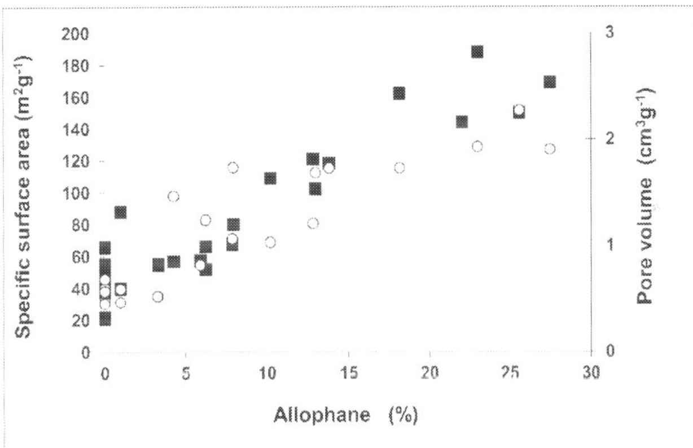

Figure 2. Pore volume (empty circles, ○) and specific surface area (black squares, ■) versus allophane content

Figure 3 shows the structure of amorphous clay in comparison with classical phyllosilicate clay: kaolinite or halloysite. These micrographs confirm the spongy structure of allophane clay.

The morphology of the allophane aggregates is peculiar [21]. Allophane has a very open structure made up of aggregated small particles (3-5 nm) that form clusters of around 10-20 nm. The clusters can stick together and form larger and larger aggregates up to ~ 100 nm in size. In comparison, the plate-like particles of phyllosilicate clay are 300-1000 nm in size (Figure 3). This aggregation mechanism is in agreement with results in the literature [22, 24].

Phyllosilicate clay (halloysite or kaolinite)   Allophanic clay

Figure 3. Scanning electronic micrographs showing the structure of phyllosilicate clay and amorphous clay from [23]

Wada [22] describes allophane particles as nearly spherical, with diameters ranging from 3-5 nm. The aggregation mechanism corresponds to a fractal morphology. Several authors assumed that allophane could have a fractal structure [20, 22]. The structure of allophane can be studied at nanoscale using scattering experiments to quantify the fractal features of the soil samples. Small angle X-ray scattering (SAXS) experiments make it possible to calculate the fractal dimension D_f (which expresses the compactness and tortuosity of the clusters) and the extent of the fractal aggregates (ξ) [25]. The fractal extent can be considered as the size of the tortuous "nano-labyrinth". SAXS curves [26] show that the fractal dimension D_f is constant (2.5-2.7). Table 1 lists changes in ξ versus allophane content. Our results showed that the size of the fractal labyrinth increased with an increase in allophane content (P=0.001 and r^2=0.71).

Allophane (%)	0	3	5	8	10	12	13	15	18	22
ξ (nm)	0	12	23	22-32	18	23-35	23	35	34-45	42-60

Table 1. Fractal range (ξ) versus allophane content

These data (high specific surface area and pore volume, and fractal features) describe a highly tortuous microstructure and small mesopores, suggesting that accessibility inside the clay microstructure is reduced.

2.2. Pesticide sequestration in allophane

Allophane clay has a spongy structure comprising aggregated small particles that form a tortuous network with small pores. This peculiar structure influences the concentration of pesticide in the soil.

Figure 4 shows the marked increase in CLD concentration in soils with increasing allophane content (P<0.0001 and r^2=0.807). This finding confirms previous data in the literature [1, 16, 27] stating that allophanic soils are more contaminated than other kinds of tropical soils,

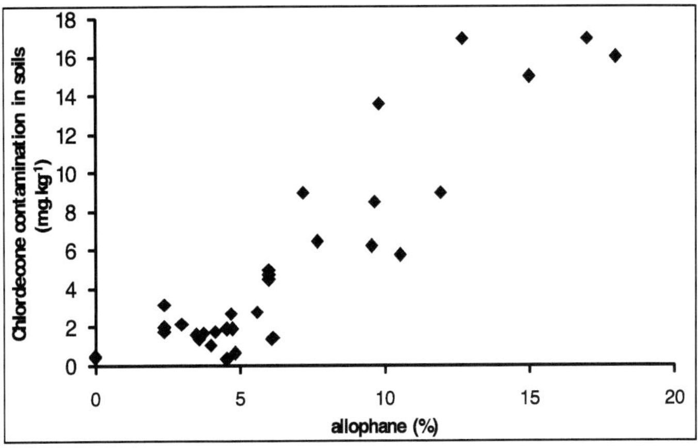

Figure 4. Soil CLD contamination versus allophane content. CLD data have a confidence interval of 30%

which contain the usual crystalline clays (like halloysite and kaolinite). Figure 4 also clearly shows the link between CLD contamination of the soil and the allophane content of the soil and one can thus assume that the pesticide retention properties of the soil are partially dependent on the features of allophane.

Another interesting result of our previous study was that for similar soil CLD contents, crops cultivated on allophanic soils were much less contaminated than the same crops cultivated on soils containing classical clays, [26]. We calculated the mean soil to plant transfer (expressed in μg.kg^{-1} of fresh matter / μg.kg^{-1} of dry soil), for different crops. Table 2 confirms that, whatever the crop concerned [26], CLD soil to plant transfer was always higher in halloysite soils than in allophanic soils. For the three crops studied (lettuce, yam, and dasheen) the ratio was close to 3.

Crops	Soil to crop transfer (halloysite soils)	Soil to crop transfer (allophanic soils)	Ratio (halloysite: allophanic)
Lettuce	0.64 (0.14)	0.23 (0.15)	2.83
Yam	3.2 (2.1)	1.04 (0.75)	3.13
Dasheen	12.2 (5.8)	3.26 (2.56)	3.73

Table 2. Mean CLD transfers from halloysite and allophanic soils to lettuce, yam and dasheen, expressed in μg.kg^{-1} of fresh matter / μg.kg^{-1} of dry soil (standard deviation in brackets) [26]

2.3. Trapping mechanism in allophane

The results shown in Figure 4 and Table 2 may appear contradictory because one would expect allophanic soils, which are more contaminated, to strongly pollute cultivated vegetables. One explanation for the observed effect is that CLD is trapped in the microstructure of the allophane clay, thus reducing its transfer from the soil to the plant. The influence of soil allophane content on CLD retention is the signature of the peculiar microstructure of the allophane aggregate. The spongy structure influences transport inside the allophane aggregates. The SAXS data made it possible for us to propose a mechanism for the retention of pesticides in allophanic soils and also for the limited release of CLD to crops and water resources [1, 19]. At the scale of the allophane, accessibility is difficult because of the fractal structure and small pore size of allophane clay. CLD transfers within the soil depend on hydraulic conductivity (*K*) and diffusion processes (*Di*) in the porous microstructure. The fractal structure allows an approximation of *K* and *Di* at the aggregates scale *l* [26] through the following equations:

$$K(l) \infty \left[1-\left[l/a\right]^{Df-3}\right]/l^2 \text{ and } Di \infty \left(1-(l/a)^{Df-3}\right)\left[1-2/3\left(2-(l/a)^{Df-3}\right)\left((l/a)^{Df-3}\right)^{3/2}\right]. \quad (1)$$

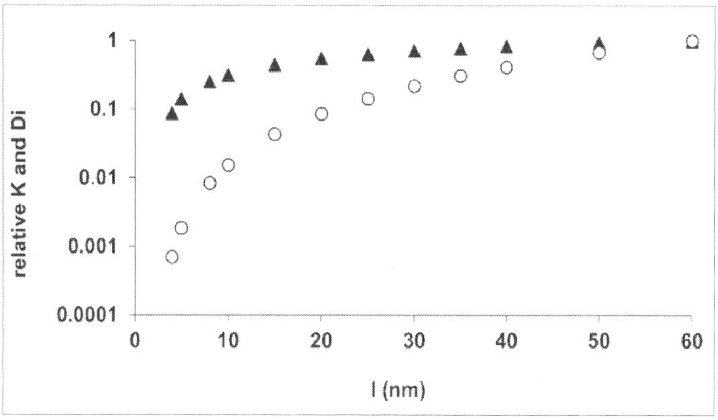

Figure 5. Relative K (o) and Di (▲) versus the scale length, l (nm)

We calculated the transport properties (hydraulic conductivity and diffusion, inside the allophane fractal aggregate, i.e. between 3 and 60 nm (Table 2). Figure 5 shows changes in *K* and *Di* normalized to *K* and *Di* at *l* = 60 nm. Hydraulic conductivity decreased by 4 orders of magnitude and *Di* decreased by 20 orders of magnitude when *l* decreased from 60 to 4 nm.

The very low calculated *Di* and *K* suggest that CLD is trapped in the porosity. The trapping mechanism is favored by the large size of the fractal labyrinth. The higher the allophane content, the bigger the labyrinth and the higher the retention. Like nanoporous materials [28], the paradox of allophane clay is that it has large porosity but poor transport properties. In these fractal structures, possible reactions with chemical or biological species that could extract the pesticides are thus hindered; the pesticide remains trapped inside allophane clay and cannot be extracted.

2.4. Influence of allophane on the decontamination process

Soil features influence pollutant availability and hence natural decontamination of the soil. In the case of CLD and other organochlorine pesticides, natural decontamination is very slow, and pollution management and remediation have to take into account any soil properties that could influence the efficiency of the treatment considered [1].

Although different strategies are proposed in the literature to remediate diffuse pollution of CLD including phytoextraction [29] and microbial degradation [30, 31], to date, these approaches have not been very successful. Recently "In situ chemical reduction" was tested for the reduction of the pesticide in the soil [32], but the technique was clearly less efficient in soils containing allophane clay. Six months of in situ chemical reduction resulted in an 88% decrease in CLD content in halloysite soils but only in a 47% decrease in allophanic soils. The efficiency of the in situ chemical reduction technique was probably affected by poor accessibility to the allophane clusters. Trapping thus likely reduces the degradation of pesticides in andosols. Whatever the soil decontamination process used to desorb or degrade the CLD, the confining structure of the clay has to be taken into account to ensure the process is effective.

3. Key farming practices

Soil type determines the potential for CLD sequestration but, in the analysis of risk and in the diagnosis of soil pollution, farming practices are key factors that determine the level of the pollutant stock, its potential availability, and its potential transfer to different environmental compartments [1, 10, 33]. In this section, we examine the effect of three main types of practices characteristic of past and present farming practices. The first is the level of intensification of the cropping system, which, in the case of CLD, determines the initial input [1, 10, 33]. The second is soil tillage, which determines the depth (and hence the volume) of soil affected by the pollution [1, 10, 33]. The third is organic matter amendment, which affects the availability of CLD [34].

3.1. Level of intensification of the cropping system

It is a truism to say that soils are polluted because they have received pollutants. Farmers use a wide range of strategies based on pesticides to protect their crops. These strategies mainly depend on the intensification of the cropping system [35-37]. Intensive systems produce a higher cash flow and more profit than small diversified systems [38, 39] and farmers tend to minimize the risk of yield loss by intensifying chemical pest control [10]. These farmers have to deal with higher pest pressure than systems that include crop rotation and diversification, and consequently require more frequent treatments [40, 41]. In agro-industrial banana plantations in the French West Indies, agronomic and economic conditions led to the intensive use of agrochemicals [39] i.e. frequent applications and/or high doses of the persistent molecule CLD over large areas. In 2013, twenty years after the treatments ended, the ubiquitous presence of the CLD in different environmental compartments raises the question of the impact of these past applications on soil contamination and its true extent.

To answer this question, the link between CLD supply and soil contamination was first examined using a simple CLD leaching model called WISORCH [1]. This model predicts soil CLD content based on the history of CLD applications under different farming systems, different soil types, and different average annual rainfalls. The model accounts for andosol, nitisol and ferralsol through their main characteristics, notably the soil-water partition coefficient relative to organic C content (K_{oc} in m^3 kg^{-1}), soil organic carbon content, depth of tillage and soil bulk density.

Simulation results first showed that the schedule of CLD applications, i.e. CLD loads, had the most impact. Long after the application of CLD, the effect of CLD loads on soil CLD content meant that soil decontamination was extremely slow. By exploring different assumptions, the WISORCH model provided two main explanations for the slowness of decontamination: 1) the absence of degradation; 2) only lixiviation by percolation water can slowly reduce soil contamination. These assumptions are consistent with the lack of evidence for natural degradation of CLD reported in the literature [9]; given the very low volatility of CLD, water is the only vector of CLD dispersion.

In addition to the main effect of CLD applications, WISORCH simulations identified tillage depth as the second factor that influences soil CLD content. This factor depends on cropping systems and is discussed below. Here we simply note that physical factors like soil gravimetric carbon content, bulk density and average annual rainfall have less impact. Consequently, human activity was the first determinant of soil contamination.

The WISORCH model explained why large areas are still contaminated even though treatments were halted long ago. As the model first identified the schedule of pesticide applications, this suggested that tracing the history of these applications would help assess the level and distribution of soil contamination at a regional scale. A historical analysis was performed in Guadeloupe using maps of banana plantations at different dates. The results revealed three gradients of land use for banana in 1 145 plots (1 376 ha) that were analyzed for the presence of CLD [10].

Overall, plot contamination increased with the duration of land use in andosols and ferralsols. Table 3 shows that on average, plots with short term banana land use were less contaminated than those with medium and long term use. The lack of a significant difference between medium and long term banana land use in CLD stocks and concentrations was consistent with the widespread use of CLD in the 1980s and the 1990s, but this was not the case in the 1970s, when CLD was used frequently but not systematically. The effect of banana land use duration was significant for andosols and ferralsols but not for nitisols. This difference in behavior between soils could be explained by their ability to retain CLD. In this case, unlike andosols and ferralsols, the lower retention capacity of nitisols could mask the impact of variations in CLD inputs with respect to the length of time the land was used for banana cultivation.

	Land use duration			
log([CLD])	**Short**	**Medium**	**Long**	***Mean soil***
Andosol	0.87 a	1.14 b	1.16 b	*1.11 c*
	(1.39)	(2.13)	(2.19)	*(2.05)*
Ferralsol	0.25 a	0.52 b	0.38 ab	*0.45 a*
	(0.28)	(0.69)	(0.46)	*(0.57)*
Nitisol	0.62 a	0.66 a	0.64 a	*0.65 b*
	(0.86)	(0.94)	(0.89)	*(0.91)*
Mean duration	*0.64 a*	*0.91 b*	*0.91 b*	
	(0.9)	*(1.48)*	*(1.48)*	

Table 3. Effect of soil type and banana land use duration (short, medium and long term) on the log of mean soil CLD concentration (corresponding values in mg kg-1 in brackets). Differences in means were assessed at two levels: globally, for the "mean soil" column and "mean duration" rows (in italics); for each soil, i.e. each row (in normal font). For each case, the letters a, b and c indicate significant differences between factor levels at p <0.05 (Kruskal-Wallis test). [10]

The soil sorption capacity was first assessed by inverting the WISORCH model [1] to determine K_{oc}. Results showed that andosols had a higher sorption capacity (K_{oc} of 12-25) than ferralsols (K_{oc} of 7.5-12) and nitisols (K_{oc} of 2-3).

This is consistent with our previous results concerning CLD sorption. Indeed, these differences in K_{oc} do not stem from the chemical composition of the clays, since the allophane in andosols does not significantly differ from phyllosilicate clays (nitisols and ferralsols) [42]. The higher apparent K_{oc} in andosols could be the result of the allophane microstructure, leading to lower CLD availability.

Moreover, large farms were more contaminated than small farms [10]. We selected farms with more than 10 plots to assess the intra-farm variation in CLD concentrations. Figure 6 shows that inter-farm variation in CLD concentration was far higher than the intra-farm variation whatever the soil type, meaning the farm factor was decisive in explaining the distribution of CLD concentration. Regardless of the type of soil, the treatment strategy used on a given farm was an essential component of soil contamination.

Concerning the type of soil and other physical factors of CLD retention in soils, at regional scale, the effect of the type of soil was clear. Analysis of CLD concentrations (Table 2) revealed significant differences between soils (p<0.05 Kruskal-Wallis test) showing that andosols were the most contaminated. Values of CLD concentrations in andosols were 2.3-fold higher than in nitisols, and 3.6-fold higher than in ferralsols. Organic carbon (OC) content was only calculated for andosols to avoid possible interactions with other physical variables (notably, allophane). Results showed that andosols with high OC content tended to retain CLD better. However, the relationship was weak. A probable explanation is that OC had less impact due to the variability of inputs. Finally, although carbon is considered to deter-

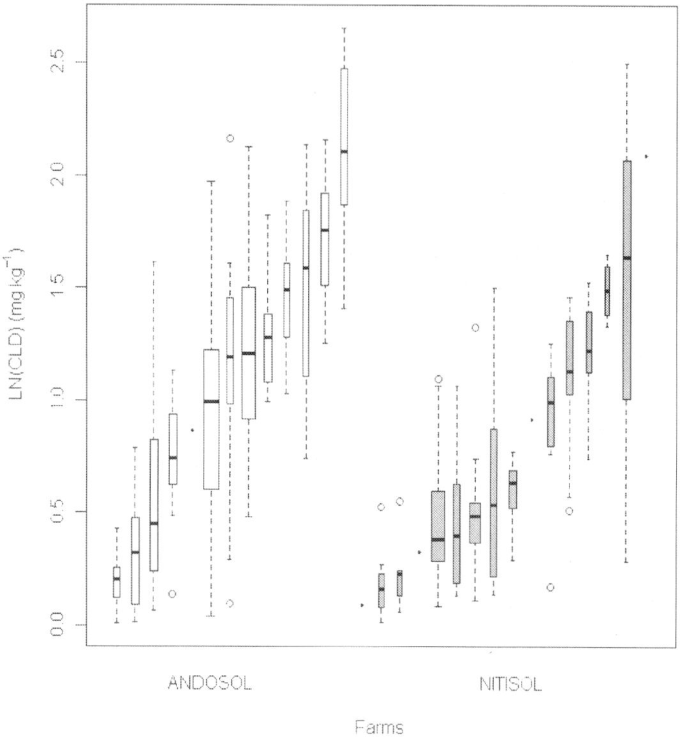

Figure 6. Distribution of CLD concentration on farms located on andosols and nitisols with more than 10 banana plots. Each box represents a farm. The bottoms and tops of the boxes represent the 25th and 75th percentile; the band inside the box is the median; the whiskers extend to the most extreme data point, which is no more than 1.5 times the interquartile range of the box [10].

mine sorption of CLD, at the regional scale, organic carbon was not a prime factor in explaining variations in CLD concentrations.

To summarize, findings at the regional scale were consistent with the findings of the WISORCH model at the plot scale. Farming systems mainly explained soil CLD contamination, and physical factors like soil carbon content had less influence. Concerning the type of soil, although the findings were the result of observations at different scales - region, plot, microstructure – they all highlighted the specific sorption capacity of andosols for CLD.

These important observations made it possible to draw maps of areas with a risk of CLD contamination in Guadeloupe and Martinique based on soil type and on an historical analysis of the supply of CLD [43, 44]. They also identified a pollution system characterized by remarkable inertia comprising (i) a persistent molecule – CLD; (ii) intensive large scale applications; (iii) andosols with high organic matter content and an allophanic microstructure that can trap pollutants. This inertia was assessed by simulating changes in soil CLD contamination using the WISORCH model. Figure 7 shows that one hundred years will be needed to clean up nitisols, and six hundred years to clean up andosols (Cabidoche et al., 2009).

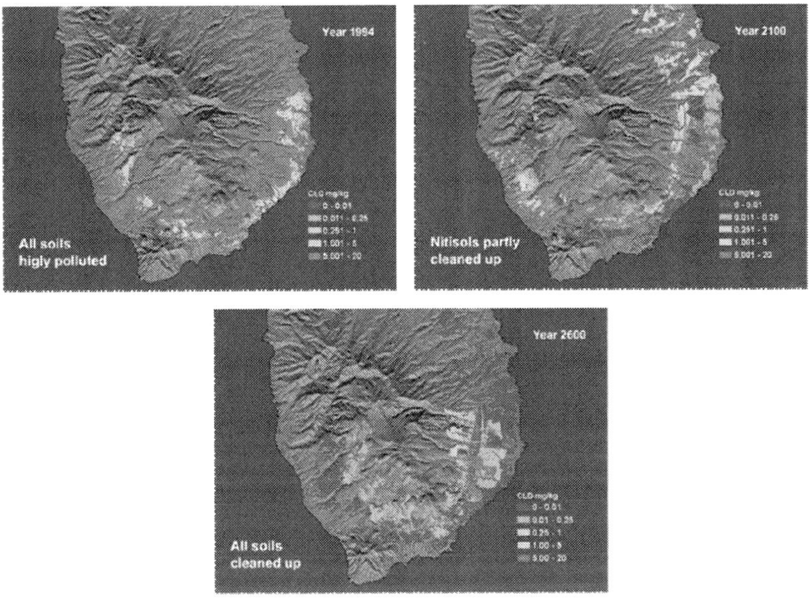

Figure 7. Simulation of changes in soil CLD contamination in Guadeloupe

Figure 7 highlights the need to use different temporal scales to assess pollution: a long term scale when soil retention properties are probably the main explanation for the spatial variability of soil contamination; short and medium term scales when applications of the pollutant explain most spatial variability. From a management point of view, authorities are thus justified in focusing on reducing pollutant loads (the frequency of application and the quantities of pollutant used) on cropped areas, even in the case of less persistent molecules. In the case of CLD, the persistence of pollution calls for further research on soil decontamination. However, in the meantime, different agricultural practices can help manage the risk of contamination, this being the case of soil tillage and organic matter amendment.

3.2. Tillage practices

A second step of the diagnosis is the analyses of soil tillage The heterogeneity of CLD content at field scale and the effect of tillage were investigated [33].

In the French West Indies, different tillage practices are used on banana plantations. These range from no tillage, especially on sloping plots, to regular deep tillage to a depth of 60 cm or more, every four years [10, 45]. The mode of application of the pesticide (powder spread on the ground around the foot of the banana tree), and the semi-perennial arrangement of trees [46] account for the high heterogeneity observed at field scale. Indeed in our study, in the same plot, CLD contamination of the upper soil layer (0-30 cm) could range from 0.2 to 2.7 mg kg^{-1} and, in plots of less than 1 ha, from 2.9 to 17.6 mg kg^{-1}.

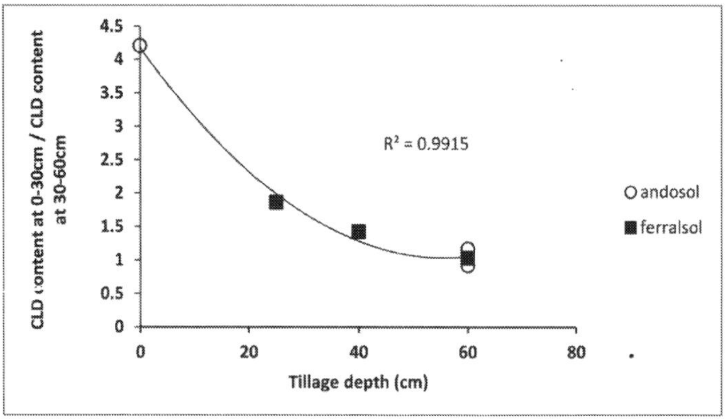

Figure 8. Ratio of CLD contents in the 0-30 cm and 30-60 cm soil layers as a function of the depth of tillage, adapted from [33].

Figure 8 shows that, whatever the soil type, the upper soil layer (0-30 cm) was generally more contaminated than the lower layer (30-60 cm). With no tillage, the 0-30 cm layer was four times more contaminated than the 30-60 cm layer (21 mg kg^{-1} and 5 mg kg^{-1}). We also showed that tillage had a significant effect ($P < 0.0001$) on the horizontal distribution of the contaminant, leading to pesticide dilution in the soil profile. With deep tillage (60 cm and deeper), whatever the type of soil and mean CLD content, CLD content was similar at the two sampling depths, with mean values of 11 vs. 12 mg kg^{-1}, 12 vs. 11 mg kg^{-1} and 2 vs. 2 mg kg^{-1}. Thus CLD content was homogenized in the 0-60 cm layer. In plots where tillage was shallower, the upper layer was still significantly more contaminated than the lower layer although the proportion depended on the tillage depth. This result is in accordance with results observed for DDT [47].

Likewise, tillage tended to reduce CLD horizontal heterogeneity. For this reason, when sampling soil, it is important to take such heterogeneity into account at intra-field scale for accurate assessment of CLD content. For sampling, it is recommended to subdivide all plots of more than 2 ha and, more generally, to use an appropriate sampling procedure that takes into account landscape (slope and resulting erosion), field history (tillage, cropping system, former inter-row distance, etc.) but also the reasons for sampling: cropping system management or analysis of overall risk.

Tillage is of major concern because it modifies the vertical distribution of the pollutant within the soil profile, and hence the volume of soil that is contaminated and the level of contamination. The risk of the pollutant being transferred to the crop depends on the type of soil and soil CLD content [19, 26]. Thus, in some cases (mainly low soil CLD content), this will determine the range of crops that can be cultivated while respecting regulatory thresholds, in particular the maximum residue limit of 20 μg kg^{-1} fresh matter for food products [23].

It is also important to keep in mind that tillage practices may negatively affect CLD sequestration and distribution. Indeed tillage during the dry season can cause surface desiccation and reduce pore volume, which will irreversibly alter the micro-structure of allophane [45, 48], thus possibly modifying soil CLD sequestration potential. Moreover, inappropriate tillage practices increase the risk of erosion, driving the top soil layer downslope [49] thereby modifying CLD distribution at plot scale, with higher CLD content at the bottom of the slope than at the top [33].

In conclusion, analyzing past and present farming practices provides insight into CLD content and distribution at intra-field scale. These practices can also affect CLD availability by modifying soil retention properties.

3.3. Pesticide sequestration by compost addition

The use of soil organic matter to control the environmental mobility and fate of pesticides has already been reported in the literature [50-53]. Here, we propose an alternative strategy which is quite the opposite of total soil decontamination: CLD sequestration enhanced by soil organic amendment. Because CLD is tightly trapped in the soil, an alternative solution to decontamination may be to further increase its sequestration in the soil thereby reducing pesticide diffusion into the environment. This could be a way to reduce further release of CLD from contaminated soils towards other environmental compartments until efficient remediation techniques become available. We now examine the hypothesis that adding organic matter to contaminated soils improves their CLD sequestration ability with the objective not of removing the pesticide from the soil but rather of controlling its release into the environment.

As detailed above, combined with high organic matter content, the microstructure and the large specific surface area of clay favor the accumulation of pollutants in the soil. In allophanic soil, the high CLD content is the result of the combination of CLD's high affinity for the soil organic content and the poor accessibility of CLD into the mesopore structure.

With the aim of preventing consumer exposure, we tested the incorporation of compost in soils as a possible way to reduce plant contamination [34], based on the hypothesis that adding organic matter would improve CLD trapping and thus reduce its bioavailability for crops. We characterized the transfer of CLD from soil to radish, a crop belonging to roots and tubers, a CLD sensitive group. Two months after incorporation of the compost, the contamination of the different plants organs was 3, 15 and 5 times lower in small roots, tubers and leaves, respectively than without added compost (Figure 9.).

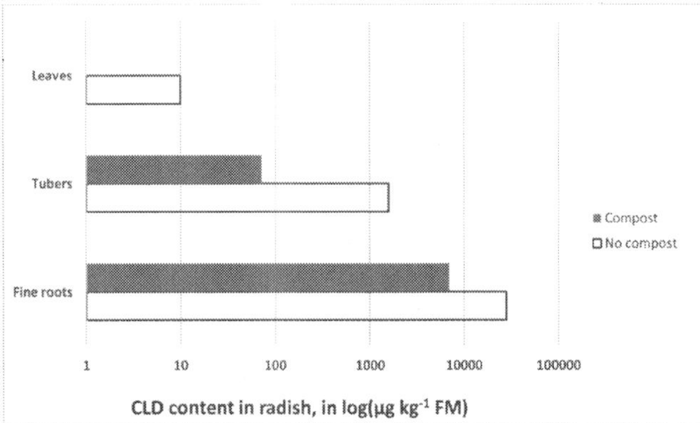

Figure 9. CLD transfer in radish organs (leaves, tubers and fine roots) with added compost (in grey) and without (in white).

These experiments also showed that adding compost closed the microstructure of allophane clays, thus favouring CLD retention in allophanic soils. Adding compost altered the porosity of the allophane clay in the size range 10 to 60 nm, while the intensity of this effect varied with the allophane content. We suggest that these pore changes are the consequence of capillary stress and of the low mechanical properties of the fractal structure [54].

Organic matter (OM) amendments aim to modify chemical conditions within the soil profile. As mentioned above, the OM soil status influences the availability of the pollutant, a factor of primary importance for both pollutant transfer and degradation [55]. At field scale, OM amendments modify the potential sequestration of persistent organic pollutants in the soil by enhancing the soil's sorption capacity for CLD [56]. This sorption capacity depends on the quantity of OM supplied, the type of OM (stable vs. labile) and the frequency of the application. This practice needs to be studied over time, as OM degrades and could modify the OM – pollutant relationships.

Finally, all these practices depend on farm strategies. When the aim is to modify these practices, both the scale and the type of farm need to be taken into account. In the case of CLD pollution, analyses revealed a strong "farm effect" [10]. A typology should be built including the farms' overall strategy and objectives, the types of crops grown, practices (more or less intensive) and the farms' specific field orientation (how often the land is used for each crop).

4. Discussion

When dealing with agricultural soil pollution, two major tasks must be included in the risk and management analysis: the characterization and analysis of soil physical properties and farmers' past and present practices (Figure 10). These two domains provide information on the level of soil pollution, on the behavior and fate of the pollutant in the soil, and on its remediation potential.

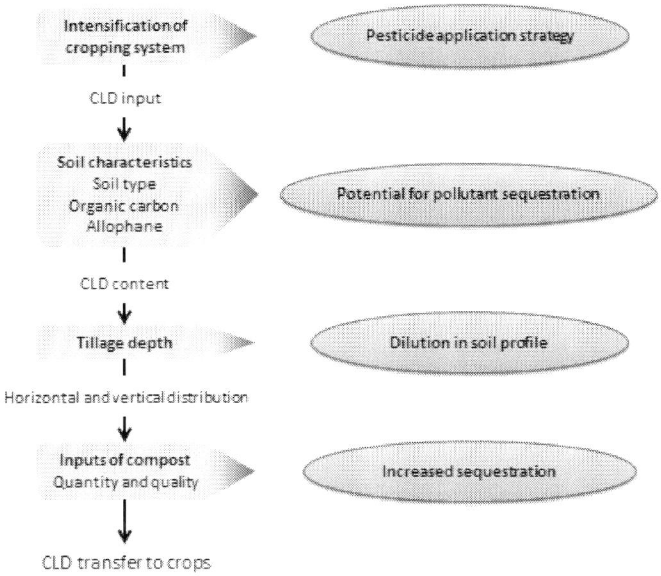

Figure 10. Key determinants of CLD soil pollution.

For soil physical properties, porosity can range from micro-, to meso- and macro-porous in volcanic soils. Fluids containing inorganic and organic solutes and gaseous species can occupy the pores and several factors (size, shape, distribution and connectivity of the pore geometries) determine how fluids migrate into and through the porosity and ultimately adsorb and react with the solid surfaces. The low hydraulic conductivity calculated for fractal allophane aggregates thus explains the high pesticide content of these materials. Because of the resulting low hydraulic conductivity, fluid exchange is slow and pesticide bioavailability is consequently reduced. This leads to the accumulation of pesticides that are not easily chemically or biologically transformed. Future studies on pesticide degrading microorganisms [57] and other bioremediation tools to clean up polluted soils should take the high soil organic carbon content and CLD accessibility in volcanic soils into account.

At field scale, past pesticide application practices account for the potential stock of pollutant in the soil, and an historical analysis will improve the initial diagnosis in terms of the quantity (doses) of pollutant and its horizontal distribution. Modeling and mapping are appropriate tools to roughly simulate current levels of pollution in the field and to identify the parts of a field that contribute to environmental pollution (water, food, animals). Current practices explain the diffusion of pollution from the soil reservoir: tillage accounts for the vertical distribution of the pollutant, i.e. dilution of the pollutant in the soil profile, and organic matter amendments account partially for the availability of the pollutant. In this way, current practices modify the level and volume of polluted soil and the environmental state of the field (surface conditions, OM content, soil profile, etc.), which in turn, influence the fate of the pollutant in the environment. Variability among and within farms also needs to be taken into account by building a farm typology including the general orientations and strategies of the farm and farm practices.

Finally, at the regional scale, all these factors combine to determine soil pollution. In the case of CLD, in agreement with the results of physical analyses, andosols, which contain allophanic clay and have high organic matter content, are the most polluted. In addition, the effect of agricultural practices was evidenced by a strong farm effect and an effect of how long the land was used for banana cultivation.

These two aspects (agricultural practices and soil physical properties) proved to be effective in the diagnosis of CLD pollution and in management analysis and could make sense for other forms of agricultural pollution, especially in the case of persistent pesticides. To illustrate this point, the WISORCH model, which accounts for agricultural loads as well as soil and climate characteristics, can simulate changes in soil pollution over time. In the case of CLD contamination, simulation results showed that soils will remain contaminated for six hundred years, with andosols decontaminating at a slower rate than other soils.

5. Conclusion

Wide use of chlorinated pesticides such as chlordecone has led to severe contamination of the environment in the French West Indies. Chlordecone is an organochlorine insecticide that was regularly applied to banana crops more than 20 years ago and is now on the list of POPs prohibited by the Stockholm Convention. Chlordecone is a hydrophobic, non-biodegradable pesticide that is strongly bound to the organic soil matrix which is why soils are still contaminated today. The level of soil pollution and the fate of the pollutant in the ecosystems depend on two main determinants at two different scales: agricultural practices and the soil physical properties, i.e. the clay microstructure.

Obviously, the history of agricultural practices (land-use changes and the intensity of pesticide use) partly explains soil contamination. The stocks of pesticide still present in soil are important indicators, which calls for the reinforcement of databases on agricultural practices, especially on pesticide use. However, soil type mainly explains the concentration of pollutant in the soil. Our experimental data showed that clay microstructure controls the accumulation of pollutants in the soil and their release into the environment. In the case of andosols, this study revealed the importance of the fractal microstructure and the associated tortuous porosity in trapping the pollutant. Moreover, as chlordecone tends to bind with soil organic matter, the high organic content of tropical volcanic soils also favors its retention.

More generally, it is worth noting that since pesticide retention depends on soil type, the concentration of pesticide in the soil cannot provide reliable information about the actual risk of contamination from the soil to the ecosystem. This makes mapping contamination hazards problematic because there is no univocal relation between soil pollution and its capacity to contaminate crops. Additional studies are thus necessary to characterize the behavior of pesticides in soils and to identify the link with plant absorption. This will increase the relevance of mapping pesticide contamination and availability

Concerning possible soil decontamination procedures, the efficiency of chemical or biological species able to degrade the pesticide will be controlled by the microstructure of the clays. Future research into pesticide degrading microorganisms and other remediation tools that could be used to clean up polluted soils should take the accessibility of the pesticide in the soil microstructure into account.

Acknowledgements

Funding was provided by the French Chlordecone National Plan ("JAFA" project), the French National Research Agency ("Chlordexco" project) and the French Ministry for Overseas development (MOM).

Author details

T. Woignier[1,2*], F. Clostre[3,4], P. Cattan[5], J. Levillain[5], Y.M. Cabidoche[6] and M. Lesueur-Jannoyer[3,4]

*Address all correspondence to: thierry.woignier@imbe.fr

1 CNRS UMR 7263- Institut Méditerranéen de Biodiversité et d'Ecologie marine et continentale (IMBE), Aix-Marseille Université, Faculté de St-Jérôme, Marseille, France

2 IRD UMR 237- Campus Agro Environnemental Caraïbes, Le Lamentin, Martinique

3 Cirad UR HortSys, TA B-103/PS4, Montpellier, France

4 Cirad/PRAM, UPR HortSys Agroecological Functioning and Performances of Horticultural Cropping Systems, Le Lamentin, Martinique

5 Cirad UPR Banana, Plantain and Pineapple Cropping Systems, Guadeloupe, France

6 Inra ASTRO (TROpicalAgroSystems) Research Unit, Domaine Duclos, Guadeloupe, France

References

[1] Cabidoche YM, Achard R, Cattan P, Clermont-Dauphin C, Massat F, Sansoulet J. Long-term pollution by chlordecone of tropical volcanic soils in the French West Indies: A simple leaching model accounts for current residue. Environmental Pollution. 2009;157(5):1697-705.

[2] Coat S, Monti D, Legendre P, Bouchon C, Massat F, Lepoint G. Organochlorine pollution in tropical rivers (Guadeloupe): Role of ecological factors in food web bioaccumulation. Environmental Pollution. 2011;159(6):1692-701.

[3] Luellen DR, Vadas GG, Unger MA. Kepone in James River fish: 1976–2002. Science of the Total Environment. 2006;358(1–3):286-97.

[4] Roche H, Salvat B, Ramade F. Assessment of the pesticides pollution of coral reefs communities from French Polynesia. Revue d'écologie. 2011;66(1):3-10.

[5] Wei S, Lau RKF, Fung CN, Zheng GJ, Lam JCW, Connell DW, et al. Trace organic contamination in biota collected from the Pearl River Estuary, China: A preliminary risk assessment. Marine Pollution Bulletin. 2006;52(12):1682-94.

[6] Epstein SS. Kepone-Hazard evaluation. Science of the Total Environment. 1978;9(1): 1-62.

[7] Dawson GW, Weimer WC, Shupe SJ. Kepone: a case study of a persistent material. Water American Institute of Chemical Engineers. 1979;Symposium Series 75(190): 366-74.

[8] Kenaga EE. Predicted bioconcentration factors and soil sorption coefficients of pesticides and other chemicals. Ecotoxicology and Environmental Safety. 1980;4(1):26 - 38.

[9] Jablonski PE, Pheasant DJ, Ferry JG. Conversion of Kepone by Methanosarcina thermophila. FEMS Microbiology Letters. 1996;139(2–3):169-73.

[10] Levillain J, Cattan P, Colin F, Voltz M, Cabidoche Y-M. Analysis of environmental and farming factors of soil contamination by a persistent organic pollutant, chlordecone, in a banana production area of French West Indies. Agriculture, Ecosystems & Environment. 2012;159(0):123-32.

[11] Gourcy L, Baran N, Vittecoq B. Improving the knowledge of pesticide and nitrate transfer processes using age-dating tools (CFC, SF6, 3H) in a volcanic island (Martinique, French West Indies). Journal of Contaminant Hydrology. 2009;108(3-4):p. 107-17.

[12] Jondreville C, Bouveret C, Lesueur-Jannoyer M, Rychen G, Feidt C. Relative bioavailability of tropical volcanic soil-bound chlordecone in laying hens (Gallus domesticus). Environ Sci Pollut Res Int. 2012.

[13] Boucher O, Simard M-N, Muckle G, Rouget F, Kadhel P, Bataille H, et al. Exposure to an organochlorine pesticide (chlordecone) and development of 18-month-old infants. NeuroToxicology. 2013;35(0):162-8.

[14] Dallaire R, Muckle G, Rouget F, Kadhel P, Bataille H, Guldner L, et al. Cognitive, visual, and motor development of 7-month-old Guadeloupean infants exposed to chlordecone. Environmental Research. 2012;118(0):79-85.

[15] Multigner L, Ndong JR, Giusti A, Romana M, Delacroix-Maillard H, Cordier S, et al. Chlordecone Exposure and Risk of Prostate Cancer. Journal of Clinical Oncology. 2010;28(21):3457-62.

[16] Brunet D, Woignier T, Lesueur-Jannoyer M, Achard R, Rangon L, Barthès BG. Determination of soil content in chlordecone (organochlorine pesticide) using near infrared reflectance spectroscopy (NIRS). Environmental Pollution. 2009;157(11):3120-5.

[17] Kumar M, Philip L. Adsorption and desorption characteristics of hydrophobic pesticide endosulfan in four Indian soils. Chemosphere. 2006;62(7):1064-77.

[18] Wang T-Y, Lu Y-L, Dawson RW, Shi Y-J, Zhang H, Xing Y. Effects of Environmental Factors on Organochlorine Pesticide Residues in Soils of the Guanting Reservoir Area, China. Journal of Environmental Science and Health, Part B. 2006;41(3):309-21.

[19] Cabidoche YM, Lesueur-Jannoyer M. Contamination of Harvested Organs in Root Crops Grown on Chlordecone-Polluted Soils. Pedosphere. 2012;22(4):562-71.

[20] Adachi Y, Karube J. Application of a scaling law to the analysis of allophane aggregates. Colloids and Surfaces A Physicochemical and Engineering Aspects. 1999;151(1-2):43-7.

[21] Chevallier T, Woignier T, Toucet J, Blanchart E, Dieudonné P. Fractal structure in natural gels: effect on carbon sequestration in volcanic soils. Journal of Sol-Gel Science and Technology. 2008;48(1):231-8.

[22] Wada K. The distinctive properties of Andosol. Advances in soil science. 1985.

[23] Lesueur-Jannoyer M, Cattan P, Monti D, Saison C, Voltz M, Woignier T, et al. Chlordecone in French West Indies: cropping system changes and their incidence on pollution dispersion. Agronomie Environnement & Sociétés. 2012;2(1):45-58.

[24] Denaix L, Lamy I, Bottero JY. Structure and affinity towards Cd2+, Cu2+, Pb2+ of synthetic colloidal amorphous aluminosilicates and their precursors. Colloids and Surfaces A: Physicochemical and Engineering Aspects. 1999;158(3):315-25.

[25] Teixeira J. Small angle scattering by fractal systems. Journal of Applied Crystallography. 1988;21:781-5.

[26] Woignier T, Clostre F, Macarie H, Jannoyer M. Chlordecone retention in the fractal structure of volcanic clay. Journal of Hazardous Materials. 2012;241–242(0):224-30.

[27] Martin-Laurent F, Sahnoun M, Merlin C, Vollmer G, Lübke M. Detection and quantification of chlordecone in contaminated soils from the French West Indies by GC-MS using the 13C10-chlordecone stable isotope as a tracer. Environmental Science and Pollution Research. 2013:1-6.

[28] Brinker CJ, Scherer GW. Sol-gel science: the physics and chemistry of sol-gel processing: Academic Press; 1990. 908 p.

[29] Topp E, I. S, Attar A, Korte F. Factors affecting the uptake of 14C-labeled organic chemicals by plants from soil. Ecotoxicology and Environmental Safety. 1986;11(2): 219 - 28.

[30] George SE, Claxton LD. Biotransformation of chlordecone by Pseudomonas species. Xenobiotica the fate of foreign compounds in biological systems. 1988;18(4):407-16.

[31] Orndorff SA, Colwell RR. Microbial transformation of kepone. Applied and Environmental Microbiology. 1980;39(2):398-406.

[32] Mouvet C, Bristeau S, Amalric L, Dictor MC, Mercier A, Thannberger L, et al., editors. In Situ Chemical Reduction (ISCR) for Removal of Persistent Pesticides; focus on kepone in tropical soils. International Symposium on Bioremediation and sustainable Environmental Technologies; 2011 Jun; Reno, États-Unis.

[33] Clostre F, Lesueur-Jannoyer M, Achard R, Letourmy P, Cabidoche Y-M, Cattan P. Decision support tool for soil sampling of heterogeneous pesticide (chlordecone) pollution. Environmental Science and Pollution Research. 2013:1-13.

[34] Woignier T, Fernandes P, Jannoyer-Lesueur M, Soler A. Sequestration of chlordecone in the porous structure of an andosol and effects of added organic matter: an alternative to decontamination. European Journal of Soil Science. 2012;63(5):717-23.

[35] Biarnès A, Rio P, Hocheux A. Analyzing the determinants of spatial distribution of weed control practices in a Languedoc vineyard catchment. Agronomie. 2004;24(4): 187-96.

[36] Houdart M. Spatial organisation of agricultural activities and water pollution by pesticides Modelling in Capot's watershed, Martinique (French West Indies). Martinique: UAG; 2006.

[37] Nkamleu GB, Adesina AA. Determinants of chemical input use in peri-urban lowland systems: bivariate probit analysis in Cameroon. Agricultural Systems. 2000;63(2):111-21.

[38] Blazy J-M, Ozier-Lafontaine H, Doré T, Thomas A, Wery J. A methodological framework that accounts for farm diversity in the prototyping of crop management systems. Application to banana-based systems in Guadeloupe. Agricultural Systems. 2009;101(1–2):30-41.

[39] de Barros I, Blazy JM, Rodrigues GS, Tournebize R, Cinna JP. Emergy evaluation and economic performance of banana cropping systems in Guadeloupe (French West Indies). Agriculture, Ecosystems & Environment. 2009;129(4):437-49.

[40] Duyck P-F, Lavigne A, Vinatier F, Achard R, Okolle JN, Tixier P. Addition of a new resource in agroecosystems: Do cover crops alter the trophic positions of generalist predators? Basic and Applied Ecology. 2011;12(1):47-55.

[41] Mollot G, Tixier P, Lescourret F, Quilici S, Duyck P-F. New primary resource increases predation on a pest in a banana agroecosystem. Agricultural and Forest Entomology. 2012;14(3):317-23.

[42] Woignier T, Braudeau E, Doumenc H, Rangon L. Supercritical Drying Applied to Natural "Gels": Allophanic Soils. Journal of Sol-Gel Science and Technology. 2005;36(1):61-8.

[43] Desprat J-F, Comte J-P, Chabrier C. Cartography of soil pollution risk through organochlorine in Martinique. Technical report Phase3: Synthesis. 2004 Contract No.: BRGM/RP53262-FR.

[44] Tillieut O, Cabidoche Y-M. Cartography of soils CLD pollution in Guadeloupe. Technical report. Abymes, France: DAAF-SA & INRA-ASTRO, 2006.

[45] Clermont-Dauphin C, Cabidoche YM, Meynard JM. Effects of intensive monocropping of bananas on properties of volcanic soils in the uplands of the French West Indies. Soil Use and Management. 2004;20(2):105-13.

[46] Cattan P, Voltz M, Cabidoche YM, Lacas JG, Sansoulet J. Spatial and temporal variations in percolation fluxes in a tropical Andosol influenced by banana cropping patterns. Journal of Hydrology. 2007;335(1–2):157-69.

[47] Spencer WF, Singh G, Taylor CD, LeMert RA, Cliath MM, Farmer WJ. DDT Persistence and Volatility as Affected by Management Practices after 23 Years. J Environ Qual. 1996;25(4):815-21.

[48] Dorel M, Roger-Estrade J, Manichon H, Delvaux B. Porosity and soil water properties of Caribbean volcanic ash soils. Soil Use and Management. 2000;16(2):133-40.

[49] Perret S, Dorel M. Relationships between land use, fertility and Andisol behaviour: examples from volcanic islands. Soil Use and Management. 1999;15(3):144-9.

[50] Ahmad R, Kookana RS, Alston AM, Skjemstad JO. The nature of soil organic matter affects sorption of pesticides. 1. Relationships with carbon chemistry as determined by C-13 CPMAS NMR spectroscopy. Environmental Science & Technology. 2001;35(5):878-84.

[51] Fushiwaki Y, Urano K. Adsorption of pesticides and their biodegraded products on clay minerals and soils. Journal of Health Science. 2001;47(4):429-32.

[52] Kanazawa J. Relationship between the soil sorption constants for pesticides and their physicochemical properties. Environmental Toxicology and Chemistry. 1989;8(6): 477-84.

[53] Li H, Sheng G, Teppen BJ, Johnston CT, Boyd SA. Sorption and Desorption of Pesticides by Clay Minerals and Humic Acid-Clay Complexes. Soil Sci Soc Am J. 2003;67(1):122-31.

[54] Fernandes P, Jannoyer-Lesueur M, Soler A, Achard R, Woignier T. Effects of clay microstructure and compost quality on chlordecone retention in volcanic tropical soils: consequences on pesticide lability and plant contamination. 19th World Congress of Soil Science, Soil Solutions for a Changing World2010. p. 50-3.

[55] AFNOR. Amendements organiques et supports de culture - Caractérisation de la matière organique par fractionnement biochimique et estimation de sa stabilité biologique - Soil improvers and growing media - Characterization of organic matter by biochemical fractioning and estimation of its biological stability. Saint Denis la Plaine, France: AFNOR; 2009. p. 19.

[56] Dolfing J, Novak I, Archelas A, Macarie H. Gibbs free energy of formation of chlordecone and potential degradation products: implications for remediation strategies and environmental fate. Environ Sci Technol. 2012;46(15):8131-9. Epub 2012/07/12.

[57] Dolfing J, Novak I, Archelas A, Macarie H. Gibbs free energy of formation of chlordecone and potential degradation products: implications for remediation strategies and environmental fate. Environ Sci Technol. 2012;46(15):8131-9. Epub 2012/07/12.

Chapter 22

Chemical and Hydraulic Behavior of a Tropical Soil Compacted Submitted to the Flow of Gasoline Hydrocarbons

Wisley Moreira Farias, Geraldo Resende Boaventura,
Éder de Souza Martins,
Fabrício Bueno da Fonseca Cardoso,
José Camapum de Carvalho and
Edi Mendes Guimarães

Additional information is available at the end of the chapter

http://dx.doi.org/10.5772/57234

1. Introduction

Gasoline is a fuel comprised basically of hydrocarbons such as aromatic, olefinic and saturated compounds of a carbon chain comprised of 4 to 12 atoms. The aromatic compounds such as benzene, toluene, ethylbenzene, *o*-, *m*-, *p*-xylene (BTEX) are harmful to human health (Cairney et. al., 2002). As these compounds are harmful to health, the legislation becomes restrictive. The U.S. Environmental Protection Agency for drinking water (US EPA) establishes the maximum concentration of benzene in 5μg.L^{-1}. In Brazil the Ordinance of the Ministry of health number 2,914 in 12th December 2011, stipulates that the maximum allowable concentration of benzene is 5 μg.L^{-1} regulation of drinking water contaminant. Soil in contaminated residential areas, Brazil has been adopting as intervention guide value, the concentration of benzene 0.08 mg.kg $^{-1}$ set by the State of São Paulo in 2001. This value indicates the intervention limit of contamination where there is potential risk to human health.

Brazil produces type-C gasoline which is different than other types due to its anhydrous alcohol content (ethanol), in the proportion of 25% (Farias, 2003). The alcohols are soluble in water, and have a significant mobility potential to percolate through the soil until reaching underground water (Ulrich, 1999; Corseuil and Fernandes, 1999). The alcohol in gasoline in an aqueous medium promotes co-solvency which is the increase in the solubility of the hydrocarbons in the gasoline in an aqueous solution (Banerjee and Yalkowsky, 1988; Cline *et al.*, 1991).

Solubility is generally controlled by the polarity effect, which decreases in size for molecules with the same organic function. Non-polar or weakly polar substances dissolve in similar solvents. Thus, highly polar compounds dissolve in polar solvents such as water. The polarity or dipolar moment is proportional to the dielectric constant, and therefore high dielectric constant compounds (values of 80 for water and 34 for methanol) dissolve ions through hydration of the disassociated types (Fernandez and Quigley, 1985).

On the surface of clay-minerals, the absorbed water forms a double layer, which reduces the strength of interaction between the negatively charged clay particles and the cations in the colloidal solution. The hydrophobic hydrocarbons in the gasoline have low dielectric constant values, thus provoking the collapse of the double layer. This collapse is due to the contraction of the double layer through the attraction of the contra-ions which are closer to the superficial charge of the clay-minerals, favoring flocculation, and consequently the increase in permeability due to the increase in pore space (Mesri and Olson, 1971; Fernandez and Quigley, 1985 and 1988).

The co-solvêncy is responsible for the partition of BTXs to the aqueous phase, promoting the reduction of density of colloidal solution of soil, providing increased viscosity and a reduction of the surface tension (Mcdowell and Powers, 2003). This reduction in surface tension and generated by the collapse of the electrical double layer that there was between the soil and water (Farias, 2003).

1.1. Aspects of the transport of pollutants in soils

The transport of pollutants in the soil can occur through the porous medium and saturated or unsaturated fractured media. This transportation occurs through physical or chemical processes, or through both processes. The chemical process becomes evident when the velocity of the fluid is not sufficiently high (i.e., less than 10^{-6} cm/s), generating a gradient due to the flow of the solute (contaminating agent) from the more concentrated medium to the less concentrated one. This process is called molecular diffusion (Rowe, 1988; Pastore and Mioto, 2000). This type of flow has been widely studied with metals and organic compounds in solid waste landfill leachate contaminants, for application in compacted soil layer, also called liners (Shackelford and Daniel, 1991; Rowe, 1988; Barone *et al.*, 1988).

Fernandez and Quigley (1985) developed an experimental research program to evaluate the hydraulic behavior of clayey-like soil (Sarnia, Ontario), permeated with liquid substances such as benzene, xylene, cyclohexane, aniline, propanol, acetone, alcohol and water. The results have shown that Hydraulic conductivity increased from 5×10^{-9} to 1×10^{-4} $cm.s^{-1}$ along with a decrease in the dielectric constant from 80 (water) to 2 (benzene).

When there is a hydraulic gradient, the velocity of the solvent is relatively high and the transportation of the solute is practically managed by the velocity of the solvent, a mechanism which is known as an advection process. In this process, the velocity of the fluid is governed by Darcy's Law, which considers not only the characteristics of the soil, but also those of the fluid (Fernandez and Quigley, 1988).

In order to have good performance, the compacted clay liners must have a hydraulic conductivity less than 10^{-8} cm/s. However Daniel and Koerner (1995) defined that the hydraulic conductivity of clay liners must be less than or equal to 10^{-7} cm/s. This low flow is normally associated with the presence of clay-minerals, and at least 15 to 20% of particles with sizes under 2 mm, as well as a minimum plasticity greater than 7%, activity greater than 0.3, and cation exchange capacity (CEC) greater than 100 $mmol_c/dm^3$ of soil (Rowe *et al.*,1995).

The natural organic material of the soils have proven to be efficient in the retarding process through the sorption of hydrophobic hydrocarbons, which are also found in gasoline (Chiou *et al.*, 1983; Karickhoff *et al.*, 1979; Schwarzenbach *et al.*, 1993).

1.2. Importance of research

The aim of this study is to evaluate the behavior of a tropical soil, and their performance as liner against the flow of hydrocarbons from gasoline, by interpreting transportation according to physical and chemical parameters, as well as micromorphological aspects. For this characterized the mineralogy of the soil and the influence of his organic matter (OM), considering the adsorption processes of hydrocarbons from gasoline and hydraulic behavior in the laboratory by variation of the hydraulic gradient in front of the gasoline flow through compacted soil. This study also aims to contribute to the understanding of the dynamics of the flow through the soil of specific groups of compounds: aromatic, olephine, saturated hydrocarbons and the ethanol found in Brazilian type-C gasoline (a complex mixture of organic compounds).

2. Location and Soil classification

The soil sample was collected indisturbed block in depth of 4 m in the experimental field of foundations and test field of the Civil Engineering Department of the University of Brasília, located on the University campus in the city of Brasília, Brazil with coordinates 15° 56 ' 45 "S, 47° 52 ' 20" W (Fig. 1).

The sample of lateritic soil typical of the Brazilian *cerrado* region was studied. According to the Brazilian Soil Classification System (Embrapa, 1999), the soil was classified as Red Latossoil, considered as Ustic Rhodic Oxisol according to the U.S. Soil Taxonomy and Geric Ferralsol Ferric (FAO, IUSS Working Group WRB, 2007). It possesses a silt-clay-like texture, a large quantity of granular aggregates, and small pores. Visually, it is homogeneous and isotropic, without the presence of discontinuities.

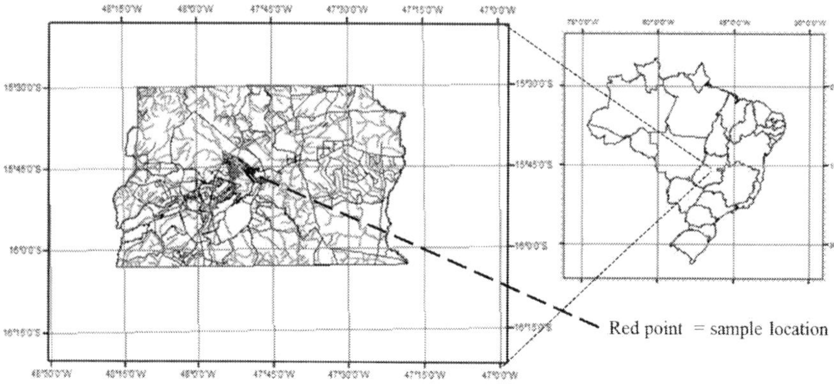

Figure 1. Map of location of the soil collection.

3. Methodology

The characterization of the soil involved the use of physical, chemical and mineralogical analysis.

3.1. Physical tests

Geotechnical tests of physical properties of soils were performed following the Brazilian Association of Technical Standards (ABNT) procedures: test of limits of consistency called Atterberg limits following the ABNT NBR 7180/84 plastic limit; 6459/84 liquid limit following the Casagrande method. Before the grain-size determination, the real density was determined according the ABNT NBR 6508/84 method. The grain-size distribution curve was determined using a grain-size digital meter Malvern Mastersizer with lens de 300Rf for grain size range of 0.05 μm to 900μm at 25 ° C. For this analysis, the sample was previously passed through a No. 40 sieve. The analyses of the samples were done either with or without ultrasound dispersion. Ultrasonic condition was 5 minutes of dispersion in distilled water with ultrasonic level set at 5. The grain size fractions were classified following the Brazilian standard NBR 6502/93.

The degree of flocculation and dispersion of soil particles was determined comparing the results of grain size determinations before and after ultrasonic dispersion.

3.1.1. Hydraulic conductivity

The test of hydraulic conductivity in compacted soil in standard Proctor energy were performed in a conventional manner with water using the variable charge and special form for gasoline (Fig. 2 and Fig. 3).

http://dx.doi.org/10.5772/57234

Figure 2. Hydraulic Conductivity cell.

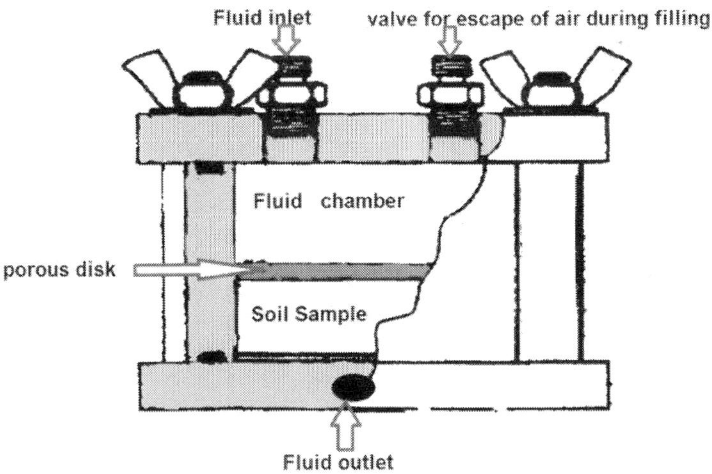

Figure 3. Schematic of permeameter Cell.

The gasoline hydrocarbons, for possessing volatile and low-density compounds require a special sealed cell to avoid losses due to evaporation and leakage and to support the applied tensions. The material selected for the construction of the special cells was stainless steel, to avoid reaction and adsorption problems in the walls, which is the case of plastics and acrylics (Doanhue *et al.*, 1999).

The system used to perform the gasoline's hydraulic conductivity was similar to that applied by Fernandes (1989). The special cell may be disassembled, and is made up of three parts. The first part is a cylinder, where the test material and reservoir are found. This part is 5 mm thick, 110 mm long and has an internal diameter of 77.2 mm. The other two parts are the upper and lower lids. Both have cavities filled with rubber rings which are able to prevent the reaction of the hydrocarbons in the gasoline and act as a seal when the cylinder is assembled. The upper lid has two openings, one for the entry of fluid and the other for the application of vertical tension with compressed air. The lower lid is made up of an outgoing flow register which is connected to a collecting container. The two lids are 120 x 120 mm^2 square, 10 mm thick. The connections were made out of aluminum, due to its low cost and flexibility; the connecting joints were sealed with 3M automotive glue and winding sealing thread, in order to prevent leaks and to make the system more secure.

The conductivity test was performed with test material 5 cm long, compacted at normal Proctor energy at optimal water content condition, in the cylinder of the hydraulic conductivity cell. Then, a thin disk of porous ceramic was placed top of the sample. The small space between the disk and the cylinder wall was filled with 3M glue to prevent preferential flows along the wall, and to ensure that the gasoline only passed through the porous ceramic disk. The cell was then assembled, and the upper and lower lids were connected to the cylinder. The cylinder is 11 cm high, of which the remaining 6 cm were filled with type C gasoline. After the cell was totally sealed and connected to the compressed air system, with pressure controlled by a manometer, it was connected with plastic tubes able to support high pressure. The conductivity tests were performed for various applied vertical pressures. For each pressure applied, the hydraulic conductivity was measured. The pressures were varied to see how the soil sample behaved with an increase in hydraulic gradient upon the flow of gasoline. The hydraulic conductivity was measured in the laboratory at static tensions σ_v of 50, 100, 150, 200, and 300 kPa, with respective hydraulic gradients of 75, 150, 225, 300, and 450.

The residual water of the soil pores mixed with gasoline collected in the test was previously run through a separating funnel to remove the aqueous phase to later take a reading of hydrocarbons of the gasoline through infrared technique.

The test material of the lateritic soil sample, before and after the hydraulic conductivity test conducted with water, and the other with the flow of gasoline, were dried at room temperature. Micromorphological analyses were performed on Thin Lamina (TL) in vertical sections, prepared by impregnating the sample with plastic resin (Cardoso, 1995; Martins, 2002). The instrumental technique used for the microscopic views of the TL was Optical Microscopy.

3.2. Mineralogical characterization

The identification and quantification of minerals in the sample were carried out by the method developed by Martins (2000). This method involves the use of X-ray diffraction (XRD) technique for identifying the minerals, chemical analysis for the determination of major elements (Al, Fe, Si, Ca, Mg and Ti), thermogravimetric analysis (TGA), and the use of Munsell color code (Munsell color company Inc., 1975). The determination of major chemical elements was performed by ICP-AES after the fusion of samples with alkaline NaOH as fondant at a temperature of 450 ° C for 40 minutes using the nickel crucible. Determinations of elements by ICP / AES (atomic emission spectrometry of Plasma Induced Coupling) were performed with Thermo Jarrell ASH equipment, model Iris / AP.

The thermogravimetric analysis were applied to quantify the kaolinite and gibbsite. For this used the TGA Shimadzu equipment with temperature ramp of 20 °C to 1500 °C, with speeds ranging from 0.2 to 60 ° C / min, using the software TAS60WS for the treatment of data. The Munsell code was used for determining the ratio of hematite and goethite in the soil samples. The CEC of soil was determined using the principle of the simple as the sum total of the exchangeable cations that a soil can adsorb. The determination of the organic matter content was done prior to extraction using wet oxidation method with potassium dichromate in sulfuric medium. The excess of dichromate after oxidation was titrated with standard solution of ferrous ammonium sulfate (Mohr salt).

3.3. Chemical characterization

The pH was measured in the soils samples in distilled water medium using a combined glass electrode Ag/AgCl (potentiometric method).

In order to study the influence of OM and mineralogy in the gasoline sorption process, an experiment was performed with samples treated with H_2O_2 and another without treatment.

The extraction of the OM used 15 g of soil in a porcelain capsule, with 10 mL of H_2O_2 volume 30% and with agitation in a 50 mL Becker cup. After agitation, there was an effervescent reaction, when the capsule was covered with clock glass for one night. The process was repeated until the complete disappearance of the reaction. It was then washed 3 to 5 times in distilled water, using a Büchner funnel with filtering under reduced pressure. Then, for the gasoline sorption test, the sample was allowed to dry at room temperature.

The sorption test procedure used 2 g of soil with 25 mL of gasoline placed in an amber glass jar under agitation for 24 hours at a temperature of 22°C. After this, the samples were centrifuged as in the processes described above, with the removal of a 15 mL portion for analysis.

The hydrocarbons content of the gasoline samples was determined at the National Petroleum Agency (ANP) laboratory, in Brasilia, with a (FTIR = Fourier Transform Infrared), manufactured by Grabner Instruments, model IROX 2000. This instrument qualified and quantified the compounds, generating the mass and volume percentages of the ethanol, aromatic, olephine and saturated compounds.

4. Results

Tab. 1 and 2 present data of the physical, chemical and mineralogical Brazilian soil and constituents of the gasoline type C studied.

Test	Lateritic
Atterberg Limits	
Liquid limit-W_L (%)	41
Plastic limit-W_P (%)	29
Plastic Index-I_P (%)	12
Activity	0,18
Grain size distribution*	
Clay (%)	65
Silt (%)	34
Sand(%)	1
Degree of flocculation (%)	92
Degree of dispersion (%)	8
Chemical Parameters	
pH	5,70
Organic Matter content (%)	0,41
CEC ($mmol_c/dm^3$)	6,4
Mineralogy	
Quartz (%)	30,2
Anatase (%)	1,57
Kaolinite(%)	24,6
Gibbsite (%)	25,5
Goethite (%)	4,6
Hematite (%)	7,5
Illite (%)	2,2
Vermiculite (%)	3,7
Hydraulic Conductivity in water (cm/s)	3,7.E-07

*Grain size data obtained by ultra-sound waves using a laser beam grain size analyser.

Table 1. Characteristics of the soil (Farias, 2003).

Tab. 2 presents the composition of the Brazilian type-C gasoline, according to Farias (2003).

Compounds	Mass (%)
Aromatics	20,8
Olefins	22,4
Saturated	31,4
Ethanol	25,4

Table 2. Brazilian Type C gasoline data.

Fig. 4 presents the increase in hydraulic conductivity with an increase in the hydraulic gradient. At a gradient of approximately 210, conductivity becomes practically constant. Fig. 5 presents the intrinsic permeability, which considers the characteristics of the soil, but does not consider the chemical and physical properties of the fluid. Intrinsic permeability reaches values close to $10^{-13}m^2$. However, as the hydraulic gradient increases, stability reaches approximately $10^{-11}m^2$.

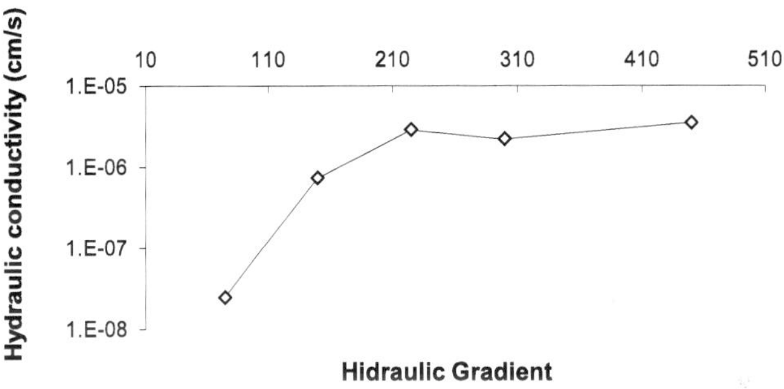

Figure 4. Behavior of hydraulic conductivity and hydraulic gradient of laterite soil on the gasoline flow.

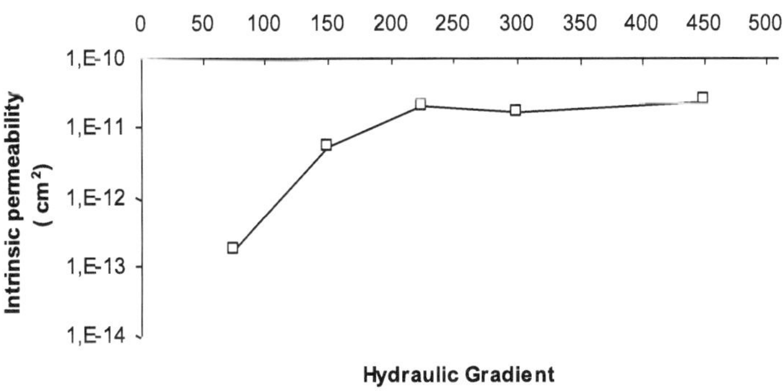

Figure 5. Behavior of the intrinsic permeability and hydraulic gradient of laterite soil on the gasoline flow.

Fig. 6 depicts the behavior of the hydraulic conductivity relative to the volume of pores while undergoing saturation in the test material with gasoline at a tension of σ_v of 50 kPa. The saturation process takes place with the expulsion of the interstitial water accumulated in the pores due to optimal compacting moisture content (w_{opt} = 26%) is the test material at normal Proctor energy. It may be observed that as the volume of pores in the gasoline flow increases, conductivity decreases from 4 to 2 x 10^{8} $cm.s^{-1}$. This suggests that the behavior of the reduction may be represented by a second-order equation.

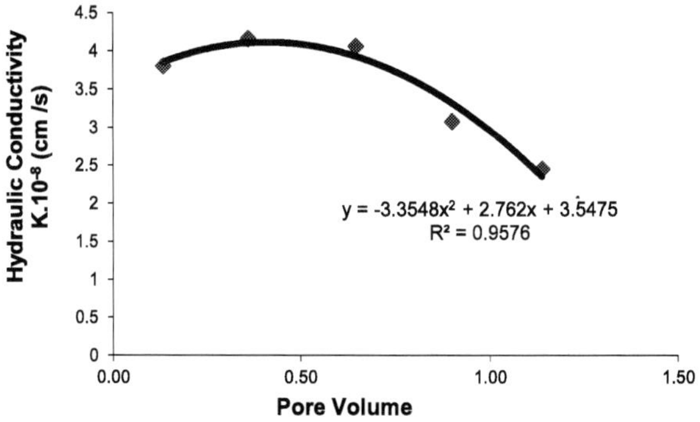

Figure 6. Behavior of the lateritic soil saturated with gasoline at 50 kPa.

Fig. 7 presents the saturation process at a σ_v tension of 50 kPa, based on the ratio between the concentration (C) of the gasoline hydrocarbons passing through the soil sample, and the initial concentration (C_o) added to the reservoir, in relative to the volume of pores. The hydrocarbons concentration data are from the Light Non-aqueous Liquid Phase (LNALP), after the flow through the soil sample in the hydraulic conductivity test.

Sample	w (%)	γ ($kN.m^{-3}$)	γ_{dmax} ($kN.m^{-3}$)	γ_s ($kN.m^{-3}$)	e	n	Sr (%)	Vv cm^3
lateritic*	1,7	17,7	17,4	27,5	0,58	0,4	8,1	134,3
lateritic**	1,7	15,8	15,6	27,5	0,77	0,4	6,2	178,5
lateritic***	1,8	14,7	14,5	27,5	0,90	0,5	5,3	210,0

*Dry soil sample before the hydraulic conductivity test

**Dry soil sample after the hydraulic conductivity test with the water flow

*** Dry Soil sample after the hydraulic conductivity test with the gasoline flow

Table 3. Result of the physical parameters of the test material.

The results in Tab. 3 present the physical parameters of the compacted test materials dried at room temperature before and after the hydraulic conductivity test. Highlights the volume of voids (**Vv**), which changes substantially when there is a flow of gasoline. The degree of saturation also decreases after the flow of gasoline.

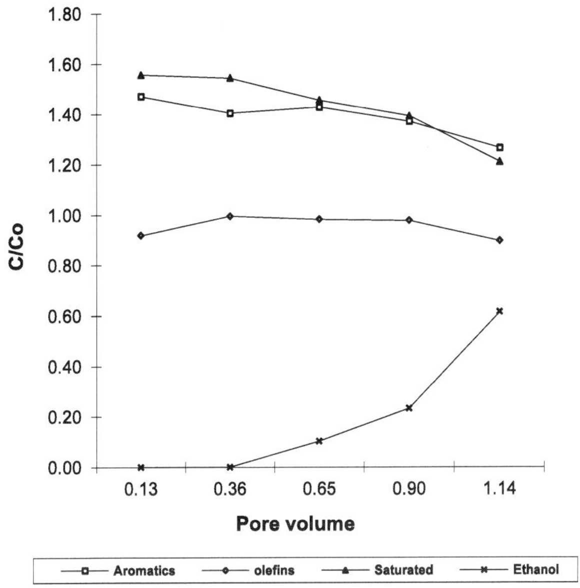

Figure 7. Light non-aqueous liquid phase ratio of the gasoline relative to the volume of pores of the lateritic soil in a saturation process at 50 kPa.

The micromorphology of the three compacted soil samples was important in order to visualize the behavior of the test material before the hydraulic flow (Fig. 8), after the hydraulic flow with water, and after the flow with gasoline. It must be noted that the grains of quartz make up approximately 40% of the total solid material; variable in size, 0.12 mm on average; and overall, are sub-rounded to angular. They are highly fractured, without orientation and their contours present corrosion. In spite of the compacting, the structure of this soil is not totally dispersed, for microaggregations of oxyhydroxides of Fe and Al remain, forming micropores. The compacted soil sample submitted to percolation in water showed a single micro-structural difference relative to the one performed on the LT of the compacted soil sample. Actually, there was an increase in small canal-type voids, generated by the flow of water (Fig. 9). The micromorphology regarding the LT of the compacted soil submitted to the flow of gasoline also showed only a quantitative increase in canal-type voids (Fig. 10). However, this variation was greater than that registered in the previous sample with the water flow.

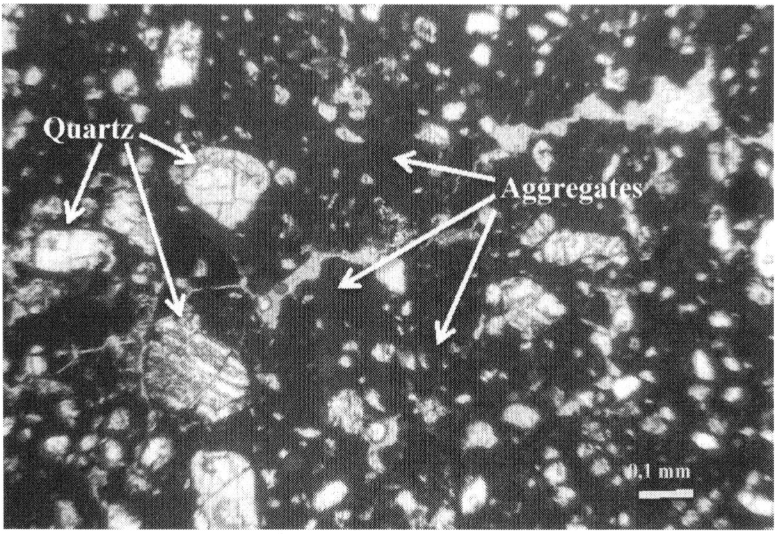

Figure 8. Photomicrography of the porfirosquelic APE, aggregates, and quartz grains of the compacted lateritic soil. Parallel nichols (N//).

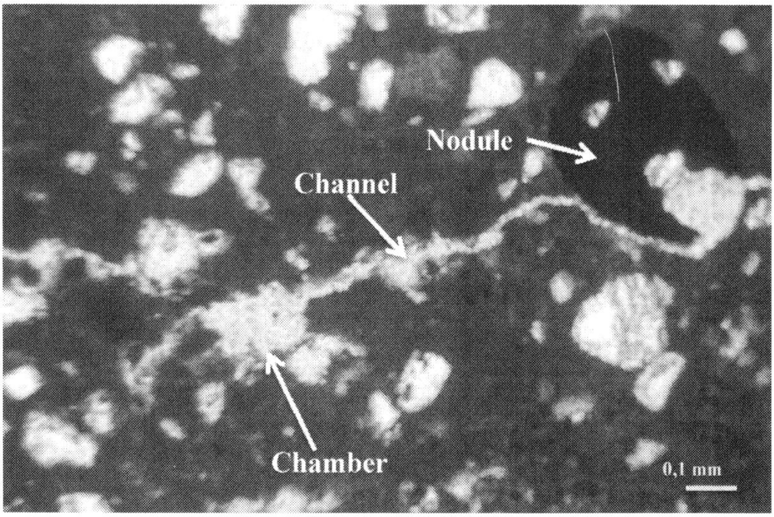

Figure 9. Photomicrography showing the nodules and canal- and chamber-type voids of the compacted lateritic soil submitted to percolation with water. Parallel nichols (N//).

http://dx.doi.org/10.5772/57234

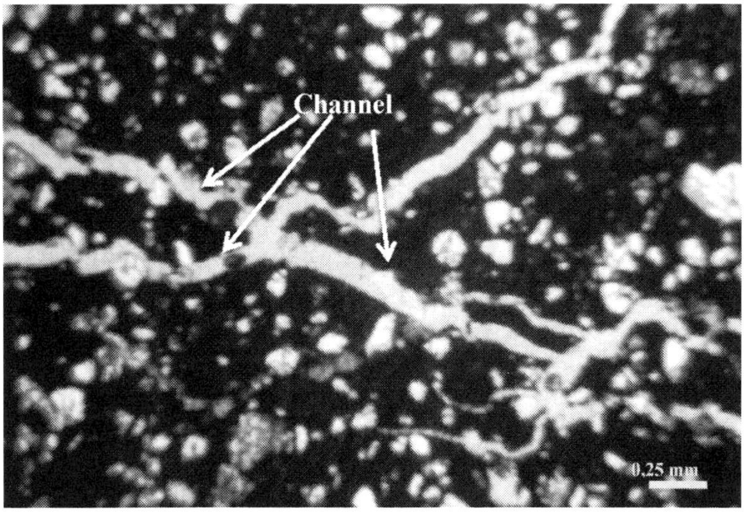

Figure 10. Photomicrography showing the canal-type voids of the compacted lateritic soil submitted to percolation with gasoline. Parallel nichols (N//).

Fig. 11 presents the results of the adsorption of the ethanol and aromatic substances in the samples with and without the extraction of organic matter with the use of hydrogen peroxide. Note that the samples treated with extractor presented low adsorption. Aromatic compounds showed no adsorption after extraction of organic matter contained in the soil.

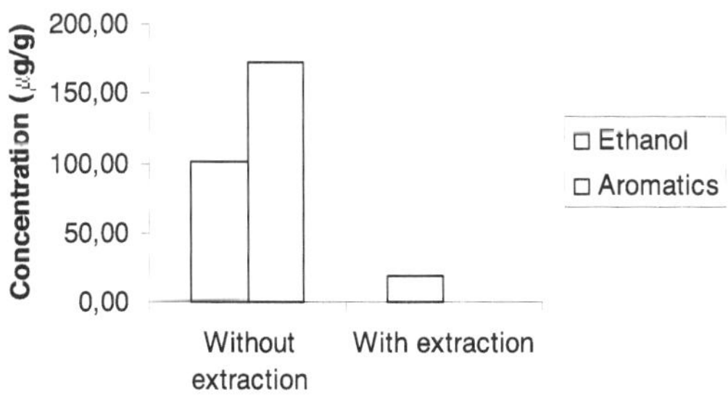

Figure 11. Results of the adsorption of the gasoline hydrocarbons in the soils with and without the extraction of the soil organic matter.

Gasoline ethanol can be adsorbed on the sites of hydroxyls of the octahedron of Al, exposed by fractures, Scrubs or crystalline lattice imperfections, or by interactions with the Fe oxides and hydroxides and Al amorphous. This occurs from adsorption of hydrogen bonds, which can also occur with water strongly adsorbed on the surface of the clay minerals (Fig. 12).

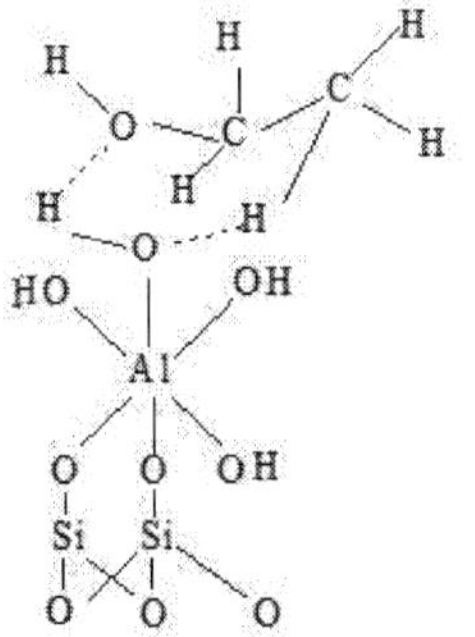

Figure 12. Coordination of interaction of hydrogen and hydroxyl ethanol exposed in the clay mineral (1:1).

5. Discussion of the results

The discussion of the results is focused on three main aspects. The first considers mineralogical, chemical and physical characteristics of the material with potential for liners. The second aspect is assessed the performance of Laterite soil on gasoline hydrocarbon flow subjected to high hydraulic gradients, causing an acceleration of the process of formation of flow channels for compressed soil in power of Proctor. The last aspect to be evaluated is the power of gasoline hydrocarbon adsorption by soil with OM and no OM.

5.1. The delimiters criteria material with potential for liners

Evaluating the criteria prescribed by Rowe et al., (1995) the soil presents considerable levels of Fe oxides and hydroxides and Al (hematite, goethite and gibbsite) and kaolinite with only 30.2% of quartz. As the mineral vermiculite is low soil activity levels was 0.18, less than the 0.3 suggested by the literature. However this value of activity indicates that the material is not expandable, being a good quality for liners. The cationic exchange capacity (CEC) also presented low value (6.4 $mmol_c/dm^3$) comparing with value defined in literature. Tropical soils lateritic in general are highly weathered with low or no mineral content of 2:1, which are typical of temperate climate. Therefore, the activity and CEC are low. The granulometry performed 65% clay fraction indicated more than 20% of particles less than 2 mm confirming the material rich in clay fraction indicating low permeability material when compressed. Thus, the hydraulic conductivity parameter value introduced into water in the order 10^{-7} cm/sec subjected to a pressure of 20 kPa (Tab. 1). These results of the hydraulic conductivity characterize the material with great potential for liner according to predefined values in the literature.

5.2. Lateritic soil performance as Liner

The hydraulic conductivity of gasoline type C brazilian obtained values between 10^{-8} to 10^{-7} cm/s to a gradient of 75 with a pressure of 50 kPa, which corresponds to 5 m of the water column (Fig. 4). Such a result of hydraulic conductivity defines the material as great for barrier on gasoline hydrocarbon flow according to predefined values in literature (Rowe et al., 1995; Daniel and Koerner, 1995). With the increase of the hydraulic gradient there was an increase in hydraulic conductivity until it reaches a level of stabilization in gradient greater than 210 (Fig. 4). Although it does not occur to the destruction of the liner to the 210 to 450 gradients suggests avoid gradients greater than 100 in the projects of protection of underground fuel tanks, ensuring in this way a hydraulic conductivity around 10^{-7} cm/s for liners according to the literature. The intrinsic permeability or specific considers simply the porous medium, not considering the characteristics of fluid. The values found for intrinsic permeability compacted laterite soil is similar to that found in the literature to clay (Freeze and Cherry, 1979).

The compacted soil voids indexes suffered increased 0.58 before tests to 0.77 with water flow and 0.90 with gasoline flow in hydraulic gradient of 75. The empty volume also increased from 134.3 to 210.0 (Tab. 3). The soils studied presented a high degree of flocculation due to the aggregates of the oxyhydroxides of Fe and Al. Even when compacted, they contain micro-aggregates which are not destroyed. When a flow is established through the soil, the micro-aggregates may interconnect, forming flow channels. The physical behavior provoked by the flow may be visualized in the micromorphology of the samples in Fig. 8, 9 and 10. However even with these micros channels formed in the compacted soil hydraulic conductivity limit of 10^{-7} cm/s (Daniel and Koerner, 1995) is not affected considering a gradient of 100.

5.2.1. *Adsorption performance for hydrocarbons of gasoline*

The performance of laterite soil to a gasoline hydrocarbon flow subjected to a pressure of 50 kPa with 75 gradient was evaluated for pore volume and the ratio C/Co in the process of saturation of the compacted clay liner for gasoline, in other words, there was the expulsion of the water contained in the soil by the process of compression to achieve the optimum water content of compaction. The reason indicates that values above 1 there is an LNAPL phase concentration of groups of substances evaluated. The groups were evaluated for aromatic, olefins, saturated and ethanol.

In Fig. 7, the aromatic compounds appear as constants in the saturation process. Since they are hydrophobic, their polarity is low, and are more easily transported in the soil. The olefines and saturates have a greater C/Co ratio in the LNAPL due to their low solubility in water, being lower than the aromatic compounds, which are more affected by ethanol through co-solvency. In the 0.13 a 0.36 pore volume range, the ethanol is partitioned to the aqueous phase and, as the saturation of the pores with gasoline increases, the C/Co ratio for ethanol in the LNAPL also increases. The partitioning of the ethanol for the aqueous phase is natural and is due its polarity, which makes it mixable in water. Thus, the ethanol, along with the other hydrophobic compounds in the gasoline, favors the collapse of the double layer, as well as the increase in micropores (Rowe et al., 1995)

The results in Fig. 11 show that the soil organic matter, although in low quantities, has an influence of almost 0.41 %, in the sorption process. The soil studied was collected at a depth of 4 meters, thus contained evolved organic matter, possibly fulvic acid. The removal of the organic material with hydrogen peroxide showed low ethanol sorption. The aromatic compounds, which are hydrophobic, were not absorbed.

Evaluating the transportation of gasoline compounds by soil with application of 50 kPa of pressure, indicates low retention and greater mobility, because mostly they are hydrophobic compounds that do not bind the soil particles. Another aspect of this experiment is that has not been evaluated by diffusion flux, which occurs at speeds equal to or less than 10^{-10} cm/s. The test of sorption to organic matter proved to be important in the retention process of ethanol. In view of the low adsorption of gasoline compounds by soil suggests considering projects of liners gradients below 75 and pressures less than 50 kPa ensuring a hydraulic conductivity greater than 10^{-8} cm/s and use clayey material rich in organic matter to promote greater retention of ethanol and avoid or reduce the effect of co-solvency.

6. Conclusion

Since the lateritic soil studied possesses a high aggregation capacity, even when compacted at normal Proctor energy, micropores remain which, in high hydraulic gradient situations, are interconnected, forming flow channels. However, even under higher hydraulic gradients in the gasoline percolation tests, this soil presents good material of liners. This is due to the stabilizing of the flow channels formed, favoring also the stabilizing hydraulic conductivity.

The measure that gasoline occupies the pores in the process of saturation the concentration of ethanol increases. This is due to the polarity of the ethanol. The aromatic compounds maintain a $\mathbf{C/C_0}$ ration close to 1, as the volume of pores increases, indicating that these are tracers due to their low dielectric constant and polarity. Due to their low solubility in water, the olephines and saturates are more present in the LNAPL phase. These hydrocarbons may form emulsions, favoring transportation through the soil in the aqueous phase.

Regarding the retarding potential of the lateritic soil, evaluated by the sorption parameter, it is not directly correlated with the mineralogy, because the aromatic compounds are not absorbed when the organic material is extracted. Actually, this sorption may be correlated with a certain type of humic substance, which may be interacting with the poly-amorphs of the oxyhydroxides of Fe and Al in the soil, favoring interaction with the aromatic compounds.

Finally, a low hydraulic gradient context (< 75), hydraulic conductivity < 10^{-8} and organic matter, in lateritic soil can improve the performance of liner.

Acknowledgements

The authors are gratefully to the Conselho Nacional de Desenvolvimento Científico e Tecnológico – CNPq, CAPES and ANP for the fellowships and financial support granted to the accomplishment of this research.

http://dx.doi.org/10.5772/57234

Author details

Wisley Moreira Farias[1], Geraldo Resende Boaventura[1], Éder de Souza Martins[2], Fabrício Bueno da Fonseca Cardoso[3], José Camapum de Carvalho[1] and Edi Mendes Guimarães[1]

1 Universidade de Brasília, Brasília-DF, Brazil

2 Embrapa/ Cerrados, Planaltina -DF, Brazil

3 ANA-Agência Nacional de Águas -Brasília-DF, Brazil

References

[1] ABNT (1984) Solo - Determinação do limite de liquidez -NBR 6459/84.Associação Brasileira de Normas Técnicas, Rio de Janeiro, RJ, 6p.

[2] ABNT (1984) Solo - Determinação do limite de plasticidade - NBR7180/84. Associação Brasileira de Normas Técnicas, Rio de Janeiro, RJ, 3p.

[3] ABNT (1984) Solo - Grãos de solo que passam na peneira de 4,8 mm - Determinação da massa específica (método de ensaio) - NBR 6508/84. Associação Brasileira de Normas Técnicas, Rio de Janeiro, RJ, 8p.

[4] ABNT (1993) Rochas e solos - Terminologia - NBR 6502/93. Associação Brasileira de Normas Técnicas, Rio de Janeiro, RJ, 19p.

[5] Banerjee, S. and Yalkowsky, S.H. (1988). Cosolvent-Induced Solubilization of Hidrophobic Compounds into Water. Analytical Chemistry, v.60, p. 2153-2155.

[6] Barone, F.; Yanful, E.K.; Quigley, R.M. and Rowe, R.K. (1988). Effect of multiple contaminant migration of diffusion and adsorption of some domestic waste contaminants in a natural clayey soil. Geotechnical Research Report GEOT -5-88, Geotechnical Research Centre. The University of Westem Ontario, London, Ont.

[7] Cairney, S., Maruff, P., Burns C., Currie B (2002) The neurobehavioural consequences of petrol (gasoline) sniffing. Neuroscience and Biobehavioral Reviews, v.26:1, p. 81-89.

[8] Cardoso, F. B.F (1995) Análise Química, Mineralógica e Micromorfológica de Solos Tropicais Colapsíveis e o Estudo da Dinâmica do Colapso. Dissertação de Mestrado, Departamento de Engenharia Civil, Universidade de Brasília, Brasília, DF, 142p.

[9] Chiou, C.T., Porter, P.E. and Schmedding, D.W. (1983). Partition equilibrium of nonionic organic compounds between soil organic matter and water. Environmental Science and Technology, 17(4):227-231.

[10] Cline, P.V., Delfino, J.J. and Rao, P.S.C. (1991). Partitioning of aromatic constituents into water from gasoline and other complex solvent mixtures. Environmental Science and Tecnology, 25(5):914-920.

[11] Corseuil, H. X. ; Fernandes, M.(1999). Efeito do Etanol no Aumento da Solubilização de Compostos Aromáticos Presentes na Gasolina Brasileira. Revista Engenharia Sanitária e Ambiental, v. 4, n. 1 e 2, p. 71-75.

[12] Daniel, D.E and Koerner, R.M. (1995) Waste containment facilities –Guidance for construction, Quality Assurance and Quality Control of Liner and Cover Systems, American Society of Civil Engineers, ASCE Press, New York, 354 pp.

[13] Donahue, R.B., Borbour, S.L. and Headley, J.V. (1999). Diffusion and adsorption of benzene in Regina clay. Canadian Geotechnical Journal, 36(3):430-442.

[14] FAO, IUSS Working Group WRB (2007). World Reference Base for Soil Resources 2006, first update 2007. World Soil Resources Reports No. 103, Rome.

[15] Farias, W.M. (2003). Condutividade Hidráulica de Solos Tropicais Compactados a Hidrocarbonetos da Gasolina. M.Sc. thesis, Department of Civil and Environmental Engineering, University of Brasilia, Brasilia DF, 152 pp.

[16] Fernandez, F. and Quigley, R.M. (1985). Hydraulic conductivity of natural clays permeated with simple liquid hydrocarbons. Canadian Geotechnical Journal, 22:205-214.

[17] Fernandez, F. and Quigley, R.M. (1988). Viscosity and dieletric constant controls on the hydraulic conductivity of clayey soils permeated with water – soluble organics. Canadian Geotechnical Journal, 25:582-589.

[18] Freeze, R. A. and Cherry, J. A (1979). Groundwater. Prentice-Hall, New Jersey, USA, 604p.

[19] Karickhoff, S.W., Brown, D.S. and Scott, T.A. (1979). Sorption of hydrophobic pollutants on natural sediments. Water Research, 13:241-248.

[20] Martins, E.S., (2000). Petrografia, Mineralogia e Geomorfologia de Rególitos lateríticos do Distrito Federal. Tese de Doutorado, Instituto de Geociências, Universidade deBrasília, Brasília, DF, 196 p.

[21] McDowell, C.J. and Powers, S.E. (2003). Mechanisms Affecting the infiltration and distribution of ethanol-blended gasoline in the vadose zone. Environmental Science and Technology, 37:1803-1810.

[22] Mesry, G. and Olson, R.E. (1971). Mechanisms controlling the permeability of clays. Clay and Clay Minerals, 19(3):151-158.

[23] Mussell Soil Color Company (1975) Mussell soil color charts. Baltimore, 1V, 117p.

[24] Pastore, E.L. and Mioto, J.A. (2000). Impactos ambientais em mineração com ênfase à drenagem mineira ácida e transporte de contaminantes. Solos e Rochas. São Paulo, SP, 23(1):33-53.

[25] Rowe, R.K. (1988). Contaminant migration though groundwater: The role of modeling in the design of barriers. Canadian Geotechnical Journal, 25(4):778-798.

[26] Rowe, R.K., Quigley, R.M. and Booker, J.R. (1995). Clayey Barrier Systems for Waste Disposal Facilities. E and FN Spon, London, England, 390p.

[27] Scharzenbach, R.P., Gschwnd, P.M. and Imboden, D.M. (1993). Environmental Organic Chemistry. Wiley Interscience, New York, NY, 1313p.

[28] Shackelford, C.D. and Daniel, D.E. (1991). Diffusion in saturated soil. II: Results for compacted clay. Journal of Geotechnical Engineering, 117(3): 485-506.

[29] Ulrich, G. (1999). The Fate and Transport of Ethanol-Blended Gasoline in the Environment. Governors' Ethanol Coalition, Lincoln, Nebraska, 88p.

Chapter 23

Exposure and Hazard Identification of Sulphonamides in the Terrestrial Environment

Anna Białk-Bielińska, Joanna Maszkowska,
Alan Puckowski and Piotr Stepnowski

Additional information is available at the end of the chapter

http://dx.doi.org/10.5772/57265

1. Introduction

For ten years now, interest has been increasing in research focused on pharmaceutical residues in the environment. Special attention has been given to the residues of antimicrobials, since it has been demonstrated, that due to the formation of the dangerous phenomenon of bacterial resistance, these substances could pose a real threat not just to ecosystems, but also to human health. Most antimicrobial agents are used in large quantities for different purposes in veterinary medicine.

Various antibiotics are commonly used in this field, but we shall concentrate on sulphonamides (SAs). The physicochemical properties and chemical structures of selected SAs are presented in Table 1.

Having been used for more than fifty years, SAs are among the oldest groups of veterinary chemotherapeutic substances. They are inexpensive and offer a broad spectrum of activity for the prevention and treatment of bacterial infections. After tetracyclines, they are the most commonly consumed veterinary antibiotics in the European Union [1,10]. However, as animals do not completely metabolize these compounds, a large fraction of them is being excreted unchanged in faeces and urine. Therefore, both the unmetabolized drugs and their metabolites are discharged to the environment, mainly to the soil component, either directly by grazing animals or indirectly during the application of manure [11]. Once in the soil, they and their transformation products are distributed among its different compartments and can be transported to the surface and ground waters. The physicochemical properties of these compounds, the applied dosage and the nature of the environmental compartment where they are released and further interact, comply the whole process.

Substance [CAS] **Abbreviation**	**Chemical structure**	**Selected physicochemical properties**
Sulphachloropyridazine [80-32-0] SCP	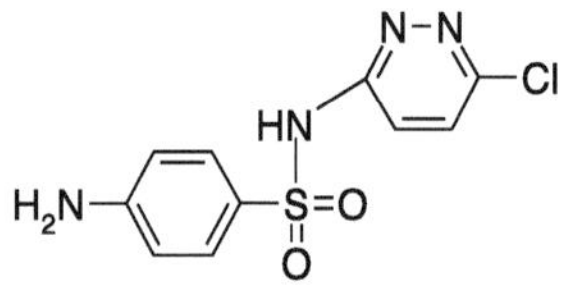	M = 284.7 g mol^{-1} pK_{a2} = 1.72 pK_{a3} = 6.39 logP = 0.71
Sulphadiazine [68-35-9] SDZ	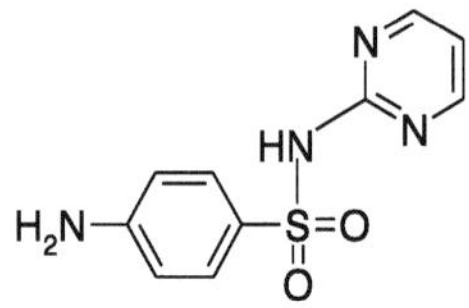	M = 250.3 g mol^{-1} pK_{a2} = 1.98 pK_{a3} = 6.01 logP = -0.09
Sulphadimethoxine [122-11-2] SDM	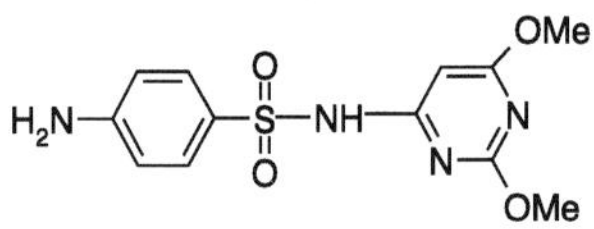	M = 310.3 g mol^{-1} pK_{a2} = 2.5 pK_{a3} = 6.0 logP = 1.63
Sulphadimidine (sulphamethazine) [57-68-1] SDMD (SMZ)	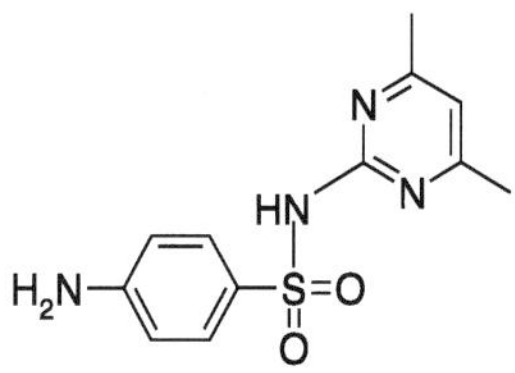	M = 278.3 g mol^{-1} pK_{a2} = 2.46 pK_{a3} = 7.45 logP = 0.27
Sulphaguanidine [57-67-0] SGD	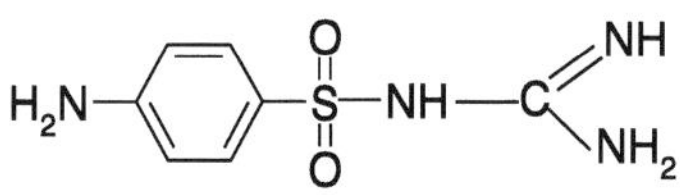	M = 214.2 g mol^{-1} pK_{a2} = 2.8 pK_{a3}= 12.0 logP = -1.22
Sulphamerazine [127-79-7] SMR	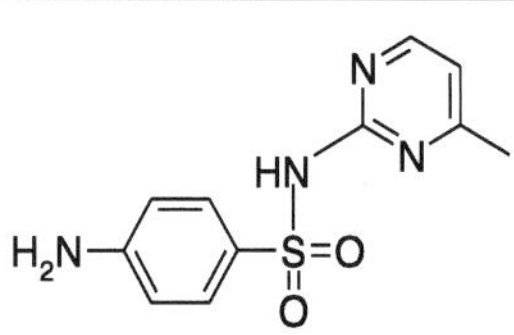	M = 264.3 g mol^{-1} pK_{a2} = 2.16 pK_{a3} = 6.80 logP = 0.11
Sulphamethoxazole [723-46-6] SMX	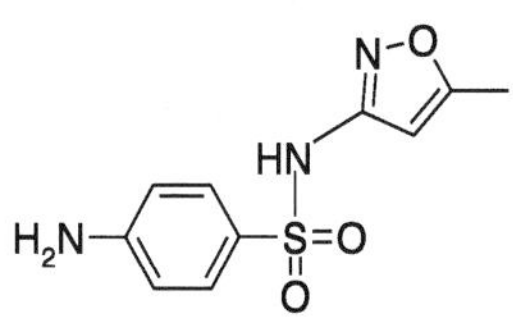	M = 253.3 g mol-1 pK_{a2} = 1.81 pK_{a3} = 5.46 logP = 0.89

Substance [CAS] Abbreviation	Chemical structure	Selected physicochemical properties
Sulphamonomethoxine [1220-83-3] SMM	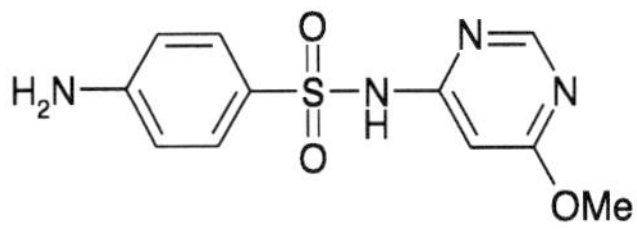	M = 280.3 g mol^{-1} pK_{a2} = 1.73 pK_{a3} = 6.22 logP = -0.04
Sulphapyridine [144-83-2] SPY	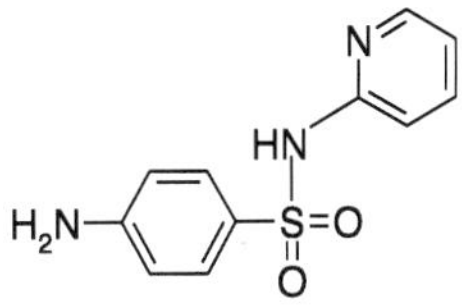	M = 249.2 g mol^{-1} pK_{a2} = 2.37 pK_{a3} = 7.48 logP = 0.03
Sulphathiazole [72-14-0] STZ	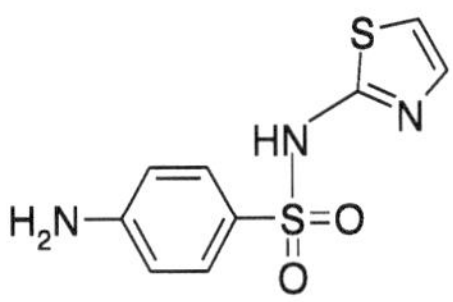	M = 255.3 g mol^{-1} pK_{a2} = 2.06 pK_{a3}= 7.07 logP = -0.04
Sulphisoxazole [127-69-5] SSX	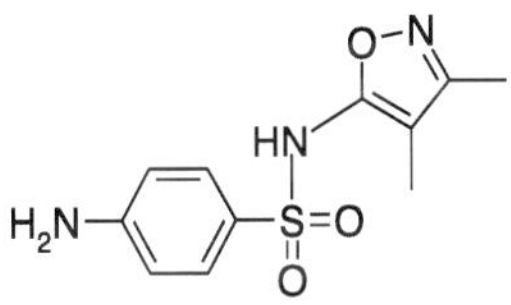	M = 267.3 g mol^{-1} pK_{a2} = 2.15 pK_{a3}= 5.00 logP = 1.01

Table 1. The structures and physicochemical properties of the sulphonamides discussed in this chapter (according to [1-9])

SAs are fairly water-soluble, polar compounds, however quite persistent - resistant to biodegradation [10,12-13] and hydrolysis [3]. This goes a long way to explaining why they have been regularly detected in both aquatic and terrestrial environments in the last ten years [1,10-11]. Although SA concentrations in environmental samples are relatively low (at the ppb or ppt level), they are continuously being released, so ultimately they may pose an elevated risk. SAs are designed to specifically target the biosynthetic pathway of folate (an essential molecule required by all living organisms) by competitively inhibiting the conversion of *p*-aminobenzoic acid (pABA); hence, they not only target bacteria but can also elicit hitherto unknown effects in environmentally relevant, non-target organisms like invertebrates and plants [14-16]. As they belong to different trophic levels, these taxonomic groups may be exposed to SAs to various extents.

So far only a small number of studies concerning the presence and effects of SAs in the soil environment have been carried out. Hence, there are a number of very pertinent questions that need to be addressed, for example: (i) What is the fate of these compounds in the terrestrial environment? (ii) What are the effects of their presence in the soil environment? (iii) Do they pose a risk to different soil organisms and also to human health?

For these reasons, the aim of this chapter is to review and summarize existing knowledge of the fate and effects of SA residues in the terrestrial environment.

Conventionally, the environmental fate of antimicrobials in the soil ecosystem is assessed with respect to their persistence and sorption onto soil. In the case of SAs, as they are very stable, only photodegradation process could have recognizable influence on their elimination from the environment [1,10]. However this process in the soil ecosystem is of lesser importance. Therefore, sorption processes influence the environmental behaviour of SAs to the greatest extent, so it is these that we shall be discussing in detail.

Although a few review papers have been published summarizing the available ecotoxicity data of pharmaceuticals, including some SAs [14-15,17-18], they focus on aquatic organisms rather than soil. In this chapter, therefore, we shall discuss the available data on SA toxicity towards different soil organisms on various trophic levels like bacteria, invertebrates and plants. These results will be discussed with respect to the existing requirements for the environmental risk assessment of veterinary pharmaceuticals (VICH, 2000 [19]; VICH, 2004 [20]; EMEA, 2007 [21]; EMEA, 2008a [22]; EMEA, 2008b [23]). In addition, we shall identify some areas where future work is warranted as well as the needs for further investigations.

2. Fate of SAs in the terrestrial environment

2.1. Basic concept of sorption modelling studies

Sorption is defined as a phenomenon during which chemicals become associated with solid phases. Immensely important, this process affects the fate of chemicals in the environment [24-25]. Adsorption/desorption experiments are useful for generating essential information on the mobility of various contaminants and their distribution in the soil, water and air. Assessing the distribution of a chemical between the soil and aqueous phases is not a straightforward process, however. It depends on various factors, such as the chemical nature of the substance and the characteristics of the soil (*e.g.* pH, organic matter (OM) content, clay fraction content). Furthermore, climatic factors such as rainfall, temperature, sunlight intensity and wind, which can affect the strength of sorption, also have to be taken into consideration. Thus, the numerous phenomena and mechanisms involved in the adsorption of a chemical by soil cannot be completely defined by simple laboratory models. Nevertheless, such investigations can provide valuable information on the environmental relevance of the adsorption of chemicals [26].

Laboratory sorption experiments can be carried out under static (batch) and dynamic (column) conditions. Static tests are commonly used when the aim of the study is to calculate the distribution coefficient (Eq. 1), in equilibrium time, which is specific to every chemical. Column tests, on the other hand, enable time-dependent monitoring of contaminant leaching from soil and waste materials; in addition, the flow-through pattern of such tests resembles actual environmental conditions. Therefore, the release of a contaminant depending on local equilibrium time and advection conditions can be evaluated based on column testing systems [27].

Both column and batch tests can be used to assess the possible leaching/release potential of contaminated materials on the soil – ground water pathway [28-30]. The release potential of water-soluble contaminants can be assessed as an expression of the source term, which gives an indication of their bioavailability. In this case a batch test provides a snapshot of a particular liquid-to-solid ratio. This type of test focuses on constituent solubility and release over a range of conditions by varying a single parameter (e.g. final extract pH, liquid-to-solid (LS) ratio) [27].

The common approach for modelling sorption results involves using only the initial linear part of the isotherm, plotted as c_w and c_s:

$$c_s = K_d \cdot c_w \tag{1}$$

where: K_d is the partition coefficient; c_s is the content of test substance adsorbed on the soil at adsorption equilibrium (mg kg^{-1}); c_w is the mass concentration of test substance in the aqueous phase at adsorption equilibrium (mg L^{-1}).

The sorption isotherm can be also mathematically described by the Freundlich or Langmuir model.

The Langmuir equation is written as:

$$c_s = \frac{c_{\max} \cdot K_L \cdot c_w}{1 + K_L \cdot c_w} \tag{2}$$

where: c_s is the content of test substance adsorbed on the soil at adsorption equilibrium (mg kg^{-1}); c_w is the mass concentration of test substance in the aqueous phase at adsorption equilibrium (mg L^{-1}); c_{max} is the maximum sorption capacity of the sorbent; K_L is the Langmuir constant.

The linear form of this equation is:

$$\frac{1}{c_s} = \frac{1}{K_L c_{\max} c_w} + \frac{1}{c_{\max}} \tag{3}$$

In the Langmuir model the mass of solute sorbed per unit mass of sorbent (c_s) increases linearly as the solute concentration increases. The model is based on three main assumptions:

- the sorption energy is the same for all sites and is independent of the degree of surface coverage;

- sorption occurs only at localized sites, with no interaction between adjoining sorbed molecules;
- the sorption maximum (c_{max}) represents a monolayer coverage.

Based on these findings, it is justified that the Langmuir isotherms are observed only very rarely in case of sorption of when organic compounds are sorbed in such a complex and heterogeneous sorbent as soils [31].

The Freundlich model takes the following form:

$$c_s = K_F \cdot c_w^{1/n} \tag{4}$$

where: c_s is the content of test substance adsorbed on the soil at adsorption equilibrium (mg kg^{-1}); c_w is the mass concentration of test substance in the aqueous phase at adsorption equilibrium (mg L^{-1}); K_F is the Freundlich adsorption coefficient; n is the regression constant

Based on the value of *1/n*, which describes the relative magnitude and diversity of energies associated with sorption, the mechanism of this process can be defined more accurately [31-36]. If $1/n = 1$, adsorption is linear, which indicates the occurrence of homogeneous energy sites on the sorbent. This type of adsorption is generally typical at very low solute concentrations with many sites on the sorbent still available for possible interaction with the adsorbate. A value of $1/n > 1$ indicates a concave, upward-curving isotherm, which is sometimes referred to as a solvent-affinity type isotherm (S-type), where the sorption energy increases with increasing surface concentration. However, phenomena unrelated to the sorption of the analyte to the sorbate surface, such as strong adsorption of the solvent, strong intermolecular attraction within the adsorbent layers, penetration of the solute in the adsorbent, and the monofunctional nature of the adsorbate, can also affect the shape of the isotherm [31]. When $1/n < 1$, the isotherm is of the convex, downward-curving Langmuir-type (L-type), where the marginal sorption energy decreases with increasing surface concentration. This may occur when the competition of solvent for sites is minimal or the absorbate is a planar molecule [36].

In order to calculate the Freundlich factors as *1/n* or K_F, Eq. 4 can be linearized by a logarithmic transformation:

$$\log c_s = 1/n \log\ c_w + \log K_F \tag{5}$$

It has been shown that the sorption of pharmaceutical compounds (PCs) to soils is influenced by solution chemistry and the type of mineral and organic sorbents [24,25,37]. Pharmaceuticals can exist as either neutral or charged species (e.g. cations, anions, zwitterions) [24], depending on the pH. Various adsorptive forces may therefore be acting. Whereas neutral molecules partition to solid matrices via relatively weak van der Waals and electron donor-acceptor interactions, charged species can interact via stronger electrostatic mechanisms, such as cation exchange, cation-bridging and complexation. The acid-base equilibrium of sulphonamides resulting from either the loss or the gain of a proton is similar for all sulphonamides apart from sulphaguanidine (see Białk-Bielińska et al. [3]). The basic properties are due to the aniline group present in all SAs and specific to each SA heterocyclic base. SAs can thus be described by three dissociation constants. Nevertheless, since the presence of the protonated form of the heterocyclic functional group is extremely unlikely, only two of the possible ionized functional groups existing in the structure of SA molecules are taken into consideration – during sorption experiments [38-45]. Hence, K_{a2} is the dissociation constant for the equilibrium between the positively charged, protonated amino group of a SA and its electrically neutral conjugate base, whereas K_{a3} refers to the equilibrium involving the loss of the SA proton to yield its negatively charged conjugate (Figure 1) [7].

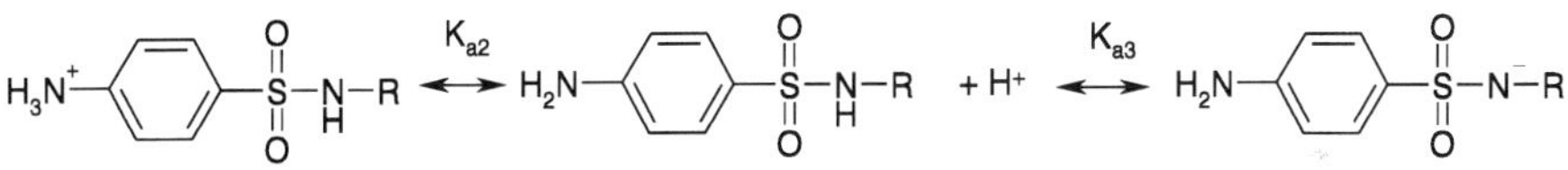

Figure 1. Dissociation equilibrium of sulphonamides [7]

Assessing the sorption of veterinary pharmaceuticals in soils is extremely important for estimating the risk of the large-scale usage of veterinary medicines to human health and environmental matrices, because this affects the fate and transport of chemicals in ground water.

2.2. Sorption potential of SAs to soils

Although SA sorption is quite a common topic of investigation, authors generally focus on just a few SAs, so that knowledge of the sorption behaviour of some of them (e.g. sulphaguanidine, sulphisoxazole) is still limited. So far, only a few review papers have been published describing the sorption of SAs to soils [16,46-49]. However, they cover a wide range of pharmaceuticals, so SAs are inadequately reviewed. Furthermore, since 2011 (when the last review on SA sorption was published), new data have been published, which are included in the present review. The available information on the sorption of the most commonly investigated compounds will therefore be discussed in depth.

The most widely investigated SA is sulphamethazine (SMZ). The level to which its undergoing sorption onto soils was investigated already more than thirty years ago by Langhammer [50]. That author calculated adsorption coefficients for four different soils, differing in pH and OM content. Based on the values of the distribution coefficients (from 1.0 to 3.1 L kg $^{-1}$), this drug can be considered as a very mobile chemical. These results are in accordance with the investigations of other researchers, such as Thurman et al. [51] and Tolls et al. [52], who reported low sorption coefficients for SMZ (0.6 L kg $^{-1}$ and 3.0 L kg^{-1} respectively) or Thiele-Bruhn et al. [53], who gave a K_d value of 2.4 L kg^{-1} for humus-rich soil. This was also supported by Lertpaitoonpan et al. [54], who examined this SA in terms of the distribution in soils varying in OM content (K_d lies between 0.2 and 3.9 L kg^{-1} depending on the physicochemical parameters of soils). However, Fan et al. [55] reported a higher sorption potential of the polar metabolite of SMZ (*N*4-acetyl-SMZ) during a miscible-displacement experiment (column test). The K_d values obtained by these authors range from 7.5 andto 206.2 L kg^{-1} and are much higher than previous data for the native compound. However, this may well be due to the polar functional group present in *N*4-acetyl-SMZ, which could enhance the association of this compound to the negatively charged soil surfaces via cation bridging or complexes. The high mobility of SMZ was also reported by Kurwadkar et al. [41], who observed a 50-90% release of SMZ from a soil column system. The most recent studies presented by Leal et al. [56] underscore the concern regarding the possible occurrence of this compound in the environment. These authors investigated a number of different Brazilian soils, finding a tendency for SMZ to leach from soil matrices.

Much attention has also been given to calculating the sorption potential of sulphadiazine (SDZ). Just recently, Doretto et al. [45] reviewed the available literature data through Freundlich sorption coefficient (K_F) for SDZ. On this basis they concluded that the potential of SDZ to interact with soil particles is relatively low and depends on the type of soil, thus on the physicochemical properties of the sorbent. In another work, these authors demonstrated the weak interaction of SDZ with binding sites on the soil surface (K_F values from 0.4 to 2.6 $\mu g^{1-1/n}(cm^3)^{1-1/n}g^{-1}$). The column studies of Wehrhan et al. [57] showed that the amount of leached SDZ depends strongly on the duration of the process. The eluted mass fraction was considerably higher in long-pulse experiments (83 and 61% respectively) than in short-pulse ones, during which only 18% was leached. Furthermore, these authors recorded the highest concentrations at the top of the column, with concentrations steadily decreasing towards the bottom. In the column with the short pulse application, solute concentrations were relatively uniformly distributed. Environmental conditions like rainfall can therefore affect the distribution of contaminants in soil.

It was also observed that SDZ exhibits, for example, a lower tendency to be retained in solid matrices than SMZ [53,56], with respective K_d values for SDZ and SMZ varying from 2.0 to 2.4 L kg^{-1} as reported by Thiele-Bruhn et al. [53], and from 5.2 to 10.5 L kg $^{-1}$, as obtained by Leal et al. [56].

Although sulphachloropyridazine (SCP) is not as widely studied a sulphonamide as SMZ or SDZ, this drug has been extensively examined using various tests besides batch or laboratory column tests. The data available in the literature show a sorption potential in soil similar to that of other SAs. For example, Boxall et al. [58] reported low sorption coefficients for SCP in soil and a soil/slurry mixture ranging from 0.9 to 1.8 L kg^{-1}. They also confirmed the high mobility of SCP in field studies, demonstrating the rapid transport of SCP to surface waters for concentrations as high as 590 μg L^{-1}. Other studies are also consistent with this statement [38,44,52,56], giving K_d values from 0.7 to 70.1 L kg^{-1}.

Blackwell et al. [59], who also examined the leaching of SCP under field conditions, detected this compound in surface run-off samples even at a concentration of 25.9 μg L^{-1} following application at a rate of 1.18 kg ha^{-1}. These authors reported the occurrence of SCP in soil water samples at a concentration of 0.8 μg L^{-1} at 40 cm depth as long as 20 days after treatment. On the basis of their results the authors concluded that SCP poses a moderate risk of contaminating ground or surface water but that its potential to accumulate in soils is low. Further lysimeter-based studies by Blackwell et al. [60] sporadically detected SCP in leachate at levels from 0.7 to 8.5 μg L^{-1}, depending on the irrigation conditions. SCP was applied in slurry (197 mL per lysimeter), which corresponds to a SCP application rate of 5.2 mg (1.18 kg ha^{-1}). The authors concluded that this compound has the potential to reach ground and surface waters. On the basis of a lysimeter study, field investigations and laboratory column tests, Kay et al. [61,62,63] pointed out that soil tillage prior to slurry application can significantly reduce losses of SCP to tile drainage systems, thereby reducing the risk of surface water contamination by SCP residues in the slurry. The observed losses of SCP in a soil column with a soil surface broken as a result of tillage fell from 54.6 % of the applied amount to zero [62].

Studies of sulphamethoxazole (SMX) have reported a similar sorption potential to that of SMZ [40]. With some exceptions, Leal et al. [56] recorded similar values of the distribution coefficient K_d for both SMX and SMZ in an examination of thirteen soils. The results obtained by Yu et al. [64] are in agreement with that. These authors calculated a K_d of 18.9 L kg^{-1} for one of three investigated soils. Their aim was to assess the suitability, inter alia, of SMX as a wastewater indicator. However, owing to the formation of non-extractable residues, such an application of SMX was regarded as limited. On the basis of a few investigations into the sorption of SMX to activated sludge, we can state that the sorption potential of sulphonamides to this sorbent is much greater than to soils [65-69]. Hrsing et al. [65] presented K_d values for SMX ranging from 280 to 370 L kg^{-1}, depending on the type of activated sludge. The results are consistent with the investigations of Hyland et al. [66], who studied the sorption of 75 pharmaceuticals onto activated sludge, obtaining a K_d value of 269 L kg^{-1}, or with those of Göbel et al. [67], who obtained a similar value of K_d for SMX. In contrast, Yang et al. [68] reported a lower sorption of SMX to activated sludge (K_d = 28.6 L kg^{-1}). However, these differences may have arisen, for example, from the different methodologies used in the tests. Nevertheless, Yang et al. [68, 69] concluded that sorption of SAs to activated sludge is highly reversible (the amount retained after desorption is 0.9% of the original dose of 100 μg L^{-1}). Therefore, the use of sewage sludge as fertilizer may constitute an additional source of SAs in ground and surface waters. Moreover, since SAs may be taken up by farmland crops, as demonstrated by Li et al. [70], the use of sewage sludge as fertilizer poses a serious risk to human health as well.

Knowledge of the sorption potential of other SAs like sulphathiazole (STZ), sulphapyridine (SPY), sulphamerazine (SMR), sulphadimethoxine (SDM), sulphamonomethoxine (SMM), sulphaguanidine (SGD) or sulphisoxazole (SSX) is very limited. In the literature there are only a few reports dealing with the sorption of these pharmaceuticals. K_d for STZ adsorption onto soil particles ranges from 1.0 to 62.5 L kg^{-1} depending on soil properties [56]. In the case of SDM, Sanders et al. [71] pointed out that the linear sorption coefficient for SDM differs somewhat, ranging from 0.4 to 25.8 L kg^{-1} as a single solute and from 2.5 to 22.1 L kg^{-1} as a co-solute with ormetoprim, another antimicrobial. Moreover, the sorption of SDM was less linear in combination with ormetoprim. In turn, Maszkowska et al. [72] did not determine the influence of the co-solute on sulphonamide release. SDM exhibited a similar leaching behaviour from the soil when it was tested alone or in a mixture with SGD. Nevertheless, these authors also reported the considerable mobility of three SAs (SDM, SGD, SSX) in three different soils; SDM was released the slowest from the soil column. These results are consistent with those published previously by Białk-Bielińska et al. [43], who showed that SDM had a greater sorption potential than SGD (K_d = 0.3 – 107.5 L kg^{-1} for SDM, 1.0 – 31.0L kg^{-1} for SGD). SDM was also found to have the strongest tendency of all the SAs investigated to interact with activated sludge [68,69]. In addition, these authors investigated the sorption strength of sulphamonomethoxine (SMM) on activated sludge, finding a lower affinity of SMM than of SDM for activated sludge particles. Jin et al. [73] demonstrated the relatively high mobility of SDM in soil, obtaining a K_d of 18.9 L kg^{-1}. These authors also highlighted the influence of different co-contaminants on adsorption. They concluded that anionic surfactants and urea could adversely affect the sorption potential of SDM, whereas cationic surfactants could improve the retention of SDM on soil particles. Figueroa-Diva et al. [40] found that SMR exhibited the lowest level of sorption of the four SAs (SDM, SMX, SMZ and SMR) that they examined. According to Thiele-Bruhn et al. [53], SPY was the most strongly retained SA in the soil matrix, with K_d higher than that of SMZ, SDZ and SDM.

Summing up, the available data indicate that determining the environmental fate of SAs in soils is not an easy task, as this depends largely on the physicochemical properties of soils and the chemical structures of the SAs. Nevertheless, one can infer from these results that these pharmaceuticals will tend to leach into ground or surface water rather than persist in soils. These data also show, however, that a certain amount of the SAs entering the soil can be retained there for quite a long time. Furthermore, their sorption to soils can increase or decrease depending on a number of different factors, which are discussed below.

2.3. Factors influencing sorption of SAs to soils

2.3.1. Influence of the organic/mineral composition of soil

Soils can be regarded as mixtures of mineral and organic fractions. The differences in their texture, structure, consistency, colour, chemical, biological and other characteristics arise from the type of parent material. Soils are therefore diverse matrices in which different sorption mechanisms can occur. The organic matter (OM) content undoubtedly plays a critical role in the sorption capacity of soils [74]. Overall, in accordance with the available literature data, it has been shown that OM positively affects sorption strength of organic compounds. Figueroa-Diva et al. [40] reported values of K_d for all examined SAs increasing in the same sequence as the organic carbon content (f_{OC}) in the soils they investigated. Białk-Bielińska et al. [43] pointed out that SAs predominantly interact with soil OM by nonbonding van der Waals interactions and hydrogen bonding. Furthermore, such weak bonding forces are susceptible to desorption, resulting in the free release of SAs following their prior surface adsorption, an observation previously made by Thiele-Bruhn et al. [53]. The authors indicated that the influence of soil OM on adsorption depends not only on the organic carbon content, but also on its composition. Sukul et al. [39] demonstrated increased sorption of SDZ in soils in the presence of manure compared to soil without manure, which greatly emphasizes the role of dissolved OM and organomineral soil particles in SDZ sorption. On the basis of original research and literature available data, Hou et al. [75] demonstrated a positive relationship between K_d and the organic carbon content (f_{OC}) for SMX sorption on soils/sediments with f_{OC} > 2%. However, for adsorbents with $f_{OC} < 2\%$, a lower f_{OC} could result in increased sorption, suggesting competition between SMX and organic matter on mineral particles. Hyland et al. [66] confirmed the positive influence of organic matter on SA sorption. The high values of K_d for SMX sorption onto activated sludge are fully justified, due in great part to the organic carbon (average f_{OC} = 44.1%) in the sorbent. Leal et al. [56] analysed the influence of the OM and clay content on the sorption of several SAs, concluding that hydrophobic partition was important in SA sorption. Nevertheless, they also found that non-hydrophobic interactions with organic and/or mineral surfaces, mainly with Al and Fe oxides and hydroxides (abundant in the investigated soils) were also important in SA retention in soils. Boxall et al. [58] determined the influence of the type of mineral fraction in soil (clay or sand) on K_d. Their results showed that clay had a greater sorption capacity for SCP than sand. The same was reported by Ter Laak et al. [38], whose K_d value for soil with greater amount of clay was twice as high as that for sandy soil. During field studies, however, Boxall et al. [58] reported faster leaching of SCP to ground water from a clay site than from a sandy site, an observation corroborated by Fan et al. [55]. The K_d values for SMZ were positively related to the OM content in case of sorbents without sand. However, the latter authors' K_d value was higher for sand (%OM=0) than for soil containing OM. They explained this as having resulted from the transport of SAs on mobile colloids (< 2 μm, dissolved organic matter and clay-sized materials) in accordance with EPA [76], which resulted in faster elution from a soil column with OM content than from sand.

2.3.2. *Influence of pH*

In the context of the acid-base equilibrium of SAs, pH can strongly affect sorption. This has been confirmed in many investigations. The overall trend presented in the literature indicates decreasing sorption of SAs with increasing pH. This is explained by the amphoteric nature of SAs, which consequently can occur in cationic, anionic or neutral form. The strongest possible interactions (ion-exchange mechanism) arise from competition for negatively charged sites on the soil surface between a cationic analyte and other cations present in the solute. Nevertheless, the existence of cationic SAs is limited due to the relatively low pK_{a2} value. Cation exchange is therefore not regarded as a favourable mechanism for SA sorption to soil matrices [39]. Although decreasing SA sorption is observed at high pH, Sukul et al. [39] achieved relatively strong adsorption in the case of one soil at a pH where the anionic form of SDZ was dominant, claiming that this was due to possible partition to the positively charged surfaces of pedogenic oxides, very abundant in the clay fraction [39,53]. Kim et al. [42] and Białk-Bielińska et al. [43] also observed a negative correlation between K_d and pH. The former authors considered that the changes were better evident for soil with a greater OM content. Pinna et al. [77], in turn, did not observe such a strong dependence on OM content. The addition of cow manure (f_{OC} = 30.58) did not significantly affect antibiotic sorption to one of the investigated soils, but did increase the extent of sorption to another soil about three times, even though larger amounts of manure had been added to the first soil than to the second one. These authors concluded that the greater sorption to the second soil prior to the addition of cow manure was most probably due to the low pH of the soil suspension rather than to its high organic carbon content. On the other hand, the high pH of the first soil suspension (7.8) could have been responsible for the reduced sorption, despite the considerable amount of OM in this amended soil.

2.3.3. *Influence of ionic strength*

Another environmentally important factor that can affect SA sorption is ionic strength. But this has not been examined extensively. Ter Laak et al. [38] carried out sorption studies of SCP, among other compounds. Generally speaking, they did not observe any significant influence of ionic strength, except in the case of one soil (clay loam), in which sorption doubled when the $CaCl_2$ concentration was raised from 0.006 to 0.2 M. The authors concluded that this increase in sorption was probably due to the neutral form of SCP increasing from 3.3 to 8.3% because of the decreasing pH. Protons are displaced from the cation-exchange sites by the addition of Ca^{2+} cations, which are ultimately responsible for the decrease in pH. Elevated cation concentrations near negatively charged soil surfaces, resulting in a decrease in the electrostatic repulsion of negatively charged sorbate molecules and soil particles, is another explanation considered by those authors. Srinivasan et al. [44] reported the different behaviour of SMX under conditions of increasing ionic strength. They explained the increasing K_d for SMX in the case of one soil as being due to cation bridging. Since positively charged calcium ions are present in the solution, bonding of anionic SMX to calcium is possible. In addition, the occurrence of a salting out effect, reducing the solubility of SMX in the salt solution so that it precipitates in the soil, was taken into consideration as a possible reason for the increase in sorption. The positive influence of ionic strength on sorption can also be attributed to the replacement of protons from the soil surface as the ionic strength increases, causing a slight reduction in pH, and shifting acidic SMX towards neutral forms that are more strongly sorbed than the anionic forms. Two other soils examined by Srinivasan et al. [44] exhibited an opposite and irregular trend in sorptive affinity of SMX, with elevated ionic strengths resulting in decreased sorption coefficients of SMX in the case of both soils. A slight decrease in sorption with increasing ionic strength of solute was also observed by Białk-Bielińska et al. [43] in the case of SDM and SGD and three soils. Srinivasan et al. [44] concluded that the ionic composition plays an important role in the sorption of ionizable organic compounds. Nevertheless, they, too, highlighted the necessity for further research in view of the conflicting results published in the literature.

2.4. Available data on the presence of SAs in soils

Although many methods have been developed in the past decade for the analysis of SAs in aqueous matrices, only a few are described in the literature for the determination of these contaminants in soil matrices. This is because the chemical analysis of pharmaceuticals from soil matrices is complicated by the need for extraction. Hence, our knowledge about the quantity of SAs in solid matrices is still limited. Nevertheless, the available literature data indicate their occurrence in agricultural soils after conventional fertilization. In a two-year monitoring study Höper et al. [78] determined SMZ at a concentration of 11 $\mu g\ kg^{-1}$. Pawelzick et al. [79] reported a maximum concentration of 4.5 $\mu g\ kg^{-1}$ for SMZ; these results are in agreement with Hu et al. [80], who demonstrated the occurrence of SMX (0.03 – 0.9 $\mu g\ kg^{-1}$) and SCP (0.18 – 2.5 $\mu g\ kg^{-1}$). Karcı et al. [81] found three SAs in agricultural soils in Turkey at concentrations even two orders of magnitude higher than those reported in previous studies: STZ (0.05 – 0.4 $mg\ kg^{-1}$), SCP (0.05 – 0.1 $mg\ kg^{-1}$) and SMX (0.05 – 0.1 $mg\ kg^{-1}$). There are some

http://dx.doi.org/10.5772/57265

discrepancies in the available literature data, which may be due to differences in the physico-chemical properties of the solid samples examined. They may also stem from the intensity of fertilization and the initial quantities of SAs applied in animal husbandry. Nevertheless, even low concentrations of SAs reported in soil samples may contaminate other environmental compartments as a result of release via desorption.

In general, concentration limits of antibiotics in the environment are not regulated, even though growing public concern has been taken into account in the prescription of environmental risk assessments of veterinary pharmaceuticals in the USA and Europe [19-23]. For these reasons, it is still necessary to develop analytical methods for the quantitation of the most important SAs in soil samples: this will help to estimate the exposure potential as well as the concentration of these substances in the terrestrial environment. Nevertheless, for a full risk assessment of these compounds not only is an exposure assessment necessary but also a hazard characterization, which addresses the question whether a substance has the potential to cause harmful effects. This will be discussed below.

3. Effects of sulphonamides in the soil environment

3.1. Introduction to soil ecotoxicology

SAs are commonly present in agricultural soils, though in fairly low concentrations (ppb and ppt levels), and are continuously being released into the environment via several routes (e.g. grazing animal faeces, manure spreading, WTP effluents). Such a state of affairs requires an Environmental Risk Assessment investigation, which should answer the basic question of whether the presence of SAs in soils poses a hazard. Ecotoxicology is the discipline that addresses this issue. It encompasses the study of organisms, populations, communities and ecosystems in terms of exposure to chemical agents, i.e. their transfer from the environment to organisms and their toxic effects. Simply put, it is the science of assessing the effects of toxic substances on ecosystems in order to protect these as a whole, rather than particular compartments, such as populations or organisms. In a practical manner, besides toxic effects, ecotoxicology explores the occurrence, distribution, accumulation and dissipation of anthropogenic toxic substances in ecosystems. The fundamental tools for this kind of research are ecotoxicological tests [82-83].

3.1.1. Ecotoxicological tests

These tests are a special group of quantitative research methods based on a thorough assessment of the impact of toxic substances (single or mixtures) on living organisms. Quantification of the results enables us to estimate the cumulative toxicity and the interactions between the introduced substances [82]. Obtaining such data enables scientists to extrapolate the results and define safe concentration levels in the ERA process (which is described in greater detail in Section 4 of this chapter).

In order to prepare valid ecotoxicity tests numerous factors need to be considered, the main ones being exposure time (acute or chronic), type of medium used, target species, toxic substance concentration range and choice of endpoint (e.g. mortality, growth inhibition, respiration).

Some 60 years ago scientists realized the need to establish uniform, standard test procedures in order to increase the repeatability and comparability of data obtained from tests. Researchers publish their own designs for tests together with results, enabling others to mimic the conditions for further experiments and allowing a better comparison of the results. Until now many standardized test procedures have been established by various environmental and governmental organizations/institutions. The best-known of these are the Organization for Economic Cooperation and Development (OECD), International Standards Organization (ISO), American Public Health Association (APHA), Environmental Protection Agency (EPA), American Society for Testing and Materials (ASTM) and International Seed Testing Association (ISTA), as well as many others from non-English speaking countries, like the the German Institute for Standardization (Deutsches Institut für Normung – DIN) and Polish Norms (Polskie Normy – PN). With this in mind it is common practice to perform tests strictly according to a chosen norm, or to modify just some aspects of a method as and when the conditions require this [82,84].

3.1.2. Soil organisms in ecotoxicological studies

The guidelines for ecotoxicological tests recommend using the best suited organisms. If the species stipulated in a guideline is unavailable, a similar one can be chosen, but it is important to select species that are extensively described in the literature. As species usually differ between ecosystems, their choice should take account of specific local conditions [82]. It is very important to realize that no single species is representative of all ecosystems; several single-species and multispecies tests have to be carried out in order to evaluate the behaviour of a toxic substance in an ecosystem.

Three main groups of organisms are evaluated in soil ecotoxicology: plants, microorganisms (microfauna) and animals (pedofauna). In the case of pedofauna, most ecotoxicological studies of soils are based on invertebrates and focus on worms, collembolans or enchytraeids because of their rapid reproduction times and easy maintenance. The most often examined endpoints here are weight change, survival, reproduction and behaviour (e.g. avoidance). Spermatophytes are the most popular plants, in which the measured effects usually relate to physiological disorders, growth inhibition and seed germination. Microorganisms constitute a very sensitive indicator of chemical stress as there are many parameters that can be evaluated: the usual ones are respiration, nitrification and growth. However (as in the case of SAs), microorganisms are also regularly evaluated for the occurrence and magnitude of increasing resistance towards pharmaceutical compounds [82-84].

3.2. Available data on sulphonamide soil ecotoxicology

An extensive literature review of data on SA soil toxicity has shown that there is a considerable gap in knowledge concerning the effects of these substances towards soil organisms. The vast majority of publications are dedicated to the analysis of microorganisms, followed by a small number of works on plants and just a few on pedofauna. Investigations involving the determination of quantified dose-response data (such as EC_{50} – the median effective concentration) are rare. However, rather more experiments have been done to detect the ecotoxicity of SAs (e.g. effects observed at a single concentration). Several investigations into the accumulation of these drugs in plants and the problem of bacterial resistance have also been done and are summarized below.

3.2.1. *Toxic effects of sulphonamides towards soil organisms*

Literature results in the form of a dose-response relationship concerning soil organisms affected by SAs are relatively scant in comparison to the numbers available for aquatic organisms. Nonetheless, such material as has been gathered does allow us to establish some basic trends and consider the potential risk posed by SAs in the soil environment. Table 2 lists effective concentrations of SAs towards soil organisms determined for different taxonomic groups. Some of the endpoints deemed less relevant have not been included in Table 2, but they will nevertheless be mentioned in the following section.

To make the results presented in Section 3.2. more transparent, all cited concentrations have been recalculated into ppm units. However, it is of paramount importance to bear in mind that each individual study was designed separately; this implies, for example, differences between the media (soil or liquid) used. Careful thought is therefore advisable in this respect as one could grossly over- or underestimate the inferences drawn from the results. On that account, for a more detailed inquiry, we recommend that the reader refer to the original versions of the cited papers.

The pedofauna is the most understudied group of soil organisms mentioned in the literature: there have been just a handful of studies. Very interesting experiments using SMX, SDZ, SPY and SMZ were conducted on the nematode *Caenorhabditis elegans*, where several kinds of effects were evaluated (behavioural – movement, and growth – body length) during 24 – 96 hours. The second generation, not exposed to SAs, was examined in the same manner. The results showed that SAs affected growth and behaviour in all the exposed nematodes in a time- and concentration-dependent way. Also, as one might expect, behavioural effects were more sensitive than growth in all cases. Interestingly though, transgenerational effects were observed: the unexposed progeny of the examined nematodes exhibited significant toxic effects. This was speculated to correspond to the ability of SMX, SDZ, SPY to penetrate the placenta and the secretion of SDZ, SPY and SMZ in maternal milk [85-86]. A different study group evaluated the effects of SCP in a multispecies soil system, where one of the examined endpoints was the mortality of earthworms *Eisenia fetida*. In this case no effect was observed for up to 21 days of exposure to SCP at concentrations reaching 100 ppm [87].

Two independent groups performed studies of the potential impact of SMX and SMZ towards plant growth and soil quality. The plants investigated were rice *Oryza sativa* L., cucumber *Cucumis sativus* L., endive *Cichorium endivia* [88], lettuce *Lactuca sativa*, alfalfa *Medicago sativa* L. and carrot *Daucus carota* [89]. There are several differences between the approaches of the two groups, such as the time of exposure, range of concentrations or the types of tests used. Despite this, the results are comparable and some conclusions are shared. In all the investigations SMX and SMZ were deemed to have the potential to affect soil organisms in environmentally relevant concentrations. Seed germination was found to be an insensitive endpoint. One of the groups [88] evaluated seed germination using the root length of seedlings in order to obtain better results. In nearly all cases SMX was found to be more effective than SMZ. Rice and carrot were found to be the most sensitive organisms with respective EC_{10} values of 0.1 ppm and 0.011 ppm [88-89]. Additionally, one group explored soil respiration and soil phosphatase activity: in these cases the respective EC_{10}s for SMX were 7 ppm and 1 ppm [88]. The inference to be drawn here is that antibiotic residues in manure and soils may affect soil microbial and enzyme activities.

A relatively original investigation was performed to assess the impacts of anthropogenic stressors (i.a. SAs) on symbiotic plant-microbe relationships [90]. The authors studied the effects of SMX on the arbuscular mycorrhizal fungus *Glomus intraradices* grown on carrot *D. carota* root-organ cultures. The assay endpoints were root length (carrot), hyphae growth and spore production (fungus). The exposure period lasted up to 28 days and the highest concentration tested was 1 ppm. SMX was found to be effective at low concentrations towards both organisms: the respective EC_{50}s for carrot and fungus (hyphae growth) were 0.0454 and 0.0451 ppm. Assessment of the endpoints was as follows: root lengths responded quickly to the presence of phytotoxic pharmaceuticals in the culture medium; hyphae length was a sensitive endpoint after 21 days' exposure; spore production required 28 days' exposure before significant differences could be detected [90].

The toxicity of STZ towards soybean (*Glycine max* (L.) Merr.) was evaluated as a potential nitrification inhibitor [91]. The effects were measured according to the growth of these plants. Fresh weight and dry weight of roots and plant tops were measured. The concentration range for STZ reached 200 ppm. STZ drastically reduced both main root elongation and lateral root development, the suppression increasing concentration-wise. Effects on soybean plants were detectable but statistically non-significant at a concentration of 10 ppm. STZ EC_{50} for dry root yield was equal to 29.5 ppm. It is worth noting that STZ inhibited root growth more than top growth.

The effects of sulphamonomethoxine sodium (SMM-Na) and sulphadiazine sodium (SDZ-Na) were investigated in three plant species: wheat *Triticum aestivum* L., Chinese cabbage *Brassica campestris* L. and tomato *Cyphomandra betacea* [92]. All of the plants exhibited linear correlations between the effects (root and shoot elongation) and SA concentrations. Seed germination was also considered, but was not sensitive to toxicity within the chosen range of SA concentrations. The tests were conducted over 2-5 days. The data acquired showed that wheat was the plant most sensitive to the toxicity of SDZ-Na with an IC_{50} = 28.1 ppm and that cabbage was the most sensitive to SMM-Na with an IC_{50} = 27.1 ppm. Worthy of note is the fact that in this study root and shoot elongation of the three crops exhibited different sensitivities, depending on the particular drug and plant species [92].

A study was developed specifically for SAs; it attempted to assess different susceptibility patterns of soil bacteria *Pantoea agglomerans* and standard antibiotic test bacteria *Pseudomonas aeruginosa,* depending on intercellular and environmental pH [93]. The results of the study revealed the effects at low concentrations (max. 20 ppm) of 8 SAs (SMX, STZ, SDZ, SDM, SMZ, SCP, SPY and SGD) at different pH values (from 5 to 8). The effects corresponding to the most sensitive pH values are listed in Table 2. The brief conclusion of this work is that the effects of SAs on microbial soil populations may depend closely on the ability of the bacteria to regulate their intercellular pH [93].

Substance	Type of test	Species	Critical effect	Time	Toxicity [ppm]	Ref.
SMX	tailored design	nematode *Caenorhabditis elegans*	body length *	24 h	$EC_{10} > 100$	[85]
				48 h	$EC_{10} = 1.02$	
				72 h	$EC_{10} = 0.0302$	
				96 h	$EC_{10} = 0.00131$	
	ISTA 1985	rice *Oryza sativa* L.	root length (seed germination)	4-5 d	$EC_{50} = 8$ NOEC = 1	[88]
		cucumber *Cucumis sativus* L.			$EC_{50} > 300$ NOEC = 1	
		endive *Cichorium endivia*			$EC_{50} = 69$ NOEC = 0.1	
	OECD 1984 (modified)	rice *Oryza sativa* L.	root length (plant growth)*	20 d	$EC_{50} = 13$ NOEC = 1	
		cucumber *Cucumis sativus* L.			$EC_{50} > 300$ NOEC = 100	
	tailored design	soil microbe	soil respiration	2 d	$EC_{50} = 7$	
	ASTM 2003	lettuce *Lactuca sativa*	root length*	5 d	$EC_{50} > 10$ $EC_{10} = 1.367$	[89]
		alfalfa *Medicago sativa* L.			$EC_{50} > 10$ $EC_{10} > 10$	
		carrot *Daucus carota*		7 d	$EC_{50} = 0.06$ $EC_{10} = 0.011$	
	DIN 58959-7	bacteria *Pseudomonas aeruginosa*	growth inhibition	24 h	$EC_{50} = 2.98$	[93]
		bacteria *Pantoea agglomerans*			$EC_{50} = 0.34$	
	tailored design	carrot *Daucus carota*	root length	14 d	$EC_{50} = 0.0627$ LOEC = 0.01	[90]
				21 d	$EC_{50} = 0.0612$ LOEC − 0.03	
				28 d	$EC_{50} = 0.0454$ LOEC = 0.01	
		mycorrhizal fungus *Glomus intraradices*	hyphae length*	14 d	$EC_{50} = 0.5029$	
				21 d	$EC_{50} = 0.0749$ LOEC = 0.3	
				28 d	$EC_{50} = 0.0451$ LOEC = 0.3	
STZ	tailored design	soybean *Glycine max* (L.) Merr.	growth reduction – fresh weight	61 d	$ED_{50} = 42$	[91]

Substance	Type of test	Species	Critical effect	Time	Toxicity [ppm]	Ref.
			growth reduction – dry yield top		ED_{50} = 33.5	
			growth reduction – dry yield root		ED_{50} = 29.5	
	DIN 58959-7	bacteria *Pseudomonas aeruginosa*	growth inhibition	24 h	EC_{50} = 5.47	[93]
		bacteria *Pantoea agglomerans*			EC_{50} = 0.77	
SDZ	DIN 58959-7	bacteria *Pseudomonas aeruginosa*	growth inhibition	24 h	EC_{50} = 2.85	[12]
		bacteria *Pantoea agglomerans*			EC_{50} = 0.5	
SDZ-Na	OECD 1984	wheat *Triticum aestivum* L.	root elongation inhibition *	2 d	IC_{50} = 28.1	[92]
		cabbage *Brassica campestris* L.		3 d	IC_{50} = 31.3	
		tomato *Cyphomandra betacea*		5 d	IC_{50} = 92.9	
SMM-Na	OECD 1984	wheat *Triticum aestivum* L.	root elongation inhibition *	2 d	IC_{50} = 120.7	[92]
		cabbage *Brassica campestris* L.		3 d	IC_{50} = 27.1	
		tomato *Cyphomandra betacea*		5 d	IC_{50} = 88.0	
SDM	DIN 58959-7	bacteria *Pseudomonas aeruginosa*	growth inhibition	24 h	EC_{50} > 20	[93]
		bacteria *Pantoea agglomerans*			EC_{50} = 2.05	
SDMD (SMZ)	ISTA 1985	rice *Oryza sativa* L.	root length (seed germination)	4-5 d	EC_{50} = 45 NOEC = 1	[88]
		cucumber *Cucumis sativus* L.			EC_{50} > 300 NOEC = 1	
		endive *Cichorium endivia*			EC_{50} = 37 NOEC = 0.1	
	OECD 1984 (modified)	rice *Oryza sativa* L.	root length (plant growth)*	20 d	EC_{50} = 43 NOEC = 1	
		cucumber *Cucumis sativus* L.			EC_{50} > 300 NOEC = 100	
	tailored design	soil microbe	soil respiration	2 d	EC_{50} = 13	

Substance	Type of test	Species	Critical effect	Time	Toxicity [ppm]	Ref.
	ASTM 2003	lettuce *Lactuca sativa*	root length*	5 d	EC_{50} >10 EC_{10} = 0.851	[98]
		alfalfa *Medicago sativa* L.		5 d	EC_{50} > 10 EC_{10} = 5.336	
		carrot *Daucus carota*		7 d	EC_{50} >10 EC_{10} = 0.065	
	DIN 58959-7	bacteria *Pseudomonas aeruginosa*	growth inhibition	24 h	EC_{50} > 20	[93]
		bacteria *Pantoea agglomerans*			EC_{50} = 1.14	
SCP	DIN 58959-7	bacteria *Pseudomonas aeruginosa*	growth inhibition	24 h	EC_{50} = 7.08	[93]
		bacteria *Pantoea agglomerans*			EC_{50} = 0.48	
SPY	DIN 58959-7	bacteria *Pseudomonas aeruginosa*	growth inhibition	24 h	EC_{50} > 20	[93]
		bacteria *Pantoea agglomerans*			EC_{50} = 2.22	
SGD	DIN 58959-7	bacteria *Pseudomonas aeruginosa*	growth inhibition	24 h	EC_{50} > 20	[93]
		bacteria *Pantoea agglomerans*			EC_{50} > 20	

* this author also examined the results for several other endpoints – see Section 3.2.1.

Table 2. Sulphonamide soil ecotoxicology – literature review

3.2.2. Other relevant soil ecotoxicology data

Several publications by Migliore et al. have shed much light on the toxicity and bioaccumulation of SDM for different terrestrial plants [94-98]. The species included in the research belonged to two groups:

- crop plants *Panicum miliaceum* L., *Pisum sativum* L., *Zea mays* L. and *Hordeum distichum* L.
- weeds *Amaranthus retroflexus* L., *Plantago major* L., *Rumex acetosella* L and *Lythrum salicaria* L.

All the plants exhibited bioaccumulation and toxicity during post-germinative development at concentrations of 300 ppm, though of course to different extents. *Lythrum salicaria* L., exposed to lower concentrations, demonstrated a hormetic response. Crop plants accumulated SDM at dissimilar rates but always higher in roots than in foliage. In order to present the versatility of these results, those of additional research using other SAs and terrestrial plants are listed in Table 3.

Substance	Organism	Observed effects	Ref.
SDM	*Panicum miliaceum* L., *Pisum sativum* L., *Zea mays* L., *Hordeum distichum* L., *Amaranthus retroflexus* L., *Plantago major* L., *Rumex acetosella* L. and *Lythrum salicaria* L.	accumulation, growth inhibition, hormesis	[94-98]
SDZ SDM SMZ	*Salix fragilis* L., *Zea mays* L., *Hordeum vulgare* L.	accumulation, toxic effects (root physiology impairment)	[101-103]
SDZ SMX	*Brassica rapa* L. and *Lumbricus terrestris* (p), activated sludge (sm)	non-extractable residues – low uptake by organisms	[104]
SDZ	soil microorganisms	respiration inhibition, adaptation	[105]
SDZ	*Eisenia fetida* (p)	no accumulation (uptake detected)	[106]
SDM	*Rhizobium etli* (sm)–*Phaseolus vulgaris* L. symbiosis	growth inhibition of both organisms	[107]
SDZ SMZ SMX	*Brassica chinensis* L. and soil microorganisms	accumulation, soil microbial biomass inhibition - higher effects for combined pollution	[70]
12 SAs	*Arthrobacter globiformis* (sm)	no effect in t = 4 h	[100]
SMX	*Salmonella typhimurium* (sm)	mutagenic activity	[108]
SDZ	*Zea mays* L. and soil microorganisms	molecular-chemical pattern changes	[109]
SDZ	soil microorganisms	soil respiration and bacterial community structure were influenced only after the addition of glucose	[110]
SDZ	soil microorganisms	soil community structure shift	[111]
SDZ	*Triticum aestivum* L.	accumulation (mainly in roots)	[112]
SMZ	*Lupinus luteus*, *Pisum sativum* L., *Lens esculenta* Medik., *Glycine max* (L.) Merr., *Vigna angularis*, *Medicago sativa* L.	root growth inhibition, necrotic changes, cytochrome c oxidase activity shifts	[113]
SDM SMX	*Pisum sativum* L., *Cucumis sativus* L.	accumulation (mainly in roots)	[114]
SMX	*Brassica campestris* L.	minimal accumulation	[115]
SMZ	*Lolium perenne* L., *Poa pratensis* L., *Poa trivialis* L., *Nasturtium officinale* R. Br.	accumulation	[116]
SDM SMR	*Solanum tuberosum* L., *Daucus carota*, *Zea mays* L., *Lycopersicon esculentum* Mill.	accumulation	[117]
SDZ	*Lactuca sativa*, *Daucus carota*	not detected in subject plants	[118]

All organisms in Table 3 belong to one of three groups: soil microorganisms **(sm)**, pedofauna **(p)** or terrestrial plants.

Table 3. Simplified list of published SA soil ecotoxicology research

As stated before soil microbiota is sensitive and easy to evaluate; hence, it is often examined for several target effects. In some cases, however, the results are closely dependent on incubation time – OECD 209 guidelines recommend relatively short exposure times – which can lead to underestimated results. This issue has been mentioned by authors working with SAs [99-100]. Moreover, microorganisms are often not the primarily evaluated body in a test design. In fact effects on microbe communities are sought as additional results, helpful in monitoring the conditions in the test environment. Some research papers covering the soil ecotoxicity of SAs (not mentioned hitherto) are very briefly summarized in Table 3.

3.2.3. *Development of resistance of soil microorganisms to sulphonamides*

The current increasing interest in research into pharmaceutical residues in the environment has drawn the attention of scientists to the causation of bacterial resistance by antimicrobial residues. As the case of SA residues in terrestrial compartments is no exception, relevant research results have been published by several authors.

The effect of SCP together with pig slurry was examined during a three-week exposure, using Biolog® multiwall plates to determine pollution-induced community tolerance (PICT). Several physiological processes were monitored as well as community-level physiological profiling (CLPP) [119]. As a result of the tests it was established that the soil microbial community's tolerance increased as soon as 7 days following exposure. Indeed, a SCP concentration of 7 ppm was sufficient to trigger the first effects. An increase in tolerance has been reported for a procedure comparable to normal agricultural practice [119].

Several investigations have been conducted with SDZ in manure. It was established that such a combination synergistically increased antibiotic resistance in bacteria. Some of the explored variations of the tests included the use of a multispecies system (microcosm) with soil bacteria, the preparation of tests on different kinds of soil and multiple amendment with pig manure. SA resistance genes were detected using hybridization and the polymerase chain reaction (PCR). In all cases there was a significant increase in resistance, though differing in extent depending on the test design. Following amendment, the bacterial populations carrying the SA resistance genes introduced to the soil declined strongly in the first weeks; nonetheless, they have the potential to be present for several months [120-122].

Sulphamethoxazole (SMX) was also examined with respect to its effect on soil bacteria. Two methods were used to assess PICT: leucine incorporation and Biolog® plates. Community structure was assessed using phospholipid fatty acid (PLFA) analysis, CLPP and bacterial growth. No effect was seen at 1 ppm SMX. At higher concentrations (20 500 ppm) effects were significant but relatively small (a ca twofold increase in community tolerance). Nonetheless, the impact of SMX on soil reflected both the direct inhibition of bacterial growth rates and changes in community structure [123].

4. Environmental risk assessment of veterinary pharmaceuticals (VPCs) in soil ecosystems

Guidelines describing how the environmental risks of veterinary products should be assessed for a range of countries have been published [22]. The approach used in Europe is based on the recommendations of the International Co-operation on Harmonization of Technical Requirements for Registration of Veterinary Products (VICH), which has attempted to harmonize the environmental risk assessment requirements of veterinary products in the USA, Europe and Japan. The approach is a two-phase process [19-20,22]. According to these guidelines for the environmental risk assessment (ERA) of VPCs, the ERA process starts with an initial exposure assessment (Phase I). With some exceptions, a fate and effects analysis (Phase II) is only required when exposure-based thresholds, the so-called action limits, are exceeded in different environmental compartments. Thus, risk assessment, described by the Risk Quotient (RQ), is carried out by calculating the ratio of the predicted (or measured) environmental concentration (PEC or MEC respectively) and the predicted biological non-effective concentrations (PNEC) on non-target organisms. The PNEC is ultimately derived from the toxicity data by applying an assessment factor (AF), usually calculated as the ratio of EC_{50} or NOEC to AF. The assessment factor (in the range from 10 to 1000) takes into account interspecies variation, acute/subchronic to chronic extrapolation and laboratory data to field impact extrapolation. For example, for acute toxicity tests its value is 1000. Nevertheless, if RQ is less than one, no further testing is recommended. Calculations of environmental concentrations rely on information on treatment dosage and intensity along with default values for standard husbandry practices and are based on a total residue approach reflecting worst-case assumptions. No fate and effects analysis is required for VPCs if the predicted environmental concentration in soil (PEC_{soil}) is < 100 μg/kg dry weight of soil. In this case the ERA is brought to a close. However, if PEC_{soil} is higher than the action limit, then Phase II, divided into two parts, comes into play: Tier A, in which the possible fate of the pharmaceutical or its metabolites and its effects on earthworms (mortality) and plants (germination and growth) as well as the effects of the test compound on the rate of nitrate mineralization in soil are determined; and Tier B, in which only effect studies are recommended for affected taxonomic levels (when RQ > 1 or in the case of soil microorganisms an effect > 25 %); no further guidance on Tier B testing is provided. There are no requirements regarding the species of organisms that should be used at this stage apart from the statement that the study using terrestrial plants should be repeated on two additional species from the most sensitive species category in the Tier A study, in addition to repeating the study on the most sensitive species. However, if after Tier B testing RQ is still > 1, more studies may be needed in order to further elucidate the effects on terrestrial ecosystems [19-23].

The main problem associated with this approach is the fact that the no actual sales figures or measured environmental concentrations are at hand when a risk assessment is conducted. Therefore, only crude PEC calculations are performed [124]. Moreover, the (eco)toxicity tests in Phase II are carried out only for single compounds and on a limited range of species. As these compounds occur in natural media not as single, isolated drugs but usually together with other compounds of the same family or the same type, accumulated concentrations or synergistic-antagonistic effects need to be considered. For these reasons, Schmitt et al. suggested checking whether the current action limits are sufficiently protective for aquatic and terrestrial organisms by at least performing limit tests to confirm the absence of effects due to pharmaceuticals with PECs below the action limits [125]. Therefore, it seems to very important to check whether other pharmaceuticals do not also pose a threat to such organisms. In addition, it must be highlighted that contaminants in ecosystems can cause adverse effects, and that the severity of such effects is not dependent on the total substance concentration but on the bioavailable concentration of the xenobiotics. Therefore, the central concept for an assessment of soils should be the bioavailability representing the link between exposure and effect of contaminants. As a consequence a "bioavailability concept" was developed, which links environmental chemistry and ecotoxicology. According to Frische et al. [126] "bioavailability describes the complex processes of mass transfer and uptake of contaminants into soil-living organisms which are determined by substance properties, soil properties, the biology of organisms and climatic influence. The bioavailable contaminant fraction in soil represents the relevant exposure concentration for soil organisms". For these reasons, risk assessments and subsequent regulatory measures (permission to use chemicals, threshold values and redemption goals in soils) are particularly subject to considerable uncertainties as long as nominal and measured total concentrations are the basis of exposure assessment [126]. Owing to necessary extrapolations, the present non-consideration of bioavailability in practice can result in both over- and underestimating the risks posed by chemicals to soils, which is why it should be included in such an ecotoxicological test. To date, only a few authors have investigated the ecotoxicity potential of pharmaceuticals with respect to their bioavailability in soil samples (e.g. [127]).

5. Risk assessment of sulphonamide residues in soil ecosystems

The approach to risk evaluation of sulphonamides in the terrestrial environment, based on the discussed issues, is presented in Figure 2.

Although, as stated in Section 2, available studies have demonstrated that the sorption potential of SAs is low, this feature, along with their mobility, is strongly influenced by the physicochemical properties of soils, the ionic strength of soil solutions as well as the physicochemical properties of the drug itself. Hence, even a slight change in these parameters can greatly increase the immobilization of SAs in soil. Furthermore, it must be emphasized that SAs are continuously being released into terrestrial ecosystems; therefore, the kind of exposure of soil organisms may be subjected to will resemble that of traditional pollutants (e.g. pesticides, detergents), even those of limited persistence. Consequently, SAs (like other pharmaceuticals) may be considered pseudo-persistent, which explains why they are detected in soils, even at such high concentrations. For all the above reasons, the ERA of SAs is a very complex and difficult process. The simple approach defined in the available guidelines, while undoubtedly useful, may not be sufficient. In accordance with existing requirements [19-23] we present a calculation of RQ for three SAs, assuming the worst case scenario; maximum reported

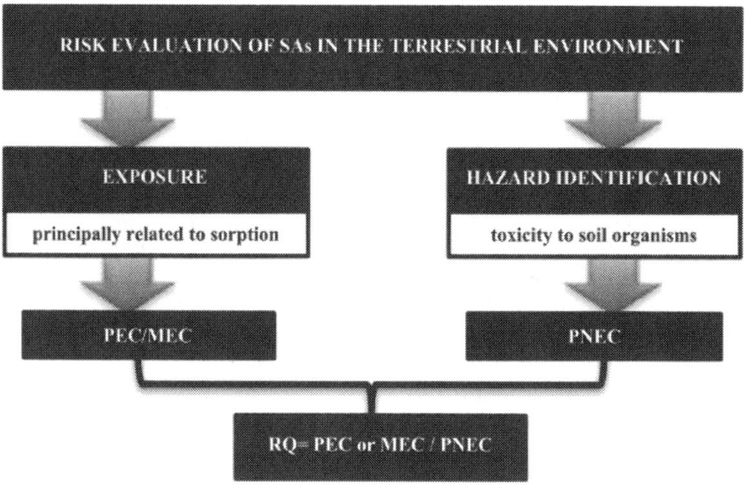

Figure 2. Risk evaluation of sulphonamides in the terrestrial environment

concentrations (Section 2.4) were used for the MEC value, PNEC was calculated on the basis of the data presented in Table 2, and AF was set at 1000 to ensure reliability (see Table 4).

Substance	MEC_{soil} [(μg kg^{-1}]	Organism	$PNEC=EC_{50}/AF$ (AF=1000) [μg kg^{-1}]	RQ=MEC/PNEC
SMX	100	rice *Oryza sativa* L.	13	**7.7**
		soil microbe	7	**14.3**
	0.9	rice *Oryza sativa* L.	13	0.07
		soil microbe	7	0.13
STZ	400	soybean *Glycine max* (L.) Merr.	42	**9.5**
SMZ	11	rice *Oryza sativa* L.	43	0.25
		soil microbe	13	0.84
	4.5	rice *Oryza sativa* L.	43	0.10
		soil microbe	13	0.35

Table 4. RQ calculations for three SAs (RQs >1 are presented in bold)

The data presented in Table 4 support our above statement, as different RQs were obtained, depending on the data used in this evaluation. It seems that a single environmental concentration, which can differ in time and place, as well as the PEC values, which may also differ in different countries, can lead to the over- or underestimation of the risk posed by these compounds. Hence, in our opinion, more realistic (reliable) approaches should be incorporated that are based on data obtained from longer term monitoring studies in each country.

6. Conclusions

Our knowledge of the presence of SAs in soils is increasing, but information in the peer-reviewed literature regarding the fate and ecotoxicological effects is still limited.

As sorption to the soil matrix governs the transport, persistence and (bio)availability of these chemicals in the environment, it can be assumed that low K_d values, together with the physicochemical properties of these compounds, indicate that they are highly mobile, readily bioavailable and easily transported from soil surfaces to aquifers, causing surface- and groundwater contamination. Being readily bioavailable to micro-organisms, plants and animals, SAs can affect these directly; indeed, they have the potential to affect entire terrestrial ecosystems. The literature records the effects of many SAs on soil organisms, although these are mainly microorganisms and plants; as there are few data on pedofauna, it is impossible to form any clear judgment in this respect. SAs have been detected in soils, and the evidence points to possible effects on soil organisms at environmentally relevant concentrations. Furthermore, SAs can be accumulated by several terrestrial plants, such as the willow *Salix fragilis* L., which could be employed for the phytoremediation of SA-contaminated soils. However, some vegetables are also reported to accumulate SAs, which could lead to adverse effects along the food chain, ultimately affecting human health. Nevertheless, research into bioaccumulation as well as the phytoremediation of these compounds is still needed.

The most and least sensitive endpoints in plant studies are root length and seed germination respectively. The effects of SAs on microorganisms have been studied in many ways, e.g. with single species and multispecies designs, and different endpoints. Most of the available data show a strongly dose-dependent relationship for the explored endpoints. Moreover, their toxicity can be strongly influenced by the pH in the environment and organisms. Furthermore the issue of microorganisms developing antibiotic resistance is related to SAs. Especially when SA-contaminated manure is used, there is a noticeable increase in resistance genes.

Hardly any information has been found concerning the toxicity of SA mixtures in soils. Since these compounds are almost always present in the form of mixtures in the environment, this issue is one to be addressed in the future. Furthermore, there is a lack of data relating to the long-term exposure of non-target organisms, and especially how continuous exposure for several generations may affect a whole population.

In conclusion, the presented data on the fate and potential effects of SAs in the terrestrial environment appear to indicate a possible negative impact on soil ecosystems and imply a threat to public health. However, further studies are necessary to characterize the risk completely.

Abbreviations

Abbreviation	Full name	Abbreviation	Full name
AF	Assessment Factor	**PICT**	Pollution-induced Community Tolerance
APHA	American Public Health Association	**PN**	Polish Norms (Polskie Normy)
ASTM	American Society for Testing and Materials	**PNEC**	Predicted Non-Effective Concentrations
CLPP	Community-level Physiological Profiling	**RQ**	Risk Quotient
DIN	German Institute for Standardization (Deutsches Institut für Normung)	**SA(s)**	Sulphonamide(s)
EC_{50}	Effective Concentration	**SCP**	Sulphachloropyridazine
ED_{50}	Effective Dose	**SDM**	Sulphadimethoxine
EMEA	European Medicines Evaluation Agency	**SDMD (SMZ)**	Sulphadimidine (Sulphamethazine)
EPA	Environmental Protection Agency	**SDZ**	Sulphadiazine
ERA	Environmental Risk Assessment	**SGD**	Sulphaguanidine
IC_{50}	Inhibitory Concentration	**SMM**	Sulphamonomethoxine
ISO	International Organization for Standardization	**SMR**	Sulphamerazine
ISTA	International Seed Testing Association	**SMX**	Sulphamethoxazole
LOEC	Lowest Observed Effect Concentration	**SPY**	Sulphapyridine
MEC	Measured Environmental Concentration	**SSX**	Sulphisoxazole
NOEC	No Observable Effect Concentrations	**STZ**	Sulphathiazole
OECD	Organization for Economic Cooperation and Development	**VICH**	Veterinary International Conference on Harmonization
OM	Organic Matter	**VPCs**	Veterinary pharmaceuticals
pABA	*p*-aminobenzoic acid	**WTP**	Wastewater Treatment Plant
PEC	Predicted Environmental Concentration	f_{oc}	Organic Carbon Content

Table 5. List of abbreviations used in the text

Acknowledgements

Financial support was provided by the Polish National Science Centre under grants DEC-2011/03/B/NZ8/03009 and DEC-2011/03/B/NZ8/03010 and DS 30-8110-D195-13.

Author details

Anna Białk-Bielińska, Joanna Maszkowska, Alan Puckowski and Piotr Stepnowski

Department of Environmental Analysis, Faculty of Chemistry, University of Gdańsk, Gdańsk, Poland

References

[1] Sukul P, Spiteller M. Sulphonamides in the environment as veterinary drugs. Reviews of Environmental Contamination and Toxicology 2006;187 67-101.

[2] Şanli S, Altun Y, Şanli N, Alsancak G, Baltran JL. Solvent Effects on pKa values of Some Substituted Sulphonamides in Acetonitrile-Water Binary Mixtures by the UV-Spectroscopy Method. Journal of Chemical and Engineering Data 2009;54 3014-3021.

[3] Białk-Bielińska A, Stolte S, Matzke M, Fabiańska A, Maszkowska J, Kołodziejska M, Liberek B, Stepnowski P, Kumirska J. Hydrolysis of sulphonamides in aqueous solutions. Journal of Hazardous Materials 2012;221-222 264-274.

[4] Stoob K. Veterinary sulphonamide antibiotics in the environment: fate in grassland soils and transport to surface waters. PhD thesis. Swiss Federal Institute of Technology Zurich, Zurich; 2005.

[5] http://www.vcclab.org/lab/alogps/start.html (ALOGPS 2.1 program)

[6] http://www.syrres.com/what-we-do/databaseforms (SRC PhysProp Database)

[7] Babić S, Horvat AJM, Mutavdžić Pavlović D, Kaštelan-Macan M. Determination of pK_a values of active pharmaceutical ingredients. Trends in Analytical Chemistry 2007;26 1043-1061.

[8] Carda-Broch S, Berthod A. Countercurrent chromatography for the measurement of the hydrophobicity of sulphonamide amphoteric compounds. Chromatographia 2004;59 79-87.

[9] Ruiz-Angel MJ, Carda-Broch S, García-Alvarez-Coque MC, Berthod A. Effect of ionization and the nature of the mobile phase in quantitative structure-retention relationship studies. Journal of Chromatography A 2005;1063 25-34.

[10] García-Galán MJ, Díaz-Cruz MS, Barceló D. Identification and determination of metabolites and degradation products of sulphonamide antibiotics. Trends in Analytical Chemistry 2008;27 1008-1022.

[11] Reemtsma T, Jekel M. (Eds) Organic Pollutants in the Water Cycle. Wiley-VCh Verlag GmbH & Co. KGaA, Weinheim, Germany, 2006.

[12] Ingerslev F, Halling-Sørensen B. Biodegradability properties of sulphonamides in activated sludge. Environmental Toxicology and Chemistry 2000;19 2467-2473.

[13] Kümmerer K. Antibiotics in the aquatic environment – A review – Part I, II. Chemosphere 2009;75 417-434.

[14] Park S, Choi K. Hazard assessment of commonly used agricultural antibiotics on aquatic ecosystems. Ecotoxicology 2008;17 526-538.

[15] García-Galán MJ, Díaz-Cruz MS, Barceló D. Combining chemical analysis and ecotoxicity to determine environmental exposure and assess risk from sulphonamides. Trends in Analytical Chemistry 2009;28 804-819.

[16] Schauss K, Focks A, Heuer H, Kotzerke A, Schmitt H, Thiele-Bruhn S, Smalla K, Wilke BM, Matthies M, Amelung W, Klasmeier J, Schloter M. Analysis, fate and effects of the antibiotic sulphadiazine in soil ecosystems. Trends in Analytical Chemistry 2009;28 612-618.

[17] Carlsson C, Johansson AK, Alvan G, Bergman K, Kühler T. Are pharmaceuticals potent environmental pollutants? Part II: Environmental risk assessments of selected pharmaceutical excipients. Science of the Total Environment 2006;364 67-87.

[18] Santos LHLM, Araújo AN, Fachini A, Pena A, Delerue-Matos C, Montenegro MCBSM. Ecotoxicological aspects related to the presence of pharmaceuticals in the aquatic environment. Journal of Hazardous Materials 2010;175 45-95.

[19] VICH, 2000. Guideline on environmental impact assessment (EIAS) for veterinary medicinal products – Phase I. VICH Topic GL6. International cooperation on harmonisation of technical requirements for registration of veterinary medicinal products.

[20] VICH, 2004. Guideline on environmental impact assessment for veterinary medicinal products – Phase II. VICH Topic GL38. International cooperation on harmonisation of technical requirements for registration of veterinary medicinal products.

[21] EMEA, 2007. Guideline on environmental impact assessment for veterinary medicinal products in support of the VICH guidelines GL6 and GL38. European Medicines Agency. Committee for medicinal products for veterinary use (CVMP). EMEA/CVMP/ERA/418282/2005-Corr.

[22] EMEA, 2008a. Revised Guideline on environmental impact assessment for veterinary medicinal products in support of the VICH guidelines GL6 and GL38. European

Medicines Agency. Committee for Medicinal Products for Veterinary Use (CVMP). EMEA/CVMP/ERA/418282/2005-Rev.1, 17 Nov. 2008.

[23] EMEA, 2008b. Reflection paper on the implementation of directive 2001/82/EC, as amended, in respect to the assessment of environmental risks of veterinary medicinal products. Committee for medicinal products for veterinary use (CVMP). Available: Doc. Ref. EMEA/CVMP/18112/2006-Consultation.

[24] Aga DS. (Ed) Fate of Pharmaceuticals in the Environment and in Water Treatment Systems. CRC Press Taylor & Francis Group, Boca Raton- London-New York, 2008.

[25] Liu P, Zhu D, Zhang H, Shi X, Sun H, Dang F. Sorption of polar and nonpolar aromatic compounds to four surface soils of eastern China. Environmental Pollution 2008;156 1053–1060.

[26] OECD, 2000. Adsorption-desorption using a batch equilibrium method. OECD Guideline for the Testing of Chemicals 106. Organization for economic Cooperation and Development, Paris, France.

[27] Lopez Meza S, Garrabrants AC, van der Sloot H, Kosson DS. Comparison of the release of constituents from granular materials under batch and column testing. Waste Managment 2008;28 1853–1867.

[28] Delay M, Lager T, Schulz HD, Frimmel FH. Comparison of leaching tests to determine and quantify the release of inorganic contaminants in demolition waste. Waste Managment 2007;27 248–255.

[29] Grathwohl P, van der Sloot HA. Groundwater risk assessment at contaminated sites (GRACOS)®: Test methods and modeling approaches. In: Quevauville P. (ed.) Groundwater Science and Policy, Cambridge: RSC; 2007.

[30] Grathwohl P, Susset B. Comparison of percolation to batch and sequential leaching tests: theory and data. Waste Managment 2009;29 2681–8.

[31] Delle A. Factors Affecting Sorption of Organic Compounds in Natural Sorbent/Water Systems and Sorption Coefficients for Selected Pollutants. A Review. Journal of Physical and Chemical Reference Data 2001;30(1) 187-439.

[32] Schwarzenbach RP, Gschwend PM, Imboden DM, Environmental Organic Chemistry, second ed. New Jersey, Hoboken: John Wiley & Sons, Inc.; 2003.

[33] Weber WJ, McGinley PM, Katz LE. A Distributed Reactivity Model for Sorption by Soils and Sediments. 1. Conceptual Basis and Equilibrium Assessments. Environmental Science and Technology 1992;26, 1955-1962.

[34] Giles CH, Smith D, Huitson J. A general treatment and classification of the solute adsorption isotherm. I. Theoretical. Journal of Colloid and Interface Science 1974;47 755-765.

[35] Farrell J, Reinhard M. Desorption of Halogenated Organics from Model Solids, Sediments, and Soil under Unsaturated Conditions. 1. Isotherms. Environmental Science and Technology 1994;28 53-62.

[36] Haque R, Coshow WR. Adsorption of Isocil and Bromacil from Aqueous Solution onto Some Mineral Surfaces. Environmental Science and Technology 1971;5(2) 139-141.

[37] Williams MC, Saison CLA, Williams D, Kookana R. Can aquatic distribution of human pharmaceuticals be related to pharmacological data? Chemosphere 2006; 65 2253-2259.

[38] Ter Laak TL, Gebbink WG, Tolls J. The effect of pH and ionic strength on the sorption of sulfachloropyridazine, tylosin, and oxytetracycline to soil. Environmental Toxicology and Chemistry 2006;25 904–911.

[39] Sukul P, Lamshöft M, Zühlke S, Spiteller M. Sorption and desorption of sulfadiazine in soil and soil-manure systems. Chemosphere 2008;73 1344–1350.

[40] Figueroa-Diva R, Vasudevan D, MacKay A. Trends in soil sorption coefficients within common antimicrobial families. Chemosphere 2010;79:786–793.

[41] Kurwadkar ST, Adams CD, Meyer MT, Kolpin DW. Comparative mobility of sulfonamides and bromide tracer in three soils. Journal of Environmental Management 2011;92 1874–1881.

[42] Kim Y, Lim S, Han M, Cho J. Sorption characteristics of oxytetracycline, amoxicillin, and sulfathiazole in two different soil types. Geoderma 2012;185-186 97–101.

[43] Białk-Bielińska A, Maszkowska J, Mrozik W, Bielawska A, Kołodziejska M, Palavinskas R, Stepnowski P, Kumirska J. Sulfadimethoxine and sulfaguanidine: their sorption potential on natural soils. Chemosphere 2012;86 1059–1065.

[44] Srinivasan P, Sarmah AK, Manley-Harris M. Co-contaminants and factors affecting the sorption behaviour of two sulfonamides in pasture soils. Environmental Pollution 2013;180 165–172.

[45] Doretto KM, Rath S. Sorption of sulfadiazine on Brazilian soils. Chemosphere 2013;90 2027–2034.

[46] Thiele-Bruhn S. Pharmaceutical antibiotic compounds in soils – a review. Journal of Plant Nutrition and Soil Science 2003;166 145–167.

[47] Beausse J. Selected drugs in solid matrices: a review of environmental determination, occurrence and properties of principal substances. Trends in Analytical Chemistry 2004;23 753–761.

[48] Tolls J. Critical Review Sorption of Veterinary Pharmaceuticals in Soils®: A Review. Environmental Science and Technology 2001;35(17) 3397-3406.

[49] Baran W, Adamek E, Ziemiańska J, Sobczak A. Effects of the presence of sulfonamides in the environment and their influence on human health. Journal of Hazardous Materials 2011;196 1–15.

[50] Langhammer J-P. Untersuchungen zum Verbleib antimikrobiell wirksamer Arzneistoffe als Rückstände in Gülle und im landwirtschaftlichen Umfeld. PhD thesis. University of Bonn, Germany. 1989.

[51] Thurman E, Lidsey M. Transport of antibiotics in soil and their potential for groundwater contamination: conference proceedings, May 21–25, 2000. 3rd SETAC World Meeting, Brighton, UK; 2000.

[52] Tolls J, Gebbink W, Cavallo R. pH-dependence of sulphonamide antibiotic sorption: data and model evaluation: conference proceedings: 12–16 May, 2002, SETAC Europe 12th Annual Meeting, Vienna, Austria; 2002.

[53] Thiele-Bruhn S, Seibicke T, Schulten H-R, Leinweber P. Sorption of Sulfonamide Pharmaceutical Antibiotics on Whole Soils and Particle-Size Fractions. Journal of Environment Quality 2004;33 1331–1342.

[54] Lertpaitoonpan W, Ong SK, Moorman TB. Effect of organic carbon and pH on soil sorption of sulfamethazine. Chemosphere 2009;76 558–564.

[55] Fan Z, Casey FXM, Hakk H, Larsen GL, Khan E. Sorption, Fate, and Mobility of Sulfonamides in Soils. Water Air and Soil Pollution 2010;218 49–61.

[56] Leal RMP, Alleoni LRF, Tornisielo VL, Regitano JB. Sorption of fluoroquinolones and sulfonamides in 13 Brazilian soils. Chemosphere (2013), http://dx.d oi.org/10. 1016/ j.ch emosphere.2013.03.01 8

[57] Wehrhan A, Kasteel R, Simunek J, Groeneweg J, Vereecken H. Transport of sulfadiazine in soil columns: experiments and modelling approaches. Journal of Contaminant Hydrology 2007;89 107–135.

[58] Boxall ABA, Blackwell PA, Cavallo R, Kay P, Tolls J. The sorption and transport of a sulphonamide antibiotic in soil systems. Toxicology Letters 2002;131 19–28.

[59] Blackwell PA, Kay P, Boxall ABA. The dissipation and transport of veterinary antibiotics in a sandy loam soil. Chemosphere 2007;67 292–299.

[60] Blackwell PA, Kay P, Ashauer R, Boxall ABA. Effects of agricultural conditions on the leaching behaviour of veterinary antibiotics in soils. Chemosphere 2009;75 13–19.

[61] Kay, P, Blackwell, PA, Boxall, ABA. Fate of veterinary antibiotics in a macroporous tile drained clay soils. Environmental Toxicology and Chemistry 2004;23 1136–1144.

[62] Kay P, Blackwell PA, Boxall ABA. Column studies to investigate the fate of veterinary antibiotics in clay soils following slurry application to agricultural land. Chemosphere 2005;60 497–507.

[63] Kay P, Blackwell PA, Boxall ABA. A lysimeter experiment to investigate the leaching of veterinary antibiotics through a clay soil and comparison with field data. Environmental Pollution 2005;134 333–341.

[64] Yu L, Fink G, Wintgens T, Melin T, Ternes TA. Sorption behavior of potential organic wastewater indicators with soils. Water Research 2009;43 951–960.

[65] Hörsing M, Ledin A, Grabic R, Fick J, Tysklind M, la Cour Jansen J,Andersen HR. Determination of sorption of seventy-five pharmaceuticals in sewage sludge. Water Research 2011;45 4470–82.

[66] Hyland KC, Dickenson ER V, Drewes JE, Higgins CP. Sorption of ionized and neutral emerging trace organic compounds onto activated sludge from different wastewater treatment configurations. Water Research 2012;46 1958–1968.

[67] Göbel A, Thomsen A, McArdell CS, Joss A, Giger W. Occurrence and sorption behavior of sulfonamides, macrolides, and trimethoprim in activated sludge treatment. Environmental Science and Technology 2005;39(11) 3981–3989.

[68] Yang S-F, Lin C-F, Lin AY-C, Hong P-KA. Sorption and biodegradation of sulfonamide antibiotics by activated sludge: experimental assessment using batch data obtained under aerobic conditions. Water Research 2011;45 3389–3397.

[69] Yang S-F, Lin C-F, Wu C-J, Ng K-K, Lin AY-C, Hong P-KA. Fate of sulfonamide antibiotics in contact with activated sludge-sorption and biodegradation. Water Research 2012;46 1301–1308.

[70] Li X, Yu H, Xu S, Hua R. Uptake of three sulfonamides from contaminated soil by pakchoi cabbage. Ecotoxicology and Environmental Safety 2013;92 297–302.

[71] Sanders S, Srivastava P, Feng Y, Dane J, Basile J, Barnett M. Sorption of the veterinary antimicrobials sulfadimethoxine and ormetoprim in soil. Journal of Environment Quality 2008;37 1510–1518.

[72] Maszkowska J, Kołodziejska M, Białk-Bielińska A, Mrozik W, Kumirska J, Stepnowski P, Krüger O, Kalbe U. Column and batch tests of sulfonamide leaching from different types of soil. Journal of Hazardous Materials 2013;260 468–474.

[73] Jin C, Pi Y, Wu C, Xue W, Chen Q, Liu J. Influence of four environmental substances on adsorption behavior of sulfamonometoxine in the soil. Acta Agriculturae Zhejiangensis 2013;25 130–134.

[74] Birkeland P. Soils and Geomorphology. Third ed. New York: Oxford University Press; 1999.

[75] Hou J, Pan B, Niu X, Chen J, Xing B. Sulfamethoxazole sorption by sediment fractions in comparison to pyrene and bisphenol A. Environmental Pollution 2010;158 2826–2832.

[76] EPA; 1999. Understanding variation in partition coefficient, Kd, values, volume I: the Kd model, methods of measurement, and application of chemical reaction codes. Washington, D.C.: Office of Air and Radiation, Environmental Protection Agency.

[77] Pinna MV, Castaldi P, Deiana P, Pusino A, Garau G. Sorption behavior of sulfamethazine on unamended and manure-amended soils and short-term impact on soil microbial community. Ecotoxicology and Environmental Safety 2012;84 234–242.

[78] Höper H, Kues J, Nau H, Hamscher G. Eintrag und Berbleib von Tierarzneimittelwirkstoffen in Böden. Bodenschutz 2002;4 141–148.

[79] Pawelzick HT, Höper H, Nau H, Hamscher G. A survey of the occurrence of various tetracyclines and sulfamethazine in sandy soils in northwestern Germany fertilized with liquid manure; conference proccedings: 18–22 April, 2004, SETAC Europe 14th Annual Meeting,, Prague, Czech Republic; 2004.

[80] Hu X, Zhou Q, Luo Y. Occurrence and source analysis of typical veterinary antibiotics in manure, soil, vegetables and groundwater from organic vegetable bases, northern China. Environmental Pollution 2010;158 2992–2998.

[81] Karci A, Balcioğlu IA. Investigation of the tetracycline, sulfonamide, and fluoroquinolone antimicrobial compounds in animal manure and agricultural soils in Turkey. The Science of the Total Environment 2009;407 4652–4664.

[82] Traczewska TM. Biologiczne metody oceny skażenia środowiska. Wrocław: Oficyna Wydawnicza Politechniki Wrocławskiej; 2011.

[83] Cardoso EJBN, Alves PRL. Soil Ecotoxicology. In: Begum G, (ed.). Ecotoxicology, InTech; 2012, p27–50.

[84] Van Gestel CAM. Soil ecotoxicology: state of the art and future directions. ZooKeys 2012;176 275–296.

[85] Yu Z, Jiang L, Yin D. Behavior toxicity to *Caenorhabditis elegans* transferred to the progeny after exposure to sulphamethoxazole at environmentally relevant concentrations. Journal of Environmental Sciences 2011;23(2) 294–300.

[86] Yu Z, Zhang J, Chen X, Yin D, Deng H. Inhibitions on the behavior and growth of the nematode progeny after prenatal exposure to sulphonamides at micromolar concentrations. Journal of Hazardous Materials 2013;250-251 198–203.

[87] Boleas S, Alonso C, Pro J, Babin M, Fernández C, Carbonell G, Tarazona JV. Effects of Sulphachloropyridazine in MS 3-Arable Land: A Multispecies Soil System for Assessing the Environmental Fate and Effects of Veterinary Medicines. Environmental Toxicology and Chemistry 2005;24(4) 811–819.

[88] Liu F, Ying G-G, Tao R, Zhao J-L, Yang J-F, Zhao L-F. Effects of six selected antibiotics on plant growth and soil microbial and enzymatic activities. Environmental Pollution 2009;157 1636–1642.

[89] Hillis DG, Fletcher J, Solomon KR, Sibley PK. Effects of Ten Antibiotics on Seed Germination and Root Elongation in Three Plant Species. Archives of Environmental Contamination and Toxicology 2011;60 220–232.

[90] Hillis DG, Antunes P, Sibley PK, Klironomos JN, Solomon KR. Structural responses of *Daucus carota* root-organ cultures and the arbuscular mycorrhizal fungus, *Glomus intraradices*, to 12 pharmaceuticals. Chemosphere 2008;73 344–352.

[91] Maftoun M, Sheibany B. Comparative Phytotoxicity of Several Nitrification Inhibitors to Soybean Plants. Journal of Agricultural and Food Chemistry 1979;27 1365–1368.

[92] Jin C, Chen Q, Sun R, Zhou Q, Liu J. Eco-toxic effects of sulphadiazine sodium, sulphamonomethoxine sodium and enrofloxacin on wheat, Chinese cabbage and tomato. Ecotoxicology 2009;18 878–885.

[93] Tappe W, Zarfl C, Kummer S, Burauel P, Vereecken H, Groeneweg J. Growth-inhibitory effects of sulphonamides at different pH: Dissimilar susceptibility patterns of a soil bacterium and a test bacterium used for antibiotic assays. Chemosphere 2008;72 836–843.

[94] Migliore L, Brambilla G, Cozzolino S, Gaudio L. Effect on plants of sulphadimethoxine used in intensive farming (*Panicum miliaceum, Pisum sativum* and *Zea mays*). Agriculture, Ecosystems and Environment 1995;52 103–110.

[95] Migliore L, Brambilla G, Civitareale C, Cozzolino S, Gaudio L. Effect of sulphadimethoxine contamination on barley (*Hordeum distichum* L., Poaceae, Liliopsida). Agriculture, Ecosystems and Environment 1996;60 121–128.

[96] Migliore L, Civitareale C, Brambilla G, Cozzolino S, Casoria P, Gaudio L. Effects of sulphadimethoxine on cosmopolitan weeds (*Amaranthus retroflexus* L., *Plantago major* L. and *Rumex acetosella* L.). Agriculture, Ecosystems and Environment 1997;65 163–168.

[97] Migliore L, Civitareale C, Cozzolino S, Casoria P, Brambilla G, Gaudio L. Laboratory models to evaluate phytotoxicity of sulphadimethoxine on terrestrial plants. Chemosphere 1998;37 2957–2961.

[98] Migliore L, Rotini A, Cerioli NL, Cozzolino S, Fiori M. Phytotoxic Antibiotic Sulphadimethoxine Elicits a Complex Hermetic Response in The Weed *Lythrum salicaria* L. Dose-response 2010;8 414–427.

[99] Kümmerer K, Alexy R, Hüttig J, Schöll A. Standardized tests fail to assess the effects of antibiotics on environmental bacteria. Water Research 2004;38 2111–2116.

[100] Białk-Bielińska A, Stolte S, Arning J, Uebers U, Böschen A, Stepnowski P, Matzke, M. Ecotoxicity evaluation of selected sulphonamides. Chemosphere 2011;85 928–933.

[101] Michelini L, Meggio F, La Rocca N, Ferro S, Ghisi R. Accumulation and Effects of Sulphadimethoxine in *Salix fragilis* L. Plants: A Preliminary Study to Phytoremediation Purposes. International Journal of Phytoremediation 2012;14(4) 388–402.

[102] Michelini L, Reichel R, Werner W, Ghisi R, Thiele-Bruhn S. Sulphadiazine Uptake and Effects on *Salix fragilis* L. and *Zea mays* L. Plants. Water, Air, & Soil Pollution 2012;223 5243–5257.

[103] Michelini L, La Rocca N, Rascio N, Ghisi R. Structural and functional alterations induced by two sulphonamide antibiotics on barley plants. Plant Physiology and Biochemistry 2013;67 55–62.

[104] Heise J, Höltge S, Schrader S, Kreuzig R. Chemical and biological characterization of non-extractable sulphonamide residues in soil. Chemosphere 2006;65 2352–2357.

[105] Kotzerke A, Sharma S, Schauss K, Heuer H, Thiele-Bruhn S, Smalla K, Wilke BM, Scholter M. Alterations in soil microbial activity and *N*-transformation processes due to sulphadiazine loads in pig-manure. Environmental Pollution 2008;153 315–322.

[106] Norr C, Riepert F. Bioaccumulation Studies with *Eisenia fetida* Using an Established Degradation Test System. Journal of Soils and Sediments 2007;7(6) 393–397.

[107] Sartorius M, Riccio A, Cermola M, Casoria P, Patriarca EJ, Taté R. Sulphadimethoxine inhibits *Phaseolus vulgaris* root growth and development of N-fixing nodules. Chemosphere 2009;76 306–312.

[108] Isidori M, Lavorgna M, Nardelli A, Pascarella L, Parrella A. Toxic and genotoxic evaluation of six antibiotics on non-target organisms. Science of the Total Environment 2005;346 87–98.

[109] Reichel R, Rosendahl I, Peeters ETHM, Focks A, Groeneweg J, Bierl R, Schlichting A, Amelung W, Thiele-Bruhn S. Effects of slurry from sulphadiazine- (SDZ) and difloxacin- (DIF) medicated pigs on the structural diversity of microorganisms in bulk and rhizosphere soil. Soil Biology and Biochemistry 2013;62 82–91.

[110] Zielezny Y, Groeneweg J, Vereecken H, Tappe W. Impact of sulphadiazine and chlorotetracycline on soil bacterial community structure and respiratory activity. Soil Biology and Biochemistry 2006;38 2372–2380.

[111] Hammesfahr U, Heuer H, Manzke B, Smalla K, Thiele-Bruhn S. Impact of the antibiotic sulphadiazine and pig manure on the microbial community structure in agricultural soils. Soil Biology and Biochemistry 2008;40 1583–1591.

[112] Grote M, Schwake-Anduschus C, Michel R, Stevens H, Heyser W, Langenkamper G, Betsche T, Freitag M. Incorporation of veterinary antibiotics into crops from manured soil. Landbauforschung Völkenrode 2007;57 25–32.

[113] Piotrowicz-Cieślak AI, Adomas B, Nałecz-Jawecki G, Michalczyk DJ. Phytotoxicity of Sulphamethazine Soil Pollutant to Six Legume Plant Species. Journal of Toxicology and Environmental Health Part A 2010;73(17-18) 1220–1229.

[114] Tanoue R, Sato Y, Motoyama M, Nakagawa S, Shinohara R, Nomiyama K. Plant Uptake of Pharmaceutical Chemicals Detected in Recycled Organic Manure and Reclaimed Wastewater. Journal of Agricultural and Food Chemistry 2012;60 10203–10211.

[115] Holling CS, Bailey JL, Heuvel B Vanden, Kinney CA. Uptake of human pharmaceuticals and personal care products by cabbage (*Brassica campestris*) from fortified and biosolids-amended soils. Journal of Environmental Monitoring 2012;14 3029–3036.

[116] Chitescu CL, Nicolau AI, Stolker AAM. Uptake of oxytetracycline, sulphamethoxazole and ketoconazole from fertilised soils by plants. Food Additives & Contaminants: Part A 2012;DOI: 10.1080/19440049.2012.725479 1–9.

[117] Sabourin L, Duenk P, Bonte-Gelok S, Payne M, Lapen DR, Topp E. Uptake of pharmaceuticals, hormones and parabens into vegetables grown in soil fertilized with municipal biosolids. Science of the Total Environment 2012;431 233–236.

[118] Boxall ABA., Johnson P, Smith EJ, Sinclair CJ, Stutt E, Levy LS. Uptake of Veterinary Medicines from Soils into Plants. Journal of Agricultural and Food Chemistry 2006;54 2288–2297.

[119] Schmitt H, Haapakangas H, Van Beelen P. Effects of antibiotics on soil microorganisms: time and nutrients influence pollution-induced community tolerance. Soil Biology and Biochemistry 2005;37 1882–1892.

[120] Heuer H, Smalla K. Manure and sulphadiazine synergistically increased bacterial antibiotic resistance in soil over at least two months. Environmental Microbiology 2007;9 657–666.

[121] Heuer H, Focks A, Lamshöft M, Smalla K, Matthies M, Spiteller M. Fate of sulphadiazine administered to pigs and its quantitative effect on the dynamics of bacterial resistance genes in manure and manured soil. Soil Biology and Biochemistry 2008;40 1892–1900.

[122] Heuer H, Solehati Q, Zimmerling U, Kleineidam K, Schloter M, Müller T, Focks A, Thiele-Bruhn S, Smalla K. Accumulation of Sulphonamide Resistance Genes in Arable Soils Due to Repeated Application of Manure Containing Sulphadiazine. Applied and Environmental Microbiology 2011;77 2527–2530.

[123] Demoling LA, Bååth E, Greve G, Wouterse M, Schmitt H. Effects of sulphamethoxazole on soil microbial communities after adding substrate. Soil Biology and Biochemistry 2009;41 840–848.

[124] Ankley GT, Brooks BW, Huggett DB, Sumpter JP. Repeating history: Pharmaceuticals in the environment. Environmental Science and Technology 2007;15 8211-8217.

[125] Schmitt H, Boucard T, Garric J, Jensen J, Parrott J, Péry A, Römbke J, Straub JO, Hutchinson TH, Sánchez-Argüello P, WennmalmÅ, Duis K. Recommendations on the Environmental Risk Assessment of Pharmaceuticals – effect characterization. Integrated Environmental Assessment and Management 2010;6 588-602.

[126] Frische T, Mebes KH, Filser J. Assessing the bioavailability of contaminants in soils: a review on a recent concepts Berlin. Research Report 2003;201 64 214.

[127] Hammesfahr U, Kotzerke A, Lamshft M, Wilke BM, Kandeler E, Thiele-Bruhn S. Effects of sulphadiazine-contaminated fresh and stored manure on a soil microbial community, European Journal of Soil Biology 2011;47 61-68.

Section 5

Advances in the Assessment of Heavy Metal Contamination - Case Studies

Chapter 24

Heavy Metal Deposition in Soils and Plants Impacted by Anthropogenic Modification of Two Sites in the Sudan Savanna of North Western Nigeria

Sunday Paul Bako,
Augustine Uwanekwu Ezealor and Yahuza Tanimu

Additional information is available at the end of the chapter

http://dx.doi.org/10.5772/57299

1. Introduction

Heavy metals occur naturally in the ecosystem, most of them in trace quantities [1]. Depending on their concentration, some of them like Zn, Fe, Mn and Cu are essential plant nutrients [2, 3, 4], others like Pb, Hg and Cr do not have any known use to plants.

Deposition of heavy metals in soils from anthropogenic activities have been implicated for an increase in heavy metal concentration above background and recommended levels [5, 6, 7]. Heavy metals are important components of agro-allied products such as pesticides, herbicides, fertilizers; manufacturing and other synthetic products such as paints and batteries [8]. Mining activities, industrial, municipal and domestic wastes have been reported to be important sources of heavy metal pollution to the environment [9].

Combustion of fuel from petroleum, abrasion of tyres, brake lining, corrosion of the body work of vehicles and engine wear have been associated with elevated concentrations of Cd, Cu, Mn, Ni, V and Zn [10, 11, 12, 13].

Excessive concentration of heavy metals in the environment is of great concern because of their non-biodegradability. Therefore, their persistence in the environment portends health hazard to plants and animals and consequently trigger ecological imbalance in the ecosystem [14]. Another concern that high concentrations of heavy metals raise is their ability to bioaccumulate across the food chain, with members that are high up the food chain having concentration of such metals several times higher than what is obtainable in the environment [15, 16, 17].

The effect of metals in living organisms could be chronic, due to exposures over a long period of time as a result of food chain transfer or acute poisoning due to ingestion or dermal contact [18]. The concentration of heavy metals in the shoot of plants may vary with season as a result of inherent growth dynamics of the plant; metal concentration and its bio-availability in the environment. Data on the response of plants to anthropogenic modification of the environment, particularly in relation to soil and air pollution by heavy metals in northern Nigeria is limited and includes earlier assessments [3, 6, 15, 16] and more recently [19], in which an attempt was made to document the visual symptoms expressed by some ruderal plant species in relation to air pollution as a step towards developing a reference for field identification of pollution events [19].

This study was carried out to determine the spatial and seasonal variations of heavy metal deposition in soils and plants in Nigeria's Sudan Savanna in order to assess the extent of pollution and to identify indigenous plant species that may be pollution tolerant and thus have potential for use in phytorememediation of heavy metal polluted sites.

2. Materials and methods

2.1. Study area

The two sampling sites were located in Katsina State, a state that falls within the Sudan Savanna eco-region of northern Nigeria (See Fig 1).

The Katsina Steel Rolling Mill (See Fig 2) was established in the mid 1970s with an installed capacity of 207, 000 mts per annum [20]

Zobe dam was constructed in 1977 to provide Portable water and for irrigation. It has a storage capacity of 170 Million cubic meters, a surface area of 39.6 km^2 and supports 8137 ha of irrigated land (See Fig 3).

Location/Site	Sampling Stations	Longitude	Latitude	Altitude
Katsina Steel Rolling Mill environs	1	7 37 06. 46 E	12 57 24.72 N	557m
	2	7 37 10. 23 E	12 57 10.23 N	553m
	3	7 37 23.06E	12 57 23.07 N	556m
Zobe Dam catchment	1	7 27 56.40 E	12 21 42. 18 N	510 m
	2	7 27 4.8 E	12 21 19. 12 N	520 m
	3	7 27 52.31 E	12 22 28. 19 N	520 m

Table 1. Geographical Positioning Coordinates of sampling stations at Katsina Steel Rolling Mill environs and Zobe Dam catchment, Katsina state, north Western Nigeria.

http://dx.doi.org/10.5772/57299

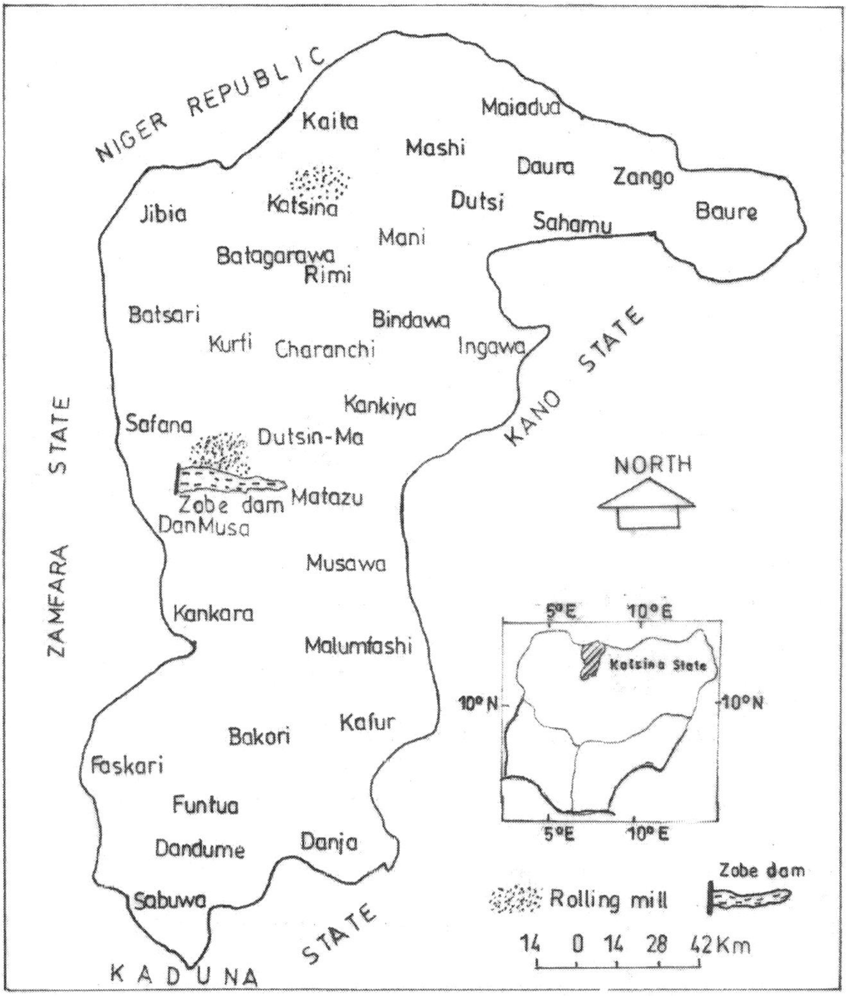

Figure 1. Katsina State, Nigeria showing approximate location of the Steel Rolling Mill and Zobe dam.

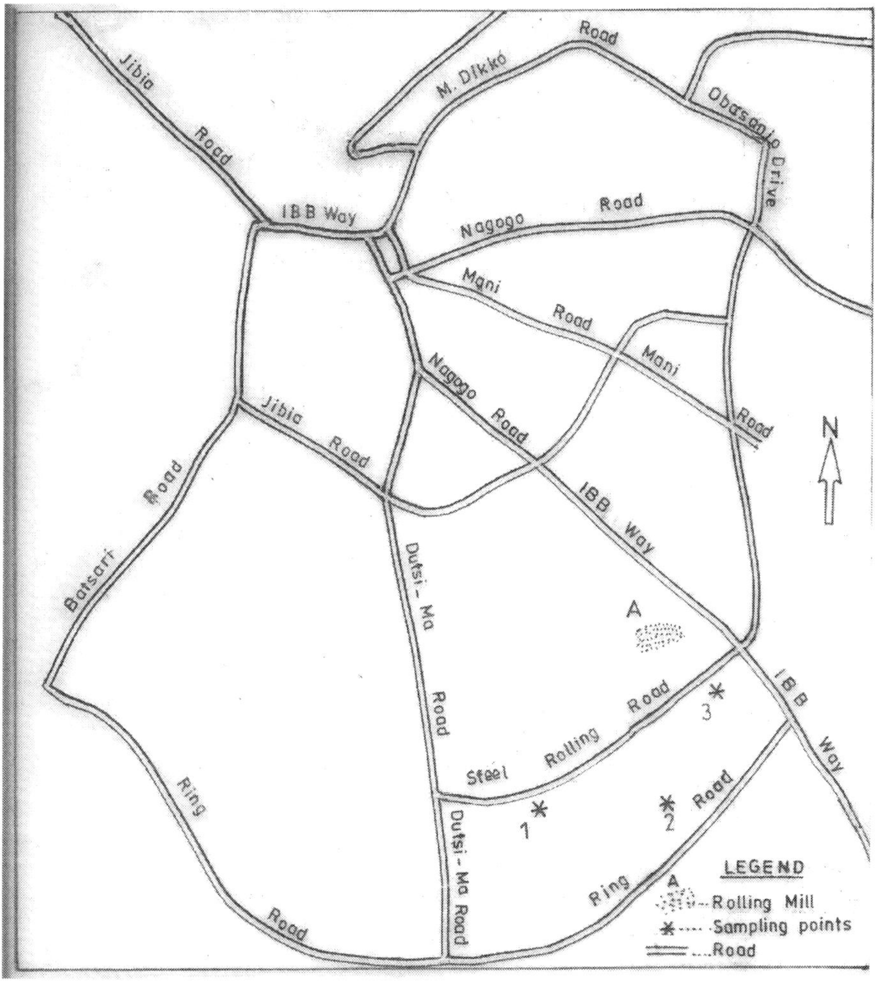

Figure 2. Environs of the Katsina Steel Rolling Mill Showing approximate sampling points.

http://dx.doi.org/10.5772/57299

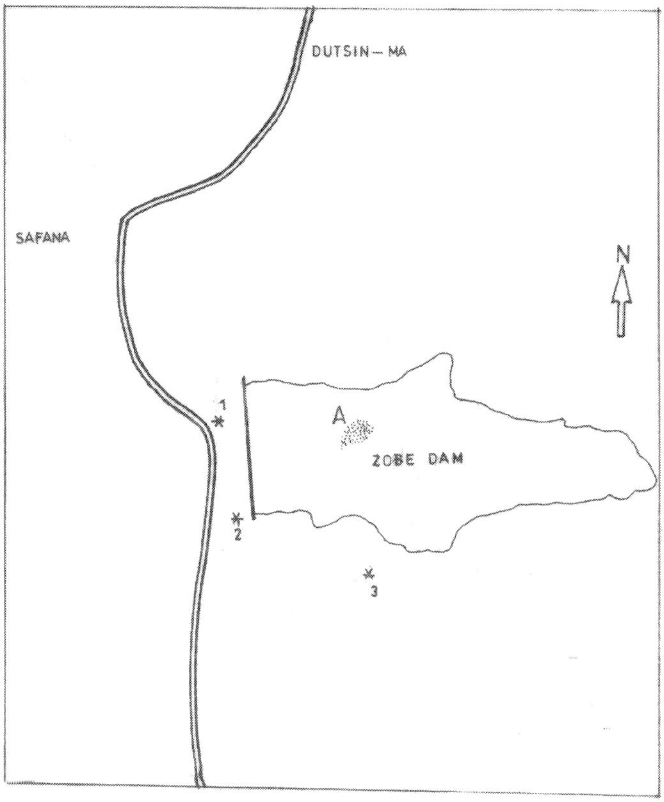

Figure 3. Environs of Zobe Dam showing approximate sampling points

2.2. Sample collection

Soil and plant samples were collected in both dry and wet seasons from three sampling stations per location in 5m by 5m quadrats. Soil samples were collected at 0-15cm depth using a soil auger. The shoot of herbaceous plants with relative abundance greater than five per quadrat, were clipped with a pair of secateurs for heavy metal analysis. Identification of plant samples were confirmed in the herbarium of the Department of Biological Sciences, Ahmadu Bello University, Zaria.

2.3. Sample preparation

Plant samples were washed with tap water and then with distilled water to remove debris and surface contamination. Samples were then bulked and air dried to remove excess moisture. Similarly, samples of the soils collected were bulked into composite samples and air dried for three days (72 hours).

Dried plant and soil samples were ground using a porcelain mortar and pestle and sieved to attain a uniform particle size. Each sample was put in a small transparent polythene bag and labeled.

2.4. Metal analysis

Analysis of the elemental content of the samples was done using the Energy dispersive X-ray fluorescence Spectroscopy (EDXRF) method [21].

The samples were ground manually to powder with an agate mortar and pestle to grain size of less than 125μm. Pellets of 19mm diameter were prepared from 0.3-0.5kg powder mixed with three drops of organic liquid binder and pressed at 10 tons of pressure in a hydraulic press.

Measurements were performed using an annular 25mCi ^{109}Cd as the excitation source, that emits Ag – K rays (22.1 KeV) in which case all elements with lower characteristic excitation energies are accessible for detection in the samples. The system consists furthermore of Si (Li) detector, with a resolution of 170eV for the 5.90KeV line, coupled to a computer controlled ADC – card.

Quantitative analysis of the sample was done using the Emission Transmission (E-T) method and that involves the use of pure target material (Mo) to measure the absorption factors in the sample.

The Mo target served as a source of monochromatic X-rays, which are excited through the sample by primary radiation and then penetrate the samples on the way to the detector. In this way, the absorption factor is experimentally determined which the program uses in the quantification of concentration of the elements. In addition, the contribution to the Mo-K peak intensity by the Zr-K is subtracted for each sample.

Sensitivity calibration of the system was performed using thick pure metal foils (Ti, Fe, Co, N, Zn, Nb, Zr, Mo, Sn, Ta and Pb) and stable chemical compounds (K_2CO_3), $CaCO_3$, Ce_2O_3), WO_3, ThO_2, U_3O_3). The spectra for the samples were collected for 3000s with the ^{109}Cd source and the spectra were then evaluated using the AXIL-QXAS program [22]. ^{109}Cd source was used for the analysis of K, Th, Y, Zr, Nb and Mo.

The accuracy and precision of the measurements was confirmed though an analysis of IAE – V10 (hay powder) and IAEA – 259 (cabbage) certified reference material, distributed by International Atomic Energy Agency (IAEA).

http://dx.doi.org/10.5772/57299

2.5. Bioconcentration factor (Enrichment Coefficient) [23]

This estimates the capacity of plants to accumulate metals, and was computed for each species as:

$$BCF=\frac{\text{Mean concentration of metal in the plant}}{\text{Mean concentration of metal in the soil}} \tag{1}$$

3. Results

3.1. Metal concentration in soils and plant samples

In the soils around the Zobe dam catchment, the mean concentration of metals generally followed the order Fe>Mn>Zn>Cr>Cu>Ni>Cd and Fe>Mn>Zn>Ni>Cr>Cu>Cd in dry and wet seasons respectively. In soils in the environs of the Katsina Steel Rolling Mill (KTSRM), the mean wet season concentration trend followed the order Fe>Mn>Zn>Cr>Ni>Cu>Cd and Fe>Mn>Zn>Cu>Cr>Ni>Cd in the dry season (Table 2; Fig 4). Differences in soil concentrations of all the metals between seasons were not statistically significant (P=0.05)

3.2. Chromium

The soil around the Zobe dam catchment had a mean Cr concentration of 66.00 mg/kg and 70.53 mg/kg, in dry and wet seasons respectively (Table 2; Fig 4). The mean Cr concentration in the environs of the Katsina Steel Rolling Mill was 46.67 mg/kg in the dry season and 70.53mg/kg in the wet season. The wet season concentrations were observed to be higher than acceptable limits for soils in Canada and the Netherlands (Table 3).

Sampling Locations	Seasons and P Values	Metals						
		Cr	Cu	Ni	Cd	Fe	Mn	Zn
Zobe Dam	DRY	66.00	29.66	26.87	0.15	22700.00	360.58	263.33
	WET	70.53	7.92	184.57	1.05	55516.93	891.66	428.45
	P value	0.13ns	0.33ns	0.64ns	0.29ns	0.25ns	0.33ns	0.34ns
KTSRM	DRY	46.67	56.91	46.31	0.33	18266.67	277.22	693.67
	WET	69.58	12.05	84.50	0.00	58867.00	745.75	385.00
	P value	0.23ns	0.60ns	0.12ns		0.32ns	0.17ns	0.94ns

Differences in soil heavy metal content were not significant (ns) between seasons in both locations.

Table 2. Mean Seasonal Heavy Metal Concentration (mg/kg) and T-test P-values of Soils collected from Zobe dam catchment and the environs of Katsina Steel Rolling Mill (KTSRM)

Recommending Agency/Country		Cr	Cu	Ni	Cd	Fe	Mn	Zn
WHO [24]		100	30	80	NA	NA	NA	200
Mexico (Agricultural soils) [25]		NA	NA	1600.00	37.00	NA	NA	NA
USA [26]		3, 000	4, 300	75.00		NA	NA	7500
Japanese MOE [27]		250	NA	NA	150	NA	NA	NA
Canadian CCME [28]	**Agriculture**	64	63	50	14	NA	NA	NA
	Residential/Park	64	63	50	10	NA	NA	NA
	Commercial	87	91	50	22	NA	NA	NA
	Industrial	87	91	50	22	NA	NA	NA
Austria [29]		100	60 to 100	50 t0 70	1 to 2	NA	NA	NA
Germany [29]		60	40	50	1	NA	NA	NA
France [29]		150	100	50	2	NA	NA	NA
Luxembourg [29]		100-200	50 to 140	30 to 75	1 to3	NA	NA	NA
Netherlands [29]		30	40	15	0.5	NA	NA	NA
Sweden [29]		60	40	30	0.4	NA	NA	NA
United Kingdom [29]		400	135	75	3	NA	NA	NA

NA — Not available

Table 3. International Recommended Levels of Heavy Metals in Soils (mg/kg)

In the plant samples, the highest concentration of Cr was recorded in *Senna siberiana* (2.64 mg/kg) in the dry season and *Englerina gralicinus* and *Terminalia mollis* (3.30 mg/kg) in the wet season (Fig 5). "Normal" concentration of Cr in plants is described as being from 0.1 to 0.5 mg/kg, while concentrations between 5 to 30 mg/kg as "Excessive" and 75 to 100 mg/kg, "Phytotoxic" (Table 4).

3.3. Copper

Mean concentrations of Cu in the soil samples collected around the Zobe dam catchment was 29.66mg/kg in the dry season and 7.92 mg/kg in the wet season. Around the environs of KTSRM, the mean Cu concentration was observed to be 56.91 mg/kg in the dry season and 12.05 mg/kg in the wet season (Table 2; Fig 4). Only soil samples collected from the environs of KTSRM in the dry season were observed to have mean Cu concentration above the acceptable limits of the WHO (30 mg/kg), Germany, the Netherlands and Sweden (40 mg/kg) Table 3).

In plant samples collected, the highest concentration of Cu was observed in members of the genus *Diospyros, D. piscatoria,* 10.41 mg/kg and *D. mespiliformis* 10.40 mg/kg in the dry season. In the wet season, the concentration of Cu was generally higher in the plants, with the highest concentration recorded in *Combretum mucronatum* (34.58 mg/kg) (Fig 5). Concentrations between 20 to 100 mg/kg in plants have been described as being excessive (Table 4).

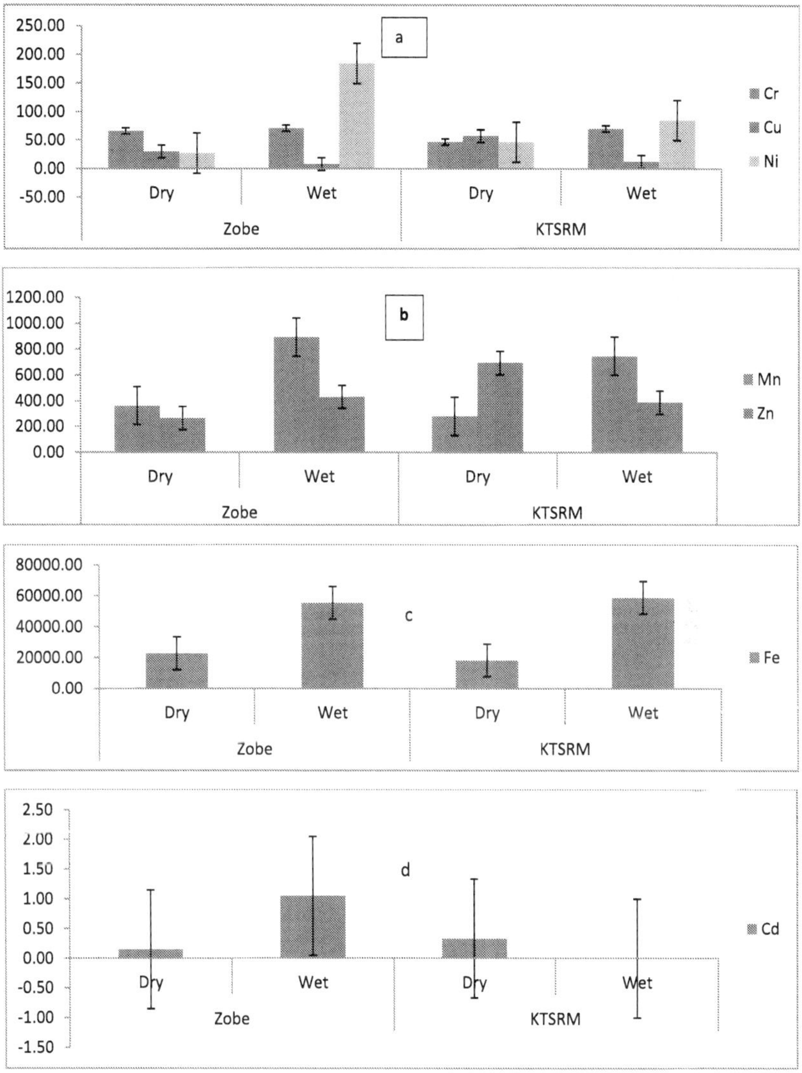

Figure 4. Seasonal variation in heavy metal concentration (mg/kg) in soils of Zobe dam catchment and the environs of Katsina Steel Rolling Mill (KTSRM)(a, Cu, Cr and Ni; b, Mn and Zn; c, Fe; and d, Cd).

	NORMAL	DEFICIENCY	EXCESSIVE	PHYTOTOXIC
Cd	NA	NA	NA	5 to 30
Zn	27-150	10 to20	100 to 400	70 to 400
Cr	0.1- 0.5	NA	5 to 30	75-100
Cu	5.1-30	2 to 5	20-100	60-125
Ni	0.1 -5	NA	10-100	100

Table 4. Range Values of Heavy Metals in Vegetation (mg/kg)

3.4. Nickel

In soils around the Zobe dam catchment, the mean Ni concentration was 184.57 mg/kg in the wet season and 26.87 mg/kg in the dry season. Whereas mean concentrations of Ni of 46.31 mg/kg and 84.50. mg/kg were observed in dry and wet seasons respectively, in soils around the environs of the KTSRM (Table 2; Fig 4). The mean concentration of Ni in both sites were observed to be higher than the acceptable limits of WHO (80 mg/kg), USA (75 mg/kg), Canada, France and Germany (50 mg/kg), Austria (50 to 70 mg/kg), Luxembourg (30-75 mg/kg); and Netherlands, Sweden and the UK (Table 3).

In plant samples, *Senna siberiana* had the highest Ni concentrations of 3.69 mg/kg and 13.77 mg/kg in dry and wet seasons respectively in the environs of KSTRM. *Diospyros piscatoria* was observed to have the highest concentration of 2.20 mg/kg for Ni in the dry season while *Ipomoea ascarafolia* had the highest concentration of 3.63 mg/kg among plants collected in the wet season around the Zobe dam catchment (Fig 5). Normal concentration of Ni in plants has been described to be between 0.1 to 5 mg/kg, whereas concentrations between 10 to 100 mg/kg as excessive (Table 4).

3.5. Cadmium

Soils around the Zobe dam catchment had a mean Cd concentration of 0.15 mg/kg in the dry season and 0.33 mg/kg in the wet season. In the environs of KTSRM, a mean concentration of 1.05 mg/kg was observed in the dry season. In the wet season Cd concentrations were below detectable limits in the wet season (Table 2; Fig 4). The above mean concentrations were found to be within the acceptable limits of the WHO and the Countries listed in Table 3.

Ipomoeaascarafolia had the highest concentration of cadmium in dry (3.90 mg/kg) and wet (1.03 mg/kg) seasons among plants sampled around the Zobe dam catchment. In the environs of KTSRM, *Ziziphus abyssinica* was observed to have the highest concentration of 4.5 mg/kg of Cd in the dry season, while in the wet season *Cadaba farinosa* was observed to have the highest concentration of 0.54 mg/kg (Fig 5).

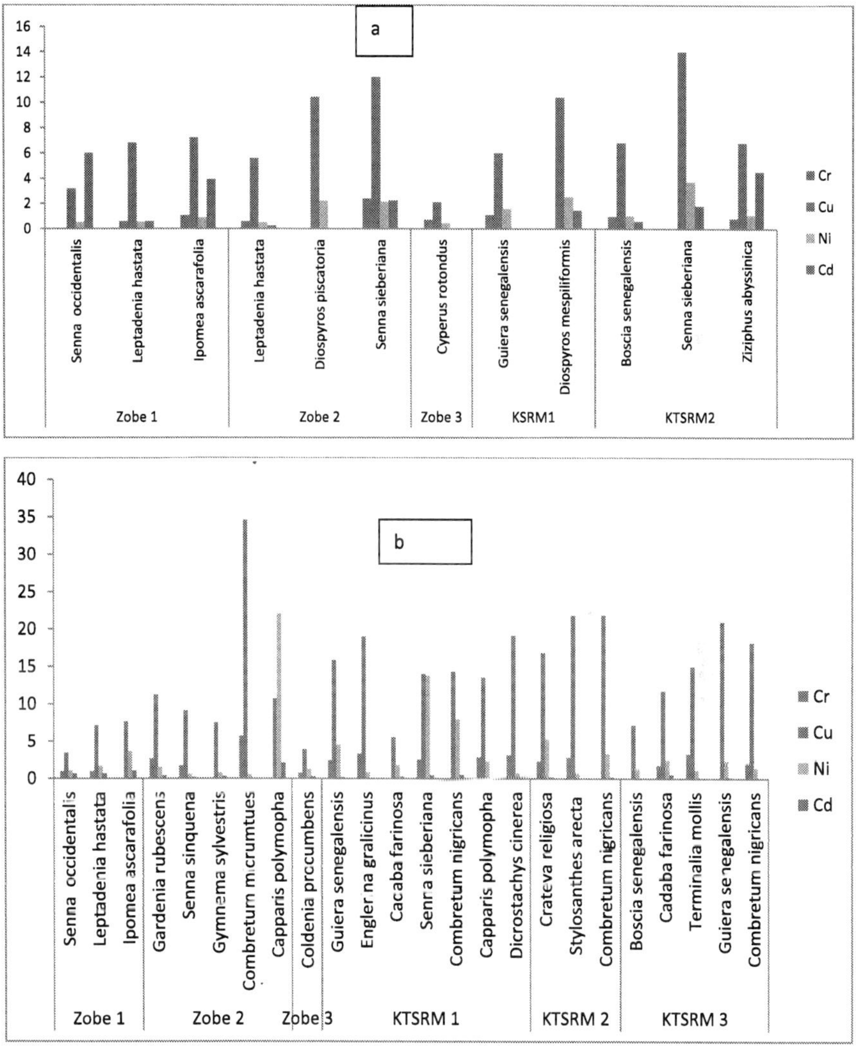

Figure 5. Concentration (mg/kg) of Cr, Cu, Ni and Cd in plants collected from Zobe Dam catchment and the environs of Katsina Steel Rolling Mill (KTSRM) during the Dry (a) and Wet (b) Seasons

3.6. Iron

Soils in the two study sites were observed to have a higher concentration of Fe in the dry season than in the wet season. Soils from Zobe dam catchment and the environs of KTSRM had means of 22, 700 and 555, 16.93 mg/kg; and 18, 266.67 and 58867 mg/kg in dry and wet seasons, respectively (Table 2; Fig 4).

Soil samples collected in the wet season were observed to have a higher concentration of Fe (891.66 and 745.75 mg/kg in Zobe dam catchment and the environs of KTSRM, respectively) than the dry season (360.58 and 277.22 mg/kg in Zobe dam catchment and environs of KTSRM, respectively (Table 2; Fig 4).

Plant species with the highest concentration of Fe in the wet season were *Ipomoea ascarafolia* (115.50 mg/kg) and *Guiera senegalensis* (264.10 mg/kg) for Zobe dam catchment and environs of KTSRM, respectively. In the dry season, *Coldenia procumbens* (184.00 mg/kg) and *Guiera senegalensis* (176.84 mg/kg) were observed to have the highest concentration of Fe Zobe dam catchment and the environs of KTSRM, respectively (Fig 6).

3.7. Manganese

The mean concentration of Manganese in soils was observed to be higher in the wet season (891.66 and 745.75mg/kg) than in the dry season (360.66 and 277mg/kg) for Zobe dam catchment and KTSRM respectively.

In plants, *Diospyros piscatoria* (36.5 mg/kg) and *Boscia senegalensis* (102.87 mg/kg) were observed to have the highest Mn concentration during the dry season in Zobe dam catchment and environs of KTSRM, respectively. In the wet season, *Capparis polymopha* (syn. *C. tomentosa)* (80.72 mg/kg) and *Senna siberiana* (190.60 mg/kg) were observed to have the highest concentration of Mn in Zobe reservoir and KTSRM, respectively (Fig 6).

3.8. Zinc

Zn concentrations in soils showed a similar pattern with Mn and Fe in Zobe dam catchment during the dry season, being higher in the wet (428.45 mg/kg) than the dry season (263.67 mg/kg). A reversed trend was observed in the environs of KTSRM, with the concentration being higher in the dry season (693.67 mg/kg) than the wet season (385.00 mg/kg) (Table 2; Fig 4). These observed mean concentrations of Zn were higher than the 200 mg/kg acceptable limit of the WHO.

Leptadeniahastata (38.5 mg/kg), *Senna siberiana* (54.84 mg/kg) were observed to have the highest concentration of Zn in the dry season for Zobe dam catchment and the environs of KTSRM, respectively. In the wet season, the highest concentrations of Zn in plant samples were presented by *Capparis polymopha* (syn. *C. tomentosa)* (158.50 mg/kg) and *Senna siberiana* (132.60 mg/kg) (Fig 6). Normal concentrations of Zinc in plants fall between 27 to 150 mg/kg. However, concentrations between 100 to 400 mg/kg may be considered Excessive, while from 70 to 400 mg/kg, as Phytotoxic (Table 4). This is dependent on the plant species in question.

http://dx.doi.org/10.5772/57299

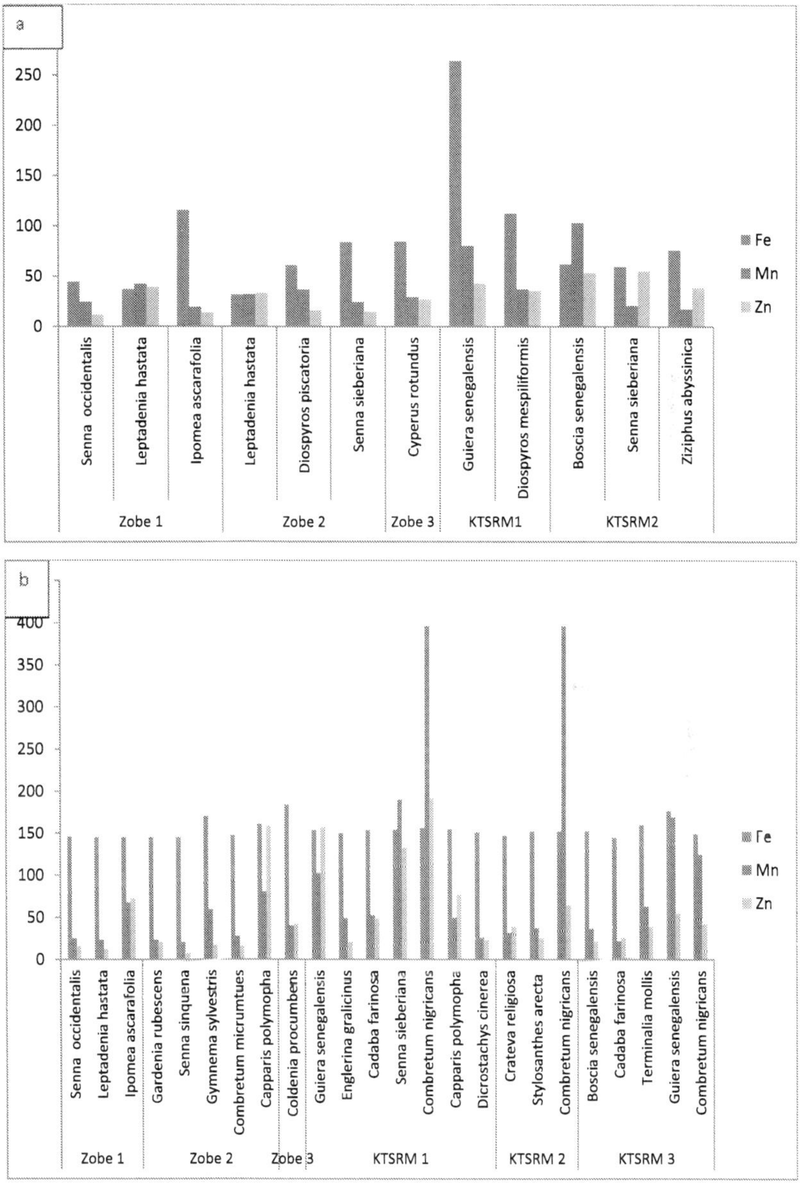

Figure 6. Concentration (mg/kg) of Fe, Mn and Zn in plants collected from Zobe dam catchment and the environs of Katsina Steel Rolling Mill (KTSRM) and the Dry (a) and Wet (b) Seasons

4. Bioconcentration Factor (BCF) of heavy metals in plants

Chromium: BCF for Cr was generally low across seasons and site. In the two study areas, the BCF of Cr ranged between 0.00 to 0.03 in the dry season, while in the wet season it varied between 0.00 and 0.07. *Combretummucronatum* was observed to have the highest BF value (Fig. 7and 8).

Copper: BCF for Cu (Fig. 7 and 8) was generally higher in the wet season than in the dry season, with *Combretum mucronatum* having the highest value of 6.16. *Ipomoea ascarafolia* was observed to have the highest BF in the dry season. Across seasons and sites, other plant species that had BCF >1.0, were: *Gardenia rubescens, Senna singuena, Gymnema sylvestris, Capparis polymopha, Guiera senegalensis, Englerina gracilinus, Senna siberiana, Combretum nigricans, Dicrostachys cinerea, Crateva religiosa, Stylosanthes arectalea* and *Terminalia mollis* (Fig. 7 and 8).

Nickel: Highest BCF of Ni was observed in *Combretumnigricans* (1.61) and *Capparis polymopha (*syn. *C. tomentosa)* (0.61). Most plants show a higher BF during the wet season in comparison to the dry season (Fig. 7 and 8).

Iron: BCF of Fe was generally low, with the highest value of 0.07 observed in *Sennasiberiana,* across seasons and sites (Fig. 7 and 8).

Manganese: For Manganese, BCF was generally low (<1.0) (Fig. 9) across seasons and sites. In the wet season, *Combretum nigricans* was observed to show a high BF of 0.53 and 0.41 of Mn in two different replicate plots. *Guiera senegalensis* was observed to accumulate Mn to a BF of 0.38. *Boscia senegalensis* was observed to have a BF of 0.62 during the dry season (Fig. 7 and 8).

Zinc: For Zinc, BCF was also generally low (<1.0) (Fig. 9) across seasons and sites.The highest BCF of Zn were observed in Ipomoea ascarafolia (0.70), Capparis polymopha (syn. C. tomentosa) (0.59), Guiera senegalensis (0.41) and Combretum nigricans (0.50) (Fig. 7 and 8).

Cadmium: *Sennaoccidentalis* and *Ipomoea ascarafolia* were observed to have considerably high BCF of 20.69 and 13.45 respectively. Most of the other plants were observed to have recorded higher BF in the wet season than the dry season. Across sites and seasons, other species that had BCF >1.0, were *Leptadenia hastata, Capparis polymopha, Senna siberiana* and *Ziziphus abyssinica* (Fig. 9)

Table 5 presents a summary of the most frequently accumulated metal in all species encountered at the sampling locations. Eighteen plant species were found to bioaccumulate metals (BCF >1) in their above ground part to varying degrees. Cu was the most frequently accumulated, being found to occur in all of the 18 species (100%) across sites and seasons. This was followed by Cd in 6 species (33.33%) and Ni in only one (05.6%) species.

http://dx.doi.org/10.5772/57299

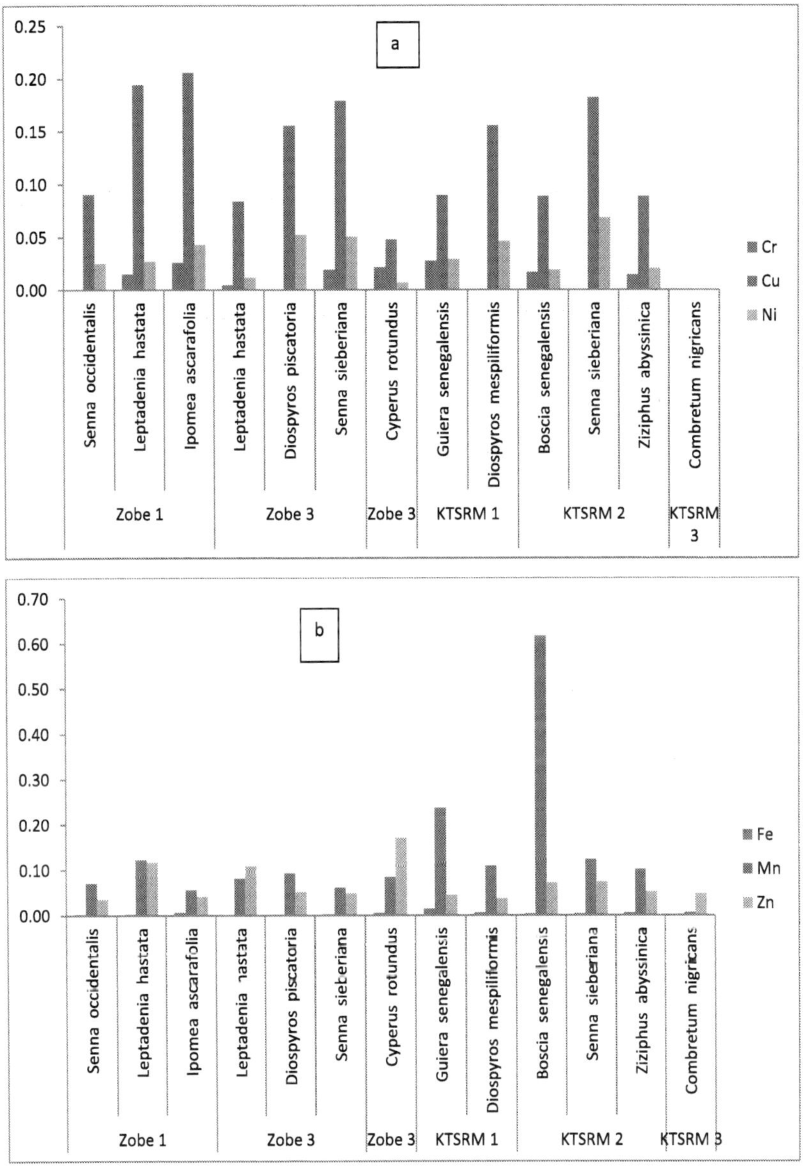

Figure 7. Bioconcentration of (a) Cr, Cu and Ni, (b) Fe, Mn and Zn in plants collected from Zobe dam catchment and the environs of Katsina Steel Rolling Mill during the Dry Season

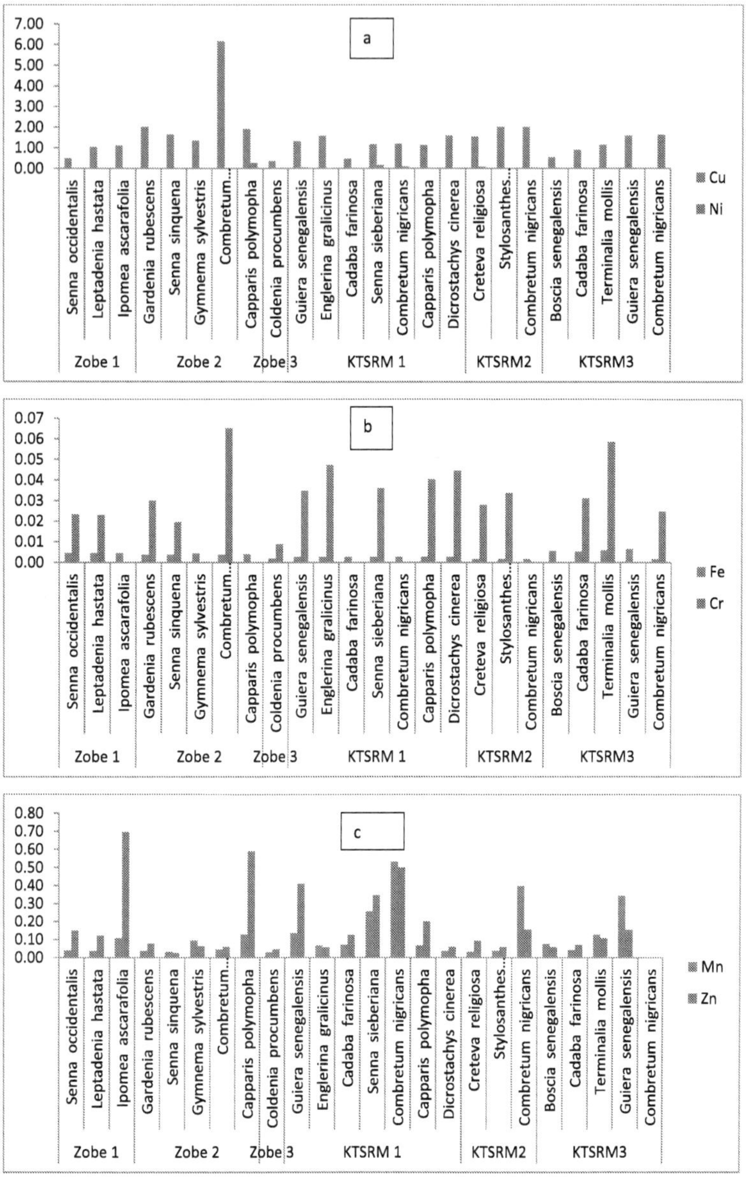

Figure 8. Bioconcentration of (a) Cu and Ni, (b) Fe and Cr, and (c) Mn and Zn in plants collected from Zobe dam catchment and the environs of Katsina Steel Rolling Mill during the Wet Season

http://dx.doi.org/10.5772/57299

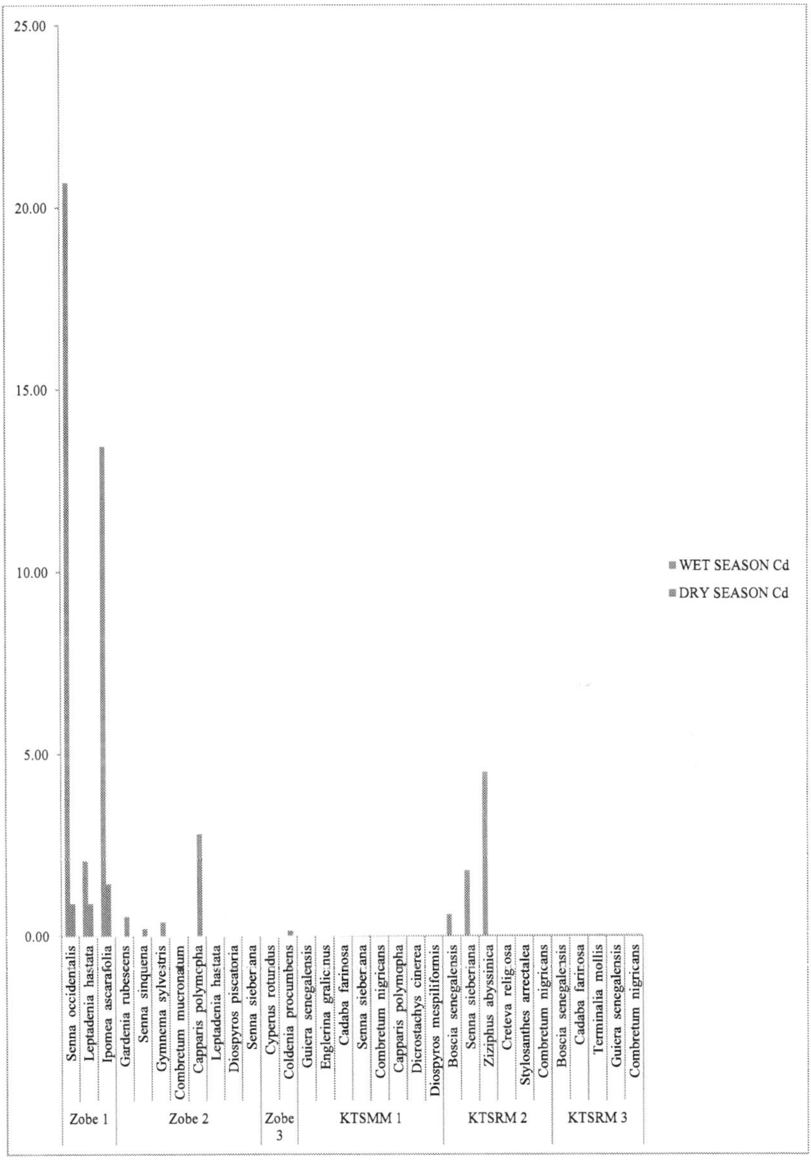

Figure 9. Bioconcentration of Cd in plants collected from Zobe dam catchment and the environs of Katsina Steel Rolling Mill during the Wet and Dry Seasons

S/No.	Family	Plant Species	Frequency of Occurrence (%)				Phtoextractive potential (as function of BCF)					
			Zobe Dam		KTSRM		Cu		Ni		Cd	
			Wet	Dry	Wet	Dry	Wet	Dry	Wet	Dry	Wet	Dry
1	Caesalpiniaceae	*Sennaoccidentalis*	0	33	0	0	-	-	-	-	-	+
2	**Asclepiadaceae**	*Leptadeniahastata*	33	67	0	0	+	-	-	-	-	+
3	Convolvulaceae	*Ipomeaascarafolia*	33	33	0	0	+	-	-	-	+	+
4	Rubiaceae	*Gardeniarubescens*	33	0	0	0	+	-	-	-	-	-
5	Caesalpiniaceae	*Sennasinguena*	0	33	0	0	+	-	-	-	-	-
6	**Asclepiadaceae**	*Gymnemasylvestre*	33	0	0	0	+	-	-	-	-	-
7	Combretaceae	*Combretummucronatum*	33	0	0	0	+	-	-	-	-	-
8	Capparidaceae	*Capparispolymopha*	33	0	0	0	+	-	-	-	+	-
9	Caesalpiniaceae	*Sennasieberiana*	0	33	0	0	+	-	-	-	-	-
10	Combretaceae	*Guierasenegalensis*	0	0	66	33	+	-	-	-	-	-
11	Loranthaceae	*Englerinagralicinus*	0	0	33	0	+	-	-	-	-	-
12	Combretaceae	*Combretumnigricans*	0	0	100	0	+	-	+	-	-	-
13	Mimosaceae	*Dicrostachyscinerea*	0	0	33	0	+	-	-	-	-	-
14	Caesalpiniaceae	*Sennasieberiana*	0	0	33	33	+	-	-	-	-	+
15	**Rhamnaceae**	*Ziziphusabyssinica*	0	0	0	33	-	-	-	-	-	+
16	Capparidaceae	*Cratevareligiosa*	0	0	33	0	+	-	-	-	-	-
17	Papilionaceae	*Stylosanthesarectalea*	0	0	33	0	+	-	-	-	-	-
18	Combretaceae	*Terminaliamollis*	0	0	33	0	+	-	-	-	-	-

(+) = BCF > 1; (-) = BCF < 1

Table 5. Seasonal occurrence of plant species with potential for phytoextraction (BCF) of Cu; Ni and Cd in the catchment of Zobe Dam and the environs of Katsina Steel Rolling Mill.

5. Discussion

Concentrations of heavy metals in soils were generally observed to be higher during the wet season in both locations although the differences were not statistically significant (P=0.05). This differs somewhat with the findings of [30], who observed a higher concentration of these metals in the dry season than the wet season. Seasonal variations in patterns of metal deposition could be related to the intensity and duration of climatic variables such as precipitation, temperature etc., that interface with topography, drainage, soil structure/texture etc., to determine the physicochemical properties of the soil in a particular location. Soil physicochemical properties have complex, interdependent effects on metal solubility, with the most important of these including solution composition (inorganic and organic solubles), Eh, and pH; type and density of charge on soil colloids; and reactive surface area, that interact with factors like metal concentration and form, particle size distribution, quantity and reactivity of hydrous oxides, mineralogy, degree of aeration and microbial activity [31]. The aggregate effects of these complex interactions determine the bioavailabilty of metals to plants.

The above acceptable limits of the observed concentrations of Cr, Cd, Cu, Ni and Zn may be attributed to some of the human activities (mainly agriculture and industrial) going on around the sites. This presents health risks to humans and other animals as the metals contaminate both aquatic and terrestrial ecosystems. Above background values of these metals may have resulted from metal smelting and electroplating activities, burning of fossil fuels, application of phosphate fertilizers, disposal of solid wastes, and quarry activities [32, 33, 34, 35]. High levels of heavy metals in terrestrial or aquatic ecosystems ultimately end up being transmitted and accumulated in the food chain. Health risks to humans arise when metal polluted water is used as drinking water or when animals that have consumed vegetative materials in which metals have accumulated, are used for food. Furthermore, cultivation of crops on metal polluted soils indicates a possibility of consuming crops in which metals have accumulated. Although, specific effects of the various metals on human health have been discussed by several authors, the biotoxic effects of metals to humans have generally been outlined as ranging from gastrointestinal disorders, diarrhoea, stomatitis, tremor, hemoglobinuria, ataxia, paralysis, vomiting and convulsion, depression, coughing and wheezing, respiratory inflammation, dermatitis, leukocytis, low blood pressure, jaundice haemolytic anemia pneumonia and coma to death (Cd, Pb, As, Hg, Zn, Cu, Cr, Ni and Al). The nature of these effects could be toxic (acute, chronic or sub-chronic), neurotoxic, carcinogenic, mutagenic or teratogenic. For example, see [15, 16, 40, 41, 42].

Concentration of Cr in all the plants observed was found to be below the 5 to 30 mg/kg described as phytotoxic to plants [35, 36]. The excessive to phytotoxic concentrations of Zn in *Capparis polymopha* (syn. *C. tomentosa*) and *Guiera senegalensis*; Cd in *Senna occidentalis;* Ni in *Senna siberiana* and Cu in *Combretum mucronatum,* may be a consequence of the high values observed in the soils or direct deposition from the atmosphere. In addition to soil concentrations of metals, other factors that determine the uptake, translocation and accumulation of metals in plants include soil pH, cation exchange capacity, organic matter content, soil texture and interaction with other metals, as well as translocation factor (rate of movement of metals between root and shoot tissues) for the particular metal [43, 44]. Heavy metals in toxic concentrations within the plant have inhibitory effects on enzymatic activity, stomatal function, photosynthesis and nutrient uptake, which may be expressed visually as chlorosis, reduced/stunted growth and yield depression. Plants vary widely in their ability to tolerate high concentrations of metals in their tissues. This variation is usually natural and dependent on inherent genetic factors. The genetic disposition confers the ability to employ a range of avoidance/exclusion or detoxification mechanisms that enable the plants cope with high metal loads. These may include the binding of metals (e.g. Ni and Cr) with amino acids, peptides and organic acids to form low molecular weight compounds, formation of phytochelatins, by binding (e.g. Cu and Pb) with sulphur-rich proteins and cellular adaptations. Other strategies may involve roles for mychorrhizas, the cell wall, extra-cellular exudates, efflux pumping mechanisms in the plasma membrane and formation of stress proteins etc [3, 45, 46, 47].

Plants with BCF of metals >1.0, have been described as suitable for phytoextraction [37, 38]. Some of the plants observed in this study with this potential include; *Combretum mucronatum, Ipomoea ascarafolia, Gardenia rubescens, Senna singuena, Gymnema sylvestre, Capparis polymopha* (syn. *C. tomentosa*), *Guiera senegalensis, Englerina gracilinus, Senna siberiana, Combretum nigricans, Dicrostachys cinerea, Crateva religiosa, Stylosanthes arectalea* and *Terminalia mollis* for Cu; *Senna occidentalis, Ipomoea ascarafolia, Leptadenia hastata, Capparis polymopha* (syn. *C. tomentosa*), *Senna siberiana* and *Ziziphus abyssinica* for Cd; *Combretum nigricans* for Ni. The ability of these plants to concentrate high levels of these metals suggests that they may have a good potential for phytoremediation.

No hyper-accumulator was observed in this study. Hyper-accumulators are plants that can accumulate at least 0.1% wt of Cu, Cd, Cr, Pb, Ni and Co or 1% wt of Zn and Mn [39]. There are possibilities for genetic modification of plants to enhance their capacity for metal tolerance [48].

6. Recommendations for further research

There is a great need to establish environmentally safe limits for metals in plant and soils of the various eco-regions in Nigeria. This need is emphasized by the observed variations in published background values from one country to another and even within the same country. These background values are often dependent on the geological history of the area. A comparison of observed field values with national recommended levels for heavy metals, developed from the background values will give a more realistic assessment of the pollution status. Furthermore, the search for alternative green technology that can be employed in remediation of pollution events must necessarily be a continuous one, due to the relative low cost and environmental friendliness of this option as compared to others. In this regard, ruderal species rather than those with agricultural value must be the candidates of choice for avoidance of obvious conflicts. The species that have indicated potentials for phytoextraction of Cu, Cd and Ni in this research may therefore be further evaluated.

http://dx.doi.org/10.5772/57299

7. Conclusion

The concentrations of Cr, Cu, Ni and Zn in soils around the Zobe dam catchment and the environs of Katsina Steel Rolling Mill were found to be above the acceptable limits. This presents health risks to humans and other animals as the metals contaminate both aquatic and terrestrial ecosystems.

Although no hyper accumulator plant species was encountered in this study, eighteen (18) plant species were identified to have high bioconcentration of metals, which indicated tolerance to excessive or phytotoxic metal concentrations. In addition, they generally produce high above ground biomass, due to rapid vegetative growth. These plants include: *Combretum mucronatum, Ipomoea ascarafolia, Gardenia rubescens, Senna singuena, Gymnema sylvestre, Capparis polymopha* (syn. *C. tomentosa*), Guiera senegalensis, Englerina gracilinus, Senna siberiana, Combretum nigricans, Dicrostachys cinerea, Crateva religiosa, Stylosanthes arectalea, Terminalia mollis, Senna occidentalis, Leptadenia hastata, and Ziziphus abyssinica. These present further possibilities for evaluating metal tolerance in relation to their potential use in phytoremediation programmes in metal polluted sites.

Acknowledgements

We wish to appreciate the University Board of Research of the Ahmadu Bello University, Zaria, Nigeria for providing the funds for this research. We also thank Mallam Gallah S. Umar who helped with the field identification of plant species and collection of plant/soil samples.

Author details

Sunday Paul Bako[1], Augustine Uwanekwu Ezealor[2] and Yahuza Tanimu[1]

*Address all correspondence to: spbako2002@yahoo.com

1 Department of Biological Sciences, Ahmadu Bello University, Zaria, Nigeria

2 Department of Forestry and Environmental Management, Michael Okpara University of Agriculture, Umudike, Nigeria

References

[1] Lar, U. A. Trace elements and health: an environmental risk in Nigeria. *Earth Science*, 2013; 2(3): 66-72.

[2] An'ongo, M.C., Bako, S.P., Ezealor, A.U. Trace metal content in relation to populations of micro-organisms in soils along some highways in Nigeria's Guinea Savanna. *Journal of Biological Sciences*. 2005; 5(6): 703-706.

[3] Bako, S.P., Funtua, I.I., Ijachi, M. Heavy metal content of some Savanna plant species in relation to air pollution. *Water, Air and Soil Pollution*. 2005;161(1-4): 125-136.

[4] Tanimu, J., Uyovbisere, E. O., Lyocks, S.W.J., Tanimu, Y. Cowdung management on the calcium and magnesium content and total microbial population in the Northern Guinea Savanna of Nigeria. *Global Journal of Biology, Agriculture and Health Sciences*, 2013; 2 (2): 7-11.

[5] Maine, M. A., Sune, N. L., Lagger, C. Chromium bioaccumulation: comparison of the capacity of floating aquatic macrophytes. *Water Research*, 2004; 38: 1494-1501.

[6] Bako, S.P., Odiwo, J.E., Ezealor, A.U. Plant diversity in relation to anthropogenic Trace Metals in soils of selected sites in Nigeria's Guinea Savanna. *International Journal of Environment and Pollution*: Special issue on Biogeochemistry of toxic trace metals in Water, Soil and Plant, 2008; 33 (2/3):185-194.

[7] Tanimu, Y., Bako, S. P., Tiseer, F. A. Effects of sewage pollution on water quality of Samaru strea, Zaria, Nigeria. *In*: Fernando Sebastian Garcia Einschlag and Luciano Carlos (Eds). *Waste Water: Treatment Technologies and Recent Analytical Developments*, INTECH Publishers, Rijeka, Croatia. 2013; Pp 189-195.

[8] USDA (United States department of Agriculture). Heavy metal soil contamination. *Soil Quality – Urban Technical Note*, 2000; No. 3, 1-7.

[9] Mathews-Amune, O. C., Kakulu, S. Investigation of heavy metal levels in roadside agricultural soil and plant samples in Adogo, Nigeria. *Academic Journal of Environmental Sciences*, 2013; 1 (2):31-35.

[10] Albasel N., Cotteine A. *In*: Iyaka, Y. A. and Kakulu, S. E. Lead, Copper and Zinc Concentrations in Roadside Topsoil of Niger State, Nigeria. *Journal of Emerging Trends in Engineering and Applied Sciences* (JETEAS), 2011; 2 (5): 754-758

[11] Nyagababo, J. T., Hamya, J. W. *In* : Iyaka, Y. A., Kakulu, S. E. Lead, Copper and Zinc Concentrations in Roadside Topsoil of Niger State, Nigeria. *Journal of Emerging Trends in Engineering and Applied Sciences* (JETEAS), 2011; 2 (5): 754-758

[12] Munch, D. *In* : Iyaka, Y. A., Kakulu, S. E. Lead, Copper and Zinc Concentrations in Roadside Topsoil of Niger State, Nigeria. *Journal of Emerging Trends in Engineering and Applied Sciences* (JETEAS), 2011; 2 (5): 754-758

[13] Menaci, F., Bargagli, R. *In* : Iyaka, Y. A., Kakulu, S. E. Lead, Copper and Zinc Concentrations in Roadside Topsoil of Niger State, Nigeria. *Journal of Emerging Trends in Engineering and Applied Sciences* (JETEAS), 2011; 2 (5): 754-758

[14] Ekmekyapar, F., Sabudak, T., Seren, G. Assessment of heavy metal contamination in soil and wheat (*Triticum aestivum* L.) plant around the Corlu-Cerkezkoy highway in Thrace region. *Global Nest Journal*, 2012; 14 (4): 496-504.

[15] Bako, S.P., Bhwankot, Edna S., Ezealor, A.U., Chia, A.M., Funtua, I.I. Human health implications of trace metal contents in parts of Maize (*Zea mays* L.) plants cultivated along highways in Nigeria's Guinea Savanna. *Soil Remediation Series: Pollution Science, Technology and Abatement*. Lukas Aachen and Paul Eichmann (Eds). Nova Science Publishers, Inc. Hauppauge, New York, United States of America, 2009; pp. 345-356 (Available on-line at www.novapublishers.com).

[16] Bako, S.P., Bhwankot, Edna S., Ezealor, A.U., Chia, A.M., Funtua, I.I. Human health implications of trace metal contents in parts of Cowpea (*Vigna unguiculata* L.Walp.) plants cultivated along highways in Nigeria's Guinea Savanna. *Soil Remediation Series: Pollution Science, Technology and Abatement*. Lukas Aachen and Paul Eichmann (Eds). Nova Science Publishers, Inc. Hauppauge, New York, United States of America 2009; pp.357-368(Available on-lineat www.novapublishers.com).

[17] Megateli, S., Semsari, S., Couderchet, M. Toxicity and removal of heavy metals (Cadmium, Copper, and Zinc) by *Lemna gibba*. *Ecotoxicology and Environmental Safety*, 2009; 72: 1774-1780.

[18] Oti Wilberforce, J. O., Nwabue, F. I. Uptake of heavy metal by *Discorea rotundata* (white yam) and *Ipomoea batatas* (Sweet potato) from Enyingba lead-zinc derelict. *Environment and Pollution*, 2013; 2 (2): 79-84.

[19] Bako, S.P., Odiwo, J. E., Bako, Lois S.P. Visual symptoms expressed by some plant species in response to anthropogenic air pollution in Nigeria's Guinea Savanna. *Mambilla: Journal of Sciences and the Environment*, 2013; (in press)

[20] Dana Group. www.danagroup.com/Businesses/Dana-Group-Of-Companies/Dana-Steel-Mill.aspx. 2013

[21] Funtua, I.I. Application of the transmission emission method in ED-XRF for the determination of trace element in geological and biological materials. *Journal of Trace and Microprobe Techniques*, 1999; 17, 293-297.

[22] Bernasconi, G.B. AXIL-QXAS, *Instruction Manual*. IAEA, Vienna, 1996; pp7-11

[23] Zu, Y. Q., Li, Y., Chen, J. J., Chen, H. Y., Qin, L., Schvartz, C. Hyperaccumulation of Pb, Zn, and Cd in herbaceous plants grown on lead–zinc mining area in Yunnan, China. *Environment International*, 2005; 31, 755–762.

[24] World Health Organization, (WHO). *World Health Organization Guidelines for Drinking-Water Quality*. 1996; 2nd Ed., Vol. 2, Health Criteria and Supporting Information, WHO, Geneva.

[25] Dávila, O. G., Gómez-Berna, J. M.l., Ruíz-Huerta, E. S. Plants and Soil Contamination with Heavy Metals in Agricultural Areas of Guadalupe, Zacatecas, Mexico *Environ-*

mental Contamination, Dr. Jatin Srivastava (Ed.), 2012; ISBN: 978-953-51-0120-8, IN-TECH Publishers, Available from: http://www.intechopen.com/books/environmental-contamination/plants-and-soil-contamination-with-heavymetals- in-agricultural-areas-of-guadalupe-zacatecas-Mexico Osiel González Dávila, 1 Juan Miguel Gómez-Bernal2

[26] United States Environmental Protection Agency (USEPA). *Clean Water Act*, Section, 503: 1993; 58(32). www.epa.gov.

[27] MOE, Japan. *Enforcement status of Agricultural Land-Soil Pollution Prevention Law in 2005 fiscal year*. MOE, Japan, 2006; Available at: http:/ / w w w. e n v. g o. j p / w a t e r / d o j o / n o u y o / jokyo_h17.pdf

[28] Canadian Council of Ministers of the Environment. *Canada environment quality guidelines.* Published by the Canadian Council of Ministers of the Environment. Winipeg, Canada 2009; pp. 8.

[29] European Commission Director General, Environment, (ECDGE). *Heavy metals and organic compounds from wastes used as organic fertilizers. Final Report*, July 2010. WPA Consulting Engineers Inc. Ref. Nr. TEND/AML/2001/07/20, pp73-74. http;//ec.europa.eu/environment/waste/compost/pdf/hm_finalreport.pdf

[30] Najib, N. W. A.Z., Mohammed, S. A., Ismail, S. A., Amiza, W., Ahmad, W. A. Assessment of Heavy Metals in Soil due to Human Activities in Kangar, Perlis, Malaysia. *IJCEE-IJENS*, 2012; 12 (6): 28-33.

[31] Cataldo, D.A., Widung, R.E., soil and plant factors influencing the accumulation of heavy metals by plants. *Environmental Health Perspectives*, 1978; 27, 149-159.

[32] Alloway, B.J., The origins of heavy metals in soils. *In*: Alloway, B.J. (Ed.), *Heavy Metals in Soils*, Second Ed. Blackie Academic & Professional Publishers, 1995.

[33] Knox, A.S., Gamerdinger, A.P., Adriano, D.C., Kolka, R.K., Kaplan, D.I. Sources and Practices Contributing to Soil Contamination. *In*: Adriano, D.C., Bollag, J.M., Frankenberg, W.T.Jr, Sims, R.C. (Eds.), *Bioremediation of the Contaminated Soils. Agronomy Series No. 37*, ASA, CSSA, SSSA, Madison, Wisconson, USA, 1999; p.53-87.

[34] Liu, X.M., Wu, Q.T., Banks, M.K., Effect of simultaneous establishment of *Sedum alfridii* and *Zea mays* on heavy metal accumulation in plants. *International Journal of Phytoremediation*, 2005; 7(1):43-53. [doi:10.1080/16226510590915800

[35] Kabata-Pendias A., Pendias, H. *Trace elements in soils and plants*, Published by CRC Press, Boca Raton, USA. 1992; 365 p.

[36] Galfata, I., Bilal, E. Beji Sassi, A., Abdallah, H., Zaier, A. Accumulation of heavy metals in native plants growing near the phosphate treatment industry, Tunisia. *Journal of Earth and Environmental Sciences*, 2013; 6 (2): 85-100.

[37] Fitz, W.J., Wenzel W.W. Arsenic transformation in the soil–rhizosphere–plant system, fundamentals and potential application of phytoremediation, *Journal of Biotechnology*, 2002; 99: 259-78.

[38] Yoon, Y., Cao X., Zhou Q., Ma L.Q. Accumulation of Pb, Cu, and Zn in native plants growing on a contaminated Florida site, *Science of Total Environment*, 2006; 368: 456-464.

[39] Baker, A.J.M., Brooks, R.R. Terrestrial higher plants which hyperaccumulate metallic elements - a Review of their distribution, ecology and phytochemistry, *Biorecovery*, 1989; 1: 81-126.

[40] Duruibe, J.O., Ogwuegbu, M.O.C., Egwurugwu, J.N. Heavy metal pollution and human toxic effects. *International Journal of Physical Sciences*, 2007; 2(5): 112-118.

[41] Lars, J. Hazards of heavy metal contamination. *British Medical Bulletin*, 2003; 68: 167-182.

[42] Friends of Nature. The IT industry has a critical duty to prevent heavy metal pollution. 2010 Study of heavy metal pollution by IT brand supply chain. Institute of Public and Environmental Affairs, Green Beagle, April, 2010, 17pp

[43] Jung, M. C. Heavy metal concentrations in soils and factors affecting metal uptake by plants in the vicinity of a Korean Cu-W mine. *Sensors*, 2008; 8: 2413-2423.

[44] Nwachukwu, O.I., Agbede, O.O. Plant bioaccumulation and root-to-shoot transport of metals in a field soil contaminated by mining activities. *PAT* 2010; 5(2): 309-319.

[45] Guala, S.D., Vega, F. A., Covelo, E. F. The dynamics of heavy metals in plant-soil interaction. *Ecological Modelling*, 2010; 221: 1148-1152.

[46] Addo, M.A., Darko, E.O., Gordon, C., Nyarko, B.J.B., Gbadago, J.K., Nyarko, E., Affum, H.A., Botwe, B.O. Evaluation of heavy metals contamination of soil and vegetation in the vicinity of a cement factory in the Volta Region, Ghana. *International Journal of Science and Technology*, 2(1): 40-50.

[47] Nouri, J., Khorasani, N., Lorestani, B., Karami, M., Hassani, A.H., Yousefi, N. Accumulation of heavy metals in soil and uptake by plant species with phytoremediation potential. *Environmental and Earth Sciences*, 2009; 59, 315-323.

[48] Kotrba, P., Najmanova, J., Macek, T., Rumi, T., Mackova, M. Genetically modified plants in phytoremediation of heavy metal and metalloid soil and sediment pollution. *Biotechnology Advances*, 2009; doi:10.1016/j.biotechadv.2009.06.003. 12ppwww.elsevier.com/locate/biotechadv

Chapter 25

Geochemical Speciation and Risk Assessment of Heavy Metals in Soils and Sediments

Santosh Kumar Sarkar, Paulo J.C. Favas,
Dibyendu Rakshit and K.K. Satpathy

Additional information is available at the end of the chapter

http://dx.doi.org/10.5772/57295

1. Introduction

Heavy metal pollution is a serious and widely environmental problem due to the persistent and non-biodegradable properties of these contaminants. Sediments serve as the ultimate sink of heavy metals in the marine environment and they play an important role in the transport and storage of potentially hazardous metals. They are introduced into the aquatic system as a result of weathering of soil and rocks, from volcanic eruptions and from a variety of human activities involving mining, dredging, processing and use of metals and/or substances containing metal contaminants. Heavy metals entering natural water become part of the water-sediment system and their distribution processes are controlled by a dynamic set of physico-chemical interactions and equilibria. The properties of metals in soils and sediments depend on the physiochemical form in which they occur [1]. Heavy metals are distributed throughout soil and sediment components and associated with them in various ways, including adsorption, ion exchange, precipitation and complexation and so on [2]. Changes in environmental conditions, such as temperature, pH, redox potential and organic ligand concentrations, can cause metals to be released from solid to liquid phase and sometimes cause contamination of surrounding waters in aquatic systems [3]. They are not permanently fixed by soil or sediment. Therefore, it cannot provide sufficient information about mobility, bioavailability and toxicity of metals if their total contents are studied alone.

Natural and anthropogenic activities have the capacity to cause changes in environment conditions, such as acidification, redox potential, or organic ligand concentrations, which can remobilize contaminated soils and sediments releasing the elements from soils and sediments and pore water to the water column resulting contamination of surrounding waters. Daily tidal currents, wind energies, and storms in coastal and estuarine systems can cause periodical remobilization of surface sediments [4]. More turbulent conditions, such as seasonal flooding or storms, or bioturbation, due to feeding and movement of benthic organisms, can expose anoxic sediments to oxidant conditions. In addition, activities such as dredging result in major sediment disturbances, leading to changes in chemical properties of sediment [5].

The remediation of heavy metal pollution is often problematic due to their persistence and non-degradability in the environment. As a sink and source, soils and sediments constitute a reservoir of bioavailable heavy metals and play a significant role in the remobilization of contaminants in the aquatic systems under favorable conditions. Such potential of sediment for being a sink as well as a source of contaminant can make sediment chemistry and toxicity key components of the quality of aquatic system. Much concern has been focused on the investigation of the total element contents in soils and sediments. However, it cannot provide sufficient information about mobility, bioavailability and toxicity of elements and thus may not be able to provide information about the exact dimension of pollution. The data on total contents of metals are quite insufficient to estimate the possible risk of remobilization of total metals under changing environmental conditions and potential uptake of liberated metals by biota and thus the determination of different fractions assume great importance. This has been described as "speciation" [6]. Since each form have different bioavailability and toxicity, the environmentalists are rightly concerned about the exact forms of metal present in the aquatic environment.

The concept of speciation dates back to 1954 when Goldberg introduced the concept of speciation to improve the understanding of the biogeochemical cycling of trace elements in seawater. Kinetic and thermodynamic information together with the analytical data made it possible to differentiate between oxidized versus reduced, complexed or chelated versus free metal ions in solution and dissolved between particulate species. Florence [7] has defined the term speciation analysis as the determination of the individual physicochemical forms of the element, which together make up its total concentration in a sample. According to Lung [8], speciation analysis involves the use of analytical methods that can provide information about the physicochemical forms of the elements. Schroeder [9] distinguishes physical speciation, which involves differentiation of the physical size or the physical properties of the metal, and chemical speciation, which entails differentiation among the various chemical forms. The main objective of measuring metal species relates to their relative toxicities to aquatic biota. The second and long term aim of speciation studies is to advance an understanding of metal interactions between water and bed sediments in an aquatic ecosystem. In the last decade researchers have followed different sequential extraction techniques for the fractionation of metals in sediments of different river systems. Rauret et al. [10] studied the speciation of copper and lead in the sediments of River Tenes (Spain) while Pardo et al. [11] studies the speciation of zinc, cadmium, lead, copper, nickel and cobalt in the sediments of Pisuerga River, Spain, in order to establish the extent to which these are polluted and their capacity to remobilization. Jardo and Nickless [12] investigated the chemical association of zinc, cadmium, lead and copper in soils and sediments of England and Wales. In most samples, these four metals were associated with all the chemical fractions. Tessier et al. [13] studied speciation of cadmium, cobalt, copper, nickel, lead, zinc, iron and manganese in water and sediments of St. Fransois River, Quebec, Canada. Elsokkary and Muller [14] studied speciation of chromium, nickel, lead and cadmium in the sediments of Nile River, Egypt, reporting that a high proportion of chromium, nickel and lead is bound to organic material and sulphides, while cadmium is bound to carbonate fraction. Ure [15] and Rauret [16] have reviewed the chemical extraction procedures used for heavy metal determinations in contaminated soils and sediments. Owing to the need for validation of extraction schemes, the EC Measurement and Testing Programme (formerly BCR) has organized a project for improving the quality of determinations of extractable heavy metals, where development and validation of extraction procedures has been discussed [17, 18].

The present article aims to summarize the potentials of sequential extraction technique adopting different analytical protocols for gaining information on the mobility and dynamics of operationally determined chemical forms of heavy metals in soils and sediments. The BCR (Community Bureau of Reference, now superseded by the Standards, Measurement and Testing Programme of the European Community) procedure has been illustrated considering the case study of Ganges (Hugli) River Estuary and adjacent Indian Sundarban mangrove wetland (a UNESCO World Heritage Site), northeastern part of the Bay of Bengal. In addition, the authors also evaluate the modified BCR sequential extraction technique as devised by various scientists, the risk assessment code (RAC) as well as assessment of toxicity comparing with sediment quality guidelines. The RAC classification is based on the strength of the bond between the metals and the different geochemical fractions in sediments or soils and the ability of metals to be released and enter into the food chain.

2. Sequential extraction: Merits and demerits

The sequential extraction provides more or less detailed information concerning the origin, mode of occurrence, biological and physicochemical availabilities, mobilization and transport of heavy metals. The procedure stimulates the mobilization and retention of these species in the natural environment using changes in environmental condition such as pH, redox potential and degradation of organic matter [16]. A series of reagents is applied to the sample, increasing the strength of the extraction at each step, in order to dissolve the trace metal present in different sediment phases. The extractants are inert electrolytes, weak acids, reducing agent, oxidizing agents and strong mineral acids [19].

The 3-stage sequential extraction procedure proposed by the European Community Bureau of Reference (BCR) was developed in an attempt to standardize the various schemes described in the literature [2, 20, 21], since the use of different procedures, varying in the number of steps, types of reagents and extraction condition. Hindered comparison of results obtained in the many studies of heavy metals chemical fractionation in environmental samples [22].

The BCR methods has been widely adapted by various authors, and applied to a range of type of solid sample including fresh water sediment [23-25], salt water sediment [26-28], sewage sludge and particulate matter [29-31]. This scheme enables us to associate the meals with one of the following four geochemical phases:

I. Acid-soluble phase: This phase is made up of exchangeable metals and others bound to carbonates that are able to pass easily into the water column, for example, when the pH drops. It is the fraction with the most labile bond to the soil/sediment and, therefore, the most dangerous for the environment.

II. Reducible phase: This phase consists of metals bound to iron and manganese oxides that can be released if the sediment changes from the oxic to the anoxic state, which could be caused, for example, by the activity of microorganisms present in the soils/sediments.

III. Oxidisable phase: This shows the amount of metal bound to organic matter and sulphides, which can be released under oxidizing conditions. Such conditions can occur, for example, if the sediment is resuspended (by dredging, currents, flooding, tides, etc.) and the sediment participles come into contact with oxygen-rich water.

IV. Residual phase : Lithogenous and inert (Non-bioavailable).

The heavy metals in the soils and sediments are bound to different fractions with different strengths, the value can, therefore, give a clear indication of soil and sediment reactivity, which in turn assess the risk connected with the presence of heavy metals in a terrestrial or aquatic environment. The rationale of the sequential extraction procedure is that each successive reagent dissolves a different component, which can content heavy metals within their crystalline structures. Under natural conditions, metals in minerals are unlikely to experience significant release over the time frames of interest [32, 33].

3. Analytical protocols for sequential extraction

In recent years a great number of papers have been published on various analytical techniques proposed for the fractionation analysis of trace elements in various environmental samples (soils, sediments, etc.). An approach that has been found to be preferable is the fractionation of heavy metal into operationally defined forms under the sequential action of different extractants [2]. Selective extractants, used in sequential extraction procedures, are aimed at the simulation of natural conditions whereby metals associated with certain soil (sediment) components can be released. For example, changes in the ionic composition affecting adsorption–desorption reactions or a decrease in pH may lead to the release of metals, retained on a matrix by weak electrostatic interactions or co-precipitated with carbonates ("exchangeable" and "acid soluble" forms). Decreasing the redox potential can result in dissolution of oxides, unstable under reducing conditions, and liberation of scavenged metals ("reducible" forms). Changes in oxidizing conditions may cause the degradation of organic matter and release of complexed metals ("oxidizable" forms). Finally, the destruction of primary and secondary mineral lattice releases heavy metal retained within the crystal structure, e.g., due to isomorphous substitution ("residual" forms) [2]. The nominal "forms" determined by operational fractionation can help to estimate the amounts of total metals in different reservoirs which

http://dx.doi.org/10.5772/57295

could be mobilized under changes in the chemical properties of the soil [34]. Since the 1970s a considerable number of extraction procedures have been proposed for determining the forms of heavy metal [2, 35-39]. Most of these procedures are based on the scheme of Tessier et al. [2]. Although most of the extracting reagents were originally used in the chemical analysis of soils, the procedures proposed have been tested on a wide variety of contaminated environmental samples—sediments, road dust, sewage sludge, etc.

Sequential extraction can be useful to have an operational classification of metals in different geochemical fractions [2] which is the most reliable criteria to quantify the potential effect of soil/sediment contamination by heavy metals. This can provide information about the identification of the main binding sites, the strength of element binding to the particulates and the phase associations of trace elements in soil/sediment. Following this basic scheme, some modified procedures with different sequences of reagents or operational conditions have been developed [40-43]. Considering the diversity of procedures and the lack of uniformity in different protocols, a European Community Bureau of Reference (BCR, now the European Community Standards Measurement and Testing Program) method was proposed [6] and was applied by a large group of researchers [31, 44-47]. In this study, we followed the sequential extraction procedure proposed by the European Union's Standards, Measurements and Testing program [3].

4. Modified BCR Sequential extraction process

As discussed above it is evident that sequential extraction provides valuable information regarding identification of main binding site, the strength of the element binding to the particulates and the phase associations of heavy metals in sediments. However, various complicated sequential extraction procedures were experimented to provide more detailed information regarding different metal phase associations [2, 48, 49]. A wide range of techniques is available whereby various extraction reagents and experimental conditions are used. These techniques involve a 5-step [2], 4-step (BCR, Bureau Commune de Reference of the European Commission), 6-step [50] and 7-step [51, 52] extraction, and are thus becoming popular methods to be used for sequential extraction [53, 54]. Following this basic scheme, some modified procedures with different sequences of reagents or operational conditions have been developed [40-43].

Several sophisticated instruments have been used for the determination of heavy metals contents in marine environments. These include; flame AAS [55, 56], atomic fluorescence spectrometry [57], anodic stripping voltametry [58, 59], ICP-AES [60] and ICP-MS [61, 62].

Heavy metal mobility and bioavailability depend strongly on their chemical and mineralogical forms in which they occur [63]. Several speciation studies have been conducted to determine study different forms of heavy metals rather their total metal content. These studies reveal the level of bioavailability of metals in harbour sediments and also confirm that sediments are indicators of heavy metal pollution in marine environment [64-67].

Since the early 1980s and 1990s sequential extraction methodology has been developed to determine speciation of metals in sediments [2, 68] due to the fact that the total concentration of metals often does not accurately represent their characteristics and toxicity. In order to overcome the above mentioned obstacles it is helpful to evaluate the individual fractions of the metals to fully understand their actual and potential environmental effects [2]. To date, strong acid digestion is used often for the determination of total heavy metals in the sediments. However, this method can be misleading when assessing environmental effects due to the potential for an overestimation of exposure risk. Moreover, in order to determine the mobility of heavy metals in sediments, various sequential extraction procedures have been developed [69-71].

Among a range of available techniques using various extraction reagents and experimental conditions to investigate the distribution of heavy metals in sediments and soils, the 5-step Tessier et al. [2] and the 6-step extraction method, Kersten and Fronstier [50] were mostly widely used. Following these two basic schemes, some modified procedures with different sequences of reagents or experimental conditions have been developed [40-43]. Considering the diversity of procedures and lack of uniformity in different protocols, a BCR, Bureau Commun de Recherche (now called the European Community (EC) Standards Measurement and Testing Programme) method was proposed [6]. It harmonized differential extraction schemes for sediment analysis. The method has been validated using a sediment certified reference material BCR-701 with certified and indicative extractable concentration of Cd, Cr, Cu, Ni, Pb and Zn [72]. This method was applied and accepted by a large group of specialists [31, 44, 45, 47, 73, 74] despite some shortcoming in the sequential extraction steps [75, 76].

Wang et al. [77] used a modified Tessier sequential extraction method to investigate the distribution and speciation of Cd, Cu, Pb, Fe, and Mn in the shallow sediments of Jinzhou Bay, Northeast China. This site was heavily contaminated by nonferrous smelting activities. They found out that the concentrations of Cd, Cu and Pb in sediments was to be 100, 73, 13 and 7 times, respectively, higher than the National guidelines (GB 18668-2002). The sequential extraction tests revealed that 39%-61% of Cd was found in exchangeable fractions. This shows that Cd in the sediments posed a high risk to the local environment. Copper and Pb were found to be at moderate risk levels. According to the relationships between percentage of metal speciation and total metal concentration, it was concluded that the distributions of Cd, Cu and Pb in some geochemical fractions were dynamic in the process of pollutants migration and stability of metals in marine sediments from Jinzhor Bay decrease in the order Pb>Cu>Cd.

Yuan et al. [78] applied BCR-sequential extraction protocol to obtain metal distribution patterns in marine sediments from the East China Sea. The results showed that both the total contents and the most dangerous non-residual fractions of Cd and Pb were extremely high. More than 90% of the total concentration of V, Cr, Mo and Sn existed in the residual fraction while more than 60% of Fe, Co, Ni, Cu, and Zn were mainly present in the residual fraction. Manganese, Pb, and Cd were dominantly present in the non-residual fractions in the top sediments.

Jones and Turki [79] worked on distribution and speciation of heavy metals in surface sediments from the Tees estuary, North East England. Tessier et al. [2] metal speciation scheme modified by Ajay and van Loon [80] was used for the study. They observed out that the sediments were largely organic-rich clayey silts in which metal concentrations exceed background levels, and which attain peak values in the upper and middle reaches of the estuary. Chromium, Pb and Zn were associated with the reducible, residual, and oxidizable fractions. Cobalt and Ni were not highly enriched while Cu is associated with the oxidizable and residual fractions. Cadmium is associated with the exchangeable fractions.

Pempkowlak et al. [81] investigated the speciation of heavy metals in sediments and their bioaccumulation by mussels. They used a 4-step sequential extraction procedure adapted from Forstner and Watmann [82]. Their investigation which was characterized by varying metal bioavailability was aimed at revealing differences in the accumulation pattern of heavy metals in mussel inhabiting that inhabit in sediments. The bioavailabilities of metals were measured using the contents of metals adsorbed to sediments and associated with Fe and Mn hydroxides. The biovailable fraction of heavy metals contents in sediments collected from Spitsbergen represented a small proportion (0.37% adsorbed metals and 0.11%, are associated with metals hydroxides). It was also revealed that the percentages of metals adsorbed and bound to hydroxides of the sediments ranged from 1 to 46% and 1 to 13%, respectively.

Wepener and Vermeulen [66] investigated on the concentration and bioavailability of selected metals in sediments of Richards Bay harbor, South Africa. Sequential extraction of sediments was carried out according to Tessier et al. [2] method. The following metals were investigated: Al, Cr, Fe, Mn, and Zn. Their studies revealed that metals concentrations in sediments samples varied only slightly between seasons, but showed significant spatial variation, which was significantly correlated to sediment particle size composition. Highest metal concentration was recorded in sites with substrates dominated by fine mud. Manganese and Zn had more than 50% of this concentration in reducible fraction while more than 70% of the Cr was associated with the inert fractions and the concentration recorded at some sites were still above action levels when considering only the bioavailable fractions. They also concluded that the concentration of Zn recorded was not elevated their results were compared with the historic data.

Coung and Obbard [54] used a modified 3-step sequential extraction procedure to investigate metal speciation in coastal marine sediments from Singapore as described by the European Community Bureau of Reference (ECBR). Highest percentages of Cr, Ni, and Pb were found in residual fractions in both Kranji (78.9%, 54.7% and 55.9% respectively) and Pulang Tokong (82.8%, 77.3% and 62.2% respectively). This means that these metals were strongly bound to sediments. In sediments from Kranji, the mobility order of heavy metals studied were Cd>Ni>Zn>Cu>Pb>Cr while sediments from Pulan Tekong showed the same order for Cd, Ni, Pb and Cr, but had a reverse order for Cu and Zn (Cu>Zn). The sum of the 4-steps (acid soluble + reducible + oxidizable + residual) was in good agreement with the total metal content, which confirmed the accuracy of the microwave extraction procedure in conjunction with the GFASS analytical method.

Fedotov et al. [83] applied a modified technique for accelerated fractionation of heavy metals in contaminated soils and sediments using rotating coiled columns. Rotating coiled columns

(RCC) is valuable for the continuous-flow sequential extraction and can be successfully applied to the dynamic leaching of heavy metals from soil and sediments. This is a fluoroplastic or steel coil wound around a rigid cylindrical drum, which revolves about its axis and, at the same time, revolves around the central axis of the device called planet centrifuge. The stationary (liquid, solid, or heterogeneous) phase is retained in the column because of the centrifugal force field, and the mobile liquid phase is continuously pumped through the column. A solid sample was retrieved in the rotating column as the stationary phase under the action of centrifugal forces while different elements (aqueous solution of complexing reagents, mineral salts and acids) were continuously pumped through. This procedure developed is time saving and requires only 4-5 hr instead of the several days needed for individual sequential extraction. Losses of solid sample are minimal. Further studies are needed to better estimate the reproducibility of the technique.

Nemati et al. [84] used a modified BCR sequential extraction procedure (SEP) in combination with ICP-MS to obtain the metal distribution patterns in different depths of sediments from Sungai Buloh, Selangor, Malaysia. The results showed that heavy metal contaminations at Sungai Buloh River sediments were more severe than at other sampling sites, especially for Zn, Cu, Ni and Pb. Nevertheless, the element concentrations from top to bottom layers decreased predominantly.

Mossop et al. [85] compared of original and modified BCR sequential extraction procedures for the fractionation of Cu, Fe, Pb, Mn and Zn in soils and sediments. The procedures were applied to five soil and sediment substrates: a sewage sludge-amended soil, two different industrially contaminated soils, river sediment and intertidal sediment. Extractable Fe and Mn concentrations were measured to assess the effects of the procedural modifications on dissolution of the reducible matrix components. Statistical analyses (two-tailed t-tests at 95% confidence interval) indicated that recovery of Fe in step 2 was not markedly enhanced when the intermediate protocol was used. However, significantly greater amounts were isolated with the revised BCR scheme than with the original procedure. Copper behaved similarly to Fe. Lead recoveries were increased by use of both modified protocols, with the greatest effect occurring for the revised BCR extraction. In contrast, Mn and Zn extraction did not vary markedly between procedures. The work indicates that the revised BCR sequential extraction proves better attack on the Fe-based components of the reducible matrix for a wide range of soils and sediments.

5. Sequential extraction of metals in sediments of the Hugli River Estuary and Indian Sundarban wetland: A case study

5.1. Materials and methods

5.1.1. Sample collection and sediment quality analysis

The delta region formed by Hugli (Ganges) River Estuary (HRE) and is famous for its luxuriant mangrove vegetation, known as Sundarban wetland, acclaimed as UNESCO World Heritage

Site for its capacity of sustaining an excellent biodiversity. The wetland is characterized by a complex network of tidal creeks, which surrounds hundreds of tidal islands exposed to different elevations at high and low semi-diurnal tides. This is one of the most sensitive and vulnerable ecosystems in the world and suffers from environmental degradation due to rapid human settlement, tourism and port activities, operation of mechanized boats, deforestation, and increasing agricultural and aquaculture practices. The ongoing degradation is also related to huge siltation, flooding, storm runoff, atmospheric deposition, and other stresses resulting in changes in water quality, depletion of fishery resources, choking of river mouth and inlets, and overall loss of biodiversity. Moreover, the rapid economic development in this deltaic region has caused highly dense areas of human activity and led to serious contamination including heavy metals and persistent organic pollutants (POPs).

Nine sampling sites, namely Barrackpur (S_1), Dakshineswar (S_2), Babughat (S_3), Budge budge (S_4), Ulubaria (S_5), Diamond Harbor (S_6), Frezergunge (S_7), Gangasagar (S_8), and Haribhanga (S_9) were selected considering the existence of typical sediment dispersal patterns along the drainage network systems (as shown in Figure 1) and their position was fixed by a global positioning system (GPS). The stations are representative of the variable environmental and energy regimes that cover a wide range of substrate behavior, wave–tide climate, and intensity of bioturbation (animal–sediment interaction), geomorphological–hydrodynamic regimes and distances from the sea (Bay of Bengal). The sites are exposed to a variable level of heavy metal contamination mainly from anthropogenic sources as mentioned earlier. Six sampling sites (S_1 to S_6) have been chosen along the lower stretch of Hugli River Estuary, while residual three sites (S_7 to S_9) were taken into account in the coastal regions of Sundarban wetland. All sampling sites together with the main stresses to which they are subjected are presented in Table 1.

During winter months (January–March 2009) surface sediment samples weighting 10 g were randomly collected in triplicate from the top 3–5 cm of the surface at each sampling site during low tide using a grab sampler, pooled and thoroughly mixed. Immediately after collection, the samples were placed in sterilized plastic bags in the ice box and transported to the laboratory. Samples were oven dried at 50°C, most gently disaggregated, transferred into precleaned inert polypropylene bags and stored in deep freeze prior to analyses. Each sample was divided into two aliquots: one unsieved (for the determination of sediment quality parameters) and the other sieved through 63 μm metallic sieves (for elemental analyses). Organic carbon content was determined following a rapid titration method [86] and pH with the help of a deluxe pH meter (model no. 101E) using combination glass electrode manufactured by M.S. Electronics Pvt. Ltd. (India). Mechanical analyses of sediment were done by sieving in a Ro-Tap Shaker manufactured by W.S. Tyler Company, Cleveland, Ohio.

5.1.2. Analytical procedure

To determine the total element concentration, sediment samples were digested in polytetrafluoroethylene vessels with aqua regia (HCl/HNO_3, 3:1) and HF neutralized with H_3BO_3 in a 650 W microwave oven (CEM MDS 2000) with a program consisting of a 20-min ramp and a 30-min hold at 100% power in pressure and temperature controlled conditions (150 psi and

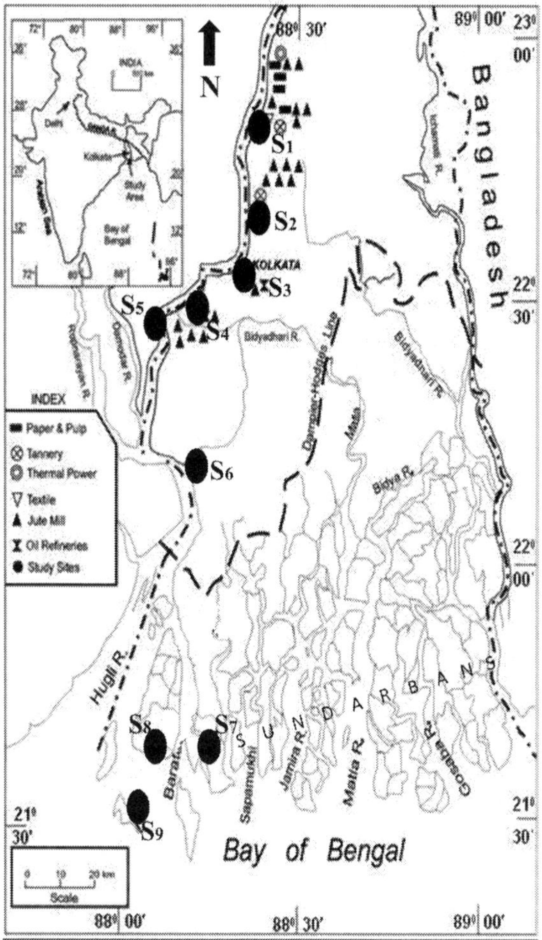

Figure 1. Map showing the location of the monitoring sites (S_1 to S_9) covering Hugli River Estuary and Sundarban mangrove wetland along with the location of the major industries.

180°C). The digested samples were filtered, transferred to polyethylene containers and stored at +4°C until analysis. All reagents were Suprapur® grade (Merck). Reagent blank was processed with the samples and did not show any significant contamination. Accuracy of the procedure was checked using two different certified reference materials (CRM): MESS-2 and PACS-2, which are both marine sediments certified by the National Research Council of Canada for the element content. The MESS-2 recovery ranged between 91% and 116% for all the elements (Table 2). Precision, calculated as relative standard deviation (RSD%), resulted always lower than 5%.

http://dx.doi.org/10.5772/57295

Station number	Site	Main stresses
S_1	Barrackpur	Industrial effluents, domestic sewage disposal, boating, bathing.
S_2	Dakshineswar	Industrial and domestic effluents, boating, bathing, idol immersion site.
S_3	Babughat	Power plant discharges, domestic sewage, boating, idol immersion site.
S_4	Budge budge	Domestic and industrial effluents, bathing, boating.
S_5	Ulubaria	Domestic and industrial effluents
S_6	Diamond Harbour	Boating, recreational activities, bathing, fishing, jetties for fishing trawlers
S_7	Frezergunge	Tourist activities, ferry services, fishing
S_8	Gangasagar	Boating, tourist activities, dredging, fishing, agricultural, domestic and aquaculture practices
S_9	Haribhanga	Boating, fishing and ferrying

Table 1. Details of the nine sampling sites and the main stresses to which they are subjected.

	Al	As	Cd	Co	Cr	Cu	Fe	Mn	Ni	Pb	Zn
Found MESS-2	86,613± 17,773	24.0± 2.4	0.230± 0.010	15.6± 1.5	112± 12	38.4± 6.1	47,385± 3,668	372± 42	54.1± 3.8	20.0± 1.4	170±12
Found PACS-2	70,190± 3,784	29.2± 1.3	1.59± 0.80	12.8± 0.6	94.9± 4.4	307± 22	46,630± 1,411	465± 23	44.3± 2.5	184± 10	398±16
Certified MESS-2	85,698± 2,600	20.7± 0.8	0.240± 0.010	13.8± 1.4	106± 8	39.3± 2.0	43,504± 2,266	365± 21	49.3± 1.8	21.9± 1.2	172±16
Certified PACS-2	66,125± 3,184	26.2± 1.5	2.11± 0.15	11.5± 0.3	90.7± 4.6	310± 12	43,738± 585	440± 19	39.5± 2.3	183±8	364±23
Recovery MESS-2	101%	116%	95.8%	113%	106%	97.8%	109%	102%	110%	91.2%	99.0%
Recovery PACS-2	106%	111%	75.4%	111%	105%	99.1%	107%	106%	112%	100%	109%

Table 2. Results of certified reference materials MESS-2 and PACS-2 as well as the observed values. All the values are expressed in µg/g of dry weight. MESS-2 and PACS-2 recovery rates are also reported.

In this study, we followed the sequential extraction procedure proposed by the European Union's Standards, Measurements and Testing program [3]. Selective extraction is based on the procedure used by Tessier et al. [2] with improvements made according to the BCR, which examined and finally eliminated irreproducibility sources. It is made up of three steps, which dissolve the following phases, respectively: exchangeable and bound to carbonate, bound to Fe and Mn oxides and hydroxides, bound to organic matter and sulphides. Exchangeable and bound to carbonate phase (phase 1) is extracted with 0.11 M acetic acid, while the fraction bound to Fe–Mn oxides (phase 2) with 0.5 M hydroxylamine hydrochloride, adjusted to pH 2 with nitric acid (65%). The phase bound to organic and sulphides (phase 3) is extracted with 8.8 M hydrogen peroxide (stabilized at a pH included between 2 and 3), treated at 80°C in a microwave oven using a program consisting of a 30-min ramp and a 60-min hold at 50% power in pressure and temperature controlled conditions (80 psi and 85°C), and 2 M ammonium acetate adjusted to pH 2 with nitric acid (65%). Each extraction was carried out overnight (16 h) at room temperature. All the reagents employed were Tracepur® grade (Merck Eurolab, Italy). After each extraction, the samples were separated from the aqueous phase by centrifuging at 4, 000 rpm for 15 min. The sediments were washed with Milli-Q water and centrifuged again. The wash water was added to supernatants. The element content of the residual phase was obtained from the difference between the total content and the sum of phases 1, 2 and 3, according to Ianni et al. [37, 38], Ramirez et al. [39], and Mester et al. [27]. Sequential extraction reagent blanks showed no detectable contamination. Accuracy of the procedure was checked with BCR-701 (SM&T). The recovery rates for trace elements in the standard reference material ranged between 77% and 118% (Table 3). Precision, calculated as RSD%, resulted generally lower than 5%, except As and Cr in the phase 1 (~20%).

	Al	As	Cd	Co	Cr	Cu	Fe	Mn	Ni	Pb	Zn
Found BCR-701 step 1	198±1	2.57± 0.28	6.09± 0.09	2.06± 0.08	2.41± 0.51	47.7± 1.7	43.8± 5.8	180±1	14.5± 0.3	3.38± 0.35	185±4
Found BCR-701 step 2	3,451± 46	16.5± 0.3	3.37± 0.08	3.22± 0.03	39.2± 0.4	100±2	7,042± 106	128±3	24.5± 0.4	111±2	102±1
Found BCR-701 step 3	1,912± 74	3.09± 0.20	0.28± 0.01	1.86± 0.17	169±4	64.8± 1.5	1,147± 56	31.9± 2.6	17.4± 1.7	7.15± 0.12	58.4± 5.0
Certified BCR-701 step 1	n.a.	n.a.	7.34± 0.35	n.a.	2.26± 0.16	49.3± 1.7	n.a.	n.a.	15.4± 0.9	3.18± 0.21	205±6
Certified BCR-701 step 2	n.a.	n.a.	3.77± 0.28	n.a.	45.7±2	124±3	n.a.	n.a.	26.6± 1.3	126±3	114±5
Certified BCR-701 step 3	n.a.	n.a.	0.27± 0.06	n.a.	143±7	55.2± 4.0	n.a.	n.a.	15.3± 0.9	9.3±2.0	54.2± 2.0
Recovery step 1	n.a.	n.a.	83.0%	n.a.	107%	96.8%	n.a.	n.a.	94.4%	106%	90.3%
Recovery step 2	n.a.	n.a.	89.4%	n.a.	85.9%	80.6%	n.a.	n.a.	92.3%	87.9%	89.4%
Recovery step 3	n.a.	n.a.	104%	n.a.	118%	117%	n.a.	n.a.	114%	76.9%	108%

Table 3. Results of certified reference materials BCR-701 as well as the observed values (expressed in µg/g of dry weight) together with recovery rates for each step. n.a.= not available.

http://dx.doi.org/10.5772/57295

The elemental concentrations were determined with an inductively coupled plasma atomic emission spectrometer Vista Pro (Varian), with the internal standard method. Cadmium was determined by electrothermal atomization atomic absorption spectrometry. A Varian Spectra A300 spectrometer with Zeeman effect background correction and autosampler Varian Model 96 was used employing the standard addition method for calibration. All the metal analyses were performed at the Department of Chemistry and Industrial Chemistry of the University of Genoa (Genoa, Italy).

5.1.3. *Statistical analyses*

Principal component analysis (PCA) was used to characterize the metal composition in sediments, and cluster analysis was used for grouping the sampling stations. Principal component analysis (PCA) is a multivariate statistical technique used for data reduction and for deciphering patterns within large sets of data. With PCA, a large data matrix is reduced to two smaller ones that consist of principal component (PC) scores and loadings. PC loadings are eigenvectors of the correlation or covariance matrix depending on which is used for the analysis. The PC scores contain information on all of the variables combined into a single number, with the loadings indicating the relative contribution of each variable to that score [87]. Hierarchical cluster analysis (HCA) characterizes similarities among samples by examining interpoint distances representing all possible sample pairs in high-dimensional space. The sample similarities are represented on two dimensional diagrams call dendrograms [88]. All statistical analyses were performed using the computer software STATISTICA (StatSoft, Inc. 2001).

5.2. Results and discussion

5.2.1. *Sediment geochemistry*

Table 4 shows values of pH; organic carbon (%); and percentage of sand, silt, and clay in sediments of the nine sampling sites. Organic carbon values, ranging from 0.22% (in station S_8) to 1.02% (in station S_2), are low in comparison with values found in sediments from other Indian coastal areas, such as Gulf of Mannar [89], Cochin [90], and Muthupet mangroves [91]. The low organic carbon values might be related with the poor absorbability of organics on negatively charged quartz grains, which predominate in sediments in this estuarine environment [92]. In addition, the constant flushing activity by tides along with the impact of waves can support the low percentage of organic carbon in the sediments. The sediments of the studied stations are characterized by slightly basic pH (7.50–8.36) with maximum values recorded in the stations closest to the sea (stations S_6, S_8, and S_9) and minimum in station S_7.

These were different from the low pH values in most of the mangrove swamps in Hong Kong [93], where sediments were not frequently flooded by the tide and become acidic in reducible conditions. With respect to texture, the sediment samples show a variable admixture of sand, silt, and clay. Clay fractions dominate in low-energy areas of suspensional deposits. On the contrary, silt, and sand dominates where the energy level is high. Sediments from station S_7

Stations	Latitude and Longitude	Salinity	pH	Organic carbon (%)	Sand (%)	Silt (%)	Clay (%)
S1	22°43′ 16″ N 88°21′ 20″ E	0	7.86	0.35	4	87.1	8.9
S2	22°39′ 17″ N 88°12′ 25″ E	0	7.80	1.02	1	76.5	22.5
S3	22°33′ 53″ N 88°20′ 19″ E	0	7.90	0.52	2.24	41.97	55.79
S4	22°30′ 10″ N 88°11′ 48″ E	0–2.5	7.60	0.74	18.25	47.42	34.33
S5	22°28′ 06″ N 88°06′ 54″ E	0–1	7.90	0.91	16.7	69.6	13.7
S6	22°11′ 14″ N 88°11′ 15″ E	0–5.6	8.36	0.56	3.15	41.13	55.71
S7	21°34′ 44″ N 88°15′ 03″ E	30–34.3	7.50	0.36	98.02	0.18	0
S8	21°38′ 15″ N 88°03′ 53″ E	32–35	8.14	0.22	32.85	58.45	8.7
S9	21°34′ 20″ N 88°01′ 25″ E	35	8.10	0.46	39.3	44.25	16.45

Table 4. Geographical position, physicochemical and textural properties of sediment samples of 9 sampling sites.

contain higher percentage of sand (98%) compared to the others, while sediments from S_1, S_2, S_5, and S_8 contains higher percentage of silt (more than 50%) compared to the others. A variable mixture of sand, silt, and clay is present in the other stations and reflect a variable amount of erosion and deposition.

5.2.2. *Total element concentrations*

Total element concentrations in the investigated stations varied in a narrow range of values (Table 5) and were comparable with data obtained for other Indian coastal areas [94, 95]. Datta and Subramanian [96] found very similar trace element concentrations throughout the Bengal Basin, where anthropogenic perturbation is low and river channel may receive a several centimeter-thick sediment layer in a single event during peak flow, preventing to bear the signature of an accumulation of trace elements. The highest concentrations for As, Cu, Fe, Mn, and Ni were measured at station S_9 while for Cd and Pb at station S_3, close to Calcutta city (about 4.5 million residents, but about 14.2 million including suburbs). An anthropogenic input from vehicular traffic and in-dustrial activities may cause high Cd and Pb con-centrations measured in samples collected in the Calcutta urban area. The lowest element concentrations were found at station S_5. Very low (close to the detection limit) Cd concentration was found in the coastal stations (S_7, S_8, and S_9).

Stations	Al	As	Cd	Co	Cr	Cu	Fe	Mn	Ni	Pb	Zn
S1	70,289	8.81	0.165	13.0	67.6	27.8	37,737	591	31.9	20.4	86.6
S2	70,879	8.44	0.452	14.0	74.8	36.8	39,405	625	34.2	22.3	90.7
S3	72,134	8.65	1.79	14.5	73.5	32.3	40,070	712	35.0	33.2	83.1
S4	72,613	8.49	0.492	14.9	76.8	27.9	40,303	726	35.0	19.6	80.4
S5	62,044	6.41	0.220	12.1	58.2	21.1	33,428	597	27.5	17.0	64.1
S6	64,325	6.79	0.106	12.0	64.8	32.0	34,273	613	31.3	17.9	69.6
S7	77,529	7.77	0.044	14.0	75.1	28.4	40,084	389	38.1	20.5	74.4
S8	68,146	8.08	0.027	12.7	62.5	22.2	36,786	511	34.3	19.7	61.4
S9	72,666	9.40	0.044	14.1	74.2	36.6	40,838	785	40.1	22.9	74.9

Table 5. Total element concentrations (µg/g) in sediments of 9 sampling sites (instrumental precision, calculated as RSD%, resulted lower than 5% for each element in all samples).

The geoaccumulation index (I_{geo}) of Muller [97] has been calculated for the analyzed elements, by comparing current concentrations with pre-industrial levels, in order to estimate the metal contamination in sediments. The equation used for the calculation of I_{geo} is: $\log_2 (C_n/1.5\ B_n)$, where C_n is the measured content of element "n" and B_n the element's content in "average shale" [98]. Factor 1.5 is used because of possible variations in background values for a given element in the environment, as well as very small anthropogenic influences [99]. As shown in Figure 2, all sediments fall in class 0 for Al, Co, Cr, Cu, Fe, Mn, Ni, Pb, and Zn, therefore the area is not contaminated for these elements. Unlike the Hugli river, in other rivers of the Bengal Basin, such as Meghna and Brahmaputra, Cr exhibits higher I_{geo} values respect to the other elements [96]. For Cd, two stations fall in class 1 and three in class 2 exhibiting a moderate contamination for this element. In all stations, As falls in class 2 (moderate pollution). In this area, As contamination was already observed in previous studies and it is probably due to groundwater contamination [100]. This contamination can have natural origin, such as coal seams in Rajmahal basin and arsenic mineral in mineral rocks in the upper reaches of the Ganges river system. The highly reducing nature of groundwater would reduce As, causing the possible desorption of As [101].

5.2.3. Speciation patterns

The potential environmental risk of trace elements in sediments is associated with both their total content and their speciation. The chemical partitioning of the considered elements (Al, As, Cd, Co, Cr, Cu, Fe, Mn, Ni, Pb, and Zn) from each extraction step has been described. Aluminum, Cr, and Fe are present mainly in the residual phase, representing 95.8–96.8%, 88.9–91%, and 83.0 94.7% of the total concentration, respectively, which implies that these elements are strongly linked to the inert fraction of the sediments. This result was in good agreement with data reported by several studies carried out worldwide in marine coastal areas [45, 46, 78, 102]. The high percentage of Fe in the residual phase indicates that most of the Fe exists as crystalline Fe peroxides (goethite, limonite, magnetite, etc.). The remaining Fe is associated with the reducible phase (mean, 11.25%). Large amounts of Fe accumulate in the residual phase probably because it is basically of natural origin (it is the most common element in the earth's crust).

Concentrations of Al, Fe, and Cr are very low in exchangeable phase (0.08%, 0.26%, and 1.72% as mean values, respectively), limiting their potential toxicity as pollutants. It should be noted that sediments always act as reservoir for elements; therefore, their potential risk of pollution to environment has always to be considered.

Arsenic, Co, Ni, and to some extent Zn, are found mainly in the residue (~50% of the total concentration). Nickel and Co are associated to the residue respectively for 56% and 74% of the total concentration, with a speciation similar in all the samples. A mean of 23% of Co is present in the phase 2. The highest percentage of labile Co (~13%) was measured in S_6 (Diamond Harbour) and S_8 (Gangasagar) and can be due to a recent input of this element. The dominant proportion of Ni in the residual phase is in agreement with the results of other studies [27, 46]. Nickel is present, apart from the residue, in phases 2 and 3 (about 10% in each phase). Arsenic is distributed mainly between the residual (mean 47%), the reducible, and the oxidizable phases (mean 19% and 22%, respectively). Acharyya et al. [101] observed that As is adsorbed to iron-hydroxide-coated sand grains and to clay minerals in the sediments of the Ganges delta from West Bengal. Among the studied elements, As is found with the greatest proportion in the oxidizable phase coinciding with organic and sulfur compounds. Arsenic is present in the phase 1 for about 10% of the total content, in station S_7 phase 1 percentage rises up to 16%. The lower land alluvial basin of the Ganges River is recognized as an arsenic-affected area. Arsenic in solution probably is easily entrapped in the fine grained organic-rich sediments deposited in the Ganges delta [101]. The percentage of silt (lower than 70% except in S_1 and S_2) may have contributed to a low retention of dissolved As since coarse sediments are less efficient at retaining As.

Cadmium was mainly present in the labile phase (more than 60%) in all the stations with the exception of station S_7, where the Cd labile percentage represents only 25% of the total concentration. Cadmium concentrations were negligible in phases 2 and 3. The highest labile Cd concentration was measured at station S_3, the closest to the city of Calcutta. Datta and Subramanian [96] found that the concentrations of elements in the non-detrital phases were higher in stations sampled in the Hugli river around Calcutta than in samples collected along Brahmaputra and Meghna rivers. The petroleum refinery, industrial, and mining effluents carried by the Hugli river may be responsible for this higher concentrations of non detrital fractions.

About 40% of the total Cu concentration is associated to the residue, while 33% of Cu is bound to Fe-Mn oxide and hydroxide (phase 2). The high percentage of Cu in the residue is likely due to the fact that Cu is easily chemisorbed on or incorporated in clay minerals [103]. All the samples showed lower Cu concentrations in exchangeable phase, with percentage ranging from 7% (S_7) to 22% (S_5), with a mean of 15%. Copper is characterized by high complex constant with organic matter thus it can be hypothesized that Cu is bound to labile organic matter such as lipids, proteins, and carbohydrates. On the other hand, high-element concentration in labile phase could be related to recent coastal input [39].

Manganese was found in all the four sediment phases, as observed by other researchers [45, 104]. Manganese is the most mobile element since it is present with the highest percentage (a mean of 42%) in the labile phase. This is probably because of the known close association of Mn with carbonates [105] as endorsed by other workers [69, 106]. In this phase, weakly sorbed Mn retained on sediment surface by relatively weak electrostatic interactions may be released by ion exchange processes and dissociation of Mn-carbonate phase [2]. The result indicates that considerable amount of Mn may be released into environment if conditions become more acidic [107]. The highest Mn labile percentage was measured in S_6 (57%). Differently, in S_7, Mn in the residue represents 65% of the total concentration, while the labile Mn is only 15%. A substantial Mn percentage was also found in the residue (mean 37.8%), followed by the reducible phase (14.7%), in which Mn exists as oxides and may be released if the sediment is subjected to more reducing conditions [108].

The major geochemical phase for Pb in these sediments was the Fe-Mn oxides phase (mean 55.7%) followed by the residual phase (mean 30.2%) while lower percentage of the total Pb are bound to exchangeable-labile (mean 5.3%) and oxidizable phases (mean 6.8%). At S_3 (Babughat), the reducible part is as high as 65% and only 19.9% of the total is associated with the residue. Atmospheric input as fallout from vehicular emission can be probably the major input of Pb for this station. The relatively high percentage of Pb in reducible phase is in agreement with the known ability of amorphous Fe–Mn oxides to scavenge Pb from solution [109, 110]. Caille et al. [111] observed that resuspension of anoxic sediment results in a rapid desorption of Pb and Cu adsorbed to sulphides. Thus, a high element percentage in the reducible fraction is a hazard for the aquatic environment because Fe and Mn species can be reduced into the porewaters during early diagenesis by microbially mediated redox reactions [112]. Dissolution will also release Pb associated with oxide phases to the porewater possibly to the overlying water column [113] and to benthic biota [79]. The major sources of Pb are from intensive human activities, including agriculture in the drainage basin [114], auto exhaust emission together with atmospheric deposition [115]. In addition, a substantial contribution from the factories located in the upstream of the Hugli river dealing with Pb producing lead ingots and lead alloys play a vital role as referred by Sarkar et al. [116].

The percentage of Zn in residue is highly variable (38.5–70%) and the distribution pattern in each fraction showed the following order: residual>reducible>oxidizable> exchangeable and bound to carbonates. There was some difference in Zn speciation among the sampling sites: in stations S_1, S_2, and S_3 about 40% of Zn is present in the residue, while in the other stations this percentage increases to more than 60%. In station S_1, the exchangeable and oxidisable phases shared over 22% of the total Zn, whereas labile Zn was as low as 4.6% at S_7. A major part of Zn (16.3%) is associated with Fe–Mn oxide phase, because of the high stability constants of Zn oxides. Iron oxides adsorb considerable quantities of Zn and these oxides may also occlude Zn in the lattice structures [117].

The BCR procedure as discussed above showed satisfactory recoveries, detection limits, and standard deviations for determinations of heavy metals/metalloid in the sediments. It is evident from the present results of the fractionation studies that the metals/metalloids in the sediments are bound to different fractions with different strengths leading to variations in mobility and availability and some of them show significant spatial variations subject to diverse environmental stresses. This type of association between metals and the sediments can be understood in detail by sequential extraction techniques. Hence the application of sequential extraction is fully justified as the quantification of different forms of metal is more meaningful than the estimation of its total metal concentrations. The strength values can, therefore, give a clear indication of sediment reactivity, which in turn assess the risk connected with the presence of metals in this wetland environment. The results obtained suggest the need for corrective remediation measures due to the higher accumulation of potentially dangerous metals/metalloids, which in most cases exceed the limits established by certain legislation.

5.2.4. Comparison with Sediment Quality Guidelines

Results obtained after total and sequential extraction are compared with Sediment Quality Guidelines (SQGs). Table 6 reports consensus-based values, such as TEC (concentration below which harmful effects on sediment-dwelling organisms were not expected) and PEC (concentration above which harmful effects on sediment-dwelling organisms were expected to occur frequently), and effect range-low and range-medium, such as ERL (concentrations below which adverse biological effects were observed in less than 10% of studies) and ERM (concentrations above which effects were more frequently observed in more than 75% of studies).

Element	Phase	Si<TEC	TEC	Si<TEC<PEC	PEC	Si<ERL	ERL	ERL<Si<ERM	ERM
As	Total	All	9.79	None	33	S5,S6,S7,S8	8.2	S1,S2,S3,S4,S9	70
	Labile	All		None		All		None	
Cd	Total	S1,S2,S4,S5, S6,S7,S8,S9	0.99	S3	4.98	S1,S2,S4,S5, S6,S7,S8,S9	1.2	S3	9.6
	Labile	S1,S2,S4,S5, S6,S7,S8,S9		S3		S1,S2,S4,S5, S6,S7,S8,S9		S3	
Cr	Total	None	43.4	All	111	All	81	None	370
	Labile	All		None		All		None	
Cu	Total	S1,S4,S5,S7,S8	31.6	S2,S3,S6,S9	149	S1,S3,S4,S5, S6,S7,S8	34	S2,S9	270
	Labile	All		None		All		None	
Ni	Total	None	22.7	All	48.6	None	20.9	All	51.6
	Labile	All		None		All		None	
Pb	Total	All	35.8	None	128	All	46.7	None	218
	Labile	All		None		All		None	
Zn	Total	All	121	None	459	All	150	None	410
	Labile	All		None		All		None	

Table 6. Sediment Quality Guidelines concentrations with respect to total and labile element concentrations found in the analyzed samples (expressed as µg/g of dry weight).

http://dx.doi.org/10.5772/57295

Comparing our results with the SQGs, it is revealed that for Pb and Zn in all the stations the measured concentrations are lower than both TEC and ERL. As regards Cd, concentration measured in station S_3 is higher than TEC and ERL but lower than PEC and ERM both in term of total and labile concentration. For this station, some possible toxic effect on benthic organism can be hypothesized, in particular because of the large amount of element bound to the most labile phase of the sediment. Considering Cu, some stations (S_2, S_3, S_6, and S_9) exhibit total concentrations higher than TEC but lower than PEC. Concentrations of Cu are higher than ERL but lower than ERM only in stations S_2 and S_9. Since only 7–22% of total Cu is bound to the labile phase, in all stations Cu labile concentrations are lower than TEC and ERL. Total As concentrations in stations S_1, S_2, S_3, S_4, and S_9 are higher than ERL value but lower than TEC value. Since more than 50% of total As is not found in the residue, attention should be paid to a change in the environment conditions which could induce a release of As from the sediments. Total Ni and Cr concentrations are higher than TEC (Ni is also higher than ERL) but lower than PEC (and ERM in the case of Ni) in all the stations. Nevertheless, more than 70% of Ni as well as 90% of Cr are present in the residual fractions, therefore adverse impacts on organisms is very much negligible.

Mean sediment quality guidelines quotients (mSQGQ) have been developed for assessing the potential effects of contaminant mixtures in sediments [118]: they are determined by calculating the arithmetic mean of the quotients derived by dividing the concentrations of chemicals in sediments by their respective SQGs. The probability of observing sediment toxicity can be estimated by comparing the mSQGQ in a sample to previously published probability tables. It is important to keep in mind that mSQGQs cannot be used to accurately predict the uptake and bioaccumulation of sediment-bound chemicals by fish, wildlife, and humans, even if there is considerable evidence that this assessment tool can be predictive of the presence or absence of toxic effects [118].

SQGQs are calculated for seven elements considering ERM as sediment quality guidelines (Table 7). The mean quotient values ranges from 0.16 in station S_5 to 0.24 in station S_3. Using PEC values instead of ERM, the mean SQGQ ranges from 0.25 in station S_5 to 0.38 in station S_3 (Table 7).

Stations	$SQGQ_{PEC}$	$SQGQ_{ERM}$
S1	0.30	0.19
S2	0.33	0.21
S3	0.38	0.24
S4	0.33	0.21
S5	0.25	0.16
S6	0.28	0.18
S7	0.32	0.21
S8	0.28	0.18
S9	0.34	0.22

Table 7. Mean Sediment Quality Guidelines Quotients calculated for the nine stations using PEC and ERM as SQGs.

Compiled data from multiple data sets reporting 10-day toxicity test conducted on amphipod species in saltwater showed that the incidence of toxicity for a range of SQGQ of 0.25–0.5 is ~35%, while for a mean SQGQ range from 0.1 to 0.25, the incidence of toxicity lowers to ~20%. Measures recorded in a survey of Biscayne Bay (port of Miami and the adjoining saltwater reaches of the lower Miami River, FL, USA) showed that the average amphipod survival (*Ampelisca abdita*) decreased slightly from the least contaminated (ERMQ <0.03) to the intermediate category, (ERMQ included in 0.03–0.2 range) then decreased greatly in the most contaminated sediments (ERMQ included in 0.2–2 range). Therefore, we can presume a low toxicity of sediments sampled in the nine stations for benthic organisms. It is important to note that the benthic response to contaminants covaried among stations with both the mean ERM quotients and the effect of natural factors, such as the sediment texture, TOC, and salinity [118].

5.2.5. Statistical analyses

The relationships between variables and the differences between stations were evaluated by PCA. The analysis was performed on 36 objects (four sediment phases for nine stations) and 11 variables (Al, As, Cd, Co, Cr, Cu, Fe, Mn, Ni, Pb, Zn). Two significant components were identified explaining 68.3% and 14.5% of the total variance, respectively. By studying the loadings of the variables (Figure 2a) on the components it can be seen that all the elements except Cd, Mn, and Pb are significantly correlated.

Unlike the other elements, most of Cd and Mn is present in the first phase: labile Cd and Mn represent more than 60% and 40% of the total concentration, respectively, except in station S_7. Cadmium and Mn speciation can be ascribed to their considerable affinity for carbonates. Lead is the only element which is bound to the reducible phase for more than 50%. Lead is a very reactive element in water column and, having scavenging type behavior, is easily bound to hydroxy- and oxyligands. Copper is positively and significantly correlated with all elements except Cd and Mn, but with lower correlation coefficients (0.66–0.81).

In the score plot (Figure 2b) phases 1, 2, 3, and 4 (corresponding to labile, reducible, oxidizable, and residual phases respectively) are identifiable by 1, 2, 3, and 4 suffix, respectively. In all stations, residue concentrations were characterized by negative values of PC1 and consequently by high concentrations of Al, As, Co, Cr, Fe, Ni, and Zn. Conversely, in the positive PC1 semi-axis labile and oxidizable metal concentrations, which represent a small percentage of the total elements, are distributed. For all stations, reducible concentrations are distributed along the positive PC2 semi-axis, i.e., high Pb concentrations, with a maximum for station S_3 and a minimum for S_5. The group formed by elements bound to organic matter and sulphides (phase 3) is characterized by low values of both PC1 and PC2. Therefore, a low percentage of elements (higher than 20% exclusively for all As data and for Zn in stations S_1 and S_3) is bound to the oxidizable phase, suggesting the presence of an oxidant environment. High Mn and Cd concentrations are associated with negative values of PC2, therefore a relatively high concentration of labile Mn and Cd is present in all samples (in particular in S_3), except station S_7. Samples are prevalently grouped in relation to the sediment geochemical phase, suggesting a similar element speciation among the stations. Station S_7 represents an exception, in fact the labile fraction is closely associated to the oxidizable phase group.

http://dx.doi.org/10.5772/57295

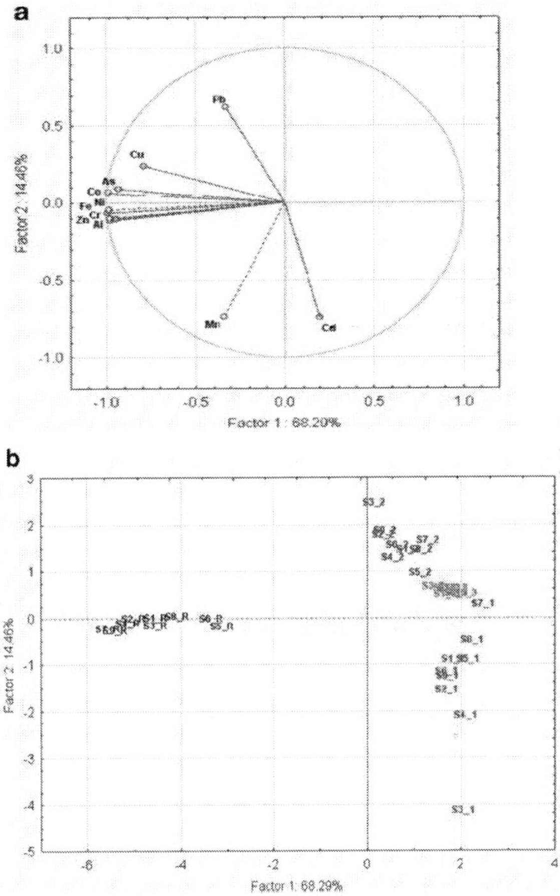

Figure 2. Principal component analysis: a) variable plot; b) score plot (phase 1, 2, 3 and 4, corresponding to labile, reducible, oxidizable and residue, are identifiable by 1, 2, 3 and 4 suffix and different colors in the score plot).

A HCA was carried out by applying Euclidean distances to quantitatively identify specific groups of similar stations. In the dendrogram of the sampling stations (Figure 3), we can note two main clusters: the first represented by station S_7, characterized by the highest element percentage bound to residue, and the second constituted by all the remainder stations. In the second group, a subgroup formed by station S_5 and S_6 can be individuated.

Station S_7 was sampled in a marine coastal environment; it is characterized by a peculiar grain size percentage respect to the other stations, being the sand percentage as high as 98.6%. In general, the concentrations of elements are much higher in fine than in coarse fraction because the fine fraction larger specific surface facilitates absorption processes. As previously noted

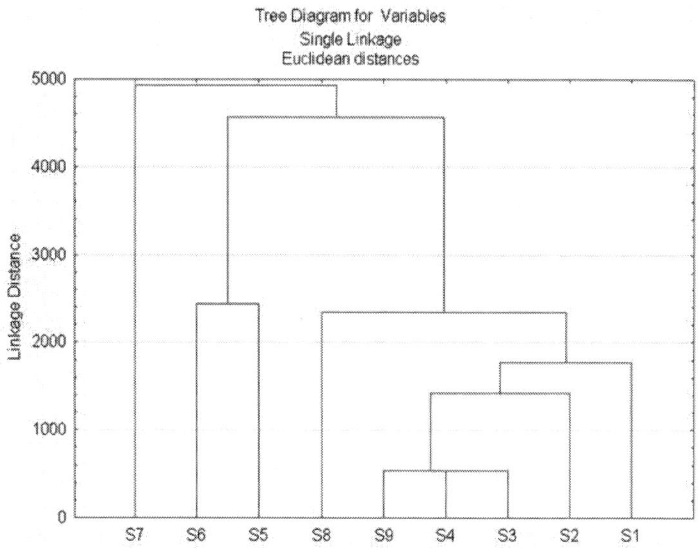

Figure 3. Dendrogram indicating linkage of sites on the basis of element concentrations.

by Ramirez et al. [39], this pattern is particularly evident for Cd. It is interesting to note that the marine coastal stations S_8 and S_9 are more similar to river stations than to station S_7. Both the stations S_8 and S_9 are, in fact, located in front of the Hugli river runoff, while station S_7 is located easternmost and probably is less influenced by the Hugli river discharge.

5.3. Conclusion and recommendation

The study provides valuable information on the potential mobility of trace elements in sediments collected along the stretch of Hugli River and in the Sundarban mangrove wetland (northeastern part of the Bay of Bengal). The results obtained adopting BCR sequential extraction method provided the following important information: (i) Al, Cr, and Fe were found mostly in the residual phase while the other elements were found in the four phases sharing different proportions; (ii) the dominant Cd, Mn, and Pb proportion was found in the non-residual fractions and (iii) Mn had the highest percentage in the labile phase. This is worthwhile to mention that coastal environment of West Bengal is considerably constrained due to implementation of dredging, construction of port/ harbor and other industrial activities. The authors strongly recommend for periodical monitoring on the bioavailability and mobility of trace elements, control the mixing of effluent of the concentration of heavy metals in the region, environmental remediation, treatment of industrial effluent and municipal wastewater for effective management of this estuarine system. It is wisely suggested that an environmental recovery framework should be urgently implemented to avoid extension of heavy metal contamination (especially As).

http://dx.doi.org/10.5772/57295

6. Case studies from other coastal regions in India

Although the importance of metal speciation and fractionation has been realized in developed countries, the subject has not really taken off in India and only few references are available on the speciation of metals in Indian rivers. Speciation of selected heavy metals geochemistry in surface sediments (n=10 was studied by Venkatramanan et. al. [119] from Tirumalairajan river estuary, east coast of India. The results obtained from sequential extraction showed that a larger portion of the metals were associated with the residual phase, although they are available in other fractions too.

Trace metal fractionation in the Pichavaram mangrove–estuarine sediments in southeast coast of India was studied by Ranjan et. al. [120] considering the pronounced changes due to occurrence of tsunami (2004). A 5-step sequential extraction procedure was applied to assess the effects of tsunami on mobility and redistribution of selected elements (Cd, Cr, Cu, Fe, Mn, Ni, Pb, and Zn) in coastal sediments revealed that metals in the residual fraction (lattice bound) had the highest concentration suggesting their non-availability and limited biological uptake in the system. Majority of the metals (except Mn) do not constitute a risk based on the different geochemical indices.

Fractionation of selected metals in the sediments of Cochin estuary and Periyar River (southwest coast of India) was studied by Mohan et. al. [121]. The results reveal that remobilization potential of metals bound is in the range of low to medium risk to various sedimentary phases is different and is based on bond strength. Therefore, the strength values can give a clear indication of sediment reactivity that can be used to assess the risk related with metals to the aquatic organisms.

7. Risk Assessment Code (RAC)

The risk assessment code (RAC) mainly applies the sum of the exchangeable and carbonate bound fractions for assessing the availability of metals in sediments. These fractions are considered to be weakly bonded metals which may equilibrate with the aqueous phase and thus become more rapidly bioavailable [11, 33]. This is important because the fractions introduced by anthropogenic activities, such as agricultural runoff and tourism, are typified by the adsorptive, exchangeable, and bound to carbonate fractions, which are weakly bonded metals that could equilibrate with the aqueous phase and thus become more rapidly bioavailable [122]. According to RAC guideline (Table 8), for any metal, soil/sediment which can release in exchangeable and carbonate fractions, less than 1% of the total metal will be considered safe for the environment and soil/sediment with 11-30% carbonate and exchangeable fractions will be at medium risk to the environment. On the contrary, soil/sediment releasing in the above fractions more than 50% of the total metal has considered being highly dangerous, which can be easily enter the food chain [123].

Grade	Exchangeable and bounded to carbonate metal (%)	Risk
I	<1	No risk
II	1 – 10	Low risk
III	11 – 30	Medium risk
IV	31 – 50	High risk
V	>50	Very high risk

Table 8. Criteria of Risk Assessment Code [123].

Heavy-Metal Fractionation in surface sediments was studied by Dhanakumar et. al. [124] in the Cauvery river estuarine region, southeastern coast of India. The results revealed that most of the samples fall under the category from low- to high-risk class and from low-risk to very high-risk class in terms of labile fractions of Pb as well as Zn and Cu, respectively.

8. Conclusion

From the above discussion it is revealed that geochemical fractionation approach to the chemical speciation has provided a useful tool and opens a new dimension in assessing the potential mobility/bioavailability of heavy metals and metalloids in soils/sediments and opens a new dimension in the field of ecology and environmental chemistry. More efficient, non-laborious and time saving processes techniques in this field of chemical speciation are also coming up to get valid information regarding geochemical behavior of soils/sediments. Besides geochemical fractionation, Dezileau et al. [125] opined that total Fe or Fe/Al may be used to infer millennial-scales climate changes in the south eastern pacific while performing sequential extraction of Fe in marine sediments from the Chileau continental margin. However, the chemical partitioning should be carefully used in the assessment of environmental pollution as large amount of metals may naturally occur as anthropogenic fractions (including loosely bonded ions, sulfide ions and metals associated with sediments).

Acknowledgements

The authors gratefully acknowledge full support and cooperation of the Springer press, UK for extending permission in publishing the research paper of the journal Environmental Monitoring & Assessment, vol. 184(12), pp:7561-77, 2012.

This study was partially supported by the European Fund for Economic and Regional Development (FEDER) through the Program Operational Factors of Competitiveness (COMPETE) and National Funds through the Portuguese Foundation for Science and Technology (PEST-C/MAR/UI 0284/2011, FCOMP 01 0124 FEDER 022689).

Author details

Santosh Kumar Sarkar[1], Paulo J.C. Favas[2,3*], Dibyendu Rakshit[1] and K.K. Satpathy[4]

*Address all correspondence to: pjcf@utad.pt

1 Department of Marine Science, University of Calcutta, Calcutta, West Bengal, India

2 Department of Geology, School of Life Sciences and the Environment, University of Trás-os-Montes e Alto Douro, Vila Real, Portugal

3 IMAR-CMA Marine and Environmental Research Centre, Faculty of Sciences and Technology, University of Coimbra, Coimbra, Portugal

4 Indira Gandhi Centre for Atomic Research, Environment and Safety Division, Kalpakkam, Tamil Nadu, India

References

[1] Gleyzes C, Tellier S, Astruc M. Fractionation studies of trace elements in contaminated soils and sediments: a review of sequential extraction procedures. Trends in Analytical Chemistry 2002; 21 (6) 451-467.

[2] Tessier A, Campbell PGC, Bisson M. Sequential extraction procedure for the speciation of particulate trace metals. Analytical Chemistry 1979; 51 844-851.

[3] Sahuquillo A, Rigol A, Rauret G. Overview of the use of leaching/extraction tests for risk assessment of trace metals in contaminated soils and sediments. Trends in Analytical Chemistry 2003; 22(3) 152-159.

[4] Calmano W, Hong J, Förstner U. Binding and mobilization of heavy metals in contaminated sediments affected by pH and redox potential. Water Science and Technology 1993; 28(8/9) 53-58.

[5] Eggleton J, Thomas KV. A review of factors affecting the release and bioavailability of contaminants during sediment disturbance events. Environment International 2004; 30 973-980.

[6] Ure AM, Quevauviller P, Muntau H, Griepink B. Speciation of heavy metals in solids and harmonization of extraction techniques undertaken under the auspices of the BCR of the Commission of the European Communities. International Journal of Environmental Analytical Chemistry 1993; 51 135-151.

[7] Florence TM. The speciation of trace elements in waters. Talanta 1982; 29 345-69.

[8] Lung W. Speciation analysis – why and how? Fresenius' Journal of Analytical Chemistry 1990; 337 557-564.

[9] Schroeder WH. Development in the speciation of mercury in natural waters. Trends in Analytical Chemistry 1989; 8 339-342.

[10] Rauret G, Rubio R, Lopez-Sanchez JF, Cassassas E. Determination and speciation of copper and lead in sediments of a Mediterranean River (River Tenes, Catalonia, Spain). Water Research 1988; 22(4) 449-51.

[11] Pardo R, Barrado E, Perez L, Vega M. Determination and association of heavy metals in sediments of the Pisuerga River. Water Research 1990; 24(3) 373-379.

[12] Jardo CP, Nickless G. Chemical association of Zn, Cd, Pb and Cu in soils and sediments determined by the sequential extraction technique. Environmental science & Technology Letters 1989; 10 743-52.

[13] Tessier A, Campbell PGC, Bisson M. Heavy metal speciation in the Yamaska and St. Francois rivers (Quebec). Canadian Journal of Earth Sciences 1980; 17 90-105.

[14] Elsokkary LH, Muller G. Assessment and speciation of chromium, nickel, lead and chromium in the sediments of the river Nile, Egypt. Science of the Total Environment 1990; 97/98 455-63.

[15] Ure AM. Single extraction schemes for soil analysis and related applications. Science of the Total Environment 1996; 178(1/3) 3-10.

[16] Rauret G. Extraction procedures for the determination of heavy metals in contaminated soil and sediment. Talanta 1998; 46(3) 449-455.

[17] Quevauviller PH, Lachica M, Barahona E, Rauret G, Ure A, Gomez A, Muntau H. Interlaboratory comparision of EDTA and DTPA procedures prior to certification of extractable trace elements in calcareous soil. Science of the Total Environment 1996; 178(1/3) 127-132.

[18] Quevauviller PH, van der Slootr HA, Ure A, Muntau H, Gomez A, Rauret G. Conclusions of the workshop: harmonization of leaching/extraction tests for environmental risk assessment. Science of the Total Environment 1996; 178(1/3) 133-139.

[19] Bacon JR, Davidson CM. Is there a future for Sequential chemical extraction? Analyst 2008; 133 25-46

[20] Förstner U, Lechsber R, Davis RA, Hermitte PL. (eds.), Chemical methods for assessing bioavailable Metals in Sludges. Elsevier, London, 1985.

[21] Meguellati M, Robbe D, Marchandise P, Astruc M, Proceedings International Conference on Heavy Metals in the Environment, Heidelberg CEP Consultants, Edinburgh, 1983, p. 1090.

[22] Filgueiras AV, Lavilla I, Bendicho C. Chemextraction for metal partitioning in environmental solid samples, Journal of Environmental Monitoring 2002; 4 832-857.

[23] Lopez-sanchez JF, Rubio R, Rauret G. Comparison of two sequential extraction procedures for trace metal partitioning in sediments. International Journal of Environmental Analytical Chemistry 1993; 51 113-121.

[24] Hlavay J, Polyak K. Chemical speciation of elements in sediment samples collected at Lake Balaton. Microchemical Journal 1998; 58 281-290.

[25] Tokalioglu S, Kartal S, Elçi L. Determination of heavy metals and their speciation in lake sediments by flame atomic absorption spectrometry after a four-stage sequential extraction procedure, Analytical Chimica Acta 2000; 413 33-40.

[26] Belazi AU, Davidson CM, Keating GE, Littlejohn D. Determination and speciation of heavy metals in sediments from the Cumbrian coast, NW England, UK. Journal of Analytical Atomic Spectrometry 1995; 10 233-240.

[27] Mester Z, Cremisini C, Ghiara E, Morabito R. Comparison od two sequential extraction procedures for metal fractionation in sediment samples. Analytical Chimica Acta 1998; 259 133-142.

[28] Gomez-Ariza JL, Giraldez I, Sanchez-Rodas D, Morales E. Comparison of the feasibility of three extraction procedures for trace metal partitioning in sediments from southwest Spain, Science of Total Environment 2000; 246 271-283.

[29] Zhang T, Shan X, Li F. Comparison of two sequential extraction procedures for speciation analysis of metals in soils and plant availability. Communications in Soil Science and Plant Analysis 1998; 29 1023-1034.

[30] Albores AF, Cid BP, Gomez P, Lopez EF. Comparison between sequential extraction procedures and single extrations for metal partitioning in sewage sludge samples. Analyst 2000; 125 1353-1357.

[31] Ho MD, Evans GJ. Operational speciation of cadmium, copper, lead and zinc in the NIST standard reference materials 2710 nad 2711 (Montana soil) by the BCR sequential extraction procedure and flame atomic spectrometry. Analytical Communications 1997; 34 353-364.

[32] Zemberyova M, Bartekov J, Hagarov I. The utilization of modified BCR three-step sequential extration procedure for the fractionation of Cd, Cr, Cu, Ni, Pb and Zn in soil reference materials of different origins, Talanta 2006; 70 973-978.

[33] Singh KP, Mohan D, Singh VK, Malik A. Studies on distribution and fractionation of heavy metals in Gomti river sediments- a tributary of the Ganges India. Journal of Hydrology 2005; 312 14-27.

[34] Davidson CM, Duncan AL, Littlejohn D, Ure AM, Garden LM. A critical evaluation of the three-stage BCR sequential extraction procedure to assess the potential mobili-

ty and toxicity of heavy metals in industrially-contaminated land. Analytica Chimica Acta 1998; 363(1) 45-55.

[35] McLaren RG, Crawford D. Studies on soil copper I. The fractionation of copper in soils. Journal of Soil Science 1973; 24(2), 172-181.

[36] Kersten M, Förstner U. Speciation of trace elements in of water-soluble organic components. sediments and combustion waste.In Ure AM, Davidson CM (ed.) Chemical speciation in the environment. Blackie Academic and Professional, Glasgow, UK. 1995; 234-275.

[37] Ianni C, Magi E, Rivaro P, Ruggieri N. Trace metals in Adriatic coastal sediments: distribution and speciation pattern. Toxicological and Environmental Chemistry 2000; 78 73-92.

[38] Ianni C, Ruggieri N, Rivaro P, Frache R. Evaluation and comparison of two selective extraction procedures for heavy metal speciation in sediments. Analytical Sciences 2001; 17 1273-1278.

[39] Ramirez M, Massolo S, Frache R, Correa J. Metal speciation and environmental impact on sandy beaches due to El Salvador copper mine, Chile. Marine Pollution Bulletin 2005; 50 62-72.

[40] Borovec Z, Tolar V, Mraz L. Distribution of some metals in sediments of the central part of the Labe (Elbe) River, Czech Republic. Ambio 1993; 22 200-205.

[41] Campanella L, D'Orazio D, Petronio BM, Pietrantonio E. Proposal for a metal speciation study in sediments. Analytica Chimica Acta 1995; 309 387-393.

[42] Zdenek B. Evaluation of the concentrations of trace elements in stream sediments by factor and cluster analysis and the sequential extraction procedure. Science of the Total Environment 1996; 177 237-250.

[43] Gomez-Ariza JL, Giraldez I, Sanchez-Rodas D, Moralesm E. Metal sequential extraction procedure optimized for heavily polluted and iron Oxide rich sediments. Analytica Chimica Acta 2000; 414 151-164.

[44] Lopez-Sanchez JF, Sahuquillo A, Fiedler HD, Rubio R, Rauret G, Muntau H, Quevauviller P. CRM 601, a stable material for its extractable content of heavy metals. Analyst 1998; 123 1675-1677.

[45] Usero J, Gamero M, Morillo J, Gracia I. Comparative study of three sequential extraction procedures for metals in marine sediments. Environment International 1998; 24 478-496.

[46] Martin R, Sanchez DM, Gutierrez AM. Sequential extraction of U, Th, Ce, La and some heavy metals in sediments from Ortigas River, Spain. Talanta 1998; 46 1115-1121.

[47] Agnieszka S, Wieslaw Z. Application of sequential extraction and the ICPAES method for study of the partitioning of metals in fly ashes. Microchemical Journal 2002; 72 9-16.

[48] Templeton DM, Ariese F, Cornels R. IUPAC guidelines for terms related to chemical Speciation and Fractionation of elements. Pure and Applied Chemistry 2001; 72 1453-1470.

[49] Bordas F, Bourg ACM. A critical evaluation of sample for storage of Contaminated Sediments to Be investigated for the potential mobility of their heavy metal load. Water. Air. Soil. Pollution 1998; 103 137-149.

[50] Kersten M, Frostner U. Chemical fractionation of heavy metals in anoxic estuarine and coastal sediments. Water Science and Technology 1986; 18 121-130.

[51] Dold B. Speciation of the most soluble phases in a sequential extraction procedure adapted for geochemical studies of copper sulfide mine waste. Journal of Geochemical Exploration 2003; 80 55-68.

[52] Favas PJC, Pratas J, Gomes MEP, Cala V. Selective chemical extraction of heavy metals in tailings and soils contaminated by mining activity: Environmental implications. Journal of Geochemical Exploration 2011; 111 160-171.

[53] Pardo R, Vega M, Debán L, Cazurro C, Carretero C. Modelling of chemical fractionation patterns of metals in soils by two-way and three-way principal component analysis. Analytica Chimica Acta 2008; 606 26-36.

[54] Cuong TD, Obbard JP. Metal speciation in coastal marine sediments from Singapore from Singapore using a modified BCR-sequential extraction procedure. Applied Geochemistry 2006; 21 1335-1346.

[55] Dapaah RK, Takano N, Ayame A. Solvent extraction of Pb (II) from acid medium with zinc Hexamethylenedithiocarbamate followed by back-extraction and subsequent determination by FAAS. Analytica Chimica Acta 1999; 386 281-286.

[56] Gomez-Ariza JL, Giraldez I, Sanchez-Rhodes D, Morales E. Metal readsorption and re-distribution during analytical fractionation of trace elements in toxic estuarine sediments. Analytica Chimica Acta 1999; 399 295-307.

[57] Cheam V, Lechner J, Sekerka I, Desrosiers R, Nriagu J. Development of laser- excited atomic fluorescence spectrometer and a method for the direct determination of lead in Great Lake waters. Analytica Chimica Acta 1992; 269 129-136.

[58] Fischer E, Van D, Berg CMG. Anodic Stripping Voltammetry of Pb and Cd using a Hg film electrode and thiocyanate. Analytica Chimica Acta 1999; 385 273-280.

[59] Morales MM, Mart P, Llopis A, Compos L, Sagrado S. An environmental study by factor Analysis of surface sea waters in the Gulf of Valencia (Western Mediterranean). Analytica Chimica Acta 1999; 394 109-117.

[60] Hirade M, Chen Z, Sugimoto K, Kawaguchi H. Co precipitation with tin (IV) hydroxide followed by removal of tin carrier for the Determination of trace heavy metals by graphite-furnace atomic absorption Spectrometry. Analytica Chimica Acta 1980; 302 103-107.

[61] Ridout PS, Jones HR, Williams JG. Determination of trace elements in a marine reference material of lobster hepatopancreas (TORT-1) using inductively coupled plasma mass spectrometry. Analyst 1988; 113 1383-1386.

[62] Sakao SY, OgawaY, Uchida H. Determination of trace elements in seaweed samples by inductively coupled plasma mass spectrometry. Analytica Chimica Acta 1999; 355 121-127.

[63] Baeyens W, Monteny F, Leermakers M, Bouillon S. Evalution of sequential extractions on dry and wet sediments. Analytical and Bioanalytical Chemistry 2003; 376 890- 901.

[64] Guevara-Riba A, Sahuquillo A, Rubio R, Rauret G. Assessment of metal mobility in dredged harbour sediments from Barcelona, Spain. Science of the Total Environment 2004; 321 241-255.

[65] Idris AM, Eltayeb MAH, Potgieter-Vermaak SS, Grieken R, Potgieter JH. Assessment of heavy metal pollution in Sudanese harbours along the Red Sea coast. Microchemical. Journal 2007; 87 104-112.

[66] Wepener V, Vermeulen LA. A note on the concentrations and bioavailability of selected metals in sediments of Richards Bay Harbour, South Africa. Water SA 2005; 31 589-595.

[67] Esslemont G. Heavy metals in seawater, marine sediments and corals from the Townsville section, Great Barrier Reef Marine Park, Queensland. Marine Chemistry 2000; 71 215- 231.

[68] Coetzee PP. Determination and speciation of heavy metals in sediments of the Hartebeespoort Dam By sequential extraction. Water SA 1993; 19 291-300.

[69] Salmons W, Förstner U. Trace metal analysis on polluted sediments. Part II:evaluation of Environmental impact. Environmental science and Technology letters 1980; 1 14-24.

[70] Li X, Thornton I. Chemical partitioning of trace and major elements in soils contaminated by mining and smelting activities. Applied geochemistry 2000; 16 1693-1706.

[71] Kiratli N, Ergin M. Partitioning of heavy metals in surface Black Sea sediments. Applied Geochemistry 1996; 11 775-788.

[72] Rauret G, Lopez-Sanchez JF. New sediment and soil CRMs for extractable Trace metal content. International Journal of Environmental Analytical Chemistry 2001; 79 81-95.

[73] Salmons W. Adoption of common schemes for single and sequential extractions of Trace metals in soil and sediments. International Journal of Environmental Analytical Chemistry 1993; 51 3-4.

[74] Fiedler HD, Lopez-Sanchez JF, Rubio R, Rauret G, Quevauviller PH. Study of the stability of extractable trace metal contents in a river sediment using Sequential extraction. Analyst 1994; 119 1109-1114.

[75] Ramos L, Hernandez LM, Gonzalez MJ. Sequential fraction of copper, lead, copper, Cadmium and zinc in soils from or near Donana National Park. Journal of Environmental Quality 1994; 23 7-50.

[76] Tu Q, Shan XZ, Ni Z. Evaluation of a sequential extraction procedure for the Fractionation of amorphous iron and manganese oxides and organic matter in soils. The Science of The total Environment 1994; 151 159-165.

[77] Wang S, Jia Y, Wang S, Wang X, Wang H. Fractionation of heavy metals in shallow marine sediments from Jinzhou Bay, China. Journal of Environmental Science (China) 2010 22 23-31.

[78] Yuan CG, Shi JB, He B, Liu JF, Liang LN. Speciation of heavy metals in marine sediments from the East China Sea by ICP-MS with sequential extraction. Environment International 2004; 30 769-783.

[79] Jones B, Turki A. Distribution and Speciation of heavy metals in surficial sediments from the Tees Estuary, North – East England. Marine Pollution Bulletin 1997; 34 768-779.

[80] Ajay SO, Van Loon GW. Studies on redistribution during the analytical fractionation of metals in sediments. The Science of the Total Environment 1989; 87 171-187.

[81] Pempkowiak J, Sikora A, Biernacka E. Speciation of heavy metals in marine sediments vs their bioaccumulation by mussels. Chemosphere 1999; 39 313- 321.

[82] Forstner U, Wittmann GTW. Metal Pollution in the Aquatic Environment, Springer-Verlag, Berlin. Springer-Verlag, Heidelberg; 1981.

[83] Fedotov PS, Zavarzina, a AG, Spivakov BYa, Wennrich, b R, Mattusch J, De K, Titzeb PC, Demin VV. Accelerated fractionation of heavy metals in contaminated soils and sediments using rotating coiled columns, Journal of Environmental Monitoring 2002; 4 318-324.

[84] Nemati K, Kartini N, Bakar A, Abas MR, Sobhanzadeh E. Speciation of heavy metals by modified BCR sequential extraction procedure in different depths of sediments from Sungai Buloh, Selangor, Malaysia, Journal of Hazardous Materials 2011; 192(1) 402-410.

[85] Mossop KF, Davidson CM. Comparison of original and modified BCR sequential extraction procedures for the fractionation of copper, iron, lead, manganese and zinc in soils and sediments. Analytica Chimica Acta 2003; 478 (1) 111-118.

[86] Walkey A, Black TA. An examination of the Dugtijaraff method for determining soil organic matter and proposed modification of the chronic and titration method. Soil Science 1934; 37 23-38.

[87] Farnham IM, Johannesson KH, Singh AK, Hodge VF, Stetzenbach KJ. Factor analytical approaches for evaluating groundwater trace element chemistry data. Analytica Chimica Acta 2003; 490 123-138.

[88] Ragno G, De Luca M, Ioele G. An application of cluster analysis and multivariate cassification methods to spring water monitoring data. Microchemical Journal 2007; 87 119-127.

[89] Jonathan MP, Ram Mohan V. Heavy metals in sediments of the inner shelf off the Gulf of Mannar, Southeast coast of India. Marine Pollution Bulletin 2003; 46 263-268.

[90] Sunil Kumar R. Distribution of organic carbon in the sediments of Cochin mangroves, south west coast of India. Indian Journal of Marine Science 1996; 25 274-276.

[91] Janaki-Raman D, Jonathan MP, Srinivasalu S, Armstrong-Altrin J S, Mohan SP, Ram-Mohan V. Trace metal enrichments in core sediments in Muthupet mangroves, SE coast of India: Application of acid leachable technique. Environmental Pollution 2007; 145 245-257.

[92] Sarkar SK, Bilinski SF, Bhattacharya A, Saha M, Bilinski H. Levels of elements in the surficial estuarine sediments of the Hugli river, northeast India and their environmental implications. Environment International 2004; 30 1089-1098.

[93] Tam NFY, Wrong YS. Spatial variation of heavy metalsin surfe sediments of Hong, Kong mangrove swamps. Environmental Pollution 2002; 110 195-205.

[94] Subramanian V, Mohanachandran G. Heavy metals distribution and enrichment in the sediments of southern east coast of India. Marine Pollution Bulletin 1990; 21 324-330.

[95] Chatterjee M, Massolo S, Sarkar SK, Bhattacharya AK, Bhattacharya BD, Satpathy KK, Saha S. An assessment of trace element contamination in intertidal sediment cores of Sunderban mangrove wetland, India for evaluating sediment quality guidelines. Environmental Monitoring and Assessment 2009; 150 307-322.

[96] Datta DK, Subramanian V. Distribution and fractionation of heavy metals in the surface sediments of the Ganges–Brahmaputra–Meghna river system in the Bengal Basin. Environmental Geology 1998; 36 93-101.

[97] Müller G. Schwermetalle in den sedimenten des Rheins-Veranderungen seit. Umschan Verlag 1979; 79 133-149.

[98] Salomon W, Förstner U. Metals in the hydrocycle. Berlin: Springer; 1984.

[99] Buccolieri A, Buccolieri G, Cardellicchio N, Dell'atti A, Leo AD, Maci A. Heavy metals in marine sediments of Taranto Gulf (Ionian Sea, Southern Italy). Marine Chemistry 2006; 99 227-235.

[100] Dowling CB, Poreda RJ, Basu AR, Aggarwal PK. Geochemical study of arsenic release mechanisms in the Bengal Basin groundwater. Water Resources Research 2002; 38(9) 1173-1190.

[101] Acharyya SK, Lahiri S, Raymahashay BC, Bhowmilk A. Arsenic toxicity of groundwater in parts of the Bengal basin in India and Bangladesh: the role of Quaternary stratigraphy and Holocene sea-level fluctuation. Environmental Geology 2000; 39 231-238.

[102] Takarina ND, Browne DR, Risk MJ. Speciation of heavy metals in coastal sediments of Semarang, Indonesia. Marine Pollution Bulletin 2004; 49 854-874.

[103] Pickering WF. Metal ion speciation—soil and sediments (a review). Ore Geology Reviews 1986; 1 83-146.

[104] Ngiam LS, Lim PE. Speciation patterns of heavy metals in tropical estuarine anoxic and oxidized sediments by different sequential extraction schemes. Science of the Total Environment, 2001; 275 53-61.

[105] Dassenakis M, Adrianos H, Depiazi G, Konstantas A, Karabela M, Sakellari A, Scoullos M. The use of various methods for the study of metal pollution in marine sediments, the case of Euvoikos Gulf, Greece. Applied Geochemistry 2003; 18 781-794.

[106] Morillo J, Usero J, Gracia I. Heavy metal distri-bution in marine sediments from the southwest coast of Spain. Chemosphere 2004; 55 431-442.

[107] Thomas RP, Ure AM, Davidson CM, Littlejoh D, Rauret G, Rubio R, López-Sánchez JF. Three-stage sequential extraction procedure for the deter-mination of metals in river sediments. Analytica Chimica Acta 1994; 286 423-429.

[108] Panda D, Subramanian V, Panigrahy RC. Geochemical fractionation of heavy metals in Chilka Lake (east coast of India) – a tropical coastal lagoon. Environmental Geology 1995; 26 199-210.

[109] Dawson EJ, Macklin MG. Speciation of heavy metals in floodplain and flood sediments: a reconnaissance survey of the Aire Valley, West Yorkshire, Great Britain Environmental Geochemistry and Health 1998; 20 67-76.

[110] Ramos L, González M, Hernández L. Sequential extraction of copper, lead, cadmium, and zinc in sediments from Ebro River (Spain): relationship with levels detected in earthworms. Bulletin of Environmental Contamination and Toxicology 1999; 62 301-308.

[111] Caille N, Tiffreau C, Leyval C, Morel JL. Solubility of metals in an anoxic sediment during prolonged aeration. Science of the Total Environment 2003; 301 239-250.

[112] Canfield DE. Reactive iron in marine sediments. Geochimica et Cosmochimica Acta 1989; 53 619-632.

[113] Petersen W, Wallman K, Li PL, Schroeder F, Knauth HD. Exchange of trace elements at the sediment– water interface during early diagenesis processes. Marine and Freshwater Research 1995; 46 19-26.

[114] Monbet P. Mass balance of lead through a small macrotidal estuary: the Morlaix River estuary (Brittany, France). Marine Chemistry 2006; 98 59-80.

[115] Adriano DC. Trace elements in terrestrial environments. New York: Springer; 1986.

[116] Sarkar SK, Saha M, Takada H, Bhattacharya A, Mishra P, Bhattacharya B. Water quality management in the lower stretch of the river Ganges, east coast of India: an approach through environmental education. Journal of Cleaner Production 2007; 15 1559-1567.

[117] Banerjee ADK. Heavy metal levels and solid phase speciation in street dusts of Delhi, India. Environmental Pollution 2003; 123(1) 95-105.

[118] Long ER, Ingersoll CG, MacDonald DD. Calculation and uses of mean sediment quality guideline quotients: a critical review. Environmental Science and Technology 2006; 40 1726-1736.

[119] Venkatramanan S, Ramkumar T, Anithamary I, Jonathan MP. Speciation of selected heavy metals geochemistry in surface sediments from Tirumalairajan river estuary, east coast of India. Environmental Monitoring and Assessment 2013; 185(8) 6563-6578.

[120] Ranjan RK, Singh G, Routh J, Ramanathan AL. Trace metal fractionation in the Pichavaram mangrove–estuarine sediments in southeast India after the tsunami of 2004. Environmental Monitoring and Assessment 2013 (article in press).

[121] Mohan M, Augustine T, Jayasooryan KK, Chandran MSS, Ramasamy EV. Fractionation of selected metals in the sediments of Cochin estuary and Periyar River, southwest coast of India, Environmentalist 2012; 32 383-393.

[122] Hseu ZY. Extractability and bioavailability of zinc over time in three tropical soils incubated with biosolids. Chemosphere 2006; 63 762-771.

[123] Perin G, Craboledda L, Lucchese M, Cirillo R, Dotta L, Zanette ML, Orio AA. Heavy metal speciation in the sediments of Northern Adriatic Sea- a new approach for environmental toxicity determination, in: T.D. Lekkas (Ed.), Heavy Metal in the Environment 1985; 2 454-456.

http://dx.doi.org/10.5772/57295

[124] Dhanakumar S, Murthy KR, Solaraj G, Mohanraj R. Heavy-Metal Fractionation in Surface Sediments of the Cauvery River Estuarine Region, Southeastern Coast of India, Arch Environmental Contamination Toxicology 2013; 65 14-23.

[125] Dezileau L, Pizarro C Rubio MA. Sequential extraction f iron in marine sediments from the Chilen continental margin. Marine Geology 2007; 241 111-116.

Chapter 26

The Effect of Industrial Heavy Metal Pollution on Microbial Abundance and Diversity in Soils — A Review

Anna Lenart-Boroń and Piotr Boroń

Additional information is available at the end of the chapter

http://dx.doi.org/10.5772/57406

1. Introduction

Metals are essential components of the ecosystem, whose biologically available concentrations depend mainly on geological and biological processes [1]. There are several definitions of heavy metals, and some of them are based on the mass density of these elements. Authors of numerous publications use different limits to define the threshold density for a "heavy metal", ranging from 3.5 to 7 g×cm^{-3}, however, the majority of authors suggests that the mass density of heavy metals should be greater than 4.5 g×cm^{-3} [2]. Within the group of heavy metals one can distinguish both elements that are essential for living organisms (microelements) and the elements whose physiological role is unknown and thus they are "inactive" towards organisms. The metals that serve as microelements in living organisms usually occur in trace amounts, precisely defined for each species and both their deficiency and excess badly affect living organisms [3]. The term "heavy metal" is linked in many people's minds to metals that are toxic. However, this is not always the truth. The effect of any substance on a living system is always dependent on its available concentration to cells. Also, several heavy metal ions are crucial in metabolic processes at low concentrations but are toxic at high concentrations [2]. Nevertheless, locally elevated levels of these elements can create significant environmental and health problems when the release of metals through various biological, geological and anthropogenic processes far exceeds its natural content resulting from processes of metal cycling. Heavy metal pollution of terrestrial environments is of great concern, due to the persistence of metals in the ecosystem and their threat to all living organisms [4].

Given the importance of the subject of soil heavy metal pollution and its effect on soil microorganisms, this chapter gives an overview of the severity of the problem when it comes to the reaction of soil microbial community to the environmental pollution. The first part of this chapter deals with the abundance of microorganisms in soils and their role in this environment.

The next part concerns major sources of heavy metals in soils with particular emphasis on the most important source of soil pollution, i.e. human activity (and more precisely – industry and mining). The following part discusses the effects that toxic levels of heavy metals may have on the microbial population in soils. The last two parts of this chapter describe the ways of dealing with heavy metal pollution – one introduces the term of phytoremediation (soil remediation with the use of plants) and the other one focuses on the use of microorganisms resistant to heavy metals in the process of soil remediation.

2. The complexity of microbial community in soils

Except for occasional insects or earthworms, once visible traces of plant biomass are removed, soil appears as a lifeless mass, that is composed of mineral particles and organic residues. However, even desert soils are abundant source of living microorganisms. This seemingly lifeless matter contains complex microbial community, including bacteria, fungi, protozoa and viruses. The integrity of the aboveground and belowground ecosystems depends on the stability, resilience and function of the soil microbial community [5].

Soil is an interesting medium for growing microorganisms, as it contains various nutrients that the microbes need for their metabolism. Unfortunately, nutrients are not always readily available [6]. However, it is one of the richest reservoirs of microorganisms, i.e. 1 gram of agricultural soil may contain even several billion colony forming units (CFUs) of microorganisms belonging to thousands of different species [7], and even though microorganisms constitute less than 0.5% of the soil mass, they have a major impact on soil properties and processes [5]. Table 1 presents the average numbers of soil microorganisms in a "typical" temperate soil. Destruction of the soil microbiota through mismanagement or environmental pollution causes decline or even death of the aboveground plant and animal populations.

Organisms	Numbers		Biomass [wet kgxha^{-1}]
	Per m2	Per g	
Bacteria	10^{13}-10^{14}	10^{8}-10^{9}	300-3000
Actinomycetes	10^{12}-10^{13}	10^{7}-10^{8}	300-3000
Fungi	10^{10}-10^{11}	10^{5}-10^{6}	500-5000
Microalgae	10^{9}-10^{10}	10^{3}-10^{6}	10-1500

Table 1. Relative numbers and approximate biomass of the soil microbiota in a fertile soil [8].

The most characteristic feature of microbial habitats is the great micro-spatial variability of environmental parameters, like temperature or nutrient availability. Many basic requirements of heterogeneous microorganisms are satisfied by various soil microhabitats. This is the reason why, in ecological terms, a number of varying microbial niches can be described. Therefore, the microbial community is composed of diverse taxa with different nutritional demands

http://dx.doi.org/10.5772/57406

within small microenvironments [9]. Analysis of the spatial distribution of bacteria at micro-habitat levels showed that in soils subjected to different fertilization treatments, the majority of bacteria were located in micropores of stable soil micro-aggregates (2 – 20 μm), as they contained over 80% of cells [10]. Such microhabitats offer the most favorable conditions for microbial growth in terms of water and substrate availability, gas diffusion and protection against predation. The microhabitat-adapted groups of microorganisms form so-called consortia which are held together by mutually facilitating metabolic processes. The consortia are characterized by more or less sharp boundaries, and variable level of interaction with each other and with other parts of the soil biota. Numerous investigations emphasize the impact of soil structure and spatial isolation on microbial diversity and community structure [11]. Some studies indicate that the soil particle size affects the diversity of microorganisms and community structure to a greater extent than other factors such as bulk pH and the type or amount of available organic compounds [12]. Other investigations show that the type and amount of available organic substrates strongly affect the abundance of microbial groups and their functional diversity in soils [13]. Fierer and Jackson [14] claim that the structure of soil bacterial communities is not random also at continental scale and that the diversity and composition of soil bacterial communities at large spatial scales can be predicted to a large extent by a single variable, that is soil pH. The diversity of soil microorganisms comprises different levels of biological organization. It includes genetic variability among taxa (species), number (richness), relative abundance (evenness) of taxa and functional groups within communities [11]. The overall biodiversity of soil microflora comprises bacteria, fungi, actinomycetes and photosynthetic microorganisms [6].

Bacteria constitute the most numerous group of soil microbes – a teaspoon of productive soil contains between 100 million and 1 billion bacterial cells. As soil environment changes rather drastically, spore-forming bacteria tend to be the most common. When environmental conditions become too difficult for normal growth, the bacteria form spores and remain dormant until the environment returns to proper conditions [6]. They facilitate various processes in soils, e.g. those related to water dynamics, nutrient cycling or disease suppression [15]. Soil-dwelling bacteria may be divided into different groups based on:

- Shapes: rods (also called bacilli), sphere (also called cocci) and spiral (also called spirilla)
- Their reaction to oxygen: aerobic (bacteria that need oxygen for their survival) and anaerobic (the ones that do not require oxygen and in most cases cannot bear oxygen that is deadly for them)
- Result of Gram staining: Gram negative (stain pink and have thinner cell walls, they are the smallest ones and tend to be more sensitive to water stress) or Gram positive (stain violet, have thicker cell walls, are larger in size and tend to resist water stress)
- Source of carbon they use: autotrophs (obtain carbon from carbon dioxide – some autotrophic bacteria directly use sunlight in order to produce sugar from carbon dioxide, while others depend on various chemical reactions) or heterotrophs (they obtain carbohydrates from their environment)

- Classification based on phyla: based on morphology, barcode DNA sequences, physiological requirements and biochemical characteristics, bacteria have been classified into 12 phyla. Each phylum corresponds to a number of bacterial species and genera [15].

Tate [5] lists the most commonly encountered soil bacterial genera as: *Acinetobacter, Agrobacterium, Alcaligenes, Arthrobacter, Bacillus, Brevibacterium, Caulobacter, Cellulomonas, Clostridium, Corynebacterium, Flavobacterium, Hyphomicrobium, Metallogenium, Micrococcus, Mycobacterium, Pseudomonas, Sarcina, Streptococcus* and *Xanthomonas*. These are the heterotrophic bacteria that are augmented in soil by autotrophic and mixotrophic representatives, including nitrifiers, *Thiobacillus* species and iron bacteria.

Bacteria facilitate a number of physical and biochemical alterations or reactions in soils and thereby directly or indirectly support the development of higher plants. Their performance is vital for a variety of processes that include: decomposition of cellulose or other carbohydrates (e.g. *Bacillus, Achromobacter, Cellulomonas, Clostridium, Methanococcus*), ammonification (*Bacillus, Pseudomonas*), nitrification (*Nitrosomonas, Nitrobacter*), denitrification (*Achromobacter, Pseudomonas, Bacillus, Micrococcus*) and nitrogen fixation (symbiotic *Rhizobium, Bradyrhizobium* etc., non-symbiotic *Azotobacter, Beijerinckia*) [16].

On the other hand, soil fungi form three functional groups: decomposers, mutualists and pathogens. Fungi, along with bacteria, are important decomposers of hard to digest organic matter and they increase nutrient uptake of phosphorus. Mycorrhizal fungi support plants by promoting root branching and increasing nitrogen, phosphorus and water uptake. They improve plant resilience to pests, diseases or drought and improve soil structure, as fungal hyphae binds soil particles together to create water-stable aggregates. They in turn create the pore spaces in the soil that enhance water retention and drainage [17]. The most common fungi found in soil belong to the *Penicillium* and *Aspergillus* genera together with the representatives of the Zygomycetes and the mycorrhizae-associated Ascomycetes and Basidiomycetes [5].

Actinomycetes are a large group of microorganisms, systematically identified as bacteria, that grow as hyphae. They decompose a wide range of substances, but they are particularly important in degrading recalcitrant (difficult to degrade) compounds such as chitin, lignin, keratin and cellulose. Moreover, they produce a number of secondary metabolites such as antibiotics i.e. streptomycin [18] or geosmine which is responsible for "earthy" smell after soil plowing [15]. Actinomycetes are important in forming stable humus, which enhances soil structure, improves soil nutrient storage and increases water retention in soils. According to Tate [5], the most commonly encountered soil actinomycetes belong to *Nocardia* and *Streptomyces* genera.

Algae are the most common among photosynthetic microorganisms found in soil. They are found only near soil surface, where light is readily available [6]. The most common genera of green algae found in soil are: *Chlorella, Chlamydomonas, Chlorococcum, Protosiphon* etc. and that of diatoms are *Navicula, Pinnularia. Synedra, Frangilaria.* Their functions include the maintenance of soil fertility, increasing water retention capacity of soil, prevention of soil erosion due to the fact that they act as cementing agents in binding soil particles. They add organic matter to soil after the cell death and thus increase the amount of organic carbon, while their photosynthetic activity release large quantity of oxygen that facilitate the aeration in submerged soils or oxygenate the soil environment. They also take part in weathering rocks, thus building up the soil structure [19]

Although biomass of all microorganisms living in soil constitutes only several percent of organic matter content, they play an important role in the functioning of entire ecosystems [20]. They take part in soil formation, mineralize organic substances, provide plants with bioavailable compounds, cooperate with plants or may be used as a source of insecticidal substances [21]. One of the most important and most widely studied microbial groups in terms of beneficial effects to soil and plants is the group of Plant Growth Promoting Rhizobacteria (PGPR) [22]. This group includes bacterial species from genera such as *Azotobacter, Azospirillum, Bacillus, Burkholderia, Enterobacter, Erwinia, Flavobacterium, Pseudomonas* and *Rhizobium* [23]. Activity of these bacteria significantly increases plant growth and yield due to a variety of mechanisms, such as phytohormone production, symbiotic and asymbiotic N_2 fixation, production of siderophores, activity against phytopathogenic microorganisms, synthesis of antibiotics, enzymes and/or fungicidal compounds, as well as solubilization of mineral phosphates and other nutrients [24]. PGPR may improve plant growth, salinity and metal toxicity stress tolerance, as well as they are able to produce phytohormones such as indole-3-acetic acid (IAA) [25]. Some PGPR produce the enzyme 1-aminocyclopropane-1-carboxylate (ACC) deaminase, which hydrolyses ACC, the immediate precursor of ethylene in plants. By decreasing its concentration in seedlings and thus its inhibitory effect, these PGPR stimulate seedlings' root length [26]. Figure 1. shows the ways how Plant Growth Promoting Rhizobacteria can stimulate plants. Bacteria from the genus *Rhizobium* form symbiotic associations with roots of leguminous plants like clovers, peas or alfalfa. These Gram-negative, rod-shaped bacteria infect growing root hairs, forming visible nodules. In this form of symbiosis, plants supply simple carbohydrates to bacteria while bacteria convert nitrogen (N_2) from air into the forms (NO_3^- or NH_4^+) that plant can use. When leaves or roots from the plant decompose, nitrogen content increases in soil [15]. Some microbial species are capable of detergent decomposition, taking part in self-purification process of soils. Decomposers are particularly important in immobilizing or retaining nutrients in their cells, thus preventing the loss of nutrients, such as nitrogen, from the rooting zone.

Despite beneficial effects of numerous soil microbes on plant growth or development, soil structure and functioning, some soil-dwelling microorganisms may cause plant, animal and human diseases. Similarly to the beneficial soil microflora, soil pathogens include bacteria, fungi and viruses. One of the example of the most important or best known plant pathogens include *Agrobacterium tumefaciens* (whose updated scientific name is now *Rhizobium radiobacter*) [27] which is the causal agent of crown gall disease of walnuts, grape vines, stone fruits and many others. These bacteria infect plant roots and induce cells to divide (due to overproduction of auxin and cytokinin), causing a tumor-like swellings that contain infected cells [28]. *Erwinia carotovora* (or now called *Pectobacterium carotovorum*) and *Erwinia amylovora*, the Gram-negative plant pathogens with a diverse host range cause infections of numerous agriculturally and scientifically important plant species, such as potato, apple, pear and some members of the family *Rosaceae* [29]. Soil is also an abundant source of fungal pathogens. Among them we may

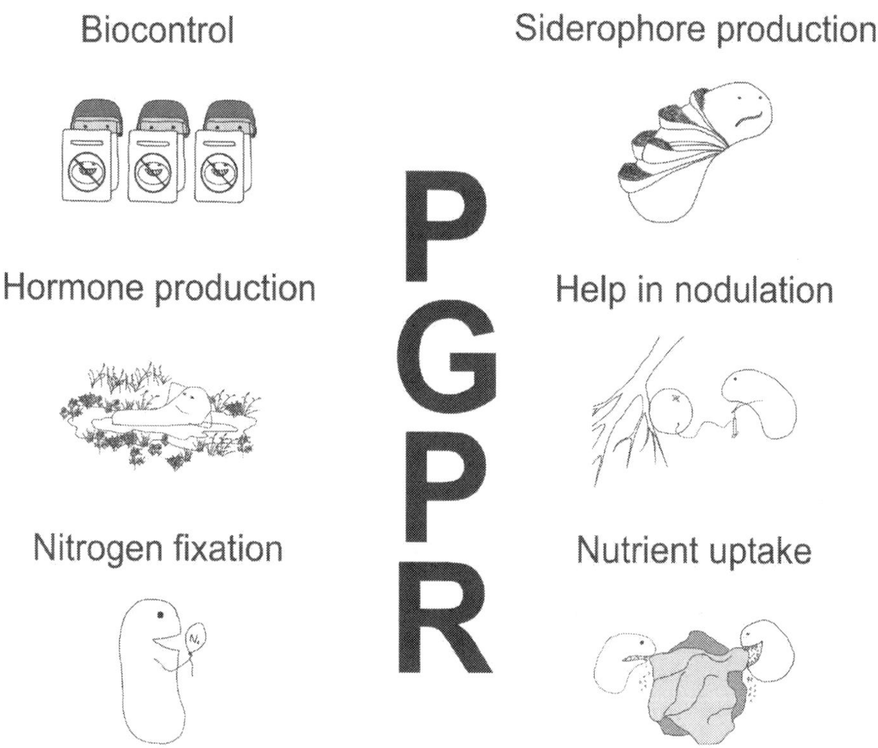

Figure 1. Summary of mechanisms employed by Plant Growth Promoting Rhizobacteria to stimulate plant development.

distinguish *Rhizoctonia solani*, a plant pathogenic fungus with a wide host range and worldwide distribution. It causes collar rot, crown rot, root rot, damping off and wire stem [30]. It mainly attacks plant seeds below the soil surface, but may also infect leaves and stems. Due to a variety of hosts that this pathogen attacks, it is of great importance and is detrimental to a variety of crops. The *Armillaria* root rot, caused by several species of basidoimycete genus *Armillaria* – the honey fungus is, on the other hand, one of the greatest threat for woody plants [31]. Another example of soil-borne plant pathogens is an important genus of fungi – *Fusarium*, which contains a number of, worldwide distributed, phytopathogenic species [32]. Moreover, *Fusarium* has also been more recently reported as an emerging human pathogen for immuno-compromised patients [33]. *Clostridium tetani* is an example of one of the most dangerous soil-borne human pathogens. It is a tetanus-causing Gram-positive bacterium, whose transmission occurs through the contamination of wounds with soil carrying its spores [34]. Generally, soil is a typical carrier of human bacterial and fungal pathogens. Another example of them is *Bacillus anthracis*, the causative agent of anthrax, which is found worldwide in a variety of soil environments. Inhalation of *B. anthracis* spores can be fatal. Nevertheless, the incidence of both of these fatal diseases has been largely controlled in developed countries due to the development of vaccines [35].

Undoubtedly, soil is an inexhaustible reservoir of microorganisms, both beneficial and pathogenic ones. Causing the imbalance between groups of soil macro- and microorganisms may be irreversible and result in a variety of effects, sometimes unpredictable. Such imbalance may be caused by soil pollution resulting from developing industry, therefore understanding the sources and effects of industrial soil pollution is an important element in preventing the environmental degradation.

3. Sources of soil heavy metal pollution

Chemical compounds, entering the ecosystem as a result of different human activities, may accumulate in soil and water environments. Therefore, soil may be regarded as a long-term reservoir of pollutants, from which these compounds may be introduced to food chains or groundwater [36]. Inappropriate and careless disposal of industrial waste often results in environmental pollution. The pollution includes point sources such as emission, effluents and solid discharge from industry, vehicle exhaustion and metal smelting or mining, as well as nonpoint sources (e.g. the use of pesticides or excessive use of fertilizers) [37]. Each of the sources have their own damaging effects on plant, animal and human health, but those that add heavy metals to soils are of serious concern due to the persistence of these elements in the environment. They cannot be destroyed, but are only transformed from one state to another [38].

Soil pollution may be defined as presence of xenobiotics (e.g. chemical compounds, radioactive elements) that alters the soil properties – both chemical, physical and biological. Soil pollution, including heavy metals, may be of natural origin, like volcanic eruptions, animal excrements or ore leaching. Nevertheless, human activity and mostly chemical industry, mining and metallurgy, as well as municipal management and traffic emissions are the main source of environmental pollution. Some authors also mention that waste disposal, waste incineration, fertilizer application and long-term application of wastewater in agricultural lands may result in heavy metal pollution of soils [39].

Heavy metals occur naturally in soils due to pedogenetic processes of weathering parent materials, however concentrations of these metals are regarded as trace (<1000 $mg \times kg^{-1}$) and rarely toxic [40]. Due to the disturbance and acceleration of the natural slow geochemical cycles of metals by man, most soils of rural and urban environments accumulate one or more heavy metals above the defined background levels, high enough to cause risks to ecosystems [41]. Nevertheless, heavy metals occurring in soils from anthropogenic sources tend to be more mobile, therefore more bioavailable than pedogenic or lithogenic ones [42].

Communication routes, such as roads, railways etc., are an important source of soil pollution, especially in the case of lead and zinc. Despite restricted use of leaded gasoline adopted in most countries, lead remains one of the most serious automotive-originating metal pollutant. The areas located nearby roads, particularly in urban sites, are the most vulnerable to automotive pollution. Apart from lead and zinc, chromium, cadmium, nickel and platinum are among the pollutants emitted by combustion engine-powered vehicles [43]. Heavy metals enter the environment as a result of tire wear and damage of vehicle parts. Moreover, grease used in vehicles may also be the source of cadmium pollution along roads [44]. Nickel emission results from this metal being added in gasoline and atmospheric abrasion of nickel-containing parts of automobiles [45]. The changes in the concentrations of lead, nickel, cadmium, copper and zinc in roadside soils are frequently attributed to traffic density [46].

Standard agricultural practices are also a significant source of heavy metals in soils, as application of fertilizers and pesticides has contributed to a continuous accumulation of these elements. Heavy metals can accumulate in soils due to the application of liquid and solid manure, as well as inorganic fertilizers [47]. The application of numerous biosolids, such as livestock manures, composts and municipal sewage sludge on agricultural soils leads to the accumulation of various heavy metals, such as, Cd, Cr, Cu, Hg, Mo, Ni, and Zn [48]. Lime and superphosphate fertilizers contain not only major elements necessary for plant nutrition and growth but also trace metal impurities such as cadmium. The presence of high concentrations of Cd in some fertilizers (particularly in phosphatic fertilizers) is of most concern due to the toxicity of this metal and its ability to accumulate in soils as well as due to its bioaccumulation in plant and consequently in animal tissues [49, 50]. Additionally, copper-containing compounds have been widely used in agricultural practice as pesticides. Copper oxychloride is annually applied on vineyards as a fungicide to control a significant number of plant diseases. Inevitably, this Cu ends up in the agricultural soil and adjacent pristine natural vegetation [51]. Lead arsenate was used in fruit orchards for many years to control some of the parasitic insects. Arsenic-containing compounds were also extensively used to control pests in banana plantations in new Zealand and Australia [52]. High fertilizer applications and acid atmospheric deposition, combined with insufficient liming, may also cause a decrease in pH and thus increase heavy metal bioavailability, aggravating the problem of deteriorating food quality, metal leaching and impact on soil organisms [53]. The application of municipal wastewater or industrial waste as fertilizers and liming agents in agriculture is a separate issue. Application of this type of waste requires constant monitoring of the amount and proportion of harmful factors, including heavy metals. The high risk of soil pollution with Cd, Zn, Ni and Pb as a result of industrial waste application as fertilizers was also evidenced [50].

Airborne sources of heavy metals include stack emissions or fugitive emissions such as dust from storage areas or waste heaps. Stack emissions can be distributed over a wide area by natural air currents, while fugitive emissions are often distributed over much smaller areas. In general, concentrations of pollutants are much lower in fugitive emissions compared to stack emissions. The type and concentration of metals emitted from both types of sources depend on site-specific conditions. All solid particles in smoke from fires and other emissions from factory chimneys are deposited on land or sea. Most forms of fossil fuels contain some heavy metals and this form of environmental pollution has been increasing since the industrial revolution began. For instance, very high concentrations of Cd, Pb and

http://dx.doi.org/10.5772/57406

Zn have been found in plants and soils adjacent to smelting plants. Another major source of soil pollution is the aerial emission of lead from combustion of petrol containing tetraethyl lead; this contributes substantially to the content of Pb in soils in urban areas and in those adjacent to major roads [52].

Another, and one of the most significant sources of heavy metal pollution of soils, includes heavy industry, e.g. mining and metallurgy. Industrial airborne heavy metal contamination of the nonferrous smelters surrounding landscapes is a well-known and widely occurring phenomenon. Emissions of metallurgical dust are spread according to the wind direction and particle size while soil is the main receiver of heavy metals in dry land. Dust emissions from smelters using sulfide copper-nickel ores are similar, regardless of their location, owing to the fact that the same raw materials are used in metallurgical processes. The following major metal-containing compounds are deposited onto the landscape in the form of dust emissions from smelters: pentlandite (Ni,Fe)9S8, pyrrotite Fe7Sg(Nix), chalcopyrite CuFeS2, chalcosite Cu2S, covellite CuS, cuprite Cu20, tenorite CuO, and metal copper and nickel [54]. Surface soil layers in the mining or metallurgy areas are often heavily polluted with copper. In the vicinities of steel plants the concentration of this element exceeds several thousand ppm and the pollution remains for a long time, even after the operation of mines or steel plants had been stopped [50]. The fine fractions of dust are enriched with lead, arsenic, and zinc. The quantity and composition of dust derived from different sources (metallurgical processes) varies according to the raw materials and the condition of the gas cleaning systems [54]. The cause for the frequently widely dispersed metal pollution in habitats of mining areas was found in the formation of acid mine drainage (AMD). The runoff from mining heaps of active and abandoned mines can be extremely acidic, with pH values reaching as low as pH 2 [9]. Chemical and biological oxidation of the abundant mineral pyrite (FeS_2) occurs after the unearthing of pyrite-containing rock formations and results in an acidification of the dump material [55]. Under acidic conditions, the majority of heavy metals is leached from the waste dump and they are transported as AMD in streamwaters [9]. Galvanization industry may cause soil pollution with silver as well as other industrial facilities that use silver salts. Additionally, the increased amount of silver may by introduced to soils with municipal sewage. Municipal sewage contains also large amounts of highly soluble forms of zinc, which may then easily contaminate soil environment [50]. Zinc is also extensively used in metallurgical industry, as an anti-corrosion agent in alloys and in galvanization. It is frequently used in paint industry [50]. The concentration of cadmium highly increases in soils polluted with emissions from nonferrous metal plants, which constitute over 60% of all anthropogenic sources of this element in soils. Municipal sewage contains on average 10 – 40 ppm of cadmium, while industrial sewage may contain over 1000 ppm. This is also a case of large amounts of lead that may be introduced into soils from municipal sewage and waste, as they contain mobile forms of this element. This may result in large increase in the concentration of lead in soils that may exceed several times the admissible limits. Additionally, dust emissions from landfills of nonferrous metal plants may become dangerous sources of lead in soils [50]. Table 2 shortly summarizes the major sources of different heavy metals in soil.

Heavy metals	Sources
As	Semiconductors, petroleum refining, wood preservatives, animal feed additives, coal power plants, herbicides, volcanoes, mining and smelting
Cu	Electroplating industry, smelting and refining, mining, biosolids
Cd	Geogenic sources, anthropogenic activities, metal smelting and refining, fossil fuel burning, application of phosphate fertilizers, sewage sludge
Cr	Electroplating industry, sludge, solid waste, tanneries
Pb	Mining and smelting of metalliferous ores, burning of leaded gasoline, municipal sewage, industrial wastes enriched in Pb, paints
Hg	Volcano eruptions, forest fire, emissions from industries producing caustic soda, coal, peat and wood burning
Se	Coal mining, oil refining, combustion of fossil fuels, glass manufacturing industry, chemical synthesis (e.g., varnish, pigment formulation)
Ni	Volcanic eruptions, land fill, forest fire, bubble bursting and gas exchange in ocean, weathering of soils and geological materials
Zn	Electroplating industry, smelting and refining, mining, biosolids

Table 2. Different sources of heavy metals in soils [56]

4. The effects of heavy metals on soil microorganisms

Metals without biological function are generally tolerated only in minute concentrations, whereas essential metals with biological functions, are usually tolerated in higher concentrations [9]. They have either metabolic functions as constituents of enzymes or meet structural demands, e.g. by supporting the cell envelope. Frequently the concentration and the speciation of metal determine whether it is useful or harmful to microbial cells [9].

Microorganisms are the first biota that undergoes direct and indirect impacts of heavy metals. Some metals (e.g. Fe, Zn, Cu, Ni, Co) are of vital importance for many microbial activities when occur at low concentrations. These metals are often involved in the metabolism and redox processes. Metals facilitate secondary metabolism in bacteria, actinomycetes and fungi [9; 57]. E.g. chromium is known to have stimulatory effect on both actinorhodin production and growth yield of the model actinomycete *S. coelicolor* [58]. However, high concentrations of heavy metals may have inhibitory or even toxic effects on living organisms [59]. Adverse effects of metals on soil microbes result in decreased decomposition of organic matter, reduced soil respiration, decreased diversity and declined activity of several soil enzymes [60]. Some of the general changes in morphology, the disruption of the life cycle and the increase or decrease of pigmentation are easy to observe and evaluate [9]. Rajapaksha et al. [61] compared the reactions of bacteria and fungi to toxic metals in soils (Zn and Cu). They concluded, that bacterial community is more sensitive to increased concentrations of heavy metals in soils than the fungal community. The relative fungal/bacterial ratio increased with increasing metal levels. Those authors also noticed the varying effect of soil pH on the microbial reaction to soil pollution, i.e. that lower pH in contaminated soils enhanced the negative effect on bacteria, but not on fungi.

The toxic concentration of heavy metals may cause enzyme damage and consequently their inactivation, as the enzymes-associated metals can be displaced by toxic metals with similar structure [59]. Moreover, heavy metals alter the conformational structures of nucleic acids and proteins, and consequently form complexes with protein molecules which render them inactive. Those effects result in disruption of microbial cell membrane integrity or destruction of entire cell [62]. Heavy metals also form precipitates or chelates with essential metabolites [63].

Various metals may affect different microbial populations and the resulting impact may vary depending on the metal whose limit concentrations in soils were exceeded. For instance, the pollution of soils with copper affects microorganisms that take part in nitrification and mineralization of protein compounds [50]. Silver is one of the most toxic metals to heterotrophic bacteria. This effect is used for the production of antiseptic preparations. However, there are some silver-resistant bacteria, both in clinical and natural conditions. Some strains of *Thiobacillus ferrooxidans* are able to accumulate particularly large amounts of silver [50]. About 100 ppm of zinc in soils may inhibit nitrification processes and about 1000 ppm inhibits the majority of microbiological processes in soils [64]. Microorganisms play vital role in circulation and transformation of mercury compounds in the environment. Numerous bacteria and fungi show high tolerance (also acquired) to increased concentrations of mercury in soils. However, some microorganisms are sensitive to excess mercury, e.g. the concentration of <10 ppm Hg may have toxic effects on nitrifiers in soils [50]. Increased concentrations of lead in surface soil layers negatively affect soil microflora. Processes of organic matter decomposition, particularly cellulose, are inhibited as a result of decreased enzymatic activity of microorganisms. This results in soil degradation. Biosorption of lead by soil microorganisms reaches on average 0.2% of this metal, but in some cases it may reach even 40% of biomass and may be used for biological remediation [50]. Some studies indicate that long-term contamination of soils with heavy metals has adverse effects on soil microbial activity. For instance, Juwarkar et al. [65] while researching the remediation strategies for cadmium and lead contaminated soils, compared the numbers of the selected groups of microorganisms in natural and heavy metal spiked soils. The results that they obtained indicated that the examined microbial groups were much less abundant in contaminated soils than in natural ones [Table 3]. On the other hand, Lenart and Wolny-Koładka [66] recorded significantly variable numbers of the selected microbial groups while analyzing the uncontaminated and heavy metal contaminated soils of ArcelorMittal steelworks in Cracow. Similarly, their results indicated that except for fungi, the soil-dwelling microorganisms were much less abundant in heavy metal polluted soils than in uncontaminated soils (Table 4). Heavy metal contamination results in reduction of microbial biomass and even if they do not cause the reduction in their number – they reduce biodiversity or disturb the community structure [64].

Analyzed microorganisms [CFUxg^{-1}]	Natural soil	Heavy metal contaminated soil
Total number of bacteria	17x10^5	58x10^4
Total number of fungi	26x10^3	42x10^2
Actinomycetes	43x10^3	18x10^1
Azotobacter spp.	23x10^3	17x10^1
Rhizobium spp.	21x10^3	16x10^2

Table 3. Microbiological characteristics of natural and heavy metal spiked spoil samples in Nagpur (India) [65]

Analyzed microorganisms [CFUxg^{-1}]	Uncontaminated soil	Heavy metal contaminated soil
Total number of mesophilic bacteria	22.50x10^2 – 10.44x10^6	0 – 13.15x10^5
Total number of fungi	84.00x10^1 – 21.03x10^3	0 – 57.90x10^3
Actinomycetes	62 – 99.50x10^3	0 – 20.26x10^3
Azotobacter spp.	0 – 28.90x10^2	0 – 57.00x10^1

Table 4. Ranges of the selected microbial groups in heavy metal contaminated and uncontaminated soils of ArcelorMittal steelworks in Cracow, Poland [66].

However, one of the reasons of decreasing biodiversity of microorganisms in heavy metal polluted soils is the selection for tolerant species or strains. Metal exposure may lead to the establishment of tolerant microbial populations, that are often represented by several Gram-positive genera such as *Bacillus, Arthrobacter* and *Corynebacterium* or Gram-negatives, e.g. *Pseudomonas, Alcaligenes, Ralstonia* or *Burkholderia* [68]. It was shown that the impact of heavy metals on the bacterial metabolism depends on the growth form. The resistance towards metals seems higher in consortia than in pure cultures [69]. A great number of heavy metal-resistant bacteria, such as e.g. *Cupriavidus metallidurans* possess efflux transporters that excrete toxic or overconcentrated metals outside the cell [70]. Efflux transporters have high substrate affinity and can therefore maintain low cytosolic concentration of metals [9]. Alternatively, microbial cells may prevent the intoxication by the release of metal-binding compounds into the extracellular surroundings. In that case, metals are chelated outside the cell and thus blocked from entering the cell through the membrane transporters that otherwise facilitate the influx [9]. Some fungal and bacterial species are able to keep metals outside their cells by the extracellularly active melanin [71]. It is a secondary metabolite that has strong cation chelating properties through the anionic function such as carboxyl and deprotonated hydroxyl groups [9]. A substantial number of soil microorganisms, such as widespread fungus *Aspergillus niger,* solubilize metals by the release of organic acids or by the immobilization of metals through excretion of different compounds, such as oxalates [72]. Some microorganisms possess the abilities to protect their cells by a cytosolic sequestration mechanisms. These mechanisms are activated once the metal enters the cell and cannot be excreted. In this case internal inclusion bodies, e.g. polyphosphate granules (volutin) bind large amounts of metal cations [73].

Investigation and understanding of microbial resistance mechanisms towards heavy metals are crucial for the potential application of microorganisms for remediation of polluted soils.

5. General outline of soil remediation strategies

The overall objective of any soil remediation approach is to create a final solution that is protective both for human health and the environment [74]. For heavy metal-polluted soils, the physical and chemical form of the heavy metal contaminant in soil strongly influences the selection of the appropriate remediation treatment approach. Details on the physical characteristics of polluted soils, type and level of the pollution at the site must be known to enable accurate assessment of the problem severity and adjustment of remedial measures [52].

Remediation of heavy metal-polluted sites is very expensive and difficult, therefore the best method to protect the environment from contamination is to prevent it. Nevertheless, it is not always possible and once metals are introduced and pollute the soil, they will remain there. Unlike carbon-based organic pollutants, heavy metals cannot be degraded or eliminated completely, therefore the traditional treatments for heavy metal pollution of soils are complicated and cost-intensive.

There are several technologies for remediation of heavy metal-polluted soils. One of the classifications divides the methods into *in situ* and *ex situ* treatment technologies. *In situ* (in place) means that the polluted soil is treated in its original location, i.e. it remains at the site or in the subsurface. Such technologies remove the pollutant from soil without excavation or removal of the soil. In this case fixing agents are applied on the unexcavated soil. This technique's advantages may include low invasiveness, simplicity and rapidity. Moreover, it is fairly inexpensive and generates relatively low amount of waste. However, it is only a temporary solution. This is due to the fact that when physicochemical properties of soil change, the pollutants may again become active. Moreover, the reclamation process is applied only to the surface layer of soil [75]. *Ex situ* means that the treated soil is removed or excavated from the site [52]. It is applied in areas where heavily polluted soil must be removed from its place of origin and its storage is associated with high ecological risk. Fast and easy applicability, relatively low costs of investment and operation are the advantages of this method. On the other hand, it is highly invasive to the environment, generates a significant amount of solid wastes, and it is necessary to control the stored waste permanently. Evanko and Dzombak [76] divide *in situ* remediation strategies into solidification/stabilization, vitrification, soil flushing, electrokinetic extraction and biological treatment. *Ex situ* treatment technologies are divided by these authors into: solidification/stabilization, soil washing, vitrification and pyrometallurgical separation. Another classification of remedial strategies divides the technologies under five categories of general approaches to remediation: isolation, immobilization, toxicity reduction, physical separation, and extraction. There are several physicochemical techniques that include excavation and burial of soil at a hazardous waste site, chemical processing of soil to immobilize metals, leaching by using acid solutions or appropriate leachants to desorb metals from soil followed by the return of clean soil to the site [77], precipitation or flocculation followed by sedimentation, ion exchange, reverse osmosis and microfiltration [78]. Nevertheless, physicochemical techniques for heavy metal remediation are generally costly and have side effects [37]. Therefore, continuous efforts have been made to develop techniques that are easy to use, sustainable and economically feasible.

6. The use of plants for biological remediation of heavy metal polluted soils

Phytoremediation is one of the best techniques for treatment of heavy metal-polluted sites. It is an *in situ* strategy that uses vegetation and associated microbiota together with agronomic practices to aid in metal remediation [79]. It is based on the use of special type of plants to decontaminate soil by inactivating metals in the rhizosphere or translocating them in the aerial parts [56]. Some plants developed mechanisms to remove ions selectively from the soil to regulate the uptake and distribution of metals. Potentially useful phytoremediation technologies for heavy metal-polluted sites include phytoextraction, phytostabilization and rhizofiltration [75].

Phytoextraction uses hyperaccumulating plants to remove metals from soil by absorption into the roots and shoots of the plant. The aboveground shoots can be then harvested to remove metals from the site and subsequently stored as hazardous waste or employed for the recovery of metals. The ideal plant for phytoextraction should grow rapidly, produce high amount of biomass and be able to tolerate and accumulate high metal concentrations in shoots [80]. Hyperaccumulating plants belong to the families of *Brassicaceae, Fabaceae, Euphorbiaceae, Asterraceae, Lamiaceae,* and *Scrophulariaceae* [77]. Studies indicate that many *Brassica* species, such as *B. juncea, B. napus* or *B. rapa* exhibit enhanced accumulation of Zn and Cd [81] In comparison to conventional methods like e.g. soil excavation (*ex situ* remediation), phytoextraction is time consuming, but on the other hand it is cost-effective and less labor-intensive [9].

Phytostabilization is based on the use of plants to limit the mobility and bioavailability of metals in soil. Plants used in this method are characterized by high tolerance of metals in surrounding soils together with their low accumulation. Phytostabilization can be carried out through the process of sorption, precipitation, complexation, or metal valence reduction. This technique is useful for the removal of Pb, As, Cd, Cr, Cu, and Zn [82]. This process is advantageous because in this case disposal of hazardous material/biomass is not required, and it is very effective when rapid immobilization is needed to preserve soils or ground and surface waters [82].

Rhizofiltration (or phytofiltration) removes metals from contaminated soil via absorption, concentration and precipitation by plant roots. This technique is used to remove pollutants from groundwater and aqueous-waste streams rather than for the remediation of polluted soils [76]. Apart from the above described phytoremediation methods, some authors [83] include also phytovolatization and phytodegradation.

Phytovolatization involves the use of plants to volatilize pollutants from their foliage such as Se and Hg, while phytodegradation uses plants and associated microorganisms to degrade

http://dx.doi.org/10.5772/57406

organic pollutants. Even though phytoremediation strategies are inexpensive, effective, environmentally friendly and can be implemented *in situ*, a substantial proportion of metal pollutants are unavailable for root uptake by field grown plants [84]. Therefore, methods of increasing phytoavailability of heavy metal pollutants in soil and their transport to plant roots are vital to the success of *in situ* phytoremediation. In this case it is useful to apply microbial populations that are able to affect trace metal mobility and availability to plants, through the release of chelators, acidification and redox changes [85]. It was proved that the presence of rhizosphere bacteria increases the available concentrations of various heavy metals to hyperaccumulative plants [80]. Microbial populations may be used not only for increasing metal bioavailability to plants, but also for the promotion of hyperaccumulative plant growth through N_2 fixation, production of phytohormones and siderophores, and transformation of nutrients [26]. Figure 2 summarizes the mechanisms of plant-mediated remediation of contaminated soils.

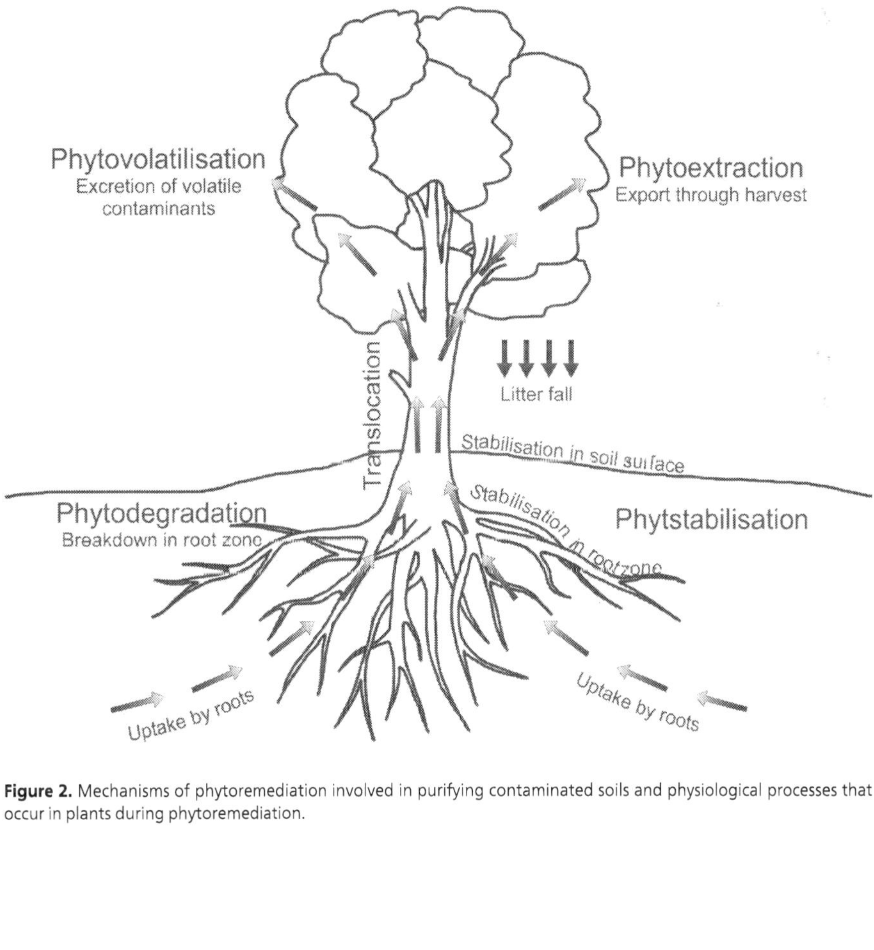

Figure 2. Mechanisms of phytoremediation involved in purifying contaminated soils and physiological processes that occur in plants during phytoremediation.

7. Application of microorganisms to remediate heavy metal-polluted soils

Another approach for biological remediation of heavy metal-polluted soils includes the use of microorganisms to detoxify metals by valence transformation, extracellular chemical precipitation or volatilization etc. [56]. Bioleaching is the method that uses microorganisms to solubilize heavy metal pollutants either by direct bacterial processes, or as a result of interactions with metabolic products, or both [76]. It can be used *in situ* or *ex situ* to help to remove the pollutants from soils. This process is based on mobilization of metal cations from insoluble ores by biological oxidation and complexation. This process was adapted from mining industry for the use in soil remediation and a general term covering both bioleaching and biooxidation techniques could be "biomining". This technique is mainly employed for copper, cobalt, nickel, zinc and uranium, which are extracted either from insoluble sulfides or (in the case of uranium) from oxides [86]. The classical bioleaching bacteria belong to the genus *Acidithiobacillus* (*A. thiooxidans* and *A. ferrooxidans*), *Acidiphilium*, *Acidimicrobium*, *Ferromicrobium* or *Sulfobacillus* [86].

Another solution for soil bioremediation using microorganisms is to apply microbially-mediated biochemical processes, such as oxidation/reduction or methylation reactions [87]. Often, biostimulation and bioaugmentation are the components of bioremediation strategies. Biostimulation is a form of *in situ* bioremediation which uses growth rate stimulation nutrients, electron donors or acceptors to encourage the growth of site-specific indigenous microorganisms capable of degrading environmental pollutants. Common electron donors and acceptors used in biostimulation include: acetate, sulfate, nitrate and ethanol [88]. Bioaugmentation is the introduction of specific competent microorganisms to the local microbial population in order to increase the metabolic capacities needed for remediation [89]. Biosorption is a physicochemical process that occurs naturally and allows to passively concentrate and bind contaminants onto the microbial cell structure [90]. Metal biosorption by living organisms is a complicated process that consists of two steps. In the first step, metal ions are adsorbed on the cell surface by interactions between metals and cell surface functional groups. Biosorption of metal ions occurs primarily on the outer surface of microbial cells and is the first step in the interactions between metals and microbial cell walls [4]. The cell wall consists of a variety of polysaccharides and proteins, and hence offers a number of active sites capable of binding metal ions [91]. Differences in the cell wall composition among various microbial groups, i.e. algae, bacteria, cyanobacteria and fungi, cause significant differences in the type and amount of metal ions binding to them [91]. Physical adsorption via electrostatic or van der Waals forces allow to retain metal ions on the outer surfaces of bacterial cells. In addition to physical adsorption, ion exchange and complexation are believed to be the dominant mechanisms involved in metal biosorption [4]. The first step, passive biosorption, is metabolism-independent and proceeds rapidly by any one or a combination of metal binding mechanisms. In the second step, due to active biosorption, metal ions penetrate the cell membrane and enter into the cells. This is, however, a slowly occurring process. Active mode is metabolism-dependent and related to metal transport and deposition [91]. There are several microbial genera and species capable of metal biosorption. Fungi were found to be efficient biosorbent organisms, as their cells are characterized by a high percentage of cell wall material, which shows excellent metal binding properties [92]. *Aureobasidium pullulans, Cladosporium resinae, Aspergillus niger, Aspergillus versicolor* or *Rhizopus nigricans* are the fungal species proved to be effective in heavy metal biosorption [91]. Numerous studies also identified several species of bacteria as efficient metal accumulating microorganisms. For instance, *Bacillus* spp. has been reported to have a high potential of metal sequestration and has been used in commercial biosorbent preparation [91]. Other bacterial species capable of metal transformation include, among others: *Escherichia coli, Pseudomonas maltophilia, Shewanella putrefaciens, Pseudomonas aeruginosa, Enterobacter cloacae* [4].

Mechanisms involved in biochemical interactions between bacteria and metal ions involve specific enzymes that catalyze the oxidation, reduction, methylation, dealkylation and precipitation reactions. Microorganisms transform a substantial number of metals and metalloids by reducing or oxidizing them directly to a lower or higher redox state. Additionally, indirect oxidation or reduction is an alternative for immobilization of toxic metals in the environment. Methylation is an important process involved in geochemical cycling of metals and the removal of metal pollutants from soils. Methylation processes derive the methyl group from methylocarbolamine (CH_3B_{12}) which is implicated in the methylation of multiple metals and metalloids, such as Pb, Sn, Pd, Pt, Au, Ti, As, Se and Te [93]. Methylation of Hg, Sn and Pb can be mediated by a range of microbes, including *Clostridium* spp., methanogens and sulfate-reducing bacteria under anaerobic conditions and principally by fungi (e.g. *Penicillium* spp. and *Alternaria* spp.) under aerobic conditions. Methyl groups are enzymatically transferred to metals and a given species may transform a number of different metals [94]. Methyl-metal compounds are generally highly volatile and available to plants [50]. Another mechanism that has the potential for the application in heavy metal-polluted sites is the production of siderophores by different microbial genera. Siderophores are the largest class of compounds that can bind and transport Fe. They are highly specific Fe(III) ligands and are excreted by a wide variety of fungi and bacteria to aid Fe assimilation [94].

Microorganisms play an important role in the environmental biogeochemical cycle of metals and their properties are of significant interest in the remediation of contaminated sites. The microbial ability to absorb and transform metals is a promising aspect in respect of solving the pollution problems [4]. The potential of numerous microbial metal transformations in treatment of environmental pollution may be employed and some processes are already in commercial operation. However, many processes are still at the laboratory scale and yet to be tested in a rigorous applied and/or commercial context [94]. Another interesting aspect of the microbial community is their ability to multiply even under undesirable environmental conditions. These microorganisms sometimes affect soil environment more quickly than abiotic processes can. Therefore, the structure of soil microbial populations may be useful as a highly sensitive bioindicator of soil disturbance and progress of remediation [95].

Facing the increasing heavy metal pollution severity accompanied by rising land prices the communities around the world need to struggle for available investment grounds. This is mostly the problem of big cities, especially those with limited opportunities for development due to geographical barriers such as seashores, mountain ranges or desert areas. In such situations the polluted industrial areas cannot be left unused for long time to recover naturally.

This creates a need for the development of various remedial procedures adjusted to changing contamination level, environmental conditions, available time and funding. Thus, remedial measures need to be almost always modified in order to meet those criteria. This makes that the continuous effort should be made to increase the effectiveness, flexibility and decrease the cost and side effects of the procedures available today. Although a number of measures was developed to remove the even toxic level of contamination, there are many degenerated areas that still cannot be successfully treated now. Those cases involve sites where remediation would be too expensive, time consuming or even technically disputable with currently available treatment procedures.

8. Conclusion

Heavy metals pose a significant threat towards the soil environment and the rapid industrialization will result in increasing problems of environmental pollution. Therefore, it is necessary to carry out the continuous monitoring of both industrial areas and their vicinities for possible transgressions of the limits given by the authorities. When necessary, the remedial measures should be applied as soon as possible by all available means. On the other hand, research should be promoted to understand the mechanisms of microbial response to heavy metal pollution and to enable screening for possible resistant microorganisms that could be used for both remediation and restoration of soil environment fertility.

Author details

Anna Lenart-Boroń[1] and Piotr Boroń[2]

*Address all correspondence to: annalenart82@gmail.com

1 Department of Microbiology, Faculty of Agriculture and Economics, University of Agriculture in Cracow, Cracow, Poland

2 Department of Forest Pathology, Faculty of Forestry, University of Agriculture, Cracow, Poland

References

[1] Ehrlich H.L. Geomicrobiology. 4th ed. New York: Marcel Dekker; 2002.

[2] Appenroth K.J. Definition of "heavy metals" and their role in biological systems. In: Sherameti I., Varma A. (eds.) Soil heavy Metals, Soil Biology. Berlin: Springer; 2010. P. 19: 19-29.

[3] Szyczewski P., Siepak J., Niedzielski P., Sobczyński T. Research on heavy metals in Poland. Polish Journal of Environmental Studies 2009; 18: 755-768.

[4] Han X., Gu J.D. Sorption and transformation of toxic metals by microorganisms. In: Mitchell R., Gu J.D. (eds.) Environmental Microbiology 2nd ed. New Jersey: Wiley-Blackwell; 2010. p. 153–175.

[5] Tate R.L. Soil Microbiology. 2nd ed. Hoboken: John Wiley and Sons Inc. USA; 2000.

[6] McKinney R. E. Environmental Pollution Control Microbiology. New York: Marcel Dekker, Inc.; 2004.

[7] Roselló-Mora R., Amann R. The species concept for prokaryotes. FEMS Microbiology Reviews 2001; 25: 39-67.

[8] Metting F.B. Soil Microbial Ecology. New York: Marcel Dekker Inc.; 1993.

[9] Hafeburg G., Kothe E. Microbes and metals: interactions in the environment. Journal of Basic Microbiology 2007; 47:453-467.

[10] Ranjard L., Richaume A. Quantitative and qualitative microscale distribution of bacteria in soil. Research in Microbiology 2001; 152: 707-716.

[11] Torsvik V., Øvreås L. Microbial diversity and function in soil: from genes to ecosystems. Current Opinion in Microbiology 2002; 5:240-245.

[12] Sessitsch A., Weilhalter A., Gerzabek M.H., Kirchmann H., Kandeler E. Microbial population structures in soil particle size fractions of a long-term fertilizer field experiment. Applied Environmental Microbiology 2001; 67: 4215-4224.

[13] Grayston S.J., Griffith G.S., Mawdsley J.L., Campbell C.D., Bardgett R.D. Accounting for variability in soil microbial communities of temperate upland grassland ecosystems. Soil Biology and Biochemistry 2001; 33: 533-551.

[14] Fierer N., Jackson R.B. The diversity and biogeography of soil bacterial communities. Proceedings of the National Academy of Sciences 2006; 103: 626-631.

[15] Hoorman J.J. The Role of Soil Bacteria. Fact Sheet – Agriculture and Natural Resources. The Ohio State University; 2011.

[16] Agriinfo. Soil Microorganism: Bacteria. http://agriinfo.in/default.aspx?page=topic&superid=5&topicid=147 (accessed 10 July 2013)

[17] Jenkins A. Soil fungi. In: Soil biology basics. State of New South Wales, Department of Primary Industries; 2005. http://www.dpi.nsw.gov.au/__data/assets/pdf_file/0020/41645/Soil_fungi.pdf (accessed 20 June 2013).

[18] Schlegel H.G. General microbiology. Cambridge: Cambridge University Press; 1993.

[19] Agriinfo. Soil Microorganism: Algae. http://agriinfo.in/?page=topic&superid=5&topicid=150 (accessed 10 July 2013).

[20] Barabasz W., Albińska D., Jaśkowska M., Lipiec J. Biological effects of mineral nitrogen fertilization on soil microorganisms. Polish Journal of Environmental Studies 2002; 11: 193-198.

[21] Lynch J.M. The terrestrial environment. In: Lynch J.M., Hobbie J.E., (eds.) Microorganisms in Action: Concepts and Applications in Microbial Ecology. Oxford: Blackwell; 1988. p. 67-91.

[22] Kloepper J.W., Leong J., Teintze M., Schroth M.N. Enhanced plant growth by siderophores produced by plant growth-promoting rhizobacteria. Nature 1980; 286: 885-886

[23] Rodriguez H., Fraga R. Phosphate solubilizing bacteria and their role in plant growth promotion. Biotechnology Advances 1999; 17: 319-339.

[24] Gholami A., Shahsavani S., Nezarat S. The effect of Plant Growth Promoting Rhizobacteria (PGPR) on germination, seedling growth and yield of maize. World Academy of Science, Engineering and Technology 2009; 25: 19-24.

[25] Do Vale Barreto Figueiredo M., Seldin L., de Araujo F.F., de Lima Ramos Mariano R. Plant Growth Promoting Rhizobacteria: fundamentals and applications. In: Maheshwari D.K. (ed.) Plant Growth Promoting Bacteria, Microbiology Monographs 18, Berlin Heidelberg: Springer-Verlag; 2010. p. 21-43.

[26] Glick B.R., Patten C.L., Holgin G., Penrose D.M. Biochemical and genetic mechanisms used by plant growth promoting bacteria. London: Imperial College Press; 1999.

[27] Young J.M., Kuykendall L.D., Martínez-Romero E., Kerr A., Sawada H. A revision of *Rhizobium* Frank 1889, with an emended description of the genus, and the inclusion of all species of *Agrobacterium* Conn 1942 and *Allorhizobium undicola* de Lajudie et al. 1998 as new combinations: *Rhizobium radiobacter, R. rhizogenes, R. rubi, R. undicola* and *R. vitis*. International Journal of Systematic and Evolutionary Microbiology 2001; 51:89-103.

[28] Viss W.J., Pitrak J., Humann J., Cook M., Driver J., Ream W. Crown-gall-resistant transgenic apple trees that silence Agrobacterium tumefaciens oncogenes. Molecular Breeding 2003; 12: 283-295.

[29] Toth I.K., Bell K.S., Holeva M.C., Birch P.R. Soft rot erwiniae: from genes to genomes. Molecular Plant Pathology 2003; 4: 17-30.

[30] Koike S.T., Subbarao K.V., Davis R.M., Turini T.A. Vegetable diseases caused by soilborne pathogens. Publication 8099 of the regents of the University of California, Division of Agriculture and Natural Resources, US; 2003. http://anrcatalog.ucdavis.edu/pdf/8099.pdf (accessed 21 June 2013).

[31] Williams R.E., Shaw III C.G., Wargo P.M., Sites W.H. Armillaria Root Disease. Forest Insect and Disease Leaflet 78. U.S. Department of Agriculture Forest Service; 1989. http://na.fs.fed.us/spfo/pubs/fidls/armillaria/armillaria.htm (accessed 16 July 2013).

[32] Moss M.O., Smith J.E. The applied mycology of Fusarium. Cambridge: Cambridge University Press; 1984.

[33] Vartivarian S.E., Anaissie E.J., Bodey G.P. Emerging fungal pathogens in immunocompromised patient: classification, diagnosis and management. Clinical Infection and Disease 17 (Suppl 2) 1993; S487-491.

[34] Ryan J.R. Clostridium, Peptostreptococcus, Bacteroids, and other Anaerobes. In: Ryan K. J., Ray C. G. (eds.) Sherris Medical Microbiology: An Introduction to Infectious Diseases 4th ed.. Columbus: McGraw-Hill; 2004. pp. 309-326.

[35] Maier R.M., Pepper I.L., Gerba C.P. Environmental Microbiology. Philadephia: Elsevier Inc.; 2009.

[36] Pečiulytė D., Dirginčiutė-Volodkienė V. Effect of long-term industrial pollution on soil microorganisms in deciduous forests situated along a pollution gradient next to a fertilizer factory. Ekologija 2009; 55: 67-77.

[37] McGrath S.P., Zhao F.J., Lombi E. Plant and rhizosphere process involved in phytoremediation of metal-contaminated soils. Plant and Soil 2001; 232: 207-214.

[38] Gisbert C., Ros R., de Haro A., Walker D.J., Pilar Bernal M., Serrano R., Avino J.N. A plant genetically modified that accumulates Pb is especially promising for phytoremediation. Biochemical and Biophysical Research Communications 2003; 303: 440-445.

[39] Bilos C., Colombo J.C., Skorupka C.N., Rodriguez Presa M.J.. Sources, distribution and variability of airborne trace metals in La Plata City area, Argentina. Environmental Pollution 2001, 111: 149-158.

[40] Pierzyński G.M., Sims J.T., Vance G.F. Soils and Environmental Quality. London: CRC Press; 2000.

[41] D'Amore J.J., Al-Abed S.R., Scheckel K.G., Ryan J.A. Methods for speciation of metals in soils: a review. Journal of Environmental Quality 2005; 34: 1707-1745.

[42] Kaasalainen M., Yli-Halla M. Use of sequential extraction to assess metal partitioning in soils. Environmental Pollution 2003; 126: 225-233.

[43] Indeka L., Karczun Z. Accumulation of selected heavy metals in soils along busy traffic routes. Ecology and Technology 1999; 6: 174-180.

[44] Antonkiewicz J., Macuda J. Levels of heavy metals and hydrocarbons in grounds surrounding some petrol stations in Kraków. Acta Scientiarum Polonorum, Formatio Circumiestus 2005; 4 (2): 31-36.

[45] Voegborlo R.B., Chirgawi M.B. Heavy metals accumulation in roadside soil and vegetation along major highway in Libiya. Journal of Science and Technology 2007; 27: 1-12.

[46] Arslan H., Gizir A.M. Heavy-metal content of roadside in Mesin, Turkey. Fresenius Environmental Bulletin 2006; 15: 15-20.

[47] Atafar Z., Mesdaghinia A., Nouri J., Homaee M., Yunesian M., Ahmadimoghaddam M., Mahvi A.H. Effect of fertilizer application on soil heavy metal concentration. Environmental Monitoring Assessment 2010; 160: 83-89.

[48] Basta N.T., Ryan J.A., Chaney R.L. Trace element chemistry in residual-treated soil: key concepts and metal bioavailability. Journal of Environmental Quality 2005; 34: 49-63.

[49] Alloway B.J. Soil processes and the behavior of metals., New York: Wiley; 1995.

[50] Kabata-Pendias A., Pendias H. Biogeochemistry of trace elements. Warsaw: PWN Scientific Publishing House; 1999.

[51] Du Plessis K.R., Botha A., Joubert L., bester R., Conradie W.J., Wolfaardt G.M. Response of the microbial community to copper oxychloride in acidic sandy loam soil. Journal of Applied Microbiology 2005; 98: 901-909.

[52] Wuana R.A., Okieimen F.E. Heavy metals in contaminated soils: a review of sources, chemistry, risks and best available strategies for remediation. ISRN Ecology 2011. doi:10.5402/2011/402647.

[53] De Vries W.R., Römkens P.F.A.M., Van Leeuwen T., Bronswijk J.J.B. Heavy metals. In: Haygarth P.M., Jarvis S.C. (eds.) Agriculture, hydrology and water quality. Nosworthy Way: CABI ; 2002. p. 107-132.

[54] Barcan V. Leaching of nickel and copper from soil contaminated by metallurgical dust. Environment International 2002; 28: 63-68.

[55] Colmer A.R., Hinkel M.E. The role of microorganisms in acid mine drainage: a preliminary report. Science 1947; 106: 253-256.

[56] Lone M.I., He Z., Stoffella P.J., Yang X.. Phytoremediation of heavy metal polluted soils and water: progresses and perspectives. Journal of Zhejiang University Science B (Biomedicine & Biotechnology) 2008: 9: 210-220

[57] Weinberg E.D. Roles of trace metals in transcriptional control of microbial secondary metabolism. Biology of Metals 1990; 2: 191-196.

[58] Abbas A.S., Edwards C. Effects of metals on Streptomyces coelicolor growth and actinorhodin production. Applied Environmental Microbiology 1990; 56: 675-680.

[59] Bruins M.R., Kapil S., Oehme F.W. Microbial resistance to metals in the environment. Ecotoxicology and Environmental Safety 2000; 45: 198-207.

[60] Tyler G. Heavy metal pollution and soil enzymatic activity. Plant and Soil 1974; 41: 303-311.

[61] Rajapaksha R.M.C.P. Tabor-Kapłon M.A., Bååth E. Metal toxicity affects fungal and bacterial activities in soil differently. Applied and Environmental Microbiology 2004; 70: 2966-2973

[62] Bong C.W., Malfatti f., Azam F., Obayashi Y., Suzuki S. The effect of zinc exposure on the bacteria abundance and proteolytic activity in seawater. In: Hamamura N., Suzuki S., Mendo S., Barroso C.M., Iwata H., Tanabe S. (eds.) Interdisciplinary Studies on Environmental Chemistry – Biological Responses to Contaminants. Tokyo: Terrapub; 2010. p. 57-63.

[63] Sobolev D., Begonia M.F.T. Effects of heavy metal contamination upon soil microbes: lead-induced changes in general and denitrifying microbial communities as evidenced by molecular markers. International Journal of Environmental Research and Public Health 2008; 5: 450-456.

[64] Doelman P. Resistance of soil microbial communities to heavy metals. In: Jensen V., Kioller A., Sorensen C.H. (eds.) Microbial communities in soil. London: Elsevier Applied Science Publishers; 1986. p. 369–384.

[65] Juwarkar A.A., Nair A., Dubey K.V., Singh S.K., Devotta S. Biosurfactant technology for remediation of cadmium and lead contaminated soils. Chemosphere 2007; 68: 1996-2002

[66] Lenart A., Wolny-Koładka K. The effect of heavy metal concentration and soil pH on the abundance of selected microbial groups within ArcelorMittal Poland steelworks in Cracow. Bulletin of Environmental Contamination and Toxicology 2013; 90: 85-90

[67] Wyszkowska J., Kucharski J., Borowik A., Boros E. Response of bacteria to soil contamination with heavy metals. Journal of Elementology 2008; 13: 443-453.

[68] Piotrowska-Seget Z., Cycoń M., Kozdrój J. Metal-tolerant bacteria occurring in heavily polluted soil and mine spoil. Applied Soil Ecology 2005; 28: 237-246.

[69] Sprocati A.R., Alisi C., Segre L., Tasso F., Galletti M., Cremisini C. Investigating heavy metal resistance, bioaccumulation and metabolic profile of a metallophile mi crobial consortium native to abandoned mine. Science of the Total Environment 2006; 366; 649-658.

[70] Nies D.H. Efflux-mediated heavy metal resistance in prokaryotes. FEMS Microbiology Reviews 2003; 27: 313-339.

[71] Fogarty R.V, Tobin J.M. Fungal melanins and their interactions with metals. Enzyme and Microbial Technology 1996; 19: 311-317.

[72] Gadd G.M. Fungal production of citric and oxalic acid: importance in metal speciation, physiology and biogeochemical processes. Advances in Microbial Physiology 1999; 41: 47-92.

[73] Gonzalez H., Jensen T.E. Nickel sequestering by polyphosphate bodies in Staphylococcus aureus. Microbios 1998; 93: 179-185.

[74] Martin T.A., Ruby M.V. Review of in situ remediation technologies for lead, zinc and cadmium in soil. Remediation 2004; 14: 35-53.

[75] USEPA. Recent developments for in situ treatment of metal contaminated soils. Tech. Rep. EPA-542-R-97-004, Washington DC: USEPA; 1997. http://www.clu-in.org/download/remed/metals2.pdf (accessed 15 June 2013).

[76] Evanko C.R., Dzombak D.A. Remediation of metals-contaminated soils and groundwater. Technology Evaluation Report. Pittsburgh: Ground-Water Remediation Technologies Analysis Center; 1997.

[77] Salt D.E., Smith R.D., Raskin I. Phytoremediation. Annual Reviews in Plant Physiology & Plant Molecular Biology 1998; 49: 643-668.

[78] Raskin I., Smith R.D., Salt D.E. Using plant seedlings to remove heavy metals from water. Plant Physiology 1997; 111: 552-552.

[79] Cunningham S.D., Ow D.W. Promises and prospects of phytoremediation. Plant Physiology 1996; 110: 715-719.

[80] Jing Y., He Z., Yang X. Role of soil rhizobacteria in phytoremediation of heavy metal contaminated soils. Journal of Zhejiang University Science B (Biomedicine & Biotechnology) 2007; 8: 192-207.

[81] Ebbs S.D., Lasat M.M., Brady D.J., Cornish J., Gordon R., Kochian I.V. Phytoextraction of cadmium and zinc from a contaminated soil. Journal of Environmental Quality 1997; 26: 1424-1430.

[82] Jadia C.D., Fulekar M.H. Phytoremediation of heavy metals: recent techniques. African Journal of Biotechnology 2009; 8: 921-928.

[83] Garbisu C, Alkorta I. Phytoextraction: A cost effective plant-based technology for the removal of metals from the environment. Bioresource Technology 2001; 77: 229-236.

[84] Ajaz Haja Mohideen R., Thirumalai Arasu V., Narayanan K.R., Zahir Hussain M.I. Bioremediation of heavy metal contaminated soil by the Exigobacterium and accumulation of Cd, Ni, Zn and Cu from soil environment. International Jornal of Biological Technology 2010; 1: 94-101.

[85] Smith S.E., Read D.J. Mycorrhizal symbiosis. San Diego: Academic Press Inc.; 1997.

[86] Rohwerder T., Gehrke T., Kinzler K., Sand W. Bioleaching review part A: Progress in bioleaching: fundamentals and mechanisms of bacterial metal sulfide oxidation. Applied Microbiology and Biotechnology 2003; 63: 239-248.

[87] Means J.L., Hinchee R.E. Emerging technology for bioremediation of metals. Boca Raton: Lewis Publishers; 1994.

[88] Miller H. Biostimulation as a form of bioremediation of soil pollutans. Basic Biotechnology eJournal 2010 http://ejournal.vudat.msu.edu/index.php/mmg445/article/viewArticle/MMG445.4543071/395 (accessed 10 July 2013).

[89] Gentry T.J., Rensing C., Pepper I.L. New approaches for bioaugmentation as a remediation technology. Critical Reviews in Environmental Science and Technology 2004; 34: 447-494.

[90] Volesky B. Biosorption of heavy metals. Florida: CRC Press; 1990.

[91] Das N., Vimala R., Karthika P. Biosorption of heavy metals – an overview. Indian Journal of Biotechnology 2008; 7: 159-169.

[92] Horikoshi T., Nakajima A., Sakaguchi T. Studies on the accumulation of heavy metal elements in biological systems: accumulation of uranium by microorganisms. European Journal of Applied Microbiology and Biotechnology 1981; 12: 90-96.

[93] Ridley W.P. Dizikes L.J., Wood J.M. Biomethylation of toxic elements in environment. Science 1977; 197: 329-332.

[94] Gadd G.M. Metals, minerals and microbes: geomicrobiology and bioremediation. Microbiology 2009; 156: 609-643.

[95] Gremion F., Chatzinotas A., Kaufmann K., Sigler W.V., Harms H. Impacts of heavy metal contamination and phytoremediation on a microbial community during a twelve-month microcosm experiment. FEMS Microbiology Ecology 2004; 48: 273-283.

Chapter 27

Pollution of Pb in Soils Affected by Pyrite Tailings: Influence of Soil Properties

F. Martín, M. Simón, I. García, A. Romero and V. González

Additional information is available at the end of the chapter

http://dx.doi.org/10.5772/57270

1. Introduction

Pb is considered one of the most common contaminant in the environment [1] and it is considered neither essential nor beneficial for plants or animals but highly toxic. Pb is one of the less mobile heavy metals so its compounds tend to accumulate in soils and sediments, and it is slightly affected by microbial degradation [2-4]. It has a strong affinity by the sulphur, so is frequently associated to sulphides, being the galena (PbS) the most frequent ore mineral. Pb has two stable oxidation states: Pb(II) and Pb(IV), but the most dominant species in nature is the plumbous one. Geochemically, Pb^{2+} can replace isomorphously K^+ in silicate lattices or Ca^{2+} in carbonates and apatites, due to the similar ionic radius with these both elements [2].

In crustal rocks, the average Pb content is approximately 16 mg kg^{-1} [5]. In igneous rocks trend to increase from ultrabasic (gabbro 1.9) to acid (granite 22.7), meanwhile shales and mudstones have an average Pb content of 23 mg kg^{-1} and sandstones 10 mg kg^{-1}, being limestones and dolomites the rocks with higher average Pb content (71 mg kg^{-1}). The average concentration in uncontaminated soils is estimated to be 40 mg kg^{-1}, ranging from 2 to 200 mg kg^{-1} [6, 7].

On 25 April 1998, a breach in the dam of the holding pond of tailings from a pyrite mine in Aznalcóllar (S Spain) spilled around 45 x 10^5 m^3 of acidic waters and toxic tailings containing high concentrations of Zn, Cu, Cd, As, Pb, Sb, Bi and Tl into the Agrio and Guadiamar river basins, affecting around 43 km^2 [8]. The soils were polluted in two stages [9]. First, the toxic water (liquid phase) and tailings (solid phase) covered the soils and penetrated in depth through macro and micropores; the acidic water had far lower heavy-metal content (Pb concentration ranging from 2.05 to 2.60 mg dm^{-3}) than the tailings had (Pb concentration ranging from 4352.3 to 9635.9 mg kg^{-1}). In the second stage, the oxidation of the pyrite tailings causes sulphides to oxidize to sulphates, a marked fall in pH and a solubilization of heavy metals [10]. The lowest pH values were reached 25 days after the spill, and the total Pb concentration at 88 days doubled or tripled (depending the soil type) the initial value of contamination in the upper 10 cm of the soils [11].

The present study provides an evaluation of the contamination processes occurring just after the removal of the tailings and part of the upper soil, to assess the behaviour of the residual Pb concentrations in the affected area at this early stage. The aim of the present work is to analyse the mobility and availability of Pb in the initial steps of the contamination process, to obtain useful information in the planning of different strategies for future remediation activities in other affected areas based on a detailed study of the soil properties.

2. Material and methods

After the removal of the tailings, the basin was gridded into 91 sectors of 400 x 400 m, and in the centre of each sector a plot 10 x 10 m was laid out. In every plot, soils samples were taken at the four corners and centre of this plot, at depths of 0-10, 10-30 and 30-50 cm. For each depth, 250 g of soil from the five sampling points per plot were mixed and homogenized, providing three samples per plot, for a total of 273 soil samples for the entire basin. All the samples were air dried and screened to 2 mm, and the percentages of gravels (>2mm) and fine earth (<2mm) were determined. The laboratory analyses were made with the fine-earth fraction.

Particle-size distribution was measured by the pipette method after eliminating organic matter with H_2O_2 and dispersion with sodium hexametaphosphate [12]. The pH was measured potentiometrically in a 1:2.5 soil:water suspension, as recommended by the International Society of Soil Science, in a CRISON Digit 501 instrument. The $CaCO_3$ equivalent was determined manometrically by the Barahona method [13]. Total carbon and total sulphur were measured by dry combustion with a LECO mod. SC-144DR instrument. Organic carbon was calculated as the difference between total carbon and inorganic carbon from $CaCO_3$. The cation-exchange capacity (CEC) was determined with 1N Na-acetate at pH 8.2 [14], measuring the sodium in a METEOR NAK-II flame-photometer. The total concentration of iron (Fe_t) was measured by X-ray fluorescence in a Philips PW-1404 instrument, from a disc of soil and lithium tetraborate in a ratio of 0.6:5.5. Amorphous and poorly crystallized iron (Fe_o) and aluminium (Al_o) oxides were extracted with oxalic-oxalate [15] and measured by atomic-absorption spectroscopy.

Samples of air-dried soils, ground to < 0.05 mm, were digested in open flasks by strong acids (HNO_3:HF in a ratio 2:3) at 160 ºC till dry, and 100 ml HNO_3 at 4% were added. In the digested samples, total Pb (Pb_t) was measured by ICP-MS with a PE SCIEX-ELAN 5000A spectrometer. A multi-element calibration standard 4 (Perkin-Elmer) was used with Rh as the internal standard. The detection limit for Pb was 0.001 μg L^{-1}, and the accuracy of the method was corroborated by analyses (six replicates) of a standard reference material: SRM 2711 (soil with moderately elevated trace-element concentrations [16]). For Pb, the mean certified value was

1162.0 mg kg^{-1} with a standard deviation of 31.0; the mean experimental value was 1138.1 mg kg^{-1} with a standard deviation of 11.0.

Soluble Pb (Pbl_w) was measured in the extract from a saturated soil paste. Pb extracted by calcium chloride 0.01 M (Pb_c) using the method [17], Pb extractable by acetic acid 0.43 M (Pb_a) following [18], and Pb extractable by oxalic-oxalate pH 3 (Pb_o) as in [15]. For the EDTA 0.05 M extractable fraction (Pb_e) method [19] were followed. The Pb of all the extracts was measured by ICP-MS.

For the statistical analysis, the SPSS 15.0 program was used. The areal distribution of the contamination was established by the algorithm of the least Euclidean, using the software Surfer 7.0.

The climate of this area is typically Mediterranean (hot, dry summers; cold, wet winters; temperate autumns and springs with variable rainfall). The mean annual rainfall of 613 mm, the mean temperature of 17.7 °C, and the potential evapotranspiration of 900 mm (taken from 7 weather stations near the spill, over 27 years). The affected soils were Typic Xerofluvents and Typic Xerothents [20] developed on alluvial deposits coming from the erosion of sedimentary and metamorphic materials of the Agrio and Guadiamar River basin, mainly gravels, sands, silts and clays from the Miocene and Plio-Quaternary.

3. Results and discussion

We assigned five different soil types, according to the soil properties between 0 and 50 cm in depth, with significant differences ($p < 0.05$) by a cluster analysis via the k-mean method (Table 1). Soil types 1 and 2 (hereafter Group A), located in the upper part of the basin, were acidic, with little or no carbonate; type 1 had a loam texture while type 2 was dominated by gravel and sand. Soil types 3, 4 and 5 (hereafter Group B) were predominantly neutral or slightly alkaline and carbonated, the main differences between them being texture (type 3, clay loam; type 4, loam; and type 5, silty clay).

Soil type	n	Gravel (%)	Sand (%)	Silt (%)	Clay (%)	pH	$CaCO_3$ (%)	OC (%)	CEC ($cmol_c\ kg^{-1}$)
1	36	5.2 (3.5)	50.7 (18.4)	30.8 (14.1)	18.5 (6.9)	5.8 (1.9)	1.7 (3.5)	1.0 (0.4)	15.5 (5.8)
2	33	37.2 (24.5)	76.2 (17.1)	13.6 (9.3)	10.2 (8.6)	5.9 (1.8)	0.8 (1.7)	0.4 (0.3)	7.7 (3.7)
3	63	6.9 (9.4)	19.6 (11.2)	49.4 (9.7)	31.0 (11.2)	7.7 (0.5)	10.0 (5.0)	1.1 (0.5)	18.1 (5.2)
4	87	4.8 (8.5)	48.8 (12.6)	32.5 (9.4)	18.7 (6.0)	7.8 (0.7)	6.9 (6.3)	0.9 (0.4)	12.1 (4.2)
5	54	8.2 (9.6)	4.9 (7.1)	47.7 (11.2)	47.4 (14.0)	8.0 (0.3)	13.6 (4.9)	1.3 (0.5)	22.9 (4.9)

Table 1. Mean and (standard deviation) of the particle size, pH, $CaCO_3$, organic-carbon (OC) content, and cation exchange capacity (CEC) in the different soil types (n = number of samples in each soil type).

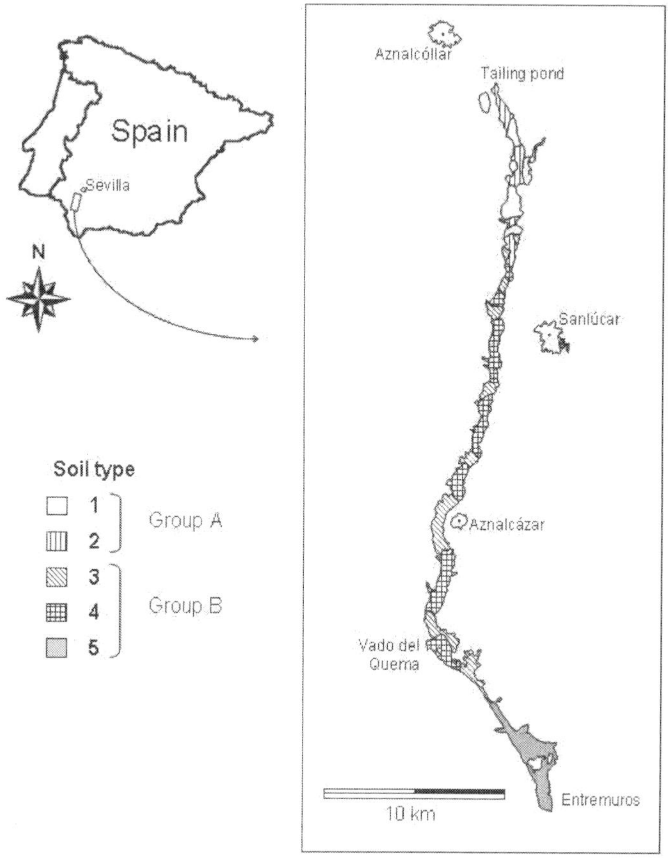

Figure 1. Location of the study area, including the distribution of the different soil types analysed in Table 1. Group A include non-carbonate soils and Group B consist of carbonate soils.

As mentioned above, these soils were quite homogeneous in depth, so the variation in the soil properties with the depth were not significant in the cases of texture, calcium-carbonate content, organic-carbon content and cation-exchange capacity. Nevertheless, total sulphur content (S_t), related to the contamination, registered its highest values in the upper 10 cm in all soil types, with values declining progressively and significantly ($p < 0.001$) in depth (Fig. 1). The difference in S_t between the first 10 cm and the interval 10-30 cm was greater in Group A soils than in Group B. In addition, the contamination also affected the pH increasing the acidity towards the soil surface, and with significant differences in depth. In the Group B soils, where differences between depths were less pronounced for the presence of $CaCO_3$, the pH remained higher, rarely falling below 6.0 (Fig. 1).

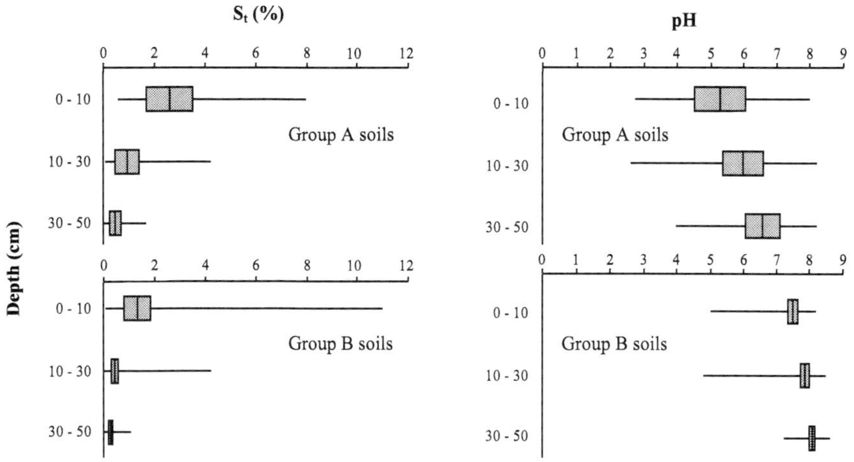

Figure 2. Total sulphur content (S_t) and pH in the two soil groups (horizontal lines range from minimum to maximum values; boxes show the confidence interval at 95% for the mean; vertical lines mark the mean values).

Total Pb content

In uncontaminated soils next to the affected area the concentrations of the total Pb (Pb_t) were between 22.0 and 74.6 mg kg^{-1}, with a mean value of 41.8 mg kg^{-1} [9]. Meanwhile, in contaminated soils, the mean Pb content in the uppermost 10 cm was 385.8 mg kg^{-1}, ranging between 35.8 and 3231.0 mg kg^{-1}. This mean value significantly declined with depth in all soil types (Fig. 2), with a distribution similar to that of S_t (Fig.1). Consequently, in all soils, a highly significant linear relationship ($p < 0.001$) was found between Pb_t and S_t (Table 2).

		Pb_t	
variable	value	std. Error	P value
	r^2 = 0.832		< 0.001
Intercept	27.58	10.95	0.012
S_t	22.45	0.65	< 0.001

Table 2. Multivariable regression model of the Pb_t (mg kg^{-1}) and S_t (g kg^{-1}) in all soils.

In addition, the Pb_t concentration in the 10-30 cm layer did not significantly differ from that of the 30-50 cm layer, and had values higher than unaffected soils (doubled or tripled the mean value for uncontaminated soils), which could be associated with a previous contamination in the area [21].

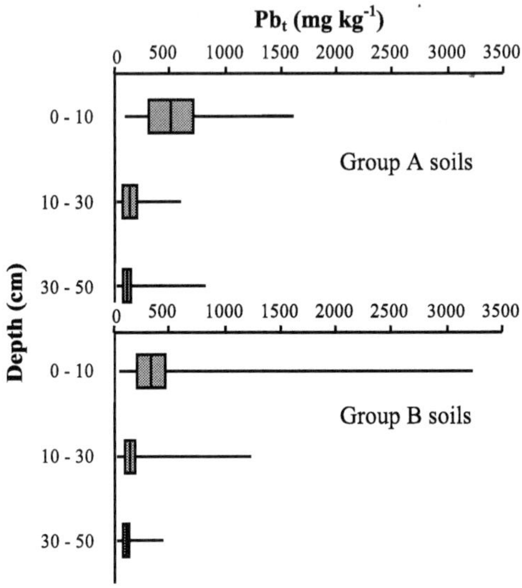

Figure 3. Total Pb content (Pb_t) in the two soil (horizontal lines range from minimum to maximum values; boxes show the confidence interval at 95% for the mean; vertical lines mark the mean values).

In any case, the pH range as well as the S_t and Pb_t values in the upper 10 cm of the soils were very broad, indicating a highly heterogeneous contamination, as previously reported in [9]. To quantify the degree of Pb_t pollution, we used the intervention levels proposed by the Andalusian Regional Government (350 and 500 mg kg^{-1} for agricultural soils with pH < 7 and > 7, respectively; and 1000 mg kg^{-1} for natural parks). The soil use in the area was mainly for agriculture and grazing and, according to these levels, 15.0% of the area proved to be uncontaminated, with values below the reference level for the area (< 75 mg kg^{-1}), 23.7% was contaminated for agricultural use, and 10.6% was seriously contaminated (> 1000 mg kg^{-1}). The remaining 50.7% had anomalous values (above the reference level and below the intervention levels).

Pb extractions

To assess the speciation of Pb in the polluted soils, selective extractions using reagents with different strengths were used, although most of them are not specific to a given soil phase [22], they provide useful information in relation to mobility and availability of the pollutant element. The average concentration and standard deviation of the different extracted forms of Pb are shown in Table 3.

In general, water extracted forms constitute the soluble fraction [23, 24]; forms extracted with $CaCl_2$ is considered the exchangeable fraction [25, 26]; forms extracted with acetic acid is

Soil type	Depth (cm)	Pb_e		Pb_a		Pb_o		Pb_c		Pb_w	
		Mean	SD	Mean	SD	Mean	SD	Mean	SD	Mean	SD
1	0-10	149.29	155.50	41.16	49.52	17.92	28.14	4.00	5.94	0.32	0.24
	10-30	29.74	54.56	21.30	48.65	7.21	8.83	0.86	2.72	0.30	0.57
	30-50	13.39	17.94	7.80	10.59	4.12	5.47	0.70	1.92	0.40	0.63
2	0-10	117.96	137.29	43.48	63.77	5.61	12.60	7.74	15.42	0.39	0.35
	10-30	25.61	22.38	9.21	9.89	4.35	6.12	1.31	1.52	0.29	0.33
	30-50	19.58	28.56	18.79	32.71	4.61	4.13	1.47	3.69	0.21	0.20
3	0-10	83.00	68.52	31.59	28.93	5.36	23.32	0.49	0.49	0.58	0.98
	10-30	31.79	22.31	9.88	6.66	0.79	1.63	0.20	0.36	0.45	0.78
	30-50	29.53	20.67	8.66	9.79	1.49	5.36	0.73	1.27	0.69	1.49
4	0-10	55.21	68.16	20.29	38.82	2.11	3.23	1.61	5.46	0.74	1.32
	10-30	37.39	69.96	8.87	14.87	4.55	9.14	1.18	3.46	0.65	1.45
	30-50	26.55	27.98	7.11	9.91	4.01	8.41	0.48	0.60	1.10	2.05
5	0-10	37.86	35.56	11.56	13.09	5.65	13.49	0.80	0.76	0.17	0.50
	10-30	20.58	13.32	4.61	2.35	3.41	5.80	0.89	0.99	0.11	0.11
	30-50	15.94	14.61	4.20	3.21	2.21	4.47	0.83	0.80	0.14	0.20

Table 3. Mean content (mg kg^{-1}) and standard deviation (SD) in the five soil types for Pb extracted with EDTA (Pb_e), acetic acid (Pb_a), oxalic-oxalate (Pb_o), calcium chloride (Pb_c) and water (Pb_w).

used as the bio-available fraction associated with carbonates and also specifically adsorbed by oxides as well as organic matter [25, 27-29]; forms extracted with EDTA is also considered to be the bio-available fraction associated with carbonates, inorganic precipitates, amorphous oxides and organic matter [24, 27-30]; and the forms extracted with oxalic oxalate is held on, or occluded in, amorphous or poorly crystallized iron, aluminium or manganese oxides [22, 31].

The reagent that extracted the highest concentrations of Pb was EDTA (around 20 - 25 % of the total Pb), indicating that a very high proportion of the Pb forms inorganic precipitates easily complexed by this reagent, but not dissolved by acetic acid or oxalic-oxalate. When all soils were pooled, the Pb extracted with EDTA (Pb_e) was related to total Pb (Pb_t), sulphates (SO_4^{2-}), and amorphous iron oxides (Fe_o) with highly significant ($p < 0.001$) relationship (Table 4). Therefore, apart from the relation with the total contamination (Pb_t), we can suppose that the main forms in which Pb is retained in the soils could be iron oxy-hydroxysulphates. The presence of iron compounds promote the sorption of Pb in the soil [32]; and these results are consistent with the obtained by [33] in alluvium contaminated by metal in an adjacent mining area (Tinto river basin). These authors pointed that the highest amounts of Pb of all contaminant metal-bearing minerals in the area were characterised by Fe oxy-hydroxysulphate minerals.

		Pb_e		
	value	std. Error	P value	VIF[a]
All soils	**$r^2 = 0.726$**		< 0.001	
Pb_t	0.121	0.008	< 0.001	1.689
SO_4^{2-}	0.014	0.002	< 0.001	1.552
Fe_o	0.003	0.001	0.019	1.736
Group A soils		Pb_e		
	$r^2 = 0.867$		< 0.001	
Pb_t	0.207	0.017	< 0.001	1.966
SO_4^{2-}	0.008	0.003	0.004	1.966
Group B soils		Pb_e		
	$r^2 = 0.651$		< 0.001	
Pb_t	0.100	0.008	< 0.001	1.416
Fe_o	0.006	0.001	< 0.001	1.416

[a] The variance inflation factor (VIF) is the reciprocal of tolerance. Values next to 1 indicate the independent variables have no redundant information.

Table 4. Multivariable Regression Model of the Pb_e (mg kg^{-1}), Pb_t (mg kg^{-1}), SO_4^{2-} (mg kg^{-1}) and Fe_o (g kg^{-1}) in the different soil groups.

Nevertheless, when the soils were separated into acidic (Group A) and neutral alkaline (Group B), the former correlated significantly Pb_e with Pb_t and SO_4^2, meanwhile the latter correlated Pb_e with Pb_t and Fe_o (Table 4), indicating that the soil properties of each group are essential in the forms of Pb retention. In the case of acid soils, the relationship between Pb_e and SO_4^{2-} is significant, but the one with the Fe_o disappear. In these soils, the pH is maintained in acidic values, so the iron released from the pyrite oxidation (Fe^{2+}) is mobile, being able to rise and accumulate in the soil surface by the effect of the semi-arid climate of the area. This process occurs under Mediterranean climatic conditions, where the precipitation are lower than the evapotranspiration and the predominant upward movement of the soil solutions generate the formation of hard surface crusts, of scarce mm thickness and high iron concentration, observed by us in the acid soils of the affected area [34]. According to this behaviour, the Pb retention in acid soils seems to be related mainly to the precipitation as Pb sulphate ($PbSO_4$); this retention has been already reported as the main process in the Pb immobilization from acid mine waters [35].

In the case of basic soils, the relationship between Pb_e and Fe_o is statistically significant, but the one with the SO_4^{2-} disappears. In these soils, the presence of calcium carbonate produce the releasing of Ca^{2+} by the acidification caused by the oxidation of sulphides, and the precipitation of gypsum ($CaSO_4$ $2H_2O$) is dominant. These reactions could be responsible in the disappearance of the statistical relation between Pb_e and sulphates, but solubility of gypsum allows to maintain certain amounts of soluble sulphates (1.51 10^{-2} M, at 25 ºC), so the formation of other types of sulphates is possible, being the presence of jarosite $[KFe_3(SO_4)_2(OH)_6]$ common in this type of contamination [36]. In addition, the basic values of the pH encourage the precipitation of iron oxy-hydroxydes, but the Pb retention by poorly crystallized iron oxides in our basic soils, seems to be less important than the one produced by iron oxy-hydroxysulphates [37]. In fact, the concentration of Pb extracted by oxalic-oxalate (Pb_o) is very low (between 1 and 3% of the total Pb). According to these results, we suppose that the main Pb-bearing mineral in basic soils should be plumbojarosite $[PbFe_6(SO_4)_4(OH)_{12}]$. The formation of this mineral has also been described in others sulphide mine contamination of the area [33, 38].

The Pb extracted with calcium chloride (Pb_c) in Group A soils was significantly related only to the percentage of clay fraction (Table 5). This implies that, in acid soils, a little fraction of Pb is retained by the clay fraction as easily exchangeable forms. On the contrary, in Group B soils, the Pb_c was nor significantly related to any of the properties studied, but only to the Pb soluble in water (Pb_w). In addition, if we compare the amounts of Pb_c to the amounts of Pb_w in basic soils, we note that there are no statistical differences between the two forms.

		Pb_c	
Group A soils	value	std. Error	P value
	r^2 = 0.836		0.011
Intercept	0.807	0.066	0.029
clay	0.058	0.013	0.011
		Pb_w	
Group B soils			
	r^2 = 0.871		< 0.001
Intercept	1.304	0.122	< 0.001
clay	- 0.024	0.004	< 0.001

Table 5. Multivariable Regression Model of the Pb_c (mg kg^{-1}) and clay (%) in acid soils (Group A); and Pb_w (mg kg^{-1}) and clay (%) basic soils (Group B).

Soluble in water Pb (Pb_w) was not significantly related to any of the soil properties in Group A soils, while in Group B soils, Pb_w proved to be negatively and significantly related to the clay fraction (Table 5), implying the retention of the soluble Pb forms by this soil fraction. This decrease in availability of Pb in relation to clay content is in agreement with the described by [39], although the type of clay should be studied in more detail [40]; in this way, high illite content has been described as an efficient mineral in the adsorption of Pb in soils [41]. Finally, we found that the concentrations of Pb_w was generally higher in basic soils in relation to the acid ones (Fig. 3), being this difference between groups statistically significant ($p < 0.05$). This finding contradicts the usual increase in Pb solubility in acidic conditions [42, 43]. Therefore, this variation could be related to the apparent solubility and stability of the iron oxy-hydroxides precipitated at different values of pH [44, 45], but more detailed studies are carrying out to verify this point.

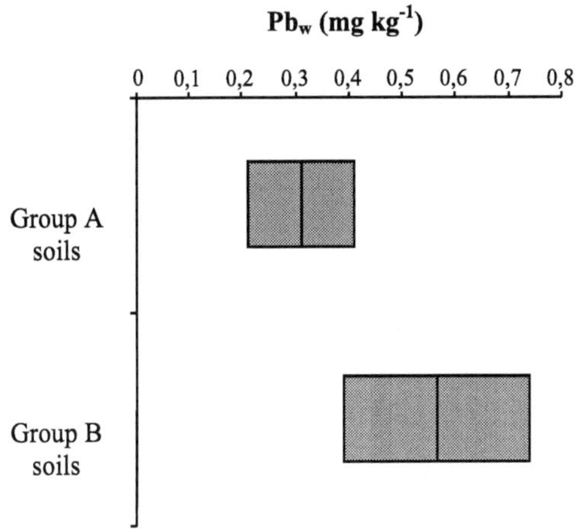

Figure 4. Soluble Pb content (Pb_w) in the two Group of soils (A: acid, B: basic). (boxes show the confidence interval at 95% for the mean; vertical lines mark the mean values).

4. Conclusions

The mobility and solubility of Pb in the contaminated soils of this study is mainly controlled by the soil properties and ageing (oxidation) of the pyrite tailings. Regardless of the soil type, the Pb accumulates mainly in the top of the soils without contaminating the subsoil or groundwater. The climatic conditions (semi-arid Mediterranean climate) produce the upward movement of the soil solutions, which cause the formation of surface crusts but only in acidic soils (where the iron and contaminants remain soluble in higher proportion). The main Pb-bearing mineral in acid soils should be Pb-sulphates, and in basic ones should be iron hydroxysulphates (mainly plumbojarosite). The amounts of Pb retained by iron oxyhydroxides are very little in all soil types. The soluble Pb is significantly higher in basic soils than in acid ones, which contradict the normal behaviour of this contaminant in soils. According to the obtained results, more detailed studies in relation to the solubility and stability of the different phases precipitated at the early stages of the contamination are required, to improve the applications in the future of adequate remediation measurements in similar affected areas.

Author details

F. Martín[1], M. Simón[2], I. García[2], A. Romero[1] and V. González[2]

1 Departamento de Edafología y Química Agrícola, Facultad de Ciencias, Universidad de Granada, Granada, Spain

2 Departamento de Agronomía, Área de Edafología y Química Agrícola, Universidad de Almería, Almería, Spain

References

[1] Adriano DC. Trace elements in terrestrial environments. Biogeochemistry, bioavailability and risk of metals. Second Edition. New York: Springer-Verlag; 2001.

[2] Davies BE. Lead, In: Alloway BJ. (ed.), Heavy metals in soils (2nd edition). Glasgow: Blackie Academic and Profesional; 1995. p206-223.

[3] Kabata-Pendias A., Pendias, H. Trace elements in soils and plants (3rd edition). Boca Raton: CRC Press; 2001.

[4] Pais I., Benton J. (Jr). The handbook of trace elements. Florida: St. Lucie Press. CRC press LLC; 1997.

[5] Nriagu JO. The biogeochemistry of lead. Amsterdam: Elsevier Biomedical Press; 1978.

[6] Davies BE. A graphical estimation of the normal lead content of some British soils. Geoderma 1983;29 67-75.

[7] Temmerman LO., Hoening M., Scokart PO. Determination of "normal" levels and upper limit values of trace elements in soils. Z. Pflanzen, Bodenk 1984;147 687-694.

[8] Grimalt JO., Ferrer M., Macpherson, E. The mine tailing accident in Aznalcóllar. The Science of the Total Environment 1999;242(1-3) 3-11.

[9] Simón M., Ortiz I., García I., Fernández E., Fernández J., Dorronsoro C., Aguilar J. Pollution of soils by the toxic spill of a pyrite mine (Aznalcóllar, Spain). The Science of the Total Environment 1999;242(1-3) 105-115.

[10] Aguilar J., Dorronsoro C., Fernández E., Fernández J., García I., Martín F., Simón M. Remediation of Pb-contaminated soils in the Guadiamar river basin (SW Spain). Water, Air, and Soil Pollution 2004;151 323–333.

[11] Simón M., Martín F., Ortiz I., García I., Fernández J., Fernández E., Dorronsoro C., Aguilar J. Soil pollution by oxidation of tailings from toxic spill of a pyrite mine. The Science of the Total Environment 2001;279(1-3) 63-74.

[12] Loveland PJ., Whalley WR. Particle size analysis. In: Smith KA., Mullis ChE. (eds.) Soil analysis: physical methods. New York: Marcel Dekker; 1991. p271-328.

[13] Barahona E., Iriarte A. Determinaciones analíticas en suelos. Normalización de métodos IV. Determinación de carbonatos totales y caliza activa In: Grupo de trabajo de normalización de métodos analíticos. Actas del I Congreso de la Ciencia del Suelo, Madrid; 1984. p53-69.

[14] Rhoades JD. Cation exchange capacity. In: Page L., Miller RH., Keeney DR (eds.) Methods of soil analysis. Part 2. Agron. Monogr. 9A., Madison: Am. Soc. Agron.; 1982. p149-157.

[15] Schwertmann U., Taylor RM. Iron oxides. In: Dixon JB., Webb SB. (eds.) Mineral in Environments. Madison: Soil Sci. Soc. Am.; 1977. p148-180.

[16] Gills TE., Kane JS. Certificate of analysis. Standard reference material 2711. Gaithersburg: National Institute of Standards and Technology; 1993.

[17] Novozamsky I., Lexmond TH., Houba VJ. A single extraction procedure of soil for evaluation of uptake of some heavy metals by plants. Int Environ Anal Chem. 1993;51 47-58.

[18] Ure AM., Quevanviller Ph., Muntau H., Griepink B. Especiation of heavy metal in soils and sediments. An account of the improvement and harmonisation of extration techniques undertaken under the auspices of the BCR of the Commision of the European Communities. Int. J. Environ. Anal. Chem. 1993;51 135-151.

[19] Quevanviller Ph., Lachica M., Barahona E., Gómez A., Rauret G., Ure A., Muntan H. Certified reference material for the quality control of EDTA and DTPA extractable trace metal contents in calcareous soils (CRM 6000) Presenius J. Anal. Chem. 1998;360 505-511.

[20] Soil Survey Staff. Soil Taxonomy. Handbook 436. 2nd edition. Washington: USDA. Natural Resources Conservation Service;1999.

[21] Cabrera F., Soldevilla M., Cordón R., Arambarri P. Heavy metal pollution in the Guadiamar river and the Guadalquivir estuary (SW Spain). Chemosphere 1987;16 463-468.

[22] Ure AM. Methods of analysis for heavy metals in soils. In: Alloway BJ. (ed.) Heavy metals in soils (2nd edition) Glasgow: Blackie Academic and Professional; 1995. p58-102.

[23] Viets FG. Chemistry and availability of micro-nutrients in soils. J. Agric. Food Chem. 1962;10 174-178.

[24] Sposito G., Lund LJ., Chang AC. Trace metal chemistry in arid-zone field soils amended with sewage sludge. I: Fractionation of Ni, Cu, Zn, Cd and Pb in solid phases. Soil Sci. Soc. Am. J. 1982;46 260-264.

[25] McLaren RG., Crawford DV. Studies on soil copper. I. The fractionation of Cu in soils. Journal of Soil Science 1973;24 172-181.

[26] Iyengar SS., Martens DC., Miller WP. Distribution and plant availability of soil zinc fractions. Soil Sci. Sco. Am. J.1981;45 735-739.

[27] Louma SN., Jenne EA. Estimating bio-availability of sediments-bound trace metals with chemical extractants. Trace Subst. Environ. Ealth. 1976;10 343-351.

[28] Stover RC., Sommers E., Silviera DJ. Evaluation of metals in waste-waters sludge. Journal Water Poll. Control Fed. 1976;48 2165-2175.

[29] Beckett PHT. The use of extractants in studies on trace metals in soils, sewage sludge, and sludge-treated soils. In: Stewart BA. (ed.) Advances in Soil Science. vol.9, New York: Springer-Verlag; 1989. p143-176.

[30] Rendell PA., Batley GE., Cameron AJ. Adsorption as a control of metal concentrations in sediment extracts. Environm. Sci. Technol. 1980;14 314-318.

[31] Chao TT., Zhou L. Extraction techniques for selectives dissolution of amorphous iron oxides from soil and sediments. Soil Sci. Soc. Am. Proc. 1983;36 764-768.

[32] Sipos P., Németh T., Mohai I., Dódony I. Effect of soil composition on adsorption of lead as reflected by a study on a natural forest soil profile. Geoderma 2005;124 363–374.

[33] Hudson-Edwards KA., Macklin MG., Taylor MP. 2000 years of sediment- borne heavy metal storage in the Yorkshire Ouse basin, NE England, U.K. Hydrological Processes 1999;13 1087-1102.

[34] Martín FJ. Pollution of soils by the spill of a pyrite mine (Aznalcóllar, Spain) PhD Thesis, University of Granada, Spain 2001.

[35] Monterroso MC., Macías F. Procesos de inmovilización de elementos traza en aguas ácidas de mina. Edafología 1998;5 59-70.

[36] Nordstrom DK., Alpers CN. (1999). Geochemistry of acid mine waters. In: Plumlee GS., Logsdon MJ. (eds) The Environmental Geochemistry of Mineral Deposits. Part A. Processes, methods and health issues. Rev Econ Geol. 1999; 6A 133-160.

[37] Simón M., Diez M., González V., García I., Martín F., de Haro S. Use of liming in the remediation of soils polluted by sulphide oxidation: A leaching-column study. Journal of Hazardous Materials 2010;180 241–246.

[38] Domènech C., de Pablo J., Ayora C. Oxidative dissolution of pyritic sludge from the Aznalcóllar mine (SW Spain). Chemical Geology 2002;190 339-353.

[39] Bur T., Crouau Y., Bianco A., Gandois L., Probst A. Toxicity of Pb and of Pb/Cd combination on the springtail Folsomia candida in natural soils: Reproduction, growth

and bioaccumulation as indicators. Science of the Total Environment 2012;414 187–197.

[40] Appel, C., Ma, L. Concentration, pH, and surface charge effects on cadmium and lead sorption in three tropical soils. J. Environ. Qual. 2002;31 581-589.

[41] Yang JY; Yang XE., He ZL., Li TQ., Shentu JL., Stoffella PJ. Effects of pH, organic acids, and inorganic ions on lead desorption from soils. Environmental Pollution 2006;143 9-15.

[42] Ming H., He WX, Lamb DT., Megharaj M., Naidu R. Bioavailability of lead in contaminated soil depends on the nature of bioreceptor Ecotoxicology and Environmental Safety 2012;78 344–350.

[43] Cheyns K., Peeters S., Delcourt D., Smolders E. Lead phytotoxicity in soils and nutrient solutions is related to lead induced phosphorus deficiency. Environmental Pollution 2012;164 242-247.

[44] Bigham J M., Schwertmann U., Traina SJ., Winland RL., Woolf M. Schwertmannite and the chemical modelling of iron in acid sulfate waters. Geochim. Cosmochim. Acta 1996;60 2111–2121.

[45] Yu J-Y., Heo B., Cho J-P., Chang H-W. Apparent solubilities of schwertmannite and ferrihydrite in natural stream waters polluted by mine drainage. Geochim. Cosmochim. Acta 1999;63 3407–3416.

Chapter 28

Copper Accumulation in Vineyard Soils: Distribution, Fractionation and Bioavailability Assessment

Marija Romić, Lana Matijević, Helena Bakić and Davor Romić

Additional information is available at the end of the chapter

http://dx.doi.org/10.5772/57266

1. Introduction

Metals are chemically very reactive in the environment, which results in their mobility and bioavailability to living organisms. Metals can be present in all environmental compartments as different species, with the TMs associated with different ligands, but never being irreversibly transformed or metabolized, and in those meaning metals are different from organic compounds. People can be exposed to high levels of toxic metal by breathing air, drinking water, or eating food that contains it. As a consequence, metals get into the human body by different routes - by inhaling, over skin, and ingestion of contaminated food. The issue of toxicity is usually merely a matter of quantity, with the range varying for each element.

1.1. Why we need to study trace metals in soils?

Soil is an important compartment of the environment in which anthropogenic loading of trace metals puts ecosystems and their inhabitants at a health risk. Repeated use of metal-enriched chemicals, fertilizers, and organic amendments such as sewage sludge as well as wastewater may cause contamination at a large scale. So far, it is believed that most soils in Europe have not been significantly enriched in trace metals by anthropogenic activity. This is changing as livestock production expands, fertilizer application increases, and biosolids and effluent applications to agricultural soils become more common. Accumulation of trace metals in soil has potential to restrain the soil functions, cause toxicity to plants, and enter the food chain.

Many chemical reactions are responsible for the behaviour of TMs in soils, but the most important processes that control their bioavailability and mobility are precipitation-dissolution, adsorption-desorption, and complexation. The ability of soils to adsorb metal ions from aqueous solution is of special interest and has consequences for both agricultural issues such as soil fertility and environmental questions such as remediation of polluted soils and waste deposition [1].

Metal-soil interaction is such that when metals are introduced at the soil surface, their mobilisation does not occur to any great extent unless the metal retention capacity of the soil is overloaded, or metal interaction with the associated waste matrix enhances mobility [2]. The most important interfaces involved in TMs transformation in soils are mineral groups commonly found in soil: aluminosilicates, oxides and organic matter. Through their surface electrochemical properties, these soil minerals control adsorption, transformation, and release behaviour of chemical constituents (e.g. nutrients and contaminants) to water and soil solution [3]. Furthermore, soil-surface electrochemical properties vary between soil types and depend on factors such as parent material, climate, and vegetation.

So that, full understanding and prediction of chemical behaviour of an element in the terrestrial environment is possible only by identification of all forms in which that element can be found in soil under different environmental conditions. Copper is one of the major toxic metals, and a highly reactive one, as well. Elevated levels of Cu in agricultural soils result from the use of Cu-containing compounds to control plant diseases and from application of manure or sewage sludge. These applications may lead to gradual accumulation of Cu in the soil and thereby increase Cu toxicity toward crop and beneficial microorganisms. In this article, the actual risk of high concentrations of copper and its mobility in vineyard soils is reviewed considering sources, chemical processes in soil and biogeochemical behaviour of copper as well as impact on agroecosystem and environment in general.

1.2. Sources and behaviour of copper in soils

Copper occurs in the Earth's crust at concentrations between 25-75 mg kg^{-1}, with the abundance pattern that shows the tendency for the concentration in mafic igneous rocks (60-120 mg kg^{-1}) and argillaceous sediments (40-60 mg kg^{-1}), but it is rather excluded from the carbonate rocks (2-10 mg kg^{-1}) [4]. Values for soil contents generally range worldwide from 1 to 140 mg kg^{-1} depending on the nature of the soil parent material.

In soil solids and solution copper occurs almost exclusively as the divalent cation Cu^{2+}, and the reduction of Cu^{2+} to Cu^{+} and Cu^{o} is possible under reducing conditions. As a chacophile, copper associates with with sulfide in the very soluble minerals, Cu_2S and CuS. Being very reactive in soil, copper is found in all matrix components. Most of the colloidal soil material (clay minerals, oxides of Mn, Al i Fe, and organic matter) adsorb copper strongly, and increasingly so as the pH is raised [5]. For copper, specific adsorption, which is not significant for the most of metal ions, seems to play a more important then nonspecific adsorption. Amorphous and cristalline oxides of Fe and Al easy adsorb Cu^{2+}, regardless the excess of alkali metals in the solution [6]. However, the most important sink for Cu is soil organic matter, and its complexation with organic matter is one of the most efficient mechanisms of Cu^{2+}

retention in soil [7]. This restricts Cu bioavailability, but also considerably reduces the risks of phytotoxicity of the accumulated anthropogenic input and its vertical migration. Organically complexed Cu^{2+} is bound more tightly than any other divalent transition metal and of low lability these complexes results in limiting copper bioavailability. This prevents copper mobility and its transport through soil to underground, and reduces substantially the risk of the groundwater contamination.

1.3. Anthropogenic inputs and copper contamination of cultivated soils

Trace elements in general enter an agroecosystem through both natural and anthropogenic processes. The latest includes TEs inputs through use of agrochemicals, farm manure, biosolids and composts, industrial and municipal waste, irrigation, and wet and/or dry deposits. Being widely used, copper is a common metal pollutant released to environment as a result of man's activities. Copper is an essential nutrient, but in excess in soils it becomes toxic to plants and some micro-organisms, disrupting nutrient-cycling and inhibiting the mineralisation of essential nutrients such as nitrogen and phosphorus. Some species accumulate copper. Toxic effects on fish and other aquatic organisms have also been observed. For humans, excess amount of this trace metal can have serious health effects.

Elevated levels of copper in agricultural soils result from the use of Cu-containing compounds to control plant diseases and from application of manure or sewage sludge. Increased concentration of Cu in soils under long-term production of grapevine, citrus and other fruit crops have been recorded in numerous studies. The Bordeaux mixture, an efficient agent for prevention of vine «Downy Mildew», *Plasmopara viticola,* has been routinely used in Europe since the end of the 19th century with its concentrations and the number of treatments depending on weather conditions, infection intensity and vineyard management. The century-old practice of using Cu-sulphates and other copper containing fungicides to protect grapevine, but also other agricultural crops, in temperate and tropic climatic regions, resulted in significant Cu accumulation in soils [8]. Most of the copper accumulated in leaves and soil by spraying will be retained in topsoil through the biological cycle and tillage [9-11]. Comparison of copper contents of 110 to 1500 mg kg^{-1} with its usual content in agricultural soils (20 – 30 mg kg^{-1}) points to their connection with such practice [12]. Copper can be either a micro nutrient or a toxic element which depends on the copper concentration. Determination of the total content of metals in soils is an important step in estimating the hazards to the vital roles of soil in the ecosystem, and also in comparison with the quality standards in terms of the effects of pollution and sustainability of the system.

From the ecotoxicological point of view, it is equally important to determine the bioavailability of Cu accumulated in vineyards, i.e., the fraction of the total metal content in soil that can be utilized by biota [13]. It depends on the soils properties, temperature, water content and aeration, and also on plant species. The toxicity of copper is essentially observed in acid soils, but not in calcareous soils and for copper contents as high as those reported in vineyard soils. A number of authors have found positive correlation between copper retention and pH [14, 15] and sum of bases or exchangeable calcium [16]. The bioavailability of copper has also been reported to decrease when the cation exchange capacity or the level of organic matter increases [5, 15]. Various electrolytes such as water, buffered or unbuffered salt solutions, chelating agents, diluted acids or a mix of these reagents can be used to estimate the biavailability of copper in soils (reviews given by [17, 18]).

1.4. Fractionation and bioavailability of copper in soils

The concepts of «bio-availability» and «bio-accessibility» were introduced to express whether the actual concentration of a toxic element would have effects on organisms [19]. The main challenge that comes out from the assessment of loads of trace and toxic metals is the methodology of determination or prediction of the trace element content in a soil that results in toxicity [20]. Trace element mobility and bioavailability is determined by their transfer between the soil solid phase and the soil solution [21], and trace element in soils can be divided into inert and the potentially toxic labile fraction [22]. Thus, the impact of trace metals on soil and the surrounding environment in most cases cannot be predicted simply by measuring their total concentration. This is because only soluble and mobile fractions have the potential to leach or to be taken up by plants, and enter the food chain. Furthermore, Cu content in the plants usually does not well correspond to the total soil copper content [23].

Water-soluble and exchangeable copper fractions are considered to be bioavailable; copper complexed with oxide, carbonate and organic matter are potentially bioavailable fractions; and mineral fraction is considered to be non-bioavailable [24]. However, fractionation does not provide information about species of metals in soil. Metal speciation is one of the most important properties that determine the behavior and toxicity of metals in the environment. Chemical speciation of an element refers to its specific form characterized by a different isotopic composition, molecular structure, and electronic or oxidation state [25]. Speciation is the process of identification and determination of different chemical and physical forms of elements present in a sample [26]. TMs speciation is determined by their reactivity and physical and chemical properties of the soil. Metals are very reactive in the environment and can relatively easy change form in soil, making their speciation non stable. Among many chemical processes that are involved in the transformation of TEs in soils, precipitation-dissolution, adsorption-desorption and complexation are the most important in controlling their bioavailability and mobility. Metals that occur in cationic forms have a higher ability of binding to negatively charged soil colloids, and are thus less bioavailable, but more easily accumulate in soil, unlike the anionic forms that are mainly present in soil solution and are more bioavailable, but are more readily leached from the soil. Formation of complexes with soil organic matter, adsorption onto the surfaces of clays and Fe and Mn oxides regulate the behavior of copper in the soil [23]. Thus, copper bioavailability in soil depends on soil pH, redox potential, CEC, amount and nature of organic matter and soil minerals [27].

Chemical extractions are used for soil copper bioavailability predictions, often in comparison with its content in plants [23]. Copper is not readily mobile in plants and root concentration is considered to be a good indicator of the plant copper content. It has been assumed that the factors affecting metal fractionation and bioavailability in soil include root-induced pH changes, metal binding by root exudates, root-induced microbial activities and root depletion as a consequence of plant uptake [28]. In the root developing zone, rhizosphere,

http://dx.doi.org/10.5772/57266

processes that control the mobility, transformation and toxicity of metals in soil are under direct influence of plant roots and may differ from those in bulk soil. Thus, root activities can considerably modify TM speciation in the rhizosphere. Metal plant root adsorption is determined by ionization of negatively charged binding sites for metal on root surfaces. Cupric ions bind to a specific carrier on the root cell plasmalemma surface [29] and plant uptake depends on the available copper in the soil and the nutritional status of a plant [30]. Soil pH determines aqueous metal speciation, affecting metal sorption and desorption on the solid phase [31], but plant roots as well. As the degree of biotic ligands ionization increases with pH, the metal ion root adsorption increases. Vice versa, the adsorption capacity of the plant root for cupric ion decreases with decreasing solution pH. Rhizosphere pH modifications by plant are a known occurrence and root zone alkalization may decrease the exposure of plant roots to copper by promoting formation of organic complexes and reducing copper solubility throughout rhizosphere area [32]. Furthermore, plant may directly decrease copper bioavailability near roots by excretion of metal-binding compounds that can complex the free cupric ions [33].

In this article, the actual risk of high concentrations of copper and its mobility in vineyard soils is reviewed considering sources, chemical processes in soil and biogeochemical behaviour of copper as well as impact on agroecosystem and environment in general.

2. Case study: Spatial distribution of copper concentrations in vineyard soils of Croatia: Wine–growing subregion of Plesivica

Elements inherited from the bedrock are partitioned within the soil through specific processes. Besides the parent material characteristics, geomorphology and landscape features contribute greatly to the variability of elements distribution. The topography is especially important, since it affects water infiltration and drainage on the one hand and soil erosion on the other. This variability is additionally enhanced in cultivated soils, especially in mountain regions and on sloping terrains where erosion processes are more expressed [34].

Steep southern slopes of the mountains in north-western Croatia have been used for centuries to grow vine and produce wine (Figure 1). Wine quality is almost always associated with the location, which means with the specific natural viticultural environment where soil is one of the major factors: firstly in terms of soil physical properties, and secondly in terms of soil inorganic chemistry [35, 36].

Although the systems of vineyard planting and maintenance in north-western Croatia have been changing with time and modern plantations prevail nowadays, parcels under traditional cultivation and old, almost forgotten, cultivars can still be found. Therefore, the aim of this study was to explore the distribution and retention pattern of copper concentrations in vineyard soil and, and to study the copper speciation and its distribution within five operationally defined fractions to assess its bioavailability and possible downward movement.

Figure 1. Grapevine plantation in the study region (Site Lokosin dol, Plesivica wine-producing subregion, Croatia, photo M. Romic)

2.1. Study area

Research was carried out on different wine-producing subregions in Croatia, in the wider Zagreb region: the wine-growing subregion of Plesivica (approximate coordinates: latitude, 45° 42′ and longitude, 15° 37′). The area of Žumberak, which is a mountainous territory located to the west of Zagreb, is bordered to the north and west by Slovenia, and to the south by Kupa River (Figure 3). Soils are developed on Pannonian sediments. This sediments consist of limy marls, sands, sandstones, conglomerates and breccias [37-39]. Week consistency of those sediments, solid Triassic dolomite as bedrock and periodic streams led to fast erosion and filling in valleys and formation of amphitheater-shaped valleys. Namely, those landscape shapes were formed by the distinct climatic oscillations and pulsation between glacial and stadial periods, along with constant tectonic activity and elevation of terrain [39]. The landscape reflects the features of the Dinaric, so numerous formations characteristic of the karst are found on highly dissected limestone terrains. As the Zumberacka Mt. piedmont spreads perpendicularly to the mountains, some slopes are firmly interlinked by ridges, forming well protected, amphitheatre shaped, vineyard areas in the wine-growing subregion of Plesivica.

Anthropogenic vineyard soils, classified as Aric Anthrosols [40], have been developed on Tertiary sediments and Pleistocene loams. Owing to exceptional geomorphological and agro-ecological conditions, these locations have been occupied almost exclusively by vineyards for many decades.

http://dx.doi.org/10.5772/57266

The climate of the wider area is humid and the average annual rainfall is 836 mm. The mean annual temperature is 10.3 °C, ranging from –0.6 °C (January) to 20.4 °C (July).

To investigate the spatial variability of surface soils, 67 soil samples were taken at the nodes of a square grid at intervals of 1 km (Figure 3). The samples were defined as composite samples made up of 10 increments collected from the soil upper 10 cm in a cross pattern, with a 5 m distance between increments (Eijkelkamp soil sampling kit used). Site descriptions were registered at the time of sampling to record the sample location in relation to vineyard characteristics and major environmental features.

Three soil profiles were then located according to the landscape feature: profile 1 (Aric Anthrosols) was dug in the vineyard plot down to the parent material (80 cm depth) at the 258 m asl, profile 2 (Aric Anthrosols) was located in vineyard plot at 231 m asl, and profile 3 (Colluvic soil) was located in the meadow at the foot of the hill 200 m asl, with the altitude difference of 28 % between P1 and P2, and 13 % between P-2 and P-3. Table 1 reports the selected physical and analytical features of soil profiles, and Figure 2 shows a sequence of horizons.

Profile / Horizon		Depth cm	pH_{H2O}	$CaCO_3$ %	Corg %	Exchangeable cations ($cmol_{(+)}kg^{-1}$)				CEC $cmol_+kg^{-1}$	Sand	Silt	Clay
						Ca_{ex}	K_{ex}	Mg_{ex}	Na_{ex}		%		
P1	P	0 - 45	8.2	48.8	2.6	24.1	0.63	2.14	0.09	24.0	11	64	26
	C	45 - 100	8.5	52.1	0.9	15.5	0.16	1.32	0.03	13.2	14	65	21
P2	P I	0 - 35	8.2	47.7	3.1	25.9	0.72	2.22	1.02	26.1	11	63	27
	P II	35 - 90	8.4	47.3	1.8	26.9	0.31	2.30	0.11	26.6	8	59	33
	C	90 - 130	8.5	51.0	0.5								
P3	Aa	0 - 10	7.9	41.0	3.6	32.6	0.18	2.30	0.05	31.1	15	68	17
	Ap	10 - 30	8.2	46.2	2.4	29.1	0.16	1.64	0.03	28.6	13	64	23
	I	30 - 50	8.2	33.9	1.8	28.2	0.19	1.64	0.05	30.5	9	59	32
	II	50 80	8.3	34.2	0.8	25.2	0.17	1.64	0.09	26.6	9	56	35
	Cg	80 - 150	8.4	44.2		22.0	0.20	1.64	0.25	20.9	11	58	31

Table 1. Selected physical and chemical features of soil profiles

2.2. Chemical analysis

Soil samples were air-dried, sieved at 2-mm, and subjected to the following analyses: pH in a 1:5 soil/water ratio (MettlerToledo MPC 227 pH- meter), soil organic carbon (SOC) by sulfochromic oxidation [41], calcium carbonate ($CaCO_3$) by the volumetric calcimeter method after HCl attack, and effective cation exchange capacity (CEC) using $BaCl_2$ solution. Particle

Profile 1	
Coordinates (m) (Gaus-Krueger projection)	y 5 063 845, x 5 551 878
Altitude	258 m
Slope	28.1 %
Land use	Vineyard, >60 years old
Soil type	Aric Anthrosol (FAO, 1990)
Horizons	Ap 0-45 cm, heavy silt loam, C 45-100 cm, silt loam
Profile 2	
Coordinates (m) (Gaus-Krueger projection)	y 5 063 758, x 5 551 884
Altitude	231 m
Slope	13.2 %
Land use	Vineyard, recently abandoned
Soil type	Aric Anthrosol (FAO, 1990)
Horizons	Ap I 0-40 cm, heavy silt loam, Ap II 40-90 cm, heavy silt loam, C 90-130 cm, silt loam
Profile 3	
Coordinates (m) (Gaus-Krueger projection)	y 5 063 523, x 5 551 872
Altitude	200 m
Slope	level
Land use	Meadow
Soil type	Calcaric Regosol (FAO, 1990)
Horizons	Aa 0-10 cm, silt loam, Ap 10-30 cm, heavy silt loam, A I 30-50 cm, heavy silt loam, A II 50-80 cm heavy silt loam, Cg 80-120 cm, heavy silt loam

Figure 2. Soil profiles description (Site Lokosin dol, Plesivica wine-producing subregion, Croatia, photo M. Romic)

size distribution was determined by the pipette method after disaggregation in sodium pyrophosphate (HRN ISO 11277:2004). Soil samples were also digested in aqua regia [42] with the microwave technique on a MARSXpress system (CEM).

Copper concentrations in soil digests were determined by inductivey coupled plasma optical emission spectroscopy (ICP-OES) on a Vista MPX AX (Varian). All concentrations were calculated on the basis of dry weight of samples (105 °C, 24 h). Quality control procedure consisted of reagent blanks, duplicate samples and several referenced soil and sediment samples with similar matrix from the inter-laboratory calibration program [43]. Maximum allowable relative standard deviation between replicates was set to 10 %.

Two other methods were used for evaluating soil available copper: DTPA extraction [44] and calcium chloride extraction [45].

DTPA extraction: 10 g of soil were extracted with 20 ml DTPA 0.005 M + TEA 0.1 M + $CaCl_2$ 0.01 M for 2 h at 20 °C under stirring (Heidolph PROMAX 2200 used), prior to being filtered.

Calcium chloride extraction: 0.5 g of soil were extracted with 50 ml of $CaCl_2$ 0.01 M for 2 h at 20 °C under stirring, prior to being filtered.

2.3. Metal fractionation

The selective sequential dissolution procedure was employed to divide metals into five solid-phase fractions [46, 47]. Chemical reagents and the experimental conditions applied are

http://dx.doi.org/10.5772/57266

summarized in Table 2. In this procedure, 0.5 g of each sample was weighted into 50 ml polyethylene centrifuge tube and the extractions were carried out directly in the tubes, except in the last step where Teflon tubes for microwave digestion were used. At each extraction step, after shaking and equilibration, solid-liquid separation was achieved by centrifugation at 3500 rpm (2000 x g) for 10 min (centrifuge Sigma 3-15 used). The supernatant was filtered through S&S 583 filter paper and placed to an acid-washed polyethylene tube. The solid residue was washed three times successively with 5 ml of ethanol and the liquid was discarded leaving the residue soil for the next step. The supernatant obtained at each step was analysed for metals using inductively coupled plasma optical emission spectroscopy (ICP-OES) on a Vista MPX AX (Varian). Single element standards were prepared for each extraction in the same solution as the extracting agent to minimise matrix effects. Blanks were used for background correction and other sources of error.

Fraction	Abbreviation	Procedure
Exchangeable	EXCH (F1)	0.1 M $Mg(NO_3)_2$ pH=7, 10 ml, room temperature, shake for 2 h
Associated with carbonates	CARB (F2)	1M CH_3COONa + CH_3COOH (pH = 5.0), 10 ml, room temperature, shake for 6 h
Associated with Fe/Mn oxides	Fe/Mn OX (F3)	0.04 M $NH_2OH.HCl$ in 25 % HOAc, 25 ml, water-bath at 90° for 3 h
Associated with organic matter	ORG (F4)	0.1 M $K_4P_2O_7$ 25 ml, shake for 24 h
Residual	RESID (F5)	4 M HNO_3, 50 ml, microwave digestion

Table 2. The selective sequential dissolution procedure; chemical reagents and the experimental conditions.

2.4. Metal statistical analysis and data management

Linear regression: Relations between extractible copper in soils and soil properties were analyzed by simple and multiple linear regression [48]. General conditions for model application were verified after parameter determination. Thus, residuals should follow normal distribution, the assumption of homoscedasticity of variable variances should be proven, and they should be independent of one another. Distribution normality was tested using the Shapiro-Wilk test, while the other hypotheses were checked visually after the graphs were drawn. Details given by [49]. All soil data were incorporated into the GIS database.

3. Copper concentrations in surface vineyard soil

Total copper concentrations in vineyard soils ranged from 30 to 700 mg kg^{-1}, while total copper in 88 % samples exceeded the maximum tolerant concentration under the Croatian regulation of 100 mg kg^{-1} [50] (Figure 3).

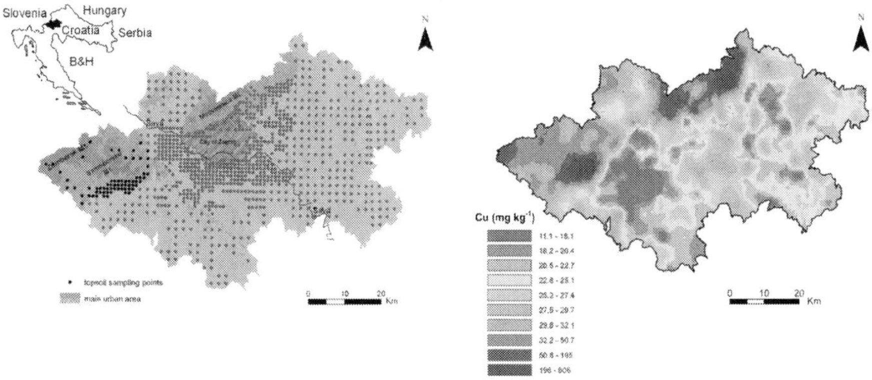

Figure 3. Study area, surface soil sampling scheme and interpolated map of total surface soil copper concentrations [51]

The amount of copper built up in the surface vineyard soil was estimated on a *per hectare* basis taking into account the total copper concentration and the weight of 10 cm thick surface layer of soil, assuming a bulk density of 1.5 (Table 3). Approximate plantation age (with within-decade precision) was estimated for most plots, and a detailed history for some plots: time of the first and possibly second deep ploughing, time of vineyard restoration or supplementary planting, common growing practice in the past, and thereby also approximate frequency of copper fungicide applications. More than 64 % vineyards are more than 40 years old and it is assumed that the same percent of all plots received similar annual amounts of copper, ranging from 2 to 5 kg ha^{-1}. This certainly allows only an estimation of the overall copper input into soil throughout the vineyard history. The upper layer of plots planted with about a century old vines exhibited high copper contents.

Location[1]	Cu_{TOT}(mg kg^{-1})	Cu_{DTPA} (mg kg^{-1})	Cu_{CaCl2} (mg kg^{-1})	Cu_{TOT}/ha (kg ha^{-1})[2]
351	363	136	1.15	491
352	154	48	0.63	208
353	380	142	1.17	514
354	389	121	0.74	525
355	181	56	0.73	244
356	166	54	0.55	224
357	655	368	1.45	884
358	586	296	1.96	790
359	369	135	0.94	498
360	565	296	1.35	763
361	633	340	2.08	855
362	641	228	1.68	866

Location[1]	Cu_{TOT}(mg kg^{-1})	Cu_{DTPA} (mg kg^{-1})	Cu_{CaCl2} (mg kg^{-1})	Cu_{TOT}/ha (kg ha^{-1})[2]
363	336	196	1.24	454
364	385	152	1.15	519
365	263	101	0.82	355
366	667	316	1.36	901
367	615	332	1.56	830
368	419	143	1.19	565
369	494	166	1.47	667
370	244	95	0.93	330
371	519	282	1.15	701
372	134	47	0.42	181
373	199	71	0.52	268
374	623	348	1.84	841
375	35	4.32	0.10	47
376	389	140	1.26	525
377	442	138	0.96	597
379	252	104	0.92	340
380	30	808	0.93	40
381	531	198	1.17	717
382	305	117	1.04	412
383	453	199	1.35	612
384	360	131	1.2	486
385	79	21.2	0.21	107
386	398	146	1.36	537
387	324	120	0.94	437
390	51	8.28	0.21	69
391	74	11.8	0.11	100
392	228	89	0.83	307
395	533	170	1.46	720
396	627	254	1.66	847
397	32	12.4	0.10	43
398	43	10.6	0.21	58
399	610	180	1.98	823
401	691	278	1.93	933
402	427	154	1.35	576

[1]Location: All the data available from the GIS database

[2]Cu/ha (kg ha^{-1}) Calculated from the data on total Cu (mg kg^{-1}) in the top 10 cm assuming a bulk density of 1.5

Table 3. Total Cu, extractable Cu and total amount of Cu per ha in the upper layer (10 cm) of vineyard soils (n=67) (Romic et al., 2004)

Determination of the total metal content in soils is an important step in assessing the hazards to the vital roles of soils in the ecosystem, but also in comparing them with quality standards referring to the effects of contamination and system sustainability. However, the mobility and availability of soil copper are governed by the processes of dynamic equilibrium, and not only by its total concentration [52]. This study shows that the mobile ($CaCl_2$-extractable) fraction of copper in vineyard soils amounted only to 0.2 – 3.1 % of its total content. As Cu accumulation in the studied soils is restricted to surface layer the risk of Cu phytotoxicity for grapevine is small, since grapevine develops most of its roots at a depth >30 cm, depending on the soil type and the profile depth. Material suspended by erosion, however, carries away also a part of the applied copper, and redistribution of this material depends on a number of factors (relief, size and shape of the drainage basin, etc.). Halamic et al. [53] applied factor analysis in geochemical investigations of stream sediments in drainage basins in the Mt. Medvednica region, which also includes part of the studied area, without determining correlation of copper concentrations with any lithological unit, so they assumed anthropogenic influence, mostly grapevine production. Ribolzi [54] carried out the research in the Mediterranean drainage basins of wine-growing regions in France with the aim to characterize copper forms in suspended material and recorded an average total concentration as high as 245 mg Cu kg^{-1}, but Brun et al. [55] reported the maximum of 250 mg kg^{-1} of copper in vineyard soils of the Mediterranean part of France.

3.1. Correlation between total, extractable Cu and soil properties

In the total copper concentration extracted with aqua regia, DTPA-extractable copper amounted to 12-81 % and $CaCl_2$- extractable to 0.2-3.1 %. Both DTPA- and $CaCl_2$-extractable copper were largely explained by the total copper concentrations, as confirmed by their high correlation coefficient (R = 0.899 and R = 0.896). They were also highly correlated to each other (R = 0.763) (Table 4).

	Cu_{total}	Cu_{DTPA}	Cu_{CaCl2}	pH_{H2O}	Org. C	CEC	$CaCO_3$
Cu_{total}	1	0.899^{***}	0.896^{***}	0.388^{**}	0.601^{***}	0.277^{*}	0.314^{**}
Cu_{DTPA}		1	0.763^{***}	0.321^{**}	0.395^{***}	0.084^{ns}	0.236^{ns}
Cu_{CaCl2}			1	0.392^{***}	0.734^{***}	0.257^{*}	0.341^{**}
pH_{H2O}				1	0.121^{ns}	0.315^{**}	0.587^{***}
Org. C					1	0.339^{**}	0.173^{ns}
CEC						1	0.130^{ns}
$CaCO_3$							1

ns - not significant

* Correlation significant at $p < 0.05$

** Correlation significant at $p < 0.01$

*** Correlation significant at $p < 0.001$

Table 4. Correlation matrix, upper triangle

http://dx.doi.org/10.5772/57266

Highly significant positive correlation was also determined between total copper and organic matter in soil, and a weaker correlation, but still significant, between total copper and cation exchange capacity and carbonate content (Table 4). In the case of extractable copper, a significant correlation was determined between Cu_{DTPA} and organic carbon content and pH, but not between cation exchange capacity and carbonate content. Significant correlation was recorded between all the analyzed soil properties and Cu_{CaCl2}.

3.2. Linear regression of total and extractable copper and selected soil properties

To establish the relation between copper fractions after particular extractions (aqua regia, DTPA and $CaCl_2$) and soil properties that may affect their behavior in soil and availability to plants, the multiple linear regression analysis was done. The model included those variables for which correlation probability $p<0.05$ was determined [48]. For the regression model of aqua regia extracted copper, this condition was met by the following properties: Cu_{DTPA}, Cu_{CaCl2} and cation exchange capacity (CEC).

For total copper, the regression model explains 92 % of total variance (Table 5). The largest contribution to the variance in regression was that of Cu_{DTPA}.

Variable explained: Cu_{TOT}

Source of variation	Degree of freedom	Sum of squares	F	Pr > F
CuDTPA	1	301395	101,8	0.0000
$CuCaCl_2$	1	192403	65,0	0.0000
CEC	1	30384	10,3	0.0021
Total	63	2319596		

Regression equation: Cu_{TOT} = -63.42 + 0.956* Cu_{DTPA} + 170.9*Cu_{CaCl2} + 3.761*CEC

R^2 = 0.92

Table 5. Linear regression of total copper (Cu_{TOT}) as a function of DTPA-extractable (Cu_{DTPA}), $CaCl_2$-extractable (Cu_{CaCl2}) and cation exchange capacity (CEC).

Concentration of total copper in vineyard topsoil went up with an increase in the cation exchange capacity. Square root transformation (SQRT) was applied tothe regression model of DTPA-extractable copper, whereby the model conditions were satisfied, and the transformed variables SQRT(Cu_{DTPA}) were predominantly dependent on Cu_{tot}. As this relation was already determined, it was omitted from the model. Two other properties met the condition of correlation probability $p<0.05$: organic matter content (Org-C) and CaC$_l$2-extractable copper (Cu_{CaCl2}), and they were included into the model. For Cu_{DTPA}, the regression model explains 85 % of total variance (Table 6).

Variable explained: SQRT (Cu_{DTPA})				
Source of variation	**Degree of freedom**	**Sum of squares**	**F**	**Pr > F**
$CuCaCl_2$	1	685.7	240.0	0.0000
Org-C	1	44.54	15.59	0.0002
Total	63	1224		
Regression equation: $SQRT(Cu_{DTPA}) = 5.534 + 9.400*Cu_{CaCl2} - 1.326*$ Org-C				
$R^2 = 0.853$				

Table 6. Linear regression of DTPA-extractable (Cu_{DTPA}), as a function of $CaCl_2$-extractable (Cu_{CaCl2}) and soil organic matter (Org. C)

Two parameters were included into the regression model of $CaCl_2$-extractable copper: organic matter content (Org-C) and pH, and the model explains 62 % of total variance (Table 7). Concentrations of $CaCl_2$- extractable copper mainly depend on pH, which relation was also confirmed by this investigation. However, since these are predominantly alkaline soils, this relation is not as strong as in the case of soils with a more varying pH [55].

Variable explained: Cu_{CaCl2}				
Source of variation	**Degree of freedom**	**Sum of squares**	**F**	**Pr > F**
Org-C	1	8.30	83.37	0.0000
pH	1	1.62	16.23	0.0002
Total	64	17.34		
Regression equation: $Cu_{CaCl2} = -2.673 + 0.379*$ Org-C $+ 0.337*$ pH				
$R^2 = 0.621$				

Table 7. Linear regression of $CaCl_2$-extractable copper as a function of soil organic matter (Org. C) and pH

Over 65 % of vineyard plots under study were more than 40 years old, and some have been continuously cultivated for more than 100 years. It is assumed that the same percent of parcels received a similar annual amount of copper, ranging from 2 to 5 kg ha^{-1}. This, naturally, does not allow an exact estimate of the overall copper input into soil throughout the vineyard history. Vineyard age parameter was not therefore included in the multiple linear regression model. There is, however, strong statistical evidence that an increase in vineyard age is related to the increase of expected total copper content (Figure 4). According to historical documents, the vine downy mildew infection started spreading in the vineyards of northwestern Croatian in 1882, and the Bordeaux mixture application became indispensable during the wine boom period at the end of the 19th century. Recognizing the benefits, winegrowers often did not observe the recommended concentrations and application times, and a large number of treatments, as many as 8 to 14, were often applied at positions exposed to disease attacks. However, numerous other factors, such as scattering during applications, washing off the leaves by rain, input of treated plant residues into soil, tillage and erosion, make it difficult to establish the relation between vineyard age and accumulated copper. In France, for example, the Bordeaux mixture has been used since 1855, and it was found that after several decades of its continuous application the soil total copper reached a concentration of as much as 1.0 g kg^{-1} [56]. Research done by Deluisa et al. [11] on 43 plots in a humid region of northern Italy revealed an average copper accumulation in topsoil of 297 mg kg^{-1}. Moolenaar and Beltrami [57] have calculated that organic protection of grapevine, which implies exclusive use of the Bordeaux mixture, can result in an increase of soil copper concentration up to 600 mg kg^{-1} after 100 years.

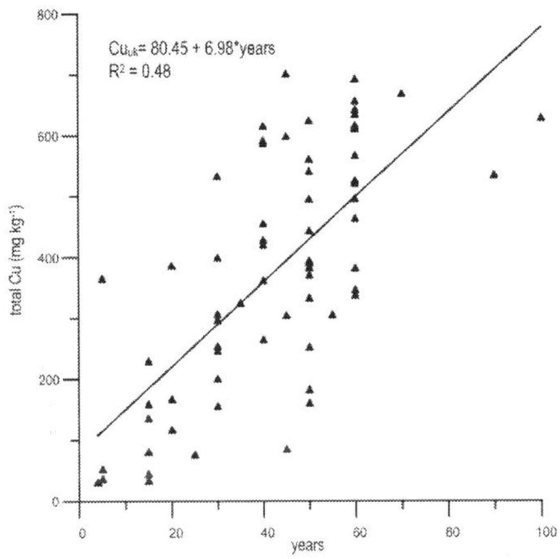

Figure 4. Vine age as a function of total soil copper

Diethylenetriaminepentaacetic acid (DTPA) is a potent synthetic chelating agent, and the method of extraction with DTPA was developed for the purpose of determining zinc, iron, manganese or copper deficiency in neutral and carbonate soils [58]. Haq and Miller [59] reported negative results of the DTPA test, which they explained by their failure to determine sufficiently significant relations between concentrations of metals (copper and manganese) extracted from soil and those found in the tested plants. Mention should be also made of the research done by O'Connor [60], who gave a number of comments on the DTPA test, based also on non-significant correlation between DTPA-extractable metals in soil and their concentrations in plants. Regardless of the above considerations, DTPA is the most widely used agent for extraction of "available" cadmium, copper, nickel and zinc, and thereby also the most standardized one [44, 61, 62]. Starting from the fact that the data on total copper content reveals very little about its bio-availability, such strong correlation between copper extracted with aqua regia and DTPA actually indicates that neither the latter extraction method is suitable for assessing copper availability to plants.

Merry et al. [63] in vineyard soils recorded 25-35 times higher contents of copper, lead and arsenic, originating from plant protection agents, than their common values in uncontaminated soils; they also determined a strong correlation between total and DTPA-extractable copper ($0.93 < r < 0.96$). Also Brun et al. [55] found that the regression model Cu_{DTPA} explained 90 % of total variance in vineyard soils, its largest part referring to total copper content.

When cation exchange capacity was included into the model, it was found that the DTPA-extractable copper decreased with increasing cation exchange capacity.

Soil extraction 0.01 M $CaCl_2$ is the method that was increasingly used in the last decade for soil testing to determine soil fertility and the behavior of nutrients and contaminants in the soil. The capabilities of instrumental chemical analysis have improved to such an extent, even in the last few years, that it become possible to determine very low concentrations of nutrients and pollutants in soil extracts [45]. The advantage of this method for determining metal concentrations in soil is that the concentration of electrolytes stays practically constant and metal concentrations reflect the difference in binding strength or solubility between soils. The extractant is an unbuffered solution and therefore the measured metals reflect their availability at the pH of the soil.

The best criterion of the efficiency of the method for determining the soil bioavailable fraction is the high correlation between the Cu content observed in plants grown in situ, at least for neutral to acid soils [55].

3.3. Vertical distribution of total copper in soil profiles

High copper concentrations were found in vineyard soils down to 20 cm depth (to 800 mg kg^{-1} in profile 1, and to 500 mg kg^{-1} in profile 2) (Figure 5). Profile 1 is situated at a higher altitude, erosion is more pronounced, and the anthropogenic horizon is less thick. Marl appears already at 45 cm depth, so that the root zone extends into horizon C as well, thus opening the transport routes of water, dissolved substances and solid particles deeper into the profile. In profile 2, the anthropogenic horizon is much thicker, while total copper concentrations are lower down to 30 cm depth. Erosion material was deposited at the base of the slope, so that as much as 100 mg kg^{-1} of copper was found in the topsoil of colluvial soil, to which no copper agents for plant protection had ever been directly applied. Accumulation of copper in colluvial soil (profile 3) was recorded down to 30 cm depth, that is, over the entire depth of the humus-accumulative horizon. Uniform copper concentrations of <25 mg kg^{-1} were found at greater depths.

Land use for agriculture causes great changes in the natural properties of soil. Translocation of soil by tillage may be the key reason for redistribution of soil particles within the profile and over the entire site, while erosion due to tillage is especially present in hilly landscapes [64]. Tillage and homogenization of several natural horizons alter both morphological and physicochemical characteristics of soil.

3.4. Sequential extraction

Results of the five-step sequential extraction are presented in Figure 6, which shows the distribution of exchangeable (Exch), carbonate-associated (Carb), Fe-Mn oxides-bound (Fe-Mn

http://dx.doi.org/10.5772/57266

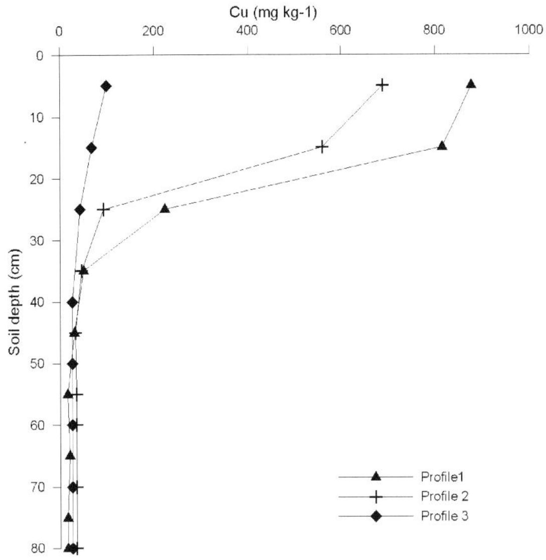

Figure 5. Verical distribution of total copper within the soil profiles

ox), organic-bound (Org) and residual (Resid) copper found in the soil samples. These five metal fractions were separated in soil samples from 10-cm profile layers of Aric Anthrosols and master horizons of Calcaric Regosols. The data were used to calculate the relative error (RE %), which for most metals amounted to ±10 %.

In the surface 10 cm of profile 1, 877 mg kg^{-1} of copper was determined by extraction in *aqua regia*. The sum of copper fractions separated by selective sequential extraction is slightly lower and amounts to 816 kg^{-1} (RE = 7 %). As much as 47 % of total copper was bound in the organic fraction, 20 % in the residual fraction, and 18 % and 16 % in the reductive and carbonate fractions, respectively. Down to 30 cm depth, the share of copper in exchangeable fraction was <1 %, whereas it was not detected in deeper layers (Figure 6). However, copper distribution per fractions changes with the profile depth. As the total concentration decreases to background values, the share of copper in residual fractions increases, since complexing ability decreases with the reduced amount of organic matter.

Considering the high content of the organic fraction (54 % in the 0–10 cm layer and 27 % at 60–80 cm depth), a very probable mechanism is translocation through complexation with soil organic matter. However, copper is certainly not translocated in the same way with surface eroded material as vertically through the profile depth. Eroded material deposited at the base of hillsides under vineyards is richer in silt, and copper is mostly strongly bound in the residual fraction. In the surface 10 cm of soil of the profile 1, 20 % of total copper content is bound in the residual fraction. In the profile 2, copper content in the residual fraction was increased to 21 %, while it amounted to 50.3 % in colluvial soil (profile 3).

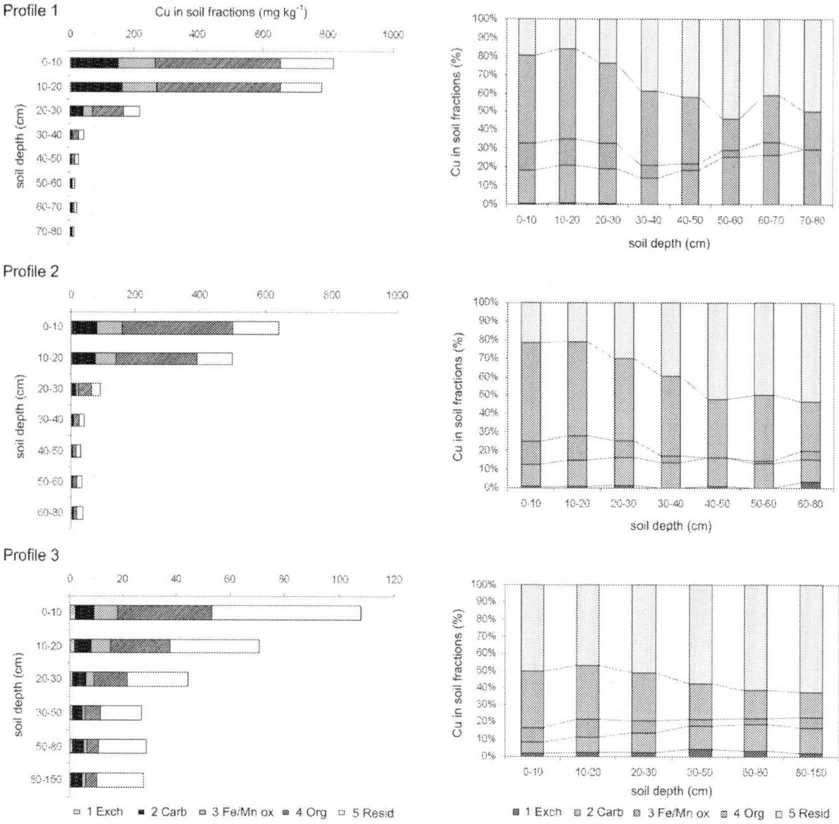

Figure 6. The distribution of exchangeable (Exch), carbonate-associated (Carb), Fe-Mn oxides-bound (Fe-Mn ox), organic-bound (Org) and residual (Resid) copper in soils

Copper distribution per fractions in profile 2 resembled that in the preceding profile. High copper concentrations decrease with depth, the profile being deeper as well. Copper content of <1 % was found in the exchangeable fraction, and that of 3 % at 60–80 cm depth. It is obvious that copper translocation occurs within the profile, but this is also the zone where most of the vine roots develop.

Although no copper fungicides had ever been directly applied to the area on which profile 3 was dug, as much as 100 mg Cu kg^{-1} was determined in topsoil by extraction in aqua regia. Background concentrations were reached at 30 cm depth, and copper was predominantly bound in the residual fraction (47 %-63 %). There was a significant share of the organic fraction, but it decreased with depth from 33 % to 15 %. The content of exchangeable copper was higher than in vineyard soils – 2 % to 5 %.

Research has shown that soil type is the main factor of accumulation and distribution of both natural and anthropogenic concentrations of heavy metals. In natural profiles, the indigenous element distribution is generated by long-term pedogenesis [65]. In the case of cultivated soils, their characteristics in time and space change in dependence on the ecological environment and land use and management, whereby also their production capacity and environmental impact get changed [66, 67]. When growing woody crops, and thus also grapevine, soil is homogenized to a greater depth, first by deep ploughing to xx cm depth before the setting up of the plantation, and then by regular tillage. This somewhat disrupts the morphogenetic soil properties, changes the sequence of genetic horizons, and often deepens the active part of the profile. Such changes are naturally reflected in the distribution of elements, changes in their mobility and bioavailability. Selective sequential extraction was used in this research to determine the way and strength of binding and retention of heavy metals in soil, which under certain conditions enables the estimation of potential mobility and bioavailability.

The spatial variability of trace metals in agricultural surface soils of the wider Zagreb area has shown that the application of agrochemicals has caused high accumulation of copper and zinc [51, 68]. This especially applies to vineyard soils, but also to orchard and vegetable garden soils. The presence of a buffer material, such as carbonate, can be particularly important in the retention of heavy metals. The trace metal retention capacity of silty soils with high carbonate content can be as high as, or higher than, the retention capacity of certain clayey soils [69].

In soils and sediments that were receiving high concentrations of copper (along with other metals) for at least 6 years, Hickey and Kittrick [70] established that about 28 % copper was bound in the organic fraction. It is also in the soils treated with waste sludge or stable manure rich in copper that most of this metal is bound in the organic fraction [71, 72]. Affinity of humic and fulvo acids to copper sorption was, among others, reported by Senesi et al. [73] while McLaren et al. [74], pointed to the importance of soluble copper chelates in soil solution. In soils of lighter texture, poorer in organic matter, the added copper is initially retained in the exchangeable fraction, whereafter it is translocated into the carbonate fraction [75]. Incubation in the laboratory experiment revealed that translocation of copper to more stabile fractions was much slower in texturally light, but acid, vineyard soil, also poor in organic matter [76]. However, Flores-Velez et al. [10] report that in the case of sandy acid vineyard soils the selective sequential extraction procedure was not selective enough to specify the form of copper. The same authors report that copper in anthropogenic vineyard soil, originating from Cu-fungicides, was concentrated in the coarse organic fraction (plant residues) and in the mineral colloid fraction. In soils or sediments deficient in organic matter, a larger part of copper was bound to Fe and Mn oxides that it was found in this research. Thus, Szarek–Gwiazda and Mazurkiewicz–Boron [77] found that 40.2-54.1 % of total copper in fluvial sediment was bound to Mn oxides and amorphous Fe hydroxides, and only about 10 % to the organic fraction. These authors maintain that Fe(III) and Mn(IV)-oxides can occur either as coats on detritus particles, as cement between them or as pure concretions.

Their ability to adsorb and control heavy metal distribution between the solution and the matrix has been thoroughly explained in scientific literature (e.g., [78]).

Romic et al. [49] did not establish, either by individual correlations or by factorial analysis in the vineyard soils of NW Croatia, the importance of the contents of clay, Fe and Mn oxides or cation exchange capacity for copper sorption in soil, though some authors stress the importance of these fractions [74]. In their investigations, the relation between the content of metals and soil properties was assessed on the basis of their total contents. In this research, correlations were determined between copper fractions and the selected soil properties: significant correlation was recorded between the contents of organic C and ORG and RESID fractions, or its total content, whereas no significant correlation was found between copper in the said fractions and cation exchange capacity, or total carbonates. This corroborates the reports that the distribution of copper of anthropogenic origin among fractions depends prevailingly on soil organic matter.

Finally, the use of copper containing fungicides is allowed in the organic agriculture by the European Union regulation, and the official guidelines for soil copper content are usually derived from the total soil copper content. However, these guidelines should be modified according to the soil properties, such as pH and organic matter content, which will affect the Cu solubility, and consequently its bioavailability. Furthermore, soil copper thresholds should be confirmed with toxicological data obtained for biota (e.g. plants, microorganisms, invertebrates). Above mentioned implies that site-specific guidelines should developed for the risk assessment of soil copper toxicity.

4. Summary

Increased anthropogenic inputs of trace metals in soils have received considerable attention since they can enter the food chain by different ways. Soils receiving repeated applications of fungicides, pesticides or manure exhibit high concentrations of extractable metals, especially copper. From the commercial aspect, wine-growers are now showing increasing interest in the effects of soil composition, its fertility and texture upon wine quality. Special importance is laid on the influence of soil geochemical characteristics, including accumulation of certain toxic elements, on grape and wine quality. A variety of factors, both spatial and temporal, affect the grape quality, many of them being specific exactly to the given wine-growing site. Grape growing conditions and enological potential have been created, among other factors, also by landscape characteristics: soil, climate and topography. These factors are much less changeable than biological (cultivar, stock) or human (ampelotechnics, vinification) factors and for this reason the concept of *viticultural terroir* is based on the simple relationship between soil and wine. Since *terroir* is defined as an interactive ecosystem, it is very difficult to evaluate scientifically its contribution to plant capacity to accumulate bioactive phytochemicals good for human health.

Soil is a factor of the natural environment and its effect on wine quality and grape composition is highly complex, since it affects mineral nutrition of grapevine, water uptake, as well

http://dx.doi.org/10.5772/57266

as rooting depth and rhisosphere temperature. Land use for agriculture causes great changes in the natural properties of soil. Translocation of soil by tillage may be the key reason for redistribution of soil particles within the profile and over the entire site, while erosion due to tillage is especially present in hilly landscapes. Tillage and homogenization of several natural horizons alter both morphological and physicochemical characteristics of soil, which affect metal behaviour in soil – plant system.

The mobility and bioavailability of metals depends not only on the its total concentration in soil but also on soil properties, metal characteristics and environmental conditions. Trace metals are present in soil in various forms: water soluble, exchangeable, carbonate associated, oxide associated, bound on organic matter and residual forms. Obviously, various metal forms have different mobility. Water soluble and exchangeable fractions are readily released to the environment, whereas the residual fractions are immobile under natural conditions. The most common observation in majority of studies dealing with sequential or partial extraction of soil trace elements, including copper as well, is that they are not completely specific to metals or chemical phases. The complexity of bioavailability phenomenon comes out from an arry of matrix-related, species-related and metal-related issues.

Plant copper uptake occurs predominantly from the soil solution, indicating that soil characteristics that determine copper solubility, will also determine the copper phytoavailability. However, roots can alter the chemical mobility and thus the bioavailability of copper in the rhizosphere. Remediation of contaminated vineyard soils is an issue of debate and it should be focused on growing practice and vineyard management more than application of the remediation techniques per se (i.e. phytoextraction). Periodical monitoring of soil quality indicators is a required wine-growing practice. If systematically collected information is associated to a location, then it can be useful to all participants in the complex process of grape and wine production, wine marketing and consumption. Use of advanced analytical and information technologies may significantly improve the production and contribute to rational utilization of resources.

Author details

Marija Romić, Lana Matijević, Helena Bakić and Davor Romić

University of Zagreb, Faculty of Agriculture, Croatia

References

[1] Bradl HB. Adsorption of metal ions on soils and soils constituents. Journal of Colloid and Interface Science 2004; 277: 1-18.

[2] McLean JE, Bledsoe BE. Behavior of materials in soils. U.S. EPA Ground Water Issue EPA 540-S-92-018:25; 1992.

[3] Evangelou VP. Environmental soil and water chemistry: principles and applications. New York: John Wiley & Sons Inc; 1998.

[4] Kabata-Pendias A, Mukherjee AB. Trace elements from soil to human. Berlin: Springer-Verlag; 2007.

[5] McBride MB. Forms and distribution of copper in solid and solution phases in soil. In: Lonergan JF, Robson AD, Graham RD. (eds.) Copper in soils and plants: proceedings of the Golden Jubilee International Symposium, 7 – 9 May 1981, Perth, Western Australia. Academic Press Australia 1981; 25-45.

[6] Kinniburgh DG, Jackson ML, Syers JK. Adsorption of alkaline earth, transition and heavy metal cations by hydrous gels of iron and aluminium. Soil Science Society of America Journal 1976; 40: 796–799.

[7] McBride MB. Environmental chemistry of soils. New York: Oxford University Press; 1994.

[8] Komarek M, Cadkova E, Chrastny V, Bordas F, Bollinger J-C. Contamination of vineyard soils with fungicides: A review of environmental and toxicological aspects. Environment International 2010; 36: 138–151.

[9] Walsh LM, Erhardt WH, Seibel HD. Copper toxicity in snap beans (Phaseolus vulgaris L). Journal of Environmental Quality 1972; 1: 197-200.

[10] Flores-Velez LM, Ducaroir J, Jaunet AM, Robert M. Study of the distribution of copper in an acid sandy vineyard soil by three different methods. European Journal of Soil Science 1996; 47(4): 523-532.

[11] Deluisa A, Giandon P, Aichner M, Bortolami P, Bruna L, Lupetti A, Nardelli F, Stringari G. Copper pollution in Italian vineyard soils. Communications in Soil Science and Plant Analysis 1996; 27(5-8): 1537-1548.

[12] Baker DE. Copper. In: Alloway BJ. (ed.) Heavy Metals in Soils. Glasgow: Blackie and Son Ltd; 1990. p151-175.

[13] Lepp N. Bioavailability – A concept driven by science or legisation: conference proceedings of the 7th International conference on the biogeochemistry of trace elements. Uppsala Sweden: June 15-19 2003; pp. 12-12.

[14] Tyler LD, McBride MB. Mobility and extractability of cadmium, copper, nickel, and zinc in organic and mineral soil columns. Soil Science 1982; 134: 198-205.

[15] Gupta SK, Aten C. Comparison and evaluation of extraction media and their suitability in a simple-model to predict the biological relevance of heavy-metal concentrations in contaminated soils. International Journal Of Environmental Analytical Chemistry 1993; 51(1-4): 25-46.

[16] Harter RD. Adsorption of copper and lead by Ap and B2 horizons of several north-eastern United-States Soils. Soil Science Society of America Journal 1979; 43(4): 679-683.

[17] Lake DL, Kirk PWW, Lester JN. Fractionation, characterization, andsSpeciation of heavy metals in sewage sludge and sludge-amended soils: A review. Journal of Environmental Quality 1984; 13: 175-183.

[18] Ross SM. Toxic Metals in Soil-Plant System. New York: John Wiley & Sons Inc; 1994.

[19] Peijnenburg WJGM, Jager T. Monitoring approaches to assess bioaccessibility and bioavailability of metals: Matrix issues. Ecotoxicology and Environmental Safety 2003; 56: 63–77.

[20] Romić M. Bioavailability of trace metals in terrestrial environment: Methodological issues. European Chemical Bulletin 2012; 1(11): 489-493.

[21] Kabata-Pendias A. Soil-plant transfer of trace elements-an environmental issue. Geoderma 2004; 122: 143–149.

[22] Yobouet YA, Adouby K, Trokourey A, Yao B. Cadmium, copper, lead and zinc speciation in contaminated soils. International Journal of Engineering Science and Technology 2010; 2(5): 802-812.

[23] Chaignon V, Hinsinger P. Heavy metals in the environment: A biotest for evaluating copper bioavailability to plants in a contaminated soil. Journal of Environmental Quality 2003; 32: 824–833.

[24] Pakula K, Kalembasa D. Copper fractionation from Cambisols and Luvisols using the BCR procedure. Polish Journal of Environmental Studies 2013; 22(3): 809-817.

[25] Manouchehri N, Besancon S, Bermond A. Major and trace metal extraction from soil by EDTA: Equilibrium and kinetic studies. Analytica Chimica Acta 2006; 559: 105-112.

[26] Wang G, Su M, Chen Y, Lin F, Luo D, Gao S. Transfer characteristics of cadmium and lead from soil to the edible parts of six vegetable species in southeastern China. Environmental Pollution 2006; 144: 127–135.

[27] Cattani I, Fragoulis G, Boccelli R, Capri E. Copper bioavailability in the rhizosphere of maize (Zea mays L.) grown in two Italian soils. Chemosphere 2006; 64: 1972–1979.

[28] Ernst WHO. Bioavailability of heavy metals and decontamination of soils by plants. Applied Geochemistry 1996; 11(1–2): 163-167.

[29] Jiang W, Liu D, Liu X. Effects of copper on root growth, cell division, and nucleolus of Zea mays. Biologia Plantarum 2001; 44(1): 105-109.

[30] Tang S, Wilke B-M, Huang C. The uptake of copper by plants dominantly growing on copper mining spoils along the Yangtze River, the People's Republic of China. Plant and Soil 1999; 209: 225–232.

[31] Yin Y, Impellitteri CA, You S-J, Allen HE. The importance of organic matter distribution and extract soil: Solution ratio on the desorption of heavy metals from soils. The Science of the Total Environment 2002; 287: 107-119.

[32] Youssef RA, Chino M. Root-induced changes in the rhizosphere of plants. II. distribution of heavy metals across the rhizosphere in soils. Soil Science and Plant Nutrition 1989; 35(4): 609-621.

[33] Bruus Pedersen M, Kjær C, Elmegaard N. Toxicity and bioaccumulation of copper to Black bindweed (Fallopia convolvulus) in relation to bioavailability and the age of soil contamination. Archives of Environmental Contamination and Toxicology 2000; 39: 431–439.

[34] Hesterberg D. Biogeochemical cycles and processes leading to changes in mobility of chemicals in soils. Agriculture Ecosystems and Environment 1998; 67: 121–133.

[35] Northcote KH. Soils and Australian viticulture. In: Coombe BG, Dry PR. (eds.) Viticulture Volume 1: Resources. Adelaide South Australia: Winetitles; 1988.

[36] Tesic D, Woolley EW, Hewett EW, Martin DJ. Environmental effects on cv Cabernet Sauvignon (Vitis vinifera L.) grown in Hawke's Bay, New Zeland. 1. Phenology and characterisation of viticultural environments. Australian Journal of Grape and Wine Research 2001; 8: 27-35.

[37] Šikić K, Basch O, Šimunić A. Osnovna geološka karta 1:100.000, list Zagreb, L 33-80. Beograd: Institute for geological research Zagreb (1972) Federal gological department; 1978.

[38] Šikić K, Basch O, Šimunić A. Tumač za OGK 1:100.000, list Zagreb, L 33-80. Beograd: Institute for geological research Zagreb (1972) Federal gological department; 1979.

[39] Gregl Z. Žumberak–od prapovijesti do kasne antike. Archaeological museum in Zagreb and Zagreb City Museum 2002, 143.

[40] FAO/UNESCO/ISRIC. Soil Map of the World, Revised Legend. FAO Rome Italy: World Soil Resources Reports 1990; 60.

[41] HRN ISO 14235:1998. Soil Quality–Determination of organic carbon by sulfochromic oxidation. International standard. Zagreb Croatia: Croatian Standards Institute; 1998.

[42] HRN ISO 11466:2004. Soil quality–Extraction of trace elements soluble in aqua regia. International standard. Zagreb Croatia: Croatian Standards Institute; 2004.

[43] Houba VJG, Uittenbogaard J, Pellen P. Wageningen evaluating programmes for analytical laboratories (WEPAL) organization and purpose. Communications in Soil Science and Plant Analysis 1996; 27(3-4): 421-431.

[44] ISO/DIS 14870. Soil quality: Extraction of trace elements by buffered DTPA solution. International Organisation for Standardisation 1997.

[45] Houba VJG, Temminghoff EJM, Gaikhorst GA, Van Vark W. Soil analysis procedures extraction with 0.01 M $CaCl_2$. Wageningen Agricultural University Netherlands 1999.

[46] Tessier A, Campbell PGC, Bisson M. Sequential extraction procedure for the speciation of particulate trace metals. Analytical Chemistry 1979; 51(7): 844–851.

[47] Miller WP, Martens DC, Zelazny LW. Effect of sequence in extraction of trace metals from soils. Soil Science Society of America Journal 1986; 50: 598–601.

[48] SAS Institute. SAS Procedures Guide for Personal Computers Version 8th Edition. Cary North Carolina: SAS Institute Inc; 1999.

[49] Romic M, Romic D, Ondrasek G. Heavy metals accumulation in topsoils from the wine-growing regions. Part 2. Relationships between soil properties and extractable copper contents. Agriculturae Conspectus Scientificus 2004; 69(2–3): 35-41.

[50] Croatian Official Gazette (2010). Regulation on protection of agricultural land in the Republic of Croatia. Government of the Republic of Croatia Zagreb: vol 32/10.

[51] Sollitto D, Romic M, Castrignano A, Romic D, Bakic H. Assessing heavy metal contamination in soils of the Zagreb region (Northwest Croatia) using multivariate geostatistics. Catena 2010; 80(3): 182-194.

[52] Kuo S, Heilman PE, Baker AS. Distribution and forms of copper, zinc, cadmium, iron, and manganese in soils near a copper smelter. Soil Science 1983; 135(2): 101-109.

[53] Halamic J, Peh Z, Bukovec D, Miko S, Galovic L. A factor model of the relationship between stream sediment geochemistry and adjacent drainage basin lithology, Medvednica Mt., Croatia. Geologia Croatica 2001; 54(1): 37-51.

[54] Ribolzi O, Valles V, Gomez L, Voltz M. Speciation and origin of particulate copper in runoff water from a Mediterranean vineyard catchment. Environmental Pollution 2002; 117(2): 261-271.

[55] Brun LA, Maillet J, Richarte J, Herrmann P, Remy JC. Relationships between extractable copper, soil properties and copper uptake by wild plants in vineyard soils. Environmental Pollution 1998; 102(2–3): 151-161.

[56] Delas J, Dartigues A. Exemples des problèmes régionaux. II Le sud-ouest. Annales Agronomiques 1970; 21: 603-615.

[57] Moolenaar SW, Beltrami P. Heavy-metal balances of an Italian soil as affected by sewage sludge and Bordeaux mixture applications. Journal of Environmental Quality 1998; 27: 828-835.

[58] Lindsay WL, Norvell WA. Development of a DTPA soil test for zinc, iron, manganese and copper. Soil Science Society of America Journal 1978; 42: 421-428.

[59] Haq AU, Miller MH. Prediction of available soil Zn, Cu, and Mn using chemical extractans. Agronomy Journal 1972; 64: 779-782.

[60] O'Connor GA. Use and misuse of the DTPA soil test. Journal of Environmental Quality 1988; 17: 715-718.

[61] Singh BR, Myhr K. Cadmium uptake by barley as affected by Cd sources and pH levels. Geoderma 1998; 84(1-3): 185-194.

[62] Amacher MC. Nickel, cadmuim and lead. In: Methods of Soil Analysis. Part 3-Chemical Methods. Madison USA: SSSA-ASA Publications; 1996. p739-768.

[63] Merry RH, Tiller KG, Alston AM. Accumulation of copper, lead and arsenic in some Australian orchard soils. Australian Journal of Soil Research 1983; 21(4): 549-561.

[64] Bažon I, Bakić H, Romić M. Soil geochemistry as a component of terroir of the winegrowing station Jazbina, Zagreb. Agriculturae Conspectus Scientificus 2013; 78(2): 95-106.

[65] Fujikawa Y, Fukui M, Kudo A. Vertical distribution of trace metals in natural soil horizons from Japan. Part 1. Effect of soil types. Water Air and Soil Pollution 2000; 124: 1-21.

[66] Paz-Gonzalez A, Vieira SR, Castro MTT. The effect of cultivation on the spatial variability of selected properties of an umbric horizon. Geoderma 2000; 97(3-4): 273-292.

[67] Makela-Kurtto R, Sippola J. Monitoring of Finnish arable land: changes in soil quality between 1987 and 1998. Agricultural and Food Science Finland 2002; 11: 273-284.

[68] Romic M, Romic D. Heavy metals distribution in agricultural topsoils in urban area. Environmental Geology 2003; 43(7): 795-805.

[69] Cabral AR, Lefebvre G. Use of sequential extraction in the study of heavy metal retention by silty soils. Water Air and Soil Pollution 1998; 102: 329-344.

[70] Hickey MG, Kittrick JA. Chemical partitioning of cadmium, copper, nickel and zinc in soils and sediments containing high levels of heavy metals. Journal of Environmental Quality 1984; 13(3): 372–376.

[71] Emmerich WE, Lund LJ, Page AL, Chang AC. Solid-phase forms of heavy metals in sewage sludge-treated soils. Journal of Environmental Quality 1982; 11(2): 178–181.

[72] McGrath SP, Cegarra J. Chemical extractability of heavy metals during and after long-term applications of sewage sludge to soil. Journal of Soil Science 1992; 43(2): 313–321.

[73] Senesi N, Sposito G, Holtzclaw KM, Bradford GR. Chemical properties of metal-humic acid fractions of a sewage sludgeamended aridisol. Journal of Environmental Quality 1989; 18: 186–194.

[74] McLaren RG, Swift RS, Williams JG. The adsorption of copper by soil materials at low equilibrium solution concentrations. Journal of Soil Science 1981; 32(2): 247-256.

[75] Han FX, Banin A. Long-term transformations and redistribution of potentially toxic heavy metals in arid-zone soils. I. Incubation under saturated conditions. Water Air and Soil Pollution 1997; 95: 399–423.

[76] Romic M. Heavy metal contents, forms and redistribution in agricultural soils of the Zagreb region. PhD Thesis. University of Zagreb Faculty of Agriculture; 2002.

[77] Szarek-Gwiazda E, Mazurkiewicz-Boron G. Deposition of copper in the eutrophic, submontane Dobczyce Dam reservoir (southern Poland) - Role of speciation. Water Air and Soil Pollution 2002; 140(1-4): 203-18.

[78] Ariza JLG, Giraldez I, Sanchez-Rodas D, Morales E. Metal sequential extraction procedure optimized for heavily polluted and iron oxide rich sediments. Analytica Chimica Acta 2000; 414: 151–164.

Chapter 29

Characterization and Remediation of Soils and Sediments Polluted with Mercury: Occurrence, Transformations, Environmental Considerations and San Joaquin's Sierra Gorda Case

I. Robles, J. Lakatos, P. Scharek, Z. Planck,
G. Hernández, S. Solís and E. Bustos

Additional information is available at the end of the chapter

http://dx.doi.org/10.5772/57284

1. Introduction

Soil as important part of the ecosystems which must be protected in the environment context, and it is necessary be studied the possible overall impact of measures for protection, with a very special attention from mining activities. The soil resource occupies a fundamental part of the ecosystems; when a soil is degraded, the others components of the ecosystems are degraded too.

The fate of the heavy metal in soils depends upon many soil processes that are governed by several soils properties of which soil pH and redox potential are known to be the most important parameters. Thus, the solubility of trace elements is often shown as a function of pH affected by amount and kind of organic matter. Trace elements are known to be accumulated in surface soils as a result of contamination from point sources as mining activities. An appreciable amount of the soils has been made unusable because of pollution. Highly contaminated soils belong to a high healthy risk to human being and their environmentally harmful effects. That is why soil should be correctly understood and underestimated long range lethal effects that can have irreversible consequences. The improvement of soils damaged and contaminated by pollutants need of the particular soils, requires a full understanding of soil properties and of the deteriorating factors.

Mercury is one of the most toxic elements to human health and ecosystem; because of all mercury species are toxic. A wide variety of mercury species exist in the environment and its various chemical forms can differ in bioavailability, transport, persistence, and toxicity. Still, every mercury species is toxic with methyl mercury being the most toxic species. The World Health Organization (WHO) recommends a maximum methyl mercury intake of 1.6 μg Kg^{-1} per week, while the Environmental Protection Agency (EPA) lists a maximum recommended intake of 0.1 μg Kg^{-1} of body weight per day for adults. Due to high bioaccumulation, mercury is found on many levels of the food chain (Hinton and Veiga, 2001; Bengtsson, 2008). Any form of mercury in the environment may evolve into a more toxic species (methyl mercury) under biogeochemical transformation processes (Figure 1). Due to these processes and the high mobility of mercury species, a good understanding of how mercury species transform and accurate monitoring are essential for assessing the risk of mercury in the environment.

The impact of mercury depends strongly on its chemical species; understanding mercury transformations and the impact of its various chemical forms are vital to preventing harmful effects on humans and the environment. Nevertheless, the physicochemical characteristics of mercury are either useful or necessary for many industrial and agricultural applications, and mercury may be scattered over large area, depending on the source (Leopold et al, 2010).

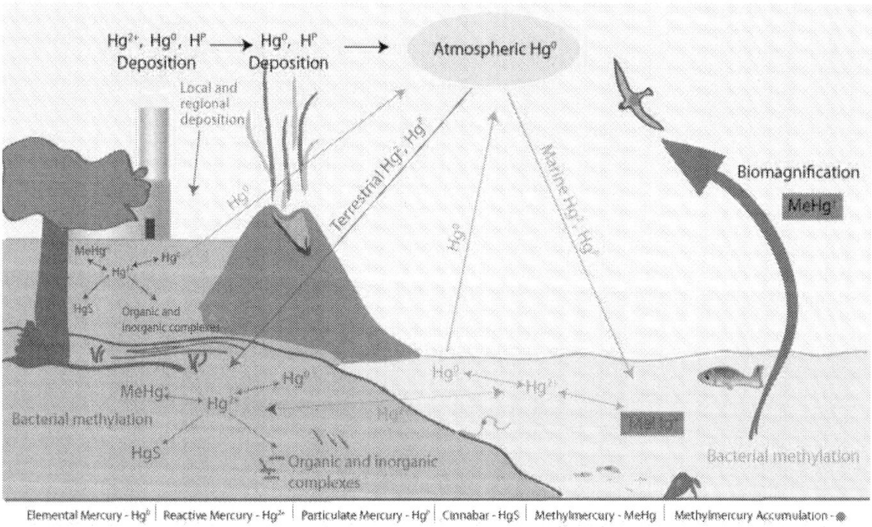

Figure 1. Biogeochemical transformation processes of mercury (Leopold et al, 2010).

Mercury concentrations in ground water indicate that the highest concentration of mercury in groundwater comes from the soil and from aquifers. While simulating mercury predictions can often be difficult, mercury can be estimated in experiments conducted in batch mode or in columns. The percentage of Hg that can potentially leach from the soil was previously estimated in batch experiments. Distribution analyses of species in leachate confirmed the presence of inorganic species (Hg^{2+} and Hg^0) ranging from 90 – 100 % (Bollen, 2008; Harvey, 2002).

The most commonly used techniques for the remediation of mercury contaminated soils have been classified as either excavation techniques or containment techniques, and are grouped as follows (Hinton and Veiga, 2001): (a) *ex situ* treatments: physical separation, thermal treatments, hydrometallurgical treatments; (b) *In situ* recuperation: vapor extraction coupled with evaporation (soil), permeable reactive barriers; (c) *In situ* leaching and extraction: electrokinetic separation, interceptor systems, phytoremediation, passive remediation; (d) *containment:* pump and treat impermeable barriers, sealed surfaces and drainage, stabilization and solidification, sediment covering.

2. Physicochemical properties of mercury

Mercury (Hg) is a chemical element with an atomic number of 80. Mercury is a silver plated heavy metal, liquid and odorless at normal conditions. It easily alloys with many other metals like gold or silver producing amalgams, is insoluble in water and soluble in nitric acid. The main source of Hg is cinnabar or mercury sulfide (HgS), a stable compound and insoluble usually recovered as a byproduct of ore processing. Mercury in this form is found in the earth's crust average concentrations of 0.5 ppm (Hinton and Veiga, 2001).

Mercury is one of the most toxic elements to human health and ecosystem. At temperatures above 40 °C mercury produces toxic and corrosive fumes. It is harmful by inhalation, ingestion and contact, is a very irritating to skin, eyes and respiratory tract, even to nervous system, its gaseous form is absorbed by lung tissues (Hinton and Veiga, 2001, Bengtsson, 2008).

A wide variety of mercury species exist in the environment and its various chemical forms can differ in bioavailability, transport, persistence, and toxicity. Still, every mercury species is toxic with methyl mercury being the most toxic species. This element can exist in the environment as elemental (Hg^0), oxidized inorganic (Hg^{2+} -mercuric, Hg_2^{2+}-mercurous) or oxidized organic (methyl/ethyl mercury) forms. Mercuric and mercurous forms are more stable under oxidizing conditions. In moderately reducing conditions, the organic or inorganic mercury can be reduced to its elemental form and be converted to forms leased by biotic or abiotic processes: these are the most toxic forms of mercury, as well as being soluble and volatile. Hg (II) forms strong soluble complexes with a variety of organic and inorganic ligands oxidized in aqueous systems. Hg sorption in soil, sediment and humic materials is an important mechanism for the removal of mercury from solutions, another mechanism, a high pH is their co-precipitation sulfide (HgS) (Leopold et al, 2010).

Any form of mercury in the environment may evolve into a more toxic species (methyl mercury) under biogeochemical transformation processes. Due to these processes and the high mobility of mercury species, a good understanding of how mercury species transform and accurate monitoring are essential for assessing the risk of mercury in the environment. The impact of mercury depends strongly on its chemical species; understanding mercury transformations and the impact of its various chemical forms are vital to preventing harmful effects on humans and the environment. Nevertheless, physicochemical characteristics of mercury are either useful or necessary for many industrial and agricultural applications, and mercury may be scattered over large area, depending on the source (Leopold et al, 2010; Nick, 2012).

The metal mercury (Hg^0) is mainly used to produce chlorine gas and caustic soda, and is part of some types of alkaline batteries, fluorescent lamps, electrical contacts, and instruments such as pressure gauges and thermometers, among others. Hg salts are used in antiseptic ointments and creams and skin lightening. Among the activities that generate the most pollution by Hg, is the burning of coal and chlor-alkali plants: other important sources are mining and metallurgy and the burning of municipal solid waste, which may contain instruments such as pressure gauges, thermometers, alkaline batteries and fluorescent lamps. The mercury released into the air tends to settle and adhere to soil organic matter (Hinton and Veiga, 2001; Nick, 2012).

Natural and anthropogenic mercury emissions are mainly in the form of elemental mercury (Hg^0), which makes up about 99 % of total atmospheric mercury. However, biogeochemical transformations can oxidize it, forming Hg^+ and Hg^{2+}. Most inorganic Hg-compounds are water soluble in small doses, and can be found in soil and sediments. In contrast, the presence of inorganic forms of Hg^{2+} bonded to organic and/or inorganic species ($[HgCl_x]^{2-x}$; $[Hg^{II}\text{-DOC}]$; [HgS]) depends on the local chemical environment. The life time of these compounds in air is very short (on the scale of minutes) and they are rapidly removed by deposition processes because of high water solubility and surface activity. Figure 2 shows the main mercury species in the atmosphere, hydrosphere and sediment (Leopold et al, 2010; Nik, 2012; Slowey et al, 2005; Wartel et al, 1999; Shi et al, 2005).

Over 90 % of surface water mercury is from atmospheric deposition. Hg^{2+} usually undergoes a biomethylation process that forms methylmercury (MeHg, CH_3Hg^+) and dimethyl mercury (DMeHg, $(CH_3)_2Hg$), though these reactions can be reversed using microorganisms and/or photolytic decomposition. All these species are highly mobile.

Three main forms of mercury are found in natural waters: elemental mercury (Hg^0), inorganic Hg^{2+} ($Hg^{2+}$and its complexes) and organic mercury (MeHg, MeHg complexes and DMeHg). With solubility (at 25°C) of 0.08 mg L^{-1}, Hg^0 can be found at all depths. Inorganic mercury (Hg^{2+}) and MeHg forms complexes with other dissolved compounds in fresh water, but for the most part, only forms complexes with chlorine in sea water. DMeHg is found in the deep sea. (Wartel et al, 1999, Shi et al, 2005, Slowey et al, 2005).

3. Edaphology properties of soil related with mercury

Soil is a collection of natural bodies on the Earth' surface, in places that most of them have been modified by man in its quality and containing living matter and supporting or capable of supporting plants. Soil grades at its lower margin to hard rock or to earthy materials virtually devoid of roots, animals or marks of other biologic activity.

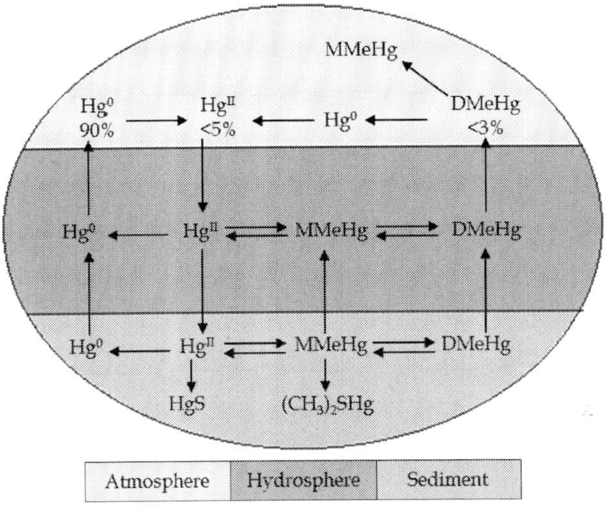

Figure 2. Distribution of mercury species in atmosphere, hydrosphere and sediment (Leopold et al, 2010).

In general, physical, chemical and biological soil characteristics are highly correlated parameters that are necessary to understand. The specific elemental composition of each particular soil reflects, to a degree modified over time by weathering and the chemical composition of the parent material from which the soil is formed. For instance, the extractability of the different elements depends on the soil properties.

Soil as important part of the ecosystems which must be protected in the environment context, and it is necessary be studied the possible overall impact of measures for protection, with a very special attention from mining activities. The soil resource occupies a fundamental part of the ecosystems; when a soil is degraded, the others components of the ecosystems are degraded too.

The fate of the heavy metal in soils depends upon many soil processes that are governed by several soils properties of which soil pH and redox potential are known to be the most important parameters. Thus, the solubility of trace elements is often shown as a function of pH affected by amount and kind of organic matter. Trace elements are known to be accumulated in surface soils as a result of contamination from point sources as mining activities.

An appreciable amount of the soils has been made unusable because of pollution. Highly contaminated soils belong to a high healthy risk to human being and their environmentally harmful effects. That is why soil should be correctly understood and underestimated long range lethal effects that can have irreversible consequences. The improvement of soils damaged and contaminated by pollutants need of the particular soils, requires a full understanding of soil properties and of the deteriorating factors.

Mercury is a microelement: its Clark value in the Earth's crust is 56 μg Kg^{-1} (Fügedi et al, 2011). It is characterized by a dual geochemical behavior: it is liable to extreme concentration and to dispersion, the latter resulting in an approximately entirely even concentration. It is found either as a native metal (near to 80 % in hydrothermal and vapors) or in cinnabar, corderoite, livingstonite and other minerals. Cinnabar (HgS) is the most common ore. Mercury ores usually occur in very young orogenic belts where rock of high density on upper mantle is forced to the crust of the Earth (Ozerova, 1996).

Given that mercury is enriched by an extremely wide variety of geological processes (Fergusson, 1990) from the formation of hydrocarbon to hydrothermal mineral occurrences, it can be regarded as an universal geochemical indicator of young geological effects; its dispersion halos are more extensive than that of any other element (Fügedi et al, 2011).

Historically there were two main registered Mercury mines: Almaden (Spain) and Idrija (Slovenia) in Europe. Later new occurrences were found in California and worldwide. It was used in gold separation. In Mexico we know mercury mines from the Pre-Hispanic era (Scharek et al., 2010) and a usage in cultic fests (Figure 3). In 2005, China was the top producer of mercury with almost two-thirds global share followed by Kyrgyzstan. Several other countries are believed to have unrecorded production of mercury from copper electro winning processes and by recovery from effluents.

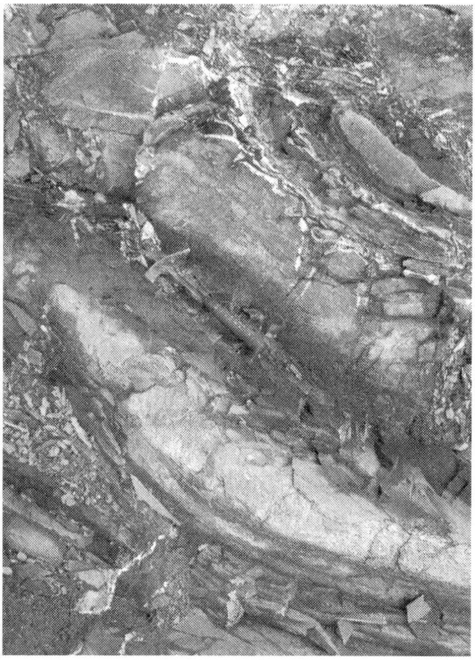

Figure 3. Typical cinnabar occurrences in a limestone system (Formation Las Trancas, San Joaquin, Querétaro, Mexico, photo by P. Scharek)

4. Mobilization and loadings the mobilized mercury

The mobilization of the different mercury forms can due to by evaporation and dissolution. Here we do not deal with the erosion which able to mobilize all form of mercury if it has attached to the solid particle, however in case of soils and fly as particles in the flue gas this particle associated mobilization mechanism can play important role in the mercury transport.

4.1. Evaporation

The evaporation governed by the vapor pressure depends on the volatility of the compound and the temperature. Concerning the volatility of the different mercury compounds evaporation at ambient temperature can be significant in case of elemental and the organic mercury cases, however the volatilization of the other mercury forms (the inorganic mercury compounds) can become considerable if the temperature reaches a couple hundred degree centigrade. All of the mercury compounds have relatively low boiling points (Table 1), some of them decompose before melting, others can sublimate. Based on these data it is obvious the vaporization can play important role of the mercury compound transport and mobilization.

Hg	**Melting T, ˚C**	**Boiling T, ˚C**
	-38.9	356.5
HgS (cinnabar, α)	580 sublimate	
HgS (metacinnabar, β)	446 sublimate	
Hg_2O	100 decompose	
HgO	500 decompose	
Hg SO_4	450 decompose	
Hg_2Cl_2	302	384
$HgCl_2$	277	304
$HgBr_2$	237	322
HgI_2	259	354
CH_3HgCl	-	92
$(CH_3)_2Hg$	-43	94

Table 1. Melting and boiling points of the mercury and mercury compounds

The partial pressure of the elemental mercury (Hg^0) reach 1 Pa at 42 ˚C and enhances exponentially till the boiling point (T_b = 356.5 ˚C). At 20 ˚C the Hg vapor pressure is 0.18 Pa the Hg concentration in the air saturated with the mercury is 7.64 10^{-8}mol dm^{-3} = 15.3 μg m^{-3}. Due to this high volatility the elemental mercury evaporate if stored and processed an open container. Elemental mercury can escape from solution if the oxidized mercury is able to reduce. The analytical data will be inaccurate if the sample is not preserved agents the elemental mercury formation.

Concerning the global mercury contamination till the middle of the past century evaporation of elemental mercury used to extract silver and gold was the main source of the mercury emission, (Nriahu, 1994). All the once produced and recently available elemental mercury stock (in the past five century one million tons was produced from cinnabar and from other ores) if in used either evaporation or after transformation can contribute to the mercury contamination worldwide (Hylanderand Meili, 2003). This is the reason why the elemental mercury use is banned. Recently one of the most significant sources of mercury emission by evaporation is the coal firing. During the coal burning the mercury associated with pyrite and be organically bonded to the coal minerals are released in the combustion flame as elemental mercury, which is partially oxidized to Hg(II) in homogeneous and heterogeneous catalytic reaction governed by the chlorine and the ash content of the combustion gases (Sondreal et al, 2004).

Generally accepted view is that the evaporated oxidized forms of mercury contaminate the environment locally, close to the emission source. However they can transform to elemental mercury and depend on this transformation rate it can become part of the global mercury cycle.

The transformation of the oxidized to reduce the reduced to oxidized forms can happen both in gas and aquatic environment according to the circumstances. There is similar transformation between the inorganic and organic forms.

Different species of macro algae from the dissolved mercury can produce different methylated mercury compounds in the ocean. Because of these methylated mercury compounds have high volatility and at the dimethylated form has low solubility in ocean water they are easily emitted into the atmosphere and can contribute significantly to the global atmospheric mercury (Pongratz and Heuman, 1998). Beside this different bacteria (e.g. sulfate reducing) and in case of abiotic route the tin- alkyls and the humic acids also can transform the dissolved mercury (II) to methyl mercury form (Weber, 1993). A quite detailed set of possible transformation in gas and aquatic media and the Henry constants which inform about the dissolved compound volatility are collected by Shon at al, 2005.

During heating mercury compounds can transforms directly or via oxides to elemental mercury. Beside the elemental mercury only the halogenides and the sulfides since last have a tendency to sublimate can occur in evaporated forms. The sulfides at presence of oxygen at 600 °C transforms to Hg and SO_2, however in presence of Fe and CaO the HgS also will decompose to Hg and Fe- or Ca- sulfides. Using the temperature programmed evaporation technique based on the volatility difference of mercury compounds the compound forms can be distinguished and can use for mercury speciation in solids (Lopez-Anton et al, 2010, 2011).

In a high temperature process since the mercury compounds decompose the original speciation of mercury does not preserve a new speciation can formed which determined by the gas composition. In the high temperature gases high portion of mercury exists in elemental and just a small portion in oxidized form. This is the reason why these technologies such as coal fired energy production, the cement kiln, the incineration has difficulty in the mercury capture.

Focusing to the soil, the heating comes from sunlight can mobilize only the weakly sorbed elemental and organic mercury but the fire on the soil surface, for example the forest fire can evaporate the less volatile mercury forms as well. This case the contamination level of the fired soil decreases but, due to the transport, at other places the contamination becomes higher (Caldwel et al, 2000).

The volatilization can be the cause of contamination but can use for decontamination as well. Based on the volatilization of mercury compounds, mercury removal process was established from coal cleaning by mild pyrolysis (Wang et al., 2000) and for the soil cleaning by thermal treatment. In case of coals the speciation of mercury determines the maximum efficiency of the mercury removal. The efficiency of the process generally remains below 100 %, (bituminous coal case at 500 °C it was aprox. 75 %). Since the efficiency remains below 100 % the rest of mercury still remain in the process and pass to the flue gas after the coal burning. The speciation of the mercury in the contaminated soil also has influence on the efficiency of the thermal remediation, see more details later.

Concerning that the different mercury forms exhibit different volatility the actual distribution of the mercury species in a medium the rate of the transformation process which able to modify it together govern the mercury mobilization by evaporation.

It is well known, if elemental mercury forms in the water this elemental mercury can easily escape to the gas phase. It is quite intensive if gas bubbling through the water or the water surface is disturbed (Okouchi and Saaski, 1984). Sunlight induced H_2O_2 formation in alkaline condition can result reduction of the oxidized mercury forms to elemental mercury. This can explains that the Hg concentration above the lake water surface can be higher day time than night. The fulvic and humic compounds are able to complex the mercury (II) ion in aquatic media but these compounds can take part in the mercury alkylations, further at a suitable pH can work as a reducing agent. The redox potential at 0 pH for Hg (II) reduction is 0.85 V(Allard and Arsenie, 1991).

This type of mercury transformation between oxidized and reduced forms together with the alkylation will generate not only a modification between the concentrations of the mercury species in the aquatic phase but will modify the mercury transport between the phases. The mercury transformation processes are important in the technological processes used for the mercury removal since can effect they efficiency (Somoano et al., 2007)

4.2. Dissolution

The mobilization by dissolution can arrange two groups: (a) dissolutions ways exist in the nature (b) dissolution way can be applied in the laboratory and in the remediation technology to determine the loading forms or remove the mercury from the contaminated media.

4.2.1. Dissolutions ways exist in the nature

The solubility of elemental mercury and the ore of mercury can find in the nature (cinnabar etc.) are very low in water. This low solubility result low mercury concentration level in aquatic phase and restricts the transport between phases by dissolution. However the oxidation both cases enhances these mercury forms solubility. The elemental mercury can be oxidized by ozone, halogens, some components of acid rains, or by oxy-acids in laboratory (HNO_3 and the hot H_2SO_4) resulting a soluble form. The ozone in air if does not consumed by the other more reactive air contaminants can oxidize Hg to Hg(II) (Iverfeld and Linquist, 1986; Shonet al, 2005).

In aquatic media oxidation can occur at acidic conditions if the sunlight produces oxidative radicals OH, or peroxides. This process can play role in the trap of the physically dissolved elemental mercury in water, and also can hinder the transformation of the oxidized mercury forms towards the reduced elemental mercury direction. The oxidative transformation of elemental mercury is essential in case of many mercury capture process since the oxidized forms of mercury has higher tendency to sorb and dissolve, therefore different oxidation procedures are available and applied in the demercurysation technologies (Ko et al, 2008; Lakatos et al, 2009; Sondreal et al, 2004).

However the mercury in the natural minerals is in the oxidized forms these minerals luckily due to the very low solubility can be considered not a mobile occurrence of the mercury. The environmental risk improves if the natural processes can transform the minerals a more soluble form. One of the most significant ore transformations which effect the mercury mobilization is the sulfide ore oxidation in the air. The oxidations of sulfides to sulfate a considerable enhancement ensue in mercury solubility (Holley et al, 2007). This process, the oxidation of the tailings, can accused for the mercury contamination all around the abandoned mercury ore mines.

Compound	**Solubility in Water c, ppm**
Hg	0.049*
HgS (cinnabar, α)	0.01
HgS (metacinnabar, β)	-
$HgSO_4$	-
$Hg(NO_3)_2$	soluble
Hg_2O	51
HgO	51
Hg_2Cl_2	10
$HgCl_2$	66 000*
$HgBr_2$	5 100
HgI_2	51
CH_3HgCl	5 780
$(CH_3)_2Hg$	-

*Solubility from paper of Ko et al, 2008.

Table 2. Solubility of different mercury compounds.

Among the mercury compounds (Table 2) the mercury-chloride and nitrates are those which have the highest solubility in water. The simple cationic form of Hg(II) is not the common form in the aquatic media, it exist only in acidic solutions, at less acidic condition the dissolved mercury appears as $HgOH^+$, HgOHCl, $Hg(OH)_2$ and $HgCl_2$ molecules and complex anions $HgCl_4^{2-}$ at high chloride concentration. It means that the sea water contains the oxidized mercury mainly in this chlor- complex form. Beside the chlor- complex the mercury -fulvo and -humic complexes also exists in aquatic environment. The speciation in the solution, the molecular forms govern the mercury loadings and play important role at the way and efficiency of the removal.

4.2.2. Dissolution for leaching mercury from different medium

Beside the thermal way and the application of the species sensitive analytical methods for mercury analysis (XPS, EXAFS etc.) the sequential extraction is often applied technique to specify the mercury chemical form and associations in solids. The thermal methods and the species sensitive elemental analysis can distinguish the elemental Hg, the HgS forms and the organically bonded mercury forms which generally exist in the soil. However these techniques do not allow doing any estimation about mobility and bioavailability of mercury. To get the loading specific information in soil for mercury, beside the classical Tessier six step extraction used generally, different modified procedure are available for Hg which able to distinguish better the mercury forms than the Tesssier method can do (Orecchio and Polizzotto, 2013; Han et al,2006). For example it can determine mercury bounded to amorphous iron oxides (by NH_4 oxalate-oxalic acid extraction), mercury bonded to crystalline iron oxides (by $NH_2OH \bullet HCl$- 25 % acetic acid extraction), non-cinnabar mercury (elemental mercury, organic bounded, humine bounded (by 4 M HNO_3 extraction), cinnabar mercury (by extracted with saturated Na_2S,Han et al, 2006). The advantage of this protocol is the ability of the separation of humic and sulfide bounded mercury which important in the soil case. Two set of sequential extraction regime can compared at Table 3 and 4.

Mobilization of mercury can occur through complex formation, ligand exchange reactions with chloride and sulfur-containing ligands which leading to enhanced Hg solubility in soil solutions. The sulfur containing ligands: tiosulfates ($S_2O_3^{2}$), thyocyanites (SCN^-) can mobilize the mercury efficiently and could improve the phytoextraction efficiency (Moreno et al, 2004).

Removal mercury by phytoextraction from soils and others soil like materials eg. waste water plants biosolids often need additives which improve the solubility of the mercury. These mobilizations agents are used in accelerate phytoextraction. One type is the chelating agents: citrate, oxalate, malate, succinate, tartarate, salicilate, acetate, and amino-poly-carboxylic acids: EDTA (Lomonte et al, 2011). Since the EDTA is persistent compound, recently the biodegradable ethylendiamine-disuccinate (EDDS) or nitrilotriacetic acid (NTA) suggested as alternative chelator instead of EDTA (Evangelu et al, 2007).

Specific compounds used the mercury extraction from tissues: they can pay role in case of poisoning for detoxification : EDTA was tested for detoxification by Aposhian (Aposhian et al,1995), 2,3- dimercapto- 1- propansulfonate was used to extract mercury from tissues of rats exposed to different mercury compounds (Buchet and Lauwerys, 1989). The EDTA was not

Mobilization Protocol	Determination of Association of Mercury to Soil Component
Loading form of mercury	Extraction procedure
Water soluble mercury	NH_4-acetate
Exchangable mercury	(1 M NH_4-acetate pH 7 set with NH_4OH: solid:liquid 1:25, 30 min 25 ºC)
Carbonate bounded mercury	Hidroxylamine – HCl
Easily reducible oxides bounded mercury (Mn-oxides)	(0.1 M NH_2OH.HCl +0.01 M HCl solid:liquid1:25, 30 min 25 ºC)
Elemental and organic bounded mercury	H_2O_2 (3 mL 0.1 M HNO_3+5 mL 30 % H_2O_2 80 ºC 2 h; 2 mL H_2O_2 80 ºC 1 h; 50 mL 1M NH_4- acetate)
Amorphous iron- oxide bounded mercury	NH4-oxalate –oxalic acid (0.2 M oxalate buffer 1:1, pH 3.25 solid:liquid1:25)
Crystalline iron oxide bounded mercury	Hydroxilamine – HCl- acetic acid (hot) (0.04 M NH_2OH.HCl in 25 % acetic acid 97-100 ºC 3 h solid:liquid 1:25)
Non cinnabar bounded mercury (Hg, organically bounded, humin bounded) TOT (non cinnabar mercury)	4 M HNO_3 (4 M HNO_3 80 ºC16 h solid:liquid1:25)
Cinnabar bounded mercury	Na_2S (4 mL saturated Na_2S 12 h repeated twice)

Table 3. Mobilization protocol for determination of association of mercury to soil component (Han et al, 2006).

Mobilization Protocol	Determination of Association of Mercury to Soil Component
Loading form of mercury	Extraction procedure
Water soluble mercury	H_2O100ºC, 1h stirred, solid:liquid 1:8
Exchangable mercury	Na-acetate (1M Na-acetate, 1 h stirred, solid:liquid1:8
Carbonate bounded mercury	Na-acetate –aceticacidpH 5 (1M Na-acetate –acetic acid pH 5, 4 h stirred, solid:liquid1:8)
Fe, Mn-oxide bounded mercury	Hydroxylamin HCl - acetic acid 0,04 M NH_2OH•HCl in 25 % Acetic acid, 96 °C, 6 h, Solid:liquid 1:8
Elemental and organic bounded mercury	Mineralisation by HNO_3-H_2O_2 (a)heated previously 180 ºC - organic bounded - microwave digestion in cc HNO_3-H_2O_2 mixture; (b) no heat - elemental + organic bounded - microwave digestion in cc HNO_3-H_2O_2 mixture)
Sulfid bounded mercury	Aqua regia (HCl:HNO_3 3:1)

Table 4. Mobilization protocol for determination of association of mercury to soil component (Orecchio and Polizzotto, 2013).

found the best for mercury removal in human application since does not the best chelator for Hg and has side effects, however there are two other compounds which suggested to keep in stock in any poison control center as mercury chelator DMPS (2,3-dimercapto-1-propane sulfonic acid, (unithiol)) and DMSA (meso-2,3 dimercapto succinic acid, succimer, Guzzi et al, 2010).

4.3. Loadings of mercury

The mercury contaminations are in the environment can exist different phases: as vapor in the air (Hg, and compounds, particle associated), dissolved in aquatic media (Hg^{2+}, $Hg(OH)_2$, $HgCl_2$, $HgCl_4^{2-}$, different complexes, particle associated) and solid as precipitates or minerals and in associated forms bonded to different manner to the component of different solids (soil, fly ash, waste water sludge, etc). In the previous section it was demonstrated that how the associations can be identified.

The loadings of mercury to solid can be considered as positive or negative phenomenon. It can restrict the dispersion of the contamination one side it is positive, the negative this way it can preserve the contamination. Since the loadings depends on the character of the collector and the speciation of the mercury, difficult to establish general rules for this process. However it can state that the elemental mercury has low sorption ability, the cationic sorbs better than the anionic forms on clays, and negatively charged carbon surfaces (coals, activated carbons, humic materials) the loading is more effective to that surfaces which have contain sulfides. It was interesting findings after the cinnabar oxidation a part of liberated mercury could load to the cinnabar surface this way it can be not just the source but the collector of mercury ions. Unfortunately the most toxic forms (alkyls) have the highest ability for bioaccumulation.

The nature works against the mercury contamination. Except the alkylation it transforms the mercury toward the most stable less soluble form. Near the chlor-alkali plant the total mercury sometimes reach the four order of magnitude higher level, than the background concentration, luckily it found a non-volatile and non-soluble associations since transforms to sulfides (Bernaus et al, 2006).

The history of mercury contamination is recorded by loadings. The dept profile of mercury concentration on peat can provide a clear picture how the mercury contamination changed during the mankind history (Barraclough et al, 2002).

The loadings play important role in the environmental technologies used for decrease the mercury emission or clean the contaminated medium. Sulfur and halogen containing carbons, oxidative inorganic sorbents (Lakatos et al, 2009) were developed for elemental mercury removal from flue gas. Beside a range of, classical, functionalized sorbents, sulphur containing carbon nanotubes widen the collection one can chose among for eliminate the mercury contamination in aquatic media (Pillay et al, 2013).

The coals especially the low rank and the oxidized coals are very good mercury ion collectors. Due to this feature we must face that the coal-firing are the main source of the anthropogenic mercury contamination nowadays. However this material offers us an application for cure a slice of the mercury problem: remove the mercury from aquatic media. It can use in batch mode or dynamic systems as the reactive barrier material, by the high mercury capture coal are able to retard or remove the aquatic mercury contamination (Lakatos et al,1999).

5. Remediation of polluted soil with mercury

The most commonly used techniques for the remediation of mercury contaminated soils have been classified as either excavation techniques or containment techniques, and are grouped as follows (Hinton and Veiga, 2001): excavation and *ex-situ* treatments, containment and *in situ* chemical treatment (Figure 4).

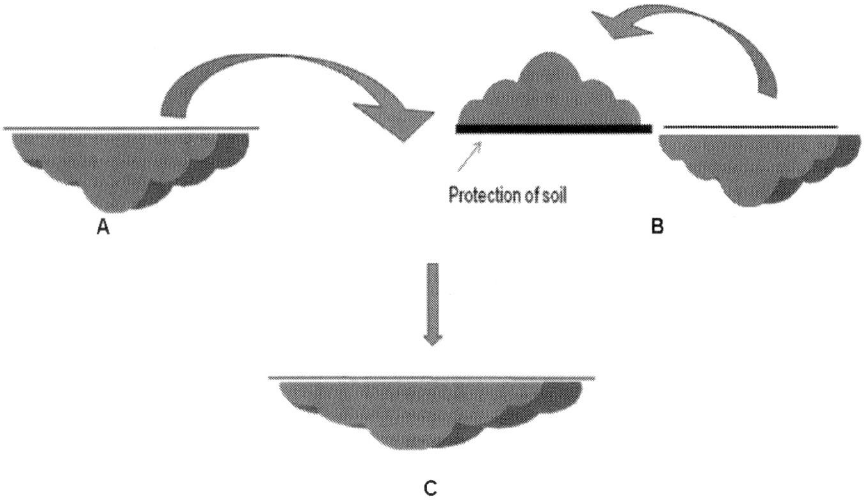

Figure 4. Representation of the techniques for the remediation of mercury contaminated soils: (A) excavation and *ex-situ* treatments, (B) containment and (C) *in situ* chemical treatment.

5.1. Excavation and *ex situ* treatments

They will be treated off-site soil contaminated when removing soil or contaminated soil-like materials to a place outside the place in which they are located, for submission to authorized treatment fixtures. (Hinton and Veiga, 2001):

1. *Physical separation.* Mercury has affinity for the smallest particles of soil.
2. *Thermal treatment.* The volatility of mercury increases with increasing temperature; therefore, a heat treatment technique of excavated soil is a potentially effective technique for the removal of mercury in soils.

3. *Hydrometallugical treatment.* Chemical extraction of mercury in contaminated soils can be induced by four mechanisms: desorption of adsorbed species, oxidation of metallic mercury, use of strong complexing agents, and dissolution of Hg precipitate. The efficiency and mechanism employed decrease with respect to time due to recomplexation, readsorption and removal of the soluble fraction. Two of the two most promising hydrometallurgical techniques are electrokinetics and electroleaching / methods of leaching.

4. *In situ recuperation. In situ* techniques have not been studied as much as *ex situ* techniques due to surface heterogeneity and longer treatment times. However, *in situ* techniques may be promising due to better cost-effectiveness and practicality.

5. *Vapor extraction coupled with evaporation.* Vacuums are used in unsaturated zone to remove volatile and semi-volatile contaminants.

6. *Permeable reactive barriers.* Dissolved compounds react with compounds found on the walls and then precipitate out. This technique has been employed for the treatment of sites contaminated with organic compounds and metals. These barriers are placed perpendicularly to the flow of contaminants.

7. *In situ leaching and extraction.* Used together with pumps in treatment, this method uses chemicals injections to improve the solubility of mercury in groundwater.

8. *Electrokinetic separation.* This process involves the generation of an electric field by applying a potential difference or current into a soil matrix. Metals such as mercury migrate towards electrodes placed in the soil where they accumulate and can be removed at a lower cost by excavating the affected area.

9. *Interceptor systems.* Interceptor systems such as ditches and drains are simple and effective for the recovery of mercury as free product, but these treatments are limited by site topography and stratigraphy.

10. *Phytoremediation.* Some plants have the ability to assimilate and concentrate metals in soil. The recovery of these metals occurs after collecting and incinerating the plants.

11. *Passive remediation of the wetlands.* Using wetlands to immobilize mercury is a controversial topic as some wetlands contain microorganisms that can convert mercury into even more toxic species.

5.2. Containment

In the containment treatment the soils are treated on one side of the contaminated site, where the processing is performed on an area adjacent to the contaminated site or an area within the contaminated site upon removal of soil or soil-like materials. In this classification are (Hinton and Veiga, 2001):

1. *Pump and treat.* With certain contaminants or systems, pollution removal is not possible and it is necessary to protect hydraulic content. When the contaminant mass remains in the subsurface, pump and treat systems can prevent site contamination.

2. *Impermeable barriers (sealed surfaces and drainage).* Mud barriers are slightly permeable barriers made of bentonite or cement-bentonite mixtures. Generally, these barriers are between 0.5 and 2 m thick and have a maximum depth of 50 m. There are other types of barriers that are constructed by injection molding or by vibratory forces. On the other hand, surface seals and drainage are used to controlling filtration and limit pollutant movement towards groundwater.

3. *Stabilization and solidification.* Stabilization and solidification techniques use both *in situ* or *ex situ* conditions by mixing impacted sites. Stabilization attaches contaminants to the soil structure, which usually decreases soil permeability. Moreover, solidification improves the physical characteristics of materials such as mudor sediments; they can be excavated and transported more easily.

4. *Sediment covering. In situ* covering involves placing an insulating layer over the contaminated material.

5.3. *In situ* chemical treatment

Another option is the *in situ* chemical treatment option, which is the name of all treatments that involve the injection of a chemical reagent into an aquifer source upstream of the contaminated site. This chemical agent reacts with the contaminant, transforming it into an innocuous form; eventually, it can pump through a given volume of water which can later be recycled for injection. The following actions must be considered:

1. Increase the output rate of the ground water through the contaminated zone by increasing the hydraulic gradient through injection and extraction.

2. Transform the contaminant using chemical reaction within the aquifer.

5.4. Electroremediation of polluted soil with mercury

Electroremediation has been successfully applied in a variety of soil restoration studies, this methodology having the advantage of exhibiting simultaneous chemical, hydraulic and electrical gradients. Indeed, for efficient mercury removal from a saturated soil with electroremediation, application of either an electric field or direct current through two electrodes (anode and cathode) is required. These are usually inserted in wells containing a supporting electrolyte made from inert salts, leading to improved electric field conductive properties (Rajeshwar et al, 1994; Huang et al, 2001; Acar and Alshawabkeh, 1993).

Furthermore, since electroremediation is a physicochemical technique based on ion transport, it is an excellent tool for the removal of inorganic species, such as Hg^{+2} (Rajeshwar et al, 1994;

Bustos, 2013). The main advantages of electroremediation, as compared with other soil treatment procedures, are (Huang et al, 2001; Acar and Alshawabkeh, 1993; Ibañez et al, 1998; Segall and Bruell, 1992; Cabrera – Guzmán et al, 1990): (1) electroosmotic flow is not dependent on either pore or particle size, (2) hydraulic gradient is enhanced by electromigration, (3) treatment can be applied *in situ*, (4) it can be applied to low permeability soils, (5) there is minimal disruption of normal activities at the site, (6) the required investment is usually lower than that for other conventional treatments, and (7) it can be applied in conjunction with techniques such as pumping, vacuum extraction or bioremediation.

The processes taking place during electroremediation can be classified into two main categories: (a) processes occurring as a consequence of the applied electric potential. These processes include electromigration (ion transport), electroosmosis (mass transport), and electrophoresis (charged particle transport); (b) processes occurring in the absence of an electric potential. This includes concentration induced processes like diffusion, sorption, complexation, precipitation and acid - base reactions (Reed et al, 1995; Bustos, 2013).

Specifically, for mercury polluted soil electroremediation, the use of complexing agents like ethylendiaminetetraacetic acid (EDTA), KI, and NaCl under a constant potential gradient has been reported (Reddy et al, 2003). Based on the above precedents, the electroremediation was developed aided by extracting agents for mercury removal from San Joaquin's Sierra Gorda soil samples (Figure 5, Robles et al, 2012).

Electroremediation of mercury polluted soil, facilitated by the use of complexing agents, proved to be an attractive alternative treatment for the removal of mercury from polluted soil in mining areas located at Sierra Gorda in Queretaro, Mexico (Figure 5A and 5B). Implementation of this remediation protocol is expected to improve the living conditions and general health of the population in the Mine "El Rincón" in San Joaquin (Figure 5C). Experimental observations suggest that it is possible to remove up to 75 % of metal contaminants in mercury polluted soil samples by wetting them with 0.1 M EDTA, placing them in an experimental cell equipped with Ti electrodes, and then applying a 5 V electric field for 6 hours (Figure 5D, Robles et al, 2012). When we followed the electrochemical removal of mercury in a batch reactor (Figure 6A), it was removed around 87 % of Hg^{2+} in a time of 9 hours close to the anode side by the presence of EDTA (Figure 6B). The pH remains nearly constant at 4 and conductivity showed values close to 10 mS cm^{-1} by the ionic species.

The efficient removal of mercury contaminants observed under these conditions is attributed to electromigration of the coordination complexes that form between the terminal hydroxyl groups in EDTA and divalent mercury (Hg^{+2}), which is probably strengthened by supramolecular interactions between unshared electrons at EDTA's tertiary amino nitrogens and Hg^{+2}. These interactions are particularly effective with the presence of potassium ions. This observation is supported by molecular modeling of several possible interactions in the proposed complex using the Density Functional Theory method (B3LYP LANL2DZ, Robles et al, 2012, Figure 7).

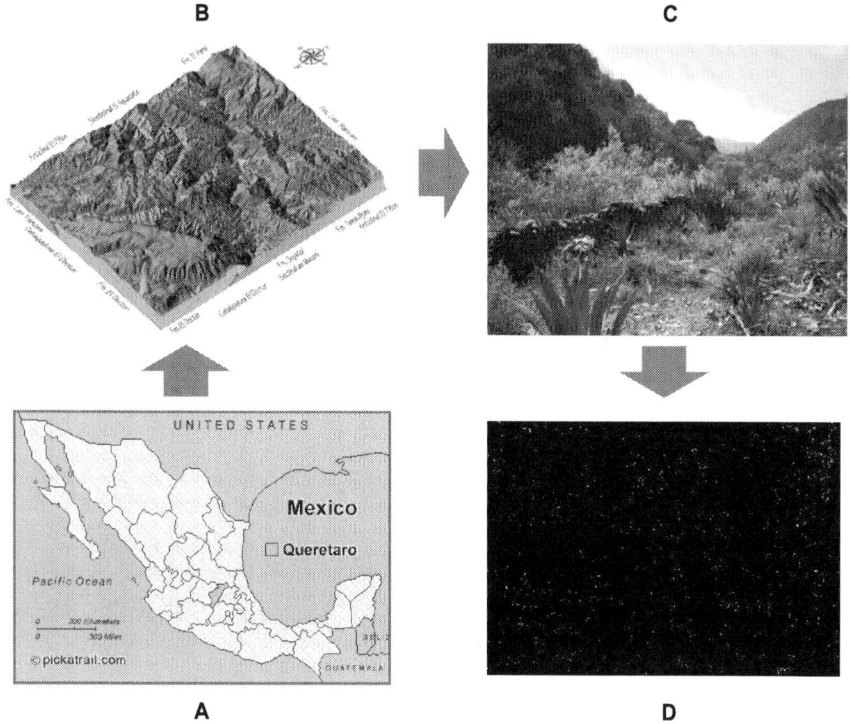

Figure 5. Localization of Queretaro in Mexico (A) with satellite image from San Joaquin's Sierra Gorda, Queretaro (B) where there is the Mine "El Rincón" (C) with high concentration of Hg^{2+}, which was removed with electroremediation process in continues flow in presence of EDTA (D).

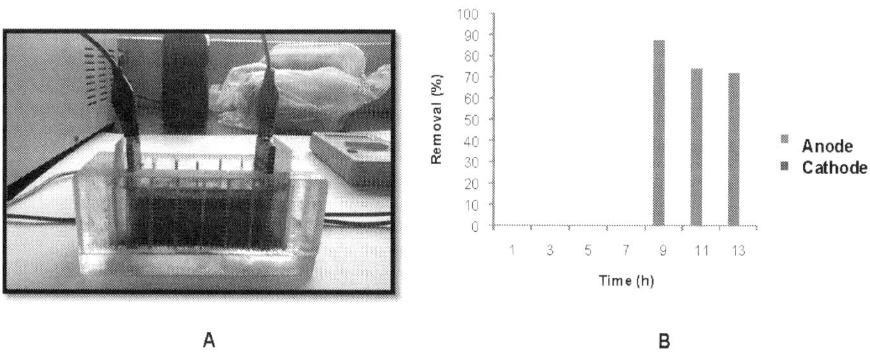

Figure 6. Electroremediation process in batch reactor assisted by EDTA (A), and its corresponding removal percentage of Hg^{2+} followed during 13 h of treatment, close to anode and cathode.

http://dx.doi.org/10.5772/57284

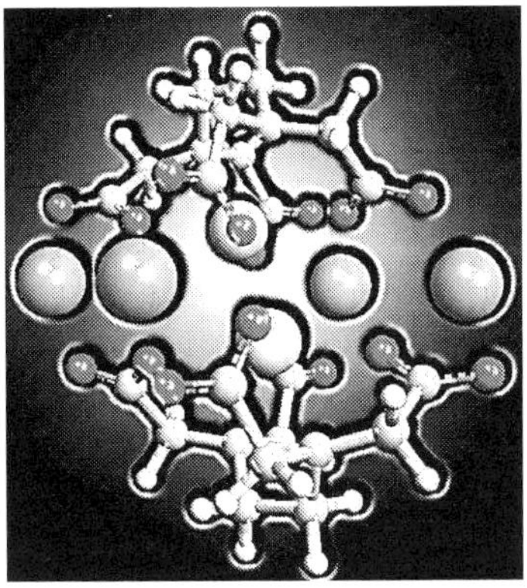

Figure 7. Optimized conformation and molecular structure of the proposed 2 Hg^{+2} / 2 EDTA / 4 Na^{+} complexes (B3LYP LANL2DZ, Robles et al, 2012).

6. Conclusions

Mercury is a non-essential metal that can bioacumulated in living organisms, causing toxic effects of various kinds. Therefore, it is vital to understand how this metal is transmitted through the environment and the changes that occur due to contact with living organisms, or environmental conditions such as high temperatures or strong winds. A study of the reactions that form organic or inorganic compounds, which are even more toxic, is also necessary to limit mercury toxicity. This research gives a brief overview of the techniques commonly used for treatment of sediment and soil contaminated with mercury, mobilization and loadings the mobilized mercury across different matrixes of environment, in specially electroremediation of mercury polluted soil, facilitated by the use of complexing agents as EDTA, proved to be an attractive alternative treatment for the removal of mercury from polluted soil in mining areas.

Acknowledgements

The authors would like to thank to Consejo Nacional de Ciencia y Tecnología de los Estados Unidos Mexicanos (CONACyT), L'Oreal, Academia Mexicana de Ciencias (AMC), Fundación México – Estados Unidos para la Ciencia (FUMEC) and the International Cooperation Program across Bilateral Cooperation Mexico – Hungary for the funding of this research. The contribution of J. L to this research was (partially) carried out in the framework of the Center of Excellence of Sustainable Resource Management / Applied Materials Science and Nano-Technology / Mechatronics and Logistics / Innovative Engineering Design and Technologies at the University of Miskolc. I. Robles is grateful to CONACyT for her scholarship.

Author details

I. Robles[1], J. Lakatos[2], P. Scharek[3], Z. Planck[3], G. Hernández[4], S. Solís[4] and E. Bustos[1*]

*Address all correspondence to: ebustos@cideteq.mx

1 Centro de Investigación y Desarrollo Tecnológico en Electroquímica, S. C. Parque Tecnológico Querétaro S/N, Sanfandila, Pedro Escobedo, Querétaro, Mexico

2 University of Miskolc, Faculty of Material Sciences and Engineering, Institute of Chemistry, Miskolc, Hungary

3 Geological and Geophysical Institute of Hungary, Budapest, Hungary

4 Centro de Geociencias, Campus UNAM-Juriquilla, Querétaro, Qro., Mexico

References

[1] Acar, Y. B., Alshawabkeh, A. N. (1993). Principles of Electrokinetic Remediation. *Environmental Science and Technoogy,*27 (13), 2638 - 2647.

[2] Allard, B., Arsenie I. (1991). Abiortic reduction of mercury by humic substances in aquatic system- an important process for mercury cycle. *Water Soil Air Pollution,* 56, 457-464.

[3] Aposhian H. V., Maiorino R. M., Ramirez D. G., Charles M. Z., Xu Z., Hurlbut K. M., Munoz J.P., Dart. R. C. ,Aposhian M. M. (1995). Mobilisation of heavy metals by newer, therapeutically useful chelating agents. *Toxicology,* 97, 23-38.

[4] Barraclough, F. R., Givelet, N.,Cortizas, A. M, Goodsite, M. E., Biester, H. Shotyk, W. (2002).An analytical protocol for determination of total mercury concentrations in solid peat samples. *The Science of Total Energy,* 292, 129-139.

[5] Bengtsson, G. (2008). Mercury sorption to sediments: dependence on grain size, dissolved organic carbon, and suspended bacteria. *Chemosphere, 73,* 526–531.

[6] Bernaus A., Gaona X., Ree D., Valiente M. (2006) Determination of mercury in polluted soils surrounding a chlor-alkali plant. Direct speciation X-ray absorption spectroscopy techniques and preliminary geochemical characterisation of the area. *Analytica Chimica Acta,* 565, 73-80.

[7] Bollen, A. (2008). Mercury speciation analyses in $HgCl_2$-contaminated soils and groundwater-implications for rosk assessment and remediation strategies. *Water Research, 42,* 91–100.

[8] Buchet J. P., Lauwerys R. R. (1989). Influence of 2,3 dimercaptopropane-1-sulfonate and dimercaptosuccinic acid on the mobilisation of mercury from tissues of rats pre-

treated with mercuric chlorid, phenylmercury acetate or mercury vapours. *Toxicology*, 54, 323-333.

[9] Bustos, E. (2013). Remediación Electro – Cinética de Suelos Contaminados con Hidrocarburo. Chapter 5: Remediación de Suelos y Acuíferos Contaminados en México: Bases Teóricas y Experiencias Reales. Luis G. Torres Bustillos y Erick R. Bandala González. FUNDAP / CONCyTEQ / GTZ, México.

[10] Cabrera - Guzmán, D., Swartzbaugh, J. T., Weisman, A. W. (1990). The Use of Microscale in Hazardous Waste Site Remediation. *Journal of Air and Waste Management Association*, 40, 1670 – 1676.

[11] Caldwell C. A., Canavan C. M., Blom N. S. (2000).Potential effect of forest fire and storm flow on total mercury and methyl mercury in sediments of an arid-lands reservoir. *The Science of the Total Environmental*, 260, 125-133.

[12] Evangelu M.W.H., Ebel, M., Schaefer, A. (2007). Chelate assisted phytoextraction of heavy metals from soil. Effect, mechanism, toxicity and fate of chelating agents. *Chemosphere*, 68, 889-1003.

[13] Fergusson, J. E. (1990). The Heavy Elements: Chemistry, Environmental Inpact and Health Effects, Pergamon Press

[14] Fügedi, U, J. Vatai and L. Kuti (2011). Mercury Content in the Superficial Geological Formations of Hungary, Central European Geology Vol. 52 (3-4), 287-298

[15] Guzzi G., Minoia C., Pigatto P.: Fatal mercury poisioning and chelating agents. Letter to the Editor. Forensic Science Intern. 202, (2010), 61.

[16] Han, F. X., Su, Y., Monts, D. L., Waggoner, C.A., Plodinec, M.J. (2006).Binding, distribution and plant uptake of mercury in a soil Oak Ridge, Tenessee, USA.*Science of the Total Environment*, 368, 753-768.

[17] Harvey, J. W. (2002). Interactions between surface water and ground water and effects on mercury transport in the north-central everglades. Reston Virginia: USGS, Science for changing world.

[18] Hinton, J.,Veiga, M. (2001). Mercury contaminated sites: a review of remedial solutions. National Institute for Minamata Disease.In: Proceedings of the NIMD (National. Institute for Minamata Disease)

[19] Holley, E.A, Mc. Quillan, A. J., Craw, D., Kim, J. P, Sander S. G. (2007).Mercury mobilization by oxidative dissolution of cinnabar and metacinnabar. *Chemical Geology*, 240, 313-325.

[20] Hylander, L. D.,Meili, M. (2003). 500 years of mercury production: global annual inventory by region until 2000 and associated emissions. *The Science of Total Environment*, 304,13-27.

[21] Huang, C. P., Cha, D., Chang, J. –H, Qiang, Z. (2001).Electrochemical Process for in-situ Treatment of Contaminated Soils, Newark, Delaware.

[22] Ibanez, J. G.; Singh, M. M.; Szafran, Z.; Pike, R. M. (1998).Laboratory Experiments on Electrochemical Remediation of the Environment.Part 3.Microscale Electrokinetic Processing of Soils. *Journal of Chemical Education*, 75 (5), 634 - 635.

[23] Iverfeld, A., Lindquist, O. (1986).Atmospheric oxidation of elemental mercury by ozone in the aqueous phase.*Atmospheric Energv.* 20, 1567-1573.

[24] Ko, K. B., Byun, Y., Cho, M., Namkung, W., Shim, D. N.., Koh, D. J., Kim, K. T. (2008). Influence of HCl on oxidation of gaseous elemental mercury by dielectric barrier discharge process. *Chemosphere*, 71, 1674-1682.

[25] Lakatos, J., Brown, S. D.,Snape, C. E. (1999). Influence of coal properties on mercury uptake from aqueous solution. *Energy and Fuel*, 13, 1046-1050.

[26] Lakatos, J., Akcin, G., Brown, S. D., Snape, C.E. (1999). Application of coal and biomass type sorbents for Hg(II) and Cr(VI) removal in the environmental protection technology. Challenges of an inderdisciplinary Science, AcademiaiKiado, Budapest, 327-336.

[27] Lakatos, J., Cheng-gong, S. C., Perry, R., Kennedy, M., Snape, C. E. (2009). Ultra high capacity co-precipitated manganese oxide sorbents for oxidative mercury capture. 237 th ACS meeting Spring, Salt Lake City.

[28] Leopold, K., Foulkes, M., Worsfold, P. (2010). Methods for the determination and speciation of mercury in natural waters- a review. *Analytica Chimica Acta*, 663 (2), 127–138.

[29] Lomonte, C., Doronila, A.,Gregory, A., Baker, A.J.M., Kolev, S.D. (2011). Chelate –assisted phytoextraction of mercury in biosolids. *Science of Total Environment*, 409, 2685-2692.

[30] Lopez –Anton, M. A., Yuan, Y., Perry, R., Morato – Valer, M. (2010).Analysis of mercury species present during coal combustion by thermal desorption. *Fuel* 89, 629-634.

[31] Lopez –Anton, M., Perry, R., Abad – Valle, P.,Diaz – Somano, M., Martinez –Tarazona, M. R., Morato-Valer, M. (2011).Speciation of mercury in fly ashes by temperature programmed desorption. *Fuel Processing Technology*, 92, 707-711.

[32] Moreno, F. N., Anderson, C.W.N., Stewart, R.B., Robinson, B.H. (2004). Phytoremediation of mercury contaminated mine tailings by induced plant-mercury accumulation. *Environmental Practice*, 6, 165-175.

[33] Nik, M. G. (2012). The estudy of mercury pollution distribution aroun a chlor-alkali petrochemical complex, Bandar Iman, southern Iran. *Environmental Earth Science*, 67 (5), 1485 – 1492.

[34] Nriahu J. O. (1994).Mercury pollution from the past mining of gold and silver in Americas. *The Science of Total Environment*, 149, 167-181.

[35] Orecchio, S., Polizzotto (2013). Fractionation of mercury in sediments during draining of Augusta (Italy) coastal area by modified Tessier method. *Microchemical Journal*, 110, 452-457.

[36] Okouchi, S., Sasaki, S. (1984). Volatility in mercury in water. *Journal of Hazardous Materials*, 8, 341-348.

[37] Ozerova, N. A. (1996). Mercury in Geological Systems In: W. Baeyens, R. Ebinghaus and O. Vasiliev (eds) (1996). Global and Regiopnal Mercury Cycles: Sources, Fluxes and Mass Balances, NATO ASI Series 2. Environment Vol. 21 Kluwer Academic Publishers Dondrecht/Boston/London, 463-474

[38] Pongratz, R., Heumann, K. G. (1988).Production of methylated mercury and lead by polar macroalgae. A significant natural source for atmospheric heavy metals in clean room compartments. *Chemosphere*, 36, 1935-1946.

[39] Pillay, K., Cukrowska, E. M., Coville N.J. (2013).Improved uptake of mercury by sulphur-containing carbon nanotubes.*Microchemical Journal*, 108, 124-130.

[40] Rajeshwar, K., Ibanez, J. G., Swain, G. M. (1994). Electrochemistry and the environment, *Journal ofApplied Electrochemistry*, 24,1077–1091.

[41] Reed, B. E., Berg, M. T., Thompson, J. C., Hatfield, J. H. (1995). Chemical conditioning of electrode reservoirs during electrokinetic (EK) soil flushing of a Pb- contaminated silt loam, *Journal of Environmental Engineering* ASCE, 121 (11)805 - 815.

[42] Reddy, K. R., Chaparro, C., Saichek, R. E. (2003). Removal of Mercury from Clayey Soils Using Electrokinetics, *Journal of Environmental Science and* Health, Part A – Toxic / Hazardous Substances & Environmental Engineering, A38 (2), 307 - 338.

[43] Reddy, K. R., Chaparro, C., Saichek, R. E. (2003). Iodide – Enhanced Electrokinetic Remediation of Mercury – Contaminated Soils, *Journal of Environmental Engineering*, ASCE, 129 (12),1137 – 1148.

[44] Robles, I., García, M. G., Solís, S., Hernández, G., Bandala, Y., Juaristi, E. and Bustos E. (2012). Electroremediation of mercury polluted soil facilitated by complexing agents, *International Journal of Electrochemistry Science*, 7, 2276 – 2287.

[45] Robles – Gutiérrez, I., Solís – Valdéz, S., Hernández – Silva, G., Bustos, E. (2012). Electroremediation of Mercury Polluted Soil by Complexing Agents. Environmental Influences of Mercury Ore Processing: Case Studies Selected at Slovenian, Mexican, Hungarian Group Meeting in Idrija, Editors: MatejaGosar, Tatjana Dizdarevič, Miloš Miler, UNESCO, Slovenian National Committee of the International Geoscience Programme, Idrija, Slovenia.

[46] Segall, B. A.; Bruell, C. J. (1992). Electroosmotic Contaminant Removal Processes, *Journal of Environmental Engineering*, 118 (I), 84 - 100.

[47] Scharek, P. Hernández–Silva, G. Solorio–Munguia, G. Vassallo–Morales, L. Bartha, A. Soliz–Valdez, S. Tullner, T. (2012). Total Mercury Content In Soils, Sediments and Tailings in San Joaquin, Querétaro, Mexico Annual Report of the Geological Institute of Hungary, 2010, 125–129

[48] Shi, J. B., Liang, L. N., Jiang, G. B., Jin, X. L. (2005). The speciation and bioavailability of mercury in sediments of Haihe River, China. *Environment International, 31,* 357–365.

[49] Shon Z. H., Kim K.H., Kim M.Y, Lee M. (2005). Modelling study of reactive gaseous mercury in urban air.*Atmospheric Environment*, 39, 749-761.

[50] Slowey, A. J., Rytuba, J. J., Brown, J. G. E. (2005). Speciation of mercury and mode of transprot from placer gold mine tailings. *Enviromental Science and Technology*, 39 (6), 1547-1554.

[51] Somoano M. D., Unterberger S., Hein K. R. G. (2007). Mercury emission control in coal-fired plants: The role of wet scrubbers. *Fuel Processing Technology*, 88, 259-263.

[52] Sondreal, E.A., Benson, S.A., Pavlish, J.H., Ralston N.V.C. (2004). An overview of air quality III. Mercury, trace element and particulate matter. *Fuel Processing Technology*, 85, 425-440.

[53] Wang M., Keener T. C., Khang S.J. (2000).The effect of coal volatility on mercury removal from bituminous coal during mild pyrolysis. *Fuel Process Technology*, 67,147-161.

[54] Wartel, M., Mikac, N., Ouddane, B., Niessen, S. (1999). Speciation of mercury in sediments of the Seine estuary (France). *Applied Organometallic Chemistry*, 13 (10), 715 – 725.

[55] Weber J. H. (1993). Review of possible path for abiotic methylation of mercury(II) in the aquatic environment. *Chemosphere*, 26, 2063-2077.

Chapter 30

Conducted and Investigate Arsenic (As), Cobalt (Co) Copper (Cu), Manganese (Mn), Lead (Pb), and Zinc (Zn) Contamination in Agricultural' Soils and Wheat Crops Projects at Three Regions in Libya

Youssef F. Lawgali, Andy A. Meharg and
Yousef M.T. Elgimati

Additional information is available at the end of the chapter

http://dx.doi.org/10.5772/57429

1. Introduction

Libya is one of the driest countries in the world [2]. The rainfall in the northern part of the country varies between 100–500 mm/year, while the southern region receives only 10 mm/year and some parts receive no precipitation at all [17]. Average evaporation rates are also high, ranging from 1.7 mm/day in the north to 6.9 mm/day in the south, much higher than the average annual rainfall. Irrigation water use has increased as part of the green revolution program to make a Libya self-sufficient in terms of food production, e.g. for wheat and barley. Agriculture in Libya developed remarkably during the last two decades, as can be seen from the large increase in the irrigated areas, which has doubled during the period 1975–2005 [11,15]. The cultivated area has also increased, from 282,250 ha in 1975 to 470,000 ha in 2000 [15]. Increased agricultural activities have resulted in higher demands for groundwater, irrigation water and applications of fertilizer and have also resulted in higher salinity (nitrate, chloride, Na, Ca, Mg, ammonia) and increases in some trace elements in soil [3]. The development in agriculture has occurred at great cost to the environment, and concern is now being expressed about the impact of agricultural chemicals. Whereas agricultural production increased from 1137 thousand tons in 1975 to 1779 thousand tons in 2000, the agrarian productivity per hectare decreased from 4 tons/hectare in 1975 to 3.5 tons/hectare in 1980 and 1990, and then showed a slight increase to 3.7 tons/hectare in 2000. Libya's annual import of fertilizer is about 2 × 10^8 metric tons, including fertilizer nutrients N, P205, and K20.1. [5,6-7]. [6] Observed that application of fertilizers to dry land not only increases plant shoot and stem growth, but also increases evapotranspiration through development of a larger root system and greater extraction of stored water. [4.14] found that the amount of N and P fertilizer used in the Montanan North African soils ranged from 20–120 kg/ha and 0–30 kg/ha, respectively. However, under irrigated conditions, critical P levels are high (i.e. 12–15 mg/kg). The high pH affects the availability of many crop nutrients in the southern Libyan soil [13].

The objective of the present study was to establish levels of As and other trace elements in the grains and soils of the Libyan arable belt. Soil and wheat samples were collected from 3 Libyan agricultural regions, East (ESR), West (WSR) and South (SSR). Each region included several sampling districts (Fig.1).

2. Materials and methods

2.1. Sample preparation

Seventy-nine samples each of soils, shoots and wheat grains from arable land were collected for the study, representing major soil types and climatic regions of agronomic cultures in the Eastern, Western and Southern soil regions of Libya (ESR, WSR and SSR, respectively). Each region included several sampling districts (52, 10 and 17, respectively). Soil samples were taken from the root zone of each sample, and within the soil horizon (0–100 mm), using a spiral auger of 20–50 mm diameter. Three samples were taken around the plant to make composite soil samples and stored in airtight bags. The sampling was carried out between May 26th and June 30th in 2007. The sampling points are shown in Fig. 1.

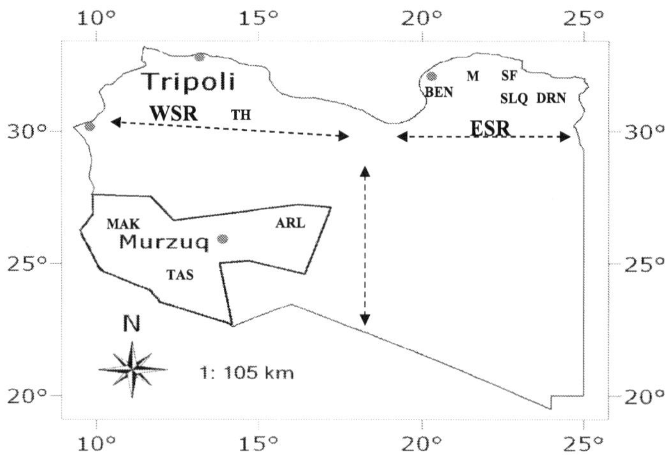

Figure 1. Soil survey sampling locations in Libya.

All samples were air-dried and homogenized by grinding and sieving through a 2 mm mesh. A soil sample of 0.2 g was placed in a digestion tube, 2 ml concentrated HNO_3 added and the sample pre-digested (i.e. left to stand overnight). 2 ml of H_2O_2 was added immediately before digestion on a digest block at 100 °C for 1 hour, 120 °Cfor 1 hour and then at 140 °C until samples were clear (*c.* 2 hours). Samples were then heated to 160 °C until the sample digest appeared clear or a pale soil color.

Quality control and assurance

Each batch of 40 samples digested included a blank, a spike and a soil reference material (NCS ZC 73007) (Table 1). The samples were allowed to cool, decanted into well-labelled 15 ml centrifuge tubes and made up to 10 ml with ultra pure decant deionised water (18.2 M) obtained from a Milli-Q system (Millipore) and analysed by ICP-MS. If the outcome of the SRM NCS ZC 73007 was above 25 % of the certified values, the analysis of the batch was rejected.

Soil sample analysis by ICP-MS

Cd, Pb, Cu, P, Zn, Co, Mn and As concentrations were measured in all soil sample digestions by An ICP-MS Agilent™ 7500c (Agilent Technologies, Tokyo, Japan) octopole reaction system (ORS)-ICP-MS with a Meinhard nebulizer was used to measure the elements Ti (m/z 49), Mn (m/z 55), Co (m/z 59), Cu (m/z 63), Zn (m/z 64, 66, 67, 68), As (m/z 75), Se (m/z 77, 78, 82) and In (m/z 115). Hydrogen (H) was employed as the reaction gas for the ORS, set at a flow rate of 1.8 ml/min. With this method, low detection thresholds can be obtained for elements such as metalloids that have high ionization energies (e.g. As).

Soil As extraction

Soil extraction to determine As fractionation was performed. Sample material was oven-dried at 90 °C for 48 hrs. Extracts were obtained by shaking 1 g of soil with 10 ml of 0.1 % HNO_3, as described by [10]. The soil was mixed with 10 ml of 0.1 % HNO_3 in 15 ml plastic centrifuge tubes shaken at a speed of 60 rpm for 24 hrs. After shaking, soil solids were separated from the solution by centrifugation at 2958 *g* for 15 min. The concentrations of As were determined as described above for the soil digestion. Quality controls (spikes and blanks) were run with each digest set of 40 samples. Concentrations were determined using five point calibrations.

Soil As extraction and analysis by hydride generation (HG)

The 0.1 % $NHNO_3$ water-soluble As fraction was determined by hydride generation (HG, Perkin-Elmer A. Analyst 300) [10]; in order to achieve this, the system must first be calibrated using a standard curve. As^V was reduced to As^{III} by a mixture of hydrochloric acid (HCL) (30 % v/w), potassium iodide (KI) (10 % w/v) and ascorbic acid (5 % w/v). The solution was pumped into a gas-liquid separator and reacted with sodium borohydride and sodium hydroxide ($NaBH_4$) (0.2 % w/v solution in 0.1 M NaOH) to from arsenic hydride generation [10]. Sample concentrations were calibrated using a standard curve of 0.5, 1, 5, 10, 15 and 20 μg/l As, in 10 % HCl 1 % KI and further diluted where necessary. All As concentrations were below the limit of detection (LOD).

Soil As and other element extraction and analysis by ICP-MS

Since the analysis by hydride generation of all Asconcentrations were below the LOD, As and other elements (Cd, Pb, Ni, Cu, P, Zn, Cu, Co and Mn) were analysed with ICP-MS.

The analytical procedures used ICP-MS to extract As and other elements extraction from the soil are identical to those outlined in section 3.2.4 (Soil As extraction), as are the aristar grade reagents used.

Minitab v. 15 was used for all statistical analyses.

3. Results

The certified reference material (CRM) as well as the LOD and spike recoveries for each of the trace elements analysed are presented in Table 1.

	Co	Zn	Cu	As	Cd	Pb	Mn
CRM (%) n=4	87 ±0.1	66 ±1.7	87 ±1.3	103 ±0.2	101 ± 0.01	90 ±1.2	88 ±5.1
Spike (%) n=4	97 ±0.7	96 ±7.4	112 ± 3.1	123 ±2.5	81 ±0.1	97 ±1.2	104 ± 5.4
LOD (mg/kg)	0.01	0.30	0.20	0.01	0.01	0.03	0.14

Table 1. Quality control measurements for analysis of Co, Zn, Cu, As, Cd, Pb and Mn concentration in Libyan agricultural soils. (CRM = NCS ZC 73007).

It should be noted that the concentration of Cd in all samples was below the LOD. The descriptive statistics for trace element concentrations in agricultural soil investigated in this study. They are provided for each region, with a summary of the As levels and other trace element concentrations in soils sampled from ESR, WSR and SSR. A one-way analysis of variance (ANOVA) showed that the differences in As levels between location means are highly significant (P<0.001,). ESR recorded the highest mean As the level in Libyan soil (8.10 ± 0.48 mg/kg), followed by WSR (1.97± 0.41mg/kg) and SSR (1.70 ± 0.58 mg/kg), i.e. As mean concentrations in ESR are nearly 4.8 and 4.1 times higher than in SSR and WSR, respectively. Conversely, the average As levels from the three regions were below 10 mg/g dry weight (dwt). The concentration of As in the agricultural soil, therefore, ranged from 0.01–18.94 mg/kg in ESR, 1.19–2.48 mg/kg in WSR and 0.73–2.92 mg/kg in SSR.The underlying rock structure in ESR is composed of Lower Palaeozoic rocks. The clay agricultural soils of ESR recorded the highest mean As the level in Libyan soil (8.10 mg/kg); this may have some relation to the parent material.

http://dx.doi.org/10.5772/57429

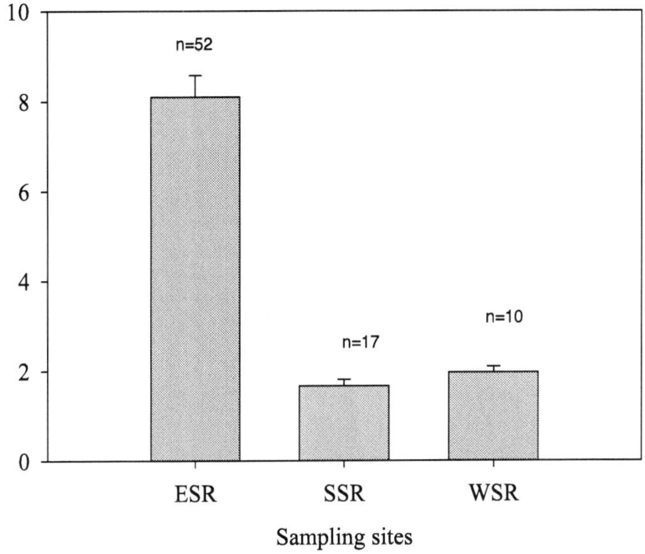

Figure 2. Mean As concentrations in agricultural soils from Libya (Error bars represent ± std. error).

The descriptive statistics for trace element concentrations in the Libyan agricultural soil investigated in this study are shown in Appendix.1, together with a summary of the levels of As and other trace elements.

3.1. Southern soil regions (SSR)

In SSR, the sites used for sampling were ARL, JAF and TAS (sites are shown in Fig.1). The mean As levels in sites in SSR decreased from TAS > JAF >ARL. In general, As levels were low in SSR, ranging from 1–2.9 mg/g, 0.7–1.8 mg/g and 1.1–1.7 mg/g for the TAS, JAF and ARL sites, respectively.

3.2. Eastern soil regions (ESR)

The sampling sites in ESR were BEN, DRN, HEN, MR, SF and SLQ (sites are shown in Fig.1). The mean As levels in these sites decreased from SF > DRN > BAY > MAR > HEN > BEN >SLQ. In general, As levels in ESR were high, ranging from 0.01 to 18.9 mg/kg.

Western soil regions (WSR)

Samples in WSR were taken from two sites, THR and SIRT (Fig.1). WSR recorded the second highest mean total As level compared with ESR and SSR (2 ± 0.1 mg/kg vs.8.1± 0.50 and 1.7± 0.14, respectively). The mean As levels in sites in WSR decreased from THR > SIRT (range 2–2.5 mg/g and 1.2–2.1mg/g for THR and SIRT, respectively).

Results of analysis of variance (ANOVA)

A one-way analysis of variance (ANOVA) was performed on each of the three regions studied. For SSR, the ANOVA showed that that the mean As levels between sites were not significantly different (SSR: P=0.398), but were for ESR and WSR (P<0.0016, P<0.015, respectively).

Multivariate analysis results

Principal component analysis (PCA)

PCA was applied in each region to assist in identifying sources of elements. 3-D plots of the PCA loadings are presented in Fig.3, where the relationships among the six trace elements are readily seen. As expected, four factors were obtained, accounting for 97.4, 98.0 and 99.4 % of the total variance in ESR, SSR and WSR, respectively. In ESR, Factor 1 is dominated by Co, Cu, Pb and Mn, while in SSR it was dominated by Co, Zn, As and Pb and in WSR by Co, Zn, As and Mn (accounting for 67.4, 69.9 and 90.7 % of the total variance, respectively). In each area, the loadings of certain minerals were lower than the others, which may, therefore, imply a quasi-independent behaviour within the group. These elements were Cu and Mn (0.42 and 0.4, respectively) for ESR, Co and Zn (0.5 and 0.4, respectively) in SSR and Zn and As (0.4 and 0.4, respectively) in WSR.

Factor 2 is dominated by Zn and As in ESR, by Mn in SSR and by Cu and Pb in WSR (accounting for 23.1, 15.3 and 6.0 % of the total variance, respectively). Factor 3 is dominated in ESR by As and Pb, by Zn As and Cu Pb in SSR and by Zn and As in WSR (accounting for 4.1, 7.7 and 1.4 % of the total variance, respectively). Zn is negatively correlated with Cu in SSR, and with As in WSR. In each region, certain elements (As and Zn in ESR, Mn in SSR and Cu in WSR) were separated from the others by a large distance in the 3-D PCA loading plot, suggesting that these elements are poorly correlated and may have different sources (Fig.3). Factor 4 is dominated by Cu, As and Mn (accounting for 2.7 % of the total variance; Mn is negatively correlated with Cu and As) in ESR, Zn, As and Pb (accounting for 5.4 % of the total variance; Zn and Mn are negatively correlated with As and Pb) in SSR and Pb and Mn (accounting for 1.3 % of the total variance; Mn is negatively correlated with Pb) in WSR.

Pearson's correlation matrix for the trace element concentrations in agricultural soils from Libya.

The correlations between trace element concentrations in agricultural soils showed a high, positive, linear relationship (Table 2). To determine whether geochemical associations exist between the different trace elements analysed in this study, the datasets for agricultural soils from each region and soils from ESR, WSR and SSR were subjected to simple correlation analysis. A scatter plot for the trace element correlations that were found to be significant (P<0.05) in some of the soils studied are presented in Fig.4. The correlation analysis was also performed on the composite data sets of all agricultural soils for this study. The Pearson correlation coefficients and P-values for Co, Zn, Cu, As, and Pb were significant and all the analyses are shown in Table 2.

Manganese

A summary of trace element Pearson correlation coefficients of soils from different regions of Libya are shown in Table 2, Fig.4 and Appendices 3.11, 3.12 and 3.13.Mn concentration showed

http://dx.doi.org/10.5772/57429

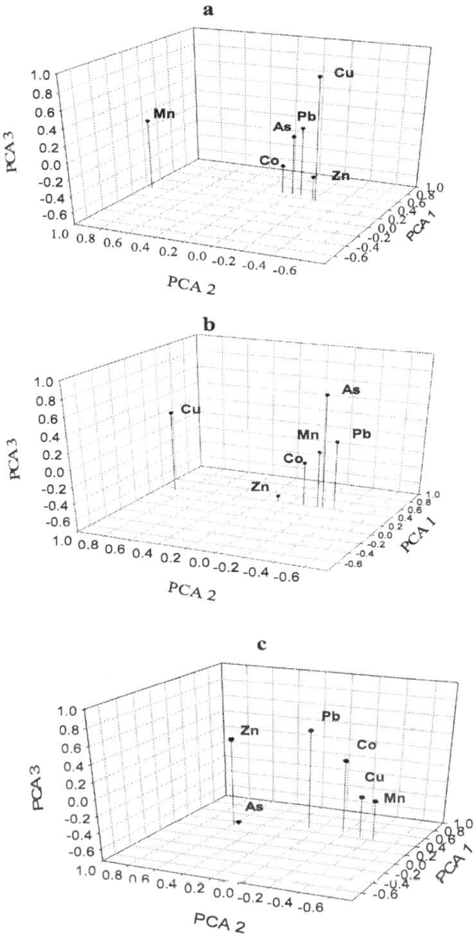

Figure 3. PCA loading plots for the three rotated components for a) SSR, b)WSR and c) ESR.

a significant correlation with Co, Zn, Cu and Pb in WSR and ESR. However, Mn is not correlated with As and Pb in ESR or SSR, respectively, or with the other metals (Co, Zn and Cu) in SSR, reflecting different sources for Mn and the other elements. Moreover, Mn is not correlated with As and not significant in ESR and SSR, which may suggest a common origin (Table 2 and Fig.4).

Lead

The Pb concentration showed a significant correlation with Co, Zn, Cu and As concentrations in all three of the agricultural soil regions (Fig.5a, b & c and Table 2).

Arsenic

As is not correlated with Co and Cu in ESR; however, As concentration showed a significant correlation with Zn concentrations in all three of the soil regions. Trace element correlations with As were slightly varied across regions. As is also positively correlated with Co, Cu and Zn in SSR and WSR (Table 2 and Fig.6a, b & c).

Zinc

Zn concentration correlated well with Co and Cu in soils from WSR and SSR sites but not with those from ESR (Table 2). The Cu-Zn and Zn-Co correlations, however, remained significant ($P<0.01$) when the analysis was performed on the composite data sets (Figs 3.7 and 3.8, respectively). [6]Reported that in Libya, depending on the crop and soil, farmers use an average of 100 kg/hectare of phosphate fertilizer annually.

	ESR	WSR	SSR
Mn-Co	(+ C *)	(+ C *)	(+ 0 NS)
Mn-Zn	(+ C *)	(+ C *)	(+ 0 NS)
Mn-Cu	(+ C *)	(+ C *)	(+ 0 NS)
Mn-As	(+ 0 NS)	(+ C *)	(+ C NS)
Mn-Pb	(+ C *)	(+ C *)	(+ 0 NS)
Pb-Co	(+ C *)	(+ C *)	(+ C *)
Pb-Zn	(+ C *)	(+ C *)	(+ C *)
Pb-Cu	(+ C *)	(+ C *)	(+ C *)
Pb-As	(+ C *)	(+ C *)	(+ C *)
As-Co	(+ 0 *)	(+ C *)	(+ C *)
As-Zn	(+ C *)	(+ C *)	(+ C *)
As-Cu	(+ 0 *)	(+ C *)	(+ C *)
Cu-Co	(+ C *)	(+ C *)	(+ C *)
Cu-Zn	(+ 0 *)	(+ C *)	(+ C *)
Zn-Co	(+ 0 *)	(+ C *)	(+ C *)

+: positive correlation (significant at $P< 0.05$)

–: negative correlation (significant at $P< 0.05$)

NS: not significant at $P< 0.05$

*:significant at $P< 0.05$

0: no correlation

C: correlation ($r> 0.50$)

Table 2. Summary of trace element correlation in soils from different regions of Libyan soils (see for correlation coefficients and *P*-values).

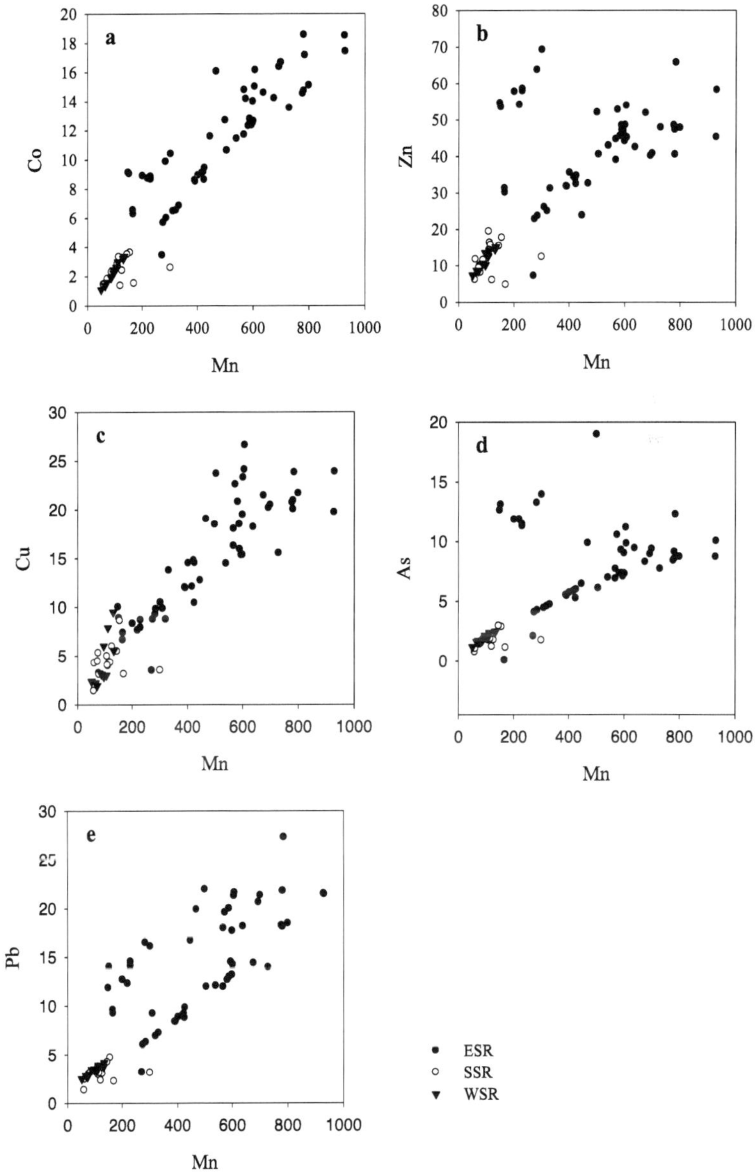

Figure 4. Mn correlations against other trace elements in agricultural soils from ESR, SSR and WSR sites.

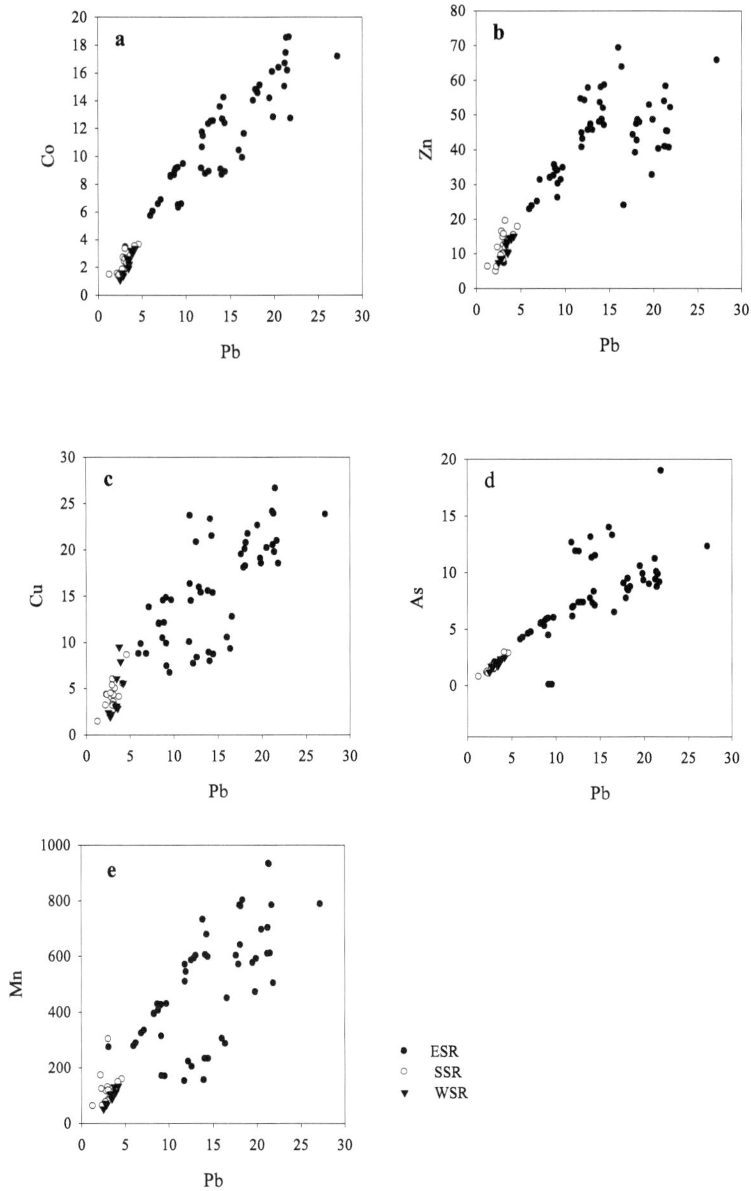

Figure 5. Pb correlations against other trace elements in agricultural soils from ESR, SSR and WSR sites.

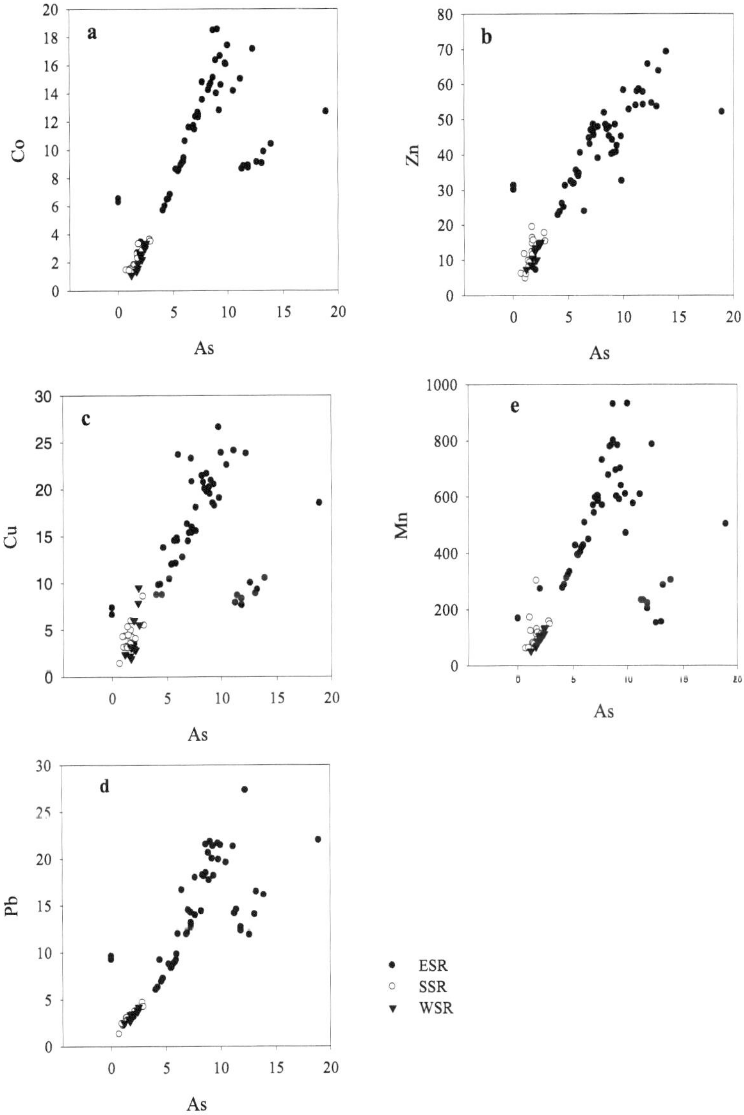

Figure 6. As correlations against other trace elements in agricultural soils from ESR, SSR and WSR sites.

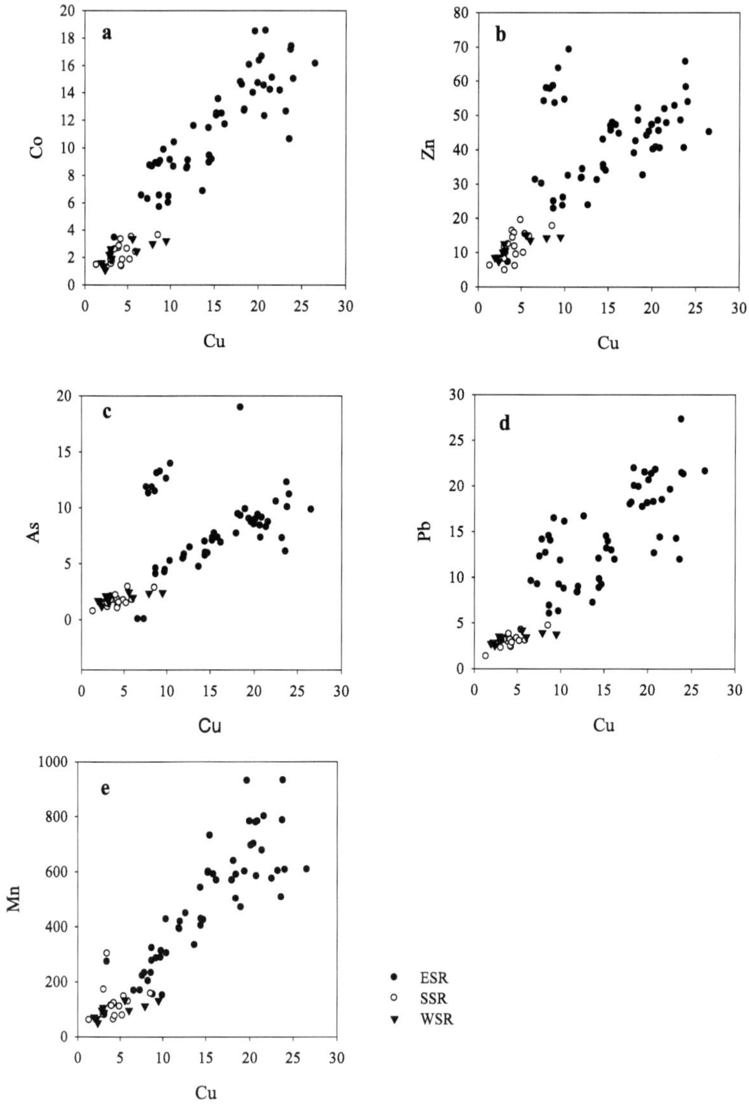

Figure 7. Cu correlations against other trace elements in agricultural soils from ESR, SSR and WSR sites.

http://dx.doi.org/10.5772/57429

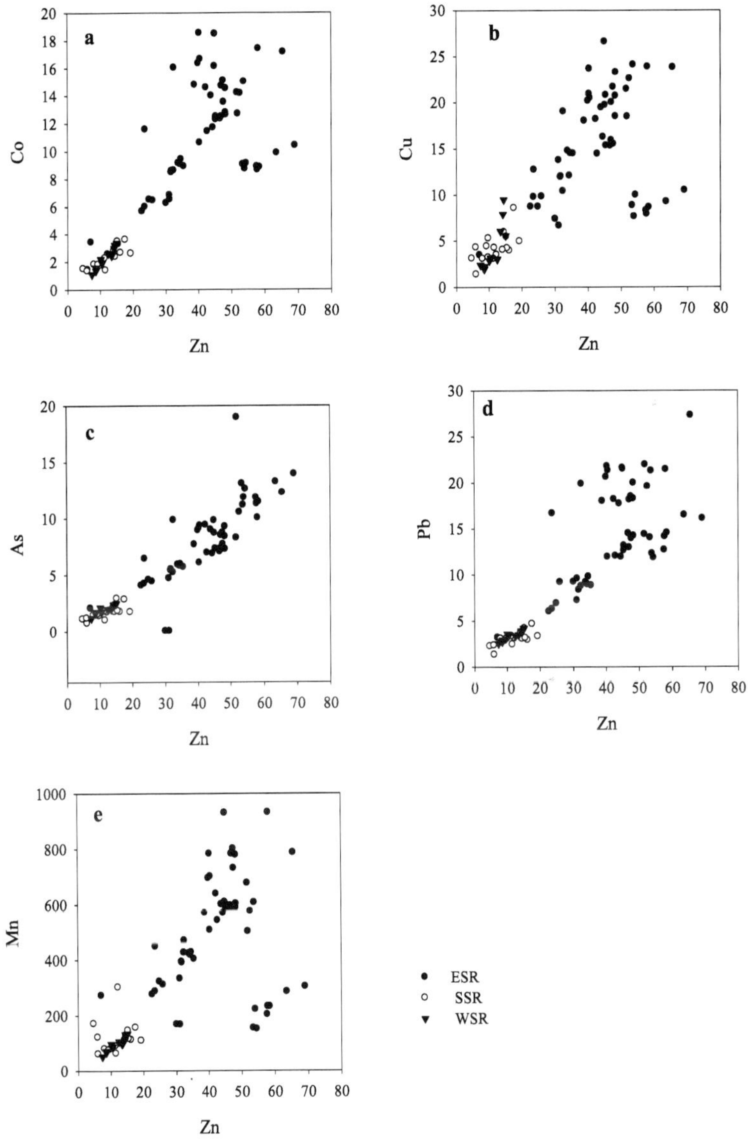

Figure 8. Zn correlations against other trace elements in agricultural soils from ESR, SSR and WSR sites.

Overview of total shoot, soil and grain As for ESR, SSR and WSR

A one-way analysis of variance (ANOVA) was performed on each of the three regions studied. The ANOVA showed that that the mean As levels for grain between sites were not significantly different (P=0.662), but that those for soil and shoot were (soil: P<0.001; shoot: P<0.002). Fig.9 shows a comparison of the distribution of As concentrations in the shoot, soil and grain from ESR, WSR and SSR.

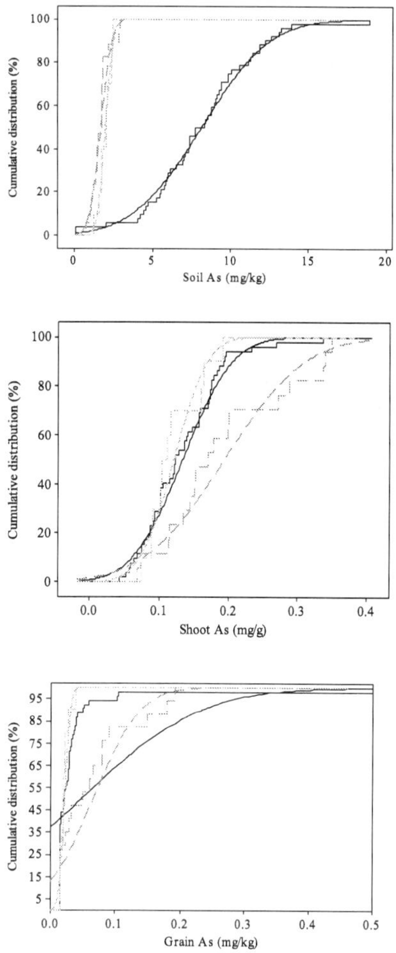

Figure 9. Cumulative ranked distribution of As concentrations in Libyan agricultural soil, shoot and grain from ESR (solid), WSR (dash) and SSR (square dot).

The mean shoot As content for SSR (range 0.07–0.35 mg/kg; mean: 0.19 ± 0.02 mg/kg) was higher than for ESR (range 0.04–0.34 mg/kg; mean: 0.13 ± < 0.01 mg/kg) and WSR (range 0.01–0.19 mg/kg; mean: 0.12 ± < 0.01mg/kg). At the 50th percentile, shoot As content for SSR (0.17 mg/kg) was more than 1.5 times that recorded for ESR (0.12 mg/kg) and WSR (0.11 mg/kg; Fig.9).

ESR recorded the highest mean As level in Libyan soil (8.10 ± 0.48 mg/kg, range 0.01–18.94 mg/kg), followed by WSR (1.97± 0.41 mg/kg, range 1.19–2.48 mg/kg) and SSR (1.70 ± 0.58 mg/kg, range 0.73–2.92 mg/kg), i.e. mean As concentrations in ESR are nearly 4.8 and 4.1 times higher than in SSR and WSR, respectively.

The mean grain As content for SSR (range 0.02–0.2 mg/kg; mean: 0.07±0.01 mg/kg) was also significantly higher (*P*< 0.01) than for ESR (range 0.02–1.1 mg/kg; mean: 0.05 ± 0.02 mg/kg) and WSR (range 0.02–0.04 mg/kg; mean: 0.02 ± < 0.01mg/kg). At the 50th percentile, grain As content for SSR (0.05 mg/kg) was over twice that recorded for ESR (0.02 mg/kg) and WSR (0.02 mg/kg; Fig.9).

The ranges of total grain, shoot and soil As concentrations over the three regions studied were relatively wide; however, the distribution was markedly skewed, with 65 % in ESR having grain, shoot and soil As concentrations of < 0.03, 0.17 and 9.9 μg, respectively, 21 % in SSR < 0.09, 0.28 and 1.8, respectively and 12 % in WSR < 0.02, 0.16 and 2.4, respectively (Fig.9).

As transfer from soil to shoot and grain

The range of As shoot/soil transfer factors (shoot/soil TFs) was considerable. Minimum and maximum shoot/soil TFs were 0.06 for ESR, 0.04–0.29 for SSR and 0.04 for WSR, while means and medians were 0.1 and 0.02, respectively, for ESR; 0.13 and 0.12 for SSR and 0.06 and 0.06 for WSR. In SSR, the median shoot/soil TF was nearly twice as high as that in ESR and WSR. In this study, transfer of As was an order of magnitude greater from shoot to grain, despite lower rates of soil to grain transfer. The differences in these transfer ratios may be due to differences in As speciation. [18],reported that, in general, higher shoot As levels were consistent with low grain/shoot transfer factors.

Estimation of bioavailable As concentrations by nitric acid extraction

Nitric acid extraction is a widely used method of estimating the bioavailable concentration of elements. The precision of the results in the present work was controlled by analysis of CRM, blank and spike samples. The correlation factors showed good precision, with the exception of two analytical bioavailable and total results (Table 3). A summary of the bioavailable As in soils sampled from ESR, WSR and SSR are shown in Table 3. A one-way analysis of variance (ANOVA) showed that the differences in As levels between location means are highly significant (*P*<0.001, Table 3). SSR recorded the highest mean levels of bioavailable soil As (0.05 ± 0.01 mg/kg) (Fig.10). In contrast, the highest total mean As level was recorded in ESR (8.10 ± 0.48 mg/kg;.Table 3 and Fig.11b). The mean bioavailable As concentrations in the three regions studied decreased from SSR > ESR > WSR (Table 3) and was in general low, ranging from 0.02–0.07 mg/g, 0.02–0.03 mg/g and 0.01–0.14 mg/g for SSR, ESR and WSR, respectively. Bioavailable concentrations of As are 1–5 % of the total concentrations. In this study, the highest median bioavailable / median total (B/T) As was recorded in SSR (Table 3, Fig.10 and 3.11b).

Regions Elements	T	T	T	B	B	B	(B/T)*100	
	mg/kg	mg/kg	mg/kg	mg/kg	mg/kg	mg/kg	r[a]	B/T[b]
	Mean	Median	Range	Mean	Median	Range		%
As ESR	7.81	8.45	24.36	0.02	0.01	0.13	0.1	0.2
SSR	4.05	2.48	6.50	0.05	0.05	0.05	–0.4	**2.0**
WSR	4.42	3.96	7.16	0.02	0.02	0.02	–0.1	0.5
Co ESR	11.49	11.63	15.11	0.76	0.02	5.59	0.4	0.2
SSR	2.27	2.27	2.26	0.10	0.02	0.33	–0.4	**0.9**
WSR	2.28	2.33	2.26	0.02	0.02	0.01	0.9*	0.7
Cu ESR	15.39	15.32	23.12	0.29	0.06	2.30	0.3*	0.4
SSR	4.28	4.22	7.21	0.08	0.05	0.33	–0.1	**1.2**
WSR	4.46	3.07	7.57	0.04	0.05	0.05	0.6	1.5
Pb ESR	14.36	14.06	24.11	0.22	0.01	2.98	0.2	0.1
SSR	3.05	3.04	3.36	0.01	0.01	0.06	–0.2	**0.3**
WSR	3.39	3.46	1.69	0.00	0.01	0.01	–0.6	0.3
Mn ESR	492.30	504.60	780.60	58.50	8.50	300.50	0.5*	**1.7**
SSR	117.80	111.40	241.50	14.17	1.34	72.57	0.7*	1.2
WSR	95.66	96.89	82.96	1.15	0.98	2.50	–0.3	1.0
Ni ESR				0.72	0.10	4.32		
SSR				0.19	0.15	0.29		
WSR				0.09	0.10	0.03		
P ESR				7.66	1.95	105.61		
SSR				52.80	33.40	154.40		
WSR				8.92	9.96	12.27		

[a]* Correlation is significant at the 0.05 level
[b]B/T= Median bioavailable/ Median total %

Table 3. The descriptive statistics for total (T) and bioavailable (B) trace element concentrations in Libyan soils (mg/kg).

There was no correlation between the mean total and the bioavailable As concentrations in ESR and WSR. However, there was a negative correlation between the mean total and the bioavailable As in SSR. [16]. reported that appreciable As can move with leaching water, especially in coarse-textured soils. The mean and median bioavailable As concentration values for SSR were close to the those for WSR and ESR (Table 3). The correlation factors between the bioavailable and the total element concentrations were: 0.073, –0.419 and –0.097 (As); 0.371, –0.349 and 0.892 (Co); 0.318, –0.124 and 0.614 (Cu); 0.24, –0.161 and –0.545 (Pb); and 0.466, 0.659 and –0.309 (Mn) for ESR, SSR and WSR, respectively (Table 3, Fig.10). Adriano (1986) reported that As mobility and phytotoxicity is usually greater in sandy than in clayey soils. The SSR samples had high B/T bioavailable As, Co, Pb and Mn contents, compared to B/T total elements (Table 3 &Fig.10). In these samples, the levels of bioavailable elements were derived from P fertilization) [12]. The low correlation value for total Cd is because the concentrations were lower than the ICP-MS detection limit. Weathering will increase bioavailability of trace elements in SSR. [1],found that at P:As ratios of 4:1 or greater, phytotoxicity on wheat was markedly reduced and deduced phytotoxicity to be a function of P concentration.

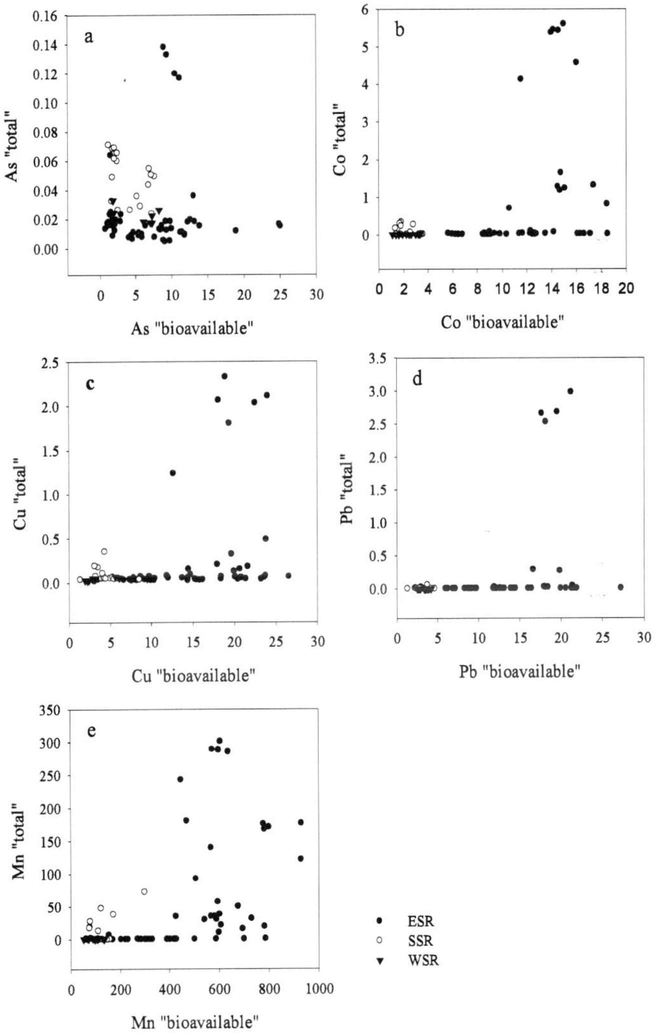

Figure 10. Scatter diagram of total vs. bioavailable trace element concentrations in agricultural soils from ESR, SSR and WSR.

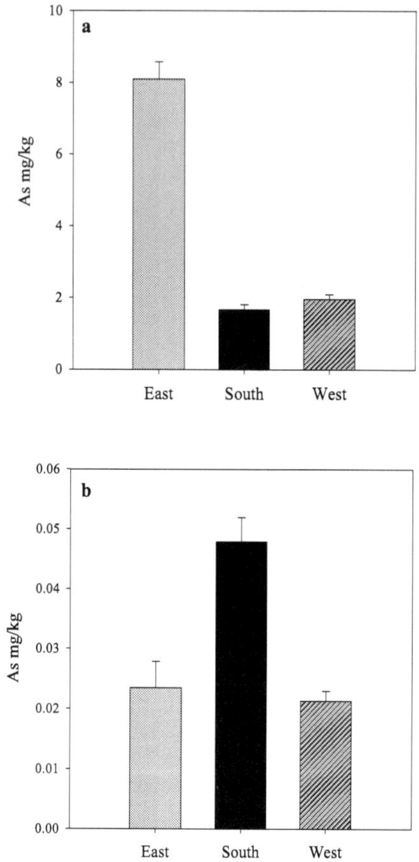

Figure 11. Comparison of (b) soil extractable of As with (a) total soil concentrations in Libya

4. Discussion

Total As mean concentrations in ESR are nearly 4.8 and 4.1 times higher than in SSR and WSR, respectively. Total As loads in ESR may be related to the soil parent material, which is comprised of subsequently covered Lower Palaeozoic rocks. This may be the reason why the clay agricultural soils of ESR have the highest background levels of As. This is in agreement with [8], who reported higher concentrations of trace elements in Eastern Libyan soil than in other areas. In the present study, the trace elements present in Eastern Libyan soil (Co, Zn, Cu,

As, Cd, Pb and Mn) were nearly five times higher than in WSR and SSR. Trace element concentrations in agricultural soils showed positive linear correlations (Table 2).

In this study, transfer of As was an order of magnitude greater from shoot to grain, despite lower rates of soil to grain transfer. In addition, this study considered samples of soil, shoot and wheat grains from arable land with a sampling design that allowed for the consideration of intra-field variations. The differences in these transfer ratios may be due to differences in As speciation. [18]reported that, in general, higher shoot As levels were consistent with low grain/shoot transfer factors. Although this similarity in trends for plants from varying geographical locations strongly suggests the involvement of plant physiological regulation, the actual mechanisms involved are still far from clear. However, the findings from the present study indicate that As transfer to grain isgoverned by multiple factors (Fig. 7.1). The bioavailability of trace elements in soils is influenced by a wide array of biophysical factors (Fig. 7.1), including pH, redox potential, organic matter content and soil texture. These factors were, however, not considered in the experimental/sampling design since the main thrust of the field studies was to establish the contribution of elevated soil As to grain As levels.

In this study, the highest median bioavailable / median total (B/T) As was recorded in SSR (Table 3, Fig.10 and 3.11b).[15]reported that appreciable amounts of As can move with leaching water, especially in coarse-textured soils. This is in agreement with the [9], which reported that As mobility and phytotoxicity is usually greater in sandy than in clayey soils. The SSR samples had high B/T bioavailable As, Co, Pb and Mn contents, compared to B/T total elements (Table 3 &Fig.10). In these samples, the levels of bioavailable elements derived from P fertilization (Hurd-Karrer 1939) and the high pH affects the availability of many crop nutrients in the southern Libyan soil [12]. Weathering will increase bioavailability of trace elements in SSR. [10]found that at P:As ratios of 4:1 or greater, phytotoxicity on wheat was markedly reduced and deduced phytotoxicity to be a function of P concentration.

The present study agrees with [16], who reported that the available soil moisture holding capacity for silty clay loams is three times higher than for loamy sands. Therefore, in the present study, total As mean concentrations in ESR are nearly 4.8 times higher than in SSR. Total As loads in Libyan agricultural soil may be related to the soil moisture holding capacity, with different types of soil having varying capacities and also different levels of As.

5. Future research

- To investigate the differential efficiency with which the main species of As found in Mediterranean conditions wheat are transferred to the grain, resulting either from remobilised shoot As, or coming direct from the soil.
- Effects of As on As uptake, speciation, and nutrient uptake by winter wheat (*Triticum aestivum* L.) under hydroponic conditions.
- Determination and comparison of total As concentrations and speciation in different parts of the wheat plant at various stages of growth and in different genotypes.

- Future research needs to focus on the balance of trace elements in an agro-ecosystem, elaboration of soil chemical and biochemical parameters that can be used to diagnose soil contamination with or deficiency in trace elements, and quantification of trace metal transport from an agro ecosystem to the environment.
- Investigate the relationship between mobility and leaching of As and soil profile depth.
- Set up baseline concentrations of As in Libya soil and wheat. This is of great importance since Libya, as a nation, consumes an enormous quantity of rice.
- The determination of the trace elements iron, zinc, phosphors, cobalt and cadmium in Libyan arable soils is essential for the management of fertilizers and pest side in that soils projects.

Acknowledgements

I wish to express my deepest gratitude to:Friends: Dr Youssef Gamtey Benghazi University,for his assistance and guidance in statistical.To Mrs Susan McCourt of the Queen Mother Library, University of Aberdeen, for her assistance and guidance in the use of the Refworks software and to Mrs Alison Sage for her copy editing. Libyan Agricultural Projects (Aril, Maknowsa, Tasswat, Bengzh, Darna Srit and Sluq) for soil and plant sampling.

Author details

Youssef F. Lawgali[1*], Andy A. Meharg[2] and Yousef M.T. Elgimati[3]

*Address all correspondence to: yogaly6@yahoo.com

1 Department of plant production, University of Sirte, Libya

2 School of Biological,Queen's University Belfast, UK

3 Benghazi University, Libya

References

[1] Adriano, D.C., Bolan, N.S., Vangronsveld, J. and Wenzel, W.W. (2005). Heavy metals. In: Daniel Hillel, ed, *Encyclopedia of Soils in the Environment.* Oxford: Elsevier, pp. 175-182

[2] Atkinson K, Waugh B. Morphology and mineralogy of red desert soils in the Libyan Sahara. Earth Surface Processes. 1979;4(2):103-15. DOI: 10.1002/esp.3290040202

[3] Beke GJ, Entz T, Graham DP. Long-term quality of shallow ground water at irrigated sites. Journal of Irrigation and Drainage Engineering, ASCE. 1993;119(1): 116-128.DOI.org/10.1061/(ASCE)0733-9437(1993)119:1(116)

[4] Cooper JR, Gilliam JW. Phosphorus redistribution from cultivated fields into riparian areas. Soil Science Society of America Journal. 1987;51(6):1600-4. DOI.10.2136/ sssaj1987.03615995005100060035x

[5] Cooper PJM, Gregory PJ. Soil water management in the rain-fed farming systems of the Mediterranean region. Soil Use and Management,. 1987;3(2):57-62.DOI.org/ 10.1111/j.1475-2743.1987.tb00711.x

[6] El-Ghawi U, Pa'tzay G, Vajda N, Bo'dizs D. Analysis of selected fertilizers imported to Libya for major, minor, trace and toxic elements using ICP-OES and INA. Journal of Radioanalytical and Nuclear Chemistry. 1999;242(3):693-701.DOI:10.1007/BF02347381

[7] El-Ghawi, U., Vajda, N. and Pa'tzay, G. (1999). Determination of some trace elements in natural and fertilised Libyan soils using INAA and ED-XRF. Journal of Radioanalytical and Nuclear Chemistry, 241(3), pp. 605-610.DOI.org/10.1007/BF02347220

[8] El-Ghawi UM, Bejey MM, Al-Fakhri SM, Al-Sadeq AA, Doubali KK. Analysis of Libyan arable soils by means of thermal and epithermal naa. Arabian Journal for Science and Engineering. 2005;30(1 A):147-53..

[9] Environment CCoMot. Arsenic. Canadian Water Quality Guidelines for the Protection of Agricultural Water Uses. 1999.

[10] Flynn HC, McMahon V, Diaz GC, Demergasso CS, Corbisier P, Meharg AA, et al. Assessment of bioavailable arsenic and copper in soils and sediments from the Antofagasta region of northern Chile. Science of the Total Environment. 2002;286(1-3): 51-9.DOI.org/10.1016/S0048-9697(01)00962-7

[11] Huda FS. Development of Rural Areas and Societies in Libya. Szent Istvan University [Online]. Available: http://phd.okm.gov.hu/disszertaciok/tezisek/2005/tz_hu1937.pdf [9/2009, 2009]. 2004.

[12] Hurd-Karrer A. Antagonism of certain elements essential to plants toward chemically related toxic elements. Plant Physiology. 1939;14:9-29Doi: 10.1104/pp.14.1.9

[13] Phosyn. Wadi Aril Prodction Project. May 2005. York - UK: Phosyn.. 2005.

[14] Ryan JaM, A.E. Fertilizer use efficiency under rainfed agriculture in West Asia and North Africa, ICARDA, ed. In: The Fourth Regional Workshop, Aleppo, Syria: ICARDA.. 1992.

[15] Shaki AA, Adeloye AJ. Mathematical modelling of effects of Irawan irrigation project water abstractions on the Murzuq aquifer systems in Libya. Journal of Arid Environments. 2007;71(2):133-56.DOI 10.1016/j.jaridenv.2007.03.013

[16] Stevens DR, Walsh LM, Keeney DR. Arsenic phytotoxicity on plain field soil as affected by ferric sulfate or aluminm sulfate. Journal of Environmental Quality. 1972;1(3):301-3. DOI.10.2134/jeq1972.13303x

[17] Wheida E, Verhoeven R. An alternative solution of the water shortage problem in Libya. Water Resources Management. 2007; 21(6):961-82. http://dx.doi.org/10.1007/s11269-006-9067-6

[18] Williams PN, Price AH, Raab A, Hossain SA, Feldmann J, Meharg AA. Variation in arsenic speciation and concentration in paddy rice related to dietary exposure. Environmental Science and Technology. 2005;39(15):5531-40. DOI: 10.1021/es0502324

Chapter 31

Risks of Heavy Metals Contamination of Soil-Pant System by Land Application of Sewage Sludge: A Review with Data from Brazil

Alysson Roberto Baizi e Silva and Fábio Camilotti

Additional information is available at the end of the chapter

http://dx.doi.org/10.5772/58384

1. Introduction

The discharge of urban sewage directly to rivers and lakes is among the principal causes of surface water contamination. Contaminated water resources threaten the water supply of cities and the ecological equilibrium of aquatic ecosystems. To minimize or avoid these negative impacts, the sewage needs to be treated.

Sewage treatment is increasing to the extent that is necessary to maintain the water quality. However, despite purifying municipal wastewater, making it suitable for discharge in receiving water bodies such as rivers and lakes, the sewage treatment process generates a large volume of sludge that needs to be destined appropriately and quickly to avoid its accumulation and consequently its transformation into environmental liabilities in wastewater treatment plants (WTPs).

The options for disposal of sewage sludge are varied. In general, they include: (i) land application, (ii) industrial reuse, (iii) disposal in landfills, (iv) incineration and (v) discharge to oceans [1]. Land application and disposal in landfills are the most widely adopted disposal methods in various parts of the world [2, 3]. The application to land is considered the most attractive option, because the sewage sludge can improve soil conditions for agricultural production, since it is rich in organic matter and plant nutrients [4, 5]. However, there are undesirable constituents in its composition.

Heavy metals, such as cadmium (Cd), lead (Pb), among others, are the most concern undesirable constituents [5-7], since they may be toxic to microorganisms, plants, animals and humans in not very high concentrations [8-10]. Therefore, soils amended with sewage sludge should be evaluated for heavy metal contamination in order to prevent its excessive entry into the food chain, thus reducing the risk of toxicity in living organisms posed by the method of disposal of sludge on land.

The evaluation of heavy metals in soil can be made in several ways. The basic and most common is to quantify the total concentration [11]. The great advantage of this measure is its robustness, that is, it does not easily change with environmental conditions, but it may be altered by the addition of metals from external sources, such as sludge, which allows its use as an indicator of soil contamination by heavy metals. However, the total concentration not satisfactorily represents the amount of metal that would be available for uptake by plants, that is, the fraction of contaminant which could cause phytotoxicity and enter the food chain, affecting animals and humans. Due to its direct relationship with potential toxic effects, the available concentration should be determined in addition to total concentration.

The availability of heavy metals to plants has been characterized using chemical extractants, some of them already employed in routine soil analysis, as Mehlich 1, Mehlich 3 and DTPA. In general, extractants are (i) acid, (ii) chelating agents, (iii) acid-chelating or (iv) saline solutions [11]. The chemical nature of the extracting solution interferes with the ability of extracting metals and, ultimately, the efficiency of the extractant to represent the available fraction. Therefore, the extractants should be systematically tested before being used in monitoring heavy metals in soils amended with sewage sludge.

Fractionation is also interesting technique to evaluate heavy metals in soils. Its principle is to separate the metals in soil fractions in which they have variable solubility [12]. With this procedure, it is possible to determine the contribution of each fraction in the availability of metals to plants [13, 14]. It also allows selecting the best chemical extractant, based on its relationship with the fractions that most contribute to uptake of metals by plants. Moreover, the redistribution of metals among fractions in response to changes in soil conditions can be studied [15]. Thus, the fractionation can indicate whether the metals added to the soil by the sludge are to be redistributed in fractions in which they are either more or less available, that is, whether they have either greater or lesser potential to cause toxicity problems respectively [7, 16].

The study of speciation is another interesting strategy to evaluate heavy metals in soils, since it enables to distinguish different chemical species in the soil solution. Each species has a particular chemical behavior in terms of availability and mobility in soil. Free ions are more relevant to the availability of metals, because they are the preferred forms of plant uptake [11]. In contrast, organo-metal complexes are more related to the mobility of metals and, consequently, to their leaching [17]. Thus, the speciation can indicate if the risk of phytotoxicity and contamination of the food chain is higher or lower than the risk of groundwater contamination, based on the proportion of the chemical species formed in response to application of sewage sludge.

Besides the contamination of soil, it is also necessary to evaluate whether the sewage sludge applied to land can contaminate crops with heavy metals. The assessment of crop contamination must include studies of differential capacity of uptake, translocation, accumulation and allocation of metals in different plant species. In the case of allocation, it is essential to evaluate the metals of interest in the harvested parts, especially those that are edible [18]. As a result of these investigations, plants can be separated by their susceptibility to contamination with heavy metals and thus the indication of using less susceptible plants in areas that receive sewage sludge can be made.

In Brazil, the sewage treatment has grown considerably in recent years. In 2001, 25.6 % of the sewage generated were treated [19]. Ten years later, in 2011, this index had risen to 37.5 % [20]. Growth should remain strong, since there is a significant amount of resources to be invested in sanitation in the country [21]. The expansion in sewage treatment causes inevitable increase in the generation of sewage sludge.

The Brazilian production of sewage sludge is estimated at 150-220 thousand tons per year (dry basis) [22] with a perspective to increase. As in other parts of the world, land application has been one of the preferred forms of sludge disposal [1]. As seen above, this option requires a careful evaluation of heavy metals in soils amended. In recent decades, many experiments have been conducted in Brazil to evaluate the effects of sewage sludge on heavy metals in soils and crops. The results of these works can help technicians to manage more safely sludge application to land.

Thus, our objective was to review the scientific literature on the impacts of sewage sludge on heavy metals in soils and plants in conditions of Brazil, attempting to evaluate the risks of contamination of the soil-plant system. We focused on studies involving field experiments because they represent better real situations of management of the sludge.

2. Heavy metals: An overview

Heavy metals comprise a class of not very well defined chemical elements. They have been commonly characterized as metals and metalloids (semimetals) with density higher than 5 g cm^{-3} associated with problems of environmental contamination and toxicity, although some of them are required in low concentrations for some organisms such as higher plants, animals and humans [23, 24]. However, this definition (attempt) has been criticized for its chemical and toxicological inadequacy and it has had no institutional support of IUPAC (International Union of Pure and Applied Chemistry) [25]. Despite the criticism, there is no more appropriate term to refer collectively to the elements known as heavy metals. Therefore, we use this terminology in this text and support its use in the definition introduced by Hawkes [26], according to which heavy metals comprise a block in the Periodic Table with all metals and metalloids in Groups 3 to 16 that are in periods 4 and greater. As some authors consider selenium (Se) it as a metalloid and therefore heavy metal [10, 27], we also consider it as metalloid in this chapter, even though it is a non-metal. Thus, we expand the definition of Hawkes [26] to include Se as a heavy metal, keeping in mind which this redefinition is simply operating, just to join in the same class specific chemical elements commonly associated to problems of environmental contamination and toxicity.

Heavy metals can be divided according to their need for different organisms. There are those which are doubtless essential and those which are not recognized as essential (Table 1). Copper

(Cu), iron (Fe), manganese (Mn), molybdenum (Mo) and zinc (Zn) are essential to plants, animals and humans. Cobalt (Co) and Se are essential only to animals and humans, while chromium (Cr) and nickel (Ni) are essential to humans and plants, respectively. In contrast, arsenic (As), Cd, Pb and mercury (Hg) are not essential to any of these organisms. However, essential or not essential metals may be toxic. For example, manganese (Mn) is an essential element (i.e., micronutrient) to plants, but its excessive uptake can cause toxicity in crops [28, 29]. On the other hand, As, Cd, Pb and mercury (Hg) are not essential to humans but in excess can also cause toxicity. People exposed to Pb or Hg develop neurological disorders, while exposure to Cd is associated with kidney damage and fragile bones, and various forms of cancer can occur due to the ingestion of food or water contaminated with As [30]. A summary of the toxicity of heavy metals commonly associated with environmental contamination is given in Table 1.

Heavy metal (Symbol)	Organism†	Essentiality‡	Toxicity§	Reference
Arsenic (As)	Plants	No	Increased oxidative stress and reduced plant growth.	[31-33]
	Animals	No	Blindness and reduced weight gain.	[9]
	Humans	No	Increased cancer risk.	[30]
Cadmium (Cd)	Plants	No	Increased oxidative stress and reduced plant growth.	[34, 35]
	Animals	No	Reduced weight gain.	[36]
	Humans	No	Kidney damage and fragile bones.	[30]
Chromium (Cr)	Plants	No	Increased oxidative stress and reduced plant growth. Cr(VI) is more toxic than Cr(III).	[9, 37]
	Animals	No	?	[38]
	Humans	Yes	Allergy and increased cancer risk. Cr(VI) is more toxic than Cr(III) or Cr(V).	[39-41]
Cobalt (Co)	Plants	No	Chlorosis in younger leaves (restricted Fe translocation) and reduced plant growth.	[42, 43]
	Animals	Yes	Body weight loss and muscular incoordination.	[9, 44]
	Humans	Yes	Cardiomyopathy and Increased cancer risk.	[45]
Copper (Cu)	Plants	Yes	Intervenial chlorosis in younger leaves, reduced branching, thickening, darkening of rootlets and reduced plant growth.	[9, 43, 46]
	Animals	Yes	Gastroenterits, liver damage and death.	[9, 47]
	Humans	Yes	Wilson's disease, hemolysis, hepatic necrosis and kidney damage.	[9, 48]
Iron (Fe)	Plants	Yes	Leaf bronzing, roots with black coating and reduced plant growth. Common in flooded rice.	[9, 49]
	Animals	Yes	Anorexia, diarrhea, metabolic acidosis, reduced body growth rate and death.	[9, 44]

Heavy metal (Symbol)	Organism†	Essentiality‡	Toxicity§	Reference
	Humans	Yes	Vomiting, diarrhea, metabolic acidosis and increased risk of atherosclerosis and Alzheimer's disease.	[9, 50]
Lead (Pb)	Plants	No	Chlorosis, root system darkening, stunted plant growth and increased oxidative stress.	[51]
	Animals	No	Appetite loss, diarrhea, anemia and body weight loss.	[9]
	Humans	No	Neurological problems (from headache to psychosis) and kidney damage.	[9, 30]
Manganese (Mn)	Plants	Yes	General chlorosis, necrotic leaf spots and stunted plant growth.	[9]
	Animals	Yes	Anemia, gastrointestinal lesions and growth retardation.	[9]
	Humans	Yes	Psychiatric disturbance and neurodegenerative disorder, including Parkinson's disease.	[9, 52]
Mercury (Hg)	Plants	No	Hypertrophic root, retarded plant growth and increased oxidative stress.	[53, 54]
	Animals	No	Vomiting, bloody diarrhea and necrosis of the alimentary mucosa.	[9]
	Humans	No	Neurological disturbances, kidney damage and decreased fertility.	[30]
Molybdenum (Mo)	Plants	Yes	Yellow or orange-yellow chlorosis, seedling injury, delayed maturity and reduced plant growth.	[9, 55]
	Animals	Yes	Diarrhea, anorexia, depigmentation of hair and wool and neurological disturbance.	[9]
	Humans	Yes	Possibly acting joints, hyperuricosuria and gout.	[9, 56]
Nickel (Ni)	Plants	Yes	Increased oxidative stress, retarded germination, stunted root growth, chlorosis, inhibited plant growth and reduced yield.	[9, 57]
	Animals	No	?	[37]
	Humans	No	Skin allergies, lung fibrosis, kidney and cardiovascular system damage and stimulation of neoplastic transformation.	[58]
Selenium (Se)	Plants	No	Chlorosis, wilted and dried leaves, reduced plant growth and plant premature death.	[59]
	Animals	Yes	Impaired vision, anemia, loss of hair, ataxia, stiffness of joints, paralysis, atrophy of heart and death.	[9, 60]
	Humans	Yes	Hypochromic anemia, damaged nails and hair loss.	[9, 61]
Zinc (Zn)	Plants	Yes	Chlorosis (Fe-deficiency-induced), stunted plant growth and reduced yield.	[9, 62]

Heavy metal (Symbol)	Organism†	Essentiality‡	Toxicity§	Reference
	Animals	Yes	Diarrhea, anorexia, jaundice, kidney and abomasum damage, arthritis and weight loss.	[9, 63]
	Humans	Yes	Diarrhea, nausea, vomiting, epigastric pain, lethargy, anemia, neutropenia, impaired immune function and decreased HDL cholesterol.	[9, 64]

†In case of animals, they are livestock animals.

‡Yes: essentiality recognized. No: essentiality unrecognized.

§Most common manifestations, symptoms and consequences of heavy metal toxic effects.

Table 1. Essentiality and toxicity of heavy metals for different organisms.

The relative importance of heavy metals toxicity was addressed by McLaughlin et al. [10] in terms of food chain contamination. According to these authors, Cd is the metal with greatest potential to contaminate plants and subsequently to be transferred to animals and humans that eat these contaminated plants or part of them. This statement is based on the fact that (i) Cd poses animal and human health risks in plant tissue concentrations that are not generally phytotoxic and (ii) Cd concentrations in agricultural soils are increasing in many parts of world due to Cd inadvertent additions through the use of fertilizers, sewage sludge and soil amendments. Due to the high risk of contaminating the food chain, the risk of Cd to cause toxicity is considered to be high as well. Despite increased concern with Cd, the toxicity risk of other heavy metals should not be neglected.

The toxicity of heavy metals in living organisms is a phenomenon somewhat complex. Toxic effects of a metal depend on a number of factors that often include (i) rate, (ii) exposure time, (iii) tolerance of the organism and (iv) environmental conditions. In recent years, the effect of the interaction between heavy metals on the expression of toxicity has been considered very intensely. As a result of the interaction, a given metal may increase or decrease the negative effects of other metal in the organism [65].

Despite the complexity, the toxicity of heavy metals in plants and in animals and humans that eat contaminated plants is primarily associated with previous environmental contamination. Soils may be contaminated with such hazardous elements by the use of sewage sludge. High concentrations of metals in the sludge increase the risks of contamination and therefore toxicity. Thus, it is important to know the chemical composition of sewage sludge.

3. Sewage sludge

Sewage sludge is generally defined as a semi-solid or liquid organic residue generated during the treatment of domestic sewage. He has also been called biosolids, but this term is more appropriate for the sludge that was conveniently selected and treated to allow its application to land (i.e., agricultural recycling) [6, 66]. Although the use of the term biosolids is becoming common, sewage sludge was used in this text because it encompasses sludges with different types and degrees of treatment and is the term adopted by the Brazilian legislation [67].

The chemical composition of sewage sludge is highly variable, particularly in terms of heavy metals. Such variability can be seen in Table 2 for sludge generated at WTPs in Brazil. The coefficients of variation (CVs) of heavy metal concentrations are somewhat high. Cd, Co, Cr and Ni concentrations had the highest CVs, above 80 %, and Zn the lowest, 43 %. Intermediate variability was observed for Cu, Fe, Mn, Mo and Pb. It was not possible to calculate CV for As, Hg and Se because of the paucity of data. The high variability is explained primarily by the origin of the sludges, since they were obtained in WTP from different cities, each with particular characteristics of generation and wastewater treatment as well as treatment of generated sludge. However, there was also considerable variation for a same city, which indicates the occurrence of temporal variability (e.g., seasonal, annual etc.) for concentration.

Despite the high variability, the concentrations of heavy metals in sewage sludge for land application may not exceed limits established by environmental institutions. In Table 2 two limits for each metal are shown, one from the Environmental Agency of the State of São Paulo (CETESB) and other from the Brazilian National Environment Council (CONAMA). There is agreement for the limits between the institutions only for the Se. For all elements, CONAMA limits are lower. As CONAMA rule is federal and more restrictive, it must prevail over the rule of CETESB, which is state and less restrictive.

Considering Resolution # 375/2006 of CONAMA, some sewage sludges presented in Table 2 had heavy metal concentrations above the established limits. Seven, six, four and three sludges were with concentrations of Zn, Ni, Cr and Pb, respectively, higher than the maximum allowable concentrations. Generally, more than one metal was in excess, but not necessarily. Improper sludges were generated in the cities of Barueri, Franca and São Paulo, however, Franca produced sludges less contaminated with heavy metals, which probably reflects its lower level of industrialization in relation to the other two cities.

The concentrations of Cd and Cu in sewage sludges were below the limits of CONAMA (Table 2). Similarly, the concentrations of As, Hg, Mo and Se were also below, but there were few values determined, making this finding unrepresentative. However, the high number of cases in which these elements were below the limit of detection (LOD) of the analytical method suggests low concentrations, probably far from reaching the maximum allowable concentrations. Although they are not legally restrictive, Co, Fe and Mn were in relatively high concentrations in the sludges analyzed.

The finding that the Cd did not exceed the limit established by CONAMA is the most positive result of this survey of sewage sludge used in experiments conducted in Brazil. This result indicates that the sludge has low potential for soil contamination with this heavy metal. Soil contaminated with Cd is a constant concern throughout the world, because this metal can easily enter the food chain and impose risks to human health [10]. On the other hand, the sludge seems to be a considerable source of other heavy metals such as Cr and Pb. Considering that

State	City	As	Cd	Co	Cr	Cu	Fe	Hg	Mn	Mo	Ni	Pb	Se	Zn	Reference
		mg kg^{-1}													
SP	Barueri		28.4		385.0	784.3					238.6	152.8		1568.5	[68]
SP	Barueri		16.0		386.0	534.0					286.0	171.0		1649.0	[68]
SP	Barueri					664.0			228.0		268.0	152.0		1800.0	[69]
SP	Barueri					551.0			294.0		595.0	371.0		3810.0	[69]
SP	Barueri					660.0			257.0		360.0	180.0		2328.0	[69]
SP	Barueri					719.0			263.0		354.0	171.0		1745.0	[69]
SP	Barueri					627.0			287.0		350.0	155.0		2354.0	[69]
SP	Barueri	<1.0	12.8	12.2	823.8	1058.0	54181		429.0	<0.01	518.4	364.4		2821.0	[70]
SP	Barueri	<1.0	9.5	5.0	1071.0	1046.0	32500		335.0	<0.10	483.0	233.0		3335.0	[70]
SP	Barueri	<1.0	9.4	9.3	1297.2	953.0	37990		418.9	<0.01	605.8	348.9		3372.0	[70]
SP	Barueri		11.0		808.0	722.0			222.0		231.0	186.0		2159.0	[71]
SP	Barueri		10.0		736.0	690.0			194.0		297.0	173.0		2930.0	[71]
SP	Barueri		8.0		798.0	998.0			206.0		299.0	169.0		2474.0	[71]
SP	Barueri		8.0		798.0	998.0			206.0		299.0	169.0		2474.0	[71]
SP	Barueri		20.3		477.7	754.1					355.3	155.8		1926.9	[72]
SP	Franca	<1.0	3.3	5.0	633.8	239.8	33793		349.3	<0.01	54.7	199.6		1230.0	[71]
SP	Franca	<1.0	2.0	4.9	1325.0	259.0	31700		267.0	<0.10	74.0	118.0		1590.0	[71]
SP	Franca	<1.0	2.1	4.8	1230.0	240.9	24176		232.5	<0.01	72.4	140.5		1198.0	[71]
SP	Franca		3.3	29.0	284.5	572.6	184		730.0	2.8	56.6	77.3		1028.3	[73]
SP	Franca		2.0		102.0	204.0	18883		243.0		69.0	100.0		1279.0	[74]
SP	Franca		1.5		41.3						34.8	43.8			[75]
SP	Franca		1.5		41.3						34.8	43.8			[75]
SP	Franca		1.5		41.0						34.3	42.4			[75]
SP	Franca		1.6		41.7						35.2	45.3			[75]
SP	Franca					98.0	42224		242.0	9.8		127.0		1868.0	[76]
SP	Jundiaí	<0.10	5.8		149.3	284.1		<0.10	676.8	<0.10	41.8	283.1	<0.10	1364.8	[77]
SP	Jundiaí	<0.01	6.6		188.5	864.8		<0.01	693.3	<0.01	35.4	206.6	<0.01	1738.1	[77]
SP	Jundiaí	4.2	14.0		277.7	304.1		1.1		9.8	65.6	201.6	1.8	1869.9	[78]
SP	Piracicaba		<0.02		207.2	192.9	30295		173.0		<0.01	107.6		943.4	[79]
RJ	Rio de Janeiro		0.8			47.0			60.0			82.0		217.0	[80]
SP	São Paulo					791.0	38000		315.0		322.0			1888.0	[81]
SP	São Paulo					975.0	44000		423.0		401.0			2904.0	[81]
SP	São Paulo					1148.0	41000		479.0		459.0			3326.0	[81]
SP	São Paulo					989.0	84000		223.0		1240.0			4035.0	[81]
Statistics															
n		1	23	7	23	30	14	1	26	3	31	30	1	30	
Minimum (mg kg^{-1})		4.2	0.8	4.8	41.0	47.0	184	1.1	60.0	2.8	34.3	42.4	1.8	217.0	
Maximum (mg kg^{-1})		4.2	28.4	29.0	1325.0	1148.0	84000	1.1	730.0	9.8	1240.0	371.0	1.8	4035.0	
Coefficient of variation (%)			91	79	81	51	51		51	54	92	53		43	
Limits (mg kg^{-1})															
By CETESB†		75	85			4300		57		75	430	840	100	7500	[82]
By CONAMA‡		41	39		1000	1500		17		50	420	300	100	2800	[67]

†Technical Norm # P4.230/1999 of CETESB (Environmental Agency of the State of São Paulo).

‡Resolution # 375/2006 of CONAMA (Brazilian National Environment Council).

Table 2. Heavy metals in sewage sludge samples used in experiments conducted in Brazil.

the sludge can introduce these metals in excess, soils amended with sewage sludge should be carefully evaluated for accumulation and availability of Cu, Cr, Mn, Pb and Zn. Additionally, the Cd should also be monitored because it is highly dangerous.

4. Heavy metals in soils amended with sewage sludge

4.1. Total concentration

The determination of the total concentration of heavy metals in soils has been performed with previous extraction of metals from samples using different mixtures of acids. In Brazil, the methods of extraction most widely used are presented in Table 3. Because of the difference in the chemical composition, some of them may produce quite contrasting results. Comparison between the methods HNO_3–H_2O_2–HCl and $HClO_4$–HF shows this contrast. It is observed in Table 4 that concentrations of Cr, Pb and Zn extracted with HNO_3–H_2O_2–HCl were 72 %, 31 % and 62 % lower than concentrations of these metals extracted with $HClO_4$–HF. For Cd, however, the results were not different between the two methods, suggesting that the difference in values between them may depend on the type of metal in question. The contrasting results between these methods are due to extraction differential capacity of the reagent mixtures employed in each of them. The HNO_3–H_2O_2–HCl mixture does not normally extract metals bound to silicates. On the other hand, hydrofluoric acid contained in the $HClO_4$–HF mixture is able to dissolve silicates and extract metals eventually present in this mineral class. Therefore, concentrations of metals extracted with HNO_3–H_2O_2–HCl tend to be smaller in relation to extraction with $HClO_4$–HF. Thus, it is necessary to standardize the method of extraction of heavy metals to assess the contamination of soil amended with sewage sludge.

Extraction method	Common name	External heating	Denomination by USEPA†	Reference
HNO_3–H_2O_2–HCl			Method 3050B	[83]
HNO_3–HCl 1+3	Aqua regia	Microwave oven	Method 3051A	[84]
HNO_3–$HClO_4$ 5+1	Nitric-perchloric			[85]
$HClO_4$–HF				[86]

†United States Environmental Protection Agency. The composition of the extraction methods 3050B and 3051A may vary slightly but it always has HNO_3.

Table 3. Methods commonly used in Brazil for extraction of heavy metals from soils.

Brazilian official institutions of environment, as CETESB [82] and CONAMA [67], recommend the use of the methods 3050B and 3051A of United States Environmental Protection Agency (USEPA) [83, 84] for extraction of heavy metals in soils amended with sewage sludge. Such methods are not designed to extract fully the metals from the soil, since they normally do not dissolve elements bound to silicates, which are not generally available in the environment,

Extraction method	Heavy metal†			
	Cd	Cr	Pb	Zn
	mg kg^{-1}			
HNO_3–H_2O_2–HCl	1.93 a	28.72 b	15.72 b	64.82 b
$HClO_4$–HF	1.99 a	104.17 a	23.05 a	171.97 a

†Means within a column followed by the same letter are not significantly different according to Tukey test ($p < 0.05$). Values are means of four sewage sludge rates.

Source: Adapted from Nogueira et al. [71].

Table 4. Cd, Cr, Pb and Zn extracted by two different methods from a Brazilian Oxisol amended with sewage sludge.

thus without implications for environmental contamination. They extract only elements which could become environmentally available, consequently harmful to living organisms. As the extraction is not fully, the terms pseudototal concentration and total recoverable concentration have recently been employed to designate the concentrations of metals extracted by these official methods, although the use of the term total concentration is still very common. In recent years, many field experiments with sewage sludge were performed in Brazil, particularly in São Paulo state, therefore there are several data on total concentrations of heavy metals in soil which allow studying the risks of contamination pose by agricultural use of the waste. A summary of these experiments is presented below.

Oliveira and Mattiazzo [68] conducted an experiment with sugarcane to test varying rates of sewage sludge applied to an Oxisol for two consecutive years, in which heavy metals were added in different amounts, including rates above the annual maximum rates allowed in São Paulo state by CETESB. The data in Tables 5 and 6 show that there was at least one rate above the annual rate allowed for Cd, Cu, Ni, Pb and Zn, but there was no rate higher than the maximum cumulative rate. In soil, the total concentrations of Cu and Zn (extraction with HCl–HNO_3 3+1 in microwave) in layer 0-0.20 m depth increased with increasing sewage sludge rate in two years especially for the higher rates in the second year, indicating a cumulative effect. Cd and Pb were below the LOD of the analytical method (Atomic Absorption Spectrometry–AAS). Ni was detected only in the second year and at the two highest rates of sludge. The addition of metals above the allowed maximum rates increased total concentrations of Cu, Ni and Zn in the second year above the natural concentrations established for São Paulo state (Tables 3 and 7), suggesting that soils receiving high loads of heavy metals by the application of excessive rates of sewage sludge could be contaminated in a short time.

In addition to the possibility of rapidly contaminating the soil with heavy metals, high rates of sewage sludge cause prolonged effect on contamination. Martins et al. [81] observed linear increases in total concentrations of Cu and Zn (extraction with HNO_3–HClO 5+1) in a clayey Oxisol in the year of application of sewage sludge single rates which reached maximum of 80 Mg ha^{-1}, being in this rate added quantities of these metals that exceeded the maximum annual limits established for São Paulo state. Four years later, increases were still linear and the concentrations were similar to the first year, indicating that these elements persist in the soil for a long time. In fact, high persistence is a characteristic of heavy metals added to soils [87].

On the other hand, low rates of sewage sludge applied to land do not increase excessively the total concentrations of heavy metals in soils. Oliveira et al. [69] found that the total concentrations of Cu, Ni, Pb and Zn (extraction with HNO_3–H_2O_2–HCl) in 0-0.20 m layer of an Oxisol cultivated with maize increased after five annual applications of sewage sludge rates up to 10 Mg ha^{-1} $year^{-1}$, but the increases did not exceed the limits of these heavy metals allowed in São Paulo state. In the study conducted by Silva et al. [88], the total concentrations of Cu, Ni and Zn (extraction with HCl–HNO_3 3+1) in 0-0.20 m layer of a clayey Oxisol cultivated with maize increased in response to rates of sewage sludge from the Barueri and Franca municipalities, but they increased less with Franca sewage sludge, which had lower concentrations of these metals and it was applied at lower rates. Thus, application of low rates of sludge with low concentrations of heavy metals in their composition seems to be a strategy to minimize the excessive accumulation of heavy metals in soils.

However, this strategy may be insufficient for Cd. Nogueira et al. [78] observed that application of 10.8 Mg ha^{-1} of sewage sludge, rate defined to supply 100 % of N required by sugarcane, increased total concentrations of As, Cd, Cu, Ni, Pb and Zn (extraction by Method 3051A – see Table 3) in 0-0.20 m layer of an Ultisol in assessments performed 360 and 720 days after application of the sludge. While the concentrations of As, Cu, Ni, Pb and Zn were well below the limits established by CETESB and CONAMA, the concentration of Cd (0.2 mg kg^{-1}) was relatively close to the limit of CETESB (< 0.5 mg kg^{-1}, Table 7). Although the concentration of Cd has more than double to reach this limit, the fact that the limit is too low cause concern in relation to any increase in Cd concentration due to new applications of sewage sludge. In the work of these authors has been reported for the first time in Brazil the effect of sewage sludge on the accumulation of Se in soil. Se concentration increased from 0.068 to 0.092 mg kg^{-1} with application of 10.8 Mg ha^{-1} of sludge, but this increase was below the limit of CETESB (0.25 mg kg^{-1}) and far below the limit of CONAMA (5 mg kg^{-1}).

The concentrations of heavy metals presented above refer only to the topsoil (0-0.20 m depth). However, deeper and stratified sampling can give an idea of how these metals are distributed among soil layers. Merlino et al. [73] evaluated the concentrations of Cd, Cr and Pb (extraction with HNO_3–H_2O_2–HCl) in the layers 0-0.10, 0.10-0.20 and 0.20-0.40 m of an Oxisol cultivated with maize after 11 years of annual application of sewage sludge rates up to 20 Mg ha^{-1}. The concentrations of Cr and Pb increased only in the 0-0.10 m layer, suggesting accumulation in the superficial layer. There was no effect on the concentration of Cd. In another study, there was also no effect on total concentrations of Cd, Cr, Ni and Pb (extraction with HNO_3–H_2O_2–HCl) in the layers 0-0.10, 0.10-0.20, 0.20-0.30, 0.30-0.40 e 0.40-0.50 m depth of a clayey Oxisol after four annual applications up to 15 Mg ha^{-1} $year^{-1}$ to supply nitrogen (N) for sugarcane [75]. In contrast, Oliveira and Mattiazzo [89] observed significant increases in total concentrations of Zn (extraction with HCl–HNO_3 3+1 in microwave) until the layer of 0.40-0.60 m of an Oxisol amended with relatively high rates of sewage sludge (> 30 Mg ha^{-1} $year^{-1}$) for two consecutive years and cultivated with sugarcane, suggesting that Zn was leached to layers below the incorporation layer of sludge (0-0.20 m depth). For Cu and Cr, there was no evidence of leaching. It was not possible to assess the mobility of Cd, Ni and Pb in soil, because their concentrations were below the LOD of the analytical method (AAS).

Sewage sludge rate		Cd		Cr		Cu		Ni		Pb		Zn	
1996/97	1997/98	1996/97	1997/98	1996/97	1997/98	1996/97	1997/98	1996/97	1997/98	1996/97	1997/98	1996/97	1997/98
Mg ha^{-1}		Rate applied to soil (kg ha^{-1})											
0	0	0	0	0	0	0	0	0	0	0	0	0	0
33	37	0.9	0.6	13	14	26	20	8	11	5	6	52	61
66	74	1.9	1.2	25	29	52	40	16	21	10	13	104	122
99	110	2.8	1.8	38	42	78	59	24	31	15	19	155	181
		Total concentration in soil† (mg kg^{-1})											
0	0	nd‡	nd	16.57	15.55	17.86	16.87	nd	nd	nd	nd	21.14	20.19
33	37	nd	nd	21.56	18.14	19.47	26.17	nd	nd	nd	nd	39.22	50.03
66	74	nd	nd	25.30	25.21	27.90	36.39	nd	10.65	nd	nd	40.41	77.11
99	110	nd	nd	25.59	27.19	30.97	42.85	nd	14.63	nd	nd	41.27	97.21

†In the 0–0.20 m soil layer. Digestion of soil samples was performed with $HCl + HNO_3$ (3:1) in a microwave oven.
‡Not detected. Concentration of the metal was below the limit of detection of the analytical method (AAS).
Source: Adapted from Oliveira and Mattiazzo [68].

Table 5. Rate and total concentration of heavy metals in a Brazilian Oxisol amended with sewage sludge and cultivated with sugarcane

http://dx.doi.org/10.5772/58384

Norm of	Abrangency	As	Cd	Cr	Cu	Hg	Mo	Ni	Pb	Se	Zn
		Annual loading rate (kg ha^{-1} 365 days^{-1})									
CETESB†	São Paulo State	2.0	1.9		75	0.85		21	15	5.0	140
		Cumulative loading rate (kg ha^{-1})									
CETESB†	São Paulo State	41	39		1500	17		420	300	100	2800
CONAMA‡	Brazil	30	4	154	137	1.2	13	74	41	13	445

†Technical Norm # P4.230/1999 of CETESB (Environmental Agency of the State of São Paulo) [82].
‡Resolution # 375/2006 of CCNAMA (Brazilian National Environment Council) [67].

Table 6. Loading rates for heavy metals applied by sewage sludge in São Paulo State and Brazil

Norm of	Abrangency	As	Cd	Co	Cr	Cu	Hg	Mo	Ni	Pb	Se	Zn
		Limits ($mg\ kg^{-1}$)										
CETESB†	São Paulo State	3.5	<0.5	13	40	35	0.05	<4	13	17	0.25	60
CONAMA‡	Brazil	15.0	1.3	25	75	60	0.50	30	30	72	5	300

†CETESB [90].

‡Resolution # 420/2009 of CONAMA [91].

Table 7. Limits for heavy metals in soils from São Paulo State and Brazil

Besides increasing or not be changed, the concentrations of heavy metals can also decrease in response to the application of sewage sludge, as shown in the work of Macedo et al. [92]. Concentrations of Cd, Cr and Pb (extraction with HNO_3–H_2O_2–HCl) were evaluated in the layers 0-0.10, 0.10-0.20 and 0.20-0.40 m of a clayey Oxisol cultivated with maize after 11 years of application of sludge sewage rates up to 20 Mg ha^{-1} $year^{-1}$. In general, the application of sewage sludge reduced the concentrations of heavy metals in the surface layers, with decreasing quite evident for Cd in layer of 0-0.10 m depth. This reduction was not expected since metals were added to the soil. Probably, the heavy metals added and part of those natives may have combined with components of sewage sludge forming highly stable compounds resistant to the attack of the extractant used.

4.2. Availability

Availability refers to the amount of a chemical (e.g. heavy metal) in the soil that would be available to be absorbed by plants or other biological receptors (e.g. microorganisms). The availability to plants is called phytoavailability, which is a specific term for this class of organisms. In recent years, the availability has also been called bioavailability. Although bioavailability not yet has a clear and accepted definition [93], it has been used to characterize the availability of heavy metals in the environment [94]. However, the uncertainty about the meaning of bioavailability restricts the expansion of its use and thus contributes to the maintenance of the term availability.

The availability of a heavy metal is given by its available concentration. Quantification of the available concentration is performed by extracting the fraction of the metal that is sufficiently soluble to be absorbed by plants. The extraction of this fraction is performed using chemical extractants that try to simulate the potential for uptake of an element of the soil by the roots of plants. There are several extractants which can be used to extract heavy metals from the soil. The extractants used to assess the availability of heavy metals added to Brazilian soils by the application of sewage sludge to land are usually the same ones used in the evaluation of cationic micronutrients, especially Mehlich 1 and DTPA as the most common. Other extractants, as 0.1 mol L^{-1} HCl and Mehlich 3, have also been used in this assessing. The concentration of a metal in the soil can only be considered available if it is closely and positively correlated with the concentration of the metal in plant tissues. In terms of assessing the contamination, the available concentration can be correlated to the concentration of the metal in raw or processed agricultural product. When there is no correlation or the correlation is poor, it is more appropriate to use the term extractable concentration.

Knowledge of available concentrations is essential for assessing the environmental impact of sewage sludge application to land, since such concentrations represent the amount of heavy metals that could be leached to reach groundwater or could be absorbed by plants and transmitted to the food chain levels until reaching the man. Assessment of the availability of heavy metals in soils amended with sewage sludge was carried out in some field experiments conducted in Brazil. The main results are summarized below.

Oliveira and Mattiazzo [68] found different effects of sewage sludge rates (Table 5) on the availability of some heavy metals evaluated. The sludge was applied for two consecutive years to an Oxisol cultivated with sugarcane, and concentrations of available Cd, Cr, Cu, Ni, Pb and Zn by extractants 0.1 mol L^{-1} HCl, Mehlich 3 and DTPA were determined in soil samples collected in 0-0.20 m depth one year after each application of the waste. Increase in rates of sewage sludge consistently increased concentrations of Cu and Zn extracted with the three extractants in both evaluations. There was increase in concentration of available Ni only at the second assessment and for sewage sludge treatments. The concentrations of Cd, Cr and Pb were below the LOD of the analytical method (AAS). Variations in concentrations of DTPA-Cu and DTPA-Zn between the control (without application of sewage sludge) and the maximum rate of the waste (110 Mg ha^{-1}) were from 0.70 to 10.70 mg kg^{-1} and 0.62 to 19.12 mg kg^{-1}, respectively. The higher values of Cu and Zn are considered high and very high, respectively, for soils of São Paulo state (Table 8), assuming a similarity between the values expressed in mg kg^{-1} and mg dm^{-3}. However, the authors did not report any negative consequence of these high concentrations to the crop. Positive correlations between the availability of Cu and Zn in soil assessed with three extractants and the concentrations of these metals in different plant components (leaf+1, stalk and juice) were significant when included data from treatments with sewage sludge (three rates), treatment with mineral fertilization and control. However, when included only the sludge treatments, there were few significant correlations, indicating generally low efficiency of extractants to assess the availability of Cu and Zn in soil amended with sewage sludge. The exceptions were significant correlations between 0.1 mol L^{-1} HCl-Zn and juice-Zn in the two years of evaluation and between Zn extracted by the three extractants and stalk-Zn and juice-Zn only in the second year.

Interpretation	Limits			
	Cu	Fe	Mn	Zn
	mg dm^{-3}			
Low	0.0-0.2	0-4	0.0-1.2	0.0-0.5
Medium	0.3-0.8	5-12	1.3-5.0	0.6-1.2
High	0.9-1.5	13-24	5.1-9.0	1.3-2.3
Very high	1.6-15	25-60	10-50	2.4-15
Toxicity				>130

Source: Adapted from Abreu et al. [93].

Table 8. Interpretation limits for concentrations of cationic micronutrients extracted by DTPA pH 7.3 in soils from São Paulo State, Brazil

In another study with sugarcane, Nogueira et al. [78] applied sewage sludge rates up to 10.8 Mg ha^{-1} to an Ultisol to supply 100 % of N required by the crop, and after 360 and 720 days they assessed the availability of heavy metals in the layer of 0-0.20 m depth using DTPA. The concentrations of Cd, Cu, Ni, Pb and Zn available increased with increasing sludge rates in the two evaluation periods, reaching maximum values of 0.112, 2.64, 0.47, 2.09 and 7.61, respectively. Concentrations of Cu and Zn are considered very high (Table 8), while the concentrations of Cd, Ni and Pb are above the normal range for soils of São Paulo state (Table

9), indicating contamination. Nevertheless, there was no toxicity in the plant. The extractant used was efficient only for Cd and Zn, since their available concentrations in the soil were highly correlated with the concentrations in leaf with top visible dewlap, stalk and juice. For concentrations of As, Cr and Se, there was no effect of sewage sludge application. All metals intentionally evaluated in this study were detected, probably by the use of inductively coupled plasma mass spectrometry (ICP-MS), which had LODs very low.

Abragency	Limits			
	Cd	Cr	Ni	Pb
	mg dm^{-3}			
São Paulo state	0.020	0.030	0.180	0.85
Other states	0.019	0.007	0.147	0.76

†Values above the limits are indicative of contamination.

Source: Adapted from Abreu et al. [93].

Table 9. Limits for heavy metals extracted by DTPA pH 7.3 in soils from São Paulo State and other states in Brazil†.

The availability of heavy metals in soils amended with sewage sludge has also been evaluated for maize crop. Martins et al. [81] found that sewage sludge rates up to 80 Mg ha^{-1} increased the concentrations of extractable Cu, Ni and Zn by DTPA and Mehlich 3 in a clayey Oxisol with or without lime after a maize cultivation. However, the effect of liming on extractability of these metals was different between the two extractants. Liming reduced the concentrations of extractable Ni and Zn by DTPA, but did not change concentrations of Cu. In the case of Mehlich 3, concentrations of extractable Cu and Zn increased with liming, which was unexpected, because it is well known the fact that the increase in soil pH due to liming generally decreases the availability of cationic micronutrients. Positive correlations between soil Cu or Zn extracted by DTPA and Mehlich 3 and concentrations of Cu and Zn in leaf+4 and shoots of maize were observed, suggesting that extractants were effective in assessing the availability of these micronutrients. In the case of Ni, it has not been possible to establish these relationships because its concentration in plant tissues was below LOD of the analytical method (AAS).

Silva et al. [88] applied sewage sludges from Franca and Barueri municipalities (São Paulo state) in three consecutive crops of maize to a clayey Oxisol. For both sludges, rates were to supply up to eight times the amount of N required by the crop, reaching maximum values of 30 and 64 Mg ha^{-1} for sludges of Franca and Barueri, respectively. Soil samples were taken from the layer 0-0.20 m depth after each waste application and before each maize sowing for assessing availability of Cu, Mn, Ni, Pb and Zn by DTPA and Mehlich 1. The concentrations of Cu, Ni and Zn extracted by both extractants increased in response to the application of the two sludges in the three maize crops. For Mn, the extractability varied with sludge, crop and extractant. In extraction with Mehlich 1, the concentration of Mn increased in the three crops for Franca sewage sludge and only in the first crop for Barueri sludge. The concentration of Mn extracted by DTPA increased for both sludges, but only in first and second crops. There were no significant changes in the extractability of Pb measured by Mehlich 1 for both sludges in all crops. In contrast, the concentration of Pb extracted by DTPA increased in all crops for Barueri sludge and in the second crop for Franca sludge. In all cases in which concentrations of extractable metals increased, increases were higher for Barueri sludge, probably because it had the highest concentrations of metals and was applied in higher rates. Regardless of sewage sludge, increases in concentrations of extractable metals by both extractants were lower in the third than in the second crop, suggesting that successive applications of sewage sludge reduced the solubility of such elements. The significance of the correlations between the concentrations of Cu, Mn, Ni, Pb and Zn in soil and leaf below the ear and grains varied with sewage sludge and crop for both extractants, indicating that they were not consistently effective in assessing the availability of these metals to maize. The element that had more cases of significance was Zn.

In long term experiments conducted by Oliveira et al. [69], sewage sludge rates up to 10 Mg ha^{-1} $year^{-1}$ were applied to two Oxisols (Typic Haplorthox and Typic Eutrorthox) for five years and in each year the soils were cultivated with maize. The concentrations of Cu, Ni and Zn extractable by Mehlich 1 in layer of 0-0.20 m depth increased in both soils after the fifth year of sludge application. The concentration of extractable Pb in this layer increased only in Typic Haplorthox. In layer of 0.20-0.40 m depth, there was an increase in extractability of Ni in both soils, whereas concentrations of extractable Cu, Pb and Zn were increased only in Typic Haplorthox. There was no change in concentration of extractable Mn in any of the layers of both soils. Although the concentrations of extractable Cu, Ni, Pb and Zn increased with sludge application, these increases were not adequately correlated with the accumulation of these metals in the shoot and grains of maize plant, indicating that the Mehlich 1 was ineffective in assessing the availability of such heavy metals in both soils. Although the extractant has been inefficient to evaluate the availability, increased extractability in layer of 0.20-0.40 m depth of Typic Haplorthox suggests that Cu, Ni, Pb and Zn were leached due application of sewage sludge.

The results presented in Galdos et al. [77] also suggest the occurrence of leaching, but not for all metals evaluated. The authors applied sewage sludge to a clayey Oxisol to supply up to double of N requirement of maize by two successive crops. The maximum rates were 21.6 Mg ha^{-1} in first crop and 20.5 Mg ha^{-1} in second crop. The sludge was incorporated into soil layer of 0-0.10 m depth. The concentrations of extractable Cu, Ni and Zn by DTPA in soil layers of 0-0.05, 0.05-0.10 e 0.10-0.20 m depth were evaluated 267 days after first application and 179 days after second application. The applications of sewage sludge increased the concentrations of Cu and Ni in the layers 0-0.05, 0.05-0.10 m only in the second evaluation, but did not change concentrations in 0.10-0.20 m layer in any of the evaluations, indicating no leaching, probably because the concentrations of these metals in sludge were low (284.1 and 864.8 mg kg^{-1} Cu and 41.8 and 35.5 mg kg^{-1} Ni). In contrast, Zn concentrations increased in all layers, including the layer of 0.10-0.20 m in both evaluations, suggesting that Zn was leached to the layer below the incorporation layer of sewage sludge. The leaching may have occurred because of high concentrations of Zn in sludge (11,364.8 and 1,738.1 mg kg^{-1}). Increased mobility of heavy metals in soils amended with sewage sludge is very concerning because the metals leached can reach groundwater, contaminating it.

4.3. Chemical fractions

In Brazil, field studies about fractionation of heavy metals in soils amended with sewage sludge are scarce. Among the few studies that exist, one of Nogueira et al. [94] can be considered comprehensive because they evaluated Cd, Pb and Zn in different soil fractions [exchangeable (Exch), organic matter (OM), amorphous Fe oxide (AFeO), crystalline Fe oxide (CFeO), and residual (Res)] and in soil organic matter (SOM) chemical fractions [fulvic acid (FA), humic acid (HA) and humin (Hum)] after nine annual applications of sewage sludge rates up to 20 Mg ha^{-1} to a clayey Oxisol cultivated with maize. The Cd concentration in the fraction Res was not altered by application of sewage sludge, and concentrations in other fractions were below the LOD of the analytical method (AAS). Similarly, there was no effect of sludge on concentration of Pb in fractions AFeO, CFeO and Res, and the concentrations in other fractions were below the LOD. In contrast, the sludge application increased the concentration of Zn in all fractions, mainly in AFeO and CFeO fractions, indicating that a considerable part of the added Zn was adsorbed on these oxides. In the fractions of SOM, Cd and Pb concentrations were not changed in response to sewage sludge application. While the concentration of Zn in Hum fraction was not changed, the concentrations in FA and HA fractions increased by sludge application, indicating that, in SOM, the added Zn is bound to fractions less stable and thus can become more easily available in the environment.

In another study, the redistribution of Ni among humic fractions of a medium-textured Oxisol cultivated with maize and amended with sewage sludge rates up to 20 Mg ha^{-1} $year^{-1}$ for six years was assessed by Melo et al. [95]. The authors observed that the application of sewage sludge increased more the proportion of Ni in the humin fraction (Figure 1). This means that the sewage sludge redistributed added Ni preferably to more stable fraction of SOM. As the humin fraction is insoluble in acid and alkaline medium, Ni associated with it can be considered unavailable to plants and not directly subject to leaching, which favors its accumulation in soil layer where sewage sludge is incorporated.

4.4. Speciation

In this work, speciation refers to the separation of heavy metals in different chemical forms and possible oxidation states (e.g., chemical species) in the soil solution. Such metals can be separated basically into free ions, complexes and ion pairs [11, 96]. The participation of each form in the total concentration of a metal in the soil solution can be estimated by using the following general procedures: (i) extraction of soil solution, (ii) determining total concentration of cations, anions and organic compounds in solution and (iii) calculation of the activity of the metals of interest, based on determined concentrations, using specific software for this purpose. Speciation is the basis of the free ion activity model (FIAM) and also other models, which relate the responses of plants to different chemical species of metals present in the soil solution [96].

Plants uptake heavy metals preferably as free ions. Thus, if these ions are increased in the soil solution, so they can be absorbed in excess, and excessive uptake of these metals can cause phytotoxicity and also contamination of the food chain. Indeed, uptake and toxicity of metals in plants generally correlate better with the activity of free ions [11, 96]. On the other hand,

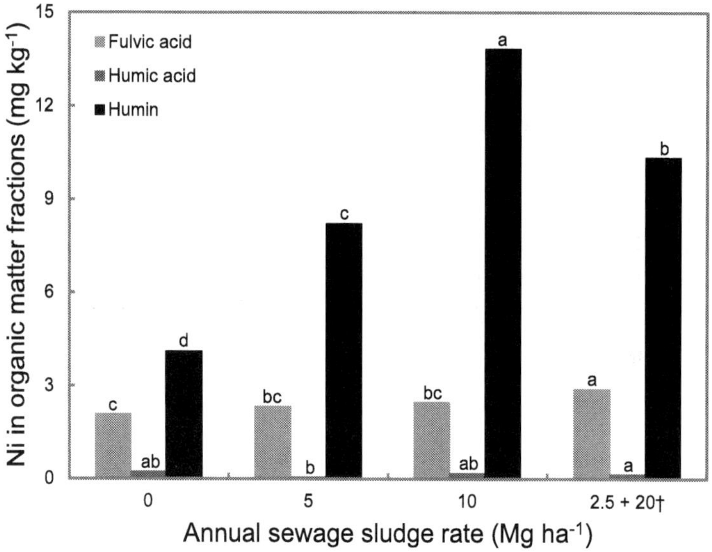

Source: Adapted from Melo et al. [95].

Figure 1. Nickel associated to soil organic matter chemical fractions in medium-textured Oxisol amended with sewage sludge applied annually during six years and cultivated with maize. Means within each fraction followed by the same letter are not significantly different according to Tukey test ($p < 0.05$). †2.5 Mg ha^{-1} in 1st, 2nd, and 3rd years and 20 Mg ha^{-1} in 4th, 5th and 6th years.

other chemical species that are not immediately relevant to the uptake of plants can also cause environmental impact. Heavy metals bound to dissolved organic matter, forming organo-metal complexes in the soil solution, may have increased their mobility in the soil profile, as demonstrated in experiments with undisturbed soil columns [17], favoring the leaching of these potentially toxic metals, which increases the risk of groundwater contamination

Addition of organic materials to the soil is able to change the availability and mobility of heavy metals in the soil solution, which can minimize some environmental impacts and maximize others. Organic materials added to the soil can decrease the activity of free ions [97] and, consequently, reduce the availability of metals to plants, minimizing the risk of phytotoxicity and contamination of the food chain. In contrast, the addition of organic materials can increase the activity of organo-metal complexes [97] and thus increasing mobility of heavy metals, which become more readily leachable, maximizing the risk of groundwater contamination. Sewage sludge, as an organic material, can also generate these contrasting environmental effects.

As occurs for fractionation, field studies about speciation of heavy metals in soils amended with sewage sludge are also rare in Brazil. The only study found with these characteristics was the Silva's MS thesis [98]. In this work, sewage sludge rates up to 20 Mg ha^{-1} $year^{-1}$ were applied for seven years to a clayey Oxisol cultivated with maize. The author evaluated chemical species of heavy metals in the soil solution from the layer 0-0.20 m depth at 22, 26, 37 and 43 months after the seventh application of sludge. Considering all evaluation times, Cd, Cr, Mo, Ni and Pb concentrations in the soil solution were generally below the LOD of the analytical method used (Inductively Coupled Plasma-Optical Emission Spectroscopy–ICP-OES). On the other hand, Cu and Zn were consistently detected. The data of speciation for Cu and Zn show that application of sewage sludge generally decreased the proportion of these metals as free ions (Cu^{2+} and Zn^{2+}) and increased their proportion as organo-metal complexes [dissolved organic carbon-Cu (DOC-Cu) and DOC-Zn]. The observed decrease in the proportion of Cu and Zn free ions in the soil solution indicates that sewage sludge can restrict the availability and consequently uptake of theses metals by plants, reducing the risk of phytotoxicity and contamination of the food chain. In contrast, increased proportions of COD-Cu and COD-Zn in the soil solution indicate that the sludge can intensify the leaching of Cu and Zn, increasing the risk of groundwater contamination. These results suggest that the risk of environmental contamination by application of sewage sludge on land should be better assessed considering the intensity these two contrasting effects.

4.5. Risks of contamination

Sewage sludge is an organic material containing heavy metals. Therefore, its application to land generates risks of soil contamination by heavy metals. Although it is difficult to quantify this risk, due to the complex interaction among the factors that determine it, it is possible to mention some steps that can minimize it. Knowledge of these measures is relevant, since they allow the adoption of a more appropriate management for the disposal of sewage sludge on land.

As noted above, research about the impact of sewage sludge agricultural use on soil contamination by heavy metals has advanced considerably in recent years in Brazil. Field experiments using the total concentration as an indicator of heavy metal contamination have shown interesting results. Sludge rates above 65 Mg ha^{-1} $year^{-1}$ can quickly contaminate (approximately two years) the soil by Cu, Ni and Zn. Once contaminated with Cu and Zn, soil may remain in this condition for several years, even without additional application of sewage sludge. Zn may be leached in soils amended with relatively high rates of sludge (> 30 Mg ha^{-1} $year^{-1}$). In contrast, rates as low as 10 Mg ha^{-1} $year^{-1}$, applied for five years, have low potential for soil contamination by Cu, Ni, Pb and Zn. Single application of sludge rate of this magnitude is unlikely to contaminate soil for As and may be insufficient to cause contamination by Cd. But as the limits for Cd in soils are very low (Table 7), any increase in its total concentration is concerning. Therefore, maximum rates for annual increase of Cd total concentration in soil could be adopted as an optional security criterion to restrict application of sewage sludge on land even though there is no legal restriction to its application.

Studies reviewed in this work also show the usefulness of available concentration to assess soils contaminated by heavy metals. Sewage sludge rates relatively low (~ 10 Mg ha^{-1} $year^{-1}$) can contaminate soil by Cd and Zn due to the excessive increase in available concentrations of these metals by DTPA. Similarly, soil contamination by Cu and Zn, indicated by increase in available concentrations by DTPA and Mehlich 3, may occur by applying high sludge rates. Liming reduces the availability of Zn and hence the potential for contamination of the soil, but only DTPA is able of detecting this reduction. Mehlich 1 extractant, which is widely used in routine soil analysis (soil testing) in Brazil, seems to be inefficient to assess whether a soil is contaminated by heavy metals. Likewise, Mehlich 3 and DTPA are not always efficient. The alternative extractant 0.1 mol L^{-1} HCl has also shown poor performance. As the factors associated to inefficiency of the extractants are variables (e.g., type of heavy metal, soil class, type of crop, cropping sequence and origin of sewage sludge) and they may act in combination, the available concentration can not yet be considered a consistent and relatively safe indicator for assessing the risk of heavy metals contamination in soils amended with sewage sludge.

Nevertheless, extractable concentrations by DTPA and Mehlich 1 may be useful to monitor the leaching of heavy metals in soils amended with sewage sludge. Relatively low sludge rates (~ 10 Mg ha^{-1} $year^{-1}$) applied for a long time (5 years) stimulate the leaching of Cu, Ni and Pb, as indicated by increase in extractable concentrations by Mehlich 1 in depth. High sludge rates (~ 20 Mg ha^{-1} $year^{-1}$) with high concentration of Zn in its composition (> 1,300 mg kg^{-1}) have the potential to leach Zn from the superficial layers of the soil in a short time (2 years), as suggested by the increase in extractable concentrations by DTPA in depth. Thus, the extractable concentration, used in the monitoring of leaching, can help in assessing the risk of groundwater contamination by these heavy metals added to the soil by sewage sludge application to land.

The results of fractionation of heavy metals, albeit very limited, have shown opposite trends for Zn and Ni. Zn added to the soil by the sludge is preferably bound to Fe and Al oxides. In this form, its solubility is limited and thus the risk of being excessively absorbed by plants or leached is reduced. On the other hand, added Zn which binds to SOM is most often associated with fulvic and humic acids. As these humic substances are poorly stable, its transformation tends to release Zn retained in soil organic matrix, thereby increasing the risk of contamination. In contrast, added Ni which binds to SOM is mainly in humin fraction. Since this fraction is very stable, it restricts the solubility of Ni, minimizing the risk of contamination of crops and groundwater.

Speciation of heavy metals in soils amended with sewage sludge is still incipient. We found only one work on this topic, which is a MS thesis. The results of this thesis, however, have shown significant trends. Applications of sewage sludge rates as low as 20 Mg ha^{-1} $year^{-1}$ for seven years declined forms of Cu and Zn (Cu^{2+} and Zn^{2+}) that are preferentially absorbed by plants, but increased forms of these metals (Cu-DOC and DOC -Zn) that are easily leached. This means that the sewage sludge, when applied in low rates over a long period, decreases the risk of phytotoxicity and contamination of the food chain, but increases the risk of groundwater contamination.

Simultaneous use of different measures to evaluate heavy metals in soils can increase security in the risk assessment of environmental contamination due to the application of sewage sludge to land. From this perspective, total concentration, available concentration, chemical fractions and chemical species of metals should be assessed together to better characterize the dynamics of possible contamination. These different measures represent a gradient of solubility with immediate and potential impact on plant uptake and leaching of heavy metals in soils. Thus, the environmental risk posed by the application of sewage sludge could be predicted for short and long term. In addition, these integrated measures could also be useful for the development of mathematical models to predict the availability of heavy metals as free ions using easily measurable input variables (e.g., total metal concentration, pH and SOM) [96]. Such models would be particularly advantageous in predicting the activity of free ions in response to changes in soil properties by the application of sewage sludge [96]. We did not find studies in Brazil using this integrated approach. The studies summarized in this work enable only superficial considerations, because the data are from independent experiments, and only for Zn, since it was the only metal with results for all measures. Total concentration is shown to be a good measure of soil contamination by Zn, but only at high rates of sewage sludge. Under these conditions, the leaching of metal can be found by the total concentration. For relatively low rates of sludge, however, this measure suggests only low potential for contamination. In both cases, it is not possible to establish a close and reliable relationship between total concentration and toxic effects. Nevertheless, availability and chemical fractions indicate environmental contamination even with application of relatively low rates of sludge. Speciation, in turn, suggests that the leaching of Zn may be more relevant than its excessive uptake by plants, which increases the risk of groundwater contamination to the detriment of phytotoxicity and contamination of the food chain.

5. Heavy metals in crops grown on soils amended with sewage sludge

5.1. Concentration in plants

Land application of sewage sludge can successively increase heavy metals availability in soil, uptake by plants and accumulation in plant tissues. Thus, concentration in plant tissue can be used as an indicator of heavy metals transfer from soils amended with sludge to plants and of the entry of hazardous elements in the food chain. It can also be used to evaluate the phytotoxicity and contamination of harvested products by heavy metals added to the soil by sewage sludge application, thus presenting important role in determining crop performance and quality of food originated in field. The concentrations of heavy metals in plants vary depending on several factors, including rate of sewage sludge, type of heavy metal, type of plant and analyzed plant part. These factors were studied in field experiments conducted in Brazil, particularly in São Paulo state. Their main results are summarized below.

Galdos et al. [77] evaluated the concentrations of Cu, Ni and Zn in the middle third of the leaf below the ear of maize grown for two years on a clayey Oxisol amended with sewage sludge to supply up to twice the N requirement of the crop. The maximum rates were 21.6 Mg ha^{-1} in the first year and 20.5 Mg ha^{-1} in the second year. Cu concentration in the leaf was changed only in the first year, increasing with application of the maximum rate of sludge. The waste had no effect on the concentration of Ni, which was measured only in the second year. For Zn, the concentration in the leaf increased progressively with sewage sludge rates in the two years. The concentrations of Cu and Zn were within the range of concentrations suitable for the crop

[99], indicating no phytotoxicity, which is consistent with the absence of adverse effects of the waste on productivity of grains.

Oliveira et al. [69] applied sewage sludge rates up to 10 Mg ha^{-1} year^{-1} to two Oxisols (Typic Haplorthox and Typic Eutrorthox) for five years and measured the concentrations of Cu, Mn, Ni, Pb and Zn in shoot and grains of maize in the 5th year. There was no effect of the sludge on concentration of Cu in any case. The Mn concentration varied only in the shoot, decreasing in Typic Haplorthox and not having clear effect on Typic Eutrorthox. The sludge did not change the concentration of Ni in shoot and grains, except for concentration in grains in the Typic Eutrorthox which was below the LOD of the analytical method (AAS). The Pb concentration was not changed in the shoot and it was below the LOD in grains. The concentration of Zn increased in shoot and grains, with the exception of the concentration in the grains in Typic Eutrorthox, which has not changed. Increased concentrations of Zn were below the tolerance limit in foods as corn (50 mg kg^{-1}) established by Brazilian Health Surveillance Agency [100].

Silva et al. [88] applied sewage sludge from Franca and Barueri municipalities to a clayey Oxisol in three successive annual crops of maize and measured the concentrations of Cu, Mn, Ni, Pb and Zn in middle third of leaf opposite and below the ear and in its grains. Sludge rates were defined to supply until eight times the amount of N required by the crop, reaching maximum values of 30 and 64 Mg ha^{-1} for sludges from Franca and Barueri, respectively. The concentration of Cu in leaf increased only in the first crop with Barueri sludge application and in third crop with Franca sludge application. Despite the increase, the values are below the toxic level (> 50 mg kg^{-1}) for maize leaf presented in Barbosa Filho et al. [99]. The concentration of Mn in the leaf increased with increasing sludge rates, except for the first crop when was used Franca sewage sludge. Suitable concentrations of Mn in maize leaf can be as high as 214 mg kg^{-1} [99]. No concentration exceeded this value suggesting that there was no toxicity of Mn. The concentration of Ni increased only in second crop by Franca sludge application. On the other hand, the Zn concentration increased with increasing rates of both sludges and in all crops. The only situation in which the concentration of Zn in leaf was slightly above the appropriate range of concentrations (15-100 mg kg^{-1}) presented in Barbosa Filho et al. [99] was in third crop with application of higher Barueri sludge rate. However, this excess Zn in leaf must not have been toxic to the plant, since no symptoms of toxicity were reported. In grains, there was much less cases of elevated concentrations of heavy metals when compared with the leaf, suggesting some limitation in the redistribution of these metals to the harvested plant part. Consistent increases were observed for Mn in second crop with sludge application, Ni also in second crop, but with Barueri sludge application, and Zn in the second and third crops with application of the sludges. In the first crop, metals concentrations were below the LOD of the analytical method used (ICP-OES). The concentrations of Cu, Ni, Pb and Zn in grains were below the maximum limits for cereals in general (30, 5.0, 8.0 and 50 mg kg^{-1}, respectively) by Brazilian Food Industry Association [101] and Mn concentrations were below the critical range presented in Kabata-Pendias and Pendias [102] for grains of plants grown in contaminated soils with Mn.

Oliveira et al. [71] evaluated the concentrations of Cd, Cr, Pb and Zn in stem, leaves, straw, ear husk, cobs and grains of maize grown on a clayey Oxisol after nine years of annual applications of sewage sludge rates (Table 10). The Cd concentrations were below the LOD of the analytical method (AAS) (0.06 mg kg^{-1} in grains and 0.2 mg kg^{-1} in other plant tissues). Cr concentration increased in stem and leaves in response to sludge application and was below the LOD in ear husk, cob and grains. Pb concentration increased in stem, leaves and ear husk, did not change in cob and was below the LOD in grains. Except for grains, all plant parts had increases in Zn concentration. These results show that stem and leaves were the most sensible plant tissues for expressing the effect of sewage sludge rates on Cr, Pb and Zn concentrations in maize.

In other long term experiment, Cd, Cr and Pb concentrations were evaluated in leaf apposite and below the ear, whole plant and grains of maize after 11 annual applications of sewage sludge rates up to 20 Mg ha^{-1} to an Oxisol [73]. There was no effect of sludge on Cd and Pb concentrations in leaf. For Cr in this plant part, its concentrations were below the LOD (< 0.19 mg kg^{-1}) of the analytical method (AAS). Similarly, concentrations of these heavy metals were below the LOD in grains (0.03, 0.15 and 0.19 mg kg^{-1} for Cd, Cr and Zn, respectively). When considered the whole plant, Cd, Cr and Pb concentrations were not significantly changed by sewage sludge applications.

There are also results for sugarcane. Camilotti et al. evaluated the concentrations of Cd, Cr, Ni and Pb in stem and leaves of sugarcane cultivated in a clayey Oxisol after three [103] and four [75] annual applications of sewage sludge rates, which reached a maximum of 15 Mg ha^{-1}, with the goal of supplying 100 % and 200 % of N required by the crop. There was no effect of three applications of sewage sludge in concentrations of Cr in stalk and leaves, Ni in leaves and Pb in stalk and leaves, as well as Cd concentrations in all plant parts analyzed. Cr in leaves and Ni in stalk were below the LOD of the analytical method used (AAS). Similarly, the fourth sludge application did not affect the concentrations of Pb in leaves, but all other cases the concentrations of evaluated heavy metals were below the LOD.

In an experiment conducted on Ultisol, Nogueira et al. [78] applied sewage sludge rates up to 10.8 Mg ha^{-1} at sugarcane planting for supplying N required by the crop and evaluated As, Cd, Cr, Cu, Ni, Pb, Se and Zn concentrations in leaf (with top visible dewlap), stalk and juice at the harvest time of plant cane (first year) and ratoon cane (second year). In these two evaluations, As concentration increased in leaf and was not affected in stalk and juice by sludge application. Cd concentration increased in leaf, stalk and juice in both crops. On the other hand, Pb concentration decreased in stalk and juice of plant cane and was not affected in other situations. Cu concentration decreased in stalk of plant cane and leaf of ratoon cane and was not affected in other cases. Ni concentration also was not generally affected, but it increased in stalk of plant cane and decreased in stalk of ratoon cane. For Pb concentration, there was no effect of sludge rates. Se concentration was affected only in leaf of plant cane, increasing with sludge application. Zn concentration increased in all situations. There were no cases of concentration below the LOD probably because the authors used ICP-MS.

Sewage sludge rate		Heavy metal in plant tissue		
Annual	Cumulative	Cr	Pb	Zn
Mg ha^{-1}		mg kg^{-1}		
			Stem	
0	0	0.13 c†	0.68 b	15.76 c
5	45	0.33 b	0.61 b	19.11 c
10	90	0.45 a	0.83 a	26.62 b
2.5 + 20‡	127.5	0.19 c	0.47 c	44.35 a
			Leaves	
0	0	0.74 b	1.10 c	29.35 b
5	45	1.20 a	1.43 b	28.54 b
10	90	1.48 a	1.72 a	39.34 ab
2.5 + 20‡	127.5	0.67 b	1.64 ab	50.77 a
			Ear husk	
0	0	< 0.3	4.28 ab	13.03 b
5	45	< 0.3	3.95 b	12.91 b
10	90	< 0.3	4.86 a	17.71 ab
2.5 + 20‡	127.5	< 0.3	4.70 ab	20.52 a
			Cob	
0	0	< 0.3	0.88 a	13.49 b
5	45	< 0.3	0.98 a	15.93 b
10	90	< 0.3	1.04 a	27.47 a
2.5 + 20‡	127.5	< 0.3	0.98 a	32.35 a
			Grains	
0	0	< 0.09	< 0.4	32.36 a
5	45	< 0.09	< 0.4	29.32 a
10	90	< 0.09	< 0.4	35.77 a
2.5 + 20‡	127.5	< 0.09	< 0.4	36.70 a

†Means within a column in each plant tissue followed by the same letter are not significantly different according to Tukey test ($p < 0.05$).

‡2.5 Mg ha^{-1} in 1st, 2nd, and 3rd years and 20 Mg ha^{-1} in 4th, 5th, 6th, 7th, 8th and 9th years.

Source: Nogueira et al. [71].

Table 10. Cr, Pb and Zn concentrations in stem, leaves, ear husk, cob and grains of maize after nine annual applications of sewage sludge to a clayey Oxisol from São Paulo State, Brazil.

http://dx.doi.org/10.5772/58384

5.2. Risks of contamination

The results reviewed above have shown contrasting effects of sewage sludge on the concentrations of heavy metals in maize and sugarcane. Sludge effects on Cu, Pb and Ni concentrations ranged greatly. On the other hand, its effects were more consistent for Zn. Sludge application generally increased Zn concentrations in plants, but not in phytotoxic levels. Increases occurred even in edible parts of crops, such as grains of corn, but concentrations did not exceed the limit established by Brazilian legislation. For sugarcane, in addition to Zn accumulation, Cd concentrations may also increase with sludge application. However, even increasing, Cd concentrations remained very low.

These findings support the view that the sewage sludge applied to land has low potential to contaminate maize and sugarcane with heavy metals. However, this does not mean that risks of contamination of other crops are also low. Vegetable crops, for example, can easily be contaminated by heavy metals since they have contact with sewage sludge applied to soil or bed. Because of this high risk of contamination, Brazilian law prohibits the agricultural use of sewage sludge for production of vegetable crops and also other crops [67].

6. Summary and conclusions

The expected increase in the generation of sewage sludge in Brazil should intensify its disposal on land. Although the sludge is rich in organic matter and plant nutrients, it has heavy metals in its composition. Due to the presence of potentially toxic metals, application of sludge to land requires a risk assessment of soil-plant system contamination. Knowledge of the sludge composition in terms of heavy metals is the first step in assessing risks of contamination.

We have shown in this work that concentrations of heavy metals in the sludge generated in Brazilian WTPs can vary widely depending on the metal. Cd concentration is extremely variable, while Zn concentration varies relatively little. High variability of Cd concentration may hamper the planning sludge application by the need for additional adjustment in order to avoid excessive application of Cd in soil. Zn concentrations found in the sludges analyzed were very high. Thus, the risk of contamination by Zn can be considered high as well.

Soil contamination by Cu, Ni and Zn, as determined by the total concentration, can be rapid, since high rates of sludge are applied to land. In contrast, low rates of sludge have shown low potential to contaminate soil with Cu, Ni, Pb and Zn. However, available concentrations can indicate contamination by Cu and Zn even with application of low rates of sludge. Furthermore, they can detect changes in the solubility of metals due to liming.

Nevertheless, the available concentrations depend on the efficiency of the extractant. DTPA, 0.1 mol L^{-1} HCl, Mehlich 1 and Mehlich 3 have not been as efficient as needed. DTPA has shown best performance among these extractants. DTPA and Mehlich 1 extractants can also be used to monitor the leaching of heavy metals such as Cu, Ni, Pb and Zn in soils amended with sewage sludge, whose importance is to assess the soil contamination at depth and the potential for contamination of groundwater. The fractionation and speciation data have confirmed that

Zn has a potential to be leached in soil amended with sludge, which increases the risk of groundwater contamination. Regard to contamination of the plant, the results have shown that the application of sewage sludge has low potential to contaminate maize and sugarcane with Cu, Ni, Pb and Zn.

Despite the apparent low risk of contamination of the soil-plant system by heavy metals due to the application of sewage sludge to land, this review has shown that (i) the results from the field experiments are limited to a restricted number of potentially toxic metals, (ii) the chemical extractantsaregenerally inefficient to estimate the available concentrations for these metals in a range of situations and (iii) the studies on fractionation and speciation of these hazardous elements in soils are still incipient. Furthermore, studies have not used an integrated approach involving total concentration, available concentration, chemical fractions and chemical species of heavy metals to assess more reliably the risk of phytotoxicity and contamination of the food chain and groundwater. Thus, this scenario suggests that the application of sewage sludge to land should be as restrictive as possible in Brazil.

Author details

Alysson Roberto Baizi e Silva[1*] and Fábio Camilotti[1]

*Address all correspondence to: alysson.silva@embrapa.br

1 Embrapa Eastern Amazon, Belém, PA, Brazil

2 Jaboticabal College of Technology, Jaboticabal, SP, Brazil

References

[1] Tsutiya MT. Alternativas de disposição final de biossólidos. In: Tsutiya MT, Comparini JB, Alem Sobrinho P, Hespanhol I, Carvalho PCT, Melfi AJ, Melo WJ, Marques MO. (ed.) Biossólidos na agricultura. São Paulo: SABESP; 2001. p133-180.

[2] National Research Council. Biosolids applied to land: advancing standards and practices. Washington: The National Academies Press; 2002.

[3] Kelessidis A, Stasinakis AS. Comparative study of the methods for treatment and final disposal of sewage sludge in European countries. Waste Management, 2012;32(6) 1186-1195.

[4] Wang M-J. Land application of sewage sludge in China. The Science of the Total Environment, 1997;197(1-3) 149-160.

[5] Bettiol W, Camargo OA. A disposição de lodo de esgoto em solo agrícola. In: Bettiol W, Camargo OA. (ed.) Lodo de esgoto: impactos ambientais na agricultura. Jaguariúna: Embrapa Meio Ambiente; 2006. p25-43.

[6] Tsutiya MT. Características de biossólidos gerados em estações de tratamento de esgotos. In: Tsutiya MT, Comparini JB, Alem Sobrinho P, Hespanhol I, Carvalho PCT, Melfi AJ, Melo WJ, Marques MO. (ed.) Biossólidos na agricultura. São Paulo: SABESP; 2001. p89-131.

[7] Rossi G, Pennelli B, Socciarelli S, Figliolia A. Effects of medium-term amendment with sewage sludges on heavy metal distribution in soil. Developments in Soil Science, 2002;28A 99-107.

[8] McGrath SP, Chaudri AM, Giller KE. Long-term effects of metals in sewage sludge on soils, microorganisms and plants. Journal of Industrial Microbiology, 1995;14(2) 94-104.

[9] Gupta UC, Gupta SC. Trace element toxicity relationships to crop production and livestock and human health: implications for management. Communications in Soil Science and Plant Analysis, 1998;29(11-14) 1491-1522.

[10] McLaughlin MJ, Parker DR, Clarke JM. Metals and micronutrients – food safety issues. Field Crops Research, 1999;60(1-2) 143-163.

[11] Mattiazzo ME, Berton RS, Cruz MCP. Disponibilidade e avaliação de metais pesados potencialmente tóxicos. In: Ferreira ME, Cruz MCP, Raij B van, Abreu CA. (ed.) Micronutrientes e elementos tóxicos na agricultura. Jaboticabal: CNPq/FAPESP/POTAFOS; 2001. p213-234.

[12] Shuman LM. Fractionation method for soil microelements. Soil Science, 1985;140(1) 11-22.

[13] Shuman LM. Effect of liming on the distribution of manganese, copper, iron, and zinc among soil fractions. Soil Science Society of America Journal, 1986;50(5) 1236-1240.

[14] Sims JT. Soil pH effects on the distribution and availability of manganese, copper, and zinc among soil fractions. Soil Science Society of America Journal, 1986;50(2) 367-373.

[15] Shuman LM. Zinc, manganese, and copper in soil fractions. Soil Science, 1979;127(1) 10-17.

[16] Sánchez-Martín MJ, García-Delgado M, Lorenzo LF, Rodríguez-Cruz MS, Arienzo M. Heavy metals in sewage sludge amended soils determined by sequential extractions as a function of incubation time of soils. Geoderma, 2007;142(3-4) 262-273.

[17] Camobreco VJ, Richards BK, Steenhuis TS, Peverly JH, McBride MB. Movement of heavy metals through undisturbed and homogenized soil columns. Soil Science, 1996;161(11) 740-750.

[18] Chumbley CG, Unwin RJ. Cadmium and lead content of vegetable crops grown on land with a history of sewage sludge application. Environmental Pollution Series B, Chemical and Physical, 1982;4(3) 231-237.

[19] Sistema Nacional de Informações sobre Saneamento. Diagnósticos dos serviços de água e esgotos – 2001. Brasília: SEDU-PR/IPEA; 2002.

[20] Sistema Nacional de Informações sobre Saneamento. Diagnósticos dos serviços de água e esgotos – 2011. Brasília: MCIDADES.SNSA; 2013.

[21] Leoneti AB, Prado EL, Oliveira SVWB. Saneamento básico no Brasil: considerações sobre investimento e sustentabilidade para o século XXI. Revista de Administração Pública, 2011;45(2) 331-348.

[22] Pedroza MM, Vieira GEG, Sousa JF, Pickler AC, Leal ERM, Milhomen CC. Produção e tratamento de lodo de esgoto – uma revisão. Revista Liberato, 2010;11(16) 89-188.

[23] Adriano, DC. Trace elements in terrestrial environments: biogeochemistry, bioavailability, and risks of metals. 2nd ed. New York: Springer-Verlag; 2001.

[24] Alloway BJ. Introduction. In: Alloway BJ. (ed.) Heavy metals in soils: trace metals and metalloids in soils and their bioavailability. 3rd ed. Dordrecht: Springer; 2013. p3-10.

[25]] Duffus JH. "Heavy metals" – A meaningless term? Pure and Applied Chemistry, 2002;74(5) 793-907.

[26] Hawkes SJ. What is a "heavy metal"? Journal of Chemical Education, 1997;74(11) 1374.

[27] Christophersen OA, Lyons G, Haug A, Steinnes E. Selenium. In: Alloway BJ. (ed.) Heavy metals in soils: trace metals and metalloids in soils and their bioavailability. 3rd ed. Dordrecht: Springer; 2013. p429-464.

[28] Foy CD, Chaney R, White C. The physiology of metal toxicity in plants. Annual Review of Plant Physiology, 1978;29 511-566.

[29] Fageria NK, Baligar VC, Clark RB. Micronutrients in crop production. Advances in Agronomy, 2002;77 185-268.

[30] Järup L. Hazards of heavy metal contamination. British Medical Bulletin, 2003;68(1) 167-182.

[31] Requejo R, Tena M. Proteome analysis roots reveals that oxidative stress is a main contributing factor to plant arsenic toxicity. Phytochemistry, 2005;66(13) 1519-1528.

[32] Geng C-N, Zhu Y-G, Tong Y-P, Smith SE, Smith FA. Arsenate (As) uptake by and distribution in two cultivars of winter wheat (*Triticum aestivum* L.). Chemosphere, 2006;62(4) 608-615.

[33] Castillo-Michel H, Parsons JG, Peralta-Videa JR, Martínez-Martínez A, Dokken KM, Gardea-Torresdey, JL. Use of X-ray absorption spectroscopy and biochemical techniques to characterize arsenic uptake and reduction in pea (*Pisum sativum*) plants. Plant Physiology and Biochemistry, 2007;45(6,7) 457-463.

[34] Prasad MNV. Cadmium toxicity and tolerance in vascular plants. Environmental and Experimental Botany, 1995;35(4) 525-545.

[35] Rodríguez-Serrano M, Romero-Puertas MC, Pazmiño DM, Testillano OS, Risueño MC, del Río LA, Sandalio LM. Cellular response of pea plants to cadmium toxicity: cross talk between reactive oxygen species, nitric oxide, and calcium. Plant Physiology, 2009;150(1) 229-243.

[36] Cousins RJ, Barber AK, Trout JR. Cadmium toxicity in growing swine. Journal of Nutrition, 1973;103(7) 964-972.

[37] Shanker AK, Cervantes C, Loza-Tavera, H, Avudainayagam S. Chromium toxicity in plants. Environmental International, 2005;31(5) 739-753.

[38] Sutle NF. The mineral nutrition of livestock. 4th ed. Oxfordshire: CABI; 2010

[39] Anderson RA. Chromium as an essential nutrient for humans. Regulatory Toxicology and Pharmacology, 1997;26(1) S35-S41.

[40] Gad SC. Acute and chronic systemic chromium toxicity. The Science of the Total Environment, 1989;86(1-2) 149-157.

[41] Dayan AD, Paine AJ. Mechanisms of chromium toxicity, carcinogenicity and allergenicity: review of the literature from 1985 to 2000. Human & Experimental Toxicology, 2001;20(9) 439-451.

[42] Liu J, Reid RJ, Smith FA. The mechanism of cobalt toxicity in mung beans. Physiologia Plantarum, 2000;110(1) 104-110.

[43] Chatterjee J. Chatterjee C. Phytotoxicity of cobalt, chromium and copper in cauliflower. Enviromental Pollution, 2000;109(1) 69-74.

[44] Ammerman CB, Goodrich RD. Advances in mineral nutrition in ruminants. Journal of Animal Science, 1983;57(Supplement 2) 519-533.

[45] Simonsen LO, Harbak H, Bennekou P. Cobalt metabolism and toxicology – a brief update. The Science of the Total Environment, 2012;432 210-2015.

[46] Yruela I. Copper in plants. Brazilian Journal of Plant Physiology, 2005;17(1) 145-156.

[47] Perrin DJ, Schiefer HB, Blakley BR. Chronic copper toxicity in a dairy herd. The Canadian Veterinary Journal, 1990;31(9) 629-632.

[48] Uauy R, Olivares M, Gonzalez M. Essentiality of copper in humans. The American Journal of Clinical Nutrition, 1998;67(5) 952S-959S.

[49] Fageria NK, Santos AB, Barbosa Filho MB, Guimarães CM. Iron toxicity in lowland rice. Journal of Plant Nutrition, 2008;31(9) 1676-1697.

[50] Brewer GJ. Risks of copper and iron toxicity during aging in humans. Chemical Research in Toxicology, 2010;23(2) 319-326.

[51] Sharma P, Dubey RS. Lead toxicity in plants. Brazilian Journal of Plant Physiology, 2005;17(1) 35-52.

[52] Crossgrove J, Zheng W. Manganese toxicity upon overexposure. NMR in Biomedicine, 2005;17(8) 544-553.

[53] Patra M, Sharma A. Mercury toxicity in plants. The Botanical Review, 2000;66(3) 379-422.

[54] Cargnelutti D, Tabaldi LA, Spanevello RM, Jucoski GO, Battisti V, Redin M, Linares CEB, Dressler VL, Flores EMM, Nicoloso FT, Morsch VM, Schetinger MRC. Mercury toxicity induces oxidative stress in growing cucumber seedlings. Chemosphere, 2006;65(6) 999-1006.

[55] McGrath SP, Micó C, Zhao FJ, Stroud JL, Zhang H, Fozard S. Predicting molybdenum toxicity to higher plants: estimation of toxicity threshold values. Environmental Pollution, 2010;158(10) 3085-3094.

[56] Vyskočil A, Viau C. Assesment of molybdenum toxicity in humans. Journal of Applied Toxicology, 1999;19(3) 185-192.

[57] Sreekanth TVM, Nagajyothi PC, Lee KD, Prasad TNVKV. Occurrence, physiological responses and toxicity of nickel in plants. International Journal of Environmental Science and Technology, 2013;10(5) 1129-1140.

[58] Denkhaus E, Salnikow K. Nickel essentiality, toxicity, and carcinogenicity. Critical Review in Oncology/Hematology, 2002;42(1) 35-56.

[59] Terry N, Zayed AM, Souza MP, Tarum AS. Selenium in higher plants. Annual Review of Plant Physiology and Plant Molecular Biology, 2000;51 401-432.

[60] Koller LD, Exon JH. The two faces of selenium – deficiency and toxicity – are similar in animals and man. Canadian Journal of Veterinary Research, 1986;50(3) 297-306.

[61] Tinggi U. Essentiality and toxicity of selenium and its status in Australia: a review. Toxicology Letters, 2003;137(1-2) 103-110.

[62] Broadley MR, White PJ, Hammond JP, Zelko I, Lux A. Zinc in plants. New Phytologist, 2007;173(4) 677-702.

[63] Allen JG, Masters HG, Peet RL, Mullins KR, Lewis RD, Skirrow SZ, Fry J. Zinc toxicity in ruminants. Journal of Comparative Pathology, 1983;93(3) 363-377.

[64] Fosmire GJ. Zinc toxicity. The American Journal of Clinical Nutrition. 1990;51(2) 225-227.

[65] Fränzle S, Markert B. The Biological System of the Elements (BSE) – a brief introduction into historical and applied aspects with special reference on "ecotoxicological identity cards" for different element species. (e.g. AS and Sn). Environmental Pollution, 2002;120(1) 27-45.

[66] EPA – Environmental Protection Agency. Standards for the use or disposal of sewage sludge. Washington, DC: USEPA; 1993. (40 CFR Part 257)

[67] CONAMA. Brazilian National Environment Council. Resolução CONAMA nº 375/2006. Available: <http://www.mma.gov.br/port/conama/legiabre.cfm?codlegi=506>. Accessed: 09/08/2013.

[68] Oliveira FC, Mattiazzo ME. Metais pesados em Latossolo tratado com lodo de esgoto e em plantas de cana-de-açúcar. Scientia Agricola, 2001;58(3) 581-593.

[69] Oliveira KW, Melo WJ, Pereira GT, Melo VP, Melo GMP. Heavy metals in Oxisols amended with biosolids and cropped with maize in a long-term experiment. Scientia Agricola, 2005;62(4) 381-388.

[70] Rangel OJP, Silva CA, Bettiol W, Dynia JF. Efeito de aplicações de lodos de esgoto sobre os teores de metais pesados em folhas e grãos de milho. Revista Brasileira de Ciência do Solo, 2008;30(3) 583-594.

[71] Nogueira TAR, Oliveira LRO, Melo WJ, Fonseca IM, Melo GMP, Melo VP, Marques MO. Cádmio, cromo, chumbo e zinco plantas de milho e em Latossolo após nove aplicações anuais de lodo de esgoto. Revista Brasileira de Ciência do Solo, 2008;32(5) 2195-2207.

[72] Pires AMM, Mattiazzo ME. Cinética de solubilização de metais pesados por ácidos orgânicos em solos tratados com lodo de esgoto. Revista Brasileira de Ciência do Solo, 2007;31(1) 143-151.

[73] Merlino LCS, Melo WJ, Macedo FG, Guedes ACTP, Melo VP, Melo GMP. Bário, cádmio, cromo e chumbo em plantas de milho e em Latossolo após onze aplicações anuais de lodo de esgoto. Revista Brasileira de Ciência do Solo, 2010;34(6) 2031-2039.

[74] Borges MR, Coutinho ELM. Metais pesados do solo após aplicação de biossólido. I - Fracionamento. Revista Brasileira de Ciência do Solo, 2004;28(3) 543-555.

[75] Camilotti F, Andrioli I, Marques MO, Silva AR, Tasso Júnior LC. Avaliação dos teores de metais pesados no solo e na planta de cana-de-açúcar sob adubação com lodo de esgoto e vinhaça. Bioscience Journal, 2009;25(6) 23-31.

[76] Camilotti F, Silva ARB, Marques MO. Biomass and yield of peanut grown on tropical soil amended with sewage sludge contaminated with lead. Applied and Environmental Soil Science, 2012;2012 1-6. doi:10.1155/2012/896090

[77] Galdos MV, De Maria IC, Camargo OA. Atributos químicos e produção de milho em um Latossolo Vermelho eutroférrico tratado com lodo de esgoto. Revista Brasileira de Ciência do Solo, 2004;28(3) 569-577.

[78] Nogueira TAR, Franco A, He Z, Braga VS, Firme LP, Abreu-Junior CH. Short-term usage of sewage sludge as organic fertilizer to sugarcane in a tropical soil bears little threat of heavy metal contamination. Journal of Environmental Management, 2013;114 168-177.

[79] Simonete MA, Kiehl JC. Extração e fitodisponibilidade de metais em resposta à adição de lodo de esgoto no solo. Scientia Agricola, 2002;59(3) 555-563.

[80] Oliveira C, Sobrinho NMBA, Marques VS, Mazur N. Efeito da aplicação do lodo de esgoto enriquecido com cádmio e zinco na cultura do arroz. Revista Brasileira de Ciência do Solo, 2003;29(1) 109-116.

[81] Martins ALC, Bataglia OC, Camargo OA. Copper, nickel and zinc phytoavailability in an Oxisol amended with sewage sludge and liming. Scientia Agricola, 2003;60(4) 747-754.

[82] CETESB. Environmental Agency of the State of São Paulo. Aplicação de lodos de sistemas de tratamento biológico em áreas agrícolas – critérios para projeto e operação. São Paulo: CETESB; 1999. (Norma técnica, P4.230) Available: http://www.cetesb.sp.gov.br/servicos/normas---cetesb/43-normas-tecnicas---cetesb>. Accessed: 09/08/2013.

[83] USEPA. United States Environmental Protection Agency. Method 3050B. Acid digestion of sediments, sludges, and soils. USEPA; 1996. Available: <http://www.epa.gov/osw/hazard/testmethods/sw846/pdfs/3052.pdf>. Accessed: 09/22/2013.

[84] USEPA. United States Environmental Protection Agency. Method 3051A. Microwave assisted acid digestion of sediments, sludges, soils, and oils. USEPA; 2007. Available: <http://www.epa.gov/osw/hazard/testmethods/sw846/pdfs/3051a.pdf>. Accessed: 09/22/2013.

[85] Abreu MF, Berton RS, Andrade JC. Comparison of methods to evaluate heavy metals in organic wastes. Communications in Soil Science and Plant Analysis, 1996;27(5-8) 1125-1135.

[86] Jackson ML. Soil chemical analysis. Englewood Cliffs: Prentice Hall; 1958.

[87] Alloway BJ, Jackson AP. The behavior of heavy metals in sewage sludge-amended soils. The Science of the Total Environment, 1991;100 151-176.

[88] Silva CA, Rangel, OJP, Dynia JF, Bettiol W, Manzatto, CV. Disponibilidade de metais pesados para milho cultivado em Latossolo sucessivamente tratado com lodos de esgoto. Revista Brasileira de Ciência do Solo, 2006;30(2) 353-364.

[89] Oliveira FC, Mattiazzo ME. Mobilidade de metais pesados em um Latossolo Amarelo distrófico tratado com lodo de esgoto e cultivado com cana-de-açúcar. Scientia Agricola, 2001;58(4) 807-812.

[90] CETESB. Environmental Agency of the State of São Paulo. Valores orientadores para solos e águas subterrâneas no Estado de São Paulo. CETESB; 2005. Available: <http://www.cetesb.sp.gov.br/solo/publicações-e-Relatórios/1-Publicações-/-Relatórios>. Accessed: 09/11/2013.

[91] CONAMA. Brazilian National Environment Council. Resolução CONAMA nº 420/2009. Available: <http://www.mma.gov.br/port/conama/legiabre.cfm?codlegi=620>. Accessed: 09/08/2013.

[92] Macedo FG, Melo WJ, Merlino LCS, Ribeiro MH, Melo GMP, Camacho MA. Acúmulo e disponibilidade de cromo, cádmio e chumbo em solos tratados com lodo de esgoto por onze anos consecutivos. Semina: Ciências Agrárias, 2012;33(1) 101-114.

[93] Abreu CA, Raij B van, Abreu MF, González AP. Routine soil testing to monitor heavy metals and boron. Scientia Agricola, 2005;62(6) 564-571.

[94] Nogueira TAR, Melo WJ, Fonseca IM, Marcussi AS, Melo GMP, Marques MO. Fractionation of Zn, Cd and Pb in a tropical soil after nine-year sewage sludge applications. Pedosphere, 2010;20(5) 545-556.

[95] Melo WJ, Aguiar OS, Melo GMP, Melo VP. Nickel in a tropical soil treated with sewage sludge and cropped with maize in a long-term field study. Soil Biology & Biochemistry, 2007;39(6) 1341-1347.

[96] Nolan AN, Lombi E, McLaughlin MJ. Metal bioaccumulation and toxicity in soils – Why bother with speciation? Australian Journal of Chemistry, 2003;56(3) 77-91.

[97] Hernandez-Soriano MC, Peña A, Mingorance MD. Soluble metal pool as affected by soil addition with organic inputs. Environmental Toxicology and Chemistry, 2013;32(5) 1027-1032.

[98] Silva LFM. Especiação iônica da solução do solo após sucessivas aplicações de solo de esgoto. Dissertação de Mestrado. Instituto Agronômico de Campinas; 2012.

[99] Barbosa Filho MP, Cantarella H, Wiethölter. Arroz, milho e trigo. In: Ferreira ME, Cruz MCP, Raij B van, Abreu CA. (ed.) Micronutrientes e elementos tóxicos na agricultura. Jaboticabal: CNPq/FAPESP/POTAFOS; 2001. p.285-318.

[100] ANVISA. Brazilian Health Surveillance Agency. Decreto nº 55.871/1965. Available: <http://portal.anvisa.gov.br/wps/content/Anvisa+Portal/Anvisa/Inicio/Alimentos/Assuntos+de+Interesse/Legislacao/Aditivos+Alimentares+e+Coadjuvantes+de+Tecnologia>. Accessed: 23/09/2013.

[101] ABIA. Brazilian Food Industry Association. Compêndio da legislação de alimentos. São Paulo: ABIA; 1985.

[102] Kabata-Pendias A, Pendias H. Trace elements in soils and plants. 3rd ed. Boca Raton: CRC Press; 2001.

[103] Camilotti F, Marques MO, Andrioli I, Silva AR, Tasso Júnior LC, Nobile FO. Acúmulo de metais pesados em cana-de-açúcar mediante a aplicação de lodo de esgoto e vinhaça. Engenharia Agrícola, 2007;27(1) 284-293.